The Cytogenetics of Mammalian Autosomal Rearrangements

Progress and Topics in Cytogenetics

Series Editor
Avery A. Sandberg
Roswell Park Memorial Institute, Buffalo, New York

Volume 1
Cytogenetics and Breeding of Pearl Millet and Related Species
Prem P. Jauhar

Volume 2
Sister Chromatid Exchange
Avery A. Sandberg, *Editor*

Volume 3
Cytogenetics of the Mammalian X Chromosome

Part A: Basic Mechanisms of X Chromosome Behavior
Part B: X Chromosome Anomalies and Their Clinical Manifestations

Avery A. Sandberg, *Editor*

Volume 4
Radiation-Induced Chromosome Damage in Man
Takaaki Ishihara and Masao S. Sasaki, *Editors*

Volume 5
Catalog of Chromosome Aberrations in Cancer, Second Edition
Felix Mitelman

Volume 6
The Y Chromosome

Part A: Basic Characteristics of the Y Chromosome
Part B: Clinical Aspects of Y Chromosome Abnormalities

Avery A. Sandberg, *Editor*

Volume 7
Aneuploidy

Part A: Incidence and Etiology
Part B: Induction and Model Systems

Baldev K. Vig and Avery A. Sandberg, *Editors*

Volume 8
The Cytogenetics of Mammalian Autosomal Rearrangements
Art Daniel, *Editor*

The Cytogenetics of Mammalian Autosomal Rearrangements

Editor

Art Daniel

Cytogenetics and Molecular Cell Biology Laboratory
Department of Medical Genetics
Shodair Hospital
Helena, Montana

and

Department of Biology
Montana State University
Bozeman, Montana

Alan R. Liss, Inc., New York

Address all Inquiries to the Publisher
Alan R. Liss, Inc., 41 East 11th Street, New York, NY 10003

Printed in the United States of America

Library of Congress Cataloging-in-Publication Data

The Cytogenetics of mammalian autosomal
rearrangements/editor, Art Daniel.
p. cm. — (Progress and topics in cytogenetics: v. 8)
Includes bibliographies and index.
ISBN 0-8451-2410-2
1. Human chromosome abnormalities. 2. Mammals—
Genetics.
I. Daniel, Art. II. Series.
[DNLM: 1. Chromosome Aberrations. W1 PR666EH
v. 8/QH 462.A1 C997]
RB155.C97 1987
599'.0159—dc19
DNLM/DLC
for Library of Congress 87-29886
CIP

Contents

Contributors

Judith A. Brown, Departments of Human Genetics, and Obstetrics and Gynecology, Medical College of Virginia, Richmond, VA 23298-0033 **[697]**

Ernesto Capanna, Dipartimento di Biologia Animale e dell'Uomo, Universitá La Sapienza, 00161 Roma, Italy **[315]**

Ann C. Chandley, MRC Clinical and Population Cytogenetics Unit, Western General Hospital, Edinburgh EH4 2XU, Scotland **[361]**

Cristina Cuoco, Gaslini Institute, Genova, Italy **[615]**

Art Daniel, Department of Medical Genetics, Shodair Children's Specialty Hospital, Helena, MT 59601, and Department of Biology, Montana State University, Bozeman, MT; present address: Cytogenetics Unit, Western Genetics Service, Institute of Clinical Pathology and Medical Research, Westmead Hospital, Westmead, N.S.W. 2145, Australia **[xv,73]**

John R. Davis, Departments of Pathology and Obstetrics-Gynecology, Arizona Health Sciences Center, University of Arizona, Tucson, AZ 85724 **[419]**

P. De Boer, Department of Genetics, Agricultural University, 6703 BM Wageningen, The Netherlands **[295]**

Jean de Grouchy, U. 173 INSERM - UA. 119 CNRS, Laboratoire de Cytogénétique Humaine et Comparée, Hôpital Necker-Enfants-Malades, 75743 Paris Cedex, France **[769]**

Gordon W. Dewald, Cytogenetics Laboratory, Mayo Clinic and Mayo Foundation, and Departments of Laboratory Medicine and Medical Genetics, Mayo Medical School, Rochester, MN 55905 **[533]**

Fredrick F.B. Elder, Departments of Pediatrics and Human Biological Chemistry and Genetics, The University of Texas Medical Branch, Galveston, TX 77550; present address: Institute for Molecular Genetics, Kleberg Cytogenetics Laboratory, Baylor College of Medicine, Houston, TX 77030 **[481]**

Jean-Pierre Fryns, Center for Human Genetics, University of Leuven, B-3000 Leuven, Belgium **[453]**

Giorgio Gimelli, Gaslini Institute, Genova, Italy **[615]**

Roberta M. Hagaman, Division of Computer Systems and Biostatistics, Arizona Health Sciences Center, University of Arizona, Tucson, AZ 85724 **[419]**

I. Hansmann, Institut für Humangenetik, Universität Göttingen, 3400 Göttingen, Federal Republic of Germany **[295]**

The numbers in brackets are the opening page numbers of the contributors' articles.

Gerald P. Holmquist, Department of Biology, Beckman Research Institute, City of Hope, Duarte, CA 91010 **[803]**

Ernest B. Hook, Bureau of Environmental Epidemiology and Occupational Health, New York State Department of Health, Albany, NY 12237, and Department of Pediatrics, Albany Medical College, Albany, NY 12237 **[73]**

T.C. Hsu, Section of Cellular Genetics, The University of Texas Cancer Center, M.D. Anderson Hospital and Tumor Institute, Houston, TX 77030 **[481]**

Hirotami T. Imai, Department of Cytogenetics, National Institute of Genetics, Shizuoka-ken 411, Japan **[551]**

Colleen K. Jackson-Cook, Department of Human Genetics, Medical College of Virginia, Richmond, VA 23298-0033 **[697]**

H. Jalbert, Laboratoire de Cytogénétique et de Biologie de la Reproduction, Hôpital Nord, 38043 Grenoble Cedex, France **[267]**

P. Jalbert, Laboratoire de Cytogénétique et de Biologie de la Reproduction, Hôpital Nord, 38043 Grenoble Cedex, France **[267]**

Peter Kaiser, Institut für Humangenetik und Genetische Poliklinik der Philipps-Universität, D-3550 Marburg, Federal Republic of Germany; present address: Klinische Genetik, Institut für Anthropologie und Humangenetik, D-7400 Tübingen, Federal Republic of Germany **[163]**

K. Margaret Kirk, Genetics Division, Medical Research Council Radiobiology Unit, Chilton, Didcot, Oxon OX11 ORD, England; present address: Pitton Mill Cottage, Pitton Green, Rhosili, Gower, Wales **[739]**

Alice Kleczkowska, Center for Human Genetics, University of Leuven, B-3000 Leuven, Belgium **[453]**

Elisabeth Kubień, Center for Human Genetics, University of Leuven, B-3000 Leuven, Belgium **[453]**

Susan E. Long, Department of Animal Husbandry, University of Bristol, Langford, Bristol BS18 7DU, England **[383]**

K. Madan, Institute of Human Genetics, Free University, 1081 BT Amsterdam, The Netherlands **[249]**

R. Ellen Magenis, Departments of Pediatrics and Medical Genetics, Crippled Children's Division, Clinical Cytogenetics Laboratory of University Hospitals, Oregon Health Sciences University, Portland, OR 97201 **[583,855]**

Paola Maraschio, Department of General Biology and Genetics, University of Pavia, Pavia, Italy **[615]**

Renee H. Martin, Division of Medical Genetics, Department of Pediatrics, University of Calgary, Calgary, Alberta, Canada T2N 4N1; present address: Genetics Clinic, Alberta Children's Hospital, Calgary, Alberta, Canada T2T 5C7 **[397]**

J.F. Mattei, Centre de Génétique Médicale and INSERM U. 242, Hôpital d'Enfants de la Timone, 13385 Marseille Cedex, France **[601]**

M.G. Mattei, Centre de Génétique Médicale and INSERM U. 242, Hôpital d'Enfants de la Timone, 13385 Marseille Cedex, France **[601]**

Walter E. Nance, Departments of Human Genetics and Pediatrics, Medical College of Virginia, Richmond, VA 23298-0033 **[697]**

Susan B. Olson, Department of Medical Genetics, Crippled Children's Division, Clinical Cytogenetics Laboratory of University Hospitals, Oregon Health Sciences University, Portland, OR 97201 **[583]**

Carlo Alberto Redi, Dipartimento di Biologia Animale and Centro di Studio per l'Istochimica del CNR, Universitá di Pavia, 27100 Pavia, Italy **[315]**

Beverly Barton Rogers, Department of Pathology, Baylor University, College of Medicine, Houston, TX 77030 **[419]**

Avery A. Sandberg, Department of Genetics and Endocrinology, Roswell Park Memorial Institute, Buffalo, NY 14263; and The Genetics and Cancer Centers, Southwest Biomedical Research Institute, Scottsdale, AZ 85251 **[xiii,835]**

Albert A. Schinzel, Institute of Medical Genetics, University of Zürich, CH-8001 Zürich, Switzerland **[725]**

Antony G. Searle, Genetics Division, Medical Research Council Radiobiology Unit, Chilton, Didcot, Oxon OX11 ORD, England **[739]**

Jeremy B. Searle, School of Biological Sciences, University of East Anglia, Norwich NR4 7TJ, England; present address: Department of Zoology, University of Oxford, Oxford OX1 3PS, England **[507]**

B. Sele, Laboratoire de Cytogénétique et de Biologie de la Reproduction, Hôpital Nord, 38043 Grenoble Cedex, France **[267]**

R.M. Speed, Medical Research Council Clinical and Population Cytogenetics Unit, Western General Hospital, Edinburgh EH4 2XU, Scotland **[295]**

Jon Stene, Institute of Statistics, University of Copenhagen, DK-1455 Copenhagen, Denmark **[3]**

Sabine Stengel-Rutkowski, Department of Pediatric Genetics, University of Munich, D-8000 Munich, Federal Republic of Germany **[3]**

Grant R. Sutherland, Cytogenetics Unit, Department of Histopathology, Adelaide Children's Hospital, North Adelaide, SA 5006, Australia **[471]**

Luciano Tiepolo, Department of General Biology and Genetics, University of Pavia, Pavia, Italy **[615]**

Catherine Turleau, U. 173 INSERM - UA. 119 CNRS, Laboratoire de Cytogénétique Humaine et Comparée, Hôpital Necker-Enfants-Malades, 75743 Paris Cedex, France **[769]**

Herman Van den Berghe, Center for Human Genetics, University of Leuven, B-3000 Leuven, Belgium **[453]**

Daniel L. Van Dyke, Medical Genetics and Birth Defects Center, Henry Ford Hospital, Detroit, MI 48202 **[635]**

Gary Wulf, Montana State Department of Administration, Information Services Division, Helena, MT 59601 **[73]**

Herman E. Wyandt, Department of Laboratory Medicine, Cytogenetics Laboratory, Danbury Hospital, Danbury, CT 06810 **[667]**

Orsetta Zuffardi, Department of General Biology, University of Florence, Florence, Italy **[615]**

Foreword

The recognition that chromosomes are key elements in the transmission of hereditary material and in the control of cellular life antedated most of our knowledge of the structure and function of DNA and genes. The concept of position effect and the role of "transposable elements" were postulated half a century ago and have only recently been appreciated. In the last 30 years there has been an explosion of cytogenetic information, primarily because of technical developments. Unfortunately, this information is widely dispersed among a large number of journals and publications, and even in this age of readily available computerized bibliographic services, the great volume and range of ideas and concepts cannot be critically or competently analyzed or interpreted.

Particularly exciting and challenging are the many studies and reports of the seemingly myriad chromosomal rearrangements in mammals, including man. Chromosome rearrangements in humans are responsible for prenatal risks, infertility, missed abortion, fetal and perinatal loss, congenital abnormalities, and various neoplastic states.

Thus, it was with great enthusiasm that I accepted Dr. Art Daniel's offer to edit a volume on "The Cytogenetics of Mammalian Autosomal Rearrangements." I anticipated that this would be a difficult if not herculean task, considering the extent and complexity of the field. However, Dr. Daniel accomplished exactly what he set out to do—i.e., to generate a volume in which the leading experts in all fields encompassed have stated their views and supplied information on a wide array of autosomal rearrangements and their effects. We owe Dr. Daniel thanks for this accomplishment, with the hope that readers will find this volume to be a major source of cytogenetic information on chromosomal rearrangements.

Avery A. Sandberg, MD
Buffalo, New York
Scottsdale, Arizona

Preface

Because two previous volumes in the series Progress and Topics in Cytogenetics were devoted to the mammalian X chromosome (volume 3) and Y chromosome (volume 6), for those of us interested in the diversity of chromosome structures, a volume devoted to the cytogenetics of mammalian autosomal rearrangements seemed appropriate. At a time when "cytogenetics" is beginning to encompass the use of chromosome-specific DNA libraries and fluorescent (biotin-labeled) probes to detect chromosomes 13, 18, 21, and X at interphase, such an overview and reexamination of perspectives is in order. Whereas these issues are only alluded to in the book, and in situ hybridization (of ^{3}H probes) and mapping with the use of RFLPs are addressed more directly, the real emphasis of the book is classical cytogenetics. In this connection the working interrelationships between clinical cytogeneticists and those in many other fields have always benefited by regular cross-fertilization, and this book contributes to that process. Consequently, specialists from many nations and from various scientific and medical fields in evolutionary genetics, animal husbandry, reproduction and fertility, clinical genetics, molecular genetics, cancer genetics and clinical cytogenetics have all participated.

The book is divided into four sections, the first of which (chapters 1–5) deals with empiric data and models of risk of unbalanced progeny, both at term and at 16 weeks gestation, the latter because of the impact that prenatal diagnosis (by amniocentesis) has made on genetic counseling practices. Current data on reciprocal, Robertsonian, and complex translocations, pericentric and paracentric inversions, and insertions are all considered in Part I.

The second section (chapters 6–13) addresses the enormous burden of unsuccessful pregnancies, as illustrated in mouse and man, for carriers of structural rearrangements and for normal euploid (human) individuals. For the latter, there is evidence that 60–75% of all conceptuses may be lost either before implantation, at implantation, later in pregnancy (missed or spontaneous abortion), or at term (stillbirth or neonatal death). This loss may become more evident in other mammalian species if conceptuses could be studied at the earliest stages.

Part III of this book (chapters 14–23) deals with the origin of chromosome rearrangements. This subject is approached from several perspectives, including some of the evolutionary and population genetics data from *Mus*, *Sorex*, and *Muntiacus* and *Sigmodon*; the disparity between the sexes in the meiotic origins of certain chromosomal structures; and the inclusion of several other rearrangement types more often of de novo origin than those considered in Part I, namely, ring chromosomes, isochromosomes, tandem fusions, fusion dicentrics, and (human) bisatellited marker chromosomes.

Chapters 24–29 consider some provocative pathways that are being pursued by researchers. Here several hypotheses and approaches are discussed: e.g., the two genome hypothesis where tissue specific genes are considered to be localised in G-positive (chromosome) bands and the various corollaries; the concept that lethality in chromosomally unbalanced fetuses is more the result of chromosome imbalance size than the accompanying interplay with the specific loci involved in the affected chromosome segment; the relationships between common rearrangements in neoplasias and the juxtaposition of certain loci (T-cell receptor cistrons, immunoglobulin chain loci, for example) with protooncogenes; the fact that specific congenital malformations may be induced by manifold events—both diverse genomic causes and others—but that it is the minor anomalies of the phenotype that may be related to specific duplications or deficiencies in the genome; and lastly the documentation that mendelian disorders meet cytogenetic abnormalities at the level of submicroscopic duplications and deletions as detected by RFLPs.

I thank my co-authors for their patience in the preparation of this volume. It certainly engaged all of us for much longer than I initially envisaged, but for a result that I trust will do them justice and will aid our colleagues, in the widest sense, in their understanding of the field.

Art Daniel
Helena, Montana

ASCERTAINMENT AND RISKS OF RECOMBINANT PROGENY

The Cytogenetics of Mammalian Autosomal Rearrangements, pages 3–72

Chapter 1
Genetic Risks of Familial Reciprocal and Robertsonian Translocation Carriers

Jon Stene and Sabine Stengel-Rutkowski

INTRODUCTION

In this chapter, we consider inherited translocations (or interchanges) involving two nonhomologous chromosomes in man. More specifically, the discussion is confined to *reciprocal translocations* (rcp), formed after one break in each of two nonhomologous chromosomes and a subsequent exchange of end segments resulting in two new derived (der) chromosomes.

If both chromosomes are acrocentric, i.e., chromosomes 13, 14, 15, 21 and 22, and both breakpoints are situated close to or within the centromeres, the resulting translocation is called a *Robertsonian translocation* (rob; or, a centric-fusion translocation), and this consists of a single chromosome formed by the long arms of the two acrocentric chromosomes. The other (usually acentric) translocation chromosome, formed by the short arms of both acrocentric chromosomes, will be lost.

Reciprocal and Robertsonian translocations are among the most frequent, inherited chromosome aberrations in man and may be inherited unchanged for millennia through many generations. From seven pooled cytogenetic studies of consecutive new births, 104 children with one of these structural (balanced) rearrangements were found among 54,749 new births [1]. This gives a birth prevalence of 0.19%.

In some families with an inherited rcp or rob there may be a very high risk (or probability) of having chromosomally unbalanced children who are congenitally malformed and mentally retarded. For some translocations, the risk may exceed 20% for a newborn child having such an abnormality. In the

Institute of Statistics, University of Copenhagen, DK-1455 Copenhagen K. Denmark (J.S.), Department of Pediatric Genetics, University of Munich, D-8000 Munich 2, Federal Republic of Germany (S.S.-R)

above-mentioned seven newborn studies, a further 28 children had such an abnormality [1].

In other families with such translocations, multiple spontaneous abortions will occur and few, if any, abnormal children; finally, in a third group of translocation families, the translocation may have been inherited through many generations without leaving any detectable sign of abnormal offspring, abortions, or any other indications of its presence. These facts imply that the risks for abnormal offspring vary considerably from translocation to translocation. This large variation has caused great concern for those who are involved in genetic counselling of families with inherited translocations.

The aim of this chapter is to provide means to determine risks for abnormal offspring in such families. The main sections are devoted to rcp, for which more problems have to be solved than for Robertsonian rearrangements. The general principles are illustrated by means of examples.

A more detailed account has been given in a recent paper by Stengel-Rutkowski et al. [2], which contains the results of analyses of data from 1,120 pedigrees with inherited rcp. References to this paper are made frequently throughout the chapter. The statistical problems are discussed only briefly in connection with the examples and references made to the relevant literature. The clinical aspects are not considered here; these have been treated in detail elsewhere [3–5].

In this chapter, we concentrate on the risks for chromosomally unbalanced offspring, which are of particular interest to medical geneticists. The proportion of balanced carriers in phenotypically normal offspring is treated more briefly.

A FAMILY WITH AN INHERITED RCP

As an illustration, consider the family with the rcp between chromosomes 5 and 10 illustrated in Figure 1. The pedigree of the family is given in Figure 2, and the family has been described by Stengel-Rutkowski et al. [6]. At the bottom of Figure 1, the four chromosomes involved are presented in the model form of the Paris Conference [7]. Because such translocations occur mostly in phenotypically normal individuals, it is assumed that no chromosome material has been lost or added through this exchange. Such translocations are called *balanced*, and individuals with such a chromosome constitution are termed *carriers* or *translocation heterozygotes* (since the interchange involves only one of each pair of chromosomes). The gametes produced by such a carrier may be very different. In the family represented in Figure 1, the grandmother (I–1) is a carrier, and she has transmitted the

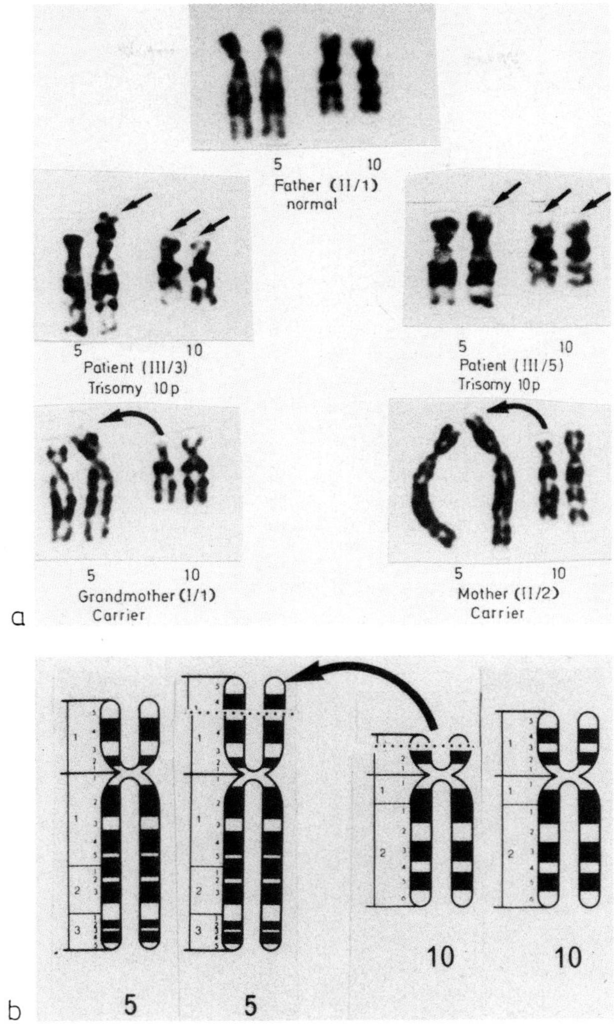

Fig. 1. **a:** Partial karyotypes for some members of a family in whom a reciprocal translocation t(5;10) (p15;p13), between chromosomes 5 and 10 is inherited [6]. **b**: Schematic representation of the balanced translocation (Paris-Nomenclature [7]). The breakpoint on chromosome 5 is in the terminal band of the short arm, denoted 5p15, where p refers to the short arm and q to the long arm. On chromosome 10, the breakpoint is in band 10p13.

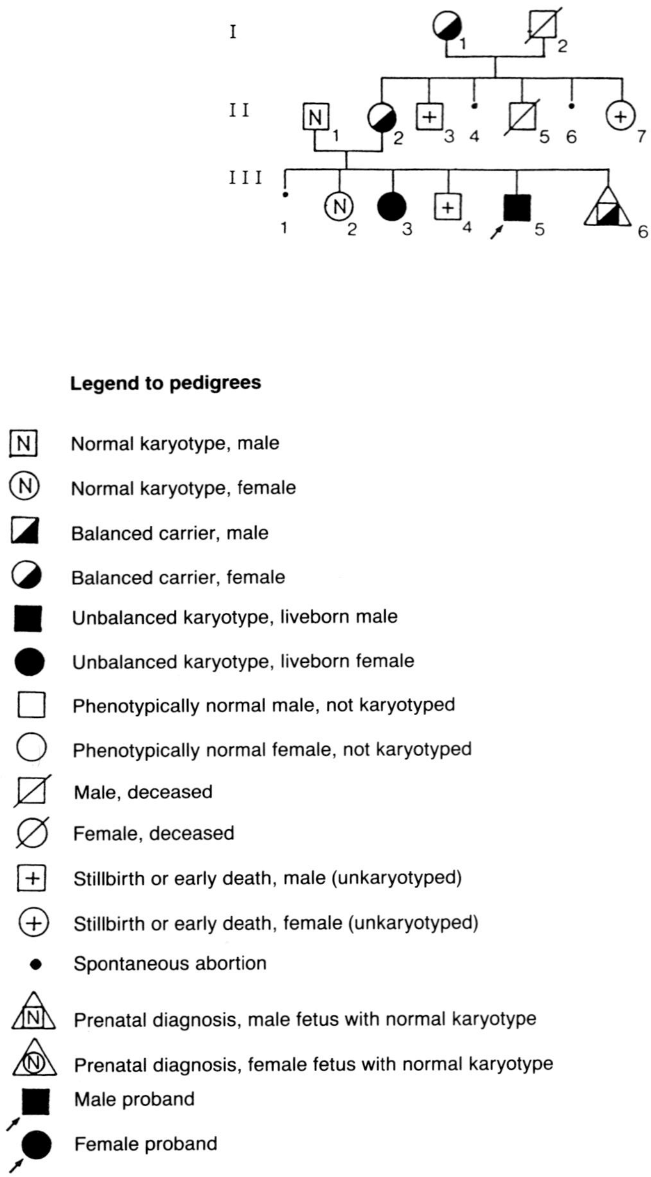

Fig. 2. Pedigree of the family with the inherited t(5;10)(p15;p13) represented in Figure 1.

balanced translocation to her daughter (II–2). This daughter herself has a child (III–2) with normal chromosomes.

Other types of gametes are possible. If the gamete contains the larger translocation chromosome and the normal chromosome 10, the resulting child will be *chromosomally unbalanced*, with trisomy for the end segment of the short arm of chromosome 10 and monosomy for the end segment of the short arm of chromosome 5. The patients (III–3 and the proband, III–5) in this family belong to just this category. They are abnormal and mentally retarded. After the detection that the mother (II–2) was a carrier, a subsequent pregnancy (III–6) was prenatally monitored, and the fetus was found to be a male carrier.

From this pedigree (Fig. 2) we can point out that five different categories of pregnancy outcome are possible: 1) Individuals with normal chromosomes, 2) Carriers of balanced translocations, 3) Abnormal liveborns of the type described above, 4) Stillbirths or early deaths (unkaryotyped), and 5) Spontaneous abortions. In addition to these, there are records of fetuses investigated prenatally with normal chromosomes, balanced rcp, or unbalanced chromosome constitutions.

To describe translocations, the international system for human cytogenetic nomenclature is used. The first version was the report of the Paris Conference [7]. This has been revised several times since then [8–10]. This particular rcp, involving chromosomes 5 and 10 with breakpoints in p15 and p13, respectively, is denoted as t(5;10)(p15;p13).

FURTHER EXAMPLES

Example 1

In Figure 3, a pedigree with an inherited translocation between the short arms of chromosome 5 and 15, i.e., t(5;15) (p13;p12) [11], is shown. In this family, there are two types of chromosomally unbalanced offspring, both with 46 chromosomes, the first type with a trisomy for 5p13 to pter and a monosomy for 15p12 to pter. Four members of the pedigree exhibit this imbalance, i.e., III-3 (indicated as proband) III-4, III-8, and by phenotype presumably also II–8. The other type has a monosomy for 5p13 to 5pter and a trisomy of 15p12 to 15pter. This latter type is to be suggested by phenotype analysis in a single member of the pedigree, II–4, who died early in childhood (unkaryotyped) and had the *cri-du-chat* syndrome [12]. Both types of abnormality arise by 2:2 disjunction with adjacent-1 segregation. The mother [II-2] sought genetic advice when she noticed the abnormal mental development of a younger cousin (III–8) similar to that known for two affected sons (III–3, III–4) and her younger brother (II–8). She brought III-3 with her.

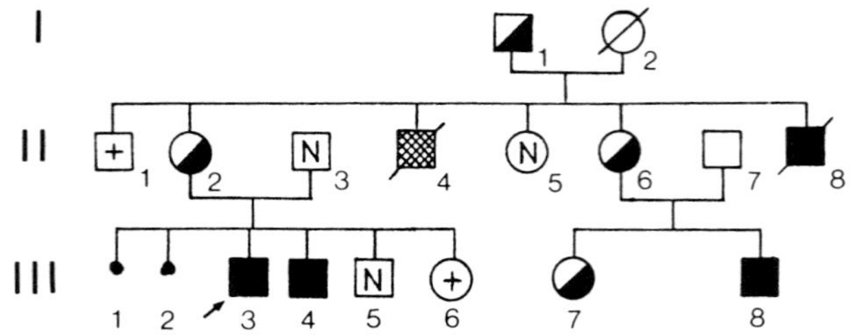

Fig. 3. Pedigree of a family with an inherited t(5;15)(p13;p12) [11]. The male II-4 phenotypically had *cri-du-chat* syndrome, presumably a partial monosomy of the segment 5pter→p13 and a trisomy of the segment 15pter→p12. The other individuals with unbalanced karyotypes had the complementary imbalance (the trisomy and the monosomy interchanged). (See Fig. 2 for pedigree symbols.)

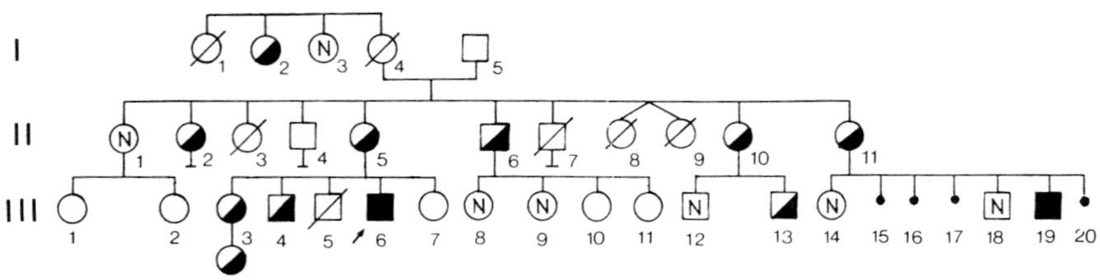

Fig. 4. Pedigree of a family with an inherited t(11;22)(q25;q13) (family 23 of Fraccaro et al. [13]). (See Fig. 2 for pedigree symbols.)

Example 2

Another example is shown in Figure 4. This is family 23 Fraccaro et al. [13]. Here, there is a translocation between chromosomes 11 and 22, with long arm breakpoints at 11q25 and 22q13. Two chromosomally unbalanced children with identical imbalances were born in this family, III–6 and III–19. They have 47 chromosomes and a trisomy for 11q25 to qter and also a trisomy for 22q13 to pter, including the centromere, resulting from meiotic 3:1 disjunction. The designation for the rcp is t(11;22)(q25;q13), and the imbalance is a *tertiary trisomy*. The family was ascertained through the abnormal child III–6 because of severe psychomotor retardation at the age of 5 years. This is a comparatively large pedigree, with a large number of pregnancies. However, only two unbalanced children have been born, indicating a small risk for such offspring. Sterility has been reported for three members of the pedigree, II–2, II–4, and II–7, but the latter two have not been karyotyped.

Example 3

A further example of a familial rcp, t(1;4)(p32;p14) [14], is shown in Figure 5. No abnormal children have been born in the family, but there are a number of spontaneous abortions. The family was detected when a carrier fetus, III–7, was prenatally diagnosed in a maternal age (42 years) referral. Neither breakpoint is terminal or occurs in an acrocentric short arm; i.e., both are in intermediate positions. For this type of rcp, it seems that none of the possible types of abnormal offspring can survive to birth.

The examples demonstrate that the segregation patterns can vary considerably from pedigree to pedigree and from rcp to rcp. Different rcp may result in different types of abnormal offspring [14]. For some rcp, all types of abnormal offspring are incompatible with survival to birth, but spontaneous abortions may occur frequently; for other rcp, phenotypically abnormal, chromosomally unbalanced offspring may be born; the abnormal offspring may have the normal number of chromosomes, 46, or a different number, e.g., 47 (or 45); for some rcp, different types of abnormal offspring may be born.

The properties of an individual rcp will depend on the chromosomes involved in the translocation and on the breakpoint positions; these determine which chromosome segments may be unbalanced as well as the possible combinations of unbalanced segments (see Breakpoint Positions, Meiotic Divisions, and Risks, below). The genetic content of the unbalanced segments in the cells of a fetus or a child determines their abnormalities in all clinical aspects. Each translocation can, in principle, be considered as unique because of the great variation in breakpoint positions.

GROUPING OF RCP FOR RISK ESTIMATION

Among the properties of a specific rcp are the *probability of survival* to a certain gestational age or to birth for a fetus with a certain type of chromo-

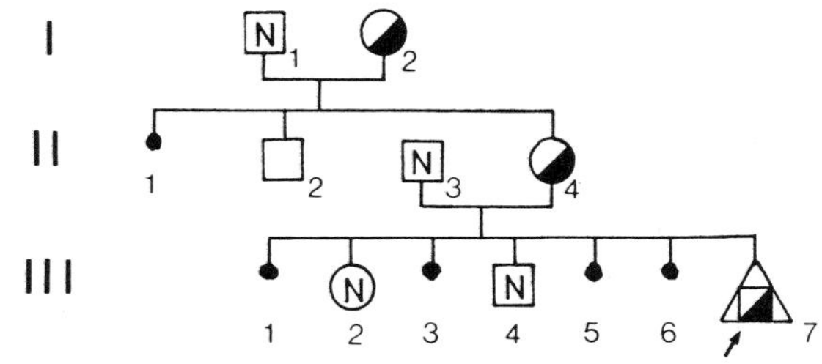

Fig. 5. Pedigree of a family with an inherited t(1;4)(p32;p14) [14]. (See Fig. 2 for pedigree symbols.)

somal imbalance. It will be shown below that the *probabilities for the production* of different types of unbalanced gametes also depend on the breakpoint positions. These probabilities should ideally be estimated on the basis of data for each rcp separately, which in practice means the data from a single pedigree. However, because most pedigrees are very small, this procedure would result in inaccurate estimates.

Instead, we shall try an alternative approach, i.e., grouping rcp that are likely to produce almost the same types of unbalanced offspring and therefore are likely to carry approximately the same risks. Many of these groups will contain sufficient observations to obtain reasonably accurate risk estimates.

To illustrate the grouping procedure, let us return to the t(5;10)(p15;p13) in Figures 1 and 2. The breakpoint in chromosome 5 is in the terminal band p15; that in chromosome 10 is in the third band from the end, p13. We have reason to assume that imbalances of the terminal band do not genetically affect the phenotype. Therefore, the abnormalities of the patients in this pedigree are determined almost exclusively by the trisomy for the distal short arm segment of chromosome 10 [6].

In our data [2], there are a few other rcp with breakpoints in the same band of chromosome 10 (p13) and with the other breakpoint in a terminal band of another chromosome or in a short arm region of an acrocentric chromosome, which are also considered to have no genes affecting the phenotype. In the families with those rcp, usually one type of unbalanced offspring is born: partial trisomy for the short arm of chromosome 10. These abnormal offspring are clinically very similar, which suggests similar survival probabilities. Rarely, offspring with monosomies of the same segment occur. The production probabilities for the nullosomic gametes leading to these types of fetuses are also likely to be approximately equal. Thus we can conclude that for carriers of this group of rcp the risks for abnormal offspring are almost equal, and we can estimate them from pedigree data for all these rcp taken together.

Consider a different breakpoint close to the centromere on 10p, e.g., p11. Translocations with this breakpoint and the other breakpoint in a terminal band of another chromosome will lead to births of abnormal children with a trisomy 10p11→pter. This is a larger segment of 10p than that in the previous group. This larger imbalance of 10p can be assumed to result in offspring who are more seriously affected than those with the smaller imbalance and who have a smaller survival probability. Assuming that the production of gametes is similar, the risks for this latter type of offspring will be smaller for carriers of the rcp with a breakpoint in p11 than for those with a breakpoint in p13.

Such translocations with a breakpoint within the arm of one chromosome and in a terminal or acrocentric short arm region of the other chromosome will be at risk for so-called *single-segment imbalances*. The clinical phenotype for such offspring is relatively similar for those with breakpoints between p11 and p13, so that one can define a "pure" trisomy 10p syndromes [3–6] (Fig. 6). (Similarly, a "pure" monosomy 10p syndrome may be defined [3–5]).

If the breakpoints in both chromosomes are neither in a terminal nor in an acrocentric short arm region, the segments distal to the breakpoints in relation to the centromere, i.e., between the breakpoint and the end of the same arm, are both assumed to contain genetic material that affects the phenotype. Such translocations carry a risk for so-called *double-segment imbalances*. The clinical pictures of those unbalanced offspring will show mixtures of the "pure" syndromes of both segment imbalances (phenotypic hybrids). This means that we can form much larger groups of rcp for single-segment imbalances than for this type of rcp, so the risks for abnormal offpring can generally be more accurately estimated for the former. In fact, reasonably accurate risk estimates could be obtained only for exceptional cases of rcp for double-segment imbalances in our data [2]. Therefore, we confine ourselves mainly to considering rcp at risk for single-segment imbalances. We return to those at risk for double-segment imbalances more briefly at a later stage.

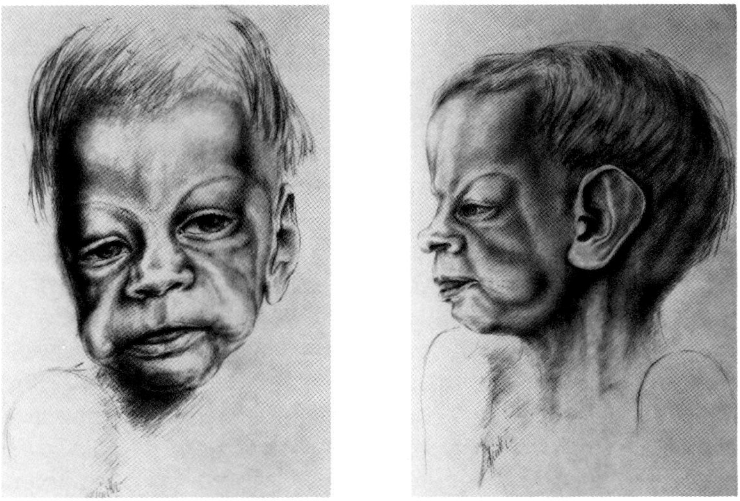

Fig. 6. Artist's rendering of the trisomy 10p syndrome based on photographs of 12 patients. (Drawings performed by Gertrud Schmidt, Munich [6].)

ASCERTAINMENT (ASC) PROBLEMS AND CORRECTIONS

The examples discussed above have demonstrated that the pedigrees can be detected (ascertained) in different ways. In all examples, the detection of some type of abnormality in the offspring led to a chromosome investigation of the parents, which revealed that one of them was a carrier of a balanced translocation causing the abnormality in the offspring. In some cases, chromosome abnormalities are found in the offspring that do not originate from the parental translocation but are results of fresh chromosome mutations. These latter, de novo cases and the risks for such spontaneously occurring chromosome abnormalities are not considered in this chapter.

In our data [2], almost all pedigrees with segregating rcp were ascertained by information about abnormal conceptuses (including some not karyotyped, such as spontaneous abortions), leading to chromosome investigations of the parents and subsequent detection of the familial rcp. Sibships that led to the detection of the familial rcp are called *index sibships*.

As soon as the first carrier parent in a pedigree has been detected, other relatives of this carrier parent will have their chromosomes investigated if they are available and cooperative. Such an investigation will unambiguously demonstrate whether a phenotypically normal individual is a carrier of the rcp (i.e., the detection probability is 100%). Other carriers in the pedigree are traced through a chain of carriers starting with the parent of the first-detected carrier parent. Thus the ascertainment of parental carriers in a pedigree is a two-stage procedure: 1) detection of the pedigree through one or more index sibships and 2) tracing of carriers within the pedigree. We shall consider these two stages separately.

Index Sibships

In our data [2] the index sibships have been detected through one of the following categories of pregnancy outcome: 1) one or more clinically abnormal individuals who have been karyotyped, 2) stillbirths or abnormal children suffering early deaths but not karyotyped, 3) spontaneous abortions (usually two or more), 4) carriers of balanced rcp found through cytogenetic population surveys (e.g., of newborn children) or by chance (independent investigations), 5) fetuses carrying a balanced or an unbalanced rcp found through prenatal diagnosis carried out for reason other than the known presence of a parental rcp (e.g., advanced parental age). The term *index sibship* may be interpreted in different ways depending on which category of ascertainment was the actual one. The asc probabilities will also vary for the different categories.

Categories 1, 2, and 3 are based primarily on clinical investigations of abnormal offspring or abnormal pregnancy outcomes. The probability that the presence of an abnormal child will lead to a chromosome investigation of this child and/or of the parents is likely to be high if the abnormality is severe and well known and if there are good diagnostic facilities available. The chances are small if the abnormality is not very conspicuous or is unknown or if diagnostic facilities are reduced or absent. In that stillbirths are usually (clinically regarded as) less conspicuous than abnormal liveborn infants, the probability will be even lower that a stillbirth will lead to an investigation of the parents. Finally, for spontaneous abortions, the asc probability may be very low. Many rcp families unfortunately incur a large number of spontaneous abortions without being investigated. They have, however, often been ascertained either through an unbalanced child, who in many of these families will have a very low recurrence risk, or through a carrier fetus detected at prenatal diagnosis performed for different reasons.

Categories 4 and 5 represent different types of investigations performed on either liveborn individuals or fetuses not suspected *a priori* of having either a balanced or an unbalanced rcp. Both types of investigation will immediately reveal if the individual or fetus has such a chromosome constitution (100% detection probability). These investigations are fundamentally different from those considered above. Often, they can be considered as population studies.

The probability for a randomly chosen, phenotypically normal carrier having his or her carrier status detected through a cytogenetic study, without any knowledge about the familial rcp, is very small because screening studies are conducted only rarely. Therefore, the asc probability for liveborn, phenotypically normal carriers is very low compared to that for liveborn abnormal offspring. Similarly, the asc of a carrier through prenatal diagnosis also has a very low probability, except perhaps when the parents are old.

The specification of the index sibship may cause some difficulties in these latter ascertainment categories if only some of the sibs have been chromosomally investigated and others not. If one or more children in a sibship have had their chromosomes investigated in a cytogenetic survey or by chance and at least one of these is found to be a carrier or to be unbalanced, we will formally define the index sibship as consisting of only these individuals. The other children in the sibship, not investigated in this way, will be found through their common parents and represent, from the asc point of view, a different sibship.

If prenatal diagnoses have been performed without any knowledge about the parental rcp and one of the fetuses is found to be a carrier or unbalanced, the index sibship will consist of just these prenatally diagnosed individuals, and the other sibs not diagnosed prenatally will be found through their common parents.

When a balanced rcp of the carrier parent has been found through the offspring belonging to one of categories 1–5 mentioned above, then the rcp has been ascertained conditional to the fact that the pregnancy outcomes have shown the actual abnormalities. Carrier parents, who, by chance, have no such abnormal offspring, will not be detected in this way and will not be presented in our data unless they have been detected in a different way. Such asc procedures lead to biased samples, with an overrepresentation of abnormal pregnancy outcomes. To obtain unbiased risk estimates from such data, corrections are necessary. For the categories mentioned above, we shall briefly describe the appropriate corrections.

We assume that asc for a certain category and for the same type of (clinical) abnormality occurred roughly in the same way for all index sibships ascertained for this category. For example, in a number of index sibships, in which the liveborn, abnormal offspring have almost the same chromosomal imbalance, and the syndrome is severe and well known, the asc through category 1 will usually be made through the oldest unbalanced child. On the other hand, if the abnormality is neither particularly conspicuous nor well known, each of the liveborn, abnormal offspring in the index sibships may have the same probability of being a *proband*, i.e., an abnormal case leading to the ascertainment of the parental translocation. (Other terms for proband are *index case* and *propositus*). Empirical studies have indicated that this assumption holds approximately.

Different types of asc will lead to different corrections. To determine the actual method of asc for a group of index sibships, the birth orders of the probands among the abnormals should be studied [15,16].

The asc of an index sibship through categories 1–5 may roughly be described formally by four asc types depending on whether there are one or two probands and whether the probands are the eldest abnormal offspring. In a few cases, there may be more than two probands. Such situations can be treated formally in a similar way. For asc types A and C below, asc corrections are given in the Appendix. Illustrations of the practical use of these corrections have been given elsewhere [17–20].

A. *Ascertainment through a single proband who is the oldest abnormal case*. The appropriate asc correction [5] is to use $A_1(m)$ and $B_1(m)$ in the Appendix and in Stene [17]. This correction should also be used if there is a single abnormal case child in the sibship. The same correction should be used if the asc criterion is the presence of at least one abnormal case disregarding the number of abnormal cases in the sibship [21].

B. *Ascertainment through a single proband; or if there are two abnormal offspring if the older and the younger are equally often the proband; or if there are three abnormal offspring the oldest, the middle, and the youngest are almost equally often the proband.* This means that the probability for an abnormal case to be a proband can be considered to be independent of the birth order among those who are abnormal. The appropriate asc correction is to omit the proband from the data [15] and to carry out the subsequent analyses without the proband.

C. *Ascertainment through two probands and these are the two oldest abnormal offspring.* The appropriate asc correction is to use $A_2(m)$ and $B_2(m)$ in the Appendix [16]. The same correction should be used if the asc criterion is the presence of at least two abnormal offspring in the sibship disregarding the total number of those who are abnormal [22]. Translocation families with no liveborn offspring and multiple abortions have been ascertained through this criterion [23].

D. *Ascertainment through two probands and, if there are three or more abnormals in the sibship, then the birth order among the abnormals and proband status are independent just as for type B.* The appropriate asc correction is to omit both probands from the data [16].

For categories 1–5 above, the asc types seem to be as follows. 1 and 2 (Unbalanced offspring, unkaryotyped stillbirths and early deaths): In our data [2] the asc types generally correspond to B and D. In that the knowledge of chromosomal syndromes and cytogenetic techniques are improving continuously, the asc types will in future include A and C. 3. (Spontaneous abortions): If a pedigree is asc through abortions, this usually is due to reports of at least two. This type of asc corresponds to C. 4 and 5 (Balanced carriers, balanced and unbalanced fetuses at prenatal diagnosis): This ascertainment corresponds to type A; the first individual with a chromosomal abnormality will lead to the ascertainment of the translocation. The index sibship (m) refers here only to the individuals who have been investigated in a cytogenetic survey or to the prenatally investigated fetuses and who have led to an investigation of the parents. When using the asc correction $A_1(m)$ and $B_1(m)$, m will refer to the number of prenatal diagnoses carried out (these pregnancies have been studied in the same, highly accurate way). Previous liveborn sibs are found through their common parents.

To simplify the analysis of our data [2], some modifications of the recommendations given above were necessary, because data processing by computer was not available. Regarding categories 3 and 5, the probands were simply omitted. This results in an overcorrection for asc and hence an underestimation, in particular for category 5, as demonstrated elsewhere

[24]. The consequences of this simplification for the estimation of the proportion of carriers among phenotypically normal fetuses will be discussed below. The estimation of risk for liveborn, unbalanced children, which was our primary aim, is hardly affected by this simplification.

As mentioned above, the asc correction applied in our analysis [2] was to omit the probands from the data. In some instances, it was not clear from the description of an index sibship with more than one abnormal case if the asc had taken place through one or more probands. In such instances, all abnormal cases were omitted in order to avoid an overestimation of the risk. This may lead to an overcorrection, in particular in families ascertained through multiple spontaneous abortions.

To compare the methods discussed above with the widely applied models and methods developed within the framework of "classical segregation analysis," one has to take into account that a basic assumption for those models and methods is *independent ascertainments of probands* [25,26]. Furthermore, in most instances, one assumes that the probability for an abnormal case being a proband is constant for all abnormal cases in a sibship and is independent of the birth order among these cases. Finally, in the way this model has often been applied, the users have often more or less implicitly assumed this asc probability for the individual abnormal cases to be the same for all families [24,26–29].

In our analysis [2], independent asc of probands has not been detected or assumed, the probands have always been reported simultaneously when there are more than one. For asc types A and C above, the individual asc probabilities are, in fact, much higher for the oldest abnormal cases than for the younger ones [15,16]. Furthermore, the methods for asc corrections discussed above do not presuppose the same asc probability for abnormal offspring in different families [15,16,21,22]. Therefore, on these grounds, the present methodology is preferable to "classical segregation analysis" [24,27–29].

Occasionally, a carrier parent is found directly through a cytogenetic study of adults or by chance, and the carrier status is determined without any reference to his or her offspring. If thereafter, the offspring are studied and not the parents of this carrier, no asc correction should be performed.

Tracing of Carriers

Carriers in a pedigree other than the carrier parents of probands are traced through a chain of carriers. The usual asc types concern the following two: 1) A carrier is traced because of a proband (or probands) in direct line of descent (index carrier); 2) A carrier is traced through the family but has no proband in direct line of descent (non-index carrier). In case 1, the appropriate ascertainment correction is to omit the index carriers from the sibships they belong to [19,30]. In case 2, no asc correction is necessary in the sibship of the nonindex carriers [19,30].

As is shown in Figure 2, as an example, II-2 is an index carrier in the progeny of her mother (I-1) and has to be omitted from her sibship (II-2–7). In Figure 3, there are in fact two index carriers (II-2 and II-6); the asc actually took place through the progeny of these two carriers. Both individuals have to be omitted from the progeny of their carrier father I-1 (II-2–8). In Figure 4, the carrier grandmother, I-4, was traced through simultaneous information from balanced sister I-2 and several children with the balanced translocation (II-2, II-5, II-6, II-10, and II-11). For asc correction, only individual II-5 has to be omitted from her sibship; in her progeny, there is a proband (III-6). The other balanced carriers in this sibship have no proband in direct line of descent and have to be retained, and the progeny of the carriers II-6, II-10, and II-11 can be used uncorrected.

If a carrier is unavailable for cytogenetic investigation and has been traced through information from relatives carrying the familial translocation, further modifications of asc corrections might be considered. As was mentioned above, the detection probability for these carriers is 100% if a chromosome analysis is carried out. The methods for pedigree analysis developed within the framework of classical segregation analysis are developed for other modes of inheritance, e.g., monogenic and polygenic modes, taking into account reduced penetrance and delayed onset, and for other modes of asc [31–34]. These methods are unnecessarily complicated for the present type of data and are also based on assumptions that are not satisfied here [24].

When the different types of asc corrections regarding the index sibships and the carriers in the pedigrees have been performed, the pedigrees can be split up into separate nuclear families each with a carrier parent. These nuclear families can be treated as independent families [19,30].

ANALYSIS OF DATA

The theoretical bases for the analyses discussed in the following section have been developed through a series of papers by the present authors and our colleagues [2,19,35–39]. After each pedigree has been corrected for asc and divided into nuclear families, as described in the preceding section, the families are grouped according to breakpoint positions as described above under Grouping of rcp for Risk Estimation. If we confine ourselves to rcp carrying risk for single-segment imbalances after a 2:2 disjunction with adjacent-1 segregation, all families with a breakpoint in the same band on the chromosome arm responsible for the abnormality will be considered as a single group. To determine if the segregation pattern is homogeneous for the different families, homogeneity tests were carried out hierarchically, as was done in our previous studies [17–20]. The risk estimations discussed below are based on such groups for which the segregation patterns are assumed to be approximately homogeneous. If, for a single band, there are too few observations to obtain reasonably accurate estimates, data for adjacent bands have been pooled.

In our study [2], the following categories of risk estimates for abnormal offspring were considered: 1) for unbalanced offspring at birth, most of them karyotyped, and some others with sufficient clinical evidence for including them in this category of unbalanced offspring; 2) for unkaryotyped, stillborn offspring and those who died early after birth; and 3) for spontaneous abortions. For all risk estimates, the total number of recorded, terminated pregnancies consists of the sum of the numbers in these three categories after asc corrections and the numbers of liveborn, phenotypically normal children, but not including the pregnancies prenatally diagnosed. As the accuracy of the different categories of observations varies considerably, objections can be raised to this procedure. The problems have been discussed elsewhere [2]. The proportions of unbalanced fetuses of prenatally investigated pregnancies are estimated separately.

In some cases, a number of abnormal offspring may have been reported in a group of families, but, if all are probands, we are left with no abnormal cases after asc corrections. For these cases, an upper limit for the risk is estimated by a formula given in the Appendix section on risk estimation.

ESTIMATION OF RISKS FOR SINGLE-SEGMENT IMBALANCES

Consider once again the group of translocations carrying risk for single-segment imbalances for the short arm of chromosome 10. These are illustrated in the upper part of Figure 7. The diagram of the chromosome 10 on the left is from the 1978 version of the standard nomenclature [8], which in its simplification of the bands and regions reflects the average quality of the breakpoint localizations in our data [2]. In Figure 7, each number at the top represents a pedigree with an rcp breakpoint in the short arm of chromosome 10. References to the authors or the laboratories that provided these rcp are given in Stengel-Rutkowski et al. [2]. The different symbols are explained in Figure 8: A white vertical bar indicates that a liveborn child with a trisomy for the actual 10p segment has been recorded; a black bar represents a child

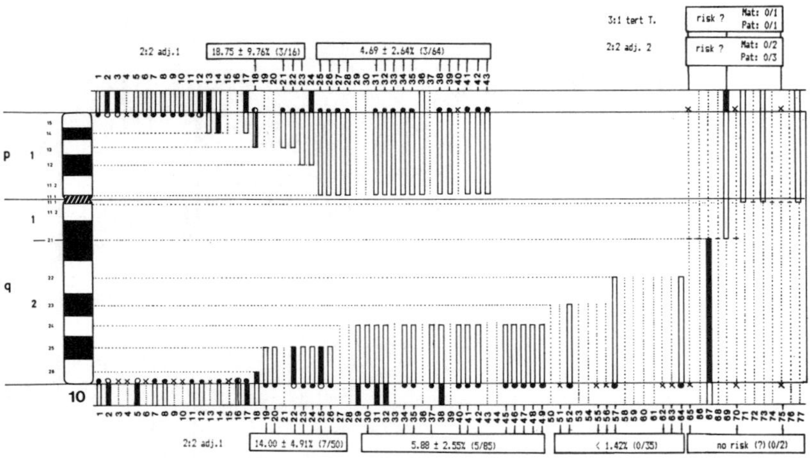

Fig. 7. Diagrammatic representation of translocations involving chromosome 10. The symbols are explained in the legend to Figure 8.

with a monosomy for the segment. Single-segment imbalances can be recognized by small circles at the top, indicating unbalanced terminal or acrocentric short arm segments from the recipient chromosomes. Open circles means trisomy, solid circles monosomy. A dotted line with a × at the top indicates a family at risk for single-segment imbalance with no records of liveborn, chromosomally unbalanced offspring. Those rcp with bars or dotted lines at the top (instead of circles or crosses) represent rcp in which the breakpoint in the recipient chromosome is in a subterminal or an interstitial band and thus carry risk for double-segment imbalance. These rcp are discussed later.

Consider rcp No. 18 at the top of Figure 7. This is a t(1;10)(q44;p13) translocation. The black and white bars indicate that both single-segment trisomy for the segment 10p13→pter and single-segment monosomy for the same segment have been observed in this family, similar to what we have seen in example 1 given in the section on further examples of rcp families.

Consider the group of rcp carrying risk for single-segment imbalances, i.e., with a circle or a × at the top of Figure 7, with breakpoints in 10p11. These are the pedigrees 25–28, 31–35, 38–43.

For this group of rcp, 64 pregnancies have been recorded after ascertainment corrections, i.e., after omission of probands and carriers with a proband in direct line of descent. Three of these 64 pregnancies were karyotyped, abnormal cases. This gives a risk estimate determined as the proportion 3/64

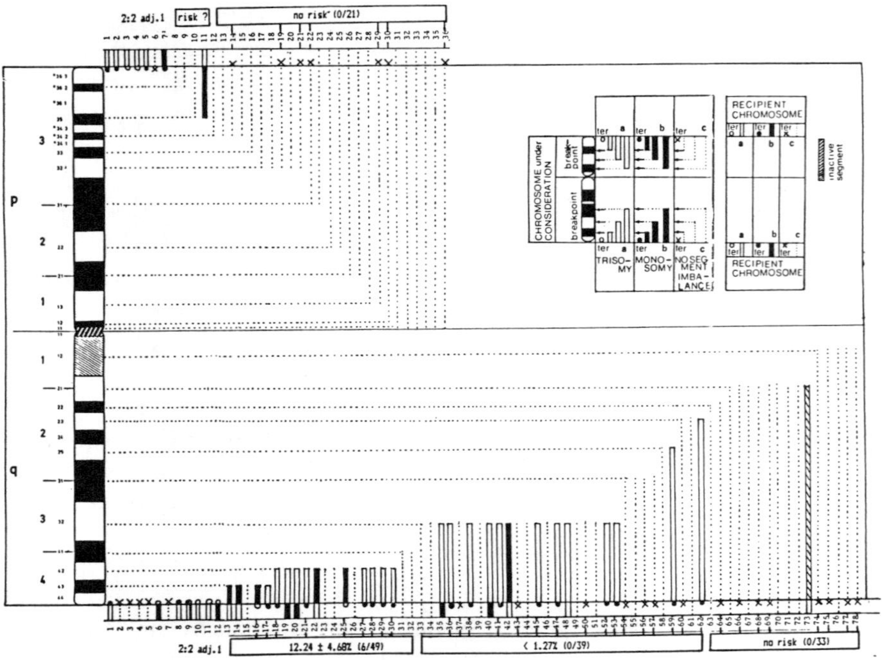

Fig. 8. Diagrammatic representation of translocations involving chromosome 1, which is shown to the left in the ISCN (1978) diagrammatic representation [8]. The arabic numbers at the top refer to translocations (listed in Stengel-Rutkowski et al. [2]) with a breakpoint in the short arm of chromosome 1, those at the bottom having a breakpoint in the long arm. Vertical white bars to the right of the chromosome refer to recorded, chromosomally unbalanced, liveborn offspring with trisomies for the indicated segments, black bars to liveborn offspring with monosomies. Hatched bars indicate that the recipient chromosome is an X and that X-chromosome inactivation attenuated the phenotypic effect of the resulting trisomy. Circles refer to imbalances for a terminal or an acrocentric short arm segment; open circles represent trisomies, solid circles monosomies. The bars and circles above the top line and below the bottom line refer to imbalances of the recipient chromosome (to which the above-mentioned segments have been translocated). Dashed lines and crosses indicate that no liveborn, imbalanced offspring have been recorded.

= 4.7%, with a standard deviation of 2.6%. Details are given in Stengel-Rutkowski et al. [2].

Now consider more distal breakpoints in the short arm of chromosome 10. For a breakpoint in p12, there is only a single pedigree with a translocation leading to abnormal offspring with single-segment imbalances, i.e., pedigree 23, t(7;10)(p22;p12). For a breakpoint in p13, there are three pedigrees leading to single-segment imbalances. These are pedigrees 18 (mentioned above), 21 (see Figs. 1 and 2), and 22 [t(10;18)(p13;q23)]. To have sufficient observations to obtain a reasonably accurate estimate of the risk for abnormal offspring, the data for these two adjacent segments, p12 and p13, are pooled. After asc corrections, we are left with 16 pregnancies, three of which are unbalanced liveborn cases, both trisomies and monosomies. This gives a risk estimate of 3/16 = 18.8%, with a standard deviation of 9.8%. Thus, for the carriers we considered in Figures 1 and 2, the risk for an unbalanced newborn child is estimated to be 18.8%. At the top of Figure 7, these risk estimates are shown in blocks with vertical lines down to the pedigrees that provided the data for the risk estimates.

For breakpoints in 10p14, there is no pedigree with a translocation for single-segment imbalance in these data. The pedigrees 1–12 have rcp with breakpoints in p15, which is considered to be terminal, i.e., distal to a syndrome-specific segment of 10p. These rcp are supposed to be without risk for any abnormality caused by a chromosome 10p segment imbalance. It can be noticed that the estimated risk is higher for the shorter segment imbalance than for the longer one, in accordance with what we had expected.

The same pattern emerges for rcp with a breakpoint in the long arm of chromosome 10 and leading to abnormal offspring with a single-segment imbalance for the segment from the breakpoint to 10qter. For those with a breakpoint in 10q25, the estimated risk is 14.0%. For those with a breakpoint in 10q24, the corresponding value is 5.9%. For those with breakpoints in q23 and q22, the data have to be pooled. After asc corrections, we are left with no liveborn, imbalanced offspring among 35 pregnancies, although some of the probands were observed in this category of offspring. The risk is estimated to be below 1.4%

Among rcp with breakpoints in 10q21 and 10q11, there are a few that theoretically could have led to abnormal offspring with a single-segment imbalance for the longarm segment of 10q from the breakpoint to 10qter. However, no such liveborn, abnormal cases have been recorded, and there may be no risk for abnormal liveborns of this type.

Consider, then, chromosome 1 in Figure 8. For the long arm, the same pattern emerges as was found for chromosome 10: A breakpoint in 1q42 or in 1q43 results in a risk estimate of 12.2% (for abnormal offspring with a single-segment imbalance, whereas, with a breakpoint between 1q32 and 1q23, the estimated risk will be below 1.3%. For breakpoints more proximal to the centromere, there seems to be no risk for this type of abnormal liveborns.

For the short arm of chromosome 1, the picture is different. With subterminal or more proximal breakpoints, no abnormal liveborns with a single-segment imbalance have been recorded, although there are a few pedigrees in which such abnormal cases theoretically could have been born. One has to conclude that there does not seem to be any associated risk.

Risk estimates for single-segment imbalances after 2:2 disjunction with adjacent-1 segregation, such as those shown for chromosomes 10 and 1 in Figures 7 and 8, respectively, have been calculated for all chromosome arms [2].

GENERALIZATIONS FOR SINGLE-SEGMENT IMBALANCES AFTER 2:2 DISJUNCTION WITH ADJACENT-1 SEGREGATION

The types of liveborn, unbalanced offspring so far considered are all the result of 2:2 disjunction with adjacent-1 segregation, and this pattern will be considered more formally below. By considering imbalances for segments in all chromosome arms as above, some general patterns emerge [2]. The magnitude of the associated risk seems to be dependent on the genetic content of the segment under consideration; unbalanced segments of the same length in different chromosomes may carry different risks. Generally though the risk for a long imbalanced segment of a chromosome is likely to be lower than the risk for a shorter part of it (see also [40]). This pattern is demonstrated in Figure 9, where risks for short and long segments have been compared for those chromosome arms for which sufficient data are available. The differences are quite substantial for many chromosome arms. The only exception is the short arm of chromosome 12, for which the inverse pattern may be a result of random variations because of limited numbers of observations.

The results may be summarized in a map of the human genome, as shown in Figure 10. Five risk groups (2:2 disjunction with adjacent-1 segregation) have been used—above 20% (very high risk), 10–20% (high risk), 5–10% (medium risk), below 5% (low risk), and no risk—to produce a picture in reasonable accordance with the accuracy of the data.

In genetic counseling, the total burden of unsuccessful pregnancies, i.e., the sum of unbalanced offspring at birth, stillbirths and spontaneous abortions, is a matter of considerable interest. In Figure 11, estimates of the total burden as a proportion of all pregnancies are shown for four groups of rcp grouped after the magnitude of the estimated risk for abnormal offspring with a single-segment imbalance after a 2:2 disjunction with adjacent-1 segregation. The total burden tends to decrease with decreasing live birth risk but is still high for no (live birth)-risk rcp, indicating that a large number of spontaneous abortions may occur in these families. It should be noted that

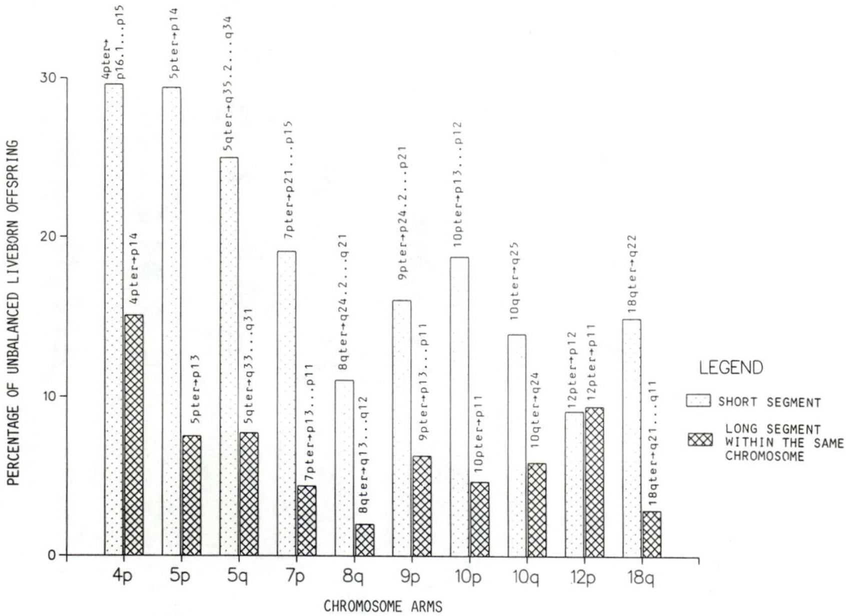

Fig. 9. Estimated risks for liveborn, unbalanced offspring after a 2:2 disjunction with adjacent-1 segregation and single-segment imbalances for short and for long segments within the same chromosome arms. Larger segments are usually associated with a smaller risk.

the data presented in Figure 11 are average values based on all translocations within the risk groups and that there are large variations in the total burden for different rcp. Figure 11 indicates only a general trend.

RISKS FOR MONOSOMIES AND FOR TRISOMIES

In the estimation of risks for liveborn abnormal cases with single-segment imbalances, no distinction has been made between the type of imbalance (whether it is trisomic or monosomic). There are three pragmatic reasons for this. First, the aim has been to estimate the risk for any type of liveborn abnormal offspring without further specifications, because this is the main concern in genetic counseling. Second, a further division of the data would have led to smaller groups and less accurate data. Third, in some families, both types of abnormal cases may occur as results of the same meiotic segregation: 2:2 disjunction with adjacent-1 segregation. Thus the risks

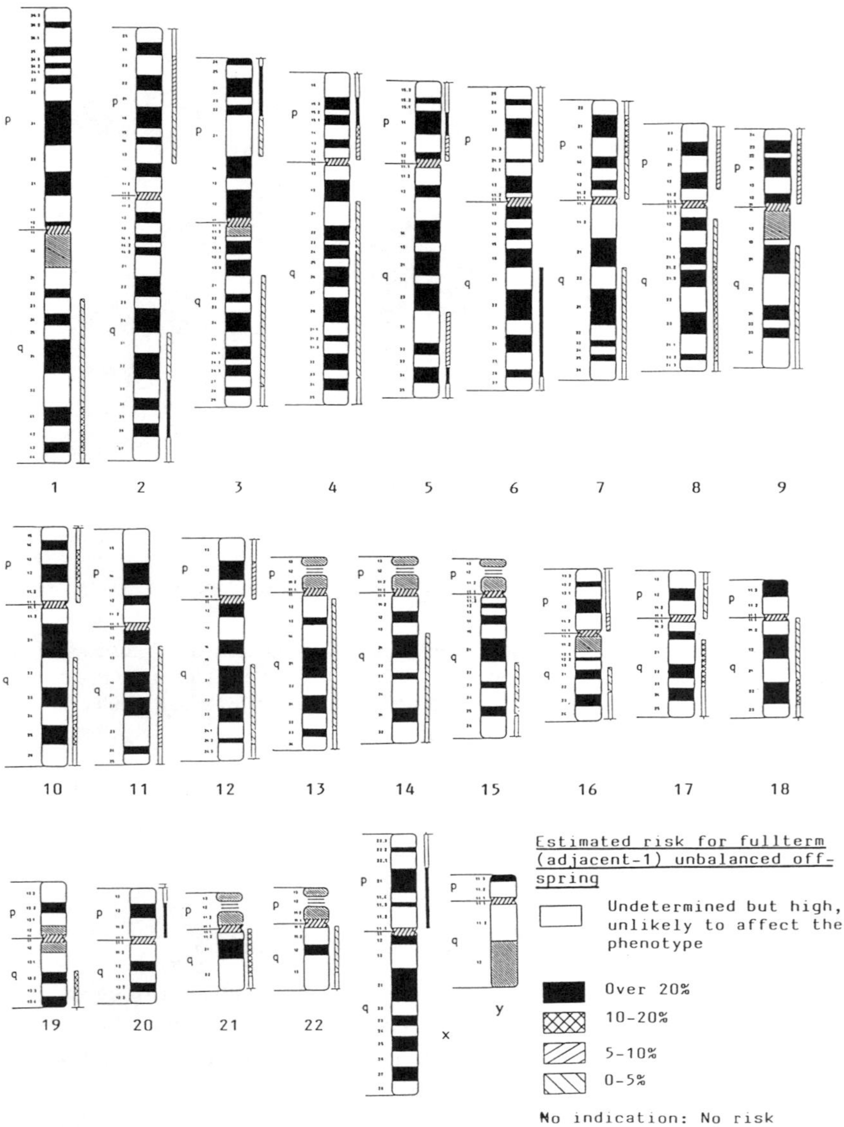

Fig. 10. Risk estimates for full-term, unbalanced progeny for different regions of the human genome. Diagram shows the connections between breakpoint positions and estimated risks for liveborn, unbalanced offspring after a 2:2 disjunction with adjacent-1 segregation and single-segment imbalances from the breakpoint to the end of the chromosome arm. White bars at the end indicate undetermined but usually high risks but with no detectable effect of the imbalance on the phenotype (no risk). This also applies to breakpoints in the short arms of the acrocentric chromosomes. The regions without any bar are no-risk regions for this type of unbalanced, liveborn offspring.

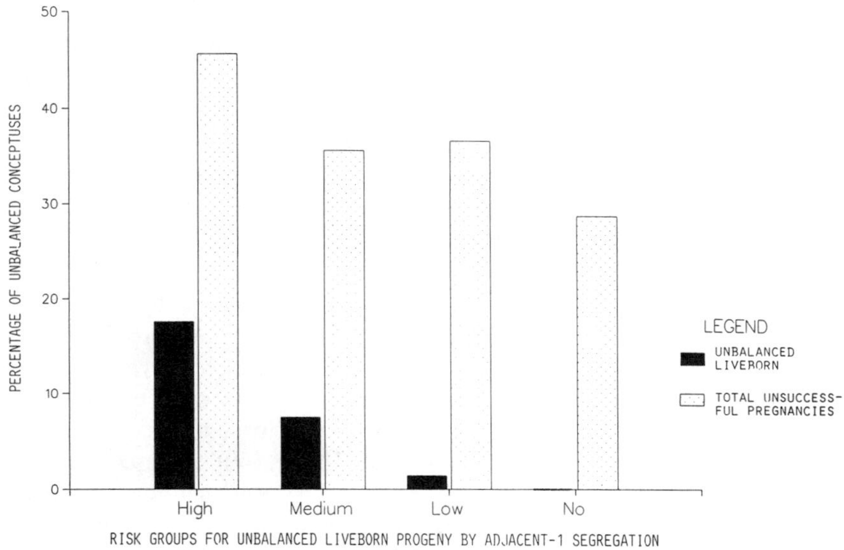

Fig. 11. Ratio of liveborn unbalanced offspring and of total abnormal conceptuses to total number of pregnancies in rcp risk groups. Rcp carrying risk for unbalanced offspring after 2:2 disjunction with adjacent-1 segregation and single-segment imbalance are considered. The rcp are grouped after the risk for full-term, unbalanced offspring: high (≥10%), medium (5–10%), low (0–5%), and no apparent risk and total percentage of unsuccessful (including unbalanced, full-term) pregnancies for families belonging to the same risk groups. Note that the proportion of (recognized) unsuccessful pregnancies is relatively similar in the four risk groups.

discussed so far are considered as risks for either type of liveborn abnormal offspring as a result of this type of disjunction and segregation.

With regard to the relative risks for an imbalance with a single-segment monosomy or the corresponding trisomy, some general patterns emerge from the data. By considering the single-segment imbalances for chromosomes 10 and 1 in Figures 7 and 8, respectively, it may be noticed that rcp for which abnormal cases with a monosomy for the actual segment occur (black vertical bar) are much more frequent when the segment is short than when it is long. From clinical experience, we know that unbalanced offspring with a monosomy for a given segment are generally more seriously affected than those with a trisomy for the same segment. If an imbalance is extended beyond a certain length within a single chromosome, the only abnormal offspring recorded have a trisomy for the segment. This means that the survival probability, and hence the birth risk for a fetus with a monosomy, is generally lower than that for one with a trisomy although both types originate from the same meiotic division.

The relatively high frequency of some pedigrees with records of partially monosomic offspring for a short segment may, to some extent, be a reflection of the fact that such offspring are more seriously affected than those with the corresponding trisomy and hence may be more conspicuous. Thus pedigrees in which offspring with such a monosomy have been born may have a higher probability of being ascertained than those in which only cases with a trisomy for the same segment have been born.

This is certainly the case for pedigrees with an rcp for single-segment imbalance of the short arm of chromosome 5. The monosomy for segments of this arm when the breakpoint is in 5p151 or is more proximal to the centromere corresponds to the well known cri-du-chat syndrome [12], whereas cases with a trisomy for the same segment may be difficult to recognize clinically [11]. The higher viability of trisomies compared with that of monosomies has been pointed out by Daniel [40] previously.

The consequence that the risk is higher for a trisomy than for a monosomy of the same segment has to be taken into account when counseling families in which only one of the types may occur at birth. An inspection of the actual diagram [2] (like Figs. 7 and 8) will reveal whether the estimated risk is based only on monosomies or only on trisomies or on a mixture of both types. In the last case, the actual risk for trisomic offspring will be higher than the estimated, average risk and lower for monosomic offspring.

SINGLE- VERSUS DOUBLE-SEGMENT IMBALANCES

So far, only rcp in which single-segment imbalances are paramount have been considered. Such rcp have a terminal breakpoint (bp) in the recipient chromosome. When the breakpoint in the recipient chromosome is neither in the terminal band nor in the short arm of an acrocentric chromosome, all types of abnormal segregants will have imbalances of two chromosome segments. In this section such translocations, with which there is a risk for abnormal offspring with double-segment imbalances after 2:2 disjunction with adjacent-1 segregation, will be considered.

Because both involved chromosomes and breakpoint positions have to be taken into account, it is generally difficult to form groups with sufficient numbers. In many cases the group consists only of a single pedigree with a rather limited number of pregnancies. Although reasonably accurate risk estimates cannot be obtained by direct estimation, an approximation can be made by utilizing the available data in different ways [2].

In these double-segment imbalances, the clinical abnormalities will be a mixture of the two "pure" abnormalities [39] but may also be the result of an interaction between the effects of the genetic content of the two separate imbalances. The variation of the phenotypic effect seems to be large. Little work has been done on interactions between different chromosome imbalances, and especially it has not yet been studied whether such interactions may result in more or less seriously affected individuals, as concerns survival, than one would expect from the effects of the two "pure" abnormalities.

A rule that seems to hold in most cases is that there is a smaller risk for a child with a double-segment imbalance than the least of the two risks for the single-segment imbalances considered separately, often considerably less. This general rule may be illustrated by the example of chromosome 4p (Fig. 12), which is the only chromosome arm for which there is sufficient data for the comparison [2]. In this figure, the risk for unbalanced offspring at birth is contrasted with the total burden of unsuccessful pregnancies for single (S)- and double (D)-segment imbalances of short, medium, and long segments of

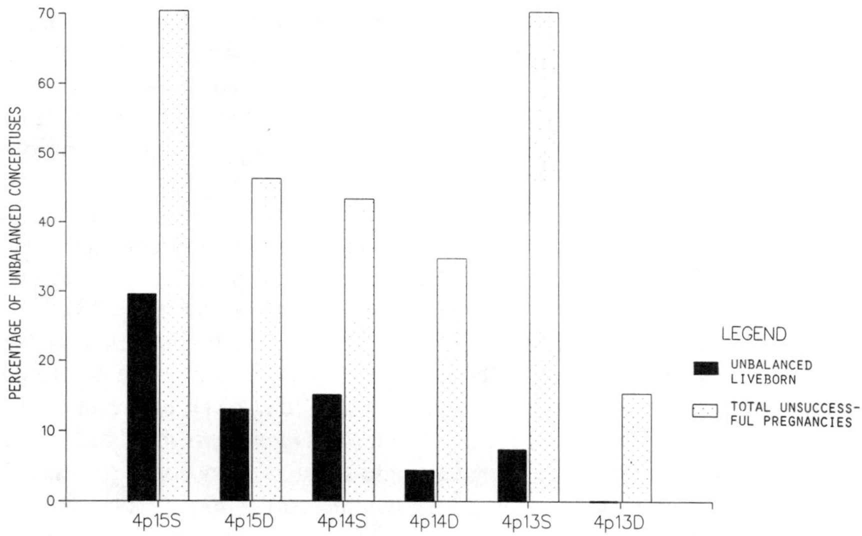

Fig. 12. Risks for unbalanced, liveborn offspring and the total percentage of unsuccessful (including unbalanced, full-term) pregnancies for translocations carrying risk for single-segment (S) and double-segment (D) imbalances after 2:2 disjunction with adjacent-1 segregation. The rcp all have a breakpoint in the short arm of chromosome 4. Risks and total burden are to be compared for S and D for translocations resulting in short (4pter→4p15), medium (4pter→4p14), and long (4pter→4p13) imbalances of chromosome 4.

4p. The estimated average risks for the double-segment imbalances are, in fact, lower than the risks for the corresponding single-segment imbalances.

In a few special cases, however, the risk for unbalanced offspring at birth seems to be higher for double-segment imbalances than for both single-segment imbalances. This points to the possibility of the above-mentioned interactions. The most important category of translocations of this type is those in which the recipient chromosome is an X. Here, because of spreading of X-chromosomal inactivation to the autosomal segment [41], a trisomy of a larger segment of an autosomal chromosome arm may be more compatible with survival until birth than the largest single-segment trisomy of the same arm. An example is shown in Figure 8 for the long arm of chromosome 1. The longest single-segment 1q imbalance compatible with life is of 1qter→1q23. However, in pedigree 73, the t(1;X)(q21;q22) is inherited, and here a trisomy of 1qter→1q21 is tolerated.

Another hypothesis not proved so far is that the reduced survival probability of an autosomal monosomy can, to a certain extent, be compensated for by an additional autosomal trisomy and vice versa; e.g., if one breakpoint is on 5p, a monosomy for a part of this arm often can be tolerated jointly with a larger imbalance of the recipient chromosome better than in the single-segment case (E. Niebuhr, personal communication). An example is the rcp t(5;15)(p14;q21) (No. 46 on 15q in [2]), which resulted in viable offspring with a monosomy for 5p14pter→pter and a trisomy for 15q21→qter. However, the largest long arm (single-segment) trisomy observed for chromosome 15 in a liveborn child is 15q22→qter. This possibly special effect of monosomies for 5p may also be demonstrated through a comparison such as that done for 4p. For rcp with a breakpoint in 5p14, a slight increase is observed both in estimated risk and in total burden for double-segment imbalances compared with those for the single-segment ones (Fig. 13).

With regard to chromosomes 10 and 1, illustrated in Figures 7 and 8, there are three cases in which a similar interaction effect might be described. Among rcp with a 10q breakpoint rcp 67, t(5;10)(q22;q21), has resulted in a liveborn infant with a trisomy for 5q22→qter and *monosomy* for 10q21→qter. Among liveborn infants, the longest single-segment monosomy recorded is 10q25→qter, and the longest single-segment trisomy is 10q22→qter. Among rcp with a 1p breakpoint, rcp 11, t(1;20)(p35;p11), has resulted in a liveborn infant with monosomy 1p35→1pter and trisomy 20p11→20pter, although, for chromosome 1p, no single-segment imbalance has been recorded. Finally, among rcp with a 1q breakpoint, rcp 42, t(1;6)(q32;q26), has resulted in a liveborn child with monosomy 1q32→qter and trisomy 6q26→qter, whereas the longest, single-segment monosomy recorded in a liveborn child is

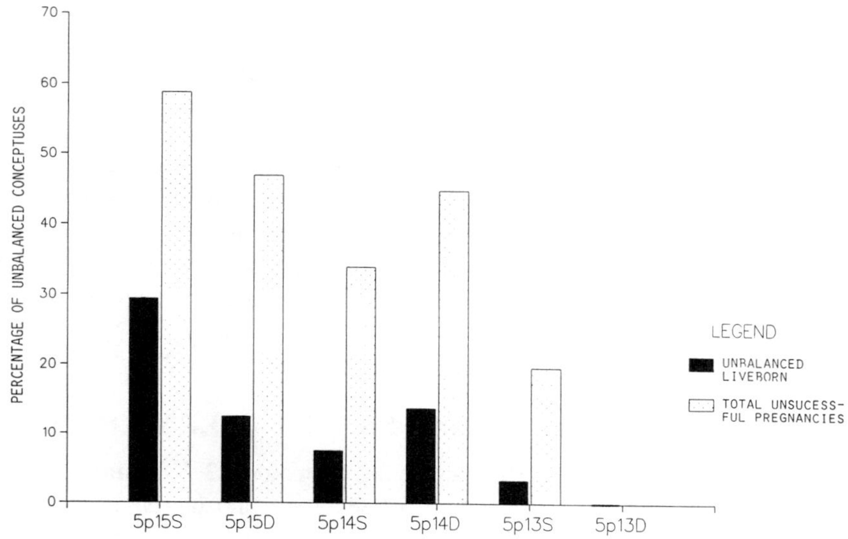

Fig. 13. Comparison (similar to Fig. 12) of translocations with one breakpoint in the short arm of chromosome 5 similar to those of chromosome 4 in Figure 12.

1q42→qter. A few similar cases can be found for other chromosome arms in our data [2], but, because no systemic study of phenotypic effects and survival in double-segment imbalances has been undertaken so far, it remains open whether such interactive genetic effects are a general phenomenon. Another explanation for this apparently discrepant pattern may be that a trisomy of a certain (small) segment is phenotypically more conspicuous when it is combined with a monosomy than when it occurs as a single-segment imbalance.

To determine the magnitude of the risk for liveborn, abnormal offspring after a 2:2 disjunction with adjacent-1 segregation with a double-segment imbalance the following guidelines may be useful. 1) Consider each of the possible single-segment imbalances separately. 2) Estimate the risk for liveborn offspring for each single-segment imbalance separately by means of the chart shown in Figure 10 or the more accurate estimates provided by Stengel-Rutkowski et al. [2]. 3) A first approximation of the risk estimate is to indicate it as being below the smaller risk of both single-segment imbalances. 4) A second approximation of the risk estimate may be to use a rule of thumb [2], i.e., to use as a risk estimate *one-half* of the lower risk of both single-segment imbalances, as is suggested by Figure 12. (A thorough study of this figure suggests a risk reduction factor between 2 and 3. Our recommendation of the factor 2 may result in an overestimation of the risk for double-segment imbalances). 5) Consider the chromosomes involved in the rcp in Figures 1–23 (corresponding to all chromosomes individually) of Stengel-Rutkowski et al. [2]. Examine these figures for the relevant single-segment imbalances and whether double-segment imbalances of the rcp under consideration have been observed and whether single-segment imbalances have been observed only for shorter segments. The latter observation could indicate interactions of the type mentioned above. The study of the diagrams may also indicate that, although both single-segment imbalances may be compatible with live birth, the actual double-segment imbalance is not. These guidelines will need to be revised when further data become available.

MEIOTIC 2:2 DISJUNCTION: DIFFERENT SEGREGATION TYPES

So far, we have considered the results of meiotic 2:2 disjunction with adjacent-1 segregation. The majority of recorded unbalanced offspring at birth are the result of this type of division, and many segments of almost all chromosome arms are involved. For the majority of these families at risk, only one type of segregation will lead to unbalanced offspring compatible with survival until birth. Therefore, the rcp can be grouped according to this "preferred" type of segregation. Finally, there are sufficient data to demonstrate a wide variation in risks for abnormal offspring after 2:2 disjunction with adjacent-1 segregation. These are the reasons we have considered some practical and clinically relevant aspects of this meiotic segregation in some depth before undertaking discussions of this and other segregation patterns and their consequences.

During gametogenesis, for the carrier of an rcp, the complete pairing of homologous segments will result in a quadrivalent, instead of the two bivalents for the normal chromosome pairs. Figure 14 shows a diagram of such a quadrivalent, with the two normal chromosomes shown as black and white. The diagram is somewhat simplified in that the chromosomes are double-stranded at this stage in the meiotic division. More detailed diagrams are shown in Stene and Stengel-Rutkowski [38].

As a result of the first meiotic division, the four chromosomes of the quadrivalent, identified by their centromeres, may be allocated to the two daughter cells in different ways. The numbers of these chromosomes allocated to the two daughter cells define the different types of disjunctions: 4:0, 3:1, and 2:2 disjunctions. In Figure 14 also, the three possible types of 2:2 disjunctions are shown. These are called *alternate segregation*, *adjacent-1 segregation*, and *adjacent-2 segregation*.

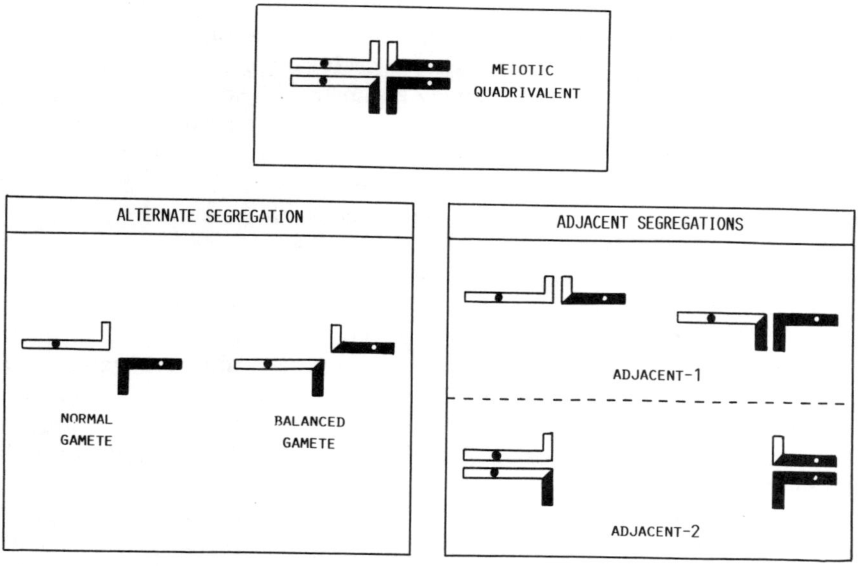

Fig. 14. Diagram of a pachytene quadrivalent for an rcp between two nonacrocentrics with breakpoints on both long arms. There are three types of segregants by 2:2 disjunction, i.e., with two chromosomes in each daughter cell at the first meiotic division for which the two types of gametes are shown for each segregation type. *Alternate segregation:* Gamete with the normal chromosomes (to the left) and the gamete with the balanced translocation (to the right). *Adjacent-1 segregation* (separation of homologous centromeres): Two types of gametes for which the imbalanced region does not include a centromere. After fertilization by a normal gamete, the gamete to the left will incur a partial trisomy for the white end segment and a partial monosomy for the black end segment, whereas the gamete to the right will incur a partial trisomy for the black and a partial monosomy for the white end segment. *Adjacent-2 segregation* (non-separation of homologous centromeres): Two types of gametes, in which both imbalanced regions include a centromere. After fertilizaton by a normal gamete, the gamete to the left will incur a partial trisomy for the complete short arm and the proximal long arm region of the white chromosome and a partial monosomy of the complete short arm and the proximal long arm of the black chromosome, and vice versa for the gamete to the right: The trisomy and the monosomy will be interchanged.

Through alternate segregation, the two daughter cells will receive either both normal chromosomes or both derivatives, i.e., the balanced rcp. After fertilization by a gamete with normal chromosomes, either zygote will develop into a phenotypically normal individual. The properties of this division will be considered later.

The two other segregation types will result in chromosomally unbalanced offspring. Adjacent-1 segregation is defined by a separation of the chromosomes with the same (homologous) centromeres. Each of the two daughter cells, and subsequently after a second meiotic division the gametes, will have one segment from a breakpoint to the end of the chromosome arm in duplicate and will lack the segment from the other breakpoint to the end of the chromosome arm. After fertilization by a gamete with normal chromosomes, the possible zygotes will have a partial trisomy and a partial monosomy as indicated in the figures. The unbalanced segments resulting from an adjacent-1 segregation do not include any centromere. Adjacent-2 segregation is defined by a nonseparation of the chromosomes with the same (homologous) centromeres. Each of the two daughter cells, and subsequently after a second meiotic division the gametes, will have in duplicate a segment from a breakpoint (usually on the long arm of the chromosome) over the centromere to the end of the opposite arm (usually the short arm) and will lack the corresponding segment from the other chromosome. After fertilization by a gamete with normal chromosomes, the resulting zygotes will have a comparatively large partial trisomy and partial monosomy as indicated in Figure 14. The unbalanced segments originating from an adjacent-2 segregation will both include a centromere.

Translocations carrying risk for an unbalanced child after an adjacent-2 segregation are usually characterized by breakpoints in the proximal long arms of both chromosomes. Furthermore, an unbalanced segment of the complete short arm including the proximal long arm has to be compatible with survival until birth. This is not possible for all chromosomes. Monosomies of the complete short arm including proximal long arm segments seem to be compatible with live birth only for acrocentric and subtelomeric chromosomes and for the X chromosome. Trisomies of the complete short arm and proximal long arm segments are compatible with live birth for only some chromosomes (see Figure 16 below). The risks for unbalanced offspring after a 2:2 disjunction with adjacent-2 segregation will be considered in the next section.

As was mentioned above, the diagram of the quadrivalent is simplified by disregarding the fact that the chromosomes are double-stranded. Furthermore, we have disregarded the possibility of crossing-over between the centromeres and the breakpoints (the interstitial regions). One result of such crossing-over is that gametes produced through an alternate or an adjacent-1 segregation cannot be distinguished [38]. Another result is that four additional gametes can be produced after adjacent-2 segregation and crossing-over in the interstitial segments [42]. These gametes however, are so heavily unbalanced that they are unlikely to develop into liveborn offspring.

RISKS FOR LIVEBORN, ABNORMAL OFFSPRING AFTER A 2:2 DISJUNCTION WITH ADJACENT-2 SEGREGATION

Liveborn offspring after a 2:2 disjunction with adjacent-2 segregation can occur in families with certain types of rcp. In most of these families, as was observed previously, only one type of abnormal liveborn child has been recorded. Consider once again rcp with a breakpoint on chromosome 10q (Fig. 7). Pedigree 69 concerns the t(10;21)(q21;q21). Here an unbalanced offspring was trisomic for the segment 10q21→pter and monosomic for the segment 21q21→pter; i.e., both unbalanced segments include the centromere. This observation demonstrates that a complete duplication of 10p together with the proximal segment of the long arm to 10q21 is compatible with survival until birth together with a monosomy for the complete short arm and the proximal long arm of chromosome 21 down to q21. The rcp in our data [2] with records of unbalanced offspring at birth after a 2:2 disjunction with adjacent-2 segregation are listed in Tables I and II. In a few cases,

TABLE I. Translocations with one nonacrocentric chromosome and recorded unbalanced offspring at birth after a 2:2 disjunction and adjacent-2 segregation

Translocation	Trisomy	Monosomy
t(4;15)(q12;q13)	4q12→pter	15q13→pter
t(9;13)(q13;q12)	9q13→pter	13q12→pter
t(9;13)(q22;q12)	9q22→pter	13q12→pter
t(9;14)(q21;q21)	9q21→pter	14q21→pter
t(9;15)(q13;q11)	9q13→pter	15q11→pter
t(9;15)(q13;q11)	9q13→pter	15q11→pter
t(9;21)(q13;q11)	9q13→pter	21q11→pter
t(9;21)(q13;q11)	9q13→pter	21q11→pter
t(9;22)(q13;q11)	9q13→pter	22q11→pter
t(10;21)(q21;q21)	10q21→pter	21q21→pter
t(12;21)(q12;q11)	12q12→pter	21q11→pter
t(17;X)(q21;q13)	17q21→pter	Xq13→pter
t(20;22)(q11;q11)	20q11→pter	22q11→pter
t(20;22)(q11;q11)	20q11→pter	22q11→pter
t(21;X)(q21/22;q22/23)	Xq22/23→pter	21q21/22→pter

TABLE II. Translocations with two acrocentric chromosomes and recorded unbalanced offspring at birth after a 2:2 disjunction and adjacent-2 segregation

Translocation	Trisomy	Monosomy
t(13;14)(q12;q22)	14q11→pter	13q12→pter
t(13;15)(q32/33;q14/15)	13q32/33→pter	15q14/15→pter
t(13;21)(q14;q11)	13q14→pter	21q11→pter
t(13;21)(q21;q21)	13q21→pter	21q21→pter
t(14;15)(q11.2;q13)	14q11.2→pter	15q13→pter
t(14;21)(q12;q22)	14q12→pter	21q22→pter
t(14;22)(q21;q11)	14q21→pter	22q11→pter
t(15;22)(q24;q11)	15q24→pter	22q11→pter

the same breakpoints are given for two rcp, although they were detected independently.

Risks for unbalanced offspring will depend on the production probabilities of unbalanced gametes and on the survival probabilities for the unbalanced zygotes and the interactions of segments. Survival probabilities for such unbalanced segments vary considerably from chromosome to chromosome; e.g., monosomy of the complete short arm of chromosome 14, including the proximal long arm segment, was observed only once after adjacent-2 segregation in our data [2], in contrast to the situation with chromosome 21, which in all cases has occurred as a monosomy for the corresponding segment. As is expected when an X-chromosome is the recipient, larger segments are tolerated than for recipient acrocentric chromosomes. For example, a liveborn, unbalanced offspring after adjacent-2 segregation for a t(17;X) translocation has been recorded for a chromosome 17 segment so large that it may otherwise not be compatible with live birth in a trisomic state (see above, Single- Versus Double-Segment Imbalances, and further discussions below, under Risks for Abnormal Offspring With Single-Segment Imbalance After a 3:1 Disjunction).

Sufficient data for risk estimation are present only for such rcp as when chromosomes 9 and 20 are involved. In Table III, risks are compared for rcp segregation by adjacent-1 and adjacent-2 modes with breakpoints in 9p11 and 9q13, respectively. A similar comparison is made for translocations with a breakpoint in 20p11 and 20q11, respectively. From Table III it may be concluded that the risks are similar for unbalanced offspring after an adjacent-2 segregation with a proximal q breakpoint as after an adjacent-1 segregation with a proximal p breakpoint in nonacrocentric chromosomes. Although the imbalance is larger in the former case, the estimated risk has not decreased. This observation is at variance with recent observations from male gametogenesis that unbalanced gametes after an adjacent-1 segregation are produced in larger quantities than after an adjacent-2 segregation [43]. However, rcp

TABLE III. Risk comparisons for translocations with breakpoints near the centromeres in chromosomes 9 and 20 and with unbalanced offspring recorded at birth after adjacent-1 or adjacent-2 segregation

Chromosome	Adjacent-1 trisomy for 9p11→pter or 20p11→pter	Adjacent-2 trisomy for 9q13→pter or 20q11→pter
9	3/21 = 14.3%	15/77 = 19.5%
20	5/25 = 20.0%	3/11 = 27.3%

predisposing to unbalanced offspring after an adjacent-2 segregation have not been investigated specifically in this way.

Furthermore, it may be important to study specifically gametogenesis in female carriers, since there are 30 female carriers of such rcp with unbalanced offspring in contrast to only 12 male carriers [2]. In this connection, adjacent-2 segregation represents a nonseparation (nondisjunction) of homologous centromeres, which from the data of primary (free) trisomies is known to occur more often in the female. However, the differences in risk between female and male carriers of such rcp suggested several times in the literature [38,44] were not supported by our recent data [2], although we believe that this question is still open because of the small amount of available data.

In summary, there are only a small number of rcp for which unbalanced offspring after an adjacent-2 segregation have been recorded. This is a consequence of the fact that compatibility with live birth for the combinations of trisomic and monosomic segments is possible only for a few pairs of chromosomes. In that the segregation type itself is not particularly rare, about one-half as frequent as alternate and adjacent-1 segregations [43], the suggestion that it occurs only occasionally [44–47] reflects the small proportion of such segregants surviving until term.

MEIOTIC 3:1 DISJUNCTION: DIFFERENT TYPES OF SEGREGATION

As was mentioned above, there are three possible types of meiotic disjunction of a quadrivalent: 4:0, 3:1, and 2:2 disjunctions. The 4:0 disjunction will lead to gametes so heavily unbalanced that they are inviable, and no such offspring have been reported. On the other hand, liveborn, chromosomally unbalanced offspring after 3:1 disjunction have been reported for a large number of rcp.

As an example, consider rcp 36, t(10;14)(p11;q11) (Fig. 7). In this pedigree, offspring has been born alive with 47 chromosomes, and this is shown diagrammatically in Figure 15, in which the white chromosome may be No. 10 and the black No. 14. One rcp derivative [the der(14)] is formed by the two segments 10pter→p11 and 14q11→cen→pter (including the 14 centromere), and this has been included in the gamete together with both normal chromosomes. The resulting unbalanced gamete has subsequently been fertilized by a gamete with normal chromosomes. Thus the unbalanced offspring is trisomic for this rcp derivative in its complete length. This is called a *tertiary trisomy*.

The other daughter cell of the same segregation event contains only the der(10), the segment (10qter→cen→p11 joined to the segment 14q11→qter)

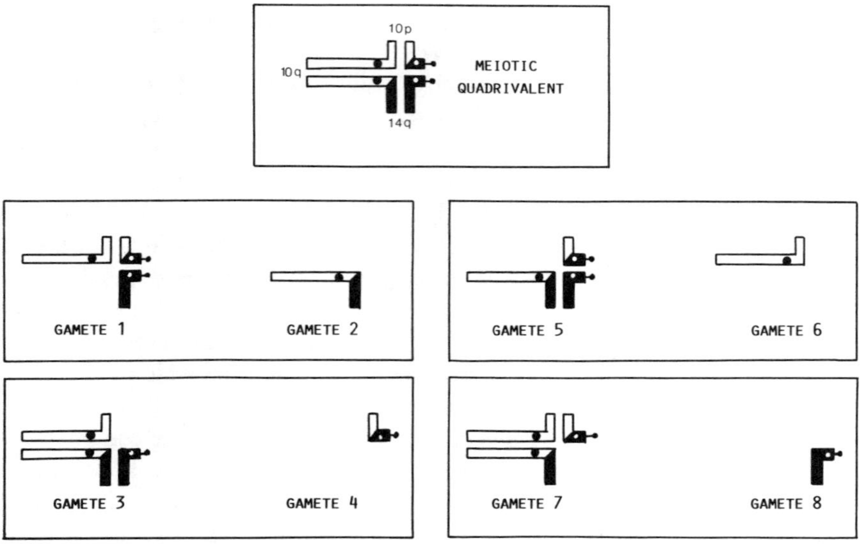

Fig. 15. Diagram of a pachytene quadrivalent for a 3:1 disjoining translocation, t(10;14)(p11;q11), and the resulting gametes. The four possible segregations are shown. After fertilization, by a normal gamete, gametes 1–8 will result in the following types of unbalanced fetuses: gamete 1 in a tertiary trisomy for the derivative (with centromere originating from) chromosome 14, and gamete 2 in a tertiary monosomy for the same derivative; gamete 3 in a tertiary trisomy for the derivative (with centromere originating from) chromosome 10, and gamete 4 in a tertiary monosomy for the same derivative; gamete 5 in an interchange trisomy i.e., an extra chromosome 14, and gamete 6 in an interchange monosomy (lacking) of chromosome 14; gamete 7 in an interchange trisomy of chromosome 10, and gamete 8 in an interchange monosomy of chromosome 10. The fetuses originating from gametes 3, 4, 6, 7, and 8 are so chromosomally unbalanced that they are unlikely to be viable.

and neither of the normal chromosomes 10 and 14. This gamete will lack the two segments composing the der(14): 10pter→p11 and 14q11→cen→pter. After fertilization by a gamete with normal chromosomes, the fetus will contain 45 chromosomes and will be monosomic for the der(14). This imbalance is called a *tertiary monosomy*. No liveborn offspring of this type have been recorded for this rcp, and it is doubtful that this imbalance is compatible with live birth.

Tertiary monosomy has, however, been recorded for a liveborn, abnormal child with 45 chromosomes in a family in which the t(5;14)(p14;q11) is inherited. This child was monosomic for the two segments 5pter→p14 and 14q11→cen→pter, which together form one of the derivatives, and had the cri-du-chat syndrome [12]. Tertiary monosomies are also recorded from several other rcp.

There are two possible segregation types leading to tertiary trisomies and monosomies, depending on which of the two derivatives is included in a daughter cell together with the two normal chromosomes.

For rcp carrying risk for unbalanced offspring after a 3:1 disjunction, it is not essential that one of the involved chromosomes is an acrocentric. As an example, consider once again Figure 7 and rcp 73, t(5;10)(p13;q11). Liveborn offspring have been born with 47 chromosomes and a trisomy for the two segments 5pter→p13 and 10q11→cen→pter, which together form the der(10). A tertiary monosomy of the same segments may be incompatible with life.

Translocations carrying risk for liveborn, abnormal offspring with a tertiary trisomy may be characterized in either of the following ways:

1) A proximal long arm breakpoint in one chromosome in which the *complete short arm and the proximal long arm segment* are compatible with survival until birth in the trisomic state, and a breakpoint in the recipient chromosome *either* in a terminal band or an acrocentric short arm region, leading to single-segment imbalance, *or* in a subterminal or interstitial region, leading to double-segment imbalance.

2) A breakpoint in a subterminal or interstitial band in the long or the short arm of any chromosome (except an acrocentric short arm), when the trisomy of the *segment distal to the breakpoint* is compatible with survival; the recipient chromosome is mostly acrocentric with a breakpoint in the proximal long arm; rarely, a chromosome 18 is also involved with the breakpoint in the proximal long arm region, and with the complete short arm and the proximal long arm and is also compatible with life in the trisomic state; the tertiary trisomies produced by this type of rcp are always double-segment imbalances.

In Figure 16, the breakpoint regions for rcp of group 1 leading to a possible imbalance after 3:1 disjunction are indicated; i.e., the proximal (or intermediate) long arm regions of chromosomes for which the complete short arm and the proximal long arm segments are compatible with survival until birth in its trisomic state (tertiary trisomy). It should be noted that compatibility with survival until birth for such a large imbalance has been recorded only for the autosomes 4, 8, 9, 10, 12, 13, 14, 15, 16, 18, 20, 21, 22. For some chromosomes, e.g., 9 and 14, the breakpoints in the long arm may be relatively distal. For chromosome 16 also the complete long arm and a proximal short arm segment are compatible with survival in its trisomic state as part of a tertiary trisomy after 3:1 disjunction.

Figure 16 also indicates the possible breakpoint positions for rcp carrying risk of liveborn, unbalanced offspring after a 2:2 disjunction with adjacent-2

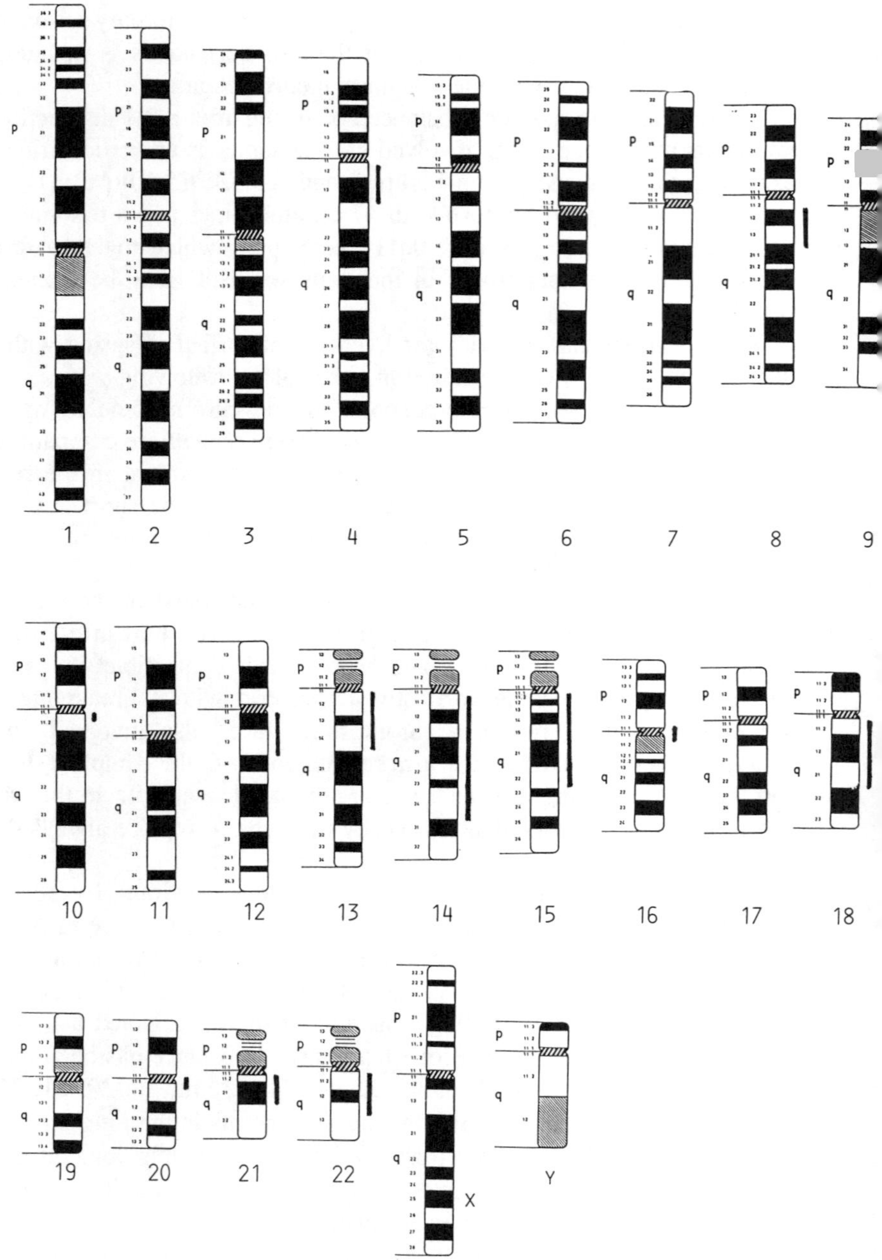
1
2
3
4
5
6
7
8
9
10
11
12
13
14
15
16
17
18
19
20
21
22
X
Y

segregation (Tables I and II). The breakpoints for trisomic segments shown in Tables I and II are all situated within the regions indicated in Figure 16, except in those cases when the recipient chromosome is an X. The regions indicated in Figure 16 are wider than those indicated in Tables I and II. One reason is that Figure 16 is based on tertiary trisomies (no complementary monosomies), whereas the imbalances in Tables I and II are all of a trisomy together with a monosomy. It should be pointed out that, after the 2:2 disjunctions with adjacent-2 segregation, the centromere of the recipient chromosome is included in the imbalance, but this is not so in the tertiary trisomies produced by the group 1 (3:1) segregating rcp. Some of the translocations in group 1 may also carry risk for tertiary monosomies but only for rcp with a breakpoint in the proximal long arm close to the centromere.

The rcp in group 2 may be compared with those carrying risk for single-segment imbalance after a 2:2 disjunction and adjacent-1 segregation with a breakpoint in a recipient acrocentric *short* arm region. In group 2, the corresponding breakpoint is situated in the proximal *long* arm region of the acrocentric chromosome, and the centromere of the acrocentric will be included in the imbalance. Also, rcp in group 2 may be carrying risk for tertiary monosomies, although the larger imbalances involved may be compatible only with live births as tertiary trisomies.

A different type of 3:1 segregation will occur when one of the normal homologues segregates with the two derivatives. After fertilization by a gamete with normal chromosomes, the fetus (with 47 chromosomes) will be trisomic for the normal homologue. This type of imbalance is called *interchange trisomy*. Interchange trisomies have occurred most frequently for rcp in which No. 21 is one of the chromosomes involved. Hence the fetus will be trisomic for chromosome 21 and will have Down syndrome. An inter-

---

Fig. 16. Diagram of all human chromosomes showing rcp breakpoints for tertiary trisomic progeny with imbalances of the complete short arm and proximal long arm segments. These imbalances have been observed in live births as part of a tertiary trisomy with an autosomal recipient chromosome after 3:1 disjunction. For chromosome 16, this also holds true for the complete long arm and a proximal short arm segment. The observed breakpoint regions are indicated with a bar to the right of the chromosome. Liveborn offspring with the corresponding tertiary monosomy and liveborn, unbalanced offspring after a 2:2 disjunction with adjacent -2 segregation may also be observed for autosomal translocations with breakpoints in the indicated regions. For chromosomes 5 and 7, liveborns with trisomies of the complete short arm have been recorded, but none with an additional, proximal long arm segment as yet have been recorded. They may also be at risk for unbalanced offspring after a 3:1 disjunction (or adjacent-2) if the breakpoint lies in the proximal long arm region.

change trisomy will be compatible with survival until birth if this is the case for the primary (standard) trisomy of the normal chromosome, such as for trisomy 21, trisomy 18, trisomy 13, and XXX. Also interchange trisomies of 14 (only an abortion) have been recorded [2,5]. Because a monosomy for any of the normal autosomes is not compatible with live birth, interchange monosomies may be seen at term only for rcp in which the X-chromosome is involved.

To be certain of the origin of the extra chromosome in a presumed interchange trisomy, a marker analysis is necessary. Such marker analyses (excluding nondisjunction in the normal parent) have not been reported in any of the possible interchange trisomies in our data [2].

RISKS FOR ABNORMAL OFFSPRING WITH SINGLE-SEGMENT IMBALANCE AFTER A 3:1 DISJUNCTION: TERTIARY TRISOMIES OR MONOSOMIES

For this group of rcp, unbalanced offspring may originate from different types of disjunction and segregation mechanisms. We will not distinguish between risks for different products of the same type of meiotic segregation; e.g., tertiary trisomies or monosomies will be considered together as unbalanced offspring.

As an example, consider case 1 reported by Rethoré et al. [48]. One breakpoint is in proximal 9q at q11 and the other in terminal 6p at p25, i.e., t(6;9)(p25;q11). In this pedigree, liveborn offspring (with 47 chromosomes) with a tertiary trisomy of the segment 9pter→cen→q11 (and the distal part of 6p25) have been reported after a 3:1 disjunction. This is a single-segment imbalance.

Another example is rcp No. 54 in Figure 17, t(5;14)(p15;q13), in which there is one breakpoint in proximal 14q, in q13, and another in terminal 5p, in p15. In this pedigree, one liveborn infant with a tertiary trisomy of 14pter→cen→q13 and the distal part of 5p (47 chromosomes) and one with a tertiary monosomy of 14pter→cen→q13 and the distal part of 5p (45 chromosomes) have been observed.

For some rcp within this group, the breakpoints in the long arm of the first chromosome may be intermediate in position. As is shown in Figure 17, this is the case for some translocations in which the first chromosome is a 14. Breakpoints as distant as 14q24 have been recorded for rcp belonging to the group. This is hardly a proximal long arm segment, but the complete 14p and the proximal 14q are included in the imbalance. Figure 17 shows these data for chromosome 14 and it appears that rcp belonging to this group more frequently involve chromosome 14 than chromosome 13 or 15.

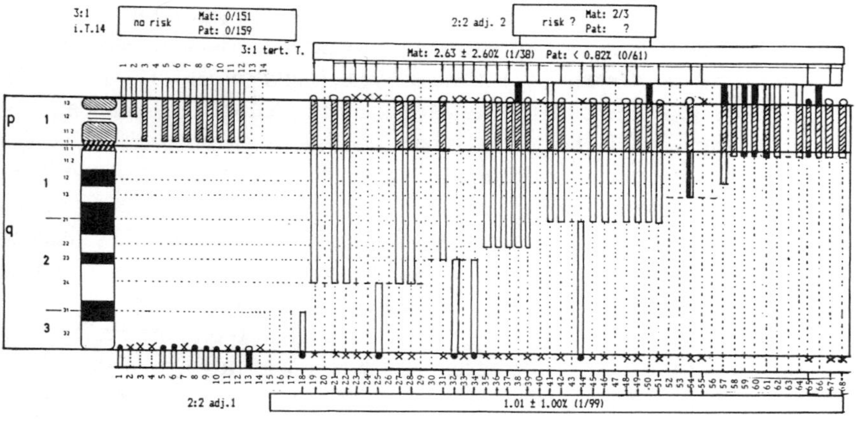

Fig. 17. Diagrammatic representation of translocations involving chromosome 14. The symbols are explained in the legend to Figure 8.

For rcp with a breakpoint in 14q carrying risk for 3:1 single-segment imbalance in the offspring of female carriers, there was only one unbalanced child with a tertiary trisomy among 38 pregnancies after ascertainment corrections. Hence we obtain the risk estimate 1/38 = 2.6%. For male carriers, no unbalanced liveborn children remained among 61 pregnancies after ascertainment corrections. Thus the corresponding risks estimate for male carriers is below 0.8%. The discrepancy between the risks for female and male carriers would be more pronounced were ascertainment corrections not considered; there were 22 tertiary trisomies and 14 tertiary monosomies among 65 pregnancies of 31 carrier mothers (21 chromosomally unbalanced probands) but only four among 80 pregnancies with carrier fathers (all probands).

The estimated risks for carriers of rcp belonging to this group are generally low, on the order of 2% or less. The exception is for rcp with a breakpoint in 21q, for which the risk estimate for female carriers is 6.9%. For male carriers, no risk estimate could be obtained in this case; no information was available about pregnancies with carrier fathers. For most chromosome arms for which liveborn, unbalanced offspring of this type have been recorded with single-segment tertiary trisomy or monosomy after a 3:1 disjunction, the numbers of observations are very limited, so the risk estimates are inaccurate.

The low risks for these rcp may be explained partly by the low frequency of production of gametes through a 3:1 disjunction and partly by the characteristically large imbalances compared to those for single-segment imbalances after a 2:2 disjunction with adjacent-1 segregation. The former have been shown in recent studies of sperm from two male carriers of balanced rcp [43]; frequencies of such (3:1) gametes were found to be 3.5% and 5.0%, respectively, although it should be noted that rcp predisposing to unbalanced offspring after a 3:1 disjunction were not investigated specifically. For chromosome 9, we can compare the estimated risks for almost the same segment, i.e., 9pter→cen→q proximal after adjacent-2 segregation or 3:1 disjunction [2]. The difference between the risk estimates of 19.6% (adjacent-2 segregation) and only 1.67% (3:1 disjunction) for female carriers may be ascribed mainly to the difference in production probability.

A specific feature of rcp predisposing to unbalanced offspring after a 3:1 disjunction seems to be an increasing risk with advancing age of the carrier parent. Such age effects have not been found for carrier parents with rcp carrying risk for abnormal offspring by 2:2 disjunction. For the former, the mean maternal age at birth of the unbalanced child and the standard deviation of these ages were significantly higher than the corresponding figures for female carriers of rcp carrying risk for unbalanced offspring after 2:2 disjunctions. Also, the very few carrier fathers who have had abnormal offspring after a 3:1 disjunction were comparatively old [2].

DOUBLE-SEGMENT IMBALANCES AFTER 3:1 DISJUNCTIONS

Translocations carrying risk for unbalanced offspring with double-segment imbalances after 3:1 disjunction may be classified into two groups, groups 1 and 2 as discussed above. Estimations of the magnitude of risks for double-segment imbalances for rcp belonging to these groups may be made with some modifications of the suggestions given above at the end of the section on Single- Versus Double-Segment Imbalances.

Translocations belonging to group 1, in contrast to the rcp discussed in the previous section, have a subterminal or an interstitial breakpoint in the recipient chromosome (not an acrocentric short arm). To determine the order of magnitude of the associated risk, we can start from the perspective of the estimated risk for single-segment imbalance after 3:1 disjunction as discussed in the previous section. This single-segment imbalance will include the complete short arm and the proximal long arm segment of the first chromosome (for a few chromosomes, e.g., 16 and 18 complete long arm and proximal short arm segments may also be involved). The magnitude of the risk for double-segment imbalance will most often be less than the risk for this single-segment imbalance. The risk reduction depends on the survival probability of the imbalance of the recipient chromosome, which may be inferred from the risk for single-segment imbalance for this segment after a 2:2 disjunction and adjacent-1 segregation. If this latter risk is low, the above-mentioned risk for single-segment imbalance after a 3:1 disjunction has to be greatly reduced. A suggestion (on the cautious side) for the magnitude of the risk for double-segment imbalance is one-half of the estimated risk for the above-mentioned single-segment imbalance after 3:1 disjunction (similar to point 4 for adjacent-1). Because possible interactions between the two unbalanced segments may be present, further adjustments of the risk may be made as suggested in point 5 given at the end of the section on Single- Versus Double-Segment Imbalances.

It is more difficult to determine the risk for rcp in group 2. A grouping is possible only for the t(11;22) rcp to be considered in the next section. To determine the magnitude of the (rcp group 2-associated) risk, we may start from the risk estimate for the single-segment imbalance after a 2:2 disjunction with adjacent-1 segregation considered above for the segment distal to the breakpoint in the first chromosome mentioned in the definition of group 2. Then, consider the corresponding group 2 rcp with a breakpoint in q11 of an acrocentric (or an 18) chromosome. This will carry risk for a double-segment imbalance: a tertiary trisomy or monosomy involving the derivative consisting of the complete short arm and the proximal long arm segment of this acrocentric and of the segment distal to the breakpoint in the first-mentioned chromosome. Although this is a double-segment imbalance, the contribution from the acrocentric on the phenotype and hence to the reduction in survival probability may be small. A comparison of the risk for this imbalance after a 3:1 disjunction with the risk for the above-mentioned single-segment imbalance after a 2:2 disjunction with adjacent-1 segregation will therefore, reflect mainly the reduction in risk for the group 2 rcp because of the lower production probability of unbalanced gametes after the 3:1 disjunction.

Assuming approximately the same risk reduction for a group 2 rcp with a breakpoint in q11 of the acrocentric as with the corresponding rcp at risk for single-segment imbalance after a 2:2 disjunction with adjacent-1 segregation, one may estimate a risk reduction factor on the basis of data from corresponding groups of rcp. On the basis of such a comparison, based on the limited available data, we have suggested a reduction factor of 3 for female carriers [2]. For male carriers, the risk will in all cases be very low.

If the q-arm breakpoint in the acrocentric chromosome is in q12 or is more distal, the double-segment imbalance will be further extended and the risk further reduced or modified through possible interactions. To determine the magnitude of the risk, one may proceed as suggested in points 3–5 given at the end of the section on Single- Versus Double-Segment Imbalances.

Some rcp may be classified in both groups. An example is the rcp 41 shown in Figure 17 with the designation t(3;14)(p25;q21). In such cases, one may carry out the risk estimation according to the suggestions for both groups separately and take the lower value.

Finally, we should note that these results and suggestions concerning risk estimates for unbalanced offspring after 3:1 disjunctions are preliminary; they are based on the very limited numbers of observations available at present. For chromosome 9, a comparison of risks for the corresponding segment after 2:2 disjunction with adjacent-2 segregation (18.5% for maternal carriers) and after 3:1 disjunction (1.7%) was possible [2]. This comparison indicates that the risk reduction factor is different for different chromosomes and breakpoints and is higher than suggested above.

t(11;22) TRANSLOCATIONS

This large group of rcp carrying risk for double-segment imbalances belongs to group 2 as discussed above. These have breakpoints in the regions 11q25subter . . . q23 and in 22q11 . . . q13. Example 2 in the section on further examples of rcp families is one such case. The liveborn, unbalanced offspring in this group have 47 chromosomes as a result of 3:1 disjunction, with a double-segment tertiary trisomy.

In our data [2], 74 rcp of this type were reported. There are three main groups of reported pairs of breakpoints (11q23;22q11), (11q24;22q12), and (11q25;22q13). It is generally recognized that exact breakpoint localizations are difficult because of the morphological similarity between 11q and 22q [2,13]. Therefore, these rcp can all be identical and will be considered as a single group. Most unbalanced individuals are probands. After ascertainment corrections in these data, there remain among the offspring of female carriers only nine abnormal cases among 241 pregnancies (3.7%), but, among the offspring of male carriers, no abnormal cases remain among 77 pregnancies, resulting in a risk estimate below 0.7%.

If one divides the rcp into two groups, assuming different point positions in 11q, the risk estimate for female carriers with a more distal 11q breakpoint, i.e., 11q25subter . . . q24 (hence more proximal in chromosome 22), is 5.5%, and for those with a more proximal 11q breakpoint, i.e., 11q23 (hence more distal on chromosome 22), it is 2.7%. This difference, although not significant, may indicate a somewhat lower risk for rcp causing a larger imbalance of chromosome 11q if the breakpoints are in fact different.

The risk estimation for the rcp (11;22) can also be considered from the perspective of rcp carrying risk for single-segment imbalances as discussed in the preceding section.

Our data [2] do not contain any rcp with a breakpoint in 22q13 carrying risk for a single-segment imbalance after a 3:1 disjunction for the complete 22p and the proximal 22q, i.e., a segregant with an extra der(22). With a breakpoint in 22q12 . . . q11, there are 13 such rcp. In this latter group, no unbalanced offspring remained among 14 pregnancies after ascertainment corrections. Thus the estimated risk is less than 3.5%, which is below the figure of 3.7% obtained for all 74 t(11;22) (with the additional imbalance from 11q).

As for the rcp with a breakpoint in 11q and carrying risk for single-segment imbalances after adjacent-1 segregation, there are only six such rcp with a breakpoint in 11q23 and none with a breakpoint in 11q24 [2]. The risk estimate for such carrier parents is found to be 7.0%. This compares well with the (2.7%) risk estimate given above for the 11q23 segment plus the proximal 22q segment after a 3:1 disjunction, with a factor 3 for risk reduction. The reduction in risk may reflect the ratio between production probabilities for gametes after a 2:2 disjunction with adjacent-1 segregation and after a 3:1 disjunction and partly the ratio between survival probabilities for single- and double-segment imbalances. Finally, it remains open whether interactions between the 11q and 22q segments play a role.

RISKS FOR INTERCHANGE TRISOMIES

As was mentioned above in the section on Meiotic 3:1 Disjunction: an interchange trisomy may be detected in newborn or spontaneous abortions if one of chromosomes 13, 14, 18, 21, or X is involved in the rcp. In our data [2], there were 428 female carriers who had, in all, 1,307 pregnancies among which 18 involved possible interchange trisomies, 16 with an extra 21, one with an extra 18, and one with an extra 13. With one exception, all patients were probands; hence the overall risk estimate for interchange trisomy is 0.1%. Of 787 pregnancies with carrier fathers, two possible interchange trisomies were recorded at birth, one with an extra 18 and one with an extra 21, both cases being probands; hence the overall estimate is much lower than 0.09%.

When an X chromosome was involved in the rcp, one interchange trisomy, in a proband, was found among 96 pregnancies of 26 female carriers. No interchange monosomy was recorded. Hence, after asc corrections, the risk estimate is below 0.8%.

ABNORMAL OFFSPRING AFTER MORE THAN ONE TYPE OF MEIOTIC DIVISION

So far, we have considered only the possibility that an rcp led to unbalanced offspring through a single type of meiotic disjunction and segregation.

This holds true for the majority of rcp. However, there are several rcp that can result in abnormal offspring after different types of meiotic divisions. This is in particular the case when at least one of the acrocentric chromosomes or an 18 or an X is involved in the rcp. Some examples are presented in Stengel-Rutkowski et al. [2]. The total risk for abnormal offspring may be estimated by adding the estimated risks for the actual types of disjunctions and segregations.

BREAKPOINT POSITIONS, MEIOTIC DIVISIONS, AND RISKS

The results considered so far have demonstrated that the risk for unbalanced offspring to be born depends 1) on the different types of meiotic disjunctions and segregations, being high for 2:2 disjunctions and low for 3:1 disjunctions, and 2) on the survival probability for the unbalanced zygotes and fetuses.

For an individual translocation, it is important to determine which types are compatible with live births and which type(s) of meiotic division may lead to them [40]. To determine the possible disjunction and segregation that will lead to a chromosomal imbalance, Tables IV and V may be utilized. These show the connections between the breakpoint localizations and the

TABLE IV. General segregation diagram: for rcp with unbalanced segments from pter to breakpoint (relationship between breakpoint positions, unbalanced segments, and type of imbalances as determined by the disjunction and segregation mechanisms)

		Recipient chromosome (breakpoint position)		
		pter→p11 (imbalance)	qter→q11 (imbalance)	
Breakpoint position	Imbalance	pter→b	pter→cen→b	qter→b
Main chromosome				
pter→p11	pter→b	2:2 Adj-1	3:1 Tert	2:2 Adj-1
qter→q11	pter→cen→b	3:1 Tert	2:2 Adj-2	3:1 Tert
qter→q11	qter→b	2:2 Adj-1	3:1 Tert	2:2 Adj-1

Breakpoint localizations: pter→p11; breakpoint in short arm; qter→q11; breakpoint in long arm. Unbalanced segments: pter→b; partial short arm imbalance down to breakpoint b (if b = p11, complete short arm imbalance); pter→cen→b, imbalance for complete short arm and proximal long arm segments down to breakpoint b; qter→b; partial long arm imbalance up to breakpoint b (if b = q11, complete long arm imbalance); qter→cen→b; imbalance for complete long arm and proximal short arm segment up to breakpoint b. Type of imbalance (determined by disjunction and segregation mechanism): 2:2 adj-1 and 2:2 adj-2; trisomy for one segment, monosomy for the other; 3:1 tert; trisomy for both segments or monosomy for both segments.

TABLE V. General segregation diagram: for rcp with unbalanced segments from qter to breakpoint.

		Recipient chromosome (breakpoint position)		
		pter→p11 (imbalance)	qter→q11 (imbalance)	
Breakpoint position	Imbalance	pter→b	pter→cen→b	qter→b
Main chromosome				
qter→q11	qter→b	2:2 Adj-1	3:1 Tert	2:2 Adj-1
pter→p11	qter→cen→b	3:1 Tert	2:2 Adj-2	3:1 Tert
pter→p11	pter→b	2:2 Adj-1	3:1 Tert	2:2 Adj-1

See Table IV for explanations.

appropriate disjunction and segregation types that determine the actual combinations of trisomic and monosomic segments.

One of the chromosomes is denoted the main and the other the recipient chromosome. Either can be considered as the main chromosome, as shown in the following examples. When using the tables, one starts from the localization of the breakpoints in the long or short arm. The next point is to consider the polarity of the unbalanced segment of the main chromosome, whether it extends from pter to the breakpoint or from qter to the breakpoint. In the former case one uses Table IV, in the latter Table V.

If the breakpoint is on the short arm of the main chromosome and the possible imbalance goes from pter to the breakpoint, the upper row of Table IV is used. If the breakpoint is on the long arm and one considers an imbalance involving the complete short arm and a proximal long arm segment, then one uses the middle row of the same table. When the breakpoint is on the long arm and one considers the imbalance from qter to the breakpoint, the lower row of Table IV can be used. With regard to the recipient chromosome, the left, the middle, and the right columns of Table IV are used in the same way that the upper, middle, and lower rows were used for the main chromosome.

Table V is used when the imbalance for the main chromosome extends from qter to the breakpoint and that for the recipient extends from pter to the breakpoint. For some combinations of breakpoint localizations and imbalances, both figures can be used, but, in others, i.e., when the actual imbalance covers one complete arm and proximal segments of the other, only one table can be used. By choosing both chromosomes in turn as the main chromosome and going systematically through the different possibilities, the two tables will cover all situations except for one unlikely case, namely, when the breakpoints in both chromosomes are situated on the short arms and complete long arm and proximal short arm imbalances are to be considered for both chromosomes. (A diagram also covering this situation can be constructed from Table IV by interchanging the terms short and long arms as well as the letters p and q at the left side and the top of the table. Then, the disjunction and segregation type with this combination of breakpoints and imbalances will be 2:2 disjunction with adjacent-2 segregation. This possibility, however, only very rarely produces offspring compatible with live birth.)

The tables can be used in genetic counseling by first eliminating those possibilities that are incompatible with live birth. We shall consider two examples and also show how the tables can be used to facilitate breakpoint determinations.

Example 1

A male carrier of the t(9;18)(p12;p11.1) requested genetic counseling. The potentially unbalanced segments of the rcp are the following: 1) 9pter→p12, 2) 9qter→cen→p12, 3) 18pter→p11.1, 4) 18qter→cen→p11.1. These segments may occur either as trisomies or as monosomies and in various combinations of pairs depending on the meiotic disjunction and segregations.

To use Tables IV and V, we begin by confining ourselves to those segments that, on the basis of our previous work [2], can be assumed to be compatible with live birth in either a trisomic or a monosomic state. In this example, the segments 1, 3, and 4 belong to this category. Then, in turn, one of the chromosomes is chosen as the main chromosome and the other as the recipient, and the possibilities are explored.

Segment 1. Let chromosome 9 be the main chromosome and, considering the imbalance 9pter→p12, refer to the left hand side of Table IV and the upper row. The imbalance to be considered in the recipient chromosome is 18pter→p11.1; refer to the upper left row of the table. The unbalanced gametes corresponding to this combination of breakpoints and unbalanced segments will be produced by 2:2 disjunction with adjacent-1 segregation, and there are two possibilities for abnormal offspring: 1A) trisomy 9pter→p12 combined with monosomy 18pter→p11.1 and 1B) monosomy 9pter→p12 combined with trisomy 18pter→p11.1. The estimated risk for the 9pter→p12 single-segment imbalance is 6.4% [2].

For the 18pter→p11.1 single-segment imbalance, no estimate is available from our data. Monosomy 18p is a well known and a common syndrome, and it is likely to reduce further the survival probability for case 1A. For case 1B, although trisomy for 18p seems to have only little effect on the phenotype [5], the survival probability for the single-segment 9pter→p12 monosomy is smaller than for the corresponding trisomy. An estimate for the total imbalanced progeny (cases 1A and 1B, together) is approximately 3.2% for male or female carriers (based on a reduction factor of 2).

Segment 3. Let chromosome 18 be the main chromosome and consider the imbalance 18pter→p11.1, referring again to the left side (upper row) of Table IV. The imbalance in the recipient chromosome is 9pter→p12. The diagram again indicates that a 2:2 disjunction with adjacent-1 segregation will lead to the unbalanced gametes, and these are the same as those considered for segment 1 above.

Segment 4. Let chromosome 18 again be the main chromosome and consider the imbalance 18qter→cen→p11.1. This is shown in the left side of Table V (middle row). In the recipient chromosome, the breakpoint is in 9p12, and we are oriented again to the left middle row of the table. For this combination of breakpoints and imbalanced segments, 3:1 disjunction will lead to unbalanced gametes, and the possible unbalanced offspring will have tertiary trisomies or monosomies: 4A) trisomy 18qter→cen→p11.1 combined with trisomy 9pter→p12 or 4B) monosomy 18qter→cen→p11.1 combined with monosomy 9pter→p12. The latter combination is unlikely to be compatible with live birth.

With regard to case 4A, this is a double-segment imbalance of group 1 discussed above in the section on Double-Segment Imbalances After 3:1 Disjunctions. There is no information in our data [2] on the risk for a single-segment trisomy of 18qter→cen→p11.1 after a 3:1 disjunction. In that the complete trisomy 18 is compatible with live birth, we may assume that this nearly complete trisomy will also allow survival. As was mentioned above in the section on Risks for Abnormal Offspring With Single-Segment Imbalance After a 3:1 Disjunction , the estimated risk for carriers of rcp carrying risk for single-segment imbalance after a 3:1 disjunction is generally 2% or less, so we can use 2% as a rough estimate for this single-segment trisomy for maternal carriers and perhaps no risk for paternal carriers. For the single-segment imbalance 9pter→p12 after 2:2 disjunction with adjacent-1 segregation, the estimated risk is 6.4%. According to suggestions given above, in the section on Double-Segment Imbalances After 3:1 Disjunctions, for estimation of risks for group 1 double-segment imbalances, the magnitude of this risk may be reduced to one-half of the above-mentioned values, i.e., 1% for female carriers and no risk for males.

Finally, we have to consider that an interchange trisomy 18 may occur. According to our data, such a risk is estimated to be less than 0.3% for female and male carriers. Thus the overall risk for an unbalanced live birth for this rcp, t(9;18)(p12;p11.1), can be estimated for female carriers to be less than 3.2 + 1.0 + 0.3 = 4.5% and for male carriers to be less than 3.2 + 0 + 0.3 = 3.5%.

Imbalance for the segment 9qter→cen→p12 was not considered, because according to our present knowledge, this will be incompatible with live birth. However, for completeness, it should be mentioned that, because the breakpoint in the recipient chromosome was in 18p11.1, by interchanging the symbols and terms in Table IV (as described above), the disjunction and segregation related to the combination of these two breakpoints and the imbalances 9pter→cen→p12 and 18qter→cen→p11.1 will be by 2:2 disjunction with adjacent-2 segregation.

Example 2

As another example, consider translocation 38 in Figure 17, with the designation t(13;14)(q12;q22). The possibly unbalanced segments are 1) 13pter→cen→q12, 2) 13qter→q12, 3) 14pter→cen→q22, 4) 14qter→q22. In our data [2], segment 1 has been seen as both as a trisomy and as a monosomy, the other segments only as trisomies. For this rcp, interchange trisomy 13 may occur, and also perhaps interchange trisomy 14. The other types of unbalanced offspring can be found by means of Table IV.

Let 13 be the main chromosome, and let the first imbalance under consideration be *13pter→cen→q12*. On the left side of Table IV, we are oriented to the middle row. In the recipient chromosome, the imbalance under consideration is 14pter→cen→q22, which leads us to the middle column in the table. The actual disjunction and segregation mechanism will be 2:2 disjunction with adjacent-2 segregation, with the following two possibilities (disregarding the four heavily unbalanced products after crossing-over in the interstitial regions): 1A) monosomy 13pter→cen→q12 combined with trisomy 14pter→cen→q22 and 1B) trisomy 13pter→cen→q12 combined with monosomy 14pter→cen→q22. The former type, 1A, has been recorded in the family. The latter is unlikely to be compatible with survival to birth. On the basis of our data [2], no risk assessment has been possible because of the very limited data. Referring to the section above on the risks for Liveborn, Abnormal Offspring After 2:2 Disjunction With Adjacent-2 Segregation; it can be seen that the risk may be relatively high.

Consider again the imbalance *13pter→cen→q12* for the main chromosome, 13. Now we consider the imbalance 14qter→q22 for the recipient chromosome. This brings us to the right column of Table IV. The middle row of this column shows that this combination of breakpoint positions and imbalances is related to 3:1 disjunction, leading to the tertiary trisomies or monosomies 2A) trisomy 13pter→cen→12 combined with trisomy 14qter→q22; 2B) monosomy 13pter→cen→12 combined with monosomy 14qter→q22. The tertiary monosomy 2B will not be compatible with live birth, whereas 2A may be. This is a double-segment imbalance and it may be classified in both groups 1 and 2, as discussed above in the section on Double-Segment Imbalances After 3:1 Disjunctions.

Consider it first from the perspective of group 1: The risk for the single-segment trisomy 13pter→cen→q12 after 3:1 disjunction is estimated as 2.6% for female carriers and as less than 0.8% for male carriers [2]. The estimated risk for the single-segment imbalance 14pter→cen→q22 after a 2:2 disjunction with adjacent-1 segregation is 1.0% for male and female carriers. From the suggestions made above in the section on Double-Segment Imbalances After 3:1 Disjunctions, a risk estimate for double-segment imbalance may be found through the reduction factor 2 to 1.3% for female carriers and to less than 0.4% for male carriers. (Because the estimated risk for the 14qter→q22 is very small, these values may overestimate the risk.)

From the point of view of group 2, we start by considering the estimated risk of 1.0% for the single-segment trisomy 14qter→q22 after a 2:2 disjunction with adjacent-1 segregation. Had the breakpoint on chromosome 13 been in 13q11, this risk should be divided by 3 to result in a risk of 0.3% for female carriers and a very low risk for male carriers. The breakpoint is in fact in 13q12, so the risk will be smaller. Thus we may conclude that the risk for type 2A will be less than 0.3% for female carriers and will be very low for males.

Consider now the imbalance *13qter→q12* for the main chromosome. This leads us to the bottom row of Table IV. With the imbalance 14pter→cen→q22 for the recipient chromosome, we use the middle column of the table and find that a 3:1 disjunction will lead to the tertiary trisomies and monosomies 3A) trisomy 13qter→q12 combined with trisomy 14pter→cen→q22, 3B) monosomy 13qter→q12 combined with monosomy 14pter→cen→q22. Here, 3B is incompatible with life, whereas the tertiary trisomy 3A may allow survival to birth. The estimation situation is similar to that discussed above for 2A since the double-segment imbalance in 3A may be considered from the point of view of both groups 1 and 2 as discussed in the section on Double-Segment Imbalances After 3:1 Disjunctions. The risk for the single-segment imbalance 14pter→cen→q22 after a 3:1 disjunction is estimated as 2.6% and that for the single-segment imbalance trisomy 13qter→q12 after a 2:2 disjunction with adjacent-1 segregation is estimated as 1.6%, so the group 1 risk estimate for 3A is 1.3% for female carriers. The group 2 estimate in case of a breakpoint in 14q11 would be 0.53% with the reduction factor 3 for female carriers. Since the breakpoint on 14 is in 14q22, and not in 14q11, the additional imbalance may suggest that we use a further reduction factor of 2 such that the group 2 risk estimate for 3A will be 0.3% for female carriers and will be very low for males. In that this is smaller than the group 1 estimate, the final risk estimate for 3A will be 0.3% for female carriers.

Consider once again the imbalance *13qter→q12* for the main chromosome but now the imbalance 14qter→q22 for the recipient chromosome. We are now oriented to the lower part of the right column in Table IV; note that 2:2 disjunction with adjacent-1 segregation will lead to 4A) trisomy 13qter→q12 combined with monosomy 14qter→q22 or 4B) monosomy 13qter→q12 combined with trisomy 14qter→q22, neither of which is compatible with life (neither of segments 2 or 4 seems to be compatible with life in a monosomic state).

In that the estimated risks for these carriers to have an unbalanced offspring of one of the types 2A and 3A is 0.3 + 0.3 = 0.6% for female carriers and is very low for males, for the rcp t(13;14)(q12;q22), the total risk for unbalanced offspring is almost equal to that for type 1A. Although this risk is unknown, it may be relatively high.

In using Table V, one is led to the types 3A, 3B, 4A, and 4B, but not to types 1A, 1B, 2A, and 2B. By going through the same steps with 14 as the main chromosome, exactly the same eight types will be found.

Example 3

Tables IV and V may also facilitate the determination of a breakpoint position in the centromeric region, where it is often difficult to decide on the basis of the banding patterns whether it is in p11 or q11.

Consider t(9;15)(p11;q11) [49,50]. This rcp has been ascertained through liveborn, unbalanced offspring with a trisomy for 9pter→p11. According to the published breakpoints, this is a complete short arm trisomy with breakpoints in 9p11 and in 15q11 for the recipient chromosome. For the latter, only the segment 15pter→ cen→q11 is compatible with life either as a trisomy or as a monosomy. This orients us to the upper row, middle column, of Table IV. Thus a 3:1 disjunction should have led to unbalanced offspring, with a tertiary trisomy and 47 chromosomes. However, the infant is reported to have had only 46 chromosomes; i.e., the imbalance must have been the result of a 2:2 disjunction with either adjacent-1 or adjacent-2 segregation. From Table IV it follows that adjacent-1 segregation corresponds to the designation t(9;15)(p11;p11) and the imbalances 9pter→p11, 15pter→p11 and adjacent-2 segregation to the designation t(9;15)(q11;q11) and the imbalances 9pter→cen→q11, 15pter→cen→q11. This means that both breakpoints should have been on the same side of the centromere, not on opposite sides. On the basis of the available information, it is not possible to determine the correct alternative. From Table III and the discussion above in the section on Risks for Liveborns, Abnormal Offspring After a 2:2 Disjunction with Adjacent-2 Segregation, the risks seem to be of the same order of magnitude for the two types of segregation for 2:2 disjunction. However, both these risks are much higher than that for a 3:1 disjunction. Therefore, it is important in the genetic counseling of a family to determine the disjunction type correctly.

RISKS FOR CATEGORIES OF ABNORMAL OFFSPRING OTHER THAN OFFSPRING UNBALANCED AT BIRTH

So far, the discussion has been confined to the risks for carriers of rcp to have unbalanced offspring at birth. This is the most important category of abnormal offspring from a counseling point of view. However, estimated risks for unkaryotyped stillbirths and early deaths and for spontaneous abortions are also of importance. Of special interest is the estimated risk of a prenatally diagnosed fetus to have an unbalanced chromosome constitution.

These problems have been discussed by Stengel-Rutkowski et al. [2], where estimates for these categories are discussed in parallel to those given in the preceding sections. One may also estimate the total burden for unsuccessful pregnancies as the sum of the risks for unbalanced offspring at birth, unkaryotyped stillbirths, and early deaths and abortions [2].

In Table VI the average risks for these categories and for unbalanced fetuses among second-trimester prenatal diagnoses are given for four groups of rcp carrying risk for abnormal offspring after a 2:2 disjunction with adjacent-1 segregation. These rcp are grouped in order of magnitude of the risk for unbalanced offspring at birth (see also Fig. 11).

Although these data are heterogeneous, some general tendencies can be seen in Table VI. The proportion of spontaneous abortions is high for all categories of rcp but is somewhat higher for rcp carrying low or no (unbalanced liveborn) risk than for rcp carrying high and medium risk. The proportion of unkaryotyped stillbirths and early deaths tends to be higher for

TABLE VI. Average estimated risks (%) for different categories for unbalanced segregants after 2:2 disjunction with adjacent-1 segregation

Risk category for unbalanced live births	Unbalanced live births	Unkaryotyped stillbirths	Abortions	Unbalanced fetuses among second-trisemester prenatal diagnoses
High risk ⩾ 10%	17.0	5.1	23.1	28.6
Medium risk 5–10%	7.4	8.0	20.3	34.6
Low risk <5%	1.4	7.9	27.4	13.8
No risk 0%	0	1.3	27.4	0

medium and low risks than for high risks whereas for "no-risk" (at birth) rcp, this proportion is low. Both these results indicate that fetuses with a lower live birth risk tend to have a lower survival probability, which will result in a higher number of abortions, stillbirths, or early deaths.

With regard to the proportion of chromosomally unbalanced fetuses at second-trimester amniocentesis, this is found to be substantially higher than the estimated risk for unbalanced children at birth for all rcp groups in Table VI except for those without risk. The difference indicates a proportion of abnormal fetuses detected at amniocentesis would have been aborted spontaneously later in the pregnancy. The difference may also indicate that prenatally diagnosed, abnormal fetuses have a better rate of detection than abnormal liveborn offspring.

DIFFERENT APPROACHES TO CHARACTERIZING AND ASSESSING RISKS FOR RCP

The approach presented in the previous sections is based on groupings of the rcp utilizing the identity of chromosomes, their breakpoint positions, the presence of single- or double-segment imbalances, and the meiotic disjunction and segregation mechanisms. The approach has been developed through a series of papers [2,19,35–39], and the necessary asc corrections for this type of data have been developed as discussed above.

Other approaches to this problem of risk assessment have also been described by other authors over the past 2 decades [40,42,44–47,51–56]. In many of these papers, geometric configurations of the pachytene quadrivalent have been discussed, in particular the relative lengths of exchanged segments and the interstitial segments. The involved chromosomes and the genetic content of the possible unbalanced segments have not been considered separately. Differentiated risk estimates, as in our approach, have not been provided, nor have means to calculate risks for individual rcp, with one exception [46].

The point of departure for some of these studies is the strong association [40,55] between risk and ascertainment, which was discussed above. This has been interpreted as if "the mode of ascertainment determines the risk" [47] (indicating a causality). Translocations have been classified according to type of ascertainment: through abnormal liveborns, balanced carriers (or carrier fetuses), spontaneous abortions, or incidental means. In some studies, even the different sibships within a pedigree transmitting a single rcp have been classified in different groups depending on the way the rcp has been traced [51]. For each group, a separate risk estimate has been calculated.

The basic ideas for these approaches are very different from those of our own study [2] and the present paper, in which we have tried to show that the main determinants of the risk are the involved chromosomes, the breakpoint positions, and the imbalances of the fetus and that there is no causal relationship between risk and asc in either direction. Furthermore, we have assumed that risks for carriers of the same translocation and the same sex are equal, and no investigations have indicated that this assumption should be questioned.

We have argued above that the asc probability is far higher for a liveborn, unbalanced child, in particular when the abnormality is severe and well known, than the ascertainment probability for spontaneous abortions and the asc probability for phenotypically normal carriers and carrier fetuses except perhaps for parents of advanced age. Thus rcp families with a high risk for liveborn, unbalanced offspring with an easily recognizable abnormality will almost always be ascertained through these liveborn abnormal individuals and not in other ways.

On the other hand, families with a low risk for liveborn, unbalanced offspring may have been ascertained in different ways, although asc through an occasional, liveborn abnormal will often occur because of the much higher asc probability for this offspring category. One may compare low-risk families at risk for the same single-segment imbalance after the same disjunction and segregation mechanism either ascertained through abnormal liveborns or ascertained in a different way. Then, after asc corrections, one will find that the risks are equal except for random variations for these two groups [2]. From a biological point of view, it is difficult to argue that, given two pedigrees with almost the same rcp, the one that by chance was ascertained through an abnormal liveborn has a higher risk than the other, ascertained in a different way.

OFFSPRING OF CHROMOSOMALLY UNBALANCED (PHENOTYPICALLY NORMAL) INDIVIDUALS WITH TERTIARY MONOSOMIES

It has been assumed implicitly that the unbalanced offspring of the rcp considered so far are so abnormal and mentally retarded that they are unlikely to reproduce. If, however, the chromosomal imbalance has little or no effect on the phenotype, this assumption is not correct.

In this section, we consider a group of families among whom some of the parents have 45 chromosomes and an unbalanced chromosome consitution that is a tertiary monosomy arising from a 3:1 disjunction. In some cases, these carriers have transmitted their abnormal chromosome constitution unchanged to their offspring, others have offspring with several types of imbalances, and some offspring have normal chromosomes.

The translocations to be considered are the following: t(11;22)(q25;q11) [57], t(19;21)(p13?;q21) [58] and t(X;22)(q24;q11) [59], for which breakpoint positions have been reported, and a translocation between chromosomes 6 and 21, for which the designation may be t(6;21)(q25?;q11?) [60]. All these translocations are characterized by one breakpoint in a terminal band of a nonacrocentric chromosome and a breakpoint in a proximal long arm band of a small acrocentric chromosome.

The phenotypically normal carriers have a derivative that is formed by translocating almost the complete long arm of the small acrocentric chromosome onto the terminal segment, conserving virtually all the nonacrocentric one. The derivative that is lacking is formed by the complete acrocentric short arm and the proximal long arm segment and perhaps a terminal segment from the nonacrocentric chromosome.

Of the four translocations, the loss of the derivative does not seem to have had any effect of the phenotype for individuals with the tertiary monosomies of the t(19;21) and the t(6;21) translocations. For the former the breakpoint on chromosome 21 might have been in 21q11 instead of the reported 21q21. For the t(11;22) translocation, both the carrier mother and the carrier children are somewhat intellectually handicapped because of the missing chromosomal material. For the t(X;22) translocation, the carrier parent was a mosaic, and one carrier child with 45 chromosomes was retarded because of lack of chromosome material for chromosome 22.

In contrast to carriers of balanced rcp, the carriers of these unbalanced products have three chromosomes involved, and these form a *trivalent* during meiosis (Fig. 18).

Alternate segregation will lead to one daughter cell with only the derivative and the other daughter cell with the two normal chromosomes. The former will lead to offspring with the parental chromosome constitution, the latter to an individual with normal chromosomes. Adjacent segregations will produce more heavily unbalanced offspring. Unbalanced liveborns may be produced if a trisomy for either the long arm of the acrocentric chromosome or the complete nonacrocentric chromosome is compatible with live birth. This will occur when the translocation chromosome is included in a daughter cell together with one of the normal chromosomes. These unbalanced offspring have 46 chromosomes in contrast to 45 for the carrier parent. Nothing is known about the probabilities for these different types of gametes.

In the families with t(6;21) and t(19;21), unbalanced offspring with Down syndrome have been born; in the t(X;22) family, two unbalanced offspring

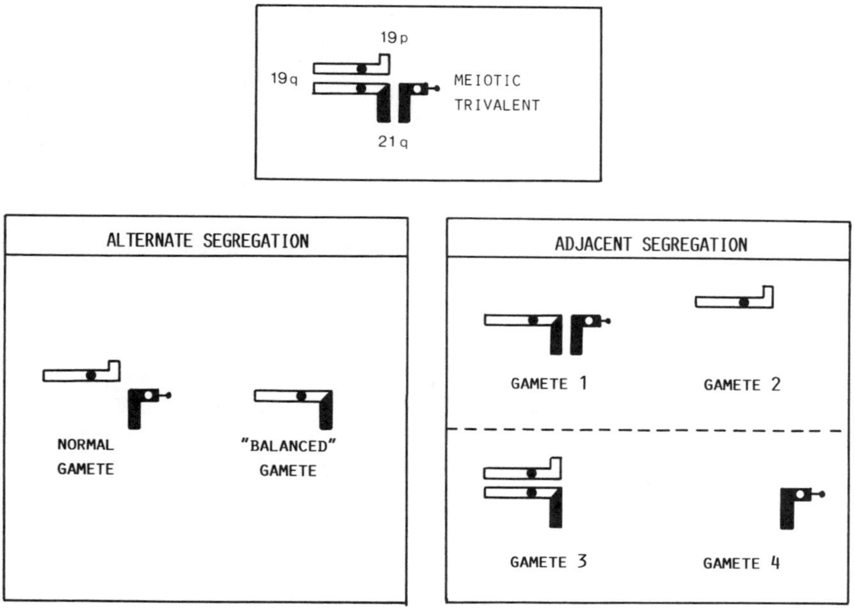

Fig. 18. Segregation in an unbalanced rcp carrier parent. Parent with a tertiary monosomy that forms a meiotic trivalent at pachytene with three possible types of segregation. The translocation is der(19), t(19;21)(p13;q21) [58]. The parent has a virtually normal phenotype; only the heterochromatic der(21) is missing. *Alternate segregation:* After fertilization by a normal gamete, the normal gamete produced from the trivalent will lead to a fetus with normal chromosomes and the "balanced" gamete to a fetus with a tertiary monosomy with 45 chromosomes like the carrier parent. *Adjacent segregations:* After fertilization by a normal gamete, the following types of unbalanced fetuses may arise: Gamete 1 will lead to a fetus with the der(19) and two copies of chromosome 21. He or she will have 46 chromosomes and Down syndrome. Gamete 2 will lead to a monosomy 21, in fact an interchange monosomy. Gamete 3 will lead to a fetus with the der(19) and two normal chromosome 19s, i.e., corresponding to trisomy 19. Gamete 4 will lead to a monosomy 19. The fetuses originating from the gametes 2–4 are unlikely to be viable.

with an extra X chromosome are recorded. The risks for such unbalanced offspring in these two families may be medium to high; both types of unbalanced offspring have high survival probabilities.

With regard to t(11;22), the only possibility for liveborn offspring, other than with a tertiary monosomy, is a trisomy for the long arm of chromosome 22. The risk for this abnormality is likely to be small in that the estimated risk for single-segment imbalance after a 2:2 disjunction with adjacent-1 segregation is estimated to be less than 2.6%, which indicates that this imbalance carries only a small survival probability.

The individuals in this t(11;22) family are the only tertiary monosomics reported from the large group of t(11;22) families considered above. A possible ancestor, having a balanced t(11;22) translocation, has not been traced. In that the carriers in this family are lacking the derivative that led to tertiary trisomies in the t(11;22) families considered above this, family should not be included in such grouped data, as has been done previously [13].

In each of these four families, only the offspring of one carrier, a female, have been studied. An open question concerning such tertiary monosomics is whether the risk will be the same for male and for female carriers or is higher for female carriers, as for 3:1 disjoining rcp or the analogous Robertsonian translocations. For both these groups of translocations, unbalanced offspring will have one chromosome more than the carrier parent, and that is also the case for the families considered in this section. It is assumed that the extra X or the extra chromosome 21 found in the cells of the more unbalanced offspring in these families originates from the carrier parent, but a nondisjunction in the normal parent has not been excluded.

ROBERTSONIAN TRANSLOCATIONS

Robertsonian translocations (rob) consist of a single derivative formed from the long arms of two acrocentric chromosomes with breakpoints close to or at the centromeres. The acrocentric chromosomes may belong to the same or to different pairs. A rob between the long arms of chromosomes 14 and 21 is denoted t(14q21q). The other derivative, formed from the short arms of the two acrocentric chromosomes, has been lost, just as for the carriers considered in the previous section. Formally, one can consider carriers of rob also as tertiary monosomics for the short arms of the two involved acrocentric chromosomes. Again, this imbalance will usually have no effect on the phenotype. Phenotypically normal carriers of the retained derivative, which is termed the Robertsonian translocation, will have 45 chromosomes. They are regarded as balanced carriers, although this does not hold true exactly.

Many rob are dicentric, with one centromere suppressed as shown by C-banding [61]. In other cases, it is claimed they are formed through a fusion of the two centromeres. This has led to the term *centric fusions,* often used as a synonym for rob.

Some rob are formed from the long arms of two homologous acrocentrics, which may happen at a prezygotic stage although not to produce carriers. Parents with such balanced, homologous rob cannot have normal offspring, and some female carriers have long histories of unsuccessful pregnancies. In one case, a phenotypically normal female carrier of a t(21q21q) had 12 pregnancies, eight Down syndrome children and four spontaneous abortions [62]. For male carriers, this has not been observed (reduced fertility?). Only in t(21q21q) and in one instance of t(13q13q) [5] have liveborn, unbalanced offspring been observed. Thus female carriers of these translocations will have a risk of 100% for liveborn offspring to be unbalanced.

For rob involving two different acrocentrics, there are ten possible pairs, each pair representing a separate risk group. Before banding techniques were available, chromosomes 13, 14, and 15 were called the D group, and 21 and 22 the G group [63,64]. Thus there were three main groups of rob D/D, D/G, and G/G translocations, the last being the t(21q22q) group.

Because a rob consists of a single chromosome, the segregation patterns are different from those we considered for rcp. Instead of the quadrivalent, there is a pachytene trivalent formed by the translocation chromosome and both normal homologues. As is shown in Figure 19 for t(14q21q), there are three possible segregation types that can lead to viable offspring. One type is alternate segregation in which the rob is included in one daughter cell and the two normal chromosomes in the other, which after fertilization will produce a carrier or an individual with normal chromosomes, respectively. In the two other types of segregation, one normal chromosome together with the rob will be included in one daughter cell and the remaining normal chromosome will be included in the other. All these segregation products will lead to unbalanced fetuses. Only those trisomic for the long arm of an acrocentric chromosome may be viable.

RISKS FOR ROB

In contrast to rcp, a study of the risks for rob carriers taking the involved chromosomes into account has not been carried out. However, by combining the evidence from different studies, some general risk patterns emerge. We shall first consider rob involving chromosome 21.

Before banding techniques were available, rob(D;21) could be identified as such if any of the carriers had a Down syndrome child with 46 chromosomes. A segregation analysis of 38 such pedigrees was carried out [19] with specially developed methods [17–19,30]. The formal basis of these analyses has also been considered. [35]. Because banding techniques were unavailable at that time, it was not possible to distinguish between t(13q21q), t(14q21q), and t(15q21q), nor to determine whether they were dicentric or had different origins and centromeric structures. The risk estimate for female carriers to have liveborn children with translocation trisomy 21 was found to be 10.1%; that for male carriers was very low, but the risks were heterogeneous [19].

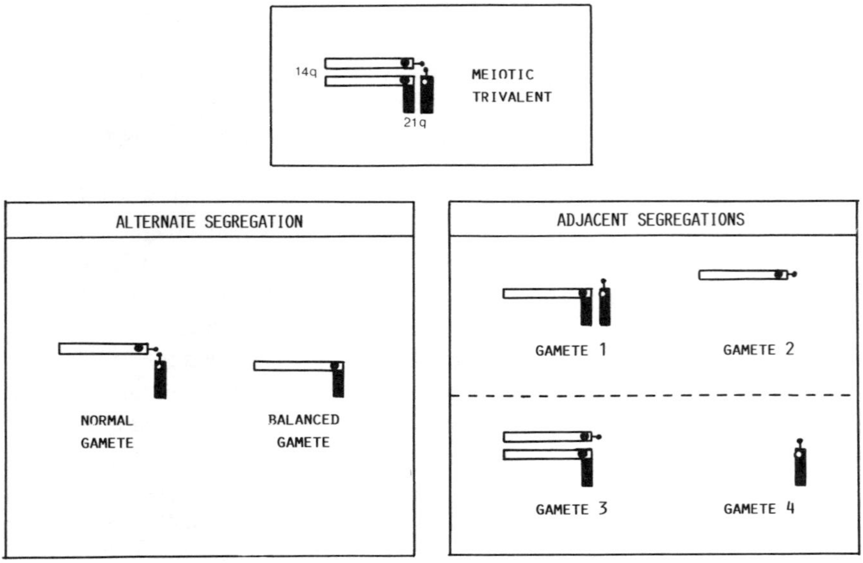

Fig. 19. Pachytene trivalent diagram of a Robertsonian translocation, t(14q21q). Robertsonian translocation showing three different types of segregation. *Alternate segregation:* After fertilization by a normal gamete, the normal gamete will lead to a fetus with normal chromosomes and the gamete with the balanced translocation to a carrier fetus with 45 chromosomes. *Adjacent segregations:* After fertilization by a normal gamete, the following types of unbalanced fetuses may arise: Gamete 1 will lead to a fetus with the translocation and two copies of chromosome 21. Because the long arm of this chromosome occurs in triplicate, the fetus will have Down syndrome and 46 chromosomes. Gamete 2 will lead to a fetus with a monosomy for chromosome 21. Gamete 3 will lead to a fetus with the translocation and two copies of chromosome 14. The fetus will be trisomic for the long arm of this chromosome. Gamete 4 will lead to a fetus with a monosomy for chromosome 14. The fetuses originating from gametes 2 and 4 are unlikely to survive.

In the G/G groups, the t(21q22q) could be identified without banding techniques if the carriers in a pedigree had both children with 46 chromosomes and Down syndrome and phenotypically normal offspring.A segregation analysis of seven such pedigrees produced a risk estimate of 8.9% for female carriers and again a very low risk estimate for male carriers [20].

No further segregation analyses of these rob groups seem to have been carried out on data from liveborns. Some information about the relative prevalence and segregation patterns can be obtained from two European collaborative studies on prenatal diagnosis data [49,50].

A rough estimate of the relative prevalence of the different groups of rob with chromosome 21 involved can be obtained by considering the proportions of recorded carriers in the different groups in these European studies [50]. These are ascertained in roughly the same way, usually through a previous child with Down syndrome. Some carriers are related but in approximately the same proportion in all groups. In these data [50], there were 208 carriers with such a translocation, 11.5% with t(13q21q), 75.0% with t(14q21q), 5.8% with t(15q21q), and 7.7% with t(21q22q); thus t(14q21q) is by far the most frequent in this group.

In second-trimester prenatal diagnoses for these rob carriers, there were only a few analyses performed for translocations other than t(14q21q). Thus no comparison regarding risks for abnormal fetuses could be made between the groups. For t(13q21q), t(14q21q), and t(15q21q) taken together, the estimated risk for unbalanced fetuses was 14.5%, and for t(21q22q), it was 15.8% (in both cases for female carriers). Taking into account that these data originate from about week 16 of pregnancy and that a proportion of the fetuses with Down syndrome or other trisomies would have been aborted spontaneously later in the pregnancy, these figures are in accordance with those from segregation analyses of data from liveborns [19,20]. For paternal carriers, the agreement between liveborn and prenatal data is also good: No fetus with a carrier father was unbalanced [50].

The risk figures discussed so far have to be considered as averages for the different rob groups. The groups may be heterogeneous with regard to the risk for Down syndrome. In a number of pedigrees from Scotland, ascertained through a balanced carrier instead of through a Down syndrome child, no Down syndrome case was reported [65]. On the other hand some dicentric rob may carry an elevated risk [61,66]. In connection with the previous segregation analyses [19,35], it was suggested that the risk may depend on the origin and structure of the centromeric region of the rob. If the centromere in a t(14q21q) translocation originates from chromosome 14 or in a dicentric dic(14;21)(p11;p11), the 21 centromere is suppressed; a segregation in which the rob and chromosome 21 are included in a daughter cell corresponds to adjacent-1 segregation with separation of homologous centromeres. If, instead, the centromere originated from chromosome 21 (or the 14 centromere was suppressed), there would have been adjacent-2 segregation, i.e, nonseparation (or nondisjunction) of homologous centromeres. A subgrouping of, e.g., the t(14q21q) based on origin and centromeric structure should be undertaken and segregation analyses carried out for each subgroup separately. For the above-mentioned pedigrees from Scotland [64], no information is available regarding the centromeric structure.

Let us turn now to the rob in which chromosome 21 is not involved. The D/D rob for which a segregation analysis has been carried out, i.e., the t(13q14q) [67], may be the largest homogeneous group of translocations in man. All abnormal segregants in this study [67], which in most cases entailed Patau syndrome (corresponding to trisomy 13), were probands and so were omitted from the data through asc corrections. The estimated risk both for male and female carriers was very low, below 1%. This result agrees fully with the data from European collaborative studies, in which none of 230 prenatal diagnoses for carriers of t(13q14q) translocations produced an abnormal fetus [50].

For other rob, i.e., t(13q15q), t(14q15q), t(13q22q), t(14q22q), and t(15q22q), few data are available. In the European collaborative studies, only 32 prenatal diagnoses were carried out for all these. One unbalanced fetus, with translocation trisomy 13, was diagnosed in a carrier mother of a t(13q22q) [50].

PROPORTION OF BALANCED ROB AND RCP CARRIERS IN THE PHENOTYPICALLY NORMAL OFFSPRING

So far, only risks for *abnormal* offspring have been considered in the progeny of rcp and rob carriers. These are the most important figures in a medical context. However, from the point of view of population genetics, the transmission rate of the balanced translocation among phenotypically normal offspring is significant, particularly since the introduction of prenatal diagnosis. With the latter, induced abortion of most unbalanced fetuses is the norm, whereas chromosomally balanced or normal pregnancies are continued; therefore, the prevalence of balanced carriers may increase in the population.

In the segregation analysis of D/G translocations, the proportion of carriers among phenotypically normal individuals was estimated as 0.57 ± 0.03 [19]. For t(13q14q) translocations, the corresponding figure was found to be 0.55 ± 0.04 [67]. In both cases, the figure is above the theoretical value of 50%, in the latter case not significantly. In the European collaborative study [50], similar patterns emerge. For the two major rob groups there were 96 normal fetuses and 134 balanced carriers for t(13q14q) and 68 normal fetuses and 99 balanced carriers for t(14q21q). Some of the carrier fetuses had been found through a prenatal diagnosis carried out without knowledge of the familial rob but for a different reason, e.g., advanced parental ages. This may lead to an overrepresentation of balanced carriers, i.e., an ascertainment bias. Such was the case in the European study for three fetuses carrying a balanced translocation with chromosome 21 involved and for 18 with chromosome 21 not involved. However, the translocations were not further specified regarding chromosomes involved [50]. If we make the assumption that all three fetuses with a rob involving chromosome 21 had t(14q21q), we are left with 96 carrier fetuses instead of 99 after asc correction. Furthermore, if we assume that all 18 fetuses carrying a rob without 21 had a t(13q14q), we get 116 carrier fetuses after this asc correction. Even in this extreme situation, we find (by a Fisher omnibus test considering both groups separately) that the carrier proportion is significantly above 50% ($P = 0.01$). In all those rob included in the test a chromosome 14 was involved.

In prenatal diagnoses of fetuses of rob carriers with chromosome 14 not involved, the difference between the numbers of balanced carriers and normals is considerably smaller. There were 41 fetuses with normal chromosomes and 45 with the balanced translocation [50]. If some of these carrier fetuses belong to the 3+18 above-mentioned carrier fetuses found without previous knowledge of the familial translocation, the difference will be even smaller after an asc correction.

The segregation distortion indicated above may therefore be connected with the involvement of chromosome 14, but the problem needs to be investigated further. It could be pointed out that if the involvement of chromosome 14 causes such a segregation distortion, these rob will have a selective advantage, which might explain the relatively high prevalences of just the commonest rob in man: t(13q14q) and t(14q21q). In the same context, one might argue that the higher prevalence of t(13q14q) than of t(14q21q) can be explained by the much lower risk of unbalanced offspring for carriers of the former.

Turning then to rcp, our data [2] have shown that, after asc corrections for carriers at risk for unbalanced offspring after a 2:2 disjunction with adjacent-1 segregation, the carrier proportion among the phenotypically normal, liveborn offspring of female carriers was $341/600 = 0.568 \pm 0.020$ and of male carriers was $214/362 = 0.591 \pm 0.026$. These carrier proportions are both significantly above the theoretical value 0.50 for carrier mothers ($P = 0.00035$) and for carrier fathers ($P = 0.00020$).

On the other hand, in the phenotypically normal offspring of carriers of translocations similarly corrected and at risk after 3:1 disjunction, the carrier proportion was found to be $51/104 = 0.500 \pm 0.049$ when the mother was the carrier, and $23/44 = 0.523 \pm 0.075$ when the father was the carrier. Neither of these values deviated significantly from 0.50. These results indicate that the proportion of carriers may be different for different types of rcp.

When considering the corresponding proportion among fetuses investigated prenatally, a different picture emerges. For families at risk after a 2:2

disjunction with adjacent-1 segregation after asc corrections, we found that the proportion of carrier fetuses of maternal carriers was 85/179 = 0.475 ± 0.037 and of paternal carriers was 56/104 = 0.538 ± 0.049. These figures do not deviate from the theoretical value 0.50 but are significantly less than the corresponding figures for liveborn individuals ($P = 0.0096$). For carriers of translocations at risk after a 3:1 disjunction, there are too few prenatal diagnoses for a comparison.

There are several possible explanations for this discrepancy in carrier proportions between liveborn and prenatal diagnosis data. A possible explanation is that the asc correction usually applied to these data (simply to omit the probands that usually are balanced carriers) independently of asc category, may result in an overcorrection in prenatal diagnosis data and produce estimates that are too small. A more appropriate asc correction, which is somewhat more complicated, is given in the Appendix. This difficulty has been discussed above. For inherited inversions, a similar discrepancy between the carrier proportions in liveborn and prenatal data [24,27] is found.

A different explanation may be sought in a possible overreporting of liveborn carriers compared with chromosomally normal individuals [2]. However, this explanation implies a higher carrier ratio than observed in liveborn offspring of carriers of translocations at risk of unbalanced offspring after a 3:1 disjunction. Different manners of reporting are not very likely for families belonging to one or the other of the two translocation groups, so this explanation is more doubtful. To solve these problems, further analyses of the data are necessary.

CONCLUSIONS

This chapter has demonstrated that carriers of translocations can have very different risks of liveborn offspring being chromosomally unbalanced. The risk may vary between 0% and about 30% for rcp and may even be 100% for certain types of rob. The risk has been shown to depend heavily on the chromosomes involved and the breakpoint positions, which define the genetic content of the unbalanced segments and thereby determine the survival probability of an unbalanced fetus. The risk is not directly dependent on the actual length of these segments measured by the same scale, such that risks cannot be calculated by means of methods related to regression analysis based on segment lengths, which has been suggested [46].

The risk is not causally determined by the mode of asc which is stated in the literature several times [47,51,55]; on the other hand, the asc probability is found to be much higher for liveborn, unbalanced offspring, in particular when the abnormality is well known, than for other categories of offspring. Therefore, families with a high risk for such liveborn, abnormal offspring will frequently have been ascertained through one of these, whereas for families among whom the risk is small for liveborn, unbalanced offspring, different modes of asc will occur. The risk estimates, however, will be almost equal for groups of almost identical translocations ascertained in different ways. Formal relationships are demonstrated between breakpoint positions, possible combinations of balanced segments, and disjunction and segregation mechanisms leading to the different types of unbalanced offspring.

The number of observations available in our data [2] has been sufficient to produce a fairly comprehensive picture of risk patterns for rcp leading to offspring with single-segment imbalances after 2:2 disjunction with adjacent-1 segregation. However, for several subgroups of segments, the data are few and the estimates not very accurate. More data are needed especially for the low-risk groups, in which reasonably accurate risk estimates only can be established through large numbers of observations. The number of pedigrees has also been insufficient to study in detail the relationships between individual risks for single- and double-segment imbalances and also those for risks for specific abnormalities after different disjunction and segregation events.

The risk is clearly higher for female than for male carriers when the number of chromosomes in the unbalanced offspring is different from the chromosome number in the carrier. This holds for rob as well as for rcp carrying risk for unbalanced offspring after a 3:1 disjunction. This pattern is similar to risk patterns for ordinary trisomies, such as trisomy 21, in which the maternal risk is higher than the paternal one. A further similarity is that risks after 3:1 disjunction increase with advancing age of the carrier. For rob, this problem has not been studied in any detail as yet. For rcp carrying risk after 2:2 disjunction with adjacent-1 segregation, no sex differences have been found for the carriers. For those carrying risk after a 2:2 disjunction with adjacent-2 segregation, the question is still open.

Some problems have not been considered in this paper or in our study [2]. One is the relationship between the chromosomes involved and the sex ratio in the offspring. For rcp in which 5p or 9p is involved, it has been found that the proportion of females is exceptionally high in all categories of offspring [12,38,68], but this is not the case for some other chromosome arms, e.g. the short arms of 4, 10, and 12 [38,39]. Another problem is whether there is some connection between the sex of the carrier and the sex of the abnormal offspring, which was suggested in one study [46].

A final problem is whether the probabilities for the different types of disjunction and segregation (production of unbalanced gametes) depend on the chromosome involved or on their centromeres. This is an important problem both for rcp and for rob.

The principles and methods used to organize and analyze pedigree data for inherited translocations may be adapted to other types of inherited, structural rearrangements, e.g., inversions or insertions, for which to date they have not been used [24,27,69].

ACKNOWLEDGMENTS

The authors are indebted to Anders Milhøj for help with several of the figures and to Joan Rasmussen for skilful typing of the different versions of the manuscript, both from the Institute of Statistics, University of Copenhagen. We also thank Markus Bürgle from the Department of Pediatric Genetics, University of Munich, for diagramming the pedigrees and two figures. Finally, we thank the editor, Dr. Art Daniel, for many constructive comments and suggestions on an earlier version.

REFERENCES

1. Jacobs PA: Epidemiology of chromosome abnormalities in man. Am J Epidemiol 105:180, 1977.
2. Stengel-Rutkowski S, Stene J, Gallano P: Risk estimates in balanced parental reciprocal translocations. Ann Genet (in press).
3. de Grouchy J, Turleau C: "Atlas des Maladies Chromosomiques 2nd ed." Paris: Expansion Scientifique Francaise, 1982.
4. de Grouchy J, Turleau C: "Clinical Atlas of Human Chromosomes." New York: John Wiley & Sons, 1982.
5. Schinzel A: "Catalogue of Unbalanced Chromosome Aberrations in Man." Berlin: Walter de Gruyter, 1983.
6. Stengel-Rutkowski S, Murken J, Frankenberger R, et al: New chromosomal dysmorphic syndromes. 2. Trisomy 10p. Eur J Pediatr 126:109, 1977.
7. Paris Conference (1971): Standardization in Human Cytogenetics (Birth Defects, Original Article Series, Vol VIII No. 7)." New York: The National Foundation, 1972 (also in Cytogenetics 11:313, 1972).
8. ISCN (1978): "An International System for Human Cytogenetic Nomenclature (Birth Defects: Original Article Series, Vol XIV, No. 8)."New York: The National Foundation, 1978 (also in Cytogenet Cell Genet 21:309, 1978).
9. ISCN (1981): "An International System for Human Cytogenetic Nomenclature—High Resolution Banding (1981) (Birth Defects: Original Article Series, Vol XVII No. 5)." New York: March of Dimes Birth Defects Foundation, 1981(also in Cytogenet Cell Genet 31:1, 1981).
10. ISCN (1985): "An International System for Human Cytogenetic Nomenclature." Harden DG, Klinger HP (eds). Published in collaboration with Cytogenet Cell Genet (Basel: Karger), 1985 (also in Birth Defects:Original Article Series, Vol XXI, No. 1. New York: March of Dimes Birth Defects Foundation, 1985).

11. Blum U: "Partielle Trisomie 5p. Versuch der Definition eines neuen autosomalen Trisomie-Syndroms anhand 3 Falluntersuchungen." Thesis, University of Munich, 1981.
12. Niebuhr E: The cri du chat syndrome. Epidemiology, cytogenetics and clinical features. Hum Genet 44:227, 1978.
13. Fraccaro M, Lindsten J, Ford CE, Iselius L: The 11q;22q translocation: A European collaborative analysis of 43 cases. Hum Genet 56:21, 1980.
14. Wirtz A, Stengel-Rutkowski S, Stene J, et al: Pränatale Diagnostik bei Eltern mit einer balancierten chromosomalen Strukturaberration. 10 Fallberichte. Geburtsh Frauenheilk 40:575, 1980.
15. Stene J: Choice of ascertainment model I. Discrimination between multi-proband models by means of birth order data. Ann Hum Genet 42:219, 1978.
16. Stene J: Choice of ascertainment model II. Discrimation between multi-proband models by means of birth order data. Ann Hum Genet 42:493, 1979.
17. Stene J: Analysis of segregation patterns between sibships within families ascertained in different ways. Ann Hum Genet 33:261, 1970.
18. Stene J: Comparison of segregation ratios of families ascertained in different ways. Ann Hum Genet 33:395, 1970.
19. Stene J: Statistical inference on segregation ratios for D/G translocations, when the families are ascertained in different ways. Ann Hum Genet 34:93, 1970.
20. Stene J: A statistical segregation analysis of (21q22q) translocations. Hum Hered 20:465, 1970.
21. Stene J: Assumptions for different ascertainment models in human genetics. Biometrics 33:523, 1977.
22. Stene J: Ascertainment models for human families selected through several affected individuals. Biometrics 34:455, 1978.
23. Broustet A, Serville F, Meynie M, Mouliner J, Moretti G: Etude cytogénétique de 20 couples ayant présentés plusieurs avortements spontanés. Semin Hop Paris 51:299, 1975.
24. Stene J: Comments on methods and results in: Sherman et al., "Segregation analysis of balanced pericentric inversions in pedigree data." Clin Genet 30:95, 1986.
25. Elston RC, Bonney GE: Sampling considerations in the design and analysis of family studies. In Rao DC, Elston RC, Kuller LH, Feinleib M, Carter C, Havlik R (eds): "Genetic Epidemiology of Coronary Heart Disease: Past, Present and Future." New York: Alan R. Liss, Inc, 1984, p 349.
26. Morton NE, Rao DC, Lalouel JM: "Methods in Genetic Epidemiology." Basel: Karger, 1983.
27. Sherman SL, Iselius L, Gallano P et al.: Segregation analysis of balanced pericentric inversions in pedigree data. Clin Genet 30:87, 1986.
28. Sherman SL, Iselius L: Response to Stene's comments on the segregation of balanced pericentric inversions. Clin Genet 30:521, 1986.
29. Stene J: Reply to a response by Sherman & Iselius to comments on the segregation of balanced pericentric inversions. Clin Genet 31:355, 1987.
30. Stene J: Sampling of pedigrees. Adv Appl Prob 7:18, 1975.
31. Elston RC, Stewart J: A general model for the genetic model of pedigree data. Hum Hered 21:523, 1971.
32. Elston RC, Sobel E: Sampling considerations in the gathering and analysis of pedigree data. Am J Hum Genet 31:52, 1979.
33. Lalouel JM, Morton NE: Complex segregation analysis with pointers. Hum Hered 31:312, 1981.

34. Cannings C, Thompson EA, Skolnik M: Probability functions on complex pedigrees. Adv Appl Prob 10:26, 1978.
35. Stene J: A model for translocation inheritance in man. Segregation patterns for a single centric-fusion translocation. Theor Population Biol 9:151, 1976.
36. Stene J, Stengel-Rutkowski S: Risk for short arm 10 trisomy. A segregation analysis of eleven families with different translocations. Hum Genet 39:7, 1977.
37. Stengel-Rutkowski S, Albert A, Zahn-Messow K, et al: New chromosomal dysmorphic syndromes. 4. Trisomy 12p. Eur J Pediatr 136:249, 1981.
38. Stene J, Stengel-Rutkowski S: Genetic risks for familial reciprocal translocations with special emphasis on those leading to 9p, 10p and 12p trisomies. Ann Hum Genet 46:41, 1982.
39. Stengel-Rutkowski S, Warkotsch A, Schimanek P, Stene J: Familial Wolf's syndrome with a hidden 4p deletion of an 8p segment. Unbalanced inheritance from a maternal translocation (4;8)(p15;p22). Case report, review and risk estimates. Clin Genet 25:500, 1984.
40. Daniel, A: Structural differences in reciprocal translocations. Potential for a model of risk in rcp. Hum Genet 51:171, 1979.
41. Summit RL, Tipton RE, Wilroy RS, Martens PR, Phelan JP: "X-Autosome Translocations: A review." New York: Alan R. Liss, Inc, Birth Defects: Original Article Series XIV (6C):219, 1977.
42. Ford CE, Clegg HM: Reciprocal translocations. Br Med Bull 25:110, 1969.
43. Brandriff B, Gordon L, Ashwort LK, et al.: Cytogenetics of human sperm: Meiotic segregation in two translocation carriers. Am J Hum Genet 38:197, 1986.
44. Jalbert P, Sele B: Factors predisposing to adjacent 2 and 3:1 disjunctions: Study of 161 human reciprocal translocations. J Med Genet 16:476, 1979.
45. Duckett DP, Roberts SH: Adjacent 2 meiotic disjunction. Report of case resulting from a familial 13q;15q balanced reciprocal translocation and review of the literature. Hum Genet 58:377, 1981.
46. Schaefer MSD: "Segregation und Pathologie autosomaler familiärer Translokationen beim Menschen." Thesis, University of Kaiserslautern, 1983.
47. Vogel F, Motulsky AG: "Human genetics. Problems and Approaches, 2nd ed." Berlin: Springer Verlag, 1986, p 66.
48. Rethoré MO, Larget-Piet L, Abonyi D, et al.: Sur quatre cas de trisomie pour le bras court du chromosome 9. Individualisation d'une nouvelle entité morbide. Ann Genet 13:217, 1970.
49. Boué A: European collaborative study on structural chromosome anomalies in prenatal diagnosis. Group report. In Murken J-D, Stengel-Rutkowski S, Schwinger E (eds): "Prenatal Diagnosis." Stuttgart: Enke, 1979, p 34.
50. Boué A, Gallano P: A collaborative study of the segregation of inherited chromosome structural rearrangements in 1356 prenatal diagnoses. Prenat Diagn 4:45, (Spec Iss), 1984.
51. Lejeune J, Dutrillaux B, de Grouchy J: Reciprocal translocations in human populations. In Jacobs PA, Price WH, Law P (eds): "Human Population Cytogenetics." Edinburgh: Edinburgh University Press, 1970, p 81.
52. Hamerton JL: "Human Cytogenetics I." New York: Academic Press, 1971.
53. Lindenbaum RH, Bobrow M: Reciprocal translocations in man. 3:1 meiotic disjunction resulting in 47- or 45-chromosome offspring. J Med Genet 12:29, 1975.
54. Lurie IW, Presman H, Lazjuk GI: Risque génétique dans les cas de translocation reciproques. J Genet Hum 26:385, 1978.

55. Aurias A, Prieur M, Dutrillaux B, Lejeune J: Systematic analysis of 95 reciprocal translocations of autosomes. Hum Genet 45:259, 1978.
56. Jalbert P, Sele B, Jalbert H: Reciprocal translocations: A way to predict the mode of imbalanced segregation by pachytene-diagram drawing. A study of 151 human translocations. Hum Genet 55:209, 1980.
57. Fu W-N, Borgaonkar DS, Ladewig PP, Weaver J, Pomerance HH: Structural aberrations of the long arm of chromosome No. 22. Report of a family with translocation t(11;22)(p25;q11). Clin Genet 10:329, 1976.
58. Pfeiffer RA, Kessel EK, Soer K-H: Partial trisomies of chromosome 21 in man. Two new observations due to translocations 19;21 and 4;21. Clin Genet 11:207, 1977.
59. Jenkins MB, Davis E, Thelen TH, Boyd L: A familial X-22 translocation with an extra X chromosome. Am J Hum Genet 26:736, 1974.
60. Borgaonkar DS, Bias WB, Chase GA, et al.: Identification of a C6/G21 translocation chromosome by Q-M and Giemsa banding techniques in a patient with Down's syndrome with possible assignment of Gm locus. Clin Genet 4:53, 1973.
61. Daniel A, Lam-Po-Tang PRLC: Structure and inheritance of some heterozygous Robertsonian translocations in man. J Med Genet 13:381, 1976.
62. Furbetta M, Falorni A, Antignani P, Cao A: Sibship (21q21q) translocation Down's syndrome with maternal transmission. J Med Genet 10:371, 1973.
63. London Conference on the Normal Human Karyotype (1963): Cytogenetics 2:264, 1963. Reprinted in Chicago Conference (1966). [63]
64. Chicago Conference (1966)"Standardization in Human Cytogenetics (Birth Defects: Original Article Series Vol II-No. 2)." New York: The National Foundation, 1966.
65. MRC Clinical and Population Cytogenetics Unit, Edinburgh, 1973 (unpublished data).
66. Daniel A, Williams K, Lam-Po-Tang PRLC: Higher risk to D;G translocation carriers of tdic (13;21) as compared to tdic (14;21). J Med Genet 17:491, 1980.
67. Harris DJ, Hankins L, Begleiter ML: Reproductive risk of t(13q14q) carriers: Case report and review. Am J Med Genet 3:175, 1979.
68. Lurie IW, Lazjuk GI, Gurevich DB, Usoev SS: Genetics of the +p9 syndrome. Hum Genet 32:23, 1976.
69. Kaiser P: Pericentric inversions. Problems and significance for clincal genetics. Hum Genet 68:1, 1984.
70. Edward AWF: "Likelihood." Cambridge: Cambridge University Press, 1972, p 75.
71. Stene J: "Interval Estimates for the Binomial p When All Available Individuals Belong to a Single Category (Research Report 98)." Copenhagen: University Institute for Statistics, 1985.

APPENDIX

Ascertainment Corrections for Index Sibships Ascertained Through the Oldest Abnormal Offspring

Let m be the number of sibs and a the number of abnormal offspring in the index sibship, p the probability (risk) for abnormal offspring, and $q = 1 - p$. If there is only one proband and he or she is the oldest affected child, the appropriate asc correction is to use the quantities $A_1(m)$ and $B_1(m)$ as demonstrated below.

$$A_1(m) = \frac{m^2p^2q^{m-1}}{(1-q^m)^2}, \quad B_1(m) = \frac{m}{1-q^m} - \frac{m^2pq^{m-1}}{(1-q^m)^2}$$

If there are two probands and they are the two oldest affected children, then the quantities $A_2(m)$ and $B_2(m)$ should be used.

$$A_2(m) = \frac{m^2 (m - 1)p^3q^{m-2} (1 - q^{m-1})}{(1 - q^m - mpq^{m-1})^2} - \frac{m(m - 1)p^2q^{m-2}}{1 - q^m - mpq^{m-1}}$$

$$B_2(m) = \frac{m(1 - q^{m-1}) + m(m - 1) pq^{m-2}}{1 - q^m - mpq^{m-1}} - \frac{m^2(m - 1)p^2q^{m-2}(1 - q^{m-1})}{(1 - q^m - mpq^{m-1})^2}$$

If there are three probands and they are the three oldest affected children, the corresponding correction is given in formula 4.39 in Stene [17]. To illustrate the use of the asc corrections, we shall consider a couple of examples.

Consider a pedigree with one index sibship ascertained through a single unbalanced child, who was the oldest of the unbalanced offspring, and some other sibships with carrier parents. Let c be the total number of chromosomally unbalanced children including the proband. Let N be the total number of pregnancies in the other sibships with the exception of those carrier children with the proband in direct line of descent. On the basis of the data from the other sibships, one can calculate a preliminary value of the parameter $p = (c - a)/N$. This value of p and that of $q = 1 - p$ are used to calculate $A_1(m)$ and $B_1(m)$. Then, an improved estimate of p is found through the following formula.

$$\hat{p} = \frac{c - A_1(m)}{B_1(m) + N} \qquad (1)$$

If the difference between $\hat{p}$ and p is small, e.g., 0.01 or less, one can use $\hat{p}$ as the risk estimate. If not, $\hat{p}$ is considered as a preliminary value to calculate new values of $A_1(m)$ and $B_1(m)$, and a second improved value of p is calculated by equation 1, and so on. If there are no unbalanced offspring outside the index sibship, then $c - a = 0$. In this case, one has to guess the preliminary value of p, e.g., 0.05, and proceed as before.

Consider next three pedigrees in a group with one index sibship each. In the first, there are m_1 children in the index sibship and two probands who are the two oldest unbalanced children. In the two others, there are m_2 and m_3 children in the index sibships, respectively, and in both cases there is a single proband in each index sibship who is in both cases the oldest unbalanced child. Let the numbers of unbalanced children in the three index sibships be a_1, a_2, and a_3, respectively, and now let N be the total number of pregnancies in the sibships other than the index sibships excluding carriers children with probands in direct line of descent. Then, a preliminary estimate of $p = (c - a_1 - a_2 - a_3)/N$. This value is used to calculate $A_2(m_1)$ and $B_2(m_1)$ for the first family, $A_1(m_2)$ and $B_1(m_2)$ for the second family, and $A_1(m_3)$ and $B_1(m_3)$ for the third family. Then, an improved estimate of p is calculated as follows.

$$\hat{p} = \frac{c - A_2(m_1) - A_1(m_2) - A_1(m_3)}{B_2(m_1) + B_1(m_2) + B_1(m_3) + N}$$

If the difference between $\hat{p}$ and p is not sufficiently small, the procedure is repeated. A number of examples are given elsewhere [17–20].

When applying these formulae to the different asc categories 1–5 as given in the section on Ascertainment Problems and Corrections, the parameter p is interpreted as the risk for liveborn unbalanced offspring in category 1, as the risk for stillbirths and liveborn unkaryotyped offspring in category 2, as the risk for spontaneous abortions in category 3, as the proportion of balanced carriers among phenotypically normal offspring in category 4, and as the proportion of fetuses with a balanced translocation or an unbalanced translocation product in category 5.

On the Risk Estimation

One wants to estimate the risk p for one type of abnormal segregant in a group, consisting of n individuals or pregnancies after asc corrections. Let the number of abnormals be x after asc corrections, as was done in our study [2] (see section on Ascertainment Problems and Corrections). If $x > 0$, then the usual method is to estimate p by $\hat{p} = x/n$ and report the interval

$$\hat{p} \pm \sqrt{\hat{p}(1 - \hat{p})/n} \qquad (2)$$

where the latter term is the standard deviation of the estimate $\hat{p}$.

A common situation in our data [2] has been that several abnormal segregants are recorded in a group of translocations but are probands and so are omitted by means of the asc correction. Then, $x = 0$ and $\hat{p} = 0$, and no interval of the type given in equation 2 could be reported.

Because this situation is unsatisfactory, e.g., from a counseling point of view, one may construct a one-sided risk interval for $x = 0$ corresponding approximately to that mentioned above for $x > 0$. The argument is based on the fact that the endpoints of the interval in equation 2 can be determined approximately from the log likelihood function of p for the given values of x and n. This function is called L(p). The maximum likelihood estimate of p corresponds to the maximum value of L(p), called $L(\hat{p})$. The endpoints of the interval in equation 2 are approximately the two values of p for which L(p) is 0.5 less than the maximum value $L(\hat{p})$ [70]. This argument can be used to estimate an upper endpoint for an interval for $x = 0$ corresponding to equation 2. For $x = 0$, the log likelihood function $L(p)\text{-}n \cdot \ln(1 - p)$, where ln is the natural logarithm. For $x = 0$, the maximum of L(p) is attained for $p = 0$, and it is $L(0) = 0$. Hence, the upper endpoint p_u of the interval is found by solving the equation

$$n \cdot \ln(1 - p_u) = -½$$

or

$$p_u = 1 - e^{-½n}$$

where $e = 2.718$.

As an example, let $n = 20$ and $x = 0$; then one gets $p_u = 0.025$. More details are given elsewhere [71].

The Cytogenetics of Mammalian Autosomal Rearrangements, pages 73–162

Chapter 2
Collaborative U.S.A. Data on Prenatal Diagnosis for Parental Carriers of Chromosome Rearrangements: Risks of Unbalanced Progeny

Art Daniel, Ernest B. Hook, and Gary Wulf

Data from 832 prenatal diagnoses of familial chromosome rearrangements, reciprocal (rcp) and Robertsonian (rob) translocations, para-and pericentric inversions (inv), insertions (ins), and complex or multiple rearrangements, were contributed by 63 U.S. cytogenetics laboratories for a collaborative study. The data were examined for relationships between balanced or unbalanced result and ascertainment, carrier parent, chromosome(s) involved, and (for rcp and inv) chromosome imbalance size (percent haploid autosome length; %HAL). Unlike rob (13, 14, or 15;21) in which maternal (mat) and paternal (pat) risk rates of unbalanced segregants are dissimilar, sex of carrier parent was relatively uninfluential in rcp. Although there were very few ins reported, four of six prenatal results were of imbalanced offspring, indicating a substantial risk for this rearrangment type. In contrast, for paracentric inv (N = 30), there were no unbalanced results. There were seven instances of unrelated chromosome abnormalities [four 47,+21, one 47,+13, one 47,XXY, one 47,XX,+i(Xq)] detected among 1,428 fetuses of translocation carriers. This frequency is close to the number expected from live birth data after adjustment for the fact these abnormalities were diagnosed at amniocentesis in a group probably at least slightly biased to advanced maternal age. Therefore, these data do not

Department of Medical Genetics, Shodair Children's Specialty Hospital, Helena, Montana 59601; present address: Cytogenetics Unit, Western Genetics Service, Institute of Clinical Pathology and Medical Research, Westmead Hospital, Westmead, N.S.W. 2145, Australia (A.D.)., Bureau of Environmental Epidemiology and Occupational Health, New York State Department of Health, Albany, New York 12237 (E.B.H.), Department of Pediatrics, Albany Medical College, Albany, New York 12237 (E.B.H.), Information Services Division, Montana State Department of Administration, Helena, Montana 59601 (G.W.), Department of Biology, Montana State University, Bozeman, Montana (A.D.).

provide evidence of an interchromosomal effect. For rob(13;14) (N = 125), there were 1.6% (2/125) unbalanced results (1 pat and 1 mat carrier). Eighteen of the rob(13;14) were ascertained through offspring with translocation trisomy 13 and one through a miscarried fetus with translocation trisomy 14; there were no recurrences in either group. For pericentric inv (N = 102), there were 2.0% (2/102) unbalanced prenatal diagnoses (one mat and one pat). For rcp and pericentric inv, each rearrangement was graphed once with actual (unbalanced result) or potential (normal or balanced result) imbalances (%HAL) plotted with trisomy (ordinate) and monosomy (abscissa). The graphed data were divided into regions of 1.0% trisomy and 0.5% monosomy and the rate of unbalanced pregnancies determined for each region. The modified data (N = 596 cases) of the European Cooperative Study were added to the U.S. data for rcp and similarly analyzed. For the rcp, the maximum risk rates observed were not significantly different for carriers of either sex; i.e., for rcp carriers ascertained by a previous unbalanced child (rcp type 1), there were maximum rates of prenatally diagnosed, unbalanced progeny of 48.0% (mat carriers) and 50.0% (pat carriers), both for the graphed region with the smallest defined interchange (rcp) segment sizes. These risks are close to the maximum theoretical risk of 50% and imply a single mode of segregation for individual rcp (for those segregating by more than one mode, the risks are not additive). For pooled rcp ascertained through balanced carriers because of recurrent miscarriages (rcp type 2) or incidentally ascertained prior to this prenatal diagnosis (asc mode 3), the maximum risk rate observed was 16.1% (both carriers). This result, again for rcp with the smallest interchange segments in these groups, is much higher than the average risk rates (for asc 2, 3) of 5.8% (mat carriers) and 3.2% (pat carriers). Thus this graphing method (which on present data is applicable only to 2:2-segregating rcp) identifies the rcp carrier at highest risk within the asc groups. Of particular interest is the observation that, in asc groups 1 and 2, there is no significant difference in the mean size of the imbalanced segments for *unbalanced* progeny, although there are four to five times fewer such progeny in asc group 2. Through a comparison with mouse rcp (based on mean imbalanced segment sizes), it is suggested that most unbalanced progeny in human rcp type 3 (incidental asc mostly maternal age) manifest as preimplantation and deciduomata losses. In such families, one or two offspring in a reproductive lifetime is no real test of subfertility.

INTRODUCTION

Ps 14:1—[Only] the fool has said in his heart, there is no God. Isaiah 6:3—... the whole earth is full of His glory. Genesis 1:28—And God blessed them and said... and have dominion over every living thing.

The frequencies of chromosome reciprocal translocations (rcp) and inversions (inv) in the various types of gametes recoverable from heterozygotes are influenced by many factors [1–3]. These include, for rcp, the comparative sizes of the chromosomes involved, their interchange segments, and the positions of centromeres; for inv, the size of the distal segments; for rcp and inv, chiasma frequency and chiasma formation in the interstitial or inverted segment; and for rcp and Robertsonian translocations (rob), the formation of chain or ring multivalents and the effects of meiotic drive and allied phenomena. Although there are some models for the mode of segregation in rcp [4,5], overall the interpretation of the empiric risks of unbalanced offspring for carriers of chromosome rearrangements is rendered very difficult with the approach of the expectation of different gametic types. The regular approach to this situation is the simple empiric method of pooling segregation data from many families as a genetic counseling resource. This has been done for rcp and rob and pericentric inv by Boué and Gallano [6], who utilized reports from 71 European collaborative prenatal diagnosis centers. Similarly, by compiling published cases, data on pericentric and paracentric inv are presented in this volume (see Chapters 3 and 4). In the present study, data from many North American centers performing prenatal cytogenetic diagnosis were used to correlate balanced/unbalanced prenatal result with mode of ascertainment (asc) sex of carrier parent, and identity of the chromosome(s) involved for the several types of rearrangement studied. These include rcp and rob, paracentric (pai) and pericentric (pii) inv, insertions (ins), and complex or multiple rearrangements (crr). For simple rcp and pii, further inferences were made assuming a selective viability of conceptuses related to the size of the chromosomal imbalance [1,3,5]. With this approach, the higher risks for chromosomally unbalanced progeny (surviving 16 weeks or longer) arising from the translocation are thought to be associated with the smaller potential imbalances. With this method, risk estimates can be determined for rcp and inv ascertained in various ways when empiric data from similar family histories are not available. The rcp cases of the European Collaborative Prenatal Diagnosis Laboratories [6] were previously analyzed in this way [5], and the data are extended in the present study.

MATERIALS AND METHODS

The data (comprising 832 cases) were collected in 1986 from the 63 participating U.S. cytogenetics laboratories (Appendix VIII) and after verification (February–October, 1986) were formatted in the following appendices: rcp (Appendix I; 401 cases), pii (Appendix II; 102 cases), rob (Appendix III; 254 cases), pai (Appendix IV; 30 cases), ins (Appendix V; six cases), sex chromosome rearrangements (Appendix VI; 28 cases), and ccr (Appendix VII; 11 cases). Data of 596 cases of rcp from the European Collaborative prenatal diagnosis centers [6], recently analyzed [5] in a similar fashion, were included in this report (Appendix I).

Simple Empiric Data

The cases were recorded (and analyzed when there were sufficient data) by chromosome(s) involved, breakpoints, sex of parental carrier, mode of asc (see footnote to Appendix I), and prenatal diagnostic result [5,6].

Data on autosomal rcp and pii, in addition to this simple empirical analysis were analyzed by methods addressing potential chromosome imbalance sizes (Appendices I and II). Further explanation of these methods is outlined in Daniel et al. [5].

Correlations With asc Mode

U.S. cases for all rearrangement types were divided into five asc modes (the fourth and fifth do not apply to the European data). These are 1) through a previous full-term, unbalanced child; 2) because of recurrent miscarriage; 3) incidentally, for reasons ostensibly unrelated to the rearrangement; 4) because of infertility; and 5) through a chromosomally balanced but mentally handicapped proband. There are few data for modes 4 and 5. Asc modes 2–5 are through carriers with the balanced form of the rearrangement. Correlations of a particular rearrangement type with any asc were noted.

Abnormal Prenatal Results Unrelated to the Rearrangement

Five cases in the present study were ascertained by a presumptively unrelated trisomy. In these cases, the affected individuals were found to have a balanced rearrangement in addition to the primary trisomy. They represent a small subgroup of asc mode 3. There were (see Table IVA) two of 401 rcp (0.5%), two of 254 rob (0.8%), and one of 102 pii (1.0%) found in this way. Data were analyzed to determine whether heterozygotes for rearrangements are at an increased risk over the normal population for such unrelated chromosome abnormalities. Contributing U.S. laboratories were requested to submit *all* their data for a particular period in an attempt not to introduce a bias for unusual results.

Tables and Graphs of Imbalance Sizes for rcp and inv

The unbalanced offspring of rcp carriers usually arise by adjacent-1 meiotic segregation. Segregants therefore usually have a duplication of one interchange segment of the rcp and a deficiency of the other. Adjacent-1 segregants with the other derived chromosome have the reverse imbalance, with a deficiency of the first segment and a duplication of the last. Similarly, unbalanced offspring of pii heterozygotes arise by meiotic crossing-over within the inv segment (in a pachytene inv loop) to give offspring with a duplication of one segment distal to the inv breakpoints and a deficiency of the other distal segment. The alternative segregant similarly has the reverse imbalance. The size of these distal (inv) or interchange (rcp) segments can then be obtained from the table of prometaphase chromosome segment sizes [7] and plotted as for Figures 1–3 with trisomy (ordinate) and monosomy (abscissa). In reading from the table [7] it was assumed that breakpoints occurred in the middle of the designated band. The rcp and inv data of Appendices I and II were so plotted and the distribution divided into regions of 1% trisomy by 0.5% monosomy so that the rate of unbalanced progeny diagnosed prenatally could be calculated for each region. This calculation was made only when there were at least 20 cases in a region to exclude determinations with overtly insufficient data. [Data analysis (through 21 separate subprograms) was by the SAS program (SAS Institute, Cary, NC 27511) on an IBM 3081 mainframe (State of Montana).] Each rearrangement was plotted once with unbalanced results assigned to the graphs as its particular imbalance irrespective of what would have been predicted. Each chromosomally normal or balanced pregnancy was plotted as the "primary potential" mode of segregation chosen for each rcp. The basis of this selection was the segregation mode giving the recombinant with the smallest imbalance. In most rcp, this was an adjacent-1 segregant. In a minority of rcp, e.g., the (11;22)(q23;q11) group, the "primary potential" mode of segregation is a 3:1 segregant. In these cases, the 3:1 potential imbalance (in normal and balanced prenatal diagnostic outcomes) was plotted instead of the interchange segments of the 2:2 potential segregant. In a second, small group of rcp cases ascertained by or expected to predispose to an unbalanced adjacent-2 segregant, the potential imbalances plotted were also adjacent-2.

With a combined duplication/deficiency in rcp, usually the larger imbalanced segment occurs as the trisomy. In the *unbalanced* offspring in the European collaborative study [5], 25% (irrespective of carrier parent) had the complementary situation, i.e., with the larger segment as the monosomy. Therefore, to match this trend in the balanced/normal outcomes when potential imbalances were calculated, one in four (25%) trisomy/monosomy sets of each asc group treated independently were plotted (Figs. 1–3) with the larger segment as the monosomy. This was done by assignment of the listed cases in the Appendix, and the trisomy/monosomy sets for each case are recorded therein in the same fashion as they were plotted.

Initial analysis of the rcp data indicated that mat and pat risk rates were very similar and that the major determinants of rates of unbalanced progeny were mode of asc and potential size of the imbalances. Therefore, for asc group 1, mat and pat carriers were considered separately (Fig. 1a,b) and together (Fig. 1c). Second, the rates of unbalanced progeny were on initial analysis similar for rcp ascertained through balanced carriers either by recurrent miscarriage (asc mode 2) or incidentally (asc mode 3). Therefore, Figure 2 was prepared with plots comprising data of asc modes 2 and 3 for mat (Fig. 2a), pat (Fig. 2b), and both carriers (Fig. 2c). The risks determined from these plots assume a single mode of segregation for individual rcp. For those few rcp segregating by more than one mode, the risks are not additive but refer to the commonest mode. This model of risk in rcp from present data applies only to 2:2 segregating rcp in which the relative proportions of normal:carrier:unbalanced progeny are similar for either sex of carrier parent.

Comparative Clinical Significance of Rearrangement Types

Some rearrangements, e.g., inv (2)(p11.2q13), seem to be totally unassociated with any clinical expression, whereas others have varying degrees of association with recombinant unbalanced offspring, recurrent miscarriage, infertility, balanced rearrangment but abnormal phenotype, or expression of a mendelian disorder. Association with recombinant offspring is the most common concern and an attempt was made to rank rearrangements in this connection by correlating the proportion of asc in each mode with the rate of unbalanced results (see Table VI).

Randomness of Breakpoints: Are Some Imbalances More Viable Than Others?

It is assumed in the present analysis that unbalanced chromosomal segments of equal size are comparable in their affect on viability irrespective of their chromosomal origin. This is tested by a frequency analysis of the rcp breakpoints (see Table VII) to detect if the distribution of certain chromosomal segments is biased towards rcp of asc type 1 or rcp of asc type 2. This question is not as easy to address as it may seem; e.g., the mere preferred occurrence of 12p in rcp type 1 does not necessarily imply that a trisomy of 12p is more viable. It may reflect only the fact that 12p is a small segment. Since larger segments are more frequently found in rcp type 2 and smaller segments in rcp type 1, a small segment is likely to be more frequent in a sample of rcp type 1. It is, therefore, necessary to compare the distribution of rcp in the *same* segment size class and not between classes. Since most unbalanced offspring of rcp carriers arise by apparent adjacent-1 segregation, the two segments distal to the breakpoints of two chromosomes involved in the rcp will be found as the trisomy/monosomy or the monosomy/trisomy in unbalanced offspring. Individual breakpoints for segment size classes used for this comparison (see Table VII) were chosen mostly because a narrow range of segment size was defined by the breakpoint. This question was addressed in a previous report [5], but, since the numbers of rcp have been increased by 70%, the data were reexamined.

RESULTS

The prenatal results of all cases are listed in Appendices I–VII together with the actual/potential imbalance sizes (%HAL) in Appendix I (for rcp) and Appendix II (for pii).

Simple Empiric Data

Summaries of the prenatal results for rcp are shown in Table I, for pii in Table II, and for rob(13, 14, and 15;21) in Table III. Table I (for rcp) shows that the mat and pat rates of unbalanced progeny are similar in each asc group, i.e., asc mode 1, 19.8% and 23.3%; asc mode 2, 5.1% and 1.6%, and asc mode 3, 6.9% and 5.1%, respectively. The data carrying asterisks are those rearrangements first detected in the present prenatal diagnosis; they are, therefore, biased towards balanced/unbalanced results and are excluded from the segregation figures. Since the rates for asc groups 2 and 3 are similar (but markedly different from asc 1), these data have been pooled for graphs of imbalance size (Figs. 2a–c). For pai, ins, sex chromosome rearrangements, and ccr, the data are brief enough to be summarized in Appendices IV–VII, respectively. Though there were only a very small number of parental carriers of ins, four of six offspring were unbalanced, indicating a substantial risk for this rearrangement type. In contrast, there were no unbalanced pregnancies in the 30 cases of pai *or* for most *types* (see Chapter 3, this volume) of pii (Table II). The two cases of unbalanced pregnancies in the pii data are from nontype (unique breakpoint) inv.

Of the 125 rob(13;14), there were one of 86 (1.2%) unbalanced translocation-trisomy 13 pregnancies (i.e., asc mode 1) from mat carriers and one of 39 (2.6%) from pat carriers (Table III). Data were insufficient for separate analysis of offspring from rob(13, 14, and 15;21) carriers, but pooled data showed 12.7% translocation-trisomy 21 offspring for mat carriers and 3.3% for pat carriers. A previous, small study [8] indicated a higher risk for rob(13;21) carriers, but this could not be addressed in this small data set.

Correlations With asc Mode

For rcp, there are sufficient data (997 cases) to document (Table 1) the four to fivefold difference between asc through a previous unbalanced, full-

A

TRISOMY % HAL \ MONOSOMY % H.A.L.	0.00–0.50	0.50–1.00	1.00–1.50	1.50–2.00	2.00–2.50	2.50–3.00	3.00–3.50	3.50–4.00
5.00–6.00	T=0 U=0	T=0 U=0	T=0 U=0	T=0 U=0	T=0 U=0	T=0 U=0	T=0 U=0	T=0 U=0
4.00–5.00	T=1 U=0	T=0 U=0	T=0 U=0	T=0 U=0	T=0 U=0	T=0 U=0	T=0 U=0	T=0 U=0
3.00–4.00	T=6 U=1	T=1 U=0	T=0 U=0	T=0 U=0	T=0 U=0	T=0 U=0	T=0 U=0	T=0 U=0
2.00–3.00	T=35 U=7 20.0%	T=5 U=0	T=5 U=0	T=0 U=0	T=0 U=0	T=0 U=0	T=0 U=0	T=0 U=0
1.00–2.00	T=70 U=14 20.0%	T=18 U=2	T=3 U=0	T=1 U=0	T=0 U=0	T=0 U=0	T=0 U=0	T=0 U=0
0.00–1.00	T=25 U=12 48.0%	T=24 U=3 12.5%	T=21 U=4 19.0%	T=4 U=0	T=2 U=1	T=1 U=0	T=0 U=0	T=0 U=0

B

TRISOMY % HAL \ MONOSOMY % H.A.L.	0.00–0.50	0.50–1.00	1.00–1.50	1.50–2.00	2.00–2.50	2.50–3.00	3.00–3.50	3.50–4.00
5.00–6.00	T=0 U=0	T=0 U=0	T=0 U=0	T=0 U=0	T=0 U=0	T=0 U=0	T=0 U=0	T=0 U=0
4.00–5.00	T=0 U=0	T=0 U=0	T=0 U=0	T=0 U=0	T=0 U=0	T=0 U=0	T=0 U=0	T=0 U=0
3.00–4.00	T=5 U=0	T=0 U=0	T=0 U=0	T=0 U=0	T=0 U=0	T=0 U=0	T=0 U=0	T=0 U=0
2.00–3.00	T=17 U=4	T=9 U=2	T=1 U=0	T=1 U=0	T=0 U=0	T=0 U=0	T=0 U=0	T=0 U=0
1.00–2.00	T=57 U=11 19.2%	T=16 U=4	T=2 U=0	T=0 U=0	T=0 U=0	T=0 U=0	T=0 U=0	T=0 U=0
0.00–1.00	T=20 U=10 50.0%	T=18 U=3	T=12 U=3	T=1 U=0	T=0 U=0	T=0 U=0	T=0 U=0	T=0 U=0

Fig. 1. Ratio of unbalanced (U) to total (T) pregnancies in rcp type 1 (ascertained through a full-term child with an unbalanced form of the rearrangement). **A:** Data for maternal carriers. **B:** Data for paternal carriers. **C:** Pooled data for both maternal and paternal carriers. Note that the maximum risk is about 50% (i.e., with the smallest imbalances, closest to the origin) irrespective of sex of carrier.

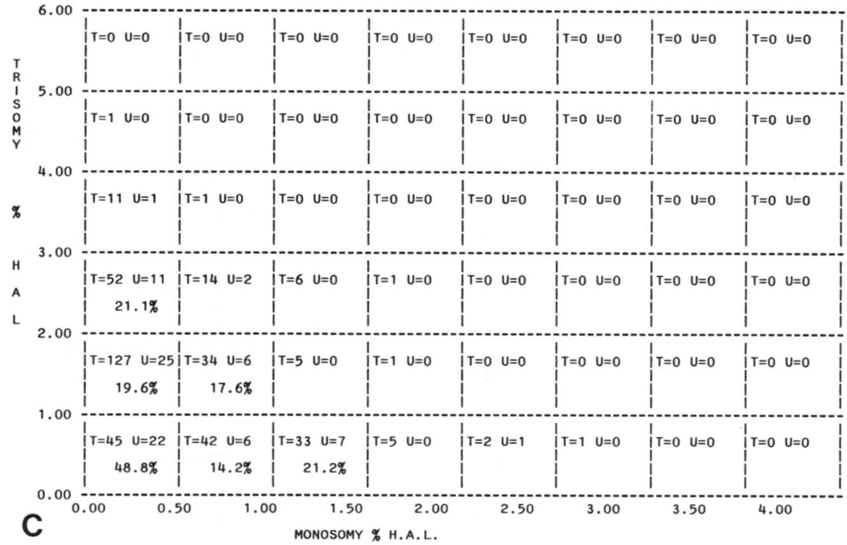

Fig. 1. ***(continued)***

term child (asc mode 1) and asc through balanced carriers (asc modes 2–5) irrespective of carrier parent. To draw attention to associations further corroborated in this study, for rob(14;15) and rob(15;22), three of five cases were ascertained by recurrent miscarriage and two of five cases of rob(13;22) were ascertained by a translocation-trisomy 13 child. In the rob(13;14) cases, three of 125 (2.4%) were ascertained because of infertility; this, of course, is a baseline figure, since most infertility cases do not proceed to prenatal diagnosis. In these cases, the fertile couple may be relatives of the proband. For the two commonest "types" of pii in this study, inv(2)(p11.2q13) and inv(10)(p11.2q 21.2), neither of which carries risk for full-term recombinant progeny, there were one of 31 and three of 17 respective asc through a couple with recurrent miscarriage. These are likely to be noncausative associations seized upon because they represented the only abnormality in the karyotype.

Abnormal Prenatal Results Unrelated to the Rearrangement

In four of 997 rcp (i.e., one of 250 = 0.4%), the fetus had trisomy 21 in addition to the regular segregation of the rcp (Table IVB). In one additional rcp, (i.e., one of 1,000), a fetus had trisomy 13. In two pii (N = 102 observations), there was only XXY fetus (i.e., one of 100) and one 47,XX,inv(12),+i(Xq) fetus (i.e., one of 100). All these figures represent (?four to tenfold) increases over the rates expected in offspring at full-term.

TRISOMY % HAL \ MONOSOMY % H.A.L.	0.00–0.50	0.50–1.00	1.00–1.50	1.50–2.00	2.00–2.50	2.50–3.00	3.00–3.50	3.50–4.00
5.00–6.00	T=1 U=0	T=0 U=0	T=0 U=0	T=0 U=0	T=0 U=0	T=0 U=0	T=0 U=0	T=0 U=0
4.00–5.00	T=8 U=0	T=3 U=0	T=1 U=0	T=1 U=0	T=0 U=0	T=3 U=0	T=1 U=0	T=0 U=0
3.00–4.00	T=14 U=1	T=2 U=0	T=10 U=0	T=3 U=0	T=5 U=0	T=6 U=0	T=1 U=0	T=1 U=0
2.00–3.00	T=27 U=2 07.4%	T=13 U=1	T=16 U=1	T=9 U=0	T=2 U=1	T=1 U=0	T=0 U=0	T=0 U=0
1.00–2.00	T=39 U=4 10.2%	T=29 U=0	T=6 U=0	T=4 U=0	T=6 U=0	T=2 U=0	T=3 U=0	T=2 U=0
0.00–1.00	T=20 U=3 15.0%	T=13 U=2	T=15 U=1	T=9 U=0	T=7 U=0	T=4 U=1	T=2 U=0	T=3 U=0

A

TRISOMY % HAL \ MONOSOMY % H.A.L.	0.00–0.50	0.50–1.00	1.00–1.50	1.50–2.00	2.00–2.50	2.50–3.00	3.00–3.50	3.50–4.00
5.00–6.00	T=1 U=0	T=0 U=0	T=0 U=0	T=0 U=0	T=0 U=0	T=0 U=0	T=0 U=0	T=0 U=0
4.00–5.00	T=5 U=0	T=2 U=0	T=4 U=0	T=1 U=0	T=1 U=0	T=0 U=0	T=1 U=0	T=0 U=0
3.00–4.00	T=7 U=0	T=0 U=0	T=5 U=0	T=1 U=0	T=1 U=0	T=3 U=0	T=1 U=0	T=0 U=0
2.00–3.00	T=28 U=2 07.1%	T=11 U=0	T=15 U=0	T=3 U=0	T=1 U=0	T=0 U=0	T=1 U=0	T=0 U=0
1.00–2.00	T=29 U=2 06.8%	T=26 U=1 03.8%	T=5 U=0	T=0 U=0	T=5 U=0	T=3 U=0	T=0 U=0	T=1 U=0
0.00–1.00	T=11 U=2	T=4 U=0	T=13 U=0	T=9 U=0	T=9 U=0	T=6 U=0	T=3 U=0	T=2 U=0

B

Fig. 2. Ratio of unbalanced (U) to total (T) pregnancies in rcp type 2 (ascertained by a couple with recurrent miscarriages) and type 3 (incidental ascertainment). **A:** Data for maternal carriers. **B:** Data for paternal carriers. **C:** Pooled data for both maternal and paternal carriers. Note that the maximum risk is about 15% (i.e., with the smallest imbalances closest to the origin) irrespective of sex of carrier.

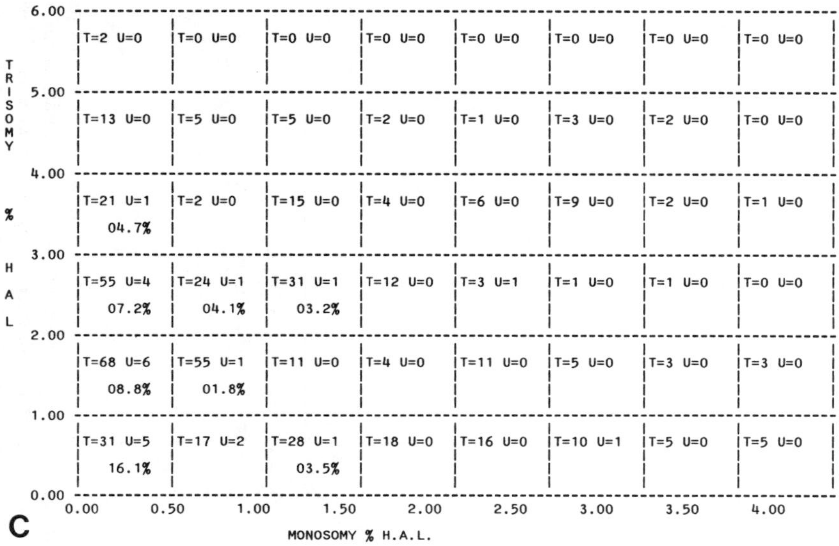

Fig. 2. ***(continued)***

Imbalanced Segment Sizes for rcp and inv

The mean imbalanced segment sizes for rcp, correlated with asc and balanced/unbalanced result, are shown in Table V. The smallest segment sizes are those for unbalanced progeny, and the largest mean segment sizes (in each asc group) are those for balanced/normal progeny; i.e., those segregants with the smallest imbalanced segment sizes are at the greatest risk. In rcp, the maximum risk rates observed were 48.0% (mat carriers; Fig. 1A) and 50.0% (pat carriers; Fig. 1B), both for the graphed region with the smallest defined interchange/imbalance segment sizes. These results were for rcp ascertained by a previous unbalanced child (rcp type 1) for which in pooled data (Table I) there was a 20–25% recurrence risk. Thus the method defined rcp within this asc group that were at a higher than average risk of unbalanced progeny, i.e., those with a risk closer to 50%. It should be noted that these figures of 48–50% are close to the maximum theoretical risk of 50% (given 50% alternate segregation) and that the risks for male and female carriers are not significantly different.

For pooled rcp either ascertained through balanced carriers for recurrent miscarriages (rcp type 2) or ascertained incidentally prior to the present prenatal diagnosis (asc mode 3), the maximum risk rate observed was 16.1% (both carriers; Fig. 2C). This result, again for the rcp with smaller inter-

TABLE I. Summary of prenatal results for rcp

Ascertainment mode	Maternal carrier [No. of cases (%)]				Paternal carrier [No. of cases (%)]				Total
	Balanced	Normal	Unbalanced	Total	Balanced	Normal	Unbalanced	Total	(N = 997)
1 (Through full term, unbalanced progeny)	90 (40.5)	88 (39.6)	44 (19.8)	222	63 (39.6)	59 (37.1)	37 (23.3)	159	381
2 (Through recurrent miscarriages)	91 (51.1)	78 (43.8)	9 (5.1)	178	66 (53.7)	55 (44.7)	2 (1.6)	123	301
3 (Incidentally found)	66 (57.4)	41 (35.7)	8 (6.9)	115	57 (58.2)	36 (36.7)	5 (5.1)	98	213
Other 1,* 2,* 3,* 4,* 5,* 4,5	51	5	3	59	42	0	1	43	102

*Rearrangement first detected in the present prenatal diagnosis.

TABLE II. Summary of prenatal results for pericentric inversions

Ascertainment mode	Maternal carrier (No. of cases)				Paternal carrier (No. of cases)				Total
	Balanced	Normal	Unbalanced	Total	Balanced	Normal	Unbalanced	Total	(N = 102) (%)
1	5	1	1	7	1	3	0	4	11 (10.8)
2	5	4	0	9	1	0	0	1	10 (9.8)
3	3	1	0	4	6	3	0	9	13 (12.8)
3*	27[a]	0	1	28	40[a]	0	0	40	68 (66.6)

[a]Most cases (81/102; 79.4%) were ascertained incidentally either in the *present* prenatal diagnosis (asc mode 3*) or by prior incidental means (asc mode 3); this implies a low clinical significance for the majority of pii. This low-risk majority of pii are mainly "type" inversions such as inv (2)(p11.2q13), with 31 cases; inv(Y)(p11.2q11.2), with 20 cases; and inv(10)(p11.2q21.2), with 17 cases. For those cases ascertained through a prior recombinant child (asc mode 1) there was one of 11 unbalanced prenatal diagnosis. A further recombinant, again to a maternal carrier, was first ascertained prenatally (mode 3*).

TABLE III. Summary of prenatal results for rob(13, 14, and 15;21)

Rob	Maternal carrier (No. of cases)				Paternal carrier (No. of cases)				Total
type	Balanced	Normal	Unbalanced (%)	Asc[a]	Balanced	Normal	Unbalanced (%)	Asc[a]	(N = 96)
(13;21)	3	2	0	2	2	4	0	0	13
(14;21)	21	14	7 (16.7)	6	9	8	1 (5.6)	3	69
(15;21)	4	4	0	0	4	2	0	0	14
Total	28	20	7 (12.7)	8	15	14	1 (3.3)	3	96

Data are insufficient for separate analysis of offspring from the three tdic (D;G) types, but pooled data show 12.7% translocation-trisomy 21 offspring to maternal carriers and 3.3% for paternal carriers.

[a]Ascertained in present fetus, thus biased to balanced/unbalanced results and so excluded from the calculation of segregation ratios.

TABLE IVA. Rearrangements ascertained through unrelated trisomies

Observation	Ascertainment	Rearrangement	Carrier	Prenatal result code
64	Trisomy 21	rcp(1;8)(p11;p23)	P	N
963	Trisomy 18	rcp(16;19)(q12;q13)	P	B
1075	Trisomy 13	inv(10)(p11q21)	P	B
1208,9	Trisomy 21	rob(13;14)(p12;p12)	M	B
1220	Trisomy 21	rob(13;14)(p12;p12)	M	B

Note that two of 401 rcp (0.5%), two of 254 rob (0.8%), and one of 102 pii (1.0%) were ascertained because of an unrelated primary trisomy; i.e., in these cases, in addition to the trisomy, a balanced rearrangement was detected. Carriers in the families had subsequent prenatal diagnoses mainly because of the latter.

change segments in those groups, is much higher than the average rates (for asc 2 or 3) of 5.8% (mat carriers) and 3.2% (pat carriers; see Table I). It should be noted that, remarkably, there is no significant difference in the mean size of the imbalanced segments (Table V) for *unbalanced* progeny between asc groups 1 and 2, although there are four to five times fewer such progeny in asc group 2.

Since there were only two of 102 unbalanced progeny from pii (both from mat carriers), there were insufficient data for an analysis of pii risk rates by carrier and asc mode. However, when the data are plotted with both carriers and all ascertainments (Fig. 3), it is noteworthy that there are one of seven (closest) and one of ten unbalanced results in two regions close to the origin. That is, there is a risk of about 10–15% for those pii with small distal segments. The two groupings of T = 31 (upper right) and T = 20 lower central) in Figure 3 represent inv(2)(p11.2q13) and inv(10)(p11.2q21.2), respectively, both of which are clinically not significant and have large distal segments.

Comparative Clinical Significance of Rearrangement Types

The various types of rearrangement discussed in this study are compared in Table VI on the basis of asc and percentage of unbalanced results. The rearrangements are ranked by the proportion of unbalanced progeny, with the highest rates at the top of the table. It should be noted that a greater proportion of cases for each rearrangement type near the top of the table has been ascertained by mode 1 and a greater proportion of cases for those types at the foot of the table by asc mode 3 (incidental asc). The rcp types of very different risk, i.e., rcp types 1 and 2, could not be entered in this table comparatively, since all rcp of simple type 1 are ascertained by mode 1 and

TABLE IVB. Abnormal results unrelated to rearrangements

Observation	Ascertainment	Rearrangement	Carrier	Result code[a]	Prenatal result
518	1	rcp(5;15)(p12;p11)	M	B	47,XX,t(5;15),+21
624	3	rcp(6;22)(q16;q13)	P	B	47,XY,t(6;22),+21
702	1	rcp(8;12)(q13;p12)	P	B	47,XX,t(8;12),+13
833	2	rcp(11;16)(p11;q11)	M	B	47,XY,t(11;16),+21
886	3	rcp(11;22)(q25;q13)	P	U	47,XY,der(22),t(11;22),+21
1060	2	inv(7)(p15.3q11.2)	M	N	47,XXY
1091	3	inv(12)(p13q23)	P	B	47,XX,inv(12),+i(Xq)

In four of 997 rcp (i.e., 1/250) the fetus is affected with trisomy 21; this represents an increase over the consecutive newborn figures but *not* over those for maternal age referrals at 35 years [10].
[a]With respect to rearrangement. B, balanced; U, unbalanced; N, normal.

all rcp of simple type 2 are ascertained by mode 2. However, comparisons are possible if rcp types are defined by their potential or actual imbalance segment sizes. Four rcp types have been defined for Table VI (see footnote) and then the data (Appendix I) were searched for cases (ascertained in any way) that correspond to these definitions, and the percentage of unbalanced results determined for each type. Three of these rcp types, I^t, II^t, and III^t, have been defined (mean imbalance segments sizes ± 0.5 standard deviations) from the total cases ascertained by modes 1, 2, and 3, respectively (Table V). The fourth rcp type (I^u) was defined (mean segment size ± 1.0 standard deviations) from the unbalanced prenatal results ascertained by mode 1 (a full-term, unbalanced child). It can be seen that the highest proportion of unbalanced progeny (17.2%) for these theoretical rcp types is seen for rcp type I^u. This should prove, along with Figures 1–3, to be a valuable counseling tool for use with rcp heterozygotes. The determination of types with a higher risk from the main bodies of data results in the helpful corollaries of defining rcp with a lesser risk, and this should be kept in mind when considering the conclusions of this study.

Randomness of Breakpoints: Are Some Imbalances More Viable Than Others?

Table VII (see footnote) compares the frequencies of a number of breakpoints in rcp of asc types 1 and 2. Departures from randomness for breakpoints are most likely the result of differences in viability for the 2:2 segregating) imbalanced segments defined by the breakpoint. No correlation with fragile sites has been found for reciprocal translocation breakpoints in the present study (see Discussion). Since smaller chromosome arms are more

TABLE V. Mean size[a] of imbalanced segments in rcp

Ascertainment mode	Data	No. of cases (%)	Potential or actual trisomy[a,b]	Potential or actual monosomy[a,b]
1	Balanced/normal cases	300 (78.7)	1.34 ± 0.88	0.51 ± 0.45
	Total cases	381 (100.0)	1.29 ± 0.85	0.48 ± 0.44
	Unbalanced cases	81 (21.3)	**1.12 ± 0.74**	**0.40 ± 0.40**
2	Balanced/normal cases	290 (96.3)	1.80 ± 1.10	1.02 ± 0.89
	Total cases	301 (100.0)	1.78 ± 1.09	1.00 ± 0.89
	Unbalanced cases	11 (3.7)	**1.11 ± 0.77**	**0.45 ± 0.62**
3	Balanced/normal cases	200 (93.9)	1.97 ± 1.26	1.21 ± 1.05
	Total cases	213 (100.0)	1.95 ± 1.25	1.19 ± 1.04
	Unbalanced cases	13 (6.1)	1.61 ± 1.05	0.57 ± 0.75

For unbalanced progeny, there is no significant difference [5] in the mean segment sizes for asc modes 1 or 2 although there are five to six times fewer such progeny with asc mode 2 (recurrent miscarriage rcp). Also, note that the unbalanced cases have smaller segments than either the balanced/normal or total cases.

From the data in this table, four types of rcp have been defined for comparative purposes (Table VI). These are three rcp types: I^t, II^t, III^t defined from the mean segments (±0.5 SD) of the *total* cases (t) of rcp ascertained by modes 1, 2, and 3, respectively. The fourth type, termed I^u, is defined from the mean segments (±1.0 SD) of the *unbalanced* cases (u) ascertained by mode 1.

The ranges of segment sizes defining these types, I^u, I^t, II^t, and III^t, are for type I^u 0.38–1.86 (potential or actual trisomy) and 0.00–0.80 (potential or actual monosomy); for type I^t, 0.86–1.72 and 0.26–0.70; for type II^t 1.23–2.33 and 0.55–1.45; and for type III^t 1.32–2.58 and 0.67–1.71.

[a]%HAL (mean ± SD).

[b]Actual for unbalanced cases and potential for balanced/normal cases.

likely to be involved in rcp type 1, breakpoints that define similar-sized segments have to be used to address the present point.

In the larger segment sets (e.g., 2.5 %HAL) the breakpoints occur in rcp type 2 more frequently, and in the smaller segment sets (e.g., the 0.3 %HAL set) the breakpoints occur in rcp type 1 more frequently. This is what would be expected on the basis only of segment size, not a difference in the viability associated with segments. Some differences do occur, however. The segments defined by bands 1p36, 8p21, 6p21, 13q14, 15q22, 15q26, 17q21, and 20q12 seem to be more frequent in rcp type 2 than expected on the basis of their size (less viable regions?), and segments defined by bands 4p14, 4p16, 4q27, 4q35, 5p12, 5p13, 5p15, 7q32, 10q24, 11q23, 18q21, and 21q22 seem to be more frequent in rcp type 1 than expected (more viable regions?).

A similar analysis was attempted on only the European portion of the data [5]. Some of the differences seen earlier have disappeared in the expanded data, but ten of 18 breakpoints with frequencies departing from expectation remain the same.

DISCUSSION

Why Should Risk in rcp Depend on Chromosome Imbalance Size?

It is patently obvious to even the cytogenetically uninitiated that, whereas human trisomies of autosomes 13, 18, 21 occur in *full-term offspring*, (non-mosaic) trisomies of 14, 15, 16, 19, and 20 have never been so observed. However, some members of the latter set of chromosomes are smaller than those of the former set. Therefore, if nothing else, this approach of measuring autosome segments should affirm that imbalance size has little (or nothing) to do with the viability of unbalanced fetuses and that the latter may rely on lethal genes or other complex ontogenic phenomena. This was approximately our expectation when it was found [1] that usually the segregation mode yielding the smallest potential imbalance was that observed in full-term, unbalanced offspring and that (recurrent miscarriage) rcp of asc type 2 had larger (mean) interchange segments than those of asc type 1. This approach of comparing imbalance sizes in rcp irrespective of autosome origin is affirmed by a singular finding [5], namely, that the mean sizes of imbalances for *unbalanced* offspring are not significantly different for rcp ascertained by recurrent miscarriages (asc type 2) or by full-term, unbalanced offspring (asc type 1) even though the proportion of such offspring is five to sixfold greater in the latter rcp group (Table V).

What, then, of differences in viability between different regions? It is probable (from the argument in the preceding paragraph) but surprisingly difficult to prove (see Table VII) that imbalances of some segments *are* more viable than others. It may be, given the overall uniqueness of rcp [except for rcp (11;22)(q23;q11), only a few similar cases have been found], that such differences are averaged out in the data or are compensated for by giving primacy to asc mode in the analysis.

If one focuses on the human autosomal monosomies found at full term or in miscarriages, then it is clearer that size is important. Monosomy for 21 (the smallest autosome) has been observed in some miscarriages and possibly (very rarely) at full term, yet, of course, the complementary product of non-disjunction for all (primary) trisomies is a monosomic gamete. Evidence for the importance of deficiency in viability was found in *Drosophila* by Lindsley et al. in 1972 [11]. In that study, the authors examined segregants from 467 Y-autosome rearrangements, involving small segments from all over the genome, to test for the presence of haplolethal loci. Surprisingly, the absence of such loci was demonstrated in 95% of the small regions. Their results foreshadowed the present results in that they concluded that the cumulative effects of small segmental deficiencies (rather than the unmasking of lethal alleles) were mainly responsible for lethality. A corollary of this is the explanation that the deleterious effects of aneuploidy are the consequence of the additive effects of genes that are slightly sensitive to abnormal dosage.

If we accept that in mammals different segments of the same size have different effects on viability, but that these differences are mostly undefined at present, what can be done to compensate for them with the present approach? Very little, apparently, but perhaps classifying data by asc mode is a partial compensation. One may expect that the segments found in recurrent miscarriage (type 2) rcp would be more lethal than those in full-term, unbalanced rcp (type 1). For example, for unbalanced fetuses in rcp types 1 or 2 (Table V), the maximum risk observed in type 1 is 50% (Fig. 1C). However, for the same region of the plot, the maximum risk observed for type 2 rcp is about 15% (Fig. 2C). This may mean that progeny with similar sized imbalanced segments in type 2 rcp are less viable than those of type 1 rcp; alternatively, data might merely be insufficient for rcp type 2.

TRISOMY % HAL \ MONOSOMY % H.A.L.	0.00–0.50	0.50–1.00	1.00–1.50	1.50–2.00	2.00–2.50	2.50–3.00	3.00–3.50	3.50–4.00
5.00–6.00	T=0 U=0	T=0 U=0	T=0 U=0	T=0 U=0	T=1 U=0	T=0 U=0	T=0 U=0	T=0 U=0
4.00–5.00	T=0 U=0	T=2 U=0	T=0 U=0	T=0 U=0	T=0 U=0	T=0 U=0	T=31 U=0	T=0 U=0
3.00–4.00	T=2 U=0	T=1 U=0	T=8 U=0	T=0 U=0	T=2 U=0	T=1 U=0	T=0 U=0	T=0 U=0
2.00–3.00	T=4 U=0	T=2 U=0	T=20 U=0	T=0 U=0	T=0 U=0	T=0 U=0	T=0 U=0	T=0 U=0
1.00–2.00	T=10 U=1	T=4 U=0	T=0 U=0	T=0 U=0	T=0 U=0	T=0 U=0	T=0 U=0	T=0 U=0
0.00–1.00	T=7 U=1	T=2 U=0	T=0 U=0	T=0 U=0	T=0 U=0	T=0 U=0	T=0 U=0	T=0 U=0

Fig. 3. Ratio of unbalanced (U) to total (T) pregnancies in pericentric inversions. Pooled ascertainment and for both carriers. Note that the only unbalanced pregnancies occur (close to the origin) with small imbalanced segments. Data are insufficient to determine rates.

TABLE VI. Ranking of rearrangements by percentage of unbalanced prenatal diagnoses

	No. of cases in ascertainment modes (%)					
Rearrangement	Mode 1	Mode 2	Mode 3	Mode 4	Mode 5	Result unbalanced (%)
Insertions	5/6	0	1/6	0	0	66.7
rcp type I^{u}[a]	183/332 (55.1)	85/332 (25.6)	62/332 (18.7)	2/332 (0.6)	0	17.2
rcp type I^{t}[a]	59/117 (50.4)	29/117 (24.8)	29/117 (24.8)	0	0	12.0
rob(13;21)	10/13	1/13	2/13	0	0	Insufficient data
rob(14;21)	50/69 (72.5)	7/69 (10.1)	12/69 (17.4)	0	0	5.6 P, 16.7 M
rob(15;21)	13/14	0/14	1/14	0	0	Insufficient data
rcp type II^{t}	28/109 (25.7)	52/109 (47.7)	26/109 (23.9)	1/109 (0.9)	2/109 (1.8)	5.5
rcp type III^{t}[a]	21/111 (18.9)	58/111 (52.3)	29/111 (26.2)	1/111 (0.9)	2/111 (1.8)	4.5
rob(21;22)	4/6	1/6	1/6	0	0	—
rob(13;22)	2/5	0	3/5	0	0	—
Pericentric inversions	11/102 (10.8)	10/102 (9.8)	81/102 (79.4)	0	0	2.0
rob(13;14)	19/125 (15.2)	39/125 (31.2)	64/125 (51.2)	3/125 (2.4)	0	1.2 M, 2.6 P
Paracentric inversions	1/30	3/30	26/30	0	0	—
rob(13;15)	0	2/9	7/9	0	0	—
rob(14;15)	0	2/3	1/3	0	0	—
rob(14;22)	0	1/7	6/7	0	0	—
rob(15;22)	0	1/2	1/2	0	0	—
inv(10)(p11.2q21.2)	0	3/17	14/17	0	0	—
inv(1)(p13q21)	0	0	5/5	0	0	—
inv(2)(p11.2q13)	0	1/31	30/31 (96.8)	0	0	—
inv(5)(p13q13)	0	0	6/6	0	0	—
inv(Y)(p11.1q11.2)	0	0	20/20 (100)	0	0	—

Note the trend that the greater percent of cases ascertained by a previously unbalanced child (asc mode 1) is associated with a higher rate of unbalanced results and the trend that the greater percent of cases ascertained incidentally (asc mode 3), the lesser the rate of unbalanced results.

[a]Since all rcp type 1 have been ascertained by mode 1 and all rcp type 2 by mode 2, the raw data defining these distinct types could not be included in this table in a comparative way. Rcp types I^{u}, I^{t}, II^{t}, and III^{t} are contrived groups defined from the raw data by their mean interchange segment sizes (see Table V). The total rcp data (997 cases) were then searched for numbers of rcp in all asc groups conforming to these definitions. The search parameters of rcp type I^{u} alone identify 52% (57/109) of *all* the unbalanced rcp prenatal results.

How does the present approach compare with the method of Jalbert and his colleagues [12] (also, this volume)? Their method seeks to predict segregation modes, whereas the present approach attempts to do the same but also to give an empiric risk. Rivera et al. [13] point out that there is a discordance using the Jalbert et al. method of about 15% [9] (also, Davis et al, 1985 and this volume) between the expected and observed segregation modes. In some cases, according to Rivera et al [13], the rcp has segregated by the mode that results in progeny with the smaller imbalance (i.e., the present model) rather than what is suggested by the relative lengths of the pachytene cross. Rivera et al. point out that, in other instances, the observed segregants have *larger* imbalances than at least one of the segregants expected from the Jalbert et al. model.

Such discrepancies in predictions from the Jalbert et al. and from the present model are for translocations of a particular type, those that mostly involve acrocentric autosomes (or a small derived chromosome), which have a propensity for 3:1 segregation and chain quadrivalents (which tend to induce male infertility because of pairing difficulties at pachytene). This means that, as for rcp (11;22)(q23;q11) and rob(13, 14, 15;21), rob (21;22), etc., there will be very different risks for male and female carriers than risk based, for example, on the size of chromosome 21. The risks of unbalanced progeny for at least male carriers will be lower than expected. The figures and tables used to interpret empiric risk in the present model are for rcp that are 2:2 segregating, which encompasses most of these data and is very clearly associated with equivalent risks to male and to female carriers for the various asc groups (Tables I, V). It will be necessary to prepare different plots (and tables) with which male and female carriers can be considered separately for translocations ascertained by or predisposing to 3:1 or 1:3 segregants. To place this point into perspective for this model, five of 29 of the interacrocentric rcp considered by Rivera et al. [13] (i.e., about 16%) produced full-term progeny by 3:1 and 2:2 segregations, which is about five times greater than the proportion (about 3%) of rcp as a whole [14].

How To Use the Chromosome Imbalance Model With rcp and pii

As discussed above, the present plots (Figs. 1, 2) are for 2:2 segregating rcp (i.e., the vast majority) or for pii (Fig. 3). The data for the latter (Fig. 3) are still less than adequate, and it is our intent to continue the collection of data to refine these models and to generate a fourth set of plots for 3:1 segregating rcp. For 2:2 rcp and pii, the size of the imbalanced *interchange* (rcp) or *distal* (pii) segments is first determined from a table of genomic segments [7]. The "prime potential" mode of segregation is evaluated by checking for the smallest imbalance that could arise by the adjacent-1 forms (either der chromosome inherited) or by the several types of adjacent-2 segregants or 3:1 segregants in doubtful cases. Some of both of the latter types can arise by crossing-over. Table III from the Paris Conference Report (1971) lists the types of segregants (same as Table IV in ISCN, 1978). Most adjacent-2 imbalances will be enormous (and can therefore be ruled out) for all but a small minority of rcp, i.e., those involving autosomes with short p arms and proximal q-arm breakpoints (e.g., a chromosome 18q11 breakpoint). This preferred mode of segregation could then be compared with that predicted by the Jalbert et al. model. The rcp or pii in question can then be correlated with the plot appropriate by asc since it is asc mode, not sex of carrier, that is the major factor in recurrence risk for these rearrangments (sex of carrier is of much greater import in 3:1 segregating rcp and rob).

Appendices I and II can also be a useful resource in considering the risks of individual rearrangments. Entries are listed in these appendices in order of breakpoint, so a close match can be made with most pii/rcp in question. The asc mode of the most comparable cases should be informative.

Which Unbalanced Human rcp Segregants Are Not Implanted or Become Deciduomata?

The rcp ascertained incidentally in the present study (asc group 3) have larger mean segments than those of asc group 1 or 2 (Table V). However, there are still 6.1% unbalanced fetuses at amniocentesis from this group, but the imbalanced segments of such progeny are larger. Most imbalanced segregants of group 3 rcp, if they occur, must not survive until the usual time of amniocentesis. The implications of these results are unknown, but some light on the question has been provided by data from Kirk and Searle (this volume). As these authors point out, the length of survival of particular trisomics and monosomics in the mouse depends to some extent on genetic background, but chromosome imbalance size is the limiting factor. Duplications of up to 2% of the haploid genome may be compatible with postnatal survival to maturity, though duplications plus deficiencies of this magnitude usually lead to embryonic death around the time of implantation, e.g., in T31H stock, some tertiary monosomics manage to survive to maturity despite a 2% deficiency. Irrespective of such variation, there seem to be two groups of murine rcp associated with different types of embryonic lethality. The first group, with larger imbalanced segments, is associated with earlier embryonic lethality of unbalanced products. These have a mean (+SD) actual or potential trisomy of 3.33 ± 1.38 %HAL and a mean actual or potential monosomy of 2.01 ± 1.04 %HAL. The second group of murine rcp, with smaller mean imbalanced segments, is associated with later embryonic lethality. These have a mean trisomy of 2.35 ± 1.70 %HAL and a mean monosomy of 1.10 ± 1.22 %HAL. These figures have been calculated from autosomal rcp data in Tables II and III, respectively, of Kirk and Searle (this volume) reversing one in four entries (entries 4, 8 and 12), as for the correction applied to human data (see footnote to Appendix I). The first mouse group has mean preimplantation losses of 11.8 ± 5.2%, mean percentage of deciduomata of 55.5 ± 14.2%, and mean later deaths of implants (very few survive birth) of 1.0 ± 1.0%. For the second murine rcp group, these figures are 21.1 ± 7.6%, 38.1 ± 13.6%, and 16.5 ± 8.1%, respectively. Most of the latter 16.5% are spontaneously aborted, few surviving birth. As for comparisons with man (Table VIII), the imbalanced segment sizes and reproductive history of carriers of human rcp type 2 closely resemble the second mouse group, whereas human rcp type 3 most closely resemble the first mouse rcp group. However, the mean segments for the mouse rcp type 1 are larger than those for human type 3; this probably reflects falsely small segment sizes for

TABLE VII. Frequency[a] of breakpoints (segments) in rcp asc groups 1 and 2

0.3% HAL set	0.5% HAL set	0.75% HAL set	1.0% HAL set
1q43 (0.66,0.17)	1q36 (0.39,1.67)[c]	1q42 (1.18,1.00)	1p35 (0.13,0.17)
2p25 (0.52,0.00)	3p25 (0.00,0.33)	3p24 (0.39,0.50)	1q41 (0.13,0.17)
2q37 (0.66,0.33)	3q27 (1.05,0.66)	4q33 (0.13,0.00)	2p23 (0.26,0.33)
3q28 (0.00,0.83)	4q34 (0.13,0.33)	5q33 (0.00,0.50)	2q35 (0.00,0.17)
4p16 (1.05,0.17)	5q34 (0.52,0.33)	8p21 (0.26,1.00)[c]	4p14 (1.18,0.33)[b]
4q35 (0.92,0.17)[b]	6p23 (0.66,0.66)	9q33 (0.52,0.14)	5p13 (1.57,0.33)[b]
5p15 (3.15,0.67)[b]	7q35 (0.39,0.17)	10q25 (0.52,0.00)	5q32 (0.00,0.17)
5q35 (0.39,0.66)	8p22 (0.26,0.66)	11p14 (0.13,0.17)	6q24 (0.00,0.17)
6p25 (0.26,0.17)	9p22 (1.31,0.83)	11q23 (3.67,1.16)[b]	7p15 (0.13,0.33)
6q26 (0.26,0.17)	13q22 (0.79,0.33)	12p12 (0.26,0.17)	7q32 (0.79,0.17)[b]
7q36 (0.39,0.66)	10p13 (0.13,0.50)	14q31 (0.00,0.33)	8q23 (0.52,0.00)
9q34 (0.26,0.00)	15q26 (0.39,1.66)[c]	16q22 (0.13,0.17)	9p12 (0.00,0.33)
10p15 (1.05,0.00)	17p12 (0.00,0.50)	18q21 (2.49,0.50)[b]	10p11 (0.79,0.50)
10q26 (1.44,0.66)	17q25 (0.79,0.83)	21q11 (0.26,0.66)	10q24 (1.05,0.33)[b]
11q25 (1.57,0.50)		22q12 (0.79,1.33)	11p13 (0.00,0.17)
12p13 (1.04,0.50)			11q22 (0.00,0.17)
13q33 (0.39,0.17)			12p11 (0.52,0.50)
18p11 (1.57,0.84)			13q22 (1.05,1.00)
18q22 (0.66,0.83)			15q23 (0.39,0.17)
21q22 (3.94,0.83)[b]			16p11 (0.92,0.50)
			16q21 (0.39,0.00)
			19p11 (0.00,0.17)
			20q12 (0.26,0.66)[c]
			22q11 (2.62,1.83)

(continued)

TABLE VII. Frequency[a] of breakpoints (segments) in rcp acs groups 1 and 2 (continued)

1.5% HAL set	2.0% HAL set	2.5% HAL set	3.0% HAL set
1p34 (0.13,0.50)	1p32 (0.00,0.83)	1p31 (0.00,0.17)	1p22 (0.00,0.17)
1q32 (1.57,1.00)	1q31 (0.26,0.33)	2p12 (0.00,0.33)	2p11 (0.00,0.33)
2p21 (0.26,0.66)	2p15 (0.26,0.33)	2q31 (0.13,0.17)	3p12 (0.00,0.33)
2q33 (0.52,1.16)	2q32 (0.13,0.33)	3p14 (0.26,0.50)	3q13 (0.26,0.33)
3q25 (0.13,0.00)	4q28 (0.26,0.33)	3q21 (0.39,1.16)	4q23 (0.26,0.00)
4p13 (0.26,0.50)	5q23 (0.26,0.17	4q27 (0.92,0.33)[b]	5q15 (0.00,0.17)
4q31 (1.18,0.66)	6q21 (0.39,1.33)	5q22 (0.13,0.17)	6q15 (0.00,0.50)
5p12 (1.57,0.00)[b]	7p11 (0.13,0.66)	9q13 (0.26,0.00)	7q11 (0.00,0.50)
5q31 (0.13,0.50)	7q22 (0.52,0.83)	10q21 (0.13,0.33)	8q11 (0.00,0.33)
6p21 (0.52,1.50)[c]	8q21 (0.00,0.00)	12q15 (0.00,0.33)	10q11 (0.13,0.83)
7p13 (0.00,0.50)	10q22 (1.05,0.83)	13q13 (0.13,0.50)	11q11 (0.00,0.33)
8q22 (0.39,0.50)	12q15 (0.00,0.33)	14q13 (0.92,0.50)	13q12 (0.00,0.66)
9p11 (0.39,0.33)	13q14 (0.52,2.49)[c]	15q13 (0.13,0.50)	
11p11 (0.26,0.17)	14q22 (0.79,0.00)		
11q21 (0.00,0.33)	15q15 (0.26,0.33)		
12q22 (0.00,0.33)			
14q23 (0.13,0.17)			
15q22 (0.52,1.99)[c]			
16q11 (0.00,0.66)			
17q21 (0.13,0.83)[c]			

An assumption of the chromosome imbalance–viability model [1,3] used in the analysis of the present data is that segments are comparable (by size) irrespective of their chromosomal origin. This is tested in the table when a frequency analysis has been performed to detect if the distribution of certain chromosomal segments is biased towards rcp of asc type 1 (unbalanced full-term offspring) or asc type 2 (recurrent miscarriage). To do this without bias the frequency of breakpoints (defining adjacent I segments) must be compared for segments of the same size (hence % HAL sets). In the larger segment sets, the breakpoints occur in rcp type 2 more frequently, and, in the smaller segment sets, the breakpoints occur in rcp type 1 more frequently. This is the distribution expected from the effect of the size of the segments (see Table V). Some minor differences do occur, however; 4p14, 4p16, 4q27, 4q35, 5p12, 5p13, 5p15, 7q32, 10q24, 11q23, 18q21, and 21q22 seem to be more frequent in rcp type 1 than expected[b] (defining more viable regions), and 1p36, 8p21, 6p21, 13q14, 15q22, 15q26, 17q21, and 20q12 seem to be more frequent in rcp type 2 (less viable regions) than expected[c] on the basis of the size of potential imbalanced segments they define (by adjacent segregation). Previous analyses [9] of this principle have referred to entire chromosome arms themselves, but such an approach is limited (invalidated?) by the many overlapping segments present therein.

[a]Frequency (% HAL set) of breakpoint in (asc groups 1 and 2): Figures on the left in parentheses in each column are the percentage of breakpoints at this site among breakpoint of asc group 1 rcp. Figures on the right are the percentage of breakpoints among those of asc group 2 rcp. There are 762 breakpoints in asc group 1 and 602 in group 2. Much more data would be necessary to demonstrate statistically significant differences in breakpoint (=chromosome segment) preference for an rcp type 1 or 2.

TABLE VIII. A comparison of mouse and human rcp

rcp	Mean imbalance segment sizes (total cases)		Reproductive history (%)			
	Trisomy	Monosomy	Pre-implantation losses	Decidu-omata	Later loss	Unbalanced full-term
Human type 1	1.29	0.48	Unknown	Unknown	Unknown	21.3
Mouse equivalent undefined						
Human type 2	1.78	1.00	Unknown	Unknown	Many	3.7
Mouse type 2[a]	2.35	1.10	21	38	17	Few survive
Human type 3[b]	1.95	1.19	Unknown	Unknown	Few	6.1
Mouse type 1[a]	3.33	2.01	12	56	1	Few survive

[a]Calculated from data in Kirk and Searle (this volume). For simplicity, SD are not shown.
[b]Falsely low segment sizes because of incomplete disclosure at referral; data contaminated by human rcp types 1 and 2.

human rcp type 3 resulting from limited disclosure to commercial cytogenetics laboratories by referring physicians. In short, the data on human rcp type 3 ("incidental" asc) are contaminated by undisclosed rcp types 1 and 2. These larger segments in both murine rcp groups than in human rcp type 1 are consistent with the very few unbalanced segregants surviving birth, as many such offspring in human type 1 do. Human rcp type 1 have smaller mean segments than human rcp type 2 or 3. The mean potential/actual trisomy and monosomy for mouse group 1 rcp are 3.33% HAL and 2.01% HAL and for human rcp type 3 they are 1.95% HAL and 1.19% HAL, respectively. This comparison for mouse rcp group 2 is 2.35% HAL and 1.10% HAL and for human rcp type 2, 1.78% HAL and 1.00% HAL, respectively. Putting aside the shortcomings of the comparison, it may be that group 3 rcp in man are associated mainly with such gross imbalances that they manifest as preimplantation and deciduomata losses, hence their "incidental" asc in the small nuclear families characteristic of industrialized society. In such families, the production of one or two full-term offspring in a reproductive lifetime is no real test of subfertility.

Pericentric Inversions

There were only two cases of recombinant fetuses (Table II) for pii heterozygotes; both were for nontype (unique breakpoint) pii. This implies that many of the type pii are not clinically significant. This category includes inv(1)(p13q21), inv(2)(p11.2q13), inv(5)(p13q13), inv(10)(p11.2q21.2), inv(Y)(p11.1q11.2), and, of course, inv(9)(p11q13) (not included in this study). Other "type" pii clearly carry risk for unbalanced recombinants; these include (Kaiser, this volume) inv(8)(p23q22), inv(3)(p25q21), inv(13)(p13q21), inv(18(p11q21), and inv(4)(p14q35). Other (as yet) nontype pii carrying risk for recombinants include inv(5)(p13q35), inv(5)(p15q33), inv(10)(p15q24), and inv(13)(p11q31). All pii with risk have small distal segments [3] (see Fig. 3).

Paracentric Inversions

No unbalanced karyotypes were found in fetuses of the 30 autosomal cases in the present study, and the overall risk of recombinant progeny to carriers of pai is probably quite low; nevertheless, some cases have been reported. The usual meiotic recombinant product in a paracentric inversion is a dicentric (with a sister acentric fragment) chromatid [3]. So far, 65 cases have been reported, and these are summarized by Madan in the present volume. Chromosomes 3, 7, and 11 are represented most frequently. Frequent breakpoints are found in 7q22 (N = 9), 7q11 (N = 5), 3p25 (N = 5), and 11q23

(N = 5), and "type" pai (as for pii) may exist. In eight of 65 of these cases, the pai was found in a parent of a child with an unbalanced (recombinant) karyotype, and, in a further ten of 65 cases (three de novo), the *balanced* pai was found in patients with congenital abnormalities. No doubt the discovery of many of the latter is due to asc bias, but a position effect apparently may exist for some pai as it does for some rcp. This phenomenon, i.e., the association of balanced pai with mental retardation and congenital anomalies, is presumably explicable in some cases (as for rcp) by the unmasking of recessives.

In two of eight of the recombinant cases (Madan, this volume), breakage of the dicentric bridge formed at meiotic anaphase I was the source of the (marker chromosome) duplications and deficiencies. This mechanism circumvents predictions made on the basis of the potential sizes of imbalance, and caution is thus necessary in counseling pai families using this method. In a further two of eight cases, a recombinant child with a recombinant deletion arising from the parental pai was observed. In the remaining four of eight cases, combined duplication/deficiencies (some quite small) of different parts of the inverted segment were seen in the recombinant offspring.

Infertility was the reason for asc in a further eight of 65 cases in Madan's study (this volume), but in five of eight there was in addition an XXY or 45,X sex chromosome complement. In two of three of the remainder, a causal mechanism for the infertility may be present. There is a similar equivocal relationship for pai ascertained by recurrent miscarriage; ten of 65 cases were ascertained in this manner, and Madan points out that many other carriers in the same families had no reproductive anomalies. However, as full-term, pai recombinant offspring are found, then, at least for some pai [14], a causal relationship with recurrent miscarriage is probable.

Familial Rearrangements and Balanced Probands with Mental Retardation or Multiple Congenital Anomalies (MR-MCA)

Whereas it has been known for some years [16–18] that there is an increase in de novo balanced rearrangements among the mentally retarded, numbers of familial rearrangements are also ascertained through a (isolated?) balanced but mentally retarded proband. Many of these cases are probably correctly interpreted as resulting from asc bias [19]; nevertheless, there are theoretical mechanisms for causal relationships. If at the breakpoints, as is suspected for the de novo examples, an interference with normal genetic mechanisms takes place (e.g., with reading frame), but there is also an allele for a disorder at the same locus on the proband's normal homologue inherited from the non-rcp parent, then a recessive disorder can be simulated or one can be unmasked [20]. Other balanced rcp family members could well be normal. This scenario could be called "recombination position effect." Likewise, genes at a breakpoint may be altered in such a way that a dominant mutation is simulated. This mechanism has been termed "unmasking of heterozygosity" by Buhler [21]. The de novo balanced rcp cases with phenotypic abnormalities seem to be simulating dominant mutations. Associated retardation may preclude reproduction and thus the demonstration that the condition is dominant. In the examples of familial rcp with a balanced/mentally retarded proband, prenatal diagnosis is sometimes offered (as in several cases in the present study; asc mode 5) simply to address the possibility of unbalanced offspring. Recording the outcome in such cases is valuable in determining the risk of further balanced rcp MR-MCA progeny. This is probably negligible in recessive simulations, but mechanisms (genetic heterogeneity?, multifactorial conditions?) are unproven, and, since cases present for prenatal diagnosis, data are necessary for appropriate counseling. Such an approach has been used in the present study, but ongoing data collection will be necessary for several more years.

Infertility

One of the striking elements of the European data [5] and the present data in relation to 2:2 segregating rcp is that the risk of unbalanced offspring is similar for male and female heterozygotes (Table I) irrespective of asc mode. This is in contrast to the conventional wisdom, by which we leap to test the mother first when a child with an unbalanced rcp is ascertained. We must conclude that, once a woman is pregnant, the child is as likely to be unbalanced or balanced/normal (in such rcp) with a carrier parent of either sex. However, there are (25–30%) fewer pat carriers of these prenatally diagnosed rcp offspring (Table I) than mat carriers for reasons best understood as the greater susceptibility of spermatogenesis to heterozygosity for a rearrangment; whereas the sex ratio of the offspring (Appendix I) is 518 males:479 females; i.e., there are at least as many males. Two theories are usually invoked to explain this subfertility effect or "germ cell maturation impairment" (Chandley, this volume) in pat heterozygotes. The first concerns the criticality of nonrandom contact between rcp quadrivalents and the transcriptionally inactive (at meiotic prophase) XY bivalent (the sex vesicle) for either X-autosome translocations [22] or as extended to male sterile autosomal translocations by Forejt and colleagues [23, 24]. The second hypothesis is that one of Miklos [25] linking difficulties in meiotic pairing (such as a constitutional trisomy or tendency to form chain quadrivalents with the chain ends unpaired) with abnormal spermatogenesis. Burgoyne and Baker [26] have argued that this could result in reactivation of unpaired sites and eventual quasimolecular suicide of the nucleus. Both of these models have some problems, and it is becoming clear that oogenesis is also not immune to some effects of aneuploidy.

Rob and 3:1 segregating rcp [e.g., (11,22)(q23;q11)] clearly demonstrate male infertility effects, for the latter, moreso than for 2:2 segregating rcp. In a sense, rob (being mostly dicentric) are 3:1 segregating structures, if chromosome arms and centromeres are counted, and like 3:1 rcp can form (male fertility-affecting) chain "quadrivalents." Chandley (this volume) points out that, among male patients attending infertility clinics, rob(13;14) is the commonest rob reported, but cases of rob(13,15), rob(14;21) rob(14;22), and rob(21;22) have also been documented. In that the present study considers only asc by cases that go on to prenatal diagnosis, it was difficult to accumulate many cases through asc mode 4 (infertility) for a comparison of rearrangment types. Martin (this volume) points out that in the two rob heterozygotes, one (13;14) and one (14;21) in whom sperm chromosomes have been studied, there were actually lower frequencies of unbalanced sperm (8%) than for all rcp similarly studied. One can infer that the progenitors of unbalanced sperm were inviable, that there were fewer sperm (subfertility) and also fewer unbalanced sperm. We note that these particular men were not infertile, but the potential for the effect is there, and, depending on genetic background, it may have a lesser or a greater effect on spermatogenesis in heterozygous individuals. This concept of how genetic background dramatically affects segregation is documented for *Mus* in rcp by Kirk and Searle (this volume) and in rob by Redi and Capanna (this volume).

Fragile Sites and Breakpoints

This association was examined in a previous study [5]. Essentially, the argument is as follows. It has been proposed [15] that there is an excess of chromosome fragile sites at the breakpoints of the human chromosome rearrangements. At least for those termed the constitutive fragile sites or c-fra [17–28], there is no such relationship in the European rcp data when the (11;22)(q23;q11) common rcp and the 11q23 c-fra or heritable fragile (h-fra) sites are excluded. There are 293 bands (including Dp and Gp as single bands) in the human haploid autosomal genome (using the region-band binary nomenclature, i.e., excluding subbands). Of these, 48 (16.4%) contain c-fra sites [28]. In the European data [5] of breakpoints in translocation heterozygotes, there are a total of 918 breakpoints, of which 169 (18.4%) occur at c-fra sites. This difference is nonsignificant ($\chi^2 = 1.08$, $P \geqslant 0.25$). The same correlation can be made for the 13 (excluding 11q23 as above) h-fra sites

[29] at 2q13, 6p23, 7p11, 8q22, 9p21, 9q32, 10q23, 10q25, 11q13, 12q12, 16p12, 16q22, and 17p12. These h-fra sites represent 4.9% of the 288 bands of the haploid autosomal genome, and, of the 918 breakpoints in the European data, they were found at 54 (5.9%). This difference is also nonsignificant ($\chi^2 = 1.97$, $P \geqslant 0.1$). Therefore, in these data, no support is found for the proposed relationship for either h-fra or c-fra sites.

Association of Rearrangements With Recurrent Miscarriage

Asc by recurrent miscarriage (asc mode 2) clearly is an important factor in many types of rearrangments (Table VI). In rcp type III[t] (see footnote to Table V), over 50% of the cases were ascertained in this way (Table VI, whereas this figure is about 30% for rob(13;14) and 10% for rob(14;21). For rob(13;14), this asc mode may be enriched because of physician concern and subsequent selective referral for prenatal diagnosis as compared to rob(13;14) incidentally ascertained. In connection with recurrent miscarriage, one study of rob(13;14) families is in apparent disagreement with such an association. Harris et al. [30] analyzed 86 sibships and found that 16.2% (32/197) of pregnancies ended in abortion for female carriers of rob(13;14) but only 9.2% for pregnancies when the father was the carrier. The figure for female carriers is not increased over the population norm.

Some common type pii, such as inv(10)p11.2q21.2) and inv(2)(p11.2q13), have been recorded in couples with recurrent miscarriage (Table VI), but such an association is probably an artifact, having been seized on because it was the only karyotypic abnormality. Several cases of less common rob, such as rob(14;15), rob(15;22), rob(13;15), and rob(14;22), were ascertained by mode 2 in the present study; given the known association of rob(14;14) and rob(15;15) with fetal wastage [31,32]; (i.e., loss later than, at, or before implantation), there is probably a causal relationship.

Multiple and Complex Rearrangements

Eleven cases of prenatal diagnosis of complex translocations (N = 3) or multiple rearrangements (N = 8) are reported in the present study. There are insufficient cases in any single category to suggest generalizations, but it may be safe to say that the presence of a common pii type *in addition to* a simple rcp or rob is little cause for increased concern. Most cases of complex or multiple *rcp* rearrangments (CCR) reported in the literature are ascertained through an infant with mental retardation and/or congenital abnormalities [33–35]. The more complex the haploid set of involved chromosomes, the less likely that they will be inherited together from a carrier parent. The few inherited cases are all from mat carriers because of probable difficulties in pairing during spermatogenesis and the resulting effects on fertility for males. These mat carriers had difficult obstetric histories, e.g., in ten pregnancies (some cytogenetically unstudied) from two families, there were four miscarriages, one stillbirth, one neonatal death, and two full-term unbalanced and two (20%) balanced/normal results. Both of these translocations [36,37] involved three breakpoints/chromosomes, whereas larger numbers of breakpoints are generally involved in the group ascertained through congenital malformations or mental retardation, e.g., cases with seven breakpoints [33] and eight breakpoints [35]. With a greater number of breakpoints in *balanced* CCR, there is probably a greater chance of an abnormal phenotype or mental retardation and crytic insertion events (and of abnormal segregants); hence the familial CCR will generally have fewer breakpoints. Presumably, if an insertion step is involved in familial CCR, then the risks of difficult obstetric histories/unbalanced outcomes will be substantial. In the three cases of CCR (all with mat carriers) for which the prenatal results are reported in the present study (Appendix VII), there was one unbalanced fetus, but all three CCR were ascertained through a full-term, unbalanced proband. Like insertions, these clearly are high-risk rearrangements.

Insertions

There were only six prenatal diagnoses of insertions reported in this study, but four of these are unbalanced, suggesting a substantial risk for carriers of such rearrangements. A maximum theoretical risk of 50%, given undiminished viability, could be expected on the basis of random chromosomal assortment. The reproductive histories of five families with autosomal insertions reported in the literature are shown in Table IX. With the removal of probands and carriers in direct descent, there are 11/36 (30.6%) unbalanced, full-term progeny including miscarriages, and, if the latter are excluded, the unbalanced, full-term progeny represent 11/27 (40.7%), i.e., close to the theoretical maximum. The average size of the insertions in Table IX is 1.11 ± 0.41 %HAL. Larger unbalanced (less viable) insertions will manifest more as miscarriages or earlier losses. The high rate of unbalanced ins progeny found in the present study is quite consistent with the previously published cases. This means that insertions involving segments of this average size are among the human autosomal rearrangements with the highest rates of unbalanced progeny (Table VI).

Putative Interchromosomal Effects

The term *interchromosomal effect* (coined by J. Lejeune) refers to the induction by an abnormal chromosome of altered meiotic behavior—particu-

TABLE IX. Reproductive outcome at full term in (direct and inverted) insertion carriers

Study	Ins	Proband's karyotype	Carrier	Size of ins (% HAL)	Carrier reproductive history excluding probands and carriers in direct descent[a]				
					M/C	SB	ND	UC	B/N
Chudley et al., 1974 [41]	(11;18)	Balanced/MCA (11;18)(p15;q11q21)	P	0.90	5	0	0	1 der18, 1 der11	0
			P		0	0	0	3 der18, 1 der11	6
			M		1	0	0	1 der11	1
			M		1	0	0	0	0
			M		0	0	0	0	3
Turleau et al., 1983 [42]	(16;13)	del(13)(q13q21.1)	P	1.0	1	0	0	0	1
			P		1	0	0	1 der13, 1 der16	0
Stoll et al., 1984 [43]	(4;1)[b]	dup(1)(q24q41)	M	1.56	0	0	0	1 der4	3
			M		0	0	0	1 ?der4	2
Orye and Laureys, 1984 [44]	?(4;3)[b]	dup(3)(p21p22)	M	0.59	0	0	0	0	0
Martin et al., 1985 [45]	(5;5)[b]	dup(5)(q22q33)	M	1.50	0	0	0	0	0
Totals (%)				1.11 ± 0.41	9	0	0	11 (40.7)	16 (59.3)

[a]Miscarriages (M/C), stillbirths (SB), neonatal deaths (ND), unbalanced full-term child (UC), balanced/normal child (B/N).
[b]Inverted insertions. Note the frequent alternative segregations 5 deficiencies:6 duplications in the unbalanced full-term offspring.

larly disjunction—of other chromosomes. In our own data, in the fetuses of 1,428 independently ascertained translocation carriers, there were seven abnormalities detected that were apparently unrelated to the parental chromosome rearrangement (Table IVB). Six of these were regular numerical aneuploidies (four 47,+21, one 47,+13, one 47,XXY), and one was an "extra structurally abnormal chromosome" (esac) 47,+i(Xq).

It is difficult to derive for comparison an "expected" number of unrelated cytogenetic abnormalities in this group of 1,428 fetuses, but at least a minimum boundary may be estimated as follows. Among about 57,000 newborns, there were 353 chromosome abnormalities detected, of which 209 (3.7 per 1,000) were numerical and 144 (2.5 per 1,000) were structural [38]. Excluding 18 Y-chromosome putative aberrations, there are 2.2 per 1,000 structural abnormalities, of which almost 25% are de novo (0.5 per 1,000) and the remainder inherited (1.7 per 1,000).

An inherited structural aberration, together with an unrelated numerical abnormality, is highly unlikely to be detected in the fetuses in our study. Therefore, our minimum expected rate of abnormalities could be derived considering only numerical and de novo structural abnormalities. This rate in newborns is about 3.7 per 1,000 numerical and 0.5 per 1,000 de novo structural, or about 4.2 per 1,000 total. This is the expected rate in *live births*. However, it is known that, because of spontaneous fetal death between the time of amniocentesis and live birth, the rate of abnormalities is higher in fetuses at 16–20 weeks than at term [39]. The excess spontaneous fetal death rate varies considerably from about 70% for 45,X and 47,+18 to close to zero for one extra X or Y chromosome or for balanced rearrangements [39]. An average death rate for all chromosome abnormality is roughly 20–30%, so the expected rate per 1,000 at amniocentesis is roughly 4.2/0.75, or 5.6 per 1,000, or eight in 1,428 fetuses. This is a minimum estimate, moreover, in that it assumes a maternal age distribution similar to that of the general population, but there probably is at least a slight bias to advanced maternal age in the fetuses studied here because of the vagaries of family planning with a structural chromosome heterozygosity; many of them were detected because of previous reproductive problems in the parents. Advanced maternal age is of course a well known risk factor for trisomies. It also appears to be a risk factor for de novo esac but not for other categories of structural abnormalities [40]. Thus the expected total number of abnormalities in the group of 1,428 fetuses studied is *at least* eight, greater than the seven actually observed.

It is more difficult to draw inferences about specific defects. The most suggestive observation is the four cases of 47,+21. The 95% confidence interval of four events in 1,428 (2.8 per 1,000) is 0.8–7.2 per 1,000, but the expected background rate of 47,+21 in 16–20-week-old fetuses is probably *at least* 1.1/07 = 1.6 per 1,000 (derived from the data described above and the minimum rate of 47,+21 among live births). The observation of four cases, although suggestive, is not nominally significant, because the expected proportion is within the 95% confidence interval of the observed proportion. It is difficult, moreover, to see why this particular defect, 47,+21, should on ad hoc grounds be analyzed separately from all abnormalities together.

There are some other considerations as well. Usually, the suggestion of an interchromosomal effect is raised for rob, especially those involving chromosome 21. There were 96 individuals with such balanced rearrangements in our data set, in *none* of whose fetuses was a 47,+21 detected. Moreover, none of those structural rearrangements that did occur with a 47,+21 involved a chromosome 21, although two of them did involve a chromosome 22, similar in size. On balance, these data provide little evidence in favor of a strong interchromosomal effect but cannot exclude a weak effect.

There is little firm evidence from other sources for interchromosomal effects with a *single* heterozygosity for a chromosome rearrangement. In the present data, the number of unrelated aneuploidies jointly ascertained with *single* rcp, rob, and pii (Table IVA) is not increased over those found later prenatally (Table IVB), but this sort of occurrence is the argument usually presented for the effect. However, there are some data in man suggesting an interchromosomal effect with *multiple* heterozygosity. Martin (this volume) compared the sperm chromosome complements in nine men heterozygous for rob, rcp, and pii. The frequency of unrelated abnormalities was significantly increased in only one man, the only double rcp heterozygote in whom there were 70% unrelated numerical abnormalities.

ACKNOWLEDGMENTS

This paper is dedicated to the memory and inspiration of Arthur Clifford Daniel (1908–1984), a good father. We particularly thank Mary Flesher for executive secretarial assistance, Ann Mann for persisting with Appendices III–VIII, and Linda Ekblom for a careful reading of the manuscript. We are in debt to the numerous investigators and their staffs (Appendix VIII) who freely contributed their data to the study.

REFERENCES

1. Daniel A: Structural differences in reciprocal translocations. Potential for a model of risk in rcp. Hum Genet 51:171, 1979.

2. Rickards GK: Orientation behavior of chromosome multiples of interchange (reciprocal translocation) heterozygotes. Annu Rev Genet 17:443, 1983.
3. Daniel A: Structural differences in pericentric inversions. Application to a model of risk of recombinants. Hum Genet 56:321, 1981.
4. DeArce MA, Grace PM, McManus S: A computer model for the study of segregation in reciprocal translocation carriers: Application to 20 new cases. Am J Med Genet 24:519, 1986.
5. Daniel A, Boué A, Gallano P: Prospective risk in reciprocal translocation heterozygotes at amniocentesis as determined by potential chromosome imbalance sizes. Data of the European Collaborative Prenatal Diagnostic Centers. Prenat Diagn 6:315, 1986.
6. Boué A, Gallano P: A collaborative study of the segregation of inherited chromosome structural rearrangements in 1356 prenatal diagnoses. Prenat Diagn 4[Special Issue]:45, 1984.
7. Daniel A: The size of prometaphase chromosome segments. Tables using percentages of haploid autosome length (750 band stage). Clin Genet 28:216, 1985.
8. Daniel A, Williams K, Lam-Po-Tang PRLC: Higher risk to D/G translocation carriers of tdic(13:21) as compared to the tdic(14;21). J Med Genet 17:491, 1980.
9. Davis JR, Barton Rogers B, Hagaman RM, Thies CA, Veomett IC: Balanced reciprocal translocations: Risk factors for aneuploid segregant viability. Clin Genet 27:1, 1985.
10. Ferguson-Smith MA, Yates JRW: Maternal age specific rates for chromosome aberrations and factors influencing them: report of a collaborative European study on 52,965 amniocenteses. Prenat Diag 4[Special Issue]:5, 1984.
11. Lindsley DL, Sandler L, et al.: Segmental aneuploidy and the genetic gross structure of the *Drosophila* genome. Genetics 71:157, 1972.
12. Jalbert P, Sele B, Jalbert H: Reciprocal translocations—A way to predict the mode of imbalanced segregation by pachytene-diagram drawing. A study of 151 human translocations. Hum Genet 55:209, 1980.
13. Rivera H, Rivas F, Cantu JM: Reciprocal translocations between acrocentrics, segregational analysis in twenty-nine families with unbalanced progeny. Ann Genet 29:93, 1986.
14. Bateman BG, Neu R, Nunley WC, Kelly TE: Pregnancy wastage associated with paracentric inversion of chromosome 13. J Med Genet 23:370, 1986.
15. Hecht F, Hecht BK: Fragile sites and chromosome breakpoints in constitutional rearrangements. II-spontaneous abortions, stillbirths and newborns. Clin Genet 26:174, 1984.
16. Jacobs PA: Correlation between euploid structural rearrangement and mental subnormality in humans. Nature 249:164, 1974.
17. Tharapel AT, Summitt RL, Wilroy RS, Martens P: Apparently balanced de novo translocations in patients with abnormal phenotypes. Report of six cases. Clin Genet 11:255, 1977.
18. Funderburk SJ, Spence MA, Sparkes RS: Mental retardation associated with "balanced" chromosome rearrangements. Am J Hum Genet 29:136, 1977.
19. Steinbach P: "Excess of mental retardation and/or congenital malformation in reciprocal translocation in man" (Fryns et al, 1986). Hum Genet 73:379, 1986.
20. Edwards JH: Chromosomal abnormalities in Mendelian disorders. Lancet 2:322, 1982.
21. Buhler EM: Unmasking of heterozygosity by inherited balanced translocations. Implications for prenatal diagnosis and gene mapping. Ann Genet 26:133, 1983.
22. Lifschytz E, Lindsley DL: The role of X-chromosome inactivation during spermatogenesis. Proc Natl Acad Sci USA 69:182, 1972.

23. Forejt J: Non-random association between a specific autosome and the X-chromosome in meiosis of the male mouse—Possible consequence of homologous centromeres separation. Cytogenet Cell Genet 13:369, 1974.
24. Forejt J, Gregorova S, Goetz P: XY pair associates with the synaptonemal complex of autosomal male-sterile translocations in pachytene spermatocytes of the mouse (Mus musculus). Chromosoma 82:41, 1981.
25. Miklos GLG: Sex chromosome pairing and male fertility. Cytogenet Cell Genet 13:558, 1974.
26. Burgoyne PS, Baker T: Meiotic pairing and gametogenic failure. In Evans CW, Dickinson HG (eds): "Controlling Events in Meiosis (38th Symp Soc Exp Biol)." Cambridge, MA: Company of Biologists, 1984, pp 349–362.
27. Daniel A, Ekblom L, Phillips S: Constitutive fragile sites 1p31, 3p14, 6q26 and 16q23 and their use as controls for false-negative results with the fragile (X). Am J Med Genet 18:483, 1984.
28. Yunis JJ, Soreng AL: Constitutive fragile sites and cancer. Science 226:1199, 1984.
29. Sutherland GR: The fragile X chromosome. Int Rev Cytol 81:107, 1983.
30. Harris DJ, Hankins L, Begleiter ML: Reproductive risk of t(13q14q) carriers: Case report and review. Am J Med Genet 3:175, 1979.
31. Maeda T, Ohno M, Takada M, Matsunobu A, Arai M: Postzygotic D/D translocation homozygosity associated with recurrent abortions. Am J Med Genet 15:389, 1983.
32. Gracias-Espinal R, Roberts SH, Duckett DP, Laurence KM: Recurrent spontaneous abortions due to a homologous translocation (14q14q). J Med Genet 19:465, 1982.
33. Pai GS, Thomas GH, Mahoney W, Migeon BR: Complex chromosome rearrangements. Report of a new case and literature review. Clin Genet 18:436, 1980.
34. Kleczkowska A, Fryns JP, Van den Berghe H: Complex chromosomal rearrangements (CCR) and their genetic consequences. J Genet Hum 30:199, 1982.
35. Fukushima Y, Kuroki Y, Ito T: Balanced double complex translocations [46,XX,t(1p;6p;7p;3q,11p)(11q;22p;21q)] in an infant with multiple congenital anomalies. Am J Med Genet 25:313, 1986.
36. Fujita H, Abe T, Yamamoto K, Furuyama J: Possible complex translocation t(9;14;13)(q12;p1?q31) in mother of a child with 9p trisomy syndrome. Hum Genet 25:83, 1974.
37. Sanchez O, Tunis JJ, Escobar JI: Partial trisomy 11 in a child resulting from a complex maternal rearrangment of chromosomes 11, 12, and 13. Hum Genet 22:59, 1974.
38. Hook EB, Hamerton, JL: The frequency of chromosome abnormalities detected in consecutive newborn studies—Differences between studies—Results by sex and by severity of phenotypic involvement. In Hook EB, Porter IH (eds): "Population Cytogenetics—Studies in Humans." New York:Academic Press, 1977, pp 63–79.
39. Hook EB: Chromosome abnormalities and spontaneous fetal deaths following amniocentesis: Further data and associations with maternal age. Am J Hum Genet 35:110, 1983.
40. Hook EB, Cross PK: Extra structurally abnormal chromosomes (e.s.a.c.) detected at amniocentesis: Frequency in approximately 75,000 prenatal cytogenetic diagnoses and associations with maternal and paternal age. Am J Hum Genet 40:83–101, 1987.
41. Chudley AE, Bauder F, Ray M, McAlpine PJ, Pena SDJ, Hamerton JL: Familial mental retardation in a family with an inherited chromosome rearrangement. J Med Genet 11,353, 1974.
42. Turleau C, de Grouchy J, Chavin-Colin F, Despoisses S, LeBlanc A: Two cases of del(13q)—retinoblastoma and two cases of partial trisomy due to a familial insertion. Ann Genet 26:158, 1983.

43. Stoll C, Roth M-P, Dott B: Trisomy 1q24—1q41 in two sibs with an insertion in an inverted chromosome 4. J Med Genet 21:133, 1984.
44. Orye E, Laureys G: Trisomy 3p syndrome. Report of a new case due to a chromosomal insertion. Ann Genet 27:245, 1984.
45. Martin NJ, Cartwright DW, Harvey PJ: Duplication 5q(5q22–5q33): From an intrachromosomal insertion. Am J Med Genet 20:57, 1985.

(See appendix page 109)

APPENDIX I. Prenatal diagnosis results for parental carriers of reciprocal translocations

		Breakpoint					Potential	Potential	
Observation	Laboratory	1	2	Carrier	Asc mode	Result code	or actual trisomy	or actual monosomy	Prenatal result
1	148	1p12	2p12	m	3	B	4.42	2.71	46,XY,t(1;2)
2	0200	1p31	2q31	m	3*	B	2.77	2.49	46,XY,t(1;2)
3	UK8	1q21	2p23	p	3	B	3.80	3.09	46,XY,t(1;2)
4	UK17	1q21	2q37	m	3	B	0.25	3.80	46,XX,t(1;2)
5	F10	1q32	2q14	p	2	B	2.17	4.46	46,XX,t(1;2)
6	0200	1q32	2q37	m	1	U	1.56	0.25	46,XY,der(2),t(1;2)
7	F9	1q44	2q32	p	1	B	2.03	0.12	46,XY,t(1;2)
8	SW2	1p11	3p11	p	3	B	4.54	3.06	46,XX,t(1;3)
9	D2	1p13	3p13	m	3	B	4.15	2.75	46,XY,t(1;3)
10	UK11	1q21	3q23	m	3	B	3.80	1.93	46,XX,t(1;3)
11	F8	1q21	3q29	p	3	B	0.11	3.80	46,XY,t(1;3)
12	0425	1q23	3q21	m	3*	B	3.25	2.55	46,XY,t(1;3)
13	319	1q25	3p23	p	4	U	2.39	1.04	46,XX,der(3),t(1;3)
14	F7	1q31	3q21	m	2	N	2.54	1.98	46,XX
15	F7	1q31	3q21	m	2	B	2.54	1.98	46,XY,t(1;3)
16	F10	1p21	4p13	m	2	N	3.78	1.47	46,XX
17	F10	1p21	4p13	m	2	B	1.47	3.78	46,XY,t(1;4)
18	F10	1p33	4p15	p	2	B	1.84	0.83	46,XX,t(1;4)
19	UK11	1p34	4p11	p	1	B	1.67	1.50	46,XY,t(1;4)
20	UK16	1p36	4p14	p	1	U	1.31	0.46	46,XY,der(1),t(1;4)
21	393	1q23	4p12	m	3*	B	1.59	3.17	46,XY,t(1;4)
22	0625	1q31	4q35	p	3	B	1.98	0.20	46,XY,t(1;4)
23	0625	1q31	4q35	p	3*	B	1.98	0.20	46,XY,t(1;4)
24	F15	1q32	4p16	m	1	B	2.17	0.22	46,XY,t(1;4)
25	F15	1q32	4p16	m	1	N	0.22	2.17	46,XY
26	F15	1q32	4p16	m	1	N	2.17	0.22	46,XY
27	F10	1q32	4q35	p	1	B	2.17	0.20	46,XY,t(1;4)
28	F10	1q32	4q35	p	1	U	2.17	0.20	46,XY,der(4),t(1;4)

(continued)

APPENDIX I. Prenatal diagnosis results for parental carriers of reciprocal translocations (continued)

Observation	Laboratory	Breakpoint 1	Breakpoint 2	Carrier	Asc mode	Result code	Potential or actual trisomy	Potential or actual monosomy	Prenatal result
29	Fin	1q42	4p15	p	1	N	0.77	0.83	46,XY
30	Fin	1q42	4p15	p	1	U	0.83	0.77	46,XY,der(1),t(1;4)
31	343	1q43	4p13	p	1	N	1.47	0.34	46,XY
32	UK11	1q43	4q25	m	1	N	2.78	0.34	46,XX
33	UK3	1q44	4q22	p	3	N	0.12	3.29	46,XX
34	IT7	1q44	4q28	p	3	N	2.10	0.12	46,XX
35	F10	1p31	5q13	p	2	N	3.81	2.76	46,XX
36	385	1p36	5q12	p	3	N	4.12	0.46	46,XY
37	1970	1p36	5q15	m	3	B	2.95	0.46	46,XY,t(1;5)
38	UK6	1q21	5q35	m	2	B	3.80	0.26	46,XX,t(1;5)
39	F8	1q31	5p14	m	3	N	0.78	1.98	46,XY
40	41	1q32	5p15.2	m	1	B	0.43	1.56	46,XX
41	UK14	1q23	5q35	m	3	N	3.17	0.26	46,XX
42	UK14	1q23	5q35	m	3	B	3.17	0.26	46,XY,t(1;5)
43	F2	1q41	5q11	p	2	B	1.19	4.44	46,XX,t(1;5)
44	IT2	1q42	5p15	m	1	U	0.18	0.77	46,XX,der(1),t(1;5)
45	F5	1q42	5q23	p	1	B	1.03	0.77	46,XX,t(1;5)
46	F5	1q42	5q23	m	1	N	1.03	0.77	46,XX
47	0270	1p32	6p21	p	3*	B	1.45	2.17	46,XX,t(1;6)
48	41	1q32	6p21	m	2	N	1.06	1.64	46,XY
49	IT2	1q32	6q26	m	1	N	2.17	0.26	46,XX
50	356	1q42	6q12	m	3*	B	3.75	0.77	46,XY,t(1;6)
51	F2	1q43	6q21	p	2	N	2.32	0.34	46,XY
52	381	1q44	6q25	m	2	U	0.13	0.12	46,XX,der(1),t(1;6)
53	F10	1p36	7p11	p	1	B	1.97	0.46	46,XX,t(1;7)
54	45	1p36	7p13	m	3*	B	1.58	0.46	46,XY,t(1;7)
55	1890	1q11.2	7q22	p	2	B	4.04	1.96	46,XX,t(1;7)
56	0020	1q21	7p22	m	4	B	3.80	0.20	46,XX,t(1;7)

57	0020	1q21	7p22	m	4	N	3.80	0.20	46,XX
58	90	1q23	7p15	p	3*	B	3.17	0.92	46,XX,t(1;7)
59	0200	1q32	7p22	p	3*	B	0.20	1.56	46,XX,t(1;7)
60	CH4	1q32	7q36	p	3	U	2.17	0.21	46,XY,der(7),t(1;7)
61	CH4	1q32	7q36	p	3	B	2.17	0.21	46,XX,t(1;7)
62	H2	1q32	7q36	p	3	U	2.17	0.21	46,XY,der(7),t(1;7)
63	1030	1q44	7q34	p	1	U	0.67	0.12	46,XY,der(1),t(1;7)
64	364	1p11	8p23	p	3	N	4.54	0.31	46,XY
65	H3	1p31	8q23	m	3	N	2.88	1.08	46,XY
66	IT3	1p34	8q24	m	2	N	0.46	1.50	46,XX
67	319	1q31.2	8p23.2	p	1	N	1.98	0.19	46,XX
68	SW4	1p22	9p23	m	3	N	3.36	0.36	46,XY
69	2070	1p36	9p21	p	2	N	2.17	0.46	46,XX
70	UK6	1p22	10q21	m	3	B	3.36	2.54	46,XX,t(1;10)
71	G7	1p36	10q24	p	3	B	0.77	2.20	46,XX,t(1;10)
72	0080	1p36	10q24	m	1	B	0.46	1.15	46,XX,t(1;10)
73	UK17	1q32	10q24	m	2	B	2.17	1.15	46,XY,t(1;10)
74	IT5	1q32	10q26	p	1	U	2.17	0.35	46,XY,der(10),t(1;10)
75	D2	1q44	10q21	m	3	U	0.12	2.53	46,XX,der(10),t(1;10)
76	UK11	1p11	11q11	m	3	B	4.54	2.96	46,XX,t(1;11)
77	319	1p13	11p11.2	p	3	N	4.15	1.33	46,XX
78	319	1p13	11p11.2	p	3	B	4.15	1.33	46,XY,t(1;11)
79	319	1p13	11p11.2	p	3	B	1.33	4.15	46,XX,t(1;11)
80	90	1p22	11p13	m	5	N	3.37	0.98	46,XX
81	1915	1p36	11q13	m	3	B	2.20	0.46	46,XY,t(1;11)
82	45	1p36	11p15.3	m	3*	B	0.43	0.46	46,XY,t(1;11)
83	0200	1q11	11p11	p	3*	B	4.04	1.44	46,XX,t(1;11)
84	UK17	1q23	11q13	p	3	B	3.17	2.20	46,XX,t(1;11)
85	F8	1p35	11q13	p	2	N	2.20	1.06	46,XX
86	CH2	1q42	11q13	p	2	N	0.77	2.20	46,XY
87	130	1p22	12q15	p	3*	B	3.37	2.27	46,XY,t(1;12)
88	0625	1p31	12p13	m	2*	B	2.77	0.30	46,XY,t(1;12)
89	UK11	1p32	12q24	p	3	B	0.53	2.16	46,XY,t(1;12)
90	1700	1q42	12q12	m	2	B	3.39	0.77	46,XY,t(1;12)

(continued)

APPENDIX I. Prenatal diagnosis results for parental carriers of reciprocal translocations (continued)

Observation	Laboratory	Breakpoint 1	Breakpoint 2	Carrier	Asc mode	Result code	Potential or actual trisomy	Potential or actual monosomy	Prenatal result
91	41	1q43	12p13	m	1	U	0.20	0.34	46,XX,der(1),t(1;12)
92	G5	1p21	13q12	m	3	B	3.78	2.84	46,XX,t(1;13)
93	148	1p36	13q12	m	2	N	2.92	1.15	46,XX
94	D1	1q31	13q34	m	1	N	1.98	0.12	46,XX
95	45	1q42	13q32	p	3*	B	0.56	0.77	46,XX,t(1;13)
96	1960	1q44	13q14	m	1	B	2.17	0.12	46,XX,t(1;13)
97	0970	1p32.3	14q32.1	p	2	B	0.59	1.99	46,XY,t(1;14)
98	F1	1q42	14q13	p	2	B	2.56	0.77	46,XX,t(1;14)
99	90	1p32	15q22	p	2	B	2.16	1.27	46,XY,t(1;15)
100	90	1p32	15q22	p	2	B	2.16	1.27	46,XX,t(1;15)
101	90	1p32	15q22	p	2	B	1.27	2.16	46,XX,t(1;15)
102	1980	1p36	15q24	m	2	B	0.87	0.46	46,XY,t(1;15)
103	130	1q11	15q26	p	2	N	4.04	0.23	46,XY
104	SW3	1q12	15p11	m	3	B	4.04	0.00	46,XX,t(1;15)
105	1178	1q32	15q26	m	3	N	1.56	0.23	46,XY
106	UK6	1p11	16q11	m	3	B	1.61	4.54	46,XY,t(1;16)
107	UK6	1p11	16q11	m	3	N	4.54	1.61	46,XX
108	1030	1p13.2	16p11.2	p	3*	B	4.23	1.00	46,XX,t(1;16)
109	UK17	1p32	16q22	m	2	N	2.16	0.78	46,XY
110	130	1p36	16p12	p	2	B	0.46	0.78	46,XX,t(1;16)
111	ISR	1q12	16p13	m	2	N	4.04	0.33	46,XY
112	D1	1q12	16q12	p	3	B	4.04	1.47	46,XY,t(1;16)
113	F10	1q21	16p11	m	2	B	3.80	1.07	46,XX,t(1;16)
114	90	1q21.2	16q22	m	3	N	0.78	3.80	46,XY
115	UK8	1q42	16q24	m	1	U	0.15	0.77	46,XY,der(1),t(1;16)
116	F10	1p34	17p11	m	2	B	1.50	0.78	46,XY,t(1;17)
117	319	1p34	17q12.3	p	3*	B	2.05	1.50	46,XY,t(1;17)
118	2070	1p36	17p11	m	2	N	0.46	0.78	46,XX

119	2070	1p36	17p11	m	2	N	0.78	0.46	46,XY
120	2070	1p36	17p11	m	2	N	0.78	0.46	46,XX
121	341	1p36	17q25	p	2	N	0.46	0.29	46,XX
122	341	1p36	17q25	p	2	B	0.29	0.46	46,XY,t(1;17)
123	45	1p36.1	17q25.2	p	3*	B	0.32	0.71	46,XY,t(1;17)
124	90	1q21.2	17p13	p	2	N	3.80	0.23	46,XX
125	D1	1q21	17q21	p	3	B	3.80	1.67	46,XX,t(1;17)
126	UK17	1q32	17q21	p	2	N	2.17	1.67	46,XY
127	F15	1q32	17q25	p	3	B	2.17	0.29	46,XX,t(1;17)
128	1960	1q41	17p13	p	1	U	1.19	0.23	46,XY,der(17),t(1;17)
129	1980	1q42	17p12	p	2	B	0.77	0.53	46,XY,t(1;17)
130	F10	1q42	17p12	m	2	U	0.77	0.53	46,XY,der(17),t(1;17)
131	F10	1q42	17p12	m	2	U	0.77	0.53	46,XY,der(17),t(1;17)
132	90	1p22	18q21	p	3*	B	0.84	3.37	46,XY,t(1;18)
133	402	1p34	18p11.3	m	2	B	0.05	1.50	46,XY,t(1;18)
134	UK8	1q32	18q21	p	2	N	2.17	0.83	46,XY
135	UK8	1q32	18q21	p	2	B	2.17	0.83	46,XY,t(1;18)
136	UK17	1q32	18q23	m	1	B	2.17	0.12	46,XY,t(1;18)
137	1030	1q44	18q21	p	1	N	0.12	0.69	46,XY
138	0200	1p11	19p11	m	3*	B	4.54	1.11	46,XX,t(1;19)
139	UK12	1q21	19q11	m	3	B	3.80	1.36	46,XX,t(1;19)
140	2955	1p36.1	19q13.4	m	2	N	0.70	0.25	46,XY
141	UK12	1q21	19q11	m	3	N	3.80	1.36	46,XX
142	F8	1p22	20p12	m	2	N	0.39	3.53	46,XY
143	UK17	1p32	20q13	m	3	B	0.42	2.16	46,XX,t(1;20)
144	UK5	1p35	20p11	m	1	N	1.06	0.90	46,XX
145	CH4	1q32	20q13	p	3	B	2.17	0.42	46,XX,t(1;20)
146	343	1p11	21p11	m	4	N	4.54	0.00	46,XX
147	UK5	1p11	21p12	m	1	N	4.54	0.00	46,XX
148	IT17	1q12	21q12	p	3	N	4.04	0.28	46,XX
149	UK4	1q32	21q22	m	1	B	2.17	0.28	46,XY,t(1;21)
150	UK4	1q32	21q22	m	1	N	2.17	0.28	46,XY
151	UK14	1q42	21q22	m	1	B	0.28	0.77	46,XX,t(1;21)
152	SW3	1q42	22q13	p	1	B	0.77	0.30	46,XY,t(1;22)

(continued)

APPENDIX I. Prenatal diagnosis results for parental carriers of reciprocal translocations (continued)

Observation	Laboratory	Breakpoint 1	Breakpoint 2	Carrier	Asc mode	Result code	Potential or actual trisomy	Potential or actual monosomy	Prenatal result
153	SW3	1q42	22q13	p	1	U	0.77	0.30	46,XY,der(22),t(1;22)
154	H1	1q43	22q13	m	1	U	0.30	0.34	46,XX,der(1),t(1;22)
155	H1	1q43	22q13	m	1	U	0.34	0.30	46,XY,der(22),t(1;22)
156	364	2p21	3p24	p	1	N	1.57	0.80	46,XX
157	UK18	2q21	3q27	m	2	B	4.02	0.52	46,XY,t(2;3)
158	130	2p21	4q35	m	2	B	1.62	0.20	46,XX,t(2;4)
159	2940	2p23	4q25	p	2	B	2.79	1.05	46,XX,t(2;4)
160	F10	2p25	4q31	p	1	B	0.35	1.50	46,XX,t(2;4)
161	F10	2p25	4q31	p	1	N	1.50	0.25	46,XX
162	0200	2q37	4q21	p	3*	B	3.72	0.25	46,XY,t(2;4)
163	2760	2q37	4q31	m	1	B	1.50	0.25	46,XX,t(2;4)
164	F11	2p11.2	5q31.1	p	2	N	1.71	2.98	46,XY
165	Fin	2p12	5q35	p	2	B	2.71	0.26	46,XX,t(2;5)
166	0200	2p21	5q35	p	2*	B	1.62	0.26	46,XX,t(2;5)
167	45	2q13	5q34	m	3*	B	0.61	4.84	46,XY,t(2;5)
168	F10	2q33	5q34	p	2	B	1.56	0.61	46,XX,t(2;5)
169	F7	2q37.3	5q21	p	2	N	0.14	2.69	46,XY
170	393	2p13	6p23	p	2	B	2.45	0.53	46,XX,t(2;6)
171	UK17	2p21	6q21	m	3	B	1.57	2.32	46,XY,t(2;6)
172	SW1	2p25	6p21	m	1	U	1.45	0.25	46,XX,der(2),t(2;6)
173	SW1	2p25	6p21	m	1	U	1.45	0.25	46,XY,der(2),t(2;6)
174	1970	2q12	6q27	m	2	N	5.10	0.10	46,XX
175	1510	2q24.1	6p11	p	3*	B	3.20	2.31	46,XX,t(2;6)
176	F10	2q35	6q21	p	2	N	2.32	0.98	46,XX
177	UK5	2q35	6q27	m	3	B	0.10	0.98	46,XX,t(2;6)
178	H3	2q36	6p21	m	3	B	1.45	0.69	46,XY,t(2;6)
179	F15	2q37	6q21	m	2	N	0.25	2.32	46,XX
180	UK15	2q21	7p15	m	2	N	4.02	0.92	46,XY

181	F11	2q31	7q36	m	1	N	2.49	0.21	46,XY
182	UK8	2q33	7p22	p	3	B	1.56	0.20	46,XY,t(2;7)
183	Fin	2q33	7p22	m	1	N	1.56	0.20	46,XX
184	UK6	2q33	7q22	m	2	N	1.96	1.56	46,XX
185	341	2q35	7p13	p	3*	B	1.58	0.98	46,XX,t(2;7)
186	1970	2q35	7q22	m	5	N	1.96	1.00	46,XX
187	1970	2q35	7q22	m	5	N	1.96	1.00	46,XY
188	UK18	2p13	8q22	m	3	B	2.45	1.45	46,XX,t(2;8)
189	F12	2p15	8q24	p	2	B	2.09	0.47	46,XX,t(2;8)
190	F12	2p15	8q24	p	2	N	0.47	2.09	46,XY
191	130	2p23	8p21	p	3*	B	0.86	1.05	46,XY,t(2;8)
192	130	2p23	8p21	p	3	N	1.05	0.86	46,XX
193	UK17	2q33	8p23	m	1	N	1.56	0.31	46,XY
194	0200	2p15	9q13	p	3*	B	2.60	2.10	46,XY,t(2;9)
195	0200	2q35	9p13	p	3	B	1.14	0.98	46,XX,t(2;9)
196	0200	2q35	9p13	p	3	N	1.14	0.98	46,XX
197	F10	2q37	9q22	m	3	B	1.54	0.25	46,XY,t(2;9)
198	395	2p23	10q26	p	1	N	0.35	1.05	46,XX
199	G1	2p25	10q22	m	3	B	0.25	2.07	46,XY,t(2;10)
200	F8	2q32	10p13	m	2	N	2.03	0.57	46,XX
201	UK17	2q33	10p13	p	2	B	1.56	0.57	46,XY,t(2;10)
202	UK17	2q33	10p13	p	2	N	0.57	1.56	46,XY
203	381	2q33	10q21	m	2	B	2.54	1.56	46,XY,t(2;10)
204	381	2q33	10q21	m	2	N	2.54	1.56	46,XY
205	H3	2q34	10q22	m	3	U	2.07	1.20	46,XX,der(2),t(2;10)
206	90	2q37	10q22	p	3*	B	2.07	0.25	46,XY,t(2;10)
207	364	2p13	11q21	p	3	B	2.45	1.43	46,XY,t(2;11)
208	1990	2p21	11q23	m	2	N	1.62	0.72	46,XX
209	130	2p25	11p11	m	3	N	0.25	1.43	46,XY
210	130	2p25	11p11	m	3*	B	1.43	0.25	46,XX,t(2;11)
211	364	2q21	11q25	m	2	B	4.02	0.09	46,XX,t(2;11)
212	G7	2q33	11q23	m	1	B	1.56	0.78	46,XY,t(2;11)
213	330	2p21	12p13	m	1	U	1.62	0.30	46,XX,der(12),t(2;12)
214	F10	2p21	12q13	p	2	B	1.57	2.91	46,XX,t(2·12)

(continued)

APPENDIX I. Prenatal diagnosis results for parental carriers of reciprocal translocations (continued)

Observation	Laboratory	Breakpoint 1	Breakpoint 2	Carrier	Asc mode	Result code	Potential or actual trisomy	Potential or actual monosomy	Prenatal result
215	F10[a]	2p21	12q13	p	2	B	2.91	1.57	46,XY,t(2;12)
216	1700	2q21	12q15	p	3	N	4.02	2.24	46,XX
217	UK17	2q23	12q13	m	3	N	3.45	2.91	46,XX
218	F16	2q31	12p13	m	2	N	2.49	0.30	46,XY
219	1970	2p23	13q11	m	1	N	0.00	1.10	46,XY
220	F15	2q14	13q34	m	2	B	4.46	0.12	46,XX,t(2;13)
221	385	2q22	13q34	p	2	B	0.12	3.72	46,XX,t(2;13)
222	385	2q33	13q12	m	3*	B	1.56	2.97	46,XX,t(2;13)
223	3060	2q33	13q34	m	1	N	1.56	0.12	46,XY
224	B1	2p23	14q31	p	2	N	3.09	1.50	46,XX
225	41	2q37	14q24	p	1	B	1.23	0.25	46,XY,t(2;14)
226	45	2q37	14q24	p	3*	B	0.25	1.23	46,XY,t(2;14)
227	1960	2p12	15q26	p	2	N	2.72	0.23	46,XY
228	F10	2q14	15q25	m	3	B	4.46	0.62	46,XX,t(2;15)
229	F10	2q21	15p11	m	2	N	4.02	0.00	46,XY
230	130	2q23	15q26	m	3	B	3.44	0.23	46,XY,t(2;15)
231	343	2q32	15q22	p	2	N	1.27	2.02	46,XX
232	F12	2p21	16p13	m	3	B	1.57	0.33	46,XY,t(2;16)
233	402	2q14	16q22	p	3	N	4.46	0.78	46,XX
234	393	2q33	16q22	p	3*	B	3.45	0.78	46,XY,t(2;16)
235	F10	2q31	16q23	p	3	N	2.49	0.44	46,XX
236	F10	2q31	16q23	p	3	N	2.49	0.44	46,XX
237	F10	2q31	16q23	m	3	N	2.49	0.44	46,XY
238	CH4	2q37	16p11	m	1	B	0.25	1.07	46,XY,t(2;16)
239	1030	2p25	17q23	m	1*	U	0.24	0.25	46,XX,der(17),t(2;17)
240	F14	2p11	17p13	m	2	B	3.06	0.22	46,XY,t(2;17)
241	G5	2q23	17q12	m	3	B	3.45	2.05	46,XY,t(2;17)
242	90	2p21	18q22.2	p	3*	B	0.36	1.62	46,XX,t(2;18)

243	F10	2q13	18q22	m	2	B	4.84	0.40	46,XY,t(2;18)
244	F10	2q13	18q22	m	2	N	4.84	0.40	46,XY
245	G5	2q33	18q23	p	2	B	0.12	1.56	46,XY,t(2;18)
246	1510	2q34	18p11.3	m	1	N	1.20	0.11	46,XY
247	G3	2q37	18q21	m	1	N	0.25	0.84	46,XX
248	319	2p11	19q13	p	3	B	0.57	3.06	46,XY,t(2;19)
249	0425	2q21	19q13.3	p	3*	B	4.02	0.60	46,XY,t(2;19)
250	UK11	2q33	19p13	p	3	B	1.56	0.66	46,XY,t(2;19)
251	ISR	2p15	20p13	m	1	U	2.10	0.12	46,XY,der(20),t(2;20)
252	ISR	2p15	20p13	m	1	N	2.10	0.12	46,XX
253	H3	2p16	20p12	p	3	B	1.88	0.39	46,XY,t(2;20)
254	SW4	2p16	20q11	p	3	B	1.88	1.16	46,XY,t(2;20)
255	37	2p11	21q11	m	2	B	2.98	1.16	46,XY,t(2;21)
256	UK11	2q12	21q22	m	1	N	1.22	0.00	46,XY
257	UK11	2q12	21q22	m	1	N	1.22	0.00	46,XX
258	F10	2q11	22p11	p	2	N	5.23	0.00	46,XY
259	1970	3p12	4q34	p	3	B	2.85	0.55	46,XX,t(3;4)
260	UK15	3p13	4p14	p	3	B	1.31	2.75	46,XY,t(3;4)
261	341	3p21	4p16	m	1	U	1.80	0.22	46,XY,der(4),t(3;4)
262	H1	3p25	4q25	p	2	N	0.41	2.78	46,XY
263	45	3q12	4p14	m	2	N	3.44	1.31	46,XY
264	45	3q12	4p14	m	2	N	3.44	1.31	46,XX
265	130	3q13	4q35	m	3*	B	3.07	0.20	46,XX,t(3;4)
266	H1	3q21	4q31	m	3	N	2.54	1.50	46,XX
267	0425	3q27	4q25	m	2	B	2.79	0.52	46,XX,t(3;4)
268	0425	3q27	4q25	m	2	B	0.52	2.79	46,XY,t(3;4)
269	364	3q27	4q31.3	p	3*	B	0.57	0.83	46,XX,t(3;4)
270	0710	3p24	5p15	m	1	U	0.80	0.32	46,XX,der(5),t(3;5)
271	0710	3p24	5p15	m	1	B	0.32	0.80	46,XX,t(3;5)
272	CH4	3q22	5p15	m	1	U	2.17	0.31	46,XY,der(5),t(3;5)
273	355	3q27	5q23	m	3*	B	2.06	0.52	46,XY,t(3;5)
274	0710	3p13	6q21	m	3*	B	2.75	2.32	46,XY,t(3;6)
275	45	3p14	6q25	p	2	B	2.47	0.13	46,XX,t(3;6)
276	3060	3p26	6p21	p	1	B	1.45	0.10	46,XY,t(3;6)

(continued)

APPENDIX I. Prenatal diagnosis results for parental carriers of reciprocal translocations (continued)

Observation	Laboratory	Breakpoint 1	Breakpoint 2	Carrier	Asc mode	Result code	Potential or actual trisomy	Potential or actual monosomy	Prenatal result
277	3060	3p26	6p21	p	1	B	1.45	0.10	46,XY,t(3;6)
278	37	3p26.3	6q25.3	m	1	B	0.03	0.50	46,XX,t(3;6)
279	37	3p26.3	6q25.3	m	1	B	0.50	0.03	46,XY,t(3;6)
280	343	3q13	6q23	p	2	N	3.07	1.25	46,XX
281	H1	3q29	6q23	p	3	B	0.11	1.25	46,XX,t(3;6)
282	90	3p12.1	7p11.2	m	2	N	3.15	1.97	46,XY
283	1960	3p14	7p22	m	1	U	2.47	0.20	46,XX,der(7),t(3;7)
284	F8	3p14	7q36	p	2	N	0.21	2.35	46,XY
285	F8	3p14	7q36	p	2	B	2.35	0.21	46,XY,t(3;7)
286	F11	3p24	7q31	p	2	N	1.42	0.80	46,XY
287	F11	3p24	7q31	p	2	B	1.42	0.80	46,XX,t(3;7)
288	UK4	3p26	7q21	m	3	N	2.55	0.10	46,XX
289	1950	3q21	7q36	p	2	B	0.21	2.55	46,XY,t(3;7)
290	45	3q25	7q31.2	m	3*	B	1.53	1.44	46,XY,t(3;7)
291	2150	3p14.1	8p23.1	p	1	B	2.60	0.31	46,XY,t(3;8)
292	UK3	3q13	8q24	m	3	B	2.87	0.46	46,XX,t(3;8)
293	395	3p26	9q33	p	3*	B	0.68	0.10	46,XX,t(3;9)
294	CH1	3q25	9p23	p	1	N	1.34	0.36	46,XX
295	CH4	3q26	9p24	m	1	B	0.00	0.95	46,XX,t(3;9)
296	2070	3q28	9q22	p	2	B	1.54	0.30	46,XX,t(3;9)
297	1990	3p25	10q24	p	3*	B	0.41	1.15	46,XX,t(3;10)
298	1700	3p26.3	10p11	m	1	U	1.20	0.03	46,XX,der(3),t(3;10)
299	341	3q13	10p15	m	3*	B	3.07	0.16	46,XX,t(3;10)
300	1760	3q29	10q22	m	2	B	2.07	0.11	46,XX,t(3;10)
301	130	3q24	11q23	m	2	B	1.69	0.78	46,XX,t(3;11)
302	F10	3p22	12q24	m	3	B	1.21	0.53	46,XY,t(3;12)
303	UK17	3p23	12q15	m	3	B	1.04	2.27	46,XX,t(3;12)
304	F15	3q27	12q22	m	2	N	0.52	1.47	46,XY

305	G6	3q21	13q14	m	2	U	2.54	2.17	46,XX,der(13),t(3;13)
306	F10	3q21	13q22	p	2	B	2.54	1.19	46,XX,t(3;13)
307	F10	3q21	13q22	p	2	N	2.54	1.19	46,XY
308	330	3q24	13q14	p	1	N	2.17	1.74	46,XY
309	343	3q28	13q14	m	2	B	2.17	0.18	46,XY,t(3;13)
310	343	3q28	13q14	m	2	N	0.18	2.17	46,XY
311	41	3q12	14p12	m	1	N	3.45	0.00	46,XY
312	F11	3q13	14q31	m	2	B	2.87	1.50	46,XX,t(3;14)
313	1950	3q13	14q22	p	1	N	3.07	0.34	46,XX
314	1950	3q13	14q32	p	1	B	3.07	0.34	46,XY,t(3;14)
315	1290	3q21	14q32	p	1	U	2.55	0.34	46,XY,der(14),t(3;14)
316	45	3q27	14q13	m	1	N	2.56	0.52	46,XY
317	45	3q27	14q13	m	1	N	2.56	0.52	46,XX
318	1030	3q27	14q22	p	1	B	1.95	0.52	46,XY,t(3;14)
319	1030	3q27	14q22	p	1	N	1.95	0.52	46,XY
320	1030	3q27	14q22	p	1	N	1.95	0.52	46,XY
321	H3	3p13	15q26	p	3	B	2.75	0.23	46,XX,t(3;15)
322	F7	3p24	15q23	m	2	N	1.02	0.80	46,XY
323	1178	3q21	15p13	m	3*	B	2.55	0.00	46,XX,t(3;15)
324	D2	3q21	15q22	m	3	B	2.54	1.27	46,XX,t(3;15)
325	CH4	3q23	15p12	m	3	U	1.93	0.00	46,XY,der(15),t(3;15)
326	CH4	3q23	15p12	m	3	B	1.93	0.00	46,XY,t(3;15)
327	CH4	3q23	15p12	m	3	B	1.93	0.00	46,XY,t(3;15)
328	F10	3q28	15q22	m	2	N	1.27	0.29	46,XX
329	F10	3q28	15q22	m	2	N	0.29	1.27	46,XX
330	2760	3q29	15q22	p	3*	B	0.11	1.27	46,XY,t(3;15)
331	UK9	3q21	16q12	m	2	N	2.54	1.47	46,XY
332	79	3p12.2	17p13	p	2	B	2.85	0.23	46,XX,t(3;17)
333	F12	3p13	17p13	m	2	N	2.75	0.22	46,XY
334	F12	3p25	17q21	p	2	B	0.41	1.66	46,XY,t(3;17)
335	UK6	3q21	17q25	m	1	N	2.54	0.29	46,XX
336	UK6	3q21	17q25	m	1	N	2.54	0.29	46,XX
337	G4	3q27	17p13	m	1	B	0.22	0.52	46,XX,t(3;17)
338	F10	3p13	18p11	m	3	B	2.75	0.35	46,XX,t(3;18)

(continued)

APPENDIX I. Prenatal diagnosis results for parental carriers of reciprocal translocations (continued)

Observation	Laboratory	Breakpoint 1	Breakpoint 2	Carrier	Asc mode	Result code	Potential or actual trisomy	Potential or actual monosomy	Prenatal result
339	393	3p23	18q12	p	1	N	1.35	1.05	46,XY
340	45	3q22	18q22.3	p	2	B	2.17	0.29	46,XY,t(3;18)
341	ISR	3q27	18q23	m	1	B	0.12	0.52	46,XX,t(3;18)
342	F1	3q22	20p11	p	3	B	0.74	2.17	46,XX,t(3;20)
343	H3	3q27	21q22	p	1	B	0.29	0.52	46,XY,t(3;21)
344	F10	3p14	22q13	p	3	N	2.35	0.29	46,XX
345	UK9	3p26	22q11	p	2	B	1.18	0.10	46,XY,t(3;22)
346	UK9	3p26	22q11	p	2	B	1.18	0.10	46,XY,t(3;22)
347	UK9	3p26	22q11	p	2	N	0.10	1.18	46,XY
348	79	4p11	5q11	m	2	B	1.89	1.63	46,XX,t(4;5)
349	90	4p11	5q35	m	1	B	0.26	1.67	46,XX,t(4;5)
350	IT2	4q22	5q35	m	2	N	3.29	0.26	46,XX
351	UK4	4q26	5p15	p	1	N	2.63	0.31	46,XY
352	UK5	4q31	5q13	m	3	B	3.81	1.50	46,XY,t(4;5)
353	G5	4q31	5q14	m	2	N	3.32	1.50	46,XY
354	H3	4p13	6q12	m	2	B	1.47	3.75	46,XX,t(4;6)
355	D2	4p14	6p25	p	1	B	1.31	0.00	46,XX,t(4;6)
356	G7	4p15	6p21	m	2	B	1.45	0.83	46,XX,t(4;6)
357	F10	4q12	6q14	m	3	B	4.54	3.32	46,XY,t(4;6)
358	G5	4q27	6p23	p	1	U	2.47	0.52	46,XX,der(6),t(4;6)
359	UK9	4q35	6q15	p	3	B	0.20	3.04	46,XX,t(4;6)
360	1178	4q27	7q34	p	2	B	2.48	0.67	46,XY,t(4;7)
361	335	4q31	7p13	m	2	B	1.58	1.50	46,XX,t(4;7)
362	F10	4q35	7q32	p	1	B	0.20	1.02	46,XX,t(4;7)
363	393	4p16	8p23	m	1*	U	0.31	0.23	46,XX,der(4),t(4;8)
364	D2	4p16.1	8p23.1	p	1	B	0.23	0.44	46,XX,t(4;8)
365	1030	4q21	8q22	m	3*	B	3.71	1.45	46,XY,t(4;8)
366	IT2	4q31	8p22	p	2	B	0.64	1.50	46,XX,t(4;8)

367	F10	4q32	8q21	p	3	N	1.95	0.99	46,XX
368	381	4q35	8q13	p	1	B	2.40	0.20	46,XY,t(4;8)
369	0200	4q13	9q32	m	3*	B	4.24	0.90	46,XX,t(4;9)
370	UK3	4q21	9q34	m	3	B	3.71	0.27	46,XX,t(4;9)
371	2955	4q21	9q21	m	3*	B	2.17	3.71	46,XY,t(4;9)
372	G4	4q23	9p24	p	1	B	3.09	0.00	46,XY,t(4;9)
373	UK6	4q31	9p21	p	3	B	1.50	0.84	46,XY,t(4;9)
374	B1	4q31	9q13	m	3	U	3.51	0.00	47,XY,+der(9),t(4;9)
375	F10	4q33	9q34	m	1	U	0.72	0.27	46,XX,der(9),t(4;9)
376	IT6	4q34	9q12	m	2	B	2.65	0.35	46,XX,t(4;9)
377	UK3	4q35	9p21	p	1	U	0.84	0.20	46,XY,der(4),t(4;9)
378	H3	4q35	9p13	p	1	U	1.14	0.20	46,XY,der(4),t(4;9)
379	F10	4p11	10q26	m	3	N	0.36	1.67	46,XX
380	UK6	4p13	10q23	m	3	B	1.54	1.47	46,XX,t(4;10)
381	1510	4p14	10q11.2	m	3	B	2.92	1.31	46,XY,t(4;10)
382	1510	4p14	10q11.2	p	3*	B	2.92	1.31	46,XX,t(4;10)
383	1950	4p14	10q22	m	1	B	2.07	1.31	46,XX,t(4;10)
384	1950	4p14	10q22	m	1	N	2.07	1.31	46,XY
385	SW4	4p14	10q26	p	3	N	1.31	0.36	46,XY
386	F15	4p16	10p14	p	3	N	0.23	0.40	46,XX
387	H3	4q23	10q11	m	3	B	3.18	3.09	46,XX,t(4;10)
388	F7	4q23	10q21	p	3	B	3.09	2.54	46,XY,t(4;10)
389	0425	4q25	10q22	m	3*	B	2.79	2.07	46,XX,t(4;10)
390	G3	4q25	10q26	p	3	N	0.36	2.78	46,XX
391	G4	4q31	10q22	m	2	N	2.07	1.50	46,XY
392	H3	4q31	10q26	m	1	U	1.50	0.36	46,XX,der(10),t(4;10)
393	1030	4q31	10q26	m	1	N	0.36	1.50	46,XY
394	H3	4q31	10q26	m	1	U	0.36	1.50	46,XX,der(4),t(4;10)
395	UK17	4p14	11p15	m	3	U	0.33	1.31	46,XY,der(4),t(4;11)
396	1030	4p14	11q25	p	1	U	1.31	0.09	46,XY,der(11),t(4;11)
397	B1	4q13	11p15	p	2	B	4.24	0.33	46,XX,t(4;11)
398	2070	4q22	11q11	m	3	B	3.29	2.96	46,XY,t(4;11)
399	F10	4q34	11p14	p	1	B	0.79	0.55	46,XX,t(4;11)

(continued)

APPENDIX I. Prenatal diagnosis results for parental carriers of reciprocal translocations (continued)

Observation	Laboratory	Breakpoint 1	Breakpoint 2	Carrier	Asc mode	Result code	Potential or actual trisomy	Potential or actual monosomy	Prenatal result
400	130	4q31.3	12p13.1	m	3*	B	1.25	0.55	46,XY,t(4;12)
401	130	4q31.3	12p13.1	m	3	N	1.25	0.55	46,XX
402	1830	4p11	13p11	m	2	B	0.00	1.67	46,XY,t(4;13)
403	45	4q27	13q22	m	1	B	2.48	0.50	46,XY,t(4;13)
404	341	4q21	14q32	p	3*	B	0.34	3.72	46,XY,t(4;14)
405	F15	4q26	14q21	m	2	N	2.63	2.27	46,XY
406	F11	4q34	14q23	m	2	N	1.64	0.55	46,XY
407	UK11	4p14	15q24	p	3	N	1.31	0.88	46,XY
408	F10	4q21	15q26	m	2	B	3.71	0.23	46,XY,t(4;15)
409	F10	4q22	15q21	m	3	N	3.29	1.66	46,XY
410	UK6	4q23	15q26	m	1	N	3.09	0.23	46,XX
411	F10	4q28	15q21	m	2	B	1.66	2.10	46,XX,t(4;15)
412	F10	4q28	15q21	m	2	B	2.10	1.66	46,XY,t(4;15)
413	F2	4q31	15q25	m	3	B	1.50	0.62	46,XX,t(4;15)
414	UK3-4	4p16	16q13	m	1	B	0.23	1.24	46,XX,t(4;16)
415	F11	4q22	16p13	m	3	N	0.33	3.29	46,XY
416	F11	4q22	16p13	p	3	B	3.29	0.33	46,XY,t(4;16)
417	UK5	4q22	16q23	p	3	B	3.29	0.44	46,XY,t(4;16)
418	H1	4q31	16p13	p	3	N	1.50	0.33	46,XX
419	45	4p12	17q23	m	1	B	1.09	1.59	46,XX,t(4;17)
420	45	4p13	17q25	m	1	N	0.29	1.47	46,XY
421	2070	4p15.3	17q12	m	2	B	0.65	0.53	46,XY,t(4;17)
422	2070	4p15.3	17q12	m	2	N	2.05	0.65	46,XX
423	1915	4p16	17q21	p	1	N	1.67	0.23	46,XY
424	UK1	4q27	17q21	m	2	B	1.67	2.47	46,XY,t(4;17)
425	ISR	4p14	18p11	m	1	B	1.31	0.35	46,XX,t(4;18)
426	UK11	4p14	18q21	m	1	U	0.84	1.31	46,XY,der(4),t(4;18)
427	UK11	4p14	18q21	m	1	B	1.31	0.84	46,XY,t(4;18)

428	IT7	4p15	18q23	p	1	U	0.00	0.66	46,XY,der(4),t(4;18)
429	209	4p16	18q12	m	1*	U	1.35	0.23	46,XY,t(4;18)
430	F10	4q21	18q22	m	2	N	3.71	0.40	46,XY
431	F10	4q21	18q22	m	2	N	3.71	0.40	46,XX
432	CH1	4q24	18q23	p	1	N	2.92	0.00	46,XX
433	F10	4q25	18p11	m	3	B	0.35	2.78	46,XY,t(4;18)
434	G8	4q27	18q21	p	1	U	2.47	0.84	46,XY,der(18),t(4;18)
435	G8	4q27	18q21	p	1	N	2.47	0.84	46,XY
436	G8	4q27	18q21	p	1	B	2.47	0.84	46,XY,t(4;18)
437	F7	4q28	18q23	m	1	N	2.10	0.00	46,XY
438	F10	4q31	18p11	m	1	B	1.50	0.35	46,XX,t(4;18)
439	2940	4q31	18q23	p	1	B	0.12	1.50	46,XX,t(4;18)
440	45	4q35	18q22.2	p	1	U	0.36	0.20	46,XY,der(4),t(4;18)
441	H1	4p14	20q13	m	1	N	1.31	0.42	46,XY
442	F12	4p15	20p12	p	1	B	0.83	0.39	46,XY,t(4;20)
443	F12	4p15	20p12	p	1	U	0.39	0.83	46,XX,der(4),t(4;20)
444	341	4p15	20q13	p	1	N	0.42	0.83	46,XY
445	CH4	4p16	20q11	m	2	B	1.16	0.23	46,XX,t(4;20)
446	IT6	4p16	20q13	m	3	B	0.42	0.23	46,XX,t(4;20)
447	F12	4p11	21q11	m	1	B	1.67	1.17	46,XX,t(4;21)
448	F12	4p11	21q11	m	1	B	1.67	1.17	46,XY,t(4;21)
449	Fin	4q21	21q21	m	1	N	3.71	0.85	46,XY
450	209	4q27	21p11	m	1	U	2.48	0.00	46,XY,der(21),t(4;21)
451	209	4q27	21p11	m	1	N	2.48	0.00	46,XY
452	F12	4q28	21q21	p	1	N	2.10	0.85	46,XY
453	F11	4q31	21q22	m	1	B	1.50	0.29	46,XX,t(4;21)
454	45	4p11	22p13	m	1	N	1.67	0.00	46,XX
455	364	4p16	22q11.2	p	1	N	0.00	0.63	46,XX
456	0425	4q28	22q11	p	3*	B	2.10	1.07	46,XY,t(4;22)
457	330	5p13	6q21	p	2	N	1.16	2.32	46,XY
458	330	5p13	6q21	p	2	B	2.32	1.16	46,XY,t(5;6)
459	148	5q34	6q27	m	2	B	0.61	0.10	46,XY,t(5;6)
460	45	5q35	6q13	m	3*	B	3.55	0.26	46,XX,t(5;6)
461	F10	5p13	7p21	p	3	N	0.54	1.16	46,XY

(continued)

APPENDIX I. Prenatal diagnosis results for parental carriers of reciprocal translocations (continued)

Observation	Laboratory	Breakpoint 1	Breakpoint 2	Carrier	Asc mode	Result code	Potential or actual trisomy	Potential or actual monosomy	Prenatal result
462	F2	5q14	7q31	p	3	B	3.42	1.42	46,XY,t(5;7)
463	F2	5q14	7q31	p	3	N	3.42	1.42	46,XY
464	F2	5q32	7p22	p	2	N	1.22	0.00	46,XY
465	90	5p15	8q23	m	1	B	0.32	1.08	46,XY,t(5;8)
466	0200	5q35	8q21	m	3*	B	1.95	0.26	46,XY,t(5;8)
467	1030	5q35	8q24	m	2	B	0.26	0.46	46,XX,t(5;8)
468	H3	5p11	9q33	m	2	N	1.56	0.68	46,XY
469	1915	5p12	9p24	p	1	N	1.45	0.15	46,XX
470	1915	5p12	9p24	p	1	N	1.45	0.15	46,XY
471	H3	5p15	9p13	m	1	B	1.14	0.32	46,XX,t(5;9)
472	UK4	5p15	9p24	m	1	N	0.00	0.32	46,XY
473	UK4	5p15	9p24	m	1	B	0.32	0.00	46,XX,t(5;9)
474	F16	5p12	10p15	p	1	N	1.46	0.17	46,XX
475	F10	5p13	10p15	p	1	U	0.17	1.16	46,XX,der(5),t(5;10)
476	F10	5p13	10p15	p	1	N	1.16	0.17	46,XY
477	F10	5p13	10p15	p	1	U	0.17	1.16	46,XX,der(5),t(5;10)
478	F10	5p13	10p15	p	1	U	1.16	0.17	46,XX,der(10),t(5;10)
479	H3	5p13	10q11	m	1	B	3.00	1.16	46,XY,t(5;10)
480	141	5p13	10q22	m	1	B	2.07	1.16	46,XX,t(5;10)
481	UK9	5p13	10q26	p	1	B	1.16	0.36	46,XY,t(5;10)
482	13	5p14	10p13	m	1	U	0.57	0.78	46,XX,der(10),t(5;10)
483	393	5p15	10q12	p	3*	B	0.32	2.93	46,XY,t(5;10)
484	1030	5p15	10q24	p	1	B	1.22	0.32	46,XX,t(5;10)
485	F10	5p15	10q25	p	1	B	0.32	0.88	46,XY,t(5;10)
486	D2	5q21	10p14	p	3	B	2.70	0.40	46,XX,t(5;10)
487	F10	5q31	10q11	m	2	N	3.00	1.57	46,XY
488	F10	5q31	10q11	m	2	N	3.00	1.57	46,XX
489	F10	5q33	10p14	m	2	B	0.40	0.91	46,XY,t(5;10)

490	0425	5p15	11p13	p	3*	B	0.98	0.31	46,XY,t(5;11)
491	1970	5p15	11q25	m	1	U	0.19	0.32	46,XX,der(5),t(5;11)
492	2955	5q32	11q24	m	3*	B	1.22	0.38	46,XY,t(5;11)
493	F10	5p12	11q23	p	1	B	1.46	0.78	46,XX,t(5;11)
494	F10	5p12	11q23	p	1	U	0.78	1.46	46,XX,der(5),t(5;11)
495	F10	5p12	11q23	p	1	U	1.46	0.78	46,XX,der(11),t(5;11)
496	F16	5p13	11q23	p	1	N	1.16	0.78	46,XY
497	IT2	5q15	11q23	p	3	B	0.78	2.95	46,XX,t(5;11)
498	CH4	5q31	11q25	p	3	B	1.57	0.00	46,XY,t(5;11)
499	0200	5q11	12q24	p	3	B	4.45	0.54	46,XY,t(5;12)
500	F10	5q13	12q11	m	3	B	3.81	3.55	46,XX,t(5;12)
501	IT6	5q21	12p13	m	2	B	2.70	0.31	46,XX,t(5;12)
502	381	5p13	13p12	m	1	N	0.00	1.16	46,XY
503	F11	5p15	13q11	m	3	N	0.32	3.23	46,XY
504	UK10	5q13	13q14	m	3	B	3.81	2.17	46,XY,t(5;13)
505	F8	5q14	13q22	m	3	B	3.42	1.19	46,XX,t(5;13)
506	SW2	5q15	13q12	m	2	N	2.95	2.92	46,XX
507	F10	5q35	13q22	m	1	U	1.19	0.26	46,XX,der(5),t(5;13)
508	0200	5p11	14q11	m	1	N	0.00	1.75	46,XX
509	UK2-9	5p15	14q13	m	1	N	2.56	0.32	46,XX
510	UK2-9	5p15	14q13	m	1	N	2.56	0.32	46,XX
511	UK2-9	5p15	14q13	m	1	B	2.56	0.32	46,XY,t(5;14)
512	UK2-9	5p15	14q13	p	1	B	2.56	0.32	46,XX,t(5;14)
513	UK2-9	5p15	14q13	p	1	B	2.56	0.32	46,XX,t(5;14)
514	H3	5p15	14q24	p	1	N	1.23	0.32	46,XX
515	F2	5p15	14q24	m	2	B	1.23	0.32	46,XY,t(5;14)
516	F2	5p15	14q24	m	2	B	0.32	1.23	46,XY,t(5;14)
517	Fin	5p12	15p11	m	1	B	1.46	0.00	46,XY,t(5;15)
518	Fin	5p12	15p11	m	1	B	0.00	1.46	47,XX,t(5;15), +21
519	UK5	5p12	15p12	m	1	N	1.46	0.00	46,XY
520	F2	5p15	15q22	p	1	B	1.27	0.32	46,XX,t(5;15)

(continued)

APPENDIX I. Prenatal diagnosis results for parental carriers of reciprocal translocations (continued)

Observation	Laboratory	Breakpoint 1	Breakpoint 2	Carrier	Asc mode	Result code	Potential or actual trisomy	Potential or actual monosomy	Prenatal result
521	343	5p15	15q22	m	2	N	1.27	0.31	46,XY
522	F10	5q35	15q13	m	1	N	2.48	0.26	46,XX
523	F15	5q35	15q15	p	3	N	1.99	0.26	46,XY
524	F14	5q23	16q12	m	2	B	2.06	1.47	46,XY,t(5;16)
525	41	5q34	16q24	m	1	N	0.15	0.61	46,XX
526	41	5q34	16q24	m	1	B	0.61	0.15	46,XX,t(5;16)
527	41	5q34	16q24	m	1	U	0.61	0.15	46,XY,der(16),t(5;16)
528	41	5q34	16q24	m	1	U	0.61	0.15	46,XY,der(16),t(5;16)
529	148	5q35.3	16p13.1	m	3*	B	0.12	0.49	46,XY,t(5;16)
530	SW3	5p13	17p11	m	3	N	1.16	0.79	46,XY
531	UK11	5p15	17q11	p	3	N	2.30	0.32	46,XY
532	H1	5q12	17q22	m	3	N	4.12	1.26	46,XY
533	UK13	5q22	17q23	p	3	B	1.09	2.41	46,XX,t(5;17)
534	0200	5q31	17q12	p	3	B	2.05	1.56	46,XX,t(5;17)
535	F10	5q33	17p13	p	2	B	0.91	0.23	46,XX,t(5;17)
536	2110	5q33	17q25	p	3	B	0.91	0.29	46,XY,t(5;17)
537	319	5q33.1	17q25	p	2	N	0.29	1.05	46,XX
538	219	5p12	18p11	m	1	U	0.35	1.45	46,XY,der(18),t(5;18)
539	UK6	5p13	18q22	m	3	N	1.16	0.28	46,XX
540	F10	5p15	18q21	p	1	N	0.84	0.32	46,XX
541	F10	5p15	18q21	p	1	U	0.84	0.32	46,XX,der(5),t(5;18)
542	F10	5p15	18q21	p	1	N	0.32	0.84	46,XY
543	F10	5p15	18q21	p	1	U	0.84	0.32	46,XY,der(5),t(5;18)
544	UK5	5q12	18q22	m	2	B	4.12	0.28	46,XY,t(5;18)
545	381	5q22	18p11	m	3*	B	2.41	0.34	46,XY,t(5;18)
546	233	5p11	19p12	m	3*	B	1.00	1.56	46,XX,t(5;19)
547	233	5p11	19p12	m	3	N	1.56	1.00	46,XY
548	41	5p12	19q13	p	1	B	1.45	0.25	46,XX,t(5;19)

549	UK18	5q11	19q11	p	3	B	4.45	1.36	46,XX,t(5;19)
550	2790	5q31	19q13.4	p	3*	B	1.57	0.25	46,XY,t(5;19)
551	330	5q35	19p13	p	3*	B	0.26	0.44	46,XX,t(5;19)
552	0200	5p12	20p11	p	1	B	0.74	1.45	46,XX,t(5;20)
553	79	5q22	20q13.3	m	2	B	2.41	0.14	46,XY,t(5;20)
554	F10	5p13	21q22	m	1	N	2.09	0.00	46,XY
555	F10	5p13	21q22	m	1	U	0.00	2.09	45,XX,−5,−21, +der(5),t(5;21)
556	F10	5q13	21q22	m	1	B	3.81	0.29	46,XY,t(5;21)
557	3060	5q22	21q22.1	m	1	N	1.22	0.00	46,XY
558	F10	5q31	21q22	m	1	B	1.57	0.29	46,XX,t(5;21)
559	CH4	5p13	22q13	p	1	B	1.16	0.30	46,XY,t(5;22)
560	2955	5p15.1	22q12.2	p	2	B	0.75	0.56	46,XY,t(5;22)
561	F10	6p21	7q35	p	2	B	0.51	1.45	46,XY,t(6;7)
562	UK3	6p11	7q11	m	3	B	3.19	2.32	46,XX,t(6;7)
563	0020	6q21	7q32	m	2*	B	2.32	1.02	46,XX,t(6;7)
564	UK15	6p23	8q22	m	2	B	1.45	0.53	46,XX,t(6;8)
565	F4	6p23	8q12	m	2	B	2.65	0.53	46,XY,t(6;8)
566	90	6q21	8q13	p	2	B	2.32	2.39	46,XY,t(6;8)
567	D2	6q27	8q12	m	2	B	2.65	0.00	46,XX,t(6;8)
568	2790	6p11	9p11	p	3*	B	2.31	1.55	46,XY,t(6;9)
569	41	6p12	9q33	p	3*	B	2.08	0.68	46,XY,t(6;9)
570	45	6q14	9q21	m	2	N	3.32	2.17	46,XY
571	UK18	6q27	9p22	p	1	N	0.00	0.56	46,XY
572	UK18	6q27	9p22	p	1	U	0.56	0.00	46,XX,der(6),t(6;9)
573	UK18	6q27	9p22	p	1	U	0.56	0.00	46,XX,der(6),t(6;9)
574	UK18	6q27	9p22	m	1	U	0.56	0.00	46,XX,der(6),t(6;9)
575	UK18	6q27	9p22	m	1	B	0.56	0.00	46,XX,t(6;9)
576	H1	6p12	10q26	p	3	B	0.36	2.09	46,XX,t(6;10)
577	3060	6p21	10q26	p	2	B	1.45	0.35	46,XY,t(6;10)
578	1915	6q21	10q26	m	1	B	2.32	0.35	46,XX,t(6;10)
579	0020	6q25	10q22	m	2*	B	0.62	2.07	46,XX,t(6;10)
580	G4	6q27	10q11	m	2	B	3.00	0.00	46,XY,t(6;10)

(continued)

APPENDIX I. Prenatal diagnosis results for parental carriers of reciprocal translocations (continued)

Observation	Laboratory	Breakpoint 1	Breakpoint 2	Carrier	Asc mode	Result code	Potential or actual trisomy	Potential or actual monosomy	Prenatal result
581	G5	6q25	10q24	m	3	B	1.15	0.62	46,XY,t(6;10)
582	ISR	6p23	11p15	p	3	N	0.53	0.33	46,XX
583	ISR	6p23	11p15	m	3	N	0.53	0.33	46,XX
584	355	6q15	11p15	p	2	N	3.04	0.33	46,XY
585	H2	6q26	11q22	m	2	N	1.19	0.26	46,XX
586	F5	6q23	12p12	p	3	N	0.76	1.25	46,XX
587	45	6q25	12q22	p	3*	B	1.47	0.13	46,XX,t(6;12)
588	H3	6q26	12q11	m	1	B	3.55	0.26	46,XY,t(6;12)
589	D1	6p21	13q14	p	2	B	1.45	2.17	46,XY,t(6;13)
590	H3	6p23	13q22	m	1	N	1.19	0.53	46,XX
591	1700	6p23	13q22	m	2	B	1.19	0.53	46,XY,t(6;13)
592	341	6q14	13q34	m	2	N	3.32	0.12	46,XY
593	341	6q14	13q34	m	2	B	3.32	0.12	46,XX,t(6;13)
594	F10	6q15	13q13	p	2	N	2.60	3.04	46,XY
595	UK14	6q21	13q22	m	3	B	2.32	1.19	46,XX,t(6;13)
596	F4	6q24	13q14	m	2	B	2.17	0.97	46,XX,t(6;13)
597	B1	6q25	13q13	m	1	N	2.60	0.62	46,XY
598	SW3	6p21	14q32	m	2	N	1.45	0.34	46,XY
599	Fin	6q21	14q31	p	3	N	2.32	0.83	46,XX
600	148	6q25	14q24	m	3*	B	0.62	1.23	46,XX,t(6;14)
601	F15	6p25	15q22	m	2	N	1.27	0.00	46,XX
602	41	6p25	15q24	m	1	U	0.87	0.13	46,XX,der(6),t(6;15)
603	45	6q22	15q15.1	m	3*	B	2.06	1.81	46,XX,t(6;15)
604	UK5	6q26	15q21	m	3	B	1.66	0.26	46,XY,t(6;15)
605	UK5	6q26	15q21	m	3	N	1.66	0.26	46,XX
606	2070	6p21	16p11	p	2	N	1.07	1.44	46,XX
607	F5	6q13	16q11	p	3	B	1.61	3.55	46,XX,t(6;16)

608	F12	6q21	17q25	m	2	N	2.32	0.29	46,XX
609	381	6p11	18p11	p	2	B	2.31	0.67	46,XY,t(6;18)
610	F10	6p23	18q21	m	1	N	0.53	0.84	46,XY
611	F10	6p23	18q21	m	1	N	0.84	0.53	46,XX
612	F10	6p23	18q21	m	1	B	0.84	0.53	46,XY,t(6;18)
613	45	6q16	18p11.1	m	2	N	2.72	0.70	46,XY
614	138	6q21	18q12	p	1	N	2.32	1.35	46,XY
615	IT5	6q12	21q22	p	1	B	3.75	0.45	46,XY,t(6;21)
616	F15	6q15	21q11	m	2	B	1.17	3.04	46,XX,t(6;21)
617	41	6q21	21q11	p	2	B	2.32	1.16	46,XX,t(6;21)
618	F10	6q25	21q22	m	1	N	0.62	0.29	46,XX
619	F10	6q25	21q22	m	1	N	0.29	0.62	46,XY
620	F10	6p21	22q12	p	2	B	1.45	0.74	46,XY,t(6;22)
621	F10	6p21	22q12	p	2	B	1.45	0.74	46,XY,t(6;22)
622	UK16	6q11	22p11	p	3	B	3.93	0.00	46,XX,t(6;22)
623	UK16	6q11	22p11	p	3	B	3.93	0.00	46,XY,t(6;22)
624	F10	6q16	22q13	p	3	B	2.72	0.30	47,XY,t(6;22), +21
625	1178	6q21	22p13	m	1	U	2.32	0.00	46,XY,der(22),t(6;22)
626	F10	7p21	8q23	m	3	N	0.54	1.08	46,XY
627	F7	7p22	8q12	m	1	B	2.70	0.00	46,XX,t(7;8)
628	CH1	7q11	8p11	m	2	B	1.42	3.19	46,XY,t(7;8)
629	CH2	7q22	8p21	m	2	N	1.96	0.85	46,XY
630	CH2	7q22	8p21	m	2	N	1.96	0.85	46,XX
631	B2	7q31	8q24	p	3	N	1.42	0.47	46,XX
632	F10	7q31.3	9q21	p	2	N	2.17	1.32	46,XY
633	41	7q32	9p23	p	1	N	1.02	0.36	46,XY
634	G3	7q32	9p24	m	1	N	1.02	0.00	46,XY
635	F10	7q35	9p13	m	1	N	1.14	0.51	46,XX
636	F10	7p11	10q11	m	2	B	1.97	3.00	46,XX,t(7;10)
637	SW3	7q11	10q21	m	3	B	3.19	2.54	46,XX,t(7;10)
638	F13	7q31.2	10p15.3	m	1	N	1.53	0.00	46,XY
639	238	7q32	10q22	p	3*	B	2.07	1.02	46,XY,t(7;10)
640	G3	7q35	10p14	p	1	B	0.40	0.51	46,XY,t(7;10)
641	G3	7q35	10p14	p	1	N	0.51	0.40	46,XY

(continued)

APPENDIX I. Prenatal diagnosis results for parental carriers of reciprocal translocations (continued)

Observation	Laboratory	Breakpoint 1	Breakpoint 2	Carrier	Asc mode	Result code	Potential or actual trisomy	Potential or actual monosomy	Prenatal result
642	UK6	7p21	11q21	m	3	B	0.54	1.43	46,XY,t(7;11)
643	UK15	7q36	11q13	p	1	B	1.20	0.21	46,XY,t(7;11)
644	1178	7p15	12q13.3	p	3*	B	2.62	0.92	46,XX,t(7;12)
645	D2	7q11	12q15	m	3	N	3.19	2.27	46,XX
646	F12	7q22	12p12	p	3	B	1.96	0.76	46,XY,t(7;12)
647	341	7p11	13q11	p	2	B	2.02	0.00	46,XY,t(7;13)
648	UK15	7p13	13q34	m	2	N	1.58	0.00	46,XX
649	UK3	7p15	13q14	p	2	B	2.17	0.92	46,XX,t(7;13)
650	45	7p21	13q13	m	2	B	0.54	2.60	46,XX,t(7;13)
651	G1	7p22	13q14	m	2	B	2.17	0.00	46,XX,t(7;13)
652	F1	7q11	13q12	m	2	B	3.19	2.97	46,XY,t(7;13)
653	233	7q22	13q12	m	2	B	2.97	1.96	46,XX,t(7;13)
654	F10	7q22	13q32	p	1	U	1.96	0.56	46,XX,der(13),t(7;13)
655	1915	7q32	13q14	m	2	B	1.02	2.17	46,XY,t(7;13)
656	UK19	7q32	13q22	m	1	B	1.19	1.02	46,XX,t(7;13)
657	45	7p15	14q23	m	1	N	0.92	1.64	46,XY
658	219	7q11	14q32	p	2	B	3.19	0.40	46,XX,t(7;14)
659	UK4	7q36	14q21	m	1	N	2.27	0.21	46,XX
660	UK11	7p11	15p12	m	3	N	0.00	1.97	46,XY
661	F3	7p22	15q22	p	2	N	1.27	0.00	46,XX
662	F3	7p22	15q22	p	2	N	1.27	0.20	46,XX
663	385	7p22	15q22	p	1	B	1.27	0.20	46,XY,t(7;15)
664	1178	7p22	15q24	m	1	B	0.20	0.87	46,XY,t(7;15)
665	0200	7q11	15q24	m	3	N	3.19	0.87	46,XY
666	F11	7q34	15q21	m	2	B	0.67	1.66	46,XX,t(7;15)
667	F12	7q21	16p13	p	2	N	2.56	0.33	46,XY
668	B1	7q36	16p11	p	2	N	1.07	0.21	46,XY
669	2150	7q22	17q25	m	1	N	1.96	0.29	46,XY

670	130	7p22	18q11.2	m	3*	B	1.72	0.20	46,XY,t(7;18)
671	F11	7q32	18q22	m	1	U	1.02	0.40	46,XY,der(18),t(7;18)
672	UK12	7q22	18q23	p	1	N	1.96	0.00	46,XX
673	F7	7p11	19q13	m	2	N	1.97	0.73	46,XY
674	341	7p22	19q13	m	3*	B	0.57	0.20	46,XY,t(7;19)
675	D2	7q32	19p13	m	3	B	0.73	1.02	46,XY,t(7;19)
676	F8	7p11	20p13	m	2	B	0.00	1.97	46,XX,t(7;20)
677	UK3	7p13	21p11	m	3	B	1.58	0.00	46,XX,t(7;21)
678	F11	7q21	21q21	p	1	N	2.56	0.85	46,XX
679	F15	7q22	21q2	m	1	B	1.96	0.58	46,XY,t(7;21)
680	UK17	7q32	21q22	p	1	B	0.29	1.02	46,XY,t(7;21)
681	F2	7p13	22q11	p	2	N	1.58	1.18	46,XY
682	45	8p12	9q32	m	2	B	1.16	0.90	46,XY,t(8;9)
683	F15	8p21	9p24	p	2	N	0.86	0.00	46,XY
684	IT3	8p23	9q32	m	1	U	0.90	0.31	46,XX,der(8),t(8;9)
685	F10	8q23	9p23	m	1	B	1.08	0.36	46,XX,t(8;9)
686	F10	8q23	9p23	m	1	B	1.08	0.36	46,XY,t(8;9)
687	F10	8q24	9p11	m	2	B	0.47	1.52	46,XY,t(8;9)
688	F10	8q24	9p11	m	2	N	1.52	0.47	46,XY
689	G3	8q24	9q22	m	1	N	1.54	0.47	46,XX
690	SW1	8p21	10q26	m	2	U	0.86	0.36	46,XX,der(10),t(8;10)
691	SW1	8p21	10q26	m	2	N	0.86	0.36	46,XY
692	3060	8p21	10q26	m	1	B	0.35	0.86	46,XY,t(8;10)
693	3060	8p21	10q26	m	1	B	0.86	0.35	46,XX,t(8;10)
694	1480	8q24.3	10q24	m	2	N	1.11	0.46	46,XY
695	H3	8p22	11q25	m	3	N	0.64	0.00	46,XY
696	130	8p23	11p11	p	1	U	1.43	0.31	46,XX,der(8),t(8;11)
697	2760	8p23	11q13	p	3	B	2.20	0.31	46,XX,t(8;11)
698	IT5	8p23	11q14	p	1	N	1.66	0.31	46,XX
699	F12	8p12	12p11	p	3	B	1.11	1.16	46,XX,t(8;12)
700	F11	8q11	12p11	m	2	N	1.11	3.08	46,XY
701	F11	8q11	12p11	m	2	B	3.08	1.11	46,XX,t(8;12)
702	1830	8q13	12p12	p	1	B	2.40	0.76	47,XX,t(8;12),+13
703	D1	8q24	12q12	p	1	B	3.39	0.47	46,XX,t(8;12)

(continued)

APPENDIX I. Prenatal diagnosis results for parental carriers of reciprocal translocations (continued)

Observation	Laboratory	Breakpoint 1	Breakpoint 2	Carrier	Asc mode	Result code	Potential or actual trisomy	Potential or actual monosomy	Prenatal result
704	F10	8q22	13q32	p	1	N	1.45	0.56	46,XX
705	F10	8q22	13q32	p	1	N	1.45	0.56	46,XY
706	F10	8q22	13q32	p	1	U	1.45	0.56	46,XX,der(13),t(8;13)
707	Fin	8p21	14q32	p	2	B	0.86	0.34	46,XX,t(8;14)
708	41	8p22	15q21	p	1	N	1.53	0.63	46,XX
709	41	8p22	15q21	p	1	B	0.63	1.53	46,XY,t(8;15)
710	UK6	8q11	15q11	p	3	B	3.08	2.93	46,XX,t(8;15)
711	H3	8q11	15q11	m	3	U	2.09	0.14	46,XY,−15,+der(8),t(8;15)
712	F12	8q21	15q25	m	3	B	1.95	0.62	46,XX,t(8;15)
713	F9	8p12	16q12	m	3	N	1.47	1.16	46,XX
714	UK5	8p22	16q11	m	2	N	1.61	0.64	46,XY
715	UK5	8p22	16q11	m	2	N	1.61	0.64	46,XY
716	UK5	8p22	16q11	p	2	N	0.64	1.61	46,XX
717	0200	8q22	16p13	p	3*	B	0.32	1.45	46,XY,t(8;16)
718	UK8	8q22	16q23	m	3	N	1.45	0.44	46,XY
719	UK11	8q24	17q25	m	1	N	0.29	0.47	46,XX
720	1915	8q23	18q21	m	1	B	1.08	0.84	46,XX,t(8;18)
721	1915	8q24	18q21	p	1	B	0.46	0.84	46,XY,t(8;18)
722	364	8q24	18q21	m	1	B	0.46	0.84	46,XY,t(8;18)
723	1510	8q22	20p11	m	2	B	1.45	0.75	46,XX,t(8;20)
724	1510	8q22	20p11	m	2	N	1.45	0.75	46,XY
725	UK3	9p22	10q24	p	3	B	1.15	0.56	46,XY,t(9;10)
726	F2	9p23	10q23	m	1	N	1.54	0.36	46,XY
727	IT7	9p24	10q24	p	1	U	1.15	0.00	46,XX,der(9),t(9;10)
728	41	9p13.1	11q24.2	m	1	N	1.27	0.38	46,XY
729	B2	9p22	11q25	m	3	N	0.56	0.00	46,XX
730	1178	9p22	12q13	m	3*	B	0.56	3.07	46,XX,t(9;12)
731	0270	9q12	16p13	m	3*	B	2.65	0.32	46,XY,t(9;16)

732	CH3	9p21	13q21	m	1	B	1.60	0.84	46,XX,t(9;13)
733	CH3	9p21	13q21	m	1	N	1.60	0.84	46,XY
734	Fin	9p24	13q22	m	1	N	1.19	0.00	46,XY
735	H3	9q11	13p11	m	3	N	2.65	0.00	46,XX
736	Fin	9q34	13q14	p	1	B	2.17	0.27	46,XY,t(9;13)
737	H1	9p11	14p11	m	1	B	1.52	0.00	46,XY,t(9;14)
738	IT3	9p22	14q13	p	2	B	2.56	0.56	46,XX,t(9;14)
739	UK11	9p22	14q22	m	1	N	1.95	0.56	46,XY
740	UK11	9p22	14q22	m	1	B	1.95	0.56	46,XY,t(9;14)
741	2070	9p22	14q24	m	1	B	1.23	0.56	46,XX,t(9;14)
742	0200	9q11	14p12	m	3*	B	1.61	0.00	46,XY,t(9;14)
743	UK9	9q21	14q21	m	3	U	2.64	0.97	46,XY, −14, +der(9),t(9;14)
744	0D2	9p11	15q11	m	1	U	3.22	0.19	46,XX, −15, +der(9),t(9;15)
745	F10	9p22	15q13	m	2	B	0.56	2.48	46,XY,t(9;15)
746	F10	9p22	15q13	m	2	N	2.48	0.56	46,XY
747	F10	9p22	15q13	m	2	B	2.48	0.56	46,XY,t(9;15)
748	UK9	9p23	15q15	m	3	B	0.36	1.99	46,XY,t(9;15)
749	Fin	9q33	15q12	m	1	U	0.00	1.01	45,XX, −9, −15, +der(9),t(9;15)
750	Fin	9q33	15q12	m	1	B	1.01	0.00	46,XY,t(9;15)
751	Fin	9q33	15q12	p	1	B	1.01	0.00	46,XX,t(9;15)
752	Fin	9q33	15q12	p	1	B	0.00	1.01	46,XX,t(9;15)
753	UK17	9q34	15q24	m	3	B	0.83	0.27	46,XX,t(9;15)
754	0200	9p22	16q24	m	1	B	0.15	0.56	46,XY,t(9;16)
755	393	9p13	17p12	m	3*	B	1.14	0.53	46,XX,t(9;17)
756	1960	9p22	17p11	m	2	B	0.78	0.56	46,XX,t(9;17)
757	395	9p11	18p11	p	1	N	1.55	0.34	46,XX
758	H3	9p13	18p11	p	1	N	1.27	0.35	46,XX
759	3060	9p22	18q11	p	1	U	1.86	0.56	46,XY,der(9),t(9;18)
760	41	9p23	18q23	m	1	N	0.12	0.36	46,XY
761	Fin	9p24	18q21	p	1	N	0.84	0.00	46,XX
762	Uk10	9q21	18q22	m	1	N	2.17	0.40	46,XX
763	F16	9q32	18p11	m	3	B	0.90	0.35	46,XX,t(9;18)
764	D1	9q21	19q13	p	3	B	0.57	2.17	46,XY,t(9;19)
765	F7	9q34.3	19p11	p	2	B	0.00	1.11	46,XY,t(9;19)

(continued)

APPENDIX I. Prenatal diagnosis results for parental carriers of reciprocal translocations (continued)

Observation	Laboratory	Breakpoint 1	Breakpoint 2	Carrier	Asc mode	Result code	Potential or actual trisomy	Potential or actual monosomy	Prenatal result
766	UK3	9p12	20q12	m	2	N	1.37	0.90	46,XX
767	UK3	9p12	20q12	m	2	B	1.37	0.90	46,XX,t(9;20)
768	F10	9q11	20p13	m	1	N	2.65	0.00	46,XX
769	341	9q22	20q12	p	2	N	1.54	0.90	46,XY
770	341	9q22	20q12	p	2	N	0.90	1.54	46,XX
771	45	9q22	21q21	m	1	B	2.60	0.67	46,XY,t(9;21)
772	343	9q11	22q11	p	1	N	1.60	0.29	46,XX
773	UK14	9q11	22q12	p	2	U	1.59	0.72	46,XY,−22,+der(9),t(9;22)
774	UK14	9q11	22q12	p	2	B	1.59	0.72	46,XY,t(9;22)
775	UK14	9q11	22q12	p	2	N	1.59	0.72	46,XX
776	F1	9q13	22q13	m	1	U	2.51	0.00	47,XX,+der(9),t(9;22)
777	F1	9q13	22q13	m	1	U	2.51	0.00	47,XX,+der(9),t(9;22)
778	0425	10p11	11q11	p	2	B	2.93	1.20	46,XY,t(10;11)
779	SW1	10p15	11q23	p	3	U	0.78	0.17	46,XY,der(10),t(10;11)
780	UK15	10q22	11q23	p	2	B	0.78	2.07	46,XX,t(10;11)
781	364	10q22	11q25	p	1	U	2.07	0.09	46,XY,der(11),t(10;11)
782	H1	10q24	11q23	p	1	N	1.15	0.78	46,XY
783	F10	10q26	11q13	p	1	B	2.20	0.36	46,XY,t(10;11)
784	F11	10q26	11q13	m	2	B	2.20	0.36	46,XY,t(10;11)
785	F15	10p11	12q24	p	1	N	1.26	0.54	46,XY
786	141	10p11	12q24.3	m	3*	B	1.26	0.30	46,XX,t(10;12)
787	F8	10p15	12p11	m	1	B	0.17	1.50	46,XY,t(10;12)
788	130	10q11	12p11	p	2	N	3.00	1.10	46,XX
789	319	10q22	12p13.3	m	1	N	2.07	0.20	46,XY
790	319	10q22	12p13.3	p	1	B	2.07	0.20	46,XX,t(10;12)
791	319	10q22	12p13.3	p	1	N	2.07	0.20	46,XX
792	319	10q22	12p13.3	m	1	N	2.07	0.20	46,XY
793	1970	10q26	12p11	m	1	B	0.35	1.11	46,XY,t(10;12)

794	319	10q21	13q34	m	1	B	0.12	2.54	46,XY,t(10;13)
795	393	10q25	13q14	m	3*	B	2.17	0.87	46,XX,t(10;13)
796	H3	10p11	14p11	m	3	N	1.26	0.00	46,XY
797	IT7	10p15	14q12	m	1	N	2.77	0.17	46,XY
798	D2	10q25	15q21	p	3	B	0.88	1.66	46,XX,t(10;15)
799	393	10q21.2	16q13	m	3*	B	2.53	1.24	46,XY,t(10;16)
800	UK17	10q24	16q22	p	3	B	1.15	0.78	46,XY,t(10;16)
801	UK8	10q22	17q25	p	3	B	2.07	0.29	46,XX,t(10;17)
802	69	10q24	17p13	p	1	N	1.11	0.23	46,XX
803	69	10q24	17p13	p	1	B	1.11	0.23	46,XY,t(10;17)
804	69	10q24	17p13	p	1	U	1.11	0.23	46,XY,der(17),t(10;17)
805	69	10q24	17p13	p	1	U	1.11	0.23	46,XY,der(17),t(10;17)
806	Fin	10p15	18q22	p	1	U	0.17	0.40	46,XX,der(18),t(10;18)
807	1178	10q21.2	18q11	m	3*	B	1.74	2.53	46,XX,t(10;18)
808	D1	10q25	18q23	m	1	N	0.88	0.00	46,XX
809	D1	10q25	18q23	m	1	B	0.00	0.88	46,XX,t(10;18)
810	IT2	10q25	18q23	p	1	N	0.88	0.00	46,XY
811	F10	10p11	20p11	m	2	N	1.26	0.74	46,XY
812	F10	10p11	20p11	m	2	B	0.74	1.26	46,XY,t(10;20)
813	G6	10q23	20q13	m	3	B	1.54	0.42	46,XX,t(10;20)
814	41	10p11	21p12	m	1	B	1.38	0.00	46,XY,t(10;21)
815	1178	10p11	21q22	p	1	B	1.25	0.29	46,XX,t(10;21)
816	H1	10p11	22p11	p	1	N	0.00	1.20	46,XX
817	393	10p11	22p11	m	1	B	1.20	0.00	46,XX,t(10;22)
818	330	10q22	22q13	m	2	B	2.07	0.30	46,XX,t(10;22)
819	330	10q22	22q13	m	2	N	2.07	0.30	46,XY
820	UK9	11p11	12q13	p	3	N	2.91	1.34	46,XX
821	1510	11p11.2	12q15	m	3*	B	2.27	1.33	46,XY,t(11;12)
822	F5	11q13	12q24	m	2	B	2.20	0.54	46,XX,t(11;12)
823	343	11q25	12q24	m	2	B	0.19	0.54	46,XX,t(11;12)
824	343	11q25	12q24	m	2	N	0.54	0.19	46,XX
825	H3	11q14	13q33	p	1	U	1.66	0.31	46,XX,der(13),t(11;13)
826	H3	11q14	13q33	p	1	B	1.66	0.31	46,XX,t(11;13)
827	45	11p11.2	14q11.2	m	3*	B	1.33	2.97	46,XY,t(11;14)

(continued)

APPENDIX I. Prenatal diagnosis results for parental carriers of reciprocal translocations (continued)

		Breakpoint							
Observation	Laboratory	1	2	Carrier	Asc mode	Result code	Potential or actual trisomy	Potential or actual monosomy	Prenatal result
828	UK11	11p12	14q11	p	3	N	3.05	1.15	46,XY
829	138	11p12	14q12	p	2	N	2.77	1.15	46,XY
830	3060	11q21	14q13	m	2	B	2.56	1.43	46,XY,t(11;14)
831	H2	11q21	14q32	m	3	N	1.43	0.34	46,XY
832	219	11q13	15q26	m	3*	B	2.20	0.23	46,XX,t(11;15)
833	0625	11p11	16q11	m	2	B	1.43	1.61	47,XY,t(11;16),+21
834	D1	11q13	16p11	m	3	B	1.07	2.20	46,XX,t(11;16)
835	F10	11q13	16p13	m	3	N	2.20	0.29	46,XX
836	UK14	11q21	16q13	p	2	B	1.43	1.24	46,XX,t(11;16)
837	UK9	11q25	16q22	m	1	U	0.78	0.00	46,XX,der(11),t(11;16)
838	1960	11q14	18p11.3	p	1	N	1.66	0.11	46,XY
839	F7	11q14	18q23	p	2	N	1.66	0.00	46,XX
840	F12	11q23	18p11	p	1	U	0.78	0.35	46,XY,der(18),t(11;18)
841	UK15	11p13	18q23	m	2	N	0.98	0.00	46,XX
842	90	11q21	20p13	p	3*	B	1.42	0.12	46,XX,t(11;20)
843	45	11q25	20q12	p	1	N	0.90	0.19	46,XY
844	IT7	11p11	21q22	p	1	N	1.34	0.29	46,XY
845	D2	11q23	21q22	p	1	B	0.29	0.78	46,XX,t(11;21)
846	D2	11q23	21q22	p	1	N	0.78	0.29	46,XX
847	138	11q11	22q13.1	p	2	B	0.51	2.93	46,XX,t(11;22)
848	UK16	11p14	22q13	m	2	B	0.79	0.30	46,XY,t(11;22)
849	2150	11q23	22q11	m	1	B	1.18	0.00	46,XY,t(11;22)
850	F10	11q23	22q11	m	1	B	1.18	0.00	46,XX,t(11;22)
851	69	11q23	22q11	m	2	B	1.18	0.00	46,XX,t(11;22)
852	1830	11q23	22q11	m	1	B	0.00	1.18	46,XX,t(11;22)
853	1960	11q23	22q11	m	1	B	1.18	0.00	46,XX,t(11;22)
854	F10	11q23	22q11	m	1	N	1.18	0.00	46,XY
855	148	11q23	22q11	m	4	B	1.18	0.00	46,XY,t(11;22)

856	F10	11q23	22q11	m	1	B	1.18	0.00	46,XY,t(11;22)
857	F10	11q23	22q11	m	1	B	0.00	1.18	46,XX,t(11;22)
858	1970	11q23	22q11	m	1	B	1.18	0.00	46,XY,t(11;22)
859	H3	11q23	22q11	p	2	N	1.18	0.00	46,XY
860	H3	11q23	22q11	p	2	B	0.00	1.18	46,XX,t(11;22)
861	CH4	11q23	22q11	m	3	B	1.18	0.00	46,XX,t(11;22)
862	IT3	11q23	22q11	m	1	N	1.18	0.00	46,XY
863	D2	11q23	22q11	m	1	B	1.18	0.00	46,XX,t(11;22)
864	1950	11q23	22q11	m	1	N	0.00	1.18	46,XY
865	CH4	11q23	22q11	p	1	B	1.18	0.00	46,XY,t(11;22)
866	335	11q23	22q11	m	3*	B	0.00	1.18	46,XX,t(11;22)
867	UK9	11q23	22q11	m	1	B	1.18	0.00	46,XX,t(11;22)
868	364	11q23	22q11	m	3*	B	1.18	0.00	46,XY,t(11;22)
869	F5	11q23	22q11	p	3	N	1.18	0.00	46,XX
870	CH4	11q23	22q11	m	1	B	1.18	0.00	46,XY,t(11;22)
871	45	11q23	22q11.2	m	4*	B	1.18	0.00	46,XY,t(11;22)
872	393	11q23	22q11.2	m	3*	B	1.18	0.00	46,XY,t(11;22)
873	364	11q23	22q11.2	p	1	N	0.00	1.18	46,XX
874	364	11q23	22q11.2	p	1	U	1.18	0.00	47,XY, +der(22),t(11;22)
875	SW3	11q23	22q12	m	1	B	0.78	0.75	46,XX,t(11;22)
876	SW3	11q23	22q12	m	1	B	0.78	0.75	46,XX,t(11;22)
877	SW3	11q23	22q13	m	2	B	0.78	0.30	46,XY,t(11;22)
878	2070	11q25	22q12	p	1	B	0.00	0.82	46,XX,t(11;22)
879	ISR	11q25	22q13	m	1	N	1.17	0.00	46,XY
880	ISR	11q25	22q13	m	1	B	1.17	0.00	46,XY,t(11;22)
881	ISR	11q25	22q13	m	1	U	1.17	0.00	47,XX, +der(22),t(11;22)
882	ISR	11q25	22q13	m	1	N	1.17	0.00	46,XX
883	1950	11q25	22q13	m	1	B	1.17	0.00	46,XY,t(11;22)
884	IT3	11q23	22q11	m	3	N	0.00	1.18	46,XY
885	UK5	11q23	22q11	m	1	B	0.00	1.18	46,XX,t(11;22)
886	UK5	11q25	22q13	p	3	U	0.00	0.30	47,XY,der(22),t(11;22), +21
887	Fin	11q25	22q13	m	1	B	1.17	0.00	46,XX,t(11;22)
888	H3	12q22	13q22	m	2	N	1.47	1.19	46,XX
889	45	12q24	13q14	m	2	B	2.17	0.54	46,XX

(continued)

APPENDIX I. Prenatal diagnosis results for parental carriers of reciprocal translocations (continued)

Observation	Laboratory	Breakpoint 1	Breakpoint 2	Carrier	Asc mode	Result code	Potential or actual trisomy	Potential or actual monosomy	Prenatal result
890	45	12q24	13q14	m	2	B	2.17	0.54	46,XX,t(12;13)
891	IT5	12p11	15q11	m	1	N	2.93	1.50	46,XY
892	1510	12p13	15q24	m	2	N	0.30	0.87	46,XY
893	UK3	12q13	15q15	m	3	B	2.91	1.99	46,XY,t(12;15)
894	2070	12q24	15q23	m	1	U	1.02	0.54	46,XX,der(12),t(12;15)
895	2070	12q24	15q23	m	1	B	1.02	0.54	46,XY,t(12;15)
896	3060	12p11	18p11	m	1	U	1.10	0.34	46,XX,der(18),t(12;18)
897	3060	12q23	18q23	m	1	B	1.23	0.12	46,XY,t(12;18)
898	0080	12p12	19p12	m	2	N	0.99	0.76	46,XY
899	319	12p13.1	20p11.2	p	1	B	0.55	0.71	46,XY,t(12;20)
900	319	12p13.1	20p11.2	p	1	B	0.71	0.55	46,XX,t(12;20)
901	SW3	12p12	21p12	m	1	B	0.00	0.76	46,XY,t(12;21)
902	SW3	12q11	21q11	m	3	U	1.22	0.00	47,XY,−12, +der(12),+der(21)
903	Fin	12q13	21q11	m	3	B	1.17	2.91	46,XY,t(12;21)
904	Fin	12q15	21q22	p	2	N	2.27	0.29	46,XY
905	Fin	12q15	21q22	m	2	N	0.29	2.27	46,XX
906	UK17	12q24	22q11	p	3	N	1.19	0.54	46,XX
907	UK11	13q34	14q11	m	1	N	3.05	0.00	46,XY
908	381	13q13	15q26	p	2	N	2.60	0.23	46,XY
909	F5	13q14	15q26	m	2	B	2.17	0.23	46,XX,t(13;15)
910	343	13q14	15q26	m	2	B	2.17	0.23	46,XY,t(13;15)
911	343	13q14	15q26	m	2	U	2.17	0.23	46,XY,der(15),t(13;15)
912	343	13q14	15q26	m	2	B	0.23	2.17	46,XY,t(13;15)
913	343	13q14	15q26	m	2	B	2.17	0.23	46,XX,t(13;15)
914	37	13q22	15q21	m	2	N	1.66	1.20	46,XX

915	1990	13q22	15q26	p	1	N	1.19	0.23	46,XY
916	1990	13q22	15q26	p	1	B	1.19	0.23	46,XY,t(13;15)
917	37	13q32	15q21	p	2	B	1.66	0.56	46,XY,t(13;15)
918	37	13q32	15q21	m	2	N	0.56	1.66	46,XX
919	UK6	13q32	15q14	p	1	B	2.22	0.56	46,XX,t(13;15)
920	UK6	13q32	15q14	p	1	B	2.22	0.56	46,XX,t(13;15)
921	F10	13q33	15q21	m	2	N	1.56	0.31	46,XY
922	2955	13q12	16q12.1	p	3*	B	1.88	0.00	46,XY,t(13;16)
923	1030	13q31	16p13	p	2	N	1.90	0.33	46,XX
924	UK15	13q32	16q13	m	3	B	0.56	1.24	46,XX,t(13;16)
925	H1	13q14	17q23	p	3	N	2.17	1.09	46,XX
926	A	13q22	18p11	m	1	N	0.35	1.19	46,XX
927	H1	13q33	18q12	p	1	N	1.35	0.31	46,XX
928	330	13q14	20q13	p	1	N	2.17	0.42	46,XY
929	335	13q22	21q22	p	2	B	0.28	1.20	46,XX,t(13;21)
930	H3	13q21	22q11	m	2	B	1.60	1.19	46,XX,t(13;22)
931	1030	14q24	18q22	p	1	B	1.23	0.40	46,XY,t(14;18)
932	1030	14q24	18q22	p	1	N	1.23	0.40	46,XX
933	0020	14q24	18q23	p	1	B	1.23	0.12	46,XY,t(14;18)
934	F15	14p11	20q11	p	2	U	1.16	0.00	46,XY,der(14),t(14;20)
935	Fin	14q24	20q12	m	3	B	1.23	0.90	46,XY,t(14;20)
936	0200	14q24	20q13	m	2	B	1.23	0.42	46,XY,t(14;20)
937	45	14q13	21q11	m	3	B	2.56	1.16	46,XY,t(14;21)
938	343	14q32	21q11	m	2	U	1.04	0.00	47,XY,+der(21),t(14;21)
939	SW1	14q22	22q13	p	3	U	1.95	0.30	46,XX,der(22),t(14;22)
940	SW1	14q22	22q13	p	3	N	1.95	0.30	46,XY
941	SW1	14q22	22q13	p	3	B	1.95	0.30	46,XX,t(14;22)
942	1960	14q32	22q12	m	1	N	0.34	0.74	46,XX
943	F15	15p12	17q21	m	2	N	1.67	0.00	46,XX
944	UK13	15q22	17q25	p	3	N	0.29	1.27	46,XY
945	152	15q24	17q21	p	2	N	1.67	0.87	46,XY
946	364	15q15	18p11	p	1	B	1.99	0.35	46,XX,t(15;18)
947	364	15q15	18p11	p	1	B	1.99	0.35	46,XX,t(15;18)
948	1970	15q21	18q23	p	1	N	1.66	0.12	46,XY

(continued)

APPENDIX I. Prenatal diagnosis results for parental carriers of reciprocal translocations (continued)

		Breakpoint					Potential	Potential	
Observation	Laboratory	1	2	Carrier	Asc mode	Result code	or actual trisomy	or actual monosomy	Prenatal result
949	SW3	15q15	19q13	p	2	B	0.57	1.99	46,XY,t(15;19)
950	SW3	15q15	19q13	p	2	N	1.99	0.57	46,XY
951	1960	15q22	20p13	p	2	B	1.27	0.12	46,XX,t(15;20)
952	Fin	15q22	20q13	p	3	N	0.42	1.27	46,XY
953	385	15q22	20q13	p	2	N	1.27	0.42	46,XX
954	219	15q23	20q12	m	1	U	1.02	0.90	46,XY,der(20),t(15;20)
955	ISR	15q14	21q21	m	1	N	2.22	0.85	46,XX
956	F10	15q22	21q22	m	1	B	1.27	0.29	46,XX,t(15;21)
957	F10	15q22	21q22	m	1	N	1.27	0.29	46,XY
958	UK4	15q24	22q11	m	1	N	1.19	0.88	46,XY
959	F15	15q26	22q12	m	2	B	0.23	0.75	46,XY,t(15;22)
960	UK5	16q13	17q25	m	3	B	0.29	1.24	46,XX,t(16;17)
961	UK5	16q13	17q25	m	3	N	1.24	0.29	46,XY
962	UK5	16q13	17q25	m	3	N	1.24	0.29	46,XX
963	1950	16q12	19q13	p	3	B	1.61	0.77	46,XX,t(16;19)
964	F14	16q24	19q11	m	3	B	0.00	1.36	46,XY,t(16;19)
965	F8	16q11	20p13	m	3	N	1.61	0.00	46,XY
966	UK11	16q13	20q13	p	2	B	1.24	0.00	46,XY,t(16;20)
967	UK6	16p11	21q22	m	3	N	1.07	0.29	46,XX
968	UK6	16p11	21q22	m	1	N	0.00	1.07	46,XX
969	H1	16p11	21q22	m	1	U	1.07	0.29	46,XY,der(21),t(16;21)
970	H1	16p11	21q22	m	1	N	1.07	0.29	46,XX
971	H1	16p11	21q22	m	1	B	1.07	0.29	46,XX,t(16;21)
972	H1	16p11	21q22	m	1	U	1.07	0.29	46,XY,der(21),t(16;21)
973	H1	16p11	21q22	m	1	N	1.07	0.29	46,XX
974	UK11	16q21	21p11	m	1	U	1.07	0.00	46,XX,der(21),t(16;21)
975	UK11	16q21	21p11	m	1	N	1.07	0.00	46,XX
976	IT2	16q21	22q12	m	1	N	1.07	0.75	46,XX
977	2110	17q12	19q13	p	4*	B	2.05	0.57	46,XY,t(17;19)
978	41	17q23	21q22	p	1	N	1.09	0.29	46,XX
979	45	17q23	21q22.1	m	1	B	1.09	0.44	46,XY,t(17;21)

980	F14	17p13	22q11	p	2	N	1.19	0.23	46,XY
981	UK17	17q25	22q11	p	2	N	1.19	0.29	46,XX
982	UK16	17q25	22q12	m	1	B	0.29	0.75	46,XX,t(17;22)
983	0710	18p11.2	19q13.1	m	2	B	0.11	0.95	46,XY,t(18;19)
984	2940	18p11.1	20p11.1	m	2	B	0.87	0.67	46,XX,t(18;20)
985	UK5	18p11	20p12	p	3	N	0.39	0.35	46,XX
986	F10	18q12	20p13	m	2	B	1.35	0.00	46,XY,t(18;20)
987	395	18q11.2	21q22	m	2	N	1.72	0.29	46,XY
988	1480	18q21	21q22	m	2	U	1.22	0.00	47,XY,+21,t(18;21)
989	402	18q11	22q13	p	2	N	0.30	1.74	46,XX
990	UK9	18q21	22q13	m	1	N	0.84	0.30	46,XY
991	UK15	19p13	20q11	p	3	B	0.44	1.16	46,XY,t(19;20)
992	CH4	19q13	20p13	m	1	B	0.60	0.00	46,XY,t(19;20)
993	1950	19q13	21q22	m	1	B	1.22	0.00	46,XY,t(19;21)
994	1950	19q13	21q22	m	1	N	1.22	0.00	46,XY
995	F7	19q13	22q12	p	2	B	0.75	0.57	46,XY,t(19;22)
996	364	20q13	22q13	p	1	B	0.30	0.42	46,XX,t(20;22)
997	F8	21p13	22q11	m	2	U	0.00	0.28	45,XX,−21,−22, +der(21),t(21;22)

U.S. cases for all rearrangement types were divided into five ascertainment modes (4, 5 do not apply to the 596 rcp cases from the European data included in Appendix I). These are: asc 1, through a previous full-term infant with an unbalanced form of the rearrangement; asc 2–5, through a balanced carrier; because of recurrent miscarriage (asc 2); incidentally, for reasons ostensibly unrelated to the rearrangement (asc 3); because of infertility (asc 4); or through a chromosomally balanced but intellectually handicapped proband (asc 5). The data are those of the U.S. contributing laboratories (Appendix VIII) with the addition of the European rcp cases [5,6]. The potential or actual imbalances for rcp are derived usually from the sizes (percentage haploid autosome length) [% HAL] of the interchange segments; see table in Daniel [7]. This is so for rcp segregating by adjacent-1. When a 3:1 or adjacent −2 segregation occurs or is predicted, the actual/potential imbalances have been calculated similarly from the sizes of the segments lost and gained. Balanced/normal fetuses with identical rcp have usually been assigned a potential imbalance by the same segregation mode as their sibs/"cousins," but when this is not known (or can be inferred) it is assumed to be adjacent-1. A departure from these principles has been made for some cases as follows. This concerns the finding [5], that usually the larger interchange segment of an rcp occurs in imbalanced offspring (arising by adjacent −1 segregation) as the duplication. In the *unbalanced* offspring in the European rcp data [5], 25% (irrespective of carrier parent) had the complementary situation, i.e., with the larger segment as the deficiency. Therefore, to match this trend in the balanced/normal outcomes, when only potential imbalances were plotted, 1/4 (25 percent) of the trisomy/monosomy sets in this Appendix (each asc mode corrected separately) were represented with the larger segment as the deficiency. This serves to act as a correction (the cases reversed are so entered) and does not imply that this is the segregation mode expected from the particular rcp. See footnotes to Tables II and IV B for abbreviations.

APPENDIX II. Prenatal diagnosis results for parental carriers of pericentric inversions

		Breakpoint							
Observation	Laboratory	1	2	Carrier	Asc mode	Result code	Potential or actual trisomy	Potential or actual monosomy	Prenatal result
998	45	1p12	1q21.1	p	3	N	4.42	4.04	46,XY
999	45	1p12	1q21.1	p	3*	B	4.42	4.04	46,XY,inv(1)
1000	90	1p13	1q21	p	3*	B	4.42	4.04	46,XX,inv(1)
1001	364	1p13	1q21	m	3*	B	4.42	4.04	46,XY,inv(1)
1002	393	1p13	1q21	m	3*	B	4.42	4.04	46,XY,inv(1)
1003	0200	1p36.3	1q42	m	1	B	0.89	0.20	46,XY,inv(1)
1004	148	2p11	2q12	p	3*	B	4.79	3.11	46,XY,inv(2)
1005	1510	2p11	2q13	m	3*	B	4.79	3.11	46,XX,inv(2)
1006	393	2p11	2q13	p	3*	B	4.79	3.11	46,XX,inv(2)
1007	0710	2p11	2q13	p	3*	B	4.79	3.11	46,XY,inv(2)
1008	341	2p11	2q13	m	3	B	4.79	3.11	46,XY,inv(2)
1009	45	2p11	2q13	p	3	B	4.79	3.11	46,XY,inv(2)
1010	2760	2p11	2q13	p	3*	B	4.79	3.11	46,XX,inv(2)
1011	341	2p11	2q13	p	3	B	4.79	3.11	46,XY,inv(2)
1012	341	2p11	2q13	m	3*	B	4.79	3.11	46,XX,inv(2)
1013	45	2p11	2q13	p	3*	B	4.79	3.11	46,XX,inv(2)
1014	1178	2p11	2q13	p	3*	B	4.79	3.11	46,XX,inv(2)
1015	319	2p11	2q13	m	3*	B	4.79	3.11	46,XY,inv(2)
1016	45	2p11	2q13	p	3	B	4.79	3.11	46,XX,inv(2)
1017	364	2p11	2q13	p	3*	B	4.79	3.11	46,XY,inv(2)
1018	355	2p11	2q13	p	3*	B	4.79	3.11	46,XY,inv(2)
1019	0325	2p11	2q13	m	3*	B	4.79	3.11	46,XX,inv(2)
1020	1990	2p11.2	2q13	p	3*	B	4.79	3.11	46,XY,inv(2)
1021	364	2p11.2	2q13	p	3*	B	4.79	3.11	46,XX,inv(2)
1022	45	2p11.2	2q13	p	3*	B	4.79	3.11	46,XY,inv(2)
1023	45	2p11.2	2q13	p	3*	B	4.79	3.11	46,XY,inv(2)
1024	0970	2p11.2	2q13	p	3*	B	4.79	3.11	46,XY,inv(2)
1025	393	2p11.2	2q13	p	3*	B	4.79	3.11	46,XX,inv(2)

1026	393	2p11.2	2q13	p	3*	B	4.79	3.11	46,XX,inv(2)
1027	393	2p11.2	2q13	p	3*	B	4.79	3.11	46,XY,inv(2)
1028	393	2p11.2	2q13	m	3*	B	4.79	3.11	46,XX,inv(2)
1029	398	2p11.2	2q13	p	3*	B	4.79	3.11	46,XY,inv(2)
1030	45	2p11.2	2q13	m	3*	B	4.79	3.11	46,XX,inv(2)
1031	385	2p11.2	2q13	m	2	B	4.79	3.11	46,XY,inv(2)
1032	45	2p11.2	2q13	p	3*	B	4.79	3.11	46,XY,inv(2)
1033	130	2p11.2	2q13	p	3*	B	4.79	3.11	46,XY,inv(2)
1034	45	2p12	2q13	m	3*	B	4.79	3.11	46,XX,inv(2)
1035	364	2p13	2q11	m	3*	B	5.42	2.45	46,XX,inv(2)
1036	90	3p13	3q11	m	3*	B	3.52	2.75	46,XY,inv(3)
1037	90	3p22.3	3q21	p	3*	B	2.55	1.15	46,XY,inv(3)
1038	79	3p25	3q25	m	1	N	1.44	0.41	46,XY
1039	0200	3p26	3q24	m	1	U	1.74	0.07	46,XX,rec(3), dup q,inv(3)
1040	90	4p12	4q23	m	1	B	3.10	1.50	46,XX,inv(4)
1041	45	4p15.3	4q12	p	3*	B	4.54	0.65	46,XX,inv(4)
1042	1700	4p15.3	4q12	m	3	B	4.54	0.65	46,XX,inv(4)
1043	393	4p16	4q21	m	3*	B	3.72	0.24	46,XX,inv(4)
1044	219	4p15	4q31	m	1	B	1.50	0.24	46,XY,inv(4)
1045	1030	4p16	4q31	m	2	N	1.50	0.24	46,XX
1046	1030	4p16	4q31	m	2	N	1.50	0.24	46,XY
1047	335	5p13	5q13	m	3*	B	3.81	1.16	46,XY,inv(5)
1048	364	5p13	5q13	m	3*	B	3.81	1.16	46,XY,inv(5)
1049	45	5p13	5q13	p	3	N	3.81	1.16	46,XX
1050	90	5p13	5q13	m	3*	B	3.81	1.16	46,XX,inv(5)
1051	90	5p13	5q13	m	3*	B	3.81	1.16	46,XX,inv(5)
1052	141	5p13	5q13	p	3	B	3.81	1.16	46,XX,inv(5)
1053	90	5p15	5q33	m	3*	B	0.91	0.32	46,XX,inv(5)
1054	2760	6p12	6q15	p	3*	B	3.04	2.10	46,XY,inv(6)
1055	45	6p12	6q15	p	3*	B	3.04	2.10	46,XY,inv(6)
1056	45	6p21.1	6q21	m	3	B	2.32	1.42	46,XY,inv(6)
1057	1510	6p21	6q27	p	1	N	1.83	0.10	46,XX
1058	37	6p21.3	6q15	m	2	B	3.04	1.19	46,XY,inv(6)
1059	1915	7p15.3	7q36	m	2	N	0.78	0.21	46,XX

(continued)

APPENDIX II. Prenatal diagnosis results for parental carriers of pericentric inversions (continued)

		Breakpoint					Potential	Potential	
Observation	Laboratory	1	2	Carrier	Asc mode	Result code	or actual trisomy	or actual monosomy	Prenatal result
1060	1830	7p15.3	7q11.2	m	2	N	3.17	0.78	47,XXY
1061	402	8p11	8q13	p	3	N	2.70	1.37	46,XX
1062	0200	8p21	8q22	p	1	B	1.45	0.86	46,XY,inv(8)
1063	598	8p22	8q13	p	3*	B	2.70	0.31	46,XY,inv(8)
1064	0200	8p23	8q12	p	3*	B	2.70	0.31	46,XY,inv(8)
1065	0200	8p23	8q22	p	3*	B	1.45	0.31	46,XY,inv(8)
1066	3060	8p23	8q22	m	1	B	1.45	0.31	46,XY,inv(8)
1067	41	8p23.2	8q22	m	1	B	1.23	0.31	46,XY,inv(8)
1068	0425	10p11	10q21	m	3*	B	2.66	1.38	46,XY,inv(10)
1069	364	10p11	10q21	m	3*	B	2.66	1.38	46,XY,inv(10)
1070	1700	10p11	10q21	m	3*	B	2.66	1.38	46,XY,inv(10)
1071	364	10p11	10q21	m	3*	B	2.66	1.38	46,XY,inv(10)
1072	90	10p11	10q21	p	3*	B	2.66	1.38	46,XY,inv(10)
1073	335	10p11	10q21	p	3*	B	2.66	1.38	46,XY,inv(10)
1074	152	10p11	10q21	m	2	B	2.66	1.38	46,XX,inv(10)
1075	0200	10p11	10q21	p	3	B	2.66	1.38	46,XY,inv(10)
1076	341	10p11	10q21	p	3*	B	2.66	1.38	46,XX,inv(10)
1077	90	10p11	10q21.2	m	2	B	2.66	1.38	46,XX,inv(10)
1078	90	10p11	10q21.2	m	2	B	2.66	1.38	46,XX,inv(10)
1079	1510	10p11.1	10q21.1	m	3*	B	2.66	1.38	46,XY,inv(10)
1080	385	10p11.1	10q21.2	p	3*	B	2.66	1.38	46,XX,inv(10)
1081	90	10p11.2	10q21.2	p	3*	B	2.66	1.38	46,XX,inv(10)
1082	0200	10p11.2	10q21	m	3*	B	2.66	1.38	46,XY,inv(10)
1083	393	10p11.2	10q21.2	m	3*	B	2.66	1.38	46,XY,inv(10)
1084	90	10p11.2	10q21.2	p	3*	B	2.66	1.38	46,XX,inv(10)
1085	0425	10p13	10q22	p	3*	B	2.07	0.57	46,XY,inv(10)
1086	0325	10p15	10q22	m	3*	B	2.07	0.16	46,XX,inv(10)
1087	1030	11p13	11q23	p	1	N	0.98	0.78	46,XX

1088	335	11p13	11q25	p	3*	B	0.98	0.09	46,XX,inv(11)
1089	0200	12p12	12q14	m	3	N	2.46	0.76	46,XY
1090	0325	12p13	12q11	p	3*	B	3.53	0.30	46,XX,inv(12)
1091	148	12p13	12q23	p	3	B	1.23	0.30	46,XX,inv(12)/ 47,XX,+i(Xq),inv(12)
1092	402	16p11.2	16q23	m	3*	U	1.00	0.44	46,XX,rec(16), dup p,inv(16)
1093	41	16p12.2	16q25	p	1	N	0.80	0.44	46,XY
1094	45	17p13	17q11	m	3*	B	2.30	0.15	46,XX,inv(17)
1095	319	18p11	18q12	m	3*	B	1.35	0.67	46,XY,inv(18)
1096	0200	18p11	18q21	p	3*	B	0.84	0.67	46,XY,inv(18)
1097	219	19p11	19q12	p	3*	B	1.25	1.00	46,XX,inv(19)
1098	152	19p12	19q13	p	2	B	1.25	1.00	46,XX,inv(19)
1099	2760	21p11	21q21	p	3*	B	0.85	0.00	46,XX,inv(21)

See footnote to Appendix I.

APPENDIX III. Prenatal diagnosis results for parental carriers of Robertsonian translocations

Observation	Laboratory	Breakpoint 1	Breakpoint 2	Carrier	Asc mode	Result code	Prenatal result
1100	2955	13p12	14p12	p	3*	B	45,XY,rob(13;14)
1101	2955	13p12	14p12	m	2	B	45,XY,rob(13;14)
1102	37	13p12	14p12	m	2	B	45,XX,rob(13;14)
1103	1030	13p12	14p12	m	3*	B	45,XX,rob(13;14)
1104	1030	13p12	14p12	m	1	N	46,XY
1105	1510	13p12	14p12	m	2	B	45,XY,rob(13;14)
1106	2760	13p12	14p12	m	3*	B	45,XX,rob(13;14)
1107	0020	13p12	14p12	m	2*	B	45,XY,rob(13;14)
1108	0020	13p12	14p12	p	3*	B	45,XX,rob(13;14)
1109	393	13p12	14p12	m	2	N	46,XX
1110	41	13p12	14p12	m	2	N	46,XY
1111	41	13p12	14p12	m	2	N	46,XX
1112	41	13p12	14p12	m	3*	U	46,XX,−14,+rob(13;14)
1113	41	13p12	14p12	m	1	B	45,XY,rob(13;14)
1114	41	13p12	14p12	m	2	B	45,XY,rob(13;14)
1115	41	13p12	14p12	m	1	B	45,XY,rob(13;14)
1116	41	13p12	14p12	m	1	N	46,XY
1117	41	13p12	14p12	m	1	B	45,XX,rob(13;14)
1118	41	13p12	14p12	m	3	N	46,XY
1119	0080	13p12	14p12	p	3	B	45,XY,rob(13;14)
1120	2110	13p12	14p12	m	3	B	45,XX,rob(13;14)
1121	2110	13p12	14p12	m	3	B	45,XX,rob(13;14)
1122	0710	13p12	14p12	m	3*	B	45,XY,rob(13;14)
1123	0710	13p12	14p12	m	2	N	46,XY
1124	0425	13p12	14p12	p	1	B	45,XX,rob(13;14)
1125	0425	13p12	14p12	p	3*	B	45,XX,rob(13;14)
1126	0425	13p12	14p12	m	3*	B	45,XX,rob(13;14)
1127	0425	13p12	14p12	m	2	B	45,XY,rob(13;14)

1128	341	13p12	14p12	m	3*	B	45,XX,rob(13;14)
1129	341	13p12	14p12	m	3*	B	45,XX,rob(13;14)
1130	138	13p12	14p12	m	1	B	45, XY,rob(13;14)
1131	138	13p12	14p12	m	2	B	45,XY,rob(13;14)
1132	138	13p12	14p12	m	1	B	45,XX,rob(13;14)
1133	138	13p12	14p12	m	2	N	46,XY
1134	138	13p12	14p12	m	2	N	46,XX
1135	1480	13p12	14p12	m	1	N	46,XY
1136	1480	13p12	14p12	m	2	N	46,XX
1137	1480	13p12	14p12	m	2	N	46,XY
1138	45	13p12	14p12	p	3*	U	46,XX, −14, +rob(13;14)
1139	45	13p12	14p12	m	2	N	46,XX
1140	45	13p12	14p12	p	4*	B	45,XX,rob(13;14)
1141	45	13p12	14p12	m	3*	B	45,XX,rob(13;14)
1142	3100	13p12	14p12	p	3*	B	45,XY,rob(13;14)
1143	3100	13p12	14p12	p	2	N	46,XX
1144	330	13p12	14p12	p	4	B	45,XX,rob(13;14)
1145	330	13p12	14p12	p	4	N	46,XX
1146	385	13p12	14p12	m	3*	B	45,XY,rob(13;14)
1147	385	13p12	14p12	m	2	N	46,XX
1148	385	13p12	14p12	m	2	N	46,XY
1149	0325	13p12	14p12	m	3	B	45,XY,rob(13;14)
1150	0200	13p12	14p12	m	2	N	46,XX
1151	356	13p12	14p12	m	3	B	45,XY,rob(13;14)
1152	356	13p12	14p12	m	2*	B	45,XX,rob(13;14)
1153	356	13p12	14p12	p	2	B	45,XY,rob(13;14)
1154	381	13p12	14p12	m	3*	B	45,XX,rob(13;14)
1155	381	13p12	14p12	p	3*	B	45,XY,rob(13;14)
1156	381	13p12	14p12	m	2	B	45,XY,rob(13;14)
1157	381	13p12	14p12	m	2	B	45,XY,rob(13;14)
1158	2790	13p12	14p12	p	2	N	46,XX
1159	1178	13p12	14p12	m	1	B	45,XY,rob(13;14)
1160	1178	13p12	14p12	m	3*	B	45,XX,rob(13;14)
1161	1178	13p12	14p12	p	3*	B	45,XY,rob(13;14)

(continued)

APPENDIX III. Prenatal diagnosis results for parental carriers of Robertsonian translocations (continued)

Observation	Laboratory	Breakpoint 1	Breakpoint 2	Carrier	Asc mode	Result code	Prenatal result
1162	319	13p12	14p12	p	3*	B	45,XY,rob(13;14)
1163	319	13p12	14p12	p	1	N	46,XY
1164	319	13p12	14p12	m	3*	B	45,XX,rob(13;14)
1165	319	13p12	14p12	p	1	B	45,XY,rob(13;14)
1166	141	13p12	14p12	m	1	B	45,XY,rob(13;14)
1167	130	13p12	14p12	m	3*	B	45,XX,rob(13;14)
1168	130	13p12	14p12	p	2	B	45,XY,rob(13;14)
1169	130	13p12	14p12	m	1	B	45,XX,rob(13;14)
1170	130	13p12	14p12	p	1	N	46,XY
1171	355	13p12	14p12	m	2	N	46,XY
1172	355	13p12	14p12	p	3*	B	45,XX,rob(13;14)
1173	398	13p12	14p12	m	1	N	46,XY
1174	335	13p12	14p12	m	3*	B	45,XX,rob(13;14)
1175	219	13p12	14p12	m	2	B	45,XY,rob(13;14)
1176	219	13p12	14p12	m	2	B	45,XY,rob(13;14)
1177	219	13p12	14p12	m	2	B	45,XX,rob(13;14)
1178	219	13p12	14p12	m	2	B	45,XX,rob(13;14)
1179	69	13p12	14p12	p	3*	B	45,XX,rob(13;14)
1180	69	13p12	14p12	p	3*	B	45,XX,rob(13;14)
1181	1915	13p12	14p12	m	1	B	45,XY,rob(13;14)
1182	1915	13p12	14p12	m	2	B	45,XY,rob(13;14)
1183	1915	13p12	14p12	m	1	B	45,XY,rob(13;14)
1184	1760	13p12	14p12	m	2	B	45,XY,rob(13;14)
1185	1990	13p12	14p12	m	2	B	45,XX,rob(13;14)
1186	1830	13p12	14p12	p	3*	B	45,XX,rob(13;14)
1187	343	13p12	14p12	m	3*	B	45,XY,rob(13;14)
1188	343	13p12	14p12	p	3	N	46,XY
1189	343	13p12	14p12	p	3	N	46,XX

1190	45	13p12	14p12	m	3*	B	45,XY,rob(13;14)
1191	45	13p12	14p12	m	3*	B	45,XX,rob(13;14)
1192	45	13p12	14p12	m	3*	N	46,XX
1193	45	13p12	14p12	m	3*	B	45,XX,rob(13;14)
1194	45	13p12	14p12	m	3*	B	45,XY,rob(13;14)
1195	45	13p12	14p12	m	2	B	45,XX,rob(13;14)
1196	90	13p12	14p12	m	2	N	46,XX
1197	90	13p12	14p12	p	3*	B	45,XX,rob(13;14)
1198	90	13p12	14p12	m	2	B	45,XX,rob(13;14)
1199	90	13p12	14p12	m	2	N	46,XX
1200	90	13p12	14p12	m	3*	B	45,XY,rob(13;14)
1201	90	13p12	14p12	m	3*	B	45,XX,rob(13;14)
1202	393	13p12	14p12	p	3*	B	45,XY,rob(13;14)
1203	393	13p12	14p12	m	3*	B	45,XY,rob(13;14)
1204	364	13p12	14p12	m	3*	B	45,XY,rob(13;14)
1205	364	13p12	14p12	p	2	B	45,XY,rob(13;14)
1206	364	13p12	14p12	p	2	N	46,XY
1207	364	13p12	14p12	m	3	B	45,XY,rob(13;14)
1208	364	13p12	14p12	m	3	B	45,XX,rob(13;14)
1209	364	13p12	14p12	m	3	N	46,XX
1210	1700	13p12	14p12	m	3*	B	45,XX,rob(13;14)
1211	1700	13p12	14p12	p	3*	B	45,XY,rob(13;14)
1212	0625	13p12	14p12	p	3*	B	45,XY,rob(13;14)
1213	0200	13p12	14p12	p	3*	B	45,XX,rob(13;14)
1214	0200	13p12	14p12	p	3	B	45,XY,rob(13;14)
1215	0200	13p12	14p12	p	3	N	46,XY
1216	0200	13p12	14p12	m	3	N	46,XY
1217	0200	13p12	14p12	p	3*	B	45,XY,rob(13;14)
1218	0200	13p12	14p12	p	3*	B	45,XX,rob(13;14)
1219	0200	13p12	14p12	p	3*	B	45,XY,rob(13;14)
1220	0200	13p12	14p12	m	3	B	45,XX,rob(13;14)

(continued)

APPENDIX III. Prenatal diagnosis results for parental carriers of Robertsonian translocations (continued)

Observation	Laboratory	Breakpoint 1	Breakpoint 2	Carrier	Asc mode	Result code	Prenatal result
1221	0200	13p12	14p12	p	3*	B	45,XY,rob(13;14)
1222	0200	13p12	14p12	m	3	B	45,XX,rob(13;14)
1223	1950	13p12	14p12	m	3	B	45,XY,rob(13;14)
1224	1950	13p12	14p12	m	3	N	46,XX
1225	1030	13p12	15p12	p	3*	B	45,XY,rob(13;15)
1226	0710	13p12	15p12	p	3*	B	45,XX,rob(13;15)
1227	0425	13p12	15p12	p	3*	B	45,XY,rob(13;15)
1228	0425	13p12	15p12	m	3*	B	45,XX,rob(13;15)
1229	341	13p12	15p12	p	3*	B	45,XX,rob(13;15)
1230	341	13p12	15p12	p	3	N	46,XY
1231	141	13p12	15p12	m	2	B	45,XY,rob(13;15)
1232	393	13p12	15p12	m	3	N	46,XY
1233	0200	13p12	15p12	p	2	B	45,XY,rob(13;15)
1234	1030	13p12	21p12	m	3*	U	46,XY,-13,+rob(13;21)
1235	1510	13p12	21p12	p	1	N	46,XX
1236	2760	13p12	21p12	m	1	N	46,XX
1237	2760	13p12	21p12	m	1	B	45,XX,rob(13;21)
1238	395	13p12	21p12	p	1	N	46,XY
1239	41	13p12	21p12	p	1	N	46,XY
1240	138	13p12	21p12	p	1	N	46,XY
1241	319	13p12	21p12	m	2	N	46,XY
1242	90	13p12	21p12	m	1	B	45,XY,rob(13;21)
1243	90	13p12	21p12	m	1	B	45,XY,rob(13;21)
1244	90	13p12	21p12	p	1	B	45,XY,rob(13;21)
1245	1700	13p12	21p12	p	1	B	45,XX,rob(13;21)
1246	0200	13p12	21p12	m	3*	N	46,XY
1247	2955	13p12	22p12	p	1	B	45,XX,rob(13;22)
1248	1480	13p12	22p12	m	3	B	45,XX,rob(13;22)

1249	1178	13p12	22p12	p	3*	B	45,XX,rob(13;22)
1250	364	13p12	22p12	p	1	N	46,XY
1251	0200	13p12	22p12	m	3*	B	45,XX,rob(13;22)
1252	1030	14p12	15p12	m	2	N	46,XX
1253	0710	14p12	15p12	p	2	N	46,XY
1254	0200	14p12	15p12	m	3*	B	45,XX,rob(14;21)
1255	2955	14p12	21p12	m	1	B	45,XY,rob(14;21)
1256	1030	14p12	21p12	m	1	N	46,XX
1257	1030	14p12	21p12	m	1	N	46,XX
1258	1030	14p12	21p12	p	1	N	46,XX
1259	1030	14p12	21p12	m	1	N	46,XX
1260	79	14p12	21p12	p	1	N	46,XY
1261	79	14p12	21p12	m	1	N	46,XY
1262	238	14p12	21p12	p	3*	B	45,XY,rob(14;21)
1263	1510	14p12	21p12	p	1	B	45,XY,rob(14;21)
1264	2760	14p12	21p12	m	1	U	46,XX,−14,+rob(14;21)
1265	0020	14p12	21p12	m	2	U	46,XY,−14,+rob(14;12)
1266	393	14p12	21p12	p	1	N	46,XX
1267	393	14p12	21p12	p	1	U	46,XX,−14,+rob(14;21)
1268	3060	14p12	21p12	m	1	B	45,XY,rob(14;21)
1269	395	14p12	21p12	m	1	N	46,XX
1270	395	14p12	21p12	m	1	N	46,XX
1271	41	14p12	21p12	m	1	B	45,XX,rob(14;21)
1272	41	14p12	21p12	m	2	B	45,XX,rob(14;21)
1273	41	14p12	21p12	m	1	B	45,XY,rob(14;21)
1274	41	14p12	21p12	m	1	B	45,XX,rob(14;21)
1275	2110	14p12	21p12	m	1	B	45,XY,rob(14;21)
1276	2110	14p12	21p12	p	3	B	45,XX,rob(14;21)
1277	0710	14p12	21p12	m	3*	B	45,XX,rob(14;21)
1278	0710	14p12	21p12	m	3*	B	45,XY,rob(14;21)
1279	0710	14p12	21p12	m	2	B	45,XX,rob(14;21)

(continued)

APPENDIX III. Prenatal diagnosis results for parental carriers of Robertsonian translocations (continued)

Observation	Laboratory	Breakpoint 1	Breakpoint 2	Carrier	Asc mode	Result code	Prenatal result
1280	0425	14q11.2	21p12	m	1	B	45,XY,−14,−21,+der(21),t(14;21)[a]
1281	0425	14p12	21p12	m	1	U	46,XX,−14,+rob(14;21)
1282	341	14p12	21p12	m	1	B	45,XY,rob(14;21)
1283	138	14p12	21p12	m	1	N	46,XX
1284	138	14p12	21p12	p	1	B	45,XX,rob(14;21)
1285	1480	14p12	21p12	p	1	N	46,XY
1286	385	14p12	21p12	p	3*	B	45,XX,rob(14;21)
1287	356	14p12	21p12	m	3*	B	45,XX,rob(14;21)
1288	381	14p12	21p12	m	3*	B	45,XY,rob(14;21)
1289	2940	14p12	21p12	m	1	B	45,XX,rob(14;21)
1290	1178	14p12	21p12	m	3*	B	45,XY,rob(14;21)
1291	319	14p12	21p12	m	1	N	46,XX
1292	319	14p12	21p12	m	1	N	46,XY
1293	319	14p12	21p12	m	1	N	46,XY
1294	141	14p12	21p12	m	1	N	46,XY
1295	130	14p12	21p12	p	2	B	45,XY,rob(14;21)
1296	130	14p12	21p12	p	1	N	46,XY
1297	69	14p12	21p12	m	1	B	45,XX,rob(14;21)
1298	255	14p12	21p12	m	1	B	45,XX,rob(14;21)
1299	1760	14p12	21p12	m	1	U	46,XY,−14,+rob(14;21)
1300	1960	14p12	21p12	m	1	B	45,XX,rob(14;21)
1301	1990	14p12	21p12	p	2	B	45,XX,rob(14;21)
1302	1830	14p12	21p12	m	1	B	45,XY,rob(14;21)
1303	1830	14p12	21p12	m	1	B	45,XX,rob(14;12)
1304	1830	14p12	21p12	p	1	B	45,XX,rob(14;21)
1305	1830	14p12	21p12	p	1	B	45,XY,rob(14;21)
1306	1830	14p12	21p12	m	1	B	45,XY,rob(14;21)

1307	1990	14p12	21p12	p	3	N	46,XY
1308	1990	14p12	21p12	p	3	N	46,XX
1309	2070	14p12	21p12	p	1	B	45,XY,rob(14;21)
1310	1970	14p12	21p12	m	1	N	46,XX
1311	1970	14p12	21p12	p	1	B	45,XX,rob(14;21)
1312	343	14p12	21p12	m	2	U	46,XX,−14,+rob(14;21)
1313	45	14p12	21p12	p	1	N	46,XY
1314	90	14p12	21p12	p	3*	B	45,XY,rob(14;21)
1315	90	14p12	21p12	m	2	B	45,XY,rob(14;21)
1316	90	14p12	21p12	m	3*	B	45,XY,rob(14;21)
1317	364	14p12	21p12	m	1	U	46,XY,−14,+rob(14;21)
1318	364	14p12	21p12	m	1	B	45,XY,rob(14;21)
1319	1700	14p12	21p12	m	1	B	45,XY,rob(14;21)
1320	0200	14p12	21p12	m	1	B	45,XY,rob(14;21)
1321	0200	14p12	21p12	m	1	N	46,XY
1322	1950	14p12	21p12	m	1	N	46,XY
1323	1950	14p12	21p12	m	1	U	46,XY,−14,+rob(14;21)
1324	41	14p12	22p12	p	3	N	46,XX
1325	41	14p12	22p12	p	3	N	46,XX
1326	1178	14p12	22p12	p	3*	B	45,XY,rob(14;22)
1327	319	14p12	22p12	p	3*	B	45,XX,rob(14;22)
1328	2070	14p12	22p12	p	2	N	46,XY
1329	45	14p12	22p12	m	3*	B	45,XY,rob(14;22)
1330	0200	14p12	22p12	p	3*	B	45,XY,rob(14;22)
1331	1990	15p12	21p12	m	1	B	45,XX,rob(15;21)
1332	1990	15p12	21p12	m	1	N	46,XY
1333	41	15p12	21p12	m	1	B	45,XY,rob(15;21)
1334	385	15p12	21p12	m	1	N	46,XY

[a]Maternal karyotype 46,XX,t(14;21)(q11.2;p12); i.e., bisatellited marker.
See footnote to Appendix I.

(continued)

APPENDIX III. Prenatal diagnosis results for parental carriers of Robertsonian translocations (continued)

Observation	Laboratory	Breakpoint 1	Breakpoint 2	Carrier	Asc mode	Result code	Prenatal result
1335	385	15p12	21p12	m	1	B	45,XY,rob(15;21)
1336	343	15p12	21p12	m	1	N	46,XX
1337	343	15p12	21p12	m	1	N	46,XY
1338	45	15p12	21p12	m	1	B	45,XX,rob(15;21)
1339	0200	15p12	21p12	p	1	N	46,XY
1340	0200	15p12	21p12	p	1	B	45,XX,rob(15;21)
1341	0200	15p12	21p12	p	1	N	46,XX
1342	0200	15p12	21p12	p	1	B	45,XX,rob(15;21)
1343	0200	15p12	21p12	p	1	B	45,XX,rob(15;21)
1344	0200	15p12	21p12	p	3	B	45,XX,rob(15;21)
1345	393	15p12	22p12	m	2	N	46,XX
1346	0425	15p12	22p12	m	3*	B	45,XX,rob(15;22)
1347	343	17p13.3	22p1	m	2	N	46,XY
1348	41	21p12	22p12	m	1	B	45,XX,rob(21;22)
1349	341	21p12	22p12	p	1	B	45,XX,rob(21;22)
1350	2790	21p12	22p12	m	3*	B	45,XX,rob(21;22)
1351	319	21p12	22p12	m	1	N	46,XY
1352	402	21p12	22p12	m	2	N	46,XY
1353	90	21p12	22p12	m	1	B	45,XX,rob(21;22)

APPENDIX IV. Prenatal diagnosis results for parental carriers of paracentric inversions

Observation	Laboratory	Breakpoint 1	Breakpoint 2	Carrier	Asc mode	Result code	Prenatal result
1354	130	1p31.4	1p36.3	m	3*	B	46,XX,inv(1)
1355	0425	1q21	1q31	m	3	B	46,XY,inv(1)
1356	0425	2q11	2q23	p	3	B	46,XX,inv(2)
1357	90	2q21	2q33	m	3*	B	46,XY,inv(2)
1358	0200	3q13	3q21	m	3	B	46,XX,inv(5)
1359	90	3q13.3	3q26.2	m	2*	B	46,XY,inv(3)
1360	0200	3q21	3q29	m	3*	B	46,XY,inv(3)
1361	2790	5q13	5q35	m	3*	B	46,XX,inv(5)
1362	0710	5q15	5q32	p	3*	B	46,XX,inv(5)
1363	1510	5q22	5q31	m	3*	B	46,XX,inv(5)
1364	1030	6p11.2	6p25	m	3*	B	46,XX,inv(6)
1365	2790	6p12	6p23	p	2	B	46,XY,inv(6)
1366	355	6p12	6p25	p	3*	B	46,XY,inv(6)
1367	0710	6p12.2	6p22.2	m	3*	B	46,XY,inv(6)
1368	385	6p21	6p23	m	3*	B	46,XX,inv(6)
1369	0200	7q12	7q21	m	3*	B	46,XX,inv(7)
1370	364	7q22	7q31.2	p	3*	B	46,XX,inv(7)
1371	393	7q22	7q36	p	3*	B	46,XX,inv(7)
1372	1700	8q21.2	8q22.3	m	3*	B	46,XY,inv(8)
1373	45	10q11	10q23	p	3*	B	46,XY,inv(10)
1374	1990	10q11	10q26	m	2	B	46,XX,inv(10)
1375	45	10q11.2	10q21.1	m	3*	B	46,XX,inv(10)
1376	0425	11q11	11q23	p	3*	B	46,XX,inv(11)
1377	393	11q21	11q23	m	3*	B	46,XY,inv(11)
1378	45	11q22	11q24	m	3	B	46,XX,inv(11)
1379	45	11q22	11q24	m	3*	B	46,XX,inv(11)
1380	45	12q12	12q15	m	3*	B	46,XY,inv(12)
1381	0200	12q15	12q24	p	3*	B	46,XX,inv(12)
1382	1178	14q24.2	14q32.3	m	1[a]	B	46,XX,inv(14)
1383	2955	17p11.2	17p13.3	p	3*	B	46,XX,inv(17)

[a]Ascertained by child with inv dup(14).
See footnote to Appendix I.

APPENDIX V. Prenatal diagnosis results for parental carriers of insertions

Observation	Laboratory	Breakpoint 1	Breakpoint 2	Breakpoint 3	Carrier	Asc mode	Result code	Potential or actual trisomy	Potential or actual monosomy	Prenatal result[a]
1384	395	11q22.1	3q22.1	3q24	p	1	N	0.00	0.57	46,XX
1385	395	11q22.1	3q22.1	3q24	p	1	U	0.57	0.00	46,XX,der(11),ins(11;3)
1386	395	11q22.1	3q22.1	3q24	p	1	N	0.00	0.57	46,XX
1387	3060	15q21	2q21	2q24	m	3*	U	1.40	0.00	46,XX,der(15),ins(15;2)
1388	1830	18p11.2	18q23.1	18q12.3	m	1	U	0.00	1.06	46,XX,rec(18),def q,inv ins(18)
1389	402	16q13	16p13.3	16p11.2	m	1	U	0.90	0.00	46,XY,rec(16),dup p,ins(16)

[a]In 4 of 6 cases, the result was an unbalanced fetus, indicating a substantial risk for carriers of intra- or interchromosomal insertions.
See footnote to Appendix I.

APPENDIX VI. Prenatal diagnosis results for parental carriers of sex chromosome rearrangements

Observation	Laboratory	Breakpoint 1	Breakpoint 2	Parental structural type	Carrier	Asc mode	Result code	Prenatal result
1390	335	Xq13	Xq26	inv	p	3*	B	46,X,inv(X)
1391	1030	Xp11.4	Xp22.3	inv	m	3	B	46,inv(X),Y
1392	1030	Xp11.4	Xp22.3	inv	m	3	B	46,inv(X),Y
1393	130	Xp11	Xq22	inv	p	3*	B	46,X,inv(X)
1394	0200	Xp22	Xq12	inv	m	3*	B	46,X,inv(X)
1395	393	Xp21	Xp22.3	del	m	3*	U	46,X,del(X)(pter→ p22.3::p21→qter)
1396	148	Xq25	3q26	rcp	m	5*	B	46,XX,t(X;3)
1397	138	Xq21	5q35	rcp	m	2	N	46,XX
1398	2150	Yp11.2	Yq11.23	inv	p	3	B	46,X,inv(Y)
1399	45	Yp11	Yq11.2	inv	p	3*	B	46,X,inv(Y)
1400	45	Yp11	Yq11.2	inv	p	3*	B	46,X,inv(Y)
1401	0325	Yp11.1	Yq11.2	inv	p	3*	B	46,X,inv(Y)
1402	1178	Yp11.1	Yp11.2	inv	p	3*	B	46,X,inv(Y)
1403	141	Yp11	Yq11.2	inv	p	3*	B	46,X,inv(Y)
1404	45	Yp11	Yq11.2	inv	p	3*	B	46,X,inv(Y)
1405	45	Yp11	Yq11.2	inv	p	3*	B	46,X,inv(Y)
1406	45	Yp11	Yq11.2	inv	p	3	N	46,XX
1407	45	Yp11	Yq11.2	inv	p	3*	B	46,X,inv(Y)
1408	45	Yp11	Yq11.2	inv	p	3*	B	46,X,inv(Y)
1409	45	Yp11	Yq11.2	inv	p	3*	B	46,X,inv(Y)
1410	90	Yp11	Yq11.2	inv	p	3*	B	46,X,inv(Y)
1411	90	Yp11	Yq11.2	inv	p	3	B	46,X,inv(Y)
1412	393	Yp11	Yq11.2	inv	p	3*	B	46,X,inv(Y)
1413	364	Yp11.2	Yq11.2	inv	p	3*	B	46,X,inv(Y)
1414	1700	Yp11.2	Yq11.2	inv	p	3*	B	46,X,inv(Y)
1415	1700	Yp11.2	Yq11.2	inv	p	3*	B	46,X,inv(Y)
1416	0200	Yp11	Yq11.2	inv	p	3*	B	46,X,inv(Y)
1417	0200	Yp11	Yq11.2	inv	p	3*	B	46,X,inv(Y)[a]

[a]Ambiguous genitalia at term (presumptively unrelated to the inversion). Note the high proportion of incidental asc for inv(Y)(p11q11.2), indicating the lack of clinical significance of this rearrangement.

See footnote to Appendix I.

APPENDIX VII. Prenatal diagnosis results for parental carriers of multiple or complex rearrangements

Observation	Laboratory	Paternal karyotype	Asc mode	Result code	Maternal karyotype	Asc mode	Result code	Prenatal result
1418	1700	46,XY,inv(1)(p13q21)	3*	B	46,XX,inv(9)(q32q34)	3*	B	46,XY,inv(1),inv(9)
1419	3060	45,XY,rob(13;14), inv(10)(p11q21)	3	B	46,XX	—	—	45,XY,rob(13;14), inv(10)
1420	341	46,XY	—	—	46,XX,inv(2)(p11q13),inv(5)(p13q11)	3	B	46,XX,inv(5)
1421	138	46,XY	—	—	45,XX,rob(13;13)(p12;p12)	1	N	46,XX
1422	1480	46,XY	—	—	45,XX,rob(13;14)(p12;p12)	2	—	45,XY,rob(13;13) (p12;p12)
1423	385	46,XY,inv(10)(p11q21)	3*	B	45,XX,rob(13;14)	2	N	46,XX,inv(10)
1424	385	46,XY	—	—	46,XX,t(9;14)(q11;p11)	2	B	46,XX,t(9;14),t(9;12) (p24;q13) de novo
1425	45	46,XY,t(1;14)(q43;q32.1)	3*	B	46,XX,inv(10)(p11q21)	3*	N	46,XY,t(1;14)
1426	130	46,XY	—	—	46,XX,t(3;11;20)(p13;p11;q13)	1	B	46,XY,t(3;11;20)
1427	130	46,XY	—	—	46,XX,t(3;11;20)(p13;p11;q13)	1	B	46,XY,t(3;11;20)
1428	1830	46,XY	—	—	46,XX,t(1;5;18)(q42;q15;q11)	1	U	46,XX,−1,−5,+der(1), +der(5),t(1;5;18)

See footnote to Appendix I.

APPENDIX VIII. Contributors to the North American collaborative data

Laboratory[a]	Contributor(s)	Institution
0020	Wladimir Wertelecki, Cheryl Dumont	University of South Alabama, Department of Medical Genetics, Mobile, AL 36617
37	James T. Mascarello	Genetics Laboratory, Children's Hospital of San Diego, CA 92123
41	Steven Schonberg, Mitchell Golbus	Laboratory of Cytogenetics, UCSF, San Francisco, CA 94143
45	John Mann, Karen Wcislo	Regional Genetics Laboratory, Permanente Medical Group, San Jose, CA 95119
69	Digamber Borgaonkar	Medical Center of Delaware, Cytogenetics Laboratory, Newark, DE 19718
79	Eduardo Cantu, Jaime Frias	University of Florida College of Medicine, Pediatric Genetics, Gainesville, FL 32702
0080	Susanne M. Gollin	Arkansas Children's Hospital Cytogenetics Laboratory, Little Rock, AR 72202
90	Richard L. Neu, Karen Miller	Genetics Laboratory, University of South Florida College of Medicine, Tampa, FL 33612
130	Catherine Palmer	Indiana University Medical Center, Medical Genetics, Indianapolis, IN 46223
138	Charles R. King	Cytogenetics Laboratory, University of Kansas Medical Center, Kansas City, KS 66103
141	Sechin Cho	Genetics Laboratory, Wesley Medical Center, UKSM, Wichita, KS 67214
148	Maria Varela	Tulane University Medical Center, Human Genetics Program, New Orleans, LA 70112
152	Carmen Zapata	Louisiana State University, School of Medicine, Shreveport, LA 71130
0200	Barbara Crandall	Amniocentesis Laboratory, UCLA, Los Angeles, CA 92024
209	Malcolm Jenkins	Human Genetics Unit, Minnesota Department of Health, Minneapolis, MN 55440
219	John F. Jackson	Medical Genetics, University of Mississippi Medical Center, Jackson, MS 39216
233	Doris Wurster-Hill, Eileen Rawnsley	Cytogenetics Laboratory, Dartmouth-Hitchcock Medical Center, Hanover, NH 03756
238	Alan R. Rushton	Pediatrics, Hunterdon Medical Center, Flemington, NJ 08822
255	Edward J. Schutta	Department of Genetics, Brookdale Hospital, Brooklyn, NY 11212
0270	Daisy M. Carr	Cytogenetics Laboratory, Charles R. Drew Medical School, Los Angeles, CA 90059
319	Peter Dignan, Lori Krouskop	Research Foundation Chromosome Laboratory, Cincinnati Children's Hospital, Cincinnati, OH 45229
0325	Stephen N. Bauer	Sutter Memorial Hospital, Cytogenetics Laboratory, Sacramento, CA 95819
330	Maureen McCorquodale	Genetics Center of Northwest Ohio, Medical College at Toledo, OH 43699
335	Nancy J. Carpenter	Chapman Research Institute of Medical Genetics, Children's Medical Center, Tulsa, OK 74135
341	Linda A. Cannizzaro, Roger L. Ladda	Genetics, Milton S. Hershey Medical Center, Hershey, PA 17033

(continued)

APPENDIX VIII. Contributors to the North American collaborative data (continued)

Laboratory[a]	Contributor(s)	Institution
343	Laird G. Jackson	Jefferson Medical College Cytogenetics Laboratory, Philadelphia, PA 19107
355	Mary C. Phelan	Greenwood Genetic Center, Greenwood, SC 29646
356	Virginia Johnson, Carol Christianson	Birth Defects Genetics, University of South Dakota Medical School, Vermillion, SD 57069
364	Patricia Howard-Peebles	Cytogenetics Laboratory, University of Texas Health Science Center at Dallas, Dallas, TX 75235
381	Gabriel Khodr	Southwest Genetics, P.A., San Antonio, TX 78229
385	Joan F. Atkin	Department of Pediatrics, Medical Genetics, University of Virginia, Charlottesville, VA 22908
393	Lawrence Karp, Robert Resta	Swedish Hospital Medical Center, Cytogenetics Laboratory, Seattle, WA 98104
395	Cynthia Dolan, Norman V. Vigfusson	Inland Empire Genetics Counseling Service, Spokane, WA 99210
398	Susie Ball	Central Washington Genetics Program, Yakima Memorial Hospital, Yakima, WA 98902
402	Renata Laxova, Gurbax Sekon	Clinical Genetics Center, University of Wisconsin, Madison, WI 53705-2280
0425	T.K. Mohandas, Michael M. Kaback	Harbor-UCLA Medical Center Cytogenetics Laboratory, Torrance, CA 90509
0625	John W. Larsen	Wilson Genetics Unit, George Washington University Medical Center, DS 20037
0710	Roger P. Donahue	Pediatric Cytogenetics Laboratory, University of Miami School Medicine, Miami, FL 33136
0760	Jean H. Priest	Medical Genetics, Department of Pediatrics, Emory University, Atlanta, GA 30322
0810	J. Rogers Byrd	Clinical Cytogenetics, Medical College of Georgia, Augusta, GA 30912-3395
0970	Richard C. Juberg	Caylor-Nickel Medical Center, Bluffton, IN 46714
1030	Shivanand R. Patil	Cytogenetics Laboratory, University of Iowa Hospital, Iowa City, IA 52242
1178	Gail Stetten	Prenatal Diagnostic Center, Cytogenetics Laboratory, Johns Hopkins University, Baltimore, MD 21205
1290	Leonard Atkins, Wayne Miller	Massachusetts General Hospital Cytogenetics Laboratory, Boston, MA 02114

1480	Daniel Van Dyke, V. Ramesh Babu	Cytogenetics Laboratory, Henry Ford Hospital, Detroit, MI 48202
1510	James V. Higgins, Joyce Center	Pediatrics, Michigan State University Genetics Laboratory, East Lansing, MI 48824
1700	Art Daniel	Montana State Medical Genetics Program, Shodair Hospital, Helena MT 59602
1760	Dorothy Warburton, Steven Gersen	Cytogenetics Laboratory, Morristown Memorial Hospital, Morristown, NJ 07960
1800	Rina Schmidt	MetPath Cytogenetics Laboratory, Teterboro, NJ 07608
1820	Kathleen E. Richkind	Vivigen Inc, St. Michael's Drive, Santa Fe, NM 87501
1830	Ann M. Willey	Human Genetics Laboratory, Wadsworth Research Center, Albany, NY 12201
1890	R.S. Verma	Division of Cytogenetics, Interfaith Medical Center, Brooklyn, NY 11238
1915	Daniel G. Stetka	Cytogenetics Laboratory, Human Genetics, Children's Hospital, Buffalo, NY 14222
1950	Harshad Q. Shah	North Shore University Hospital, Cytogenetics Laboratory, Manhasset, NY 11030
1960	Judith Stamberg, Steven Keiles	Human Genetics, Long Island Jewish—Hillside Medical Center, NY 11042
1970	Lilian Hsu	Prenatal Diagnosis Laboratory, NYC, New York, NY 10016
1980	Sandra R. Wolman	Cytogenetics Laboratory, Department of Pathology, New York University Medical Center, New York, NY 10016
1990	Dorothy Warburton	Genetics Diagnostic Laboratory, Columbia-Presbyterian Medical Center, New York, NY 10032
2070	Lawrence R. Shapiro	Cytogenetics Laboratory, Letchworth Village Develop Center, Thiells, NY 10984
2110	Frank S. Grass	Cytogenetics, Pediatrics, Charlotte Memorial Hospital, Charlotte, NC 28232
2150	John E. Wiley	Cytogenetics Laboratory, East Carolina University School Medicine, Greenville, NC 27834
2570	Ellen Magenis, Susan Olson	Medical Genetics, Oregon Health Sciences University, Portland, OR 97207
2730	Gail Barsel	Genetic Counselling Center, Rhode Island Hospital, Providence, RI 02903
2760	S. Robert Young	Clinical Genetics, University of South Carolina School of Medicine, Columbia, SC 29203
2790	Charles W. Shears	Clinical Genetics Center, Johnson City, TN 37601
2940	Frederick F. Elder	Pediatric Cytogenetics Laboratory, University of Texas Medical Branch, Galveston, TX 77550

(continued)

APPENDIX VIII. Contributors to the North American collaborative data (continued)

Laboratory[a]	Contributor(s)	Institution
2955	Janice L. Smith	Laboratories for Genetic Services, Houston, TX 77030
3060	Robert Fineman, Martha Morgan	Cytogenetics Laboratory, University of Utah Medical Center, Salt Lake City, UT 84113
3100	Shelly Kerman	Cytogenetics Laboratory, American Medical Laboratories, Fairfax, VA 22030
3260	David M. Cox David I. Hoar	Alberta Hereditary Diseases Program, Alberta Childrens Hospital, Calgary, Alberta T2T-5C7
3300	Dagmar K. Kalousek	Cytogenetics and Embryopathology, British Columbia Children's Hospital, Vancouver, British Columbia V6H-3V4
3305	Salma M. Styles J.S. Popkin	Cytogenetics Laboratory, Victoria General Hospital, Victoria, British Columbia V8Z-6R5
3340	Irene Uchida, Viola Freeman	Cytogenetics Laboratory, McMaster University, Hamilton, Canada L8N 3Z5
3360	Alessandra Duncan, Ellen Price	Cytogenetics Laboratory, Kingston General Hospital, Kingston, Ontario K7L-2V7
3380	F.R. Sergovich	Victoria Hospital Cytogenetics Laboratory, London, Ontario N6A-4G5
3407	Ikuko Teshima	Medical Genetics, Hospital for Sick Children, Toronto, Ontario M5G-1X8
3480	M. Vekemans, Lola Cartier	McGill Prenatal Diagnosis Program, Montreal Children's Hospital, Montreal, Quebec H3H-1P3
3500	Ernst Pinno	Cytogenetics Laboratory, Regina General Hospital, Regina, Saskatchewan S4P-OW5
3510	T. Pantzar	Cytogenetics Laboratory, Vancouver General Hospital, Vancouver, British Columbia V5Z-1M9

[a] Four-digit laboratory numbers are from the International Cytogenetic Laboratory Directory, 1986, Association of Cytogenetic Technologists, Pasadena, California. One- to three-digit laboratory numbers are from the Birth Defects Genetics Services International Directory, 1983, March of Dimes Birth Defects Foundation. Letter/digit laboratory numbers (Appendix I) are from the European Collaborative Prenatal Diagnosis Laboratories listed in Boué and Gallano [6].

The Cytogenetics of Mammalian Autosomal Rearrangements, pages 163–247

Chapter 3
Pericentric Inversions: Their Problems and Clinical Significance

Peter Kaiser

The findings pertaining to 352 inversions from the literature and from our own laboratory are analyzed in cytogenetic and clinical terms. Of these, there are 22 familial pericentric inversions (pii) that occur repeatedly in unrelated families. These common pii are given a type nomenclature as a shorthand means of referring to them. The problems of the classification and clinical significance of pii in the heterochromatic regions of chromosomes 1, 3, and 9 are delineated, and the required investigations are pointed out. The significance of the individual pii is evaluated with regard to their behavior in meiosis and their phenotypic relevance. An approximate 1:1 segregation is found. Fertility, stillbirth, and rates of abortion are not statistically altered. Gonadal findings are reported. The occurrence of aneusomic recombinants among the live offspring of carriers shows a marked, direct dependence on the length of the relative inversion segments. Since these are usually small in pii types, they result in recombinants only in exceptional cases. Other chromosomal alterations are not increased in families with pii. A correlation between pii and clinical abnormalities likewise cannot be proved.

INTRODUCTION

The application of modern banding techniques has shown that pericentric inversions (pii), with a frequency in clinical material of 1–2 % [83,189], are among the most frequent chromosomal rearrangements in man. They merit attention because of their importance, for clinical genetics as well as their role in primate evolution [91]. There are several problems with regard to the clinical significance of these rearrangements for genetic counseling and prenatal diagnosis; e.g., it is uncertain whether 1) pii with apparently identical breakpoints are genetically identical, 2) all heterozygotes of identical pii have the same risk of unbalanced recombinants, 3) an interchromosomal effect is induced by pii, and 4) pii, especially de novo rearrangements, are

Institut für Anthropologie und Humangenetik, Abteilung Klinische Genetik der Universität, D-7400 Tübingen, Federal Republic of Germany causally associated with phenotypic abnormalities. All these questions force us to pursue the clinical relevance of each inversion to decide whether it is to be assigned as a minor variant to chromosomal polymorphisms or as a major variant to the structural rearrangements with phenotypic or gametogenetic relevance or whether it requires further clarification as an unsolved variant.

In this analysis, data from 352 inversions (inv) reported in the literature and from our own family investigations are presented. The sole criterion of literature selection was a definite band representation permitting a precise identification of the chromosome number and the breakpoints (bp; inv breakpoints, ip). Both reports of isolated and familial cases are included in the study. The published data were frequently inadequate to classify the pii, especially when the report did not clearly involve a recombinant or where familial data were not recorded. Therefore, in the following reflections, an exact statistical analysis was not possible for many points.

CYTOGENETICS

The cytogenetic definition of an inv requires two breaks, the rotation of the intercalary segment by 180° around a transverse axis and the reincorporation of this region into the same chromosome [314,363]. They are thus structural rearrangements of an intrachromosomal type. Inv in combination with other rearrangements of the same chromosome are designated as complex inv. When they are present as the sole structural alteration, they are designated as single inv. In contrast to the first group, single inv are found relatively frequently in screening even clinically normal persons.

However, it is still difficult to determine their frequency in specific populations. This is due on the one hand to difficulties of collating data from reports utilizing dissimilar staining methods, when the identity of pii are in doubt, and on the other to different interpretations of some findings on certain chromosomes. Thus the frequency data on pii in heterochromatic regions of chromosome 1 are variously reported as 0.01% [104] to 3.2% [263], with a mean value of 0.3. To an even greater extent, this difference in interpretation applies to the reports on chromosome 9. The data vary from 0.2% [19] to 22.5% [263]. The frequency of the inv of the Y chromosome is specified fairly consistently as 0.1–0.2% among male neonates [126,138]. Similarly, the frequencies of inv of the remaining chromosomes in normal subjects and in material sent to the laboratory are reported as between 0.02% and 0.06%, with a mean value of 0.05% [189]. In the material sent to our laboratory, we found pii of all chromosomes in 1.35%. This agrees almost exactly with the values of de La Chapelle et al. [83] of 9/600–700, ~1.38%, found evidently under the same conditions and also generalized for population studies.

Because of the very different possible consequences in meiosis, inv must be strictly separated conceptually into paracentric inversions (pai), with both breaks in the same arm, and pii, with one break in the short and one in the long arm (Fig. 1a,b). The former can be recognized only with banding techniques because of the lack of an alteration in form (arm ratio), and they are probably overlooked in some cases, especially in screening investigations. On the other hand, pii can lead to an appreciable alteration of form because of different distances of breakpoints from the centromere (Fig. 2).

According to recent biochemical concepts, a DNA sequence can be incorporated into another polynucleotide only when its polarity is maintained. To ensure this in a double helix, the coding and complementary strand must be exchanged in the reincorporation of the inv segment (Fig. 3). Transcription and reduplication at the inverted segment are conceivable when an entire replicon or an entire multiple replicon is within the inv and there is a firm correlation between these points and the fixed points at the nuclear membrane in the interphase nucleus; i.e., genetically functional inv would have to have a certain predilection for breakpoint sites. There is no more precise knowledge in regard to the mechanism of origin of inv than there is for other chromosomal rearrangements [103,312,313,356]. Their formation in the meiotic stages is most likely; however, observations of mosaicism in C-band

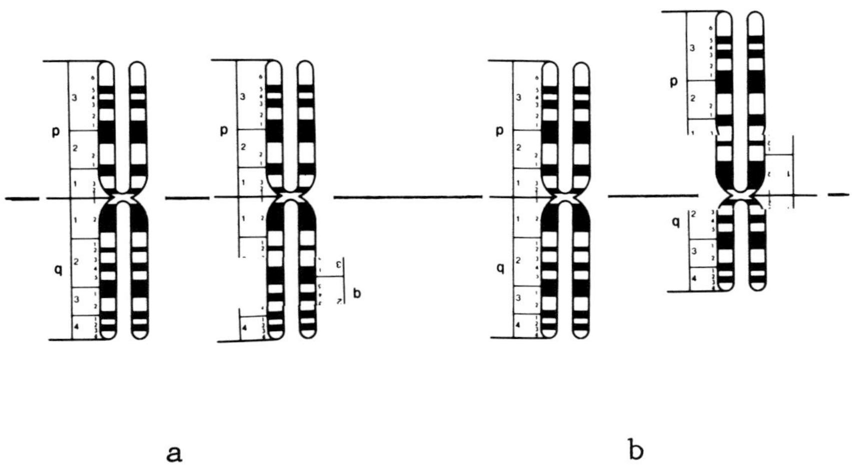

Fig. 1. Schematic presentation of inversions. **a:** Paracentric. **b:** Pericentric.

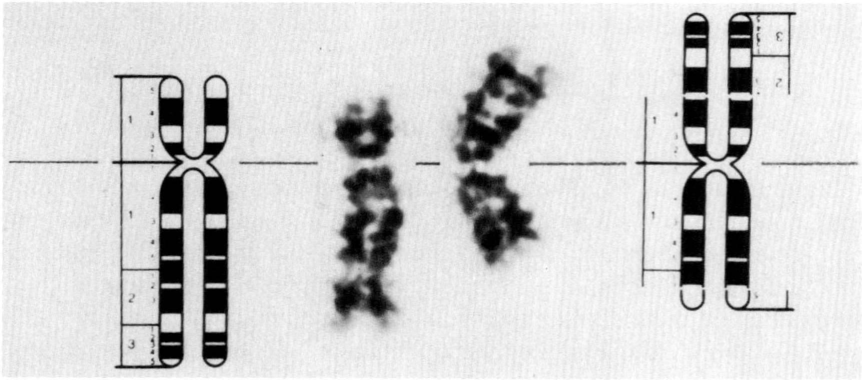

Fig. 2. Example of the change in form in a pericentric inversion. **Left:** Normal chromosome 5. **Right:** Inv(5)(p15q22); GTG bands.

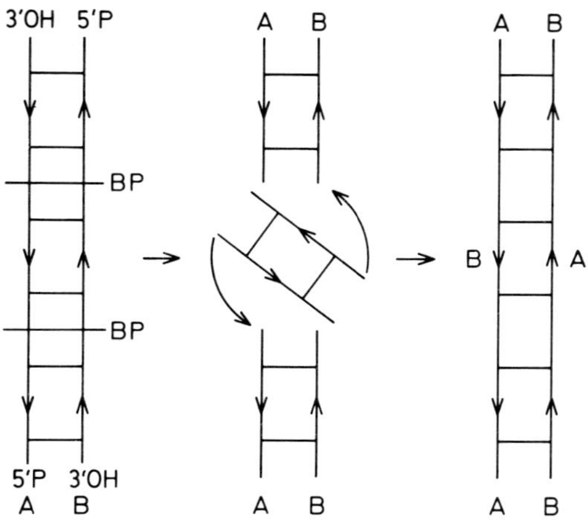

Fig. 3. Schematic presentation of an inversion on the double-stranded DNA to clarify the importance of polarity. BP, breakpoint.

polymorphisms [72] and the fact of somatic crossing-over [378] makes a postzygotic genesis of inv appear possible [129,267].

Inv are uniformly designated according to ISCN: a pii of chromosome 9 with a break in region 1, band 1, of the short arm (p) and a break in region 1, band 3, of the long arm (q) is stated as follows: inv(9) (pter→ p11::q13→p11::q13→qter), or in the abbreviated form with the breakpoints specified in parentheses: inv(9)(p11/q13).

Inv of Heterochromatic Regions

The most frequent pii observed in man is that of the heterochromatic region of chromosome 9. With conventional C staining, there are three qualitative forms: 1) Less than one-third of the total heterochromatic block is sited in the short arm, as is observed in 47% of chromosomes 9 [263]; 2) one third of the heterochromatin is located in the short arm, the "partial" inv or type II inv which occur in 10% of chromosomes; 3) the total block of heterochromatin is located in the short arm, the "total" inv or type I inv. Although this latter type is found less frequently [263] in chromosome polymorphism studies (i.e., in 0.6%), it constitutes the largest group of inv in individual publications. (The type designation does not refer to ascertainment and is only a shorthand means of referring to a common inv of certain breakpoints. For consistency in referring to inv of heterochromatin, partial inv have been referred to as type II see below.) Boué et al. [38] pointed out that a similar cytogenetic picture appears in various reports of pii 9 type I even though the breaks are recorded as occurring in different bands: p11q13 (segment size 20%) and p13q21 (segment size 37%).Those authors who have reported only the region p1 but no band as the location of the break in the short arm [83,295] obviously had the same difficulties of interpretation. This latter breakpoint localization measures as an average segment size of about 28%. These different interpretations are important for two reason. 1) Whereas in one proposal only heterochromatic material is included, the breakpoints of the other interpretations are outside this region, i.e., within the euchromatin or, most importantly, at the heterochromatic border. Heterochromatization might be possible [68]. 2) The various sizes of the segments (20%, 28%, 37% in relation to the length of chromosome 9) might lead to a different behavior of the inv heterozygotes in meiosis (see section on Recombination Aneusomy below).

In chromosome 1, the displacement of the entire heterochromatic region into the short arm (breakpoint p13q21) has been described quite often. In an analogy to inv(9), we designated it as type I. The cases of this inv tabulated here (see Appendix) are associated with pathologic manifestations in the respective propositi but not of a uniform kind. Information about other disease in these families or reports on recombinations are not available. This inv type has been found very infrequently in screening studies [47,104,227, 263,386], so it is not possible to evaluate its biological significance at present. Further investigations are necessary also on the structure of the heterochromatin. In contrast, the displacement of only a part of the heterochromatic block into the short arm of chromosome 1, designated by us as type II, is a frequent finding in polymorphism studies. Atkin and Baker [7] found this latter type more frequently in combination with carcinoma, but that has not been independently confirmed.

In chromosome 3, there is, apart from an inv type with a breakpoint in p25 and q21, a finding frequently described in screening in which the strongly fluorescing band of the proximal long arm is displaced to the short arm. Despite all difficulties, it should be designated as inv type II in an analogy to similar findings in chromosomes 1 and 9 [189,350,351].

The question arises of what extent these inv types with their high frequency are to be regarded as polymorphisms: Resulting aneusomic recombinants have not been described. Other chromosomal aberrations apart from those in the propositi (for review, see [189] have been described in only two cases [108,291]. Four trisomy 9 mosaics (out of 11 found in the literature), however, have been observed in connection with pii 9, two in a type II [40,127] and one in a type I [330].

New light might be shed on this problem by the work of Gosden et al. [143] and Mattei et al. [240]. According to these authors, the heterochromatic segment of chromosome 9 consists of two biochemically different subunits: the heterochromatin of the centromere (cen HC) itself, which stains only with C banding, and the heterochromatin of the secondary constriction (sc HC), which can be stained both with C-banding techniques and with G11 and DA/DAPI. Cen HC is normally localized as a narrow block only in the centromere or as a large bulk in the short arm, in the long arm, or in both arms. Sc HC is normally localized in the long arm or, as the result of an inv, in the short arm. Gosden et al [143] and Mattei et al [240] suspect that partial inversions of sc HC are extremely rare. It is important that screening results of pii 9 are qualified in the future with such techniques. Different clinical significance may be revealed. Until this clarification, the type classification mentioned above should be retained.

Preferred Chromosomes and Breakpoints; Inv Types

Pii have been described in every chromosome (for details see, Appendix and Table I). However, it is evident that there are both preferred chromosomes and preferred breakpoints of certain chromosomes as well as problematic cytogenic findings that need further clarification. Preferred breakpoints have been observed on chromosomes 1–5, 8–3, 18, X, and Y; at these chromosomes, identical inv have been found in several unrelated families. In Table I, inv types are designated if they have been described in three or more ndependent cases or if they have previously received a type designation from other authors [83,154]. The type designation is indicated by a Roman numeral after the inversion symbol. They are listed singly in Table I, which contains details.

TABLE I. Preferred chromosomes and breakpoints in pii

		Identical findings				
Chromosome No.	No. of all inversions observed	No.	Breakpoint in p	Breakpoint in q	Length of inversion segment in relation to chromosome length (%)	Inversion type
1	17	7	13 (11)	21	21 (15)	I
			11	12		II
2	48	27	11	13	11	I
		8	12	14	17	II
3	21	7	25	21	61	I
		3	13	12	18	II
4	12	3	14	35	81	I
5	16	4	13	13	30	I
6	11	Only two identical inv preferred bp:p25				
7	7	No identical inv				
8	21	4	23	11	33	I
		10	23	22	74	II
9	88	68	11	13	20	I[a]
			13	21	37	
			1	13	28	
		8	11	12	14	II
		3	13	13	30	III
10	16	5	11	21	21	I
		1	11	11	7	II[b]
11	14	3	15	12	37	I
12	5	2	13	11	17	I
13	11	3	13	21	60	I
14	4	No identical inv				
15	2	No identical inv				
16	1	No identical inv				
17	2	No identical inv				
18	12	8	11	21	62	I
19	6	No exact comparison possible				
20	1	No identical inv				
21	1	No identical inv				
22	6	No exact comparison possible				
X	11	5	22	13	47	I
		2	11	28	64	II
Y	19	8	11	11	33	I

[a]According to Hansmann [154] and Boué et al. [38].
[b]According to de la Chapelle et al. [83].

The length of the inverted segment in relation to the whole chromosome is on average less than one-third (32%) in the different inv types. This means that the inv segments of the types are smaller than the average of segment sizes of all inv taken together (38.4%) and especially of those that are unclassified as to type (i.e., of quasi unique breakpoints (47%). Almost all breakpoints of the inv types are in G-negative bands.

Assuming that the reasons for publication have been the same for all authors, a cautious evaluation with respect to the distribution of autosomal inv in the genome can be made (Table II): Chromosome 9, with 27.33%, is the most frequent. The figures of the Danish Central Register are even higher (45.7%) [272]. The next most frequent pii chromosomes are 2 (14.9%), 8 (6.5%), 3 (6.5%), 1 (5.28%), and 5 (4.97%). The relative frequencies, if taken as a percentage of the observed and the expected statistics with regard

TABLE II. Different distributions of pii and autosomal chromosomes in the genome*

Chromosome	Observed N	Observed Percent	Expected (%)	Observed/ expected
1	17	5.28	8.44	0.63
2	48	14.91	8.02	1.86
3	21	6.52	6.83	0.95
4	12	3.73	6.30	0.59
5	16	4.97	6.08	0.82
6	11	3.42	5.90	0.58
7	7	2.17	5.36	0.41
8	21	6.52	4.93	1.32
9	88	27.33	4.8	5.69
10	16	4.97	4.59	1.08
11	14	4.35	4.61	0.94
12	5	1.55	4.66	0.33
13	11	3.42	3.74	0.91
14	4	1.24	3.56	0.35
15	2	0.62	3.46	0.18
16	1	0.31	3.36	0.09
17	2	0.62	3.25	0.19
18	12	3.73	2.93	1.27
19	6	1.86	2.67	0.7
20	1	0.31	2.56	0.12
21	1	0.31	1.9	0.16
22	6	1.86	2.04	0.91
Total	322	100	100	

*From our own observations and published cases with given breakpoints but without polymorphism studies.

to the length of the single autosomes (according to Table 5D of the Paris Conference [287]), follow a sequence of chromosomes 9 (observed to expected: 5.69), 2 (1.86%), 8 (1.32%), 18 (1.27%); i.e., these are in part overrepresented. The following chromosomes with less than 0.5 are markedly underrepresented: 7, 12, 14–17, and 19–21. A nonrandom distribution has been found in other chromosomal rearrangements too [182,205,266] and was interpreted as common ancestry of carriers of the same aberration, increased vulnerability of specific regions, differences in the repair mechanisms, or different status of the inv forms leading to a higher frequency of specific types. In a fashion similar to pii Robertsonian translocations (rob) recur in a number of types, but this is due to the involvement of the ribosomal cistrons. However, rcp, which at first glance are a closer analogy to pii, have very few similar forms, and there may be a specific mechanism for the types in pii.

It is important to point out a few special features. There may be a different geographic frequency in inv(3)II. Soudek and Sroka [351] observed it in 4% in Canada; Mikelsaar et al. [250] observed it in 2–6% in Estonia, and Müller et al. [263] observed it in 1.2% in Bronx, New York. In their polymorphism studies, Buckton et al. [47] also found significant differences in the frequencies of inv(9) between neonates in Edingburgh on the one hand and 14-year-old children and inhabitants over 65 years old from an island of the Outer Hebrides on the other hand. As an explanation for this difference in Q-band polymorphisms, they regarded an age effect as being more likely than a population effect. Otherwise, Lubs and Ruddle [226], Walzer and Gerald [400], and Lubs et al. [227] had observed pii(9) up to 20 times more frequently in blacks than in whites. This difference is not found in pii(1) [227]. Hypothetical sex differences such as those claimed by Zankl and Zang [420] and Mikelsaar et al. [249] for chromosomal polymorphisms, are unknown for pii.

Familial Occurrence and New Mutations

Family investigations have been reported in 272 cases of this compilation (Table III). In 249, the inv were found to be familial in at least one case involving each chromosome with the exception of chromosome 17 (inadequate data). It is remarkable that each inv type was also observed to be familial.

New mutations were referred to in 23 cases (8.4%). They were observed in chromosomes 1–9, 11–14, 18, X, and Y. In only five of them are the breakpoints identical with those of corresponding types: inv(13)I [148], inv(18)I [189,395], inv(X)II [106], inv(Y)I [302]. In six, the same break-

TABLE III. Cases including information about familial occurrence or de novo mutations of pii

		Familial inv					New mutations						
												Proof of paternity	
Chromosome No.	No. of all inv observed	Inversion points p	q	Inv segment in relation to chromosome length %	Inv types	No. (=families)	Inversion points p	q	Inv segment in relation to chromosome length %	Inv types	No. (=families)	Cytogenetic	Serologic
1	17	13	21	21	I	4							
		13	23	26		2							
		13	25	34		1	13	25	34		1	+	–
		32	42	74		1							
							33	32	73		1	+	–
		34	23	60		1							
		35	21.3	55		1							
		36	32	80		1							
2	48	11	13	11	I	19							
		13	11	13		1							
		11	14	14		1							
		13	21	15		3	13	21	15		1	+	–
		12	14	17		1							
		15	21	31		1							
		12	36	64		1							
		21	11	30		1							
		11	11	8		1							
		12	14	17	II	1							
		23	11	32		1	?	?	?		1	+	–
3	21	12	11	11		1							
		13	12	19	II	1							
		12	24	37		1							
		25	21	61	I	6							
		27	21	66		1							

		25	25	73		1						
							25	11	42	1	+	+
		14	24	44		1						
		25	23	67		1						
		11	21	22		1						
4	12	14	25	38		1						
		13	35	75		1						
		14	35	81	I	3						
							16	24	58	1	+	+
		15	11	20	I	3						
		15	12	24		1						
5	16	15	11	27		2						
		13	13	30	I	1						
		14	14	43		1						
		14	15	49		1						
		15	22	68		1						
		13	33	73		1						
		13	35	79		2						
		15	31	82		1	15	31	82	1	+	–
		15	33	78		2						
6	11						12	15	26	1	+	–
		12	16	31		1						
		21	27	74		2						
		23	23	77		1						
							25	21	68	1	+	–
		25	22	79		1						
		25	25	89		1						
7	7	22	32	83		1						
							15	11	29	1	+	–
							15	22	51	1	+	–
8	21	23	11	31	1	4						
		11	21	37		1						

(continued)

TABLE III. Cases including information about familial occurrence or de novo mutations of pii (continued)

Chromosome No.	No. of all inv observed	Familial inv: Inversion points p	Familial inv: Inversion points q	Familial inv: Inv segment in relation to chromosome length %	Familial inv: Inv types	Familial inv: No. (=families)	New mutations: Inversion points p	New mutations: Inversion points q	New mutations: Inv segment in relation to chromosome length %	New mutations: Inv types	New mutations: No. (=families)	Proof of paternity: Cytogenetic	Proof of paternity: Serologic
		11	22	47		1							
		11	24	65		1							
		23	22	74	II	9							
							23	23	84		1	+	–
		23	12	41		1							
		12	13	25		1							
		23	24	88		1							
9	88	11	12	14	II	5							
		12	12	19		1							
		11	13	20	I	46							
		12	13			3							
		13	13	30	III	3							
		12	21	35		1							
		22	32	74		1							
							24	12	44		1	+	+
		13	11	16		1							
10	16	11	11	7	II	1							
		11	21	21	I	4							
		11	25	57		1							
		11	26	67		1							
		15	24	76		2							
		?	?	?		1							
		15	11	30		1							
		12	21	27		1							

11	14	11	11	5		1							
		11	13	13		1							
		15	13	45		2							
		11	25	63		1							
		15	23	86		1							
		?	?	?		1							
							?	?		?	1	+	–
		12	25	67		1							
		15	14	60		1							
		15	12	37	I	2							
		11	12	7		1							
12	5	13	11	17	I	2							
		11	13	20		1	11	13	20		1	+	–
		13	13	47		1							
13	11	11	14	26		1							
		11	21	46		2							
		11	22	54		1							
		13	21	60	I	2	13	21	60	I	1	+	–
		12	13	25		1							
		13	22	71		1							
14	4	11	23	42		1							
		1	24	67		1	1	24	67		1	+	–
		12	24	65		1							
		11	21	20		1							
15	2	12	22	73		1							
		13	13	26		1							
16	1	11	12	26		1							
17	2												
18	12	11	12	37		1							
		11	21	62	I	6	11	21	62	I	2	+	–

(continued)

TABLE III. Cases including information about familial occurrence or de novo mutations of pii (continued)

Chromosome No.	No. of all inv observed	Familial inv: Inversion points p	Familial inv: Inversion points q	Familial inv: Inv segment in relation to chromosome length %	Familial inv: Inv types	Familial inv: No. (=families)	New mutations: Inversion points p	New mutations: Inversion points q	New mutations: Inv segment in relation to chromosome length %	New mutations: Inv types	New mutations: No. (=families)	Proof of paternity: Cytogenetic	Proof of paternity: Serologic
		?	?	?		1							
							?	?	?		1	+	–
		11	11	30		1							
19	6	13	13	50?		5							
20	1	13	11			1							
21	1	12	22	70		1							
22	6	12	12	40		1							
		13	12	62		2							
		?	?	?		1							
		11	12	39		2							
X	11	22	11	35		1							
		22	13	47	I	5							
		11	28	64	II	1	11	28	64	II	1	+	–
		22	24	70		1	22	24	70		1	+	–
Y	19	11	11	33	I	6	11	11	33	I	1	+	–
		?	?	?		7							

points were found only in one other observation. The remaining 12 de novo inv are single findings. The inconspicuous transmission of a de novo inv to the next generation was observed three times: inv(2)(p13q21) [165], inv(14)(p1q24) [384], and inv(18) [406]. In two further cases, the de novo balanced inv was not transmitted to the next generation, but two recombinants occurred in one family [395] and two severely malformed children not investigated cytogenetically in the other family [189].

In seven of the 23 cases observed with new mutations, the person in whom the inv had appeared for the first time had clinical anomalies for which there was no other explanation. Significantly, in six of them, the inv forms were single observations: inv(3)(p25q11) [51], inv(6)(p12q15) [333], inv(6)(p25q21) [89], inv(7)(p15q11) [90], inv(8)(p23q23) [196], inv(9)(p24q12) [155]. These figures, however, do not allow a final evaluation of the clinical relevance of de novo inv: In the asymptomatic cases, there is no reason for a genetic investigation, and, in routine investigations, they can be fully detected only with banding techniques. Even if these are very rare events, it can be assumed that they were published mainly because of the abnormal phenotype. However, with regard to the evaluation of such findings in amniotic fluid, it is exceedingly important to address the question of clinical relevance. The data presented here must be given serious consideration until they are definitely disproved by further investigations.

MEIOTIC BEHAVIOR

Pairing at zygotene varies in inv heterozygotes. Depending on the length of the inv segment in relation to the corresponding whole chromosome, there are three possibilities (Fig. 4): 1) formation of a retrograde loop in the normal chromosome and apposition of the inv segment, 2) absent pairing in the region of the inv segment [364,409] if it is extremely small, and 3) absent pairing of the noninverted segments in extremely large inv [93]. With the exception of the second possibility, one crossover or an uneven number of crossover events in pachytene within the inv segment between two nonsister chromatids leads to two genetically unbalanced chromatids: In pai (Fig. 5), the result is both a dicentric and an acentric; in pii (Fig. 6), two even different but monocentric reciprocal duplication-deficiency chromatids are formed. Offspring with either of these duplication deficiencies are aneusomic because of recombination of a balanced chromosome rearrangement. One therefore terms them recombination aneusomies or simply recombinants. The two different duplication-deficient chromosomes are designated depending on the duplication present as short-arm recombinant [rec dup(p)] or long-arm recombinant [rec dup(q)].

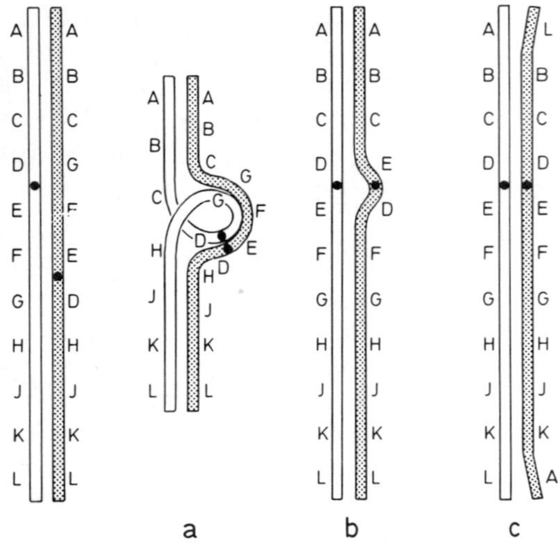

Fig. 4. Simplified scheme to demonstrate the various pairing possibilities of the inversion heterozygotes in the zygotene of prophase I of meiosis.

Recombination Aneusomy

In the present material, there are 83 recombinant offspring resulting from pii (Table IV). These concerned chromosomes 3–11, 13–15, 18, 20–22, and X. Sixty-five cases involved long-arm recombinants and only 18 involved short-arm recombinants. This corresponds to a ratio of 3.6:1. Both recombinants are extremely rare as the result of the same inv form: inv(5)(p13q35), inv(13)(p11q22), and inv(18)(p11q21). In 36 of the recombinant offspring, the fathers were the inv carriers, in 46 of them the mothers. That suggests the conclusion of Sutherland et al. [369] that female carriers have a higher risk of recombinant offspring. Only in one case did the recombinant itself constitute a de novo mutation, inv(13) [148]. In one further family, the mutation of the inv was probable in the grandparents' generation, inv(14) [84]. In 42 families, the inverted chromosome could also be demonstrated in the grandparents' generation; corresponding information is lacking in the remaining 39 publications.

The average relative segment size of all of the inv with recombinants (i.e., 64.8%) is far in excess of the average size of all inv (38.4%). Only four exceptions were observed: inv(3)(p15q11), 23%, inv(8)(p12q13), 25%; inv(13)(p11q14), 26%, and inv(13)(p12q13), 25%. Similarly, recombinants resulting from inv *types* (see above) are usually found in those with inv

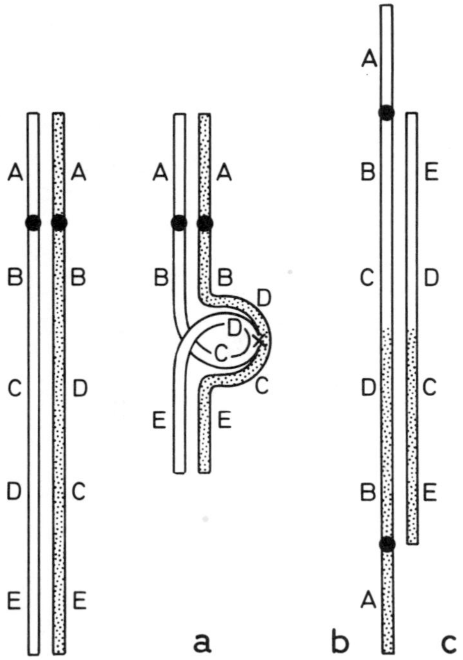

Fig. 5. Schematic presentation of the pairing of an inversion heterozygote in a paracentric inversion with a crossing-over within the inversion segment **(a)**, dicentric recombinant **(b)**, and acentric recombinant **(c)**, each with different duplication deficiencies.

segments larger than one-third. There are again a few exceptions: inv(3)I, inv(4)I, inv(8)II, inv(13)I, inv(18)I, and inv(X)I. In this material, no differences are found in regard to the relative lengths of the inv segments leading to rec dup(q) and rec dup(p).

The correlation between the size of an inv and the occurrence of recombinants among offspring has been variously observed [189,192,384,409]. However, this is not a linear correlation with the absolute size alone but a complex process that depends on the probability with which a crossing-over occurs within a certain chromosome segment and on the size of the genetic imbalance of a recombinant of a certain inv. If one assumes the same reduction of the crossing-over in the heterochromatin [231,301,318] and the same frequency in the euchromatin [179], then there is an absolute segment size at which the probability of *one* crossover is greatest. Smaller segments have no crossovers, and larger segments have two crossovers with the possibility of double crossover between the same chromatids, i.e., without

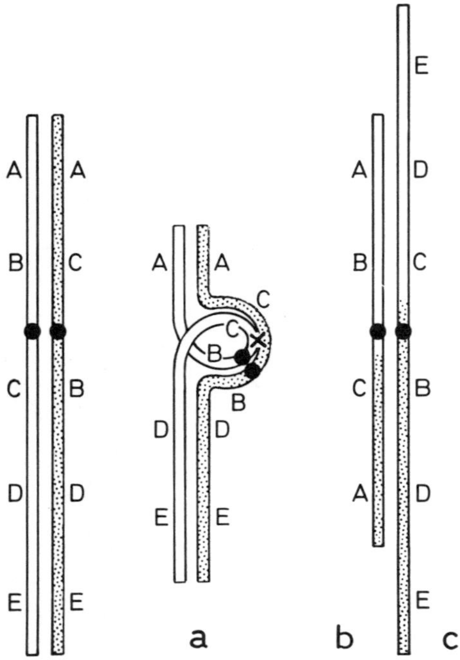

Fig. 6. Schematic presentation of the meiotic pairing of an inversion heterozygote in a pericentric inversion with a crossing-over within the inverted segment **(a)**. The two resulting recombinants are also shown: rec dup(p)del(q) **(b)** and rec dup(q)del(p) **(c)**.

pathologic recombinants too. Even larger segments would in turn have three crossovers and thus aneusomic recombinants too. If one assumes 50 chiasmata per cell [114], a chromosome of medium length has two crossovers, an inv segment thus probably has zero or one crossover; i.e., the larger an inv segment, the greater the probability of a crossover, with formation of a recombinant. However, the extent of the genetic imbalance of these recombinants depends on the relative size of the inv segment. With increasing size, the imbalance decreases (Fig. 7). This means that the duplication deficiencies of relatively large inv lead to relatively small imbalances, which are compatible with life. The recombinant chromosomes greatly resemble the initial chromosome in their form. Medium-sized and smaller inv lead to recombinants with extensive imbalance, which are more incompatible with life and frequently lead to gametic- or embryonic-lethal factors. Moreover, in these inv- the position of the centromere plays a crucial role. The percentages of the two kinds of aberrations (duplication and deficiency) in the two recom-

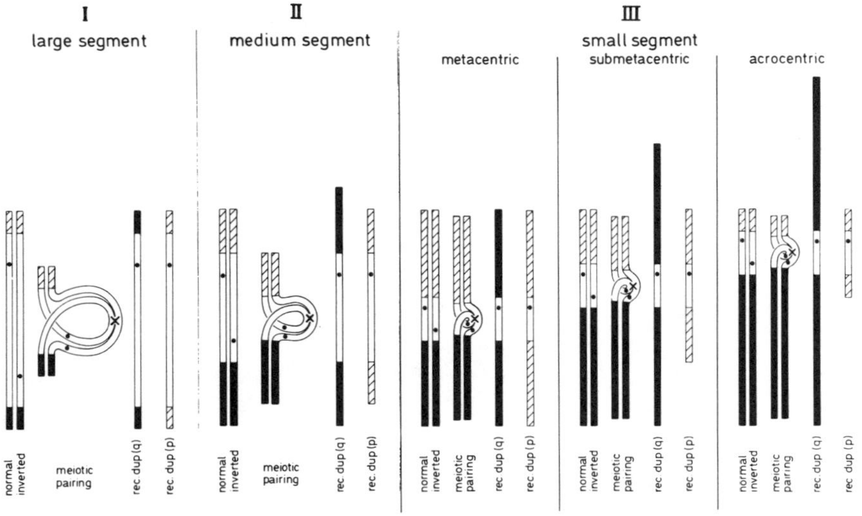

Fig. 7. Schematic presentation of pericentric inversions in relation to inversion sizes and centromere position. Meiotic pairing with one crossing-over within the inversion segment, both resulting recombinants. **I:** A large inversion segment. Both recombinants with relatively small duplications and deficiencies. **II:** A medium inversion segment. Both recombinants with larger duplications and deficiencies. **III:** A small segment. Both recombinants with even larger duplications and deficiencies, which are also different because of the position of the centromere: metacentric, submetacentric, acrocentric.

binants are altered as the inv segment deviates from the middle of the chromosome (Fig. 7). These considerations also explain the preponderance of long-arm recombinants in our results: It is known with regard to translocations that partial trisomies are more readily compatible with life than partial monosomies [328]. Since, in a not absolutely metacentrically situated inv, each rec dup(q) leads to smaller partial monosomies (and larger partial trisomies), the former are also more frequent among live births.

For the evaluation of the genetic risk, an observation of Sutherland et al. [369] is also important: Carriers whose inv was discovered via a recombinant have risks of 5% and 10% (males and females, respectively) for recombinant offspring, whereas carriers whose inv was ascertained through an inv heterozygote do not appear to have any risk. The European Collaborative Study in Prenatal Diagnosis for Chromosomal disorders (1981) confirms this trend but reports risk figures of 50% for the former and 0.8% for the latter [Boué in 16]. Daniel [80] has combined both methods. However, he measured not the inverted segments but the two terminal ("potential duplication-deletion seg-

TABLE IV. Pii with observed recombinants in offspring

Chromosome	Recombinants			Breakpoints		Segment		Carrier		De novo mutation in the patients		
No.	No.	dup(q)	dup(p)	p	q	size (%)	Type	Father	Mother	Yes	No	?
1	—											
2	—											
3	13	9		25	21	61	I	5	4		6	3
		1		15(!)	11	23			1			1
		1		25	25	73			1			1
		2		25	23	67		1	1		2	
4	5	1		14	25	38		1			1	
			1	13	35	75		1			1	
			3	14	35	81	I	1	2		3	
5	8	1		13	33	73			1		1	
		2	2	13	35	79		2	2		2	2
		3		15	33	78		3				3
6	4	1		25	22	79			1			1
		1		23	23	77		1			1	
		1		25	25	89			1		1	
			1	21	27	74		1			1	
7	2	2		22	32	83		1	1			2
8	13	11		23	22	74	II	7	4		3	8
		1		12	13	25		1				1
		1		23	24	88			1		1	
9	1	1		22	32	74						
10	5	2		15	24	76		1	1		1	1
			1	11	25	57			1		1	
			1	11	26	67			1			1
			1	?	?	?					1	
11	1		1	11	25	63		1				1

12	—											
13	13	3		13	21	60	I		2[a]			1
		2		11	31	66			2		2	
		1		11	14	26			1		1	
		1		12	22	66			1			1
		2			11	21			1			2
		1	1	11	22	54		2			2	1
		1		12	13	25			1			
		1		13	22	71			1			1
14	3	1		1	24	67				1		
		1		12	24	65			1			1
15	1	11		11	23	42			1			1
16	—											
17	—											
18	6	2	2	11	21	62	I		4		1	3
			1	11	12	37		1				
			1	?	?	?		1				1
19	—											
20	1	1		13	11	66		1				
21	1	1		12	22	70			1		1	
22	2	1		13	12	62						
			1	11	13	39			1			1
X	4	2		22	13	47	I		2		1	1
		1		22	11	35			1			1
			1	22	24	70			1			1
Y	—										1	
Total	83	65	18			64.8		36		46[a]	42	39

[a]In one case, there were de novo inv and recombination in the propositus.

ments") in relation to the haploid autosomal length. They were, according to our results, significantly shorter in the inv found via recombinants.

Gonadal Findings in Human pii Heterozygotes

The processes of meiotic behavior of inv heterozygotes are well documented in animals and plants [42,308,391]. In humans, on the other hand, there are only very isolated histologic or cytogenetic gonadal studies, most often carried out only because of the coincidental occurrence of inv with other indications that require a biopsy. For this reason, the information available from these results is limited, especially with regard to the findings of azoo- or oligospermia [247,316,374,380]. Tubules lined only by Sertoli cells [316,380] and bilateral atresia of the vas deferens [296] have been found as well as normal configurations [83,107,390]. In cytogenetic investigations, McIlree [244] found in a propositus with inv(Y) no peculiarities in pairing of X and Y. Hauksdottir et al. [157] analyzed 45 cells at diakinesis in a phenotypically normal carrier of inv(13) without finding pathologic bivalents. de La Chapelle et al. [83] were able to carry out cytogenetic and histologic investigations on two male propositi with inv(9)(p1q13). The findings were normal. van der Linden et al. [390] concluded that one or two crossovers within the loop of an inv(4)(p14q25) occurred in 52% of cells, which leads us to expect one of the two duplication deficiencies in about 26% of sperm. Page and Ferguson-Smith [cited in 409] postulated in their findings a normal pairing of inv(9) heterozygotes with the exception of the inv segment. Winsor et al. [409] could find no evidence of loop formation in a patient with inv(7)(p22q32), 83%, from which they concluded that there must be normal pairing of extremely large inv segments with the exception of the terminal regions. Faed et al. [109] found normal chiasma counts and no abnormalities of pairing in an inv(3) heterozygote and in two with inv(9). Histologic or cytogenetic studies on ovaries have not been found in the literature.

Fertility

Fertility could give us an indirect indication of the meiotic behavior of inv heterozygotes, but it is difficult to arrive at definite values in reviewing the literature. Morton et al. [259] and Jacobs et al. [183] calculated fertility in persons with structural chromosomal aberrations and found a slight but insignificant reduction of the reproductive fitness in male and in female carriers in families among whom the inv was ascertained through a euploid propositus. Liveborn children who reached their 18th year were found significantly less often for male carriers.

We attempted to compile a list of families in the literature and found 1,094 pregnancies in 349 carriers of pii (not counting index persons), that is, 3.1/

carrier. There were no differences relating to the sex of the carrier. The abortion rate in relation to all pregnancies was 9.1% (without index persons). This is within the range of the average abortion rate in Europe, about 10–35% [392]. A significant variation according to sex was not found here either.

The live birth rate (without recombinants) was 2.6/carrier. The rate of stillbirth (stillbirth in relation to the total stillbirths and live births together) was 2.4%. This is clearly higher than the present stillborn rate of 0.77% (1975), but one must remember that the number of stillbirths has dropped sharply in recent years. In 1952, it was 2.1%, which was in accordance with our average values (Health Service Data, 1977).

Subdividing this material according to individual chromosomes or inv types mostly gives numbers that are too small to be of use with respect to the question of fertility. The supposed decreased fertility in male inv(9) [38,83] cannot be substantiated (for details, see [190]. Ford et al. [121] found a significant difference in rate of inv(9) between women with recurrent miscarriages and a test group. The abortion rate (8.7%) in partners of male carriers in our material is clearly lower than in female carriers (17%) but this difference is only slightly significant. de La Chapelle et al. [83] saw a similar trend in inv(10)II, and this can be seen in all collected cases of inv(10) but has no bearing on the general statistics. Finally, Dutrillaux et al. [92] found five inv in 271 subfertile males out of 57 chromosomal aberrations. This is higher than the frequencies of inv in the population as a whole mentioned above.

Segregation

If one assumes neither decreased fertility nor increased abortion rates in pii carriers, family analyses can give an indication of the segregation rate of the heterozygotic chromosome pair. Analyzing informative families from the literature according to the method used by de la Chapelle et al [83] (for details, see [189]), we found, among all autosomal inv a segregation rate of 0.52, which, with a χ^2 of 0.73, does not differ from mendelian distribution. A significant segregation advantage of 0.63 was found only for the Y chromosome as a result of the preponderance of one very large family investigated by Grace et al. [144].

The other results on the individual pii do not differ from the 1:1 distribution. This also includes pii(9), which has a high prevalence and could be supposed to have segregation advantages. These values are in accordance with the findings of de la Chapelle et al. [83], Madan and Bobrow [231], and Robinson et al. [320]. With respect to the other chromosomes, there are only

TABLE V. Cases with clinical symptoms of unknown origin and pii

Chromosome No.	Cases with symptoms of unknown origin: Case numbers in Appendix	No.	Breakpoints p	q	Inv. type	Inv segment in relation to chromosome length %	Carrier: Father	Mother	No information	De novo inversions: Yes	No information
1	1.7	1	13	21	I	21					+
	1.6	1	13	21	I	21	+				
	1.1	1	?	?		?					+
	1.3	1	13	21	I	21			+		
	1.14	1	13	21	I	21			+		
2	2.6	1	11	13	I	11	+				
	2.18	1	12	14		17					+
	2.20	1	12	14		17					+
	2.23	1	12	14		17		+			
	2.24	1	11	13	I	11	+				
	2.27	1	11	13	I	11					
	2.28	1	11	13	I	11					+
	2.29	1	12	36		64		+			
	2.30	1	13	21		15		+			
	2.34	1	13	11 (Mosaic)		13				+	
	2.36	1	13	21		15		+			
3	3.6	3	13	12	II	19					+
	3.7						+				
	3.8									+	
	3.9	1	12	24		37		+			
	3.10	1	25	11		42				+	
	3.16	1	14	24		44	+				
	3.21	1	11	21		22	+				
4	4.8	1	15	11		20			+		
	4.10	1	15	12		24		+			
5	5.3	3	15	31		82	+ (3x)	One family			
	5.4	1	13	13	I	30					+
	5.10	1	15	22		68	+				
	5.14	1	15	11		27			+		
6	6.4	1	12	15		26				+	
	6.8	1	12	16		[illegible]					

	6.10	1	25	21		68				+	
7	7.1	1	15	11		29				+	
8	8.2	3	23	11	I	33	+				
	8.3						+				
	8.4						+				
	8.10	1	23	23		84				+	
	8.7	1	11	21		37	+				
9	9.7,9.9,9.6, 9.37,9.38, 9.39,9.43, 9.46,9.47 9.48,9.52,9.70 9.75,9.84,9.87	15			I	20–37	+(10×)[a]	+(6×)[a]			
	9.30	1	24	12		44				+	
	9.34	1	12	21		35		+			
	9.61	1	12	12		19					+
10	—										
11	11.9	1	15	14		60	+				
	11.10	1	15	13		45		+			
	11.12	1	15	12	I	37		+			
12	12.5	1	11	13		27				+	
13	13.5	1	11	21		46	+				
14	14.4	1	11	21		26		+			
15	15.2	1	11	13		26	+				
16	16.1	1	11	12		26	+				
				13							
17	—										
18	18.2	2	11	21	I	62		+			
	18.6										
19	19.2	2	13	13		50	+				
	19.3							+			
21	—										
22	—										
X	X.10	1	11	22		34			+		
Y	Y.16	1	11	11	I	33	+				
Total		72				33.5	31[a]	20[a]	6	8	9

[a]Difference from total is caused by two cases with heterozygous father and mother.

a few single investigations [16,83,183,250,320]. They too do not differ. In contrast, among animals, in the onion fly (*Hylemya antiqua*) [391] and in male *Drosophila* [326], a preferred segregation of the inverted chromosome has been observed.

Other Chromosomal Aberrations and pii

A connection between pii and the occurrence of other chromosomal aberrations would be very important for genetic counseling. We have therefore checked the entire material on the basis of two criteria: 1) coincidence of pii and other chromosomal aberrations (double aberrations) and 2) occurrence of other chromosomal aberrations in families with inv (familial heteroaberrations). Additional chromosome aberrations are found together with inv of chromosomes 1–12, 18, 19, 21, 22, and Y. Familial heteroaberrations have been recorded for two trisomies 21, one trisomy 8, one dup(10)(q22→q25), one deletion (Xp) and one i(Xq), five in a family with inv(9) and one in a family with inv(8) (for details see [190].

Conclusions that the coincidence of inv and other chromosomal aberrations are increased cannot be proved in this material; all cases were cytogenetically investigated because of their overt clinical symptoms (ascertainment bias). Nevertheless, one observation in this context may be noteworthy: In the cases of inv(9), four patients were found with a mosaic trisomy 9 and one with a partial trisomy 9. Similarly, three published cases of inv(22) have trisomy 22 or an additional r(22) in three family members. Both trisomies, 9 and 22, are extremely rare, so a nonrandom connection is not impossible [40,81,100,127,235,330,403]. In this connection, however, we have to remember that our present knowledge about the mechanisms of postzygotic nondisjunction is insufficient.

PII AND PHENOTYPE

In the present data, there are 72 patients with inv and an abnormal phenotype, the occurrence of which in several organ systems is suggestive of a chromosomal aberration (Table V). A particular clustering together with one chromosome or one inv type is not found. The average inv segment sizes in these cases (33.5%) are smaller than the average for all inv. Family studies are available in 63 cases. In eight (12.6%) of these, the inv was a de novo rearrangement. This figure is somewhat higher than in the total percentage of all inv in this collection (8.4%). It is important to note that, among the parents of these patients, the fathers were more frequently the carriers, the ratio of paternal to maternal carriers being 31:20.

Two additional comments seem to be necessary in connection with the question of pii and phenotype: 1) The cases of pii(3)II with a pathologic phenotype reported by Soudek et al. [350] are contradicted by the normal phenotype found in polymorphism studies of this inversion by Müller et al. [263] and especially by the findings of Mikelsaar et al. [250]. 2) The above-mentioned cases of trisomy 9/mosaics and trisomy 22 together with inv(9) and inv(22), respectively, should be kept in mind. The apparently balanced inv in some cases might be trisomy mosaics. The many recent observations of apparently balanced translocations in mentally retarded persons also point to such an association [43,134,182,188,375]. It is therefore important to consider theoretically plausible mechanisms. In de novo inv, the following mechanisms are possible: 1) A finding that appears to be an inv under the microscope has in reality an additional microdeletion or microduplication; 2) one of the breaks leading to an inv lies within a cistron, thus disturbing the realization of genetic information; 3) displacement of genetically active DNA to a position close to the heterochromatin results in heterochromatinization and genetic inactivation [59,262,363].

In familial inv, the following mechanisms are possible: 1) A three break rearrangement as described by de Grouchy et al. [84]; 2) unequal crossing over (not to be confused with uneven crossing over) in meiosis, resulting in a microdeletion or microduplication; this event in the area of repetitive DNA has been discussed as a mechanism for formation of qh+ regions [70] and is known as the cause of gene duplication in euchromatin [101,208,399]; 3) undetected trisomy mosaics of the inverted chromosomes.

The following findings may also be significant in this connection. Atkin and Baker [7] found an inv(1)II in lymphocytes of the peripheral blood in 15% of 76 carcinoma patients as compared to a 0.4% in the control group. This result is also confirmed in relation to pii(9) in recent publications [28,341]. Recent clonal findings of inv(7) in ataxia telangiectasia [329], inv(1) in chronic myelocytic leukemia [256], and inv(16) in acute nonlymphocytic leukosis (AMM-M4) [27,168,413,418] are also of interest.

CONSEQUENCES FOR GENETIC COUNSELING AND PRENATAL DIAGNOSIS

The conclusions to be drawn from these results can be summarized as follows. First, to clarify a possible causal connection in the simultaneous presence of a phenotype typical of aneuploidy and a balanced pii, mosaicism should be excluded. This also applies to the specific inversion types. Second, for genetic counseling: 1) Family studies are not necessary in inv(1)II, inv(3)II, inv(9)I or/-II (special staining applied), and inv(Y)I. 2) In all other inv cytogenetic family studies are indicated to detect carriers at an early stage. 3) Amniocentesis should be carried out when one of the partners is an inv heterozygote if he or she is not precluded under 1 above. The likelihood of recombinant offspring rises with the relative increase in size of an inv segment. If an inv diagnosed in amniocytes is a new mutation (with reasonable certainty of paternity), then a situation exists similar to that pertaining to de novo translocations.

REFERENCES

1. Allderdice PW: Identification of the location of a cross-over in a pericentric inversion heterozygote which resulted in a duplication deficient chromosome. Am J Hum Genet 25:11a, 1973.
2. Allderdice PW: A large kindred with segregation of pericentric inversion heterozygotes inv(3)(p25q21) and inv(8)(p23q22). Am J Hum Genet 26:9a, 1974.
3. Allderdice PW: Chromosome 3 duplication q21→pter deletion p25→pter syndrome in children of carriers of pericentric inversion inv(3)(p25q21). Am J Hum Genet 27:699–718, 1975.
4. Andreev A, Kincheva V, Khadzhitodorova E: Pericentric inversion of chromosome 6 in women with spontaneous abortions. Akush Ginekol (Sofiia) 22:14–16, 1983.
5. Andrews T, Gardiner AC, Boon AR: Recombinant chromosome 18 in two offspring of a chromosome 18 inversion heterozygote. Ann Genet 25:185–188, 1982.
6. Arrhigi FE, Hsu TC: Localisation of heterochromatin in human chromosomes. Cytogenetics 10:81–86, 1971.
7. Atkin NB, Baker MC: Pericentric inversion of chromosome 1: Frequency and possible association with cancer. Cytogenet Cell Genet 19:180–184, 1977.
8. Aurias A: Cytogenetic analysis of 21 cases of ataxia teleaniectasia. J Genet Hum 29:235–247, 1981.
9. Aurias A, Dutrillaux B, Buriot D, Lejeune J: High frequencies of inversions and translocations of chromosome 7 and 14 in ataxia telangiectasia. Mutat Res 69:369–374, 1980.
10. Autio-Harmainen H, de la Chapelle A: High resolution of a small pericentric inversion of chromosome 1. J Med Genet 17:44–47, 1980.
11. Ayme S, Mattei JF, Mattei MG, Aurran Y, Giraud F: Nonrandom distribution of chromosome breaks in cultured lymphocytes of normal subjects. Hum Genet 31:161–175, 1976.
12. Ayme S, Mattei MG, Mattei JF, Giraud F: Abnormal childhood phenotypes associated with the same balanced chromosome rearrangements as in the parents. Hum Genet 48:7–12, 1979.
13. Ayral D, Raudrant D, Charleux JP, Noel B: Duplication of the long arm of chromosome 3 (dup 3q) in a newborn infant whose father is carrier of pericentric inversion of chromosome 9. Pediatrie 39:681–690, 1984.
14. Babu KA, Verma RS, Rodriguez J, Rosenfeld W, Jhaveri RC: A possible clinical implication of homozygous inversions of 9qh regions with Cornelia de Lange syndrome (CLS). Hum Hered 35:265–267, 1985.

15. Baccichetti C, Lenzini E, Peserico A, Tenconi R: Study on segregation and risk for abnormal offspring in carriers of pericentric inversion of the (p11-q13) segment of chromosome 2. Clin Genet 18:402–407, 1980.
16. Baccichetti C, Tenconi R, Caufin D, Bortotto L: Study on segregation of the inversion of chromosome 4 (p15.2q11) in two unrelated families. Hum Genet 62:117–120, 1982.
17. Baheux G, Emerit I, Roux Ch: Trisomie 21 et anomalie de structures de l'Y. Ann Genet (Paris) 13:191–194, 1970.
18. Balkan W, Burns K, Martin RH: (1983) Sperm chromosome analysis of a man heterozygous for a pericentric inversion of chromosome 3, Cytogenet Cell Genet 35:295–297, 1983.
19. Ballesta FM, Serra M: Chromosomal polymorphism: Frequency and clinical manifestations. Clin Genet 10:349, 1976.
20. Barbata G, Granata G, Carbone P: Inherited pericentric inversion of human chromosome 5. Hum Genet 65:216, 1983.
21. Barnes IC, Kumar D, Bell RJ: A child with a recombinant of chromosome 8 inherited from her carrier mother. J Med Genet 22:67–70, 1985.
22. Bartsch-Sandhoff M, Terinde R, Wiegelmann W, Scholz W: Karyotyp-Phenotyp-Korrelation bei einem 46,X,del(X)(p22)-Befund. Hum Genet 31:263–270, 1976.
23. Bauld R, Sutherland GR, Bain DA: Chromosome studies in investigation of stillbirths and neonatal deaths. Arch Dis Child 49:782, 1974.
24. Baumann W, Zabel B, Holl M: Familial pericentric inversion of the X chromosome [inv(X)(p11q28)] Ann Genet (Paris) 27:106–108, 1984.
25. Bec P, Arne JL, Bourrouillou G, Colombies P: Rétinoblastome et transmission de remaniements chromosomiques balancés. Mod Probl Ophtalmol 18:160–164, 1977.
26. Beemer FA, de France HF, Rosina-Angelista IJ, Gerards LJ, Cats BP, Guyt R: Familial partial monosomy 5p and trisomy 5q; three cases due to paternal pericentric inversion 5(p151q333). Clin Genet 26:209–215, 1984.
27. Berger R, Bernheim A, Daniel MT, Valensi F, Sigaux F, Flandrin G: Cytogenetic studies on acute myelomonocytic leukaemia (M4) with eosinophilia. Leuk Res 9:279–288, 1985a.
28. Berger R, Bernheim A, Kristoffersson U, Mitelman F, Olson H: C-band heteromorphism in breast cancer patients. Cancer Genet Cytogenet 18:37–42, 1985b.
29. Betz A, Turleau C, de Grouchy J: Hétérozygotie et homozygotie pour une inversion percentrique de 3 humain. Ann Genet (Paris) 17:77–80, 1974.
30. Bhasin MK, Foerster W, Fuhrman W: A cytogenetic study of recurrent abortion. Hum Genet 18:139–148, 1973.
31. Bobrow M, Pearson PL, Pike MC, El-Alfi OS: Length variation in the quinacrine binding segment of human Y chromosomes of different size. Cytogenetics 10:190, 1971.
32. Bocchini V, Armellini R, Mariucci G, Migliorini G, Gambogi G, Stinchi T: Inversione pericentrica del cromosoma 4 in tre successive generazioni. Riv Biol 76:103–105, 1983.
33. Bochkov NP, Kuleshov NP, Chebotarev AN, Alekhin VI, Midian SA: Population cytogenetic investigation of newborns in Moscow. Humangenetik 22:139–152, 1974.
34. Borgaonkar DS: Application of new technics of chromosome identification to cytogenetic problems. Birth Defects IX:171–182, 1973.
35. Boué J, Boué A: L'intérêt en diagnostic prénatal des techniques nouvelles d'identification chromosomique dans des translocations et une aneusomie de recombinaison. Nouv Press Med 2:3097–3102, 1973.

36. Boué JG, Boué A: Chromosomal anomalies in early spontaneous abortion. In Gropp A, Benirschke K (eds): "Current Topics in Pathology, Vol. 62, Developmental Biology and Pathology." Heidelberg: Springer, 1976, p 193.
37. Boué J, Boué A, Lazar Ph: Retrospective and prospective epidemiological studies of 1500 karyotyped spontaneous human abortions. Teratology 12:11–26, 1975a.
38. Boué J, Taillemite L, Hazael-Massieux P, Leonhard C, Boué A: Association of pericentric inversion of chromosome 9 and reproductive failure in ten unrelated families. Humangenetik 30:217–224, 1975b.
39. Boué J, Hazael-Massieux P, Larget-Piet L, Leonard C: Inversions pericentriques et echecs de la reproduction. J Genet Human (Suppl) 23:81,1975c.
40. Bowen P, Ying KL, Chung GSH: Trisomy 9 mosaicism in a newborn infant with multiple malformations. J Pediatr 85:95, 1974.
41. Bozzola M, Larizza D, Lorini R, Cisternino M, Bolis PF, Severi F: Pregnancy in cases of Turner's syndrome. Int J Biol Res Pregnancy 1:133–135, 1980.
42. Brandham PE, Johnson MAT: The meiotic behaviour of inversions in polyploid aloineae. Chromosoma 62:85–91, 1977.
43. Breg WR, Miller DA, Allderdice PW, Miller OJ: Identification of translocation chromosomes by quinacrine fluorescence. Am J Dis Child 123:561, 1972.
44. Bross K, Krone W: On the number of ribosomal RNA genes in man. Hum Genet 14:137–141, 1972.
45. Broughton WL, Rosenbaum KN, Beauchamp GR: Congenital glaucoma and other ocular abnormalities associated with pericentric inversion of chromosome 11. Arch Ophthalmol 101:594–597, 1983.
46. Brown SW: Heterochromatin. Science 151:417–425, 1966.
47. Buckton KE, O'Riordan ML, Jacobs PA, Robinson JA, Hill R, Evans HJ: C- and Q-band polymorphisms in the chromosomes of three human populations. Ann Hum Genet 40:99, 1976.
48. Buckton KE, Newton MS, Collyer S, Lee M, Spowart G: Phenotypically normal individuals with an inversion (X) (p22q13) and the recombinant (X)dup(q). Am Hum Genet 45:159–168, 1981.
49. Bui TH, Sichong Z, Castro I: A familial pericentric inversion of chromosome 8 analysed with a high resolution chromosome banding technique. Clin Genet 21:266–271, 1982.
50. Bullerdiek J, Bartnitzke S, Pahlke R: Perizentrische Inversion des Y-Chromosomes als Ausschlußkonstellation Anthrop Anz 42:67–72, 1984.
51. Butler LJ, Hall ME, Wharton BA: A retarded child with a 46,XX,3p -q+ chromosome karyotype. J Ment Defic Res 18:41, 1974.
52. Cabrol C, Werner-Favre C, Wyss M, Pitmon D, Engel E: Acute lymphoblastic leukemia in two children with a congenital chromosome anomaly: Familial inv(11)(p15q13) in one and ring chromosome No. 21 in the other. Cancer Genet Cytogenet 8:67–74, 1983.
53. Cantu JM, Hernandez A, Vaca G: Trisomy 22q12→qter: Aneusomie de recombinaison of a pericentric inversion. Ann Genet (Paris) 24:37–40, 1981.
54. Carpenter NJ, Say B, Barber ND: A homozygote for pericentric inversion of chromosome 4. J Med Genet 19:469–471, 1982.
55. Carr DH: Chromosomal abnormalities and their relation to disease. Trans Am Assoc Genitourin Surg 54:9–13, 1962.
56. Carr DH, Gedeon M: Population cytogenetics of human abortuses. In Hook EB, Porter IH (eds): "Population Cytogenetics Studies in Humans." New York: Academic Press. 1977, pp 1–9.

57. Carrano AV, Wolff SH: Distribution of sister chromatid exchanges in the euchromatin and heterochromatin of the Indian muntjac. Chromosoma 53:361–369, 1975.
58. Carter CO, Evans KA: Risk of parents who have had one child with Down's syndrome. Lancet 2:785, 1961.
59. Cattanach BM: Position effect variegation in the mouse. Genet Res 23:291–306, 1974.
60. Catti A: Inversions familiales et autres aberrations chromosomiques concomitantes. J Genet Hum 23:92–95, 1975.
61. Catti A, Schmidt W: A pericentric inversion, 5p-q+, and additional complex rearrangements in a case of cri-du-chat syndrome. Cytogenetics 10:50–60,1971.
62. Chen ATL, Frederick RS, Mc Kim JS, Barr ML, Gruber D: Chromosome studies in full-term, low-birth-weight mentally retarded patients. Pediatrics 76:393–398, 1970.
63. Coco R, Penchaszadeh VB: Inherited partial duplication deficiency of chromosome 15(p12; q22). J Genet Hum 26:203–210, 1978.
64. Cohen MM, Capraro VJ, Takagi N: Pericentric inversion in a group D chromosome (13–15) associated with amenorrhea and gonadal dysgenesis. Ann Hum Genet 30:313, 1967.
65. Cohen MM, Rosenmann A, Hacham-Zadeh SH, Dahan S: Dicentric X-isochromosome (Xqi dic) and pericentric inversion of No. 2 [inv(2)(p15q21)] in a patient with gonadal dygenesis. Clin Genet 8:11–17, 1975.
66. Cohen MM, Dagan J, Shahan M, Cohen R: A cytogenetic study of 700 Jerusalem newborn infants. V. International Congress of Human Genetics, Mexico D.F., October 10–15, 1976. In Armendares S, Lisker R (eds): "International Congress Series No. 397," Amsterdam: Excerpta Medica, 1976, p 117.
67. Cook PJL, Robson EB, Buckton KE, Slaughter CA, Gray JE, Blank CE, James FE, Riler AC, Insley J, Hutlén M: Segregation of ABO, AK_1 and $ACON_s$ in families with abnormalities of chromosome 9. Ann Hum Genet 41:365, 1978.
68. Cooper KW: Cytogenetic analysis of major heterochromatic elements (especially Xh and Y) in Drosophila melanogaster and the theory of "heterochromatin." Chromosoma 10:535, 1959.
69. Côté GB, Tsomi K, Papadakon-Lagoyanni S, Petmezaki S: Oligohydramnion syndrome and XYY karyotype. Ann Genet (Paris) 21:226–228.
70. Craig-Holmes AP, Shaw MW: Polymorphism of human constitutive heterochromatin. Science 174:702, 1971.
71. Craig-Holmes AP, Moore FB, Shaw MW: Polymorphism of human C-band heterochromatin. I. Frequency of variants. Am J Hum Genet 25:181–192, 1973.
72. Craig-Holmes AP, Moore FB, Shaw MW: Polymorphism of human C-band heterochromatin. II. Family studies with suggestive evidence of somatic crossing over. Am J Hum Genet 27:178–189, 1975.
73. Crandall BF, Sparkes RS: Pericentric inversion of a number 15 chromosome in nine members of one family. Cytogenetics 9:307–316, 1970.
74. Crandall BF, Carrell RE, Sparkes RS: Chromosome findings in 700 children referred to a psychiatric clinic. J Pediatr 80:62–63, 1972.
75. Crawfurd M d'a, Mason MK: A pericentric inversion in three generations of a family. Bull Eur Soc Hum Genet 46–48, 1973.
76. Creasy MR, Crolla JA, Alverman ED: A cytogenetic study of human spontaneous abortions using banding techniques. Hum Genet 31:177–196, 1976.
77. Crippa L, Ferrier S: Etude cytogénétique d'un cas de syndrome de Fanconi avec inversion péricentrique (1) familiale. J Genet Hum 23:7–16, 1975.

78. Croquette MF, Couturier J, Dutrillaux B, Turleau C, de Grouchy J: Présence, chez un homme stérile, de deux inversions, péricentriques des chromosomes (9). Ann Genet 22:53–56, 1979.
79. Dallapiccola B, Capra L, Petro G, Covic M, Dutrillaux B: Inversion péricentrique du 4: inv(4)(p13q35) et trisomie du bras court du 4 par aneusomie de combinaison Ann Genet (Paris) 17:115–118, 1974.
80. Daniel A: Structural differences in pericentric inversions. Application to a model of risk of recombinants. Hum Genet 56:321–328, 1981.
81. Degenhardt K-H, Geisler M, Weisse K, Grubisic A: The mental retardation syndrome G 22. Am J Hum Genet 27:32A, 1975.
82. de la Chapelle A, Schröder J, Kokkonen J: Cytogenetics of recurrent abortion or unsuccessful pregnancy. Int J Fertil 18:215–219, 1973.
83. de la Chapelle A, Schröder J, Stenstrand K, Fellmann J, Herva R, Saarni M, Anttolainen I, Tallila I, Tervilä L, Husa L, Tallquist G, Robson EB, Cook PJL, Sanger R: Pericentric inversions of human chromosomes 9 and 10. Am J Hum Genet 26:746–766, 1974.
84. de Grouchy J, Aussannaire M, Brissaud HE, Lamy M: Aneusomie de recombinaison: Three further examples. Am J Hum Genet 18:467–484, 1966.
85. de Grouchy J, Plachot M, Sebaoun M, Bouchard R: Chromosome F en anneau (46,XY,Fr) chez un garcon multimalformé. Ann Genet (Paris) 15:121–126, 1972a.
86. de Grouchy J, Turleau C, Roubin M, Klein M: Evolutions caryotypiques de l'homme et du chimpanzé. Etude comparative de topographies de bandes après dénaturation ménagée. Ann Genet (Paris) 15:79–84, 1972b.
87. de Grouchy J, Turleau C: "Atlas des maladies chromosomiques." Paris: Expansion scientifique francaise, 1977.
88. Delprado WJ, Baird PJ: Cephalothoracopagus syncephallus: A case report with previously unreported anatomical abnormalities and chromosomal analysis. Teratology 29:1–9, 1984.
89. Duckett DP, Roberts SH: A new pericentric inversion of chromosome 6 in an abnormal infant. Ann Genet (Paris) 23:117–118, 1980.
90. Dutrillaux B: Chromosomal aspects of human male sterility. Nobel Symp 23:205–208, 1973.
91. Dutrillaux B: Chromosomal evolution in primates: Tentative phylogeny from Microcebus murinus (Prosimian) to man. Hum Genet 48:251–314, 1979.
92. Dutrillaux B, Laurent C, Couturier J, Lejeune J: Coloration des chromosomes humains par l'acridine orange après traitement par le 5 bromodéoxyuridin. C R Acad Sci [D] (Paris) 276:3179, 1973a.
93. Dutrillaux B, Laurent C, Robert JM, Lejeune J: Inversion péricentrique, inv(10), chez la mère et aneusomie de recombinaison, inv(10), chez son fils. Cytogenet Cell Genet 12:245–253, 1973b.
94. Dutrillaux B, Réthore MO, Lejeune J: Analyse du caryotype de Pan paniscus. Comparaison avec les autres Pongidae et l'homme. Humangenetik 28:113–119, 1975a.
95. Dutrillaux B, Réthore MO, Lejeune J: Comparaison du carotype de l'orang outang à celui de l'homme, du chimpanzé et du gorille. Ann Genet (Paris) 18:153–161, 1975b.
96. Eberle F, Kaiser P, Wehr M: 17-jährige Patientin mit dilativer Cardiomyopathie; familiäre, pericentrische Inversion (5). Internist 23:509–513, 1982.
97. Egozeùe J, Aragonés JM, Caballin RM, Goday C: Banding patterns of the chromosomes of man and gorilla. Ann Genet (Paris) 16:207–210, 1973a.
98. Egozeùe J, Caballin R, Goday C: Banding patterns of the chromosomes of man and the chimpanzee. Hum Genet 18:77–80, 1973b.

99. Einfeld SL, Smith A: Pericentric inversion of chromosome 11 in one of two similar retarded brothers J Med Genet 19:312, 1982.
100. Emanuel BS, Zackai EH, Aronson MM, Mellmann WJ, Moorhead PS: Abnormal chromosome 22 and recurrence of trisomy-22 syndrome. J Med Genet 13:501–506, 1976.
101. Engel W, Schmidtke J: Die Bedeutung von Genduplikationen für die Evolution der Wirbeltiere. In Becker PE (ed): "Humangenetik, ein kurzes Handbuch, Volume I/3". Stuttgart: Thieme, 1975, pp 618–649.
102. Escobar JI, Sanchez O, Yunis JI: Trisomy for the distal segment of chromosome 13. Am J Dis Child 128:217–220, 1974.
103. Evans HJ: Chromosome aberrations induced by ionizing radiation. Int Rev Cytol 13:221, 1962.
104. Evans HJ: Chromosome anomalies among livebirths. J Med Genet 14:309–312, 1977.
105. Evans HJ, Buckton KE, Sumner AT: Cytological mapping of human chromosomes: Results obtained with quinacrine fluorescence and acetic-saline Giemsa techniques. Chromosoma 35:310, 1971.
106. Evans MI, Simpson JL, Larson JW, Martin AO, Sarto GE, Schulman JD: Pericentric X-chromosome ascertained during antenatal diagnosis. Clin Genet 18:30–33, 1980.
107. Faed MJW, Marrian VJ, Robertson J, Robson EB, Cook PJL: Inherited pericentric inversion of chromosome 5: A family with history of neonatal death and a case of the "cri du chat" syndrome. Cytogenetics 11:400–411, 1972.
108. Faed MJW, Lamont M, Morton HG, Robertson J, Smail P: An XYY boy with short stature and a case of Kline-felter's syndrome (XXY) in a family with inv.9. Clin Genet 14:241–245,1978.
109. Faed MJW, Lamont MA, Baxby K: Cytogenetic and histological studies of testicular biopsies from subfertile men with chromosome anomaly. J Med Gent 19:49–56, 1982.
110. Fällstrom SP, Wählström J: A pericentric inversion of chromosome 9 and rearrangement involving chromosomes 9 and 10, observed in two generations. Clinical description of chromosome 9(p12-p21) deletion syndrome. Clin Genet 15:480–486, 1979.
111. Feingold M, Atkins L: A case of trisomy 9. J Med Genet 10:184–187, 1973.
112. Ferguson-Smith MA: Clinical cytogenetics. In Crow JF, Neel JV (eds): "Proceedings of the 3rd International Congress of Human Genetics, Chicago, 1966." Baltimore: Johns Hopkins Press, 1967, pp 69–75.
113. Ferguson-Smith MA: Autosomal polymorphisms. Birth Defects X:10–19, 1974a.
114. Ferguson-Smith MA: Meiosis in the human male. In Pearson PL, Lewis KR (eds): "Chromosomes Today," New York: John Wiley & Sons, 1974b, Vol 5, p 235.
115. Ferguson-Smith MA, Handmaker SD: Observations on the satellited human chromosomes. Lancet 1:638, 1961a.
116. Ferguson-Smith MA, Handmaker SD: Observations on the satellited human chromosomes. Lancet 2: 1362, 1961b.
117. Ferguson-Smith MA, Ferguson-Smith ME, Ellis PM, Dickson M: The sites and relative frequencies of secondary constructions in human somatic chromosomes. Cytogenetics 1:325, 1962.
118. Fitzgerald PH: The nature and inheritance of an elongated secondary constriction on chromosome 9 of man. Cytogenet Cell Genet 12:404–413, 1973.
119. Fogle TA, McKenzie WHM: Cytogenetic study of a large black kindred: Inversions, heteromorphisms, and segregation analysis. Hum Genet 55:345–352, 1980.
120. Fonatsch C: New chromosome polymorphism: Inv(16) (p11q12)or 13). Cytogenet Cell Genet 18:106–107, 1977.

121. Ford JH, Callen DF, Jahnke AB, Roberts CG: Within pair differences of human chromosome 9 C-bands associated with reproductive loss. Hum Genet 61:360–363, 1982.
122. Fraccaro M: "DNA Replication and Sex Chromosome Function in the First Stages of Gametogenesis." Wien: Vortrag, gehalten auf der 14, Tagung der Gesellschaft für Anthropologie und Humangenetik, 1975.
123. Fraisse J: Inversion péricentrique d'un chromosome 21. J Genet Hum 23:107–111, 1975.
124. Francke U, Benirschke K, Jones O: Prenatal diagnosis of trisomy 9. Humangenetik 29:243–250, 1975.
125. Friedrich U, Nielsen J: Chromosome studies in 5,049 consecutive newborn children. Clin Genet 4:333–343, 1973.
126. Friedrich U, Nielsen J: Pericentric inversion Y in a population of newborn boys. Hereditas 76:147–152, 1974.
127. Frydman M, Shabtai F, Halbrecht J, Elian E: Normal psychomotor development in a child with mosaic trisomy and pericentric inversion of chromosome 9. J Med Genet 18:390–392, 1981.
128. Fryns JP, Haspeslagh M, Goddoeris P, van Aerde J, Eggermont E, van den Berghe H: Balanced and unbalanced pericentric inversion of chromosome 11. Ann Genet 24(3): 182–183, 1981.
129. Fryns JP, Petit P, Heffinck R, van den Berghe H: Mosaic pericentric inversion of chromosome 2. J Genet Hum 31(2):157–161, 1983.
130. Fryns JP, Kleczkowska A, Londers L, van den Berghe H: Unusual chromosome 9 polymorphism and reproductive failure. Ann Genet (Paris) 28(1):49–51, 1985.
131. Fuenmayor HM, Roldan-Paris L, Bermudez H: Ectodermal dysplasia in females and inversion of chromosome 9. J Med Genet 18:214–217, 1981.
132. Fujimoto A, Wilson MG, Towner JW: Familial inversion of chromosome No. 8. Humangenetik 27:67–73, 1975.
133. Fujimoto A, Wilson MG, Towner JW: Duplication of the segment q12.2 leads to qter of chromosome 22 due to paternal inversion 22(p13q12.2). Hum Genet 63(1):82–84, 1983.
134. Funderburk STJ, Spence MA, Sparkes RS: Mental retardation associated with "balanced" chromosome rearrangements. Am J Hum Genet 29:136–141, 1977.
135. Gagne R, Laberge C: Specific cytological recognition of the heterochromatic segment of number 9 chromosome in man. Exp Cell Res 73:239–242, 1972.
136. Gardner RJM, McCreanor HR, Parslow MI, Veale AMO: Are 1q+ chromosomes harmless? Clin Genet 6:383–393, 1974.
137. Geraedts JPM, Pearson PL: Fluorescent chromosome polymorphisms: Frequencies and segregations in a Dutch population. Clin Genet 6:247–257, 1974.
138. Gerald PS, Walzer S: Newborn infants. In Jacobs PA, Price WH, Law P (eds): "Human Population Cytogenetics." Pfizer Medical Monograph 5. Edinburgh: Edinburgh University Press, 1970, p 134.
139. German J, New MI, Archibald RM, Chaganti RSK: An inherited pericentric X-chromosome associated with male pseudohermaphroditism. V. International Congress of Human Genetics, Mexico, D.F., October 10–15, 1976. In Armendares S, Lisker R (eds): "International Congress Series No. 397." Amsterdam: Excerpta Medica, 1976, p 126.
140. Ghosh PK, Singh IP: Morphologic variability of human chromosomes: Polymorphism of constitutive heterochromatin. Human Genet 32:149–154, 1976.
141. Giraldo A, Silva E, Martinez I, Campos C, Guzman J: Pericentric inversion of chromosome (1) in three sterile brothers. Hum Genet 58:226–227, 1981.

142. Giraud F, Mattei JF, Mattei MG, Ayme, S: Les inversions péricentrique. J Genet Hum 27:109–122, 1979.
143. Gosden JR, Spowart G, Lawrie SS: Satellite DNA and cytological staining patterns in heterochromatic inversions of human chromosome 9. Hum Genet 58:276–278, 1981.
144. Grace HJ, Ally FE, Paruk MA: 46, X inv (Yp+q−) on four generations of an Indian family. J Med Genet 9:293, 1972.
145. Gray JE, Mutton DE, Ashby DW: Pericentric inversion of chromosome 21. A possible further cytogenetic mechanism in mongolism. Lancet 1:21–23, 1962.
146. Grote W, Gödde E: Total perizentrische Inversion am Chromosom 9 als positiver Hinweis bei einer Vaterschaftsbegutachtung. Anthropol Anz 37:278–281, 1980.
147. Haas B, Wirtz A, Kraus CH, Stengel-Rutkowski S, Murken JD, Rehder H, Schneider ST, Lamerz R: Fallvorstellung: Fall II. In Murken J, Stengel-Rutkowski S (eds): 10. Informationsblatt über die Dokumentation der Untersuchungen im Rahmen des Schwerpunktprogramms. In "Pränatale Diagnostik genetisch bedingter Defekte" der Deutschen Forschungsgemeinschaft. München: Kinderpoliklinik der Universität, 1976, p 52.
148. Habedank M: Partial trisomy 13q21→qter de novo due to a recombinant chromosome rec(13)dupq. Hum Genet 52:91–99, 1979.
149. Habedank M: Familial pericentric inversion of chromosome 13 resulting in duplication 13q22→qter J Med Genet 19:227–229, 1982.
150. Halbrecht I, Shabtay F: Human chromosome polymorphism and congenital malformations. Clin Genet 10:113–122, 1976.
151. Hamerton JL, Ray M, Abbott J, Willianson C, Ducasse GC: Chromosome studies in a neonatal population. Can Med Assoc J 106:776–779, 1972.
152. Hamerton JL, Ray M, Douglas GR: Chromosome banding techniques in clinical cytogenetics. Nobel Symp 23:209–213, 1973.
153. Hamerton JL, Canning N, Ray M, Smith S: A cytogenetic survey of 1469 newborn infants. Clin Genet 8:223–243, 1975.
154. Hansmann I: Structural variability of human chromosome 9 in relation to its evolution. Hum Genet 31:247–262, 1976.
155. Hansmann I, Keutel J: A subtelocentric chromosome 9 in a dysplastic 18-year-old boy with dissociated mental development. Humangenetik 30:287–289, 1975.
156. Hauge M, Paulsen H, Halberg A, Mikkelsen M: The value of fluorescence markers in the distinction between maternal and fetal chromosomes. Humangenetik 26:187–191, 1975.
157. Hauksdottir H, Halldorsson S, Jensson O, Mikkelsen M, McDermott A: Pericentric inversion of chromosome No. 13 in a large family leading to duplication deficiency causing congenital malformations in three individuals. J Med Genet 9:413, 1972.
158. Hauksdottir H, Arnardottir A, Steinarsdottir M, Gudmundsdottir E, Halldorsson S, Gunnarsson A, Mikkelsen M: Pericentric inversion of chromosome 13. J Med Genet 20:154–166, 1983.
159. Hecht F: The nonrandomness of human chromosome abnormalities. In Hook EB, Porter IH (eds): "Population Cytogenetics." New York: Academic Press, 1977, p 237.
160. Heinemann M-H, Breg R, Cotlier E: Rieger's syndrome with pericentric inversion of chromosome 6. Br J Ophtalmol 63:40–44, 1979.
161. Henderson AS, Warburton D, Atwood KC: Location of ribosomal DNA in the human chromosome complement. Proc Natl Acad Sci USA 69:3394–3398, 1972.
162. Herva R: Inv(7) as a recurrent aberration in human lymphocyte cultures. Hereditas 95(1):163–164, 1981.

163. Herva R, de la Chapelle A: A large pericentric inversion of human chromosome 8. Am J Hum Genet 28:208–212, 1976.
164. Hernandez A, Rivera H, Jimenez-Sainz M, Cantu JM: Familial trisomy and pericentric inversion (9) (q.11q13). Ann Genet 22:40–42, 1979.
165. Hesselbjerg U, Friedrich U: Pericentric inversion in chromosome No. 2 as a de novo mutation. Hum Genet 53:117–119, 1979.
166. Hirschhorn K, Lucas M, Wallace I: Precise identification of various chromosomal abnormalities. Ann J Hum Genet 36:375, 1973.
167. Holbeck S, Friedrich U, Lauritsen JG, Therkelsen AJ: Marker chromosomes in parents of spontaneous abortuses. Humangenetik 25:61–64, 1974.
168. Holmes R, Keating MJ, Cork A, Broach Y, Trujillo J, Dalton WT Jr, McCredie KB, Freireich EJ: A unique pattern of central nervous system leukemia in acute myelomonocytic leukemia associated with inv(16)(p13q22). Blood 65: 1071–1078, 1985.
169. Honea KL: Paternal pericentric inversion of chromosome 9 with a 9p+ offspring. Ala J Med Sci 14:434, 1978.
170. Jongel K: Trisomy 9q- in family with a pericentric inversion of C9. In Armendares S, Lisker R (eds): "International Congress Series No. 397." Amsterdam: Excerpta Medica, 1976, p 130.
171. Hoo JJ, Parslow MI, Slaw RL, Veale AMO: Complex de novo rearrangement of chromosome 9 with clinical features of monosomy 9p syndrome. Clin Genet 16:151–1555, 1979.
172. Hooft C, Coetsier H, Orye E: Syndrome de Turner et inversion péricentrique probable du chromosome No. 2 [45,X,2(p+q−)]. Ann Genet (Paris) 11:181–183, 1968.
173. Howard-Peebles PN: Familial pericentric inversion of chromosome 1 with a note on reproductive risks. Hum Genet 45:123–125, 1978.
174. Howard-Peebles PN: X-linked mental retardation with a fragile site in Xq and an inversion of chromosome No. 9. Hum Hered 32:139–141, 1982.
175. Howard-Peebles PN, Stoddard GR: A satellited Yq chromosome associated with trisomy 21 and an inversion of chromosome 9, Hum Genet 34:223–225, 1976a.
176. Howard-Peebles PN, Stoddard GR: Pericentric inversion of chromsome 9: Benign or harmful? V. International Congress of Human Genetics, October 10–15, 1976. In Armendares S, Lisker R (eds): "International Congress Series Nos. 339, 397", 1. Amsterdam: Excerpta Medica, 1976b, p 131.
177. Howard-Peebles PN, Stoddard GR: Pericentric inversion of chromosome No. 9, benign or harmful? Hum Hered 29:111–117, 1979.
178. Hsu TC, Pathak S: Differential rates of sister chromatid exchanges between euchromatin and heterochromatin. Chromosoma 58:269–273, 1976.
179. Hultén M: Chiasma distribution at diakinesis in the normal human male. Hereditas 76:55–78, 1974.
180. Husslein P, Schnedl W, Wagenbichler P: Zytogenetische Untersuchungen bei wiederholten Aborten. Wiener Klin Wochenschr 92 (16):575–578, 1980.
181. Husslein P, Huber J, Wagenbichler P, Schnedl W: Chromosome abnormalities in 150 couples with multiple spontaneous abortions. Fertil Steril 37(3):379–383, 1982.
182. Jacobs PA, Buckton KE, Cunningham C, Newton M: An analysis of the breakpoints of structural rearrangements in man. J Med Genet 11:50, 1974.
183. Jacobs PA, Frackiewicz A, Law P, Hilditch CJ, Morton NE: The effect of structural aberrations of the chromosomes on reproductive fitness in man. Clin Genet 8:169–178, 1975.

184. Johnson DD, Dobyns WB, Dewald GW, Gordon H: Family studies of pericentric and paracentric inversions. Am J Hum Genet 37, No. 4 1985, Supplement p A 100, No. 292.
185. Jonasson J, Lindstein J, Lundberg R, Kissmeyer-Nielsen F, Lamm LU, Patersen GB, Therkelsen AJ: HL-A antigens and heteromorphic fluorescence characters of chromosomes in prenatal paternity investigation. Nature 236:312–313, 1972.
186. Jones KW: Satellite DNA. J Med Genet 10:273–281, 1973.
187. Jones KW, Prosser J, Ginelli CG: The chromosomal location of human satellite DNA III. Chromosoma 42:445–451, 1973.
188. Jongbloet PH, van Erkelens-Zwets JHJ: Letter to the editor: Apparently balanced de novo translocations in patients with abnormal phenotypes. Clin Genet 12:188–190, 1977.
189. Kaiser P: "Perizentrische Inversionen menschlicher Chromosomen." Stuttgart: Georg Thieme Verlag, 1980.
190. Kaiser P: Pericentric inversions. Hum Genet 68:1–47, 1984.
191. Kaiser P, Klein TH, Steuber ED, Gerhardt-Ratschow: "Familienuntersuchungen bei pericentrischer Inversion 9." Hamburg: Vortrag, gehalten auf der Tagung der Gesellschaft für Humangenetik und Anthropologie, 1977.
192. Kaiser P, Zabel B, Hansen S, Daume E: Short arm deletion of an X chromosome (46,XX,p-). Hum Genet 32:89–100, 1976.
193. Kaiser P, Foerster W, Steuernagel P, Hilling U, Herberg KP: Familial pericentric inversion (14)(p11; q24), with a rec dup(q) in one offspring. Clin Genet 26:73–76, 1984.
194. Kajir T, Ohama K, Niikawa N, Ferrier A, Avirachan S: Banding analysis of abnormal karyotype in spontaneous abortion Hum Genet 25:539–547, 1973.
195. Kassam G, Chen AT, Goldberg MF, Trusler S, Oakley GP Jr: Prenatal diagnosis of pericentric inversion of chromosome No. 17 in a twin pregnancy. Prenat Diagn 4:213–216, 1984.
196. Kawana S, Watanabe G, Asami T, Okada T, Yamamoto M: Pericentric inversion of chromosome No. 8. Tohoku J Exp Med 119(1):65–70, 1976.
197. Kawashima H, Maruyama S: A case of chromosome 3 duplication q deletion p syndrome born to the mother with a pericentric inversion inv(3) (p25q21). Jpn J Hum Genet 24:9–12, 1979.
198. Keitges EA, Palmer CG, Weaver DD: Pericentric X inversion in dizygotic twins who differ in X chromosome inactivation and menstrual cycle function. Hum Genet 62(3):210–213, 1982.
199. Khan F, Verma RS, Dosik H, Warman J: Primary amenorrhea in a black female with duplication and inversion of the secondary constriction regions of chromosome 9. J Clin Endocrinol Metab 47:280–283, 1978.
200. Kim HJ, Levy J, Roguero W: Prenatal detection of pericentric inversion of chromosome 12. Diagn Gynecol Obstet 2:231–234, 1980.
201. Kim HS, Hsu LYF, Hirschorn K: Prenatal detection of pericentric inversion of chromosome 11. Am J Hum Genet 27:55A, 1975.
202. Kistner G: Personal communication, 1985.
203. Klein TH: Familienuntersuchungen bei pericentrischer Inversion des Chromosomes 9. Marburg, 1979. MD Dissertation, Philipps Universität.
204. Körner H, Degen B, Roese I, Metschkarski S: Pränatale Diagnosen bei einer Familie mit perizentrischer Inversion am Chromosom Nr. 5. Zentralbl Gynakol 105:934–939, 1983.
205. Korenberg JR, Therman E, Denniston C: Hot spots and functional organization of human chromosomes. Hum Genet 43:13–22, 1978.

206. Koske-Westphal T, Pruszak-Seel RE, Niss R, Passarge E: Partial trisomy 13 presumably due to recombination in an inversion heterozygote and by unequal crossing-over. Ann Hum Genet 41:315, 1978.
207. Krone W, Wolf U: Chromosome variation and gene action. Hereditas 86:31–36, 1977.
208. Kruger J, Vogel F: Population genetics of unequal crossing over. J Mol Evol 4:201–247, 1975.
209. Kukolich MK, Althaus BW, Sears JW, Mankinen CB, Lewandowski RC: Abnormalities resulting from a familial pericentric inversion of chromosome 18. Clin Genet 14:98–104, 1978.
210. Kunze J, Mau G: A_1 and C_1 marker chromosomes in children with combined minor and major malformations. Lancet 1:273, 1975.
211. Lamm LU, Friedrich U, Petersen GB, Jorgensen J, Nielsen J, Therkelsen AJ, Kissmeyer-Nielsen F: Assignment of the major histocompatibility complex to chromosome No. 6 in a family with a pericentric inversion. Hum Hered 24:273–284, 1974.
212. Lansky-Shafer S, Daniel W, Ruiz L: Trisomy 10p produced by recombination involving maternal inversion inv(10)(p11q26). J Med Genet 18:59–75, 1981.
213. Leal-Garza C, Riojas-Valdes V, Montes de Oca-Luna R, Garza-Chapa R: Frequency and association of NOR'S in a family carrying a pericentric inversion of chromosome 14. Rev Invest Clin 36:283–286, 1984.
214. Lee CSN, Ying KL, Bowen P: Position of the duffy locus on chromosome 1 in relation to breakpoints for structural rearrangements. Am J Hum Genet 26:93–102, 1974.
215. Lee M, Schneider J, Wasant P, Yu CY, Trpis L, Liang YW, Morgan Lewis B, Borkowf S, Borgaonkar DS: Supernumerary small chromosomal anomaly: Report of three cases including one with a familial inversion of chromosome 5. J Genet Hum 26:275–285, 1978.
216. Leeksma CH, den Nijs J, Havik F, Kerkhofs H, Geraedts JP: Clones with 5q− and inversion of chromosome 7 in PHA stimulated peripheral blood. Scand J Haematol 27:19–24, 1981.
217. Lejeune J: Les conséquences meiotiques des remaniements chromosomiques. Ann Genet (Paris) 8:9, 1965.
218. Lejeune J, Dutrillaux B, Rhetoré MO, Prieur M: Comparaison de la structure fine des chromatides d'Homo sapiens et de Pan troglodytes. Chromosoma 43:423–444,1973.
219. Lele KP: Chromosome studies in 5 cases of coloboma of the iris. Lancet 1:576–578, 1965.
220. Leonard C, Hazael-Massieux P, Bocquet L, Larget-Piet L, Boue J: Inversion péricentrique inv(2)(p11q13) dans des familles non apparentées. Humangenetik 28:121–128, 1975.
221. Lewandowski RC, Yunis JJ: Trisomy 9 mosaicism. Clin Genet 11:306–310, 1977.
222. Li L, Xia J, Dai H, He X, Xu F, Xu J, Xiao G, He H, Hu X, Lu H: Genetic counseling clinic chromosome analysis in 522 cases. Clin Med J (Peking) 95:793–804, 1982.
223. Lin CC, Gedeon MM, Griffith P, Smink WK, Newton DR, Wilkil L, Sewell LM: Chromosome analyses on 930 consecutive newborn children using quinacrine fluorescent banding technique. Hum Genet 31:315–328, 1976.
224. Lozzio CB, Klepper MB: Chromosome aberrations identified by the new banding techniques. Am J Hum Genet 26:55A, 1974.
225. Lubs HA, Ruddle FJ: Application of quantitative karyotype to chromosome variations in 4000 consecutive newborns. In Jacobs PA, Price WH, Law P (eds): Human Population Cytogenetics." Pfizer Medical Monographs 5. Edinburgh: Edinburgh University Press, 1970, pp 119–142.

226. Lubs HA, Ruddle FH: Chromosome polymorphism in American negro and white populations. Nature 233:134–136, 1971.
227. Lubs HA, Patil SA, Kimberling WJ, Brown J, Cohen M, Gerald P, Hecht F, Myrianthopoulos N, Summitt RL: Q- and C-banding polymorphisms in 7- and 8-year-old children: Racial differences and clinical significance. In Hook EB, Porter IH: (eds): "Population Cytogenetics." New York: Academic Press, pp 133–159, 1977.
228. Lucas J, Le M'ee F, Picard F, Le Marec B, Junien C: Trisomie 13q13 leads to qter pure par recombinaison aneusomique d'une inversion p'pericentrique maternelle. Ann Genet (Paris) 26:187–190, 1983.
229. MacDonald IM, Cox DM: Inversion of chromosome 2 (p11p13): Frequency and implications for genetic counselling. Hum Genet 69:281–283, 1985.
230. Mace MA: Segregation of ACP_1 and MNSs in families with structural rearrangement involving chromosome 2. Ann Hum Genet 38:479, 1975.
231. Madan K, Bobrow M: Structural variation in chromosome No. 9. Ann Genet (Paris) 17:81–86, 1974.
232. Maeda T, Ohno M, Takada M, Nishida M, Tsukioka K, Tomita H: Turner's syndrome with a duplication-deficiency X-chromosome derived from a pericentric inversion X-chromosome. Clin Genet 15:259–266, 1979.
233. Magenis RE, Chamberlin J, Overton K, Lovrien E: Linkage relationship of HLA and a familial chromosome 6 inversion (pter-p23::q23-p23::q23-qter): Lack of dose effect in duplication-deficient offspring. Cytogenet Cell Genet 22:418–420, 1978.
234. Maltby EL: Familial pericentric inversion (13) detected by antenatal diagnosis. J Med Genet 21:149–151, 1984.
235. Martin AO, Turk KB, Macintyre MN: An analysis of chromosome 9 inversions in four families. Am J Hum Genet 26:58A, 1974.
236. Martin MJ, Abrisqueta JA: Unusual pericentric inversion inv(9) (p13q11) in a phenotypically normal family. Hum Genet 64:100, 1983.
237. Martin AO, Simpson JL, Deddish RB, Elias S: Clinical implications of chromosomal inversions. A pericentric inversion in No. 18 segregating in a family ascertained through an abnormal proband. Am J Perinatol 1:81–88, 1983.
238. Mattei MG, Mattei JF, Ayme S, Mapuech G, Giraud F: A dynamic study in two new cases of X-chromosome translocation. Hum Genet 41:251–257, 1978.
239. Mattei JF, Mattei MG, Adissone JP, Taramasco H, Giraud F: Pericentric inversion, inv(9) (p22q32), in the father of a child with a duplication-deletion of chromosome 9 and gene dosage effect for adenylate kinase-1. Clin Genet 17:129–136, 1980.
240. Mattei MG, Mattei JF, Guichaoua M, Giraud F: Partial inversion of the secondary constriction of chromosome 9. Does it exist? Hum Genet 59:310–316, 1981.
241. Mattei JF, Mattei MG, Balestrazzi P, Giraud F: Familial pericentric inversion of chromosome 9, inv(9) (p22q32) with recurrent duplication-deletion (letter) Clin Genet 24(3):220–222, 1983.
242. Mattevi MS, Pinheiro CEA, Erdtmann B, Flores RZ, Salzano FM: Familial pericentric inversion of chromosome 2. J Genet Hum 29:161–169, 1981.
243. McDermott A, Parrington JM: Elucidation of a pericentric inversion of a D-group chromosome in the mother of a child with Patau's syndrome. Ann Hum Genet 38:305, 1975.
244. McIlree ME: Chromosome studies on testicular cells from 50 subfertile men. Lancet 2:69–71, 1966.
245. McKenzie WH, Lubs HA: Human Q and C chromosomal variations: Distribution and incidence. Cytogenet Cell Genet 14:97–115, 1975.

246. Meyer E, Navon D, Aloni U, Gdal-On M, Zonis S, Dar H, Perlman I: Bilateral optic atrophy associated with abnormality of chromosome No. 2. Metab Pediatr Syst Ophthalmol 7:207–210, 1984.
247. Mićić M, Mićić S, Diklić V: Chromosomal constitution of infertile men. Clin Genet 25:33–36, 1984.
248. Migliori V, Ferrari F, Giamagli CA, Di Stefano P, Galanti E, Guastaferro N: Un caso di sindrome da trisomia 3q21→qter e monosomia 3p25→pter. Pediatr Med Chir 5:237–239, 1983.
249. Mikelsaar AVN, Tüür SJ, Kaosav ME: Human karyotype polymorphism. Hum Genet 20:89–101, 1973.
250. Mikelsaar A-V, Ilus T, Kivi S: Variant chromosome 3 (inv3) in normal newborns and their parents, and in children with mental retardation. Hum Genet 41:109–113, 1978.
251. Mikkelsen M, Halberg A, Paulsen H: Maternal and paternal origin of extra chromosome in trisomy 21. Hum Genet 32:17–21, 1976.
252. Miller DA, Firschein JC, Dev VG, Tantravahi R, Miller OJ: The gorilla karyotype, chromosome lengths and polymorphisms. Cytogenet Cell Genet 13:536–550, 1974.
253. Moedjono SJ, Sparken RS: Familial pericentric inversion of chromosome 8. Is the breakpoint p23q22 important in the formation of imbalanced recombinants? Ann Genet (Paris) 23:235–237, 1980.
254. Moore DE, Weise K, Zawydiwski R, Thompson EB: The karyotype of the glucocorticoid-sensitive, lymphoblastic human T-cell line CCRF-CEM shows a unique deleted and inverted chromosome 9. Cancer Genet Cytogenet 14:89–94, 1985.
255. Moorhead PS: A closer look at chromosomal inversions. Am J Hum Genet 28:294–296, 1976.
256. Morgan R, Stebbins RD, Hecht F, Sandberg AA: The Philadelphia (Ph) chromosome in leukemia. III. Complex Ph translocation plus inversion in chronic myelocytic leukemia. Cancer Genet Cytogenet 14:23–30, 1985.
257. Morichon-Delvallez N, Vendrely C: Découverte d'une inversion péricentrque du chromosome 4 chez un couple présentant des avortements à répétition. Rev Fr Gynécol Obstét 76:543–546, 1981.
258. Morishima A, Morishima A, Liu N, Grumbach MM: Multiple congenital anomalies associated with probable pericentric inversion of chromosome No. 4 and mosaicism in an asymptomatic mother. J Pediatr 65:1096–1097, 1964.
259. Morton NE, Jacobs PA, Frackewicz A, Law P, Hilditch CJ: The effect of structural aberrations of the chromosomes on reproductive fitness in man. Clin Genet 8:159–168, 1975.
260. Mosher GA, Sanger WG: Familial pericentric inversion of chromosome 20. Am J Hum Genet 37, No. 4 1985, Supplement p A 108, No. 316.
261. Mounoud RL: A clinical and cytogenetic investigation carried out in a special institution for mentally retarded patients. J Genet Hum 24:297–335, 1976.
262. Müller HJ: Radiation and genetics. Am Nat 64:220, 1930.
263. Müller HJ, Klinger HP, Glasser M: Chromosome polymorphism in a human newborn population. Cytogenet Cell Genet 15:239–255, 1975.
264. Murray J, Patil, S, Fatemi C, Aschbacher A, Buetow K, Ferrell R, Miles J, Smith M, Carlock L: Effect of a large pericentric inversion of chromosome 4 on meiotic recombination. Am J Hum Genet 37, No. 4, 1985, Supplement p A 109, No. 321.
265. Mutton DE, Daker MG: Pericentric inversion of chromosome 9. Nature 241:80, 1973.
266. Nakagome Y, Chiyo H: Nonrandom distribution of exchange points in patients with structural rearrangements. Am J. Hum Genet 28:31–41, 1976.

267. Nakagome Y, Oka S, Matsunaga E: LBA technique in the detection of chromosome variants. Hum Genet 38:307–314, 1977.
268. Natarajan AT, Ahnström G: Induced chromosomal aberrations and heterochromatin. In Pfeiffer RA (ed):"Modern Aspects of Cytogenetics: Constitutive Heterochromatin in Man." Stuttgart: Schattauer Verlag, 1973, pp 215–223.
269. Neri G, Tedeschi B, Sanfilippo S: Partial inversion of the secondary constriction of chromosome 9. It exists. Hum Genet 61:80–81, 1982.
270. Niebuhr E, Seemanova E, Losan F: Familial occurrence of pericentric inversion of X chromosomes. CS Pediatr 29:245–247, 1974.
271. Nielsen J: Chromosome abnormalities and variants in the Danish cytogenetic central register. In Becker PE, Vogel F, Lenz W (eds): "Topics in Human Genetics." Stuttgart: Georg Thieme, 1976.
272. Nielsen J, Rasmussen, J: Pericentric Y-inversion. J Genet Hum 22:21–29, 1974.
273. Nielsen J, Friedrich U, Hreidarsson AB: Frequency and genetic effect of 1qh+. Humangenetik 21:193–196, 1974a.
274. Nielsen J, Friedrich U, Hreidarsson AB: Frequency and genetic effect of 1qh+. Humangenetik 21:193–196, 1974b.
275. Nielsen J, Friedrich U, Hreidarsson AB, Zeuthen E: Frequency of 9qh+ and risk of chromosome aberrations in the progeny of individuals with 9qh+. Humangenetik 21:211–216, 1974c.
276. Nielsen J, Sillesen I: Incidence of chromosome aberrations among 11,148 newborn children. Humangenetik 30:1–12, 1975.
277. Nielsen J, Homma A, Holm V: Inversion 19 and isochromosome short arm 17 or 18. Hum Genet 37:347–350, 1977.
278. Nielsen J, Homma A, Rasmussen K, Ried E, Sørensen K, Saldano-Garcia P: Deletion 14q and pericentric inversion 14. J Med Genet 15:236, 1978.
279. Nikolis J, Stolevic E: Recombinant chromosome as a result of pericentric inversion of X-chromosome. Hum Genet 45:115–122, 1978.
280. Ohyashiki K, Ohsumi A, Ito H, Umehara S, Ohyashiki J, Yamazaki T, Yunokawa K, Osamura S, Oshimura M: A case of chronic myelogenous leukemia with a pericentric inversion of the heterochromatic region of chromosome No. 1. Rinsho Ketsueki 23:199–205, 1982.
281. Orye E, Coetsier H, Hooft E: A probable pericentric inversion of a G G translocation chromosome in a mentally retarded child with mongoloid trait. Hum Hered 19:288–298, 1969.
282. Orye E, Benoit Y, van Mele B: Complete trisomy 5p owing to de novo translocation t(5;22) (q11; p11) with isochromosome 5p associated with a familial pericentric inversion of chromosome 2, inv2(p21q11). J Med Genet 20:394–396, 1983.
282a. Orye E, Verhaaren H: A peculiar t(21;14) translocation, associated with a pericentric inversion of the Y-chromosome in a mongol boy. Ann Genet (Paris) 17(3):217–220, 1974.
283. Osztovics M, T'oth S, Wilhelm O: Unusual chromosome aberrations in 3 children with Down syndrome. Acta Paediatr Acad Sci Hung 23:283–289, 1982.
284. Page C, Ferguson-Smith MA: Gonadenbef. bei Inversionen. Zit bei Winsor 1978.
285. Palka GD, Forcucci-Zulli M, Gaglione R, Stuppia L, Calabrese G, Ramenghi M: Amenorrhoea in a girl with pericentric inversion of chromosome 11. Pathologica 76:623–625, 1984.
286. Pardue MS, Gall JG: Chromosomal localization of mouse satellite DNA. Science 168:1356–1358, 1970.

287. Paris Conference: Standardization in human cytogenetics. Birth Defects VIII:7, 1971.
288. Paris Conference: Supplement Cytogenetics 15:203, 1975.
289. Patil SR, Lubs HA, Brown J, Cohen M, Gerald P, Hecht F, Kimberling W, Myrianthopoulos N, Summitt RL: Incidence of major chromosome abnormalities in children. Cytogenet Cell Genet 18:302–306, 1977a.
290. Patil SR, Lubs HA, Kimberling WJ, Brown J, Cohen M, Gerald P, Hecht F, Moorhead P, Myrianthopoulos N, Nummitt RL: Chromosomal abnormalities ascertained in a collaborative survey of 4342 7- and 8-year-old children: Frequency, phenotype and epidemiology. In Hook EB, Porter H (eds): "Population Cytogenetics of Humans." New York: Academic Press, 1977b, pp 103–131.
291. Pauli RM, Pagon RA, Hall JG: Trisomy 18 in sibs and maternal chromosome 9 variant. Birth Defects XIV:297–301, 1978.
292. Pawlowitzki IH: Frequency of chromosome abnormalities in abortions. Humangenetik 16:131–136, 1972.
293. Pearson G, Mann JD, Bensen J, Bull RW: Inversion duplication of chromosome 6 with trisomic codominant expression of HLA antigens. Am J Hum Genet 31:29–34, 1979.
294. Perissel B, Geneix A, Charbonn'e F, Jaffray JY, Grosjean I, Malet P: Inversion péricentrique du chromosome Y. Etude familiale. Pathologica 74:563–567, 1982.
295. Pescia G, Gaide A-C, Juillard E: Trois familles avec inversion péricentrique du chromosome 9 (invC9). J Genet Hum 25:121–134, 1977.
296. Petit P, Fryns JP: Two pericentric inversions inv(7)(p15; q32) and inv(9)(p11; q13) in a male with absence of vas deferens. Hum Genet 64:303, 1983.
297. Pfeiffer RA, Kessel E: Balanced and unbalanced pericentric inversion of a chromosome 14. Hum Genet 43:103–106, 1978.
298. Phillips RB: Pericentric inversions inv(2)(p11q13) and inv(2)(p13q11) in 2 unrelated families. J Med Genet 15:388–390, 1978.
299. Pison H, Sele B, Salvat J, Pasquier D, Jalbert H, Halbert P: Mécanismes de formation des duplications à propos d'une observation de duplication 10q22→25. Ann Genet 25:137–140, 1982.
300. Podugolnikova OA, Blumina MG: Pericentric inversion of the chromosome 2 in a girl with a congenital mental deficiency and in her mother. Genetika (Moskva) VIII:156–161, 1972.
301. Polani (1974) Cited in Madan and Bobrow 1974.
302. Ponzio G, Carozzi F, Dragone E, Spada A, Brignone S, De Marchi M, Carbonara A: Descrizione di un caso di cromosoma 21 ad anello ed inversione pericentrica del cromosoma Y. Pathologica 75:276–279, 1983.
303. Pope IS, Thuline HC, Aronson MM, Bozarth B, Greene AE, Coriell LL: Duplication of a segment of chromosome 3 in a subject with multiple congenital anomalies and a 47,XXY father, inversion of chromosomes 3 and 9 in the mother, and inversion of chromosome 9 in a brother. Cytogenetics 24:127–128, 1979.
304. Poulsen H, Mikkelsen M, Holmgren G: Peri- and paracentric inversions in chromosome 12: Prenatal diagnosis and family study. Prenatal Diagn 1:35–42, 1981.
305. Prieto F, Badia L, Asensi F, Moya A, Figuera MJ: Pericentric inversions of chromosome 12 in two families. Hum Genet 57:131–133, 1981.
306. Punnett HH, Kistenmacher ML, Toro-Sola MA, Kohn G: Quinacrine fluorescence and Giemsa banding in trisomy 22. Theor Appl Genet 43:134–138, 1973.
307. Quazi OH, Masakawa A, Madahar C, Ehrlich R: Trisomy 9 syndrome. Clin Genet 12:221–226, 1977.

308. Rabbani MG, Seawright JA, Kitzmiller JB: X-ray induced pericentric inversions in anopheles albimanus. Can J Genet Cytol 19:67–74, 1977.
309. Rasmussen K, Nielsen J, Dahl D: The prevalence of chromosome abnormalities among mentally retarded persons in a geographically delimited area of Denmark. Clin Genet 22:244–255, 1982.
310. Renwick JH: Assignment and map-positioning of human loci using chromosomal variation. Ann Hum Genet 35:79, 1971.
311. Réthore M-O, Dutrillaux B, Job J-C, Lejeune J: Trisomie 4p par aneusomie de recombinaison d'une inv(4)(p14q35). Ann Genet (Paris) 17:109–114, 1974.
312. Revell SH: The accurate estimation of chromatid breaks, and its relevance to a new interpretation of chromatid aberrations induces by ionizing radiations. Proc R Soc Lond [Biol] 150:563, 1959.
313. Rieger R: Alternative Interpretationsversuche zur Entstehung chromosomaler Strukturumbauten: Bruch-Reunions- und Austausch-Hypothese. Biol Zbl 85:29, 1966.
314. Rieger R, Michaelis A, Green MM: "Glossary of Genetics and Cytogenetics," 4th ed. Heidelberg: Springer, 1976.
315. Rivera H, Hernandez A, Rolon A, Ibarra B, Alcaraz A, Cantu JM: Familial pericentric inversion (3)(p14q24). Ann Genet 22:43–46, 1979.
316. Rivera H, Alvarez-Arratia MC, Moller M, Diaz M, Cantu JM: Familial inv(1)(p3500q21.2) associated with azoospermia. Hum Genet 66:165–167, 1984a.
317. Rivera H, Moeller M, Hernandez A, Enriquez-Guerra MA, Arreola R, Cantu JM: Tetrasomy 18p: A distinctive syndrome. Ann Genet (Paris) 27:187–189, 1984b.
318. Roberts PA: Difference in the behaviour of eu- and hetero-chromatin: Crossing-over. Nature 205:725–726, 1965.
319. Robinson JA: Origin of extra chromosome in trisomy 21. Lancet 1:131–133, 1973.
320. Robinson JA, Buckton KE, Spowart G, Newton M, Jacobs PA, Evans HJ, Hill R: The segregation of human chromosome polymorphisms. Ann Hum Genet 40:113, 1976.
321. Rodriguez MT, Martin MJ, Abrisqueta JA: Familial pericentric inversion (10) and its effect on two offspring. J Med Genet 21:317–319, 1984.
322. Romain DR, Chapman CJ, Columbano-Green L, Msythe RH, Gebbie O: Two pericentric inversions, inv(2)(p11q13) and inv(5)(p13q13), in a patient referred for psychiatric problems. J Med Genet 19:153–155, 1982.
323. Russell MH, Miller P, Killam A: Familial pericentric inversion of chromosome 11 detected prenatally. J Med Genet 19:151–152, 1982.
324. Sagredo G, Sagredo G, Moran-Cabre A, San Roman-Cos-Gayon C: Familial pericentric inversion of Y chromosome and Down syndrome. An Esp Pediatr 8:68–71, 1975.
325. Sandig KR, Mücke J, Veit H: Trisomy 9p resulting from de novo 9/15 translocation and a 9p isochromosome. Hum Genet 52:175–178, 1979.
326. Sandler L, Hiraizumi Y: Meiotic drive in natural populations of Drosophila melanogaster. Can J Genet Cytol 3:34, 1961.
327. Saura R, Longy M, Sautarael M, Renouil M, Sandler B: Double trisomie et inversion pericentrique transmise [48,XXY,+21,inv(22)]. Effet interchromosomique. Ann Genet (Paris) 26:180–182, 1983.
328. Schaefer MSD: "Segregation and Pathologie autosomaler familiärer Translokationen beim Menschen." Dissertation Fachbereich Biologie, Universität Kaiserslautern, 1983.
329. Scheres JMJC, Hustinx TWJ, Weemaes CMR: Chromosome 7 in ataxia telangiectasia. J Pediatr 97:440–441, 1980.
330. Schinzel A, Hayashi K, Schmid W: Mosaic-trisomy and pericentric inversion of chromosome 9 in a malformed boy. Humangenetik 25:171–177, 1974.

331. Schmid W: Pericentric inversions. J Genet Hum 16:89–96, 1967.
332. Schmidt A: The phenotype in de novo and familial pericentric inversion 6. A problem in karyotype-phenotype correlation. Hum Genet 61:167–170, 1982.
333. Schmidt A, Samson GV: 1. "Tagung der AG-klinische Genetik." Gießen, 1978.
334. Schroer RJ, Culp M, Stevenson RE, Pott, WE, Taylor HA, Simensen RJ: Duplication-deletion syndrome in a family with pericentric inversions of chromosome 6. Clin Genet 18:83–87, 1980.
335. Schwanitz G, Schmid P, Hägele CH, Daffner HW, Grosse K-P: Partial deletion of 1q following a pericentric inversion in a boy with multiple minor morphologic anomalies and mental retardation. Acta Genet Med Gemellol (Roma) 26:173–175, 1977.
336. Seabright M, Gregson N, Mould S: Trisomy 9 associated with an enlarged 9qh segment in a liveborn. Hum Genet 34:323–325, 1976.
337. Sekhon GS, Sly WS: Trisomy 10p due to familial pericentric inversion. Birth Defects XIV (6c), 1978, p 422.
338. Selezner V, Dzenis IG: Differentiation along human mitotic chromosomes. Genetika VIII:115–119, 1972.
339. Serville F, Broustet A: Pericentric inversion and partial monosomy 4q associated with congenital anomalies. Hum Genet 39:239–242, 1977.
340. Seuanez HN: The phylogeny of human chromosomes. Heidelberg: Springer, 1979.
341. Shabtai F, Antebi E, Klar D, Kimchi D, Hart J, Halbrecht I: Cytogenetic study of patients with carcinoma of the colon and rectum: Particular C-band variants as possible markers for cancer proneness. Cancer Genet Cytogenet 14:235–245, 1985.
342. Shapiro LR, Pettersen RO, Wilmot PL, Warburton D, Benn PA, Hsu LY: Pericentric inversion of the Y chromosome and prenatal diagnosis. Prenat Diagn 4:463–465, 1984.
343. Simola K, Karli P, de la Chapelle A: Two pericentric inversions of human chromosome 11. J Med Genet 14:371–374, 1977.
344. Simpson JL, Elias S, Martin AO: Parental chromosomal rearrangements associated with repetitive spontaneous abortions. Fertil Steril 36:584–590, 1981.
345. Singh RP, Jaco NT: Pierre Robin syndrome. Am J Dis Child 120:560–561, 1970.
346. Sofuni T, Tanabe K, Naruto J, Awa AA: A new type of inversion of a human Y chromosome. Humangenetik 30:91–97, 1975.
347. Sole-Pujol MT, Rivera H, Castro-Felix LP, Moller M, Cantu JM: Familial inv(2)(p2300q11.2). J Genet Hum 33:57–62, 1985.
348. Soler A, Salami C, Balmes I, Carrio A, Tejada I, Farguell T, Cols N, Cararach J, Fortuny A: Pericentric X chromosome in a family. Clin Genet 20:234–235, 1981.
349. Soudek D, Laxova R, Adamek R: Pericentric inversion in a family with a 21/22 translocation. Cytogenetics 7:108–117, 1968.
350. Soudek DS, O'Shaughnessy S, Laraya P, McCreary BD: Pericentric inversion of "fluorescent" segment in chromosome no. 3. Humangenetik 22:343–346, 1974.
351. Soudek D, Sroka H: Inversion of "fluorescent" segment in chromosome 3. A polymorphic trait. Hum Genet 44:109–115, 1978.
352. Soukup SW, Yarema W, Robinow M: A pericentric inversion of a chromosome 4 with a t(4q+10p−) and a filial t(DqDq) in a mentally retarded girl. Humangenetik 25:69–78, 1974.
353. Sparkes RS, Muller HM, Veomett IC: Inherited pericentric inversion of a human Y chromosome in trisomic Down's syndrome. J Med Genet 7:59–62, 1970.
354. Spedicato FS, Dicomite A, Gaudio R: Pericentric inversion inv(3)(p11q21). J Med Genet 21:396, 1984.

355. Sperling K, Kaden R, Gillert KE, Weise W: Ein Fall von XX/XY-Chimärismus mit normalem männlichen Habitus. Humangenetik 21:237–244, 1974.
356. Stadler LJ: On the genetic nature of induced mutations in plants. Proc VI Int Congr Genet 1:274, 1932.
357. Stahl-Maugé C, Weiss-Wichert P, Propping P: Familial pericentric inversion of chromosome 1 in a boy with Goldenhar's syndrome. Hum Genet 61:78, 1982.
358. Stanley RG, Dev VG, Butler MG, Phillips JA: Partial 8q trisomy and 8p monosomy resulting from inversion in paternal chromosome 8. Am J Hum Genet 37:4, 1985.
359. Steinbach P: Three cases of familial pericentric inversion 2. Cases observed. Hum Genet 58:444–445, 1981.
360. Stengel-Rutkowski S, Frankenberger R: Inv(G) als Nebenbefund bei Klinefelter 47,XXY. Bericht über die funkt. In Murken JD, Stengel-Rutkowski S (eds): "Beratung und pränatale Diagnostik." München: Kinderpoliklinik der Universität, 1975, p 31.
361. Stine SB, Clark CE, Telfer MA, Casey PA, Cowell HR: Ullrich-Turner syndrome [45,X/46,X,i(Xq)] in a child with a familial inversion of chromosome 3. Am J Med Genet 12:57–62, 1982.
362. Stuehrenberg M: De novo inversion 1, in amniotic fluid cell cultures. Hum Genet 64:412, 1983.
363. Sturtevant AH: A cross-over reducer in Drosophila melanogaster due to inversion of a section of the third chromosome. Biol Zbl 46:697, 1926.
364. Sturtevant AH, Beadle GW: The relations of inversions in the X-chromosome of Drosophila melanogaster to crossing over and disjunction. Genetics 21:554, 1936.
365. Subrt I, Kozak J, Hnikova O: Microdensitometric identification of the pericentric inversion of chromosome No. 2 and of duplication of the short arm of chromosome No. 7 in a reexamined case. Hum Hered 23:331–337, 1973.
366. Sugansky E, Smith ACM, Peakman DC, McConnell ThS, Baca P, Robinson A: Familial pericentric inversion of chromosome 8. Am J Med Genet 10:229–235, 1981.
367. Sumner AT: A simple technique demonstrating centromeric heterochromatin. Exp Cell Res 75:302, 1972.
368. Sutherland GR, Carter RF, Morris LL: Partial and complete trisomy 9: Delineation of a trisomy 9 syndrome. Hum Genet 32:133–140, 1976a
369. Sutherland GR, Gardner AJ, Carter RF: Familial pericentric inversion of chromosome 19, inv(19)(p13q13) with a note on genetic counselling of pericentric inversion carriers. Clin Genet 10:54–59, 1976b.
370. Sutherland GR, Mulley JC, Goldblatt E: A large kindred with an inv(3)(p25q23): Clinical, cytogenetic and genetic marker studies. Ann Genet (Paris) 24:202–205, 1981.
371. Taysi K, Bobrow M, Balci S, Madan K, Atasu M, Say B: Duplication/deficiency product of a pericentric inversion in man: A cause of D1 trisomy syndrome. J Pediatr 82:263–268, 1973.
372. Testa JR, Rowley JD, Hawkins C, Yu RL, Aronson MM, Mulivor RA, Green AE, Coriell LL: A balanced translocation (17:22) and a pericentric inversion of chromosome 5. Cytogenet Cell Genet 27:270, 1980.
373. Teyssier JR, Bajolie F: Duplication-deficiency of chromosome 18, resulting from recombination of a paternal pericentric inversion, with a note for genetic counselling. Hum Genet 53:195–200, 1980.
374. Teyssier M, Moreau N: Inversion pericentrique familiale du chromosome 10. Deux nouveaux cas. Ann Genet (Paris) 26:183–186, 1983.

375. Tharapel AT, Summit RL, Wilroy RS, Martens P: Apparently balanced de novo translocations in patients with abnormal phenotypes: Report of 6 cases. Clin Genet 11:255–269, 1977.
376. Tharapel AT, Summitt RL: Minor chromosome variations and selected heteromorphisms in 200 unclassifiable mentally retarded patients and 200 normal controls. Hum Genet 41:121–130, 1978.
377. Therman E, Pätau K: Abnormal X chromosomes in man: Origin, behaviour and effects. Humangenetik 25:1–16, 1974.
378. Therman E, Meyer-Kuhn E: Cytological demonstration of mitotic crossing-over in man. Cytogenet Cell Genet 17:254–267, 1976.
379. Toomey KE, Rotter JI, Mohandas I, Kaback MM: Pericentric inversion of chromosome 7 and a balanced 10:15 translocation in skin and blood of an individual with acute myelogenous leukemia (AML). Birth Defects XIV(6c):424, 1978.
380. Toth A, Ga'al M, S'ara G, L'aszlo J: Inherited pericentric inversion of the chromosome 1 associated with male infertility. Orv Hetil 122:2423–2425, 1981.
381. Toth A, Gaal M, Laszlo J: Familial pericentric inversion of the Y chromosome. Ann Genet (Paris) 27:60–61, 1984.
382. Trochet-Royer C, Sele B, Jalbert P, Pison H, Racinet C, Bernard P: Avortements spontanés et anomalies chromosomiques parentales. Rev Fr Gynecol Obstet 76:195–199, 1981.
383. Tropp MR, Currie M: Mosaic trisomy 9: Two additional cases. Hum Genet 38:131–135, 1977.
384. Trunca C, Opitz JM: Pericentric inversion of chromosome 14 and the risk of partial duplication of 14q (14q31→14qter). Am J Med Genet 1:217–228, 1977.
385. Tsenghi C, Kalpini-Mavrou A, Panagou M, Metaxotou-Stavridaki C: Polymorphisms of chromosome No. 9 in 400 Greek subjects. Clin Genet 10:373, 1976.
386. Tüür S, Kaosaar M, Mikelsaar A-V: 1q+ variants in a normal adult population (one with a pericentric inversion). Humangenetik 24:217–220, 1974.
387. Turleau C, de Grouchy J, Klein M: Phylogénie chromosomique de l'homme et des primates hominiens. Essai de reconstitution du caryotype de l'ancetre commun. Ann Genet (Paris) 15:225–240, 1972.
388. Turleau C, Plachot M, Chavin-Colin F, Roubin M, Langmaid H, Cochet C, Blanc-Brude M, Lety MA, de Grouchy J: Distribution des points de cassure chromosomique dans les remaniements constitutionnels. Lyon Med 233:329–335, 1975.
389. Vamos E, Pratola D, van Regemorter N, Freund M, Flament-Durand J, Rodesch F: Prenatal diagnosis and fetal pathology of partial trisomy 20p-monosomy 4p resulting from paternal translocation. Prenatal Diagn 5:209–214, 1985.
390. van der Linden, AGJM, Pearson PL, van de Kamp JJP: Cytological assessment of meiotic exchange in a human male with a pericentric inversion of chromosome No. 4. Cytogenet Cell Genet 14:126–139, 1975.
391. van Heemert C: Somatic pairing and meiotic nonrandom disjunction in a pericentric inversion of Hylemya antiqua (Meigen). Chromosoma 59:193–206, 1977.
392. Vara P: In Schwalm H, Döderlein G (eds): "Klinik der Frauenheilkunde und Geburtshilfe," Bk 8. Munich: Urban & Schwarzenberg, 1972, p 212.
393. Vejerslev L, Friedrich U: Experiences with unexpected structural chromosome aberrations in prenatal diagnosis in a danish series. Prenatal Diagn 4:181–186, 1984.
394. Verma RS, Dosik H, Wexler IB: Inherited pericentric inversion of chromosome No. 2 with Robertsonian translocation (13q14q) resulting in trisomy for chromosome 13q. J Genet Hum 25:295–301, 1977.

395. Vianna-Morgante AM, Nozaki, MJ, Ortega CC, Coates V, Yamamura Y: Partial monosomy and partial trisomy 18 in two offspring of carrier of pericentric inversion of chromosome 18. J Med Genet 13:366–370, 1976.
396. Vigi V, Maraschio P, Bosi G, Guerine P, Fraccaro M: Risk of recombinants in pericentric inversions of the (p11-q21) region of chromosome 18. Hum Genet 37:1–5, 1977.
397. Vine DT, Yarkoni S, Cohen MM: Inversion homozygosity of chromosome No. 9 in a highly inbred kindred. Am J Hum Genet 28:203–207, 1976.
398. Vogel F: Moderne Aspekte der Evolutionsgenetik. Z Morphol Anthropol 69:110–116, 1978.
399. Vogel F, Motulsky AG: Human genetics: Problems and approaches. Heidelberg: Springer, 1979.
400. Walzer S, Gerald PS: A chromosome survey of 13,751 male newborns. In Hook EB, Porter IH (eds): "Population Cytogenetics Studies in Humans." New York: Academic Press, 1977.
401. Wang HC, Fedoroff S: Banding in human chromosomes treated with trypsin. Nature 235:52–54, 1972.
402. Warter St, Ruch J-V, Lehmann M: Karyotype with chromosomal abnormality with various inherited defects in the offspring (recombination aneusomy). Humangenetik 20:355–359, 1973.
403. Watt JL, Olson IA, Johnston AW, Ross HS, Couzin DA, Stephen GS: A familial pericentric inversion of chromosome 22 with a recombinant subject illustrating a 'pure' partial monosomy syndrome. J Med Genet 22:283–287, 1985.
404. Weitkamp LR, Janzen MK, Guttormsen SA, Gershowitz H: Inherited pericentric inversion of chromosome number two: A linkage study. Ann Hum Genet 33:53, 1969.
405. Wenger SL, Steele MW: Meiotic consequences of pericentric inversions of chromosome 13. Am J Med Genet 9:275–283, 1981.
406. Wertelecki W, Gerald PS: Clinical and chromosomal studies of the 18q− syndrome. J Pediatr 78:44–52, 1971.
407. Wikramanayake E, Renwick JH, Ferguson-Smith MA: Chromosomal heteromorphisms in the assignment of loci to particular autosomes: A study of four pedigrees. Ann Genet (Paris) 14:245–256, 1971.
408. Williamson EM, Miller, JF, Seabright M: Pericentric inversion (13) with two different recombinants in the same family. J Med Genet 17:309–312, 1980.
409. Winsor EJT, Palmer CG, Ellis PM, Hunter JLP, Ferguson-Smith MA: Meiotic analysis of a pericentric inversion, inv(7)(p22q32), in the father of a child with a duplication-deletion of chromosome 7. Cytogenet Cell Genet 20:169–184, 1978.
410. Winsor EJ, Welch JP: Prader-Willi syndrome associated with inversion of chromosome 15. Clin Genet 24:456–461, 1983.
411. Wirtz A, Stengel-Rutkowski S, Krauss C, Murken JD: Pericentrische Inv(6)13, Informationsblatt. In Murken JD, Stengel-Rutowski S (eds): "Pränatale Diagnostik genetischer Defekte." Munich: Kinderklinik der Universität, 1977, p 90.
412. Wirtz A, Stengel-Rutkowski S, Stene J, Frankenberger R, Krauss C, Albert A, Götz-Sothmann M, Murken J: Pränatale Diagnostik bei Eltern mit einer balanzierten chromosomalen Strukturaberration. Geburtsh Frauenheilkd 40:575–592, 1980.
413. Woessner S, Lafuente R, Florensa L, Sans-Sabrafen J, Antich J: Acute myelomonocytic leukaemia with abnormal eosinophils and inversion of chromosome 16. Blut 50:45–50, 1985.

414. Yamamoto M, Fujimori R, Ito T, Kamimura K, Watanabe G: Chromosome studies in 500 induced abortions. Humangenetik 29:9–14, 1975.
415. Yamamoto M, Ito T, Watanabe G: Eine weitere Chromosomenanalyse bei künstlichen Aborten. Hum Genet 34:69–71, 1976.
416. Yang-Feng TL, Finley SC, Finley WH, Francke U: High resolution cytogenetic evaluation of couples with recurring fetal wastage. Hum Genet 69:246–249, 1985.
417. Yunis E, de Caballero OT: Duplication deficiency as a result of meiotic segregation of a maternal inv(10). Hum Genet 57:71–74, 1981.
418. Yunis JJ: Recurrent chromosomal defects are found in most patients with acute nonlymphocytic leukemia. Cancer Genet Cytogenet 11:125–137, 1984.
419. Zabel B, Jovanovic V: "Informationsblatt Pränatal-diagnostik Genetische Defekte." Munich: Kinderklinik der Universität, 1979.
420. Zankl H, Zang KD: Structural variability of the normal human karyotype. Hum Genet 13:160–162, 1971.
421. Zeuthen E, Nielsen J: Pericentric Y inversion in the general population. Humangenetik 19:265–270, 1973.

(See appendix page 211)

Appendix. Summary of Cases With Inversions

APPENDIX. Summary of cases with inversions

Nos.	Author	Family specification	Chromosome	Breakpoints p	Breakpoints q	Inv type	Inv segment/chromosome length (%)	Inv segment/haploid autosomal length (%)	Familial	De novo	Recombinants
1.1	Selezner and Dzenis (1972)	[338]	1								
1.2	Bhasin et al. (1973) [30]	J.531/72	1	11	21	I	15	1.3			
1.3	Jacobs et al. (1974) [182]	K212/171/72	1	32	42		74	6.2	+		
1.4	Lee et al. (1974) [214]	V12CL	1	13	23		26	2.2	+		
1.5	Crippa and Ferrier (1975) [77]	DeRe	1	13	21	I	21	1.8	+		
1.6	Our case	Wa.10/75	1	13	21	I	21	1.8	+		
1.7	Mounoud (1976)	[261]	1	13	21	I	21	1.8			
1.8	Schwanitz et al. (1977)	[335]	1	13	25		34	2.9		+	
1.9	Our case	Treu92/77	1	13	21	I	21	1.8	+		
1.10	Howard-Peebles (1978) [173]		1	13	23		26	2.2	+		
1.11	Giraldo et al. (1981) [141]		1	13	25		34	2.9	+		
1.12	Toth et al. (1981) [380]		1	34	23		60	5.1	+		
1.13	Stahl-Maugé et al. (1982) [357]		1	13	21	I	21	1.8	+		
1.14	Ohyashiki et al. (1982) [280]		1			I					
1.15	Stuehrenberg (1983) [362]		1	33	32		73	6.3		+	
1.16	Rivera et al. (1984a) [316]		1	35	21.3		55	4.8	+		
1.17	Johnson et al. (1985) [184]		1	36.1	32		80	6.9	+		
2.1	Subrt et al. (1973) [365]		2	25	13		46	3.7			
2.2	Jacobs et al. (1974) [182]	207/161/72	2	11	13	I	11	0.9	+		
2.3	Our case	Wo230/74	2	11	13	I	11	0.9	+		
2.4	Leonard et al. (1975) [220]	Yves D.	2	11	13	I	11	0.9			
2.5		Genev. L.	2	11	13	I	11	0.9	+		

Inv + aberration	Aberration only	Inv + symptoms	Chronic abortions	Sterility	Screening	Malformed child without cytogenetic findings	Carrier: Father	Carrier: Mother	Carrier: Grandfather	Carrier: Grandmother	Remarks
		+									Disturbed sexual development
			+								
					+			+			
		+						+	+	+	Mucopolysaccharidosis
		+					+			+	Fanconi syndrome
		+					+				Mental retardation + malformations
		+								+	Mental retardation
del(1)											
			+				+			+	Chronic abortions
					+			+			Two mentally retarded persons not undergoing cytogenetic investigations
				+							Three infertile brothers (oligospermia. rsp. azoospermia)
				+				+			Infertility
47,XYY		+*									Two different carriers; *Goldenhar's syndrome, male
		+									CML; acustic neurinoma, primary aldosteronism, female
											Amniotic fluid finding, no paterinity test
				+				+			Azoospermia
			+								Three miscarriages; 20 other carriers normal
dup(7) (pl3–14)											
					+		+				
der(5)								+			Facial dysmorphia + pulmonary stenosis
			+								
						+		+		+	Spina bifida; myelomeningocele

(continued)

APPENDIX. Summary of cases with inversions (continued)

Nos.	Author	Family specification	Chromosome	Breakpoints p	Breakpoints q	Inv type	Inv segment/chromosome length (%)	Inv segment/haploid autosomal length (%)	Familial	De novo	Recombinants
2.6		Jean Paul M.	2	11	13	I	11	0.9			
2.7		Fam. 4	2	11	13	I	11	0.9	+		
2.8	Cohen et al. (1975) [65]		2	15	21		31	2.5	+		
2.9	Our case	Neu341/75	2	11	13	I	11	0.9	+		
2.10		Hö 192/76	2	11	13	I	11	0.9	+		
2.11	Verma et al. (1977) [394]		2	11	14		14	1.1			
2.12	Phillips (1978) [198]	Fam. 1	2	11	13	I	11	0.9	+		
2.13		Fam. 2	2	13	11		13	1.0	+		
2.14	Our case	Lie269/78	2	11	13	I	11	0.9			
2.15	Giraud et al. (1979) [142]	Case 1	2	12	14	II	17	1.4			
2.16		Case 2	2	11	13	I	11	0.9			
2.17		Case 3	2	12	14	II	17	1.4			
2.18		Case 4	2	12	14	II	17	1.4			
2.19		Case 5	2	12	14	II	17	1.4			
2.20		Case 6	2	12	14	II	17	1.4			
2.21		Case 7	2	12	14	II	17	1.4			
2.22	Hesselbjerg and Friedrich (1979) [165]		2	13	21		15	1.2	+	+	
2.23	Ayme et al. (1979) [12]	Case 3	2	12	14	II	17	1.4	+		
2.24	Baccichetti et al. (1980) [15]	Case 1	2	11	13	I	11	0.9	+		
2.25		Case 2	2	11	13	I	11	0.9			
2.26		Case 3	2	11	13	I	11	0.9	+		
2.27		Case 4	2	11	13	I	11	0.9	+		
2.28	Daniel (1981) [80]	Case 290/78	2	11	13	I	11	0.9			

Inv + aberration	Aberration only	Inv + symptoms	Chronic abortions	Sterility	Screening	Malformed child without cytogenetic findings	Carrier: Father	Carrier: Mother	Carrier: Grandfather	Carrier: Grandmother	Remarks
						+	+				Stillbirth; chromosomes normal, but without banding
XO								+			Turner syndrome
XO/Xi(Xq)							+				Turner syndrome
		+						+			+Acrocephalosyndactylia
						+	+		+		Heart malformation (1 child); encephalorrhagia, hydrocele (2 children)
tr13/14, tris 13								+			Unbalanced t13/14 de novo + inv(2)mat
			+			+		+			Chronic abortions; sickle cell anemia
						+		+			Sickle cell anemia
						+		+			Microgyria; hydrocephalus
		+	+								Encephalocele + dysmorphies
											Bilan system IAD
											Bilan system IAD
		+									Encephalocele + dysmorphies
			+								Chronic abortions
		+									Ectopie testiculaire
				+							Oligospermia
					+		+		−(!)	−(!)	Found in amniotic fluid
		+						+		+	Two children mentally and physically retarded, facial dysmorphia
		+					+		+		Dolychocephaly, congenital heart disease, muscular hypotonia
			+					+			Two abortions
			+					+	+		Two abortions
		+						+		+	Ichthyosis; 3 stillbirths
		+									Epileptic male, short stature

(continued)

APPENDIX. Summary of cases with inversions (continued)

Nos.	Author	Family specification	Chromosome	Breakpoints p	Breakpoints q	Inv type	Inv segment/chromosome length (%)	Inv segment/haploid autosomal length (%)	Familial	De novo	Recombinants
2.29	Mattevi et al. (1981) [242]		2	12	36		64	5.1	+		
2.30	Steinbach (1981) [354]	Case 1	2	13	21		15	1.2	+		
2.31		Case 2	2	11	13	I	11	0.9	+		
2.32		Case 3	2	11	13	I	11	0.9	+		
2.33	Romain et al. (1982) [322]		2	11	13	I	11	0.9	+		
2.34	Fryns et al. (1983) [129]	C.A.	2	13	11		13	1		+*	
2.35	Orye et al. (1983) [282]		2	21	11		30	2.3	+		
2.36	Meyer et al. (1984) [246]		2	13	21		15	1.2	+		
2.37	Vejerslev and Friedrich (1984) [393]	Case 18	2	11	13	I	11	0.9	+		
2.38		Case 19	2	11	13	I	11	0.9	+		
2.39		Case 20	2	11	11		8	0.7	+		
2.40		Case 21	2	*	*					+	
2.41		Case 22	2	11	13	I	11	0.9	+		
2.42		Case 23	2	12	14	II	17	1.4	+		
2.43	MacDonald and Cox (1985) [229]	Case 1	2	11	13	I	11	0.9			
2.44		Case 2	2	11	13	I	11	0.9			
2.45		Case 3	2	11	13	I	11	0.9	+		

Inv + aberration	Aberration only	Inv + symptoms	Chronic abortions	Sterility	Screening	Malformed child without cytogenetic findings	Carrier: Father	Carrier: Mother	Carrier: Grandfather	Carrier: Grandmother	Remarks
		+						+		+	Micrognatism, microphthalmia, syndactyly, jejunal atresia, retardation
		+						+			Defect of the ventricular septum
					+			+			Found in amniotic fluid
					+		+				Found in amniotic fluid
							+				Inv(5)(p13q13) pat in the same patient: "psychiatric problems, alcoholic"
		+									Mosaicism*: 46,XY/ 46,XYinv(2) 88:24, only blood investigated; psychomotor retardation, facial stigmata.
t(5;22), i(5p)								+			
		+						+			Bilateral optic atrophy
							+				Found in amniotic fluid; normal boy
							+				Found in amniotic fluid; normal boy
								+			Found in amniotic fluid; normal girl
											No breakpoints given*, found in amniotic fluid; abortion induced; autopsy: without malformations
*	*						+				Found in amniotic fluid; normal boy; "father's sister with Down syndrome"
								+			Found in amniotic fluid; normal girl; mental retardation in the family
			+								Only short communication
			+								Only short communication
			+					+			Found in amniotic fluid

(continued)

APPENDIX. Summary of cases with inversions (continued)

Nos.	Author	Family specification	Chromosome	Breakpoints p	Breakpoints q	Inv type	Inv segment/chromosome length (%)	Inv segment/haploid autosomal length (%)	Familial	De novo	Recombinants
2.46		Case 4	2	11	13	I	11	0.9	+		
2.47		Case 5	2	11	13	I	11	0.9	+		
2.48	Sole-Pujol et al. (1985) [347]		2	23	11.2		32	2.6	+		
3.1	Boué and Boué (1973) [35]	2822	3	25	21	I	61	4.2	+		dup(q)2x
3.2	Jacobs et al. (1974) [182]	66/201/69	3	14	27		61	4.2			
3.3	Hirschhorn et al. (1973) [166]	B(GC829)	3	25	21	I	61	4.2	+		dup(q)2x
3.4	Allderdice (1973) [1]	Case 1	3	15!	12		29	2			
3.5		Case 2	3	15!	11		23	1.6			dup(q)
3.6	Soudek et al. (1974) [350]	A.C.	3	13	12	II	19	1.3			
3.7		S.L.	3	13	12	II	19	1.3	+		
3.8		L.F.	3	13	12	II	19	1.3			
3.9	Betz et al. (1974) [29]		3	12	24		37	2.5	+		
3.10	Butler et al. (1974) [51]		3	25	11		42	2.9		+	
3.11	Catti (1975) [60]		3	27	21		66	4.5	+		
3.12	Allderdice (1974, 1975) [2,3]		3	25	21	I	61	4.2	+		1×dup(q) 2×dup(q
3.13	Kawashima and Maruyama (1979) [197]		3	25	21	I	61	4.1	+		dup(q)
3.14	Pope et al. (1979) [303]		3	25	25		73	5.0	+		dup(q)
3.15	Fogle and McKenzie (1980) [119]		3	12	11		11	0.8	+		
3.16	Rivera et al. (1979) [315]		3	14	24		44	3	+		
3.17	Sutherland et al. (1981) [370]		3	25	23		67	4.6	+		2× dup(q)
3.18	Faed et al. (1982) [109]		3	25	21	I	61	4.1			

Inv + aberration	Aberration only	Inv + symptoms	Chronic abortions	Sterility	Screening	Malformed child without cytogenetic findings	Carrier: Father	Carrier: Mother	Carrier: Grandfather	Carrier: Grandmother	Remarks
								+			Found in amniotic fluid
								+			Found in amniotic fluid
						+	+		+		Holoprosencephaly
							+			+	
					+						Mental retardation
							+				
						+	+				Stillbirth; polycystic; kidneys; two abortions
								+			Several malformations in the family
		+									Mental retardation, short stature, strabismus
		+					+		+		Mental retardation, petit mal epilepsy
		+									Mental retardation, strabismus, epicanthus, clinodactyly
		+						+	+		Two sisters mentally retarded, parents consanguineous
		+					–	–	–	–	Mental and motor retardation, facial dysmorphia
	tris 18							+		+	
							+		+		Three cases in one family
								+		+	
								+			
								+			Additional inv(9)
						+					
		+					+				Macrocephaly and minor malformations; IQ 110
							+	+			In two related families
				+							Subfertile male

(continued)

APPENDIX. Summary of cases with inversions (continued)

Nos.	Author	Family specification	Chromosome	Breakpoints p	Breakpoints q	Inv type	Inv segment/chromosome length (%)	Inv segment/haploid autosomal length (%)	Familial	De novo	Recombinants
3.19	Stine et al. (1982) [361]		3	25	21	I	61	4.1	+		*
3.20	Migliori et al. (1983) [248]	F.S.	3	25	21	I	61	4.1	+		dup(q)
3.21	Spedicato et al. (1984) [354]		3	11	21		22	1.6	+		
4.1	Dallapiccola et al. (1974) [79]		4	13	35		75	4.7	+		dup(p)
4.2	Rethoré et al. (1974) [311]		4	14	35	I	81	5.1	+		dup(p)
4.3	Soukup et al. (1974) [352]		4	16	21		44	2.7		+	
4.4	Lozzio and Klepper (1974) [224]		4	12	34		65	4.1			
4.5	van der Linden et al. (1975) [390]		4	14	25		38	2.4	+		
4.6	Serville and Broustet (1977) [339]		4	16	24		58	3.7		+	
4.7	Morichon-Delvallez P. and Vendrely (1981) [257]		4	16	11		27	1.7			
4.8	Baccichetti et al. (1982) [16]	Family A	4	15.2	11		20	1.3	+		
4.9		Family B	4	15.2	11		20	1.3	+		
4.10	Carpenter et al. (1982) [54]		4	15.2	12		24	1.6	+		
4.11	Bocchini et al. (1983) [32]		4	14	35	I	81	5.1	+		dup(p)
4.12	Murray et al. (1985) [264]		4	14	35	I	81	5.1	+		2×dup(p)
5.1	Faed et al. (1972) [107]		5	13	33		73	4.4	+		dup(q)
5.2	Warter et al. (1973) [402]		5	13	35		79	4.8	+		dup(q)/dup(p)
5.3	Crawfurd and Mason (1973) [75]		5	15	31		82	5	+		
5.4	Jacobs et al. (1974) [182]	K96/82/69	5	13	13	I	30	1.8			
5.5	Our case	Wo.230/74	5	15	31					+	
					0.33		82	5	(+)		
5.6	Turleau et al. (1975) [387]		5	14	14		43	2.6	+		

Inv + aberration	Aberration only	Inv + symptoms	Chronic abortions	Sterility	Screening	Malformed child without cytogenetic findings	Carrier: Father	Mother	Grandfather	Grandmother	Remarks
45,XO/ 46,Xi(Xq)							+			+	Five family members with the clinical signs of dup.-def.inv(3)q*
								+			Female
		+					+				Female; facial dysmorphism
							+		+		
							+		+		
	inv(4)+ tr4/10										Familial t (13;14); paternity excluded
											No information
							+			+	Testes investigation
	inv(4)+, del(4q)						–	–			Paternity not tested
			+				+				Grandparents not tested
		+	+					+		+	Two cases in one family
	+*							+	+		46,XY,−21,+der21t (inv4;21)*
		+						+			Homozygous for inv(4)
								+		+	
											* Carrier not given
			+					+		+	No banding
							+				
		+					+		+		Three mentally retarded patients
		+									Mental retardation
	inv(5)+						–	–			
	del(5)							(+)			
	inv(2)										
			+								No information on whether father or mother was carrier

(continued)

APPENDIX. Summary of cases with inversions (continued)

Nos.	Author	Family specification	Chromosome	Breakpoints p	Breakpoints q	Inv type	Inv segment/chromosome length (%)	Inv segment/haploid autosomal length (%)	Familial	De novo	Recombinants
5.7	Lin et al. (1976) [223]		5	15	11		27	1.6	+		
5.8	Lee et al. (1978) [215]	Case II	5	13	13	I	30	1.8	+		
5.9	Testa et al. (1980) [372]		5	13	13	I	30	1.8			
5.10	Eberle et al. (1982) [96]	Dal78/80	5	15	22		68	4.1	+		
5.11	Daniel (1981) [80]		5	14	15		49	2.9	+		
5.12	Li et al. (1982) [222]	Case 4	5	13	33		70	4.1	+		
5.13	Romain et al. (1982) [322]		5	13	13	I	30	1.8	+		
5.14	Barbata et al. (1983) [20]		5	15	11		27	1.6	+		
5.15	Körner et al. (1983) [204]		5	13	35		79	4.8	+		dup(p)
										+	dup(q)
5.16	Beemer et al. (1984) [26]		5	151	333		78	4.7	+		dup(q)
6.1	Lamm et al. (1974) [211]		6	21–25	22–25		ca. 74	4.4	+		
6.2	Stengel-Rutkowski and Frankenberger (1975) [360]		6								
6.3	Nielsen and Sillesen (1975) [276]		6								
6.4	Schmidt and Samson (1978) [333], Schmidt (1982) [332]		6	12	15		26	1.5		+	
6.5	Wirtz et al. (1977, 1980) [411, 412]		6	25	22		79	4.6	+		dup(q)
6.6	Magenis et al. (1978) [233]	3410	6	23	23		77	4.5	+		dup(q)
6.7	Pearson et al. (1979) [293]		6	21	27		74	4.3	+		dup(q)
6.8	Heinemann et al. (1979) [160]		6	12*	16*		31	1.8	+		
6.9	Schroer et al. (1980) [334]		6	25	25		89	5.2	+		dup(q)
6.10	Duckett and Roberts (1980) [89]		6	25	21		68	4.0		+	
6.11	Andreev et al. (1983) [4]		6	21	27		74	4.8	+		
7.1	Dutrillaux (1973) [90]		7	15	11		29	1.6		+	
7.2	Winsor et al. (1978) [409]	GIIIWH	7	22	32		83	4.4	+		dup(q)2×

Inv + aberration	Aberration only	Inv + symptoms	Chronic abortions	Sterility	Screening	Malformed child without cytogenetic findings	Carrier: Father	Carrier: Mother	Carrier: Grandfather	Carrier: Grandmother	Remarks
					+			+			Finding in amniotic fluid
metacentr. fragment								+		+	Atypical recombinant?
+t(17;22)			+					+			No family investigation
		+					+			+	Cardiomyopathy
					+		+			+	Finding in amniotic fluid
								+	+		
							+				Inv(2)(p11q13) in the same patient; ''psychiatric problems, alcoholic''
		+									Myelofibrosis
								+	+		
			+				+				
					+			+		+	Newborn screening
47,XXY,inv(6)											Secondary finding; Klinefelter syndrome
					+						Newborn screening
		+					–	–			Retardation, hydrocephalus, facial dysmorphia; no paternity tests
			+					+			
							+		+		HLA studies
			+				+		+		HLA studies
		+					+				Breakpoint estimated*; Rieger syndrome
			+					+		+	
		+									No detailed information with respect to paternity
			+					+			
		+*					–	–			Oligospermia
							+	+			Two cases in one family

(continued)

APPENDIX. Summary of cases with inversions (continued)

Nos.	Author	Family specification	Chromosome	Breakpoints p	Breakpoints q	Inv type	Inv segment/chromosome length (%)	Inv segment/haploid autosomal length (%)	Familial	De novo	Recombinants
7.3	Toomey et al. (1978) [379]		7	15	22		51	3.1		+*	
7.4	Scheres et al. (1980) [329]	Case 2	7	13	32		57	3.4			
7.5		Case 3	7	13	32		57	3.4			
7.6		Case 1	7	13	32		57	3.4			i[inv(7)q]
7.7	Petit and Fryns (1983) [296]		7	15	32		69	4.1			
8.1	Ferguson-Smith (1967) [112]		8	23	11	I	33	1.6	+		
8.2	Jacobs et al. (1974) [182]	R4/155/61	8	23	11	I	33	1.6	+		
8.3		K12/92/64	8	23	11	I	33	1.6	+		
8.4		K14/14/65	8	23	11	I	33	1.6	+		
8.5	Allderdice (1974) [2]		8	23	22	II	74	3.6			
8.6	Fujimoto et al. (1975) [132]		8	23	22	II	74	3.6	+		dup(q)2×
8.7	Our case	Fi.136/75	8	11	21		37	1.8	+		
8.8	Kaiser et al. (1976) [191]	J.Sch.11/75	8	11	22		47	2.3	+		
8.9	Herva and de la Chapelle (1976) [163]		8	11	24		65	3.2	+		
8.10	Kawana et al. (1976) [196]		8	23	23		84	4.1		+	
8.11	Moedjono and Sparken (1980) [253]		8	23	22	II	74	3.6	+		dup(q)
8.12	Sugansky et al. (1981) [366]	Case 1	8	23	22	II	74	3.6	+		dup(q)
8.13		Case 2	8	23	22	II	74	3.6	+		dup(q)
8.14		Case 3	8	23	22	II	74	3.6	+		dup(q)
8.15		Case 4	8	23	22	II	74	3.6	+		dup(q)
8.16		Case 5	8	23	22	II	74	3.6	+		dup(q)
8.17		Case 6	8	23	22	II	74	3.6	+		dup(q)2×
8.18		Case 7	8	23	22	II	74	3.6	+		dup(q)
8.19	Bui et al. (1982) [49]		8	23108/ 12100			41	2	+		
8.20	Stanley et al. (1985) [358]		8	12	13		25	1.2	+		dup(q)
8.21	Barns et al. (1985) [21]		8	23	24		88	4.2	+		dup(q)

Inv + aberration	Aberration only	Inv + symptoms	Chronic abortions	Sterility	Screening	Malformed child without cytogenetic findings	Carrier: Father	Carrier: Mother	Carrier: Grandfather	Carrier: Grandmother	Remarks
+							–	–			*+t(10;15)de novo, retardation, facial dysmorphia, acute myeloid leukemia
											Ataxia telangiectasia
											Ataxia telangiectasia
											Iso-inv(7q)
				+							Atresia of vas deferens; +inv(9)(p11q13)
						+		+		+	
		+						+			Secondary amenorrhoea
		+						+			"Congenital abnormalities"
		+						+			"Congenital abnormalities"
											No detailed information
								+			Two cases in a large family
							+				
		+				+	+				Multiple malformations, exitus
	Xp-							+			Turner-syndrome
			+				+				Male carrier; spermiogram normal
		+									Multiple malformations; mikrodeletion supposed
							+		+		
							+				
							+				Common origin suggested
							+		+		
								+			
								+			Common origin suggested
							+				
								+			
								+			Found in amniotic fluid
			+				+				
			+					+			

(continued)

APPENDIX. Summary of cases with inversions (continued)

Nos.	Author	Family specification	Chromosome	Breakpoints p	Breakpoints q	Inv type	Inv segment/chromosome length (%)	Inv segment/haploid autosomal length (%)	Familial	De novo	Recombinants
9.1	Mutton and Daker (1973) [265]	PRU6615	9	11	13	I	20	1	+		t18/21
9.2	Bowen et al. (1974) [40]		9	11	12	II	14	0.7	+		Trisomy 9- mos.
9.3	Schinzel et al. (1974) [330]		9	12	13	I	26	1.2	+		Trisomy 9- mos.
9.4	de la Chapelle et al. (1974) [83]	P	9	1	13	I	28	1.3	+		
9.5		Ma	9	1	13	I	28	1.3			
9.6		V	9	1	13	I	28	1.3	+		
			Homocygot								
9.7		S	9	1	13	I	28	1.3	+		
9.8		T	9	1	13	I	28	1.3	+		45,XO
9.9		J	9	1	13	I	28	1.3	+		
9.10	Boué (1975b) [38]	1	9	13	21		37	1.8	+		
				11	13	I	20	1			
9.11		2	9	13	21		37	1.8	+		
				11	13	I	20	1			
9.12		3	9	13	21		37	1.8			
				11	13	I	20	1			
9.13		4	9	13	21		37	1.8			
				11	13	I	20	1			
9.14		5	9	13	21		37	1.8	+		
				11	13	I	20	1			
9.15		6	9	13	21		37	1.8	+		
				11	13	I	20	1			
9.16		7	9	13	21		37	1.8	+		
				11	13	I	20				
			Homocygot								
9.17		8	9	13	21		37	1.8			
				11	13	I	20	1			
9.18		9	9	13	21		37	1.8	+		
				11	13	I	20	1			

Inv + aberration	Aberration only	Inv + symptoms	Chronic abortions	Sterility	Screening	Malformed child without cytogenetic findings	Carrier				Remarks
							Father	Mother	Grandfather	Grandmother	
										+	Grandmother:inv(9) and t18/21; propositus unbalanced
								+			
								+			
						+	+			+	Two children with malformations; cytogenetics not studied
			+								Chronic abortions; carrier: husband
		+					+	+			Consanguinity; cleft lip, physical retardation
		+						+			Turner signs, 46,XX,inv(9)
								+			
		+					+		+		Malformations
			+			+	+				+Malformations, cytogenetics normal by conventional staining
				+			+				Primary sterility
				+			+				Primary sterility
		+					+				Oligoasthenospermia; one abortion
	Trisomy 21						+				
	Trisomy 21						+				
	Trisomy 8		+				+				
			+					+			One abortion; one stillbirth
			+					+			One stillbirth

(continued)

APPENDIX. Summary of cases with inversions (continued)

Nos.	Author	Family specification	Chromosome	Breakpoints p	Breakpoints q	Inv type	Inv segment/chromosome length (%)	Inv segment/haploid autosomal length (%)	Familial	De novo	Recombinants
9.19		10	9	13	21		37	1.8	+		
				11	13	I	20	1			
9.20	Turleau et al. (1975) [388]	N1721	9	11	13	I	20	1			
9.21		N1733	9	11	13	I	20	1			
9.22		N1285	9	11	13	I	20	1			
9.23		N1737	9	11	13	I	20	1			
9.24		N1249	9	11	13	I	20	1			
9.25		N1014	9	11	13	I	20	1			
9.26		N1298	9	11	13	I	20	1			
9.27		N1203	9	11	13	I	20	1			
9.28		G6063	9	11	13	I	20	1			
9.29		N1194	9	11	13	I	20	1			
9.30	Hansmann and Keutel (1975) [155]		9	24	12		44	2.1		+	
9.31	Hongell (1976) [170]		9	11	13	I	20	1	+		
9.32	Haas et al. (1976) [147]	Case II	9	11	13	I	20	1	+		
				13	21		37	1.8			
9.33	Howard-Peebles and Stoddard (1976b, 1979) [176, 177]	Case 4 Fam. T.	9	12	13	I	26	1.2	+		
9.34		Case 3 Fam. R.	9	12	21	I	35	1.7	+		
9.35		Case 2 Fam. D.	9	13	13	III	30	1.4	+		
9.36		Case 1 Fam. N.	9	13	21	I	37	1.8	+		
9.37	Vine et al. (1976) [397]		9	11	13	I	20	1	+		
			Homocentric								
9.38	Bec et al. (1977) [25]	Case 1	9	11	13	I	20	1	+		
9.39	Pescia et al. (1977) [295]	Fam. 1	9	1	13	I	28	1.3	+		

Inv + aberration	Aberration only	Inv + symptoms	Chronic abortions	Sterility	Screening	Malformed child without cytogenetic findings	Carrier				Remarks
							Father	Mother	Grandfather	Grandmother	
Balanced tl/7								+			Malformations; tentative diagnosis: hyperphosphatasia
		+									Amenorrhoea
		+									Amenorrhoea
		+									Amenorrhoea
				+							Sterility, male
				+							Sterility, male
			+								Chronic abortions; carrier: husband
			+								Oligospermia
		+									Transsexualism
Trisomy 21								+			
						+	+				Malformations; cytogenetics normal
		+					−	−			Malformations
Partial trisomy 9							+				
			+								Habitual abortions; carrier: wife
						+		+			Daughter: cleft lip; cytogenetics normal
		+	+					+			Multiple malformations
Trisomy 21								+			
Trisomy 21								+			Yps (a satellited Y chromosome in the short arm)
		+					+	+			Ambiguous genitalia
	Homocentric										
		+					+				Retinoblastoma
		+					+				

(continued)

APPENDIX. Summary of cases with inversions (continued)

Nos.	Author	Family specification	Chromosome	Breakpoints p	Breakpoints q	Inv type	Inv segment/chromosome length (%)	Inv segment/haploid autosomal length (%)	Familial	De novo	Recombinants
9.40		Fam. 2	9	1	13	I	28	1.3	+		
9.41		Fam. 3	9	1	13	I	28	1.3	+		
9.42	Honea (1978) [169]		9						+		rec
9.43	Our observation N. 6	Eit.56/75	9	11	13	I	20	1	+		
9.44	N. 7	Hau.106/75	9	11	13	I	20	1	+		
9.45	N.8	Hec.300/76	9	11	12 or 13	II	14	0.7	+		
9.46	N.9	Hen.374/75	9	11	13	I	20	1	+		
9.47	N.10	Heu.397/76	9	11	13	I	20	1	+		
9.48	N.11	Kap.415/75	9	11	13	I	20	1	+		
9.49	N.12	Mat. 88/75	9	13	13	III	30	1.4	+		
9.50	N.13	Mue.9/75	9	13	13	III	30	1.4	+		
9.51	N.14	Pfe.336/75	9	11	13	I	20	1	+		
9.52	N.15	Rin.46/76	9	11	13	I	20	1	+		
			Homocentric								
9.53	N.16	Sal.167/76	9	11	13	I	20	1	+		
9.54	N.17	Sch.35/77	9	11 13	13	I	20	1	+		
9.55	N.18	Sch.424/74	9	11	13	I	20	1	+		
9.56	N.19	Spi.354/75	9	11	13	I	20	1	+		
9.57	N.20	Ste.196/75	9	11	13	I	20	1	+		
9.58	N.21	Sti.206/75	9	11	13	I	20	1	+		
9.59	Pauli et al. (1978) [291]	H.	9	11	13	I	20	1	+		
9.60	Mattei et al. (1978) [238]	Case 2	9						+		

Inv + aberration	Aberration only	Inv + symptoms	Chronic abortions	Sterility	Screening	Malformed child without cytogenetic findings	Carrier: Father	Carrier: Mother	Carrier: Grandfather	Carrier: Grandmother	Remarks
		+						+			Uterus agenesis
			+				+				Chronic abortions; carrier: wife
							+				Breakpoints not given; + 9qh+
		+					+				
			+					+			Chronic abortions; carrier; wife
XO/rX							+			+	
		+					+		+		
		+					+			+	+9qh+
		+					+		+		
	i(Xq)										Brother: inv (9)
del(11q)								+			
		+						+		+	
		+						+		+	Two cases of heart malformations
								Homocentric			
			+			+		+	+		+9qh+
		+						+	−		Neurotic disturbances
			+					+		+	Chronic abortions; carrier: wife
		+						+			+9qh+ of the homologous chromosome
Trisomy 21							+				9qh+ of the homologous chromosome
Trisomy 21							+				+9qh+ of the homologous chromosome
Trisomy 18			+					+			
	t9/ Xq12: p22						+		+		Breakpoints of the inv not given

(continued)

APPENDIX. Summary of cases with inversions (continued)

Nos.	Author	Family specification	Chromosome	Breakpoints p	Breakpoints q	Inv type	Inv segment/chromosome length (%)	Inv segment/haploid autosomal length (%)	Familial	De novo	Recombinants
9.61	Khan et al. (1978) [199]		9	12	12.1		19	0.9			
9.62	Faed et al. (1978) [108]		9						+		
9.63	Côté et al. (1978) [69]		9	11	13	I	20	1	+		
9.64	Sandig et al. (1979) [325]		9	12	12		19	0.9	+		
9.65	Hoo et al. (1979) [171]		9	11	12	II	14	0.7	+		
9.66	Fällström and Wahlström (1979) [110]		9	12	13	I	26	1.2	+		
9.67	Croquette et al. (1979) [78]		9	11	1209	II	14	0.7	+		
				pter	q129		18	2.2	?		
9.68	Hernandez et al. (1979) [164]		9	11	13	I	20	1	+		
9.69	Mattei et al. (1980, 1983) [239, 241]		9	22	32		74	3.4	+		dup (q) 2×
9.70	Fuenmayor et al. (1981) [131]		9	13	21	I	37	1.8	+		
9.71	Frydman et al. (1981) [127]		9	11	12	II	14	0.7	+		
9.72	Grote and Godde [146]		9	11	13	I	20	1	+		
9.73	Husslein et al. (1980)	No. 6	9	13	21	I	37	1.8			
9.74	No. 7 (180)		9	13	21	I	37	1.8			
9.75	Our observation	381/79	9	11	13	I	20	1			
9.76		12/80	9	11	13	I	20	1	+		
9.77	Howard-Peebles (1982) [174]		9	13	21	21	27	1.8			
9.78	Faed et al. (1982) [109]	695/78	9	12	12		19	0.9			
9.79		518/75	9	11	12	11	14	0.7			

Inv + aberration	Aberration only	Inv + symptoms	Chronic abortions	Sterility	Screening	Malformed child without cytogenetic findings	Carrier: Father	Carrier: Mother	Carrier: Grandfather	Carrier: Grandmother	Remarks
		+									Turner signs; (9qh+; parents not examined)
1XXY/ 1XYY							+		+		Breakpoints not given
XYY		+					+				Potter syndrome type II; encephalocele
t9;15							+				
(q11;p12) Complete rearrangment 9							+				
+t9/10								+			No detailed information
				+			(+)	+			Two different inversions of chromosome 9 in the propositus
Trisomy 21								+			One sister: Down's syndrome; deceased
							+			+	Unusual breakpoints
		+					+				Ectodermal dysplasia in two females and three males
+trisomy 9mos.								+			46,XX,inv(9)/ +inv(9)
							+				Paternity report
			+								Male carrier
			+								Male carrier
				+							Testes agenesia left
							+				Found in amniotic fluid
fra(X)											No family investigation possible
				+							Depression of spermatogenesis; testicular biopsy normal
				+							Depression of spermatogenesis; testicular biopsy normal

(continued)

APPENDIX. Summary of cases with inversions (continued)

Nos.	Author	Family specification	Chromosome	Breakpoints p	Breakpoints q	Inv type	Inv segment/chromosome length (%)	Inv segment/haploid autosomal length (%)	Familial	De novo	Recombinants
9.80	Pison et al. (1982) [299]		9	11	13	I	20	1	+		
9.81	Martin et al. (1983) [237]		9	13	11		16	0.8	+		
9.82	Petit and Fryns (1983) [296]		9	11	13	I	20	1			
9.83	Ayral et al. (1984) [13]		9			I			+		
9.84	Delprado and Baird (1984) [88]		9	11	13	I	20	1	+		
9.85	Our observation	412/84	9	11	12	II	14	0.7	+		
9.86	Rivera et al. (1984b) [317]		9	11	13	I	20	1	+		
9.87	Babu et al. (1985) [14]		9			I	20	1	+		
9.88	Fryns et al. (1985) [130]		9	11	12*	II				+	
10.1	Dutrillaux et al. (1973b) [93]		10	15	24		76	3.5	+		dup(q)
10.2	de la Chapelle et al. (1974) [83]	E	10	11	21	I	21	1	+		
10.3		Mo	10	11	21	I	21	1	+		
10.4		U	10	11	11	II	7	0.3	+		
10.5	Sekhon and Sly (1978) [737]		10								dup(p)
10.6	Yunis and de la Caballero (1981) [417]		10	11	25		57	2.6	+		dup(p)
10.7	Lansky-Shafer et al. (1981) [212]		10	11	26		67	3.1	+		dup(p)
10.8	Simpson et al. (1981) [344]		10	12	21		27	1.3			
10.9	Ostovics et al. (1982) [285]		10	13	22		43	2.1	+		
10.10	Husslein et al. (1982) [813]	Case N.6	10	12	21		27	1.3			
10.11	Teyssier and Moreau (1983) [374]	Case 1	10	15	11		30	1.4	+		

Inv + aberration	Aberration only	Inv + symptoms	Chronic abortions	Sterility	Screening	Malformed child without cytogenetic findings	Carrier: Father	Carrier: Mother	Carrier: Grandfather	Carrier: Grandmother	Remarks
							+				Mother: inv(9)+t(14;21); amniotic finding: dup(10)(q22 → qter)
t(14;21)	dup(10)q22–qter										
							+			+	
				+							Atresia of the vas deferens; +inv(7)(p15q32)
+							+				+dup(3q)t(3;14)(q21;q32)
		+						+			Cephalothoracopagus conjoined twins; mother homozygous inv (9)
t(3;17)								+			de novo translocation; phenotypically normal
i(18p)							+				
		+						+	+		Propositus: homozygous inv(9); Cornelia de Lange syndrome
			+					+		+	+9qh+*
								+			
		+					+				Corpus callosum agenesis
		+					+		+		Hypogonadotrophic hypogonadism
		+					+			+	Ambigous external genitalia
							+		+		Breakpoints not given
								+		+	+inv(7) father and propositus
								+			
			+								Female carrier
	+tris21							+			
			+								Male carrier
				+				+			Azoospermia

(continued)

APPENDIX. Summary of cases with inversions (continued)

Nos.	Author	Family specification	Chromosome	Breakpoints p	Breakpoints q	Inv type	Inv segment/chromosome length (%)	Inv segment/haploid autosomal length (%)	Familial	De novo	Recombinants
10.12		Case 2	10	11	21	I	21	1	+		
10.13	Vejerslev and Friedrich (1984) [393]	Case 24	10	11	21	I	21	1	+		
10.14	Rodriguez et al. (1984) [321]		10	15	24		76	3.5	+		dup(q)*
10.15	Our observation	260/82	10	12	21		27	1.3	+		
10.16		154/85	10	11	21	I	21	1	+		
11.1	Catti (1975) [60]		11	15	13		45	2.1	+		
11.2	Kim et al. (1975) [201]	Case 1	11							+	
11.3		Case 2	11						+		
11.4	Simola et al. (1977) [343]	Fam. 1	11	15	23		86	4	+		
11.5		Fam. 2	11	11	11		Ca.5	0.2	+		
11.6	Autio-Harmainen and de la Chapelle (1980) [10]		11	11	13.3		13	0.6	+		
11.7	Fryns et al. (1981) [128]	M.K.	11	11	25		63	2.9	+		dup(p)
11.8	Russell et al. (1982) [323]		11	12	25		67	3.1	+		
11.9	Einfeld and Smith (1982) [99]		11	15	14		60	2.8	+		
11.10	Cabrol et al. (1983) [52]		11	15	13		45	2.1	+		
11.11	Broughton et al. (1983) [45]		11	15	I		37	1.7	+		
11.12	Palka et al. (1984) [285]	DiM.A.	11	15	I		37	1.7			
11.13	Yang-Feng et al. (1985) [416]		11	11.13	12.3		7	0.3	+		
11.14	Kistner (1985) [202]	Fam. M.	11	15	I		37	1.7	+		

Inv + aberration	Aberration only	Inv + symptoms	Chronic abortions	Sterility	Screening	Malformed child without cytogenetic findings	Carrier: Father	Carrier: Mother	Carrier: Grandfather	Carrier: Grandmother	Remarks
				+			+				Oligoasthenospermia
							+				Found in amniotic fluid
Trisomy +r/ 8**								+	+		*Male **Female
							+				Found in amniotic fluid
								+		+	Found in amniotic fluid
Trisomy 21							+			+	
											Incidental finding in amniotic fluid
											Incidental finding in amniotic fluid
			+					+	+		Female carrier
					+						Family investigations; myotonic dystrophy; +inv(9)(p1q13)
				+				+			High testosterone level
							+				
							+				Found in amniotic fluid
		+					+				One brother with the same symptoms but without inv(11)
		+						+			ALL
		+						+			Congenital glaucoma + inv(11) in three family members
			+								Secondary amenorrhoea
				+				+		+	
								+		+	No symptoms

(continued)

APPENDIX. Summary of cases with inversions (continued)

Nos.	Author	Family specification	Chromosome	Breakpoints p	Breakpoints q	Inv type	Inv segment/chromosome length (%)	Inv segment/haploid autosomal length (%)	Familial	De novo	Recombinants
12.1	Kim et al. (1980) [200]		12	13	11	I	17	0.8	+		
12.2	Prieto et al. (1981) [305]	Case 1	12	13	13		47	2.2	+		
12.3		Case 2	12	13	11	I	17	0.8	+		
12.4	Poulsen et al. (1981) [304]		12	112	131		20	0.9	+		
12.5	Li et al. (1982) [222]	B373	12	1110	1310		27	1.2		+	
13.1	Hauksdottir et al. (1972, 1983) [157,158]		13	11	31		66	2.5	+		dup(q)2x
13.2	Taysi et al. (1973) [371]		13	11	14		26	1	+		dup(q)
13.3	Escobar et al. (1974) [102]		13	12	22		66	2.5	+		dup(q)
13.4	McDermott and Parrington (1975) [243]		13	13	21	I	60	2.2	+		dup(q)
13.5	Koske-Westphal et al. (1978) [206]	Pat. 1	13	11	21		46	1.7	+		dup(q)
13.6	Habedank (1979) [148]		13	13	21	I	60	2.2		+	dup(q)
13.7	Williamson et al. (1980) [408]		13	11	22		54	2	+		dup(q) dup(p)
13.8	Wenger and Steele (1981) [405]		13	13	21	I	60	2.2	+		dup(q)
13.9	Habedank (1982) [149]		13	11	21		46	1.7	+		dup(q)
13.10	Lucas et al. (1983) [228]		13	12	13		25	0.9	+		dup(q)
13.11	Maltby (1984) [234]		13	13	22		71	2.6	+		dup(q)
14.1	Trunca and Opitz (1977) [384]		14	1	24		67	2.4	+	+	dup(q)
14.2	Pfeiffer and Kessel (1978) [297]		14	11	23		42	1.5	+		dup(q)
14.3	Kaiser et al. (1984) [193]	263/83	14	12	24		65	2.3	+		dup(q)
14.4	Leal-Garza et al. (1984) [213]		14	11	21		26	0.9	+		

							Carrier				
Inv + aberration	Aberration only	Inv + symptoms	Chronic abortions	Sterility	Screening	Malformed child without cytogenetic findings	Father	Mother	Grandfather	Grandmother	Remarks
					+		+				Amniocentesis
+21			+				+				
								+			Physical retardation
						+	+			+	Amniocentesis; previous stillborn child and paracentric inv(12)
		+									Mentally retarded; highly arched palate; male
			+					+	+		Two cases in one family; testis biopsy
								+	+		
			+					+		+	
								+			
			+					+			Two abortions; 1 stillbirth
		+					+			+	+Hydrocephalus
											No paternity test
							+		+		
								+	+		
								+			Male
								+		+	Female
			+					+			Female
								+			De novo rearrangement
								+			
							+				Female
		+						+		+	Mental retardation

(continued)

APPENDIX. Summary of cases with inversions (continued)

Nos.	Author	Family specification	Chromosome	Breakpoints p	Breakpoints q	Inv type	Inv segment/chromosome length (%)	Inv segment/haploid autosomal length (%)	Familial	De novo	Recombinants
15.1	Coco and Penchaszadeh (1978) [63]		15	12	22		73	2.6	+		dup(q)
15.2	Winsor and Welch (1983) [410]		15	13	13		26	0.9	+		
16.1	Fonatsch (1977) [120]		16	11	12 or 13		26 13	0.9	+		
17.1	Winsor et al. (1978) [409]	G120 MS	17	11	25		66				
17.2	Kassam et al. (1984) [195]		17	11	21		33				
18.1	Wertelecki and Gerald (1971) [406]	Case 6	18						+	+	dup (p)
18.2	Jacobs et al. (1974) [182]	162/70	18	11	21	I	62	1.8	+		dup (p) an tris18
18.3	Turleau et al. (1975) [388]	G3730	18	11.1	21.1	I	62	1.8	+		
18.4	Our family	Vö98/75	18	11	21	I	62	1.8		+	
18.5	Vianna-Morgante et al. (1976) [395]		18	11	21	I	62	1.8		+	dup (q) dup (p)
18.6	Vigi et al. (1977) [396]		18	11	21	I	62	1.8	+		
18.7	Kukolich et al. (1978) [209]		18	11.2	21	I	62	1.8	+		dup (q)
18.8	Teyssier and Bajolie (1980) [373]		18	11	12		37	1.1	+		dup (p)
18.9	Andrews et al. (1982) [5]		18	11	11		30	0.9			dup (q) 2×
18.10	Martin and Abrisqueta (1983) [236]		18	11.2	21.3	I	62	1.8	+		dup (p)
18.11	Vejerslev and Friedrich (1984) [393]		18	11.3	21	I	62	1.8	+		
18.12	Vamos et al. (1985) [389]		18	11	11		30	0.9	+		

Inv + aberration	Aberration only	Inv + symptoms	Chronic abortions	Sterility	Screening	Malformed child without cytogenetic findings	Carrier				Remarks
							Father	Mother	Grandfather	Grandmother	
								+			
		+					+				Male; Prader-Willi Syndrome
		+					+				Hypogenitalism
											"Imbalanced child"
											Found in amniotic fluid
		+						+			
											Familial, but no exact data
						+		+			Two malformed children
								+			
		+						+	+		Peptidase-A reduced
								+			Father: inv(y)(p11q11)
						+			+		
						+					Male
			+					+		+	Male
								+			Found in amniotic fluid
t(4;20) female						+					

(continued)

APPENDIX. Summary of cases with inversions (continued)

Nos.	Author	Family specification	Chromosome	Breakpoints p	Breakpoints q	Inv type	Inv segment/chromosome length (%)	Inv segment/haploid autosomal length (%)	Familial	De novo	Recombinants
19.1	Hamerton et al. (1973) [152]	ACO90472	19	13	13		>50	1.3	+		
19.2	Jacobs et al. (1974) [182]	K82/54/70	19	13	12 or 13		>50	1.3	+		
19.3	Sutherland et al. (1976b) [364]		19	13			>50	1.3	+		
19.4	Nielsen et al. (1977) [277]		19	13	13		>50	1.3	+		
19.5	Husslein et al. (1980) [180]	Fam. 4	19	12	12		>40				
19.6	Bozzola et al. (1980) [41]		19	13	13						
20.1	Mosher and Sanger (1985) [260]		20	13.2	11.2		66		+		rec?
21.1	Fraisse (1975) [123]		21	12	22		70	1.3	+		dup (q)
22.1	Degenhardt et al. (1975) [81]		22						+		
22.2	Emanuel et al. (1976) [100]		22	12	12			0.8	+		
22.3	Cantu et al. (1981) [53]		22	13	12		62	1.2	+		dup (q)
22.4	Fujimoto et al. (1983) [133]		22	13	12.2		62	1.2	+		dup (q)
22.5	Saura et al. (1983) [327]		22	11	12		39	0.8	+		
22.6	Watt et al. (1985) [403]		22	11	12		39	0.8	+		dup (p)

Inv + aberration	Aberration only	Inv + symptoms	Chronic abortions	Sterility	Screening	Malformed child without cytogenetic findings	Carrier: Father	Carrier: Mother	Carrier: Grandfather	Carrier: Grandmother	Remarks
Trisomy 21								+			Mother: age 41 years
		+					+				Pubertas tarda
		+								+	Mental retardation and multiple malformations
+i (17p)							+				
			+					+			
+ Turner-mosaic											Mother: 45, XO, inv (19)/ 46,XX,inv (19); daughter: 46,XX, inv (19)
									+		
+trisomy 21								+		+	Mother: age 42 years
Trisomy 222×								+			
Trisomy 222×								+			Mother: age 34 years
								+			
							+				Male
48,XXY, +21, inv(22)								+			
+r(22)								+			Male recdup(q); female and male r(22)

(continued)

APPENDIX. Summary of cases with inversions (continued)

Nos.	Author	Family specification	Chromosome	Breakpoints		Inv type	Inv segment/chromosome length (%)	Inv segment/haploid autosomal length (%)	Familial	De novo	
				p	q						
X.1	Niebuhr et al. (1974) [270]		X	22	13	I	47	2.6	+		
X.2	Nikolis and Stolevic DR (1978) [279]		X	22	24		70	3.9	+		dup (p)
X.3	Maeda et al. (1979) [232]		X	22	11		35	1.9	+		dup (q)
X.4	Zabel and Jovanović (1979) [419] = Baumann et al. (1984) [24]	139/79	X	11	28	II	64	3.5	+		
X.5	Evans et al. (1980) [106]		X	11	28	II	64	3.5		+	
X.6	Soler et al. (1981) [348]		X	22	13	I	47	2.6	+		
X.7	Buckton et al. (1981) [48]	Fam. 1 K482	X	22	13	I	47	2.6	+		
X.8		Fam. 2 K491	X	22	13	I	47	2.6	+		dup(q)2
X.9	Soler et al. (1981) [348]		X	22	13	I	47	2.6	+		
X.10	Keitges et al. (1982) [198]		X	11	22		34	1.8			
X.11	Vejerslev and Friedrich (1984) [393]		X	22	24		70	3.9		+	
Y.1	Baheux et al. (1970) [17]		Y						+		
Y.2	Sparkes et al. (1970) [353]		Y								
Y.3	Grace et al. (1972) [144]		Y						+		
Y.4	Zeuthen and Nielsen (1973) [421]	No. 2494	Y						+		
Y.5		4406	Y								
Y.6	Borgaonkar (1973) [34]		Y								

Inv + aberration	Aberration only	Inv + symptoms	Chronic abortions	Sterility	Screening	Malformed child without cytogenetic findings	Carrier: Father	Mother	Grandfather	Grandmother	Remarks
							+			+	No pathologic phenotype; no abortions
								+			Turnerlike symptoms; mother premature menopause
								+			Growth retardation; streak gonads; mammae and external genitalia normal
					+		+			+	Amniocentesis; no symptoms
					+						Amniocentesis; no paternity tests
				+			+				Premature menopause resp. sterility in two patients
					+		+			+	Screening
					+				+ +	+	Amniocentesis and family investigation
				+			+				Female proband with sterility
		+									Dizygotic twins; only one with symptoms; no family investigations possible
											No symptoms; female
+Trisomy 21							+				
+Trisomy 21							+				
							+		+		Family investigation
					+		+				
					+						
					+						

(continued)

APPENDIX. Summary of cases with inversions (continued)

Nos.	Author	Family specification	Chromosome	Breakpoints p	Breakpoints q	Inv type	Inv segment/chromosome length (%)	Inv segment/haploid autosomal length (%)	Familial	De novo	Recombinants
Y.7	Nielsen and Rasmussen (1974) [272]		Y						+		
Y.8	Orye et al. (1964) [282a]		Y	11	11	I	33				
Y.9	Friedrich and Nielsen (1974) [126]	7890	Y	11	11	I	33		+		
Y.10		9748	Y	11	11	I	33		+		
Y.11	Bochkov et al. (1974) [33]		Y						+		
Y.12	Sagredo et al. (1975) [324]		Y						+		
Y.13	Sofuni et al. (1975) [346]		Y	11	12		65				
Y.14	Nielsen and Sillesen (1975) [276]	16653	Y	11	11	I	33		+		
Y.15		15732	Y	11	11	I	33		+		
Y.16	Perissell et al. (1982) [294]		Y	11	11	I	33		+		
Y.17	Ponzio et al. (1983) [302]		Y	11	11	I	33			+	
Y.18	Bullerdik et al. (1984) [50]		Y						+		
Y.19	Toth et al. (1984) [381]		Y	11	11	I	33		+		

Inv + aberration	Aberration only	Inv + symptoms	Chronic abortions	Sterility	Screening	Malformed child without cytogenetic findings	Carrier				Remarks
							Father	Mother	Grandfather	Grandmother	
					+		+				Screening in psychiatric clinic
+Trisomy 14/21							+		+		Accidental coincidence
					+		+		+		
+Ring											Family investigation
							+				
+Trisomy 21							+				
					+						
					+		+				
					+		+				
							+				Four sons with the same inv(Y); one with encephalopathy
+r(21) Mosaic											
							+				Found in amniotic fluid
			+				+				

The Cytogenetics of Mammalian Autosomal
Rearrangements, pages 249–266

Chapter 4
Paracentric Inversions and Their Clinical Implications

K. Madan

INTRODUCTION

An inversion is a chromosome structural rearrangement in which a segment of a chromosome has been inverted. It involves two breaks and reinsertion of the chromosome segment after a rotation through 180°, so that the order of genes in the inverted segment is reversed as compared to the original chromosome.

There are two types of inversions [1]. A *pericentric* inversion is one in which the centromere is included in the inverted segment. A *paracentric* inversion is one in which a segment in only one arm of the chromosome is involved.

Since the first detection in *Drosophila* by Sturtevant [2] in 1926 from an altered order of genes in a linkage group, inversions have been found to be very widely spread in natural populations of both plants and animals [3,4]. In general the presence of a chromosome inversion does not of itself give rise to phenotypic abnormality. However, meiotic crossing-over in the inverted segment in inversion heterozygotes generates chromosomally unbalanced gametes. The risk of chromosome imbalance depends on the chance of pairing and cross-over occurring within the inverted segment, which in turn is proportional to the length of the inverted segment. Pericentric inversions have been discussed in chapter 3 of this volume.

PARACENTRIC INVERSIONS

Paracentric inversions are one of the most common forms of chromosome rearrangements found in nature [3]. This is partly a consequence of the fact

Medical Faculty, Institute of Human Genetics, Free University, 1081 BT Amsterdam, The Netherlands that heterozygosis for a paracentric inversion is less prone to make its carriers semisterile than a pericentric inversion or a translocation. The role of paracentric inversions in evolution and speciation has long been recognized [5–8].

The presence of a paracentric inversion, unlike most pericentric inversions, does not alter the arm ratio of a chromosome; and it has therefore not been possible to detect paracentric inversions in mitotic cells of most animals and plants. Although the first inversions were detected through genetic linkage studies in *Drosophila* [2], they can be directly observed in the giant polytene chromosomes in the salivary gland cells of the fly larvae.

In plants and other organisms without polyteny, paracentric inversions have been studied in meiosis [9,10] as pairing in the inverted segment is achieved by a loop formation at pachytene.

Also, chiasma formation in the inverted segment leads to the formation of (dicentric) anaphase bridges and acentric fragments. The pachytene loops and the anaphase configurations can be observed in preparations of meiotic divisions and the chiasma frequency can be used to estimate the position and the size of the inverted segment [11]. Of course, short paracentric inversions which fail to form a pachytene loop or which suppress crossing-over within the inverted segment will remain undetected in meiotic studies.

Although they have been described in squirrels [12], paracentric inversions are less commonly found in natural populations of mammals. However, in the mouse, mutagenically induced paracentric inversions [13] have been extensively studied in linkage experiments [14] and in meiotic [15,16] as well as in banded mitotic chromosome preparations [17,18].

Also in man paracentric inversions were undetectable in mitotic chromosomes until the advent of banding pattern techniques in 1970. Since the first report in 1974 [19], paracentric inversions in man have appeared with a steadily increasing frequency in published reports and are a regular finding among patients referred to cytogenetic laboratories throughout the world. It is therefore of interest for human cytogeneticists and clinical geneticists to review the 65 cases published to date and to examine the clinical implications for carriers of paracentric inversions.

MEIOSIS IN PARACENTRIC INVERSION HETEROZYGOTES

The mechanism by which chromosomal imbalance is generated at meiosis in heterozygotes for paracentric inversions has been well described in both plants and animals [9,10]. If no pairing (and thus, no crossing over) takes place within the inverted segment between the homologous chromosomes, then the normal course of meiosis remains unaltered.

Result of Crossing-Over Within the Inverted Segment

During pachytene, if the inverted segment is long enough it pairs with its homologue by forming a loop (Fig. 1). Depending on the number and position of the chiasmata there are the following possibilities:

i. A chiasma formation only outside the inverted segment will not affect the course of meiosis.

ii. A crossover within the inversion loop (Fig. 1a) results in the formation of a dicentric chromatid and an acentric fragment, a normal chromatid and a chromatid with the inversion. The dicentric chromatid forms a bridge between the two poles at anaphase I. As a result there are two normal products of meiosis (a normal and an inversion chromatid) and two abnormal products with duplication and/or deficiency depending on the fate of the dicentric chromatid and the acentric fragment (see below).

iii. An additional crossover in the interstitial segment (that is, the segment between the centromere and the inversion) does not affect the bridge if two different chromatids are involved in the interstitial and the loop chiasma or if the same two chromatids are involved. If, however, one of the two chromatids involved in the interstitial chiasma is also involved in the loop chiasma (Fig. 1b), there is a loop instead of a bridge at anaphase I. This loop will appear as a bridge at anaphase II. Also in this case there will be two normal products of meiosis (one with a normal and one with an inverted chromatid) and two chromosomally abnormal products.

iv. Two chiastmata in the loop involving two different chromatids would lead to the formation of two bridges and two acentric fragments (Fig. 1c). Thus, a four-strand double crossover in the inverted segment leads to all four products of meiosis being chromosomally unbalanced.

Depending on the number of crossovers and the chromatids involved there may be single or double fragments and single or double bridges or loops at anaphase I.

Fate of the Acentric Fragments and the Dicentric Bridge

The acentric fragments, which have no capacity for movement at anaphase, are either excluded from the gametes or passively included in one of them. In the latter case they may replicate so that there are one or more per cell. In any case they are likely to be lost in subsequent cell divisions of the zygote. Cells with up to three acentric fragments have been observed in early mouse zygotes produced by paracentric inversion heterozygotes [20].

The fate of the dicentric bridge can be one of the following:

i. It may break, leading to a varying degree of deficiency or duplication depending on the position of the break. The products of broken anaphase I

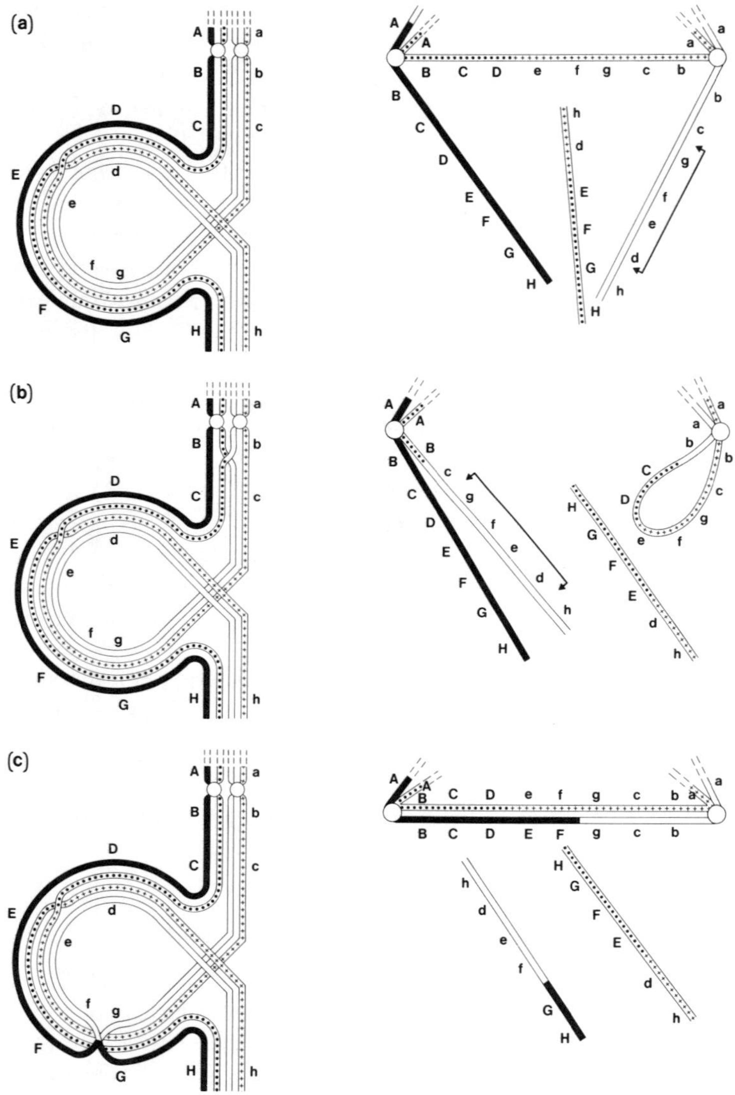

Fig. 1. Paracentric inversion loops at pachytene and anaphase I configurations showing **a)** a single crossover in the inversion loop; **b)** two cross-overs, one in the inverted and the other in the interstitial segment involving the same chromatid, and **c)** a four-strand double crossover in the inverted segment.

bridges have been seen at metaphase II in male mice carrying paracentric inversions as unequal-armed chromosomes with different degree of inequality [15].

ii. The dicentric bridge may hold the two groups of disjoining chromosomes together, so that a diploid restitution nucleus is formed. Meiotic studies from paracentric inversion heterozygote male mice show an increase in frequency of diploid second metaphases [15].

The large diploid spermatids may fail to develop into functional sperm, as is the case in species of *Chironomus* [21]. However, in the mouse there is the evidence that they develop into double-headed or large-sized sperm [22].

iii. The dicentric bridge may occasionally be excluded from both polar groups of chromosomes leading to deficiency of the whole chromatid in the two abnormal products.

iv. The dicentric chromatid may be included in one of the two telophase nuclei. After fertilization the dicentric chromosome can do one of the following:

a. It can begin a breakage-fusion-bridge cycle [23] in the zygote. Double-sized dicentrics which are a product of such a cycle have been seen in 2–8-cell embryos from paracentric inversion heterozygote mice [20].

b. The dicentric bridge may form an impediment to cytokinesis during mitotic divisions of the zygote leading to tetraploidy. Mosaic (2n/4n) 2–8-cell embryos have also been found in the above-mentioned study [20].

c. Following inactivation or deletion of one of its centromeres, a dicentric chromosome could form a stable structure, a pseudodicentric [24].

Effect on Fertility and the Fate of Chromosomally Unbalanced Gametes

From the above it is clear that crossing-over within the inverted segment can lead to either disruption of meiosis or to the production of chromosomally unbalanced gametes.

The negative effect of paracentric inversions on fertility of the carrier is either completely eliminated or considerably reduced in some species in both the male and female.

In male *Drosophila* [25] and in *Sciara* [26] there is complete absence of crossing-over in the male so that dicentric chromatids are formed only in the female. Thus there is no effect on the fertility in the male in these two genera. In male *Chironomus* the presence of a dicentric bridge leads to the formation of diploid spermatids which fail to develop into functional sperm [21]. Nonhomologous pairing reduces the frequency of crossover in the inverted segment and is a further means of reducing the production of unbalanced gametes in *Chironomus* [27].

Fertility in male carriers of two mutagenically induced inversions [13]:In(2)5RK[1] and In(5)9RK have been studied in the mouse [17]. These two inversions cover approximately 50% and 90% of the chromosome arm in chromosome 2 and 5, respectively. It was found that whereas the fertility in In(2)5RK male heterozygotes was not affected, that in In(5)9RK male heterozygotes was considerably reduced. Meiotic studies on the two types of males showed that the anaphase bridges in In(5)9RK were more breakable than in In(2)5RK. Nearly all the anaphase bridges in In(2)5RK heterozygotes resulted in diploid restitution nuclei which could be seen as diploid second metaphases with the dicentric and the associated fragment [15]. Similar diploid second metaphases were also seen in In(5)9RK heterozygotes, but haploid second metaphases with unequal-armed chromosomes with different degree of inequality (products of broken anaphase I bridges) were also seen [15]. Investigation of spermatozoa from In(5)9RK heterozygotes showed an increased frequency of double-headed or abnormally large-sized sperm as compared to the frequency in normal mice or in inversion homozygotes. This suggests that diploid spermatids develop into diploid spermatozoa in the mouse. However, there appears to be a selective barrier against diploid spermatozoa at the uterotubal junction [28]. Even if all the anaphase bridges resulted in the formation of diploid sperm as is the case for the mouse heterozygotes of In(2)5RK, there would be no noticeable effect on the fertility. There would always be sufficient sperm to affect fertilization unless bridge frequency exceeds 80% [29].

In female *Drosophila* [25] and *Sciara* [26], there is a special mechanism for the elimination of the dicentric bridge and the associated fragment during meiosis. The dicentric chromatid remains as a link between two inner nuclei and passes into the polar body so that only the non-crossover chromatids are included in the functional egg. A similar mechanism is known in *Zea mays* [30] and in the tulip [31].

Fertility in female mice heterozygotes for In(2)5RK and In(5)9RK is significantly reduced [17]. In female mice heterozygotes for a long paracentric inversion (covering about 85% of the X chromosome)—i.e., In(X)1H—there is evidence that the dicentric bridge is eliminated from the egg nucleus [14]. Whether it is left on the spindle or incorporated into the polar body nucleus is not known. In any case female heterozygotes crossed with normal males produce normal individuals, inversion heterozygotes, and XO daughters in equal frequency (the X in this last group being derived from the father). This is as expected if the anaphase bridge is always formed and eliminated, then 50% of the eggs would have no X at all. Hence, the progeny

[1]In (2)5RK is an inversion in chromosome 2, the fifth identified by Roderick et al.[13].

would be normal, inversion heterozygotes, XO and YO individuals in equal frequency, the last being incompatible with life.

Chromosomally unbalanced gametes from both sexes carrying broken bridges and/or whole dicentric chromosomes are capable of affecting fertilization in the mouse [20,32]. As mentioned already, acentric fragments and double-sized dicentrics (products of breakage-fusion-bridge cycle [23]) have been seen in 2–8-cell preimplantation embryos [20]. Also 2n/4n mosaic embryos have been observed. The failure of cytokinesis during mitosis because of the presence of the dicentric, led to the formation of the tetraploid cells. Thus unbalanced gametes can lead to the formation of unbalanced embryos. The survival of these embryos to implantation and beyond would depend on the degree of imbalance.

HUMAN CHROMOSOMES INVOLVED IN PARACENTRIC INVERSIONS

In the 65 published cases of paracentric inversions in man (details in Table I), 17 different chromosomes are involved covering the whole range of size from the longest to the shortest chromosome (Table II). Chromosomes 3, 7, and 11 are the most frequently represented, the most frequent breakpoints being in bands 7q22(9), 7q11(5), 3p25(5), and 11q23(5). Although relationships may exist, none was found between the three Belgian families with inv(3)(p13p25) [40,41] or the three patients with inv(11)(q21q23) found in the north of the Netherlands [37]. Chromosomes 3, 7, and 11 may be particularly prone to paracentric inversions. Alternatively, any shift in the position of bands in these three chromosomes which have very distinct bands may be easy to recognize and could lead to an overrepresentation of these cases in published reports.

CLINICAL IMPLICATIONS

The details of the inversion, the reason for referral for cytogenetic studies and the number of phenotypically normal family members are presented in Table I. One report of the paracentric inversion inv(12)(p12·3p13·1) is noteworthy [62]. It was found during an investigation of an eight-generation pedigree with a pericentric inversion of 12, inv(12)(p11·2q13·1). In this family four types of phenotypically normal individuals were identified: (i) with a pericentric inversion of 12, (ii) with a paracentric inversion of 12, (iii) with one chromosome with a paracentric and the other with a pericentric inversion, and (iv) with a pericentric and a paracentric inversion in the same

TABLE I. Details of 65 published reports of paracentric inversions

Case No.	First author and year	Sex	Inversion	Reason for referral	Inversion carriers in family
1.	Deroover, 1979 [33]	F	1(p22p36)	MR	1
2.	Morichon-Delvallez, 1985 [34]	M	1(p22p36)	RA	1
3.	Romain, 1983 [35]	M	1(p31.2p36.22)	MR (product brother-sister incest)	6
4.	Haapala, 1983 [36]	F	1(p31.2p35.2)	Unbalanced karyotype in child	
5.	Madan, 1984 (p.1) [37]	F	1(q25q42)	Floppy baby, cerebral palsy	2
6.	Speevak, 1985 [38]	F	1(q42q44)	Unbalanced karyotype in child	5
7.	Schmid, 1985 [39]	F	2(q31q36)	RA	7
8.	Madan, 1984 (p.2) [37]	M	3(p11p21)	PD for increased serum AFP	2
9.	Fryns, 1979 [40]	M	3(p13p25)	MR	3
10.	Fryns, 1980 (p.K) [41]	M	3(p13p25)	MR	1
11.	Fryns, 1980 (p.VK) [41]	M	3(p13p25)	PD for NTD	1
12.	Peters-Slough, 1982 [42]	F	3(p13p25)	Growth retardation	1
13.	Kelly, 1979 [43]	M	3(p14p22)	Unbalanced karyotype in child	
14.	Callen, 1985 [44]	M	3(p211.1p25)	congenital abnormalities	1
15.	Kasai, 1985 (p.2) [45]	M	3(q12q29)	Unbalanced karyotype in child	
16.	Bass, 1978 [46]	F	3(q13q26)	trisomy 10q (independent familial translocation)	2
17.	Djalali, 1984 [47]	F	3(q21q25.1)	PD for maternal age	1
18.	Crandall, 1980 [48]	M	3(q21q28)	PD for maternal age	1
19.	Kasai, 1985 (p.1) [45]	F	3(q25q27)	RA	
20.	Van Dyke, 1983 [49]	F	4(p14p16.3)	PD for maternal age	1
21.	Madan, 1984 (p.3) [37]	F	5(p13p15)	RA	9
22.	Valcárcel, 1983 [50]	M	5(p13pter)	Unbalanced karyotype in child	5
23.	Morichon-Delvallez, 1985 [34]	F	5(q12q34)	RA	4
24.	Fryns, 1980 (p.B) [41]	M	5(q13q34)	Father of trisomy 21	
25.	Canki, 1979 [51]	M	5(q21q32)	Klinefelter syndrome 47,XXY	1
26.	Stetten, 1983 [52]	F	7(p15p22)	RA	1
27.	Madan, 1984 (p.4) [37]	M	7(q11q22)	Undescended testis	4
28.	Canki, 1979 [51]	F	7(q11.3q22.3)	Turner syndrome 45,X	2
29.	Hoo, 1982 [53]	F	7(q11q22)	Unbalanced karyotype in child	
30.	Ridler, 1981 [54]	F	7(q11q22)	PD for maternal age	1
31.	Orye, 1983 [55]	F	7(q11q22)	Acute leukemia	1

32.	Madan, 1984 (p.5) [37]	M	7(q21q32)	PD for maternal age	1
33.	Shimba, 1976 [56]	M	7(q22q31)	Survey bomb survivors	
34.	Madan, 1984 (p.6) [37]	M	7(q22q31)	Klinefelter syndrome 47,XXY	1
35.	Faed, 1979 [57]	M	7(q22q34)	Subfertility	1
36.	Callen, 1985 [44]	M	7(q22.3q36.1)	One intrauterine death in wife	3
37.	Shabati, 1985 [58]	M	8(q11.23q21.1)	Some Langer-Giedion features	
38.	Madan, 1984 (p.7) [37]	F	8(q212q22)	PD for NTD	1
39.	Madan, 1984 (p.8) [37]	F	8(q22q24)	Oligomenorrhea, hirsutism	
40.	Madan, 1984 (p.9) [37]	M	9(q32q34)	PD Down syndrome in family of other parent	1
41.	Venter, 1984 [59]	F	10(q11q26)	PD for maternal age	19
42.	Trunca, 1978 [60]		11q	RA	Familial
43.	Van Dyke, 1983 [49]	M	11(q11q23)	PD for maternal age	1
44.	Verjaal, 1979 [61]	M	11(q13q21)	PD for maternal age	1
45.	Orye, 1983 [55]	M	11(q13q23)	Hypoparathyroidism	2
46.	Madan, 1984 (p.10) [37]	M	11(q21q23)	Klinefelter syndrome 47,XXY	
47.	Madan, 1984 (p.11) [37]	F	11(q21q23)	RA after AID	
48.	Madan, 1984 (p.12) [37])	M	11(q21q23)	azoospermia	1
49.	Poulsen, 1981 [62]	M	12(p123p131)	Family investigation	4
50.	Fryns, 1980 [41]	F	12(q12q24)	RA	
51.	Madan, 1984 (p.13) [37]	M	12(q15q24)	healthy volunteer for AID	
52.	Singh, 1981 [63]	M	12(q15q24)	Klinefelter syndrome 47,XXY	
53.	Riccardi, 1979 [64]	M	13(q12q22)	MR	De novo
54.	Yang-Feng, 1985 [65]	F	13(q13q22)	RA	1
55.	Sparkes, 1979 [66]	F	13(q14q22)	Unbalanced karyotype in child	
56.	Madan, 1984 (p.14) [37]	M	13(q141q22)	PD for NTD	De novo
57.	Jaeken, 1980 [67]	M	14(q13q24)	Microcephaly	De novo
58.	Madan, 1984 (p.15) [37]	M	14(q241q32)	PD for increased serum AFP	1
59.	Mules, 1984 [68]		14(q24.2q32.2)	Unbalanced karyotype in child	1
60.	Madan, 1984 (p.16) [37]	M	15(q13q24)	PD for maternal age	3
61.	Del Porto, 1984 [69]	M	15(q15q24)	Father of two dead infants	2
62.	Del Solar, 1974 [19]	M	16(q11q22)	Hypotonic newborn	1
63.	Madan, 1984 (p.17) [37]	M	16(q11q13)	Small for age	De novo
64.	Madan, 1984 (p.18) [37]	F	21(q21q22)	Down syndrome (+21)	2
65.	Herr, 1985 [70]	F	X(q13.1q26.1)	Learning and skeletal abnormalities	De novo

p. = patient; MR = mental retardation; RA = repeated abortions; PD = prenatal diagnosis; NTD = neural tube defect; AFP = alpha-fetoprotein; AID = artificial insemination by donor

TABLE II. Chromosomes involved in 65 cases of paracentric inversions

Chromosome	No. of cases
1	6
2	1
3	12
4	1
5	5
7	11
8	3
9	1
10	1
11	7
12	4
13	4
14	3
15	2
16	2
21	1
X	1

chromosome. This last had arisen as a result of crossing between the two chromosomes 12.

The 65 cases in Table I were referred for widely varying reasons. However, there were four major groups of subjects.

Normal Subjects

Out of the 65 cases, 45 were shown to be familial with one or more phenotypically normal carriers of the paracentric inversion in the family (a total of 110 normal relatives). Twenty-one of these were ascertained fortuitously during prenatal diagnosis for advanced maternal age or neural tube defect or during a survey or other investigation unrelated to the presence of the paracentric inversion.

This is as expected. As mentioned above, paracentric inversions are the most common form of chromosome polymorphism found in nature. They act as crossover suppressors holding together co-adapted groups of genes acting as "super genes" [71], which in certain circumstances can acquire a selective advantage [3,7,72].

The existence of short inversions acting as "super genes" has been predicted in all organisms including man [72]. Although in the last years steadily increasing numbers of paracentric inversions have been reported in man, it is possible that there exist many more short inversions in which crossing over is completely suppressed and which may involve single bands and therefore remain undetected even with the current banding techniques.

While short paracentric inversions are likely to be harmless, most of the abnormal zygotes resulting from unbalanced gametes produced by carriers of long inversions are likely to be so grossly abnormal that they would be lost early, even before implantation. The majority of paracentric inversions would therefore have no apparent deleterious effect on their carriers. It is therefore not surprising that so many paracentric inversions have been detected in normal individuals.

Patients With Infertility

Eight individuals (six males and two females) were referred for chromosome studies because of infertility. However, five of these (cases 25, 28, 34, 46, and 52, Table I) had another overt reason for infertility. Four males had Klinefelter syndrome with a 47,XXY karyotype and one female had Turner syndrome with a 45,X karyotype. This, of course, raises another question (considering that there was also a patient with a trisomy 21 in addition to the paracentric inversion, case 64 in Table I) of whether the presence of a paracentric inversion increases the risk of non-disjunction. However, the finding of six cases of aneuploidy out of 65 is not different from the proportion of Klinefelter, Turner, or Down syndrome patients referred to a cytogenetic laboratory.

The infertility problems in the remaining two males (cases 35 and 48, Table I) and one female (case 39) may or may not be causally related to the presence of the paracentric inversion. It should be noted that the inversion in one male (case 48, Table I) was inherited from his normal fertile father, while in the second (case 35, Table I) it was inherited from his mother.

In general, long paracentric inversions are more likely to affect fertility than short ones. As discussed above, there exist efficient mechanisms in some species to counteract the negative effects of paracentric inversions on fertility. It is not known if any such mechanisms exist in man. It is possible that fertility in heterozygotes for specific paracentric inversions (as is the case in the mouse) may be reduced.

Couples With Recurrent Abortions

Ten cases were referred because of repeated abortions (cases 2, 7, 19, 21, 23, 26, 42, 47, 50, and 54, Table I) and one because of an unexplained intrauterine death (case 36, Table I). In one couple (case 2, Table I) with three spontaneous abortions and two healthy children, while the husband had a paracentric inversion inv(1)(p22p36), a balanced reciprocal translocation t(2;4)(q14;q26) was found in the wife. The translocation could easily have been the cause of the recurrent abortions. In another family (case 21, Table I), although the inversion was initially identified in a female with repeated abortions, there was no history of abortions or abnormal children in nine other carriers of the inversion. Similarly, cases 5 and 36 in Table I had, respectively, four and three phenotypically normal (carrier) relatives with no reproductive problems.

It is expected that also in man, as in the case of the mouse, unbalanced gametes are capable of affecting fertilization, leading to the formation of chromosomally unbalanced zygotes. It is possible that such zygotes arising from certain paracentric inversions could lead to a recognized pregnancy but one which, depending on the degree of imbalance, would end in a spontaneous abortion or in the birth of an abnormal child. Occasionally, therefore, we expect to find paracentric inversions among couples with recurrent abortions or with abnormal children.

Parents of Abnormal Children

In eight cases (4, 6, 13, 15, 22, 29, 55 and 59, Table I) a paracentric inversion was found in a parent of a child with an unbalanced karyotype arising from the inversion. In addition a paracentric inversion was found in ten patients with congenital abnormalities (1, 3, 9, 10, 14, 37, 53, 57, 62, and 65, Table I). In this latter group the inversion had arisen de novo in three cases (53, 57, 65), while in the remaining seven it was familial.

The unbalanced karyotypes had arisen from the eight parental paracentric inversions in various ways.

In two families (cases 6 and 22, Table I) breakage of the dicentric bridge was the cause of the chromosome imbalance in the child. In the first family reported (case 22, Table I) inv(5)(p13pter) was present in five normal members of the family in three generations. One female carrier had had three abortions and had produced two sons with a duplication and one daughter (who had *cri-du-chat* syndrome) with a deletion of 5p. These resulted from the breakage of the dicentric bridge in different places. The paracentric inversion, the chromosome with the duplication, and the chromosome with the deletion could be easily distinguished in this family. In the second family (case 6, Table I) inv(1)(q42q44) was present in six normal members in three generations. One mentally retarded boy had had a del(1)(q42) derived from the maternal inversion.

There is one example of the formation of a stable dicentric structure. A female carrier (case 59, Table I) of a paracentric inversion inv(14)(q24.2q32.3) had an abnormal son with an inverted tandem duplication, inv dup(14)(pter→q32.3 :: q24.2→pter). Although the bisatellited dicentric showed positive NOR (nucleolus organizing region) [73,74] and C-band [75] staining at both ends, one end was only loosely constricted suggesting that the second centromere was not functional. Such a pseudodicentric [24] is stable during cell division, as shown by the fact that it was seen unchanged in all cells examined from both blood and skin of the child.

There are two reports of abnormal children with an interstitial deletion. In one case (13, Table I), the deleted segment is exactly the same as the inverted segment in the father, inv(3)(p14p22). In the other case (55, Table I), the deleted segment 13q14q22 in the child is shorter than the inversion in the mother inv(13)(q12q22). The suggestion that the abnormal chromosome in the child may have been derived from the inversion in the parent by excision of the inversion loop is easier to envisage in the former case than in the latter. An alternative hypothesis [76] offered for the second case that the mother has an inverted insertion instead of a paracentric inversion is more convincing.

In three cases (4, 15, and 29, Table I) high-resolution banding was required to demonstrate a minute difference between apparently similar inversions in the abnormal child and the phenotypically normal parent. In one case (4, Table I) there was a tiny extra band 1p13.3 as well as a minute deletion of band 1p35.1 in the child with a maternally inherited inversion inv(1)(p31.2p35.2). In the second report (case 29, Table I) the extra tiny band 7q11.22→q11.23 was found in addition to the maternally inherited inv(7)(q11.23q22). In the third case (15, Table I), an extra band 3q26.3 was present in the child, within the paternally inherited inv(3)(q12q29). In the first two reports the extra or the missing bands are at the edge of the inversion, and their presence or absence has been attributed to an unequal crossing-over at the base of the loop. In the third case the extra band lies in the middle of the inverted segment. The authors have suggested an unusual U-type exchange or an even number of crossovers with one unequal one within the inversion loop.

In two of the seven familial cases (with the same inversion in the parent and child), it is difficult to correlate the patients' abnormality with the paracentric inversion. The mental retardation in the child (case 3, Table I) with inv(1)(p31.2p36.22) may have been a result of the incest. In the case of the mentally normal patient with some signs of Langer-Giedion syndrome (case 37, Table I), the authors express caution in both the interpretation of the paracentric inversion as well as the causal relationship between the chromosome abnormality and the clinical symptoms. This is because high-resolution banding studies were not performed, and other alternative explanations could not be excluded. In the five remaining cases (1, 9, 10, 14 and 62, Table I) in which the same inversion was found in the normal parent, the child's abnormalities may be coincidental to the inversion. Alternatively, there may be minute karyotypic differences between the inversion in the parent and the child.

In at least some of these cases it is clear that the parental paracentric inversion has resulted in chromosome imbalance in the child. The expected abnormal karyotypes with the dicentric (case 59, Table I) and the duplication or deletion owing to the broken bridge (cases 6 and 22, Table I) as well as the interstitial deletions (cases 13 and 55, Table I) are easy to recognize. However, in cases (4, 15, and 29, Table I) which required very high resolution banding with 850 or more bands per karyotype [77], the detection of the abnormal karyotype and its interpretation is more difficult.

CONCLUSIONS

From this review of 65 published cases in man and from our knowledge of other species the following can be concluded about the clinical implications of paracentric inversions:

• The vast majority of paracentric inversions are likely to be harmless.

• The risk of production of unbalanced gametes by a paracentric inversion heterozygote is expected to be small.

• Most of the zygotes resulting from the unbalanced gametes are likely to be so grossly abnormal that they would be lost very early, even before implantation.

• The risk of having abnormal children is expected to be low. This risk increases with the finding of repeated abortions or abnormal children or both in carrier members of the family.

• Couples with an increased risk should be offered prenatal diagnosis. However, this should be done bearing in mind the difficulty in interpreting the results of the prenatal diagnosis. The difficulty is caused by two factors: i) there is a large variety of unpredictable unbalanced chromosome products that can result from a paracentric inversion; and (ii) although deletions and duplications involving whole bands in a standard 400-band karyotype [77] are easy to recognize, it may be very difficult if not impossible to detect with confidence minute differences requiring a 850-band [77] karyotype between apparently similar parental and fetal inversions.

ACKNOWLEDGMENTS

I am grateful to Professor Dr. M. Bobrow for reading the manuscript and to Mr. D. Bakker for the drawing of Figure 1.

REFERENCES

1. Muller HJ: Bearings of the 'Drosophila' work on systematics. In Huxley JS (ed): "The New Systematics." Oxford: Oxford University Press, 1940, p 185.
2. Sturtevant AH: A crossover reducer in Drosophila melanogaster due to inversion of a section of the third chromosome. Biol Zbl 46:697, 1926.
3. Dobzhansky T: "Genetics and the Origin of Species," 3rd Ed., revised. New York, London: Columbia University Press, 1951.
4. Darlington CD: "Cytology." London: J and A Churchill Ltd., 1965.
5. Dobzhansky T, Sturtevant AH: Inversions in the chromosomes of Drosophila pseudoobscura. Genetics 23:28, 1938.
6. Swanson CR: The distribution of inversions in Tradescantia. Genetics 25:438, 1940.
7. Darlington CD: "Evolution of Genetic Systems." Edinburgh, London: Oliver and Boyd, 1958.
8. White MJD: "Animal Cytology and Evolution," 3rd Ed. Cambridge: Cambridge University Press, 1973.
9. Swanson CP, Merz T, Young WJ: "Cytogenetics: The Chromosome in Division, Inheritance and Evolution," 2nd Ed. Englewood Clifs, New Jersey, Prentice Hall, 1981.
10. Sybenga J: "Meiotic Configurations - A Source of Information for Estimating Genetic Parameters." Berlin, Heidelberg, New York: Springer-Verlag, 1975.
11. Brandham PE: Inversion heterozygosity and sub-chromatid exchange in Agava stricta. Chromosoma 26:270, 1969.
12. Koller PC: Cytological studies on the reproductive organs. Chromosome behavior in the male grey squirrel (Sciurus carolinensis leucopus). Proc R Soc Edinburgh, 56:196, 1936.
13. Roderick TH, Hawes NL: Nineteen paracentric chromosomal inversions in mice. Genetics 76:109, 1974.
14. Evans EP, Phillips RJS: Inversion heterozygosity and origin of XO daughters of Bpa/+ female mice. Nature 256:40, 1975.
15. Evans EP, Ford CE: Some cytological properties of paracentric inversions in the mouse. In: Jones K, Brandham PE (eds): "Current Chromosome Research." Amsterdam: Elsevier, North-Holland, Biomedical Press, 1976, p 216.
16. Chandley AC: A pachytene analysis of two male-fertile paracentric inversions in chromosome 1 of the mouse and in the male-sterile double heterozygote. Chromosoma 85:127, 1982.
17. Ford CE, Evans EP, Burtenshaw MD: Meiosis and fertility in mice heterozygous for paracentric inversions. In Jones K, Brandham PE (eds): "Current Chromosome Research." Amsterdam Elsevier, North-Holland, Biomedical Press 1976, p 123.
18. Davisson MT, Roderick TH: Chromosomal banding patterns of two paracentric inversions in mice. Cytogenet Cell Genet 12:398, 1973.
19. Del Solar C, Uchida IA: Identification of chromosomal abnormalities by quinacine-staining technique in patients with normal karyotypes by conventional analysis. J Pediatr 84:534, 1974.
20. Evans EP, Burgoyne PB: The chromosome unbalance observed in early pre-implantation embryos from *mice* heterozygous for the paracentric inversion of the X chromosome, In(X)1H. (Abstract). In Bennett MD, Gropp A, Wolf U (eds): "Chromosomes Today Volume 8." London: George Allen and Unwin, 1984, p 305.
21. Beermann W: Inversions-Heterozygotie und Fertilität der Männchen von Chironomus. Chromosoma, Bd 8, S.1, 1956.

22. Hugenholtz AP, Bruce WR: Sperm size abnormalities in homozygous and heterozygous In(5)9RK mice. Can J Genet Cytol 21:115, 1979.
23. McClintock B: The stability of broken ends of chromosomes in Zea Mays. Genetics 26:234, 1941.
24. ISCN 1978: An international system for human cytogenetic nomenclature. Cytogenet Cell Genet 21:311, 1978.
25. Sturtevant AH, Beadle GW: The relations of inversions in the X chromosome of Drosophila Melanogaster to crossing over and disjunction. Genetics 21:554, 1936.
26. Carson HL: The selective elimination of inversion dicentric chromatids during meiosis in the eggs of Sciara impatiens. Genetics 31:95, 1946.
27. Martin J: Meiosis in inversion heterozygotes in Chironomidae. Can J Genet Cytol 9:255, 1967.
28. Krzanowska H: The passage of abnormal spermatozoa through the uterotubal junction of the mouse. J Reprod Fertil 38:81, 1974.
29. De Boer P: Chromosomal causes for fertility reduction in mammals. In Ferres FJ de (ed): "Chemical Mutagenesis," Volume 10. New York, London: Plenum Press, 1986, p 427.
30. Schulz-Schaeffer J: "Cytogenetics: Plants. Animals. Humans." New York, Heidelberg, Berlin: Springer-Verlag, 1980, p 208.
31. Darlington CD, La Cour L: The genetics of embryosac development. Ann Bot NS 5:547, 1941.
32. Ford CE: Gross genome unbalance in mouse spermatozoa: does it influence the capacity to fertilize? In Beatty RA, Glueckshon-Waelsch S (eds): "Edinburgh Symposium on the Genetics of the spermatozoön. Edinburgh, New York: published by the editors, 1972.
33. Deroover J, Fryns JP, Haegeman J, Van den Berghe H: Paracentric inversion in the short arm of chromosome 1. Hum Genet 49:117, 1979.
34. Morichon-Delvallez N, Marchand F: Les inversions paracentriques chez l'homme. Ann Genet (Paris) 2:119, 1985.
35. Romain DR, Columbano-Green LM, Whyte S, Smythe RH, Parfitt RG, Gebbie OB, Chapman CJ: Familial paracentric inversion of 1p. Am J Med Genet 14:629, 1983.
36. Haapala K, Herva R, Leisti J: Paracentric inversion of chromosome 1 in a mother with unbalanced progeny. Abstracts of the 8th International Chromosome Conference, Lübeck 1983.
37. Madan K, Seabright M, Lindenbaum RH, Bobrow M: Paracentric inversions in man. J Med Genet 21:407, 1984.
38. Speevak M, Hunter AGW, Hughes H, Cox DM: A familial paracentric inv(1)(q42q44) resulting in a child with a del (1)(q42) karyotype. Ann Genet (Paris) 28:177, 1985.
39. Schmid M, Hofmann R, Köhler J, Jannek U: Familial paracentric inversion inv(2)(q31q36). Hum Genet 71:270, 1985.
40. Fryns JP, Van Den Berghe H: Familial paracentric inversion of the short arm of chromosome 3. Ann Genet (Paris) 22:163, 1979.
41. Fryns JP, Van Den Berghe H: Paracentric inversion in man: personal experience and review of the literature. Hum Genet 54:413, 1980.
42. Peters-Slough MF, Planteyat HT, Timmerman MJ, Vooren MJ vd: A familial paracentric inversion in the short arm of chromosome 3: A case report. Clin Genet 22:102, 1982.
43. Kelly TE, Wyandt H, Kasprzak J, Ennis J, Willson K, Koch V, Schnatterly P: Paracentric inversion: Probable mechanism for an interstitial 3p deletion in a patient with multiple anomalies. Am J Hum Genet 31:100 A, 1979.
44. Callen DF, Woollatt E, Sutherland GR: Paracentric inversions in man. Clin Genet 27:87, 1985.

45. Kasai R, Narahara K, Kikkawa K, Takahashi Y, Wakita Y, Kimura S, Kataoka N, Kimoto H: Reproductive risk of paracentric inversion carriers: Report of two unrelated cases with paracentric inversion of the long arm of chromosome 3. Jpn J Hum Genet 30:57, 1985.
46. Bass HN, Sparkes RS, Crandall BF, Tannenbaum SM: Familial partial trisomy 10q(q23-qter)syndrome and paracentric inversion 3(q13q26) in the same patient. Ann Genet (Paris) 21:74, 1978.
47. Djalali M, Steinbach P, Barbi G: Familial paracentric inversion inv(3)(q21q25.1). Ann Genet (Paris) 27:41, 1984.
48. Crandall BF, Lebherz TB, Rubinstein L, Robertson RD, Sample WF, Sarti D, Howard J: Chromosome findings in 2500 second trimester amniocenteses. Am J Med Genet 5:345, 1980.
49. Van Dyke DL, Weiss L, Roberson JR, Babu VR: The frequency and mutation rate of balanced autosomal rearrangements in man estimated from prenatal genetic studies for advanced maternal age. Am J Med Genet 35:301, 1983.
50. Valcárcel E, Benitez J, Martinez P, Rey JA, Cascos AS: Cytogenetic recombinants from a female carrying a paracentric inversion of the short arm of chromosome number 5. Hum Genet 63:78, 1983.
51. Canki N, Dutrillaux B: Two cases of familial paracentric inversion in man associated with sex chromosome anomaly 47, XXY, inv(5)(q21, q32) and 45, X, inv(7)(q11.3q22.3). Hum Genet 47:261, 1979.
52. Stetten G, Rock JA: A paracentric chromosomal inversion associated with repeated early pregnancy wastage. Fertil Steril 40:124, 1983.
53. Hoo JJ, Lorenz R, Fischer A, Fuhrmann W: Tiny interstitial duplication of proximal 7q in association with a maternal paracentric inversion. Hum Genet 62:113, 1982.
54. Ridler MA, Sutton SD: A case of a paracentric inversion inv(7)(q11q22). Prenatal detection and counselling. Prenat Diagn 1:81, 1981.
55. Orye E, Van Bever H: Paracentric inversions: two new familial cases, inv (7)(q22q11) and inv(11)(q23q13). J Med Genet 20:231, 1983.
56. Shimba H, Ohtaki K, Tanabe K, Sofuni T: Paracentric inversion of a human chromosome 7. Hum Genet 31:1, 1976.
57. Faed MJM, Robertson J, Lamont MA, MacIntosh W, Grieve J, Baxby K, James GB, Crowder AM: A cytogenetic survey of men being investigated for subfertility. J Reprod Fertil 56:209, 1979.
58. Shabtai F, Sandowski U, Nissimov R, Klar D, Halbrecht I: Familial syndrome with some features of the Langer-Giedion syndrome and paracentric inversion of chromosome 8, inv 8(q11.23→q21.1). Clin Genet 27:600, 1985.
59. Venter PA, Dawson B, Du Toit JL, Smith EL, Kritzinger N, Landman S, Cronje AS, Hof JO: A familial paracentric inversion: A short review of the current status. Hum Genet 67:121, 1984.
60. Trunca C: Paracentric inversions: a probable cause of fetal wastage. Am J Hum Genet 30:96A, 1978.
61. Verjaal M, Leschot NJ, Treffers PE, Weduwen JJ der: Unexpected chromosomal aberrations in prenatal diagnosis in older women. In Murken JD, Stengel-Rutkowski S, Schwinger E (eds): "Prenatal Diagnosis." Stuttgart: Ferdinand Enke, 1979, p 16.
62. Poulsen H, Mikkelsen M, Holmgren G: Peri- and paracentric inversions in chromosome 12: Prenatal diagnosis and family study. Prenat Diagn 1:35, 1981.
63. Singh RP: Klinefelter's syndrome with a 47, XXY, inv(12)(q15q24) karyotype. Clin Genet 19:188, 1981.

64. Riccardi VM, Holmquist GP: De novo 13q paracentric inversion in a boy with cleft palate and mental retardation. Hum Genet 52:211, 1979.
65. Yang-Feng TL, Finley SC, Finley WH, Francke U: High resolution cytogenetic evaluation of couples with recurring fetal wastage. Hum Genet 69:246, 1985.
66. Sparkes RS, Muller H, Klisak I, Abram JA: Retinoblastoma with 13q-chromosomal deletion associated with maternal paracentric inversion of 13q. Science 203:1027, 1979.
67. Jaeken J, Fryns JP, Standaert L, De Cock P, Van den Berghe H: De novo paracentric inversion in a microcephalic boy: 46, XY, inv(14)(q13q24). Ann Genet (Paris) 23:105, 1980.
68. Mules EH, Stamberg J: Reproductive outcomes of paracentric inversion carriers. Report of a liverborn dicentric recombinant and literature review. Hum Genet 67:126, 1984.
69. Del Porto G, D'Allessandro E, Matteis C de, D'Innocenzo R, Baldi M, Pachi A, Cappa F: Familial paracentric inversion of chromosome 15(q15q24). J Med Genet 21:451, 1984.
70. Herr HM, Horton SJ, Scott CI Jr: De novo paracentric inversion in an X chromosome. J Med Genet 22:140, 1985.
71. Darlington CD, Mather K: "The Elements of Genetics." London: Allen & Unwin, 1949.
72. Ford EB: "Ecological Genetics." 3rd Ed. 1971. London: Chapman and Hall, 1964.
73. Matsui SI, Sasaki M: Differental staining of nucleolus organizers in mammalian chromosomes. Nature 246:148, 1973.
74. Howell WM, Denton TE, Diamond JR: Differential staining of the satellite regions of human acrocentric chromosomes. Experientia 31:260, 1975.
75. Paris Conference 1971: Standardization in human cytogenetics. In "Birth Defects," Original Article Series. New York: The National Foundation, 1972, Vol VIII, no 7.
76. Hoegerman SF: Chromosome 13 long arm interstitial deletion may result from maternal inverted insertion. Science 205:1035, 1979.
77. ISCN 1981. An International system for human cytogenetic nomenclature: High-resolution banding. Cytogenet Cell Genet 31:1, 1981.

The Cytogenetics of Mammalian Autosomal Rearrangements, pages 267–291

Chapter 5
Types of Imbalances in Human Reciprocal Translocations: Risks at Birth

P. Jalbert, H. Jalbert, and B. Sele

INTRODUCTION

Reciprocal translocations (rcp) occur frequently in human populations. They result from an exchange of segments between two chromosomes, any point of any chromosome may be involved, and thus almost an infinite number of different rcp is possible.

For any given rcp, a balanced carrier is potentially able to produce a large variety of different gametes through meiotic segregation—only two types without gain or loss of chromosome material, but many types with imbalances (duplications and/or deficiencies involving either the translocated segments [TS] or the centric segments [CS] or both).

These imbalances have different consequences, the main ones being abortions and birth of malformed infants. The average risks of these imbalances have been estimated from population studies: the values are dependent on the mode of detection of the rcp and on the nature and the structure of the population studied. However, these pooled values mask the fact that each particular rcp has its own risk (of producing abortions or imbalanced live births [ILB] or both). In other words, there is probably a wide range of risk from one rcp to another. Some rcp lead exclusively to abortions (those for which any imbalanced segregation would be inviable), while others produce both abortions and ILB.

Given that the true risk values for a particular rcp may deviate considerably from the estimated averages, it would be useful to have a method that addresses specific risks in order to facilitate genetic counseling (especially for couples before pregnancy to better evaluate their reproductive alternatives

Laboratoire de Cytogénétique et de Biologie de la Reproduction, Hôpital Nord, BP 217 X, 38043 Grenoble Cedex, France and for couples rejecting prenatal diagnosis). If these risk values are dependent on the viability of the different possible imbalances, the method is required to indicate for each rcp the imbalances compatible with live birth.

RISK FACTORS FOR IMBALANCED LIVE BIRTHS

Many factors are known or suspected to influence the relative risk values for abortion or ILB [1]. Some of them are purely meiotic, such as centromere position, comparative size of the interstitial segments (the segments between the centromeres and the breakpoints), chiasma frequency, chiasma position, and formation of chain or ring multivalents. Other factors are purely selective—for example, the mode of ascertainment (detection) of the translocation and the quantitative amounts of trisomic and monosomic imbalance. Finally, other factors are probably both meiotic and selective such as the comparative size of the two chromosomes involved, the position of the breakpoints, the comparative size of the exchanged segments, the sex of the carrier parent, and so on [1]. Whichever of these factors are involved two phenomena are associated with the outcome:

1. The actual risk of ILB differs greatly from one rcp to another: some rcp lead frequently to ILB, whereas for others there is no evidence that such live births have occurred [2].

2. Imbalanced offspring produced by carriers of a particular rcp are nearly always produced by the same mode of meiotic segregation.

The latter phenomenon infers that the gametes selected in pregnancies resulting in aneuploid children are generally the same for a given rcp [3,4]. The most striking example is the well known t(11;22) (q23.2; q11.2): out of 110 different families studied only one ILB type has been found, namely 47,+der(22), the product of a 3:1 tertiary meiotic segregation [5]. Whether this is due to selection against other types during gametogenesis, after fertilization, or both, is not yet clear, but preliminary results on the sperm karyotype of one carrier indicate that no apparent meiotic selection occurs [6].

Mode of Detection of the rcp

It has been repeatedly observed that the risk of ILB is higher for translocations detected through imbalanced infants than for those detected through reproductive failure [7, 8] or by chance in population surveys [9]. This difference in risk values is very large and can reach more than 15% [10] (20.8% for ILB translocations versus 4.9% for others). It is not surprising that the average risk values differ substantially from one study to another [9–

14], as they particularly depend on the proportion of translocations detected via ILB in each study.

Size of the Potential Imbalance

It has also been observed that the smaller the imbalance, the higher the risk of ILB. In man as in other species autosomal aneuploidy involving large chromosomal segments is embryonically lethal, with trisomy being generally better tolerated than the corresponding monosomy. The maximum imbalance burden (permitting viability) has been estimated to be a 5–6% trisomy or a 2–3% monosomy of the haploid autosomal complement (HAC) [15]. Daniel studying 20 different rcp observed that all of the imbalances in neonates were below 4% trisomy and 2% monosomy of the HAC [1]. Thus the abnormal segregations at term are generally those involving the least possible gain or loss of chromosomal material [1, 3, 16]. This is confirmed by the fact that imbalances observed at term are on average smaller than those expected in rcp detected through abortions [1, 10, 17, 18].

Genetic Content of the Imbalance

Lethality is dependent on the genetic content as well as the size of the imbalance. This is obvious for complete trisomies, the viable trisomies 8, 13, and 18 involve 5%, 3.8%, and 2.8% of the HAC, respectively, whereas the inviable 19 and 22 trisomies involve only 2.5% and 1.7%. As for trisomies, lethality of the partial duplications and deficiencies arising from rcp are not strictly proportional to their size, some regions being relatively insensitive to changes in dosage. It has been suggested that aneuploidy of the early replicating R-bands should be on average more lethal than that of the Q-bands [4, 17, 19]. Taken together, these differences mean that imbalances of some particular chromosomal regions are associated with viable full-term offspring and cause mental and physical handicaps, while others are inviable and lead only to more-or-less early abortions. For genetic counseling it is important to attempt to distinguish the first type of imbalance from the second and, to this end, have a schematic representation of the karyotype with the location of the regions likely to be viable in trisomy or in monosomy.

LOCATION OF IMBALANCES

The site of imbalance on the karyotype depends on the chromosomes involved, on the position of breakpoints and on the mode of meiotic segregation.

Chromosomes Involved

Some chromosomes are more frequently involved: interchromosomal distribution of breakpoints is different from that expected on the basis of their individual lengths [17, 20]. In fact breakpoints are randomly distributed for rcp detected through reproductive failure, and in contrast are not for those detected via imbalances at birth [10, 18]. With the notable exception of chromosome 22 all of the chromosomes which tend to carry an excess of breakpoints are implicated in well-known chromosome disorders [17].

Location of Breakpoints

Translocations detected through ILB show an excess of telomeric breakpoints [18, 21], while those segregating by the 3:1 mode show an excess of centromeric breakpoints [17].

With the exception of the translocation (11;22) (q23.2; q11.2), the high frequency of which remains unexplained, the majority of the translocations recorded seem to be unique [10]. In fact only a small proportion of newly ascertained translocations are likely to be identical to known rcp. The compilation of a very large number of cases possibly might allow the identification of individual translocations occurring at a higher frequency. Such a compilation would presumably also reveal that some translocations are never found, especially if only one type of detection (for example imbalances at birth) is used.

Types of Imbalanced Gametes Produced by rcp Carriers

Apart from the normal and balanced types which result from alternate segregation, every heterozygotic carrier can produce a large variety of imbalanced gametes by adjacent-1, adjacent-2 (both of these from 2:2 disjunction) or 3:1 segregation. The occurrence of these possibilities probably depends on the more-or-less complete synaptic association of the derivative chromosomes with their homologs at pachytene. The maximum synaptic association forms at this stage a cross-shaped quadrivalent (Fig. 1). The main varieties of imbalanced gametes and subsequent chromosome complements in the offspring are shown in Table I. Other possibilities exist which result either from crossing-over in the interstitial segments or from non-disjunction at anaphase II or both (for example: AB.AB., AD.AD.CB, CD.CD. BC., etc.) [12, 16, 22]. These imbalances result in extreme degrees of aneuploidy and fetal loss.

At anaphase I the different types of imbalances arise as follows:

The adjacent-1 type occurs when adjacent non-homologous centromeres reach the same pole: it results in duplication of one TS and deficiency of the other (duplication deficiency of B- and D-segments).

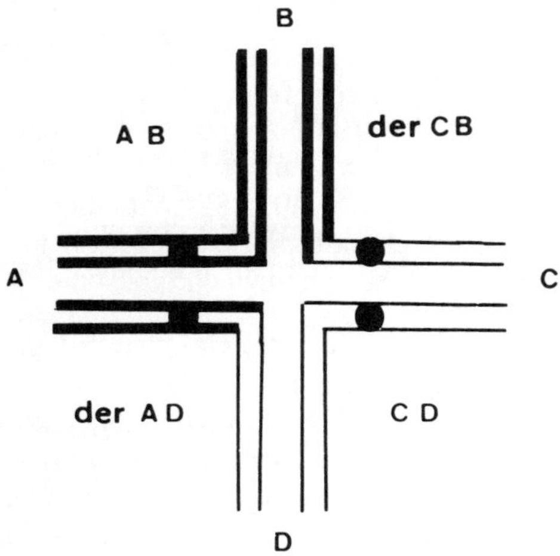

Fig. 1. Pachytene quadrivalent of a reciprocal translocation. The two centric segments (CS) are **A** and **C** and the two translocated segments (TS) **B** and **D**.

TABLE I. Main varieties of imbalanced gametes produced by a rcp carrier and subsequent chromosome complement in the offspring

Segregation	Gametes	Chromosome complement in the offspring: Nb	Trisomy	Monosomy
Adjacent-1	AB CB	46	B	D
	AD CD	46	D	B
Adjacent-2	AB AD	46	A	C
	CB CD	46	C	A
3:1				
Monosomies				
Interchange	AB	45	—	CD
Interchange	CD	45	—	AB
Tertiary	AD	45	—	CB
Tertiary	CB	45	—	AD
Trisomies				
Interchange	CB CD AD	47	CD	—
Interchange	AD AB CB	47	AB	—
Tertiary	AB CB CD	47	CB	—
Tertiary	CD AB AD	47	AD	—

The adjacent-2 type occurs when adjacent homologous centromeres reach the same pole: it results in duplication deficiency of the CS A and C (the two TS are therefore balanced).

Zygotes resulting from the two above types have 46 chromosomes.

The 3:1 type occurs when three centromeres reach the same pole: it results in trisomy (47 chromosomes) or in monosomy (45 chromosomes). Aneuploidy involves one of the two derivative chromosomes (tertiary trisomy/monosomy) or one of their normal homologs (interchange trisomy/monosomy).

Respective frequencies of the different types of gametes produced at meiosis are unknown: only some data are available on the haploid sperm karyotype from rcp carriers [6, 23]. However, frequencies of viable types of imbalance in pooled rcp are well known: the incidence decreases from adjacent-1 segregations to 3:1 types, and lastly to adjacent-2 types [12, 16]. A previous study carried out on 161 rcp showed an incidence of 66.5% for adjacent-1, 27.6% for 3:1 and 5.9% for adjacent-2 [3].

Although adjacent-1 imbalances are most frequent at term they are never observed in certain translocations despite a large number of generations: either no imbalance has occurred [2] or all of the imbalances result from another mode—3:1 for some rcp [5], adjacent 2 for others [4]. This suggests that for translocations resulting in ILB, the majority can produce at term only one type of imbalance (most of them from adjacent-1, some from 3:1, and a few from adjacent-2 segregation). However, a minority can result in both adjacent 1 and 3:1 imbalances [4].

In conclusion, each translocation shows a predisposition either not to produce any imbalance at term, or always to produce the same imbalance.

DETERMINATION OF VIABLE IMBALANCES

Factors of Choice

Adjacent-1 segregation has been very little studied, perhaps because it is the "normal" type of imbalance, also perhaps it is cytologically indistinguishable from an alternate segregation for which a crossover has occurred in one of the interstitial segments [12,24]. Since in this mode imbalances concern the two TS, it can be assumed that the shorter these are, the greater the chance of their imbalance being viable. It might therefore be expected that translocations giving rise to adjacent-1 ILB would have short TS. This is confirmed by measurements: the sum of the lengths of their two CS is always greater than that of their two TS, and the shortest CS is as long as—or longer than—the longest TS [4]. The sex of the carrier parent does not seem to influence the frequency of this type of imbalance at birth [3, 17, 18, 24]. Acrocentric chromosomes and chromosome 9 do not appear to be involved with any particular frequency in these translocations [3].

In contrast, translocations with short CS would preferentially be adjacent-2 segregants (as in this mode imbalances concern the two CS). In practice measurements do not always agree with this postulate: certain translocations with one (or even two) long CS segregate in this way. This is due to the fact that these exceptions concern translocations that occur either between the long arms of two acrocentric chromosomes or between the long arm of an acrocentric chromosome and a chromosome 9 in q13; the CS then carry short arms of acrocentric chromosomes (acro p) or 9qh. Imbalance of these regions have no influence on viability. If we exclude them from measurements the sum of the CS appears to be lower than that of the TS. Of 14 translocations [3, 18, 24], six occurred between a chromosome 9 in q13 and an acrocentric chromosome (number 13, 15, 21, or 22) in q11; a seventh was questionable (breakpoint in 9q22 or in 9q13?) [25]; three occurred between two acrocentric chromosomes with juxtacentromeric breakpoints and four between juxta-centromeric q arms of an acrocentric and another autosome (4q13, 12q13, 18q21, 20q11). Therefore the CS—acro p and 9qh excluded—are short and in the majority of cases shorter than the TS. There are departures from these rules when the chromosomes involved are of small size (breakpoints in 18q21, 20q11, 21q22); in these cases the determination of breakpoints is crucial (especially distinctions between p11 or q11).

Adjacent-2 ILB are almost exclusively maternal; and since they result from non-disjunction of homologous centromeres, it would be interesting to know the age of the mothers. No other imbalance mode has occurred at birth in these families.

Translocations leading to 3:1 imbalances at term have been well studied. Their frequency is estimated to account for at most 25% of all rcp imbalances [3, 16]. Most often it is a 3:1 tertiary segregation: the supernumerary chromosome is the smallest of the two derivative chromosomes. The monosomy is viable only if the chromosome is short. When it results from two short arms of acrocentrics its loss does not lead to any deleterious effects [26]. Rarely the imbalance results from a 3:1 interchange segregation: the supernumerary chromosome is a chromosome 21, a chromosome 13, or a chromosome 18. Theoretically a primary trisomy of these latter additional chromosomes occurring in combination with a balanced parental translocation could be misinterpreted as interchange trisomy: it would originate from non-disjunction in the normal parent and alternate segregation in the carrier parent [24]. This is only possible if the non-disjunction and the translocation involve chromosomes of the same pair, a rather unlikely coincidence. A careful study of the polymorphism of these additional chromosomes might show if this mechanism really occurs. In a retrospective study of 122 families with 3:1 segregation, 82 resulted from tertiary trisomy, 27 from tertiary monosomy, and 13 from interchange trisomy [27]. Translocations with resultant 3:1 ILB are defined by the following features [4]:

The two CS, as well as the two TS, are unequal in size; this results in an asymmetry of the meiotic quadrivalent [12].

The acrocentric and number 9 chromosomes are involved in 90% of them. As for adjacent-2 imbalances, short arms of acrocentric chromosomes and 9qh are frequently involved in the imbalance. If these regions are removed from measurements the asymmetry of the quadrivalent increases still further.

In the 3:1 tertiary mode the short TS is translocated to the short CS. The derivative aneuploid chromosome is short and carries either acrocentric satellites or the 9qh region.

In contrast, in the 3:1 interchange mode the long TS is translocated to the short CS. As the two TS are unequal in length, the two derivative chromosomes are deeply modified and their imbalance is inviable.

The 3:1 tertiary and interchange segregations are almost exclusively maternal. In the families studied by Friedmann [27] none of the 60 offspring from male carriers had an imbalanced karyotype; in contrast 18% of the 138 offspring from female carriers had imbalances. This similarity with the adjacent-2 mode of segregation could result from the fact that, in both modes, non-disjunction of homologous centromeres occurs; this meiotic error might occur more frequently in maternal than in paternal carriers [24] and might increase with maternal age [14, 16].

A Predictive Method for Genetic Counseling to rcp Heterozygotes

An attempt has been made, using a representation of the meiotic quadrivalent, to estimate the risk of imbalance at birth and to predict its type for each individual translocation [4]. To establish this "pachytene diagram" the two CS and the two TS were measured using the Paris Conference scheme of the human karyotype [22] after exclusion of acro-p and 9qh. (Although it gives imperfect renditions of length, the Paris scheme is representative of the relative size of chromosomes). Imbalances were expressed as a percentage of the haploid neutral genome (HNG) (22 autosomes + X). The shape of the quadrivalent varies from one translocation to another, as a function of the lengths of the CS and TS. Imbalances involving the shortest branches have a chance of being viable; in contrast, long branches, the imbalances of which are inviable, usually segregate normally. Therefore, it is easy to determine which segregation produces the least imbalance; it is represented by the line between the homologous segments of the two longest branches (Fig. 2). The imbalance is quantitatively assessed as a percentage of HAC (22A) or HNG (22A + X), respectively, for trisomy and for monosomy. If these values exceed the maximum tolerable imbalance burden, this imbalance is not liable to occur at birth (and *a fortiori* no other type). If the values do not exceed these thresholds, a risk of ILB exists. However, it is low if the predicted imbalance is not known to be viable, or if the carrier is the father and the imbalance arises from type 3:1 or adjacent 2 segregation.

Nevertheless these estimations should take into account certain imprecisions in the method:

Some translocations are liable to produce at term the two types adjacent-1 and 3:1 (this was the case for 5 out of 151 rcp recorded).Their pachytene diagram showed features compatible with that of adjacent 1 segregation (Σ 2 CS > Σ 2 TS) and with that of 3:1 segregation (unequal sizes of the TS and participation of acrocentric chromosomes) [4].

Furthermore, Davis et al. [18], in a study involving 202 couples with ILB, carried out a blind test of the pachytene diagram measurements for predicting the type of imbalance: in 26 cases there was discrepancy between the predicted and the observed type (87.1% predictive accuracy rate). Most of

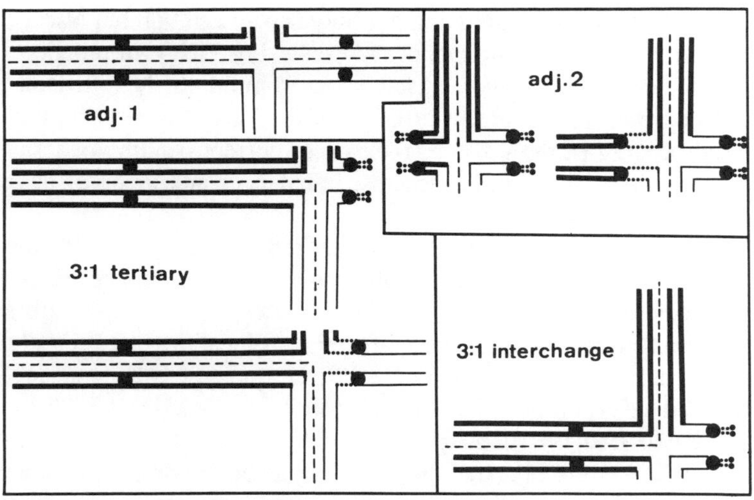

Fig. 2. Different types of pachytene diagrams. The dotted line is the segregation which defines the likely imbalance at birth. (Tertiary 3:1 and adjacent-2 segregations have two different representations according to which one of the two chromosomes is an acrocentric or a No. 9.)

them were adjacent-1 predicted and 3:1 observed or vice versa (19 cases). This suggests that there is a high frequency of translocations liable to lead to imbalance at term by either of these two modes. However, for some of these discrepant translocations, the location of the breakpoints was doubtful—two t(11;22) presented as (q25; q13) were really (q23.2; q11.2) [5]. Three translocations predicted to be 3:1 interchange segregated as 3:1 tertiary; but in fact their pachytene diagram was compatible with both modes, having asymmetric CS and two short TS. Anyhow, this relative imprecision in the method suggests that all imbalances quantitatively inferior to the maximum tolerable imbalance burden and previously observed at birth should be considered as potentially viable.

APPLICATION TO A SAMPLE OF 230 FAMILIES WITH AUTOSOMAL rcp

These observations originated from 15 laboratories in Europe. Modes of detection were (i) reproductive failures —abortions and stillbirths— (118 families, 51%); (ii) imbalanced live birth(s) (81 families, 35%); (iii) fortuitous detection (26 families, 11%); and (iv) trisomy for a chromosome not involved in the translocation (5 families, 2%).

The pachytene quadrivalent has been drawn for each translocation and the four branches measured (excluding not only acro-p and 9qh but also 1qh and 16qh). Unit used for measurements represents 1‰ HNG (in practice 1 mm of a 1-m haploid neutral genome established from the ISCN 1981 data [28]). These data are an improvement of those of the Paris Conference, and have been used because of the greater precision of current banding patterns. The 230 translocations, their measurements, imbalances and mode of detection, are grouped in Appendix I.

All of the autosomes are involved in the rcp in Appendix I. If translocations occur at random (between chromosome pairs) the number of possible binary associations should be $C^2{}_{22} = 231$, whereas 102 of them are absent from the study (for example, there are no translocations between chromosome 1 and chromosomes 8, 10, 13, 18, 19, 21, and 22). Other translocations are observed several times; there are 11 t(11;22) (q23; q11), two t(2;8) (q34; p23), and two t(10;11) (q26; q13). All others have different breakpoints. Figure 3 indicates the breakpoints according to the mode of ascertainment, and the size and region of imbalances at term, according to their original segregation. For translocations discovered by means of reproductive failure, the interchromosomal distribution of breakpoints is a function of the size of the chromosomes i.e. random, contrary to those discovered via ILB for which the distribution is not random ($P < 0.001$). Intrachromosomal localization of breakpoints is also dependent on the mode of detection; it is uniform for translocations detected by means of reproductive failures, but mainly in terminal chromosomal regions for those detected via ILB. Translocations detected by chance often have at least one breakpoint in a medium chromosomal region.

Only 83 of the 230 families have ILB (81 ascertained via ILB, 1 via abortion, 1 by chance). In total 117 imbalances are recorded (106 at term, 11 prenatally): 83 from adjacent-1, 19 from 3:1 tertiary, 9 from 3:1 interchange (giving rise to 27 trisomies and 1 monosomy) and 6 from adjacent-2. No translocation results in two different types of imbalance. Some chromosomal regions are frequently involved in imbalances (their trisomy or monosomy has generally been described at birth [29, 30]), while others seem never to be involved. ILB are more often maternal than paternal when they result

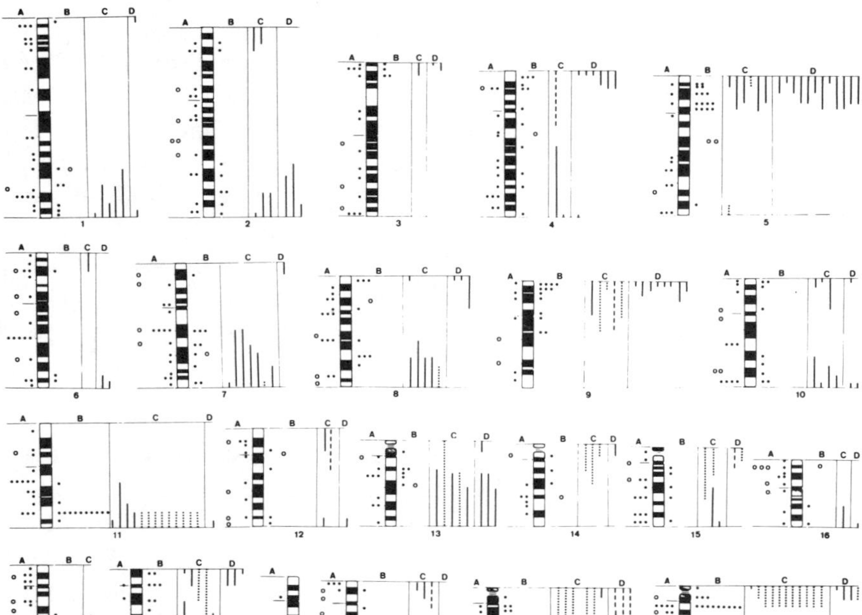

Fig. 3. Chromosomal location of breakpoints and imbalances. Breakpoints: **A:** Filled circles = rcp detected by means of reproductive failures; open circles = rcp detected by chance. **B:** Filled circles = rcp detected by ILB; open circles = rcp detected by an associated trisomy. Imbalances: **C:** In trisomy. **D:** In monosomy. Plain line = from adjacent-1 segregation; large dotted line = from adjacent 2 segregation, small dotted line = from 3:1 segregation.

from 3:1 segregation (27 mat, 1 pat), or from adjacent-2 (5 mat, 1 pat). This difference is less marked for ILB arising from adjacent-1 (50 mat, 30 pat, 3 undetermined). When the (11;22) (q23; q11) translocation results in an imbalance it is always by the 3:1 tertiary mode (ten imbalances, nine families).

The average value of the ratio Σ 2TS/ Σ 2CS differs according to the mode of detection of the translocation: 0.35 by means of ILB (n = 81), 0.52 fortuitously (n = 26), 0.41 by means of reproductive failures (n = 118).

The average value of this ratio also varies according to the type of imbalance at birth: 0.21 for adjacent 1 (n = 56), 0.52 for 3:1 (n = 23), 1.67 for adjacent 2 (n = 4).

The first two values differ significantly, the last involves too small a number of cases to be tested.

As expected the proportion of acrocentric chromosomes plus chromosome 9 varies with the mode of imbalance: 23% for adjacent-1, 43% for 3:1 and 63% for adjacent-2.

For 97 of the 106 ILB, the observed imbalance conforms with the pachytene diagram prediction (92%). In nine cases there is a discrepancy (Table II). (Six adjacent-1 expected and 3:1 were observed or vice versa; in these the unexpected imbalance often involves the 13q13 → 13qter segment. Two 3:1 expected and adjacent 2 were observed; as in the two previously reported cases [18] the unexpected monosomy involves the 21pter → 21q22 segment. One tertiary 3:1 expected and interchange 3:1 was observed.)

In Figure 4 each translocation is located according to the size of trisomy and/or monosomy at birth as previously represented by Daniel [1]; no imbalance exceeds 40‰ in trisomy and 25‰ in monosomy of the haploid neutral genome.

TABLE II. Discrepancies between observed and expected modes of segregation in live births (eight translocations, nine segregations)

Translocations	Observed segregation	Expected segregation
(2;13)(q36; q13)	Adj-1	Tertiary 3:1
(3;13)(p24; q13)	Interchange 3:1	Tertiary 3:1
(4;13)(q35; q14)	Adj-1 (2)	Tertiary 3:1
(5;21)(p13; q22)	Adj-1	Interchange 3:1
(10;13)(q26; q21)	Adj-1	Tertiary 3:1
(11;18)(q23; p11)	Interchange 3:1	Adj-1
(12;21)(q13; q22)	Adj-2	Interchange 3:1
(20;21)(q11; q22)	Adj-2	Interchange 3:1

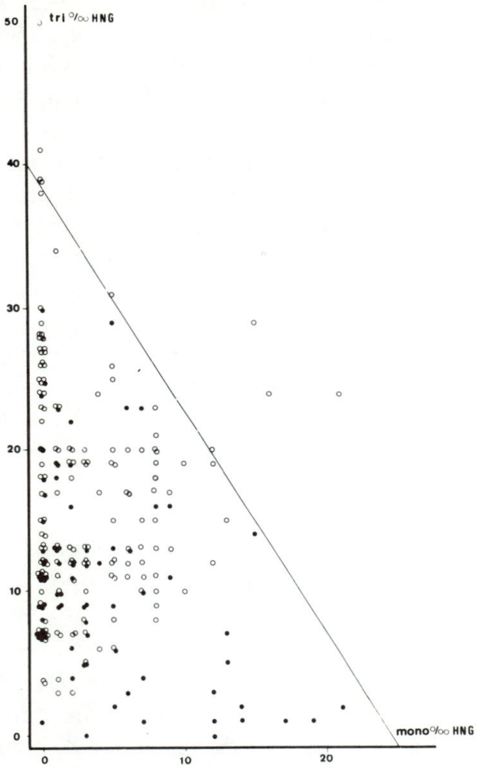

Fig. 4. Observed and potential imbalances. Filled circles: observed live birth imbalances; they occur close to the origin and to the left of the line joining 40‰ and 25‰ monosomy of the haploid neutral genome. Open circles: potential imbalances of rcp without observed imbalances (the potential imbalance is that with the minimum aneuploidy, the large aneuploidy segment in trisomy, the short in monosomy).

CONCLUSIONS

For the majority of reciprocal translocations any potential for viable imbalances and (to a certain extent) their risk of occurrence can be predicted. This should be helpful for genetic counseling.

The counseling approach should differ from one translocation to another and should be adapted to each case. Generally,—

1. If the rcp has led previously to an ILB, there is a major risk that this will occur again through the same segregation.

2. If there is no previously detected ILB the following can be undertaken: Once the breakpoints are specified, the pachytene diagram should be drawn

and its four branches measured (excluding acro-p, lqh , 9qh, and 16qh). The line between the homologous segments of the two longest branches defines the segregation leading to the minimum imbalance. When this imbalance exceeds 4% trisomy and/or 2.5% monosomy of the HNG the translocation is very unlikely to produce imbalanced offspring at term. More frequently one (or even two) potential imbalances are below these values. It may be that the viability of a certain case cannot be excluded but in practice is unlikely if it has not previously been observed at term in the literature. Care is required when considering imbalances of very small size such as those involving telomeric regions (their duplication or loss does not necessarily result in a recognized cytogenetic entity). These risk estimations need to be updated by the accumulation of new data and by more precise mapping of the viable trisomies and monosomies in the genome. It would also be useful to determine more precisely the size of chromosome bands in the normal karyotype.

SUMMARY

The risk of imbalance at birth differs from one translocation to another. It depends on the relative viability of the imbalances produced through meiosis. For a given translocation the imbalances potentially compatible with livebirth can be identified. The method consists of—:

1. The determination of the smallest possible imbalance and of its segregation mode (adj-1, adj-2, or 3:1) from a pachytene diagram. The viability of an imbalance depends both on its size (the lethality threshold is around 4% trisomy and 2.5% monosomy of the haploid genome) and on its location in the karyotype.

2. The consideration of other potentially viable imbalances (in this approach these may be longer and thus less probable but in 10% of cases the predictive choice is equivocal).

For some autosomal reciprocal translocations all potential imbalances are lethal; these translocations are uncommon or, at least, seldom recorded. For the majority of translocations only one type of resultant imbalance is viable to birth: this type can be identified by the method proposed with an accuracy rate of 90%. These data used with care could improve genetic counseling.

ACKNOWLEDGMENTS

The following colleagues were kind enough to send us translocation data: G. Bourouillou (Toulouse), J.L. Bresson (Besançon), E. Emberger (Montpellier), A. Geneix (Clermont-Ferrand), S. Girard (Paris), M. Hartung

(Marseille), J.C. Lambert (Nice), M.G. Mattei (Marseille), B. Noel (Chambéry), K. Rasmussen (Aarhus), J.V. Ruch (Strasbourg), J.L. Taillemite (Paris), and C. Turc (Dijon). We would like to thank them for their helpful collaboration. Also our continuing thanks to our secretary, Mrs. Marie-Jo Bossan.

REFERENCES

1. Daniel A: Structural differences in reciprocal translocations: Potential for a model of risk in reciprocal translocations. Hum Genet 51:171, 1979.
2. Vauhkonen AE, Sankila EM, Simola KOJ, De la Chapelle A: Segregation and fertility analysis in an autosomal reciprocal translocation t(1;8)(q41; q23.1). Am J Hum Genet 37:533, 1985.
3. Jalbert P, Sele B: Factors predisposing to adjacent 2 and 3:1 disjunctions: Study of 161 human reciprocal translocations. J Med Genet 16:467, 1979.
4. Jalbert P, Sele B, Jalbert H: Reciprocal translocations: A way to predict the mode of imbalanced segregation by pachytene-diagram drawing. A study of 151 human translocations. Hum Genet 55:209, 1980.
5. Iselius L, Lindsten J, Aurias A, Fraccaro M: The 11q;22q translocation. A collaborative study of 20 new cases and analysis of 110 families. Hum Genet 64:343, 1983.
6. Martin RH: Analysis of human sperm chromosome complements from a male heterozygous for a reciprocal translocation t(11;22) (q23; q11). Clin Genet 25:357, 1984.
7. Palmer CG, Wallace M, Breg WR, Hook EB, Magenis E, Pasztor L, Summit R: Risk in reciprocal translocation families. Am J Hum Genet 32:82A, 1980.
8. Petrosky DL, Borgaonkar DS: Segregation analysis in reciprocal translocation carriers. Am J Med Genet 19:137, 1984.
9. Jacobs PA, Aitken J, Frackiewicz A, Law P, Newton MS, Smith PG: The inheritance of translocations in man: data from families ascertained through a balanced heterozygote. Ann Hum Genet 34:119, 1970.
10. Boué A, Gallano P: A collaborative study of the segregation of inherited chromosome structural rearrangements in 1356 prenatal diagnoses. Prenat Diagn 4:45, 1984.
11. Ford CE, Clegg HM: Reciprocal translocations. Br Med Bull 25:110, 1969.
12. Hamerton JL: "Human Cytogenetics," Vol 1. New York: Academic Press, 1971, p 247.
13. Lejeune J, Dutrillaux B, Grouchy J de: Reciprocal translocations in human populations. A preliminary analysis. In Jacobs PA, Price WH, Law P (eds): "Human Population Cytogenetics." Edinburgh University Press, 1970.
14. Lurie IW, Presman H, Lazjuk GI: Risque génétique dans les translocations réciproques. J Genet Hum 26:385, 1978.
15. Sandler L, Hecht F: Genetic effects of aneuploidy. Am J Hum Genet 25:332, 1973.
16. Lindenbaum RH, Bobrow M: Reciprocal translocations in man. 3:1 meiotic disjunction resulting in 47 or 45 chromosome offspring. J Med Genet 12:29, 1975.
17. Aurias A, Prieur M, Dutrillaux B, Lejeune J: Systematic analysis of 95 reciprocal translocations of autosomes. Hum Genet 45:259, 1978.
18. Davis JR, Rogers BB, Hagaman RM, Thies AC, Veomett IC: Balanced reciprocal translocations: Risk factors for aneuploid segregant viability. Clin Genet 27:1, 1985.
19. Evans HJ: Molecular architecture of human chromosomes. Br Med Bull 29:196, 1973.

20. Stoll C: Non random distribution of exchange points in patients with reciprocal translocations. Hum Genet 56:89, 1980.
21. Francke U: Quinacrine mustard fluorescence of human chromosomes characterization of unusual translocations. Am J Hum Genet 24:189, 1972.
22. Paris Conference: Standardization in human cytogenetics. Birth Defects: Original Article Series 8, pt 7. New York: The National Foundation-March of Dimes, 1972.
23. Balkan W, Martin RH: Chromosome segregation into the spermatozoa of two men heterozygous for different reciprocal translocations. Hum Genet 63:345, 1983.
24. Stene J, Stengel-Rutkowski S: Genetic risks for familial reciprocal translocations with special emphasis on those leading to 9p, 10p, and 12p trisomies. Ann Hum Genet 46:41, 1982.
25. Howard-Peebles PN, Yarbrough KM, Stoddard GR, Rary JM: Translocation 9q/13q resulting in duplication (trisomy 9pter→9q22) and deficiency (monosomy 13pter→13q12). Clin Genet 11:46, 1977.
26. Pison H, Mattei MG, Jalbert H, Sele B, Leroux D, Augusseau S, Jalbert P: Translocation réciproque á l'origine d'une translocation robertsonienne 15;22 par perte d'un chromosome métacentrique. Conseil Génétique. J Genet Hum 32:363, 1984.
27. Freidman JM: Familial occurrence and recurrence of 3:1 segregation in reciprocal translocation heterozygotes. Am J Hum Genet 34:90A, 1982.
28. Report of the standing committee in Human Cytogenetic Nomenclature: An international system for human cytogenetic nomenclature - High resolution banding. Cytogenet Cell Genet 31:1, 1981.
29. Grouchy J de, Turleau C: "Clinical Atlas of Human Chromosomes." New York: Wiley, 1984.
30. Schinzel A: "Catalogue of Unbalanced Chromosome Aberrations in Man." Berlin/New York: De Gruyter, 1984.

(See appendix page 283)

APPENDIX Listing of the 230 rep

No.	Breakpoints	Lengths (‰ HNG) CS	TS	2 TS / 2 CS	Observed imbalances Adj-1	Adj-2	3:1	Detection
1	(1;2)(q44;q32)	(76)-62	1-17	0.13	1 pat(der2)	—	—	Imb
2	(1;3)(p36;p25)	(74)-62	3-2	0.04	—	—	—	RF
3	(1;3)(q21;q29)	(43)-62	34-1	0.33	—	—	—	RF
4	(1;4)(q32;p16)	(64)-57	13-1	0.12	1 mat(der4)°	—	—	Imb
5	(1;4)(q41;q28)	(69)-45	8-17	0.22	—	—	—	RF
6	(1;4)(q42;p16)	(72)-57	5-3	0.06	1 mat(der4) 1 mat (der4)°	—	—	Imb
7	(1;5)(p36;q23)	(74)-39	3-19	0.19	—	—	—	RF
8	(1;5)(p33;q23)	(64)-38	13-15	0.27	—	—	—	RF
9	(1;5)(p31;q21)	(56)-34	21-24	0.50	—	—	—	RF
10	(1;5)(q31;p14)	(57)-50	20-8	0.26	—	—	—	RF
11	(1;5)(q41;q11)	(69)-17	8-41	0.57	—	—	—	RF
12	(1;6)(p35;p12)	(69)-37	8-18	0.25	—	—	—	RF
13	(1;6)(p31;p22)	(56)-47	21-8	0.28	—	—	—	RF
14	(1;6)(q32;q25)	(63)-50	13-6	0.17	1 pat(der6) 1 pat(der6)°	—	—	Imb
15	(1;6)(q41;q21)	(69)-35	8-20	0.27	—	—	—	RF
16	(1;6)(q43;q21)	(74)-35	3-20	0.21	—	—	—	RF
17	(1;7)(q25;q35)	(53)-47	24-4	0.28	—	—	—	RF
18	(1;9)(p36;p12)	(75)(25)	2-12	0.15	1 pat(der1)	—	—	Imb
19	(1;9)(q31;p24)	(58)(37)	19-1	0.21	1 pat(der9)	—	—	Imb
20	(1;11)(p36;p12)	(74)-30	3-13	0.15	—	—	—	RF
21	(1;11)(p35;q14)	(70)-28	7-15	0.22	—	—	—	RF
22	(1;11)(q35;q13)	(65)-24	12-19	0.35	—	—	—	RF
23	(1;12)(q42;p12)	(77)-34	5-6	0.10	—	—	—	RF
24	(1;14)(q31;q23)	(57)(16)	20-12	0.44	—	—	—	Tri 21
25	(1;14)(q41;q24)	(69)(17)	8-11	0.22	—	—	—	RF
26	(1;15)(q24;q25)	(51)(22)	23-5	0.38	—	—	—	RF

(continued)

APPENDIX I. Listing of the 230 rep (continued)

		Lengths (‰ HNG)		2 TS	Observed imbalances			
No.	Breakpoints	CS	TS	2 CS	Adj-1	Adj-2	3:1	Detection
27	(1;16)(p34;p13)	(67)(23)	10-3	0.14	—	—	—	RF
28	(1;16)(q21;p11)	(43)(15)	34-10	0.76	—	—	—	RF
29	(1;16)(q43;q21)	(74)(17)	3-9	0.13	1 mat(der1)	—	—	Imb
30	(1;17)(p34;p12)	(66)-22	11-5	0.18	—	—	—	RF
31	(1;17)(q32;q25)	(66)-25	13-2	0.16	—	—	—	Chance
32	(1;20)(p22;p12)	(48)-16	29-4	0.52	—	—	—	RF
33	(2;3)(q24;q27)	54-59	25-5	0.27	—	—	—	Chance
34	(2;5)(p22;p14)	69-51	10-7	0.14	1 mat (der5)	—	—	Imb
35	(2;5)(p11;q24)	50-43	29-15	0.47	—	—	—	RF
36	(2;5)(q37;p15)	77-53	2-5	0.05	2 mat(der5)	—	—	Imb
37	(2;5)(q37;q31)	77-46	3-13	0.13	—	—	—	RF
38	(2;6)(q32;p22)	62-47	17-8	0.23	—	—	—	RF
39	(2;6)(q37;q21)	77-35	2-20	0.20	—	—	—	RF
40	(2;7)(q24;q21)	53-24	26-27	0.69	—	—	—	RF
41	(2;7)(q31;q36)	58-49	21-2	0.21	1 mat (der2)	—	—	Imb
42	(2;8)(p15;q24)	60-40	19-5	0.24	—	—	—	RF
43	(2;8)(q34;p23)	69-44	10-1	0.10	5 mat (der8)	—	—	Imb
44	(2;8)(q34;p23)	69-44	10-1	0.10	2 pat (der8)	—	—	Imb
45	(2;8)(q36;q11)	75-18	4-27	0.33	—	—	—	RF
46	(2;9)(p23;q34)	72(35)	7-3	0.09	—	—	—	RF
47	(2;9)(p22;p21)	69(31)	10-7	0.17	—	—	—	RF
48	(2;9)(q22;q31)	43(28)	36-10	0.65	—	—	—	Chance
49	(2;10)(q22;q25)	48-38	31-5	0.42	—	—	—	Chance
50	(2;10)(q32;p13)	62-39	17-6	0.23	—	—	—	RF
51	(2;12)(p22;q15)	69-22	10-19	0.32	—	—	—	RF
52	(2;12)(q14;q24)	35-37	44-4	0.67	—	—	—	Chance
53	(2;13)(q14;q34)	40(29)	39-1	0.58	—	—	—	RF
54	(2;13)(q36;q13)	73(6)	6-23	0.37	1 mat(der2)	—	—	Imb

55	(2;15)(q14;p11)	40-27	39(0)	0.58	—	—	—	RF
56	(2;16)(p12;p13)	53-25	26-3	0.37	—	—	—	Chance
57	(2;16)(p12;q24)	52(22)	27-2	0.39	—	—	—	RF
58	(2;17)(p11;p11)	50-20	29-7	0.51	—	—	—	RF
59	(2;18)(q13;q22)	37-21	42-4	0.79	—	—	—	RF
60	(2;22)(p23;q13)	72(10)	7-3	0.12	1 mat(der22)	—	—	Imb
61	(2;22)(q11;q11)	33(1)	46-12	1.71	—	—	—	RF
62	(3;4)(q29;q32)	63-53	1-7	0.07	—	—	—	RF
63	(3;4)(q29;q11)	62-17	1-41	0.53	—	—	—	RF
64	(3;5)(q22;q35)	44-56	19-2	0.21	—	—	—	RF
65	(3;6)(p25;q27)	61-54	3-1	0.03	—	—	—	RF
66	(3;7)(p26;q21)	63-28	1-23	0.26	1 mat(der3) 1 pat(der3)	—	—	Imb
67	(3;7)(p24;p21)	58-46	6-5	0.11	1 mat(der7) 1 ? (der7)	—	—	Imb
68	(3;7)(p24)(q31)	58-38	6-13	0.20	—	—	—	RF
69	(3;7)(p14;q36)	43-50	20-1	0.23	—	—	—	RF
70	(3;8)(p25;q22)	61-33	3-12	0.16	1 mat(der3)	—	—	Imb
71	(3;8)(q26;q21)	55-26	8-19	0.33	—	—	—	RF
72	(3;12)(p27;q22)	63-30	1-11	0.13	—	—	—	RF
73	(3;13)(p24;q13)	58(6)	6-24	0.47	—	—	1 mat interchange	Imb
74	(3;13)(q24;q31)	48(21)	15-8	0.33	—	—	—	RF
75	(3;14)(q13;q31)	37(26)	26-5	0.49	—	—	—	RF
76	(3;16)(q12;p13)	33(23)	31-3	0.61	—	—	—	Chance
77	(3;17)(p25;q21)	61-15	3-12	0.20	—	—	—	RF
78	(3;17)(p13;p13)	40-26	23-1	0.52	—	—	—	RF
79	(3;19)(q25;q13)	51-18	12-5	0.25	—	—	—	Chance
80	(4;7)(p15;p21)	53-46	8-5	0.13	—	—	—	Chance

(*continued*)

APPENDIX I. Listing of the 230 rep (continued)

No.	Breakpoints	Lengths (‰ HNG) CS	TS	2 TS / 2 CS	Observed imbalances Adj-1	Adj-2	3:1	Detection
81	(4;7)(q32;q21)	53-28	8-23	0.30	—	—	—	RF
82	(4;8)(q27;p22)	40-40	20-5	0.31	—	—	—	RF
83	(4;8)(q32;p12)	53-35	8-10	0.20	—	—	—	RF
84	(4;9)(q23;p22)	31(34)	29-5	0.52	1 mat(der9)	—	—	Imb
85	(4;10)(p16;p14)	58-44	2-4	0.06	1 pat(der4)	—	—	Imb
86	(4;11)(p15;q13)	53-24	7-20	0.35	—	—	—	RF
87	(4;11)(q21;q21)	26-30	35-13	0.86	—	—	—	RF
88	(4;11)(q34;q23)	57-38	3-5	0.08	—	—	—	RF
89	(4;12)(p15;q24)	53-40	7-1	0.09	2 mat(der4)	—	—	Imb
90	(4;13)(q35;q14)	59(10)	1-19	0.29	1 pat(der13) 1 pat(der4)	—	—	Imb
91	(4;14)(q26;q21)	37(10)	23-18	0.87	—	—	—	RF
92	(4;14)(q34;q23)	57(16)	3-12	0.21	—	—	—	RF
93	(4;15)(q13;q12)	22(2)	38-25	2.63	—	2 mat(der4)	—	Imb
94	(4;15)(q31;q25)	48(17)	12-5	0.26	—	—	—	RF
95	(4;16)(q21;p13)	26-23	34-5	0.80	—	—	—	Tri 21
96	(4;18)(p15;q12)	53-12	7-13	0.31	—	—	—	RF
97	(4;18)(q21;q22)	26-22	35-4	0.81	—	—	—	RF
98	(4;20)(p15;p12)	53-17	7-4	0.16	1 pat(der4)° 1 pat(der4)	—	—	Imb
99	(4;21)(p11;q11)	45(1)	16-10	0.55	—	—	1 mat interchange	Imb
100	(4;21)(q28;q21)	43(5)	17-8	0.52	—	—	1 mat interchange	Imb
101	(4;21)(q31;q22)	48(9)	12-3	0.26	—	—	1 mat interchange	Imb
102	(4;22)(p15;q13)	54(9)	6-3	0.14	1 mat(der4)	—	—	Imb
103	(5;6)(p15;q26)	53-52	5-3	0.08	1 mat(der6)	—	—	Imb
104	(5;7)(p14;q21)	51-27	7-23	0.38	1 mat(der5)	—	—	Imb
105	(5;7)(q14;q31)	28-37	30-13	0.66	—	—	—	Tri 21
106	(5;7)(q32;p15)	49-43	9-8	0.18	—	—	—	Chance

107	(5;8)(p11;p11)	43-31	15-14	0.39	1 mat(der5)	—	—	Imb
108	(5;10)(p13;15)	46-43	12-1	0.16	2 pat(der5)	—	—	Imb
109	(5;10)(q35;q22)	57-27	1-17	0.21	—	—	—	RF
110	(5;11)(p13;q24)	46-41	12-3	0.17	2? (der5) 1 pat(der5)° 1 pat(der11)°	—	—	Imb
111	(5;13)(p15;q11)	55(1)	3-29	0.57	—	—	—	RF
112	(5;13)(q14;q22)	28(17)	30-12	0.93	—	—	—	Tri 18
113	(5;14)(p13;q23)	46(16)	12-12	0.39	—	—	—	RF
114	(5;14)(q34;q12)	54(5)	4-23	0.46	—	—	1 mat tertiary	Imb
115	(5;15)(p15;q22)	55(14)	3-14	0.25	—	—	1 mat tertiary	Imb
116	(5;15)(p14;q21)	53(13)	5-13	0.27	1 mat(der5)	—	—	Imb
117	(5;15)(p12;q26)	44(25)	14-2	0.23	1 pat(der5)	—	—	Imb
118	(5;15)(q35;q15)	57(8)	1-19	0.31	—	—	—	RF
119	(5;18)(p15;q12)	56-13	2-12	0.20	1 pat(der5)	—	—	Imb
120	(5;18)(p11;p13)	39-24	14-1	0.24	2 mat(der5)	—	—	Imb
121	(5;18)(p11;p11)	42-17	16-8	0.41	2 mat(der18) 1 mat(der5)°	—	—	Imb
122	(5;21)(p13;q22)	45(6)	13-5	0.35	1 mat(der5)	—	—	Imb
123	(5;21)(q22;q22)	33(10)	25-1	0.60	—	—	—	RF
124	(5;22)(p13;p13)	46(12)	12(0)	0.21	2 mat(der5) 1 mat(der22)	—	—	Imb
125	(6;7)(q26;q22)	52-38	3-19	0.24	—	—	—	RF
126	(6;8)(p23;q11)	51-16	4-29	0.49	—	—	—	RF
127	(6;9)(p22;p23)	47(35)	8-3	0.13	1 mat(der9)	—	—	Imb
128	(6;11)(p12;p13)	47-30	8-13	0.27	—	—	—	Chance
129	(6;11)(q26;q13)	52-24	3-19	0.29	—	—	—	RF
130	(6;12)(q23;p12)	44-35	11-6	0.22	—	—	—	RF

(continued)

APPENDIX I. Listing of the 230 rep (continued)

No.	Breakpoints	Lengths (‰ HNG) CS	TS	2 TS / 2 CS	Observed imbalances Adj-1	Adj-2	3:1	Detection
131	(6;13)(p21;q14)	41(9)	14-21	0.70	—	—	—	RF
132	(6;15)(p25;q22)	54(14)	1-13	0.21	—	—	—	RF
133	(6;16)(q13;q11)	25(12)	30-14	1.19	—	—	—	Chance
134	(6;17)(p22;q22)	47-19	8-8	0.24	—	—	—	Chance
135	(6;17)(q21;q25)	35-26	20-1	0.34	—	—	—	RF
136	(6;18)(q21;q23)	35-24	20-2	0.37	—	—	—	RF
137	(6;20)(q23;q12)	43-15	12-7	0.33	—	—	—	Chance
138	(6;21)(q15;q11)	28(0)	27-12	1.39	—	—	—	RF
139	(7;9)(q23;p24)	38(38)	13-1	0.18	1 pat(der9)	—	—	Imb
140	(7;9)(q31;p24)	37(36)	13-1	0.19	1 mat(der9)	—	—	Imb
141	(7;9)(q35;q33)	47(32)	4-6	0.13	—	—	—	RF
142	(7;10)(p15;q22)	42-27	9-17	0.38	—	—	—	RF
143	(7;11)(q31;p15)	40-38	11-5	0.21	—	—	—	RF
144	(7;12)(p11;q11)	39-12	12-29	0.80	—	—	—	RF
145	(7;12)(q21;q21)	27-25	24-16	0.77	—	—	—	Chance
146	(7;12)(q22;p12)	32-35	17-6	0.34	—	—	—	Chance
147	(7;15)(q34;q21)	45(11)	6-15	0.38	—	—	—	RF
148	(7;15)(q36;q14)	49(6)	2-21	0.42	—	—	1 mat tertiary	Imb
149	(7;16)(q21;pter)	28(25)	23-1	0.45	—	—	—	RF
150	(7;18)(q32;q22)	42-21	9-5	0.22	1 mat(der18) 1 mat(der18)°	—	—	Imb
151	(7;20)(p11;p13)	39-19	12-2	0.24	—	—	—	RF
152	(7;21)(q21;q21)	23(4)	28-10	1.41	—	—	—	RF
153	(7;21)(q21;q22)	28(10)	23-1	0.63	—	—	3 mat interchange	Imb
154	(7;21)(q22;q22)	32(9)	19-2	0.51	—	—	1 mat interchange	Imb
155	(7;22)(q11;q13)	23(11)	28-2	0.88	—	—	—	RF
156	(8;9)(p21;p24)	38(37)	7-1	0.11	—	—	—	RF
157	(8;9)(q21;pter)	27(38)	18-1	0.29	1 mat(der9)	—	—	Imb

158	(8;9)(q22;p21)	32(32)	13-7	0.31	1 mat(der9)	—	—	Imb
159	(8;9)(q24;q11)	42-12	3(23)	0.48	—	—	—	RF
160	(8;10)(q21;q26)	27-42	18-2	0.29	—	—	—	RF
161	(8;10)(q24;q11)	43-16	2-28	0.51	—	—	—	Chance
162	(8;12)(p12;p11)	35-31	10-10	0.30	—	—	—	Tri 21
163	(8;12)(q11;p11)	16-32	29-11	0.83	—	—	—	RF
164	(8;13)(p23;q14)	43(9)	2-21	0.44	1 pat(der13)	—	1 mat tertiary	Imb
165	(8;13)(q22;p11)	33-25	12(0)	0.21	2 mat(der13)	—	—	Imb
166	(8;14)(p11;q24)	31-19	13-9	0.44	—	—	—	RF
167	(8;14)(q23;q21)	32(11)	13-8	0.72	—	—	1 mat tertiary	Imb
168	(8;15)(q21;q25)	26(22)	19-5	0.50	—	—	—	RF
169	(8;16)(p12;p12)	34(19)	11-7	0.34	—	—	—	RF
170	(8.17)(q24;p12)	44-18	1-4	0.08	—	—	—	Chance
171	(8;20)(q21;p12)	28-17	17-4	0.47	—	—	—	Chance
172	(9;10)(p23;q23)	(34)31	3-13	0.25	2 pat(der9)	—	—	Imb
173	(9;11)(p22;q14)	(33)28	5-15	0.33	—	—	—	RF
174	(9;14)(q13;p11)	(12)28	22(0)	0.55	—	—	1 mat tertiary	Imb
175	(9;18)(p23;q12)	(36)13	3-13	0.33	—	—	1 mat tertiary	Imb
176	(9;21)(q13;q11)	(12)(1)	23-12	2.69	—	2 mat(der9)	—	Imb
177	(9;22)(q11;p11)	12-13	(23)(0)	0.92	—	—	1 mat tertiary	Imb
178	(9;22)(q21;q12)	(19)(6)	20-7	1.08	—	—	—	Chance
179	(10;11)(q26;q13)	42-24	2-19	0.32	—	—	—	RF
180	(10;11)(q26;q13)	42-24	2-19	0.32	1 pat(der10)	—	—	Imb
181	(10;12)(p15;p11)	43-32	1-9	0.13	1 mat(der10)	—	—	Imb
182	(10;12)(p11;q24)	32-37	12-4	0.23	1 pat(der12)	—	—	Imb
183	(10;13)(q22;q31)	27(15)	17-13	0.71	—	—	—	RF
184	(10;13)(q25;q11)	38(1)	6-29	0.90	—	—	—	Chance
185	(10;13)(q26;q21)	42(13)	2-16	0.33	1 mat(der10) 1 mat(der13)°	—	—	Imb

(continued)

APPENDIX I. Listing of the 230 rep (continued)

		Lengths (‰ HNG)		2 TS	Observed imbalances			
No.	Breakpoints	CS	TS	2 CS	Adj-1	Adj-2	3:1	Detection
186	(10;15)(p15;q22)	43(15)	1-12	0.22	—	—	—	RF
187	(10;15)(p11;q15)	32(7)	12-20	1.10	—	—	—	Chance
188	(10;17)(q26;q11)	42-10	2-17	0.37	—	—	—	RF
189	(10;17)(q26;q21)	42-16	2-11	0.22	—	—	—	RF
190	(10;22)(p14;q11)	41(1)	3-12	0.36	—	—	—	RF
191	(10;22)(q24;p11)	35(13)	9(0)	0.19	1 mat(der22)	—	—	Imb
192	(10;22)(q25;p11)	35-12	9(0)	0.19	1 mat(der22)	—	—	Imb
193	(11;12)(q12;q24)	20-38	23-3	0.45	—	—	—	RF
194	(11;12)(q22;p12)	40-34	3-7	0.14	—	—	—	RF
195	(11;15)(p13;p12)	32(27)	11(0)	0.19	—	—	—	RF
196	(11;16)(q13;q22)	23(20)	20-6	0.60	—	—	—	RF
197	(11;16)(q22;q24)	34(24)	9-1	0.17	1 pat (der16)	—	—	Imb
198	(11;18)(q23;p11)	37-20	6-5	0.19	1 pat(der18)°	—	1 pat interchange	Imb
199	(11;22)(q23;q11)	37(1)	6-11	0.45	—	—	2 mat tertiary	Imb
200	(11;22)(q23;q11)	37(1)	6-11	0.45	—	—	1 mat tertiary	Imb
201	(11;22)(q23;q11)	37(1)	6-11	0.45	—	—	1 mat tertiary	Imb
202	(11;22)(q23;q11)	37(1)	6-11	0.45	—	—	1 mat tertiary	Imb
203	(11;22)(q23q11)	37(1)	6-11	0.45	—	—	—	RF
204	(11;22)(q23;q11)	37(1)	6-11	0.45	—	—	1 mat tertiary	Imb
205	(11;22)(q23;q11)	37(1)	6-11	0.45	—	—	1 mat tertiary	Imb
206	(11;22)(q23;q11)	37(1)	6-11	0.45	—	—	—	RF
207	(11;22)(q23;q11)	37(1)	6-11	0.45	—	—	1 mat tertiary	Imb
208	(11;22)(q23;q11)	37(1)	6-11	0.45	—	—	1 mat tertiary	Imb
209	(11;22)(q23;q11)	37(1)	6-11	0.45	—	—	1 mat tertiary	Imb
210	(12;16)(q24;p11)	40(16)	1-10	0.20	—	—	—	Chance
211	(12;21)(q13;q22)	16(9)	25-3	1.12	—	1 mat(der12)	—	Imb
212	(13;15)(q14;q26)	(10)(26)	21-2	0.64	—	—	—	RF
213	(13;15)(q21;q26)	(13)(25)	16-2	0.47	—	—	—	RF

214	(13;17)(q13;p12)	(6)22	24-5	1.04	—	—	—	RF
215	(13;18)(q14;q23)	(9)25	21-1	0.65	—	—	—	RF
216	(14;20)(p11;p11)	28-14	(0)7	0.16	1 pat(der14)°	—	—	Chance
217	(15;18)(q15;q23)	(7)25	20-1	0.66	—	—	—	RF
218	(15;20)(q22;p13)	(15)19	12-2	0.41	—	—	—	RF
219	(15;22)(q11;p11)	(1)13	26(0)	1.86	—	—	1 mat tertiary	Chance
220	(15;22)(q26;q12)	(26)(6)	2-7	0.28	—	—	—	RF
221	(16;17)(p13;q21)	(23)15	3-12	0.39	—	—	—	Chance
222	(16;17)(p12;q21)	(20)15	6-12	0.51	—	—	—	RF
223	(16;17)(p11;q11)	(16)10	10-17	1.04	—	—	—	RF
224	(16;20)(q11;p13)	13-19	(13)2	0.47	—	—	—	RF
225	(17;18)(p11;p11)	19-23	8-3	0.26	—	—	—	RF
226	(17;21)(q23;q22)	21(8)	6-2	0.28	1 pat(der21)	—	—	Imb
227	(18;20)(q21;q13)	17-18	9-3	0.34	2 mat(der20)	—	—	Imb
228	(18;21)(p11;p11)	22-12	3(0)	0.09	1 pat(der18)	—	—	Imb
229	(19;22)(p12;q12)	15(5)	8-8	0.80	—	—	—	RF
230	(20;21)(q11;q22)	11(9)	10-2	0.60	—	1 pat(der20)	—	Imb

[a]The figures in parentheses result from the subtraction of heterochromatic regions or acro-p.
[b]Segregations with a degree sign have been diagnosed from prenatal testing.
[c]Imb = imbalance; RF = reproductive failures.

INFERTILITY, GAMETIC SELECTION, AND FETAL LOSS

The Cytogenetics of Mammalian Autosomal Rearrangements, pages 295–314

Chapter 6
Aneuploidy-Related Delay of Meiotic Development in the Mouse and the Djungarian Hamster

I. Hansmann, P. De Boer, and R.M. Speed

INTRODUCTION

In this chapter we have addressed the question of what kind of evidence from experimental animal studies pertains to the relation between meiotic rate and meiotic non-disjunction.

Most experimental work has been done in the mouse (*Mus musculus*) and in the Djungarian hamster (*Phodopus sungorus*) and the present report is confined to these two species. As a second restriction, only first meiotic non-disjunction will be treated. The reasons for this are twofold.

a. In the human female, the majority of non-disjunctional events take place during the first meiotic division [33].

b. In the male with both meiotic divisions about equally vulnerable in man [33], research with respect to incidence, possible causes, etc., of second meiotic non-disjunction is almost absent in experimental animals.

As will be clear, especially from the data on females, delay is not the only aspect of meiocyte kinetics that one can attempt to relate to meiotic non-disjunction. A disharmonic follow-up of meiotic stages, i.e., catching-up after an initial delay, is equally a factor. Another impression that we gathered by writing this chapter is that, notwithstanding the principal similarities between male and female meiosis, the specific functions of the male and female gamete have consequences for the "biological boundaries" the cell can cross without genetic repercussions. In the female, cytoplasmic and

Institut für Humangenetik, Universität Göttingen, 3400 Göttingen, Federal Republic of Germany (I.H.), Department of Genetics, Agricultural University, 6703 BM Wageningen, The Netherlands (P.dB.), Medical Research Council Clinical and Population Cytogenetics Unit, Western General Hospital, Edinburgh EH4 2XU, Scotland (R.M.S.)

chromosomal processes presumably have a more vulnerable interaction than in the male, where, apparently, some kinetic tolerance exists. From a cell-biological point of view, the treatment of meiotic non-disjunction as we have executed in this chapter is necessarily superficial. On the other hand, we hope research into the cell-biological aspects of the animal models presented here will be promoted by our considerations.

MEIOTIC DELAY AND NON-DISJUNCTION IN PRIMARY SPERMATOCYTES

In a study of the meiotic behaviour of chromosome mutations in male mice it was discovered that in addition to the chromosomes involved in the mutation exhibiting meiotic irregularities, the normal bivalents often had an increased susceptibility for non-disjunction [9,10,14]. Here, we will try to discuss the aetiology of the irregularities observed for the normal bivalents (and less extensively the translocation multivalents), in terms of a deviant meiotic rate during spermatogenesis. Some hitherto unpublished data will be incorporated.

The following karyotypes are to be presented: T(1;13)70H and its translocation and tertiary trisomic derivatives, T(1;13)70H,+/+, Rb (11.13)4Bnr, T(1;11.13S)70H/+,Rb(11.13)4Bnr, T(1;13)70H/T(1;13)1Wa, and Rb(11.13) 4Bnr/+. Pachytene pairing diagrams of all these chromosome mutations are given in Figure 1.

The abnormal behaviour of translocation heterozygotes is best described by the frequency of adjacent-2 segregation and the frequency of numerical non-disjunction (3:1 segregation of the chromosomes of the potential quadrivalent). For the trivalent in Robertsonian translocation heterozygotes, the frequency of non-disjunction for the acrocentric chromosomes involved suffices to quantify abnormal meiotic behaviour. Table I gives the data available with the appropriate references. Table II gives relative data of sperm counts from the caput epididymis. Also incorporated are the results of studies of pachytene chromosome pairing in primary spermatocytes of some of the mutants. Many authors [16,20,31] have noticed that for chromosome mutants that affect male fertility, the frequency of unpaired chromosomes and chromosome segments in close contact with the normally unpaired segments of the sex chromosomes is much higher than for normal male mice. It is apparent from the literature [16] that high frequencies of incomplete meiotic pairing, visualised through high frequencies of meiotic prophase autosome-sex chromosome contact, coincide with low relative sperm counts, ultimately leading to sterility of the male. When in male mice with chromosome

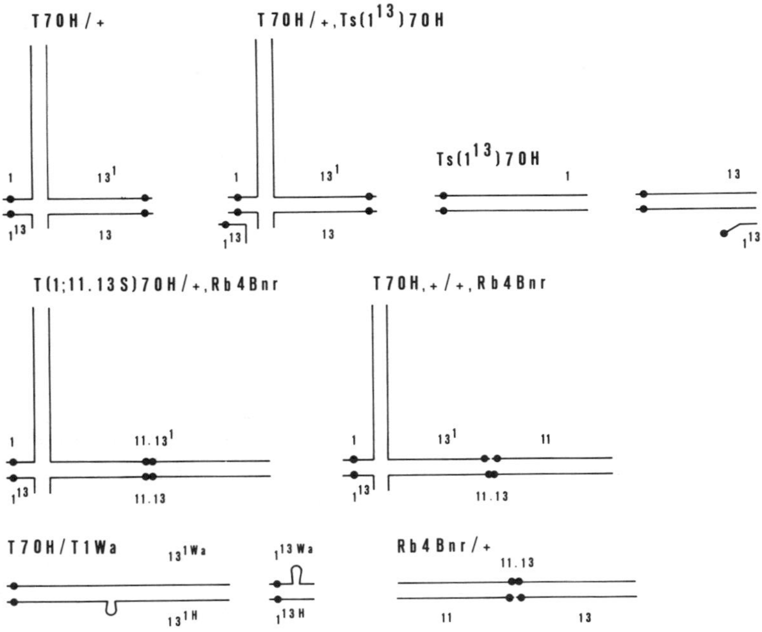

Fig. 1. Pachytene pairing diagrams of chromosome mutations presented.

mutations the sperm count is reduced by 50% or more (T70H translocation trisomics being the exception, see Table II), one consistent finding in routine air-dried preparations of spermatogenic cells is that the frequency of diakinesis-metaphase I cells with contracted and degenerating bivalents is increased [39].

This led us to investigate late diplotene-metaphase I cells in Rb4Bnr/+ males. The technique used enabled counts to be obtained per individual site along the seminiferous epithelium where, by the presence of meiotic divisions, stage XII of the cycle of the seminiferous epithelium was taking place [35]. An approximately 50% (significant) increase in late diplotene-metaphase I meiotic figures was observed. This finding prompted an autoradiographic study into the periods from premeiotic S-phase to late diplotene-diakineses and to meiotic metaphase II [41].

Four chromosome mutants have been compared with controls: T70H heterozygotes, −translocation trisomics and −tertiary trisomics, and Rb4Bnr heterozygotes. The kinetics of meiosis of T70H/+ has also been studied by tracing a small band of surviving resting primary spermatocytes after hydroxyurea-trenimon treatment, into the meiotic divisions, such that only a

TABLE I. Segregation of translocation involved chromosomes and normal bivalents during male meiosis I in seven chromosome mutants and chromosomally normal mice

	Segregation multivalent (%)		
Karyotype (N_{Males}-N_{cells})[a]	Adjacent-2[b]	Numerical non-disjunction of translocation involved chromosomes, "3:1"	Non-disjunction normal bivalents (%)
T7OH/+(7-692)[c]	26.3	8.8	3.6
T7OH/+, Ts(1^{13})7OH (6-1002)[c]	13.0	4.7	6.0
Ts(1^{13})7OH (5-141)[d]	—	8.5[i]	Low
T7OH/Rb4Bnr (5-480)[e]	23.4	23.4	14.4
T(1;11.13S)70H/Rb4Bnr (5-357)[e]	12.3	10.9	23.2
T7OH/T1Wa (3-483)[f]	—	3.9[j]	0.4
Rb4Bnr/+ (10-1400)[g]	—	26.6[k]	1.4
+/+ (26-5200)[h]	—	—	0.4

[a]The chromosome mutants are on a Swiss random-bred genetic background; the +/+ cells were from 13 different strains and stocks.
[b]Adjacent-2 denotes the failure of homologous centromeres to separate at the first meiotic division when associated in a multivalent.
[c]De Boer and Van Beek [14].
[d]De Boer and Groen [12].
[e]De Boer [10].
[f]Wauben-Penris et al. [47].
[g]The data of Nijhoff and De Boer [36] and the control data of Table IV, proving statistically not different for the "overall" non-disjunction frequency, have been pooled. Visibility of the Rb4Bnr 11.13 marker chromosome in secondary spermatocytes was 42% in both sets of data. When the Rb and not Rb carrying secondary spermatocyte classes were corrected to obtain 50% "visibility," two classes were used to estimate the frequency of non-disjunction for normal bivalents: 21 chromosomes without Rb4Bnr (four cells, obtained after correction) and 22 chromosomes with Rb4Bnr (one cell). Both have been multiplied by 4 and subtotals added to determine the frequency of non-disjunction for normal bivalents.
[h]Beatty et al. [2].
[i]Non-disjunction for chromosome 13 in a 13;13;1^{13} trivalent.
[j]Non-disjunction for heteromorphic marker bivalents, mostly the small one (1^{13}H;1^{13}Wa).
[k]Non-disjunction for the acrocentric chromosomes of the Robertsonian translocation trivalent.

few secondary spermatocytes were present for chromosome analysis of early dividing meiocytes. For Rb4Bnr/+ this technique has been repeated but omitting the trenimon treatment such that after hydroxyurea depletion of spermatogenic S-phase cells (mostly A1 to B spermatogonia and resting primary spermatocytes), the front of repopulating cells can be followed, allowing the chromosome analysis of newly developing metaphase II secondary spermatocytes.

TABLE II. First meiotic prophase delay, pachytene behaviour of unpaired chromosomes (−segments), and fertility characteristics of male chromosome mutants and Swiss controls

Karyotype	First meiotic prophase delay (h)[a]	Frequency of autosomal-sex vesicle contact during pachytene[b] (%)	Relative sperm count (caput epididymis) (%)	% Morphologically abnormal sperm (caput epididymis)
T7OH/+	0	30.3[d]	80.9[9h]	4.3[h]
T7OH/+, Ts (1^{13}) 7OH[c]	7.3	20.0[e]	50.2[h]	5.0[h]
Ts (1^{13}) 7OH[c]	7.3	Close to 100[f]	21.3[h]	77.2[h,j]
T7OH/Rb4Bnr	—	—	16.8[g]	14.6[g]
T(1;11.13S) 70H/Rb4Bnr	—	—	27.8[g]	14.0[g]
T70H/T1Wa	—	43.5[d]	13.8[d]	41.3[d]
Rb4Bnr/+	5.0	<26.7[d]	45.9[i]	14.8[g]
+/+	0	—	100	9.3[h]

[a]The average delay (relative to +/+ and calculated from the 50% labelling points, read from Figure 2 of Speed and De Boer [41]) of the period premeiotic S-phase, late diplotene-metaphase I is given.
[b]The frequency of a topologically close relation between sex chromosomes and usually unpaired autosome (segments) has been given ("+" by De Boer et al. [16]).
[c]Trisomy per se is causing a sperm count reduction of 40% of the control value [13].
[d]De Boer et al. [16].
[e]De Boer and Van Beek [14].
[f]De Boer and Branje [11].
[g]De Boer [10].
[h]De Boer and Speed [13].
[i]Speed and De Boer [41].
[j]Gives frequency of deviant uterine sperm.

In the following sections, the variation in the periods premeiotic S-phase to late diplotene-metaphase II will be discussed as to its relevance to abnormal chromosome behaviour in chromosome mutations in the mouse.

Increased Duration of the Period Premeiotic S-Phase, Late Diplotene-Metaphase II as Established by Autoradiography

Speed and De Boer [41] and Forejt [19] discovered that in chromosome mutants characterized by a diminished sperm production (<50% reduction), this period was increased for at least a proportion of the meiocytes. In the sterile Rb(16.17)7Bnr/Rb(8.17)1Iem, the delay was most spectacular (18 h for front cells [19]). From the data of Speed and De Boer [41] (see Table II as well), it can be seen that the tendency for primary spermatocytes to be delayed is greater when the resulting sperm count is lower. This observation, together with that of degenerating diakinesis-metaphase I figures, makes clear that up to the majority of the delayed meiocytes will never develop into functional spermatozoa, but die during the meiotic divisions or develop into morphologically aberrant spermatozoa (De Boer et al. [16] and Table II).

In T70H heterozygotes and −translocation trisomics, there is little cell death at diakinesis-metaphase I. T70H translocation trisomics are especially interesting in that when the translocation chromosomes unite in a chain of IV plus an univalent (CIV+I, approximately 10% of pachytene spermatocytes), meiotic prophase delay measured at late diplotene-diakinesis is greater, and moreover, the bivalents in these cells undergo precocious spiralization and are delayed in the transition to secondary spermatocytes [13]. By analogy, and supported by the autoradiography and bivalent morphology data, this can be taken to be the general course of events for the majority of surviving primary spermatocytes in T70H tertiary trisomics. In both instances, the 1^{13} chromosome forms a functional partnership with the sex chromosomes during pachytene [11,14]. For CIV+I descendants, which because of the independent segregation of two small 1^{13} chromosomes could be followed as secondary spermatocytes, no spectacular rises in either adjacent-2 segregation or normal bivalent non-disjunction could be established. On the other hand, for all cells in T70H translocation trisomics, there was a significant positive relation between adjacent-2 segregation of the multivalent and non-disjunction for the normal bivalents. In the T70H/+ males, which do not exhibit meiotic delay (see Table II), this relation was already observed, but in the translocation trisomic karyotype, it was much more evident [14].

In T70H tertiary trisomics, the non-disjunction frequency of the normal bivalents is negligible. All aneuploid secondary spermatocytes (corrected for random segregation of the extra chromosome 1^{13}) could be explained by non-disjunction for chromosome 13 in a 13;13;1^{13} trivalent that has a frequency of 22% in this karyotype [12].

For Rb4Bnr/+, the conclusion to be made from the autoradiography data is that, because primary spermatocytes are dying at a maximum of 50% of the initial spermatocyte numbers in this karyotype 1 (see Table II), among the secondary spermatocytes there must be some variation with regard to the kinetics of the transition from primary to secondary spermatocytes. Thus, the hypothesis forwarded by Nijhoff and De Boer [36] that "slower" primary spermatocytes will have a higher non-disjunction rate within the trivalent was not rejected.

Cell Kinetic Studies With Hydroxyurea- or Hydroxyurea-Trenimon-Restricted Spermatocyte Populations

When Wauben-Penris et al. [48] restricted spermatocyte populations of T70H/+ to a "bandwidth" of 10 h and followed this band until the first cells had reached the secondary spermatocyte stage (i.e., the more quickly developing ones), their main finding was that this approach delayed the arrival of the cells at late diplotene by 26 h (the band of cells is constructed at the resting primary spermatocyte stage). This approach led to clearly elevated chiasma frequencies in the translocation multivalent. In the first primary spermatocytes that developed into secondary spermatocytes, adjacent-2 frequencies had not changed. However, non-disjunction for normal bivalents was down from 5.7% to 1.9%, which was significant and could be interpreted as reflecting the effect of meiotic progress on normal bivalent dissociation.

The kinetics of the meiotic divisions in the Rb4Bnr/+ karyotype was studied by following the seminiferous epithelium after 3 injections of hydroxyurea [17,40]. Unlike the hydroxyurea-trenimon "band" method, some primary spermatocytes reach the late diplotene stage in a normal period (Table III), but the majority seems delayed by some 15–20 h. From the results of Wauben et al. [48] it can be inferred that this delay is occurring during meiotic prophase. Secondary spermatocytes in numbers high enough to be analyzed are encountered some 17 h after the first cells should have reached this stage. In the protocol followed by Nijhoff et al. [38], one testis was removed 513–514 h after the first hydroxyurea injection, whereas the second testis followed 5 h later. In a control series, it was established that the ether narcosis applied to remove the first testis did not affect the non-disjunction frequency of the second testis (Table IV). It was assumed that between testes (within animal) variance was absent with respect to the speed of repopulation after hydroxyurea treatment. Although it is a possibility that male mice do vary in repopulation characteristics, for each mouse the testis removed later contained relatively more slowly developing secondary spermatocytes. The difference between the trivalent non-disjunction rates of the testes removed first and second, when analysed by a Mann-Whitney U-test, proved to be significant ($U' = 6$, $P = 0.032$). On theoretical grounds, one would expect the control non-disjunction value to be higher than the non-disjunction value

TABLE III. First appearance of late diplotene-metaphase I (MI) and metaphase II (MII) spermatocytes that develop from G1 A1 spermatogonia after hydroxyurea treatment of Rb4Bnr/+ males

Period first injection HU to autopsy (h)	N males	N testes	Stage(s) reached
489.5, 492.5, 494.5 and 497.5	2	4	M1 (few cells)
509, 510.5	2	2	M1 (increasing Nos.)
513–514	6	12	M1, few MII

TABLE IV. Non-disjunction frequencies in secondary spermatocytes of Rb4Bnr/+ mice in control testes and during repopulation after a hydroxyurea-induced gap in the spermatogenic line

		N males	N cells	% Non-disjunction[a]
Control series	First testis	4	400	27.0
	Second testis	4	400	24.5
Experimental series	First testis	6	432	16.0
	Second testis	6	660	21.5

[a]As cell breakage is negligible in secondary spermatocytes, the non-disjunction frequency has been calculated from both secondary spermatocytes with 19 and with 21 chromosome arms.

for the second testis if primary spermatocytes with the greatest delay have not yet become secondary spermatocytes by 518–519 h after the initial hydroxyurea injection. When this suggestion was addressed with a χ^2 test to prove the significance of the increasing order of non-disjunction (testis removed first—testis removed second—control testis), a trend was established ($\chi_2{}^2 = 15.8$, $P < 0.005$). So our results are suggestive of a) heterogeneity in the kinetics of the first meiotic division in Rb4Bnr/+, as was already apparent from the autoradiography results; and b) a relationship between cell delay and non-disjunction caused for the Rb trivalent. A conclusion also to be drawn is that despite the relative high survival rate of primary spermatocytes in an "endangered spermatogenic environment," non-disjunction for normal bivalents is hardly increased.

Karyotypes Where Meiotic Delay Can Be Inferred From the Characteristics of the Chromosome Mutation

As is clear from the previous sections, meiotic delay can be inferred from such characteristics as breakpoint positions (chances for unpaired segments at first meiotic prophase), the morphology of the majority of late diplotene-metaphase I cells, and the relative sperm count. In this category, the combinations between T70H and Rb4Bnr can be grouped as well as the double heterozygotes between T(1;13)70H and T(1;13)1Wa (see Tables I, II; Fig. 1).

Most T70H/T1Wa carriers are sterile and do not yield many secondary spermatocytes for chromosome analysis. However, regularly some fertile probands occur and three of those have been used to estimate non-disjunction frequencies [47]. Their sperm counts were 35.7%, 32.8%, and 7.8% of normal, and the percentage of abnormal spermatozoa was only slightly increased. As can be seen (Table I), almost all non-disjunction found in this karyotype is due to the heteromorphic translocation bivalents, especially the small one (1^{13}H;1^{13}Wa). Non-disjunction for normal bivalents is not elevated.

In the karyotypes that combine T70H and Rb4Bnr, we find a combination of a rather severe depression of spermatogenic efficiency (but not sterility) and clearly elevated levels of non-disjunction in normal bivalents (see again Tables I, II).

For T(1;11.13S)70H/Rb4Bnr, a similar positive relationship between adjacent-2 segregation and non-disjunction was seen as in T70H translocation trisomics [37]. An interpretation of the results with the T70H,Rb4Bnr compounds is offered in the next section.

What Can Be Learned From the Study of Male Meiosis in Carriers of Chromosome Mutations That Affect Spermatogenesis With Regard to the Aetiology of First Meiotic Non-Disjunction of Normal Bivalents?

For T70H translocation heterozygotes, T70H tertiary trisomics and Rb4Bnr heterozygotes directly, but for T70H/T1Wa and the T70H/Rb4Bnr karyotypes indirectly, there is evidence that for a significant part of the cell population, the period from premeiotic S-phase to late diplotene-diakinesis is delayed. Only for Ts(1^{13})70H and for chains of IV plus a univalent (I) in T70H translocation trisomics, which show most delay over this period, do we have indications that the transition metaphase I to metaphase II is also delayed.

In the latter two cell types, although they are not as extensively investigated as they should be, we have no indication of a significant increase in normal bivalent non-disjunction. Adjacent-2 segregation is not directly influenced by delay (chain IV+I, T70H translocation trisomics) or by delay inferred from primary spermatocyte death as in the T70H, Rb4Bnr combination karyotypes. Neither is adjacent-2 segregation decreased when a front of secondary spermatocytes is studied in a restricted spermatocyte population, as has been done for T70H/+ [48].

For T70H/+ and −translocation trisomics and for T(1;11.13S)70H/Rb4Bnr [37], we have evidence that non-disjunction for normal bivalents and adjacent-2 segregation are interrelated phenomena. If we visualize the multivalent chain in the T70H cases as a more rigid structure when adjacent-2 segregation follows, with the two 13-centromeres closely apposed to the interstitial chiasma [10], a comparison can be made with the very long multivalent in the T70H,Rb4Bnr combination karyotypes, where non-disjunction for normal bivalents is highest. Thus, it seems likely that the orientation of a very large multivalent interferes with a proper orientation of the bivalents. An effect of this process on meiotic rate (or vice versa) could also contribute to abnormal bivalent behaviour. Taking all findings in male chromosome mutants together, the male meiotic divisions in the mouse are not very conducive for non-disjunction to occur. The first meiotic division is a selective barrier, prone to cell death and probably thus plays a role in the avoidance of aneuploidy. It also has to be mentioned here that spermatogenic cells operate under a system of metabolic cooperation (among themselves and with the Sertoli cells) and that also at the first meiotic division, a minority of cells with meiotic delay can be rescued without greatly increasing the number of aneuploid gametes, as has been demonstrated for cells with a pachytene univalent in T70H translocation trisomics [13,14].

MEIOTIC RATE AND NON-DISJUNCTION IN PRIMARY OOCYTES

Female meiosis in mammals is a characteristic discontinuous process regulated by several stop and go signals [44]. After pairing and recombination at prophase I during prenatal life, the oocyte nucleus becomes arrested in the dictyate stage, which may last up to several months in rodents until resumption of meiosis I is triggered by a preovulatory surge of gonadotrophin. The second arrest of meiosis at metaphase II is only circumvented via fertilization or parthenogenetic activation. Hence, both meiotic divisions, the preovulatory first division and the postovulatory second division, have to be induced from a shortly preceding phase of meiotic arrest.

In the female, as in the male, we will confine ourselves to the first meiotic division (see Introduction). The phase of first meiotic arrest may serve as a controlling factor in selecting against errors of meiotic prophase as well as meiotic delay up to this stage. However, oocytes with pairing irregularities, such as in the XO-karyotype or mice heterozygous for the structural anomalies In(X)1H, T(11;19)42H and Ts(1^{13})70H [7,8,10], seem to be prone to degeneration shortly before meiotic arrest at dictyotene and the ovary of such females is characterized by a reduced number of oocytes. At least in XO mice, atretic follicles are not more frequently encountered [6].

An experimentally delayed pre-GVB-phase[1] of oocyte maturation by the injection of anti-LHRH immunoglobulins in PMSG-primed prepubertal mice did not increase chromosome malsegregation during meiosis I [30].

Meiotic delay may thus adversely affect chromosome segregation at both meiotic divisions only when it arises during the short time available, for the

[1]GVB = germinal vesicle breakdown; LHRH = luteinizing hormone-releasing hormone; PMSG = pregnant mare serum gonadotrophin; HCG = human chorionic gonadotrophin.

first division the few hours at most from dictyotene till metaphase I, but after having received the triggering signal.

In the section on male meiosis, we have seen that mouse chromosome mutations per se can be characterized by altered meiotic rates. Therefore, we include (first division) meiotic data of the female counterparts of some of the constructs treated in the male.

As will be seen, increased non-disjunction frequencies for normal bivalents can be encountered among this group.

However, the duration of the meiotic stages from the gonadotrophin HCG surge, or in vitro, have not yet been determined. Standard gonadotrophin stimulation, the regular PMSG-HCG sequence, is given irrespective of the female's cycle stage with 2 days in between.

In the second part of this section, attention will be given to gonadotrophin-induced first meiotic abnormalities in the Djungarian hamster and in the NMRI mouse genotype. The kinetics of the first meiotic division will be approached indirectly here (colchicine sensitivity distinct periods after the HCG injection) but also directly by releasing oocytes at varying periods post-HCG from their follicles and scoring the stage of meiosis. A synopsis of these results is given.

Chromosomal Properties Causing Aneuploidy

Female mice, heterozygous for the T70H reciprocal translocation (Fig. 1), have been used to investigate the disjunction of the translocation multivalent as well as the interaction with the segregation of nontranslocation chromosomes at meiosis I (Table V) [45,46].

Presegregation of the small 1^{13} marker chromosome (unpartnered chromatid), a result of kinetochore malfunction at meiosis I, occurred significantly more often in alloxan-treated nondiabetic T70H/+ (3.2%) and in alloxan diabetic T70H/+ females (5.5%) in comparison with oocytes from various other mouse strains. Oocytes from T70H/+ females homozygous for the dwarf mutation (dw/dw) do show presegregation of the 1^{13} small marker as well, however, only at a lower incidence (1.5%). Single unpartnered 1^{13} chromatids were also detected in secondary oocytes from aging T70H/T70H females (0.7%), but hardly in young females of this karyotype (0.2%) [15]. No significant (presegregation) splitting of the large 13^1 marker or the non-translocation chromosomes occurred in oocytes from T70H hetero- and homozygotes.

Data on presegregation during oogenesis are available for another reciprocal translocation, i.e., T(2;4)1Gö [50]. Altogether 4.4% of ovulated oocytes showed equational splitting of the small 4^2 marker chromosome,

TABLE V. Influence of reciprocal translocations on the presegregation of normal and marker chromosomes in mouse primary oocytes

Karyotype	Specification of females	No. MII oocytes	% Oocytes with presegregation of: Marker chromosomes Small (1^{13})	Large (13^{1})	Normal chromosomes
T7OH/+	Non-diabetic[a]	250	3.2	0.0	0.0
	Diabetic[a]	218	5.5	0.0	0.0
	dw/dw[b]	69	1.5	0.0	0.0
T7OH/T7OH	Young[c]	545	0.2	0.0	0.0
	Old[c]	443	0.7	0.0	0.0
+/+	C3H/HeHan[d]	1,302	—	—	0.4
	(101 × C3H)F1[d]	740	—	—	0.7
	NMRI/Han[d]	1,296	—	—	0.4

[a]Wauben-Penris and Prins [45].
[b]Wauben-Penris and Van der Hoeven [46].
[c]De Boer and Van der Hoeven [15].
[d]Hansmann and El Nahass [22].

specifically after gonadotrophin treatment at low doses. In the hypergonadotrophic situation less oocytes were affected by 4^2 presegregation (0.7%; $P < 0.05$). Presegregation of other chromosomes was not recorded. Thus from the analysis of small marker chromosomes, in the homozygous or heterozygous condition, it follows that presegregation is determined mainly by individual chromosomal characteristics such as size and related effects (in a T/+ situation) like lack of/or decreased synapsis. The quality of oocyte maturation—young versus old females (T70H/T70H [15]), high versus low gonadotrophic stimulation (T1Gö/+ [50]), may determine the extent to which kinetochore malfunction is expressed. Yet, no information is available on meiotic rates in preovulatory oocytes from females with chromosome translocations.

Parallel to the results in male T70H heterozygotes, in females with the same translocation there was frequent interference with the disjunction of normal bivalents (compare Table VII with Table I). Approximately 7–9% of all oocytes from T70H/+ females showed non-disjunction of normal bivalents. The frequency of non-disjunction in oocytes from chromosomally normal female mice from various strains is otherwise far less or at most around 1% (see Hansmann [21], for review). Hence, non-disjunction seems to result from an interaction between the bivalents and the multivalent. This type of interaction is restricted not only to translocation heterozygosity (young and old T70H/T70H females had low frequencies of hyperhaploidy for non-translocation bivalents with 0.4% and 0.5%, respectively [15]), but also occurs in females doubly heterozygous for two semi-identical reciprocal translocations (see below). No effect upon the disjunction of the normal bivalents was detected in secondary oocytes from T1Gö/+ mice [50].

The T(1;13)70H/T(1;13)1Wa females (Fig. 1) have two pairs of heteromorphic chromosomes with slightly unequal length (1^{13}H vs. 1^{13}Wa and 13^{1}H vs. 13^{1}Wa, [47]). The two interstitial segments of the translocation chromosome bivalents are too small for pairing, which could result in multivalents at diakinesis/metaphase I [16]. Although T70H/T1Wa bivalents resemble very closely those from T70H/T70H, oocytes of the former karyotype demonstrate significant alterations in chromosome pairing and segregation at meiosis I (Tables VI–VIII). Chiasma frequency was significantly reduced in the large 13^{1} bivalent (2.27 in T70H/T70H versus 2.07 in T70H/T1Wa). Furthermore, true and opposite univalents occurred significantly more often for the small 1^{13} marker in the compound heterozygotes (1^{13}H;1^{13}Wa, 12.6% vs. 1.4% in 1^{13}H;1^{13}H). Non-disjunction was signifi-

TABLE VI. Influence of homozygous and semihomozygous reciprocal translocations on frequencies of univalents and chiasma number in primary oocytes

Karyotype	Age of females	% Oocytes with univalents: Small marker (1^{13})	% Oocytes with univalents: Other bivalents	Mean chiasma No. of large marker (13^{1})
T70H/T70H	Young[a]	1.4	3.3	2.27
	Old[a]	7.2	6.0	2.27
T70H/T1Wa	Young[b]	12.6	0.4	2.07

[a]De Boer and Van der Hoeven [15].
[b]Wauben-Penris et al. [47].

TABLE VII. Influence of reciprocal translocations on chromosome segregation during the first meiotic division in mouse oocytes (for references see Table VI) demonstrating significant interference of the T7OH/T1Wa and T7OH/+ karyotypes with disjunction of normal bivalents

			% Oocytes with non-disjunction			
			Marker chromosomes			
Karyotype	Age of females	No. MII oocytes	Small (1^{13})	Large (13^{1})	Normal chromosomes	Total
T70H/T70H	Young[a]	545	0.2	0.7		0.9
	Old[b]	443	0.9	0.9		1.8
T70H/T1Wa	Young[c]	421[a]	5.4	1.7	6.2	13.8
T70H/+	Young[d]	212[a]			9.0	

[a]De Boer and Van der Hoeven [15].
[b]Wauben-Penris et al. [47].
[c]Wauben-Penris and Prins [45].
[d]Cell Nos. have been corrected for the broken and diploid secondary oocytes.

TABLE VIII. Chromosome segregation during the first meiotic division in oocytes from T7OH/T1Wa mice

	Translocation bivalents	
	Normal segregation	Non-disjunction
Normal segregation	369	26
Normal bivalents		
Non-disjunction	20	6
%	5.14	18.75[a]

[a]Non-disjunction of normal bivalents occurred significantly more often in those oocytes with malsegregating translocation bivalents than in those where the $1^{13}H;1^{13}Wa$ and $13^{1}H;13^{1}Wa$ bivalents segregate normally ($\chi_1^2 = 6.60$, $P < .025$, data from Wauben-Penris et al. [47] and corrected for cell breakage and diploidy).

cantly increased not only for the translocation chromosomes (5.4%) but surprisingly also for the non-translocation bivalents (5.4%).

Hence, non-disjunction by interference is not only generated in maturing oocytes with a specific large multivalent but also in female mice with heteromorphic marker bivalents. Here, meiotic maturation may suffer from failure of or incomplete meiotic pairing and/or from the resulting consequences for meiotic rate and spindle function. Interestingly, an additional type of meiotic anomaly was recorded in oocytes from T70H/T1Wa females. After in vitro culture, 2.3% of all oocytes with metaphase II configuration remained diploid. No such diploidy was observed in oocytes recovered from T70H/T70H females [15]. As the oocytes from T70H/T70H females were collected after gonadotrophin-induced maturation in vivo, no conclusive answer as to the significance of diploidy in T70H/T1Wa oocytes is possible yet. No comparative analysis of the meiotic rate of oocytes from T70H/T70H vs. T70H/T1Wa has been performed by the authors.

F1 hybrid XO female mice from crosses of Tabby (XO) females with NMRI/Han males, the latter stock carrying the genetically determined trait "diploidy," ovulated considerable numbers of diploid oocytes (24.6%) arrested at metaphase I after gonadotrophin stimulation [5]. In XX littermates, this frequency was only 1.0%. For the parental XO stock itself and exposed to the same gonadotrophin treatment, a lower frequency of diploidy (3.1%) is recorded than for the XO F1 hybrid females, but in comparison to other mouse strains, the level is still increased. No information is available yet on the rate of meiosis in XO,NMRI/+ oocytes.

Meiotic Rate and Aneuploidy via Alterations of Follicle-Oocyte Maturation

In a study of the spontaneous frequency of non-disjunction in oocytes from several mammalian species [22,23], we noticed that treatment with the gonadotrophic hormones (GTH) PMSG and HCG caused the ovulation of significant numbers of hyperhaploid and diploid oocytes from the Djungarian hamster (*Phodopus sungorus*; n = 14 [24]). GTH-inducible genome mutations have been a repeatable observation in females from our random-bred hamster colony. The frequency of genome mutations in ovulated oocytes depends on the dosage of both gonadotrophins, injected within 48 h. Recently a significant linear increase of hyperhaploid and diploid oocytes with logdose was detected over a dose range from 1.25 IU up to 20 IU of PMSG-HCG each [49]. Up to 15% of all ovulated oocytes were heteroploid at the highest dose levels. Since prepubertal females, as a response to gonadotrophin treatment, ovulate diploid and hyperhaploid oocytes as well, the effect in adult females cannot be explained by the rescue of atretic follicle-oocyte complexes, not ovulated during a preceding cycle [42].

The diploid oocytes in GTH-treated females were of different types. Some oocytes showed metaphase I arrest, others appeared to be blocked in meiosis at any transitional stage between metaphase I and telophase I [26]. Some diploid oocytes even had metaphase II configuration chromosomes. In an analysis of approximately 200 non-disjunctional events, it became clear that not all bivalents are involved equally in gonadotrophin-induced non-disjunction [3,28,42,49]. The large metacentric and submetacentric bivalents from the groups A–C were preferentially affected. Hence, it was suggested that chromosomal properties determining early or late segregation could provide a bivalent specific risk for meiotic non-disjunction. Properties such as length, position of the centromere, chiasma frequency, and position and presence plus activity of nucleolus organizers seem to accumulate stronger pairing forces between the homologs within a bivalent.

Because of these cytogenetic observations after GTH treatment, we suggested that failures in the endocrine control of follicular maturation would cause an altered developmental program within the maturing oocyte that increased the risk for generating either non-disjunction of one or several bivalents (hyperhaploidy) or even of all bivalents (diploidy) [21, 26]. As the NMRI mouse stock responds similarily to the treatment with gonadotrophins by ovulating diploid oocytes [1,4,22] the model of "failure in the endocrine control of meiosis as a primary cause for aneuploidy" seems to be applicable to other mammalian species. The two classes of effects, ovulation of only diploid oocytes in the NMRI mouse stock but of diploid and hyperhaploid oocytes in Djungarian hamsters, have been explained by a more or less synchronous segregation of the acrocentric mouse bivalents [26], compared to a more bivalent-specific pattern of behaviour in the latter species.

We assumed that GTH-induced errors of chromosome segregation are caused by an alteration of an otherwise synchronized maturation of the oocyte nucleus and cytoplasm [21,25]. The resulting asynchrony—i.e. a relative delay of meiotic resumption—would thus be expressed as premature cytoplasmic activity that would normally block meiosis at metaphase II but now inhibits those bivalents that are preferentially late segregating (the cytostatic factor or CSF as proposed by Masui et al. [32]). Our assumption of preferential non-disjunction of late segregators was verified by cytogenetic studies with the spindle inhibitors colchicine [29] and methylbenzimidazolcarbamate (MBC), using the Djungarian hamster as a model [28]. The latter compound is known to bind specifically to β-tubulin [34]. Both compounds, injected at a rather late stage of meiosis I, caused some oocytes to be ovulated as aneuploids for those chromosomes that are also affected by the hypergonadotrophic stimulation. "Diploid ovulation" also occurred.

The first hint that oocyte maturation induced by increasing GTH doses is indeed accompanied by the assumed asynchrony of nuclear and cytoplasmic development—i.e., by a delay in the resumption of meiosis—was derived from colchicine-induced arrest of meiosis at metaphase I. Females stimulated with high GTH doses ovulated significantly more metaphase I arrested oocytes than those receiving colchicine at the same time after the HCG injection but at lower GTH doses [28,29]. The delay in nuclear progression, measured as a delay in sensitivity of oocytes for metaphase I arrest by colchicin, was estimated to be approximately 1.5 h. Wuttke [49] was able to show that in Djungarian hamsters, the hypergonadotrophic stimulation indeed caused a considerable delay in germinal vesicle breakdown (GVB). During a detailed study of spindle formation and function and rate of meiosis, Theuring [43] observed a grossly altered rate of meiosis in oocytes from female hamsters, characterized by a high frequency of genome mutations. Maturation of high-risk oocytes was accompanied by a more-or-less severe delay in germinal vesicle breakdown. But in contrast to the model suggested by Hayden et al. [27], the delayed breakdown was followed by a shortened phase up to anaphase I, which would not allow premature separation of bivalents. Thus oocyte maturation, leading to the generation of genome mutations, is characterized by a very short transitional phase from GVB till the actual segregation of bivalents occurs. Abnormalities like multipolar and morphologically irregular spindles develop under these maturation conditions [43]. Spindle malfunction seems to result from and mediate an altered developmental program of oocyte maturation. The mechanisms involved in altering oocyte maturation by causing changes in the temporal sequence of meiosis have to be studied at the molecular level.

CONCLUSIONS

The most significant aspect of this chapter is that it has enabled a male-female comparison with respect to first meiotic non-disjunction. A phenomenon that is traced through the use of chromosome mutations in the mouse is the increased likelihood for the simultaneous occurrence of non-disjunction of normal bivalents and abnormal behaviour of the translocation multivalent (T70H/+, adjacent-2, male mice) or the heteromorphic bivalents (T70H/T1Wa, non-disjunction, female mice). In the male we have argued that a rigid multivalent in a metaphase I spindle can interfere with normal bivalent non-disjunction. The fact that the same phenomenon is encountered in females with heteromorphic bivalents might plea for a common cellular denominator for both processes. In the male, we could not relate this phenomenon to meiotic delay. In the female, a study with the potential to demonstrate an effect on timing has not yet been undertaken.

An intriguing aspect of the male/female comparison is that in the female, all meiosis I products can realize themselves in a conceptus, whereas for the male, this should not necessarily be the case. Thus, spermatogenic delay and oligospermia seem to be interconnected. The research with Rb(11.13)4Bnr as reported above suggests that some meiotic prophase delay plays a role in the formation of aneuploid secondary spermatocytes caused by this karyotype. Considering other members of the Bnr family of Robertsonian translocations, no selection against aneuploid spermatids-spermatozoa could be demonstrated up to the time of conception [18]. Thus, a certain amount of kinetic variation may occur during first meiotic prophase in the male without segregational consequences for the normal bivalents. Our results with male T70H translocation trisomics point in the same direction, and a parallel can be seen with the work of Laing et al. [30] in female mice. However, it should be noted that mouse acrocentric bivalents probably are very synchronous in their behaviour in meiosis I, not allowing "time" to differentiate between them.

In the hamster, using PMSG-HCG stimulation of oocytes as the model, a different phasing of the post-HCG meiotic stages (GVB-delay) is directly related to the chances for non-disjunction of "risk bivalents," the effect being proportional to the gonadotrophin dose. A partial blockade of meiosis I in an oocyte preparing for metaphase II arrest apparently is the case. Two leads should be pursued further and are especially suitable for female mammals: (i) endocrinological control of follicle-oocyte maturation and (ii) cell-biological research at the molecular level of oocyte-granulosa cell complexes.

ACKNOWLEDGMENTS

Thanks are due to Mrs. M. Heinze-Sprotte for the editorial work and Mrs. H. D'hondt-De Jong for typing the manuscript.

The research by I.H. has been granted by the "Deutsche Forschungsgemeinschaft" and the "Bundesministerium für Forschung und Technologie (UC/CMT)."

REFERENCES

1. Bartels I, Jenderny J, Hansmann I: Control of meiosis by somatic cells in mice: inheritance of the meiosis I error "diploidy" and nonexpression in sensitive NMRI/Han oocytes ovulated from chimeras. Cell Differ 15:189, 1984.
2. Beatty RM, Lim MC, Coulter VJ: A quantitative study of the second meiotic metaphase in male mice (Mus musculus). Cytogenet Cell Genet 15:256, 1975.
3. Beermann I: Der Einflusz von Inhibitoren der Steroidsynthese auf die Follikelentwicklung und Oocytenreifung. Diplomthesis, Math. Nat. University Göttingen, 1984.
4. Beermann F, Hansmann I: Follicular maturation, luteinization and first meiotic division in oocytes after inhibiting mitochondrial function in mice with chloramphenicol. Mutat Res 160:47, 1986.
5. Beermann F, Franke U, Hansmann I: High susceptibility for diploidy in ovulated oocytes from XO mice. Hum Genet 72:323, 1986.
6. Burgoyne PS, Baker TG: Oocyte depletion in XO mice and their XX sibs from 12 to 200 days post partum. J Reprod Fertil 61:207, 1981.
7. Burgoyne PS, Baker TG: Perinatal oocyte loss in XO mice and its implications for the aetiology of gonadal dysgenesis in XO women. J Reprod Fertil 75:633, 1985.
8. Burgoyne PS, Mahadevaiah S, Mittwoch U: A reciprocal autosomal translocation which causes male sterility in the mouse also impairs oogenesis. J Reprod Fertil 75:647, 1985.
9. De Boer P: Male meiotic behaviour and male and female litter size in mice with the T(2;8)26H and T(1;13)70H reciprocal translocations. Genet Res 27:369, 1976.
10. De Boer P: Chromosomal causes for fertility reduction in mammals. In de Serres FJ (ed): "Chemical Mutagens," Vol. 10. New York and London: Plenum Press, 1986, p 427.
11. De Boer P, Branje HEB: Association of the extra chromosome of tertiary trisomic male mice with the sex chromosomes during first meiotic prophase, and its significance for impairment of spermatogenesis. Chromosoma 73:369, 1979.
12. De Boer P, Groen A: Fertility and meiotic behaviour of male T70H tertiary trisomics of the mouse (Mus musculus): A case of preferential telomeric meiotic pairing in a mammal. Cytogenet Cell Genet 13:489, 1974.
13. De Boer P, Speed RM: Meiosis of T70H translocation trisomic male mice II. Meiotic rate, spermatocyte interactions and fertility. Chromosoma 87:315, 1982.
14. De Boer P, Van Beek MEAB: Meiosis of T70H translocation trisomic male mice I. Meiotic configuration and segregation. Chromosoma 87:303, 1982.
15. De Boer P, Van der Hoeven FA: The use of translocation-derived "marker-bivalents" for studying the origin of meiotic instability in female mice. Cytogenet Cell Genet 26:49, 1980.
16. De Boer P, Searle AG, Van der Hoeven FA, De Rooy DG, Beechey CV: Male pachytene pairing in single and double translocation heterozygotes and spermatogenic impairment in the mouse. Chromosoma 93:326, 1986.
17. Dietrich AJJ, Mulder RJP: A light microscope study of the development and behaviour of the synaptonemal complex in spermatocytes of the mouse. Chromosoma 83:409, 1981.

18. Ford CE, Evans EP: Non-expression of genome unbalance in haplophase and early diplophase of the mouse and incidence of karyotypic abnormality in post-implantation embryos. In Boué A, Thibault C (eds): "Les Accidents Chromosomiques de la Reproduction." Paris: Colloque INSERM, 1973, p 271.
19. Forejt J: X-Y involvement in male sterility caused by autosome translocations—a hypothesis. In Crosignani PG, Rubin BL, Fraccaro M (eds): "Genetic Control of Gamete Production and Function." London, Toronto, Sydney: Academic Press, New York, San Francisco: Grune & Stratton, 1982, p 135.
20. Forejt J, Gregorovà S, Goetz P: XY pair associates with the synaptonemal complex of autosomal male-sterile translocations in pachytene spermatocytes of the mouse (Mus musculus). Chromosoma 82:41, 1981.
21. Hansmann I: Factors and mechanisms in non-disjunction and X-chromosome loss. In Sandberg A (ed): "Cytogenetics of the Mammalian X Chromosome, Part A: Basic Mechanisms of X Chromosome Behaviour." New York: Alan R. Liss, Inc., 1983, p 131.
22. Hansmann I, El Nahass E: Incidence of non-disjunction in mouse oocytes. Cytogenet Cell Genet 24:115, 1979.
23. Hansmann I, Probeck HD: Chromosomal unbalance in ovulated oocytes from Syrian hamsters (Mesocricetus auratus) and Chinese hamsters (Cricetulus griseus). Cytogenet Cell Genet 23:70, 1979.
24. Hansmann I, Jenderny J, Probeck HD: Mechanisms of non-disjunction: Gonadotrophin-induced aneuploidy in oocytes from Phodopus sungorus. Eur J Cell Biol 22:24, 1980.
25. Hansmann I, Jenderny J, Probeck HD: Low doses of X-rays decrease the risk of diploidy in mouse oocytes. Mutat Res 109:99, 1983.
26. Hansmann I, Bartels I, Beermann F, Caspari D, Franke U, Hummler E, Theuring F: Mechanisms of non-disjunction: Facts and perspectives. In Dellarco VL, Voytek PE, Hollaender A (eds): "Aneuploidy." New York, London: Plenum Press, 1985, p 417.
27. Hayden TL, Crowley PH, Wilson E: A model relating the incidence of meiotic trisomy to maternal age. J Theor Biol 86:123, 1980.
28. Hummler E: Über die Ursachen einer gestörten Chromosomensegregation in der Oogenese von Phodopus sungorus. Diplomthesis. Math. Nat. University Göttingen, 1984.
29. Hummler E, Hansmann I: Preferential non-disjunction of specific bivalents in oocytes from Djungarian hamsters (Phodopus sungorus) following colchicine treatment. Cytogenet Cell Genet 39:161, 1985.
30. Laing S, Gosden RG, Fraser HM: Cytogenetic analysis of mouse oocytes after experimental induction of follicular overripening. J Reprod Fertil 70:387, 1984.
31. Mahadevaiah S, Mittwoch U: Synaptonemal complex analysis in spermatocytes and oocytes of tertiary trisomic $Ts(5^{12})31H$ mice with male sterility. Cytogenet Cell Genet 41:169, 1986.
32. Masui Y, Lohka MJ, Shibuya EK: Roles of Ca ions and ooplasmic factors in the resumption of metaphase-arrested meiosis in Rana pipiens oocytes. In Evans CW, Dickinson HG (eds): "Controlling Events in Meiosis." Cambridge: University of Cambridge, 1984, p 45.
33. Mikkelsen M, Poulsen H, Grinsted J, Lange A: Non-disjunction in trisomy 21. Study of chromosomal heteromorphisms in 110 families. Ann Hum Genet 44:17, 1980.
34. Morris NR: The genetics of microtubule polymerization and the mechanism of action of antimicrotubule drugs in Aspergillus nidulans. In De Brabander M, De Mey J (eds): "Microtubules and Microtubules Inhibitors." Amsterdam: Elsevier-North Holland, 1980, p 511.

35. Nijhoff JH, De Boer P: A first exploration of a Robertsonian translocation heterozygote in the mouse for its usefulness in cytological evaluation of radiation-induced meiotic autosomal nondisjunction. Mutat Res 61:77, 1979.
36. Nijhoff JH, De Boer P: Radiation-induced meiotic autosomal non-disjunction in male mice: The effects of low doses of fission neutrons and X-rays in meiosis I and II of a Robertsonian translocation heterozygote. Mutat Res 72:431, 1980.
37. Nijhoff JH: Meiotic behaviour and spermatogenesis in male mice heterozygous for translocation types also occurring in man. Ph. D. Thesis. Agricultural University, Wageningen, The Netherlands, 1981.
38. Nijhoff JH, De Boer P, Wessels-Kaalen MCA: Private communication. Mouse News Letter 71:52, 1984.
39. Oshimura M, Tonomura A, Tanaka M: Degenerative metaphase I's and meiotic configurations in male translocation carriers in mice. Cytogenet Cell Genet 27:209, 1981.
40. Oud JL, De Jong JH, De Rooij DG: A sequental analysis of meiosis in the male mouse using a restricted spermatocyte population by a hydroxyurea/triaziquone treatment. Chromosoma 71:237, 1979.
41. Speed RM, De Boer P: Delayed meiotic development and correlated death of spermatocytes in male mice with chromosome abnormalities. Cytogenet Cell Genet 35:257, 1983.
42. Theuring F: Der Einflusz von Gonadotropinen und des LH-Releasing Hormons auf die Ovarfunktion und den Ablauf der Meiose beim russichen Hamster (Phodopus sungorus). Diplomthesis, Math. Nat. University Göttingen, 1982.
43. Theuring F, PhD-Thesis, Math. Nat. Faculty University Göttingen, 1986.
44. Tsafriri A, Pomerantz SH: Regulation of the development of meiotic competence and of the resumption of oocyte maturation in the rat. In Evans CW, Dickinson HG (eds): "Controlling events in meiosis." Cambridge: University of Cambridge, 1984, p 25.
45. Wauben-Penris PJJ, Prins JB: Meiotic behaviour of alloxan-treated diabetic and nondiabetic T(1;13)70H/+ mice. Hum Genet 63:268, 1983.
46. Wauben-Penris PJJ, Van der Hoeven FA: Meiotic non-disjunction and translocation segregation in dwarf (dw/dw) and control T(1;13)70H heterozygous mice. J Hered 76:341, 1985.
47. Wauben-Penris PJJ, Van der Hoeven FA, De Boer P: Chiasma frequency and non-disjunction in heteromorphic bivalents: meiotic behaviour in T(1;13)70H/T(1;13)1Wa mice as compared to T(1;13)70H/T(1;13)70H mice. Cytogenet Cell Genet 36:547, 1983.
48. Wauben-Penris PJJ, Van Meel GPJM, De Boer P: Spermatogenic delay and increased chiasma frequency in T70H/+ male mice with hydroxyurea-Treminon limited spermatocyte populations. Can J Genet Cytol 27:192, 1985.
49. Wuttke A: Der Einflusz von Hormonen und Wachstumsfaktoren auf Follikelentwicklung und Meiose in Oocyten des Djungarischen Hamsters (Phodopus sungorus). Diplomthesis, Math. Nat. University Göttingen, 1986.
50. Beermann F, Bartels I, Franke U, Hansmann I: Chromosome segregation of meiosis I in female T(2;4)IGö/+ mice: No evidence for a decreased crossover frequency with maternal age. Chromosoma 95:1, 1987.

The Cytogenetics of Mammalian Autosomal Rearrangements, pages 315–359

Chapter 7
Robertsonian Heterozygotes in the House Mouse and the Fate of Their Germ Cells

Carlo Alberto Redi and Ernesto Capanna

Chromosomal translocation of whole arms (Robertsonian chromosome fusion) leading to the presence of (single/multiple) structural heterozygosity occurs frequently in natural populations of *Mus domesticus* throughout the area of distribution of this species in Western Europe and the North African coast.

From the analysis of which chromosomal arms are involved in the Rb metacentric chromosomes, it seems that there is a random chance for any of the 19 acrocentric autosomes (with the exception of chromosome 19) to be involved. Some of the structural properties of the mouse genome and the undetectable DNA losses that occur during Rb exchange indicate that Rb exchange in the mouse might be favoured by the presence of the same major type of highly repetitive satellite DNA along each chromosome in conjunction with the overall acrocentric constitution. There are several consequences of the presence of Rb metacentric chromosomes in the karyotype: The chromosome number is decreased, although the fundamental number remains unchanged. The size of the genome (DNA or A-T base pair content) does not vary detectably from the standard karyotypic strain even after nine pairs of Rb fusions. Translocation homozygosity does not seem to affect any physiological process, whereas heterozygosity has detrimental effects, especially on meiosis, resulting very often in impairment of fertility of the carrier. The effects on gametogenesis of several heterozygosity situations are analyzed. Single heterozygosities mainly cause non-disjunction events, with the production of chromosomally unbalanced gametes; females very often have higher rates of non-disjunction than males and have a distorted ratio in the transmission of the Rb metacentrics to progeny with the homozygous acrocentric type being much more frequent. Multiple independent heterozygosity produces higher rates of non-disjunction events. When there

Dipartimento di Biologia Animale and Centro di Studio per l'Istochimica del CNR, Università di Pavia, 27100 Pavia (C.A.R.), Dipartimento di Biologia Animale e dell'Uomo, Università 'La Sapienza,' 00161 Roma (E.C.), Italy

This paper is dedicated to the memory of Alfred Gropp, friend and pioneer in the study of the mouse Robertsonian system.

are arm homologies between two or among several Rb metacentrics, complex meiotic pairing of multivalents causes chain or ring configurations and male gametogenesis is much more impaired than female.

A detailed analysis of the gametogenic process in Rb heterozygosity shows that germ cells can be lost at any stage of their development, and no predictions can be made about whether and/or how much Rb heterozygosity may affect gametogenesis in a single individual. Data as to why the gametogenic process is impaired are still sparse but mechanisms are discussed.

A comparison of intrinsic characteristics of male and female gametogenesis suggests that it is possible that the fewer germ cells needed by females than by males for reproduction could "buffer" the damage caused by Rb heterozygosity. It has been shown by Mittwoch and her colleagues [88,89] that this is the situation in reciprocal translocation.

A discussion of pre- and postzygotic losses in both males and in females, especially considering pre- and postimplantation losses, yields the following conclusions: preimplantation losses vary widely with male and female heterozygotes, whereas postimplantation losses are clearly higher with females. The relative merit of methods for the evaluation of the aneuploid frequencies in pre- and postimplantation stages are described.

Regarding the fate of the aneuploid germ cell produced by Rb heterozygotes, it has been shown that in both sexes once the unbalanced gametes pass the meiotic phases they are able to complete development, contributing to postzygotic losses and causing diminished fertility. The capacity of aneuploid gametes to develop into zygotes and go on to produce newborns makes it possible to use the mouse system of chromosomal variability for studies of the relationship between aneuploid conditions (monosomy and trisomy) and developmental disorders. Trisomy 16 of the mouse can be considered to be an animal model for human trisomy 21, and this view is evaluated. Finally, an overview of the role of Rb heterozygotes in evolutionary biology and speciation is presented.

INTRODUCTION

In the house mouse structural karyotypic heterozygosity owing to whole chromosomal arm translocation is very frequent. From this perspective two questions may be asked:

How do individuals (populations) with rearranged karyotypes arise and why are they limited, among the genus *Mus*, to *Mus domesticus*, as appears to be the case judging from the karyotypically rearranged populations so far described?

What are the consequences of the presence of a rearranged chromosome to a heterozygous carrier and, more specifically, what is the fate of germ cells that carry it?

These are the two main questions posed by such a heterozygote status and the data gathered to address them are relevant to evolutionary and reproductive biology.

ROBERTSONIAN TRANSLOCATIONS IN WILD HOUSE MOUSE POPULATIONS

In the Muridae, whole arm translocation, or Robertsonian fusion (Rb) [1], is a very frequent karyotypic change both inter- and intraspecies [2,3], although deletion, duplication, and inversion can also contribute in producing structural rearrangements [4]. The frequency of this sort of intraspecies chromosomal variability in *Mus domesticus* is of a much greater order of magnitude that that found in any other mammals [5].

The Poschiavo "Tobacco" Mouse

Gropp and his colleagues [6] were the first to identify a wild population of the house mouse that had a homozygous karyotype with diploid number decreased because of Robertsonian fusion. He found them in the Poschiavo Valley, an isolated valley in the Rhaetian Alps of the Grison Canton of Switzerland. These mice were 2n = 26, with 7 Rb metacentrics. One hundred years earlier (1869), the Swiss zoologist Fatio [7] had assigned the mice of the Poschiavo Valley to a separate species, *Mus poschiavinus*, with the familiar name of tobacco mouse.

Single metacentric pairs of *poschiavinus* origin were introduced into the laboratory mouse strains by an introgressive breeding procedure in order to obtain several 38-chromosome strains, each with a different pair of metacentrics of Poschiavo origin. This greatly increased the possibilities of research, which had been conducted until then with only three Rb metacentrics that had originated in laboratory strains Rb(6.15)1Ald [8], Rb(9.19)163H [9], and Rb(5.19)1Wh [10].

Rb Metacentrics in Italian Mouse Populations

The Val Poschiavo mouse is not a unique creature, however. It is just one of a vast mosaic of populations of mice with Robertsonian rearrangements of the karyotype, closely related to each other and distributed throughout the central Alps. Joint efforts of Gropp and his colleagues with our laboratory have recognized a complex system (Rhaetian system) of not only a large number of populations with karyotypes structurally homozygotic for many different Robertsonian metacentrics, from 6 to 9, 2n from 28 to 22, but also of many natural hybrids between these populations and populations with the standard 40-chromosome, all-acrocentric karyotype [11–13]. Mice with reduced diploid numbers because of Robertsonian fusion have been found in two other Italian areas, the Apennines [14] and Sicily [15]. Beyond the geographic region of the Alps and Italy, mice with Robertsonian rearranged karyotypes have been found in widely separated areas, from subarctic Orkney Islands [16] to Antarctica [17], to the recent discovery of a population from the Tunisian coast which is 2n = 22 with 9 Rb metacentrics [18].

Undoubtedly, more and more wild house mouse populations with Rb rearrangements and Rb chromosomes of laboratory origin will be discovered with new arm composition.

Chromosomes Contributing to the Metacentrics

A detailed analysis of all the Rb chromosomes so far detected revealed G-banding patterns identical with those of the acrocentrics of the standard karyotype [19]. Only minor differences in the G-banding patterns among either the Rb variant mice or related subspecies of *Mus* have been detected, although these differences may well be accounted for by the different resolving powers of the various banding procedures employed [20]. Therefore the G-banding pattern appears to be remarkably uniform among the *Mus* subspecies [21,22] and/or the Rb variants so far investigated.

Gropp and Winking [23] and Capanna [14] employed G-banding techniques and the analysis of the pairing products of meiosis I in crosses between carriers of Rb metacentrics of known arm composition in animals with undetermined Rb metacentric compositions. They clearly demonstrated the acrocentrics contributing to the Rb chromosomes (Table I) and concluded that each autosomal arm of the mouse karyotype (with the exception of No. 19) has a random chance to be involved in an Rb metacentric chromosome. This virtually random involvement is not modified by selection for chromosomes of similar size or those bearing NORs.

Apart from chromosome 19, which has never been found to be involved in Rb formations of feral origin, all the other chromosomes were involved from nine to fifteen times (Fig. 1). The exceptions were chromosomes 7, 16, and 18, found in six, seven, and five different Rb metacentrics, respectively.

By evaluating the frequencies of occurrence of the various Rb fusions, an attempt can be made to establish whether or not there are preferential combinations among the acrocentrics of the standard 40 chromosome karyotype. For this exercise a given fusion is considered to have occurred only once in each population system, while in separate systems it is considered as often as there are different populations that have it. Gropp and Winking [23] and Capanna [12] tried this approach with less extensive data than is available

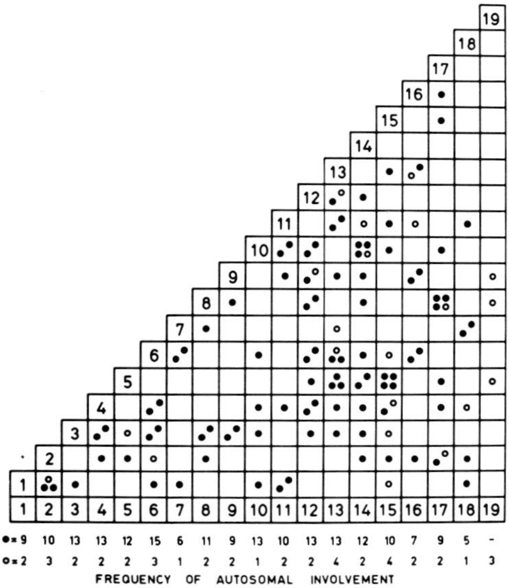

Fig. 1. Frequencies of different chromosome arm constitutions in feral (●) and laboratory-derived (○) Rb metacentrics in the house mouse *M. domesticus*. Each circle is an Rb metacentric chromosome; the autosomes of the all-acrocentrics karyotype involved in its constitution are indicated on both sides of the triangle. For each autosome, the total frequency of its involvement in feral and laboratory metacentrics is shown at the bottom.

today. At present 116 Robertsonian fusions of independent origin are known in natural populations and in laboratory strains (Table I) of the 171 possible combinations for fusion between the 19 autosomes of the standard karyotype. Among these 116 types of fusions, 51 have been found once, 22 have been found twice, 3 have been found 3 times, and only 3 have been found 4 times. No fusion has appeared more than 4 times.

Moreover, these data reveal several situations. The first concerns the Rhaetian Alps system, the populations of which have the same metacentrics to a great extent. Almost all the populations have the five fusions 5.15, 11.13, 9.14, 16.17, and 10.12; others are common to peripatric populations in contact with each other; only a few Rb metacentrics are restricted to single populations. There are still other areas, such as the Apennines, in which there are only a small number of common metacentrics, even in contiguous populations. Another situation is that geographically separated populations sometimes have considerable numbers of fusions in common. An example of this is that the population of the islands of Lipari (2n = 26) and that of the northeastern Apennines (ACR, 2n = 24), although widely separated geo-

TABLE I. Robertsonian metacentrics of the house mouse (*Mus domesticus*)

Systems and populations	Feral origin											Reference	Laboratory origin (arm composition and designation)[a]
	2n	Arm composition											
Rhaetian system													
Poschiavo	26	1.3	4.6	5.15	11.13	9.14	16.17	8.12				[6]	Rb(6.15)1 Ald
Upper Valtellina	24	1.3	4.6	5.15	11.13	9.14	16.17	2.8	10.12			[13]	Rb(5.19)1 Wh
Orobian Alps	22	1.3	4.6	5.15	11.13	9.14	16.17	2.8	10.12	7.18		[23]	Rb(9.12)1 Cam
Cremona	22	1.6	3.4	5.15	11.13	9.14	16.17	2.8	10.12	7.18		[13]	Rb(11.14)1 Dn
Milan 1st	24	7.8	2.4	5.15	11.13	9.14	16.17	3.6	10.12			[23]	Rb(3.5)1 Icg
Milan 2nd	24	6.7	3.4	5.15	11.13	9.14	16.17	2.8	10.12			[13]	Rb(10.14)1 Ad
Luino	24	3.8	2.4	5.13	11.18	9.14	16.17	6.7	10.12			[13]	Rb(7.13)1 Oxe
Mesolcina	28	1.3	7.8	13.16	10.11	2.14			4.12			[11]	Rb(13.16)1 Mpl
Bondo and Chiavenna (hybrid populations)	28–40	1.10	7.8	13.16	10.11		16.17		4.12			[11]	Rb(8.17)1 Iem Rb(2.6)4 Iem Rb(6.13)1 H Rb(11.16)2 H
Apennine system													
Cittaducale (CD)	22	1.7	3.8	6.13	10.11	4.15	2.18	5.17	12.14	9.16		[14]	Rb(4.18)3 H
Campobasso (CB)	22	6.7	1.18	2.17	5.15	4.11	3.13	8.14	10.12	9.16		[14]	Rb(2.17)4 H

Ancarano (ACR)	24	1.2	3.9	5.13	4.17	6.16	11.15	8.14	10.12		[23]	Rb(1.2)5 H
												Rb(9.19)163 H
												Rb(8.19)1 Ct
Sicily system												Rb(1.15)2 Ct
Palermo	26	3.4	2.15	5.13	6.12	8.17	10.14	9.16			[15]	Rb(12.13)3 Ct
Lipari	26	1.2	3.9	4.13	5.14	6.16	10.15	8.12			[23]	Rb(4.15)1 Rk
												Rb(3.15)2 Rk
Orkney system												
Caithness	34	9.12	4.10	6.13								
Eday	34	9.12	4.10	3.14							[16]	
Westray	36	9.12		6.14								
Other wild populations												
West Germany	28	3.6	2.5	4.12	8.17	11.13	10.14				[25]	
Yugoslavia	28	1.11	5.15	6.12	8.17	9.13	10.14				[23]	
Greece	34		5.12		15.17		3.10				[26]	
Spain	30	4.14	5.15	6.10	9.11	12.13					[25]	
Tunisia	22	1.11	4.6	3.12	2.16	13.15	10.17	5.14	7.18	8.9	[18]	

Rb metacentric chromosomes (and their chromosome arm constitution) for feral and laboratory house mice *M. domesticus*.
[a]From the list of chromosome anomalies reported in Mouse News Letters [24].

TABLE II. Taxonomy of the European house mouse

Species	Biochemical groups	2n	Ecology	Distribution
M. domesticus	Mus 1 (*M. musculus domesticus* *M. musculus brevirostris*)	40–22	Commensal	Cosmopolitan
M. musculus	Mus 2 (*M. musculus musculus* *M. musculus wagneri*?)	40	Commensal	Eastern Europe
M. spretus	Mus 3 (*M. spretus*)	40	Semicommensal	Spain and Southern France
M. hortulanus	Mus 4 (*M. spicilegus*? *M. abbotti*?)	40	Semicommensal	Southeastern Europe
M. abbotti	Mus 4 (*M. spicilegus*)	40	Semicommensal	From Macedonia to Caspian Sea

Taxonomic relationships among European house mice, from the data of Marshall and Sage [29] and of Thaler et al. [32].

graphically and quite separate and isolated from each other, have in common 3 Rb metacentrics of the 7 or 8 that characterize their karyotypes.

It is obvious that for the case exemplified by the Rhaetian Alpine populations, the simplest hypothesis is that the common Rb metacentrics were inherited from a common evolutionary ancestor or transmitted by hybridization between peripatric populations [27,28]. For the situation of the common Rb metacentric in geographically separate areas, one can only suppose that they were acquired independently and are thus the results of two different events, even though passive spreading due to interaction with man theoretically cannot be excluded.

The Taxonomic Context

On the basis of the review of Marshall and Sage [29], which takes into account the biochemical data of Hunt and Selander [30] and of Bonhomme et al. [31], five taxa can be distinguished in Europe that can be considered as species or, according to Thaler et al. [32], as semispecies. Two of these, namely, *Mus musculus* and *Mus domesticus*, are more specifically commensal, while *Mus abbotti*, *Mus hortulanus*, and *Mus spretus* live more often in the open field, rural, steppe, or Mediterranean maquis environments. Moreover, all the species have the same 40-chromosome acrocentric karyotype, corresponding to that of the Asian species of the subgenus *Mus* in both the G-banding and the distribution of the constitutive heterochromatin [33]. However, the genetic isolation of the taxa is shown by discriminant alleles that allow a subdivision into four biochemical groups, in good agreement with the morphological identification of the taxa (Table II) according to Marshall and Sage [29]. The striking finding is that it is only for *Mus domesticus* that wild populations with rearranged karyotypes have been described so far.

CYTOLOGICAL AND MOLECULAR FACTORS IN ROBERTSONIAN FUSION FORMATION

The Mouse Genome

The mouse genome consists of 6.5–6.8 pg, according to Sparrow et al. [34] and Manfredi Romanini [35], or of 8.7 pg of DNA, according to Gamperl et al. [36], 75% of which can be visualized cytologically in interphase nuclei as euchromatin. The remainder cytologically can be recognized as constitutive and facultative heterochromatin. The amount and the proportion of histones in heterochromatin is identical with that found in euchromatin [37] even though in the centromeric heterochromatin the H1 histone is modified by phosphorylation and there are specific non-histone proteins [38] in addition to the normal proteins in heterochromatin itself. The cytologically detectable constitutive heterochromatin constitutes about 25% of the mouse genome [36,39] and is limited to the pericentromeric regions, as visualized by C-banding techniques. It mainly consists of highly repetitive DNA sequences, 10% of which (with respect to the entire genome) is satellite DNA [40–42]. The satellite DNA chromatin contains all five histones, with a protein-to-DNA ratio of 1.3 to 1, i.e., similar to the chromatin associated with main band DNA, but there are differences in histone composition between the two types of chromatin.

The mouse seems to have only one major type of satellite DNA, at approximately 1 million copies per genome with a buoyant density in neutral CsCl of 1.69/g cm(−3) [Tm = 74.2, Cot = 6.6 × 10(−4)]; a base composition of A:32.2; T:33.6; G:16.4; C:17.8; and short, highly repeated sequences rich in AT [43–45]. The periodicity-range analysis performed by Horz and Zachau [46], utilizing five different restriction endonucleases, has shown, with all the nucleases tested, the presence of a major repeat unit of 245 nucleotides. It is still not known how a defined sequence, repeated in tandem within the genome, arises and is maintained, though several mechanisms have been suggested, including unequal sister chromatid crossing over [47] and saltatory amplification events [46] [reviewed in 48,49].

The two complementary strands of mouse satellite DNA exhibit lateral asymmetry owing to an unequal distribution of thymidine between the two chains: i.e., 22% versus 45% [44,50]. The centromeric lateral asymmetry that can be observed in the mouse satellite DNA could theoretically be derived by several processes, as outlined by Alves [51], including inverted insertions, centric fusion followed by paracentromeric fission, inversion of part of the centromeric satellite DNA, and, lastly, through the addition of new DNA to the centromeric heterochromatin [52]. One of these mechanisms acting independently or in conjunction with another can also produce variation in the amount of the centromeric heterochromatin [53].

Schmid [54] and Schweizer and Ehrendorfer [55] proposed that whenever C-bands have the same fluorescence properties, as in the mouse, there has been a co-evolution of highly repeated DNA on nonhomologous chromosomes. The arrangement of similar C-bands has led to the hypothesis that there are mechanisms that produce and distribute C-band material within the genome [56]. Several studies [20,57,58] have shown C-band differences in both laboratory and feral mouse populations. In four inbred strains Jonasson et al. [53] detected differences in the lateral asymmetry patterns of the heterochromatin.

However it originates, mouse satellite DNA is found in all chromosomes, with the exception of the Y, at the pericentromeric area [40–42]. Centromeric heterochromatin, moreover, does show a tendency to aggregate in one or a few neighbouring areas of interphase cells [59–62] and the basis of this may be the homogeneity of mouse satellite. This topographical clustering of the centromeric heterochromatin within the nucleus does not seem to follow any particular rules, and recently Rattner and Lin [62] argue that "this association may provide the structural framework within which Robertsonian fusion occurs."

It is generally accepted that the mechanism by which Rb fusion chromosomes are formed requires the occurrence of a breakpoint at the centromere, since no direct adhesion of chromosome ends is possible, with reunion of the chromosomal acrocentric arms in a metacentric chromosome, and the concurrent loss of the two smallest fragments [23,63,64]. If so, after chromosomal breakage occurs in closely adjacent centromeres, the presence on all mouse chromosomes of only one major type of satellite DNA could be a factor favouring random acrocentric chromosome fusion (Fig. 2). In this connection Holmquist and Dancis [70] suggested the occurrence of recombination between satellite DNA sequences of two different chromosomes during temporary telomere-telomere fusion in early S-phase.

Direct evidence that identical DNA sequences at the breakpoints can facilitate chromosome fusion was obtained from analysis by means of restriction maps of the chromosomal DNA regions surrounding the breakpoints, of a t(14;18) in man [65]. The analysis of the DNA composition at and near the breakpoints revealed identical or almost identical base sequences. An indirect test for this hypothesis would look at the chromosome rearrangement in a species or in a group of related species in which acrocentric chromosomes carry heterogeneous satellite DNA: In this situation, Rb transformation would not be favoured.

Inexplicable at present is the fact that in wild mice, Rb translocations are not randomly occurring within their worldwide distribution area, but are almost exclusively present in Western European and North African populations, which represent only one species, namely, *Mus domesticus*, of the genus *Mus*.

If similar DNA sequences do contribute to Rb fusions, then they probably constitute the "minimum requirement" for Rb chromosome formation and other factors must play a role.

NORs and Rb Fusions

To explain the high occurrence of Rb translocation in the house mouse Ohno et al. [66] suggested, as nucleoli have a great tendency to fuse, that

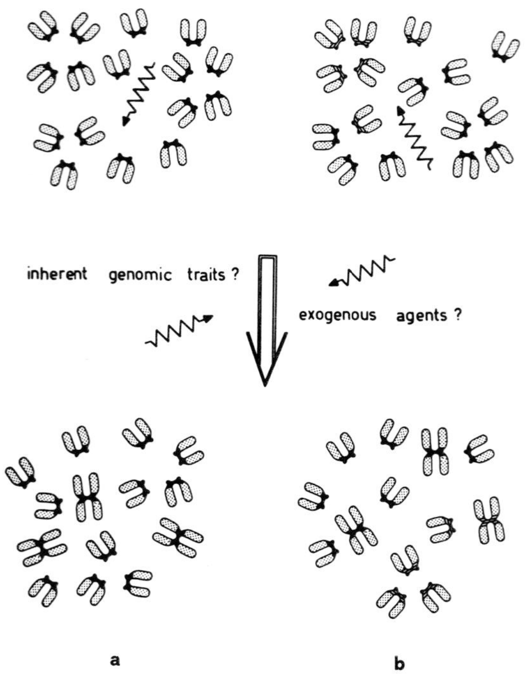

Fig. 2. Hypothesis for explaining Rb chromosome formation: When breakpoints occur as the result of inherent genome traits or of exogenous agents, high repetitiveness of the satellite DNA facilitates DNA strand reorganization, whereas its homogeneity **(a)** or heterogeneity **(b)** favours random or nonrandom acrocentric association.

this process can (by inducing topographical proximity) predispose chromosomes with NOR to centric fusions. After an analysis of NOR-bearing chromosomes involved in Rb translocations in three cell lines, Miller et al. [67] reached the conclusion that the "nucleolus plays an important role, but nucleolar fusion is relatively unimportant in the origin of Rb translocations in the mouse." These authors observed that NOR-bearing chromosomes in such lines were more likely to be involved in Rb chromosomes but only one NOR chromosome was usually present in the Rb metacentrics. Only chromosome 17 has been found fused with 15 and with 16 among NOR-bearing chromosomes 12, 15, 16, 17, 18, and 19. A concomitant analysis of a feral mouse population revealed [68] that (in the wild) fusion of nucleoli cannot be a key factor for predisposing to centric fusion, though the presence of NOR on particular chromosomes is a characteristic probably acquired prior to the Robertsonian fusion. Stahl et al. [69] document the association of homologous regions (18 and 28s RNA cistrons) on nonhomologous chromosomes in man in spermatocytes and oocytes as probable mediators of Rb fusions via a breakage-reunion event. However, in the mouse these genes are longarm in location and Rb would not be favored. Holmquist and Dancis [70] suggested the occurrence of recombination between satellite DNA sequences of two different chromosomes during temporary telomere-telomere fusion in early S phase.

CONSEQUENCES OF Rb STRUCTURAL REARRANGEMENTS

Effects on Karyotype Structure and Genome Size

Whole arm fusions decrease the 2n number, leaving the number of the chromosomal arms, which is the fundamental number (NF) of Matthey [71,72], constant.

With one homologue of an acrocentric pair involved in an Rb fusion, a single heterozygous condition arises, with the formation of a trivalent at diakinesis. The number of Rb metacentrics present in the heterozygous condition can vary and single, double, or multiple independent heterozygosities can occur. In multiple independent heterozygosities, no chromosomal arms are common to the metacentrics, and several trivalent configurations will appear at diakinesis. The most extreme example of Rb translocation is the homologue-homologue centric fusion, but in the mouse it has never been reported.

When homologous acrocentrics are shared between different metacentrics, at diakinesis they will pair in a multivalent configuration, from tetravalents to polyvalents: Multiple heterozygosities with alternating monobrachial arm homologies will result in a meiotic chain configuration, whereas a ring will be formed when the alternating arm homology involves a complete set of acrocentric pairs. Since the most plausible mode of occurrence of Rb translocation in the mouse implies the loss of some acrocentric [23] short arm material, genome size can be expected to vary whenever Rb chromosomes are present.

Several attempts to show genome size differences between Rb populations and 40 all-acrocentric strains have failed to reveal any differences. Comings and Avelino [73] used analytical ultracentrifugation to determine the differences in satellite DNA quantity between 40 all-acrocentric strains and the Poschiavo mice (with a $2n = 26$) and found no statistically significant differences. In any case, they concluded, the losses must be less than 4% of the satellite DNA.

The approach of measuring the total DNA content of interphase nuclei of different Rb populations after specific and stoichiometric reactions for DNA also did not show any differences, using either static or flow cytophotometric apparatus (Fig. 3) that could detect differences in DNA content of 1.94% and in A-T base pair content of 1.78% of the genome [74]. Finally, in our laboratory, measurement of the amounts of pericentromeric heterochromatin, either by H-33258 intensity emission or as Giemsa-positive material after the C-banding procedure, also did not reveal any difference between Rb metacentrics and the two homologous acrocentrics (involved in the Rb translocation) added together.

Effects on Mitosis and Meiosis

Rb metacentric chromosomes have only one functional centromere and either the second centromere is lost (as explained by mode No. 4 of the John and Freeman [63] scheme for Rb processes) or functionally inactivated ("latent," as suggested by Hsu et al. [75]).

Complete viability of Rb heterozygotes is indirect evidence that the mitotic process is not disturbed. Until now, only a single situation has been described that leads us to suspect an interaction of Rb heterozygosity with mitosis: this is the decreased numbers of proliferative A_0—A spermatogonia in mice with

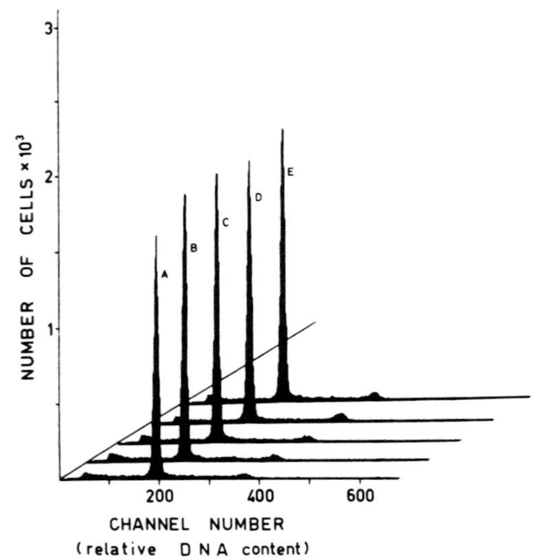

Fig. 3. Relative DNA content distributions in the thymocytes of 2n = 40 (**A**), 2n = 39 (**B**), 2n = 36 (**C**), 2n = 27 (**D**), and 2n = 22 (**E**) mice. (Flow cytometric evaluation of DAPI fluorescence intensity.) The modal channel numbers for the G0/G1 peaks and the coefficients of variation are A = 198 and 2.07; B = 196 and 2.06; C = 199 and 2.11; D = 201 and 2.07; E = 197 and 2.09. There are no significant differences in channel number ($P < 0.01$).

multiple heterozygosities with alternating arm homologies [76]. However, this cannot be attributed with certainty to interference with the spermatogonial mitotic apparatus, since other mechanisms could be responsible, such as release of an increased number of spermatogonia from the proliferative compartment in an attempt to increase the number of differentiating germ cells (see Kinetics of Rb Heterozygote Spermatogenesis).

As regards meiosis, individuals heterozygous for translocation chromosomes (apparently with unaffected viability) have several meiotic disturbances, as seen from data gathered from other mammalian species—e.g., man, cow, pig, and monkey [77,78]. Rb heterozygotes very often show impaired anaphase segregation, aneuploid gamete formation and breakdown of the gametogenesis with a wide range of effects from total arrest to a minimal disruption of the process. In analysing the meiotic consequences of Rb heterozygosity, it is useful to consider separately single, multiple independent, and multiple dependent heterozygosities, since, in general, the extent of the impairment is related to the "complexity" of the meiotic configuration and to the sex.

Single heterozygosity. Individuals with single metacentric heterozygosities show trivalent pairing configurations at diakinesis/metaphase I, instead of the usual bivalents (Fig. 4). Meiotic pairing in the presence of Rb chromosomes does not seem to be affected (under light microscopy), but the existence of minor pairing anomalies between the Rb metacentric and the homologous acrocentrics at the centromeric region is evident at a higher resolution. By electron microscopy [79], the trivalents, at mid and late pachytene, have a Y-shaped configuration with a short stump where the nonhomologous pericentric segments of the acrocentrics are located (Fig. 5).

The imperfect adjustment in the triangular centromeric area may be one possible mechanism for the effect of Rb exchange on the meiotic process, as suggested for minor (undetectable) structural rearrangements as a source of increased meiotic non-disjunction [80].

Tettenborn and Gropp [81] and Cattanach and Moseley [80] were the first to describe meiotic anaphase I non-disjunction in male and female Rb heterozygous mice. Gropp and Winking [23] and Gropp et al. [82] assessed the non-disjunction rates for a large series of single Rb metacentrics (both of feral and laboratory origin), and from these data it is clear that the non-disjunction rates ascribable to the different Rb heterozygosities show a wide range. The analysis of the chromosomal arm constitution of the Rb chromosomes reveals that there is not a preferential involvement of any standard acrocentric chromosomes in cases with a high non-disjunction rate. Female non-disjunction rates are up to several times higher than those for the male

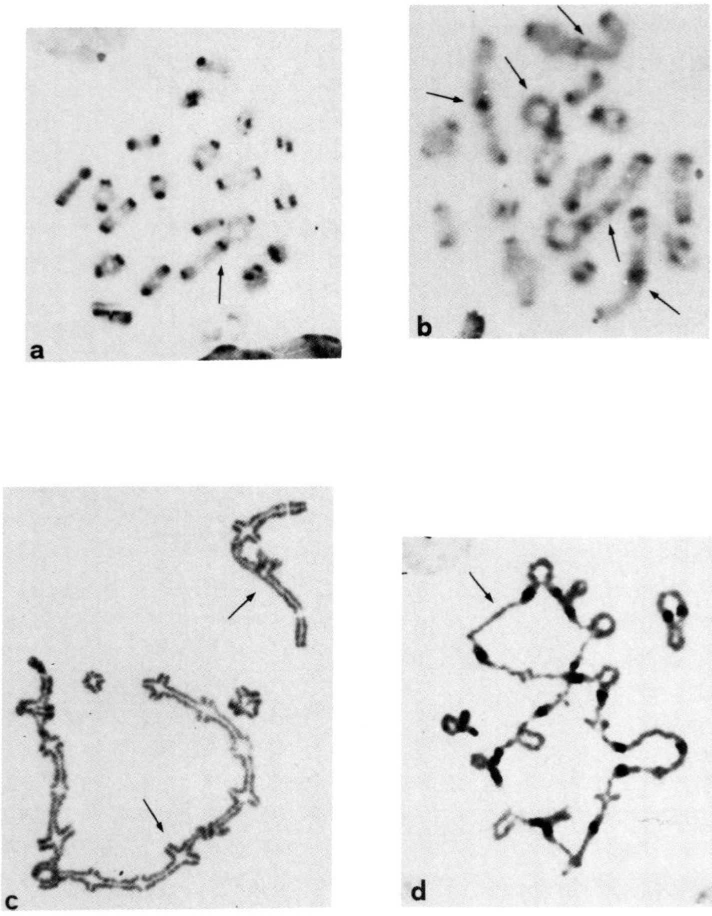

Fig. 4. Meiotic diakinesis/metaphase I of Rb heterozygous mice showing **a**) single heterozygosity (arrow); **b**) multiple independent heterozygosities (arrows); multiple dependent heterozygosities with alternating arm homologies giving rise to two long polyvalent chains (arrows) (**c**) or to a megaring (arrows) (**d**).

with the same Rb constitution (Fig. 6). There are only a few exceptions to this situation, with the non-disjunction rates almost the same in both sexes. These rates are based on chromosomal arm counts in meiotic metaphase II plates which reveal malsegregation in anaphase I (see later) and on the analysis of unbalanced versus balanced progeny. In the wild Peru-Coppock

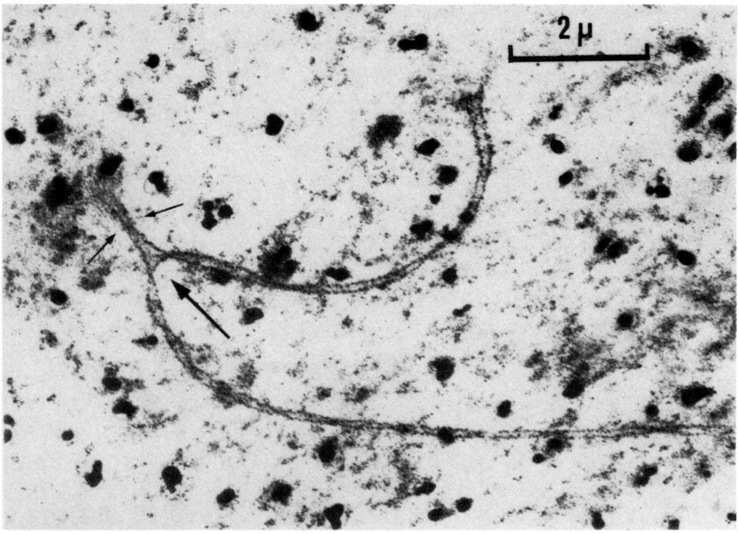

Fig. 5. Electron micrograph of an Rb-trivalent late pachytene spermatocyte from a 22 × 40 hybrid. Small arrow indicates the centromeres of acrocentrics and large arrow centromere of the Rb metacentric (preparation of M. Ponsá). The Y-shaped configuration in the centromeric area results from adjustment of the non-homologous pericentromeric segments of the acrocentrics.

mice, Harris et al. [83] estimated a 10% rate of non-disjunction for Rb(9.12)/+ males versus 12–16% for females.

The behaviour of an Rb metacentric in an oocyte differs not only in its higher likelihood to produce unbalanced gametes but also in its transmission rate to the progeny. Female heterozygotes have a distorted ratio of transmission of the Rb metacentric, and this has been suggested to be due to the more asymmetrical division than in the spermatocytes—i.e., into a large oocyte and a small polar body. It is supposed that the spindle body in the dividing oocyte has a more "open" configuration than in the spermatocyte [84] and this can lead to the inclusion of the Rb metacentric into the polar body. Thus the homozygous type (two acrocentrics) prevail among the progeny of single heterozygous females, even though Harris et al. [83] obtained an opposite result for the Rb(9.12) metacentric chromosome tested in the wild genome in which it arose.

What might account for the greater tendency of single female heterozygotes to produce a higher non-disjunction rate is still unclear, though ob-

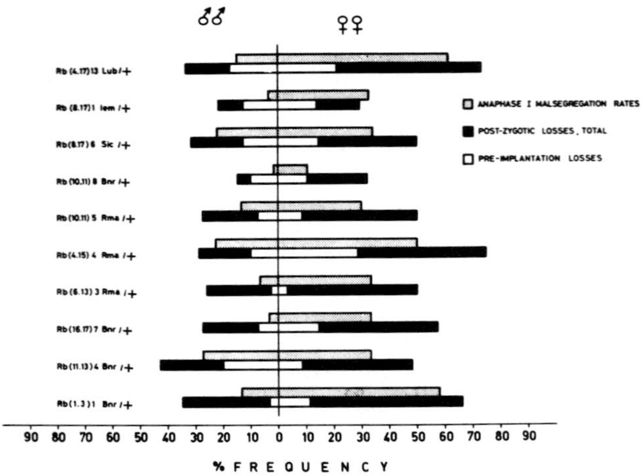

Fig. 6. Frequencies of meiotic anaphase I non-disjunction rates (▨), preimplantation losses (□), and total postzygotic losses (■) of single male and female Rb heterozygotes. The frequency of the non-disjunction rates was calculated by doubling the hypermodal value (i.e, > 20×2). The preimplantation loss values were calculated from the total number of corpora lutea and implants; the total postzygotic losses include early and late deaths plus trisomic newborns (for details see text).

viously it must be something inherent in the specific oocyte environment. This last consideration focuses attention on the importance of other factors, other than the Rb metacentric structure "per se," for non-disjunction. There are situations in which differences in the non-disjunction rates cannot be accounted for by different Rb constitution or by minor structural differences, at least any that are detectable. One such case is that of Rb metacentrics with the same chromosomal arm combinations but different geographic origins. As an example, the non-disjunction rates of males heterozygous for Rb metacentric 10.11 (or 8.17) are considerably different in different geographical origins; Rb(10.11)8Bnr/+ males have 5% (Val Bregaglia, North Italy) and Rb(10.11)5Rma/+ males have 13.8% (Central Italy) meiotic anaphase I malsegregation rates [23]. A similar situation exists for the non-disjunction rates of Rb metacentrics in the original wild genome versus the rate (that very often is much higher) for the same Rb metacentric introduced into the genome of a laboratory strain [23]. In females age is also a factor. Young female heterozygotes Rb(1.3)1Bnr/+ have lower non-disjunction rates than old females (49% vs. 73% for 2 months vs. 12 months) [23], and these values are very much higher than those of ageing nonheterozygous females.

Multiple independent heterozygosities. Mice that are carriers of multiple independent Rb metacentrics form a corresponding number of trivalents at diakinesis (Fig. 4). The non-disjunction rates are higher in both females and males than in carriers of a single Rb heterozygosity, but the increases are not additive for the single values pertaining to each Rb metacentric present in the karotype.

In general mice carriers of trivalent(s) show disturbed gametogenesis, though this never reaches the condition of a severe primary impairment of the process.

Multiple dependent heterozygosities. When there are arm homologies of two or several different Rb metacentrics, chains or rings of up to 16 chromosomal arms can be formed (Fig. 4). Capanna et al. [14] and Gropp et al. [82] have shown such meiotic configurations in laboratory mice obtained by opportune breeding, but it is a reasonable assumption that these kinds of hybrids could occur in the wild at times when two carrier populations with Rb metacentrics of suitable arm homology are in contact.

In cases examined so far there is a very high rate of non-disjunction in many of the double heterozygotes with one homologous arm. In these cases a common feature is that the extra chromosome present in a high percentage of gametes consists of the chromosomal arm sheared off one of the two Rb chromosomes. (In this case it is very difficult to ascertain the non-disjunction rate by conventional MII counting; see below).

It is interesting that in double and multiple heterozygotes the rates of unbalanced gametes produced by female and male are seemingly reversed with respect to those of single heterozygotes: males have higher rates of production of unbalanced gametes, and in certain Rb cytogenetic constitutions they can be totally sterile, whereas females are fully fertile. In addition, the ring meiotic configuration is less devastating to the gametogenic process than the chain. Electron microscopic (EM) observations of quadrivalent or multivalent chain or ring configurations [23,85] revealed that the ring configuration is, in a certain sense, a more symmetrical figure than chain pairing: i.e., there are fewer regions that remain unsynapsed at both ends of the multivalents.

The role that unsynapsed regions of chain ends have in the causation of meiotic disturbances is discussed later.

Thus, from the data so far accumulated [see also 82], two major mechanisms, acting to different extents in the two sexes, seem to be responsible for the impairment of gametogenesis due to Rb heterozygosity: a) non-disjunctional events leading to the formation of chromosomally unbalanced gametes and thus to a few pre- and more postzygotic losses; and b) a male-prevalent, non-specific depressive effect (quite similar to the "male hybrid sterility" described by Haldane by 1922 [86]) on gameteogenesis, when there are multiple dependent heterozygosities.

The consequence of such mechanisms is selective germ cell or embryo loss at many stages in development.

It should be stressed that no predictions can be made with certainty about whether or how much any Rb heterozygosity will affect gametogenesis. In addition, other factors such as genetic background, age, or the presence of other chromosome rearrangements can alter the extent of this gametogenic impairment.

PRE- AND POSTZYGOTIC LOSSES

Massive prezygotic losses cause "primary" impairment of fertility, whereas postzygotic losses cause "secondary" impairment, since in this last case the production of gametes, whatever their capacity to develop into new individuals, is achieved.

Prezygotic Losses

Prezygotic loss in females with Rb heterozygosity is still unexplored, with few data collected so far. Mittwoch described the diminished ovarian size in carriers of some chromosome anomalies, including reciprocal translocations [87–91]. The reduction of ovarian size is due to losses of oocyte cells during the prenatal stages of oocyte differentiation, but there are no data as to the cause of such losses. Moreover, nothing is known about prezygotic but postovulatory losses, although they are mainly of unfertilized eggs. These last data are difficult to collect, since only a part of the numerical discrepancy between numbers of corpora lutea and numbers of blastocysts/implants can be ascribed with certainty to unfertilized eggs. Much more is known about male prezygotic loss, probably because it is intrinsically easier to study from testicular histology and sperm analysis. The premeiotic and meiotic stages of the mouse spermatogenic process in animals with different Rb have been extensively investigated.

Kinetics of Rb Heterozygote Spermatogenesis

The study of spermatogenesis in Rb heterozygotes has shown that both premeiotic and meiotic stages are affected and that the effects are more pronounced in multiple Rb heterozygotes than in single heterozygotes. For multiple Rb heterozygotes, two cytogenetic conditions with monobrachial arm homologies, leading respectively, to ring or chain diakinetic configurations, have been better investigated than others [82,92,93], since they exemplify those cases in which Rb heterozygosities give rise to subfertility or sterility.

The ring comes from the cross of CD/CD × CB/CB strains (see Table I) and is based on the alternate arm homologies of the following Rb chromosomes: 1.7/7.6/6.13/13.3/3.8/8.14/14.12/12.10/10.11/11.4/4.15/15.5/5.17/17.2/2.18/18.1. The F_1 of the cross Rb(16.17)/Rb(16.17) × Rb(8.17)/Rb(8.17) gives rise to a chain involving two Rb chromosomes and two acrocentrics: 16/16.17/17.8/8.

Both ring and chain conditions show disturbances of all steps of spermatogenesis, and with interindividual variability.

Three main steps seem to be particularly sensitive and can account for the arrest of spermatogenesis at any stage of the process and for the disordered morphological appearance of atypical stages of the seminiferous epithelium cycle.

A first step is at the premeiotic level, when the number of the proliferative A_0—A spermatogonia is reduced [76]. This decrease might be due either to losses of spermatogonia following interference by Rb heterozygosity with the mitotic apparatus or to an increased number of spermatogonia that enter differentiation, thus impoverishing the stem compartment. The total number of differentiating spermatogonia is indeed unchanged. Evidence that tends to favour this last possibility is the increased number of gonia with positive alkaline phosphatase activity found in subfertile and (many more) in sterile individuals [92]. From comparisons with fertile mice with the standard karyotype, the APase-positive gonia could be the ones that progress from proliferation to differentiation. The trigger for more proliferating gonia to enter differentiation may be feedback from the germ cell losses that occur during the meiotic phases of spermatogenesis.

A second step occurs at the meiotic level, when most of the pachytene spermatocytes are unable to follow the normal cytodifferentiation pattern. There are several indications of that, and again chain configurations are much more severely impairing than ring configurations. A first indication comes from the morphological analysis of the pachytene cells in control and subfertile animals. The size of pachytene cells in control animals constantly increases from stage I to stage XII of the seminiferous epithelium cycle, where they reach a maximum. Instead, in cells from stage V/VI of subfertile mice, acquisition of the correct perimeter length is belated in at least 50% of the pachytene cells [76]. It is interesting to remember that the same phenomenon of belated development occurs in partially trisomic mice [94]. Even the time that cells spend in pachytene and diakinesis is lengthened, as demonstrated by Forejt [95]. That several activities of pachytene cells are disrupted is confirmed by other studies. Considering DNA metabolism, Hotta et al. [96] observed the persistence of increased nicking activity and all of the DNA repetitious classes, including the satellite were affected by this nicking activity. Probably the longer persistence of nicking activity is related to a greater ^{3}H-thymidine incorporation, after 1-h and 8-h single-pulse exposures, in subfertile and sterile than fertile mice [97]. Moreover, a deregulation in the pattern of some enzymatic activities, particularly in that of the TPPase in the Golgi area has been found [92].

A third step in the disturbance of the process is at the spermiogenesis level, when spermatids undergo morphological and physiological changes. In multiple heterozygotic mice, there is a drastic reduction in the numbers of spermatids that arise at stage I of the cycle, after the two meiotic divisions. In subfertile animals (i.e., ring carriers), this reduction halves the spermatid population, but they can continue to differentiate to stage VIII when spermiation occurs. In sterile animals (i.e., chain carriers) the few spermatids that arise at stage I degenerate soon after, revealing their defective nature (Fig. 7).

A direct consequence of the failure of spermatogonia, pachytene spermatocytes, and spermatids to follow the correct differentation pathway is the appearance of morphologically atypical stages of the seminiferous epithelium cycle. Inside the seminiferous tubules defective germ cells have a zonal distribution, supporting the concept that the disturbance of germ cell development is clonal at early stages of differentiation. Moreover, because of the particular structure of seminiferous tubules, two phenomena can induce both the loss of normal germ cells and the rescue of defective germ cells. The degeneration of cells that occupy the basal compartment of the seminiferous epithelium in the first stages of differentiation can disrupt its architecture, so that large blocks of cells of the adluminal compartment detach into the lumen and are lost for reasons other than whether they are normal or aneuploid (Fig. 8). The other phenomenon is that germ cells that have already started into disturbed development might be rescued by neighbouring cells. This means that all cells that carry the same cytogenetic constitution will not be affected to the same extent by the deregulatory factors. As a result, the morphological expression of the spermatogenic impairment is increased, and certainly both phenomena amplify the damage due to Rb heterozygosity in males versus females. The reason for postmeiotic disturbance (i.e., spermatid degeneration) is related to the high proportion of aneuploid spermatids that arise after the second meiotic division. Sertoli cells appear to be numerically untouched in all of the types of Rb variability so far investigated, and this

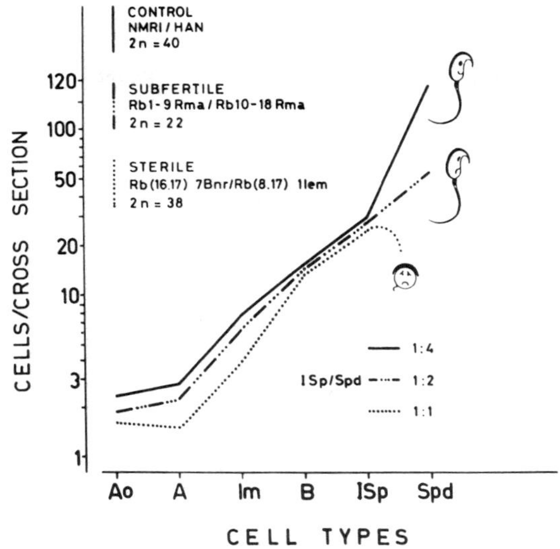

Fig. 7. Cytodifferentiation of the germ cell types in fertile, subfertile, and sterile mice. Multiple heterozygosities with alternating arm homologies giving ring or chain meiotic configurations are present in subfertile and sterile mice. Sterile mice show a decrease in the A_0—A spermatogonia compartment and a 1:1 ratio between primary spermatocytes and spermatids (which die soon after their formation), whereas subfertile mice have a 1:2 ratio (with a high proportion of aneuploid spermatozoa).

again indicates the normal viability of all the "somatic" cells in single and multiple heterozygotes.

The chromosomally determined extent of the spermatogenic impairment is related to the ring or chain configuration, with the ring being less devastating and leading to subfertility, while the chain leads to sterility since this pattern of impairment can be detected in other (*Mus domesticus*) ring and chain carriers. In any case it represents only a tendency and cannot be assumed to be a general rule, since other chain carriers are fully fertile [98].

Postzygotic Losses

Preimplantation, but postzygotic, losses can be assessed directly by counting the number of corpora lutea and then comparing it with the total number of implantations. Postimplantation losses are counted as the number of early deaths (resorptions and moles) plus late deaths (autolytic embryos) and living aneuploid embryos that sooner or later will die. Examples of such calculations are given in Figure 6 (based on unpublished data of Gropp, Winking,

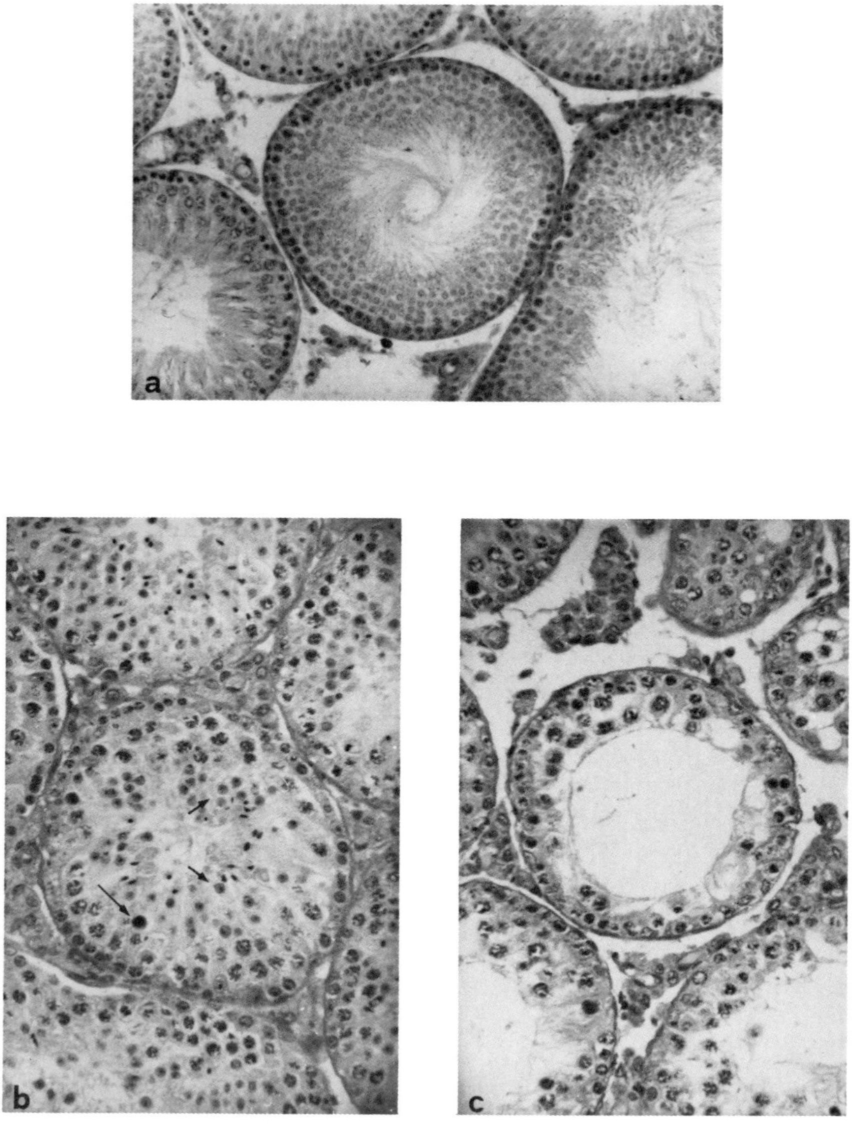

Fig. 8. Sections of mouse seminiferous tubules. The picture of normal histoarchitectural germ cell arrangement **a**) is taken from an original preparation of C. Golgi (probably 1898). Several degrees of the commonest histological abnormalities of Rb heterozygotes are shown: **b**) abnormal histological structure with degenerating pachytenes (long arrow) and spermatids (short arrow) and morphologically defective spermatids; **c**) total absence of spermatids;

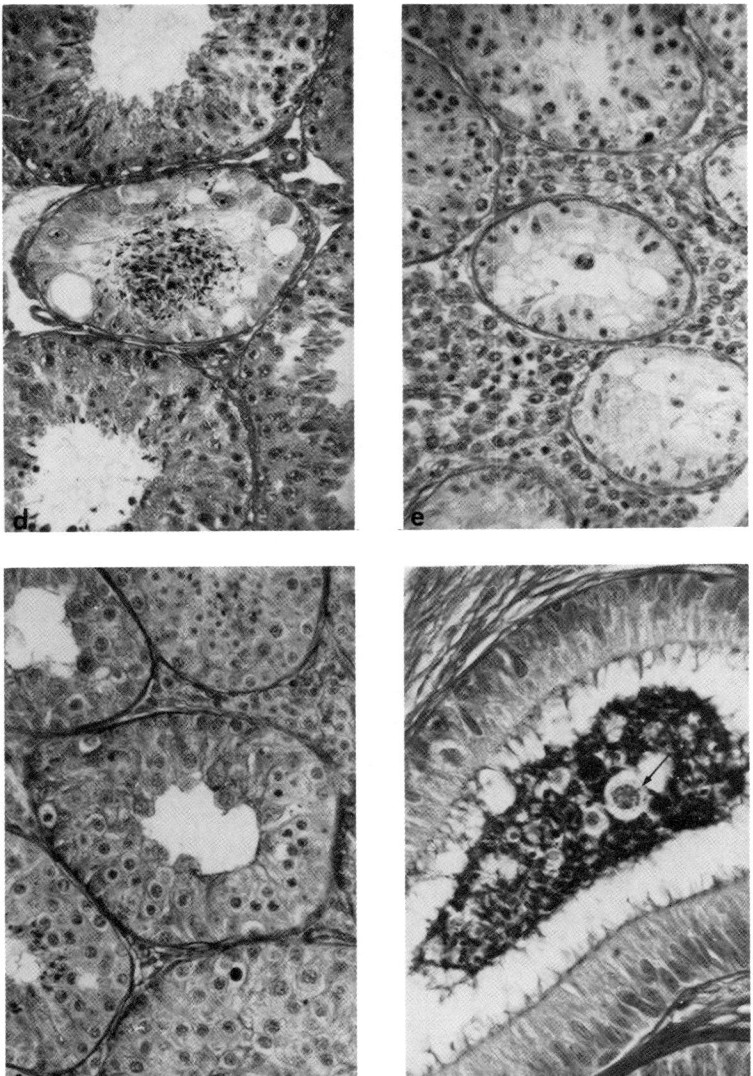

Fig. 8 ***(continued)***

d) blockage of germ cells in the lumen; **e)** totally depleted seminiferous tubules. The pictures in **f** and **g** are for an Rb heterozygote of *Spalax eherenbergi* and show an atypical stage with losses of spermatids (f) and pachytene cells in the lumen of an efferent duct (arrow) (g).

Redi, Noack, Kolbus, Louton) for some Rb/+ male or female heterozygotes mated with +/+ animals.

There is great variability in the rates of the preimplantation losses ascribable to Rb heterozygotes with different cytogenetic constitutions. Moreover, the rates for male and female heterozygotes with the same Rb constitution also vary, with some Rb chromosomes that produce higher rates in males, others with higher rates in females and still others that affect both sexes to the same degree. Rates of postimplantation abnormalities (losses) are clearly higher in female heterozygotes, sometimes triple those in males (see Rb13Lub/+). Harris et al. [83] in wild Peru-Coppock mice estimated a total postimplantation loss of 21.3% for Rb(9.12)/+ females compared to a 10% non-disjunction rate for males.

Evaluations of preimplantation loss are affected by such factors as the number of unfertilized eggs, which can account to some extent for numerical discrepancies between corpora lutea and implants. There are preimplantation losses of unspecified origin even in feral mice [99]. Detailed analysis of these discrepancies must always take into account the possibility that sources of variability of unknown nature could be active, thus restricting the validity of the data to the animals used in that experiment only. Another source of preimplantation loss is background error during second meiotic division. Non-disjunction at second meiotic division will produce double trisomies and monosomies (although nothing is known about the actual rate of second meiotic non-disjunction) that probably are subject to early preimplantation elimination. There are controversial data about whether monosomies are eliminated before or after the blastocyst stage. Ford [100] and Gropp [101] did not find any noticeable proportion of aneuploid embryo losses until this stage, whereas Epstein and Trevis [102] found that at least some monosomies were eliminated before the blastocyst stage.

There are no analytical data for multiple Rb heterozygotes, but, in general, postzygotic losses would be expected in all cases in which gametogenic disturbances create aneuploid gametes, as multiple Rb heterozygotes do.

Sole production of unbalanced gametes (nullosomic and disomic) can be expected if homologous-homologous Rb translocation occurs and in this case postzygotic losses will be 100%, since monosomies and trisomies (of any autosome) are almost all inviable (see below, Importance to Experimental Medicine). In man this kind of extremely rare translocation has been reported [103] and the carriers are totally sterile because of postzygotic loss.

Differential Effects of Rb Heterozygosity on Male and Female Gametogenesis

Gametogenesis is a complex phenomenon that requires interactions of genes distributed all over the genome and thus it can be impaired by many factors. For spermatogenesis in animal carriers of several types of chromosomal anomalies, including reciprocal translocations, several hypotheses have been proposed to explain the disruption of the process. These include altered expression of some genes [104,105] and interference with X-chromosome inactivation [106], which usually occurs at the pachytene stage in male gametogenesis and unsaturated meiotic pairing [107]. In a detailed study of single- and double-translocation heterozygote spermatogenesis, de Boer et al. [108] correlated the data on meiotic pairing with the spermatogenic impairment, since "it was hoped that an analysis of this kind would throw light on the question of which of the two main hypotheses listed above—i.e., lack of saturation of meiotic pairing [107] versus disturbance of the dormant state of the sex chromosomes during male meiotic prophase [106]—best fits the phenomenon of chromosomal male sterility in the mouse." Unfortunately, the data did not permit this distinction.

Whatever evidence one can find to support one or the other hypothesis, substantially they suggest the same general conclusion: control mechanisms specific for spermatogenesis are disturbed at the gene level. Interference with X-chromosome inactivation that leads to inappropriate gene expression (as shown by Hotta and Chandley [109] for some X-linked enzymatic activities in spermatocytes of mice with chromosomally caused sterility) could tentatively be considered as a working hypothesis to explain the limitation of the phenomenon to males and the greater facility of chains rather than rings to affect spermatogenesis. EM of chain and ring configurations [85] show unsynapsed regions (in chains much more than in rings) and these can interfere with the X-Y unsynapsed regions. As shown by Forejt [95,110] in the mouse and by Rosenmann et al. [111] in man, there are significantly more contacts between the Rb chromosomes and the X-Y bivalent in the spermatocyte.

It could be argued that these hypotheses are weakened by the finding of Mittwoch's group [87–91] that oogenesis is also affected (as judged from ovarian size and oocyte counts) in mice with chromosome anomalies, including reciprocal translocations. The lack of data about the complete period of oogenesis prevents thorough comparisons of gametogenesis in the two sexes and thus the concept of male limitation of the phenomenon cannot be considered definitive. Differences between oogenesis and spermatogenesis may corroborate this latter view. In females germ cell damage is "buffered" by the intrinsic peculiarities of oogenesis (mainly the time of occurrence of folliculogenesis and the few cells needed for it). As is well known, in mammalian females the premeiotic and the meiotic phases up to the diplotene occur before birth, when prespermatogenesis occurs in the male [112].

Secondly, taking into account how many fewer cells are needed for a female to be fully fertile than for a male, it is possible that female gametogenesis is also affected (with germ cells that die during the embryonic and fetal period of oogenesis). However, the few germ cells that escape the unfavourable conditions determined by the Rb translocations and complete the phases of ovariogenesis could provide a sufficient number of oocytes, on the order of a few thousand oocytes per mouse, to ensure ovulation to the carrier. This is exactly what Mittwoch's group has shown in females for some chromosomal anomalies known to affect spermatogenesis. The small oocyte population is markedly reduced, but the large growing oocyte population is not numerically affected. Thus, evidence is accumulating that gametogenesis can be disturbed in both sexes but, in the male, after DNA segregation ends, Rb heterozygosity can still extend its influence over a specific (male-limited) process: the genome packing and cell shaping of spermiogenesis. Thus the problem of differences in male-female fertility impairment could be regarded as reflecting different opportunities for the expression of the damage in the two gametogenic processes (Fig. 9).

Apart from these hypotheses for explaining the female's ability to overcome the interference of heterozygosity with the gametogenic process, it

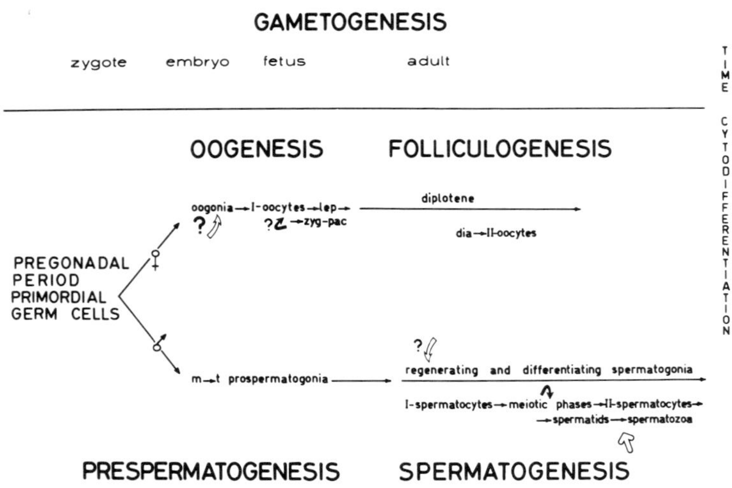

Fig. 9. Comparison of male and female mammal gametogenesis. Arrows indicate the stages at which it has been shown (or is suspected) that Rb heterozygosity actively causes germ cell loss. Germ cell cytodifferentiation stages occur at different times in males and females. Folliculogenesis and spermatogenesis take place in adult animals. If only a few germ cells can pass the critical phases affected by Rb heterozygosity, these can be enough to ensure ovulation but spermatogenesis will be drastically impaired.

must be conceded that there are no clear pieces of evidence to explain how heterozygosity interferes with the gametogenic processes. The few data we have now are for only some cytogenetic situations at a few steps in the gametogenic process, mainly for males, and more-or-less all from the laboratory strains. We have no analytical data for females nor has any of the comparative analysis of all these phenomena been done—for instance, comparing the natural Rb heterozygotes and their laboratory counterparts (i.e., changing the genome background) or comparing strains. The detrimental effects of Rb heterozygosity on the fertility of a carrier might well not occur at all if some "conditions" change. In six natural hybrid carriers of one to three independent heterozygosities Winking [113] found the non-disjunction rates to be very low. After Rb heterozygosities were introduced into the laboratory mouse genome, the non-disjunction rates rose to 20–40%!

Unfortunately, in other species in which Rb heterozygosities occur—such as man, the lemur, cattle, and *Akodon molinae*—or in species that have an Rb system similar to that of the house mouse—such as *Sorex araneus* (see elsewhere, this volume), *Acomys cahirinus*, and *Ellobius talpinus*—we have less information than for the house mouse to use as a basis for general principles by which Rb heterozygosity might cause segregational and related disorders.

Thus there is a great deal of room for further investigation in this area.

METHODS FOR EVALUATION OF ANEUPLOID FREQUENCIES IN Rb STRUCTURAL HETEROZYGOTES

Preimplantation Stages

Aneuploid frequencies in Rb heterozygotes can be assessed by direct cytogenetic study of preferably the spermatocyte [114] or oocyte [115]. Another useful technique, but applicable only to males, is to measure the DNA content of germ cells as a direct index of the "ploidy" of the cell. These methods together can provide (for males) complete profiles of the aneuploid frequencies from the gonocyte stage to the embryo (with chromosome plates prepared from the fetal membranes).

Metaphase II chromosomal arm counts should give an estimate of anaphase I non-disjunction. There is one intrinsic error in this operation, i.e., when nullisomy occurs simultaneously with disomy. The actual incidence of the occurrence of simultaneous nullisomy/disomy events is unknown, though in some multiple heterozygosities with arm homologies it may be a frequent phenomenon. Hypomodal error owing to cell breakage can be minimized by estimating the aneuploid frequency as twice the hypermodal value. The

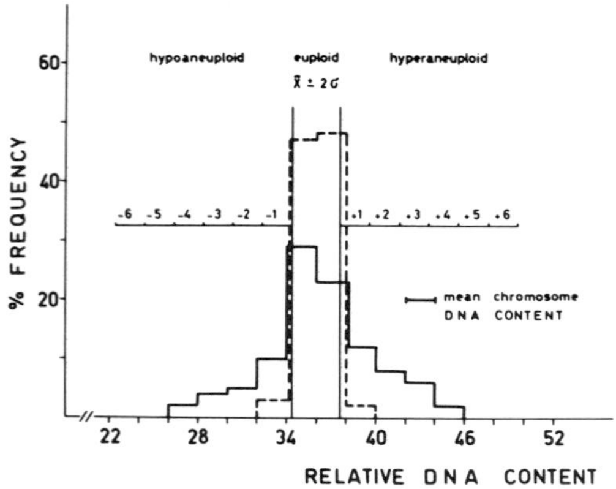

Fig. 10. Estimation of gametic aneuploidy frequency by cytophotometric DNA content evaluation. From the histogram of relative DNA content, the mean gametic DNA content ±2SD gives the euploid range for 96% of the gametes. From the mean chromosomal DNA content, it is possible to assess the frequency of hypo- and hyperaneuploid classes. - - -, the distribution curve for a +/+; –, for an Rb/+ animal.

estimates of germ cell DNA content [116–121] utilize stoichiometric reactions for DNA. Two intrinsic errors can alter the actual aneuploidy rate thus determined: 1) a pseudo-euploid DNA content could be obtained when nullisomy and disomy are present simultaneously; and 2) in some cytogenetic situations, hypo- or hyperaneuploidy for one smaller chromosome is not easy to detect because of the limits of sensitivity of the method. When used in conjunction with metaphase chromosome evaluations of spermatocytes, blastocysts, and fetal membranes, these methods offer some insight into the existence of anaphase II non-disjunction and prezygotic selection of unbalanced gametes. This last point is of particular interest since it has sometimes been claimed that prezygotic selection [122] could explain the differences between sexes in the gametic aneuploid frequencies owing to heterozygosities. Both static and flow cytophotometric determinations can be performed on sperm cells after DNA has been stoichiometrically stained. Static measurements are usually performed with the Feulgen reaction by microdensitometry, and the cell-by-cell analysis yields additional information about morphological abnormalities [93,117]. This procedure is time-consuming as compared with the flow methods. A drawback of the flow method is that the sperm head shape can constitute a difficulty for some species, especially when it is very unlike the usual ovoid shape.

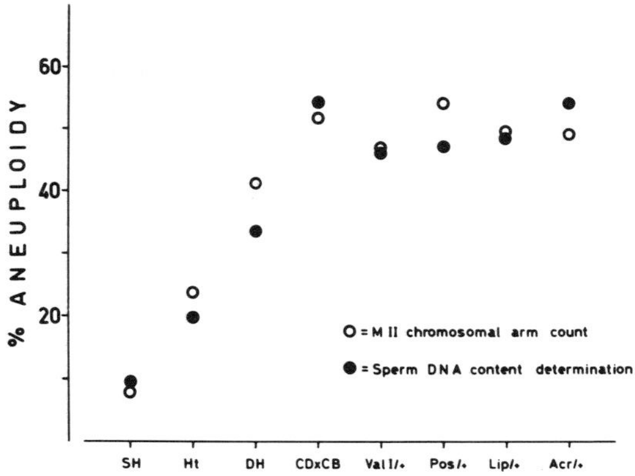

Fig. 11. Comparison of gametic aneuploidy frequency determined by metaphase II chromosomal arm count and sperm DNA content. The abbreviations refer to the three karyotypic forms of *Akodon molinae* (SH, Ht, and DH) and to different types of Rb heterozygotic mice.

Once the distribution curve of the DNA content values for control and Rb heterozygotes has been obtained, a simple comparison will give the hypo- and hypermodal values, since mathematical treatment of the statistical parameters (mean, standard deviation, and coefficient of variation) for each curve can easily be performed. It is assumed that sperm DNA values for the +/+ animals are distributed as a Gaussian curve; therefore, the limits of the range for euploidy can reasonably be chosen as the mean plus or minus twice the standard deviation that, in a Gaussian curve, includes 96% of the data. All the data outside these limits can be assumed to be "aneuploid." Other confidence limits for euploidy can be chosen, such as the mean plus or minus three times the standard deviation, which will include 98% of the data. The precise percentages of aneuploidy ascribable to each class of hypoaneuploidy (i.e., −1, −2, −3, . . . chromosomes) or hyperaneuploidy (i.e., +1, +2, +3, . . . chromosomes) can be assessed, determining what percentage of the genome is represented by a chromosome and then calculating the limits of the hypo- and hyperaneuploidy classes by substraction and addition of it to the low and high euploid limit (Fig. 10). This procedure seems to work, at least in the mouse, in which each chromosome is more or less 5% of the genome in the haploid condition. The estimation of aneuploidy frequency of a sperm population by means of sperm DNA content measurement has been applied to other nonmurine animals' heterozygosity, such as *Akodon molinae* [121] and *Sorex araneus* (see elsewhere, this volume).

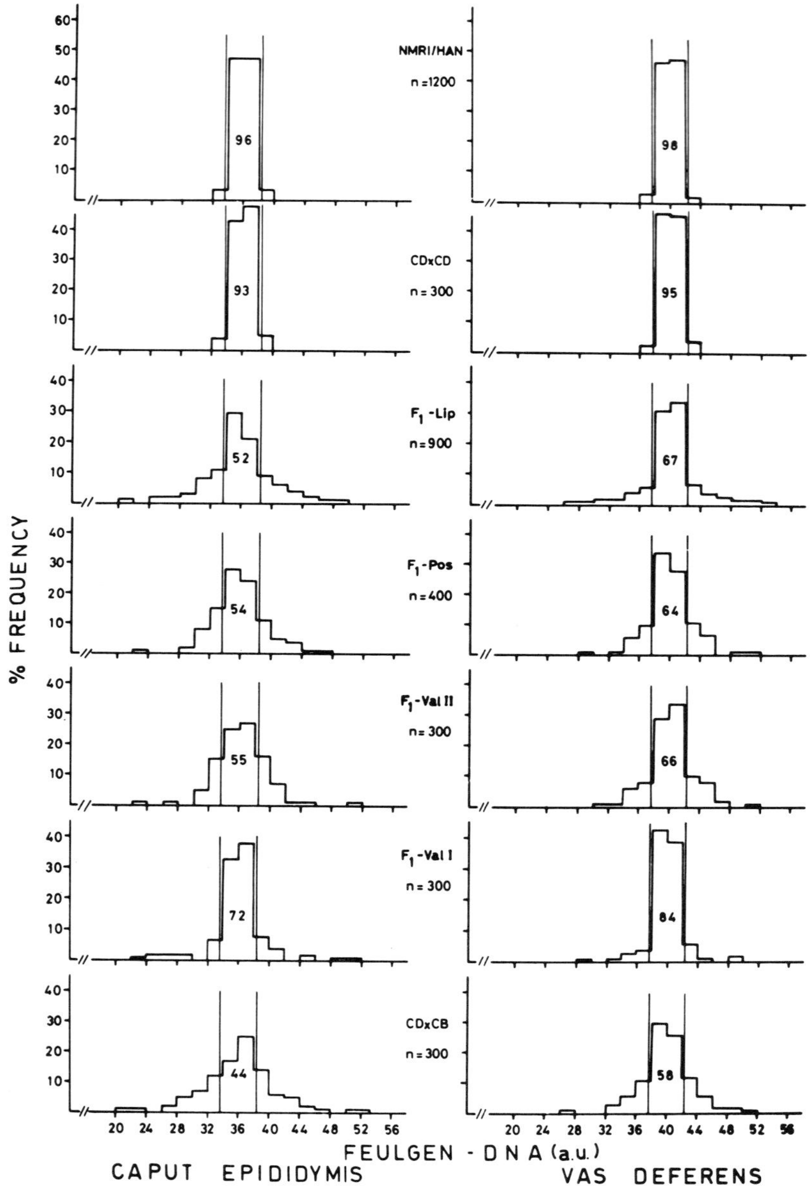

Fig. 12. Feulgen-DNA content distributions of spermatozoa from caput epididymis and vas deferens for the control group (NMRI/HAN, CD × CD), Rb heterozygotes mice with up to nine trivalents (F_1-Lip, F_1-Pos, F_1-Val II and F_1-Val I) and with polyvalent ring meiotic configuration (CD × CB). The vertical lines show the limits of the euploid range (calculated

A comparison of the cytogenetic and cytophotometric methods for the same groups of individuals shows good agreement (Fig. 11).

Postimplantation Stages

Regular cytogenetic methods can be applied to embryonic fetal membranes and living embryos. In this last case there will be a serious underestimation of the aneuploidy rate ascribable to postimplantation stages, since many fetuses with abnormalities will die before birth. Blastocyst karyotyping can also provide aneuploid frequencies, but factors other than the heterozygous condition can increase or decrease the aneuploid frequency ascertained.

Evaluation of aneuploidy in postimplantation stages requires the recording of the total number of implants.

FATE OF ANEUPLOID GERM CELLS

Primary and Secondary Impairment of Fertility

As described, Rb heterozygosities can disturb several phases of the meiotic process, leading to prezygotic losses that result in primary impairment of fertility. This is the case for some male carriers of double or multiple heterozygosities that produce chain meiotic configurations. When the Rb heterozygosities affect the gametogenic process, causing the formation of unbalanced gametes (as for many single heterozygotes, male and female, and male carriers of multiple heterozygosities with ring meiotic configuration), not only do prezygotic losses occur, but also postzygotic ones, since very often the unbalanced gametes can successfully accomplish fertilization and the unbalanced zygote will die early (pre-) or late (postimplantation). This causes secondary subfertility or sterility, since the fitness of the individual is quite reduced, even though it may be able to produce a certain number of gametes.

Fate of Aneuploid Sperm and Oocytes

The vast majority of fetal abnormalities produced by Rb heterozygote stocks are due to cytogenetic imbalances. Moreover, the MII chromosomal arm counts in both single and multiple heterozygotes of the two sexes (which

---

from the mean value of sperm DNA content of NMRI/HAN mice plus or minus two times the standard deviation), and the numbers inside them indicate the percentage frequencies of spermatozoa that fall within it. Reprinted from Redi et al. [93], with permission of the publisher. Note that the proportion of grossly aneuploid sperm is less in the vas deferens than in the caput epididymis.

give anaphase I non-disjunction rates) show values that parallel the fetal membrane rates of cytogenetic disorder. The indications are, therefore, that once a germ cell has passed the meiotic phases, it can finish development. Sperm DNA content determinations have shown that even moderate genome imbalance does not prevent the capacity to become a fully mature sperm and then to be able to fertilize [100,123]. It seems that the fate of the major proportion of aneuploid sperm is to contribute to postzygotic losses and thus to impair fertility, even though sperm with gross genome imbalance tend to disappear in the epididymis. This becomes clear when the frequency of aneuploid sperm in different localities of the male genital tract is compared: vas deferens aneuploid sperm are always fewer than those in the caput epididymis (Fig. 12). This decrease is due to the loss of sperm with DNA content equivalent to hypo- or hyperaneuploidy for four or five chromosomal arms.

The same phenomenon has been found in chromosome variability systems in nonmurine animals, such as *Akodon molinae* [124].

The disappearance is partly ascribable to epididymal phagocytic activity [125] (Fig. 13).

How the extremely aneuploid sperm are recognized and eliminated is still obscure, but morphological factors have to be excluded, since they are shaped normally and both light and heavy morphologically abnormal sperm can pass along the male genital tracts. The frequency of phagocytic activity is in any case not high enough to account for all the unbalanced sperm that disappear during transit through the male genital tract (ranging from 10% to 22%). Probably some of them die naturally because of incapacity to appropriately express some haploid genome activities. In the female genital tract aneuploid sperm can reach the upper end of the tubes and fertilize, whereas light and heavy morphologically abnormal sperm are eliminated [93*,126].

Few data are available for oogenesis, but it is quite clear from the cytogenetic analysis of Rb/+ female progeny that unbalanced oocytes can develop and be fertilized, thus contributing to postzygotic losses.

PRACTICAL AND THEORETICAL IMPLICATIONS OF THE MOUSE CHROMOSOME VARIABILITY SYSTEM

Importance to Experimental Medicine

The *Mus domesticus* system of chromosomal variability, its occurrence and distribution in nature, and its availability in the laboratory is mainly due to the efforts of Alfred Gropp, who during two decades of intensive work established an animal model suitable for studying the relationship between chromosomal disorders and development.

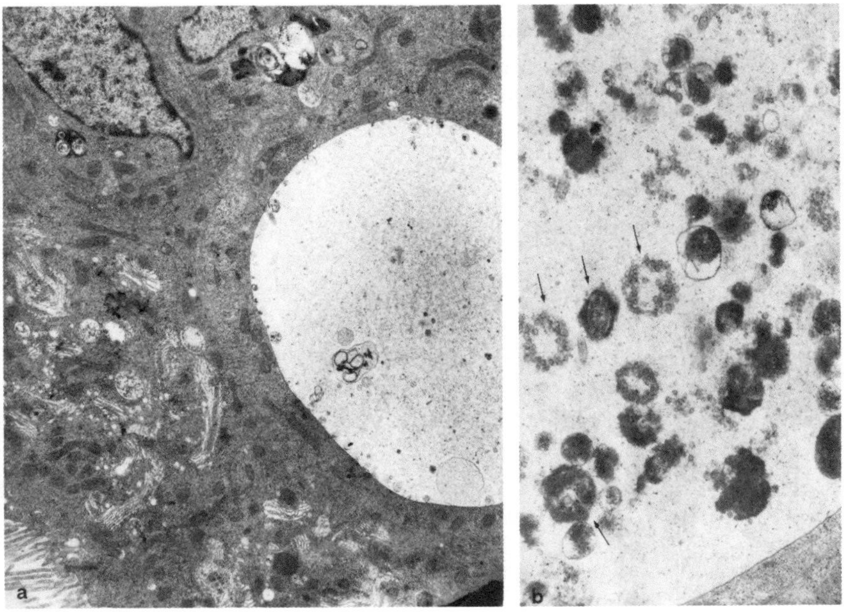

Fig. 13. Phagocytosis of presumptively grossly aneuploid sperm. Electron microscopy. **a:** The apical part of a principal cell of the cauda epididymis of an Rb heterozygous mouse (2n = 32); a large vacuole is visible, filled with digested material. ×6,000. **b:** Enlargement of the vacuole's content shows (arrow) cross-sectioned sperm flagella. ×24,000. (Preparation of C. de Martino.)

In many mammals (e.g., cow, sheep, and man), such studies are of particular interest in the perspective of, for example, the aneuploidy rate among human newborns of about 0.5% [127]. The order of magnitude becomes much greater considering the loss of embryonic and fetal life; at conception something like 15–20% is a commonly accepted loss rate. Nearly all aneuploid embryos are sooner or later aborted, but some autosomal trisomies of chromosomes 21, 18, and 13 survive until term.

With the mouse, induction of individual autosomal trisomies can be performed in a predictable manner by the cross-breeding of double heterozygotes with one arm homology with the 2n = 40 all-acrocentric animals. The production of a specific trisomy can be achieved by choosing the chromosomal arm that is shared by the two Rb chromosomes. In this way fetal development and gross phenotypic abnormalities, linked to developmental disorders, have been studied by Gropp et al. [128] for all the nineteen possible trisomies. Some trisomic animals die during organogenesis, whereas others have a much longer life span, which allows the study of developmental disorders in specific organ systems.

Of particular interest is murine trisomy 16, since it is assumed to be a suitable model of human 21 trisomy (Down syndrome) [129]. The basis for this assumption is the synteny of three gene loci (namely, superoxide dismutase, interferon receptor protein, and phosphoribosyl glycinamide synthetase) on human 21 and murine 16 chromosomes.

There are many parallels in developmental anomalies between the two systems [130,131].

Interactions between the phenotype and chromosomal imbalance are reviewed elsewhere in this volume (see Chapter 24).

At the cellular level, the fate of trisomic and monosomic cells can be studied in trisomic- or monosomic-2n aggregation chimeras produced at the preimplantation stages [132,133]. Essentially it is shown that gene dosage abnormalities can account for defective cellular processes. In addition, these studies permit the rescue of trisomic cells from early death for longer-term studies [134].

Impact on Theoretical Evolutionary Biology

The extreme chromosomal variability in the *Mus domesticus* system is accepted as an example of stasipatric speciation [135], in which chain processes of chromosomal rearrangements make it possible for postmating isolation to develop [136].

There is a basic dichotomy in theories about the role of chromosomal rearrangements in speciation. On one side, impaired fertility in heterozygotes is claimed to be a means of building a postmating barrier between chromosomally rearranged populations. On the other side, the negative heterosis of hybrids, resulting from segregational disorders of the structural heterozygotes, makes it difficult for a chromosomal rearrangement itself to become fixed in the population.

Once it appears in a natural population, a chromosomal rearrangement becomes distributed throughout it in the heterozygotic form, but this situation also impairs the relative fertility of the carriers and confers more negative heterosis than for the homozygotes. In addition, the heterozygotes with rearranged karyotypes are necessarily many fewer in number than the unchanged (acrocentric) homozygotes. It is difficult, therefore, to see how the Rb variant can outweigh the heterozygosity, overcome the barrier of negative heterosis, and become fixed in the population. The demographic conditions outlined above would seem to lead to rapid disappearance of the karyotypic variant from the population, not into its fixation.

Recent theoretical algebraic models, mostly developed by Lande [137,138] and Hedrick [139], have been used to show that the blind alley heterosis can be circumvented and the new homokaryotype can become fixed in the population. For this, the relative impairment of fertility of the structural heterozygotes must be contained within modest limits and the deme of origin must be of minimal real dimensions.

The preferential transmission of the Rb chromosome to the progeny (as shown for Rb(9.12) chromosome in wild Peru-Coppock mice) can counter the deleterious effects on gametogenesis increasing the chance of homozygosity [83].

This not only makes it possible for new homokaryotypes to become fixed but also makes it possible for Rb metacentrics to accumulate in the karyotypes of the population by two mechanisms: 1) successive appearance of new fusions in a population that already has a previously fixed Rb metacentric, and 2) hybridization of two populations that each have one pair of fixed metacentrics in the karyotype or that are still in the stage of balanced polymorphism. For this accumulation phase to continue (for the new homokaryotype to become fixed) at the larger number of Rb metacentrics a new condition becomes necessary, which is to say there must be some advantage for animals with both Rb metacentrics over animals with acrocentrics. The large number of Rb metacentrics in most mouse population karyotypes, usually seven to nine, indicates a posteriori that some advantage must exist [5,11,28]. Dover [140] felt, however, that the accumulation of Rb metacentrics is a purely stochastic phenomenon of molecular drive.

In the analysis of the role played by Rb heterozygosity in building up reproductive barriers, single and multiple Rb heterozygosities probably act to different extents. The first allow a certain degree of hybridization, since the gametogenesis of the carriers is not severely affected. Multiple Rb heterozygosities have the potential capacity of isolating populations and thus impeding gene flow, since they severely impair gametogenesis.

Corti et al. [28] have recently constructed a cladogram for the Rb metacentrics in a compatibility analysis of all the populations in the Rhaetian Alps. The branch points of this phyletic tree are all at positions where Rb metacentrics, one for each divergent lineage, with reciprocal monobrachial homology intersect.

At present we have corresponding data only for the mouse from studies of meiosis and kinetics of gametogenesis in structural heterozygotes. Therefore, for the mouse, we can propose a pathway for speciation that includes a first phase of the appearance of a pair of Rb metacentrics and their fixation in homozygotes. The fixation takes place in a small deme isolated from and peripheral to the general area of distribution, in a situation that Mayr [141] called peripatric. In this situation, postmating isolation is not involved, since the segregational impairment of fertility is relatively weak. In this way, Rb metacentrics can accumulate through hybridization between populations until Rb metacentrics with monobrachial homology appear and remain fixed in adjacent populations. At this time, postmating isolation develops and the background for complete speciation is established.

However, Mayr is probably correct when he states that speciation is always associated with allopatric or peripatric isolation, since an essential prerequisite for chromosomal evolution is a stage of fixation of the new karyotype into an isolated deme.

ACKNOWLEDGMENTS

The authors are grateful to Dr. S. Garagna for his great help in the preparation of the manuscript; to Drs. C. de Martino, M. Pansa, and H. Winking for providing part of the illustrations; to Dr. B.L. Rubin for English revision; and to Dr. M. Zuccotti, Mrs. T. Bosini, Mrs. P. Veneroni, and Mr. S. Veneroni for skillful technical assistance. This work was supported by M.P.I. 40% Progetto Differenziamento and by Wilkinson Foundation (Charitable Trust), London to CAR and by C.N.R. and M.P.I. grants to EC.

REFERENCES

1. Robertson WM, Rees B: Chromosome studies. I. Taxonomy relationship shown in the chromosomes of Tettegidae and Arcrididiae. Chromosomes and variations. J Morphol 27:179, 1916.
2. Fredga K: Chromosomal changes in vertebrate evolution. Proc R Soc Lond [Biol] 199:377, 1977.
3. Hsu TC: "Human and Mammalian Cytogenetics." New York, Heidelberg, Berlin: Springer Verlag, 1979.
4. Searle AG: Chromosomal variants. In Green MC (ed): "Genetic Variants and Strains of the Laboratory Mouse." Berlin, Heidelberg, New York, Tokyo; Springer Verlag, 1981, p 324.
5. Capanna E: Karyotype variability and chromosome transilience in rodents: The case of the genus *Mus*. In Luckett WP, Hartenberger JL (eds): "Evolutionary Relationship Among Rodents." New York: Plenum Publ Corp., 1985, p 643.
6. Gropp A, Tettenborn U, von Lehmann E: Chromsomenvariation vom Robertson'schen Typus bei der Tabakmaus, *M poschiavinus* und ihren Hybriden mit der Laboratoriumsmaus. Cytogenet Cell Genet 9:9, 1970.
7. Fatio V: "Faune des Vertébrés de la Suisse. I. Mammiferes." Geneve: H. Georg, 1869, p 410.
8. Leonard A, Deknudt GH: A new marker for chromosome studies in the mouse. Nature 214:504, 1967.

9. Evans EP, Lyon MF, Daglish M: A mouse translocation giving a metacentric marker chromosome. Cytogenet Cell Genet 6:105, 1967.
10. White BJ, Tjio JH: A mouse translocation with 38 and 39 chromosomes but normal NF. Hereditas 58:284, 1968.
11. Gropp A, Winking H, Zech L, Mueller HJ: Robertsonian chromosomal variation and identification of metacentric chromosomes in feral mice. Chromosoma 39:265, 1972.
12. Capanna E: Robertsonian numerical variation in animal speciation: *Mus musculus,* an emblematic model. In Barigozzi C (ed): "Mechanisms of Speciation." New York: Alan R. Liss, Inc., 1982, p 155.
13. Gropp A, Winking H, Redi C, Capanna E, Britton-Davidian J, Noack G: Robertsonian karyotype variation in wild house mice from Rhaeto-Lombardia. Cytogenet Cell Genet 34:67, 1982.
14. Capanna E, Gropp A, Winking H, Noack G, Civitelli MV: Robertsonian metacentrics in the mouse. Chromosoma 58:341, 1976.
15. Lehmann E, Radbruch A: Robertsonian translocations in *Mus musculus* from Sicily. Experientia 33:1025, 1977.
16. Nash HR, Brooker PC, Davis SJM: The Robertsonian translocation house-mouse populations of North East Scotland: A study of their origin and evolution. Heredity (Edinburgh) 50:303, 1983.
17. Robinson TJ: Preliminary report of a Robertsonian translocation in an isolated feral *Mus musculus* population. Mammal Chrom News Letter 19:84, 1978.
18. Said K, Jacquart T, Montgelard C, Sonjaya H, Helal AN, Britton-Davidian J: Robertsonian house mouse populations in Tunisia: A karyological and biochemical study. Genetika 68:151, 1986.
19. Committee on Standardized Genetic Nomenclature for Mice: Standard karyotype of the mouse *Mus musculus*. J Hered 63:69, 1972. Also Mouse News Letter 51:2, 1974.
20. Evans EP: Karyotype of the House Mouse. In Berry RJ (ed): "Biology of the House Mouse." New York, London: Academic Press, 1981, p 127.
21. Hsu TC, Markvong A, Marshall JT: G-band patterns of six species of mice belonging to subgenus *Mus*. Cytogenet Cell Genet 20:304, 1978.
22. Miller OJ, Miller DA: Cytogenetics. In Morse H (ed): "Origin of Inbred Mice." New York, London: Academic Press, 1978, p 591.
23. Gropp A, Winking H: Robertsonian translocations: cytology, meiosis, segregation pattern and biological consequences of heterozygosity. In Berry RJ (ed): "Biology of the House Mouse." New York, London: Academic Press, 1981, p 141.
24. Mouse News Letter 72:78, 1985.
25. Adolph S, Klein J: Robertsonian variation in *Mus musculus* from Central Europe, Spain and Scotland. J Hered 72:219, 1981.
26. Winking H, Gropp A, Bulfield G: Robertsonian chromosomes in mice from North-Eastern Greece. Mouse News Letter 64:69, 1981.
27. Capanna E, Corti M, Nascetti G: Role of contact areas in chromosomal speciation of the European long-tailed house mouse (*Mus musculus domesticus*). Boll Zool 52:97, 1985.
28. Corti M, Estabrook GF, Capanna E: Microevolutionary sequences in house mouse chromosomal speciation. Syst Zool 35:163, 1986.
29. Marshall JT, Sage RD: Taxonomy of the House Mouse. In Berry RJ (ed): "Biology of the house mouse." New York, London, 1981, p 15.
30. Hunt WG, Selander RK: Biochemical genetics of hybridization in European House Mouse. Heredity (Edinburgh) 31:11, 1973.

31. Bonhomme F, Britton-Davidian J, Thaler L, Triantaphyllidis C: Sur l'existence en Europe de quatre groupes de souris (genre *Mus* L) du rang espèce et semi-espèce, demontrée par la génétique biochimique. C R hebd Seanc Acad Sci (Paris) 287D:631, 1978.
32. Thaler L, Bonhomme F, Britton-Davidian J: Processes of speciation and semi-speciation in the house mouse. In Berry RJ (ed): "Biology of the House Mouse." New York, London: Academic Press, 1981, p 27.
33. Dev VG, Miller DA, Tantravahi RR, Schreck RR, Roderick TH, Erlanger BF, Miller OJ: Chromosome markers in *Mus musculus*: Differences in C-banding between the subspecies *M. m. musculus* and *M. m. molossinus*. Chromosoma 53:335, 1975.
34. Sparrow AH, Price HJ, Underbrink AG: A survey of DNA content per cell and per chromosome of prokaryotic and eukaryotic organisms: Some evolutionary considerations. In Smith HH (ed): "Evolution of Genetic Systems." Brookhaven Symposia in Biology, 23, New York: Gordon and Breach, 1972, p 451.
35. Manfredi Romanini MG: The nuclear content of deoxyribonucleic acid and some problems of mammalian phylogenesis. Mammalia 49:369, 1985.
36. Gamperl R, Ehmann Ch, Bachmann K: Genome size and heterochromatin variation in rodents. Genetika 58:199, 1982.
37. Comings DE, Harris DC: Nuclear proteins. I. Electrophoretic comparison of mouse nucleoli, heterochromatin, euchromatin and contractile proteins. Exp Cell Res 96:161, 1975.
38. Matsukuma S, Utakoji T: Non-histone protein associated with centromeric heterochromatin in the mouse chromosome. Exp Cell Res 105:217, 1977.
39. Hsu TC, Arrighi FE: Distribution of constitutive heterochromatin in mammalian chromosomes. Chromosoma 34:243, 1971.
40. Pardue ML, Gall JG: Chromosomal localization of mouse satellite DNA. Science 168:1356, 1970.
41. Jones KW: Chromosomal and nuclear location of mouse satellite DNA of individual cells. Nature 255:912, 1970.
42. Yunis G, Yasmineh WG: Heterochromatin, satellite DNA and cell function. Science 174:1200, 1971.
43. Marzimas JA, Balhorn R, Hatch FT: Separation of satellite DNA chromatin and main band DNA chromatin from mouse brain. Nucleic Acids Res 7:935, 1979.
44. Flam WG, McCallum M, Walker PMB: The isolation of complementary strands from a mouse DNA fraction. Proc Natl Acad Sci USA 57:1729, 1967.
45. Corneo G, Ginelli E, Soave C, Bernardi G: Isolation and characterization of mouse and guinea pig satellite deoxyribonucleic acid. Biochemistry 7:4373, 1968.
46. Hoerz W, Zachau HG: Characterization of distinct segments in mouse satellite DNA by restriction nucleases. Eur J Biochem 73:383, 1977.
47. Southern EM: Long Range Periodicites in Mouse Satellite DNA. J Mol Biol 94:51, 1975.
48. Appels R, Peacock WJ: The Arrangement and Evolution of Highly Repeated (Satellite) DNA Sequences with Special Reference to *Drosophila*. Int Rev Cytol 8:69, 1978.
49. Singer MF: Highly repeated sequences in mammalian genomes. Int Rev Cytol 76:67, 1982.
50. Lin MS, Latt SA, Davidson RL: Microfluorometric detection of asymmetry in the centromeric region of mouse chromosomes. Exp Cell Res 86:392, 1974.

51. Alves PR: The structure and position of mouse centromeric heterochromatin in BrdU-labelled chromosomes and diplochromosomes stained by the pH 10.4 method. Chromosoma 85:713, 1982.
52. Jones KW: Repetitive DNA sequences in animals, particularly primates. Chromosomes Today 5:305, 1976.
53. Jonasson J, Alves P, Stroem A: Polymorphisms in the centromeric region of the chromosomes of the laboratory mouse. Hereditas 90:111, 1979.
54. Schmid M: Chromosome banding in Amphibia. IV Differentiation of GC and AT-rich chromosome regions in Anura. Chromosoma 77:83, 1980.
55. Schweizer D, Ehrendorfer F: Evolution of C-band patterns in Asteraceae-Anthemidaeae. Biol Zentralbl 102:637, 1983.
56. John B, King M, Schweizer D, Mendelak M: Equilocality of heterochromatin distribution and heterochromatin heterogeneity in acridid grasshoppers. Chromosoma 91:185, 1985.
57. Forejt J: Centromeric heterochromatin polymorphism in the house mouse. Evidence from inbred strains and natural populations. Chromosoma 43:187, 1973.
58. Dev VG, Miller DA, Miller OJ: Chromosome makers in *Mus musculus*: Strain differences in C-banding. Genetics 75:663, 1973.
59. Jean P, Hartung M, Mirre C, Stahl A: Association of centromeric heterochromatin with the nucleolus in mouse Sertoli cells. Anat Rec 205:375, 1983.
60. Hsu TC, Cooper JEK, Mace ML Jr, Brinkeley BR: Arrangement of centromeres in mouse cells. Chromosoma 34:73, 1971.
61. Mayfield JE, Ellison JR: The fluorescence localization of mouse satellite DNA in interphase nuclei. Chromosoma 54:27, 1976.
62. Rattner JB, Lin CC: Centromere organization in chromosomes of the mouse. Chromosoma 92:325, 1985.
63. John B, Freeman M: Causes and consequences of Robertsonian exchange. Chromosoma 52:123, 1975.
64. Lau YF, Hsu TC: Variable modes of Robertsonian fusions. Cytogenet Cell Genet 19:231, 1977.
65. Tsujimoto Y, Gorham J, Cossman J, Jaffee E, Croce CM: The t(14;18) chromosome translocations involved in B-cell neoplasms result from mistakes in VDJ joining. Science 229:1390, 1985.
66. Ohno S, Trujillo JM, Kaplan WD, Kinosita R: Nucleolus-organizers in the causation of chromosome anomalies in man. Lancet ii:123, 1961.
67. Miller OJ, Miller DA, Tantravahi R, Dev VG: Nucleolus organizer activity and the origin of Robertsonian translocations. Cytogenet Cell Genet 20:40, 1978.
68. Winking H, Nielsen K, Gropp A: Variable positons of NORs in *Mus musculus*. Cytogenet Cell Genet 26:158, 1980.
69. Stahl A, Luciani JM, Hartung M, Devictor M, Bergé-Lefranc JL, Guichaoua M: Structural basis for Robertsonian translocation in man: association of ribosomal genes in the nucleolar fibrillar center in meiotic spermatocytes and oocytes. Proc Natl Acad Sci USA 80:5946, 1983.
70. Holmquist GP, Dancis B: Telomere replication, kinetochore organizers, and satellite DNA evolution. Proc Natl Acad Sci USA 76:4566, 1979.
71. Matthey R: L'évolution de la formule chromosomiale chez les vertebres. Experientia 1:50, 1945.

72. Matthey R: Le polymorphisme chromosomique des *Mus* africains du sous-genre *Leggada.* Revision générale portant sur l'analyse de 213 individus. Revue suisse Zool 73:585, 1966.
73. Comings DE, Avelino E: DNA loss during Robertsonian fusion in studies of the tobacco mouse. Nature 237:199, 1972.
74. Redi CA, Garagna S, Mazzini G, Winking H: Pericentromeric heterochromatin and A-T contents during Robertsonian fusion in the house mouse. Chromosoma 9:31, 1986.
75. Hsu TC, Pathak S, Chen TR: The possibility of latent centromeres and a proposed nomenclature system for the total chromosome and whole arm translocations. Cytogenet Cell Genet 15:41, 1975.
76. Redi CA, Garagna S, Hilscher B, Winking H: The effects of some Robertsonian chromosome combinations on the seminiferous epithelium of the mouse. J Embryol Exp Morphol 85:1, 1985.
77. Boué A, Thibault C: "Les Accidents Chromosomiques de la Reproduction." In Boué A, Thibault Ch (eds): "Proc Symp Sponsored by Institut National de la Santé et de la Recherche Médicale." Paris: 1973.
78. Chandley A: Chromosomes. In Hargreave TB (ed): "Male Infertility." Berlin, Heidelberg, New York, Tokyo: Springer-Verlag, 1983, p 144.
79. Johannisson R, Winking H: Synaptonemal complex of multivalents in mouse spermatocytes. Eur J Cell Biol 20:122, 1979.
80. Cattanach BM, Moseley M: Nondisjunction and reduced fertility caused by the tobacco mouse metacentric chromosomes. Cytogenet Cell Genet 12:264, 1973.
81. Tettenborn U, Gropp A: Meiotic non-disjunction in mice and mouse hybrid. Cytogenet Cell Genet 9:272, 1970.
82. Gropp A, Winking H, Redi CA: Consequences of Robertsonian heterozygosity: segregational impairment of fertility versus male-limited sterility. In Crosignani PG, Rubin BL, Fraccaro M (eds): "Genetic Control of Gamete Production and Function," Serono Clinical Colloquia on Reproduction, Vol. 3. London-New York: Academic Press/Grune and Stratton, 1982, p 115.
83. Harris MJ, Wallace ME, Evans EP: Aneuploidy in the embryonic progeny of females heterozygous for the Robertsonian chromosome (9.12) in genetically wild Peru-Coppock mice (*Mus musculus*). J Reprod Fertil 76:193, 1986.
84. Szollosi D, Calarco F, Donahue RP: Absence of centrioles in the first and second meiotics spingle of mouse oocytes. J Cell Sci 11:521, 1972.
85. Winking H, Johannisson R: Pattern of pachytene pairing in mouse hybrids with chain and ring multivalents. Clin Genet 17:94, 1980.
86. Haldane JBS: Sex ratio and unisexual sterility in hybrid animals. J Genet 12:101, 1922.
87. Mahadevaiah S, Mittwoch U, Moses MJ: Pachytene chromosomes in male and female mice heterozygous for the Is(7;1)40H insertion. Chromosoma 90:163, 1984.
88. Mittwoch U, Mahadevaiah S, Setterfield LA: Chromosomal anomalies that cause male sterility in the mouse also reduce ovary size. Genet Res 44:219, 1984.
89. Burgoyne PS, Mahadevaiah S, Mittwoch U: A reciprocal autosomal translocation which causes male sterility in the mouse also impairs oogenesis. J Reprod Fertil 75:647, 1985.
90. Mahadevaiah S, Mittwoch U: Synaptonemal complex analysis in spermatocytes and oocytes of tertiary trisomic Ts(5)31H mice with male sterility. Cytogenet Cell Genet 41:169, 1986.
91. Setterfield LA, Mittwoch U: Reduced oocyte numbers in tertiary trisomic mice with male sterility. Cytogenet Cell Genet 41:177, 1986.

92. Redi CA, Hilscher B, Winking H: Stage-dependent enzymatic activities in spermatogenesis of mice with the standard karyotype and of chromosomal variants with impaired fertility. Andrologia 15:322, 1983.
93. Redi CA, Garagna S, Pellicciari C, Manfredi Romanini MG, Capanna E, Winking H, Gropp A: Spermatozoa of chromosomally heterozygous mice and their fate in male and female genital tracts. Gamete Res 9:273, 1984.
94. Speed RM, de Boer P: Delayed meiotic development and correlated death of spermatocytes in male mice with chromosome abnormalities. Cytogenet Cell Genet 35:257, 1983.
95. Forejt J: X-Y involvement in male sterility caused by autosome translocations. A hypothesis. In Crosignani PG, Rubin BL, Fraccaro M (eds): "Genetic Control of Gamete Production and Function," Serono Clinical Colloquia on Reproduction, Vol. 3. London, New York: Academic Press/Grune and Stratton, 1982, p 135.
96. Hotta Y, Chandley AC, Stern H, Searle AG, Beechey CV: A disruption of pachytene DNA metabolism in male mice with chromosomally-derived sterility. Chromosoma 73:287, 1979.
97. Redi CA, Garagna S, Hilscher B, Winking H: Pachytene spermatocyte and round spermatid 3H-Thymidine incorporation in Robertsonian heterozygous mice. In IV European Workshop on Molecular and Cellular Endocrinology of the Testis. p 131, 1986.
98. Gropp A, Winking H: Fertility vs sterility in males with double Robertsonian heterozygosity and monobrachial homology. Mouse News Letter 65:33, 1981.
99. Batten CA, Berry RJ: Prenatal mortality in wild-caught mice. J Anim Ecol 36:453, 1967.
100. Ford CE: Gross genome unbalance in mouse spermatozoa: does it influence the capacity to fertilize? In Beatty RA, Gluecksohn-Waelsch S (eds): "The Genetics of the Spermatozoon." Edinburgh: Department of Genetics of the University, 1972, p 359.
101. Gropp A: Fetal mortality due to aneuploidy and irregular meiotic segregation in the mouse. In Boué A, Thibault Ch (eds): "Proc Symp Sponsored by Institut National de la Santé et de la Recherche Médicale." Paris: 1973, p 255.
102. Epstein Ch J, Travis B: Preimplantation lethality of monosomy for mouse chromosome 19. Nature 280:144, 1979.
103. Palmer CG, Schwartz S, Hodes ME: Transmission of a balanced homologous t(22q;22q) translocation from mother to normal daughter. Clin Genet 17:418, 1980.
104. Ford CE, Searle AG, Evans EP, West BJ: Differential transmission of translocations induced in spermatogonia of mice by irradiation. Cytogenet Cell Genet 8:447, 1969.
105. Cacheiro NLA, Russel LB, Swartout MS: Translocations, the predominant cause of total sterility in sons of mice treated with mutagens. Genetics 76:73, 1974.
106. Lifschytz E, Lindsley DL: The role of X-chromosome inactivation during spermatogenesis. Proc Natl Acad Sci USA 69:182, 1972.
107. Miklos GLC: Sex-chromosome pairing and male sterility. Cytogenet Cell Genet 13:558, 1974.
108. de Boer P, Searle AG, van der Hoeven FA, de Rooij DG, Beechey CV: Male pachytene pairing in single and double translocation heterozygotes and spermatogenic impairment in the mouse. Chromosoma 93:326, 1986.
109. Hotta Y, Chandley AC: Activities of X-linked enzymes in spermatocytes of mice rendered sterile by chromosomal alterations. Gamete Res 6:65, 1982.
110. Forejt J: Nonrandom association between a specific autosome and the X chromosome in meiosis of the male mouse: Possible consequence of the homologous centromeres' separation. Cytogenet Cell Genet 13:369, 1974.

111. Rosemann A, Wahrman J, Richler C, Voss R, Persitz A, Goldman B: Meiotic association between the XY chromosomes and unpaired autosomal elements as a cause of human male sterility. Cytogenet Cell Genet 39:19, 1985.
112. Hilscher W, Hilscher B: Kinetics of the male gametogenesis. Andrologia 8:105, 1976.
113. Winking H: Some aspects of Robertsonian karyotype variation in European wild mice. Curr Top Microbiol Immunol 127:68, 1986.
114. Evans EP, Breckon G, Ford CE: An air-drying method for meiotic preparations from mammalian testes. Cytogenet Cell Genet 3:289, 1964.
115. Tarkowski AK: An air drying method for chromosome preparations from mouse eggs. Cytogenet Cell Genet 5:394, 1966.
116. Doering L, Gropp A, Tettenborn U: DNA content and morphological properties of presumably aneuploid spermatozoa of tobacco mouse hybrids. J Reprod Fertil 30:335, 1972.
117. Stolla R, Gropp A: Variation of the DNA content of morphologically normal and abnormal spermatozoa in mice susceptible to irregular meiotic segregation. J Reprod Fertil 38:335, 1974.
118. Redi CA, Capanna E: DNA-content variation in the mouse spermatozoa arising from irregular meiotic segregation. Boll Zool 45:315, 1978.
119. Zante J, Schumann J, Goehde W, Hacker U: DNA-fluorometry of mammalian sperm. Histochemistry 54:1, 1977.
120. Meistrich ML, Lake S, Steinmetz LL, Gledhill BL: Increased variability in nuclear DNA content of testis cells and spermatozoa from mice with irregular meiotic segregation. Mutat Res 49:397, 1978.
121. Garner DL, Johnson LA, Lake S, Chaney N, Stephenson D, Pinkel D, Gledhill BL: Morphological and ultrastructural characterizaiton of mammalian spermatozoa processed for flow cytometric DNA analyses. Gamete Res 10:339, 1984.
122. Hamerton JL: Robertsonian translocations in man: Evidence for prezygotic selection. Cytogenet Cell Genet 7:260, 1968.
123. Ford CE: The time in development at which gross genome unbalance is expressed. In Balls M, Wild AE (eds): "The Early Development of Mammals." Cambridge, London, New York, Melbourne: Cambridge University Press, 1975, p 285.
124. Redi CA, Garagna S, Merani MS, Capanna E, Bianchi NO, Manfredi Romanini MG: Microdensitometric evaluation of DNA content, as ploidy parameter, of spermatozoa in the polymorphic chromosomal system of *Akodon molinae* Cabrera (Rodentia, Cricetidae). Gamete Res 5:345, 1982.
125. Malorni W, Mancini F, De Martino C, Redi CA, Capanna E: Spermatozoa in subfertile chromosome derived mice: Ultrastructural features of their fate in the epididymis. Ultramicroscopy 12:139, 1983.
126. Krzanowska H: The passage of abnormal spermatozoa through the utero tubal junction of the mouse. J Reprod Fertil 38:81, 1974.
127. Hamerton JL, Canning N, Ray M, Smith S: A cytogenetic survey of 14,069 newborn infants. Clin Genet 8:223, 1975.
128. Gropp A, Winking H, Herbst EW, Claussen CP: Murine trisomy: Developmental profiles of the embryo, and isolation of trisomic cellular systems. J Exp Zool 228:253, 1983.
129. Polani PE, Adinolfi M: Chromosome 21 of man, 22 of the great apes and 16 of the mouse. Dev Med Child Neurol 22:223, 1980.
130. Miyabara S, Gropp A, Winking H: Trisomy 16 in the mouse fetus associated with generalized edema, cardiovascular and urinary tract anomalies. Teratology 5:369, 1982.

131. Epstein CJ: Mouse monosomies and trisomies as experimental systems for studying mammalian aneupolidy. Trends Genet 1:129, 1985.
132. Epstein CJ, Smith SA, Zamora T, Sawicki JA, Magnuson TR, Cox DR: Production of viable adult trisomy 17 diploid mouse chimeras. Proc Natl Acad Sci USA 79:4376, 1982.
133. Magnuson TR, Smith SA, Epstein CJ: The development of monosomy 19 mouse embryos. J Embryol Exp Morphol 69:223, 1982.
134. Herbst EW, Pluznik DH, Gropp A, Uthgenannt H: Trisomic haemopoietic stem cells of fetal origin restore haemopoiesis in lethally irradiated mice. Science 211:1175, 1981.
135. White MJD: Models of speciation. Science 159:1065, 1968.
136. White MJD: "Modes of speciation." San Francisco: Freeman, 1978.
137. Lande RL: Effective deme size during long-term evolution estimated from rates of chromosomal rearrangements. Evolution 33:234, 1979.
138. Lande RL: The fixation of chromosomal rearrangements in a subdivided population with local extinction and colonization. Heredity (Edinburgh) 54:323, 1985.
139. Hedrick PW: The establishment of chromosomal variants. Evolution 35:322, 1981.
140. Dover G, Trick M, Strachan T, Coen ES, Brown SDM: DNA family turnover and coevolution of chromosomes. Chromosomes Today 8:229, 1984.
141. Mayr E: Processes of speciation in animals. In Barigozzi C (ed): "Mechanisms of Speciation." New York: Alan R. Liss, Inc., 1982, p 1.

NOTE ADDED IN PROOF

Since this manuscript was accepted for publication, discovery of new Rb variants in the natural population as well as in laboratory stocks has been reported. Rb(6.9)40Tu and Rb(10.13)42Tu, among other Rb variants already described, were detected in a mouse population from Thebes (Greece) by Tichy and Vucak (Chromosoma 95:31–36, 1987). Hubner (Mouse News Letter 77:108, 1987) recovered Rb variants in the mouse from Belgium and classified them as Rb(4.12), Rb(4.8), and Rb(5.7) even though at present they appear to be better classified as Rb(5.10) and Rb(4.12) (personal communication). Of particular interest is Adler's finding (Mouse News Letter 77:140, 1987) of an Rb(X.2) that is the first Rb translocation involving the X-chromosome recovered. This latter was a son of a female T(5.13) in an experimental group of laboratory mice.

The Cytogenetics of Mammalian Autosomal Rearrangements, pages 361–382

Chapter 8
Meiotic Studies and Fertility in Human Translocation Carriers

Ann C. Chandley

INTRODUCTION

Among all structural rearrangements in man, translocations are the most common type of abnormality, occurring among liveborn progeny with a frequency of about 1.7 per 1,000 [1]. For translocation heterozygotes of both sexes, significant increases in fetal death and in generation time between live births have been recorded, resulting in a lowering of reproductive fitness [2]. This is due to the formation of unbalanced gametes at the segregation of the chromosomes involved in the translocation at meiosis. Each translocation, however, is unique with regard to breakpoints and arm lengths and thus produces its own level of genetic imbalance [3]. Current studies employing the hamster egg-human sperm fusion technique [4] are showing just how variable these levels can be in spermatozoa from different heterozygotes, values from 12.5% to 77.0% imbalance having been found for six different translocations studied so far [5–8].

A second kind of reproductive impairment, however, has also been found associated with some translocations, and particularly in male heterozygotes. The type of effect is one in which the chromosome anomaly acts to disturb gametogenesis and thus lowers the numbers of gametes produced. Male heterozygotes carrying translocations (and other structural rearrangements) of this kind are frequently encountered among men attending infertility clinics and are ascertained through oligo- or azoospermia [9–12].

The present review will aim to provide details of this particular form of reproductive impairment in translocation heterozygotes. Details of the cytogenetic findings at meiosis will be given and several hypotheses that have been put forward to explain the findings will be discussed.

MRC Clinical and Population Cytogenetics Unit, Western General Hospital, Edinburgh EH4 2XU, Scotland

TECHNIQUES EMPLOYED IN GERM-CELL INVESTIGATION

Since the first chromosome studies were made on human male meiotic material by a squash method [13], enormous advances in technique have been made. For investigation of dividing spermatogonia and primary and secondary spermatocytes, the best available method is that of Evans et al. [14], which employs an air-drying technique following brief hypotonic treatment and fixation in 3:1 methanol:glacial acetic acid. Conventional staining with, for example, Giemsa, orcein, or carbol fuchsin can be applied to such cells in metaphase I for analysis of bivalent number, chiasma frequency, and presence or absence of a numerical or structural abnormality. If C-banding [15] or a fluorescence technique such as that employing the use of the AT specific peptide antibiotic distamycin A in combination with the AT specific fluorescent dye 4′-6′-diamidino-2-phenylindole DAPI [16] are applied, centromere positions can be revealed, and a more complete identification of individual bivalents made. Specific chromosomes such as Nos. 1, 9, 16, and the Y are then clearly recognizable by their large blocks of C-band heterochromatin showing intense fluorescence. The way in which these staining methods can help in the identification of trivalents and quadrivalents at metaphase I in translocation heterozygotes will be illustrated in the chapter.

Another air-drying technique has been devised for analysis of bivalents at the pachytene stage [17,18]. This employs a prolonged hypotonic treatment, sometimes at a raised temperature, and has the effect of bursting the pachytene nuclei and releasing the chromosome arms into a widely spread configuration. Since chromomere patterns at this stage correspond to the G-band patterning found on somatic chromosomes, individual pachytene bivalents can be identified. The sex chromosomes in such preparations occupy the dark-staining body known as the sex vesicle (Fig. 1), and bivalent No. 9 can often be recognized by the appearance at this stage of small, round, dark-staining bodies known as "parameres," which correspond to the secondary constriction seen on the long arm in metaphase of mitosis [19]. Parameres have also have described in human oocytes at the pachytene stage [20].

Perhaps the most important recent development in mammalian meiotic cell technique, however, has been the application of "microspreading" for spermatocyte investigation at the prophase stage [21,22]. In combination with appropriate staining, this method allows chromosome pairing and desynapsis to be followed in synaptonemal complexes (SC) throughout zygotene, pachytene, and diplotene, and details of structural heterozygosity and other features of anomalous chromosome behaviour can be observed. The technique allows for rapid screening of cells at the light microscope (LM) level [23] and for detailed examination by transference of cells to electron micro-

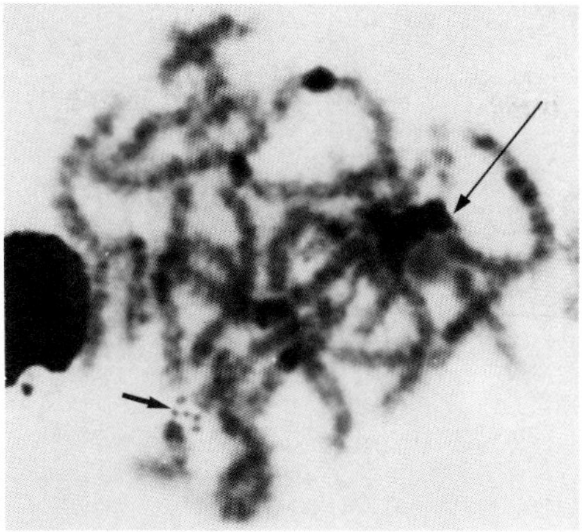

Fig. 1. Pachytene cell showing the sex vesicle (long arrow) and parameres of chromosome 9 (short arrow). Giemsa stain. Magnification ×800.

scope (EM) grids [21,22]. The spreading method has been successfully applied not only to human spermatocytes but also to oocytes [24–26]. The usefulness of the spreading technique as well as the other methods outlined above will be demonstrated in the cases described in the following sections.

X-AUTOSOME TRANSLOCATION

Reciprocal translocation involving the X chromosome and an autosome is a rare event in man, and meiotic studies on heterozygotes are even more rare. Nevertheless, some information is available. Fourteen male carriers of 13 different balanced X-autosome translocations ascertained in families have been reviewed by Madan et al. [27]. Seven carriers were prepubertal; but among the adult subjects, all but one individual showed severe oligospermia or azoospermia. In the latter case the t(X;14) reported was present in two brothers, one of whom was azoospermic and had very small atrophic testes, while the other was claimed to have fathered a child, a possibility that could not be excluded on paternity testing. His seminal fluid, however, was not examined.

Meiotic studies have been carried out by Faed et al. [28] on a t(X;21) male heterozygote who showed azoospermia. No development beyond the pachytene stage was seen in the air-dried preparations and this abrupt cessation of germ-cell maturation at the first spermatocyte stage was confirmed in the testicular histology specimens. In a case of t(X;12) studied at meiosis by

Noel and Quack (personal communication), some development to metaphase I was found and occasional dividing spermatocytes containing a chain configuration, made up of the Y, the normal No. 12, and the translocated X and No. 12 chromosomes (Fig. 2) were seen. Degeneration, however, characterized these and many other primary spermatocytes at earlier stages of development and only rare cells developed to the spermatid stage as judged by the testicular histology (Fig. 3).

This picture of germ-cell arrest principally at the primary spermatocyte level for human male X-autosome translocation heterozygotes is typical for the state of affairs found also in the mouse and a few other investigated mammalian species [29,30]. In the mouse, for example, all known reciprocal X-autosome translocations are male-sterile with severe spermatogenic im-

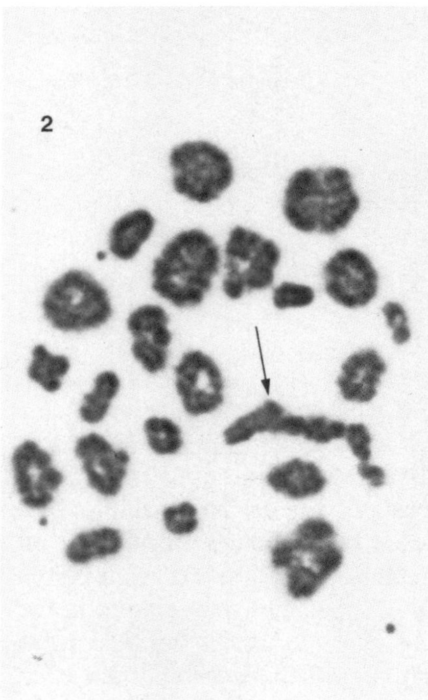

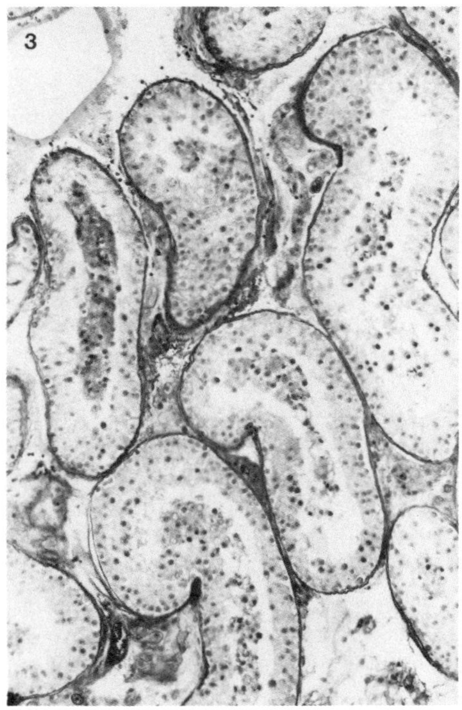

Fig. 2. Air-dried metaphase I from an X;12 translocation heterozygote. The chain quadrivalent is arrowed (kindly supplied by Dr. B. Noel). Giemsa stain. ×900.

Fig. 3. Testicular histology from an X;12 translocation heterozygote. Germ-cell arrest characterises the spermatogenic development; few spermatids and no spermatozoa are seen in the seminiferous tubules (kindly supplied by Dr. B. Noel). Haematoxylin and eosin stain. ×200.

pairment, so that metaphase I is only reached regularly in the individual translocation (T(X;16)16H. All others arrest at the pachytene stage.

For human female X-autosome heterozygotes there is evidence that the structural integrity of a critical region of the X-chromosome long arm (q) is necessary for normal ovarian function. Sarto et al. [31] first drew attention to this, pointing out that all females hitherto reported who were heterozygous for a reciprocal X-autosome translocation with a breakpoint in this region, had primary or secondary amenorrhoea. Summitt et al. [32] have reviewed the literature and found that 9 of 18 females reported to be carriers of Xq-autosome translocations exhibited either streak gonads (seven cases), ovarian hypoplasia (one case), or underdeveloped ovaries (one case). The X-chromosome breakpoints in the 13 translocations were all located within the region Xq13-Xq26. The critical region, therefore, lies in the middle of Xq and represents about two-thirds of its length. There are a few exceptions to this rule, however, and such cases retain fertility in spite of having a breakpoint in the critical region. According to Madan et al. [27], the mechanism operating may be one of variable penetrance, which allows some women to retain a few oocytes into adult life, thus enabling them to achieve pregnancy. (See Maturation Impairment in Females.)

Y-AUTOSOME TRANSLOCATION

Two types of Y-autosome translocation have been described in the human cytogenetics literature. One is generally male-sterilizing; the other is not. In the latter cases, a brightly fluorescent piece of extra chromatin of Y-chromosome origin is found attached to the short arms of an acrocentric chromosome, usually a No. 15 or No. 22, by translocation or transposition [33]. Care must be taken, however, in analysing such cases, to distinguish the extra material from simple enlargement of the acrocentric satellite itself, which can also fluoresce brightly using quinacrine dyes. A fluorescent technique such as distamycin A/DAPI, which is specific for the Yq heterochromatin, should therefore be used [16]. In these Y-autosome translocation cases, the Y chromosome itself is usually of normal length and of normal bright fluorescence, and fertility of the carrier is not disturbed. Family studies frequently show the abnormal acrocentric chromosome passing through several generations and through female as well as male carriers. The extra heterochromatin present in the karyotype does not produce phenotypic disturbances.

By contrast, a number of truly reciprocal Y-autosome translocations have been described, and these are almost universally male sterile producing severe oligospermia or azoospermia [33]. In these cases, the Y-chromosome is abnormal owing to de novo rearrangement with an autosome.

There appear to be only 14 individual cases of Y-autosome reciprocal translocation in the literature, all but one having been ascertained through infertility and/or hypogonadism. The exception is a case in which ascertainment of a Y-10 translocation in a father arose by the birth of his congenitally handicapped child [34]. The testicular histology showed all stages of spermatogenesis to be represented, and sperm counts were within normal limits. Almost all other cases reported have been azoospermic, or nearly so, with the exception of an oligospermic t(Y;16) heterozygote with a sperm count of 1×10^6 per ml reported by Faed et al. [28] and a t(Y;3) heterozygote with a sperm count of 7×10^6 per ml found by Noel and Quack (personal communication). Where information on testicular histology has been provided, spermatogenesis in human reciprocal Y-autosome translocations shows a range from near-normal to early germ-cell arrest. In most azoospermic cases germ-cell maturation arrest has occurred at the primary spermatocyte stage or before, with little development taking place beyond diakinesis/metaphase I. In at least three of the reported cases, reductions in numbers of spermatogonia were also found.

These findings in human males agree in general with those for the mouse [29,30], although in that species the data are very limited. In the mouse, however, unlike man, no case has yet been reported to be fertile. This could simply be attributable to selection bias, for in animal species karyotyping is generally only carried out on individuals selected on account of their sterility. Fertile Y-autosome translocation heterozygotes might possibly therefore be overlooked.

At meiosis in human carriers of a Y-autosome reciprocal translocation, chain quadrivalents, bivalents, and univalents and unequal bivalents have all been recorded. Figure 4 shows the chain quadrivalent observed by Laurie et al. [34] in their case of t(Y;10) when metaphase I air-dried preparations were examined. At pachytene in spread preparations at the light microscope level, Laurent et al. [35] observed the quadrivalent of their t(Y;17) heterozygote, the axes of the sex chromosomes being discernible by their dark-staining characteristics (Fig. 5). No other analyses of a Y-autosome translocation by spreading appear yet to have been carried out.

ROBERTSONIAN (Rb) TRANSLOCATION

Among all human translocations, the most common type reported is the DqDq centric fusion rearrangement brought about by Robertsonian translo-

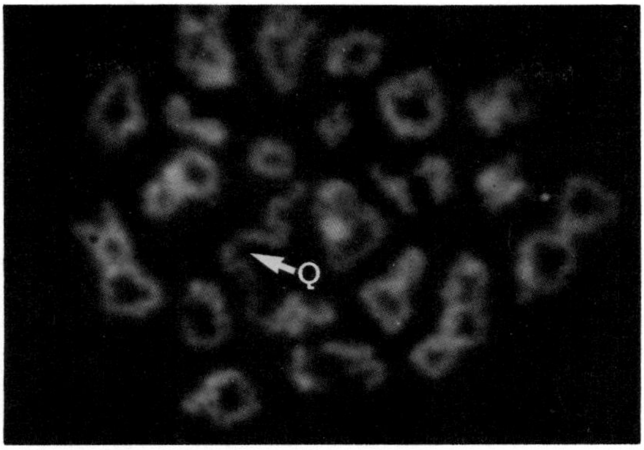

4

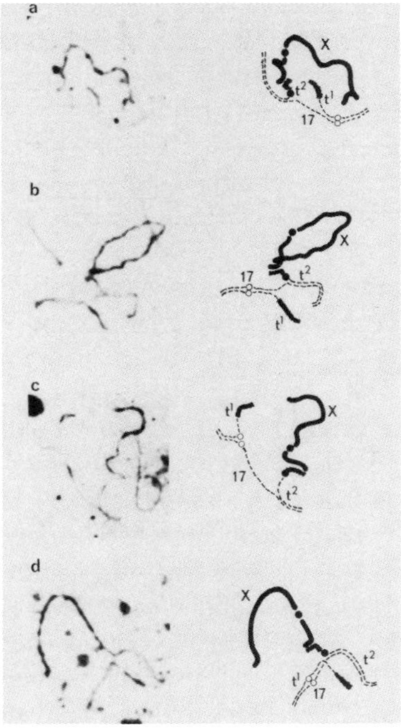

5

Fig. 4. Chain quadrivalent (Q) seen at metaphase I in a t(Y;10) heterozygote. (Reproduced from Laurie et al. [34], with permission of the publisher.) Quinacrine stain. ×800.

Fig. 5. Light microscope spread preparations showing a t(Y;17) multivalent at the pachytene stage in four different cells. On the right are diagrammatic representations of the pairing configurations in each. (Reproduced from Laurent et al. [35], with permission of the publisher, S. Karger AG, Basel.) Silver nitrate stain. ×600.

cation of two nonhomologous acrocentric D-group chromosomes, usually Nos. 13 and 14. Among male patients attending infertility clinics, the t(13q14q) translocation is also the most common type of centric fusion rearrangement [12]. Documented examples of rarer t(13q15q), t(14q21q), t(14q22q) [12], and t(21q22q) [36] cases do, nevertheless, exist.

Of interest is the fact that most of these cases ascertained in the major infertility surveys around the world are familial, and many of the infertile subjects have fertile male relatives carrying the same rearrangement. The translocation would appear, therefore, not to be the only and consistent cause of infertility. Genetic background could be an important factor, for it is known from work in the mouse that the effects on fertility of some translocations vary greatly depending on strain and genetic background [37].

Among human Rb heterozygotes ascertained through infertility, a variable testicular histology is presented, ranging from near-normality to severe spermatogenic disruption [38,39]. Increased frequencies of spermatozoa showing morphological defects have also been reported in a number of cases of DqDq and DqGq translocation [10,11,36].

At meiosis in Rb heterozygotes, a trivalent is usually found at metaphase I when standard air-dried preparations are made (Fig. 6a), and the three centromeres—belonging to the centric fusion chromosome and its two acrocentric homologues, respectively—are seen when C-banding is applied (Fig. 6b). Counts of chiasmata for Rb heterozygotes have been found generally not to differ from those recorded among normal individuals, showing values within the range 40–60 per cell [40]. The trivalent may show two, three, or four chiasmata.

The reason that spermatogenic breakdown should characterise some but by no means all male carriers of an Rb translocation still remains obscure. The various hypotheses that have been put forward to explain the germ-cell maturation arrest observed are applicable also to other forms of translocation including X-autosome, Y-autosome, and some purely autosomal reciprocal translocations, and full discussion about them will therefore be deferred until later in the chapter (see below). For the present, however, mention should be made of two recent publications in which the prophase stages of meiosis have been examined in Rb heterozygotes and possible clues to maturation impairment revealed. In the first of these, Luciani et al. [41] examined air-dried pachytene chromosome preparations from a 13q14q heterozygote. In the second, Rosenmann et al. [42] applied the spreading technique to prophase cells of two 14q21q heterozygotes. All three patients were severely oligospermic. Trivalent configurations could be readily identified in each cell analysed at the pachytene stage by the two techniques, and in a high propor-

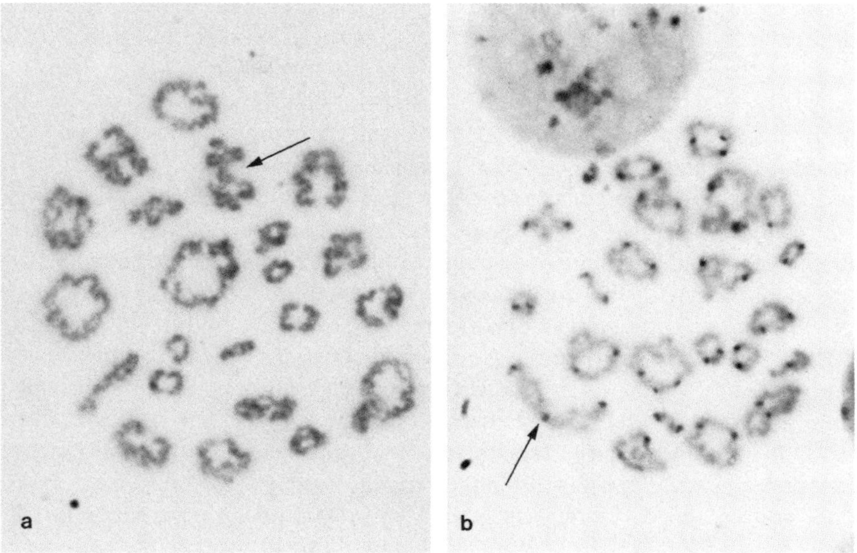

Fig. 6. **a**: Air-dried preparation showing the trivalent (arrow) formed at metaphase I in a t(13q14q) Rb heterozygote. **b**: C-banded preparation showing the three centromeres of the trivalent (arrowed) in an Rb t(13q14q) heterozygote. (Both a and b reproduced from Chandley [38], with permission of the publisher.) Giemsa stain. ×900.

tion of cells, nonrandom associations were seen between the sex vesicle or the sex bivalent and the trivalent. In all three cases, the frequency of cells showing such contacts exceeded 60%. Observation at the EM level in the 14q21q heterozygotes studied by Rosenmann et al. [42] revealed that associations were almost always made between the differential (unpaired) segments of the sex chromosomes and the unpaired ray of the trivalent constituted by the two tiny short arms of the normal No. 14 and No. 21 chromosomes. The significance of these observations will be discussed in a subsequent section.

RECIPROCAL AUTOSOMAL TRANSLOCATIONS

One of the earliest reports of a meiotic study in a reciprocal autosomal translocation in man was that of Lindsten et al. [43]. This concerned a paternal rearrangement involving two autosomes in the C-group ascertained through a severely affected girl with an unbalanced form of the translocation. A testicular biopsy was obtained from the girl's father, and squash preparations were made. Cells in diakinesis/metaphase I showed 21 bivalents, a chain (usually) quadrivalent, and the sex bivalent was normal, providing good evidence that the rearrangement was purely autosomal in nature.

Over the 20 yr that have elapsed since this investigation was carried out numerous other human reciprocal autosomal translocations have been studied at meiosis, many having been ascertained through infertility. From such studies it is known that reciprocal translocations between two autosomes also produce severe effects on spermatogenesis in some instances. Lyon and Meredith [44] first showed in the mouse that maturation impairment occurred in the germ cells of males carrying certain autosomal translocations. They also noted that such translocations tended to be ones giving a high proportion of chain configurations at meiotic metaphase I in contrast to the usual preponderance of rings found in male-fertile translocation heterozygotes. Sterile heterozygotes carrying this type of translocation in the mouse display long and short marker chromosomes in their somatic karyotypes and are likely to generate tertiary trisomic and monosomic zygotes by 3:1 disjunction at meiosis. In man, tertiary trisomic progeny carry 47 chromosomes, and in one family studied in our own laboratory, a tertiary trisomic proband who was severly handicapped both mentally and physically was found in a family where a reciprocal 9;22 translocation was segregating [45]. The proband had one fertile brother who was chromosomally normal and three sterile brothers who were all carrying the translocation in balanced form. They were all childless after many years of attempted parenthood. Cytogenetic investigation showed that the translocation was inherited from the mother, who, it seemed, was unaffected in terms of fertility, by the presence of the rearrangement. The translocation is shown in Figure 7.

Seminal analysis performed on two of the sterile brothers showed them both to be azoospermic, and a testicular biopsy from one revealed the severe effect produced on the seminiferous epithelium by the translocation (Fig. 8). Spermatogenesis was normal up to metaphase I, but secondary spermatocytes

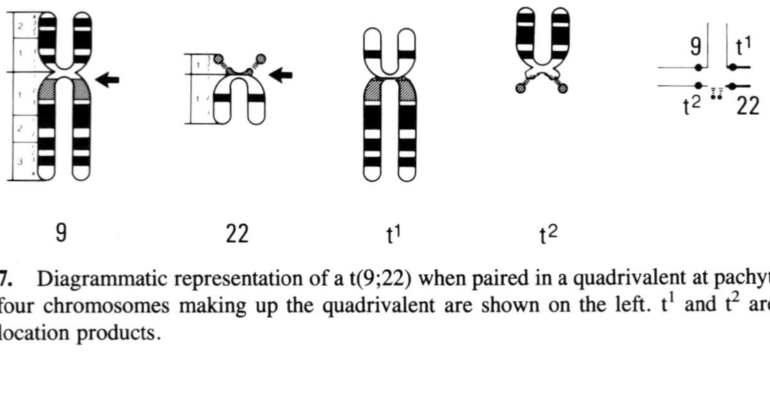

Fig. 7. Diagrammatic representation of a t(9;22) when paired in a quadrivalent at pachytene. The four chromosomes making up the quadrivalent are shown on the left. t^1 and t^2 are the translocation products.

8

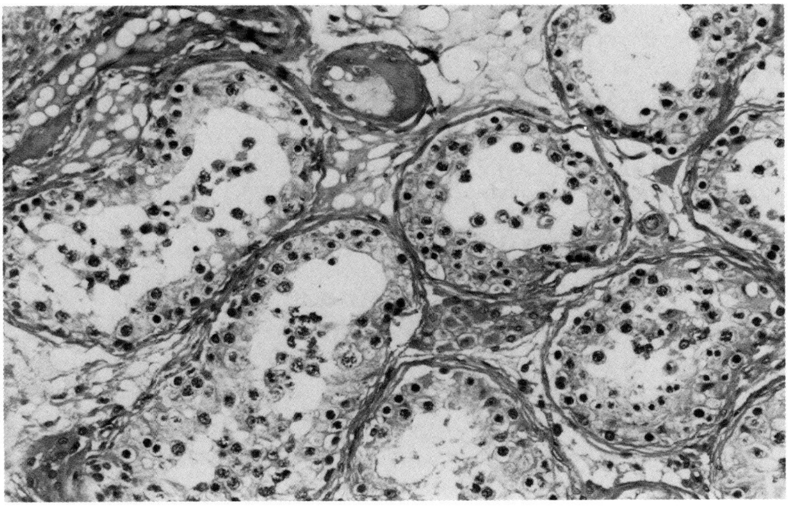

9

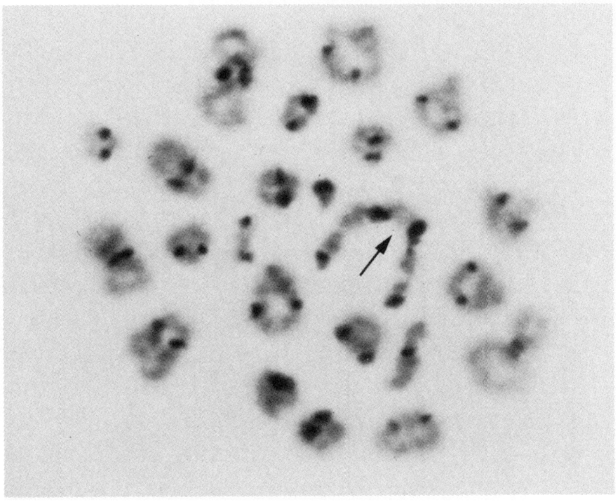

Fig. 8. Testicular histology of the t(9;22) translocation heterozygote illustrated in Figure 7. Complete spermatogenic arrest is seen and large empty spaces characterise the seminiferous tubules. Haematoxylin and eosin stain. ×200.

Fig. 9. Air-dried preparation showing the chain quadrivalent (arrow) formed in a t(9;22) translocation heterozygote. C-banding reveals the large No. 9 centromeres and the smaller No. 22 centromeres. (Reproduced from Chandley et al. [45], with permission of the publisher, S. Karger AG, Basel.) Giemsa stain. ×1000.

and spermatids were deficient, and spermatozoa were absent from the tubules. Meiotic studies showed a chain quadrivalent in every cell at diakinesis (Fig. 9). Other human translocations associated with male sterility where a preponderance of chain configurations have been reported at meiosis include those of San Roman et al. [46] and Laurent et al. [47].

A review of the available data for human autosomal reciprocal translocations [48] shows that a positive correlation exists between severity of effect on sperm count and percentage of chain configurations at meiosis. The same correlation has been found for the mouse [30].

Each (rcp) translocation giving chains at meiosis is characterized by the fact that one breakpoint at least is located very close to the end of a chromosome arm, so that a very short segment is produced at prophase when pairing occurs to form a quadrivalent (Fig. 7). Frequently, chiasma formation fails in this segment, resulting in a chain quadrivalent. If chiasma formation fails in two arms of the pachytene quadrivalent, a trivalent and univalent will result, and 3:1 disjunction may then occur. It is noteworthy that an acrocentric chromosome showing a breakpoint very close to the centromere is frequently involved in chain-forming human translocations. One tiny arm made up mostly of short arm material, satellite, and satellite stalk in which chiasma formation is generally precluded, arises in such rearrangements.

Autosomal translocations that might be expected to give chain quadrivalents or trivalent plus univalent configurations at meiosis are often accordingly ascertained by probands with unbalanced karyotypes arising by 3:1 disjunction or by infertility in a carrier male [45,49,50]. The reciprocal translocation (11;22)(q23;q11), which has been observed in an unusually large number of unrelated families [50] falls into this category. Detection of this translocation has usually been by birth of a congenitally malformed child with an unbalanced karyotype following 3:1 disjunction at meiosis. Moreover, in 44 reported cases, only 1 had inherited the translocation from the father, all the rest having received it maternally. This very striking difference in transmission can be attributed to differing conditions of meiosis in the oocyte and the spermatocyte. When ascertainment is through a balanced male proband, it is most often due to infertility and/or hypogonadism [50]. Meiotic investigations appear never to have been performed on a such a male t(11;22) balanced heterozygote, but Martin [7] has analysed the sperm chromosome complements from one individual. Evidence of all possible 2:2 segregations as well as 3:1 disjunction was obtained, these segregations occurring in approximately equal frequencies. The results indicate that the (11;22) translocation does not specifically cause 3:1 disjunction in males but that this type of segregation is more likely to result in a viable unbalanced pregnancy.

Generally speaking, the risks for spermatogenic impairment in both man and the mouse are much less for autosomal translocations making ring quadrivalents at meiosis. In this case, chiasma formation is achieved in all four arms of the quadrivalent at meiosis and effects on spermatogenesis, if they occur at all, are less severe. A few such translocations have nevertheless been ascertained among patients in attendance at infertility clinics owing to oligospermia [12,28,45], indicating that they too may be at some risk for impaired spermatogenesis. The large ring quadrivalent seen in all cells from a subfertile patient with a t(5;6) translocation, reported by Chandley et al. [45], is shown in Figure 10. In this case, the sperm count was 23×10^6, a figure only just above the minimal count of 20×10^6 per ml considered necessary for fertility.

With the advent of microspreading, a few studies have now been made at the prophase stage of meiosis in males carrying reciprocal autosomal translocations. The first such report was that of Saadallah and Hultén [51]. The proband was a man who presented to an infertility clinic following two first-trimester miscarriages in his wife. The sperm count was normal, although testis size was reported to be reduced. Karyotyping was initially performed on peripheral blood lymphocytes, and this revealed a balanced t(4;9) translocation. When meiotic investigations were performed, however, a complex rearrangement involving six chromosomes was apparent from the presence of a hexavalent, seen both at pachytene in spreads and at metaphase I. The translocation proved to be a three-way rearrangement involving chromosomes 2, 4, and 9. The remarkable hexavalent seen at pachytene (Fig. 11) is

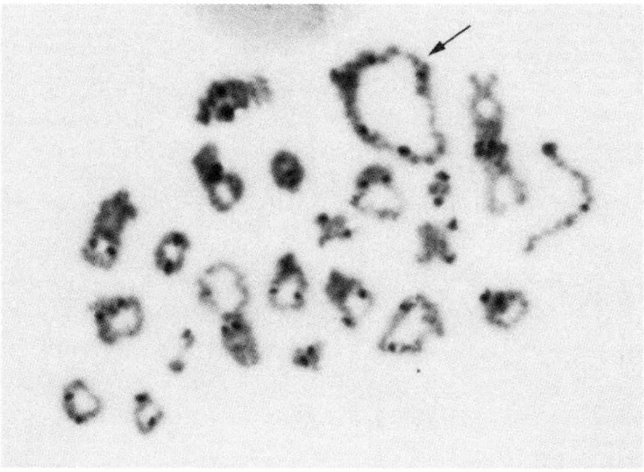

Fig. 10. Ring quadrivalent (arrow) in a t(5;6) translocation heterozygote. C-banded preparation. (Reproduced from Chandley et al. [45], with permission of the publisher, S. Karger AG, Basel.) Giemsa stain. ×950.

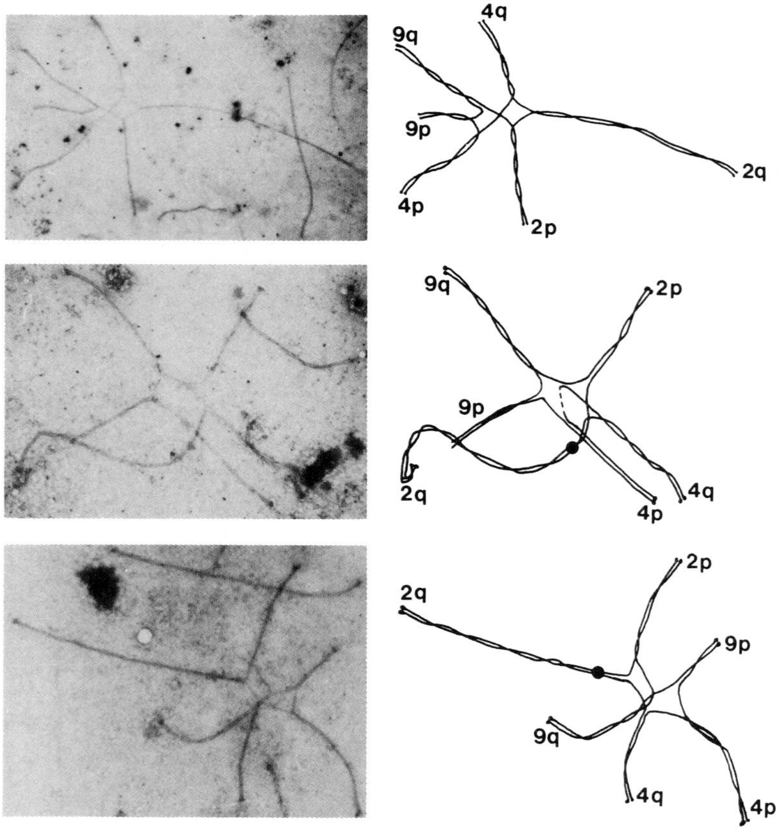

Fig. 11. Hexavalents with diagrammatic explanations in a t(2;4;9) translocation at pachytene in spread preparations. (Reproduced from Saadallah and Hultén [51] with permission of the publisher.) Stain phosphotungstic acid (PTA). Magnification not given in original paper but probably similar to Fig. 12.

illustrated. Evidence of pairing disruption around the breakpoints can be seen, but judging by the normal sperm count, this asynapsis had not resulted in impairment of spermtogenesis. Three-way translocations have, however, occasionally been ascertained through infertility and lowering of the sperm count [10,52].

In another case analysed by spreading, Chandley et al. [53] obtained clear evidence of mainly ring quadrivalent formation in an oligospermic t(9;20) heterozygote ascertained through infertility. Quadrivalents seen at pachytene in spread preparations are shown in Figure 12a, b. Pairing disruption around

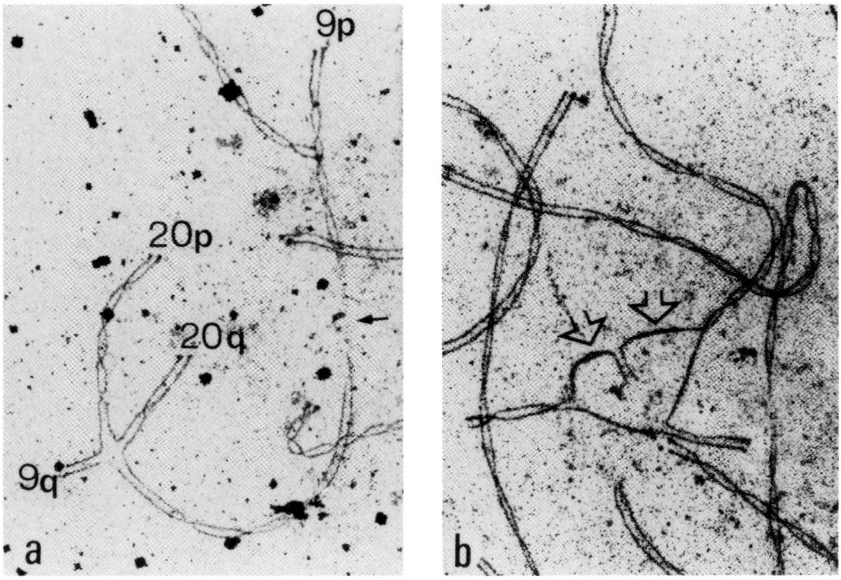

Fig. 12. Quadrivalents seen at pachytene in spread preparations from a t(9;20) heterozygote. In **a** pairing is complete. In **b** pairing disturbance is seen in the vicinity of the breakpoints (arrows). (Reproduced from Chandley et al. [53], with permission of the publisher, S. Karger, Basel.) Silver nitrate stain. Magnification ×7600.

the breakpoints was again seen in a high proportion of cells (Fig. 12b), and about 20% of cells showed association between a short unpaired arm of the quadrivalent and the XY bivalent. Additionally, there appeared to be a high degree of germ-cell degeneration among very early prophase cells before the pachytene stage was reached. The proband had inherited the translocation from his mother, whose own fertility was not impaired. (See Maturation Impairment in Females.)

HYPOTHESES TO EXPLAIN GERM-CELL MATURATION IMPAIRMENT IN MALE TRANSLOCATION HETEROZYGOTES

Several features of germ-cell development in translocation heterozygotes, and in particular those causing germ-cell maturation impairment in males, have been highlighted in the previous sections. The observations have produced a number of ideas concerning possible causes of impairment and several hypotheses have been proposed.

With regard to the total sterility and complete spermatogenic arrest encountered in X-autosome translocation carriers, Lifschytz and Lindsley [54] originally put forward the suggestion that rearrangement in which autosomal material was placed in immediate contact with sex chromosomal material might cause spermatogenic disturbance. They argued that the transcriptionally inactive state of the X chromosome during meiotic prophase in males of species such as *Drosophila*, mouse, and man, could constitute a control mechanism for normal spermatogenesis. Juxtapositioning transcriptionally active autosomal chromatin against sex chromatin might cause a reactivation of sex chromsomal material. Since the Y chromosome is also transcriptionally inactive during meiotic prophase in mammals, the same hypothesis could be applied in the case of mammalian Y-autosomal reciprocal translocation, although as mentioned earlier (see Y-Autosome Translocation), effects on spermatogenesis in such cases are generally not quite so severe as for those involving an autosome and the X chromosome.

As an extension to the Lifschytz and Lindsley [54] hypothesis, Forejt [55,56] has suggested that the breakdown in spermatogenesis recorded for many male-sterile autosomal translocations of the mouse (and for certain other autosomal anomalies, both structural and numerical), might also arise through interference with sex chromosome material. Forejt [55] and Forejt and Gregorova [57] observed nonrandom contacts between the quadrivalent and X-centromeric region of the XY bivalent at meiotic metaphase I in several male-sterile translocation stocks. The frequency of such contacts was subsequently shown to be much higher at the pachytene stage when microspread preparations were examined [58]. At metaphase I, heterochromatinization of part of the translocated autosome, apparently in phase with the allocyclic X chromosome, was also observed [55]. Contacts between quadrivalents and the XY bivalent were not, however, seen to any great extent in male-fertile translocation stocks.

For human autosomal translocations few data exist on the behavior of the quadrivalent and sex bivalent at the pachytene or metaphase I stage in male-sterile versus male-fertile cases, but both Luciani et al. [41] and Rosenmann et al. [42] reported frequencies of contact at pachytene in excess of 60% for their t(13q14q) and t(14q21q) male-sterile Robertsonian translocations, respectively. For the t(9;20) reciprocal translocation studied in our own laboratory [53], only 20% of such contacts were recorded at the pachytene stage, these being found usually when pairing failure had occurred in the shortest arm of the pachytene cross. Associations also tended to occur when there was pairing failure in the short arms of the two non-translocated chromosomes of the 14q21q trivalents studied by Rosenmann et al. [42]. For the three-way translocation studied by Saadallah and Hultén [51], no associations between the XY bivalent and hexavalent were seen. Here, however the sperm count was normal.

Whether such nonrandom contacts between a translocation multivalent and the XY pair constitutes a *cause* of germ-cell breakdown is not yet clear. It might be a secondary consequence of a primary pairing failure. Some authors have suggested that a mechanism might operate at meiosis to remove gametogenic cells in which there has been pairing failure [59]. Certainly in all the translocations studied by Forejt [56], incomplete synapsis was an additional feature of those showing high frequencies of contact and male sterility, while being rare among male-fertile heterozygotes. Based on an early hypothesis of Miklos [60], Burgoyne and Baker [59] have argued that meiotic pairing sites that need to be saturated by homologous pairing could, if unpaired, become activated, setting in motion a process that eventually leads to the destruction of the nucleus. The speed and effectiveness of the process would be related to the number of unpaired sites. In such a situation, there would be an inappropriate expression of genes in the asynapsed segments, which could eventually lead to spermatogenic death.

Meiotic pairing failures in general in spermatocytes of man and the mouse are known to be associated with breakdown in germ-cell development [48]. Simple failure of association between the X and Y chromosomes at metaphase I in these species leads to destruction of the nucleus, as can pairing failure in individual autosomes. It is interesting to note, however, that although partial asynapsis at pachytene characterised the translocation multivalents in the cases reported by Saadallah and Hultén [51] and Chandley et al. [53], only in the latter case was there a disturbance of spermatogenesis and a depression of the sperm count. XY-hexavalent associations were not recorded by Saadallah and Hultén [51], and it was noteworthy that pairing in all terminal arms of the rearrangement was complete. It could thus be that XY-multivalent associations are only formed when distal telomeric regions remain asynapsed. Chandley et al. [53] pointed out the fact that much early meiotic degeneration characterised their 9;20 translocation, seemingly unrelated to pachytene pairing failure or XY-quadrivalent association. About one-half of the prophase spermatocytes analysed showed signs of degeneration even before they had reached the pachytene stage. In this case, it looked as though the presence in the karyotype of the translocation had exerted its effect on germ-cell development at a much earlier time in the spermatogenic sequence.

It seems obvious that many facts still remain to be disclosed about human translocations and their effects on germ-cell development. Further research will undoubtedly help to clarify the situation, but when it is realised that in the mouse, where the most extensive studies have been made, there is much variability in the severity of the spermatogenic impairment, both within particular translocation stocks, and between them, and that effects are dependent on genetic background, it will be realised that the issue is very complex indeed. Moreover, it is now coming to be realised in the mouse that the female too may be prone to oogenic impairment in some instances where a chromosome anomaly exists in the genome. In oocytes, the XX bivalent is fully active and in synchrony with the autosomes [61], making the hypotheses of Lifschytz and Lindsley [54], and Forejt [56] not applicable to females. Some alternative explanation must therefore be sought.

MATURATION IMPAIRMENT IN FEMALES

There has been a tendency in the past to assume that oogenesis is unaffected by the presence of reciprocal translocations and other anomalies that sterilize the male by germ-cell maturation arrest [10,37,56]. The evidence from both man and the mouse shows that fertility is retained in female carriers of male-sterlizing anomalies, and illustrations of specific cases in which sterile male heterozygotes have inherited their translocations from their fertile mother are given in the previous sections and in the literature. Burgoyne and Baker [59] argue, however, that "fertility" is an inadequate measure of germ-cell maturation impairment in the female. They point out the marked reductions in oocyte number that have been found in the mouse for the male-sterilizing reciprocal autosomal translocation T(11;19) 42H [62] with associated reductions in ovary size within the first week of birth for female heterozygotes [63]. It has also been noted in the mouse that reproductive life span is curtailed in females heterozygous for the X-autosome translocation T(X;16)16H owing, it is assumed, to a deficiency in the pool size of oocytes available for growth and ovulation [59]. Moreover, a number of other male-sterilizing anomalies (an insertion and a tertiary trisomy) show reduced oocyte numbers, reduced ovary size, and a shorter reproductive span compared with controls [64,65]. The simple assumption that oogenesis is not affected by chromosome anomalies that sterilize the male is therefore severely challenged by these observations in the mouse. For man, data on oocyte numbers, ovary size, and reproductive life span in females heterozygous for male-sterilizing translocations and other anomalies are hard to find in the literature, but it is probable that humans too would show these effects. Shortening of reproductive span owing to depletion of oocytes might be detectable as an early menopause in some human female translocation carriers; and indeed, among fertile carriers of X-autosome rcp with breakpoints in the critical region of X_q (see X-Autosome Translocations), there have been reports of early menopause [27].

Burgoyne et al. [62] have invoked the pairing-failure hypothesis of Miklos [60] to explain the germ-cell arrests seen in females. They believe that additional damage will occur in males owing to random contact's interfering with the normal silencing of the X chromosome. It may be that even these combined explanations do not provide the full story. Meiotic spreading studies carried out on human fetal ovaries with the chromosome anomalies 45,X Speed [66] and iso-18q [26] show severe germ-cell arrests in oocytes long before they reach the zygotene-pachytene stages when synapsis occurs. In both of these anomalous situations, little meiotic progression occurs beyond the preleptotene stage. Other studies in human male fetuses show significant (testicular) premeiotic germ-cell depletions for individuals with XXY, XYY, and trisomies 13 and 18 compared with controls [67]. Also, the t(9;20) adult heterozygote cited in Reciprocal Autosomal Translocations [53] showed severe germ-cell degeneration in the very early prophase stages in a large proportion of cells. It would therefore seem that much germ-cell loss can occur prior to the pairing stage in meiosis of males or females; and such loss is not therefore explicable on the Forejt [56], Lifschytz and Lindsley [54], or Burgoyne and Baker [59] hypotheses.

The situation regarding chromosome anomaly and gametogenic impairment is thus seen to be highly complex. The conclusion is reached that individuals carrying a wide variety of chromosome anomalies can be at risk for reduced fertility, germ-cell depletion, or both. Further studies into gonadal and meiotic chromosome material in humans should contribute to the fuller understanding and elucidation of the problem.

ACKNOWLEDGMENTS

The author is extremely grateful to all of the staff of the MRC Clinical and Population Cytogenetics Unit who contributed in any way to the preparation of material cited in this review. Ann Kenmure typed the manuscript. Dr. Maj Hulten and Dr. B Noel are thanked for the use of photographs and unpublished data. Sandy Bruce prepared the illustrations.

REFERENCES

1. Jacobs PA: Population surveillance: A cytogenetic approach. In Morton NE, Chung CS (eds): "Genetic Epidemiology." New York: Academic Press, 1978, p 463.
2. Jacobs PA, Frackiewicz A, Law P, Hilditch J, Morton NE: The effect of structural aberrations of the chromosomes on reproductive fitness in man. II. Results. Clin Genet 8:169, 1975.

3. Ford CE, Clegg HM: Reciprocal translocations. Br Med Bull 25:110, 1969.
4. Rudak E, Jacobs PA, Yanagimachi R: Direct analysis of the chromosome constitution of human spermatozoa. Nature 274:911, 1978.
5. Balkan W, Martin RH: Segregation of chromosomes into the spermatozoa of a man heterozygous for a 14;21 Robertsonian translocation. Am J Med Genet 16:169, 1983.
6. Balkan W, Martin RH: Chromosome segregation into the spermatozoa of two men heterozygous for different reciprocal translocations. Hum Genet 63:345, 1983.
7. Martin RH: Analysis of human sperm chromosome complements from a male heterozygous for a reciprocal translocation t(11;22)(q23;q11). Clin Genet 25:357, 1984.
8. Brandriff B, Gordon L, Ashworth LK, Littman V, Watchmaker G, Carrano AV: Cytogenetics of human sperm: Meiotic segregation in two translocation carriers. Am J Hum Genet 38:197, 1986.
9. Chandley AC: The chromosomal basis of human infertility. Br Med Bull 35:181, 1979.
10. Chandley AC, Edmond P, Christie S, Gowans L, Fletcher J, Frackiewicz A, Newton M: Cytogenetics and infertility in man. I. Karyotype and seminal analysis. Ann Hum Genet 39:231, 1975.
11. Kjessler B: Facteurs genetique dans la subfertilite male humaine. In Durchon M (ed): "Fecondite et Sterilite du male." Paris: Masson et Cie, 1972, p 205.
12. Tiepolo L, Zuffardi O, Fraccaro M, Giarola A: Chromosome abnormalities and male infertility. In Frajese G, Hafez E, Conti C, Fabbrini A (eds): "Oligozoospermia: Recent Progress in Andrology." New York: Raven Press, 1981, p 233.
13. Ford CE, Hamerton JL: The chromosomes of man. Nature 178:1020, 1956.
14. Evans EP, Breckon G, Ford CE: An air-drying method for meiotic preparations from mammalian testes. Cytogenetics 3:289, 1964.
15. Chandley AC, Fletcher J: Centromere staining at meiosis in man. Humangenetik 18:247, 1973.
16. Schmid M: Demonstration of Y/autosomal translocations using Distamycin A. Hum Genet 53:107, 1979.
17. Hungerford DA: Chromosome structure and function in man. I. Pachytene mapping in the male, improved methods and general discussion of initial results. Cytogenetics 10:23, 1971.
18. Luciani JM, Morazzani MR, Stahl A: Identification of pachytene bivalents in human male meiosis using G-banding technique. Chromosoma 52:275, 1975.
19. Page BM: Identification of chromosome 9 in human male meiosis. Cytogenet Cell Genet 12:254, 1973.
20. Luciani JM, Devictor M, Morazzani M-R, Stahl A: Pachytene mapping of the C9 and acrocentric bivalents in the human oocyte. Hum Genet 36:197, 1977.
21. Moses MJ: Microspreading and the synaptonemal complex in cytogenetic studies. In de la Chapelle A, Sorsa M (eds): "Chromosomes Today, Vol 6." Amsterdam: Elsevier/North Holland Biomedical Press, 1977, p 71.
22. Moses MJ: New cytogenetic studies on mammalian meosis. In Serio M, Martini L (eds): "Animal Models in Human Reproduction." New York: Raven Press, 1980, p 169.
23. Fletcher JM: Light microscope analysis of meiotic prophase chromosomes by silver staining. Chromosoma 72:241, 1979.
24. Speed RM: Meiotic configurations in female trisomy 21 foetuses. Hum Genet 66:176, 1984.
25. Speed RM: The prophase stages in human foetal oocytes studied by light and electron microscopy. Hum Genet 69:69, 1985.

26. Speed RM: Prophase pairing in a mosiac 18p$^-$; iso 18q human female foetus studied by surface spreading. Hum Genet 72:256, 1986.
27. Madan K, Hompes PGA, Schoemaker J, Ford CE: X-autosome translocation with a breakpoint in Xq22 in a fertile woman and her 47,XXX infertile daughter. Hum Genet 59:290, 1981.
28. Faed MJW, Lamont MA, Baxby K: Cytogenetic and histological studies of testicular biopsies from subfertile men with chromosome anomaly. J Med Genet 19:49, 1982.
29. Searle AG: The genetics of sterility in the mouse. In Crosignani PG, Rubin BL, Fraccaro M (eds): "Genetic Control of Gamete Production and Function." New York: Academic Press, 1982, p 93.
30. Searle AG, Beechey CV, Evans EP: Meiotic effects in chromosomally derived male sterility in mice. Ann Biol Anim Biochem Biophys 18:391, 1978.
31. Sarto GE, Therman E, Patau K: X inactivation in man: A woman with t(Xq−;12q+). Am J Hum Genet 25:262, 1973.
32. Summitt RL, Tipton RE, Wilroy RS, Martens PR, Phelan JP: X-autosome translocations: A review. Birth Defects XIV(6C):219.
33. Smith A, Fraser IS, Elliott G: An infertile male with balanced Y;19 translocation. Review of Y; autosome translocations. Ann Genet 22:189, 1979.
34. Laurie DA, Palmer RW, Hulten MA: Studies on chiasma frequency and distribution in two fertile men carrying reciprocal translocations; one with a t(9;10) karyotype and one with a t(Y;10) karyotype. Hum Genet 68:235, 1984.
35. Laurent C, Chandley AC, Dutrillaux B, Speed RM: The use of surface spreading in the pachytene analysis of a human t(Y;17) reciprocal translocation. Cytogenet Cell Genet 33:312, 1982.
36. Chandley AC, Hargreave TB, Fletcher JM: Translocation 21q22q in an infertile human male. J Med Genet 19:366, 1982.
37. Forejt J: Spermatogenic failure of translocation heterozygotes affected by H-2-linked gene in mouse. Nature 260:143, 1976.
38. Chandley AC: Human meiotic studies. In Emery AEH(ed): "Modern Trends in Human Genetics." 2nd Edition. London, Boston: Butterworths, 1975, p 21.
39. Skakkebaek N, Hulten M, Philip J: Quantification of human seminiferous epithelium. IV. Histological studies in 17 men with numerical and structural autosomal aberrations. Acta Pathol Microbiol Scand Sect A 81:113, 1973.
40. Paris Conference (1971): Standardization in human cytogenetics. Birth Defects 8(7): 1972, p 38.
41. Luciani JM, Guichaoua MR, Mattei A, Morazzani MR: Pachytene analysis of a man with a 13q;14q translocation and infertility. Behaviour of the trivalent and non-random association with the sex vesicle. Cytogenet Cell Genet 38:14, 1984.
42. Rosenmann A, Wahrman J, Richler C, Voss R, Persitz A, Goldman B: Meiotic association between the XY chromosomes and unpaired autosomal elements as a cause of human male sterility. Cytogenet Cell Genet 39:19, 1985.
43. Lindsten J, Fraccaro M, Klinger HP, Zetterquist P: Meiotic and mitotic studies of a familial reciprocal translocation between two autosomes of Group 6–12. Cytogenetics 4:45, 1965.
44. Lyon MF, Meredith R: Autosomal translocations causing male sterility and viable aneuploidy in the mouse. Cytogenetics 5:335, 1966.
45. Chandley AC, Seuanez H, Fletcher J: Meiotic behaviour of five human reciprocal translocations. Cytogenet Cell Genet 17:98, 1976.

46. San Roman C, Sordo MT, Garcia-Sagredo JM: Meiosis in two human reciprocal translocations. J Med Genet 16:56, 1979.
47. Laurent C, Biemont M-Cl, Cognat M, Dutrillaux B: Studies of the meiotic behaviour of a translocation t(10;13) (q25;q11) in an oligospermic man. Hum Genet 39:123, 1977.
48. Chandley AC: Infertility and chromosome abnormality. In Clarke J(eds): "Oxford Reviews of Reproductive Biology." Vol 6. Oxford University Press, Oxford 1984, pl.
49. Lindenbaum RH, Bobrow M: Reciprocal translocations in man. 3:1 meiotic disjunction resulting in 47− or 45− chromosome offspring. J Med Genet 12:29, 1975.
50. Fraccaro M, Lindsten J, Ford CE, Iselius L, et al.: The 11q;22q translocation: A European collaborative analysis of 43 cases. Hum Genet 56:21, 1980.
51. Saadallah N, Hulteñ M: A complex three breakpoint translocation involving chromosomes 2, 4 and 9 identified by meiotic investigations of a human male ascertained for infertility. Hum Genet 71:312, 1985.
52. Joseph A, Thomas IM: A complex rearrangement involving three autosomes in a phenotypically normal male presenting with sterility. J Med Genet 19:375, 1982.
53. Chandley AC, Speed RM, McBeath S, Hargreave TB: A human 9;20 reciprocal translocation associated with male infertility analyzed at prophase and metaphase I of meiosis. Cytogenet Cell Genet 41:145, 1986.
54. Lifschytz E, Lindsley DL: The role of X-chromosome inactivation during spermatogenesis. Proc Natl Acad Sci USA 69:182, 1972.
55. Forejt J: Non-random association between specific autosome and the X chromosome in meiosis of the male mouse: Possible consequence of homologous centromeres separation. Cytogenet Cell Genet 13:369, 1974.
56. Forejt J: X-Y involvement in male sterility caused by autosome translocations—a hypothesis. In Crosignani PG, Rubin BL, Fraccaro M (eds): "Genetic Control of Gamete Production and Function." New York: Academic Press, 1982, p 135.
57. Forejt J, Gregorova S: Meiotic studies of translocations causing male sterility in the mouse. I. Autosomal reciprocal translocations. Cytogenet Cell Genet 19:159, 1977.
58. Forejt J, Gregorova S, Goetz P: XY pair associates with the synaptonemal complex of autosomal male-sterile translocations in pachytene spermatocytes of the mouse (Mus musculus). Chromosoma 82:41, 1981.
59. Burgoyne PS, Baker T: Meiotic pairing and gametogenic failure. In Evans CW, Dickinson HG (eds): "Controlling Events in Meiosis." Cambridge: Company of Biologists, 1984, p 349.
60. Miklos GLG: Sex chromosome pairing and male fertility. Cytogenet Cell Genet 13:558, 1974.
61. Epstein CJ: Mammalian oocytes: X chromosome activity. Science 163:1078, 1969.
62. Burgoyne PS, Mahadevaiah S, Mittwoch U: A reciprocal autosomal translocation which causes male sterility in the mouse also impairs oogenesis. J Reprod Fertil 75:647, 1985.
63. Mittwoch U, Mahadevaiah S, Olive MB: Retardation of ovarian growth in male-sterile mice carrying an autosomal translocation. J Med Genet 18:414, 1981.
64. Mittwoch U, Mahadevaiah S, Setterfield LA: Chromosome anomalies that cause male sterility in the mouse also reduce ovary size. Genet Res (Camb) 44:219, 1984.
65. Setterfield LA, Mittwoch U: Reduced oocyte numbers in tertiary trisomic mice with male sterility. Cytogenet Cell Genet 41:177, 1986.
66. Speed RM: Oocyte development in XO foetuses of man and mouse: The possible role of heterologous X-chromosome pairing in germ cell survival. Chromosoma 94:115, 1986.
67. Coerdt W, Rehder H, Gausmann I, Johannisson R, Gropp A: Quantitative histology of human fetal testes in chromosomal disease. Pediatr Pathol 3:245, 1985.

The Cytogenetics of Mammalian Autosomal Rearrangements, pages 383–396

Chapter 9
Segregation Patterns and Fertility of Domestic Mammals With Chromosome Translocations

Susan E. Long

INTRODUCTION

The term *translocation* was first used to denote the transference of a chromosome segment from its normal position to a position in a different chromosome [1]. In the present review the term will be expanded to include centric fusion (Robertsonian) translocations. In domestic mammals, although there have been reports of a tandem fusion [2], inversions [3–5], and X-autosome translocations [6,7], all in cattle, very little is known about their meiotic behaviour and their effects on fertility. Most have occurred as isolated cases and although each has been associated with some reduction in fertility in one form or another, this has not been investigated in any depth. Only with centric fusion translocations in sheep and cattle and reciprocal translocations in pigs is there sufficient information available to assess segregation patterns and their effects on fertility. I propose, therefore, to confine my discussion to these two types of translocations.

When one considers the variety of karyotypes in domestic mammals (Table I) it is not surprising that different translocations appear more commonly in some species than others. Their effects on fertility are expressed differently in the different species because of the differing breeding patterns. In the pig, a polytoccus species, litter size is the major yardstick of fertility. In sheep one considers the number of liveborn lambs for every hundred ewes mated (lambing percentage). Bulls are usually assessed by numbers of cows failing to return to oestrus after service (nonreturn rate), irrespective of whether the service is by artificial insemination or coitus. In the cow, conception to first service or the number of services per conception are quoted. It should be

Department of Animal Husbandry, University of Bristol, Langford House, Langford, Bristol BS18 7DU, England

TABLE I. Normal chromosome complement of domestic animals

Species	Diploid No.	X	Y	Autosomes
Cattle				
Bos taurus	60	Large sub-metacentric	Small sub-metacentric	Acrocentric
Bos indicus	60	Large sub-metacentric	Small acrocentric	Acrocentric
Sheep				
Ovis aries	54	Large acrocentric	Small metacentric	3 pairs of metacentrics; 23 pairs of acrocentrics
Goat *Capra hircus*	60	Large acrocentric	Small metacentric	Acrocentric
Pig *Sus scrofa*	38	Large metacentric	Small metacentric	5 pairs of sub-metacentrics; 2 pairs of sub-telocentrics; 5 pairs of meta-centrics; 6 pairs of acro-centrics
Horse *Equus caballus*	64	Large sub-metacentric	Acrocentric	13 pairs of meta- or submetacentrics; 18 pairs of acrocentrics
Donkey *Equus asinus*	62	Large sub-metacentric	Small sub-metacentric	24 pairs of meta- or submetacentrics; 6 pairs of acro-centrics

remembered, therefore, that because of these different methods of assessment, cross-species comparisons of fertility are not exact.

CENTRIC FUSION (ROBERTSONIAN) TRANSLOCATIONS

Centric fusions are by far the most widely distributed type of translocation in domestic mammals. They have been found in cattle, sheep, goats, dogs, and pigs (Table II)—all species with large numbers of acrocentric chromosomes in their karyotype. They are particularly common in cattle where the rob(1:29) is very widely distributed in many breeds across the world in both *Bos taurus* and *B. indicus* species (Table III).

Heterozygosity for a centric fusion translocation can be predicted to affect fertility, since the translocation and homologous acrocentric chromosomes

TABLE II. Centric fusion (Robertsonian) translocations in domestic mammals

Species	Translocation	Breed	Reference
Cattle	1:29[a]		
	2:4	Friesian	[35]
	14:20	Simmental	[13]
		(Holstein-Friesian)?	[36]
	6:16	Dexter	[37]
	27:29	Guernsey	[38]
	14:24	Podolian	[39]
	25:27	Alpine Grey	[40]
	3:4	Limousin	[41]
	7:21	Japanese Black	[42]
	5:18	Simmental	[43]
	1:25	German Piebald	[44]
	?8:?9	Brown Swiss	[45]
	7–11:20–25	Blonde d'Aquitaine cross Limousin	[46]
Sheep	5:26	New Zealand Romney	
		British Romney Marsh	
	8:11	New Zealand Romney	[47]
	7:25	New Zealand Romney	
		Drysdale	
		Perendale	
Goat	6:17	Saanen	[48]
	?	Toggenburg cross	[49]
Pig	13:17	Landrace cross	[50]
		Duroc Jersey	
Dog	?15:?35	Setter cross Retriever	[51]
	13:23	Golden Retriever	[52]
	?2:?10	Terrier	[53]
	?1:19	Miniature Poodle	[54]

[a]See Table III.

form a trivalent during meiosis (Fig. 1). Segregation of the trivalent can theoretically result in six different gamete products, four of which are unbalanced and would result in abnormal embryos. Unfortunately, data are lacking on fertility and segregation patterns for the majority of centric fusion translocations. Substantial information is only available for the three translocations in sheep and the rob(1:29) in cattle.

The most extensive and comprehensive studies have been by a group working in New Zealand on the centric fusion translocations in sheep [t_1 = rob(5:26); t_2 = rob(8:11); t_3 = rob(7:25)]. Experimental data are available on male and female meiosis, preimplantation embryos, and extensive breed-

TABLE III. Breed occurrence of the rob(1:29) translocation in cattle

Bauole	Kuri
Blonde d'Aquitaine	Limousin
British White	Marchigiana
Brown Mountain	Modicana
Brown Atlas	Montebeliard
Brown Swiss	Norwegian Red
Charolais	Ottonese
Chianina	Pisa
Czechoslovakian Red Pied	Podolian
De Lidia	Red Poll
Fleckvien	Romagnola
Gascons	Russian Black Pied
German Red Pied	Santa Gertrudis
Grauviel	Siamese
Guernsey	Simmental
Holstein Friesian	Swedish Red and White
Hungarian Grey	Vosgienne
Japanese Black	

Reproduced from Long [55].

ing trials for all three translocations. These have shown that whilst normal and translocation homozygous rams have few or no aneuploid secondary spermatocytes, translocation heterozygous rams do have raised levels of aneuploid metaphase II figures [8–11] (Table IV). In addition, although there is considerable individual ram variation [8], there are significant differences in the frequency of aneuploid secondary spermatocytes in rams, heterozygous for different translocations. In cattle, where meiotic studies have been made on three different translocations [12–14] a similar pattern exists. The heterozygous bulls have a higher aneuploid frequency compared with normal bulls, and there is a marked translocation effect, with the rob(25:27) having a very high non-disjunction rate (Table V). Data on female meiosis in sheep are too sparse to draw definite conclusions about non-disjunction [15], and no work has been done on the cow.

It is evident that heterozygosity for a centric fusion translocation is a major factor leading to meiotic non-disjunction in cattle and sheep. However, a most significant finding has been that triple heterozygous rams have a mean aneuploid frequency comparable to those levels in single heterozygous rams [11]. Their aneuploid frequencies are not the sum of the individual translocations as might have been expected. Thus the rate of non-disjunction cannot be entirely a simple function of the characteristics of the individual centric fusion translocation.

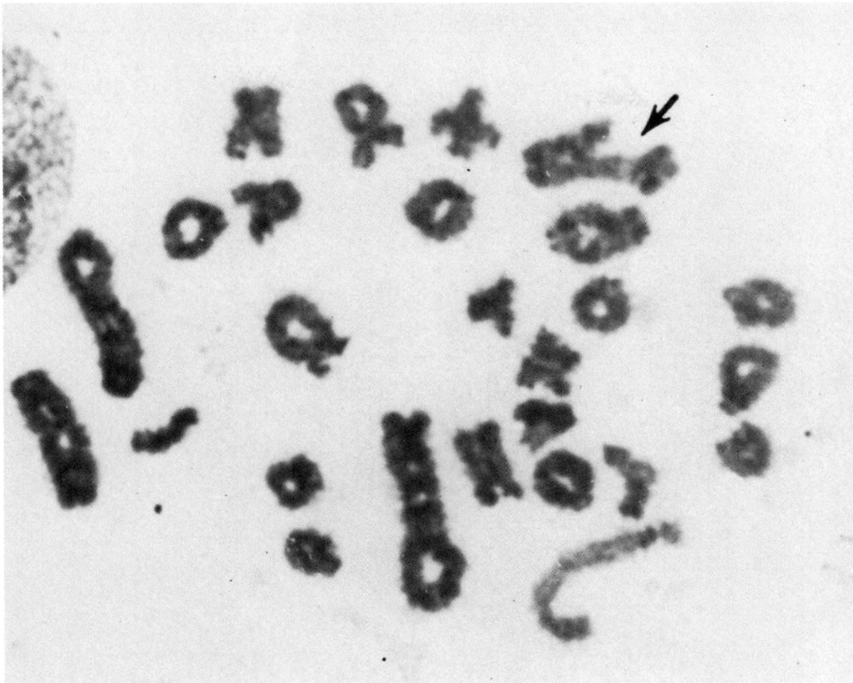

Fig. 1. Chromosome preparation from a germinal cell at diakinesis from a ram heterozygous for the rob(5:26). The arrow indicates the trivalent (conventional Giemsa stain).

When euploid secondary spermatocytes are considered, a significant excess of normal secondary spermatocytes exists in all single heterozygous rams [8,10] and bulls heterozygous for the rob(1:29) [12], although the later result was considered to be a technical artefact by the authors [12]. In the triple heterozygous rams, there was not only a significant excess of normal, but a deficit of 25,t_1, t_3 secondary spermatocytes [11].

Meiotic studies show, therefore, that centric fusions in heterozygous cattle and sheep behave the same until the secondary spermatocyte stage. However, thereafter differences arise. In sheep, extensive breeding data demonstrate that heterozygous animals show little or no impairment of reproductive competence [11,16–18]. This is true when the translocations are in the ewes, or rams, or both. Nearly 2,000 matings have been carried out; and these produced normal conception rates, normal lambing percentages, and average breeding performances. Most of the mating programmes produced the ex-

TABLE IV. Mean non-disjunction (ND) rates in normal rams and rams carrying a centric fusion (Robertsonian) translocation

Karyotype	No. rams	No. MII cells	ND%[a]	References
54XY	16	301	1.9	[8–11]
52XY,(5:26)^2 (homozygous)	2	116	0	[8,10]
52XY,(7:25)^2 (homozygous)	1	23	0	[8]
53XY,(5:26)	14	560	7.5	[8–10]
53XY,(8:11)	4	331	9	[8]
53XY,(7:25)	5	326	17.17	[8]
51XY,(5:26)(8:11)(7:25)	3	332	5.4	[11]

[a]Non-disjunction calculated as: $\frac{2 \times \text{no. of hyperhaploid MII cells}}{\text{total No. of cells counted}} \times 100.$

TABLE V. Non-disjunction (ND) rates in a normal bull and bulls heterozygous for a centric fusion (Robertsonian) translocation

Translocation	Metaphase II (ND%)	Reference
rob(1:29)	6.4	[12]
rob(14:20)	6.3	[13]
rob(25:27)	46.7	[14]
Normal bull	2.6	[13]

pected Mendelian progeny ratios of normal and balanced translocation carrying animals despite the excess of normal secondary spermatocytes. However, the small number of matings of triple heterozygous rams with normal ewes did deviate from the expected 1:1:2:2:1:1 ratio because of an excess of normal progeny and a deficit of 52, t_1,t_3 and 53 t_2 progeny [11]. No liveborn lambs have ever been found with an unbalanced complement, and a study of 102 preimplantation embryos from ewes mated to rams heterozygous for the rob(5:26) found only normal and balanced translocation carrying embryos [19]. The overwhelming evidence is that in sheep with centric fusion translocations, only balanced gametes take part in fertilisation and that aneuploid secondary spermatocytes fail to mature to aneuploid gametes.

The situation is quite different in cattle. Both monosomic and trisomic embryos have been identified from normal cows inseminated with semen from bulls heterozygous for the rob(1:29), demonstrating that unbalanced secondary spermatocytes do mature and take part in fertilisation [20–22]. Furthermore, breeding data show that both males and females heterozygous for the rob(1:29) have a reduced fertility. Cows inseminated with semen from heterozygous bulls show a lowered nonreturn rate equivalent to a reduction in fertility in the bulls of the order of 5–7% [23]. This is equivalent to the mean frequency of aneuploid secondary spermatocytes for the rob(1:29) and suggests that there is no preferential loss at all during gamete maturation. There is also a reduction in male fertility associated with the rob(25:27). Two bulls heterozygous for the rob(25:27) had poor libido and poor semen quality [14].

The fertility of heterozygous females is also reduced. For the rob(1:29) the fertility records of nearly 45,000 cows and heifers, daughters of either heterozygous or homozygous bulls, have been compared to those of nearly 619,000 cows or heifer daughters of normal bulls [24,25]. The 56-day nonreturn rate of daughters of translocation bulls was 3% less than daughters of normal bulls, and by 273 days it was 6% less. This is strong evidence that the heterozygous females were producing unbalanced embryos which were undergoing embryonic death, some of which must have been quite late in pregnancy. The lowered fertility in females was also reflected in higher culling rates in daughters of heterozygous bulls [26] and a raised incidence of heterozygous females in groups of repeat breeder heifers [27].

There is only minimal information on the fertility of heifers heterozygous for the rob(25:27) but this also indicates a reduction in fertility. Twenty-seven heterozygous heifers, daughters of one heterozygous bull, had a mean conception rate of 76% and required 1.34 services per conception compared to 80.9% and 1.18, respectively, in their 26 normal sibling heifers [28]. No liveborn calf has been identified with an unbalanced karyotype associated with any of the translocations.

It is evident, therefore, that there is a species difference in the behaviour of centric fusion translocations during gametogenesis and that the difference lies in the spermatid maturation stage and not in the meiotic process. It has been suggested that cell loss in spermatogenesis is species specific [29] and the data presented here provide evidence of prezygotic selection in the sheep.

Evidence for prezygotic selection is also found in various wild sheep (*Ovis vignei, O. ammon, O. aries,* and *O. nivicola*) where the diploid number is 58, 56, 54, and 52 with one, two, three, and four pairs of biarmed chromosomes, respectively [30]. Based upon wild sheep hybrid populations in northern Iran and experimental flocks of wild sheep at Utah State University, the progressive reduction in chromosome numbers in *Ovis* species is thought to be due to a prezygotic selection for gametes with the lower chromosome number [31,32].

That there is such a major difference in the effect of centric fusion translocations in two species as closely related as cattle and sheep is extremely interesting. It is only a pity that comparable information is not available for the other domestic species, especially the goat. What little information there is suggests that heterozygous goats do have a reduction in fertility [33,34] but non-disjunction rates have not been calculated.

RECIPROCAL TRANSLOCATIONS

In contrast to centric fusion translocations, which are most common in cattle, reciprocal translocations are found most commonly in the pig (Table VI, VII) (Fig. 2). In all species they have been discovered because of the individual's reduced fertility. At meiosis, heterozygosity for a reciprocal translocation will result in each homologous segment of the rearranged chromosomes pairing to produce a cross-configuration at pachytene and a quadrivalent at diakinesis. This may disjoin in one of three ways—alternate, adjacent-1, and adjacent-2 segregation [61]. If chiasmata form and crossovers occur, then segregation of the multivalent may result in up to ten different gametic products, assuming 2:2 segregation [62]. Since only two of these are chromosomally balanced, we can predict that heterozygosity for a reciprocal translocation could result in unbalanced embryos and embryonic death.

In the pig, heterozygosity for a reciprocal translocation has resulted in a reduction of fertility ranging from 5% to total sterility (Table VI). In cattle there has been a reduction of fertility of the order of 40–45%, and the bull with two reciprocal translocations was sterile, although he was allowed only a very few matings [59]. Sheep with the rcp($1p^-$;$20q^+$) do appear to have fewer twins and triplets, but the reduction in fertility has not been quantified.

Embryonic studies in the pig have revealed a variation in viability in embryos with unbalanced karyotypes. Five different abnormal karyotypes were found in embryos with the rcp($11q^+$;$15q^-$) and some survived until at least 80 days of gestation [63]. (Normal gestation in the pig is approximately 114 days.) In contrast, embryos with seven different unbalanced products of the rcp($13q^-$;$14q^+$) were identified and all failed to survive to the implantation stage [64]. Embryos with six different unbalanced karyotypes associated with rcp($4q^+$;$14q^-$) were present at the late preimplantation stage [65]. Unfortunately, no embryonic data are available for cattle and sheep. It would be particularly interesting to know whether there was any prezygotic selection in sheep as occurs with unbalanced products from centric fusion translocations.

The variation in viability of embryos with an unbalanced karyotype is presumably related to the size of the genome that is displaced and the

TABLE VI. Reciprocal translocations in the domestic pig and their effects on fertility

Chromosomes involved	Country	Breed	Reduction of the prolificacy (%)	Authors
trcp($11p^{+}$;$15q^{-}$)	Sweden	Landrace	56	Henricson and Backstrom, 1964 [68]
			34	Hageltorn et al., 1973 [69]
trcp($6p^{+}$;$15q^{-}$)	Belgium	Landrace	100	Bouters et al.,1974 [70]
trcp($1p^{-}$;$6q^{+}$)	Yugoslavia	Large white	26	Locniskar et al.,1976 [71]
trcp($6p^{+}$;$14q^{-}$)	G. Britian	Large white × Essex	100	Madan et al.,1978 [66]
trcp($4q^{-}$;$14p^{+}$)	France	Large white × Landrace	43	Popescu and Legault, 1979 [72]
				Bahri et al., 1984 [73]
trcp($1p^{-}$;16p+)	W. Germany	Landrace	?	Forster et al., 1981 [74]
trcp($7q^{-}$;$11q^{+}$)	Sweden	Yorkshire	50	Gustavsson et al., 1982 [75]
trcp($1p^{-}$;$8q^{+}$)	Sweden	Yorkshire	?	Gustavsson et al., 1982 [75]
trcp($1q^{+}$;$14q^{-}$)	E. Germany	Lignee synth.	5	Golisch et al., 1982 [76]
trcp($3p^{+}$;$7q^{-}$)	France	Large white	45	Popescu et al.,1983 [77]
				Bahri et al.,1984 [73]
trcp($1p^{+}$;$14q^{-}$)	Sweden	Yorkshire	?	Gustavsson (pers. comm.)
trcp($1q^{-}$;$17q^{+}$)	Sweden	Yorkshire	?	Gustavsson (pers. comm.)
trcp($5q^{-}$;$8q^{+}$)	Sweden	Yorkshire	50	Gustavsson (pers. comm.)
trcp($5q^{-}$;$14p^{+}$)	France	Hampshire × Pietrain	28	Popescu et al., 1984 [78]
trcp($16q^{+}$;$17q^{-}$)	France	English Landrace × Duroc	36	Popescu and Boscher, 1986 [56]
trcp(1;7)	Sweden		?	Gustavsson (pers. comm.)
trcp(15;16)	Sweden		?	Gustavsson (pers. comm.)

From Popescu and Boscher [56].

TABLE VII. Reciprocal translocations in cattle and sheep and their effect on fertility

Species	Translocation	Breed	60–90 day nonreturn rate[a] (%)	References
Cattle	trcp($10q^+$;$11q^-$)	Fleckvieh	30	[57,58]
	trcp($8q^-$;$15q^+$)	Alpine grey	25	[58]
	trcp($2q^-$;$20q^+$)	Dutch Friesian	0	[59]
	trcp($8q^-$;$27q^+$)			
Sheep	trcp($1p^-$;$20q^+$)	?	NA[b]	[60]

[a]Normal rates = 70%.
[b]NA = not applicable.

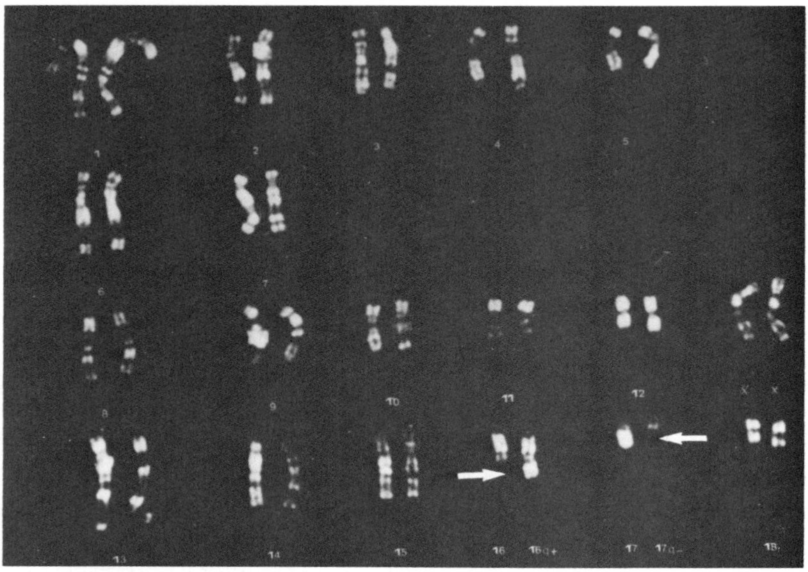

Fig. 2. The R-banded karyotype of a sow heterozygous for the rcp($16q^+$;$17q^-$). The arrows indicate the translocated segment. (Photograph supplied through the courtesy of C.P. Popescu and J. Boscher.)

importance of the genes carried in that region. In this respect it is interesting to note that the size of the segments exchanged in the rcp($11p^+$;$15q^-$) is smaller than in the rcp($13q^-$;$14q^+$) and that chromosome 11 is late replicating and, hence, relatively genetically inactive [64].

The relative frequencies of the different karyotypes in the embryos indicates the frequency of the different segregation patterns. Alternate segregation, producing normal or balanced heterozygous embryos, predominates in all three reciprocal translocations that have been studied. However, the karyotypes of the unbalanced embryos suggests that there may be an excess of adjacent-2 over adjacent-1 segregation with the rcp($11q^+;15q^-$) [66] and males heterozygous for the rcp($13q^-;14q^+$) [64]. By contrast, females heterozygous for the rcp($13q^-;14q^+$) produced a slight excess of embryos resulting from adjacent-1 segregation [64]. The slight excess of adjacent-1 segregation was also shown by direct studies on female meiosis [67].

The embryonic and meiotic studies of heterozygotes for the rcp($13q^-;14q^+$) and rcp($11q^+;15q^-$) indicate that 2:2 segregation is the norm, although total non-disjunction (4:0) did occur in one 38,XX,rcp($13q^-;14q^+$) gilt [67]. The work on the rcp($4q^+;14q^-$) suggests that some (3:1) non-disjunction occurs [65].

CONCLUSIONS

Despite the large number of different translocations described in domestic mammals, very few have been studied in any depth. This is a great pity. It is obvious from the work that has been carried out that each translocation behaves individually. Furthermore the same class of translocation may not behave in the same manner in different species. Why this should be so deserves further thought and investigation.

REFERENCES

1. Bridges CB: The translocation of a section of chromosome-II upon chromosome-III in Drosophila. Anat Rec 24:426, 1923.
2. Hansen KM: Tanden-fusion-translokation und Unfruchtbarkeit beim Rind. Geissener Beitr Erbpath Zuchthyg. [Suppl]1:115, 1970.
3. Popescu CP: Un cas possible d'inversion peŕicentrique chez les bovins. Ann Genet (Paris) 15:197, 1972.
4. Popescu CP: New data on pericentric inversion in cattle (*Bos taurus L.)* Ann Genet Sel anim 8:443, 1976.
5. McFeeley RA, Hare WCD, Biggers JD: Chromosome studies in 14 cases of intersex in domestic mammals. Cytogenet Cell Genet 6:242, 1967.
6. Gustavsson I, Fraccaro M, Tiepolo L, Lindsten J: Presumptive X-autosome translocation in a cow: Preferential inactivation of the normal X chromosme. Nature 218:183, 1968.
7. Eldridge F: X-autosome translocation in cattle. In "Proc 4th Eur Colloq Cytogenet Domest Anim." Uppsala, Sweden, 1980, p 223.
8. Chapman HM, Brueŕe AN: The frequency of aneuploidy in the secondary spermatocytes of normal and Robertsonian translocation-carrying rams. J Reprod Fertil 45:333, 1975.
9. Logue DN: Meiosis in the domestic ruminants with particular reference to Robertsonian translocations. Ann Genet Sel anim 9:493, 1977.

10. Long SE: Chiasma counts and non-disjunction frequencies in a normal ram and in rams carrying the Massey I (t_1) Robertsonian translocation. J Reprod Fertil 53:353, 1978.
11. Bruere AN, Scott IS, Henderson LM: Aneuploid spermatocyte frequency in domestic sheep heterozygous for three Robertsonian translocations. J Reprod Fertil 63:61, 1981.
12. Logue DN, Harvey MJA: Meiosis and spermatogenesis in bulls heterozygous for a presumptive 1/29 Robertsonian translocation. J Reprod Fertil 54:159, 1978.
13. Logue DN, Harvey MJA: A 14/20 Robertsonian translocation in Swiss Simmental cattle. Res Vet Sci 25:7, 1978.
14. Giovanni A, Popescu CP, Succi G, Molteni L: Meiotic study of a new autosomal translocation in the alpine Grey cattle. In ''Proc 4th Eur Colloq Domestic Anim.'' Uppsala, Sweden, 1980, p 158.
15. Logue DN, Berry JE, Bruère AN: In vitro culture and karyotyping of oocytes from 4 to 6 week old lambs. Anim Reprod Sci 1:25, 1978.
16. Bruere AN: The segregation patterns and fertility of sheep heterozygous and homozygous for three different Robertsonian translocations. J Reprod Fertil 41:453, 1974.
17. Long SE: Reproductive performance of ewes mated to rams heterozygous for the Massey I (t_1) centric fusion (Robertsonian) translocation. Vet Rec 102:399, 1978.
18. Bruère AN, Ellis PM: Cytogenetics and reproduction of sheep with multiple centric fusions (Robertsonian translocations). J Reprod Fertil 57:363, 1979.
19. Long SE: Cytogenetic examination of pre-implantation blastocysts of ewes mated to rams heterozygous for the Massey I (t_1) translocation. Cytogenet Cell Genet 18:82, 1977.
20. King WA, Linares T, Gustavsson I, Bane A: Presumptive translocation type trisomy in embryos sired by bulls heterozygous for the 1/29 translocation. Hereditas 92:167, 1980.
21. Popescu CP: Cytogenetic study on embryos sired by a bull carrier of 1/29 translocation. In "Proc 4th Colloq Cytogenet Domest Anim." Uppsala, Sweden, 1980, p 182.
22. King WA, Linares T: Cytogenetics of pre-implantation bovine embryos. Theriogenol 15:116, 1981.
23. Dyrendahl I, Gustavsson I: Sexual functions, semen characteristics and fertility of bulls carrying the 1/29 chromosome translocation. Hereditas 90:281, 1979.
24. Gustavsson I: Cytogenetics, distribution and phenotypic effects of a translocation in Swedish cattle. Hereditas 63:68, 1969.
25. Refsdal AO: Low fertility in daughters of bulls with 1/29 translocation. Acta Vet Scand 17:190, 1976.
26. Gustavsson I: Culling rates in daughters of sires with a translocation of centric fusion type. Hereditas 67:65, 1971.
27. Gustavsson I: Chromosomes of repeat-breeder heifers. Hereditas 68:331, 1971.
28. Succi G, Giovanni Macchi AM, Molteni L: Preliminary observations on the influence of 25/27 translocation on fertility and milk production of Grey Alpine cows. In "Proc 5th Eur Colloq Cytogenet Domest Anim." Milano-Garguano, Italy, 1982, p 142.
29. Roosen-Runge EC: Germinal-cell loss in normal metazoan spermatogenesis. J Reprod Fertil 35:339, 1973.
30. Bunch TD, Nadler CF: Giemsa-band patterns of the tahr and chromosomal evolution of the tribe Caprini. J Hered 71:110, 1980.
31. Bunch TD, Foote WC: Evolution of the 2n=54 karyotype of domestic sheep (*Ovis aries*). Ann Genet Sel Anim 9:509, 1977.
32. Bunch TD, Foote WC, Spillett JJ: Translocation of acrocentric chromosomes and their implications in the evolution of sheep (*Ovis*). Cytogenet Cell Genet 17:122, 1976.
33. Padeh B, Wysoki M, Soller M: Further studies on a Robertsonian translocation in the Saanen dairy goat. Cytogenet Cell Genet 10:61, 1971.

34. Ricordeau G: Observations sur les caracteres de reproduction des produits males et femelles issus d'un bouc porteur d'une fusion centrique. Ann Genet Sel Anim 4:593, 1972.
35. Pollock DL, Bowman JC: A Robertsonian translocation in British Friesian cattle. J Reprod Fertil 40:423, 1974.
36. Kovacs A, Meszaros I, Sellyei M, Vass L: Mosaic centromeric fusion in a Holstein Friesian bull. Acta Biol Hung 24:215, 1973.
37. Eldridge FE: A dicentric Robertsonian translocation in a Dexter cow. J Hered 65:353, 1974.
38. Bongso A, Basrur PK: Chromosome anomalies in Canadian Guernsey bulls. Cornell Vet 66:476, 1976.
39. Berardino D di, Iannuzzi L, Ferrara L, Matassino D: A new case of Robertsonian translocation in cattle. J Hered 70:436, 1979.
40. Giovanni A de, Succi G, Molteni L, Castiglioni M: A new autosomal translocation in Alpine Grey cattle. Ann Genet Sel Anim 11:115, 1979.
41. Popescu CP: A new type of Robertsonian translocation in cattle. J Hered 68:139, 1977.
42. Hanada H, Muramatsu S, Abe T, Fukushima T: Robertsonian chromosome polymorphism found in a local herd of the Japanese Black cattle. Ann Genet Sel Anim 13:205, 1981.
43. Papp M, Kovacs A: 5/18 dicentric translocation in a Simmental bull. In "Proc 4th Eur Colloq Cytogenet Domestic Anim." Uppsala, Sweden, 1980, p 51.
44. Stranzinger GF, Forester M: Autosomal Chromosomentranslokationen bein Fleck-und Brauvieh. Experientia 32:24, 1976.
45. Tschudi P, Zahner B, Kupfer U, Stampfli G: Chromosomenuntersuchungen an schweizerischen Rinderrassen. Schweiz Arch Tierheilkd 119:329, 1977.
46. Darré R, Berland HM, Quiennec G: Une nouvelle translocation Robertsonienne chez les bovins. Ann Génét Sél Anim 6:297, 1974.
47. Bruère AN, Zartman DL, Chapman HM: The significance of the G-bands and C-bands of three different Robertsonian translocations of domestic sheep (*Ovis aries*). Cytogenet Cell Genet 13:479, 1974.
48. Elmiger G, Stranzinger G: Identification of a centric fusion in the G-banding karyotype of a Saanen goat. In "Proc 5th Eur Colloq Cytogenet Domestic Anims." Milano-Garguano, Italy, 1982, p 407.
49. Sohrab M, McGovern PT, Hancock JL: Two anomalies of the goat karyotype. Res Vet Sci 15:77, 1973.
50. Masuda H, Okamoto A, Waide Y: Autosomal abnormality in a swine. Jpn J Zootech Sci 46:671, 1975.
51. Ma NSF, Gilmore CE: Chromosomal abnormality in a phenotypically and clinically normal dog. Cytogenet Cell Genet 10:254, 1971.
52. Larsen RE, Dias E, Flores G, Selden JR: Breeding studies reveal segregation of a canine Robertsonian translocation along Mendelian proportions. Cytogenet Cell Genet 24:95, 1979.
53. Shive RJ, Hare WCD, Patterson DF: Chromosome studies in dogs with congenital heart disease. Cytogenet Cell Genet 4:340, 1965.
54. Hare WCD, Bovee K: A chromosomal translocation in miniature poodles. Vet Rec 95:217, 1974.
55. Long SE: Centric fusion translocations in cattle. A review. Vet Rec 116:516, 1985.
56. Popescu CP, Boscher J: A new reciprocal translocation in a hypoprolific boar. Genet Sel Evol 18:123, 1986.

57. Mayr B, Themsessel H, Wöckl F, Schleger W: Reziproke Translokation 60, XY, t (10;11) (41;14) beim Rind. Z Tierzüchtg Zuchtgsbiol 96:44, 1979.
58. Mayr B, Krutzler H, Aver H, Schleger W: Reciprocal translocation 60, XY, t (18;15) (21;24) in cattle. J Reprod Fertil 69:629, 1983.
59. Schepper GG de, Aalbers JG, Te Brake JHA: Double reciprocal translocation heterozygosity in a bull. Vet Rec 110:197, 1982.
60. Glahn Luft B, Wassmuth R: G-banding and fluorescent-banding in sheep with heterozygous and homozygous translocation. Ann Genet Sel Anim 10:594, 1978.
61. McClintock B: Neurospora. 1. Preliminary observations of the chromosomes of neurospora crassa. Am J Bot 32:671, 1945.
62. Ford CE, Clegg HM: Reciprocal translocations. Br Med Bull 25:110, 1969.
63. Åkesson A, Henricson B: Embryonic death in pigs caused by unbalanced karyotype. Acta Vet Scand 13:151, 1972.
64. King WA, Gustavsson I, Popescu CP, Linares T: Gametic products transmitted by rcp ($13q^-;14q^+$) translocation heterozygous pigs and resulting embryonic loss. Hereditas 95:239, 1981.
65. Popescu CP, Boscher J: Cytogenetics of pre-implantation embryos produced by pigs heterozygous for the reciprocal translocation ($4q^+,14q^-$). Cytogenet Cell Genet 34:119, 1982.
66. Madan K, Ford CE, Polge C: A reciprocal translocation, t($6p^+;14q^-$) in the pig. J Reprod Fertil 53:395, 1978.
67. King WA: Meiotic behaviour of a rcp ($13q^-;14q^+$) translocation in heterozygous pigs. Hereditas 94:235, 1981.
68. Henricson B, Backstrom L: Translocation heterozygosity in a boar. Hereditas 52:166, 1964.
69. Hageltorn M, Gustavsson I, Zech L: The Q- and G-banding patterns of a t($11p^+$; $15q^-$) in the domestic pig. Hereditas 75:147, 1973.
70. Bouters R, Bonte P, Vandeplassche M: Anomalies chromosomiques et mortalite embryonnaire chez le porc. In "1st World Congress on Genet Applied to Livestock Production" 1974, Madrid 3:169, Minist Agric Madrid.
71. Locniskar F, Gustavsson I, Hageltorn M, Zech L: Cytological origin and points of exchange of a reciprocal chromosome translocation ($1p^-$; $6q^+$) in the domestic pig. Hereditas 83:272, 1976.
72. Popescu CP, Legault C: Une nouvelle translocation reciproque t($4q^+$; $14q^-$) chez le porc domestique (Sus scrofa domestica) Ann Genet Sel Anim 11:361, 1979.
73. Bahri I, Bonneau M, Boscher J, Popescu CP: Double heterozygote for reciprocal translocation in pigs. In "6th Eur Colloq Cytogenet Domest Anim," Zurich, Switzerland, 1984, p 275.
74. Förster M, Willeke H, Richter L: Eine autosomale, reziproke 1/16 translocation bei Deutschen Landrasse schweinen. Zuchthyg 16:54, 1981.
75. Gustavsson I, Settergren I, King A: Identification of three spontaneous reciprocal translocations in the domestic pig. In "5th Eur Colloq Cytogenet Domest Anim" Milano-Garguano, Italy, 1982, p 281.
76. Golisch D, Ritter E, Schwerin M: Zytogenetische Untersuchungen von Ebern unterschiedlicher genetischer Konstruktionen. Arch Tierz 25:337, 1982.
77. Popescu CP, Boscher J, Tixier M: Une nouvelle translocation reciproque trcp($7q^-$; $15q^+$) chez un verrat "hypoprolifique." Genet Sel Evol 15:479, 1983.
78. Popescu CP, Bonneau M, Tixier M, Bahri I, Boscher J: Reciprocal translocations in pigs. Their detection and consequences on animal performance and economic losses. J Hered 75:448, 1984.

The Cytogenetics of Mammalian Autosomal Rearrangements, pages 397–417

Chapter 10
Abnormal Spermatozoa in Human Translocation and Inversion Carriers

Renee H. Martin

INTRODUCTION

Individuals who are heterozygous for a structural chromosomal rearrangement have an increased risk for spontaneous abortions and for chromosomally unbalanced children. It is very difficult to estimate the risk of a chromosomally abnormal liveborn child because of the severe ascertainment bias inherent in families who present at a genetics clinic. Also, it is next to impossible to gain information about the meiotic segregation of translocations and inversions by studying liveborn children because lethal segregations will have already been lost as recognized or unrecognized spontaneous abortions.

One approach to overcoming some of these difficulties has been to use data from prenatal diagnosis in cases where one parent carries a structural rearrangement. In the European collaborative study of prenatal diagnosis for chromosomal disorders, the empiric risk of producing a chromosomally unbalanced fetus was 5% for inversion carriers, 4.9% for carriers of Robertsonian translocations, and 11.4% for reciprocal translocations [1]. This study provides information on the chromosome constitution of fetuses of 16–24-wk gestation, but much information has been lost by this stage. The theoretical risks of producing abnormal gametes are much higher than these empiric risks [2]. However, the theoretical chromosome segregations are based on information from plants and animals, particularly maize, the mouse, and *Drosophila*. Until recently, we have had no information in humans on the actual meiotic segregation ratios of normal and abnormal chromosome complements in heterozygotes for translocations and inversions. The advent of a technique that allows analysis of human sperm chromosomes after penetra-

Division of Medical Genetics, Department of Pediatrics, Faculty of Medicine, University of Calgary, Calgary, Alberta, Canada tion of golden hamster eggs [3–8] provides the first opportunity to study the meiotic segregation of structural chromosomal rearrangements in humans.

As well as observing the frequencies and types of chromosomally normal and abnormal sperm from men heterozygous for translocations and inversions, there are other interesting questions that can be addressed with this technique.

In reciprocal translocations, there are a number of meiotic chromosome segregations possible. Most often two of the four chromosomes move into each gamete (2:2 segregation) [9,10]. This can happen in three ways. In alternate segregation, the gametes formed will have a normal chromosome complement or a chromosomally balanced complement containing both translocated chromosomes. In adjacent-1 segregation, homologous centromeres pass to the opposite poles, and in adjacent-2 segregation, homologous centromeres pass to the same pole at anaphase I. Both types of adjacent segregation produce chromosomally unbalanced gametes. Unbalanced gametes also result from 3:1 segregation, when three chromosomes move to one pole and one chromosome moves to the other. Theoretically 4:0 segregations could also occur, but they have never been reported in humans. Studies in humans have suggested that the various types of chromosome segregations do not occur with equal frequency [11,12]. Direct analysis of the types of chromosome segregation in sperm should determine whether or not the different types of segregation occur with equal frequency. Also, some studies have suggested that among normal progeny from translocation carriers, there is an excess of balanced carriers, although theoretically there should be an equal frequency of normal and balanced carriers produced from alternate segregation [13,14]. Sperm chromosome analysis should resolve this issue as well.

A number of studies have suggested that structural chromosomal rearrangements affect the segregation of other chromosomes (interchromosomal effect). For example, Down syndrome has been reported to occur more frequently in carriers of translocations [15,16] and inversions [17], even when the structural abnormality does not involve chromosome 21. Studies of sperm chromosomes will provide information as to whether translocations and inversions affect the segregation of other chromosomes.

Because the technique for sperm chromosome analysis is so new, there have been only a small number of reports on men heterozygous for structural chromosomal abnormalities. The aim of this chapter is to summarize the information available on seven translocation carriers and two inversion carriers from three different laboratories.

MATERIALS AND METHODS

All laboratories working on analysis of human sperm chromosomes use some modification of the technique pioneered by Rudak et al. [3]. I will briefly describe the methods we employ. A more detailed description of our technique has been provided elsewhere [6].

Hamster Superovulation

The hamsters used for the experiments are 2–5 mo old. The estrous cycle is determined for each hamster by examining the vaginal discharge each morning for at least 4 days. The hamsters ovulate during the night and show a white, opaque, viscous discharge with a strong odor on the following morning. This is day 1 of their cycle. The hamsters are superovulated with an intraperitoneal injection of 25–30 IU of pregnant mare's serum gonadotropin (Gestyl, Organon, Special Chemicals Division, West Orange, NJ) on the morning of day 1 of their estrous cycle, followed by an intraperitoneal injection of 25 IU of human chorionic gonadotropin (HCG) (A.P.L., Ayerts Laboratories, New York, N.Y.) at 11 P.M. on day 3. The cumulus cell mass containing eggs is in the oviduct 15 to 17 h later (2 to 4 P.M. the next day). Each hamster yields approximately 40–60 eggs.

Media

Biggers-Whitten-Whittingham medium (BWW) [18] with some modifications is used for sperm preparation, capacitation, and egg preparation. This medium is filtered through a cellulose acetate membrane (pore size 0.22 μm) to sterilize it, and then 5% CO_2 in air is bubbled through it for 5 min. The pH is adjusted to 7.4 to 7.5 with a small amount ($\leqslant$ 0.1 ml) of acid or base HEPES solution. The medium used for egg culture is Ham's F10 (Flow Laboratories; McLean, VA) supplemented with 15% fetal bovine serum (Flow Laboratories), heat-inactivated at 60°C for 1 h; 100 IU penicillin G/ml; and 50 μg streptomycin sulfate/ml. Just before use for egg culture, 5% CO_2 in air is bubbled through the medium for 5 min, and the pH is adjusted to 7.3 to 7.4 with IN HCl.

Human Sperm Preparation

Informed written consent is obtained from all sperm donors. The semen sample is collected in a sterile container and processed as soon as liquefaction has occurred (10 to 20 min at 37°C). Approximately 10 ml of BWW working solution at 37°C is added to the semen, the components are mixed gently, and the suspension is filtered through two layers of sterile tissues (Kimwipes, Fisher). The filtrate is centrifuged at 600g for 6 min, the supernatant is decanted, and the pellet is resuspended in 10 ml BWW. The sperm are washed two more times to remove seminal fluids and debris that inhibits capacitation and therefore egg penetration. The final pellet is resuspended to yield a count of 10^7 sperm/ml. The sperm suspension is placed in tissue culture dishes (60 × 15 mm, Falcon) at 37°C. This is done by injecting drops of the suspension (0.2 ml) with a needle and syringe underneath a paraffin oil layer. The sperm drops are incubated at 37°C, 5% CO_2, and 95% humidity for 5–6 h to effect capacitation.

Hamster Egg Penetration

The hamsters are stunned with ether and killed by cervical dislocation approximately 16.5 h after HCG injection, and the oviducts are removed to a watchglass containing BWW covered with paraffin oil. Four hamsters are routinely used for one experiment. The oviducts are punctured, the cumulus cell mass is removed and transferred with a Pasteur pipet to a solution of 0.1% hyaluronidase (Type 1-S, Sigma Chemical Co., St. Louis, MO) in BWW. The hyaluronidase disperses the cumulus cells in 2–3 min, freeing the eggs, which are transferred immediately to BWW. Eggs are transferred with a micropipet attached to rubber tubing and a mouthpiece. The micropipets are made by drawing out 9-in Pasteur pipets. The eggs are put in a second BWW wash.

The eggs are then transferred to a solution of 0.1% trypsin (Type XII, Sigma) in BWW, which dissolves the zona pellucida; the trypsin acts within 1 min. The eggs are transferred to BWW as soon as the zona begins to disintegrate. The eggs are washed in a second dish of BWW and placed into the drops of capacitated sperm; 20–25 eggs are scattered throughout one drop. They are incubated for 3 h at 37°C in 5% CO_2 and at 95% humidity to allow sperm penetration to occur. Egg handling time is kept to a minimum. It takes approximately 0.5 h from dissection to incubation with sperm.

Fertilization Check

Approximately 40 oocytes are checked for evidence of sperm penetration after 2–3-h incubation. Squash mounts are prepared to detect swollen sperm heads (and tails) and/or male and female pronuclei.

Culture of Fertilized Eggs

After 3-h incubation the eggs are washed to remove excess sperm in a watchglass containing F10 at 37°C covered with paraffin oil and transferred to 0.2 ml drops of F10 under paraffin oil in tissue culture dishes. Twenty to thirty eggs are scattered throughout one drop and incubated overnight for 12–12.5 h at 37°C in 5% CO_2 and 95% humidity. The next morning the eggs are transferred directly to 0.2 ml drops of F10 containing 0.4 μg/ml Colcemid (Gibco, Grand Island, NY) under paraffin oil in tissue culture dishes and incubated for a further 4–7 h.

Slide Preparation

Approximately 15 eggs are transferred from the Colcemid solution to the hypotonic solution (1% sodium citrate) in a watchglass at room temperature for 3–6 min. Three to five eggs are transferred in a small drop of hypotonic solution (2-mm diameter) to the center of a glass slide cleaned with alcohol. Eggs are put on the slide very slowly to keep them from bursting. An automatic pipet held directly over the eggs is used to drop 20 μl of fixative (3:1 95% ethanol:glacial acetic acid) on them. The slide is observed with a dissecting microscope while three more drops of fixative are added in this manner [19].

After 2 wk chromosomes are stained with 0.5% quinacrine dihydrochloride (pH 4.5) for 15 min followed by three rinses in distilled H_2O (pH 4.5) for a total of 10 min. The slides are mounted in distilled H_2O at pH 4.5 and examined and photographed with a Zeiss fluorescent microscope.

RESULTS OF STUDIES

Reciprocal Translocations

Sperm chromosome complements have been studied in six men with reciprocal translocations. Three translocation heterozygotes—t(5;18), t(6;14) [20], and t(11;22) [21]—were studied in our laboratory; two in the laboratory of Brandriff et al.—t(8;15) and t(3;16) [22]; and one male carrying two different reciprocal translocations—t(5;11) and t(7;14)—was studied by Burns et al. [23]. The results from the studies of the individual men are presented below and a summary of the segregation patterns is shown in Table I.

Translocation heterozygote (5;18) (p15;q21). This 39-yr-old man was ascertained during investigation of his two daughters who had congenital abnormalities and developmental delay. Sperm with unbalanced forms of the

TABLE I. The segregation of sperm chromosomes in six men heterozygous for reciprocal translocations

Segregation	% of sperm resulting from the segregation						
mode	t(5;18)	t(6;14)	t(11;22)	t(8;15)	t(3;16)	t(5;11)[a]	t(7;14)[a]
Alternate	78	68	23	37	37	70	30
Adjacent-1	16	32	39	38	41	26	48
Adjacent-2	0	0	23	21	16	0	0
3:1	3	0	15	4	5	4	17
4:0	0	0	0	0	0	0	4

[a]t(5;11) and t(7;14) were both present in one man.

translocation were responsible for these abnormalities. Family studies to determine if this man inherited the translocation have not been possible.

A total of 39 sperm chromosome spreads were analyzed. Seven spreads were analyzed by Giemsa (solid) staining only (five 23,X and two 23,Y), and these were excluded from the segregation data. Fourteen of the remaining 32 sperm, including three of the abnormals, included both normal chromosomes 5 and 18. Eleven, including one abnormal (Fig. 1), carried both derived chromosomes. The ratio of normal to balanced was therefore not significantly different from 1:1.

Six (19%) of the sperm complements were chromosomally unbalanced with respect to the translocation. Two of these complements had abnormalities unrelated to the translocation as well as the unbalanced segregation of the translocation. Five of the unbalanced complements were the result of 2:2 adjacent-1 segregation; one was the result of a 3:1 segregation.

Seven of the spreads (18%) had chromosome abnormalities unrelated to the translocation. This frequency of abnormalities is somewhat higher than the mean frequency of abnormal spreads (8.5%) seen in control donors examined in our laboratory [5] but the difference was not significant (one sample test for proportions, P = .06). The ratio of X- to Y-bearing sperm

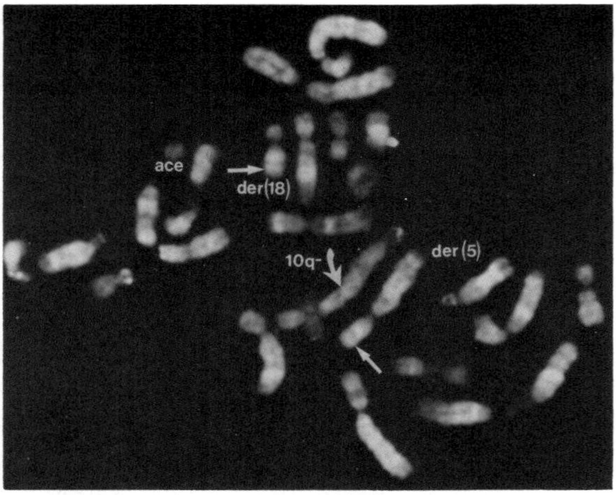

Fig. 1. Q-banded chromosomes from a sperm carrying the balanced form of the (5;18) translocation, 23,X,−5,−18,+ der(5), + der(18), t(5;18) (p15;q21), csb10. The arrow on the der(5) chromosome points to the added band. This band is 18q21 and is normally found where the arrow on the der(18) is pointing. The curved arrow indicates the point on chromosome number 10 (10q−) at which (presumably) the acentric fragment (ace) had broken off. × 2,800. (Reproduced from Balkan and Martin [20], with permission of the publisher.)

was not significantly different from the 54:46 ratio seen in a control population in our laboratory [5].

Translocation heterozygote (6;14) (p24;q22). This 34-yr-old man was ascertained after this familial translocation was discovered when his brother's wife had amniocentesis for late maternal age. This carrier inherited the translocation from his father and has passed on the balanced translocation to his only child.

Nineteen sperm chromosome spreads were karyotyped from the single experiment in which this carrier participated. With respect to the translocation, eight sperm were normal, five were balanced, and six were unbalanced. The 8:5 observation of normal to balanced sperm was not significantly different from the expected 1:1 ratio. Six (31.6%) of the sperm were unbalanced and these were all the result of adjacent-1 segregation. Two examples of chromosomally unbalanced sperm resulting from complementary events of an adjacent-1 segregation as shown in Figure 2A,B. No adjacent-2, 3:1, or 4:0 segregation was observed.

Two (11%) sperm contained abnormalities unrelated to the translocation. Both involved chromosome breaks. This was not greater than expected compared to a chromosomally normal population examined in our laboratory [20]. The ratio of X:Y-bearing sperm was not significantly different from a 1:1 segregation.

Translocation heterozygote (11;22) (q23;q11). This 30-yr-old man was ascertained during a family study after his first cousin had a child with multiple congenital abnormalities owing to an unbalanced form of the translocation. He has two children who also carry the balanced translocation.

Thirteen sperm chromosome complements were analysed from a single experiment. With respect to the translocation, one sperm was normal, two were balanced, and ten were unbalanced. The ten unbalanced sperm (77%) were the result of adjacent-1 (39%), adjacent-2 (23%), and 3:1 (15%) segregations. Two sperm (15%) had abnormalities unrelated to the translocation. This frequency of other chromosome abnormalities is not significantly higher than that seen in control donors in our laboratory [21]. The ratio of X:Y-bearing sperm was not significantly different from a 1:1 segregation.

Translocation heterozygote (8;15) (p22;q21). This 31-yr-old man was ascertained during the investigation following the birth of his daughter with multiple congenital abnormalities owing to an unbalanced form of the translocation. His other two children had a normal karyotype and a balanced form of the translocation, respectively.

Two hundred and twenty-six sperm chromosome complements were analysed from this carrier. The proportion of normal [43] and balanced [41]

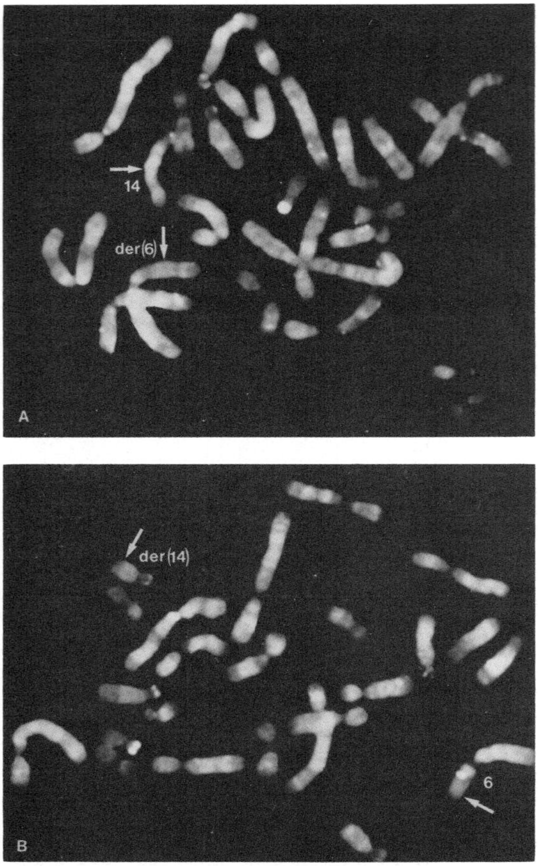

Fig. 2. Unbalanced Q-banded sperm chromosome complements produced by complementary events of an adjacent-1 segregation in the t(6;14) heterozygote. The arrows indicate the breakpoints of the translocation. **A:** 23,X,−6, +der(6), t(6;14) (p24;q22). ×2,700. **B:** 23,X,−14, +der(14), t(6;14) (p24;q22). ×2,700. (Reproduced from Balkan and Martin [20], with permission of the publisher.)

complements was not significantly different from the expected 1:1 ratio. One hundred and forty-two (63%) complements were unbalanced, and these were the result of adjacent-1 (38%), adjacent-2 (21%), and 3:1 (4%) segregations.

The frequency of chromosome abnormalities unrelated to the translocation was 5%, which was not significantly different than the control population of donors [22]. The ratio of X- and Y-bearing sperm did not differ significantly from the expected 50%.

Translocation heterozygote (3;16) (p23;q24). This 25-yr-old carrier was ascertained after his mother had amniocentesis for advanced maternal age. The fetus also carried a balanced form of this translocation.

Two hundred one sperm chromosome spreads were analysed from this carrier. The proportion of normal (40) and balanced (35) complements was not significantly different from the expected 1:1 ratio. One hundred and twenty-six complements (63%) were unbalanced, and these were the result of adjacent-1 (41%), adjacent-2 (16%), and 3:1 (5%) segregations.

The frequency of chromosome abnormalities unrelated to the translocation was 7%, which was within the normal range of control donors [22]. The proportion of X- and Y-bearing sperm did not differ significantly from the expected 50%.

Translocation double heterozygote (5;11) (p13;q23); (7;14) (q11;q24). This 34-yr-old man was heterozygous for two different reciprocal translocations. His two living children inherited the (7;14) translocation, and one child inherited a derivative 5 chromosome resulting in the *cri-du-chat* syndrome. Twenty-three sperm chromosome complements were analysed [23]. Each translocation segregated independently of the other. For the 5;11 translocation, there was a significant excess of normal (14) over balanced (2) complements (chi-square = 6.5, $P < .05$), whereas the 7;14 translocation had approximately equal proportions of normal (4) and balanced (3) spreads. Seven (30%) of the 23 sperm complements were unbalanced for the t(5;11) and these were the result of adjacent-1 (26%) and 3:1 (4%) segregations. Sixteen (70%) complements were unbalanced for the t(7;14) and these were the result of adjacent-1 (48%), 3:1 (17%), and 4:0 (4%) segregations. There were no adjacent-2 segregations for either translocation.

The frequency of chromosome abnormalities unrelated to the translocations was 70%. Although Burns et al. have not studied a population of control donors, this frequency of abnormalities is clearly higher than has been found in any normal male studied to date [3–5,7,8]. Considering chromosomal imbalance caused by both the translocations and unrelated abnormalities, there was not a single sperm that was chromosomally normal. There was a significant excess of Y-bearing sperm (17) compared to X-bearing sperm (5) (chi-square = 6.5, $P < .05$).

Robertsonian Translocation Heterozygote (14q;21q)

Only one man heterozygous for a Robertsonian translocation (14q;21q) has had the segregation of chromosomes studied in sperm [24]. This carrier was ascertained by our laboratory after his brother had a child with Down syndrome owing to an unbalanced segregation of the translocation.

Twenty-four sperm chromosome complements were analysed from this man. With respect to the translocation, seventeen (71%) were chromosomally normal and four (17%) were balanced. The difference between the expected 1:1 segregation and the observed 17:4 ratio was significant (chi-square = 7.2, $P < .01$). Three (13%) sperm complements were unbalanced.

Two sperm (8%) had abnormalities unrelated to the translocation. This was not significantly different from our control population of donors [5]. The proportion of X- and Y-bearing sperm did not differ significantly from a 1:1 ratio.

Inversions

Two men heterozygous for inversions have been studied by sperm chromosome analysis. We have studied one man with a pericentric inversion of chromosome 3 (p11q11) [25] and one man with a paracentric inversion of chromosome 7 (q11q22) [26].

Pericentric inversion (3) (p11q11). This inversion was discovered upon analysis of sperm chromosomes, when this 33-yr-old man participated in our research as a control donor. We have no information on the inheritance of this inversion, but there were no instances of abnormal children or spontaneous abortions in the family. One hundred eleven sperm chromosome spreads were analysed [25]. Sixty-four of these had the normal chromosome 3, while 47 had the inverted one. This segregation was not significantly different from the expected 1:1 ratio. None of the spreads contained an unbalanced recombinant chromosome.

Abnormalities unrelated to the inversion were observed in ten (8%) spreads. This frequency of abnormalities was similar to our results from chromosomally normal males [5]. The ratio of X- and Y-bearing sperm was not significantly different from the expected 50%.

Paracentric inversion (7) (q11q22). This 32-yr-old inversion carrier was ascertained when his wife had amniocentesis, performed for a family history of neural tube defect. Investigation of the family revealed that the inversion was present in this man, his father and the fetus.

A total of 94 sperm chromosome complements was obtained [26]. Thirty-four (37%) complements contained a normal 7, while 59 (63%) had the inverted 7. This segregation was significantly different from the expected 1:1 ratio (chi-square = 6.72, $P < .01$). None of the spreads contained an unbalanced recombinant number 7 chromosome. There were no dicentric, acentric, or duplicated/deficient chromosomes as would be expected if a crossover occurred within the inverted segment.

Abnormal sperm complements were present in 10% of the spreads. This frequency of abnormalities is similar to our results from 45 chromosomally normal control donors who have a mean frequency of 8.9% abnormal sperm [27]. The proportion of X- and Y-bearing sperm was not significantly different from the expected equal number.

Summary of All Structural Rearrangements

A total of nine men with constitutional structural rearrangements of their chromosomes have been studied. A summary of some of the salient features of the sperm chromosome complements is presented in Table II.

Most of the structural rearrangements produced an equal frequency of normal and chromosomally balanced sperm. The individuals with reciprocal translocation (5;11) and the Robertsonian translocation (14;21) had a significant excess of normal sperm. The male with paracentric inversion (7) had a significant excess of sperm with an inverted chromosome. The frequency of unbalanced sperm chromosome complements owing to the structural rearrangement varied greatly, from 0% to 77%. The frequency of abnormalities unrelated to the specific structural rearrangement was within the normal range in all men with the exception of the man heterozygous for two reciprocal translocations. The proportion of X- and Y-bearing sperm was significantly different from the expected 1:1 ratio in only one male, the carrier of two translocations.

DISCUSSION

Types of Segregations in Reciprocal Translocations

All types of 2:2, 3:1, and 4:0 segregations were observed in the sperm chromosome complements. The 4:0 segregation observed in sperm from the t(5;11) and t(7;14) heterozygote [23] is of particular interest since this segregation has never been reported in human spontaneous abortions, fetuses (from prenatal diagnosis), or liveborn offspring.

The other types of segregations that produce chromosomally unbalanced human offspring are, in decreasing frequency, 2:2 adjacent-1, 3:1, and 2:2 adjacent-2 [28]. This same order of decreasing frequency was observed in the human sperm complements. The adjacent-1 segregation was observed in all seven translocations and varied in frequency from 51% to 100% of the unbalanced sperm. In all translocations, this was the most common segregation producing unbalanced gametes. Six of the seven translocations underwent 3:1 segregation, producing 0% to 25% of unbalanced sperm. Lindenbaum and Bobrow [11] estimated that a maximum of 6–25% of chromosomally unbalanced offspring result from 3:1 disjunction. This estimate is similar to the frequencies observed in sperm complements. The

TABLE II. Sperm chromosome complements in nine men with structural rearrangements

Sperm complements	Translocations								Inversions	
	Reciprocal							Robertsonian	Pericentric	Paracentric
	t(5;18)	t(6;14)	t(11;22)	t(8;15)	t(3;16)	t(5;11)[a]	t(7;14)[a]	t(14;21)	(3)	(7)
No. normal balanced	14/11	8/5	1/2	43/41	40/35	[b]14/2	4/3	[b]16/4	64/67	[b]34/59
% unbalanced	19	32	77	63	63	30	70	13	0	0
% unrelated abnormalities	18	11	15	5	7	70	70	8	8	10

[a]t(5;11) and t(7;14) were both present in one man.
[b]Significantly different.

adjacent-2 segregation was observed in only three of the seven translocations, but these three did have a relatively high frequency of this segregation: 25–33% of the unbalanced sperm were caused by this segregation in the three translocations. Jalbert and Sele [28] have suggested that certain factors predispose to 3:1 and 2:2 adjacent-2 disjunction. Considering data from fullterm offspring, they proposed that translocations involving acrocentric chromosomes and/or a number 9 chromosome were predisposed to these rarer segregations. They also suggested that the 3:1 and adjacent-2 segregations were mutually exclusive. The evidence available from sperm chromosome analysis shows that the other possible segregation patterns do occur even if they do not result in viable offspring.

Several examples illustrating this point come to mind:

Alternate segregation is the most common segregation pattern from a translocation heterozygote observed in human liveborn offspring [28] and fetuses [1]. Similarly, studies in other mammals demonstrate that more than 90% of offspring result from alternate segregation [29,30]. This is not surprising, since this segregation produces normal progeny. However, for most of the translocations studied by sperm chromosome analysis, the frequencies of alternate and adjacent-1 segregation are approximately equal.

Furthermore translocation (11;22) (q23;q11) is the most common reciprocal translocation in humans. This translocation is of particular interest because of the propensity for imbalanced offspring by 3:1 segregation. There have been no published reports of chromosomally unbalanced offspring born as a result of adjacent-1 or -2 meiotic segregations in a heterozygote for this translocation. But there have been more than 80 cases of abnormal progeny who have tertiary trisomy (22q11q) because of a 3:1 meiotic segregation [33,34]. Various theories have been formulated as to why this translocation is predisposed to 3:1 segregation; for example, Zackai and Emanuel [33] suggested that "the pachytene configuration is extremely asymmetric and could lead to the discordant orientation of the quadrivalent on the equatorial plate and subsequent 3:1 disjunction."

Our results from sperm chromosome analysis demonstrate that all of the other segregations do occur for this translocation [21] (Table I). Therefore the explanation as to why unbalanced progeny are only produced by 3:1 segregation is not that there is a specific meiotic mechanism causing 3:1 disjunction but, rather, that the adjacent-2:2 segregations do not produce viable offspring.

Finally, some medical genetics texts explain alternate-1, adjacent-1, and adjacent-2 segregation as if these three segregations occurred with equal frequency [31]. It is clear from the data derived from sperm chromosome complements as it is from information in the mouse [32] that these segregations do not occur with equal frequency. For most translocations alternate and adjacent-1 segregations have approximately the same frequency, and adjacent-2 segregation is much rarer.

Chromosomally Unbalanced Sperm Resulting From the Structural Rearrangements

The frequency of chromosomally unbalanced sperm related to the structural aberration varied dramatically from 0% to 77%.

Inversions. There were no recombinant sperm from either inversion despite a cell sample size of approximately 100 in each case [25,26] (Table II). Recombinant chromosomes would be formed if an odd number of chiasmata occurred within the inverted segment. For the pericentric inversion of chromosome 3, this would have resulted in a large duplication/deficiency. For the paracentric inversion of chromosome 7, dicentric and acentric chromosomes would be expected. Since no unbalanced sperm were detected, this suggests a lack of pairing of the inverted segment or a suppression of crossing-over. However, specific characteristics of these two inversions might account for the fact that no recombinant sperm were found. The pericentric inversion of chromosome 3 (p11q12) may not represent a true inversion but may be a polymorphism, where the heterochromatic region normally found on the long arm, close to the centromere, is in an equivalent location on the short arm of the inverted chromosome. Various studies have found a high frequency of this inversion in normal individuals, ranging from 0.5% to 4.5% [35]. Since no recombinant chromosomes were found in our study, this might indicate that the inversion is, in fact, a polymorphism. On the other hand, it is possible that it is a true inversion and that recombinant chromosomes did not arise because crossing-over rarely occurs in the pericentric region of chromosome 3 [36]. The paracentric inversion of chromosome 7 (q11q22) is one of the most common paracentric inversions in man with six index cases reported [26,37–40]. This inversion is a type of "back mutation" since the inverted chromosome is present in the gorilla, which is presumed to be ancestral to the human chromosome 7 [40]. This may have some effect on crossing-over in this region as well.

Robertsonian translocation. The 14;21 Robertsonian translocation had only 13% unbalanced sperm [24] (Table II). Theoretically a much higher proportion is expected [41]. We must await further studies on other Robertsonian translocation carriers to determine if this low frequency of chromosomally unbalanced sperm is characteristic of Robertsonian translocation carriers or is simply a result of our small sample size.

Reciprocal translocations. The frequency of unbalanced sperm ranged from 19% to 77% in the reciprocal translocation heterozygotes. Four out of seven translocations had more than 50% chromosomally unbalanced sperm. These frequencies are all much higher than those seen in studies of human fetuses [1].

The most likely explanation for this difference is that many chromosomally unbalanced embryos are lost early in embryological development. It is also possible that sperm selection in vivo might account for the lower frequency of unbalanced fetuses. However, this is unlikely since paternal reciprocal translocation carriers were found to have the same frequency of chromosomally unbalanced fetuses as maternal carriers (11.4%) in the European collaborative study of prenatal diagnosis [1].

Ascertainment of structural rearrangement. From these studies, it appears that there is a great variation in the frequency of unbalanced gametes produced by different structural rearrangements. Many studies of liveborn offspring have demonstrated an increased risk of chromosomally unbalanced children when the rearrangement was ascertained through the birth of an abnormal child [15,42]. In all likelihood, this occurs because the rearrangements produce a chromosome segregation with a relatively small proportion that is unbalanced and thus viable. Rearrangements producing gametes with larger imbalances would lead to spontaneous abortions and not be ascertained by the birth of an abnormal child. It is not surprising that we do not find the same relationship from our studies on sperm chromosome complements. Sperm with small and large imbalances were seen in all the translocations and the proportion of unbalanced sperm had no relationship to the manner in which the translocation was ascertained.

It will be necessary to study many more translocations and inversions to discern any general principles that influence the meiotic segregation of rearranged chromosomes and the resulting frequency of unbalanced sperm.

Ratio of Normal to Balanced Sperm

In both translocations and inversions, meiosis should theoretically produce an equal number of chromosomally normal and balanced gametes. The expected equal frequency was present for most of the structural rearrangements. The reciprocal translocation (5;11) seen in the double heterozygote and the Robertsonian translocation (14;21) carrier had a significant excess of chromosomally normal sperm [23,24] (Table II). The paracentric inversion (7) carrier had a significantly increased frequency of sperm containing the inverted chromosome [26] (Table II).

Studies on human translocation carriers have suggested that more balanced carriers than normal offspring are born when the father is the carrier of a translocation [13]. Results from the European collaborative study of chromosome abnormalities during prenatal diagnosis indicated an excess of balanced carrier fetuses for paternal reciprocal translocations. For paternal Robertsonian translocations, the excess of balanced carrier fetuses was only significant for paternal 13;14 translocations and not for 14;21 translocations [1]. The majority of studies on other mammals have demonstrated an equal frequency of normal and balanced offspring born to translocation heterozygotes [43,44]. A number of studies in sheep heterozygous for Robertsonian translocations have found results similar to these; namely, a significant excess of normal secondary spermatocytes [45–47]. Researchers have suggested that this uneven distribution of meiotic segregation products might be attributed to either cell selection or degeneration during spermatogenesis [47]. Our information on humans is based on limited data and must await further studies to determine if there is an unequal representation of normal and balanced sperm in some human translocation carriers.

In the paracentric inversion (7) heterozygote we found an excess of sperm carrying the inverted chromosome [26] (Table II). A preferred segregation of the inverted chromosome has been demonstrated in the onion fly *(Hylemya antigua)* [48] and in male *Drosophila* [49], but there is no evidence in human liveborn studies of such a preferential segregation of the inverted chromosome [50]. Sperm chromosome analysis in our carrier of the inversion 3 chromosome demonstrated an equal frequency of inverted and normal chromosomes [25] (Table II). We must await further sperm chromosome studies of inversion heterozygotes to determine if preferential segregation does occur for some inversions.

Interchromosomal Effect

Cases of trisomy 21 and sex chromosomal aneuploidy have been reported in association with translocations and inversions [15–17,40]. A number of researchers have suggested that there is an increased frequency of chromosomal abnormalities unrelated to the specific structural rearrangement and have termed this an "interchromosomal effect." It is thought that the rearranged chromosomes could disrupt the normal pairing and disjunction of homologous chromosomes. According to Grell's distributive pairing hypothesis [51], two types of chromosome pairing occur at meiosis. The first involves the normal pairing of homologous chromosomes with formation of chiasmata, ensuring segregation at anaphase I. The second, distributive pairing, occurs in chromosomes in which no crossing-over has occurred. These chromosomes enter a "distributive pool" in which pairing is primarily dependent on chromosome size and is independent of homology. This would provide non-crossover homologues a second opportunity to pair and thereby segregate normally since they are closer in size to each other than a heterologue would be. However, if there were more than two chromosomes of approximately the same size within the distributive pool, this would increase the chance of nonhomologous pairing causing abnormal segregation. Thus if crossing-over did not occur within the rearranged chromosomes, these chromosomes could disrupt the disjunction of other chromosomes, particularly those of similar size.

The frequency of abnormalities unrelated to the structural rearrangement was not significantly increased for any of the translocations or inversions with the exception of the double translocation heterozygote. This man had an extremely high frequency of numerical chromosomal abnormalities, unrelated to the translocations (70%). It is possible that the presence of two reciprocal translocations greatly disrupted pairing and crossing-over at meiosis, causing the anomalous segregations that were observed.

The evidence to date does not support the concept of an interchromosomal effect for carriers of a single structural chromosomal abnormality. It would be very interesting to study other men heterozygous for two structural abnormalities to determine if these men do have an increased risk for chromosomal abnormalities unrelated to the structural abnormalities.

Sperm Selection

A central question to the utility of this technique in analysis of the segregation of chromosomes during meiosis is the possible selection of sperm by the system. Studies in mice demonstrate that chromosome constitution does not influence the ability of sperm to affect fertilization [52–54]. The fact that we have found abnormalities resulting from all theoretical segregations (even 4:0) argues against sperm selection, since one would expect sperm with less genomic imbalance if selection occurred. However, the possibility of sperm selection has not been formally tested, and future studies must be designed to answer this important question.

CONCLUSIONS

Sperm chromosome complements have been studied in nine men heterozygous for structural chromosomal rearrangements. Two men carried inversions [inv(3) (p11q12) and inv(7) (q11q22)]; one man had a Robertsonian translocation [t(14q21q)]; and six men carried reciprocal translocations, including one man who was a heterozygote for two different translocations [t(5;8) (p15;q21), t(6;14) (p24;q22), t(11;22) (q23;q11), t(8;15) (p22;q21), t(3;16) (p23;q24), t(5;11) (p13;q23), (7;14) (q11;q24)]. These men have been studied by three laboratories with the technique of hamster egg penetration by human sperm.

In the translocation carriers all theoretical types of segregations were observed in the sperm chromosome complements, including a 4:0 segregation, which has not been seen in studies of human liveborn offspring or embryos. For most of the translocations studied, the frequencies of alternate and adjacent-1 segregations were approximately equal. Adjacent-2, 3:1, and 4:0 segregations occurred more rarely. The ratio of normal to balanced sperm was equal in most of the rearrangements, with the exception of an excess of normal sperm in one of the translocations in the double translocation carrier and in the Robertsonian translocation heterozygote, and an excess of sperm carrying the inverted chromosome in the paracentric inversion carrier. There was no evidence for an interchromosomal effect in eight of the men. The double translocation heterozygote had a very high frequency of abnormalities unrelated to the translocations, perhaps indicating an effect on the segregation of other chromosomes. The frequency of unbalanced sperm varied from 0% in both inversions to 77% in a reciprocal translocation. In general, the frequencies of unbalanced sperm were much higher than those seen in studies of human fetuses. Further studies will allow us to discern any general principles that influence the meiotic segregation of rearranged chromosomes and the resulting frequency of unbalanced sperm.

ACKNOWLEDGMENTS

I would like to thank Kathy Hildebrand and Wayne Balkan for technical expertise and Dr. A. Rademaker for assistance with the statistical analysis. This research was supported by the Medical Research Council of Canada, Health and Welfare Canada, the Alberta Heritage Fund for Medical Research, and the Alberta Children's Hospital Foundation.

REFERENCES

1. Boué A: Antenatal diagnosis of chromosomal disorders from research to public health application. In Bonné-Tamin B, Cohen T, Goodman RM (eds): "Human Genetics, Part B: Medical Aspects." Proceedings of the Sixth International Congress of Human Genetics, 1981, Jerusalem. Progress in Clinical and Biological Research, Vol. 103B. New York: Alan R. Liss, Inc., 1982, p 523.
2. Hamerton JL: "General Cytogenetics." Vol. 1. New York: Academic Press, 1978.
3. Rudak E, Jacobs PA, Yanagimachi R: Direct analysis of the chromosome constitution of human spermatozoa. Nature 274:911, 1978.

4. Martin RH, Balkan W, Burns K, Lin CC: Direct chromosomal analysis of human spermatozoa. Am J Hum Genet 34:459, 1982.
5. Martin RH, Balkan W, Burns K, Rademaker AW, Lin CC, Rudd NL: The chromosome constitution of 1000 human spermatozoa. Hum Genet 63:305, 1983.
6. Martin RH: A detailed method for obtaining preparations of human sperm chromosomes. Cytogenet Cell Genet 35(4):253–256, 1984.
7. Brandriff B, Gordon L, Ashworth L, Watchmaker G, Carrano A, Wyrobek A: Chromosomal abnormalities in human sperm: Comparisons among four healthy men. Hum Genet 66:193, 1984.
8. Brandriff B, Gordon L, Ashworth L, Watchmaker G, Morre D, II, Wyrobek AJ, Carrano AV: Chromosomes of human sperm: Variability among normal individuals. Hum Genet 70:18, 1985.
9. McClintock B: Neurospora: I. Preliminary observations on the chromosomes of Neurospora crassa. Am J Bot 32:671, 1945.
10. Ford CE: Meiosis in mammals. In: Benirschke K (ed) "Comparative Mammalian Cytogenetics." New York: Springer Verlag, 1969, p 91.
11. Lindenbaum RH, Bobrow M: Reciprocal translocation in man. 3:1 meiotic disjunction resulting in 47− or 45− chromosome offspring. J Med Genet 12:29, 1975.
12. Davis JR, Rogers BB, Hagaman RM, Thies CA, Veomett IC: Balanced reciprocal translocations risk factors for aneuploid segregant viability. Clin Genet 27:1, 1985.
13. Hamerton JL: Robertsonian translocation: Evidence on segregation from family studies. In Jacobs PA, Price WH, Law P (eds): "Human Population Cytogenetics." Edinburgh: Edinburgh University Press, 1970.
14. Petrosky DL, Borgaonkar DS: Segregation analysis in reciprocal translocation carriers. Am J Med Genet 19:137, 1984.
15. Aurias A, Prieur M, Dutrillaux B, Lejeune J: Systematic analysis of 95 reciprocal translocations of autosomes. Hum Genet 45:259, 1978.
16. Okawa K, Trent M, Lebovitz R: Familial balanced translocation 4p+/17q− as a suggested cause of primary trisomy 21 Down syndrome. Arch Dis Child 52:890, 1977.
17. Fryns JP, van den Berghe: Paracentric inversion in man: Personal experience and review of the literature. Hum Genet 54:413, 1980.
18. Biggers JD, Whitten WK, Whittingham DG: The culture of mouse embryos in vitro. In Daniel JC (ed): "Methods in Mammalian Embryology." San Francisco: W.H. Freeman and Co., 1971.
19. Tarkowski AK: An air drying method for chromosome preparation from mouse eggs. Cytogenetics 5:394, 1966.
20. Balkan W, Martin RH: Chromosome segregation into the spermatozoa of two men heterozygous for different reciprocal translocations. Hum Genet 63:345, 1983.
21. Martin RH: Analysis of human sperm chromosome complements from a male heterozygous for a reciprocal translocation t(11;22) (q23;q11). Clin Genet 25:357, 1984.
22. Brandiff B, Gordon L, Ashworth LK, Littman V, Watchmaker G, Carrano AV: Cytogenetics of human sperm: Meiotic segregation in two translocation carriers. Am J Hum Genet 38:197, 1986.
23. Burns JP, Koduru PRK, Alonso ML, Chaganti RSK: Analysis of meiotic segregation in a man heterozygous for two reciprocal translocations using the hamster in vitro penetration system. Am J Hum Genet 38:954–964, 1986.
24. Balkan W, Martin RH: Segregation of chromosomes into the spermatozoa of a man heterozygous for a 14;21 Robertsonian translocation. Am J Med Genet 16:169, 1983.

25. Balkan W, Burns K, Martin RH: Sperm chromosome analysis of a man heterozygous for a pericentric inversion of chromosome number 3. Cytogenet Cell Genet 35:295–297, 1983.
26. Martin RH: Sperm chromosome analysis in a man heterozygous for a paracentric inversion of chromosome 7 (q11q22). Hum Genet 73:97–100, 1986.
27. Martin RH: Chromosomal abnormalities in human sperm. In Dellarco VL, Voytek P, Hollaender A (eds): "Aneuploidy: Etiology and Mechanisms." New York: Plenum Press, 1985, p 91.
28. Jalbert P, Sele B: Factors predisposing to adjacent 2 and 3:1 disjunctions: Study of 161 human reciprocal translocations. J Med Genet 16:467, 1979.
29. Akesson A, Henricson B: Embryonic death in pigs caused by unbalanced karyotype. Acta Vet Scand 13:151, 1972.
30. King WA, Linares T, Gustavsson I, Bane A: Presumptive translocation type trisomy in embryos sired by bulls heterozygous for the 1/29 translocation. Hereditas 92:167, 1980.
31. Connor JM, Ferguson-Smith MA: "Essential Medical Genetics." Oxford: Blackwell Scientific Publications, 1984, p 56.
32. Oshimura M, Takagi N: Meiotic disjunction in t(14;15) 6 Ca heterozygotes and fate of chromosomally unbalanced gametes in embryonic development. Cytogenet Cell Genet 15:1, 1975.
33. Zackai EH, Emanuel BS: Site-specific reciprocal translocation, t(11;22) (q23;q11), in several unrelated families with 3:1 meiotic disjunction. Am J Med Genet 7:507, 1980.
34. Narahara K, Kodama Y, Kimura S, Kamoi M, Inoue H, Kimoto H: A case of partial trisomy 22 resulting from maternal 11/22 translocation. Jpn J Hum Genet 24:253, 1979.
35. Soudek D, Sroka H: Inversion of 'fluorescent' segment in chromosome 3: A polymorphic trait. Hum Genet 44:109, 1978.
36. Hulten M: Chiasma distribution at diakinesis in the normal human male. Hereditas 76:55, 1974.
37. Ridler MA, Sutton SP: A case of a paracentric inversion inv (7) (q11q22). Prenatal detection and counselling. Prenat Diagn 1:81, 1981.
38. Orye E, van Bever H: Paracentric inversions: Two new familial cases, inv (7) (q22q11) and inv (11) (q23q13). J Med Genet 19:63, 1983.
39. Hoo JJ, Lorenz R, Fischer A, Fuhrmann W: Tiny interstitial duplication of proximal 7q in association with a maternal paracentric inversion. Hum Genet 62:113, 1982.
40. Canki N, Dutrillaux B: Two cases of familial paracentric inversion in man associated with sex chromosome anomaly 47,XY,inv (5) (q21q32) and 45,X,inv (7) (q11.3q22.3). Hum Genet 47:261, 1979.
41. Thompson JS, Thompson MW: "Genetics in Medicine." Fourth Ed. Toronto: Saunders, 1986, p 115.
42. Stene J, Stengel-Rutkowski S: Genetic risks for familial reciprocal translocations with special emphasis on those leading to 9p, 10p and 12p trisomies. Am Hum Genet 46:41, 1982.
43. Gustavsson I: Symposium: Cytogenetics of farm animals. Distribution and effects of the 1/29 Robertsonian translocation in cattle. J. Dairy Sci 62:825, 1979.
44. Larsen RE, Dias E, Flores G, Selden JR: Breeding studies reveal segregation of a canine Robertsonian translocation along Mendelian proportions. Cytogenet Cell Genet 24:95, 1979.
45. Chapman HM, Bruere AN: The frequency of aneuploidy in the secondary spermatocytes of normal and Robertsonian translocation-carrying rams. J Reprod Fertil 45:333, 1975.

46. Long SE: Chiasma counts and non-disjunction frequencies in a normal ram and in rams carrying the Massey I(t,) Robertsonian translocation. J Reprod Fertil 53:353, 1978.
47. Bruere AN, Scott IS, Henderson LM: Aneuploid spermatocyte frequency in domestic sheep heterozygous for three Robertsonian translocations. J Reprod Fertil 63:61, 1981.
48. van Heemert C: Somatic pairing and meiotic nonrandom distribution in a pericentric inversion of Hylemya antiguia (Meigen). Chromosoma 59:193, 1977.
49. Sandler L, Hiraizumi Y: Meiotic drive in natural populations of Drosophila melanogaster. Can J Genet Cytol 3:34, 1961.
50. Kaiser P: Pericentric inversions: Problems and significance for clinical genetics. Hum Genet 68:1, 1984.
51. Martin RH, Barnes M, Arthur K, Ringrose T, Douglas G: A serial study of the effects of radiotherapy on semen parameters, hamster egg penetration rates and lymphocyte chromosome abnormalities. Fertil Steril 41:415, 1984.
52. Epstein CJ, Travis B: Preimplantation lethality of monosomy for mouse chromosome 19. Nature 280:144, 1979.
53. Gropp A: Fetal mortality due to aneuploidy and irregular meiotic segregation in the mouse. In Boué A, Thibault C (eds): "Proc Symp Institut National de la Sante et de la Recherche Medicale." Paris, 1973, p 255.
54. Ford CE: Gross genome unbalance in mouse spermatozoa: Does it influence the capacity to fertilize? In Beatty KA, Gluechsohn-Waelsch S (eds): "Proc Inst Symp The Genetics of the Spermatozoan." Edinburgh: University Press, 1972, p 359.

NOTE ADDED IN PROOF

Pellestor et al. have studied a man heterozygous for a Robertsonian translocation of chromosomes 13 and 14 (Hum Genet 76:116, 1987). Similar to our 14;21 translocation heterozygote, this man had a relatively low frequency of chromosomally unbalanced sperm (8%). It is interesting that both Robertsonian translocations studied had a lower frequency of unbalanced sperm than all the reciprocal translocations. In the European collaborative study on prenatal diagnosis [1], the frequency of unbalanced fetuses from reciprocal translocation carriers was the same if the carrier was the mother or the father but the risk was significantly lower for paternal Robertsonian translocation carriers than maternal Robertsonian carriers. This difference might be explained by an actual lower frequency of unbalanced sperm produced in males heterozygous for a Robertsonian translocation as compared to a reciprocal translocation.

The Cytogenetics of Mammalian Autosomal Rearrangements, pages 419–451

Chapter 11
Factors Influencing Viability in Reciprocal Translocation Segregants in Man

John R. Davis, Beverly Barton Rogers, and Roberta M. Hagaman

INTRODUCTION

Reciprocal translocations (rcp) are a particularly interesting type of chromosomal rearrangement. A balanced carrier possesses a rcp with neither loss nor gain of genetic material and is phenotypically normal, but has the potential to have offspring suffering imbalances as a consequence of the rcp. Therefore balanced carriers pose a clinical challenge in terms of their reproductive potential.

In meiosis their rearranged chromosomes pair with normal homologous chromosomes forming a quadrivalent. The major modes of segregation include alternate, adjacent-1, and adjacent-2 pathways of gamete formation. In alternate segregation, either a chromosomally normal or a balanced rcp gamete will result. In adjacent-1 segregation, an imbalanced gamete, when fertilized by a normal gamete, results in a zygote with both partial trisomy and partial monosomy. The result of adjacent-2 segregation produces for most rcp partial trisomy and monosomy of greater magnitude. Non-disjunction of the quadrivalent results in 3:1 segregation with either full trisomy (interchange mode) or mixed partial trisomies (tertiary mode) in the zygote. In imbalanced segregations, predominantly trisomic conceptuses are observed more frequently than predominantly monosomic, suggesting a developmental advantage for duplicated autosomal burden over deficiency.

There are risks for rcp balanced carriers both in regard to occurrence of imbalanced conceptuses and in the severity of imbalance. Empirically the more severe the chromosome imbalance, the more likely a failure of the

Department of Pathology (J.R.D.) and Divison of Computer Systems and Biostatistics (R.M.H.), Arizona Health Sciences Center, University of Arizona, Tucson, Arizona 85724 and Department of Pathology, Baylor University, College of Medicine, Houston, Texas (B.B.R.)

conceptus to survive intrauterine development (implantation failure, abortion, or stillbirth). A less severe imbalance may be compatible with livebirth despite malformations and maldevelopment.

CASE COMPILATION

In order to evaluate risk factors concerning chromosome imbalance, a listing of our own and previously published cases of rcp balanced carrier couples was compiled. Table I is the AB (abortion) group and consists of 198 couples with recurrent pregnancy wastage (more than one abortion or stillbirth), with or without chromosomally normal or balanced carrier children, but without liveborn aneuploid children. Table II is the LB (liveborn) group and consists of 294 couples with at least one liveborn child with aneuploidy as a consequence of inheriting the parental rcp in an unbalanced form.

The compilation is to our knowledge the largest collection of rcp balanced carriers to date. Since rcp is not a common occurrence, no laboratory has more than a few dozen of its own cases; only 20 of these cases are from our laboratory. Therefore, the compilation is drawn predominantly from the literature. It includes a sizeable compilation by Jalbert et al. [201] of 147 families which were mostly LB. Our previous study of 297 families [33] including those of Jalbert et al. has been expanded to 492. The literature search included the major cytogenetic journals through 1985 and tabulated only cases with breakpoints identified by banding. The LB compilation is by family and segregation mode, but does not include multiple progeny in a given family with the same imbalance. Six LB aneuploids represented more than one segregation mode from the same rcp parents, and these are included in the table, bringing the total AB and LB aneuploid segregants to 498.

Information for each case was computerized (DEC-System 20) and tabulated using SPSSX. The graphs were generated using SPSSX graphics package and SPSSG post-processor.

Tables I and II include the *predicted* imbalanced segregation mode in the AB group and the *observed* modes in the LB. Predicted imbalanced modes were derived by the pachytene-diagram method of Jalbert et al. [201]. The accuracy of their method is indicated by an 87.8% concurrence of observed vs. predicted in the LB group (258 of 294 cases excluding dual segregants).

COMPARISON OF GROUPS BY MEIOTIC SEGREGATION MODE

Table III lists the predicted segregation mode in the AB group and the observed vs. predicted modes in the LB group. The frequency of predicted modes in the AB group and those observed in LB were very similar: adjacent-1, 69% each; 3:1 tertiary, 20 and 24%, respectively. The major discrepancy was in the 3:1 interchange mode: 11 and 0.7% respectively.

Table III also highlights the concordance between observed and predicted modes in the LB group. Discordant modes were adjacent-1 observed/3:1 tertiary predicted (cases 226, 312, 329, 330, 343, 344, 366, 373, 408, 464, 466, and 486); adjacent-2 observed/3:1 tertiary predicted (cases 372, 459, 479, and 483); 3:1 tertiary observed/adjacent-1 predicted (cases 206, 218, 279, 339, 354, 355, 375, 438, 439, 447, 468, and 475); and 3:1 tertiary observed/3:1 interchange predicted (cases 322, 394, 401, and 402).

FREQUENCY OF CHROMOSOMAL INVOLVEMENT

In the AB group an almost random frequency occurred relative to comparative chromosome size. Figure 1 shows the exceptions at 95% confidence intervals: chromosomes 6 and 7 were more frequent than expected, and 3, 12, and 16 were less frequent than expected.

In the LB group a non-random frequency occurred. Figure 2 shows chromosomes 9, 10, 13, 15, 21, and 22 were observed more frequently than expected and chromosomes 1, 2, 3, 6, 16, 17, and 19 less frequently.

Chromosomes 3 and 16 in both groups were involved less frequently than expected, perhaps indicating they are less prone to rcp.

Figure 3 is a histogram of frequency of chromosome involvement by percentage.

Involvement of chromosomes 1, 2, 6, 7, and 19 is predominantly in the AB group, and chromosomes 9, 21, and 22 is predominantly in the LB group. These data are not statistically significant. Combinations of chromosomes involved in a given rcp also were not significant.

SEX OF CARRIER PARENT

Maternal balanced carriers were more frequent in both groups (56% in AB, 71% in LB). Maternal carriers occurred at nearly the expected rate in 3:1 tertiary segregation mode cases predicted in the AB group (58%), but comprised 95% of the 3:1 tertiary cases observed in the LB group.

ANEUPLOID SEGMENTS

Selected chromosomes were compared regarding size and segment of imbalance in the two groups. Figure 4A represents trisomic imbalances.

TABLE I. Abortion group

Case	RCP		Parent	Mode	Ref.
1	1P34	2P23	M	ADJ1	1
2	1Q25	2Q23	M	ADJ1	2
3	1P13	2Q24	M	ADJ1	3
4	1P36	2Q37	P	ADJ1	4
5	1P36	2Q13	M	3:1T	4
6	1Q44	3P21	M	ADJ1	5
7	1P22	3P27	M	ADJ1	4
8	1Q11	4P16	P	3:1T	6
9	1Q42	4P16	P	ADJ1	7
10	1Q32	4Q31	P	ADJ1	4
11	1Q44	4Q22	M	ADJ1	4
12	1Q21	5Q35	M	ADJ1	8
13	1P32	5Q13	P	3:1T	9
14	1P34	6Q16	P	ADJ1	4
15	1Q31	7P22	M	ADJ1	5
16	1P32	7P13	M	ADJ1	10
17	1Q32	7Q36	M	ADJ1	7
18	1P22	8Q22	M	ADJ1	11
19	1Q12	10Q26	M	ADJ1	4
20	1Q42	11Q13	M	ADJ1	12
21	1Q32	11Q21	P	ADJ1	13
22	1Q25	12P13	M	ADJ1	—[a]
23	1Q11	13P11	M	3:1I	14
24	1Q25	14Q13	M	3:1T	4
25	1P21	15Q26	P	3:1I	16
26	1Q12	16P13	M	3:1I	15
27	1P36	16P13	P	ADJ1	16
28	1Q12	16Q13	M	3:1I	17
29	1P34	17P11	M	ADJ1	1
30	1P36	17P11	M	ADJ1	18
31	1Q32	18Q21	M	ADJ1	15
32	1Q22	18P11	P	ADJ1	5
33	1P11	19P11	P	3:1I	1
34	1Q24	21Q22	M	ADJ1	19
35	2P13	3P14	P	ADJ1	4
36	2Q13	4P15	M	3:1T	8
37	2P14	5P15	M	ADJ1	4
38	2Q33	6Q27	P	ADJ1	4
39	2Q21	7P15	M	ADJ1	20
40	2Q22	7Q32		ADJ1	21
41	2Q23	7Q22	M	ADJ1	9
42	2P24	7P21	P	ADJ1	4
43	2P25	7Q11	M	3:1T	4

(continued)

TABLE I. Abortion group (continued)

Case	RCP		Parent	Mode	Ref.
44	2Q14	8Q13		3:1I	15
45	2Q23	8Q24	M	ADJ1	—[a]
46	2Q23	8P11	M	ADJ1	4
47	2Q23	9P24	M	ADJ1	4
48	2Q11	10P11	P	3:1I	16
49	2Q37	10P11	M	ADJ1	22
50	2P13	10Q26	P	ADJ1	6
51	2Q21	10Q26	M	ADJ1	9
52	2Q11	14Q24	P	3:1I	4
53	2Q31	15Q25	M	ADJ1	20
54	2Q21	15Q25	M	3:1I	20
55	2P12	17P13	M	ADJ1	23
56	2Q13	18Q22		3:1I	15
57	2Q23	18Q23	M	ADJ1	5
58	2P13	18P11	M	3:1I	16
59	2P23	21Q22	M	ADJ1	9
60	3P25	4Q12	P	3:1T	15
61	3Q13	4Q28	P	ADJ1	—[a]
62	3Q25	4Q21	P	3:1T	24
63	3Q13	6Q23	P	ADJ1	10
64	3Q26	7P15	P	ADJ1	9
65	3P13	8P23	P	ADJ1	2
66	3Q21	8Q24	P	ADJ1	25
67	3Q25	8Q21	M	ADJ1	8
68	3Q25	11P15	P	ADJ1	7
69	3P21	13Q22	M	ADJ1	8
70	3Q29	13Q14	M	3:1T	10
71	3Q28	15Q22	M	ADJ1	1
72	3Q21	15Q13	M	3:1T	8
73	3P21	15Q25	M	ADJ1	4
74	3P25	19P12	M	ADJ1	1
75	3Q27	21Q22	P	ADJ1	10
76	4Q33	6Q21	M	ADJ1	19
77	4Q35	6Q23	P	ADJ1	7
78	4P15	6Q24	P	ADJ1	4
79	4Q11	9P24	P	3:1T	5
80	4Q35	9Q12	M	3:1T	26
81	4Q31	10Q22		ADJ1	15
82	4Q31	10Q22	M	ADJ1	11
83	4P14	10Q22	M	ADJ1	7
84	4P16	10Q26	P	ADJ1	4
85	4Q25	11Q13	M	ADJ1	27
86	4Q13	13Q34	P	3:1T	18
87	4P15	13Q33	P	ADJ1	9

(continued)

TABLE I. Abortion group (continued)

Case	RCP		Parent	Mode	Ref.
88	4P14	14Q13	M	3:1T	15
89	4P14	14Q32		ADJ1	21
90	4P16	15Q22	M	ADJ1	6
91	4P13	16Q24		ADJ1	21
92	4Q22	17P13	M	ADJ1	5
93	4P14	18P11	P	ADJ1	28
94	4P16	XQ13	M	3:1T	11
95	5Q22	6Q13		3:1T	15
96	5P13	7Q11	P	3:1T	5
97	5Q35	7P12	P	ADJ1	4
98	5Q35	7Q11	M	3:1T	4
99	5P11	9Q33	M	ADJ1	15
100	5Q13	9Q33	P	ADJ1	15
101	5P11	9Q33	M	ADJ1	23
102	5Q33	10Q11	M	3:1T	14
103	5Q11	10Q26	M	3:1T	1
104	5P14	14Q21	P	3:1T	29
105	5Q13	15Q25	M	3:1I	15
106	5Q15	16P11	M	3:1I	23
107	5P15	18Q21	P	ADJ1	30
108	5Q12	19P11		3:1I	15
109	5P12	19P12		ADJ1	21
110	5Q12	20P11	M	3:1I	15
111	6Q15	7P21	P	ADJ1	22
112	6P25	8Q22	M	ADJI	20
113	6Q15	8P11		ADJI	15
114	6Q27	8Q12	M	ADJ1	31
115	6Q23	8Q24	P	ADJI	19
116	6Q14	8P23		3:1T	21
117	6Q25	8Q24	P	ADJ1	24
118	6P23	8Q22	M	ADJ1	9
119	6Q15	10Q22	M	ADJ1	4
120	6Q21	11Q25	P	ADJ1	2
121	6P23	11Q22	P	ADJI	32
122	6Q14	11P13	M	ADJ1	22
123	6Q21	11Q25	P	ADJ1	4
124	6Q27	12P11	P	ADJ1	4
125	6P21	13Q21	P	ADJ1	15
126	6P21	13Q12		3:1T	15
127	6P21	13Q14	P	3:1T	15
128	6P23	13Q14		3:1T	21
129	6P23	13Q33	P	ADJ1	9
130	6Q25	14Q24	P	ADJ1	8

(continued)

TABLE I. Abortion group (continued)

Case	RCP		Parent	Mode	Ref.
131	6Q23	14Q32		ADJ1	21
132	6Q11	15P11	M	3:IT	13
133	6P21	17P13	P	ADJ1	33
134	6Q23	18P11	M	ADJ1	4
135	6Q27	19Q13	P	ADJ1	2
136	6Q25	19Q13	M	ADJ1	20
137	6Q27	19P13	M	ADJ1	7
138	6P11	21Q22	P	3:1I	16
139	6Q12	21P11	M	3:1I	7
140	7P11	8Q11	M	3:1T	20
141	7P22	8Q22	M	ADJ1	34
142	7Q36	8P12	M	ADJ1	4
143	7P15	9Q34	P	ADJ1	30
144	7Q31	9Q21	P	ADJ1	1
145	7P13	9Q22		ADJ1	21
146	7P11	11Q23	M	ADJ1	20
147	7Q11	12P13		3:1T	15
148	7Q32	12P13	P	ADJ1	8
149	7P12	12P13	M	ADJ1	35
150	7P15	12P13	M	ADJ1	4
151	7P13	13Q33	P	ADJ1	20
152	7P22	13Q14		3:1T	15
153	7Q11	13P12	M	3:1T	19
154	7P22	13Q22		ADJ1	21
155	7Q36	14Q24	M	ADJ1	36
156	7P13	18Q23	P	ADJ1	4
157	7P22	20Q11	P	ADJ1	5
158	7Q11	20P11	M	3:1I	—[a]
159	7Q11	22Q13	M	3:1I	11
160	8Q24	9Q22	P	ADJ1	28
161	8P11	11Q25	M	ADJ1	9
162	8P12	13Q33	M	ADJ1	1
163	8Q24	22Q12	P	3:1T	4
164	9P22	11P15	M	ADJ1	33
165	9P23	13Q21	P	ADJ1	15
166	9P24	18Q21	M	ADJ1	4
167	PQ22	20Q12	P	ADJ1	37
168	9Q13	XP11	M	3:1T	4
169	9P24	YQ12	P	ADJ1	30
170	10Q22	11Q25	P	ADJ1	20
171	10Q22	12Q24	M	ADJ1	12
172	10Q26	14Q24	M	ADJ1	9
173	10Q24	17Q25		ADJ1	21

(continued)

TABLE I. Abortion group (continued)

Case	RCP		Parent	Mode	Ref.
174	10Q11	18Q11	P	ADJ2	11
175	10Q11	22Q11	P	3:1I	28
176	11Q23	13Q22	P	ADJ1	33
177	11P15	18Q23	M	ADJ1	20
178	11Q23	22Q11	M	3:1T	11
179	11P11	22P11	M	3:1I	23
180	11Q23	22Q13	P	ADJ1	4
181	12Q22	13Q22	M	ADJ1	24
182	12P11	14P11	M	ADJ1	20
183	12Q24	20Q11	P	ADJ1	20
184	12Q24	21Q21	M	3:1T	38
185	13Q32	14Q22	M	ADJ1	33
186	13Q12	18Q21	P	3:1T	39
187	13Q32	18Q23	P	ADJ1	9
188	14Q24	16Q22	M	ADJ1	9
189	14Q32	18Q22	P	ADJ1	16
190	15Q15	21Q22	P	3:1T	24
191	16Q21	20Q12	P	ADJ1	19
192	17Q23	19P13	M	ADJ1	27
193	17P13	22Q11	P	3:1T	5
194	17P13	22Q11	P	3:1T	23
195	18Q12	19Q13	M	3:1T	4
196	18P11	20Q11	P	ADJ1	9
197	18Q21	21Q22	P	ADJ1	33
198	18Q12	21Q21	P	3:1I	19

M = maternal carrier, P = paternal.
[a]Cases from our laboratory.

Those seen more frequently in the LB group were 4p, 4 distal q, 9p → qh, 10p, 10 distal q, 11 distal q, 12p, 12 distal q, 21p → proximal q, and 22p → proximal q. These imbalances are not critical for intrauterine survival. In Figure 4B, monosomic imbalances of the LB group consist of 4 distal p; 5 distal p; acro-p of chromosomes 14, 15, 21, and 22; and distal q of 21 and 22. The combined imbalances also register a preponderance of trisomic imbalances over monosomic in the LB group.

POTENTIALLY SURVIVABLE ANEUPLOIDY IN THE AB GROUP

Distinction between AB and LB groups is somewhat arbitrary. The LB group demonstrates fetal survivability, but the AB group may include non-chromosomal factors responsible for or contributing to pregnancy wastage.

TABLE II. Liveborn group

Case	RCP		Parent	Mode	Ref.
199	1P36	2Q31	P	ADJ1	33
200	1Q25	3P25	M	ADJ1	15
201	1Q25	4Q35	M	ADJ1	3
202	1Q42	4P16	MP	ADJ1	40
203	1Q31	5P15		ADJ1	21
204	1Q32	7Q32	P	ADJ1	41
205	1P36	9Q21	M	3:1T	42
206	1P36	9Q22	M	3:1T	43
207	1P36	9P12	P	ADJ1	44
208	1P36	15Q13	M	3:1T	15
209	1Q21	XQ22	M	ADJ1	45
210	2P25	3Q21	M	ADJ1	46
211	2P25	4Q21	P	ADJ1	47
212	2P23	5Q31	P	ADJ1	48
213	2P23	5P15	P	ADJ1	49
214	2P21	6Q27	P	ADJ1	50
215	2P23	7Q36	M	ADJ1	51
216	2Q32	7P22	M	ADJ1	52
217	2Q34	8P23	P	ADJ1	53
218	2Q36	9Q31	M	3:1T	54
219	2Q31	9P24	M	ADJ1	55
220	2Q31	10Q24	P	ADJ1	56
221	2P24	10Q26	MP	ADJ1	57
222	2Q31	12Q24	M	ADJ1	58
223	2Q37	12Q24	P	ADJ1	59
224	2Q37	12Q24	P	ADJ1	60
225	2Q32	13Q33	M	ADJ1	61
226	2Q36	13Q13	M	ADJ1	62
227	2P22	14P11	M	ADJ1	63
228	2Q12	14Q13	M	3:1I	64
229	2P21	15Q26	P	ADJ1	65
230	2Q32	17Q25	M	ADJI	66
231	2P23	20Q13	P	ADJI	67
232	2Q32	XP22	M	ADJI	68
233	3P23	4P16	M	ADJI	69
234	3P26	4Q13	M	3:1T	70
235	3P26	5P13	M	ADJ1	71
236	3Q26	5P15	M	ADJ1	15
237	3P23	5P15	MP	ADJI	72
238	3P24	6Q22	M	ADJ1	73
239	3P26	6P21	P	ADJ1	74
240	3Q27	7Q31	M	ADJ1	75
241	3P27	7Q32	P	ADJ1	76
242	3Q27	9Q34	M	ADJ1	7

(continued)

TABLE II. Liveborn group (continued)

Case	RCP		Parent	Mode	Ref.
243	3Q28	10Q11	M	3:1T	77
244	3P26	10P11	M	ADJ1	33
245	3P21	10Q26	M	ADJ1	78
246	3P27	11Q23	M	ADJ1	79
247	3Q29	13Q22	M	ADJ1	18
248	3Q24	13Q14	P	3:1T	80
249	3Q29	13Q22	M	ADJ1	81
250	3Q23	14Q21	M	3:1T	82
251	3P23	15P11	M	ADJ1	83
252	3Q28	20P12		ADJ1	21
253	3Q25	22P11	M	ADJI	84
254	4Q32	5P14	P	ADJ1	85
255	4Q32	5P13	P	ADJ1	86
256	4P12	7Q36	P	ADJ1	87
257	4Q35	7Q32	P	ADJ1	88
258	4P12	7Q36	MP	ADJ1	89
259	4P16	8P23	P	ADJ1	90
260	4P15	8P22	M	ADJ1	91
261	4Q35	8Q13	MP	ADJ1	92
262	4Q31	9Q34	M	ADJ1	93
263	4Q34	9Q21	M	3:1T	94
264	4Q33	9Q31	M	ADJ1	95
265	4Q31	9Q13	M	3:1T	96
266	4P16	10P14	P	ADJ1	97
267	4Q35	10P11	M	ADJ1	98
268	4P16	10Q24	M	ADJ1	99
269	4P15	10P14	P	ADJ1	100
270	4P16	10Q24	P	ADJ1	101
271	4Q35	11Q21	M	ADJ1	33
272	4Q25	12P13	P	ADJ1	15
273	4Q13	12Q24	M	3:1T	102
274	4P16	12P11		ADJ1	21
275	4Q25	13Q31	M	ADJI	103
276	4Q26	13Q34		PADJ1	104
277	4Q26	13Q34	M	ADJ1	105
278	4Q33	13Q14	M	3:1T	106
279	4P16	15Q22	M	3:1T	107
280	4P11	15Q11	M	3:1T	108
281	4P12	16P13	P	ADJ1	109
282	4P14	17Q25		ADJ1	110
283	4Q26	18Q23	M	ADJ1	111
284	4Q28	18Q22	M	ADJ1	112
285	4Q31	18Q23	M	ADJ1	113
286	4Q35	18Q21	P	ADJ1	114

(continued)

TABLE II. Liveborn group (continued)

Case	RCP		Parent	Mode	Ref.
287	4Q25	20Q13	M	ADJ1	115
288	4Q32	21Q22	M	ADJ1	116
289	4P11	21P11	M	ADJ1	117
290	4Q21	21Q21	M	3:1T	118
291	4Q27	21P11	M	3:1I	119
292	4Q27	21Q21	M	3:1T	120
293	4P14	21Q22	M	ADJ1	15
294	4P11	22P11	M	ADJ1	121
295	4P14	22P11	P	ADJ1	122
296	4P16	22Q12	M	3:1T	123
297	4P12	22P11	P	ADJ1	15
298	4P16	22Q11	P	3:1T	15
299	4Q25	XQ27	M	ADJI	124
300	4P13	XP21	M	ADJI	125
301	5P13	6Q27	M	ADJI	126
302	5Q34	6Q24	M	ADJI	127
303	5Q35	7Q31	P	ADJ1	128
304	5Q33	8P23	P	ADJ1	129
305	5P14	8P12	P	ADJ1	130
306	5P12	9P13	P	ADJ1	15
307	5P13	9P22	P	ADJ1	131
308	5P15	10P11	M	ADJI	132
309	5P15	10Q11	M	3:1T	33
310	5Q31	10Q26	M	ADJI	133
311	5P13	12Q24	M	ADJI	134
312	5P13	13Q14	M	ADJI	135
313	5P11	14Q11	M	3:1T	136
314	5P15	14Q13	M	3:1T	137
315	5P15	14Q13	M	3:1T	138
316	5P11	15P11	M	ADJI	139
317	5P14	15Q21	MP	ADJ1	140
318	5P15	17Q23	P	ADJ1	141
319	5P15	17P11	P	ADJ1	142
320	5P13	18P11	M	ADJ1	143
321	5P15	20P11	P	ADJ1	144
322	5P11	21Q11	M	3:1T	145
323	5Q33	22Q13	P	ADJ1	70
324	6Q27	7Q31	M	ADJ1	15
325	6Q25	9Q11	M	3:1T	146
326	6Q21	10Q26	M	ADJ1	147
327	6Q25	11Q25	M	ADJ1	148
328	6Q27	12P11		ADJ1	21
329	6Q25	13Q13	M	ADJ1	149

(continued)

TABLE II. Liveborn group (continued)

Case	RCP		Parent	Mode	Ref.
330	6Q26	18Q11	M	ADJ1	150
331	6P21	20P13	P	ADJ1	151
332	6Q22	21P11	M	ADJ1	152
333	6Q26	22P12	M	ADJ1	153
334	6P22	22Q13	M	ADJ1	154
335	7P13	8P23	P	ADJ1	155
336	7Q31	9P24	M	ADJ1	156
337	7Q35	9P13	M	ADJ1	157
338	7Q36	9Q13	M	3:1T	158
339	7P22	9Q32	M	3:1T	159
340	7P22	10P12	P	ADJ1	160
341	7Q32	10Q25	M	ADJ1	15
342	7P22	10P11	M	ADJ1	161
343	7Q36	11Q13	P	ADJ1	20
344	7P22	13Q14	M	ADJ1	162
345	7Q31	14Q32	P	ADJ1	163
346	7P11	14P11	M	ADJ1	164
347	7Q36	14Q24	M	ADJ1	165
348	7Q35	15Q14	M	3:1T	166
349	7Q31	17Q25		ADJ1	21
350	7Q31	21P11	M	ADJ1	167
351	7P15	21P12	M	ADJ1	155
352	7P21	22Q13	P	ADJ1	168
353	8P22	9P12	M	ADJ1	169
354	8Q24	9Q22	M	3:1T	170
355	8P23	9Q22	M	3:1T	171
356	8P21	10P15	M	ADJ1	172
357	8P23	10Q23	M	ADJ1	173
358	8P21	10Q26	M	ADJ1	174
359	8Q22	10P13		ADJ1	21
360	8Q24	12Q12	P	3:1T	175
361	8Q21	13Q34	M	ADJ1	176
362	8P11	13Q34	P	ADJ1	177
363	8P23	13Q14	M	3:1T	178
364	8Q21	13Q34	M	ADJ1	179
365	8Q24	15Q13	M	3:1T	180
366	8P23	18Q12	M	ADJ1	181
367	8Q23	18Q23	P	ADJ1	174
368	9Q334	10Q24	P	ADJ1	182
369	9P24	12P11	M	ADJ1	183
370	9P21	13Q21	M	ADJ1	184
371	9P24	13Q22	P	ADJ1	185
372	9Q22	13Q12	M	ADJ2	186

(continued)

TABLE II. Liveborn group (continued)

Case	RCP		Parent	Mode	Ref.
373	9P23	13Q21	M	ADJ1	187
374	9P22	14Q32	M	ADJ1	188
375	9P22	14Q24	M	3:1T	189
376	9P22	15Q26	P	ADJ1	190
377	9Q12	15P11	M	3:1T	191
378	9P22	15Q11	M	3:1T	192
379	9Q13	15P11	M	3:1T	193
380	9Q13	15Q25	M	3:1T	157
381	9Q13	15Q11	P	ADJ2	194
382	9Q13	15Q11		ADJ2	21
383	9Q32	16Q24	P	ADJ1	195
384	9Q32	17P13	M	ADJ1	196
385	9P13	18P11	M	ADJ1	33
386	9P23	18Q12	M	3:1T	197
387	9P12	19P13	M	ADJ1	146
388	9Q12	20P13	M	3:1T	198
389	9Q13	20Q13	M	3:1T	33
390	9P23	20P13	M	ADJ1	199
391	9P13	21Q22	M	ADJ1	169
392	9Q13	21Q11	M	ADJ2	200
393	9Q13	21Q11	M	ADJ2	201
394	9P11	21Q11	M	3:1T	202
395	9P21	21Q21	M	3:1T	203
396	9P21	21Q21	M	3:1T	203
397	9P13	22P11	P	ADJ1	204
398	9P11	22P11	M	ADJ1	205
399	9P12	22P11	P	ADJ1	146
400	9Q13	22Q11	M	ADJ2	206
401	9P11	22Q11	M	3:1T	207
402	9Q13	22Q13	M	3:1T	146
403	9Q32	22P11	P	ADJ1	208
404	9P12	22P11	M	ADJ1	209
405	10Q26	11Q21	P	ADJ1	210
406	10P11	12Q24	P	ADJ1	97
407	10Q24	13Q34	M	ADJ1	211
408	10Q26	13Q14	M	ADJ1	212
409	10Q25	13P12	P	ADJ1	213
410	10P12	14P11	M	ADJ1	214
411	10P15	14Q22	M	3:1T	215
412	10Q24	14Q32	M	ADJ1	216
413	10Q26	14Q32	P	ADJ1	217
414	10Q24	15Q26	P	ADJ1	218
415	10Q25	16Q24	M	ADJ1	219

(continued)

TABLE II. Liveborn group (continued)

Case	RCP		Parent	Mode	Ref.
416	10Q24	17Q25	P	ADJ1	220
417	10Q25	18Q23	P	ADJ1	221
418	10P13	18Q23	P	ADJ1	222
419	10P12	20P12	M	ADJ1	223
420	10P11	21P11	M	ADJ1	224
421	10P11	21P11	P	ADJ1	225
422	10P11	22P11	M	ADJ1	226
423	10Q24	22Q12	M	ADJ1	227
424	11Q25	14Q23	P	ADJ1	228
425	11P15	16Q22	M	ADJ1	229
426	11Q25	16Q13	P	ADJ1	230
427	11Q23	17P13	M	ADJ1	231
428	11Q21	18P11	P	ADJ1	15
429	11Q14	18P11	P	ADJ1	232
430	11Q25	20Q11	M	3:1T	233
431	11Q23	21Q22	P	ADJ1	234
432	11Q23	22Q11	M	3:1T	235
433	11Q23	22Q11	M	3:1T	236
434	11Q23	22Q11	M	3:1T	237
435	11Q23	22Q12	M	3:1T	238
436	11Q23	22Q11	M	3:1T	231
437	11Q23	22Q11	M	3:1T	239
438	11Q25	22Q13	M	3:1T	240
439	11Q25	22Q13	M	3:1T	33
440	11Q23	22Q11	M	3:1T	241
441	11Q23	22Q11	M	3:1T	242
442	11Q23	22Q11		3:1T	21
443	12Q12	14P11	M	3:1T	243
444	12Q24	14Q21	M	3:1T	244
445	12P13	14Q13	M	3:1T	33
446	12Q24	18P11	M	ADJ1	245
447	12Q24	18Q21	M	3:1T	246
448	12P11	18Q23	M	ADJ1	247
449	12Q24	18Q23	P	ADJ1	248
450	12P11	21P11	M	ADJ1	249
451	12P11	21P11	P	ADJ1	250
452	12Q24	21P11	M	ADJ1	251
453	12P12	21P11	M	3:1T	252
454	12P12	21P11	M	ADJ1	252
455	12P11	22P11	M	ADJ1	15
456	13Q13	14Q22	M	ADJ2	253
457	13Q22	15P12	M	ADJ1	254
458	13Q22	15P12	M	3:1T	254

(continued)

TABLE II. Liveborn group (continued)

Case	RCP		Parent	Mode	Ref.
459	13Q32	15Q14	P	ADJ2	255
460	13Q22	15Q26	P	ADJ1	256
461	13P11	15Q21	M	3:1T	257
462	13Q13	17P13	M	3:1T	258
463	13Q13	17P13	M	ADJ1	258
464	13Q32	18Q11	P	ADJ1	259
465	13Q32	18Q11	P	ADJ1	260
466	13Q14	18Q23	P	ADJ1	261
467	13Q34	20P11	M	ADJ1	262
468	13P11	20Q11	M	3:1T	263
469	13Q21	21Q21	M	3:1I	187
470	13Q21	21Q21	M	ADJ2	187
471	13Q21	21Q11	M	3:1T	264
472	13Q12	22P11	M	3:1T	135
473	13Q12	22P11	M	ADJ1	135
474	13Q22	22Q12	M	3:1T	27
475	14P13	15Q22	M	3:1T	—[a]
476	14Q11	16Q24	M	3:1T	265
477	14P11	18Q21	M	ADJ1	181
478	14P11	18Q22	M	ADJ1	181
479	14Q12	21Q22	P	ADJ2	266
480	14Q21	XP22	M	3:1T	267
481	15P11	16Q11	P	3:1T	268
482	15P12	16Q11	M	ADJ1	269
483	15Q13	21Q22	M	ADJ2	270
484	15Q26	21Q21	M	ADJ1	271
485	15Q26	21Q21	M	3:1T	271
486	15Q15	XP22	M	ADJ1	272
487	16P12	18Q11	M	3:1T	273
488	16P13	20Q11	M	ADJ1	274
489	16P11	21Q22	MP	ADJ1	275
490	16Q22	21Q22	P	ADJ1	276
491	16Q24	XP11	M	ADJ1	33
492	18Q23	20P11	M	ADJ1	277
493	18Q21	21Q22	M	ADJ1	278
494	18Q12	21Q11	M	ADJ2	33
495	19P13	21Q21	M	ADJ1	120
496	19Q13	22P11	M	ADJ1	279
497	20Q13	21Q21	M	3:1T	280
498	20P11	22Q13	P	ADJ1	281

M = maternal carrier, P = paternal.
[a]Cases from our laboratory.

TABLE III. Comparison of the groups by meiotic segregation mode

	Observed	Predicted					
		Adj 1	*Adj 2*	*3:1T*	*3:1I*	*A1 and 3:1T*	*A2 and 3:1I*
Abortion Group							
Total 198		136	1	39	22		
Liveborn Group							
Adjacent-1	203	191		12			
Adjacent-2	11		7	4			
3:1 Tertiary	72	12		56	4		
3:1 Interchange	2				2		
Adj 1 and 3:1T	5	2		2		1	
Adj 2 and 3:1I	1						1
Total	294	205	7	74	6	1	1

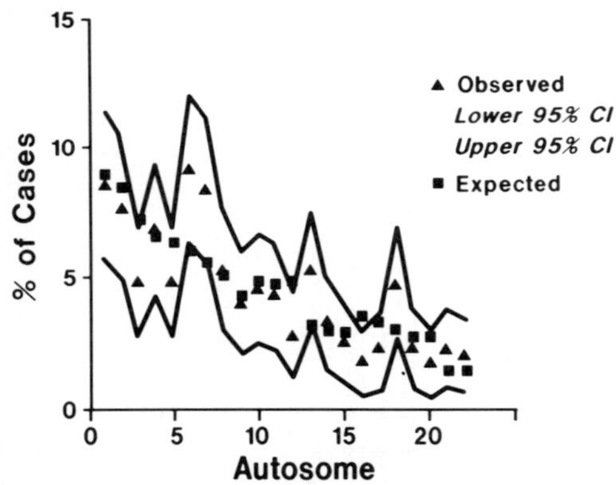

Fig. 1. Abortion group frequency of chromosomal involvement. Observed involvement is compared to involvement expected based on relative chromosome size and registered within 95% confidence intervals.

Twelve cases in the AB group (cases 9, 17, 84, 104, 124, 154, 155, 165, 173, 178, 187, and 197) had the same rcp and predicted imbalance (or the same imbalance of less magnitude) as occurred in LB cases, so must be considered potentially survivable. Ideally the AB group should be defined on the basis of fetal karyology, but this was performed in only eight instances.

CONCLUSIONS

Previous series of rcp have demonstrated that size of imbalance correlated with intrauterine survivability; the magnitude of aneuploidy in AB is greater

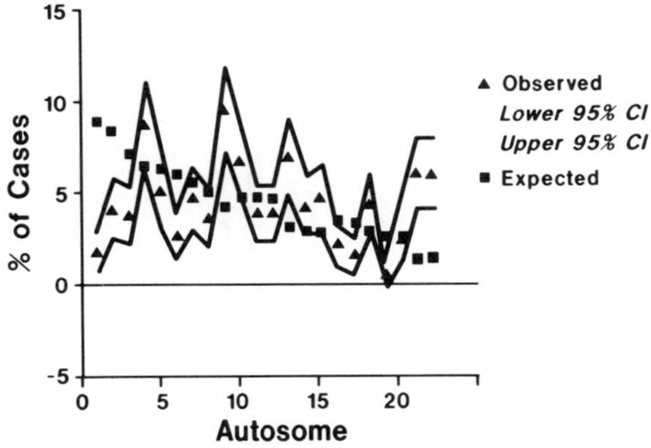

Fig. 2. Liveborn group frequency of chromosomal involvement. Observed involvement is compared to involvement expected based on relative chromosome size and registered within 95% confidence levels.

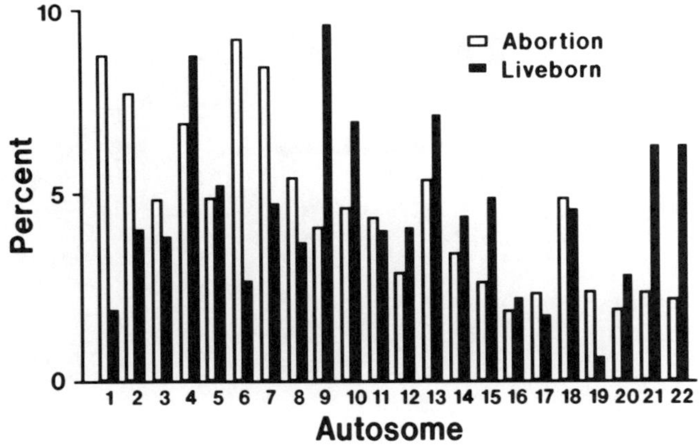

Fig. 3. Histogram of frequency of chromosomal involvement in the two groups expressed as percentage of total cases.

than that observed in LB [15,201,282,283]. Our previous study showed a mean haploid autosomal complement predicted in AB was 2.2%, and that observed in LB 1.68% [33]. Also predominantly trisomic rather than monosomic imbalance is characteristic of LB [15,33,284].

Survivable trisomic imbalances are well known clinically, and our data drawn from the literature are not new observations. In addition to full or partial LB trisomies indicated in Figure 4A, chromosomes 13 and 18 are known survivable trisomies. Survivable monosomies of distal 4p and 5p are

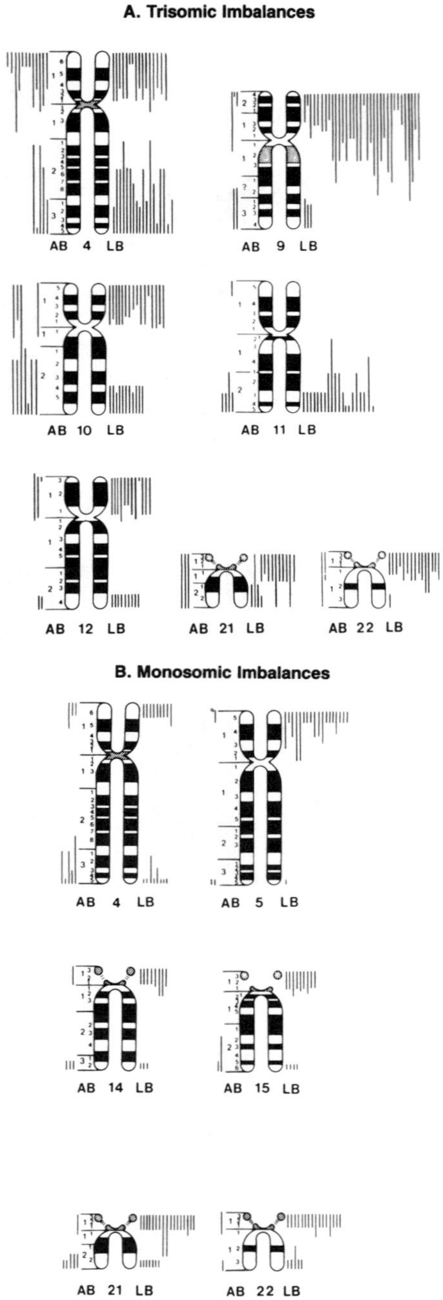

Fig. 4. Resultant trisomic **(A)** and monosomic **(B)** imbalances of selected chromosomes predicted for the AB group and observed in the LB group.

also well recognized, and the current study fails to demonstrate other significant genomic regions (acro-p segments excluded).

Other studies have identified random chromosomal involvement in AB and non-random in LB [282,283,284]. The current study was confirmatory with the exceptions that chromosomes 3 and 16 are less frequently involved relative to comparative size. In the AB group chromosomes 6 and 7 were involved more frequently (vital genomic components?), while in the LB selective tolerance in imbalance resulted in greater frequencies for chromosomes 9, 10, 13, 15, 21, and 22. Excluding the infrequently involved chromosomes 3 and 16, in the LB group chromosomes 1, 2, 6, 17, and 19 were involved less frequently than expected, suggesting that they bear vital segments poorly tolerated in imbalance. Considering the balanced carrier parents, the 71% predominance of maternal cases in the LB group would suggest that an imbalanced sperm is at a general disadvantage in procreation. Also the 95% maternal carriers in the 3:1 tertiary mode in LB group demonstrated in this study and as noted by others [285,286] reflects a distinct disadvantage for imbalanced sperm.

Despite the expanded compilation in this study, no conclusions can be drawn regarding breakpoints or the chromosome pairs involved in a given rcp. Numerous additional cases must be collected in order to gain statistical significance for these factors. Greater efforts should be made in fetal karyology of the AB cases in order to define this group more precisely.

The compilation serves as a resource in consideration of balanced carriers' risk for LB aneuploids in terms of the parental rcp itself. For a rcp carrier with untested reproductive capability, the LB (Table II) provides a checklist for potentially survivable aneuploidy. Imbalanced segregants can be accurately predicted, and in some cases compared to survivable aneuploid segments in Table IV.

ACKNOWLEDGMENTS

Assistance in manuscript preparation and arduous data and reference tabulation by Ms. Nancy Suttle and Kris Cetone is greatly appreciated.

REFERENCES

1. Turleau C, Chavin-Colon F, de Grouchy J: Cytogenetic investigation in 413 couples with spontaneous abortions. Eur J Obstet Gynecol Reprod Biol 9:65–74, 1979.
2. Mennuti MT, Jingeleski S, Schwarz RH, Mellman WJ: An evaluation of cytogenetic analysis as a primary tool in the assessment of recurrent pregnancy wastage. Obstet Gynecol 52:308–313, 1978.

3. Garrett JH, Finley SC, Finley WH: Fetal loss and familial chromosome 1 translocations. Clin Genet 8:341–348, 1975.
4. Fryns JP, Kleckowska A, Kubien E, Petit P, Van den Berghe H: Cytogenetic survey in couples with recurrent fetal wastage. Hum Genet 65:336–354, 1984.
5. Subrt I: Reciprocal translocation with special reference to reproductive failure. Hum Genet 55:303–307, 1980.
6. Schmid W: Cytogenetic results in 96 couples with repeated abortions. Clin Genet 17:85, 1980.
7. Osztovics MK, Toth SP, Wessely JA: Cytogenetic investigations in 418 couples with recurrent fetal wastage. Ann Genet 25:232–236, 1982.
8. Michels VV, Medrano C, Venne VL, Riccardi VM: Chromosome translocations in couples with multiple spontaneous abortions. Am J Hum Genet 34:507–513, 1982.
9. Brankovic S, Laca Z, Moric-Petrovic S, Arandelovic N: Types of chromosomal segregation and the risk in the families with reciprocal translocations. Clin Genet 27:298, 1985.
10. FitzSimmons J, Wapner RJ, Jackson LG: Repeated pregnancy loss. Am J Med Genet 16:7–13, 1983.
11. Diedrich U, Hansmann I, Janke D, Opitz O, Probeck H-D: Chromosome anomalies in 136 couples with a history of recurrent abortions. Hum Genet 65:48–52, 1983.
12. Kajii T, Ferrier A: Cytogenetics of aborters and abortuses. Am J Obstet Gynecol 131:33–38, 1978.
13. Robertson RD, Leddet I, Funderburk SJ, Sparkes RS: Chromosomal variants and abnormalities in couples with repeated spontaneous pregnancy loss. Clin Res 29:116A, 1981.
14. Lucas M, Wallace I, Hirschhorn K: Recurrent abortions and chromosome abnormalities. J Obstet Gynecol B Commonw 79:1119–1127, 1972.
15. Daniel A: Structural differences in reciprocal translocations. Hum Genet 51:171–182, 1979.
16. Schwartz S, Palmer CG: Chromosomal findings in 164 couples with repeated spontaneous abortions: With special consideration to prior reproductive history. Hum Genet 63:28–34, 1983.
17. Fried K, Perpinyal S, Rosenblatt M, Caspi E: Familial balanced reciprocal translocation t(1;16)(q12;q13) ascertained because of multiple abortions in a carrier. Hum Hered 27:362–365, 1977.
18. Schmidt R, Nitowsky HM, Dar H: Cytogenetic studies in reproductive loss. J Am Med Assoc 236:369–373, 1976.
19. Lippman-Hand A, Vekemans M: Balanced translocations among couples with two or more spontaneous abortions: Are males and females equally likely to be carriers? Hum Genet 63:252–257, 1983.
20. Sant-Cassia LJ, Cooke P: Chromosomal analysis of couples with repeated abortions. Brit J Obstet Gynaecol 88:52–58, 1981.
21. de Arce MA, Law E, McManus S: Outcome of pregnancy in 20 families with reciprocal translocation. Clin Genet 28:423–424, 1985
22. Lyberatou-Moraitou E, Grigori-Kostaraki P, Retzepopoulou Z, Kosmaidou-Aravidou Z: Cytogenetics of recurrent abortions. Clin Genet 23:294–297, 1983.
23. Stoll C: Cytogenetic findings in 122 couples with recurrent abortions. Hum Genet 57:101–103, 1981.
24. Geraedts JPM, Klasen EC: Chromosome studies and $\alpha 1$-antitrypsin phenotypes in recurrent abortions. Clin Genet 17:68, 1980.

25. Verma RS, Dosik H, Salazar JD: A new balanced translocation in humans. J Reprod Med 26:133–134, 1981.
26. Neri G, Serra A, Bova R, Natale MT, Tedeschi B: A balanced translocation t(4;9)(q35;q12) with a breakpoint within the heterochromatic region of chromosome 9 in a woman with recurrent abortion. Clin Genet 18:239–243, 1980.
27. Kim HJ, Hsu LYF, Paciuc S, Christian S, Quintanta A, Hirschhorn K: Cytogenetics of fetal wastage. N Engl J Med 293:844–847, 1975.
28. Wolstenholme J, Faed MJW, Robertson J, Lamont MA: Chromosome abnormality in couples with histories of multiple abortions. The outcome of pregnancies subsequent to ascertainment and a study of familial translocation carriers. Hum Genet 63:45–47, 1983.
29. Borgaonker DS, Blair SM, Lutz JB, Kelley T, Tice RR, Delaney NV, Hutchinson JR, Bias WB: Cytogenetic study of a 5/14 translocation in man. J Hered 64:299–300, 1973.
30. Tsenghi C, Metaxotou-Stavridaki C, Strataki-Benetou M, Kalpini-Mavrou A, Matsaniotis N: Chromosome studies in couples with repeated abortions. Obstet Gynecol 47:463–468, 1976.
31. Neibuhr E: A familial translocation t(6q+;8q−) identified by fluorescence microscopy. Humangenetik 18:189–192, 1973.
32. Wright Y, Clark WE, Breg WR: Craniorachischsis in a partially trisomic 11 fetus in a family with reproductive failure and a reciprocal translocation t(6p+;11q−). J Med Genet 11:69–76, 1974.
33. Davis JR, Rogers BB, Hagaman RM, Thies CA, Veomett IC: Balanced reciprocal translocations: Risk factors for aneuploid segregant viability. Clin Genet 27:1–19, 1985.
34. Hatzissevastou-Loukidou H, Bouli-Kalachani A, Malaka-Zafiriou A, Mantelenakis S: Reciprocal translocation of the long arm of chromosome 8 to the short arm of chromosome 7 in a woman with two spontaneous abortions. Hum Genet 70:379, 1985.
35. Ming PL, Parker R, Goodner DM: Familial occurrence of chromosome 7/12 translocation. Clin Genet 18:445–449, 1980.
36. De la Chapelle A, Schröder J, Kokkonen J: Cytogenetics of recurrent abortion or unsuccessful pregnancy. Int J Fertil 18:215–219, 1973.
37. Sulewski JM, Dang TP, Ferguson KA, Ward AP, Ladda RL: Chromosomal abnormalities associated with infertility. Obstet Gynecol 55:469–475, 1980.
38. Ford JH, Rofe RH, Pavy RP: Translocations involving chromosome 12. Hum Genet 58:144–148, 1981.
39. Kajii T, Meylan J, Mikamo K: Chromosome anomalies in three successive abortuses due to paternal translocation t(13q−;18q+). Cytogenet Cell Genet 13:426–436, 1974.
40. Leisti J, Aula P: Partial trisomy 1 (q42→ter). Clin Genet 18:371–378, 1980.
41. Penchaszadeh VB, Kardon NB, Davis JG: Distal trisomy 1q and monosomy 7q in two cousins due to paternal translocation (1;7) (q32;q32). Am J Hum Genet 34:138A, 1982.
42. Mason MK, Spencer DA, Rutter A: A case of partial (9p) trisomy in a family with a balanced translocation 46,XX,t(1p+;9−). J Med Genet 12:310–315, 1975.
43. Neu RL, Dennis NR, Lanman JT Jr, Bannerman RM: An infant with trisomy 9pter→9q22 resulting from 3:1 segregation in a 46,XX,t(1;9)(p36;q22) mother. Ann Genet 22:151–154, 1979.
44. Desangles F, Mourrieras P, Papouin-Rauzy M, Saliou P: Monosomie 1pter par translocation familiale (1p;9p). Ann Genet 26:53–55, 1983.
45. Wegner R: Translocation t(X;1) and the "critical region hypothesis." Hum Genet 61:79, 1982.
46. Chiyo HA, Kuroki Y, Matsui I, Niitsu M, Nakagome Y: A case of partial trisomy 3q. J Med Genet 13:525–527, 1976.

47. Biederman B, Bowen P: Partial trisomy 4q due to familial 2/4 translocation. Hum Genet 33:147–153, 1976.
48. Osztovics M, Kiss P: Familial translocation t(2;5)(p23;q31). Clin Genet 8:112–116, 1975.
49. Neu RL, Dennis NR, Fisher JE: Partial 2p trisomy in a 46,XX,der(5), t(2;5)(p23;p15)PAT infant. Ann Genet 22:33–34, 1979.
50. Mayer U, Schwanitz G, Grosse KP, Etzold R: Trisomie partielle 2p par translocation familiale 2/6. Ann Genet 21:172–176, 1978.
51. Francke U, Jones KL: The 2p partial trisomy syndrome. Am J Dis Child 130:1244–1249, 1976
52. Siffroi JP, Romani F, Viguie F, Lejeune C: Trisomie partielle pour le bras long du chromosome 2 par malsegregation d'une translocation t(2;7)(q321;p22) maternelle. Ann Genet 27:241–244, 1984.
53. Laurent C, Biemont M Cl, Guibaud P, Guillot J, Noel B, Quack B, Genevieve M, Cressens ML: Sept cas de trisomie 2q34→2qter par transmission familiale d'une translocation t(2;8)(q34;p23). Ann Genet 21:13–18, 1978.
54. Moirot H, Fessard Cl, Hemet J: Trisomie 9p par t(2:9)(q36;q31)mat. Ann Genet 20:45–51, 1977.
55. Howard-Peebles PN, Goldsmith JP: Duplication of region 2q31→2qter in a family with 2/9 translocation. Hum Hered 30:84–88, 1980.
56. Sills JA, Buckton KE, Raeburn JA: Severe mental retardation in a boy with partial trisomy 10q and partial monosomy 2q. J Med Genet 13:507–510, 1976.
57. Larson LM, Wasdahl WA, Saumur JH, Coleman ML, Hall JG, Dolan CR, Schutta CJ: Familial reciprocal translocation t(2;10)(p24;q26) resulting in duplication 2p and deletion 10q. Clin Genet 21:187–195, 1982.
58. Zabel B, Hansen S, Hartmann W: Partial trisomy 2q and familial translocation t(2:12)(q31;q24). Hum Genet 32:101–104, 1976.
59. Roberts SH, Mattina T, Laurence KM, Sorge G. Pavone L: Partial trisomy 12q: Report of a case and review. J Med Genet 18:470–473, 1981.
60. Tengstrom C, Wilska M, Kahkonen M, Autio S, Leisti J: Partial trisomy 12q: Clinical and cytogenetic observations. Clin Genet 28:112–117, 1985.
61. Forabosco A, Dutrillaux B, Toni G, Tamborino G, Cavazzutti G: Tanslocation équilibrée t(2;13)(q32;q33) familiale et trisomie 2q partielle. Ann Genet 16:255–258, 1973.
62. Giraud F, Mattei JF, Mattei MG: Trisomie 13 partielle par translocation t(2;13) maternelle. Ann Genet 20:203–208, 1977.
63. Stoll C, Messer J, Vors J: Translocation t(2;14) équilibrée chez une mere et trisomie partielle d'une partie du bras court d'un chromosome no. 2 chez deux de ses enfants. Ann Genet 17:193–196, 1974.
64. Coco R, Penchaszadeh VB: Partial trisomy 14q and familial translocation (2;14)(q12;q13). Ann Genet 20:41–44, 1977.
65. Armendares S, Salamanca-Gomez F: Partial 2p trisomy (p21→pter) in two siblings of a family with a 2p−;15q+ translocation. Clin Genet 13:17–24, 1978.
66. Giliberti P, Celona A, Della Pietra M, deMasi RV, Fioretti G, Pagano L, Renda S, Vetrella A, Ventruto V: Familial translocation 2;17 with partial trisomy 2q32→2qter. Ann Genet 23:249–250, 1980.
67. Rosenfeld W, Verma RS, Jhaveri R, Dosik H, Evans H: Partial duplication for the short arm of chromosome 2: The 2p23→pter syndrome. Ann Genet 25:28–31, 1982.

68. Turleau C, Chavin-Colon F, de Grouchy J, Repesse G, Beauvais P: Familial t(X;2)(p223;q323) with partial trisomy 2q and male and female balanced carriers. Hum Genet 37:97–104, 1977.
69. Baliesta F, Vehi L: Trisomie partielle pour la partie distale du bras court du chromosome 3. Ann Genet 17:287–290, 1974.
70. Dallapiccola B, Mastroiacovo PP, Montali E, Sommer A: Trisomy 4p: Five new observations and overview. Clin Genet 12:344–356, 1977.
71. Zabel B, Baumann W, Gehler J, Conrad G: Partial trisomy for short and long arm of chromosome no. 5. Two cases of two possible syndromes. J Med Genet 15:143–147, 1978.
72. Schwanitz G, Zerres K: Signes phenotypiques a differents ages d'une trisomie partielle 3p par translocation familiale 3/5. Ann Genet 27:167–172, 1984.
73. Fitch N: Partial trisomy 6. Clin Genet 14:181–185, 1978.
74. Pagano L, Fioretti G, Vetrella M, Risolo E, Casullo C, Celano A, Renda S, Rinaldi A, Ventruto V: Hereditary 3;6 translocation. Ann Genet 23:173–175, 1980.
75. Grace E, Sutherland GR, Stark GD, Bain AD: Partial trisomy of 7q resulting from a familial translocation. Ann Genet 16:51–54, 1973.
76. Novales MA, Fernandez-Novoa C, Hevia A, San Martin V, Galera H: Partial trisomy for the long arm of chromosome 7. Hum Genet 62:378–381, 1982.
77. Penchaszadeh VB, Coco R: Trisomy for the short arm of chromosome no. 10. J. Genet Hum 25:221–227, 1977.
78. Van Regemorter N, Vamos E, Viteux V, Hayez F, Pardou A, Flament-Durand J: Partial trisomy 3p in two siblings: Clinical and pathological findings. Clin Genet 19:531, 1981.
79. Ridler MAC, McKeown JA: 11q Aneuploidy: Partial monosomy and trisomy in the children of a mother with a t(3;11)(p27;q23) translocation. Hum Genet 52:101–106, 1979.
80. McCorquodale M, Erickson RP, Robinson M, Roszczipka K: Kleeblattschädel anomaly and partial trisomy for chromosome 13(47,XY,+der(13),t(3;13) (q24;q14). Clin Genet 17:409–414, 1980.
81. Byrd JR, Askew DE, McDonough PG: Cytogenetic findings in fifty-five couples with recurrent fetal wastage. Fertil Steril 28:246–250, 1977.
82. Simpson J, Zellweger H: Partial trisomy 14q− and parental translocation of no. 14 chromosome. Report of a case and review of the literature. J Med Genet 14:124–127, 1977.
83. Say B, Barber N, Bobrow M, Jones K, Coldwell JG: Familial translocation (3p15p) with partial trisomy for the upper arm of chromosome 3 in two sibs. J Pediatr 88:447–450, 1976.
84. Schwanitz G, Schmid RD, Grosse G, Grahn-Liebe E: Translocation familiale 3/22 mat avec trisomie partielle 3q. J Genet Hum 25:141–150, 1977.
85. McDermott A, Poulding R, Creery D: Cri-du-chat syndrome in a child with a 46,XX, der(5),t(4;5)(q32;p14)pat karyotype. Hum Genet 39:109–112, 1977.
86. Yunis E, Silva R, Egel H, Zuniga R, De Caballero OMT, Ramirez E, De Ruiz HP: Partial trisomy 5p. Hum Genet 43:231–237, 1978.
87. Andrle M, Erlach A, Killian W, Rett A: Two cases of trisomy 4p with translocation t(4p−;7q+) in several members of one family. Hum Genet 33:155–160, 1976.
88. Moric-Petrovic S, Laca Z, Krajgher A: Translocation réciproque dans la famille de deux proposants avec une trisomie partielle du chromosome 7q. Ann Genet 19:133–136, 1976.

89. Reynolds JF, Shires MA, Wyandt HE, Kelly TE: Trisomy 4p in four relatives: Variability and lack of distinctive features in phenotypic expression. Clin Genet 24:365–374, 1983.
90. Tranebjaerg L, Petersen A, Hove K, Rehder H, Mikkelsen M: Clinical and cytogenetic studies in a large (4;8) translocation family with pre- and postnatal Wolf syndrome. Ann Genet 27:224–229, 1984.
91. Stengel-Rutowski S, Warkotsch A, Schimanek P, Stene J: Familial Wolf's syndrome with a hidden 4p deletion by translocation of an 8p segment. Clin Genet 25:500–521, 1984.
92. Young RS, Hansen KL, Khodr GS: Five generations of t(4;8) (q35;ql3) leading to a case of partial 8q trisomy with consideration of potential pregnancy outcomes from translocated carriers. Clin Genet 25:522–527, 1984.
93. Issa M, Potter AM, Blank CE: Multiple congenital defects associated with trisomy for long arm of no. 4. J Med Genet 13:326–329, 1976.
94. Réthoré MOP, Ferrand J, Dutrillaux B. Lejeune J: Trisomie 9p par t(4;9) (q34;q21)mat. Ann Genet 17:157–161, 1974.
95. Turleau C, de Grouchy J, Chavin-Colon F, Roubin M, Brissaud PE, Repesse F, Safar A, Borniche P: Partial trisomy 9q: A new syndrome. Humangenetik 29:233–241, 1975.
96. Fryns JP, Azou M, Devliegher H, Eggermont E, van den Berghe H: A malformed newborn with 9p and 4q trisomy. Ann Genet 24:48–50, 1981.
97. Rolland M, Bourrouillou G, Elana G, Colombies P, Regnier C: Trisomie 10p partielle d'origine paternelle. Deux nouvelles observations dans deux familles différentes. Ann Genet 20:209–213, 1977.
98. Turleau C, Doussau de Bazignan M, Roubin M, de Grouchy J: Trisomie 10p. Une observation ancienne précisée par marquage. Ann Genet 19:61–64, 1976.
99. Cantu JM, Hernandez A, Nazara Z, Rolon A, Ramirez ML, Sanchez-Corona J, Rivera H: Simultaneous trisomy 10q24→qter and monosomy 4p16. Ann Genet 24:41–44, 1981.
100. Sanchez-Corona J, Garcia-Cruz D, Rivera H, Vaca G, Hernandez A, Rolon A, Cantu JM: Simultaneous partial trisomy 4p and monosomy 10p resulting from a paternal balanced translocation. Ann Genet 26:41–43, 1983.
101. Pauli RM, Kirkpatrick SJ, Meisner LF, Mijanovich JR, Spritz RA: Neonatal death in cousins with trisomy 10q and monosomy 4p due to a familial translocation. Clin Genet 21:340–347, 1982.
102. Tenge JR, Tang TT, Gallen WJ, Schanfield N, Hogg A: A case of 4p+ resulting from a balanced translocation in a carrier mother. Am Soc Clin Pathol ST-102:1–6, 1978.
103. Nielsen J, Rasmussen L, Lassen LB, Christiansen F: A family with a high risk of segregation for an autosomal unbalanced reciprocal translocation. Hum Genet 32:343–348, 1976.
104. Schrott HG, Sakaguchi S, Francke U, Lazzatti L, Fialkow PJ: Translocation t(4q−;13p+) in three generations resulting in partial trisomy of the long arm of chromosome 4 in the fourth generation. J Med Genet 11:201–205, 1974.
105. Stoll C, Roth MP: Partial 4q duplication due to inherited der(13), t(4;13) (q26;q34) mat in a girl with a deficiency of factor X. Hum Genet 53:303–304, 1980.
106. Fontatsch C, Flatz SD, Weitzel E: Partial trisomy 13 plus partial trisomy 4q due to unusual segregation of translocation chromosomes. Clin Genet 15:176–182, 1979.
107. Cohen MM, Ornoy A, Rosenmann A, Kohn G: An inherited translocation t(4;15)(p16;q22) leading to two cases of partial trisomy 15. Ann Genet 18:99–103, 1975.
108. Hustinx TWJ, Gabreels FJM, Kirkels VGHJ, Korten JJ, Scheres JMJC, Joosten EMG, Rutten FJ: Trisomy 4p in a family with a t(4;15). Ann Genet 18:13–19, 1975.

109. Bauknecht T, Betteken F, Vogel W: Trisomy 4p due to a paternal t(4p−;16p+) translocation. Hum Genet 34:227–230, 1976.
110. del Mazo J, Florit F, Calafell R, Abrisqueta JA: Four cases of partial trisomy 4p by preferential segregation in a familial 4p/17q balanced translocation. Hum Genet 66:370, 1984.
111. Canki N, Debevec M, Ranier S, Réthoré MO: Trisomie 4q26→4qter par translocation t(4:18)(q26;q23)mat. Ann Genet 20:195–198, 1977.
112. Dutrillaux B, Laurent C, Forabosco A, Noel B, Suerinc E, Biemont MC, Cotton JB: La trisomie 4q partielle. A propos de 3 observations. Ann Genet 18:21–27, 1975.
113. Yunis E, Giraldo A, Zuniga R, Egel H, Ramirez E: Partial trisomy 4q. Ann Genet 20:243–248, 1977.
114. Steele MW, Pan S, Mickell J, Senders V: Trisomy for the distal half of the long arm of chromosome no. 18. J Pediatr 85:827–829, 1974.
115. Sparkes RS, Francke U, Muller H, Toomey K: Partial 4q duplication due to inherited der(20),t(4;20)(q25;q13)mat. Ann Genet 20:31–35, 1977.
116. Baccichetti C, Tenconi R, Anglani F, Zacchello F: Trisomy 4q32→4qter due to a maternal 4/21 translocation. J Med Genet 12:425–427, 1975.
117. Darmady JM, Seabright M: A family with an inherited translocation involving the no. 4/no. 21 chromosomes. J Med Genet 12:408–411, 1975.
118. De la Chapelle A, Koivisto M, Schröder J: Segregating reciprocal (4:21)(q21;q21) translocation with proposita trisomic for parts of 4q and 21. J Med Genet 10:384–389, 1973.
119. Jenkins MB, Boyd L: Reciprocal translocation, 4q−;21p+, giving rise to Down's syndrome. J Med Genet 13:323–326, 1976.
120. Pfeiffer RA, Kessel EK, Soer KM: Partial trisomies of chromosomes 21 in man. Two new observations due to translocations 19;21 and 4;21. Clin Genet 11:207–213, 1977.
121. Giovannelli G, Forabosco A, Dutrillaux B: Translocation familiale t(4;22) (p11;p12) et trisomie 4p chez deux germains. Ann Genet 17:119–124, 1974.
122. Schwanitz G, Grosse KP: Partial trisomy 4p with translocation 4p-;22p+ in the father. Ann Genet 16:263–266, 1973.
123. Garlinger P, McGeary SA, Magenis E: Partial trisomy 22: A recognizable syndrome. Clin Genet 12:9–16, 1977.
124. Cervenka J, Djavadi GR, Gorlin RJ: Partial trisomy 4q syndrome: Case report and review. Hum Genet 34:1–7, 1976.
125. Morichon-Delvallez N, Couturier J, Frison B: Phenotype attenue de la trisomie 4p par translocation t(X;4) (p21.2;p13). Ann Genet 25:246–248, 1982.
126. Hashimoto T, Tsukino R, Chiyo H, Furuyama J: Reciprocal translocation t(5;6) (p13;q27) through three generations. Hum Genet 53:145–147, 1980.
127. Curry CJR, Loughman WD, Francke U, Hall BD, Golbus MS, Derstine J, Epstein CJ: Partial trisomy for the distal long arm of chromosome 5 (region q34→qter). Clin Genet 15:454–461, 1979.
128. Saadi AA, Moghadam HA: Partial trisomy of the long arm of chromosome 7. Clin Genet 9:250–254, 1976.
129. Bartsch-Sandhoff M, Liersch R: Partial duplication 5q syndrome: Phenotypic similarity in two sisters with identical karyotype (partial duplication 5q33→5qter and partial deficiency p23→pter). Ann Genet 20:281–284, 1977.
130. Rethore M, Couturier J, Villain E, Hambourg M, Lejeune J: Maladie du cri du chat et trisomie 8p par translocation paternelle t(5;8) (p1409;p12). Ann Genet 27:118–121, 1984.

131. Liberfarb RM, Atkins L, Holmes LB: A clinical chromosome associated with 5p duplication and 9p deletion. Ann Genet 23:26–30, 1980.
132. Back E, Vogel W, Hertel C, Schuchmann L: Trisomy 10p due to t(5;10)(p15;p11) segregating in a large sibship. Hum Genet 41:11–17, 1978.
133. Lazjuk GI, Lurie IW, Kirillova IA, Zaletajev DV, Gurevich DB, Shved IA, Ostrovskaya TI: Partial trisomy 5q and partial monosomy 5q within the same family. Clin Genet 28:122–129, 1985.
134. Laurent C, Biemont MC, Robert J, Dutrillaux B: Identification de deux translocations familiales. Ann Genet 17:279–281, 1974.
135. Noel B, Quack B, Réthoŕe MO: Partial deletions and trisomies of chromosome 13; mapping of bands associated with particular malformations. Clin Genet 9:593–602, 1976.
136. Bass HN, Sparkes RS, Crandall BF, Galos KJ, Howard J: A 45,XX,−5, −14,+t(5q;14q) mat cri du chat child. Ann Genet 21:56–59, 1978.
137. Abeliovich D, Yagupsky P, Bashan N: 3:1 Meiotic disjunction in a mother with a balanced translocation, 46,XX,t(5;14) (p15;q13) resulting in tertiary trisomy and tertiary monosomy offspring. Am J Med Genet 12:83–90, 1982.
138. Fried K, Tieder M, Beer S, Rosenblatt M, Krespin HI: Mental retardation with 45 chromosomes, 45,XX,−5,−14,+der(5),t(5;14)(p15;q13)mat due to familial balanced reciprocal translocation. J Med Genet 14:68–72, 1977.
139. Brimblecombe FSW, Lewis FJ, Vowles M: Complete 5p trisomy; 1 case and 19 translocations in 6 generations. J Med Genet 14:271–275, 1977.
140. Tzancheva M, Krachunova M, Damjanova Z: Trisomy 15q distal—a new chromosome syndrome? Clin Genet 19:542, 1981.
141. Turleau C, de Grouchy J, Bouveret JP: Distal trisomy 17q. Clin Genet 16:54–57, 1979.
142. Jinno Y, Matsuda I, Kajii T: Trisomy 17p due to a t(15;17)(p15;p11)pat translocation. Ann Genet 25:123–125, 1982.
143. Abrisqueta JA, Perez A, Aller V, DelMazo J, Goday C, Martin MA, DeTorres ML: Cri du chat syndrome and translocation t(5p−;18p+). J Genet Hum 24:173–182, 1976.
144. Funderburk SJ, Sparkes RS, Sparkes MC: Trisomy 20p due to a paternal reciprocal translocation. Ann Genet 26:94–97, 1983.
145. Chaganti RSK, Morillo-Cucci G, Friis L, Degnan M, German J: De novo appearance of a translocation t(5p;21q) and its transmission in both balanced and unbalanced forms to the next generation. Ann Genet 19:43–48, 1976.
146. Réthoré MO, Hoehn H, Rott HD, Couturier J, Dutrillaux B, Lejeune J: Analyse de la trisomie 9p per dénaturation ménagée. A propos d'un nouveau cas. Humangenetik 18:129–138, 1973.
147. Neu RL, Gallien JU, Steinberg-Warren N, Wynn RJ, Bannerman RM: An infant with trisomy 6q21→qter. Ann Genet 24:167–169, 1981.
148. Clark CE, Cowell HR, Telfer MA, Casey PA: Trisomy 6q25→6qter in two sisters resulting from maternal 6;11 translocation. Am J Med Genet 5:171–178, 1980.
149. Petit P, Fryns JP, van den Berghe H: Partial trisomy 13 with phenotype of Patau syndrome due to maternal reciprocal translocation t(6;13) (q25;q13). Ann Genet 23:57–59, 1980.
150. Nakagome Y, Icnuma K, Matsui I: Three translocations involving C or G group chromosomes. J Med Genet 10:174–177, 1973.
151. Breuning MH, Bijlsma JB, de France HR: Partial trisomy 6p due to familial translocation t(6;20) (p21;p13). A new syndrome? Hum Genet 38:7–13, 1977.

152. Taysi K, Chao WT, Monaghan N, Monaco MP: Trisomy 6q22→6qter due to maternal 6;21 translocation. Ann Genet 26:165–170, 1983.
153. Turleau C, de Grouchy J: Trisomy 6qter. Clin Genet 19:202–206, 1981.
154. Fryns JP, Petit P, Jaeken J, Eggermont E, Kleczkowska A, van den Berghe H: Partial distal 6p trisomy: A clinical entity. Ann Genet 26:50–52, 1983.
155. Moore CM, Pfeiffer RA, Craig-Holmes AP, Scott CI Jr, Meisel-Stosiek M: Partial trisomy 7p in two families resulting from different balanced translocations. Clin Genet 21:112–121, 1982.
156. Berger R, Turc C, Wachter H, Begue G: Partial 7q trisomy. Clin Genet 11:39–42, 1977.
157. Turleau C, de Grouchy J, Chavin-Colon F, Roubin M, Langmaid H: Trisomie 9p: Deux nouvelles observations. Ann Genet 17:167–174, 1974.
158. Penchaszadeh VB, Coco R: Partial 9 trisomy by 3:1 segregation of a balanced maternal translocation (7q+;9q−). J Med Genet 12:301–304, 1975.
159. Sutherland GR, Carter RF, Morris LL: Partial and complete trisomy 9: Delineation of a trisomy 9 syndrome. Hum Genet 32:133–140, 1976.
160. Johnson G, Bachman R, Roed T, Riddervold P: Partial trisomy 10p and familial translocation t(7;10) (p22;p12). Hum Genet 35:353;356, 1977.
161. Schleiermacher E, Schliebitz U, Steffens C: Brother and sister with trisomy 10p: A new syndrome. Humangenetik 23:163–172, 1974.
162. Martin-Lucas A, Perez-Castillo A, Abrisqueta JA, deTorres ML, Martin-Sempere MJ, del Mazo J, Aller V: Partial trisomy 13 due to maternal translocation t (7;13) (p22q14). Ann Genet 25:172–178, 1982.
163. Alfi OS, Donnell GN, Kramer SL: Partial trisomy of the long arm of chromosome no. 7. J Med Genet 10:187–189, 1973.
164. Carnevale A, Frías S, del Castillo V: Partial trisomy of the short arm of chromosome 7 due to a familial translocation rcp(7;14) (p11;p11). Clin Genet 14:202–206, 1978.
165. Markkanen A, Somer M, Nordstrom A: Distal trisomy 14q syndrome; a case report. Clin Genet 26:231–234, 1984.
166. Castel Y, Riviere D, Boucly JY, Toudic L: Trisomie 15q partielle par translocation maternelle t(7;15) (q35;q14). Ann Genet 19:75–79, 1976.
167. Vogel W, Siebers JW, Reinwein H: Partial trisomy 7q. Ann Genet 16:277–280, 1973.
168. Larson LM, Wasdahl WA, Jalal SM: Partial trisomy 7p associated with familial 7p;22q translocation. J Med Genet 14:258–261, 1977.
169. Mulcahy MT, Jenkyn J: The 9p trisomy syndrome: Two further cases arising from different familial translocations. Clin Genet 8:199–204, 1975.
170. Schwanitz G, Schamberger U, Rott HD, Wieczorec V: Partial trisomy 9 in the case of familial translocation 8/9 mat. Ann Genet 17:163–166, 1974.
171. Centerwall WR, Mayeski CA, Cha CC: Trisomy 9q−. A variant of the 9p trisomy syndrome. Humangenetik 29:91–98, 1975.
172. Clark CE, Telfer MA, Cowell HR: A case of partial trisomy 8p resulting from a maternal balanced translocation. Am J Med Genet 7:21–25, 1980.
173. Chieri P, Iolster N: Monosomy 10qter due to a balanced maternal translocation: t(10;8) (q23;p23). Clin Genet 24:147–150, 1983.
174. Fineman RM, Ablow RC, Breg WR, Wing SD, Rose JS, Rothman SLG, Warpinski J: Complete and partial trisomy of different segments of chromosome 8: Case reports and review. Clin Genet 16:390–398, 1979.
175. Nielsen J, Vetner M, Holm V, Askjaer SA, Reske-Nielsen E: A newborn child with karyotype 47,XX,+der(12)(12pter→12q12::8q24→8qter),t(8;12)(q24;q12)pat. Hum Genet 35:357–362, 1977.

176. Abuelo D, Peri DP, Henkle C, Richardson A: Partial trisomy 8 (trisomy 8q2106→8qter). J Med Genet 14:463–465, 1977.
177. Chiyo HA, Nakagome Y, Matsui I, Kuroki Y, Kobayashi H, Ono K: Two cases of 8p trisomy in one sibship. Clin Genet 7:328–333, 1975.
178. Gilgenkrantz S, Defeche C, Stehlin S, Gregoire MJ: Proximal trisomy 13. A family with balanced reciprocal translocation t(8;13) in seven members and robertsonian translocation t(13;14) in three members. Hum Genet 58:436–440, 1981.
179. Dumon J, Devos E, Leroy J, Vanden Bulcke L: Partial trisomy 8q due to 8q;13q translocation. Clin Genet 13:113, 1978.
180. Pfeiffer RA, Kessel E: Partial trisomy 15ql. Hum Genet 33:77–83, 1976.
181. Turleau C, de Grouchy J: Trisomy 18qter and trisomy mapping of chromosome 18. Clin Genet 12:361–371, 1977.
182. Fraisse J, Lauras B, Laselve A, Freycon F: Deux nouveaux cas de trisomie 10q24→10qter chez deux soeurs par translocation paternelle t(9;10) (q34;q24). Ann Genet 20:128–131, 1977.
183. Hoo JJ: 12p trisomy: A syndrome? Ann Genet 19:261–263, 1976.
184. Jotterand M, Juillard E: A new case of trisomy for the distal part of 13q due to maternal translocation t(9;13) (p21;q21). Hum Genet 33:213–222, 1976.
185. Schinzel A, Hayashi K, Schmid W: Further delineation of the clinical picture of trisomy for the distal segment of chromosome 13. Report of three cases. Hum Genet 32:1–12, 1976.
186. Howard-Peebles PN, Yarbrough KM, Stoddard GR, Rary JM: Translocation 9q/13q resulting in duplication (trisomy 9pter → 9q22) and deficiency (monosomy 13pter → 13q12). Clin Genet 11:46–52, 1977.
187. Prieto F, Badia L, Asensi F, Roques V: Two reciprocal translocations t(9p+;13q−) and t(13q−;21q+). A study of two families. Hum Genet 54:7–11, 1980.
188. Centerwall WR, Miller KS, Reeves LM: Familial "partial 9p" trisomy; six cases and four carriers in three generations. J Med Genet 13:57–61, 1976.
189. Short EM, Solitare GB, Berg WR: A case of partial 14 trisomy. 47,XY,(14q−)+ and translocation t(9p+;14q−) in mother and brother. J Med Genet 9:367–373, 1972.
190. Orye E, Verhaaren H, Van den Bogaert-Van Heesvelde AM: The 9p-deletion syndrome. Report of a patient with a 46,XX,9p− constitution due to a paternal t(9p−;15q+) translocation. Clin Genet 8:349–357, 1975.
191. Balicek P, Zizka J, Lichy J: A case of trisomy 9p in a family with translocation 9/15. Humangenetik 27:353–358, 1975.
192. Bergamo F, Crosato F, Francesconi D, Pasquali F, Zuffardi O: The 9p−deletion syndrome. A patient with a 45,XX,−9,−15,+t(9/15) constitution due to maternal 3:1 meiotic disjunction. Clin Genet 11:219–223, 1977.
193. Podruch PE, Weisskopf B: Trisomy for the short arms of chromosome 9 in two generations, with balanced translocations t(15p+;9q−) in three generations. J Pediatr 85:92–95, 1974.
194. Zaremba J, Zdzienicka E, Glowowska I, Abramowicz T, Taracha B: Four cases of 9p trisomy resulting from a balanced familial translocation (9;15) (q13;q11). J Ment Defic Res 18:153–190, 1974.
195. Soltan HC, Jung JH, Pyatt Z, Singh RP: Partial trisomy 9q resulting from a familial translocation t(9;16)(q32;q24). Clin Genet 25:449–454, 1984.
196. Aftimos SF, Hoo JJ, Parslow MI: Partial trisomy 9q due to maternal 9/17 translocation. Am J Dis Child 134:848–850, 1980.

197. Jaffray JY, Geneix A, Goumy P, Perissel B, Menut G, Malet P: Trisomie partielle 18(pter→q122) d'origine maternelle. Ann Genet 23:224–227, 1980.
198. Stoll C, Levy JM, Gardea A: Trisomy 9p in a girl whose mother has a translocation t(9;20) (q12;p13). Humangenetik 27:269–274, 1975.
199. Hoo JJ, Fischer A, Fuhrmann W: Familial tiny 9p/20p translocation. Ann Genet 25:249–252, 1982.
200. Subrt I, Blehova B, Pallova B: Trisomy 9p resulting from maternal 9/21 translocation. Hum Genet 32:217–220, 1976.
201. Jalbert P, Sele B, Jalbert H: Reciprocal translocations: A way to predict the mode of imbalanced segregation by pachytene-diagram drawing. Hum Genet 55:209–222, 1980.
202. Hadebank M, Faust J: Trisomy 9p and unusual translocation mongolism in siblings due to different 3:1 segregations of maternal translocation rcp (9;21) (p11;q11). Hum Genet 42:251–256, 1978.
203. Lewandowski RC, Yunis JJ, Lehrke R, O'Leary J, Swaiman KF, Sanchez O: Trisomy for the distal half of the short arm of chromosome 9. Am J Dis Child 130:663–667, 1976.
204. Archidiacono N, Rocchi M, De Vonderweid U, Filippi G: t(9/22) with centric fission and NOR translocation leading to a case of pure 9p trisomy in the offspring. Hum Genet 40:325–331, 1978.
205. Genest P, Roussy J, Genest FB, Dumas L: La trisomie 9p. Un cas de translocation t(9;22) (p11;p11)mat chez une fillette. Union Med Can 106:956–959, 1977.
206. Blank CE, Colver DCB, Potter AM, McHugh J, Lorber J: Physical and mental defect of chromosomal origin in four individuals of the same family. Trisomy for the short arm of 9. Clin Genet 7:261–273, 1975.
207. Philippe M, Requin C, Germain D, Licheron A, Hermier M: Trisomie 9p par translocation maternelle (9;22) (p11;q11). J Genet Hum 23:309–317, 1975.
208. Subrt I, Janovsky M, Jodl J: Partial trisomy 9q− chromosomal syndrome. Hum Genet 34:151–154, 1976.
209. Lurie IW, Lazjuk GI, Gurevich DB, Usoev SS: Genetics of the +p9 syndrome. Hum Genet 32:23–33, 1976.
210. Tusques J, Grislain JR, Andre MJ, Mainard R, Rival JM, Cadudal JL, Dutrillaux B, Lejeune J: Trisomie partielle 11q identifée grace a l'étude en "dénaturation ménagée" par la chaleur de la translocation équilibrée paternelle. Ann Genet 15:167–172, 1972.
211. Mulcahy MT, Jenkyn J, Masters PL: A familial 10/13 translocation: Partial trisomy C in an infant. Clin Genet 6:335–340, 1974.
212. Hornstein L, Soukup S: A recognizable phenotype in a child with partial duplication 13q in a family with t(10q;13q). Clin Genet 19:81–86, 1981.
213. Miro R, Templado C, Ponsa M, Serradell J, Marina S, Egozcue J: Balanced translocation (10;13) in a father, ascertained through the study of meiosis in semen, and partial trisomy 10q in his son. Hum Genet 53:179–182, 1980.
214. Hustinx TWJ, Ter Haar BGA, Scheres JMJC, Rutten FJ: Trisomy for the short arm of chromosome no. 10. Clin Genet 6:408–415, 1974.
215. Raoul O, Réthoré MO, Dutrillaux B, Michon L, Lejeune J: Trisomie 14q partielle. I. Trisomie 14q partielle par translocation maternelle t(10;14) (p152;q22). Ann Genet 18:35–39, 1975.
216. Canki N, Konjajev Z, Debevec M, Rainer S, Rethore M: Trisomie 10q24→10qter par translocation familiale t(10;14) (q24;q32). Ann Genet 23:176–178, 1980.
217. Turleau C, de Grouchy J, Chavin-Colin F, Denavit M, LeTouze P: La trisomie 14q distale. Ann Genet 26:165–170, 1983.

218. Yunis JJ, Sanchez O: A new syndrome resulting from partial trisomy for the distal third of the long arm of chromosome 10. J Pediatr 84:567–570, 1974.
219. Wegner R, Kunze J, Paust H: Monosomy 10qter due to a balanced familial translocation: t(10;16) (q25.2;q24). Clin Genet 19:130–133, 1981.
220. Prieur M, Forabosco A, Dutrillaux B, Laurent C, Bernasconi S, Lejeune J: La trisomie 10q24→10qter. Ann Genet 18:217–222, 1975.
221. Krøyer S, Niebuhr E: Partial trisomy 10q occurring in a family with a reciprocal translocation t(10;18)(q25;q23). Ann Genet 18:50–55, 1975.
222. Moric-Petrovic S, Laca Z, Krajgher A, Milosevic J: Deux cas de trisomie 10p partielle dus à une translocation paternelle t(10;18) (p13;q23). Ann Genet 19:195–197, 1976.
223. Lurie IW, Lazjuk GI, Gurevich DB, Kravtzova GI, Nedzved MK, Shved IA: Partial trisomy 10p in two generations. Hum Genet 41:235–241, 1978.
224. Cantu JM, Salamanca-Gomez F, Buentello L, Carnevale A, Armendares S: Trisomy 10p. A report of two cases due to a familial translocation rcp (10;21) (p11;p11). Ann Genet 18:5–11, 1975.
225. Yunis E, Silva R, Giraldo A: Trisomy 10p. Ann Genet 19:57–60, 1976.
226. Nakagome Y, Kobayashi H: Trisomy of the short arm of chromosome 10. J Med Genet 12:412–414, 1975.
227. Roux C, Taillemite JL, Baheux-Morlier G: Trisomie partielle 10q par translocation familiale t(10q−22q+). Ann Genet 17:59–62, 1974.
228. Fryns JP, Van Eygen M, Tanghe W, Van den Berghe H: Partial trisomy 14q due to familial t(14q−;11q+) translocation. Hum Genet 37:105–110, 1977.
229. Calva P, Frias S, Carnevale A, Reyes P: Partial trisomy 16q resulting from maternal translocation 11p/16q. Ann Genet 27:122–125, 1984.
230. Hatanaka K, Ozaki M, Suzuki M, Murata R, Fujita H: Trisomy 16q13→qter in an infant from a t(11;16) (q25;q13) translocation-carrier father. Hum Genet 65:311–315, 1984.
231. Laurent C, Biemont M C1, Bethenod M, Cret L, David M: Deux observations de trisomie 11q (q231→qter) avec la meme anomalie des organes génitaux externes. Ann Genet 18:179–184, 1975.
232. de France HF, Beemer FA, Senders RC, Gerards LG, Cats BP: Partial trisomy 11q due to paternal t(11q;18p); further delineation of the clinical picture. Clin Genet 25:295–299, 1984.
233. Delicado A, Lopez Pajares I, Vincente P, Garcia R: Partial trisomy 20. Ann Genet 24:54–56, 1981.
234. Jacobsen P, Hauge M, Henningsen K, Hobolth M, Mikkelsen M, Phillip J: An (11;21) translocation in four generations with chromosome 11 abnormalities in the offspring. Hum Hered 23:568–585, 1973.
235. Aurias A, Turc C, Michiels Y, Sinet PM, Graveleau D, Lejeune J: Deux cas de trisomie 11q (q231→qter) par translocation t(11;22) (q231;q111) dans deux familles différentes. Ann Genet 18:185–188, 1975.
236. Giraud F, Mattei JF, Mattei MG, Bernard R: Trisomie partielle 11q et translocation familiale 11-22. Humangenetik 28:323–347, 1975.
237. Ayraud N, Galiana A, Lloyd M, Deswarte M: Trisomie 11q (q231→qter) par translocation maternelle t(11;22) (q231-q111). Une nouvelle observation. Ann Genet 19:65–68, 1976.
238. Kessel E, Pfeiffer RA: 47,XY,+der(11;22) (q23;q12) following balanced translocation t(11;22) (q23;q12)mat. Remarks on the problem of trisomy 22. Hum Genet 37:111–116, 1977.

239. Noel B, Levy M, Réthoré MO: Trisomie partielle du bras long du chromosome 11 par malségrégation d'une translocation maternelle t(11;22)(q231;q111). Ann Genet 19:137–139, 1976.
240. Bofinger MK, Soukup SW: Cat eye syndrome; partial trisomy 22 due to translocation in the mother. Am J Dis Child 131:893–897, 1977.
241. Nakai H, Yamamoto Y, Kuroki Y: Partial trisomy of 11 and 22 due to familial translocation t(11;22) (q23;q11), inherited in three generations. Hum Genet 51:349–355, 1979.
242. Chauveau P, Gruchy D, Herouin C, Levesque L: A propos d'une nouvelle observation de trisomie 11q secondaire a une translocation maternelle t(11;22) (q23.1;q11.1). Ann Genet 23:213–215, 1980.
243. Réthoré MO, Kaplan JC, Junien CL, Cruveiller J, Dutrillaux B, Aurias A, Carpentier S, Lafourcade J, Lejeune J: Augmentation de l'activité de la LDH-B chez un garcon trisomique 12p par malségrégation d'une translocation maternelle t(12;14) (q12;p11). Ann Genet 18:81–87, 1975.
244. Turleau C, de Grouchy J, Bocquentin F, Roubin M, Chavin-Colon F: Trisomie 14q partielle. II. Trisomie 14q partielle par translocation maternelle (12;14) (q244;q21). Ann Genet 18:41–44, 1975.
245. Johnson G, Bachman R: A 46,XX,del(18)(pter→p1100)cebocephalic child from a 46,XX,t(12:18) (18pter→18p1100::12qter→12pter) normal parent. Hum Genet 34:103–106, 1976.
246. Fried K, Bar-Yochai A, Rosenblatt M, Mundel G: Partial 18 trisomy (with 47 chromosomes) resulting from a familial maternal translocation. J Med Genet 15:76–78, 1978.
247. Tenconi R, Giorgi PL, Tarantino E, Formica A: Trisomy 12p due to an adjacent 1 segregation of a maternal reciprocal translocation t(12;18) (p11;q23). Ann Genet 21:229–233, 1978.
248. Hemming L, Brown R: Partial trisomy 12q associated with a familial translocation. Clin Genet 16:25–28, 1979.
249. Armendares S, Salamanca-Gomez F, Nava S, Ramires S, Cantü JM: The 12p trisomy syndrome. Ann Genet 18:89–94, 1975.
250. Biederman B, Bowen P, Robertson C, Schiff D: Partial trisomy 12p due to t(12;21)pat translocation. Hum Genet 36:35–41, 1977.
251. Hobolth N, Jacobsen P, Mikkelsen M: Partial trisomy 12 in a mentally retarded boy and translocation (12;21) in his mother. J Med Genet 11:229–302, 1974.
252. Hansteen IL, Schirmer L, Hestetun S: Trisomy 12p syndrome. Evaluation of a family with a t(12;21) (p12.1p11) translocation with unbalanced offspring. Clin Genet 13:339–349, 1978.
253. Pena SDJ, Ray M, McAlpine PJ, Ducasse C, Briggs J, Hamerton JL: Tertiary trisomy 14; is there a syndrome? Birth Defects 12:113–118, 1976.
254. Zergollern L, Begovic D: Proximal and distal partial trisomy 13 in the same family. Clin Genet 19:545–546, 1981.
255. Duckett DP, Roberts SH: Adjacent 2 meiotic disjunction. Report of a case resulting from a familial 13;15q balanced reciprocal translocation and review of the literature. Hum Genet 58:377–386, 1981.
256. Rivas F, Rivera H, Plasencia ML, Ibarra B, Cantu JM: The phenotype in partial 13q trisomies, apropos of a familial (13;15) (q22;q26) translocation. Hum Genet 67:86–93, 1984.
257. Anneren G, Gustavson K-H: A boy with proximal trisomy 15 and a male foetus with distal trisomy 15 due to a familial 13p;15q translocation. Clin Genet 22:16–21, 1982.

258. Wilroy RS, Summit RL, Martens P, Manford Gooch W: Partial monosomy and partial trisomy for different segments of chromosome 13 in several individuals of the same family. Ann Genet 20:237–242, 1977.
259. Rosenmann A, Isacson M, Cohen R, Segal M, Cohen MM: Partial trisomy 18 (q11→qter) in an infant and aborted fetus resulting from a balanced paternal translocation t(13;18) (q32;q11). Ann Genet 21:60–64, 1978.
260. Murthy DSK, Patel ZM, Ambani LM: Trisomy 18q: 46,XX,13q+,t(13;18) (q32;q11) in a newborn associated with multiple congenital anomalies due to paternal reciprocal translocation. Clin Genet 18:233–238, 1980.
261. Steinbach P: Partial trisomy 13(q14→qter) due to a familial translocation t(13;18) (q14;q23). Hum Genet 58:444–445, 1981.
262. Centerwall W, Francke U: Familial trisomy 20p. Five cases and two carriers in three generations. A review. Ann Genet 20:77–83, 1977.
263. Schinzel A: Trisomy 20pter→q11 in a malformed boy from a t(13;20) (p11;q11) translocation-carrier mother. Hum Genet 53:169–172, 1980.
264. Pangalos C, Couturier J: Partial trisomy 13(q21.3→qter) resulting from a maternal translocation t(13;21). Ann Genet 24:179–181, 1981.
265. Young SR, Donovan DM, Greer HA, Burch K, Potter DC: Tertiary trisomy 47,XX,+14q resulting from maternal balanced translocation, 46,XX,t(14;16)(q11;q24). Hum Genet 33:331–334, 1976.
266. Laurent C, Dutrillaux B, Biemont M C1, Genoud J, Bethenod M: Translocation t(14q−;21q+) chez le pèṙe: Trisomie 14 et monosomie 21 partielles chez la fille. Ann Genet 16:281–284, 1973.
267. Francke U, Busby N, Shaw D, Hansen S, Brown MG: Intrachromosomal gene mapping in man. Somatic Cell Genet 2:27–40, 1976.
268. Yunis E, Gonzales JT, De Caballero OMT: Partial trisomy 16q−. Hum Genet 38:347–350, 1977.
269. Nevin NC, Coffey WW, Nevin J, Reid MM: Partial trisomy 16q in two boys resulting from a maternal translocation, t9 (15;16) (p12;q11). Clin Genet 24:375–379, 1983.
270. Réthoré MO,Dutrillaux B, Lejeune J: Translocation 46,XX,t(15;21) (q13;q22.1) chez la mère de deux enfants atteints de trisomie 15 et de monosomie 21 partielles. Ann Genet 16:271–275, 1973.
271. Raoul O, Carpentier S, Dutrillaux B, Mallet R, Lejeune J: Trisomies partielles du chromosome 21 par translocation maternelle t(15;21) (q262; q21). Ann Genet 19:187–190, 1976.
272. Zabel B, Baumann W: Trisomie partielle pour la partie distale du bras long du chromosome 15 par translocation X/15 maternelle. Ann Genet 20:285–289, 1977.
273. Stern LM, Murch AR: Pseudohermaphroditism with clinical features of trisomy 18 in an infant trisomic for parts of chromosomes 16 and 18: 47,XY,der(18),t(16;18) (p12;q11)mat. J Med Genet 12:305–308, 1975.
274. Pawlowitzki IH, Grobe H, Holzgreve W: Trisomy 20q due to maternal t(16;20) translocation. Clin Genet 15:167–170, 1979.
275. Geraedts JPM. Leschot NJ, Veenema H: Segregation analysis of a translocation (16;21) (p11;q22) in a large pedigree. Clin Genet 28:36–41, 1985.
276. Rethore M, Lafourcade J, Couturier J, Harpey JP, Hamet M, Engler R, Alcindor LG, Lejeune J: Augmentation d'activite de l'adenine phosphoribosyl transferase chez un enfante trisomique 16q22.2→16qter par translocation (16;21) (q22.2;q22.2)pat. Ann Genet 25:36–42, 1982.

277. Taylor KM, Wolfinger HL, Brown MG, Chadwick DL, Francke U: Partial trisomy 20p derived from a t(18;20) translocation. Hum Genet 34:155–162, 1976.
278. Niazi M, Coleman DV, Saldana-Garcia P: Partial trisomy 18 in a family with a translocation (18;21) (q21;q22). J Med Genet 15:148–151, 1978.
279. Lange M, Alfi OS: Trisomy 19q. Ann Genet 19:17–21, 1976.
280. Sanchez O, Mamunes P, Yunis JJ: Partial trisomy 20 (20q13) and partial trisomy 21 (21pter→21q213). J Med Genet 14:459–462, 1977.
281. Cohen MM, Davidson RG, Brown JA: A familial F/G translocation t(20p−;22q+) observed in three generations. Clin Genet 7:120–127, 1975.
282. Boué A, Boué J: Chromosome structural rearrangements and reproductive failure. Chromosomes Today 7:281–290, 1981.
283. Schwartz S, Palmer CG, Yu P-L: Evaluation of factors differentiating translocations ascertained in couples with fetal wastage and translocations ascertained through an unbalanced carrier. Am J Hum Genet 34:142A, 1982.
284. Aurias A, Prieur M, Dutrillaux B, Lejeune J: Systematic analysis of 95 reciprocal translocations of autosomes. Hum Genet 45:259–282, 1978.
285. Friedman JM: Familial occurrence and recurrence of 3:1 segregation in reciprocal translocation heterozygotes. Am J Hum Genet 34:90A, 1982.
286. Stene J, Stengel-Rutkowski S: Genetic risks for familial reciprocal translocations with special emphasis on those leading to 9p, 10p and 12p trisomies. Ann Hum Genet 46:41–74, 1982.

The Cytogenetics of Mammalian Autosomal
Rearrangements, pages 453–470

Chapter 12
Structural Chromosome Rearrangements in Couples With Recurrent Fetal Wastage: The Leuven Experience

Jean-Pierre Fryns, Alice Kleczkowska, Elisabeth Kubień, and Herman Van den Berghe

INTRODUCTION

Several years following the introduction of various banding techniques in human cytogenetics, a number of centers have now accumulated sufficient data to document the high incidence of structural chromosomal rearrangements in couples with recurrent fetal wastage (RFW) [1–12]. In this chapter, we present a review of the cytogenetic data in 1,555 couples who were examined for RFW at the Leuven Centre for Human Genetics. We included in the study all couples with at least two first trimester abortions and those in whom a spontaneous first trimester abortion was preceded or followed by a second or third trimester fetal death, particularly with multiple congenital malformations. Besides orcein staining, G- and R-banding techniques were routinely applied in all patients. After detection of a chromosome abnormality, additional banding techniques (Q-, C-, and NOR-staining) were used, if they were necessary for further identification of the translocation type and the respective breakpoints.

GENERAL DATA

The 1,555 couples (3,110 patients) were examined in the period 1970–1985. During that period a constitutional chromosome analysis (using the same banding techniques as mentioned above) was performed for different indications in a total number of 52,179 patients. Table I summarizes the number and respective incidence of different structural rearrangements de-

Centre for Human Genetics, University of Leuven, B-3000 Leuven, Belgium

TABLE I. Number and incidence of structural chromosomal anomalies

	In the total number of 52,179 chromosomally investigated patients	In 1,555 couples with recurrent fetal wastage
rcp Translocations	167 (3.2 in 1,000)	56 (36 in 1,000)
Rob translocations		
D/D	76 (1.5 in 1,000)	23 (15 in 1,000)
D/G	39 (0.7 in 1,000)	4 (2.5 in 1,000)
G/G	27 (0.5 in 1,000)	1 (0.6 in 1,000)
Inversions		
Pericentric	25 (0.5 in 1,000)	6 (3.9 in 1,000)
Paracentric	19 (0.36 in 1,000)	4 (2.5 in 1,000)
Small accessory chromosomes	32 (0.6 in 1,000)	3 (1.9 in 1,000)
Xp and Xq deletions	13 (0.2 in 1,000)	1 (0.6 in 1,000)
r(21)	1 (0.01 in 1,000)	1 (0.6 in 1,000)
Total	399 (7.6 in 1,000)	99 (63.6 in 1,000)

tected in 1,555 couples with RFW, and it shows the incidences of the same structural rearrangements in a total number of 52,179 chromosomally investigated patients. In addition to reciprocal balanced translocations and Robertsonian D/D, D/G, and G/G translocations, we also discuss para- and pericentric inversions, the small accessory chromosomes, Xp and Xq deletions, and one patient with a ring chromosome 21. Pericentric inv(9)(p11q13) was excluded on the basis that it is considered to be a population variant only. In the selected group of couples with RFW, the incidence of these types of chromosomal rearrangements was 6.36% versus 0.76% in the total population examined for various indications. The present study indicates that the frequency of chromosomal abnormalities in couples with RFW, at least with the available cytogenetic techniques at this moment, is 6–7%. Osztovics et al. [7] found 4.78% cytogenetic abnormalities in a large series of 418 couples with RFW. During the past few years, other investigators have published their results. In most reviews the number of couples studied was smaller than 200 and the frequency of cytogenetic abnormalities varied greatly from 0–4.7% [8,9,13–27]. This discrepancy is probably due to differences in selection and referral of patients for cytogenetic examination. Some studies indicated that the frequency of chromosomal aberrations increased with a higher number of early abortions (more than three) and/or in couples with late fetal wastage associated with multiple congenital malformations [4,28], but this was confirmed neither by others nor by the present study.

As summarised in Table I, more than 50% (56 of 99 index patients) of all chromosome abnormalities detected were balanced reciprocal translocations, and a further 30% constituted Robertsonian (mostly D/D) translocations. This is in agreement with a review by Tharapel et al. [12] who compiled the cytogenetic results taken from 79 published surveys of couples with two or more pregnancy losses (8,208 women and 7,834 men). In the rest of our data peri- and paracentric inversions accounted for 6 and 4% of cytogenetic abnormalities and small accessory chromosomes, an X chromosome deletion, and an autosomal ring chromosome r(21) were found in the remaining five index patients.

RECIPROCAL BALANCED TRANSLOCATION (rcp)

A total of 167 index patients with different reciprocal translocations were detected among 52,179 persons karyotyped in our centre for constitutional chromosome aberrations. The frequency of rcp translocations in this population (3.2 in 1,000, Table I) is thus obviously greater than that of the general population (1 in 800 to 1 in 1,000) [29–32]. As indicated in Table II, the probands were classified in three groups according to different indications. In the first group, the index patients presented with mental retardation and/

TABLE II. Indications for karyotyping in 167 index patients with balanced reciprocal translocations

	Origin	No.
MR/CM syndrome		
Unbalanced karyotype	Familial	26
	De novo	11
	Unknown	5
Balanced karyotype	Familial	18
	De novo	12
	Unknown	9
Normal karyotype in index patient, rcp in one parent		3
Unknown karyotype, rcp in one parent		1
Spontaneous abortions and other reproductive failure	Familial	16
	De novo	5
	Unknown	41
Other indications for karyotyping	Familial	4
	De novo	3
	Unknown	13
Total		167

or congenital malformations (MR/CM) (total of 85 index patients) [33]. In the second group, indications for karyotyping were various types of reproductive failure: repeated spontaneous abortions, stillbirths, primary and secondary infertility (total of 62 index patients but only 56 patients with RFW). In the third group of so-called fortuitous findings chromosome analysis was performed for different indications: prenatal diagnosis for advanced parental age, prenatal diagnosis for different congenital malformations and genetic counseling (total of 20 index patients). In 68/167 cases, the inheritance of the rcp translocations could not be clarified by family studies; hence the category of unknown origin (Table II).

Table III lists the type and origin of 56 rcp translocations found in 1,555 couples with RFW; more than one-third of all rcp translocations in our material were thus detected because of RFW. On the other hand, it should be stressed that the frequency of rcp translocations in those 1,555 couples with RFW was 30-fold higher than in the general population and also 10-fold higher than in the whole selected population karyotyped for constitutional chromosome aberrations in our centre (Table I). The male/female ratio of these 56 rcp heterozygotes was found to be 24:32. This slight prevalence of female carriers over male carriers of rcp translocations is very similar to that found by Lippman-Hand and Vekemans [8].

Table IV illustrates the striking similarities between different breakpoints found in the present study on RFW and those found in previous studies on constitutive chromosome breakage [34–36] and also experimentally induced constitutive fragile sites [37].

Chromosome investigation was successful in seven spontaneous abortions; in four of them a balanced karyotype was found, and in the remaining three abortuses unbalanced karyotypes were detected. Prenatal diagnosis in 37 pregnancies revealed 18 balanced reciprocal translocation fetuses and four other fetuses with unbalanced karyotype (Table V). The chromosomally unbalanced fetuses constituted 10.8% of all pregnancies monitored prenatally; this percentage does not greatly differ from that found previously by Boué et al. [38]. This frequency is lower than the approximately 25% chromosomally abnormal fetuses detected with amniocentesis after the birth of a previous child with multiple congenital anomalies and unbalanced karyotype.

ROBERTSONIAN TRANSLOCATIONS (Rob)

Up to the present, we have detected Rob translocations in 142 index patients among the total number of 52,179 chromosomally investigated pa-

TABLE III. Type and origin of balanced reciprocal translocations in 56 couples with RFW

Female carriers	Male carriers
1. 46,XX,t(1;2)(p36;q13)	1. 46,XY,t(1;2)(p22.2;q35)
2. 46,XX,t(1;3)(p22;p27),9qh+	2. 46,XY,t(1;2)(p36;q27)
3. 46,XX,t(1;3)(q32;q21)	3. 46,XY,t(1;4)(p31.2;p15.3) de novo
4. 46,XX,t(1;4)(q44;q22) mat	4. 46,XY,t(1;4)(q32;q31)
5. 46,XX,t(1;10)(q12;q26) mat	5. 46,XY,t(1;6)(p34;q16) pat
6. 46,XX,ins(1;11)(q23q24;p12), t(1;14)(q23;q13) de novo	6. 46,XY,t(1;12)(p32;q23), 1qh+
7. 46,XX,t(1;11)(q44;q23)	7. 46,XY,t(1;15)(q32.1;q23.1)
8. 46,XX,t(2;5)(p12;p15)	8. 46,XY,t(1;19)(p13;p12)
9. 46,XX,t(2;5)(p14;p15) pat	9. 46,XY,t(2;3)(p13;p14)
10. 46,XX,t(2;8)(q23;p11)	10. 46,XY,t(2;6)(q33;q27) fam
11. 46,XX,t(2;9)(q23;p24)	11. 46,XY,t(2;7)(p24;p21) pat
12. 46,XX,t(2;18)(q31;q22) de novo (fetus; parents nl)	12. 46,XY,t(2;14)(q11;q24)
13. 46,XX,t(3;4)(q25;q31)	13. 46,XY,t(3;6)(q25.3;q11.2) de novo
14. 46,XX,t(3;15)(p23;q15)	14. 46,XY,t(4;6)(p15;q24)
15. 46,XX,t(3;7)(q12.1;q35.3) mat	15. 46,XY,t(4;10)(p16;q26) mat
16. 46,XX,t(3;15)(p21;q25)	16. 46,XY,t(5;7)(q35;p12)
17. 46,XX,t(5;7)(q35;q11)	17. 46,XY,t(5;11)(p15;q11)
18. 46,XX,t(5;16)(q15;q13)	18. 46,XY,t(6;11)(q21;q25)
19. 46,XX,t(5;16)(q35.3;p11.2)	19. 46,XY,t(7;12)(q31.2;q24.3) mat
20. 46,XX,t(5;18)(q12;q11)	20. 46,XY,t(7;18)(p13;q23)
21. 46,XX,t(6;10)(q15;q22)	21. 46,XY,t(8;22)(q24;q12)
22. 46,XX,t(6;18)(q23;p11)	22. 46,XY,t(11;16)(q13;q13) mat
23. 46,XX,t(7;8)(q36;p12) pat	23. 46,XY,t(11;22)(q23;q13) fam
24. 46,XX,t(7;12)(p15;p13)	24. 46,XY,t(15;22)(q23;q12)
25. 46,XX,t(7;14)(p25;q24)	
26. 46,XX,t(8;11)(p12;p14)	
27. 46,XX,t(10;13)(q22.1;q24.1),9qh+	
28. 46,XX,t(10;17)(p11;q23) mat	
29. 46,XX,t(12;15)(q24.3;q24)	
30. 46,XX,t(13;17)(q12;q21.1)	
31. 46,XX,t(18;19)(q12;q13)	
32. 46,X,t(X;9)(p11;q13)	

tients (2.7 in 1,000). In more than half of them (76 out of 142), a D/D translocation was found (1.5 in 1,000 patients), whereas D/G and G/G translocations were found in 39 (0.7 in 1,000) and 27 (0.5 in 1,000) patients, respectively. Table VI summarizes indications for karyotyping in those 142 Rob carriers. It is interesting to note that in almost half of the D/D translocations, reproductive failure was the indication for karyotyping. This is in contrast to D/G and G/G translocations; of the 39 index patients with D/G

TABLE IV. Breakpoints found in couples with RFW (Robertsonian translocations are excluded) in comparison to previous studies on chromosome breakage

Breakpoints found in the present study	Chromosome rearrangements detected in amniocentesis [34] and in spontaneous abortions, stillbirths, and livebirths [35]	Experimentally induced constitutive chromosome breakpoints [37]	Rare autosomal fragile sites [36]
1p36	*1p36*		
1p35.3			
1p34			
1p32	*1p32*		
1p31.2		*1p31.2*	
1p22 and 1p22.1	*1p22*		
1p13			
1q12			
1q23			
1q32 and 1q32.1	*1q32*		
1q44		*1q44.1*	
2p24	*2p24*	*2p24.2*	
2p14			
2p13	*2p13*	*2p13*	
2p12			
2q11			*2q11*
2q13	*2q13*		
2q21		*2q21.3*	
2q23			
2q31	*2q31*		
2q33	*2q33*		
2q35			
2q37		*2q37.3*	
3p27			
3p25			
3p23			
3p21			
3p14		*3p14.2*	
3p13			
3p12			
3q12.1			
3q13			
3q21			
3q25 and 3q25.3			
4p16		*4p16.1*	
4p15 and **4p15.3**			
4q22			
4q31		*4q31.1*	
5p15			
5p13 and 5p13.3			

(continued)

TABLE IV. Breakpoints found in couples with RFW (Robertsonian translocations are excluded) in comparison to previous studies on chromosome breakage (continued)

Breakpoints found in the present study	Chromosome rearrangements detected in amniocentesis [34] and in spontaneous abortions, stillbirths, and livebirths [35]	Experimentally induced constitutive chromosome breakpoints [37]	Rare autosomal fragile sites [36]
5q12			
5q13 and 5q13.3			
5q15			
5q35 and 5q35.3	*5q35*		
6q11.2			
6q15			
6q16			
6q21	*6q21*		
6q23			
6q24			
6q27			
7p25			
7p21			
7p15 and 7p15.3			
7p13	*7p13*		
7p12			
7q11 and 7q11.2			
7q22.1	*7q22*		
7q31.2		*7q31.2*	
7q35.3			
7q36			
8p23			
8p12			
8p11			
8q24		*8q24.1; and 8q24.3*	
9p24			
9q13			
10p11			
10q22 and 10q22.1		*10q22.1*	
10q26		*10q26.13*	
11p14		*11p14.2*	
11p12			
11q11			
11q13	*11q13*		*11q13*
11q23	*11q23*	*11q23.3*	*11q23*
11q25			
12p13			
12q13	*12q13*		*12q13*
12q23			

(continued)

TABLE IV. Breakpoints found in couples with RFW (Robertsonian translocations are excluded) in comparison to previous studies on chromosome breakage (continued)

Breakpoints found in the present study	Chromosome rearrangements detected in amniocentesis [34] and in spontaneous abortions, stillbirths, and livebirths [35]	Experimentally induced constitutive chromosome breakpoints [37]	Rare autosomal fragile sites [36]
12q24 and 12q24.3			
13q12			
13q24.1			
14q13		*14q13*	
14q24	*14q24*	*14q24.11*	
15q15			
15q23			
15q24			
15q25			
16p11.2			
16q13			
17q21.1			
17q23		*17q23.1*	
18p11			
18q11			
18q12		*18q12.2*	
18q22			
18q23			
19p12			
19q13			
21p11			
21q22 or 23			
22q***12***	*22q12*	*22q12.2*	
22q13			
Xp21.01			
Xp11			

Breakpoints found in the present study two or more times are in bold type. Breakpoints also found in other studies are in italic type.

Rob only four (10%) were detected because of RFW, and of the 27 G/G translocations, only one was detected because of repetitive spontaneous abortions. Thus in the study of 1,555 couples with RFW, 28 spouses were found to have Rob translocations, with D/D translocations being the most prevalent (Table I).

Table VII summarizes the type and origin of Rob translocations in parents with RFW. Compared to the rcp heterozygotes (32:24—female:male car-

TABLE V. Results of 37 prenatal diagnoses in rcp translocation carriers with RFW

Balanced karyotype of the fetus	Unbalanced fetal karyotype	Normal	No.
1. 46,XX,t(1;4)(q44;q22) mat	1. 46,XY,der(11),ins(1;11)(q23q24;p12) mat	46,XY	7
2. 46,XX,t(1;4)(q44;q22) mat	2. 47,XY,+21,ins(1;11)(q23q24;p12),t(1;14)(q23;q12) mat	46,XX	8
3. 46,XY,t(1;4)(q44;q22) mat	3. 46,XY,der(5),t(2;5)(p12;p15) mat		
4. 46,XY,t(1;6)(p34;q13) pat	4. 47,XY,+18,t(5;18)(q12;q11) mat		
5. 46,XX,t(1;10)(q12;q26) mat			
6. 46,XY,t(1;5)(q32.1;q23.1) pat			
7. 46,XY,t(2;5)(p14;p15) mat			
8. 46,XY,t(3;15)(p21;q25) mat			
9. 46,XY,t(4;10)(p16;q26) pat, inv (6)(p25q21) de novo			
10. 46,XX,t(5;7)(q35;p12) pat			
11. 46,XX,t(6;11)(q21;q25) pat			
12. 46,XY,t(7;8)(q36;p12) mat			
13. 46,XX,t(7;8)(q36;p12) mat			
14. 46,XY,t(7;12)(q31.2;q24.3) pat			
15. 46,XY,t(7;12)(p15;p13) mat			
16. 46,XX,t(7;18)(p13;q23) mat			
17. 46,XY,t(11;16)(q13;q13) pat			
18. 46,Y,t(X;9)(p11;q13) mat			

riers), female carriers are more prevalent in this group (20 female carriers versus 8 male carriers). An explanation for the difference in sex distribution could be that the lower detection rate of Rob translocations in male carriers is possibly because of the lower risk of having chromosomally abnormal fetuses. This results in turn from the decreased possibility of fertilization by chromosomally unbalanced spermatozoa. On the other hand, it is known that Rob translocations in males may lead to sterility by a spermatogenic arrest [39–45]. Several studies have estimated the incidence of Robertsonian D/D translocations in the general population. Court Brown [46] has shown that t(Dq;Dq) is the commonest single cause of human chromosome heterozygosity, occurring with a frequency of about 1 in 1,000 adults. The latest data for the Canadian newborn and Danish newborn populations do not differ greatly and are 1.17 in 1,000 and 1.34 in 1,000, respectively [31,32]. In the present study, RFW D/D translocations were found 15 times more frequently than in a general population. Of these, 13/14 translocations were by far the most frequent (19 of 23; see Table VII).

Prenatal diagnosis was performed in 13 pregnancies, and normal karyotypes were detected in seven fetuses, whereas Rob translocations identical to those in parents were found in six other fetuses. No unbalanced fetal karyotype was detected in this series. Thus the risk of liveborn trisomy 13 children

TABLE VI. Indications for karyotyping in 142 index patients with Robertsonian translocations

Indication	Origin	No.
Robertsonian D/D translocations (total of 76 index patients)		
1. MR/CM syndrome		
Unbalanced karyotype	Familial	3
	De novo	3
	Unknown	1
Balanced karyotype	Familial	4
	De novo	2
	Unknown	9
Normal karyotype in index patient, Rob in one parent		4
2. Spontaneous abortions and	Familial	8
other reproductive failure	Unknown	27
3. Other indications for karyotyping	Familial	4
	Unknown	11
Robertsonian D/G translocations (total of 39 index patients)		
1. Down syndrome	Familial	9
	De novo	14
	Unknown	3
2. Spontaneous abortions and	Familial	3
other reproductive failure	Unknown	1
3. Other indications for karyotyping	Familial	3
	De novo	1
	Unknown	5
Robertsonian G/G translocations (total of 27 index patients)		
1. Down syndrome	Familial	3
	De novo	13
	Unknown	8
2. Unbalanced karyotype and MR/CM syndrome of unknown origin		1
3. Spontaneous abortions		1
4. Other indications for karyotyping		1

is relatively low, estimated as being not higher than 2% [47]. On the other hand, one may expect that for Rob D/G and particularly Rob G/G the risk of unbalanced offspring may be higher, considering the indications for karyotyping in those translocations (Table VI).

PERICENTRIC AND PARACENTRIC INVERSIONS

There have been several recent studies of peri- and paracentric inversions in man [48–53]. We detected 25 pericentric inversion carriers [excluding

TABLE VII. Type and origin of Robertsonian translocations in 28 couples with recurrent fetal wastage

Female carriers	No.	Male carriers	No.
45,XX,−13,−14,+t(13q;14q) mat	3	45,XY,−13,−14,+t(13q;14q) unknown	6
45,XX,−13,−14,+t(13q;14q) pat	1	45,XY,−13,−15,+t(13q;15q) unknown	1
45,XX,−13,−14,+t(13q;14q) unknown	9	45,XY,−15,−22,+t(15q;22q) mat	1
45,XX,−13,−15,+t(13q;15q) pat	1	Total	8
45,XX,−14,−15,+t(14q;15q) fam	1		
45,XX,−D,−D,+t(Dq;Dq) unknown	1		
45,XX,−14,−21,+t(14q;21q) unknown	1		
45,XX,−14,−22,+t(14q;22q) fam	1		
45,XX,−15,−22,+t(15q;22q) mat	1		
45,XX,−21,−22,+t(21q;22q) unknown	1		
Total	20		

TABLE VIII. Indications for karyotyping in peri- and paracentric inversions

Indication	Origin	No.
Pericentric inversions (total of 25 index patients)		
Down syndrome	Unknown	1
MR/CM syndrome (2 unbalanced)	Familial	5
	De novo	1
	Postzygotic	1
	Unknown	1
Spontaneous abortions and other reproductive failure	Familial	3
	De novo	1
	Unknown	2
Other indications for karyotyping	Familial	4
	De novo	1
	Unknown	5
Paracentric inversions (total of 19 index patients)		
Down syndrome	Postzygotic	1
MR/CM syndrome	Familial	4
	De novo	1
	Postzygotic	1
Spontaneous abortions and other reproductive failure	Familial	3
	Postzygotic	1
Other indications for karyotyping	Familial	2
	Unknown	6

inv(9)(p11q13)] and 19 paracentric inversion carriers in the total number of 52,179 chromosomally investigated patients (0.5 in 1,000 and 0.36 in 1,000 patients, respectively). Table VIII summarizes the different indications for cytogenetic examination of these patients. Six of the 25 pericentric inversions and four of the 19 paracentric inversions were detected in the selected group of 1,555 couples with RFW (3.9 in 1,000 and 2.5 in 1,000, respectively). Table IX shows the types of peri- and paracentric inversions and the numbers of male and female carriers.

The incidence of pericentric inversions in the general population is still difficult to determine because of the rarity of this type of chromosomal rearrangement and the lack of detailed studies. The data collected by Nielsen and Sillesen [54] from different studies on consecutive liveborns (for review see [54]) estimate the incidence of all autosomal abnormalities and subsequently of chromosome inversion to be 5.75 in 1,000 and 0.13 in 1,000. However, it was not stated clearly by these authors if the (normal variant) inversion of chromosome 9 [inv(9)(p11q13)] was also taken into account in this evaluation. In three large published series of newborns (a total of 36,897 liveborns) we could find mention of only five pericentric inversions, namely: 46,XX, inv(1)(p32q42); 46,XX, inv(2)(p11q13) and 47,XX,inv(18)(p11q21) + rec(18) dup p in Jacobs et al. [29]; 47,XY,+21,inv(19)(p13q13) in Hamerton et al. [55]; and an inexactly defined pericentric inv(6) in the Nielsen and Sillesen study [54]. The incidence of pericentric inversions among those 36,897 newborns was thus also 0.13 in 1,000. Different authors have reported an increased tendency to early spontaneous abortions in familial pericentric inversions [22,56–60]. Further data, however, are needed to confirm their findings. On the other hand, numerous examples of duplication/deletion syndromes resulting from familial pericentric inversions have been described [48–50,59,61–66]. The risk of duplication/deficiency syndrome for certain types of pericentric inversions has been estimated by Sutherland et al. [67] as 5–10%, and the length of the inverted chromosomal segment seems to be the most decisive factor influencing the harmful or harmless role of a pericentric inversion in the progeny of inversion carrier parents. Studies made by Trunca and Opitz [68], Winsor et al. [69], and Daniel [49] indicate that genetically unbalanced gametes have a greater chance to arise from a large inversion meiotic loop than from a small one.

TABLE IX. Type and origin of peri- and paracentric inversions in 10 patients with RFW

Female carriers	Male carriers
1. 46,XX,inv(2)(p12q21) unknown	1. 46,XY,inv(5)(p13q13) unknown
2. 46,XX,inv(3)(p12q13) (fetus; parents nl) de novo	2. 46,XY,inv(5)(p13.3q13.3) mat
3. 46,XX,inv(7)(p15.3q11.2), 1qh+, mat	3. 46,XY,inv(8)(p23q11) mat
4. 46,XX,inv(1)(p22.1p35.3) fam	4. 46,XY,inv(3)(p13p25) mat
5. 46,XX,inv(7)(q11.23q22.1) mat	
6. 46,XX/46,XX,inv(12)(q13q24) unknown	

The incidence of paracentric inversions in the general population is still unknown. We detected 19 index patients with paracentric inversions in a total of 52,179 patients (0.36 in 1,000; see Table I) referred for constitutional chromosomal analysis for different reasons. In four index patients, the indication for karyotyping was a RFW problem. In paracentric inversion carriers (as for pericentric inversions), the meiotic pairing of the homologous chromosomes occurs by the formation of an inversion loop between the normally structured and the inverted homologue. In contrast to pericentric inversions, crossing-over in the paracentric inverted loop leads to a dicentric chromosome and an acentric fragment at anaphase I, and consequently induces chromosomally unbalanced gametes. Such a process may explain the increased fetal wastage at least in certain carriers of paracentric inversions [51,52,70,71].

OTHER CHROMOSOMAL ABNORMALITIES

Small accessory chromosomes (SAC) are one of the most puzzling findings in human cytogenetics. They usually remain unidentified with the available banding techniques [72,73]. Their size and morphology are variable, even within the same family, and the presence of centrometric constitutive heterochromatin seems to be their only common feature. The incidence of SAC in different cytogenetic surveys is about 1 in 3,500 [29,54,55]. We have detected SAC in 32 index patients (0.6 in 1,000), and three of them were diagnosed among the 1,555 couples with RFW. At the present time, it is difficult to conclude definitely whether carriers of SAC have a higher risk for fetal wastage and/or MR/CM in their progeny [74].

One woman with a del(X)(p2101) and RFW was detected among the nonmosaic patients with Xq or Xp deletion. It is very probable that the two spontaneous abortions in this patient were due to an Xp nullosomy in males which is probably lethal.

A ring chromosome 21 was detected in one phenotypically normal woman. The phenotypic effect of ring chromosomes is known to be variable, sometimes even without any phenotypic manifestation [75,76]. Extensive banding techniques did not show any evidence of translocation of the deleted part of the chromosome 21 to other chromosomes, as has been demonstrated in another report on a ring chromosome [77]. Transmission of a ring chromosome through different generations has rarely been observed, and it depends on the number and kind of cross-over events between the ring chromosome and its homologue. Stoll and Roth [78] reported a family in which the ring chromosome 21 had been segregating through three generations.

CONCLUSIONS

Cytogenetic examination of a large series of 1,555 couples with recurrent fetal wastage lead to the detection of structural chromosomal rearrangements in 6.36% of this selected population. This is almost 7-fold higher than the incidence of chromosomal abnormalities in the general population.

More than 50% of the chromosomal abnormalities were reciprocal balanced translocations, and almost 30% were of various Robertsonian types; mostly D/D translocations. Peri- and paracentric inversions accounted for 6 and 4% of all chromosome abnormalities detected.

There was no significant prevalence of female carriers of balanced reciprocal translocations, but in the Robertsonian translocation carriers the female/male ratio was 2.5:1.

Striking similarities between the breakpoints found in this study and those from previous investigations seem to indicate that at least some chromosome regions are particularly sensitive for chromosome breakage.

Prenatal diagnosis in subsequent pregnancies of reciprocal translocation carriers with RFW resulted in 10.8% chromosomally unbalanced pregnancies at 15 weeks of gestation. A much lower proportion of chromosomally aneuploid offspring at prenatal diagnosis was observed in carriers of Robertsonian translocations.

Further data are needed to evaluate whether peri- and paracentric inversions, and other chromosomal rearrangements such as small accessory chromosomes, X deletion, and autosomal rings, are more frequent in patients with RFW.

REFERENCES

1. Kajii T, Ferrier A: Cytogenetics of aborters and abortuses. Am J Obstet Gynecol 131:33, 1978.
2. Turleau C, Chavin-Colin F, de Grouchy J: Cytogenetic investigation in 413 couples with spontaneous abortions. Eur J Obstet Gynecol Reprod Biol 9:65, 1979.
3. Boué A, Boué J: Chromosome structural rearrangements and reproductive failure. Chromosomes Today 7:281, 1981.
4. Tsenghi C, Metaxotou C, Kalpini-Mavrou A, Strataki-Benetou M, Matsaniotis N: Parental chromosome translocations and fetal loss. Obstet Gynecol 58:456, 1981.
5. Davis JR, Weinstein L, Veomett JC, Shenker L, Giles HR, Hauck L: Balanced translocation karyotypes in patients with repetitive abortion. Am J Obstet Gynecol 144:299, 1982.
6. Michels VV, Medrano C, Venne VL, Riccardi VM: Chromosome translocations in couples with multiple spontaneous abortions. Am J Hum Genet 34:507, 1982.
7. Osztovics MK, Tóth SP, Wessely JA: Cytogenetic investigations in 418 couples with recurrent fetal wastage. Ann Genet 25:232, 1982.

8. Lippman-Hand A, Vekemans M: Balanced translocations among couples with two or more spontaneous abortions: Are males and females equally likely to be carriers? Hum Genet 63:252, 1983.
9. Lyberatou-Moraitou E, Grigori-Kostaraki P, Retzepopoulou Z, Kosmaidou-Aravidou Z: Cytogenetics of recurrent abortions. Clin Genet 23:294, 1983.
10. Fryns JP, Kleczkowska A, Kubień E, Petit P, Van den Berghe H: Cytogenetic survey in couples with recurrent fetal wastage. Hum Genet 65:336, 1984.
11. Pantzar JT, Allanson JE, Kalousek DK, Poland BJ: Cytogenetic findings in 318 couples with repeated spontaneous abortion: A review of experience in British Columbia. Am J Med Genet 17:615, 1984.
12. Tharapel AT, Tharapel SA, Bannerman RM: Recurrent pregnancy losses and parental chromosome abnormalities: A review. Br J Obstet Gynaecol 92:899, 1985.
13. Tsenghi C, Metaxotou-Stavridaki C, Strataki-Benetou M, Kalpini-Mavrou A, Matsaniotis N: Chromosome studies in couples with repeated spontaneous abortions. Obstet Gynecol 47:463, 1976.
14. Breuker KH, Winkhaus-Schindl I, Citoler P: Chromosomenanomalien bei Ehepaaren mit wiederholten Aborten. Gebursh u Frauenheilk 38:11, 1978.
15. Duca D, Cioltei A, Ioan D, Maximilian C: The importance of cytogenetic investigation of the couples with mutliple spontaneous abortions and malformed offsprings. Endocrinology 17:17, 1979.
16. Neu RL, Entes K, Bannerman RM: Chromosome analysis in cases with repeated spontaneous abortions. Obstet Gynecol 53:373, 1979.
17. Pescia G, Jotterand M, Gaide A-C: Fausses-couches précoces itératives et caryotype. Etude cytogénétique de 80 couples consécutifs. J Gynecol Obstet Biol Reprod 8:35, 1979.
18. Kardon NB, Davis JG, Berger AL, Broekman A: Incidence of chromosomal rearrangements in couples with reproductive loss. Hum Genet 53:161, 1980.
19. Ward BE, Henry GP, Robinson A: Cytogenetic studies in 100 couples with recurrent spontaneous abortions. Am J Hum Genet 32:349, 1980.
20. Adžić S, Marković S, Mijin K, Stolević E, Sulović V, Vidaković B: Les recherches cytogénétiques chez les sujets ayant des échecs de la reproduction. J Gynecol Obstet Biol Reprod 10:371, 1981.
21. Sant-Cassia LJ, Cooke P: Chromosomal analysis of couples with repeated spontaneous abortions. Br J Obstet Gynecol 88:52, 1981.
22. Simpson JL, Elias S, Martin AO: Parental chromosomal rearrangements associated with repetitive spontaneous abortions. Fertil Steril 36:584, 1981.
23. Stoll C: Cytogenetic findings in 122 couples with recurrent abortions. Hum Genet 57:101, 1981.
24. Blumberg BD, Shulkin JD, Rotter JI, Mohandas T, Kaback MM: Minor chromosomal variants and major chromosomal anomalies in couples with recurrent abortion. Am J Hum Genet 34:948, 1982.
25. Husslein P, Huber J, Wagenbichler P, Schnedl W: Chromosome abnormalities in 150 couples with multiple spontaneous abortions. Fertil Steril 37:379, 1982.
26. Diedrich U, Hansmann I, Janke D, Opitz O, Probeck HD: Chromosome abnormalities in 136 couples with a history of recurrent abortions. Hum Genet 65:48, 1983.
27. Mameli M, Cardia S, Milia A, Aste A, Santucci S, Genazzani AR: Cytogenetic study in 50 couples with recurrent abortions. Gynecol Obstet Invest 17:84, 1984.
28. Tho PT, Byrd JR, Donough PGCM: Etiologies and subsequent reproductive performance of 100 couples with recurrent abortion. Fertil Steril 32:389, 1979.

29. Jacobs PA, Melville M, Ratcliffe S, Keay AJ, Syme J: A cytogenetic survey of 11,680 newborn infants. Ann Hum Genet (Lond) 37:359, 1974.
30. Nielsen J, Rasmussen K: Autosomal reciprocal translocations and 13/14 translocations: A populations study. Clin Genet 10:161, 1976.
31. Evans JA, Canning N, Hunter AGW, Martsolf JT, Ray M, Thompson DR, Hamerton J: A cytogenetic survey of 14,069 newborn infants. III. An analysis of the significance and cytologic behavior of the Robertsonian and reciprocal translocations. Cytogenet Cell Genet 20:96, 1978.
32. Nielsen J, Hansen KB, Sillesen I, Videbeck P: Chromosome abnormalities in newborn children. Physical aspects. Hum Genet 59:194, 1981.
33. Fryns JP, Kleczkowska A, Kubień E, Van den Berghe H: Excess of mental retardation and/or congenital malformation in reciprocal translocations in man. Hum Genet 72:1, 1986.
34. Hecht F, Hecht BK: Fragile sites and chromosome breakpoints in constitutional rearrangements. I. Amniocentesis. Clin Genet 26:169, 1984.
35. Hecht F, Hecht BK: Fragile sites and chromosome breakpoints in constitutional rearrangements. II. Spontaneous abortions, stillbirths and newborns. Clin Genet 26:174, 1984.
36. Sherman SL, Sutherland GR: Segregation analysis of rare autosomal fragile sites. Hum Genet 72:123, 1986.
37. Yunis JJ, Soreng AL: Constitutive fragile sites and cancer. Science 226:1199, 1984.
38. Boué J, Barichard F, Deluchat C: Diagnostic prénatal des anomalies de la structure chromosomique. Nouv Presse Med 10:3299, 1981.
39. Kjessler B: Meiosis in a man with a D/D translocation and clinical sterility. Lancet 1:1421, 1964.
40. Fraccaro M, Maraschio P, Pasquali F, Tiepolo L, Zuffardi O, Giarola A: Male infertility and 13/14 translocation. Lancet 1:488, 1973.
41. Plymate SR, Bremner WJ, Paulsen CA: The association of D-group chromosomal translocations and defective spermatogenesis. Fertil Steril 27:139, 1976.
42. Fraccaro M: Cytogenetics of human male infertility. Clin Genet 17:65, 1980.
43. Marmor D, Taillemite JL, Van den Akker J, Portnoi MF, Le Porrier N, Joye N, Delafontaine D, Raix C: Semen analysis in subfertile balanced-translocation carriers. Fertil Steril 34:496, 1980.
44. Mićić MD, Nicoliš JG, Mićić R: 13/14 translocation in a man with reproductive failure. Hum Genet 55:137, 1980.
45. Chandley AC: The origin of chromosomal aberrations in man and their potential for survival and reproduction in the adult human population. Ann Genet 24:5, 1981.
46. Court Brown WM: "Human Population Cytogenetics." Amsterdam: North Holland, 1967.
47. Dutrillaux B, Lejeune J: Etude de la descendance des individus porteurs d'une translocation t(Dq;Dq). Ann Genet 13:11, 1970.
48. Giraud F, Mattei JF, Mattei MG, Aymé S: Les inversions péricentriques. A propos de 47 observations. J Genet Hum 27:109, 1979.
49. Daniel A: Structural differences in pericentric inversions. Application to a model of risk of recombinants. Hum Genet 56:321, 1981.
50. Beemer FA, De France HF, Rosina-Angelista IJM, Gerards LJ, Cats BP, Guyt R: Familial partial monosomy 5p and trisomy 5q; three cases due to paternal pericentric inversion 5(p151q333). Clin Genet 26:209, 1984.
51. Madan K, Seabright M, Lindenbaum RH, Bobrow M: Paracentric inversions in man. J Med Genet 21:407, 1984.

52. Venter PA, Dawson B, Du Toit JL, Smith EL, Kritzinger N, Landman AS, Cronje AS, Hof JO: A familial paracentric inversion: A short review of the current status. Hum Genet 67:121, 1984.
53. Callen DF, Woollat E, Sutherland GR: Paracentric inversions in man. Clin Genet 27:87, 1985.
54. Nielsen J, Sillesen I: Incidence of chromosome aberrations among 11,148 newborn children. Hum Genet 30:1, 1975.
55. Hamerton JL, Canning N, Ray M, Smith J: A cytogenetic survey of 14,069 newborn infants. I. Incidence of chromosome abnormalities. Clin Genet 8:223, 1975.
56. Herva R, de la Chapelle A: A large pericentric inversion of human chromosome 8. Am J Hum Genet 28:208, 1976.
57. Simola K, Karli P, de la Chapelle A: Two pericentric inversions of human chromosome 11. J Med Genet 14:371, 1977.
58. Howard-Peebles P: Familial pericentric inversion of chromosome 1 with a note on reproductive risks. Hum Genet 45:123, 1978.
59. Schroer RJ, Culp DM, Stevenson RE, Potts WE, Taylor HA, Simensen RJ: Duplication-deletion syndrome in a family with pericentric inversion of chromosome 6. Clin Genet 18:83, 1980.
60. Stetten G, Rock JA: A pericentric chromosomal inversion associated with repeated early pregnancy wastage. Fertil Steril 40:124, 1983.
61. Fujimoto A, Towner JW, Turkel SB, Wilson MG: A fetus with recombinant of chromosome 8 inherited from her carrier father. Hum Genet 40:241, 1978.
62. Pfeiffer RA, Kessel E: Balanced and unbalanced pericentric inversion of a chromosome 14. Hum Genet 43:103, 1978.
63. Mattei JF, Mattei MG, Ardissone JP, Taramasco H, Giraud F: Pericentric inversion, inv(9)(p22q32) in the father of a child with a duplication-deletion of chromosome 9 and gene dosage effect for adenylate kinase-1. Clin Genet 17:129, 1980.
64. Yunis E, Torres de Caballero O: Duplication deficiency as the result of meiotic segregation of a maternal inv(10). Hum Genet 57:71, 1981.
65. Lucas J, Le Mée F, Picard F, Le Marec B, Junien C: Trisomie 13q13→qter pure par recombinaison aneusomique d'une inversion péricentrique maternelle. Ann Genet 26:187, 1983.
66. Watt JL, Olson IA, Johnston AW, Ross HS, Couzin DA, Stephen GS: A familial pericentric inversion of chromosome 22 with a recombinant subject illustrating a "pure" partial monosomy syndrome. J Med Genet 22:283, 1985.
67. Sutherland GR, Gardiner AJ, Carter RF: Familial pericentric inversion of chromosome 19, inv(19)(p13q13), with a note on genetic counseling of pericentric inversion carriers. Clin Genet 10:54, 1976.
68. Trunca C, Opitz JM: Pericentric inversion of chromosome 14 and the risk of partial duplication of 14q(14q31→14qter). Am J Med Genet 1:217, 1977.
69. Winsor EJT, Palmer CG, Ellis PM, Hunter JLP, Ferguson-Smith MA: Meiotic analysis of a pericentric inversion, inv(7)(p22q32), in the father of a child with a duplication-deletion of chromosome 7. Cytogenet Cell Genet 20:169, 1978.
70. Del Porto G, D'Alessandro E, de Matteis C, D'Innocenzo R, Baldi M, Cappa F: Familial paracentric inversion of chromosome 15(q15q24). J Med Genet 21:451, 1984.
71. Schmid M, Hoffmann R, Köhler J, Jannek U: Familial paracentric inversion inv(2)(q31q36). Hum Genet 71:270, 1985.
72. Fryns JP, Kleczkowska A, Van den Berghe H: Small accessory chromosomes (SAC) and their genotype-phenotype correlation. J Genet Hum 30:199, 1982.

73. Münke M, de Martinville B, Lieber E, Francke U: Minute chromosomes replacing the Y chromosome carry Y-specific sequences by restriction fragment analysis and in situ hybridization. Am J Med Genet 22:361, 1985.
74. Benn PA, Hsu LYF: Incidence and significance of supernumerary marker chromosomes in prenatal diagnosis. Am J Hum Genet 36:1092, 1984.
75. Schmid W, Tenconi R, Baccichetti C, Caufin D, Schinzel A: Ring chromosome 21 in phenotypically apparently normal persons: Report of two families from Switzerland and Italy. Am J Med Genet 16:323, 1983.
76. Kleczkowska A, Fryns JP: Ring chromosome 21 in a normal female. Ann Genet 27:126, 1984.
77. Fryns JP, De Boeck K, Jaeken J, Van den Berghe H: Malformative syndrome associated with a ring 10 chromosome and a translocated 10q/19 chromosome. Hum Genet 43:239, 1978.
78. Stoll C, Roth MP: Segregation of a 22 ring chromosome in three generations. Hum Genet 63:294, 1983.

The Cytogenetics of Mammalian Autosomal Rearrangements, pages 471–478

Chapter 13
Chromosome Rearrangements in Pediatric Autopsies

Grant R. Sutherland

Chromosome studies on 2,669 unselected pediatric autopsies are summarised. There was an increased incidence of chromosome abnormalities in all groups, ranging from 10.3% in macerated stillbirths to 3.4% in late neonatal deaths. All diagnostic groups except those with primary CNS malformation, trauma, and unclassified had higher frequencies of chromosome abnormality than seen in unselected liveborn infants. It is recommended that cytogenetic studies should be an integral part of all pediatric autopsies on stillbirths and babies aged less than 28 days except where the cause of death is a primary CNS malformation, trauma, SIDS, or a single gene defect.

INTRODUCTION

Approximately one in 20 pediatric autopsies in teaching hospitals is on an individual with a major chromosome abnormality. This is about ten times the incidence of chromosome abnormality found amongst randomly selected liveborn infants. The types of chromosome abnormality detected at autopsy are generally of the type seen among liveborn infants, with very few of the abnormalities characteristic of early spontaneous abortion being seen.

Systematic studies of chromosome abnormalities in randomly or unselected autopsies have only been carried out in four centres: Adelaide [1,2], Edinburgh [1,3], London [4], and Moscow [5]. The results from these studies will be collated and discussed with special emphasis on structural chromosome abnormalities.

DEFINITIONS

The subjects of autopsies performed in pediatric teaching hospitals range from spontaneously aborted fetuses in the first trimester of pregnancy through

Cytogenetics Unit, Adelaide Children's Hospital, North Adelaide, S.A. 5006, Australia adolescents. In the compilation of data from the above series included in this review, abortuses (spontaneous or induced) below 20 weeks' gestation are excluded, and from the London [4] and Moscow [5] series, abortuses below 28 weeks' gestation are excluded. Those included are divided into a number of categories: *Macerated stillbirths* in which there was pathological evidence of antepartum death; *non-macerated stillbirths* in which there was no evidence of antepartum death; *early neonatal deaths* which followed live birth (at any gestation) on or before the seventh day of life; *late neonatal deaths* which occurred on days 8–28 inclusive of life; *infant deaths* which occurred on days 29–365 of life; *childhood deaths* which occurred after the 365th day of life.

The first three categories enter into the perinatal period and are referred to as perinatal deaths. However it should be noted that the legal definitions of perinatal deaths and infant deaths differ from country to country.

METHODOLOGY

Chromosome results from phenotypically abnormal individuals who were born alive were often obtained from blood lymphocyte cultures established prior to death. However, in the majority of cases chromosome studies were initiated after death. These studies involved fibroblast culture for which a variety of tissues, including skin, pericardium, kidney, gonad, fascia, and amnion were used. Chromosome results obtained in this way are perhaps not as reliable as those obtained from lymphocyte culture. It is well known that in vitro chromosome changes can occur in fibroblast cultures [6,7]. Results obtained from amnion culture have special problems since, in addition to possible in vitro changes, there is the possibility of chromosomal mosaicism confined to the amnion [8] which does not reflect the chromosomal status of the fetus. These mechanisms could account for the few cases of mosaic trisomy 2 ascertained from amnion culture [2]. Furthermore, the placenta accompanying a stillborn fetus to autopsy might be the wrong one [1]. Finally, in some instances amnion culture might result in growth of maternal rather than fetal cells.

In this review, no attempt has been made to assess the validity of any results from the published series except that two probable normal variant chromosomes (Gp−, 15p+) recorded in the London study have been excluded from the calculations of incidence of chromosome abnormalities.

FINDINGS

The incidence of chromosome abnormality according to autopsy category for the published series is shown in Table I. It should be noted that the data

TABLE I. Results of published series of chromosome studies in unselected pediatric autopsies

Subjects	London [4,10]	Moscow [5]	Edinburgh [1]	Adelaide [1]	Adelaide [2]	Edinburgh [3]	Total abnormal (%)
Macerated stillbirth	34(3)[a]	22(3)	49(6)	7(1)	26(5)	115(8)	253(26) (10.3)
Non-macerated stillbirth	122(5)	61(3)	146(5)	11(0)	19(2)	57(1)	416(16) (3.8)
Early neonatal death	344(20)	92(6)	340(15)	48(0)	38(5)	185(8)	1,047(54) (5.2)
Late neonatal death	—	—	64(2)	22(3)	24(0)	37(0)	147(5) (3.4)
Infant	—	—	96(10)	107(4)	199(5)	—	402(19) (4.7)
Child	—	—	32(2)	94(7)	141(5)	—	267(14) (5.2)

Excludes abortions of less than 28 weeks' gestation.
[a] No. karyotyped (no. abnormal).

from the early part of the Edinburgh series [9] is included in Sutherland et al. [1]. This table shows that the highest incidence of chromosome abnormality is seen in macerated stillbirths and the lowest incidence is in non-macerated stillbirths and late neonatal deaths. Many of the non-macerated stillbirths were due to obstetric complications and were not a reflection of any fetal abnormality. Similarly, the late neonatal deaths were often from complications of prematurity.

The incidence of chromosome abnormality according to primary cause of death for the Adelaide, Edinburgh, and London series is given in Table II. Insufficient published data on the Moscow series prevented its inclusion in this table. There may be some minor inaccuracies in the distribution of cause of death for the London series. Data pertaining to the Adelaide [2] series differs from the originally published figures since the therapeutic abortions had not been separated from the rest of the data. The highest incidence of chromosome abnormality is seen in the macerated stillbirths with malformations, followed by babies with malformations, whether stillborn or liveborn. The frequency of chromosome abnormality is raised above that seen in unselected liveborns, approximately 0.5%, in all groups except those with primary CNS malformation, the trauma group which is heterogeneous including both birth trauma and motor vehicle accidents, and the miscellaneous group. There was an unexpected increased incidence of chromosome abnormalities in the SIDS (Sudden Infant Death syndrome) group. This was due to two cases of mosaic trisomy 8 (one also having XXXYY) in the Edinburgh

TABLE II. Chromosome abnormality according to primary cause of death

Cause of death	London [4,10]	Edinburgh [1]	Adelaide [1]	Adelaide [2]	Edinburgh [1]	Total abnormal	(%)
Macerated without malformations	27(2)	47(3)	1(0)	18(1)	114(3)	207(9)	4.3
Macerated with malformations	5(1)	3(3)	0	2(2)	21(9)	31(15)	48.4
Prematurity	124(4)	157(2)	12(0)	3(0)	119(1)	415(7)	1.7
Primary CNS malformation	50(1)	101(0)	14(0)	25(0)	21(0)	211(1)	0.47
Multiple malformations	76(16)	98(26)	28(7)	35(15)	53(7)	290(71)	24.5
Congenital heart disease	11(1)	62(2)	44(3)	56(2)	17(0)	190(8)	4.2
Primary anoxia	120(1)	143(2)	21(1)	17(0)	132(1)	433(5)	1.2
Infection	39(2)	34(1)	26(2)	25(0)	—	124(5)	4.0
SIDS	—	40(2)	66(0)	135(2)	—	241(4)	1.7
Malignancy	—	1(1)	17(1)	33(0)	—	51(2)	3.9
Trauma	25(0)	26(0)	30(1)	44(0)	—	125(1)	0.8
Other	23(0)	42(0)	31(0)	57(0)	23(0)	176(0)	0

Excludes abortions of less than 20 weeks' gestation from all series, and spontaneous abortions of less than 28 weeks from the London [4,10] and Moscow [5] series.

[1] series, one XY/XYY mosaic and an X/autosome translocation in the Adelaide [2] series.

In Table III the types of chromosome abnormality detected in the series reviewed are shown and compared with the incidence found in randomly selected liveborn infants [11]. All types of chromosome abnormality are found more commonly in pediatric autopsies than in the liveborns. This is not unexpected for autosomal trisomies, triploids, and unbalanced translocations. It is, however, a little surprising for the sex chromosome abnormalities and the balanced translocations.

The sex chromosome abnormalities are shown in Table IV. The only common sex chromosome abnormality not present in excess is 47,XXX. The large number of Turner syndrome cases [8] is attributable to death from congenital heart disease and intrauterine deaths presumably resulting from fetal lymphedema. There are four males each with an extra X or Y chromosome, and this apparent increased loss of males with sex chromosome abnormalities in the perinatal period or shortly afterwards is unexpected.

TABLE III. Types of chromosome abnormality seen in 2,669 pediatric autopsies

				Robertsonian translocations or inversions		Reciprocal translocations		
Series/Ref. no.	Autosomal trisomy or mosaic	Sex chromosome abnormality	Triploidy	Balanced	Unbalanced	Balanced	Unbalanced[a]	Other
4,10	14	5	1	2	2	1	1	2[b]
5	6	4	1	0	0	0	1	0
Edinburgh [1]	32	3[c]	1	1	1	1	1	2[c]
Adelaide [1]	6	3	1	0	0	4	1	0
2	12	4	0	1	0	1	4	0
3	13[c]	2	2	0	1	1	2	0
Total	83	21	6	4	4	8	10	2
Incidence (%)	3.11	0.788	0.225	0.150	0.150	0.300	0.375	0.075
Newborn incidence [11]	0.144	0.206	0.002	0.091	0.007	0.104	0.023	0.030

See footnote to Table II.
[a]Deletions are included here.
[b]Probable normal variant chromosomes, excluded from further consideration.
[c]One of these contained a marker chromosome. Therefore incidence of structurally abnormal karyotypes is 29/2665 (1.10%).

TABLE IV. Sex chromosome abnormalities detected in pediatric autopsies

Abnormality	Series [ref.]	Cause of death
45,X	Adelaide [1]	Congenital heart disease, death at 2 weeks
45,X	Adelaide [1]	Congenital heart disease, death at 8 days
45,X male	Adelaide [2]	Hypoplastic left heart, death at 2 days
45,X	Adelaide [2]	Macerated stillbirth at 23 weeks gestation
45,X	Moscow [5]	Congenital heart disease, death at 26 hours
45,X	London [4,10]	Congenital heart disease, talipes, neonatal death
45,X	Edinburgh [3]	Macerated stillbirth at 20–28 weeks
45,X	Edinburgh [3]	Macerated stillbirth at 20–28 weeks
45,X/46,X,i(Xq)	Edinburgh [1]	Congenital heart disease, death at 2 months
46,XY female	Adelaide [2]	Multiple malformations, death at 1 day
47,XXX	Edinburgh [1]	Stillbirth at 31 weeks
46,XX/46,XY	Moscow [5]	Female infant, stillbirth with multiple malformations
46,XX/47,XXY	Moscow [5]	Female infant, macerated stillbirth at 38 weeks
47,XYY	London [4,10]	Neonatal death, no malformations
47,XYY	Edinburgh [1]	Bilateral renal dysplasia, death at 3 hours
47,XYY	Moscow [5]	Prematurity, 30 weeks' gestation, death at 16 hours
46,XY/47,XYY	Adelaide [2]	SIDS, death at 5 months
47,XXY	Adelaide [1]	Congenital heart disease, death at 1 year
47,XXY	London [4,10]	Macerated stillbirth, no malformations
47,XXY	London [4,10]	Neonatal death, ? obstetric cause
46,XY/47,XXY	London [4,10]	Macerated stillbirth, no malformations
48,XXYY/ 49,XXYY,+8	Edinburgh [1]	SIDS at 8 months

See footnote to Table II.

Although balanced Robertsonian translocations are almost twice as frequent in the autopsy material compared with the newborn group (Table III), there is no pattern in either the chromosomes involved or the pathology seen (Table V). A similar situation exists for balanced reciprocal translocations and inversions which are more than three times as frequent in the autopsy material as in the newborns.

CONCLUSIONS

Cytogenetic investigation at the pediatric autopsy is probably more likely to reveal an abnormality than other autopsy investigation. The knowledge of a major chromosome abnormality associated with an intrauterine or pediatric death will provide a firm basis for parental counselling following the death; even knowing that the chromosomes were normal can be helpful in this situation. It has been stated [13,14] that cytogenetic studies are not indicated on perinatal deaths due to anoxia, prematurity-associated problems, or iso-

TABLE V. Balanced structural changes in pediatric autopsies

	Series [ref.]	Pathology
Robertsonian translocations		
45,XX,t(13;14)	Edinburgh [1]	Hepatic hemangioma, death at 1 hour
45,XY,t(14;21)mat	Adelaide [2]	Macerated stillbirth, no malformations
45,XY,t(14;15)pat	London [4,10]	Neonatal death, ? obstetric cause
45,XY,t(13;21)	London [4,10]	Neonatal death, ? obstetric cause
Reciprocal translocations		
46,XY,t(12;16)(q24;p11)pat	Adelaide [1]	Inhalation of peanut in 19-month-old normal child
46,XX,t(1;5)(q24;p11)	Adelaide [1]	Rhabdomyosarcoma in 12-year-old child
46,Y,t(X;5)(q13;p15)mat	Adelaide [1]	SIDS in 6-month-old boy with hypospadias [12]
46,XX,t(12;17)(24;q21)mat	London [4,10]	Stillbirth, multiple malformations
46,XX/46,XX,t(9;14)(q34;q12)	Adelaide [2]	Normal child, infectious death at 2.5 years
Inversions		
46,XY,inv(2)(p11;q13)	Adelaide [1]	Infectious death at 8 days
46,XX,inv(2)(p13;q11)	Edinburgh [3]	Stillbirth, no malformations
46,XX,inv(8)(p23;q11)	Edinburgh [1]	Stillbirth, no malformations

See footnote to Table II.

lated congenital heart defects. In the first two of these categories, the incidence of chromosome abnormalities is only 2–3 times that seen in randomly selected newborns (Table II). However, in congenital heart disease about 4% of autopsies show a chromosome abnormality, mostly of the sex chromosomes, and these are well worth finding. It has been recommended [2] that chromosome studies be carried out on all pediatric autopsies except where death is due to a primary CNS malformation, trauma, SIDS, or single gene defects where chromosome abnormalities are unlikely to be detected at an increased frequency.

This chapter is dedicated to the memory of Dr. A. Douglas Bain, a pioneer in genetic studies at pediatric autopsies, whom I thank for constructive criticism of the manuscript of this chapter, and many years of friendship. Dr. Bain died on 26 January 1987.

REFERENCES

1. Sutherland GR, Carter RF, Bauld R, Smith II, Bain AD: Chromosome studies at the paediatric necropsy. Ann Hum Genet 42:173–181, 1978.
2. Sutherland GR, Carter RF: Cytogenetic studies: An essential part of the pediatric necropsy. J Clin Pathol 36:140–142, 1983
3. Angell RR, Sandison A, Bain AD: Chromosome variation in perinatal mortality: A survey of 500 cases. J Med Genet 21:39–44, 1984.
4. Machin GA: Chromosome abnormality and perinatal death. Lancet 1:549–551, 1974.
5. Kuleshov NP: Chromosome anomalies of infants dying during the perinatal period and premature newborn. Hum Genet 31:151–160, 1976.

6. Littlefield LG, Mailhes JB: Observations of de novo clones of cytogenetically aberrant cells in primary fibroblast cell strains from phenotypically normal women. Am J Hum Genet 27:190–197, 1975.
7. Oxford JM, Harnden DG, Parrington JM, Delhanty JD: Specific chromosome aberrations in ataxia telangiectasia. J Med Genet 12:251–262, 1975.
8. Kalousek DK, Dill FJ: Chromosomal mosaicism confined to the placenta in human conceptions. Science 221:665–667, 1983.
9. Bauld R, Sutherland GR, Bain AD: Chromosome studies in investigation of stillbirths and neonatal deaths. Arch Dis Child 49:782–788, 1974.
10. Machin GA, Crolla JA: Chromosome constitution of 500 infants dying during the perinatal period. Humangenetik 23:183–198, 1974.
11. Hook EB, Hamerton JL: The frequency of chromosome abnormalites detected in consecutive newborn studies—differences between studies—results by sex and by severity of phenotypic involvement. In Hook EB, Porter IH (eds): "Population Cytogenetics. Studies in Humans." New York: Academic Press, 1977, pp 63–79.
12. Callen DF, Sutherland GR: Normal female carrier and affected male sibs with t(X;5)(q13;p15). Location of a gene determining male genital development. Clin Genet, 30:59–62, 1986.
13. Perinatal Pathology (editorial) Lancet 1:431–432, 1984.
14. Ellis PM, Bain AD: Cytogenetics in the evaluation of perinatal death. Lancet 1:630–631, 1984.

ORIGIN OF CHROMOSOME REARRANGEMENTS

The Cytogenetics of Mammalian Autosomal Rearrangements, pages 481–506

Chapter 14
Tandem Fusion in the Evolution of Mammalian Chromosomes

F.F.B. Elder and T.C. Hsu

Tandem fusions of chromosomes occur rarely in long-term tissue cultures, malignant or premalignant diseases, and as clinical cytogenetic conditions in man. Tandem fusions can also be observed in the evolution of the karyotypes of several mammalian taxa. Two of these taxa, the muntjacs (genus *Muntiacus*) and the cotton rats (genus *Sigmodon*), contain species whose karyotypes have been structured through the fixation of numerous tandem fusions. We have examined tandem fusions in these and other taxa as well as those occurring clinically in man, cell culture, and malignancy and have hypothesized that, based on the architecture of fusion chromosomes, fusions occurring in evolution are unique cytogenetic events. While tandem fusions occuring clinically in man produce dicentric or pseudodicentric chromosomes with suppression of one centromere, those occurring in evolution involve only the euchromatic blocks of the chromosome arms. Heterochromatin is lost when tandem fusions occur in evolution, and centromeres are either lost or redistributed in the process of heterochromatin elimination. Thus, radical and rapid genome reorganization in muntjacs and cotton rats has proceeded in a nonrandom way through rearrangements involving the telomeres and the junctions of heterochromatin and euchromatin. We have speculated that karyotypic destabilization may have resulted from cycles of introgressive hybridization and could have involved mobile genetic elements in a way similar to that which has been documented in *Drosophila* and other organisms. Under such conditions multiple and genetically identical rearrangements could be produced, thus facilitating rapid fixation. Karyotypic restabilization could have been achieved through loss of heterochromatin or through the generation of fusions not subject to further rearrangement.

Departments of Pediatrics and Human Biological Chemistry and Genetics, The University of Texas Medical Branch, Galveston, Texas 77550; present address: Institute for Molecular Genetics, Kleberg Cytogenetics Laboratory, Baylor College of Medicine, Texas Medical Center, Houston, TX 77030 (F.F.B.E.), Section of Cellular Genetics, The University of Texas M.D. Anderson Hospital and Tumor Institute, Houston, Texas 77030 (T.C.H.)

INTRODUCTION

There is little question that the mammalian genome is a highly organized, integrated, and regulated entity. Generally speaking, genomes are stable with little overall difference between individuals of the same species. When cytogenetic differences do occur among individuals, these are usually identifiable and can be explained in conventional cytogenetic terms. Even in an evolutionary sense, chromosomal differences are most often explained by invoking conventional cytogenetic events such as translocation, inversion, addition/deletion of heterochromatin, etc. Occasionally, however, genomes are radically altered such that numerous rearrangements separate the genomes of related species. White [1,2] noted that the rearrangements occurring in an evolutionary lineage are often of the same type. For example, one lineage might be characterized by a series of Robertsonian fusions, while another might be characterized by a series of inversions. White [1,2] termed this phenomenon "karyotypic orthoselection." More recently, Baker and his co-workers [3,4] have offered the term "karyotypic megaevolution" to describe the phenomenon of radical genome reorganization that changes many chromosomes in the genome in a variety of ways and in a relatively short evolutionary time span.

In almost every evolutionary assemblage where reasonably extensive cytogenetic investigations have been undertaken, chromosomal rearrangements of one form or another have been found. Most often, these findings have been used 1) to infer relationships among species or genera; 2) to fuel cause-and-effect arguments of chromosomal rearrangements and speciation, and 3) to provide frameworks for various models of fixation of rearrangements during speciation. Much less frequently have these studies examined the rearrangements occurring in evolution in the context of the architecture of the chromosomes involved. In this chapter we shall briefly examine the architecture of tandem fusion chromosomes as they are seen clinically in man and in cell cultures. Then we shall examine the architecture of tandem fusion chromosomes as they are seen in evolution. We shall also present hypotheses pertaining to how these rearrangements could have occurred in the evolution of the karyotypes of two mammalian groups, the muntjacs and cotton rats, in which tandem fusion has been the predominant mechanism of chromosomal rearrangement.

TYPES OF CHROMOSOMAL FUSION

As described by Hsu et al. [5], there are three basic types of fusions involving entire chromosomes or whole chromosome arms. Depending on the morphology of the participating chromosomes, these three types are 1) centromere-centromere fusions, 2) centromere-telomere fusions, and 3) telomere-telomere fusions. (This terminology describes only chromosome orientation and does not necessarily imply that centromere-centromere or centromere-telomere fusions actually involve the physical centromere. Rather, fusion may actually occur as telomere-telomere events at the centromeric ends of acrocentric chromosomes.) The morphology and behavior of a fusion chromosome will depend on the morphology and orientation of the participating chromosomes and the functional properties of their centromeres. In all of these types of fusions, centromeric activity of one of the participating chromosomes is usually suppressed such that the fusion product behaves as a monocentric chromosome in most divisions.

Centromere-centromere fusions involve acrocentric chromosomes and are often referred to as Robertsonian translocations. This type of rearrangement has classically been viewed as a reciprocal translocation with the breakpoints located in the short arm of one acrocentric chromosome and in the proximal long arm of the other. Restitution produces the familar Robertsonian translocation chromosome comprising the long arms of both chromosomes and a small accessory chromosome consisting of a centromere and the minute short arms of the participating chromosomes. The accessory chromosome is subsequently lost. The result is a monocentric chromosome and (if balanced) a reduction in the diploid number of the organism. While this sequence of events requires loss of a centromere, a growing body of evidence suggests that many Robertsonian translocations are dicentric with one suppressed centromere [6,7]. In a sense then, a Robertsonian fusion may be envisioned as a special case of tandem fusion occurring between chromosomes with extremely small short arms.

Centromere-telomere fusions may occur between two acrocentric chromosomes or between an acrocentric and a biarmed chromosome. In the former case, the two acrocentrics are oriented in such a way as to produce another longer acrocentric chromosome or a biarmed chromosome (different from the orientation in centromere-centromere fusion), depending upon which centromere remains active. In the case of an acrocentric and a biarmed chromosome, fusion would produce a larger metacentric or submetacentric chromosome. In either case, the fusion chromosome would have two centromeres and, for stability and normal function to be achieved, one of these centromeres would necessarily have to be suppressed. Examples of spontaneous centromere-telomere fusions are rare but have been reported in several instances of clinical abnormality in man [8–17].

Spontaneous telomere-telomere fusions may occur among acrocentric chromosomes such that they are joined at their distal ends or among biarmed chromosomes such that any telomeric orientation is possible (i.e., p-p, p-q, q-p, q-q). Telomere-telomere fusions are also rare occurrences, and when observed in man they are usually associated with abnormal clinical conditions. This type of rearrangement most often involves homologous chromosomes and the fusion formed represents an inverted tandem repeat (i.e., p→q::q→p or q→p::p→q). Relatively speaking, they are far more commonly observed among sex chromosomes [18,19] than among autosomes, probably owing more to differential viability than to different frequencies of occurrence. The origin of isodicentric (inverted tandem) X chromosomes has been discussed by Dewald [18] and Rivera et al. [20] and is most likely to be due to fusion of sister chromatids at their telomeres during S or G_2. Other mechanisms, however, have been proposed [21]. Among human autosomes, tandem inverted repeats are best known for chromosome 21 in cases of the Down syndrome [22–26] but involvement of other autosomes is known [27]. Telomere-telomere fusion occurring among nonhomologous biarmed chromosomes seems to be very rare clinically [28,29], but may be somewhat more common in cultured cells [30], and in certain malignancies [31,32].

From the preceding account of tandem fusion, the following generalities can be made:

1. Fusion of entire chromosomes results in a dicentric fusion product. Stability of this dicentric is achieved through suppression of the activity of one of the two centromeres, and in most cases the same centromere is suppressed in all cells. Exceptions for some dicentric chromosomes, however, are known [16,33]. Centromere suppression of one centromere in a dicentric chromosome appears to be irreversible [34], and reactivation does not occur even if the fusion chromosome is prone to dissociation [29].

2. The site of the inactive centromere is marked by constitutive heterochromatin in the vast majority of reported cases.

3. Most if not all clinically ascertained human tandem fusion chromosomes produce at least some degree of phenotypic abnormality, mental retardation, or reproductive loss.

Tandem fusions as described above are useful in attempting to understand what is permissible with regard to the mechanics of rearrangement as well as what is possible with regard to centromeric function. They are also of value in determining what rearrangements are viable although not necessarily vigorous or fit. Tandem fusions having evolutionary potential, however, are further constrained. Not only must they be mechanically permissible and viable, but they also must be acted upon favorably (or at least neutrally) by natural selection if they are to persist. Our critical evaluation of tandem fusion chromosomes as they appear in the chromosomal evolution of several mammalian taxa has led us to think that these fusions are, in most instances, fundamentally different from those described above, as we shall attempt to show in the following section.

TANDEM FUSIONS IN EVOLUTION

Tandem fusion as a mechanism of karyotypic evolution has been documented in a variety of mammalian groups, including man [35,36]. Generally speaking, tandem fusions during evolution are relatively rare occurrences, and the presence of a single tandem fusion chromosome in one or several species often defines an evolutionary lineage [36–38]. The implication of these studies is that a tandem fusion occurs, becomes fixed in a population, subsequently characterizes the species, and in turn is transmitted to daughter species as the process of evolution continues. A few genera, however, have utilized tandem fusion to an extreme in the restructuring of the genomes of some of their representative species. Perhaps the best examples of this phenomenon are seen in the muntjacs of Asia (genus *Muntiacus*) and in the cotton rats (genus *Sigmodon*) of North America. While these are not the only examples, they are the ones with which we are most familiar and in which extensive documentation of tandem fusion exists. We therefore will confine our observations primarily to them.

The Muntjacs

In the genus *Muntiacus*, diploid numbers range from a high of 2n=46 in the Chinese muntjac (*M. reevesi*) to a low of 2n=6♀/7♂ in the Indian muntjac (*M. muntjak vaginalis*) [39,40], and other low diploid numbers are known in *M. m. muntjak* (2n=8♀/9♂) [41], *M. feae* (2n=13♀; possible range 12–14 owing to Robertsonian translocation heterozygosity) [42], and *M. rooseveltorum* (2n=6♀) [43]. These deer are apparently closely related and viable F_1 hybrids of *M. m. vaginalis* and *M. reevesi* are known [44]. In their study of *M. m. vaginalis*, *M. reevesi*, and their F_1 hybrids, Shi et al. [44] compared the G-banding patterns of the chromosomes of the two species and concluded that a great deal of banding homology still exists between the two karyotypes even though the genomes differ by 20 chromosomes (Fig. 1). The simplest interpretation of these data would indicate that at least 20 fusions have been fixed in the evolution of the *M. m. vaginalis* karyotype from the more primitive *M. reevesi* karyotype. It is significant that as many as 17 of these fusions may have been tandem fusions, while as few as three fusions may have been of the Robertsonian type. Whether or not the somewhat intermediate karyotype of *M. feae* (2n=13♀) represents an isolated

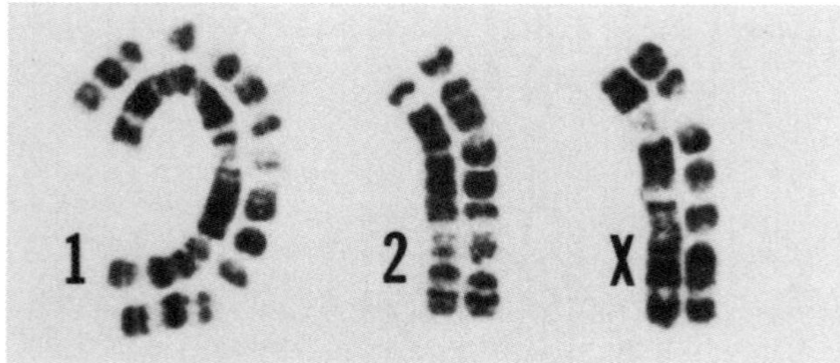

Fig. 1. Karyotype of Indian muntjac (n=3) × Chinese muntjac (n=23) hybrid. Chromosomes of the Indian muntjac are numbered and are at the left in each pairing. Chinese muntjac chromosomes are at the right in each grouping and are paired according to presumed G-band homology. Twenty fusions are inferred from this arrangement. (Photograph courtesy of Dr. Shi Liming.)

branch in this evolutionary pathway, or a derivation of the 2n=6♀/7♂ karyotype (through a process of tandem fission) is unknown.

The Cotton Rats

In the genus *Sigmodon*, diploid numbers range from a high of 2n=52 in *S. hispidus* to lows of 2n=28 in *S. mascotensis* and 2n=22 in *S. arizonae*. Chromosomal evolution of the seven currently recognized North American species has been well documented [37,45,46]. Despite the difference in diploid numbers, *S. hispidus* and *S. arizonae* appear to be closely related both morphologically [45] and biochemically [47]. Elder [37] proposed a scheme of chromosomal evolution from the primitively high, and presumed ancestral 2n=52 of *S. hispidus* to an intermediate ancestral form of (probably) 2n=36 through a process of tandem fusion. This process of tandem fusion continued from the intermediate ancestral form to modern *S. mascotensis*, whose genome is separated from *S. hispidus* by nine tandem fusions. *Sigmodon arizonae* diverged from the intermediate ancestral form through fixation of one additional tandem fusion and six Robertsonian fusions. In all, nine tandem fusions separate *S. mascotensis* from *S. hispidus* and ten tandem fusions (nine are the same as in *S. mascotensis*) and six Robertsonian fusions separate *S. arizonae* from *S. hispidus* (a few other structural rearrangements are also identifiable in this phylogeny).

Several salient points have emerged from studies of the chromosomes of muntjacs and cotton rats, as well as several other mammalian groups [48–52], that provide insight into the general mechanics of tandem fusion in chromosomal evolution.

1. Among tandem fusion chromosomes occurring in evolution, there is a singular lack of evidence for either persistence of interstitial C-band heterochromatin at the fusion site or for persistence of an interstitial (latent) centromere in any species thus far reported. This situation is quite different from that found clinically in man, in cell cultures, and in malignancies, where tandem fusion chromosomes (both centromere-telomere and telomere-telomere orientations) typically show interstitial heterochromatin as well as a second, often latent, centromere. These observations suggest that the mechanisms producing tandem fusion in evolution and those producing tandem fusion in man, cell cultures, and malignancies may be fundamentally different and that the end products are only superficially similar.

2. A critical examination of muntjac and cotton rat chromosomes shows that tandem fusion seems to involve only the large euchromatic blocks of the chromosome arms, and that these euchromatic blocks are usually retained unaltered by other forms of rearrangement (reciprocal translocations, inversions, etc.). Furthermore, all fusion orientations are apparently permitted mechanically, i.e., centromere-telomere, telomere-telomere, and centromere-centromere (in the sense that proximal euchromatic regions are joined, but not in the Robertsonian sense that one or both centromeres and centromeric heterochromatin are retained in the fusion). Figure 2 illustrates these fusion orientations. Intuitively, it seems likely that certain fusions would not be permissible evolutionarily in individual instances because the genetic linkages that they represent would be deleterious.

3. Tandem fusions occurring in evolution most often involve acrocentric chromosomes (the formation of human chromosome 2 is an exception to this statement [35,53]), and are more common among genera in which karyotypes with numerous acrocentric chromosomes are ancestral.

4. Evolutionarily stable tandem fusions occur strictly among nonhomologous chromosomes. This is another important difference from tandem fusions occurring in cell cultures, and as clinical conditions in man, in which tandem fusion chromosomes are often found to be isodicentric inverted repeats as has been previously noted.

5. There are apparently no populations in which tandem fusions exist as stable polymorphisms (unlike Robertsonian fusions for which balanced polymorphisms are not infrequently encountered [54–56]). This suggests that when these rearrangements occur, they either proceed rapidly to fixation or are rapidly lost from the population [57].

The foregoing observations lead to the following generalizations about tandem fusions as evolutionary events. First, centromeres and centromeric

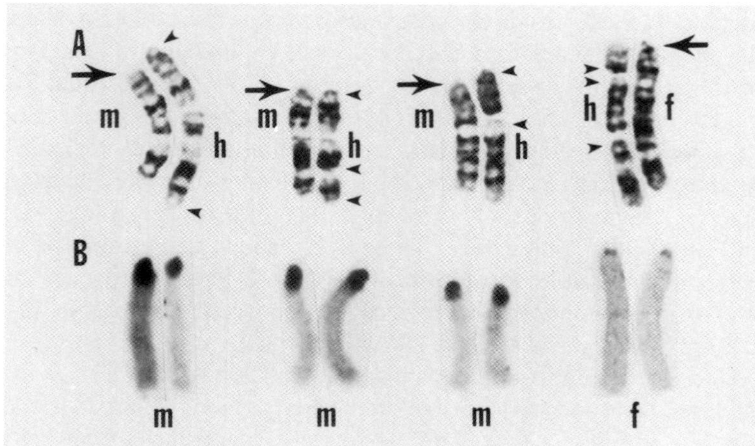

Fig. 2. A: Four G-banded tandem fusion chromosomes from the cotton rats, genus *Sigmodon*. Left elements of the first three constructs are chromosomes from *S. mascotensis* (m), 2n=28. Elements at the right are corresponding (unfused) chromosomes from *S. hispidus* (h), 2n=52, aligned to match the G-banding patterns of the *S. mascotensis* chromosomes. The construct at the far right is a chromosome from *S. fulviventer* (f), 2n=28, 29, 30, aligned with the corresponding chromosomes from *S. hispidus* (h). Arrowheads in this series indicate centromere position of *S. hispidus* chromosomes. Arrows indicate the centromeric end of the tandem fusion products found in *S. mascotensis* and *S. fulviventer*. Telomere-telomere, telomere-centromere, centromere-centromere orientations, and combinations thereof are represented in these four fusion chromosomes. **B:** C-banding of *S. mascotensis* chromosomes (m) and *S. fulviventer* chromosomes (f) as depicted in A, showing complete lack of interstitial heterochromatin at the fusion sites in these chromosomes.

heterochromatin must either be eliminated or redistributed in the genome such that they do not occur at multiple, dispersed loci on the same chromosome. Second, the fusions produce functionally as well as structurally monocentric chromosomes. Third, the euchromatic arms participating in fusions remain intact and are capable of a variety of orientations but not homologous orientations. Fourth, the notable lack of tandem fusion heterozygosity in natural populations indicates that, when they occur, they must be either rapidly fixed or lost.

Human chromosome 2 represents an exception to our generalizations regarding the mechanics of tandem fusion as an evolutionary event and deserves brief comment here. Human chromosome 2 has been formed by telomeric fusion of biarmed ancestral chromosomes represented by chimpanzee chromosomes 12 and 13 [35,36,53,58]. In this fusion, one centromere of the dicentric is thought to have been inactivated (chimpanzee chromosome 13) and the surrounding constitutive heterochromatin is thought to have been either lost or changed to a C-band negative state. The centromeric region of the ancestral chromosome is, however, retained in the human chromosome as band 2q21.2 [53], suggesting that not all centromerically associated chromatin has been lost from this locus. Apparent loss of C-banding from the site of an (interstitial) inactive centromere has been reported in the case of a pseudodicentric X chromosome [59], but the frequency of occurrence of this phenomenon as well as its underlying mechanism remain unknown. We recognize that tandem fusions such as that forming human chromosome 2 occur in evolution. Nevertheless, for now at least, we regard these as more specialized events which are mechanically different from those observed in the chromosomal evolution of species such as the muntjacs and cotton rats, and therefore we shall not elaborate upon them in detail here.

CENTROMERES AND KINETOCHORES

Hsu and his co-workers [5,60] proposed that chromosomal evolution in species such as the Indian muntjac may have proceeded through total chromosome translocation followed by centromere inactivation. This proposal was based on the existence of karyotypes evolved through numerous fusion events and observations of centromere inactivation and various drug-induced fusions in tissue cultures. They further proposed that when a centromere is inactivated in a fusion chromosome, if fission occurs at some later time, the latent centromere might be reactivated (recent evidence suggests that reactivation in cases of tandem fusion may not be a reasonable hypothesis [29]). It was envisioned that this sequence of fusion/centromere inactivation and fission/centromere reactivation might be an explanation for the variety of numerical and structural changes often seen among related species. In addition, this hypothesis circumvents the problem of centromere loss which many cytogeneticists believe would place an obligatory unidirectional restriction on chromosomal evolution. Since chromosomal evolution apparently does proceed by fission in some groups [61,62], Hsu's latent centromere hypothesis has remained an attractive one, although it does not explain the fate of the resulting interstitial centromeric heterochromatin, which we here have proposed is lost when tandem fusions occur as evolutionary events.

Implicit in the latent centromere hypothesis is the premise that the centromeres are retained interstitially at the points of fusion. The advent of immunofluorescence techniques using kinetochore-specific antiserum derived from patients with the CREST form of scleroderma [63] has allowed for the initial biochemical characterization of the mammalian kinetochore and associated centromeric structures [64]. Using these techniques, Merry et al. [65] showed that interstitial kinetochore regions (inactive centromeres) are detectable in some tandem tandem fusion "dicentric" chromosomes that function monocentrically. Earnshaw and Rothfield [66] have identified a family of human centromere proteins of molecular weights 17, 80, and 140 kilodaltons, respectively, using serum from the same scleroderma patient. In contrast to the findings of Merry et al., Earnshaw and Migeon [67] have demonstrated by combined Q-banding and immunofluorescence that all three of these centromere proteins were absent from the inactive centromere of an isodicentric X chromosome. These authors resolved the apparent conflict between their results and those of Merry et al. by noting that the translocation studied by Merry et al. retains a constriction at the site of the inactive centromere while their isodicentric X does not. They speculated 1) that one of the family of centromeric antigens is involved in establishment of the primary constriction but is not involved in microtubule-kinetochore interactions, and 2) that the possibility of gradations of centromere inactivation (as suggested by Hsu) may exist [67]. These phenomena have recently been examined in more detail by Peretti et al. [68].

If complete centromere inactivation occurred in the evolutionary processes forming a tandem fusion chromosome, then the results of Earnshaw and Migeon [67] would suggest that an interstitial latent centromere, even if present, would be undetectable. It is important to note that a tandem fusion identified today as an evolutionary event in a given lineage may have occurred thousands or millions of years in the past. This would allow ample time for progressive gradations toward complete inactivation to occur even if the initial fusion resulted in the kind of situation documented by Merry et al. [65]. It is also important to note that the dicentric chromosomes studied by Earnshaw and Migeon [67] and Merry et al. [65] represent new mutations, suggesting that fundamental differences in centromere inactivation may exist at the time of the initial rearrangement [68]. Furthermore, mammalian centromeres exist in regions of constitutive heterochromatin and these regions were noted to be present and apparently intact in the above studies [65,67]. If, in an evolutionary sense, a latent centromere were to remain interstitial, then it would have to give up the heterochromatin which surrounds it either through loss or through conversion to chromatin which is not C-band positive as has apparently been the case with human chromosome 2.

While interesting hypotheses regarding the interrelationship of centromeres, telomeres, and satellite DNA evolution have been presented [69], the fate of centromeres in chromosomal evolution is still a major point of conjecture. The recent study by Brinkley et al. [70] on muntjac kinetochores provides the best and perhaps only experimental evidence on this long-standing problem. Again, using immunofluorescence techniques, Brinkley et al. [70] found that at interphase, the single, dotlike kinetochores of the Chinese muntjac clustered nonrandomly into small aggregates similar in appearance to the beadlike arrays of the somewhat dispersed compound kinetochore subunits seen in interphase nuclei of Indian muntjac cells. These authors then provided a fascinating model for linear fusion of centromeric regions (kinetochores). In essence, the Brinkley model suggests that the unit centromeres (as observed in the Chinese muntjac) aggregate and that this aggregation is possibly mediated by centromeric heterochromatin. These aggregations may have been the targets for clastogenic agents which would have allowed for fusion of telomeres and centromeres producing translocations involving entire chromosome arms. The translocations could have been either tandem, with all chromosomal orientations possible, or Robertsonian, but the result would have been accumulation of centromeres into a compound structure at one locus along the length of the chromosome accompanied by fusion of chromosome arms. In this model, the centromeres are neither inactivated at interstitial loci nor lost. They are accumulated at one compound locus—a form of storage. This model also provides for the elimination of centromeric heterochromatin by placing the presumed breaks at (or very near) the junction between euchromatin and heterochromatin. While the Brinkley model does not specifically address the issue of chromosomal fission, we see no reason why it could not be thought of as bidirectional, reducing chromosome numbers in some groups through fusion and increasing numbers in other groups or at other times by fission.

Critical to the Brinkley model is the fact that the kinetochores of Indian muntjac chromosomes are compound. In our other example, the cotton rats, there is no evidence either for or against compound kinetochores. Nevertheless, the nonrandom aggregation of centromeres at interphase provides a mechanism for orientation of chromosomes that would again facilitate the multiple fusions which have occurred in some representatives in this genus. Whether or not the kinetochores are stored as compound units or lost in the evolutionary process remains to be seen.

The Brinkley model has been devised primarily to account for redistribution of centromeres, which has accompanied chromosome fusion during the evolution of the Indian muntjac karyotype. It seems clear, however, that two other chromosomal regions, the telomere and the pericentromeric heterochromatin, must also play an important role in the formation of evolutionarily stable tandem translocations.

TELOMERES

Telomeres form the ends of chromosomes and it is well known that chromosomes that have lost their telomeres are generally unstable. The molecular structure of the telomere has recently been reviewed [71], and a variety of models concerning structure, interaction, replication, and behavior of telomeres have been proposed [69,72–74]. Because the cytological function of telomeres appears to be capping of the ends of chromosomes, telomeric fusions are relatively rare events. On the other hand, telomeres display intricate and nonrandom behavior at different times in the cell cyle and often exhibit nonrandom associations (reviewed in [71,74]).

Dancis and Holmquist [74] have proposed a fusion model of telomere replication wherein telomeres are envisioned to pair and fuse before replication, covalently linking the DNA from each pair of telomeres (and therefore their associated chromosomes) into a continuous DNA molecule. After replication the fused ends separate through specific enzymatic cleavage and telomeric integrity and function are restored. This model assumes that telomeric fusion is nonspecific (any telomere can fuse with any other telomere) and implies that all telomeres in any given genome should be similar. Since telomere fusion occurs prior to replication, this model predicts nonhomologous chromosome linking to be more common than homologous linking. Dancis and Holmquist [74] stated that dysfunction in the replication process could result in permanently fused chromosomes, and documented the fact that when this occurs in culture, nonhomologous chromosomes are the primary contributors to the dicentrics that are formed. While the accuracy of the telomere fusion model remains an open question, the fact that telomeric DNA sequences are similar on many (nonhomologous) chromosomes within the karyotypes of several species is established [75–78].

Aberrant telomeric associations resulting in permanent fusion are well known (reviewed in [71]) and are sometimes invoked to explain specific Robertsonian fusions or Robertsonian fusion/fission events [71,74] in karyotypic evolution. Since the hypothesis of palindromic base sequences at the telomeres of eukaryotic chromosomes was proposed [72,73], their existence has been documented in some organisms (reviewed in [71]), and long tandem telomeric repeats capable of elongation in cell free extracts have been identified in others [79]. The possible palindromic nature of telomeres has also been used to account for unusual translocations involving the telomeres of some chromosomes and presumed interstitial complementary sequences in euchromatic regions of other chromosomes [80–82]. One possible explanation for these putative interstitial palindromes is that they are evolutionary remnants of telomeres from previous fusion events [77].

PERICENTROMERIC CONSTITUTIVE HETEROCHROMATIN

Constitutive heterochromatin has been lost in the evolution of tandem fusion chromosomes [37,41,83]. In the case of telomere-telomere translocations involving acrocentric chromosomes, the centromeric heterochromatin is lost from one or the other end of the translocation chromosome, yet we must assume that the new telomeric end functions normally. In other words, normal telomeric function appears at what was originally the centromeric heterochromatin-euchromatin junction of one of the participating chromosomes as if a "latent" telomere were present at that site [84]. In centromere-telomere fusions involving acrocentric chromosomes, the translocation occurs between the telomere of one chromosome and the centromeric heterochromatin-euchromatin junction of the other, eliminating the centromere (or, as in the case of the muntjacs, redistributing the kinetochores) and centromeric heterochromatin. Holmquist and Dancis [69] presented an hypothesis whereby satellite DNA sequences (in centromeric heterochromatin) are interchromosomally communicated, thus distributing the same satellite among many different chromosomes of the karyotype. Jabs et al. [85] have cloned a highly repeated sequence present in the centromeric heterochromatin of all human chromosomes. Young et al. [75] and Renkawitz-Pohl and Bialojan [76] have demonstrated what appears to be complex sequence homology between telomeric regions and beta heterochromatin in *Drosophila melanogaster*. While direct evidence is lacking for mammals, taken together, the above studies suggest the possibility that some homology may exist between telomeric DNA sequences and DNA sequences at the junction of heterochromatin and euchromatin on some or many chromosomes in a given genome. Schmidtke et al. [83] speculated that the fusions which have occurred in the evolution of the Indian muntjac karyotype may have been mediated by a specific subset of repetitive sequences located at the centromeric and telomeric ends of ancestral chromosomes.

We, therefore, can envision a eukaryotic chromosome as a structure in which the euchromatin comprising the arms might be capped at both ends by similar DNA sequences ("capping" sequences), one end representing a telomere and the other end representing the junction of euchromatin with centromeric constitutive heterochromatin. It is noteworthy that chromatin at this junction has different cytological properties as is apparent from sister chromatid exchange (SCE) studies in the Indian muntjac [86] and in the kangaroo rat, genus *Dipodomys* [87], where the frequency of spontaneous SCEs was found to be higher at this site than elsewhere in the genome. Painter [88] has attributed this increase to the difference in replication timing between euchromatin and heterochromatin.

Similar capping sequences on both ends of the same euchromatic block would facilitate whole arm (paracentric) inversions, which have been identified in the cotton rat, *Sigmodon fulviventer* [46], and in the meadow voles, genus *Microtus* [38], both genera also showing numerous tandem fusions. Similar capping sequences distributed over nonhomologous chromosomes would facilitate rearrangements in the form of tandem translocations connecting euchromatic blocks together in various orientations. If, in a given genome, similar capping sequences were not randomly distributed, then nonrandom tandem translocations could result. Tandem translocation could then be viewed as a phenomenon in which fusion occurs based on similarity of the DNA sequences at the ends of euchromatic blocks. Under such circumstances, some fusions would be favored over others at the cellular level, and particular (and genetically "identical") fusions could reoccur at a frequency which might be much higher than expected via random mutational processes.

If similarities of telomeric sequences do play a role in facilitating these rearrangements, then it can be argued that telomeric fusions producing tandem translocations should be as likely, or even more likely to occur than rearrangements between telomeres and heterochromatin-euchromatin junctions. Cytogenetic information available from human clinical material, cell cultures, and malignancies tends to support this argument. In these cases, however, dicentric or pseudodicentric chromosomes are formed and heterochromatin is retained at the centromeric loci. We have no reason to doubt that terminal rearrangments occur in other organisms as they do clinically in man, but because these rearrangements seem virtually always to be deleterious to their possessors, we have no reason to suspect that they are the kinds of chromosomal changes that would persist in evolution. Therefore, we think that the rearrangements that have produced tandem fusions during evolution, as exemplified in the karyotypes of muntjacs and cotton rats, represent a unique type of cytogenetic event and as such are probably much rarer than those occurring clinically or in culture.

THE EVOLUTION OF RADICAL CHROMOSOMAL REARRANGEMENTS

In the examples of the muntjacs and cotton rats, we are witness to the dynamics of the genome. The fact is that radical and presumably rapid reorganization has occurred in the evolution of the karyotypes of some (but not all) of these species. We propose that the answer as to how this reorganization was possible lies, at least in part, in the architecture of the chromosomes themselves. Based on this architecture, we hypothesize that rearrangements have taken place through interactions of telomeres and heterochromatin-euchromatin junctions with centromeres and centromeric heterochromatin being lost or redistributed in the process. Two other critical questions, however, must also be addressed—namely, 1) what factors could have precipitated such extensive reorganization in some species of muntjacs and cotton rats but not in others? and 2) how have so many rearrangements transcended the heterozygous stage to become fixed such that these species are typified by karyotypic uniformity?

A fundamental problem of chromosomal evolution by any mechanism, be it tandem fusion, centric fusion, inversion, or reciprocal translocation, is how any rearrangement (which must be heterozygous as a new mutation) transcends the heterozygous stage and becomes fixed in the homozygous condition in the population. Loss of fertility in the heterozygote has been widely documented in many different kinds of organisms, and strong arguments have been presented concerning why heterozygosity is deleterious [89]. The fact that tandem fusion polymorphisms are virtually unknown in mammals coupled with the relative rarity of tandem fusion in most taxa suggest that these events are highly deleterious in the heterozygous state and are probably not subject to the same meiotic regulation as is seen in heterozygotes for centric fusion [55,90,91]. White [1: pp. 225,226] states that tandem fusion heterozygotes may produce 50% aneuploid gametes and that selection against these heterozygotes will be severe.

An assumption of chromosomal evolution, which we shall state very simply here, is that if a new genetic combination is to become fixed, it must provide in the homozygous state some selective advantage that outweighs both the disadvantage of being heterozygous and the former advantage of being homozygous for the ancestral condition [92]. If these qualifications are met, then the rearrangement can proceed to fixation through selection. Alternatively, if both homozygotes are equally fit, but the rearrangement is deleterious in the heterozygous state, then fixation of the new rearrangement must occur by genetic drift. In the case of genetic drift, Lande [93] has reemphasized the long-held view that a chromosomal rearrangement has a much better chance of becoming fixed if it occurs in a small, isolated deme of perhaps tens or at most a few hundreds of individuals. Disruption of isolation through migration of even a few individuals per generation could prevent the establishment of a new rearrangement. Spreading of the rearrangement in homozygous form after fixation is envisioned to occur through random local extinctions and colonizations. Undoubtedly, both selection and genetic drift are critical to the evolutionary process depending on the particular rearrangement and on the population structure of the organisms in which it occurs.

The nature of a selective advantage conferred by a given rearrangement in a given evolutionary instance is impossible to ascertain specifically but is usually ascribed to adaptation to environment. If, for example, a translocation between chromosomes A and B (AB) provides a "better" genetic combination than A and B provided as separate chromosomes, then AB will have an advantage (i.e., increases fitness) under a particular set of environmental constraints and will be more likely to persist (become fixed) in a given population. In organisms that show numerous chromosome rearrangements and, as is thought in the cotton rats, have fixed these rearrangements in a short evolutionary time span, the environmentally adaptive nature of each and every rearrangement is hard to fathom. On the other hand, the uniformity and discreteness of these rapidly evolved karyotypes makes it equally difficult to understand how genetic drift alone could be responsible for fixation of each and every rearrangement which has occurred while leaving no peripheral isolates [94]. Since establishment of a rearrangement is necessarily a populational phenomenon, it is instructive to examine certain aspects of population biology for clues which might be helpful in understanding how establishment could occur. The population biology and ecology of the cotton rats are reasonably well known so these animals will serve as our example.

Cotton rats are, under optimal ecological conditions, restricted to grassy habitats. These rodents are highly cyclical, exhibiting both annual and long-term population fluctuations and occasionally reaching extremely high and low densities. The species that have undergone numerous tandem fusions (as well as centric fusions) are found primarily in the semiarid regions of the southwestern United States and western Mexico. These regions are characterized by highly disjunct, patchy stands of grass surrounded by desert or arid scrub habitat. The small populations (demes) therefore usually exist in strict isolation. The geographic area where these species occur is also subject to climatic cycles of moisture and drought. This creates environmental conditions favorable for occasional expansion of grassland, thus allowing contact between isolated populations followed by contraction of habitat and reestablishment of isolation. Several models of speciation are compatible with this extremely brief account of cotton rat biology, most notably the peripatric model of Mayr [95], and the stasipatric model of White [96,97]. We do not, however, exclude the potential applicability of other models of speciation [92,98–100]. In our view, this account also contains two necessary but seemingly contradictory conditions for the uniform establishment of numerous chromosome rearrangements in these species: isolation of small populational subunits, and occasional introgression.

If fixation were to occur by genetic drift in these demes, then periodic introgression would create a significant problem in that it could either retard or abort the ultimate fixation of a particular rearrangement [93]. If, on the other hand, a particular rearrangement provided a selective advantage, then introgression could serve to spread it to other demes. The most intriguing possibility, however, is that introgression may itself have played an active role in generating novel genetic combinations [97], thus providing the raw material on which evolutionary forces have acted in these populations.

The role of introgression in generating novel genetic combinations is well established in a variety of organisms. In *Drosophila*, hybridization of individuals from different populations has been shown to release mutator factors that induce frequent and nonrandom chromosome breakage [101–104]. It is thought that this mutational activity is caused by mobile genetic elements that under appropriate conditions can subject the genome to unusually high degrees of stress [105]. The potential importance of these elements in rapid genome reorganization and evolution has been suggested [97,110–111], and it is certainly possible that mobile genetic elments could be the clastogenic agents proposed by Brinkley et al. [70] in the linear fusion model of kinetochore redistribution in the karyotype of the Indian muntjac. These mobile elements are widespread among organisms and their presence in such diverse groups as bacteria, yeast, plants, and animals is established. In *Drosophila melanogaster*, half of the moderately repetitive DNA or 5–10% of the entire genome consists of transposable elements [112,113]. Introgression of strains, subspecies, or sibling species of *Drosophila* known to differ in certain transposable elements produces a syndrome of hybrid dysgenesis that is often characterized by a high frequency of nonrandom chromosome breakage and rearrangement. Likewise, Shaw et al. [114] observed repeated nonrandom chromosome fusions as a result of introgression between two subspecies of Australian grasshoppers. In that study, 12% of backcross progeny carried novel chromosome rearrangements, and it was noted that some of the same rearrangements occurred repeatedly in different backcrosses. Shaw et al. [114] also observed recurrence of the same rearrangement in several offspring of a single female and different rearrangements produced from the same germline. While Shaw et al. [114] studied backcrosses of laboratory-produced F_1 hybrids, these authors were also able to identify some of the same fusions in natural hybrids occurring in a narrow zone of overlap between the two subspecies.

The examples of fruit flies and grasshoppers show that introgression, among other things, may act to destabilize the complex organization of a genome and that this destabilization can, and often does, take the form of chromosomal rearrangements. Certain chromosomal rearrangements may themselves act to initiate destabilization or function to nonrandomly affect destabilization as the process of reorganization continues [105,106].

The presence of transposable elements in mammalian cells has been inferred from DNA sequence analysis, but the degree of their mobility is not yet known [111]. Movement or intercommunication of satellite DNA sequences [69], which are highly repeated, implies a form of transposable genetic element which could be associated with both centromeric heterochromatin and telomeres. In this regard, it is noteworthy that Young et al. [75], who found complex DNA sequence homology between telomeric regions and pericentric heterochromatin in *Drosophila*, identified an apparent mobile element in their probe. This mobile element also hybridized (in situ) at interstitial loci that varied from stock to stock [75]. The repetitive nature of many mobile element sequences in *Drosophila* led Rubin [115] to speculate that scrambled, clustered repeats found on chromosome arms and in centric heterochromatin may be the graveyards of dead transposable elements. Rubin [115] also speculated that new elements may emerge from these "graveyards" by duplicative transposition mechanisms reestablishing movement within the genome.

While introgression followed by chromosomal reorganization is a hypothesis that we think is clearly applicable to our example of the cotton rats, the obvious problem created in this and in any similar scenario is how to stop the process and return to stability with a reorganized genome. It is possible that restoration of stability results in part from the elimination of heterochromatin (Rubin's mobile element "graveyards") during chromosomal rearrangement, a characteristic of tandem translocations occurring in evolution. It seems to us that under circumstances where reorganization has been promoted, such as through introgression, new combinations which were either stable or which promoted stability would be highly advantageous. Thus, stability could be restored to the genome through generation not of environmentally adaptive chromosomal rearrangements but through rearrangements which were adaptive in the sense that they were not subject to further high frequency reorganization. Since these rearrangements are nonrandom, one could envision the same rearrangements occurring in multiple individuals in the same population (or possibly even in different populations) and stability being restored to the genome with independently derived but genetically identical translocations. Through a number of cycles of introgression-destabilization/rearrangement-isolation-restabilization, highly derived yet uniform karyotypes could rapidly evolve. This view is different from the conventional evolutionary doctrine that chromosomal rearrangements arise as unique, sequential events [96]. It is also different from the view of King [94] that multiple rearrangements may arise simultaneously in a single individual and eventually spread throughout the species. On the other hand, it has been established that repeated and nonrandom rearrangements occur [114], and we should expect the notion that similar or identical chromosomal rearrangements may have multiple origins to continue to gain acceptance in particular instances of chromosomal evolution [97,116].

It is undeniable that selection for environmentally adaptive rearrangements exists. Nevertheless, selection in the sense that rearrangements are adaptive to an organism in its environment may not necessarily be as important to some species as rearrangements which form newly stable combinations. Multiple copies of the same rearrangement produced in the same brood [114] would provide multiple and identical heterozygotes and would facilitate rapid fixation [116]. If selection were to favor stability, and if these new rearrangements were stable structurally, then they would have a distinct selective advantage simply because of the stability they provide. Under these conditions, rapid genome reorganization could occur without appreciable reorganization at the genic level, which, in fact, has been the case in the cotton rats [47] and muntjacs [83].

We do not know how applicable a model of introgression might be to the case of the muntjacs. It is worth noting, however, that the Indian muntjac and the Chinese muntjac, despite their karyotypic disparity, can produce viable but sterile F_1 hybrids [44]. It is not unreasonable to think that in evolving the Indian muntjac karyotype, ancestral populations having little or no karyotypic disparity could have been subjected to introgression, initiating cycles of genome reorganization and eventually producing the karyotype of modern *M. muntjak vaginalis*.

In our examples of muntjacs and cotton rats, an obvious question is why some species have undergone radical genome reorganization while some of their near relatives have maintained what must be regarded as ancestral karyotypes. Certainly, differences in population structure, isolation, selection, and genetic drift—the classical conditions of chromosomal evolution—may all be applicable to greater or lesser degrees. It has been argued that if a karyotype is adaptive to an organism in a particular environment, the retention of an ancestral karyotype is simply the maintenance of an optimum karyotype for that environment. Similarly, a derived karyotype may be maintained as the optimum under different environmental circumstances [117]. The apparent answer to the above question is that in some ancestral populations, both genomic and populational conditions were right for reorganization to occur. In other populations these conditions were not right or at least have not been right to this point in their evolution. The biology of transposable genetic elements has been instructive and important in that it has added a new dimension to our interpretations of how chromosomal evolution involving rapid and radical genome reorganization could occur. It is easy to imagine the nonrandom destabilization of the genome induced by transposable elements as a basis for the initial steps of karyotypic orthoselection in the sense of White [1,2], inducing particular types of rearrangements that may both stem from, and be affected by, the architecture of the chromosomes themselves. In those cotton rats and muntjacs that have maintained the ancestral karyotype, perhaps the populations have not yet been subjected to genomic stress sufficient to initiate destabilization.

Obviously, we have no direct evidence that mobile genetic elements have been involved in the reorganization of the karyotypes of cotton rats and muntjacs or even that mobile elements exist in these species. On the other hand, no evidence exists to the contrary, and certainly there is no direct evidence to support the contention that any of the rearrangements have been environmentally adaptive or fixed by random genetic drift. What we have documented is that radical genome reorganization has occurred in a nonrandom way, primarily through tandem fusion. We propose that these tandem translocations have been facilitated by the architecture or the chromosomes involved, with fusions occurring between telomeres and heterochromatin-euchromatin junctions, and with centromeres and centromeric heterochromatin being lost or redistributed in the process. We also propose that introgressive hybridization involving ancestral populations may have acted to destabilize the genome in much the same way as has been shown in *Drosophila* and other organisms. Restabilization follows from the generation of new and stable translocations. Our hypothesis differs somewhat from the conventional in that we view the reorganized karyotypes as adaptive not to the environment but because of their stability. Secondly, we think that similar or identical rearrangements may have arisen multiple times. We do not exclude the possibility that many aspects of cotton rat and muntjac biology may have influenced the evolution of the karyotypes in these groups or that factors other than introgression may have provided the genomic stress necessary to initiate destabilization. Nevertheless, radical and rapid genome reorganization potentially mediated by mobile genetic elements seems to us an attractive hypothesis and one well worth further investigation in mammals.

REFERENCES

1. White MJD: "Animal Cytology and Evolution." Third Edition. London: Cambridge University Press, 1973, Ch 7, 12.
2. White MJD: Chromosomal repatterning—regularities and restrictions. Genetics [Suppl] 79:63–72, 1975.

3. Baker RJ, Bickham JW: Karyotypic evolution in bats: Evidence of extensive and conservative chromosomal evolution in closely related taxa. Syst Zool 29:239–253, 1980.
4. Haiduk MW, Baker RJ: Cladistical analysis of G-banded chromosomes of nectar feeding bats (Glossophaginae:Phyllostomidae). Syst Zool 31:252–265, 1982.
5. Hsu TC, Pathak S, Chen TR: The possibility of latent centromeres and a proposed nomenclature system for total and whole arm translocations. Cytogenet Cell Genet 15:41–49, 1975.
6. Daniel A, Lam-Po-Tang PRLC: Structure and inheritance of some heterozygous Robertsonian translocations in man. J Med Genet 13:381–388, 1976.
7. Daniel A: Single Cd band in dicentric translocations with one suppressed centromere. Hum Genet 48:85–92, 1979.
8. Warburton D, Henderson AS, Shapiro LR, Hsu LYF: A stable human dicentric chromosome, t dic (12;14)(p13; p13) including an intercalary satellite region between centromeres. Am J Hum Genet 25:439–445, 1973.
9. Roberts SH, Howell RT, Laurence KM, Heathcote ME: Stable dicentric autosome, t dic (8:22)(p23:p13), in a mentally retarded girl. J Med Genet 14:66–68, 1977.
10. Daniel A, Perel ID, Clarke AJ, Saville T: Familial dicentric translocation t(13;18)(p13; p11.2) ascertained by recurrent miscarriages. J Med Genet 16:73–75, 1979.
11. Herva R, Saarinen I, Savikurki H, de la Chapelle A: Dicentric Y chromosome arising via tandem translocation. Am J Med Genet 7:115–122, 1980.
12. Howard PJ, Berry AC: Familial transmission of a non-Robertsonian translocation dicentric. Clin Genet 29:246–250, 1986.
13. Hayashi K, Schmid W: Tandem duplication q14 and dicentric formation by end-to-end chromosome fusions in ataxia telangiectasia (AT): Clinical and cytogenetic findings in 5 patients. Humangenetik 30:135–141, 1975.
14. Aurias A, Dutrillaux B, Griscelli C: Tandem translocation t(14;14) in isolated and clonal cells in ataxia telangiectasia are different. Hum Genet 63:320–322, 1983.
15. Aurias A, Croquette MF, Nuyts JP, Griscelli C, Dutrillaux B: New data on clonal anomalies of chromosome 14 in ataxia telangiectasia: tct(14;14) and inv(14). Hum Genet 72:22–24, 1986.
16. Ing PS, Smith SD: Cytogenetic studies of a patient with mosaicism of isochromosome 13q and a dicentric (Y;13) translocation showing differential centromeric activity. Clin Genet 24:194–199, 1983.
17. Badalyan LO, Arkhipov BA, Malygina NA, Petrukhin AS, Lysov VLA: Clinical case of tandem translocation between chromosomes 13 and 15 (in Russian). Tsitol Genet 19:304–308, 1985.
18. Dewald GW: Isodicentric X chromosomes in humans: Origin, segregation behavior, and replication band patterns. In Sandberg AA (ed): "Cytogenetics of the Mammalian X Chromosome, Part A." New York: Alan R. Liss, Inc., 1983, pp 405–426.
19. Zakharov AF, Baranovskaya LI: X-X chromosome tranlocations and their karyotype-phenotype correlations. In Sandberg AA (ed): "Cytogenetics of the Mammalian X Chromosome, Part B." New York: Alan R. Liss, Inc., 1983, pp 261–279.
20. Rivera H, Sole MT, Garcia-Cruz D, Martinez-Wilson M, Cantu JM: On telomere replication and fusion in eukaryotes: Apropos of a case of 45,X/46,X,ter rea(X;X)(P22.3; p22.3). Cytogenet Cell Genet 38:23–28, 1984.
21. Therman E, Trunca C, Kuhn EM, Sarto GE: Dicentric chromosomes and the inactivation of the centromere. Hum Genet 72:191–195, 1986.

22. Jernigan D, Curl N, Keeler C: Milledgeville mongoloid: A rare karyotype of Down's syndrome. J Hered 65:254–257, 1974.
23. Harvey J, Wiener S, Birner R: "Mirror image" chromosome No. 21. J Med Genet 14:152–153, 1977.
24. Cantu JM, Hernandez A, Plascencia L, Vaca G, Moller M, Rivera H: Partial trisomy and monosomy 21 in an infant with an unusual de novo 21/21 translocation. Ann Genet (Paris) 23:183–186, 1980.
25. Turleau C, Roubin M, Chavin-Colin F, Grouchy J de: Trisomie 21 par duplication en miroir 46,XX,psu dic(21) ter rea(21q21q). Ann Genet (Paris) 23:187–189, 1980.
26. Berg JM, Gardner HA, Gardner RJM, Goh EG, Markovic VD, Simpson NE, Worton RG: Dic (21;21) in a Down's syndrome child with an unusual chromosome 9 variant in the mother. J Med Genet 17:144–155, 1980.
27. Schwartz S, Palmer CG, Weaver DD, Priest J: Dicentric chromosome 13 and centromere inactivation. Hum Genet 63:332–337, 1983.
28. Dutrillaux B, Aurias A, Couturier J, Croquette MF, Viegas-Pequignot E: Multiple telomeric fusions and chain configurations in human somatic chromosomes. In de la Chapelle A, Sorsa M (eds): "Chromosomes Today, Vol 6." Amsterdam: Elsevier/North-Holland, 1977, pp 37–50.
29. Drets ME, Therman E: Human telomeric 6;19 translocation chromosome with a tendency to break at the fusion point. Chromosoma 88:139–144, 1983.
30. Chen TR: A simple method to sequentially reveal Q- and C-bands on the same metaphase chromosomes. Chromosoma 47:147–156, 1974.
31. Fitzgerald PH, Morris CM: Telomeric association of chromosomes in B-cell lymphoid leukemia. Hum Genet 67:385–390, 1984.
32. Mandahl N, Heim S, Kristoffersson U, Mitelman F, Rooser B, Rydholm A, Willen H: Telomeric association in malignant fibrous histiocytoma. Hum Genet 71:321–324, 1985.
33. Dewald GW, Boros SJ, Conroy MM, Dahl RJ, Spurbeck JL, Vitek HA: A tdic (5;15)(p13; p11) chromosome showing variation for constriction in the centromeric regions in a patient with cri du chat syndrome. Cytogenet Cell Genet 24:15–26, 1979.
34. Zuffardi O, Tiepolo L: Latent centromere and chromosome function in man. Clin Genet 19:547A, 1981.
35. Lejeune J, Dutrillaux B, Rethore MO, Prieur M: Comparison de la structure fine des chromatides d' *Homo sapiens* et de *Pan troglodytes*. Chromosoma 43:423–444, 1973.
36. Seuanez HN: "The Phylogeny of Human Chromosomes." Berlin: Springer-Verlag, 1979, pp 70–71.
37. Elder FFB: Tandem fusion, centric fusion, and chromosomal evolution in the cotton rats, genus *Sigmodon*. Cytogenet Cell Genet 26:199–210, 1980.
38. Modi WS: Phylogenetic analyses of chromosomal banding patterns among Nearctic Arvicolidae (Mammalia:Rodentia). Syst Zool (in press).
39. Wurster DH, Benirschke K: Chromosome studies in some deer, the springbok and the pronghorn, with notes on placentation in deer. Cytologia 32:273–285, 1967.
40. Wurster DH, Benirschke K: Indian muntjac, *Muntiacus muntjak*: A deer with a low diploid chromosome number. Science 168:1364–1366, 1970.
41. Wurster DH, Atkin NB: Muntjac chromosomes: A new karyotype for *Muntiacus muntjak*. Experientia 28:972–973, 1972.
42. Soma H, Kada H, Mtayoshi K, Suzuki Y, Meckvichal C, Mahannop A, Vatanaromya B: The chromosomes of *Muntiacus feae*. Cytogenet Cell Genet 35:156–158, 1983.
43. Wurster-Hill DH, Seidel B: The G-banded chromosomes of Roosevelt's muntjac, *Muntiacus rooseveltorum*. Cytogenet Cell Genet 39:75–76, 1985.

44. Shi L, Ye Y, Duan X: Comparative cytogenetic studies on the red muntjac, Chinese muntjac, and their F_1 hybrids. Cytogenet Cell Genet 26:22–27, 1980.
45. Zimmerman EG: Karyology, systematics and chromosomal evolution in the rodent genus, *Sigmodon*. Publ Mus Biol Ser Michigan State Univ 4:385–454, 1970.
46. Elder FFB, Lee MR: The chromosomes of *Sigmodon ochrognathus* and *S. fulviventer* suggest a realignment of *Sigmodon* species groups. J Mammal 66:511–518, 1985.
47. Johnson WE, Selander RK, Smith MH, Kim YJ: Biochemical genetics of sibling species of the cotton rat (*Sigmodon*). In Wheeler MR (ed): "Studies in Genetics VII." Univ Texas Publ No 7213, pp 297–305, 1972.
48. Buckland RA, Evans HJ: Cytogenetic aspects of phylogeny in the Bovidae. I. G-banding. Cytogenet Cell Genet 21:42–63, 1978.
49. Buckland RA, Evans HJ: Cytogenetic aspects of phylogeny in the Bovidae. II. C-banding. Cytogenet Cell Genet 21:64–71, 1978.
50. Di Beradino D, Iannuzzi L: Chromosome banding homologies in swamp and murrah buffalo. J Hered 72:183–188, 1981.
51. Yosida TH: Chromosome differentiation and species evolution in rodents. In Sharma AK, Sharma A (eds): "Chromosomes in Evolution of Eukaryotic Groups. Vol I." Boca Raton: CRC Press, 1983, pp 147–176.
52. Bickham JW: Chromosomal variation and evolutionary relationships of vespertilionid bats. J Mammal 60:350–363, 1979.
53. Yunis JJ, Prakash O: The origin of man: A chromosomal pictorial legacy. Science 215:1525–1530, 1982.
54. Lee MR, Zimmerman EG: Robertsonian polymorphism in the cotton rat, *Sigmodon fulviventer*. J Mammal 50:333–339, 1969.
55. Elder FFB, Pathak S: Light microscopic observations on the behavior of silver-stained trivalents in pachytene cells of *Sigmodon fulviventer* (Rodentia, Muridae) heterozygous for centric fusion. Cytogenet Cell Genet 27:31–38, 1980.
56. Koop BF, Baker RJ, Genoways HH: Numerous chromosomal polymorphisms in a natural population of rice rats (*Oryzomys*, Cricetidae). Cytogenet Cell Genet 35:131–135, 1983.
57. White MJD: Chromosomal rearrangements and speciation in animals. Annu Rev Genet 3:75–98, 1969.
58. ISCN (1985): "An International System for Human Cytogenetic Nomenclature." Harnden DG, Klinger HP (eds); published in collaboration with Cytogenet Cell Genet (Basel: Karger, 1985); also in Birth Defects Original Article Series, Vol 21, No 1 (New York: March of Dimes Birth Defects Foundation, 1985).
59. Nakagome Y, Nakahori Y, Mitani K, Matsumoto M: The loss of centromeric heterochromatin from an inactivated centromere of a dicentric chromosome. Jpn J Hum Genet 31:21–26, 1986.
60. Hsu TC, Pathak S, Basen BM, Stark GJ: Induced Robertsonian fusions and tandem translocations in mammalian cell cultures. Cytogenet Cell Genet 21:86–98, 1978.
61. Todd NB: Karyotypic fissioning and canid phylogeny. J Theor Biol 26:445–480, 1970.
62. Dutrillaux B: Chromosomal evolution in primates: Tentative phylogeny from *Microcebus murinus* (Prosimian) to man. Hum Genet 48:251–314, 1979.
63. Moroi Y, Peebles C, Fritzler MJ, Steigerwald J, Tan EM: Autoantibody to centromere (kinetochore) in scleroderma sera. Proc Natl Acad Sci USA 77:1627–1631, 1980.
64. Valdivia MM, Brinkley BR: Fractionation and initial characterization of the kinetochore from mammalian metaphase chromosomes. J Cell Biol 101:1124–1134, 1985.

65. Merry DE, Pathak S, Hsu TC, Brinkley BR: Anti-kinetochore antibodies: Use as probes for inactive centromeres. Am J Hum Genet 37:425–430, 1985.
66. Earnshaw WC, Rothfield N: Identification of a family of human centromere proteins using autoimmune sera from patients with scleroderma. Chromosoma 91:313–321, 1985.
67. Earnshaw WC, Migeon BR: Three related centromere proteins are absent from the inactive centromere of a stable isodicentric chromosome. Chromosoma 92:290–296, 1985.
68. Peretti D, Maraschio P, Lambiase S, Lo Curto F, Zuffardi O: Indirect immunofluorescence of inactive centromeres as an indicator of centromeric function. Hum Genet 73:12–16, 1986.
69. Holmquist GP, Dancis B: Telomere replication, kinetochore organizers, and satellite DNA evolution. Proc Natl Acad Sci USA 76:4566–4570, 1979.
70. Brinkley BR, Valdivia MM, Tousson A, Brenner SL: Compound kinetochores of the Indian muntjac. Evolution by linear fusion of unit kinetochores. Chromosoma 91:1–11, 1984.
71. Blackburn EH, Szostak JW: The molecular structure of centromeres and telomeres. Annu Rev Biochem 53:163–194, 1984.
72. Cavalier-Smith T: Palindromic base sequences and replication of eukaryote chromosome ends. Nature 250:467–470, 1974.
73. Bateman AJ: Simplification of palindromic telomere theory. Nature 253:379–380, 1975.
74. Dancis BM, Holmquist GP: Telomere replication and fusion in eukaryotes. J Theor Biol 78:211–224, 1979.
75. Young BS, Pession A, Traverse KL, French C, Pardue ML: Telomere regions in Drosophila share complex DNA sequences with pericentric heterochromatin. Cell 34:85–94, 1983.
76. Renkawitz-Pohl R, Bialojan S: A DNA Sequence of *Drosophila melanogaster* with a differential telomeric distribution. Chromosoma 89:206–211, 1984.
77. Steinemann M: Telomere repeats within the neo-Y-chromosome of *Drosophila miranda*. Chromosoma 90:1–5, 1984.
78. Carmona MJ, Morcillo G, Galler R, Martinez-Salas E, de la Campa AG, Diez JL, Edstrom JE: Cloning and molecular characterization of a telomeric sequence from a temperature-induced Balbiani ring. Chromosoma 92:108–115, 1985.
79. Greider CW, Blackburn EH: Identification of specific telomere terminal transferase activity in Tetrahymena extracts. Cell 43:405–413, 1985.
80. Lejeune J, Maunoury C, Prieur M, Van den Akker J: Translocation sauteuse (5p;15q), (8q;15q), (12q;15q). Ann Genet (Paris) 22:210–213, 1979.
81. Mattei MG, Souiah N, Mattei JF: Chromosome 15 anomalies and the Prader-Willi syndrome: Cytogenetic analysis. Hum Genet 66:313–334, 1984.
82. Elder FFB, Nichols MM, Hood OJ, Harrison WR, III: Unbalanced translocation (15;17)(q13; p13.3) with apparent Prader-Willi syndrome but without Miller-Dieker syndrome. Am J Med Genet 20:519–524, 1985.
83. Schmidtke J, Brennecke H, Schmid M, Neitzel H, Sperling K: Evolution of muntjac DNA. Chromosoma 84:187–193, 1981.
84. Hsu TC: Longitudinal differentiation of chromosomes and the possibility of interstitial telomeres. Exp Cell Res [Suppl] 9:73–85, 1963.
85. Jabs EW, Wolf SF, Migeon BR: Characterization of a cloned DNA sequence that is present at centromeres of all human autosomes and the X chromosome and shows polymorphic variation. Proc Natl Acad Sci USA 81:4884–4888, 1984.

86. Carrano AV, Wolff S: Distribution of sister chromatid exchanges in the euchromatin and heterochromatin of the Indian muntjac. Chromosoma 53:361–369, 1975.
87. Bostock CJ, Christie S: Analysis of the frequency of sister chromatid exchange in different regions of chromosomes of the kangaroo rat (*Dipodomys ordii*). Chromosoma 56:275–287, 1976.
88. Painter RB: A replication model for sister chromatid exchange. In Sandberg AA (ed): "Sister Chromatid Exchange." New York: Alan R. Liss, Inc., 1982, pp 115–121.
89. White MJD: "Modes of Speciation." San Francisco: WH Freeman and Company, 1978, Ch 3, 6.
90. Vidal F, Templado C, Navarro J, Marina S, Egozcue J: Meiotic and synaptonemal complex studies in a 14/21 translocation carrier. Int J Androl 5:21–26, 1982.
91. Luciani JM, Guichaoua MR, Mattei A, Morazzani MR: Pachytene analysis of a man with a 13q;14q translocation and infertility. Cytogenet Cell Genet 38:14–22, 1984.
92. Lewis H: Speciation in flowering plants. Science 152:167–172, 1966.
93. Lande R: Effective deme size during long-term evolution estimated from rates of chromosomal rearrangement. Evolution 33:234–251, 1979.
94. King M: A case for simultaneous multiple chromosome rearrangements. Genetica 59:53–60, 1982.
95. Mayr E: Processes of speciation in animals. In Barigozzi C (ed): "Mechanisms of Speciation." New York: Alan R. Liss, Inc., 1982, pp 1–19.
96. White MJD: Models of speciation. Science 159:1065–1070, 1968.
97. Bush GL: Stasipatric speciation and rapid evolution in animals. In Atchley WR, Woodruff DS (eds): "Evolution and Speciation: Essays in Honor of MJD White." New York: Cambridge University Press, 1981, pp 201–218.
98. Mayr E: "Populations, Species, and Evolution. An Abridgment of Animal Species and Evolution." Cambridge: The Belknap Press of Harvard University Press, 1970, Ch 17.
99. Lewis H: The origin of diploid neospecies in *Clarkia*. Am Nat 107:161–170, 1973.
100. Bush GL: Modes of animal speciation. Annu Rev Ecol Syst 6:339–364, 1975.
101. Thompson JN Jr, Woodruff RC: Mutator genes—pacemakers of evolution. Nature 274:317–321, 1978.
102. Woodruff RC, Thompson JN Jr, Lyman RF: Intraspecific hybridization and the release of mutator activity. Nature 278:277–279, 1979.
103. Woodruff RC, Thompson JN Jr: Hybrid release of mutator activity and the genetic structure of natural populations. In Hecht MK, Steere WC, Wallace B (eds): "Evolutionary Biology, Vol 12," New York: Plenum Press, 1980, pp 129–162.
104. Naveira H, Fontdevila A: The evolutionary history of *Drosophila buzzatii* IX. High frequencies of new chromosome rearrangements induced by introgressive hybridization. Chromosoma 91:87–94, 1985.
105. McClintock B: The significance of responses of the genome to challenge. Science 226:792–801, 1984.
106. McClintock B: Mechanisms that rapidly reorganize the genome. Stadler Symp 10:25–47, 1978.
107. Dover G, Brown S, Coen E, Dallas J, Strachan T, Trick M: The dynamics of genome evolution and species differentiation. In Dover GA, Flavell RB (eds): "Genome Evolution." New York: Academic Press, 1982, pp 343–372.
108. Dover G: A role for the genome in the origin of species? In Barigozzi C (ed): "Mechanisms of Speciation." New York: Alan R. Liss, Inc., 1982, pp 435–459.
109. Rose MR, Doolittle WF: Molecular biological mechanisms of speciation. Science 220:157–162, 1983.

110. Ginzburg LR, Bingham PM, Yoo S: On the theory of speciation induced by transposable elements. Genetics 107:331–341, 1984.
111. Syvanen M: The evolutionary implications of mobile genetic elements. Annu Rev Genet 18:271–293, 1984.
112. Rubin GM, Brorein WJ Jr, Dunsmuir P, Flavell AJ, Levis R, Strobel E, Toole JJ, Young E: *copia*-like transposable elements in the *Drosophila* genome. Cold Spring Harbor Symp Quant Biol 45:619–628, 1981.
113. Finnegan DJ, Will BH, Bayev AE, Bowcock AM, Brown L: Transposable DNA sequences in eukaryotes. In Dover GA, Flavell RB (eds): "Genome Evolution." New York: Academic Press, 1982, pp 29–40.
114. Shaw DD, Wilkinson P, Coates DJ: Increased chromosomal mutation rate after hybridization between two subspecies of grasshoppers. Science 220:1165–1167, 1983.
115. Rubin GM: Dispersed repetitive DNA in *Drosophila*. In Shapiro JA (ed): "Mobile Genetic Elements." New York: Academic Press, 1983, pp 329–361.
116. Chesser RK, Baker RJ: On factors affecting the fixation of chromosomal rearrangements and neutral genes: Computer simulations. Evolution 40:625–632, 1986.
117. Bickham JW, Baker RJ: Canalization model of chromosomal evolution. Bull Carnegie Mus Nat Hist 13:70–84, 1979.

The Cytogenetics of Mammalian Autosomal
Rearrangements, pages 507–531

Chapter 15
Selection and Robertsonian Variation in Nature: The Case of the Common Shrew

Jeremy B. Searle

Intraspecific Robertsonian variation (polymorphism and polytypy) is commonly observed in mammals. The presence of a Robertsonian rearrangement in the karyotype of an individual may affect the well-being or fertility, i.e., the fitness, of that individual and thus Natural Selection may influence the degree of Robertsonian variation. In this chapter particular effects of Robertsonian rearrangements which may affect fitness are examined. The possibility is entertained that there may be direct physiological effects associated with Robertsonian mutations and that there may be disruption at cell division due to centromeric abnormality, but there is no evidence that such effects are important. However, there is evidence for disruption of meiosis in Robertsonian heterozygotes associated with the formation of chain and ring configurations at prophase I. Chain trivalents are formed in simple Robertsonian heterozygotes and longer chains and rings are formed in double and multiple Robertsonian heterozygotes with monobrachial homology. Presence of long chain configurations may cause sterility in males, apparently as a result of interference of the chains with the X-chromosome at pachytene (thereby disrupting its inactive state). A more general effect associated with the presence of Robertsonian configurations is an increased frequency of anaphase I non-disjunction. There is also evidence for cross-over suppression around the centromere of the trivalent of simple Robertsonian heterozygotes; this could promote the development of Robertsonian polymorphism. The other meiotic effects are likely to result in heterozygote disadvantage that may help maintain hybrid zones between Robertsonian races. Unequal transmission of the rearranged and standard chromosomes from Robertsonian heterozygotes will help promote the spread or loss of a rearrangement. As an example of Robertsonian polymorphism and polytypy, the common shrew in the Oxford area (U.K.) is considered. The arm combinations that differ between the two races of common shrew in the Oxford area are also polymorphic. In the hybrid zone between the races, the acrocentric state predominates for all arm combinations. It is proposed that this is the result of selection against hybrids that are double

Department of Zoology, University of Oxford, Oxford OX1 3PS, England

Robertsonian heterozygotes with monobrachial homology. Such hybrids are likely to suffer a high frequency of anaphase I non-disjunction and perhaps male sterility. Away from the hybrid zone, Robertsonian metacentrics occur at high frequency, perhaps as a result of unequal transmission from heterozygotes. The polymorphism is a consequence of the meeting of areas where acrocentrics and metacentrics predominate and are apparently favoured by selection.

INTRODUCTION

Intraspecific karyotypic variation due to presence or absence of Robertsonian rearrangements is commonly observed in mammals, both between individuals within populations (polymorphism) and between races within species (polytypy). Robertsonian polymorphism has been recorded in a number of species, including the shrew *Blarina brevicauda* [1], the antelope *Aepyceros melampus* [2], and the vole *Microtus oeconomus* [3]. Examples of Robertsonian polytypy in mammals include the mole rat *Spalax ehrenbergi* [4], the mouse *Mus musculus* [5], and the lemur *Lemur fulvus* [6]. The mutations involved in the variation may be either Robertsonian fusions (= centric fusions, Robertsonian translocations) or Robertsonian fissions (= centric fissions, dissociations). Robertsonian fusions are mutations in which two chromosomes with terminal or near-terminal centromeres (acrocentrics) combine at their centromeric ends to form a single chromosome with a median centromere (a metacentric), and Robertsonian fission is the reverse process in which two acrocentrics are propagated from a single metacentric. Although it is often unclear whether the metacentric or the acrocentric state is ancestral, the mutations that generated the polytypy of *Mus musculus* were certainly Robertsonian fusions, while the polymorphism in *Microtus oeconomus* is most probably the consequence of a Robertsonian fission.

Robertsonian variation is necessarily dependent on the occurrence of Robertsonian mutations, but the degree and geographic pattern of Robertsonian variation observed in a particular species is also dependent on many other factors. Selection may influence the frequency of a Robertsonian rearrangement if 1) the well-being or fertility of individuals (i.e., their fitness) is affected by presence of the rearrangement in the karyotype (selection at diplophase) or 2) there is some inequality in the likelihood of the rearranged and standard chromosomes being transmitted from heterozygotes (selection at haplophase). Stochastic processes, such as the founder effect and genetic drift, may cause the frequencies of a rearrangement in a particular area to deviate markedly from the expectation on the basis of selective factors. The geographic pattern of Robertsonian variation in a species will also be dependent on a number of properties of that species that are present before the origin of the rearrangement and/or which continue independently of karyotype, such as the geographic range (and any changes in that range) and dispersal characteristics of the species.

This chapter will examine how selection, in particular, may be involved in the persistence and loss of Robertsonian variation. As an example, the possible influence of selection on the Robertsonian polytypy and polymorphism in the common shrew (*Sorex araneus*) as observed in the area around Oxford, U.K., will be considered.

EFFECTS OF ROBERTSONIAN REARRANGEMENTS ON THE FITNESS OF INDIVIDUALS

Robertsonian rearrangements may affect the well-being or fertility of an individual through direct physiological effects and effects at mitosis and meiosis. The theoretical basis and experimental evidence for these various effects are now considered.

Despite considerable debate about the nature of the mutational event that results in a Robertsonian rearrangement [7,8], it is generally agreed that chromosome breakage is involved, and there is the possibility that genetic material is lost or gained (in fusion and fission, respectively). Furthermore, a fusion mutation will bring together previously separate genetic material, while fission has the opposite effect, thereby affecting linkage groups. Any of these characteristics of Robertsonian mutation could have a physiological effect on an individual. Thus, if chromosome breakage occurs within an important structural or regulatory locus or if such a gene is lost, physiological function could be impaired. Also, when previously unlinked loci are brought together, gene activity at either locus could be affected (this appears to be the case for certain reciprocal translocations associated with cancer [9]).

On the other hand, it can be argued that there are unlikely to be significant direct physiological effects resulting from Robertsonian mutations because the breakages, additions, and deletions associated with these mutations occur in the centromeric region that consists primarily of highly repetitive DNA rather than genes of physiological significance. In support of this view, studies in man indicate that chromosomal (structural) mutations with breakpoints in interstitial regions (inversions, reciprocal translocations) are associated with an increased incidence of mental retardation [10,11], while Robertsonian mutations show no such correlation. Similarly, in the laboratory mouse there is evidence that heterozygotes for reciprocal translocations and inversions may be malformed and homozygotes may be lethal, but no such defects appear to have been observed in association with Robertsonian rearrangements that have arisen [12,13]. Genes may be lost during Robertsonian mutation; Robertsonian fusions in man have been shown to be associated with loss of rDNA [14,15]. However, such loss is probably of limited physiological significance [16].

According to some models of Robertsonian mutation [7,8], breakage could occur within the centromere. Clearly, it is possible that this may detrimentally affect centromere function of the rearranged chromosome. Likewise, if a Robertsonian fusion occurs by breakage in both short arms of the acrocentrics, a dicentric product will be formed. This could be unstable at division. However, as the centromeres of a dicentric product would be in close proximity, they may be able to function in unison, or alternatively, one centromere may become inactivated (for review, see [17]). There is evidence that inherited Robertsonian rearrangements in man are usually close dicentrics [15,18,19].

Even if chromosomal (structural) mutations are stable at division and have no direct physiological effect, one may expect disruption of other sorts at meiosis. In normal mammalian meiosis, homologous chromosomes pair up as and disjoin from bivalents at the first division. In homozygotes for Robertsonian rearrangements, such bivalents would be formed. However, in heterozygotes more complex configurations are expected. In "simple Robertsonian heterozygotes" where two acrocentrics have a metacentric as a homologue, trivalents are expected. In double or multiple "Robertsonian heterozygotes with monobrachial homology," where the karyotype includes two or more metacentrics that differ from each other but that share a common chromosome arm, chains of four or more chromosomes or rings of three or more chromosomes are anticipated. As one would expect the mammalian meiotic process to be adapted to the pairing and disjunction of bivalents, the formation and disjunction of longer chains and rings may be a problem. In addition, even if homologous regions do pair up as expected in Robertsonian heterozygotes, there may be a partial lack of homology between "homologous" chromosomes in the centromeric region, due to loss or gain of genetic material. This could influence pairing and chiasma formation in this region.

There is evidence that meiosis may indeed be disrupted in Robertsonian heterozygotes. As regards chromosome pairing, analyses of synaptonemal complexes in simple Robertsonian heterozygotes of lemurs (*Lemur fulvus*), mice (*Mus musculus*), and cotton rats (*Sigmodon fulviventer*) indicate that while the acrocentrics pair up normally with the homologous metacentric along most of their length, there is a degree of non-homologous pairing between the acrocentrics in the centromeric region [20–23]. In the chains formed in mice that are double or multiple Robertsonian heterozygotes with monobrachial homology, unsynapsed homologous regions are also observed [23,24]. Correct pairing is a prerequisite for chiasma formation and recombination. The peculiarity of pairing around the centromere in Robertsonian heterozygotes may lead to suppression of crossing-over in that region. Such cross-over suppression has been recorded in simple Robertsonian heterozygotes of the mouse [25]. As will become clear below, such cross-over suppression could ultimately have a profound effect on fitness.

It has been proposed that incomplete pairing at pachytene may also result in cell death at that stage or later in gametogenesis, perhaps because lack of pairing permits inappropriate gene expression [26,27]. Data from male mice that are simple Robertsonian heterozygotes or double or multiple Robertsonian heterozygotes with monobrachial homology indicate that there may be cell death at meiotic or post-meiotic stages, perhaps at least partly as a result of this effect [28,29]. Similar gametogenic disruption has been recorded in human males who are simple Robertsonian heterozygotes (reviewed in [30]). In mice, the extent of cell death is most extreme in males that are multiple Robertsonian heterozygotes with monobrachial homology in which long *chain* configurations (pentivalents or longer) are formed at prophase I. Such individuals are almost always sterile due to complete breakdown of spermatogenesis at late meiotic or post-meiotic stages [29]. Sterility or sub-fertility due to similar gametogenic breakdown has also been recorded in male mice that are double Robertsonian heterozygotes with monobrachial homology for certain combinations of chromosomes [29,31,32]. In heterozygous female mice and in males where *ring* configurations are expected, sterility is not observed [29]. The sterility in male Robertsonian heterozygotes with long chain configurations, but not ring configurations, does not appear to be due to a greater lack of synapsis at pachytene in chain configurations versus ring configurations. It has been observed that where there are long chain configurations at prophase I in males, for some reason as yet not fully explained, the chains associate with the sex vesicle [33,34]. Forejt [33,34] has proposed that this interaction may disrupt the inactive state of the X-Y pair that is necessary for normal spermatogenesis [35].

A further stage at which meiosis may proceed abnormally in Robertsonian heterozygotes, leading to reduced fertility, is anaphase I. A failure of chain trivalents, longer chains, or rings to orientate correctly at anaphase I can lead to the production of chromosomally unbalanced secondary spermatocytes (nullisomic or disomic for specific chromosomes). This "anaphase I non-disjunction" can be detected by metaphase II counts. Also, as aneuploid gametes apparently survive and fertilise as well as euploid gametes [36], anaphase I non-disjunction can be detected from the DNA content of spermatozoa and from cytogenetic studies of embryos [37]. It is because anaphase I non-disjunction ultimately results in aneuploid embryos that fertility may be reduced when it occurs. The embryos that ultimately derive from anaphase I non-disjunction of Robertsonian configurations are monosomic or trisomic for specific chromosomes. Based on studies in the mouse, monosomic zygotes die at or around the time of implantation, while trisomics tend to die later in gestation [38].

Table I presents anaphase I non-disjunction frequencies in Robertsonian heterozygotes and homozygotes for various species of mammal. Clearly, there is a great deal of variation between categories. Although bivalents, as well as Robertsonian chain and ring configurations, may undergo anaphase I non-disjunction, in all cases, except the field mouse, *Akodon molinae*, non-disjunction frequencies are greater in Robertsonian heterozygotes than in homozygotes for the standard or rearranged condition. In *A. molinae* there is a very high anaphase I non-disjunction frequency in individuals homozygous for the fission products. The reasons for this are unclear (see below). The best-studied species by far is the house mouse. The anaphase I non-disjunction frequencies for all categories of heterozygotes in this species vary with the chromosomes involved and with investigators. Frequencies in homozygotes tend to be low. In heterozygotes, frequencies are higher for feral-derived metacentrics than for metacentrics that arise spontaneously in the laboratory. They are higher in females than in males, and they are high when metacentrics in the karyotype have monobrachial homology and where an individual is a multiple simple Robertsonian heterozygote.

In summary, studies to date, particularly in the mouse, indicate that there may be considerable meiotic disruption expressed in Robertsonian heterozygotes (but not homozygotes) in particular cross-over suppression around the centromere in the heteromorphic configurations, anaphase I non-disjunction, and in males even complete meiotic interruption when long chain configurations are present. Such effects, in the main, may be expected to reduce the fertility (and hence fitness) of Robertsonian heterozygotes.

OVERALL FITNESS OF ROBERTSONIAN HETEROZYGOTES

Attempts to measure the overall fitness of individuals with Robertsonian rearrangements in their karyotypes have only been made in man, from data collected in newborn surveys and followup family studies. Robertsonian heterozygotes in man appear to be less fit than standard homozygotes due to higher levels of anaphase I non-disjunction and, to a certain extent, male sterility (reviewed in references [29,30]). On the assumption that Robertsonian variation in man is at equilibrium, reduction in fitness among heterozygotes relative to standard homozygotes can be calculated; it is that proportion of newborn heterozygotes who have a "new" rather than an inherited mutation product [39,40]. For D/D and D/G Robertsonian heterozygotes, Jacobs [40] estimates a reduction in fitness of 5.5 and 12.5%, respectively.

As the fitness of Robertsonian heterozygotes is likely to be reduced more through loss of fertility than by a diminution of viability, an approximation to overall fitness may also be obtained from breeding records. Particularly extensive records are available for domestic cattle that are heterozygous for the 1/29 Robertsonian fusion product. These records consist of frequencies at which individuals had to be remated as a result of failure of fertilisation or embryonic loss. From these records, Gustavsson [41] has demonstrated that the "56-day non-return rates" of male and female 1/29 Robertsonian heterozygotes is lowered by 4.5 and 6%, respectively, relative to standard homozygotes.

UNEQUAL TRANSMISSION FROM ROBERTSONIAN HETEROZYGOTES

In simple Robertsonian heterozygotes one would expect equal transmission of the metacentric and two homologous acrocentrics. This pattern of transmission has been well documented, for example, in rams heterozygous for the Robertsonian fusion product t_1 [42]. However, studies in man and mouse suggest that transmission may be unequal in Robertsonian heterozygotes for at least certain arm combinations. Prenatal diagnosis and family studies in man suggested that the 13q14q Robertsonian fusion product is transmitted more frequently than expected by chance in both male and female heterozygotes [43,44], while studies of spermatozoa from a 14q21q Robertsonian heterozygote man suggest unequal transmission in favour of the acrocentric chromosomes 14 and 21 [45]. In female laboratory mice that are Robertsonian heterozygotes, there is evidence that pairs of acrocentrics are transmitted at a greater frequency than a variety of homologous feral-derived metacentrics [23], but Harris et al. [46] report that in female genetically feral Peruvian mice, Robertsonian metacentric 9.12 is transmitted from heterozygotes at a significantly greater frequency than acrocentrics 9 and 12. Unequal transmission in female heterozygotes presumably results from preferential segregation of one chromosome type into the first polar body. Unequal transmission in male Robertsonian heterozygotes could come about by differences in maturation of spermatozoa carrying the two types of chromosome

TABLE I. Anaphase I non-disjunction frequencies in Robertsonian heterozygotes compared with homozygotes for different mammals

Species	Robertsonian rearrangements	Type of heterozygote	Genetic background	Sex	Ascertainment method	AI non-disjunction (%)			Reference
						Hom. for standard	Het.	Hom. for rearrangement	
House mouse *Mus musculus*	Fusions (various)								
	Lab	1 Simple	Lab	M	MII counts	0–2	3–21	—	23,69,70
	Lab	2–3 Simple	Lab	M,F	MII counts	—	16–24	0–2	70,71
	Lab	1 Simple	Feral	M,F	MII counts, Embryo studies	0	10–16	—	46
	Feral	1 Simple	Mostly lab	M	MII counts, Embryo studies	0	2–39	0–6	23,72,73
	Feral	1 Simple	Mostly lab	F	MII counts	—	11–61	—	Gropp et al., unpublished

	Feral	7–9 Simple	Lab/feral	M,F	MII counts, Sperm DNA	0–2	48–77	0	69,74,75
	Feral	2 Mono-brachial	Mostly lab	M	MII counts	—	26–44	—	31
	Feral	16 Mono-brachial	Feral	M	Sperm DNA	< 10	60–70	< 10	74
South American field mouse *Akodon molinae*	1 Fission (1a,1b)	1 Simple	Lab (recently feral)	M	MII counts, Sperm DNA	7–9	19–23	37–41	49,76
Cattle *Bos taurus*	1 Fusion (1/29)	1 Simple	Domestic	M	MII counts	3	6	—	77
		1 Simple	Domestic	M	Embryo studies	0	5–11	—	78
Sheep *Ovis aries*	3 Fusions (t_1,t_2,t_3)	1 Simple	Domestic	M	MII counts	0–2	6–18	0	79,80
		1 Simple	Domestic	M	Embryo studies	—	0	—	81
		3 Simple	Domestic	M	MII counts	0	5	—	82

that ultimately affects the ability to fertilise (as observed in mice heterozygous for a *t* haplotype [47,48]). The physiological basis of unequal transmision in Robertsonian heterozygotes is unknown for any of the cases documented above.

SELECTION AND ROBERTSONIAN VARIATION IN NATURE

Table II summarises the effects of Robertsonian rearrangements, the manner in which the fitness of individuals is affected, and how those selective factors will ultimately influence the persistence of Robertsonian variation in nature.

Let us consider the first of the selective factors in Table II. Direct physiological effects of a Robertsonian mutation, due to loss or alteration of activity of important genes, are likely to be disadvantageous. Where such effects occur, selection against homozygotes for the rearrangement is likely to be greater than the selection against heterozygotes; the effects may operate in a recessive manner. If there are such direct physiological effects they will reduce the likelihood that the rearrangement will persist. Thus, one would not expect the rearrangements that we observe in nature to have direct physiological effects. A possible exception is found in the field mouse

TABLE II. Effects of Robertsonian rearrangements in relation to the persistence of Robertsonian variation in nature

Effects of Robertsonian rearrangement	Relative fitness: Hom. for standard	Relative fitness: Het.	Relative fitness: Hom. for rearrangement	Persistence or loss of Robertsonian variation
Direct physiological effects of mutation	0	–	--	Rearrangement less likely to persist (applies to both rows above)
Dicentric formation	0	–	--	
Anaphase I non-disjunction in heterozygotes	0	–	0	Reduces probability of spread of rearrangement but once races formed, could contribute to maintenance of hybrid zone (applies to both rows)
Male sterility in heterozygotes	0	–	0	
Cross-over suppression in heterozygotes	+	0	+	Could promote polymorphism, perhaps leading to polytypy
	0	+	0	Promotes balanced polymorphism
Unequal transmission in heterozygotes	In favour of standard			Rearrangement less likely to persist
	In favour of rearrangement			Promotes spread of rearrangement

Akodon molinae, in which heterozygotes and more particularly homozygotes for the Robertsonian fission products 1a and 1b (which occur in nature) have very low fertility, substantially, but perhaps not completely, the result of anaphase I non-disjunction (see Table I); [49,50]. The unexpected high anaphase I non-disjunction and any other factors resulting in reduced fertility in the homozygotes for the rearrangement could be a reflection of direct physiological effects of the mutation on meiosis and perhaps other stages of gametogenesis.

If the product of a Robertsonian fusion mutation is a dicentric, it is conceivable that it will be unstable. Clearly, any adverse effect will be expressed more extremely when there are two copies of the aberrant chromosomes (homozygotes for the rearrangement) than when there is only one (in heterozygotes). By contrast, Robertsonian heterozygotes appear more prone to anaphase I non-disjunction than either type of homozygote. Such heterozygote disadvantage will reduce the probability that a new rearrangement will spread when it first arises. However, as a result of other factors, such as genetic drift and range changes, the new rearrangement may spread to form a new Robertsonian race. If this happens and if the new race makes contact with the ancestral race or another Robertsonian race, any hybrids formed will be simple Robertsonian heterozygotes (in the case of contact with the ancestral race and some types of Robertsonian race) or double or multiple Robertsonian heterozygotes with monobrachial homology (in the case of contact with other types of Robertsonian race). Now, if the hybrids (as Robertsonian heterozygotes) suffer from higher frequencies of anaphase I non-disjunction than pure race individuals (as homozygotes) or if they suffer male sterility (as double or multiple Robertsonian heterozygotes with monobrachial homology), then one would expect hybrids to be limited to the immediate area of contact, forming a hybrid zone, the width of which would be determined by the degree of hybrid unfitness and the extent of dispersal [51]. Thus heterozygote disadvantage due to anaphase I non-disjunction or male sterility could help maintain a hybrid zone. If there were no such heterozygote disadvantage one would expect a broad area of Robertsonian polymorphism to develop, the extent of which would be determined by factors such as dispersal and the degree and direction of genetic drift.

Cross-over suppression around the centromeres of Robertsonian chromosomes in heterozygotes will assist these regions of the chromosome to evolve independently in the two types of homozygotes. This may aid the two types of homozygote to become adapted to different ecological niches within the same environment. Taking a non-mammalian example, this could explain the finding that the most metacentric karyotype of the dog whelk *Nucella lapillus* is apparently favoured in exposed intertidal habitats, while the most acrocentric karyotype is apparently favoured in more sheltered habitats at the same locality [52,53]. If the two homozygotes do become adapted to different ecological niches in this manner, the heterozygote could be well adapted to intermediate habitats and even be at advantage there (this has been proposed for the dog whelk). Alternatively, the two homozygous forms could evolve into distinct races with the heterozygote less fit than either homozygote even in intermediate habitats (see reference [54] for a discussion of the origin and build-up of genetic differences between races).

If cross-over suppression does help maintain genic differences between homozygotes, then in addition to the scenarios discussed above, the heterozygotes may become fitter than either homozygote in all habitats, as a result of heterosis (for a detailed discussion of how such heterosis may evolve see Wasserman [55]). This may promote a balanced polymorphism in the manner proposed by E.B. Ford [56]. Heterosis arising from cross-over suppression is thought to be involved in the maintenance of inversion polymorphisms of *Drosophila pseudoobscura* [57]. Heterosis may result in an excess of Robertsonian heterozygotes over the Hardy-Weinberg expectation in a natural population. Such an excess has been recorded for Robertsonian polymorphisms in a population of the gopher *Geomys bursarius* [58].

Unequal transmission from Robertsonian heterozygotes in favour of a rearrangement will clearly promote its spread, while unequal transmission in favour of the ancestral condition will reduce the probability that such a rearrangement will persist.

THE CASE OF THE COMMON SHREW

A species that demonstrates Robertsonian variation and serves as an example for many of the preceding arguments is the common shrew (*Sorex araneus*), a small, terrestrial insectivore with an extensive distribution over the Palaearctic region. As shown in Table III, there are 14 distinct karyotypic races which differ in sets of metacentrics formed by Robertsonian fusion mutations. Robertsonian polymorphism has been recorded in most of these races, often involving several arm combinations (Table III). This extensive Robertsonian variation is unparalleled in mammals; thus the common shrew is an appropriate model for the study of this phenomenon. Further, the species is easily collected from nature so that wild-caught animals may be studied directly. This is clearly an advantage for the assessment of the selective factors involved in the persistence of Robertsonian variation under natural conditions.

TABLE III. Karyotypic races of common shrew

Site described	Known range	Karyotype[a]
Aberdeen, U.K.	N.E. Scotland	*hi, gm, jl*, ko, np, qr*
Oxford, U.K.	Central Britain	*hi, gm, jl*, kq*, no*, pr**
Hermitage, Berkshire, U.K.	S. England	*hi, gm, jl*, ko*, n, p, q, r*
Valais, Switzerland	S. Switzerland	*hj, gi, lo*, kn, m, p, q, r*
Ulm, West Germany	Central Europe	*hi, gm*, jl*, k, n, o, p, q, r*
Drnholec, S. Moravia, Czechoslovakia	Czechoslovakia	*hi, gm, jl*, ko*, nr*, p, q*
Bialowieza, Poland	E. Poland	*hn*, gr*, jl*, ik, mp*, q, o*
Lake Druzno, Poland	N. Poland	*hi, jl, ko*, g, m, n, p, q, r*
Åkarp, Skåne, Sweden	S. Sweden	*hi, gm, jl, ko, nq, pr*
Stugun, Jämtland, Sweden	Central Sweden	*hi*, gm*, jl*, kp*, nr*, oq**
Kuusijoki, Kuhmo, Finland	Finland, N. Sweden	*hn* gm*, jl*, kq*, ip*, or**
Novosibirsk, U.S.S.R.	Within Siberia	*hn, go*, jl*, ik, mp*, qr**
Chaldejevo, Tomsk, U.S.S.R.	Within Siberia	*hi, gk*, jl*, mn*, op*, qr**
Ermakovskoe, Krasnoyarsk, U.S.S.R.	Within Siberia	*ho, gk*, jl, mn*, iq*, p, r*

[a]Details of metacentrics (in format *xy*) and those elements that do not occur combined as metacentrics (in format *x, y*) formed from variable chromosome arms *g–r*. Asterisk indicates polymorphic arm combination.
Data from Searle [83], Wojcik and Fedyk [84], Zima and Kràl [85] and references therein, and J. Hausser, personal communication.

The present studies have been based in the vicinity of Oxford, U.K. It was in this region that Robertsonian variation was first described in the common shrew [59,60]. By distribution studies it has been possible to demonstrate that the Robertsonian polymorphism described by Ford and co-workers varies latitudinally and that this variation is intimately related to a hybrid zone between the Oxford race (to the north) and the Hermitage race [61] (see Table III for details of these races). The polymorphism involves the race-specific arm combinations *kq, no, pr* (of the Oxford race) and *ko* (of the Hermitage race). There is also a low level polymorphism of arm combination *jl*, which will not be considered further (Searle and Fredga, unpublished). The clines of polymorphism of the Oxford and Hermitage race-specific metacentrics along a north-south transect through the Oxford area are shown in Figure 1a. It can be seen that the Oxford race metacentrics occur at high frequency to the north with decreasing frequency south, while the Hermitage race metacentric shows the opposite trend. Around a latitudinal grid reference of 180 km the frequencies of all the race-specific metacentrics are low, while

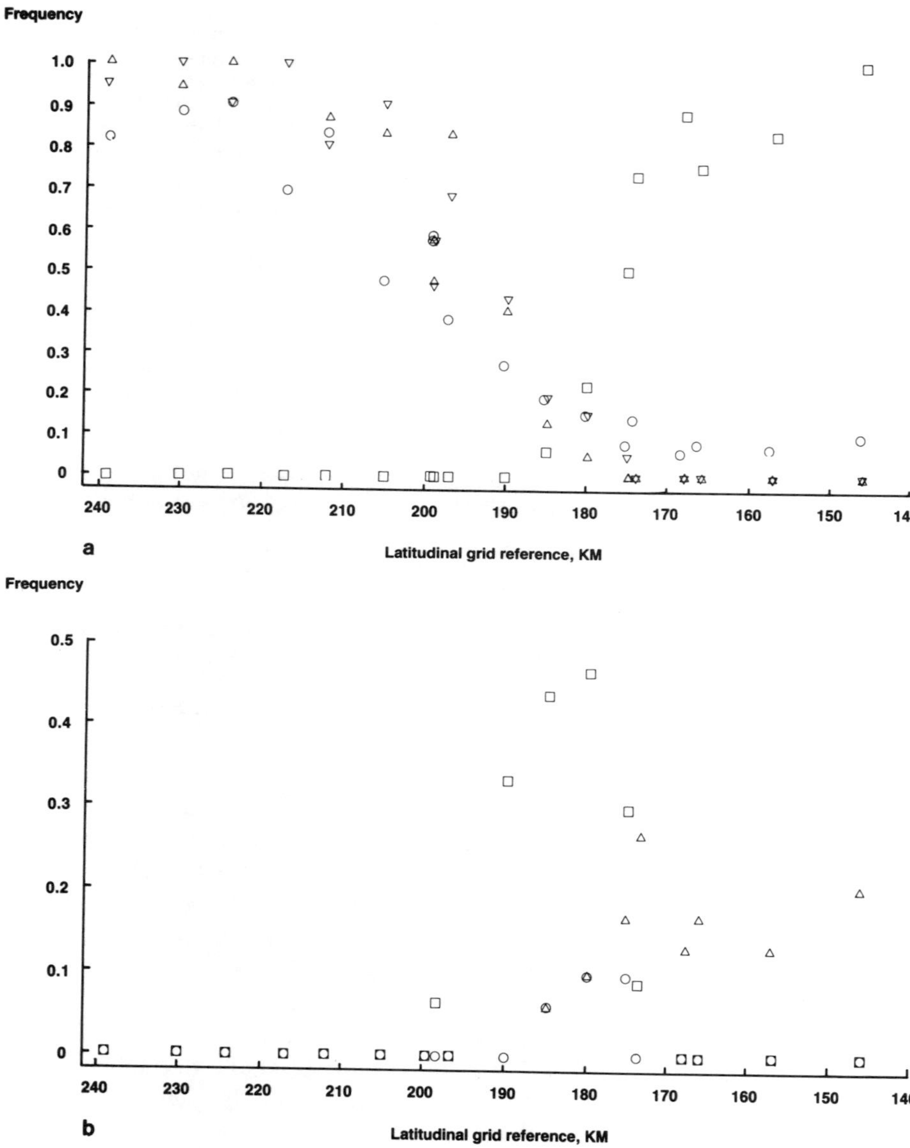

Fig. 1. **a:** Variation in frequency of metacentric morph of arm combinations *kq* (△), *no* (▽), *pr* (○), and *ko* (□) with latitude in the Oxford area. **b:** Variation in frequency of monobrachial hybrids (○), *ko/pr* hybrids (△) and individuals homozygous acrocentric for chromosomes *k, n, o, q* (□) with latitude in the Oxford area. Figures adapted from Searle [61] with permission.

acrocentrics *k, n, o, p, q, r* occur at high frequencies. The transect illustrated is perpendicular to a hybrid zone between the Oxford and Hermitage races (Searle and Reilly, unpublished). Hybrids that are double Robertsonian heterozygotes with monobrachial homology (i.e., with *ko* from the Hermitage race and *no* or *kq* from the Oxford race [23]: *ko/no* and *ko/kq* "monobrachial hybrids") were only observed in the area between latitudinal grid references 185 and 175 km, while hybrids that are simple Robertsonian heterozygotes (i.e., with *ko* from the Hermitage race and *pr* from the Oxford race: *ko/pr* hybrids), were found at all sites south of latitudinal grid reference 185 km (Fig. 1b).

Studies of laboratory mice indicate that double Robertsonian heterozygotes with monobrachial homology suffer very high frequencies of anaphase I non-disjunction (greater than 25% [31]), while non-disjunction frequencies in simple Robertsonian heterozygotes vary widely but tend to be lower, and the frequencies in homozygotes are negligible (see Table I). This suggests a model to explain the polymorphisms of arm combinations *kq* and *no* of the Oxford race and *ko* of the Hermitage race. If the hybrids that are double Robertsonian heterozygotes with monobrachial homology (i.e., monobrachial hybrids) are at a selective disadvantage relative to individuals with other karyotypes due to high frequencies of anaphase I non-disjunction, then one would expect selection in favour of individuals homozygous acrocentric for chromosome arms *k, n, o, q,* because such individuals cannot produce monobrachial hybrids whatever the mating. Individuals with all other possible karyotypes can, given the appropriate mate, produce monobrachial hybrids. If there is the expected selection in favour of individuals homozygous acrocentric for chromosome arms *k, n, o, q*, one would expect such individuals to occur at high frequency where monobrachial hybrids occur. This is indeed observed; the frequency distribution of individuals homozygous acrocentric for chromosome arms *k, n, o, q,* mirrors exactly, but is in excess of, the distribution of monobrachial hybrids (Fig. 1b).

Thus the high frequency of acrocentrics *k, n, o, q* and low frequency of metacentrics *kq, no*, and *ko* around latitudinal grid reference 180 km can be explained by selection against monobrachial hybrids. The high frequencies of metacentrics *kq, no*, and *ko* away from the hybrid zone may reflect that these metacentrics are at a selective advantage over the acrocentric condition. They could, for example, be favoured if there is unequal transmission from heterozygotes. The clines of polymorphism of *kq, no,* and *ko* and also the Oxford race arm combination *pr* could be influenced by selection against simple heterozygotes. Indeed, the cline of polymorphism for *pr* could be maintained solely by such heterozygous disadvantage. As indicated in Table

II, a possible source of disadvantage against simple Robertsonian heterozygotes is anaphase I non-disjunction (even if non-disjunction frequencies are not as high as in double Robertsonian heterozygotes with monobrachial homology).

To test these various possibilities, efforts were made to measure anaphase I non-disjunction in individuals in the Oxford area, both simple Robertsonian heterozygotes and double Robertsonian heterozygotes with monobrachial homology. Also the possibility that either class of Robertsonian heterozygote suffers disruption of spermatogenesis was investigated. Further, transmission of metacentrics from Robertsonian heterozygotes was examined. Of the other possible selective factors associated with Robertsonian rearrangements (Table II), no direct information is available on cross-over suppression, and the rearrangements do not appear dicentric down the light microscope. There is no indication of any direct physiological effects associated with the Robertsonian rearrangements. Thus, at any particular collection site the frequencies of homozygous metacentric, heterozygous, and homozygous acrocentric individuals are close to the Hardy-Weinberg expectation [61].

ANAPHASE I NON-DISJUNCTION FREQUENCIES IN COMMON SHREWS

Simple Robertsonian Heterozygotes

Anaphase I non-disjunction estimates were made for simple Robertsonian heterozygotes from the Oxford area on the basis of meiotic and fetal studies [62–64]. As the data are limited, it was not possible to make non-disjunction estimates for individual arm combinations.

In the meiotic material, anaphase I non-disjunction estimates were based on metaphase II counts of 168 spreads from five males heterozygous for either one or two arm combinations [63,64]. From the number of hypermodal metaphase II spreads, anaphase I non-disjunction in simple Robertsonian heterozygotes from the Oxford area is estimated as 1.2% per heterozygous individual and 0.8% per heterozygous arm combination (assuming only the Robertsonian trivalents undergo anaphase I non-disjunction).

In the fetal studies [62], ten pregnant females at mid-gestation were collected from nature and they and their live fetuses karyotyped (the karyotype of the male parent was inferred). The vast majority of the 78 ovulations detected by corpora lutea counts could be accounted for by a karyotypically balanced fetus (Table IV). However, one fetus in shrew 472 was a trisomic for either chromosome arm *n* or *o*. This trisomic probably resulted from anaphase I non-disjunction of the Robertsonian trivalent *n-no-o* at maternal

TABLE IV. Reproductive data for 10 pregnant common shrews in the Oxford area

	Heterozygous arm combinations			No. of ovulations accounted for by			
Shrew	Female	Male[a]	No. of corpora lutea	No visible implant	Regressing implant	Karyotypically unbalanced fetus	Karyotypically balanced fetus
463	*kq*	*no, pr*	8				8
472	*no, pr*	*pr*	7		1	1	5
489	*kq, pr*	*no, pr*	7	1			6
505	*kq, pr*	*no*	8				8
510		*ko*	7				7
717	*no*	*no, pr*	8				8
722	*pr*	*kq*	8				8
724	*pr*	*kq, pr*	10	1			9
725		*kq, pr*	8	1			7
726		*pr*	7				7

[a]Minimum heterozygosity inferred from karyotype of female and her live fetuses. Adapted from Searle [62].

meiosis (see Table IV). Another four ovulations failed to result in a live fetus, possibly due to anaphase I non-disjunction. On the basis of studies in the mouse, all monosomies and possibly some trisomies would die before mid-gestation [38]. From these fetal data, the lower limits of anaphase I non-disjunction in simple Robertsonian heterozygotes is 1.5% per heterozygous parent and 1.0% per heterozygous arm combination (assuming only the Robertsonian trivalents undergo anaphase I non-disjunction). (These frequencies are based on the expectation of an equal number of detectable trisomics and undetectable monosomics resulting from anaphase I non-disjunction.) The upper limits of anaphase I non-disjunction from the fetal data are 3.7% per heterozygous parent and 2.5% per heterozygous arm combination.

In summary, meiotic and fetal studies on a small number of single Robertsonian heterozygotes from the Oxford area indicate an anaphase I non-disjunction frequency of 0.8–2.5% per heterozygous arm combination (assuming only the Robertsonian trivalents undergo anaphase I non-disjunction) and 1.2–3.7% per heterozygous parent (single or double Robertsonian heterozygote). The non-disjunction frequencies are low compared with those recorded for simple Robertsonian heterozygotes of other species (see Table I). However, this does not necessarily mean that anaphase I non-disjunction in simple heterozygotes is not an important selective factor with regard to the Robertsonian polymorphisms in the Oxford area. Theoretical studies are required to determine what influence such non-disjunction frequencies may have.

Double Robertsonian Heterozygotes With Monobrachial Homology

Precise estimates of anaphase I non-disjunction in double Robertsonian heterozygotes with monobrachial homology from the Oxford-Hermitage hybrid zone are not available. No suitable fetal or meiotic data were collected. However, an indication of anaphase I non-disjunction in monobrachial hybrids was obtained from a preliminary microdensitometric study of DNA variation in spermatozoa (Garagna et al., unpublished). One would expect the spread of DNA content in spermatozoa to increase with the level of aneuploidy due to anaphase I non-disjunction. The spread of Feulgen-DNA content of spermatozoa from two monobrachial hybrids heterozygous for *ko* and *no* was considerably greater than in Oxford race individuals (homozygous metacentric for *kq, no*, and *pr*), Hermitage race individuals (homozygous metacentric for *ko*) and individuals homozygous acrocentric for *k, n, o, q* (also homozygous acrocentric for *p, r*) (Fig. 2). This supports the contention that the high frequency of individuals homozygous for *k, n, o, q* in the Oxford-Hermitage hybrid zone results from selection against monobrachial hybrids due to high frequencies of anaphase I non-disjunction. However, we do not consider it advisable at this stage to make a quantitative estimate of the non-disjunction frequency due to double Robertsonian heterozygosity with monobrachial homology. The data are limited and the results we have are confounded by the fact that the monobrachial hybrids are also heterozygous for arm combination *pr*. We intend to measure DNA content of sper-

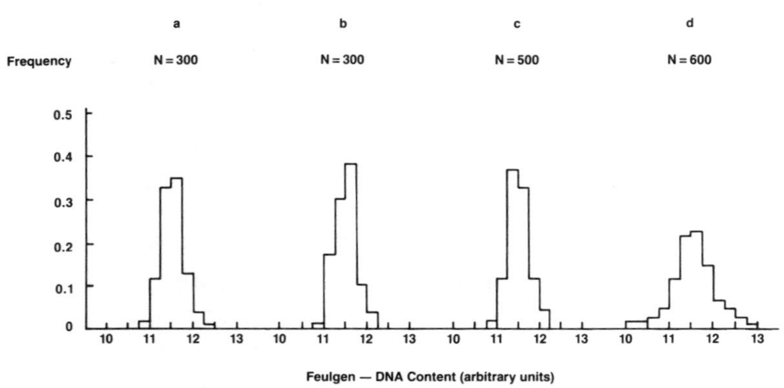

Fig. 2. Feulgen-DNA content distributions of spermatozoa from the caput epididymis of **a:** Oxford race homozygotes (two individuals); **b:** Hermitage race homozygotes (two individuals); **c:** three individuals homozygous acrocentric for *k, n, o, q*; and **d:** monobrachial hybrids (two individuals).

matozoa from a much larger series of individuals of different karyotype from the Oxford area.

SPERMATOGENESIS IN COMMON SHREWS

The testes of adult males from the Oxford area with a variety of karyotypes were analysed to assess the degree of spermatogenic impairment (Wilkinson and Searle, unpublished). The testes were collected *after* the period of growth in the spring (Table V); in a large study Brambell [65] demonstrated that testis weights were stable between mid-April and the end of June. Testis weights were determined for 75 males, including homozygotes, a variety of simple Robertsonian heterozygotes, and two monobrachial hybrids (Table V). There is no clear indication of a difference in testis weights between the different categories of homozygote and simple heterozygote. However, the mean testis weights of the two monobrachial hybrids was considerably lower than that of other categories. The two monobrachial hybrids, which were both killed on 30th May, had among the three lowest testis weights recorded in the sample of 75 males. Low testis weight is often indicative of spermatogenic impairment. Thus an XXY shrew that had no germ cells within its seminiferous tubules and an XY/XYY mosaic shrew with a depleted germ

TABLE V. Data on the testes of adult males (common shrew) from the Oxford area killed[a] between 25th April and 14th June, 1984

Karyotype category	No. of animals: Testis histology	No. of animals: Testis weight	Testis weight (mg) (mean ± S.E.)[b]
Homozygote (Oxford)	6	5	254.6 ± 12.1
Homozygote (Hermitage)	3	4	234.0 ± 13.3
Homozygous acrocentric *k, n, o, p, q, r*	15	16	257.4 ± 8.5
Single heterozygote (Oxford)	19	22	258.3 ± 6.8
Single heterozygote (Hermitage)	10	11	256.6 ± 11.5
Double heterozygote (Oxford)	11	11	274.8 ± 9.9
Triple heterozygote (Oxford)	1	1	268
ko/pr double heterozygote hybrid	3	3	234.7 ± 15.3
ko/no monobrachial hybrid	3[c]	2[d]	181.0 ± 11.0

[a]Animals were usually killed within 3 days of collection (always within 8 days).
[b]Left and right testes combined.
[c]Includes one male collected and killed during early April 1983 [86].
[d]Both animals were also heterozygous for arm combination *pr*.

cell population both had low testis weights [66,67], while Evans [32] records that sterile mice that are double Robertsonian heterozygotes with monobrachial homology have low testis weights. However histological sections of the left testes of the monobrachial hybrid shrews described above and of a third monobrachial hybrid heterozygous for *ko* and *no* did not reveal any spermatogenic abnormality. One complete transverse section from each of these hybrids was examined as well as sections from a variety of homozygotes and simple Robertsonian heterozygotes (Table V; sections from a total of 71 individuals were examined). In each case, every tubule examined displayed full spermatogenesis, with either spermatids or spermatozoa as the oldest stage within the epithelium.

TRANSMISSION OF METACENTRICS FROM SIMPLE ROBERTSONIAN HETEROZYGOTE COMMON SHREWS

Transmission of metacentrics from simple Robertsonian heterozygotes was examined in common shrews from the Oxford area using data from crosses between wild-caught individuals in captivity and from pregnant females collected in nature [68]. The frequencies of karyotypes among weanlings (in the laboratory crosses) or fetuses were compared with the Mendelian expectation on the basis of the karyotypes of the mother and father (the inferred karyotype of the father in the pregnant female data). As the data from pregnant females and laboratory crosses did not differ significantly, they were combined. Adequate data were obtained for the three Oxford race arm combinations *kq, no,* and *pr* (Table VI). For arm combination *pr,* there was, as expected, equal transmission of the metacentric and two homologous acrocentrics from heterozygotes, but for arm combinations *kq* and *no* significantly more metacentrics were transmitted than expected. Thus the high

TABLE VI. Transmission of metacentrics and the two homologous acrocentrics from simple Robertsonian heterozygotes (common shrew) collected in the Oxford area

Arm combination	No. available for karyotypic analysis[a]	Transmission frequency[b] Twin-acrocentric	Transmission frequency[b] Metacentric	χ_1^2	P
kq	71(5)	24	42	4.91	< 0.05
no	64(2)	27	51	7.38	< 0.01
pr	108(3)	77	66	0.85	n.s.

[a]Numbers in parentheses show attempted karyotypes but not obtained.
[b]When both mother and sire were Robertsonian heterozygotes, data on transmission could be obtained from both parents.
After Searle [68], with permission.

frequencies of metacentrics *kq* and *no* (and perhaps *ko*) away from the area of occurrence of monobrachial hybrids may, at least in part, be the consequence of selection, due to unequal transmission from heterozygotes.

CONCLUSIONS

The aim of this chapter was to demonstrate that it should be possible to identify and assess the selective factors involved in the persistence of Robertsonian variation in nature. It has been possible to construct a working hypothesis of the selective factors involved in the Robertsonian polymorphism of the common shrew around the Oxford area and to make the first attempts to test this hypothesis.

ACKNOWLEDGMENTS

I acknowledge support from the Natural Environment Research Council (grant GR3/5413 to Dr. G.M. Hewitt). I am grateful to my colleagues Ms. A.J. Reilly (University of East Anglia), Ms. P.J. Wilkinson (University of Oxford), and Drs. C.A. Redi and S. Garagna (University of Pavia) for unpublished results. Drs. A.E. Douglas and G.M. Hewitt (University of East Anglia) kindly read through the manuscript.

REFERENCES

1. Meylan A: Formules chromosomiques et polymorphisme robertsonien chez *Blarina brevicauda* (Say) (Mammalia: Insectivora). Can J Zool 45:1119, 1967.
2. Wallace C, Fairall N: Chromosome polymorphism in the impala (*Aepyceros melampus melampus*). S Afr J Sci 63:482, 1967.
3. Fredga K, Persson A, Stenseth NC: Centric fission in *Microtus oeconomus*. A chromosome study of isolated populations in Fennoscandia. Hereditas 92:209, 1980.
4. Wahrman J, Goitein R, Nevo E: Mole rat *Spalax*: Evolutionary significance of chromosome variation. Science 164:82, 1969.
5. Capanna E: Chromosomal rearrangement and speciation in progress in *Mus musculus*. Folia Zool 29:43, 1980.
6. Rumpler Y, Dutrillaux B: Chromosomal evolution in Malagasy lemurs. I. Chromosome banding studies in the genera *Lemur* and *Microcebus*. Cytogenet Cell Genet 17:268, 1976.
7. White, MJD: "Animal Cytology and Evolution." Cambridge: Cambridge University Press, 1973.
8. John B, Freeman M: Causes and consequences of Robertsonian exchange. Chromosoma 52:123, 1975.
9. Croce CM, Klein G: Chromosome translocations and human cancer. Scientific American 252(3):54, 1985.
10. Funderburk SJ, Spence MA, Sparkes RS: Mental retardation associated with "balanced" chromosome rearrangements. Am J Hum Genet 29:136, 1977.

11. Tierney I, Axworthy D, Smith L, Ratcliffe SG: Balanced rearrangements of the autosomes: Results of a longitudinal study of a newborn survey population. J Med Genet 21:45, 1984.
12. Searle AG: Chromosomal mutations: The genetic approach. In Serres FJ de (ed): "Chemical Mutagens," Vol. 10. New York: Plenum Press, 1986.
13. Rutledge JC, Cain KT, Cachiero NLA, Cornett CV, Wright CG, Generoso WM: A balanced translocation in mice with a neurological defect. Science 231:395, 1986.
14. Brasch JM, Smyth DR: Absence of silver bands in human Robertsonian translocation chromosomes. Cytogenet Cell Genet 24:122, 1979.
15. Gosden JR, Lawrie SS, Gosden CM: Satellite DNA sequences in the human acrocentric chromosomes: Information from translocations and heteromorphisms. Am J Hum Genet 33:243, 1981.
16. Guanti G, Zuffardi O, Tiepolo L: rDNA levels in infertile male carriers of Robertsonian translocations. Cytogenet Cell Genet 27:162, 1980.
17. Therman E, Trunca C, Kuhn EM, Sarto GE: Dicentric chromosomes and the inactivation of the centromere. Hum Genet 72:191, 1986.
18. Niebuhr E: Dicentric and monocentric Robertsonian translocations in man. Humangenetik 16:217, 1972.
19. Mattei M-G, Mattei J-F, Ayme S, Giraud F: Dicentric Robertsonian translocations in man. 17 cases studied by R, C, and N banding. Hum Genet 50:33, 1979.
20. Moses MJ, Karatsis PA, Hamilton AE: Synaptonemal complex analysis of heteromorphic trivalents in *Lemur* hybrids. Chromosoma 70: 141, 1979.
21. Elder FFB, Pathak S: Light microscopic observations on the behavior of silver-stained trivalents in pachytene cells of *Sigmodon fulviventer* (Rodentia, Muridae) heterozygous for centric fusion. Cytogenet Cell Genet 27:31, 1980.
22. Johannisson R, Winking H: Synaptonemal complex of multivalents in mouse spermatocytes. Eur J Cell Biol 20:122, 1979.
23. Gropp A, Winking H: Robertsonian translocations: Cytology, meiosis, segregation patterns and biological consequences of heterozygosity. Symp Zool Soc Lond 47:141, 1981.
24. Winking H, Johannisson R: Pattern of pachytene pairing in mouse hybrids with chain and ring multivalents. Clin Genet 17:94, 1980.
25. Cattanach BM: Crossover suppression in mice heterozygous for tobacco mouse metacentrics. Cytogenet Cell Genet 20:264, 1978.
26. Miklos, GLG: Sex-chromosome pairing and male fertility. Cytogenet Cell Genet 13:558, 1974.
27. Burgoyne, PS, Baker TG: Meiotic pairing and gametogenic failure. In Evans CW, Dickinson HG (eds): "Controlling Events in Meiosis." Cambridge: Company of Biologists, 1984, p 349.
28. Speed RM, Boer P de: Delayed meiotic development and correlated death of spermatocytes in male mice with chromosome abnormalities. Cytogenet Cell Genet 35:257, 1983.
29. Gropp A, Winking H, Redi CA: Consequences of Robertsonian heterozygosity: Segregational impairment of fertility versus male-limited sterility. In Crosignani PG, Rubin BL (eds): "Genetic Control of Gamete Production and Function." New York: Grune and Stratton, 1982, p 115.
30. Chandley AC: Infertility and chromosome abnormality. Oxford Rev Reprod Biol 6:1, 1984.
31. Gropp A, Kolbus U, Giers D: Systematic approach to the study of trisomy in the mouse. II. Cytogenet Cell Genet 14:42, 1975.

32. Evans EP: Male sterility and double heterozygosity for Robertsonian translocations in the mouse. Chromosomes Today 5:75, 1976.
33. Forejt J: Meiotic studies of the translocations causing male sterility in the mouse. II. Double heterozygotes for Robertsonian translocations. Cytogenet Cell Genet 23:163, 1979.
34. Forejt J: X-Y involvement in male sterility caused by autosome translocations—a hypothesis. In Crosignani PG, Rubin BL (eds): "Genetic control of Gamete Production and Function." New York: Grune and Stratton, 1982, p 135.
35. Lifschytz E, Lindsley DL: The role of X-chromosome inactivation during spermatogenesis. Proc Natl Acad Sci USA 69:182, 1972.
36. Ford CE: Gross genome unbalance in mouse spermatozoa: Does it influence the capacity to fertilise? In Beatty RA, Gluecksohn-Waelsch S (eds): "The Genetics of the Spermatozoon." Edinburgh: Beatty and Gluecksohn-Waelsch, 1972, p 359.
37. Nijhoff JH, Boer P de: Spontaneous nondisjunction in mammals. A study evaluating the various experimental approaches. Genetica 56:99, 1981.
38. Epstein CJ: Mouse monsomies and trisomies as experimental systems for studying mammalian aneuploidy. Trends Genet 1(5):129, 1985.
39. Bengtsson BO, Bodmer WF: The fitness of human translocation carriers. Ann Hum Genet. 40:253, 1976.
40. Jacobs PA: Structural rearrangements of the chromosomes in man. In Hook EB, Porter IH (eds): "Population Cytogenetics: Studies in Humans." New York: Academic Press, 1977, p 81.
41. Gustavsson I: Chromosome aberrations and their influence on the reproductive performance of domestic animals—a review. Z Tierzuchtg Zuchtgsbiol 97:176, 1980.
42. Long SE: Reproductive performance of ewes mated to rams heterozygous for the Massey I (t_1) centric fusion (Robertsonian) translocation. Vet Rec 102:399, 1978.
43. Hamerton JL: Robertsonian translocations. Evidence on segregation from family studies. In Jacobs PA, Price WH, Law P (eds): "Human Population Genetics." Edinburgh: Edinburgh University Press, 1970, p 63.
44. Boué A: European collaborative study on structural chromosome anomalies in prenatal diagnosis. In Murken JD, Stengel-Rutowski S, Schwinger E (eds): "Prenatal Diagnosis." Stuttgart: Enke, 1979, p 34.
45. Balkan W, Martin RH: Segregation of chromosomes into the spermatozoa of a man heterozygous for a 14;21 Robertsonian translocation. Am J Med Genet 16:169, 1983.
46. Harris MJ, Wallace ME, Evans EP: Aneuploidy in the embryonic progeny of females heterozygous for the Robertsonian chromosome (9.12) in genetically wild Peru-Coppock mice (*Mus musculus*). J Reprod Fertil 76:193, 1986.
47. Seitz AW, Bennett D: Transmission distortion of *t*-haplotypes is due to interactions between meiotic partners. Nature 313:143, 1985.
48. Olds-Clarke P, Peitz B: Fertility of sperm from *t*/+ mice: Evidence that + −bearing sperm are dysfunctional. Genet Res 47:49, 1986.
49. Merani MS, Capanna E, Bianchi NO: Cytogenetics of South American akodon rodents. VIII. Segregation of the polymorphic chromosome 1 in the testicular meiosis of *A. molinae*. Nucleus 23:54, 1980.
50. Bianchi NO, Merani S: Chromosome 1 polymorphism and viability in *Akodon* (Rodentia, Cricetidae). In Reig OA (ed): "Ecologia y Genetica de la Especiacion." Caracas: Equinoccio Press, Universidad Simon Bolivar, 1981, p 211.
51. Hewitt GM, Barton NH: The structure and maintenance of hybrid zones as exemplified by *Podisma pedestris*. Symp R Entomol Soc Lond 10:149, 1980.

52. Staiger H: Der Chromosomendimorphismus beim Prosobranchier *Purpura lapillus* in Beziehung zur Ökologie der Art. Chromosoma 6:419, 1954.
53. Bantock CR, Cockayne WC: Chromosome polymorphism in *Nucella lapillus*. Heredity 34:231, 1975.
54. Barton NH, Hewitt GM: Hybrid zones and speciation. In Atchley WR, Woodruff DS (eds): "Evolution and Speciation. Essays in Honour of MJD White." Cambridge: Cambridge University Press, 1981, p 109.
55. Wasserman M: Recombination-induced chromosomal heterosis. Genetics 58:125, 1968.
56. Ford EB: Polymorphism and taxonomy. In Huxley JS (ed): "The New Systematics." Oxford: Oxford University Press, 1940, p 493.
57. Dobzhansky T: "Genetics of the Evolutionary Process." New York: Columbia University Press, 1970.
58. Baker RJ, Chesser RK, Koop BF, Hoyt RA: Adaptive nature of chromosomal rearrangement: Differential fitness in pocket gophers. Genetica 61:161, 1983.
59. Sharman GB: Chromsomes of the common shrew. Nature 177:941, 1956.
60. Ford CE, Hamerton JL, Sharman GB: Chromosome polymorphism in the common shrew. Nature 180:392, 1957.
61. Searle JB: Factors responsible for a karyotypic polymorphism in the common shrew, *Sorex araneus*. Proc R Soc Lond B229:277, 1986.
62. Searle JB: Nondisjunction frequencies in Robertsonian heterozygotes from natural populations of the common shrew, *Sorex araneus* L. Cytogenet Cell Genet 38:265, 1984.
63. Searle JB: Nondisjunction estimates for wild Robertsonian heterozygote common shrews (*Sorex araneus* L). Genet Res 45:225, 1985.
64. Searle JB: Meiotic studies of Robertsonian heterozygotes from natural populations of the common shrew, *Sorex araneus* L. Cytogenet Cell Genet 41:154, 1986.
65. Brambell FWR: Reproduction in the common shrew (*Sorex araneus* Linnaeus). II. Seasonal changes in the reproductive organs of the male. Phil Trans R Soc Lond B 225:51, 1935.
66. Searle JB: A wild common shrew (*Sorex araneus*) with an XXY sex chromosome constitution. J Reprod Fertil 70:353, 1984.
67. Searle JB, Wilkinson PJ: The XYY condition in a wild mammal: An XY/XYY mosaic common shrew (*Sorex araneus*). Cytogenet Cell Genet 41:225, 1986.
68. Searle JB: Preferential transmission in wild common shrews (*Sorex araneus*), heterozygous for Robertsonian rearrangements. Genet Res 47:147, 1986.
69. Tettenborn U, Gropp A: Meiotic nondisjunction in mice and mouse hybrids. Cytogenetics 9:272, 1970.
70. White BJ, Crandall C, Raveché ES, Tjio J-H: Laboratory mice carrying three pairs of Robertsonian translocations: Establishment of a strain and analysis of meiotic segregation. Cytogenet Cell Genet 21:113, 1978.
71. White BJ, Tjio J-H, Water LC van de, Crandall C: Studies of mice with a balanced complement of 36 chromosome derived from F_1 hybrids with T1Wh and T1Ald translocation homozygotes. Proc Natl Acad Sci USA 69:2757, 1972.
72. Cattanach BM, Moseley H: Nondisjunction and reduced fertility caused by the tobacco mouse metacentric chromsomes. Cytogenet Cell Genet 12:264, 1973.
73. Ford CE, Evans EP: Non-expression of genome unbalance in haplophase and early diplophase of the mouse and incidence of karyotypic abnormality in post-implantation embryos. In Boué A, Thibault C (eds): "Les Accidents chromosomiques de la Reproduction." Paris: Institut National de la Santé et de la Recherche Médicale, 1973, p 271.

74. Redi CA, Capanna E: DNA-content variation in mouse spermatozoa arising from irregular meiotic segregation. Boll Zool 45:315, 1978.
75. Winking H, Gropp A. Meiotic nondisjunction of metacentric heterozygotes in oocytes versus spermatocytes. In Crosignani PG, Mitchell DR (eds): "Ovulation in the Human." New York: Academic Press, 1976, p 47.
76. Redi CA, Garagna S, Merani MS, Capanna E, Bianchi NO, Manfredi Romanini MG: Microdensitometric evaluation of the DNA content, as ploidy parameter, of spermatozoa in the polymorphic chromosomal system of *Akodon molinae* Cabrera (Rodentia, Cricetidae). Gamete Res 5:345, 1982.
77. Logue DN, Harvey MJA: Meiosis and spermatogenesis in bulls heterozygous for a presumptive 1/29 Robertsonian translocation. J Reprod Fertil 54:159, 1978.
78. King WA, Linares T, Gustavsson I: Cytogenetics of preimplantation embryos sired by bulls heterozygous for the 1/29 translocation. Hereditas 94:219, 1981.
79. Chapman HM, Bruère AN: The frequency of aneuploidy in the secondary spermatocytes of normal and Robertsonian translocation-carrying rams. J Reprod Fertil 45:333, 1975.
80. Long SE: Chiasma counts and nondisjunction frequencies in a normal ram and in rams carrying the Massey I (t_1) translocation. J Reprod Fertil 53:353, 1978.
81. Long SE: Cytogenetic examination of preimplantation blastocysts of ewes mated to rams heretozygous for the Massey I (t_1) translocation. Cytogenetic Cell Genet 18:82, 1977.
82. Bruère AN, Scott IS, Henderson LM: Aneuploid spermatocyte frequency in domestic sheep heterozygous for three Robertsonian translocations. J Reprod Fertil 63:61, 1981.
83. Searle JB: Three karyotypic races of the common shrew *Sorex araneus* (Mammalia: Insectivora) and a phylogeny. Syst Zool 33:184, 1984.
84. Wojcik JM, Fedyk S: A new chromosome race of *Sorex araneus* L. from northern Poland. Experientia 41:750, 1985.
85. Zima J, Kràl B: Karyotype variability in *Sorex araneus* in central Europe (Soricidae, Insectivora). Folia Zool 34:235, 1985.
86. Searle JB: Hybridization between Robertsonian karyotypic races of the common shrew *Sorex araneus*. Experientia 40:876, 1984.

The Cytogenetics of Mammalian Autosomal Rearrangements, pages 533–550

Chapter 16
Translocation Dicentric Chromosomes Among Human Autosomes and Centromeric Suppression

Gordon W. Dewald

Translocation dicentric chromosomes among human autosomes are occasionally found in patients with congenital anomalies. They are also sometimes observed in isolated cells of patients with chromosome breakage syndromes or persons exposed to toxic agents. In addition, they are frequently found in cells of advanced malignant processes. Translocation dicentric chromosomes usually have two C-bands. In some cases each C-band is associated with a centromeric constriction. In other cases one of the centromeric areas is less constricted or not constricted at all and its function is thought to be suppressed. When the centromeres are a considerable distance apart, the suppression of one of the centromeres becomes important to ensure normal segregation of the chromosome in cell division. Chromosomes with two functional centromeres often produce anaphase bridge formations or broken or multiple dicentric chromosomes in cell division. Theoretically, the suppression of a centromere can occur either by chromosome deletion or by molecular changes of the centromeric gene or genes. Neither Cd-banding of kinetochores nor staining with anticentromeric antibodies has shed much light on which mechanism is involved. In congenital disorders, most translocation dicentric chromosomes are Robertsonian translocations or involve at least one acrocentric chromosome: the breakpoints are generally in the satellite stalks. Thus, they are usually genetically balanced or near-balanced abnormalities. In malignant processes, the breakpoints are often interstitial, and the dicentric chromosomes are usually not genetically balanced.

INTRODUCTION

Chromosomes with two centromeres are termed "dicentric." Such chromosome abnormalities are occasionally found in patients with congenital

Cytogenetics Laboratory, Mayo Clinic and Mayo Foundation; and the Departments of Laboratory Medicine and Medical Genetics, Mayo Medical School, Rochester, Minnesota 55905 anomalies [1–22; Skovby and Niebuhr, personal communication]. They also occur in isolated cells in patients with chromosome breakage syndromes or in persons exposed to toxic agents such as chemotherapy or viruses [23–25]. They are most commonly found in chromosomally abnormal clones in advanced malignant solid tumors, leukemia, and lymphoma [26].

There are two categories of dicentric chromosomes. Those that arise from translocations between any two non-homologous chromosomes are termed "translocation dicentric chromosomes." Those that originate by sister chromatid fusion during DNA replication or repair, or translocation between homologous chromosomes to form abnormal chromosomes with symmetric band patterns, are called "isodicentric chromosomes." Most isodicentric chromosomes in humans involve the sex chromosomes. Isodicentric chromosomes have been described in an earlier volume in this series and so will not be discussed further in this chapter [27].

Most translocation dicentric chromosomes have two C-bands, but some have two centromeric constrictions and others have only one (Fig. 1). When there is only one apparent centromeric constriction, the abnormal chromosome is referred to as a pseudo translocation dicentric chromosome. When there are two centromeric constrictions, the chromosome is a true translocation dicentric chromosome. In a few patients the translocation dicentric chromosome may be a true dicentric in some cells and a pseudo dicentric in others. In cell division, true translocation dicentric chromosomes are sometimes unstable, whereas pseudo translocation dicentric chromosomes generally function normally.

There is some controversy about the centromere in the pseudo centromere region. Some investigators ask 1) whether there really is a centromere in this

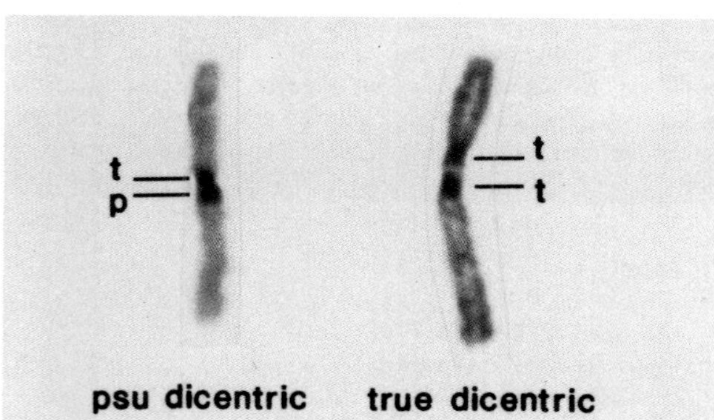

Fig. 1. Two C-banded translocation dicentric chromosomes. One is a psu tdic(13;14) (p11;p11) and the other is a tdic(5;15) (p13;p11).

area; 2) if there is a centromere, is it functional; and 3) if there is not a centromere, has it been deleted or genetically suppressed.

This chapter will discuss translocation dicentric chromosomes among human autosomes with respect to their origin, their prevalence, and the behavior of their centromeres.

ORIGIN OF TRANSLOCATION DICENTRIC CHROMOSOMES

Like most structural abnormalities, translocation dicentric chromosomes probably arise during interphase at times when the cells undergo DNA replication or repair. Figure 2 illustrates the likely sequence of DNA breakage and fusion between two chromosomes that leads to the formation of a translocation dicentric chromosome. There is ample opportunity for the DNA from different chromosomes to come in contact with one another during interphase when the chromosomal DNA is loosely distributed throughout the nucleus.

From inspection of Figure 2, it is apparent that should the broken DNA from two chromosomes become repaired incorrectly, a reciprocal translocation between them could occur. This particular translocation would result in two abnormal chromosomes: one is a dicentric chromosome and the other is an acentric chromosome. The translocation does not interfere with DNA

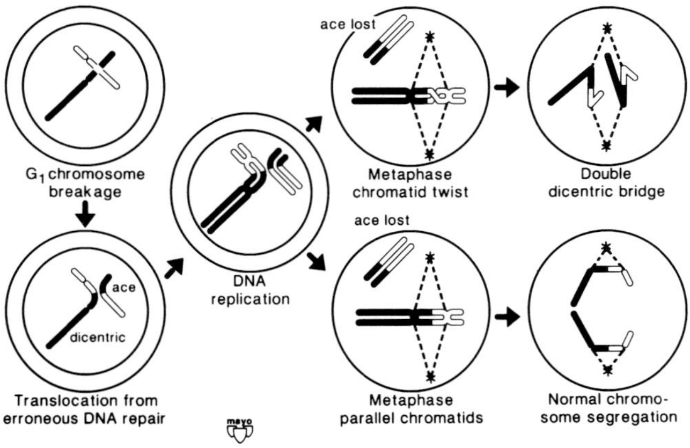

Fig. 2. Illustration of the most likely origin of a translocation dicentric chromosome. Note chromosome breakage in G_1 and erroneous DNA repair to form a dicentric and an acentric (ace) chromosome. Acentric chromosomes may become lost. Dicentric chromosomes may become aligned on the equatorial plate either so that the chromatids are twisted in the intercentromeric region or so that they are not twisted. Thus, anaphase bridges may form or there may be normal chromosome segregation.

replication, so each abnormal chromosome is capable of forming metaphase chromosomes, each consisting of two chromatids.

During cell division, the acentric chromosome may become lost in the cytoplasm or randomly incorporated into one of the daughter cells. The dicentric chromosome may become aligned on the equatorial plate in at least two ways. If the chromosome is not twisted in the intercentromeric region, then the segregation of the chromatids would be normal. It is normal because two spindle fibers from the same cell pole would attach to the two centromeres on the same chromatid. If the chromosome is twisted in the intercentromeric region, then spindle fibers from opposite poles may become attached to alternate centromeres on each chromatid. This could produce a double dicentric bridge in anaphase. The probability of a twist in the chromatids between centromeres is directly related to the size of the intercentromeric distance. Such a twist is expected to be rare when the intercentromeric distance is least and common when it is great.

A few phenomena may happen to any anaphase bridge. One or more of the spindle fibers may break so that one or both of the dicentric chromosomes become incorporated into the same daughter cell. Also, one or both of the dicentric chromosomes could break in the intercentromeric region. Thus, it would be possible to produce daughter cells with one or more monocentric portions of the dicentric chromosome. Any combination of these events may also occur.

A sister chromatid exchange in the intercentromeric region would offset or correct any chromatid twist and lead to normal anaphase segregation. A sister chromatid exchange in a dicentric chromosome in which there is no intercentromeric twist could lead to anaphase-bridge formation.

RECURRING TRANSLOCATION DICENTRIC CHROMOSOMES

In the literature dealing with congenital disorders, translocation dicentric chromosomes are only occasionally mentioned. A search of the literature revealed only 22 articles and one abstract describing any surviving patient with any kind of translocation dicentric chromosome (Table I). Although most of these anomalies were spontaneous in occurrence, these publications described 56 different families. With only five exceptions, all the chromosomes involved at least one acrocentric chromosome. Of the 58 translocation dicentric chromosomes, 41 involved two acrocentric chromosomes. One report alone describes 25 persons with translocation dicentric chromosomes involving two acrocentric chromosomes [6].

The overwhelming majority of reported translocation dicentric chromosomes have originated from short-arm breakpoints on acrocentric chromosomes. Thus, chromosome satellites have been suggested to be particularly prone to the type of breakage and fusion that leads to stable translocation dicentric chromosomes [14]. In situ hybridization of satellite DNA has shown that the breakpoints in such translocations are indeed usually within the stalks of the chromosome satellites [28]. These chromosome abnormalities would be compatible with survival because they usually are genetically balanced and permit carrier states in clinically normal individuals. In addition, the closeness of the centromeres on the dicentric chromosomes probably permits stable cell division.

There is evidence that the association of DNA in interphase chromosomes may not be entirely random. The observation of satellite association among the acrocentric chromosomes during metaphase is often attributed to the combined efforts of DNA in the stalk regions of these chromosomes to form the nucleolus during interphase. This DNA is rich in highly repetitive DNA for a short gene that makes rRNA. Also, some of the satellite association may be due to hybridization between the DNA in the nucleolar organizer regions. This association of satellite DNA in interphase may be an important factor in explaining why the proportion of translocation dicentric chromosomes that involve two acrocentric chromosomes is so great. In addition, the DNA in this area may be free from the protection of histone packing because the DNA is actively transcribing rRNA. Thus, this DNA is more vulnerable to DNA breakage.

The translocation dicentric chromosomes that involve two acrocentric chromosomes could be classified as Robertsonian translocations. Today, the term "Robertsonian translocation" is somewhat loosely applied to any translocation between two acrocentric chromosomes with break and fusion points occurring in the vicinity of their centromeres. When Robertson [29] first described such translocations, he suggested that they may be the result of centric fusion between two acrocentric chromosomes. Because human acrocentric chromosomes have considerable chromatin on their short arms, it would seem difficult to have simple fusion of centromeres without loss of other chromatin. Whether the structure of centromeres would be compatible with simple centromeric fusion as a mechanism for the origin of such translocations is unknown.

Robertsonian translocations originate by at least two mechanisms. First, a Robertsonian translocation could result from a reciprocal translocation with the breakpoint at p11 on one chromosome and at q11 on the other to form two metacentric chromosomes. One chromosome would essentially consist of the short arms of the original acrocentric chromosome and the other a metacentric chromosome made up of the long arms. Thus, the observation of

TABLE I. Reported cases of translocation dicentric chromosomes

Type	Reference	Year	Description of tdic chromosomes	No. of cases	Active centromere
AA	Schwartz et al. [1]	1983	(13;13)(q34;q34) } same	1	?
			(13;13)(q22;q11) } patient		
	Mules et al. [2]	1983	(13;15)(p11;p11)	1	?
	Weilong et al. [3]	1982	(22;22)(q11;q11)	1	?
	Daniel et al. [4]	1982	(13;21)(p11;p11)	5	?
	Berg et al. [5]	1980	(21;21)(q22.3;q22.3)	1	?
	Daniel and Lam-Po-Tang [6]	1976	(14;21)(p?;p?)	12	?
			(14;22)(p?;p?)	1	?
			(13;21)(p?;p?)	3	?
			(13;22)(p?;p?)	1	?
			(13;14)(p?;p?)	5	?
			(21;21)(p?;p?)[a]	2	?
			(15;21)(p?;p?)	1	?
	Skovby and Niebuhr [7]	1976	(21;21)(p11;p11)[a]	1	?
	Koulischer and Lambotte [8]	1974	(14;15)(q32;q26)	1	?
	Niebuhr [9]	1972	(13;14)(p?;p?)	1	14
	Niebuhr [10]	1972	(13;13)(p12;p12)	1	14
			(13;14)(p12;p12)	1	Both
			(15;21)(p11;p13)	1	Both
			(21;22)(p11;p12)	1	Both
			(14;21)(p?;p?)	1	?

AN	Daniel et al. [11]	1985	(9;13)(p22;p13)	1	Both, 9
	Ing and Smith [12]	1983	(13;Y)(?p11;q12)	1	13,Y
	Dewald et al. [13]	1979	(5;15)(p13;p11)	1	Both,5,15
	Roberts et al. [14]	1977	(8;22)(p23;p13)	1	8
	Nakagome et al. [15]	1976	(7;15)(p21;p11)	1	7
	Skovby and Niebuhr, personal communication	1978	(13;18)(p1;p11)	1	?
	Skovby and Niebuhr [7] Skovby and Niebuhr, personal communication	1976 1978	(5;13)(p11;p13)	1	Both,5,13
	Warburton et al. [16]	1973	(12;14)(p13;p13)	1	12
	Niebuhr [17]	1972	(5;13)(p?;p?)	1	5
	Sŭbrt et al. [18]	1971	(G;18)(p?;p?)	1	E
NN	Merry et al. [19]	1985	(9;11)(p24;p15)	1	11
	Ward et al. [20]	1981	(16;16)(q22;p11)	1	?
			(18;18)(p11.1;p11.3)	1	?
	Ray and Hunter [21]	1980	(8;8)(p21;p21)	1	?
	Pallister et al. [22]	1974	(6;19)(p?;p?)	1	19

AA = tdic between two acrocentric chromosomes; AN = tdic between an acrocentric and non-acrocentric chromosome; NN = tdic between two non-acrocentric chromosomes.

[a]Labeled as translocation dicentric by the authors but possibly could be isodicentric. Modified from Dewald et al. [13].

two abnormal metacentric chromosomes would be an important clue that a Robertsonian translocation was derived by a reciprocal translocation. It should be kept in mind that loss of the smaller, often bisatellited, chromosome might not be clinically important because it consists primarily of C-bands and DNA from the chromosome satellite. Bisatellited supernumerary chromosomes are sometimes attributed to this mechanism of origin [30].

Cases in which there are two abnormal metacentric chromosomes resulting from a reciprocal translocation are rare. Nevertheless, individuals with such chromosome abnormalities have been described [31], and one such patient was recently encountered at the Mayo Clinic. The karyotype of this patient was 47,XX,+21,t(21;21)(p11;q11), and the patient had Down syndrome. This karyotype had two abnormal metacentric chromosomes: one consisted of the long arms of a chromosome 21 and the other the short arms of a chromosome 21. Because the parents were chromosomally normal, this translocation was suspected to have taken place de novo during gametogenesis and to have led to a gamete with both metacentric chromosomes.

Second, in humans most Robertsonian translocations are probably derived by the formation of a dicentric chromosome. This would explain why most cells with a Robertsonian translocation are characterized by a single metacentric chromosome. Among the 17 most recent Robertsonian translocations identified at the Mayo Clinic, 13 had two C-bands and were classified as pseudo translocation dicentrics (Fig. 1). In four, two C-bands could not be resolved, but the breakpoints on each of the two chromosomes may have been so close to the C-bands that as a consequence of the translocation they may have merged together into one apparent larger one. Because the C-bands are highly polymorphic, it is difficult to be sure that the anomaly is a translocation dicentric chromosome when this happens.

NON-ROBERTSONIAN TRANSLOCATION DICENTRIC CHROMOSOMES

Other than Robertsonian translocations, there is no evidence of recurring translocation dicentric chromosomes among patients with birth defects. Apparently, only 15 patients with translocation dicentric chromosomes that were not Robertsonian translocations have been described (Table I). However, 10 of these cases still involved one acrocentric chromosome. In each case the breakpoint on the acrocentric chromosome was in the satellite region. There are only four reports, involving five patients, in which there is a translocation dicentric chromosome that does not involve an acrocentric chromosome (Table I).

Apparently, these rare cases of non-Robertsonian dicentric chromosomes are compatible with life when the break and fusion point on the non-acrocentric chromosome does not result in extensive deletion and when the intercentromeric distance is not too great to interfere with cell division. A patient with a tdic(5;15)(p13;p11), described in 1979 [13], had the greatest deletion of chromosomal material of any reported surviving individual. This patient essentially had the *cri-du-chat* syndrome as a consequence of loss of part of the short arm of a chromosome 5. The rest of the reported translocation dicentric chromosomes all involved break and fusion points in the telomeric area of the non-acrocentric chromosome and in the short arm of the acrocentric chromosome.

Translocation dicentric chromosomes are relatively common in abnormal clones of cells in advanced malignant tumors. The presence of more than one dicentric chromosome in the same malignant cells is not uncommon (Fig. 3). These chromosome abnormalities may be either pseudo or true dicentric chromosomes; sometimes both types occur in the same metaphases. Appar-

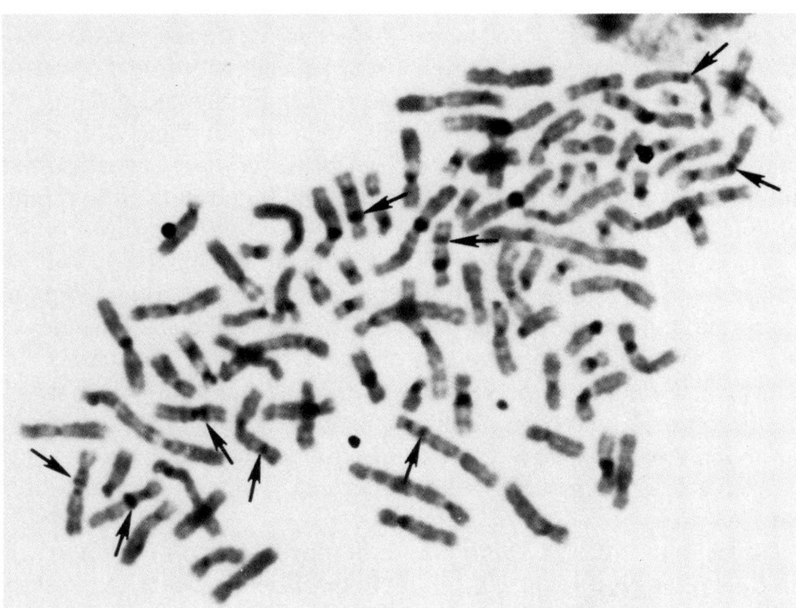

Fig. 3. C-banded metaphase from an osteogenic sarcoma. Chromosomes with more than one C-band (arrows) suggest that they have more than one centromere. Note that some chromosomes are pseudo dicentric and others are true dicentric.

ently, malignant cells can tolerate considerably more genetic imbalance than do the constitutional cells that make up an individual. Although there are a couple of recurring isodicentric chromosomes [27], no translocation dicentric chromosomes have been reported to be associated with the primary cytogenetic events of malignant neoplasms. This kind of evidence would require observing more than one patient in whom there is only one abnormality and it is a similar translocation dicentric chromosome. So far, the translocation dicentric chromosomes that have been seen in association with malignant neoplasms have been associated with chromosome evolution and advanced disease. Thus, many other chromosome abnormalities have been noted in addition to the translocation dicentric chromosomes. Consequently, it is difficult to determine the clinical significance, if any, of these translocation dicentric chromosomes.

CENTROMERIC CONSTRICTIONS

Most reports about patients with dicentric chromosomes note broken or multiple copies of the dicentric chromosome in occasional metaphases. In the absence of diacetylmethylcolchicine (Colcemid) treatment, anaphase bridges can sometimes be observed in cells from patients with true dicentric chromosomes [32]. An example of bridges formed during cell division in four somatic cells in an osteogenic sarcoma is shown in Figure 4. Broken or multiple dicentric chromosomes and anaphase bridges are all considered evidence that a dicentric chromosome is not segregating in a completely stable fashion during cell division.

In a report by Niebuhr [9], up to 4% of the metaphases in one patient showed extra or broken dicentric chromosomes. In the patient with a 5;15 translocation dicentric chromosome [13], such sporadic abnormalities of the dicentric chromosome occurred in about 1.7% of the metaphases. These sporadic abnormalities are rarely observed in patients with pseudo translocation dicentric chromosomes. Thus, some investigators think that one of the centromeres has somehow become inactivated and the apparent dicentric chromosome is permitted to behave in cell division as though it were monocentric.

The degree of constriction in the centromeric regions of dicentric chromosomes often varies from cell to cell. In the patient with a 5;15 translocation dicentric chromosome [13], variation in the degree of centromeric constriction, ranging from classic centromeric constriction to complete absence of a constriction, was observed (Fig. 5). In 15% of the metaphases, the dicentric chromosome had only one primary centromeric constriction: in 8% the

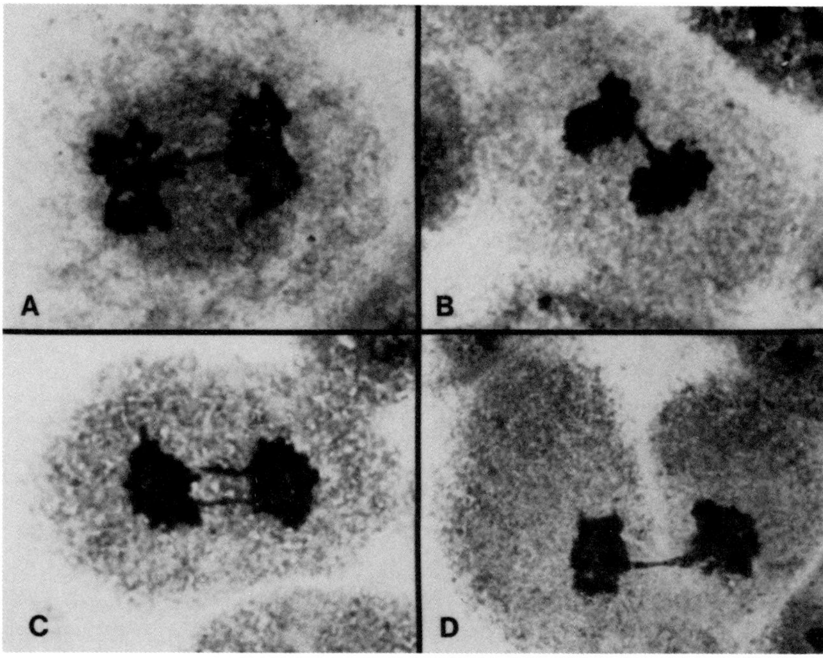

Fig. 4. Anaphase of a cell from an osteogenic sarcoma demonstrates an anaphase bridge from the abnormal segregation of a dicentric chromosome.

constriction was at the site of the chromosome 15 centromere, and in 7% it was at the chromosome 5 centromere. In the remaining 85% of metaphases, each centromeric region was associated with a constriction. The observation of differential constriction between the two centromeric regions has led some investigators to suggest that the centromere in the region of lesser constriction may be non-functional or weaker than the centromere located in the region of greater constriction [33].

KINETOCHORE STAINING

At least two reported methods can be used to stain specific parts of the centromeric region. One such method is Cd-banding, which reportedly stains the kinetochore [34,35]. The kinetochore is the site on the chromosome where chromosomal microtubules are assembled and attached, and it is thought to be a trilaminar structure that is made up of DNA, RNA, and tubulin. Another method is to use an indirect immunofluorescence technique

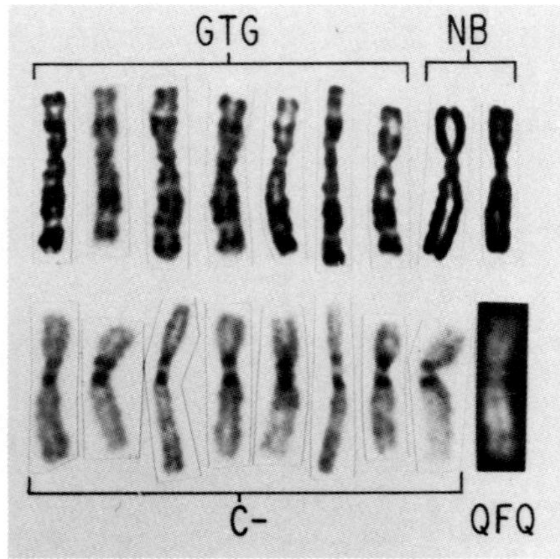

Fig. 5. Eighteen representative tdic(5;15)(p13;p11) chromosomes stained by various methods demonstrate variable appearance of centromeric constrictions. The first GTG- and first two C-banded translocation dicentric chromosomes show only one constriction, and it is on chromosome 5. The last GTG- and last two C-banded translocation dicentric chromosomes also show only one constriction, but it is on chromosome 15. The rest of the translocation dicentric chromosomes show two primary constrictions. (From Dewald GW, Boros SJ, Conroy MM, Dahl RJ, Spurbeck JL, Vitek HA: A tdic (5;15)(p13;p11) chromosome showing variation for constriction in the centromeric regions in a patient with the cri du chat syndrome. Cytogenet Cell Genet 24:15, 1979. By permission of S. Karger.)

that involves first exposing the chromosomes to an "anticentromeric antibody" derived from serum of patients with the CREST syndrome (calcinosis cutis, Raynaud phenomenon, esophageal dysfunction, sclerodactyly, and telangiectasia) and then staining the metaphases with a fluorescein-conjugated goat antihuman IgG [19,36]. This method is reported to also stain the kinetochore [19,36], but at least one investigator (A. Daniel, personal communication) suspects that it might actually stain a specific part of the C-band nucleoprotein complex in the centromeric region. A metaphase showing positive fluorescence of the centromeric region of metaphase chromosomes is shown in Figure 6.

At least one investigator has applied Cd-banding to the study of pseudo translocation dicentric chromosomes in four patients [34]. In each case, no Cd-banding was noted in the C-band pseudo centromere region. Similar findings have been reported for suppressed centromeres in isodicentric Y chromosomes [37]. Another investigator has used the anticentromeric antibody technique to study a 9;11 pseudo translocation dicentric chromosome [19]. Interestingly, the "functional centromere" had a typical intensely fluorescent anticentromeric antibody staining, whereas the anticentromeric an-

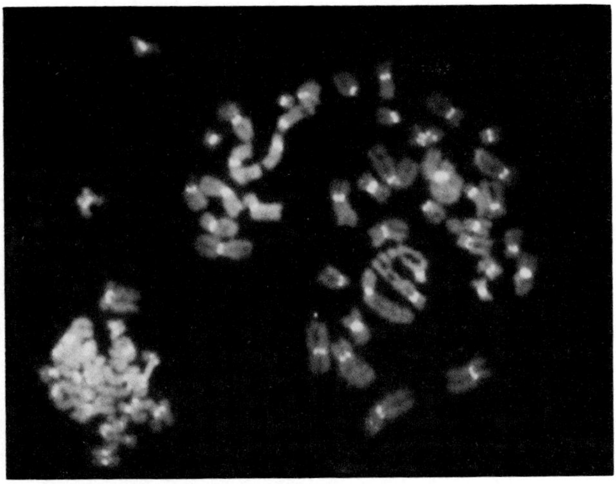

Fig. 6. Metaphase from an osteogenic sarcoma shows positive fluorescence at kinetochore regions by the anticentromeric antibody staining procedure. (From Powell FC, Winkelmann RK, Venencie-Lemarchand F, Spurbeck JL, Schroeter AL: The anticentromere antibody: disease specificity and clinical significance. Mayo Clin Proc 59:700, 1984. By permission of Mayo Foundation.)

tibody staining in the "inactive centromere" fluoresced much less intensely but was still positive. The weaker fluorescence in the pseudo centromeric area might be related to the non-functionality of the centromeres. However, the lesser fluorescence in the pseudo centromeric area could also have been caused by their relatively distant location from one another due to the separation of the chromatids because of the absence of a centromeric constriction. The brighter fluorescence of the kinetochores in the area of most centromeric constriction may have been due to the joint luminescence of the two kinetochores that were located immediately adjacent to one another.

The different results of the studies using Cd-banding and anticentromeric antibody procedures may suggest that they actually do stain different parts of the centromeric region. If so, the Cd-banding study might suggest that there is no kinetochore in the pseudo centromeric region. In contrast, the anticentromeric antibody method suggests that there is some component of the centromeric region in the pseudo centromeric region that may not be disturbed as a consequence of centromeric suppression. It is also possible that the anticentromeric antibody method does in fact stain the kinetochore but is more sensitive than Cd-banding. Alternatively, some patients may have a kinetochore, or at least part of one, in the pseudo centromeric region whereas others do not. Thus, whether the mechanism for centromeric suppression is related to an inability of the chromosome to make a normal kinetochore is unclear.

MECHANISMS OF CENTROMERIC SUPPRESSION

At least two hypotheses relate to how a chromosome may lose the service of a centromere. One hypothesis involves a chromosome deletion that leads to the loss of the centromeric gene or genes. The other hypothesis involves inactivation of the centromeric gene or genes by some internal genetic molecular process. Unfortunately, there is no effective way to prove that either of these mechanisms actually occurs.

Dewald et al. [13] tried to make a case for centromeric deletion in the loss of centromeric function in the 5;15 translocation dicentric chromosome. The evidence was circumstantial and involved the observation of a chromatid gap in the centromeric region of the dicentric chromosome in two metaphases. This was thought to be more than coincidence because no gaps or breaks were noted in any other chromosomes in any of the metaphases studied. The centromeres were speculated to be torn out of the translocation dicentric chromosomes during anaphase-bridge formation. As a consequence of their entangled chromosomal fibers, the two broken parts of the dicentric chromosome may remain together until the broken DNA is repaired without replacement of the centromeric gene or genes. The fact that in this patient some metaphases had no constriction at the chromosome 15 centromere and others at the chromosome 5 centromere suggested that the mechanisms associated with loss of centromeric function were actively occurring. In most patients this probably occurs much earlier in embryogenesis. Such random centromeric inactivation was suggested to favor the deletion hypothesis.

Merry et al. [19] suggested that the fact that their patient had a pseudo translocation dicentric chromosome that stained positive in the pseudo centromeric area with the anticentromere antibody method favors the hypothesis of molecular inactivation of centromeres. However, the loss of part of the centromere gene or genes by chromosome deletion could still leave sufficient chromatin for the anticentromere antibody to recognize, but the centromere could be left non-functional. Thus, the demonstration of anticentromere antibody staining does not really answer the question of whether centromeric inactivation occurs by chromosome deletion or gene inactivation.

Much of the problem of establishing the mechanism of centromeric suppression is that no one really understands what gene or genes are responsible for the "centromere." To some investigators the term "centromere" applies to the entire area of primary constriction on the chromosome and not only the kinetochore. The primary constriction itself is probably caused by some delay in DNA replication in the area. It seems that more than one gene must be involved in normal centromeric function. Some gene or genes probably control the time of centromeric DNA synthesis, others manufacture a kinetochore, and still others probably are involved in spindle fiber assembly and attachment. The disruption of any of these important steps in the centromeric region could render a centromere inactive.

PERMANENCE OF CENTROMERIC INACTIVATION

No one has yet determined whether centromeric inactivation is a permanent event or whether it occurs continually throughout the life of an individual in each cell cycle. Most reports of dicentric chromosomes suggest that in any given patient it is usually the same centromere that is consistently constricted from cell to cell. Only two exceptions to this pattern have been reported: one is the report in 1979 by Dewald et al. [13] of a 5;15 translocation and the other is a report by Skovby and Niebuhr in 1978 (personal communication). Thus, the literature supports the hypothesis that centromere inactivation is probably a permanent event. In the patient with the 5;15 translocation, the inactivation process was possibly also a permanent event, but it happened at least twice in this unusual patient.

The relatively few published reports of patients with translocation dicentric chromosomes seem to suggest that the same centromere is non-functional from one cell to another in certain individuals, whereas in other persons more than one cell line may be exhibiting either centromere or both centromeres functioning. The observation of mosaicism is probably related to how early in ontogeny that loss of centromeric function occurs. For dicentric chromosomes with a great intercentromeric distance, the potential for anaphase-bridge formation is high. In these cases, the survival of the embryo would depend on inactivation of one centromere early in ontogeny in order to permit stable cell division. Consequently, there may not be sufficient time to permit the formation of more than one cell line with different centromeres inactivated without affecting the survival of the embryo.

For dicentric chromosomes with a small intercentromeric distance, the loss of a centromeric function may occur at any time during ontogeny without seriously imperiling the survival of the embryo. In these cases, the probability of observing cell lines with either centromere or both centromeres functioning may be greater. Also, the intercentromeric distance may be small enough so that there is no selective advantage of forming monocentric chromosomes. These dicentric chromosomes may, in reality, remain dicentric. Some Robertsonian translocations may be examples of this phenomenon.

In 1979, Dewald et al. [13] suggested that developing clones of cells from patients such as the one in that study might provide useful clues to the mechanism of centromeric inactivation. This would permit studies that might shed some light on whether non-functional centromeres remain inactive in subsequent cell divisions or whether they can regain their function. If the mechanism is a chromosomal interstitial deletion, it would be expected to be permanent. If the inactivation process is not permanent, then quite possibly the mechanism involves some gene-regulation system. Unfortunately, contact with this patient was lost before a skin biopsy specimen could be obtained to establish a fibroblast culture, and a similar opportunity has not arisen.

REFERENCES

1. Schwartz S, Palmer CG, Weaver DD, Priest J: Dicentric chromosome 13 and centromere inactivation. Hum Genet 63:332, 1983.
2. Mules EH, Stamberg J, Jabs EW, Leonard CO: Two different structural abnormalities of chromosome 13 in offspring of chromosomally normal parents with two fragile sites. Clin Genet 23:380, 1983.
3. Weilong G, Aide Y, Hongbao F, Shumou L, Ju Y: Bisatellited dicentric chromosome: A report on a case with karyotype 47,XY,+psu dic(22)t(22;22)(p22pter → cen → 22q11::22q11→22pter). Hum Genet 61:325, 1982.
4. Daniel A, Stewart L, Saville T, Brookwell R, Paull H, Purvis-Smith S, Lam-Po-Tang PRLC: Prenatal diagnosis in 3,000 women for chromosome, X-linked, and metabolic disorders. Am J Med Genet 11:61, 1982.
5. Berg JM, Gardner HA, Gardner RJM, Goh EG, Marković VD, Simpson NE, Worten RG: Dic(21;21) in a Down's syndrome child with an unusual chromosome 9 variant in the mother. J Med Genet 17:144, 1980.
6. Daniel A, Lam-Po-Tang PRLC: Structure and inheritance of some heterozygous Robertsonian translocations in man. J Med Genet 13:381, 1976.
7. Skovby F, Niebuhr E: Centromere inactivation in dicentric human chromosomes (Abstract). Excerpta Medica Int Congr Ser No. 397:153, 1976.
8. Koulischer L, Lambotte C: Telomeric tandem translocation tan (14;15) (q32;q26) and absence of IgA in an incestuous child. Ann Genet (Paris) 17:189, 1974.
9. Niebuhr E: Unusual findings by fluorescence microscopy of a t(13q 14q). Humangenetik 15:96, 1972.
10. Niebuhr E: Dicentric and monocentric Robertsonian translocations in man. Humangenetik 16:217, 1972.
11. Daniel A, Ekblom L, Phillips S, FitzGerald JM, Opitz JM: NOR activity and centromere suppression related in a de novo fusion tdic(9;13)(p22;p13) chromosome in a child with del(9p) syndrome. Am J Med Genet 22:577, 1985.
12. Ing PS, Smith SD: Cytogenetic studies of a patient with mosaicism of isochromosome 13q and a dicentri (Y;13) translocation showing differential centromere activity. Clin Genet 24:194, 1983.

13. Dewald GW, Boros SJ, Conroy MM, Dahl RJ, Spurbeck JL, Vitek HA: A tdic(5;15)(p13;p11) chromosome showing variation for constriction in the centromeric regions in a patient with cri du chat syndrome. Cytogenet Cell Genet 24:15, 1979.
14. Roberts SH, Howell RT, Laurence KM, Heathcote ME: Stable dicentri autosome, tdic (8:22)(p23:p23), in a mentally retarded girl. J Med Genet 14:66, 1977.
15. Nakagome Y, Teramura F, Kataoka K, Hosono F: Mental retardation, malformation syndrome and partial 7p monosomy [45,XX,tdic(7;15) (p21;p11)]. Clin Genet 9:621, 1976.
16. Warburton D, Henderson AS, Shapiro LR, Hsu LYF: A stable human dicentric chromosome, t dic (12;14)(p13;p13) including an intercalary satellite region between centromeres. Am J Hum Genet 25:439, 1973.
17. Niebuhr E: A 45,XX,5 − ,13 − ,dic + karyotype in a case of cri-du-chat syndrome. Cytogenetics 11:165, 1972.
18. Sŭbrt I, Blehová B, Táborský O: Dicentric chromosome due to an unusual fusion. Humangenetik 12:136, 1971.
19. Merry DE, Pathak S, Hsu TC, Brinkley BR: Anti-kinetochore antibodies; use as probes for inactive centromeres. Am J Hum Genet 37:425, 1985.
20. Ward BE, Bradley CM, Cooper JB, Robinson A: Homodicentric chromosomes: A distinctive type of dicentric chromosome. J Med Genet 18:54, 1981.
21. Ray M, Hunter AG: Partial trisomy 8 mosaicism with 46,XX/46,XX, − 8, + dic(8). Ann Genet (Paris) 23:100, 1980.
22. Pallister PD, Patau K, Inhorn SL, Opitz JM: A woman with multiple congenital anomalies, mental retardation and mosaicism for an unusual translocation chromosome t(6;19). Clin Genet 5:188, 1974.
23. German J: "Chromosomes and Cancer." New York: John Wiley & Son, 1974, pp 601–617.
24. Miller OJ: Dicentric chromosomes. In Emery AEH, Rimoin DL (eds): "Principles and Practice of Medical Genetics." New York: Churchill Livingstone, 1983, p 57.
25. Harnden DG: Viruses, chromosomes, and tumors: the interaction between viruses and chromosomes. In German J (ed): "Chromosomes and Cancer." New York: John Wiley & Sons, 1974, pp 154–190.
26. Sandberg AA: "The Chromosomes in Human Cancer and Leukemia." New York: Elsevier North Holland, 1980.
27. Dewald GW: Isodicentric X chromosomes in humans: Origin, segregation behavior, and replication band patterns. In Sandberg AA (ed): "Cytogenetics of the Mammalian X Chromosome. Part A. Basic Mechanisms of X Chromosome Behavior." New York: Alan R. Liss, 1983, pp 405–426.
28. Gosden JR, Lawrie SS, Gosden CM: Satellite DNA sequences in human acrocentric chromosomes: Information from translocations and heteromorphisms. Am J Human Genet 33:243, 1981.
29. Robertson WRB: Chromosome studies. I. Taxonomic relationships shown in the chromosomes of Tettigidae and Acrididae: V-shaped chromosomes and their significance in Acrididae, Locustidae, and Gryllidae: Chromosomes and variation. J Morphol 27:179, 1916.
30. Steinbach P, Djalali M, Hansmann I, Kattner E, Meisel-Stosiek M, Probeck H-D, Schmidt A, Wolf M: The genetic significance of accessory bisatellited marker chromosomes. Hum Genet 65:155, 1983.
31. Larson LM, Wasdahl WA, Saumur JH, Coleman ML, Jalal SM: Trisomy 18 syndrome with an unusual karyotype: Possible double isochromosome. J Med Genet 15:73, 1978.

32. Ying KL, Ives EJ: Mitotic behavior of a human dicentric Y chromosome. Cytogenetics 10:208, 1971.
33. Sears ER, Câmara A: A transmissible dicentric chromosome. Genetics 37:125, 1952.
34. Daniel A: Single Cd band in dicentric translocations with one suppressed centromere. Hum Genet 48:85, 1979.
35. Lambiase S, Maraschio P, Zuffardi O: The Cd technique identifies a specific structure related to centromeric function. Hum Genet 67:214, 1984.
36. Powell FC, Winkelmann RK, Venencie-Lemarchand F, Spurbeck JL, Schroeter AL: The anticentromere antibody: Disease specificity and clinical significance. Mayo Clin Proc 59:700, 1984.
37. Daniel A, Lyons N, Casey JH, Gras L: Two dicentric Y isochromosomes, one without the Yqh heterochromatic segment: Review of the Y isochromosomes. Hum Genet 54:31, 1980.

The Cytogenetics of Mammalian Autosomal Rearrangements, pages 551–582

Chapter 17
Centric Fission in Man and Other Mammals

Hirotami T. Imai

INTRODUCTION: THEORETICAL BASES FOR CENTRIC FISSION

Centric fission is not a new type of chromosomal mutation that competes with centric fusion, but it provides us with a new paradigm in chromosome evolution. However, some reconciliation of the concept with classic views of chromosome architecture is necessary.

The central dogmas of chromosome structure were established long ago by Muller [1,2] and were generalized to animals by White [3,4]. These are summarized as: 1) a unitary structure for the centromere; 2) the necessity for discrete chromosome ends, i.e., telomeres; and 3) the preclusion of the de novo formation of both centromeres and telomeres (Fig. 1). Besides these structural restrictions, two break rearrangements were adopted as the guidelines of chromosomal mutations. This is known as the breakage and reunion theory. Under these principles, structural chromosomal mutations such as centric fusion, inversion, tandem fusion, centric dissociation, and reciprocal translocation are theoretically available (Fig. 2). Among them, centric fusion and inversion have long been considered to be the major chromosomal mutations that have contributed to animal karyotype evolution. This is the so-called fusion hypothesis.

If these principles are strictly applied, there is obviously no room for centric fission. The reasons are simple, i.e., centromere duplication must occur before centric fission, which is against principles (1) and (3), and telocentrics induced by centric fission are unstable because of the lack of telomeres at the naked ends (Fig. 1). The increase of chromosome number was, therefore, interpreted in the fusion hypothesis by centric dissociation, which is a kind of reciprocal translocation occurring between a metacentric and a minute donor (Y- or B-chromosome) (Fig.1).

The recent advances in cytogenetics make it possible to revise the classic principles for structural rearrangements to allow for the following phenom-

National Institute of Genetics, Mishima, Shizuoka-ken 411, Japan

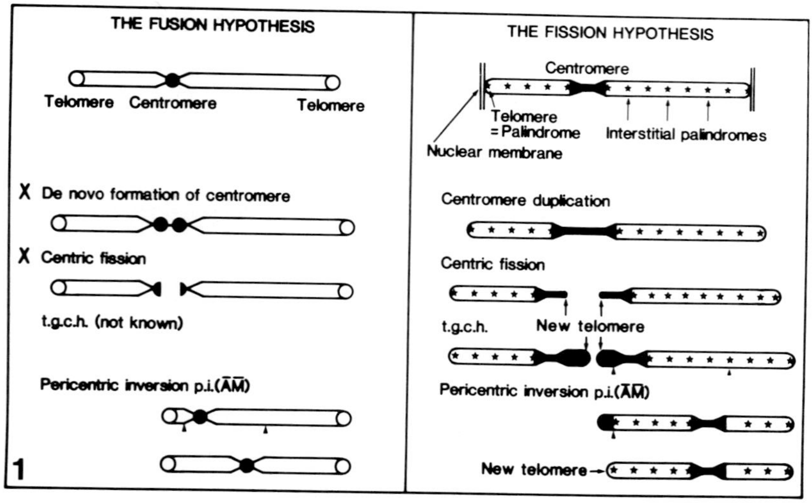

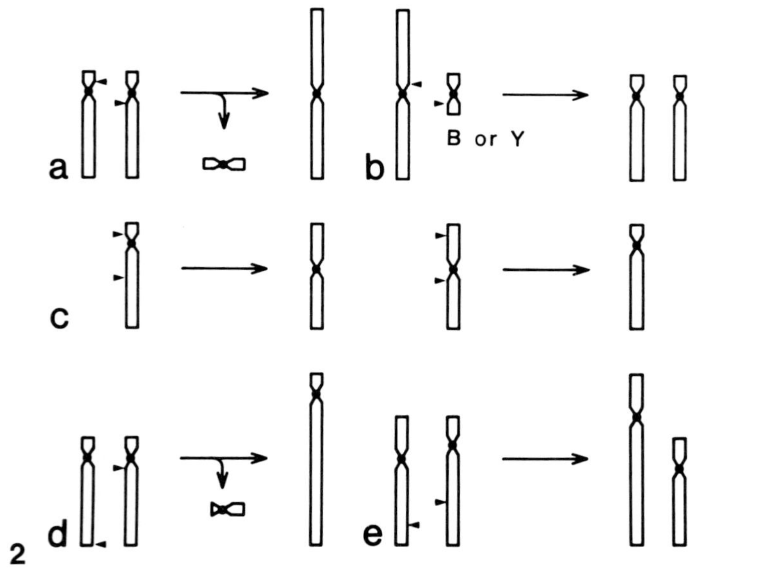

Fig. 1. Central dogma of chromosome structure in the fusion hypothesis (left) where de novo formation of centromeres and centric fission are disallowed (X). Fission hypothesis (right) with impremata for interstitial telomers and centromere duplication. t.g.c.h.: Tandem growth of constitutive heterochromatin. p.i.($\overline{\text{AM}}$) and p.i.($\overline{\text{MA}}$): Pericentric inversion changing $\overline{\text{A}}$ to $\overline{\text{M}}$ and the reverse. For definition of chromosome see Figure 8. Stars are palindromic base sequences. Black areas: Centromere with compound structure or constitutive heterochromatin. Arrow heads: breakage points or exchange sites.

Fig. 2. Spontaneous structural chromosomal mutations under the fusion hypothesis. **a:** Centric fusion. **b:** Centric dissociation. **c:** Pericentric inversions: p.i.($\overline{\text{AM}}$) (left) and p.i.($\overline{\text{MA}}$) (right). **d:** Tandem fusion. **e:** Reciprocal translocation. Arrow heads are breakage points in terms of the breakage and reunion theory. B: B-chromosome. Y: Y-chromosome.

ena: 1) a compound structure existing at least in some centromeres; 2) the sealing off of the chromosome ends by attaching palindromes to the nuclear membrane (i.e., telomeres are palindromes) and de novo formation of telomeres by interstitial palindromes; 3) tandem duplication of chromosome segments including the centromere, interstitial palindromes, and constitutive heterochromatin (Fig. 1; for details, see Imai [5]); and 4) centromeric suppression in dicentric chromosomes. These revised principles yield some theoretical bases for centric fission. As is shown in Figure 1, the naked chromosome ends caused by centric fission can be sealed with an interstitial palindrome, and thus the telocentrics induced by centric fission are theoretically stable. The compound structure of the centromere has been observed in plants [6,7], in grasshoppers [8,9], and in Indian muntjacs [10]. The palindrome model of the telomere was first proposed by Cavalier-Smith [11] and was slightly modified by Bateman [12], and the molecular bases for it refined more recently [13]. There is evidence that numerous palindromic base sequences are uniformly distributed at least in some eukaryote chromosomes [14,15]. The centric fission hypothesis is also dependent on Ohno's concept of gene duplication [16]. Indeed, gene duplication is a well-established principle in modern molecular genetics. The saltatory replication of DNA sequences was proposed by Britten and Kohne [17] and Bostock [18], by which tandem growth of constitutive heterochromatin in the short arm (t.g.c.h.) that is alien to the classic framework, can be added to the list of chromosomal mutation phenomena. T.g.c.h. is detectable cytogenetically as C-bands. The unipolarity of telomeres is easily interpreted here by attachment of the chromosome ends to the nuclear membrane. This is known as the basal knob [19] or the terminal attachment plaque [20].

By these revised principles, we can theoretically expect a new series of chromosome changes proceeding for example as follows: metacentric ($\overline{M}$) with functionally duplicated centromeres—(c. fission)→ telocentric (T)—(t.g.c.h.)→ acrocentric (A)—(p.i.(A$\overline{M}$))→ $\overline{M}$ (Fig. 1), where p.i.(A$\overline{M}$) is pericentric inversion changing A to $\overline{M}$. (For definitions of T, A, and $\overline{M}$, see Fig. 8). This process, i.e., the fission hypothesis, was proposed by Imai [5]. Chromosome mutations which are theoretically possible in the fission hypothesis are shown in Figure 3. The major chromosomal mutations which were assumed to occur by certain means in the fusion hypothesis (Fig. 2) are explained by different mechanisms in the fission hypothesis. In the latter hypothesis, new mutation phenomena, e.g., centric fission, t.g.c.h., and tandem fission, are added to the list. Note that c. fusion and c. fission, p.i.($\overline{AM}$) and p.i.($\overline{MA}$), t.g.c.h. and e.c.h., and tandem fusion and tandem fission are all paired rearrangements that are opposite in directionality. In

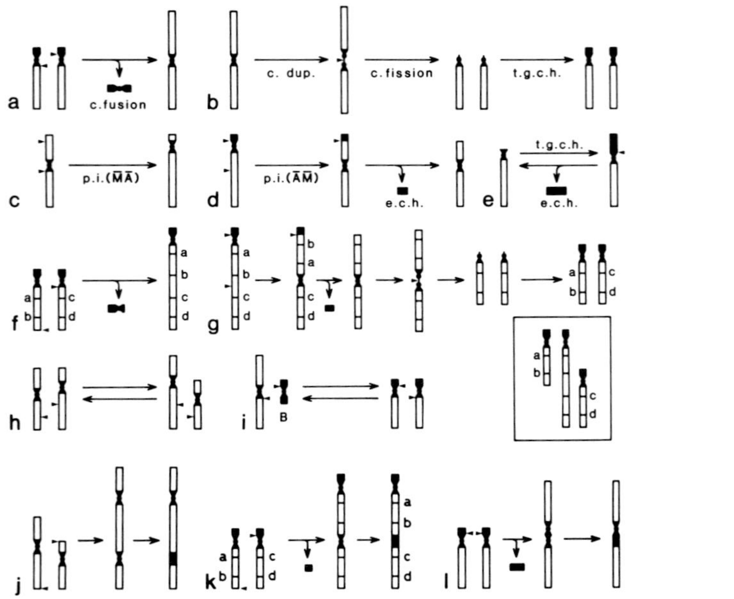

Fig. 3. Spontaneous structural chromosomal mutations under the fission hypothesis. **a:** Centric fusion in the classic sense. **b:** Centric fission in a broad sense including centromere duplication (c. dup.), centric fission in a strict sense, and t.g.c.h. **c; d:** Pericentric inversions changing $\overline{\text{M}}$ to $\overline{\text{A}}$ (c) and $\overline{\text{A}}$ to $\overline{\text{M}}$ (d). **e:** Tandem growth of constitutive heterochromatin (t.g.c.h.) and elimination of constitutive heterochromatin (e.c.h.). **f:** Tandem fusion. **g:** Tandem fission including p.i.($\overline{\text{AM}}$), centric fission, and t.g.c.h. **h:** Reciprocal translocation (centric fusion, tandem fusion, and centric dissociation are basically this type of rearrangement). **i:** Centric dissociation (= reciprocal translocation). **j:** Dicentric translocation accompanying centromeric suppression. **k:** Dicentric tandem fusion accompanying centromeric suppression and insertion of heterochromatin block. **l:** Dicentric fusion accompanying centromeric suppression. a and b, c and d, and f and g are opposite rearrangements in directionality. I use Robertsonian rearrangement as a general term to a and b. In the same manner, tandem Robertsonian rearrangement encompasses f and g. It is usually difficult to determine which of the alternatives has actually occurred in each case, i.e., they are *in a qualitative sense* bidirectional. **Inset:** A tandem arrangement of banding patterns between one large acrocentric and two small ones. Note that such a tandem arrangement of banding patterns, though it is generally believed to be a supporting evidence for tandem fusion, is also expected by tandem fission (g). Small letters on short bars of chromosomes mean G-banding. On the other hand, if the so-called tandem fusion is induced by "dicentric tandem fusion," it can be identified by the combination of tandem arrangement of G-banding pattern and insertion of heterochromatin blocks. Dicentric fusion (Fig. 3l) is more frequent than orthodox centric fusion (Fig. 3a) in human D-D, D-G, or G-G translocations. This type of fusion may be found in mammals other than man. I use centric fusion in this text in a broad sense including both fusion types. The series of theoretical approaches for karyotype evolution and especially the minimum interaction hypothesis mentioned in this text (for details see Imai [65]) is not inconsistent with dicentric fusion (see Fig. 18).

other words, these major rearrangements are *bidirectional* in the fission hypothesis. For this reason, the term "Robertsonian rearrangement" is used in the present context as a general term for fusion and fission.

A new type of chromosomal phenomenon called centromeric suppression or centromeric inactivation has been recognized in recent years in human dicentric chromosomes [21–24]. Dicentric chromosomes have long been considered in classic cytogenetics to be highly deleterious because of chromosome bridge formation at anaphase. Centric fusion has been, therefore, interpreted as a reciprocal translocation occurring in the pericentromeric heterochromatin of two nonhomologous acrocentrics (Fig. 2a). However, in human chromosomes, most metacentrics resulting from D-D, D-G, or G-G fusions have two centromeres (i.e., they are dicentric), and more surprisingly one of the two centromeres is functionally inactivated (Fig. 3). Centromeric suppression seems to be a rather general phenomenon in human dicentric chromosomes. For example, Daniel [25,26] has revealed that some special translocations denoted by him as "dicentric translocations" (Fig. 3j) and "isodicentrics" can survive by suppressing one of the two centromeres involved. Centric fusion is interpreted in this new context as a kind of dicentric translocation, different from classic reciprocal translocation in mechanism though "dicentric fusion" may be a more suitable term for this type of rearrangement. In this chapter the term centric fusion is used in a broad sense including both classic centric fusion (Fig. 3a) and dicentric fusion (Fig. 1, also see Fig. 18). This usage is chosen firstly, because centromeric suppression has the potential to circumscribe some problems of chromosomal mutation, but further accumulation of data is required from mammals other than man, and secondly because centromeric suppression does not affect any fatal damage to the present arguments.

Some cytologists believe that tandem fusion has played a significant role in karyotype evolution, e.g., in the Indian muntjac [27–30], in cotton rats *Sigmodon* [31], or in spiny mice *Mus platythrix* [32]. A tandem arrangement of G-banding patterns between two or more fragmented acrocentrics and a corresponding large acrocentric (Fig. 3, inset) has been proposed as supporting evidence of tandem fusion. There is, however, an alternative means of transforming a large sized acrocentric to small ones by p.i. ($\overline{AM}$), centric fission, and t.g.c.h. (Fig. 3g). By this means, so-called tandem fusion is theoretically bidirectional. I propose that this new type of chromosomal alteration be termed "tandem fission." Note that the small sized acrocentrics induced by tandem fission can be matched tandemly on the large sized acrocentric (Fig. 3, inset). There is another possibility, that is, tandem fusion accompanying centromeric suppression (Fig. 3k). This type of tandem fusion, possibly better termed "dicentric tandem fusion," could be recognized by the combination of tandem arrangement of G-banding patterns and insertion of heterochromatin blocks. If the karyotype of the Indian muntjac (2n = 6, 7) was derived from that of the Chinese muntjac (2n = 46) by successive dicentric tandem fusions, at least five or six interstitial heterochromatin blocks should be inserted in each chromosome arm of the Indian muntjac, as suggested in Figure 3k. Even if we neglect this criticism, the concepts of "branched chromosomes," i.e., centromeric heterochromatin branching off from the main chromosome strand, and "linear fusions" occurring selectively between centromere-centromere, centromere-telomere, or telomere-telomere, which were proposed by Brinkley et al. [30], are opposed to the basic principles of the fusion hypothesis as well as the fission hypothesis. One interpretation for the origin of compound centromeres may be to explain them simply as a result of an interstitial duplication of centromeres rather than the result of linear fusions.

Another concept I wish to introduce in subsequent sections to the mechanism of chromosomal mutations is that the exchange theory [33,34] is more consistent with the fission hypothesis than the breakage and reunion theory.

FUSION AND FISSION IN MAMMALIAN KARYOTYPE EVOLUTION

Over 1,000 mammalian species have been karyotyped during the last two decades [28,35]. These accumulated data reveal the widely divergent chromosome numbers in mammals, ranging between 2n = 6 and 92 with the mode at 2n = 48 (Fig. 4). Therefore mammals provide a considerable body of data for the consideration of karyotype evolution in perspective.

Three controversial hypotheses (the fusion, fission, and modal hypotheses), have been proposed concerning the mode of mammalian karyotype evolution (Fig. 4). The fusion hypothesis assumes that ancestral mammals had a karyotype with a large number of acrocentrics ($2K = 96\overline{A}$ or $80\overline{A}$), and further that karyotype change proceeded by reduction of chromosome numbers mainly by centric fusion and inversion changing $\overline{M}$ to $\overline{A}$ [(p.i.($\overline{MA}$) in this chapter)] [36–38].

On the other hand, the fission hypothesis assumes the opposite directionality, in which the evolutionary process starts from $2K = 6\text{–}10\overline{M}$ and centric fission and pericentric inversion changing $\overline{A}$ to $\overline{M}$ (p.i.($\overline{AM}$)) are considered to be the major rearrangements [5,39–41]. Centric fusion is adopted in the fission hypothesis as a local back eddy or a transitional step. The modal hypothesis is a compromise between the two extremes mentioned above, i.e.,

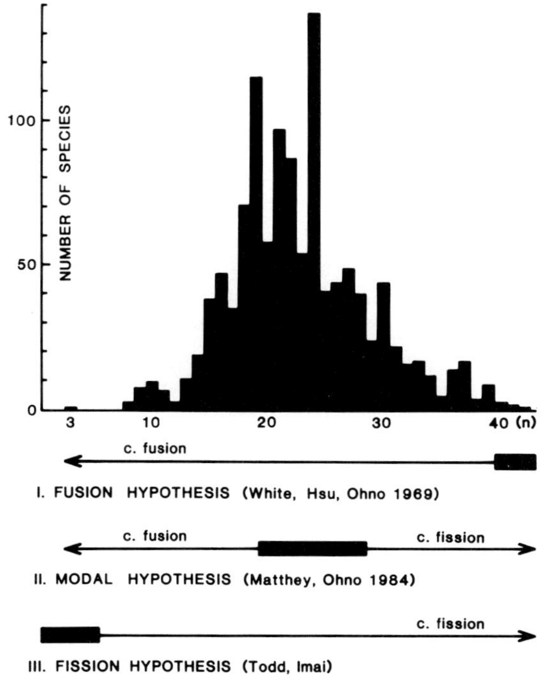

Fig. 4. Frequency distribution of haploid chromosome numbers in mammals and three hypotheses for mammalian karyotype evolution. The starting chromosome numbers expected in each hypothesis are represented by black bars with arrows. The names of the cytogeneticists who proposed or are associated with each hypothesis are shown in parentheses.

chromosome numbers disperse both upward and downward starting from 2n = 40–56 (the modal number) [28,42].

The fusion hypothesis has had the majority of support in mammals [43,44]. White [4] concluded that centric fission has no place in science. This extreme and unbalanced conclusion is, however, far from realistic, because most examples of centric fusion have been claimed without any solid evidence, with the exception of some human cases [45], *Mus musculus* in Italy and Switzerland [46], and *Rattus rattus* [32]. Indeed, it is practically difficult to determine whether fusion or fission has occurred in the majority of Robertsonian rearrangements. The actual circumstance may be that a majority of cytologists have interpreted Robertsonian rearrangement as centric fusion in the context of the fusion hypothesis. As is discussed in the previous section, centric fusion is the only available interpretation for Robertsonian rearrangement under the classic principles of chromosome structure. I wish to emphasize, however, that centric fission is not only theoretically possible (Fig. 3) but also has been documented in Chinese hamsters [47], humans [48–51], Mauritian black rats [32], Japanese racoon dogs [52], and zebras [53]. Robertsonian rearrangement is in a qualitative sense bidirectional.

The modern fusionists will argue, for example, that available evidence at least suggests that fusion is more common than fission and thus fusion, instead of fission, has played the major role in karyologic evolution in mammals (see Hsu [54]). The fusionists need to integrate the concept of dicentric fusion and accompanying centromeric suppression, but also to reconsider their kind of quantitative approach. The problem is, however, that such a conclusion (i.e., the theoretical predominance of fusion) as well as that of the modalists was deduced intuitively, based on enumerative data but not logically within a firm theoretical framework. Before I discuss a quantitative approach for karyotype evolution, I wish to mention some problems involved in the conventional approach to interpreting karyotype evolution.

The conventional procedure for analyzing karyotype evolution has two steps: 1) detection of chromosomal mutations by comparing karyotypic homologies between every two species, and 2) reconstruction of a possible pathway of karyotype differentiation based on the directionality of chromosomal mutations detected (Fig. 5). As I mentioned above, Robertsonian rearrangement is in a qualitative sense bidirectional. The same argument is

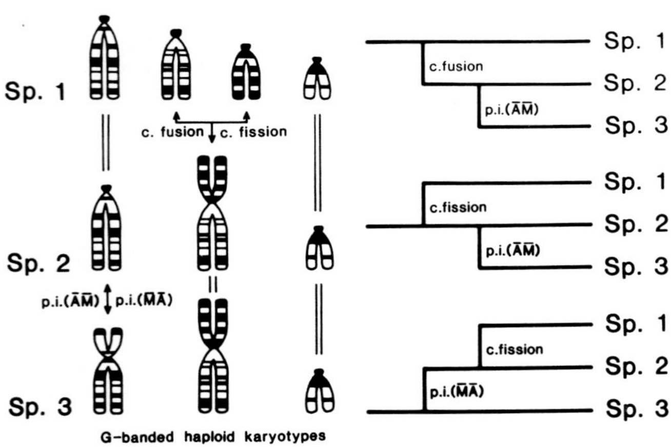

Fig. 5. Conventional method for identifying chromosomal mutations and for reconstructing phylogenetic relationships among species based on the directionality of chromosomal mutations. As Robertsonian rearrangments and pericentric inversions are bidirectional, at least three different dendrograms are possible in this case, and it is difficult to determine which of them is the genuine one.

true for pericentric inversions (Fig. 3). Therefore, it is fundamentally difficult to reconstruct a solid phylogenetic relationship of a given taxon based on only the directionality of chromosomal mutations (Fig. 5). Attempts at determining karyologic directionality by morphologic trends, e.g., by assuming that primitive species necessarily have a primitive karyotype, would seem to be counter productive for the fusion hypothesis as regards mammals. For example, marsupials and Indian muntjacs have lower chromosome numbers than eutherians or related (to muntjac) species in spite of the "primitiveness" in their anatomical morphologies [27,55]. A second technical problem concerned with the conventional method is that extrapolation of data from findings at the specific level to the family or order level is not very satisfactory. Indeed, the majority of case reports that discuss karyotype evolution at the genus level include only a few species or at most a dozen species. On the other hand, if we discuss "mammalian" karyotype evolution, we have to compare over 1,000 karyotypes most of whose karyotypic homologies (in terms of G- or C-banding) have been lost. Can we get a retrospective view of karyotype evolution at higher taxonomic levels by enumerating local experimental data at the genus level? The answer to this question may not be optimistic. What is required is a simple yet rigorous quantitative framework that will be useful even at higher taxonomic levels.

QUANTITATIVE ANALYSIS OF MAMMALIAN KARYOTYPE EVOLUTION

The crucial problem for analyzing karyotype evolution is how to decode the sequential processes of the chromosomal mutations recorded in metaphase chromosome morphologies during the course of evolution.

Four parameters (chromosome size, centromeric position, chromosome number, and arm number) are essential for this purpose. It is these parameters that are changed by Robertsonian rearrangement or pericentric inversion and have contributed to mammalian karyotype evolution (Fig. 3). Among them, the first two and the latter two are useful for a quantitative description of chromosome morphologies and karyotypes, respectively. Given these four parameters I wish to outline a quantitative approach to this problem.

Non-random Localization of the Centromere on Mammalian Chromosomes

When the size of short arms (S_w), long arms (L_w), and whole chromosomes ($C_w = S_w + L_w$) are measured (expressed as percentages) relative to the X-containing haploid set by using the photo cut-out technique [56], the centromeric position on chromosomes can be expressed quantitatively by the S_w values. As mammalian genome sizes are approximately the same except in bats [57,58], it is possible to compare quantitatively the S_w values of different chromosomes in different mammalian karyotypes without regard to the peculiarity of each individual karyotype. Using these criteria, a total of 16,817 individual chromosomes collected from 723 mammalian species were examined [59,60]. It was revealed that the frequency distribution of S_w is non-random and has a W-shaped pattern with two distinct antimodes lying at $S_w = 0.6$ and 0.1 (Fig. 6). The W-shaped pattern was found in all sized chromosomes with $C_w \geqslant 1.2$ (Fig. 7), suggesting an inherent characteristic of mammalian chromosomes.

Biologically Significant Classification of Mammalian Chromosomes

Levan et al. [61] proposed a standard nomenclature for chromosomes (i.e., M, SM, ST, A, and T) in terms of arm ratio ($r = L/S$; L and S on the length of long arm and short arms measured by the length method). This classification is, however, not always suitable for analyzing karyotype evolution quantitatively, because it is an arbitrary system, and because the length method tends to overestimate the size of short arms especially in acrocentrics [62]. These problems can be solved by the weight method combined with the photo cut-out technique [62,63]. The non-random distribution of the centromere (mentioned above), provides a new criterion for the classification of mammalian chromosomes. The W-shaped patterns shown in Figures 6 and 7 were best interpreted by assuming three structurally different short arms consisting of 1) a centromere with compound structure, 2) constitutive heterochromatin (in terms of C-banding), and 3) euchromatin (Fig. 8). Each arm type was approximately characterized by the size of short arms as $S_w < 0.1$, $0.1 \leqslant S_w \leqslant 0.6$, and $S_w > 0.6$, respectively. It was also revealed that chromosomes with $S_w \leqslant 0.6$ and $S_w > 0.6$ roughly (~ 90%) correspond, respectively, to A or T, and M, SM or ST identified intuitively by mammalian cytologists [63]. From these findings and others, mammalian chromosomes can be reclassified into three biologically significant types, T ($S_w < 0.1$), A ($0.1 \leqslant S_w \leqslant 0.6$), and $\overline{M}$ ($S_w > 0.6$, i.e., metacentric in a broad sense including conventional M, SM or ST), or more simply $\overline{A}$ (acrocentric in a broad sense; $S_w \leqslant 0.6$) and $\overline{M}$ ($S_w > 0.6$). The discrimination of T from A is important for analyzing the origin of telocentrics [5], but otherwise dispensable for the quantitative analysis of karyotype evolution. To simplify the discussion, the latter (dichotomous) classification will be frequently used in the present chapter as it is especially valuable in defining arm number strictly.

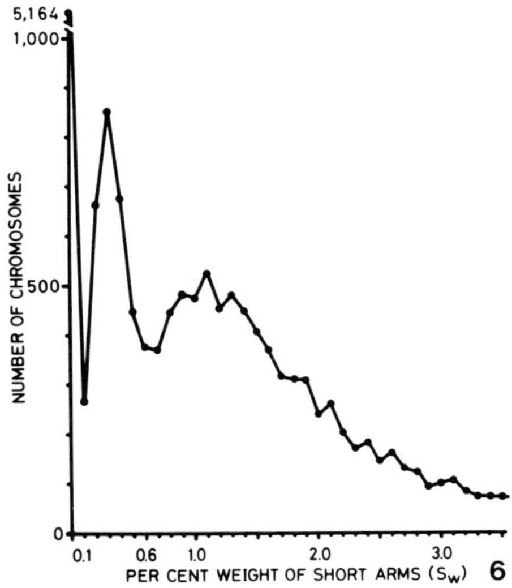

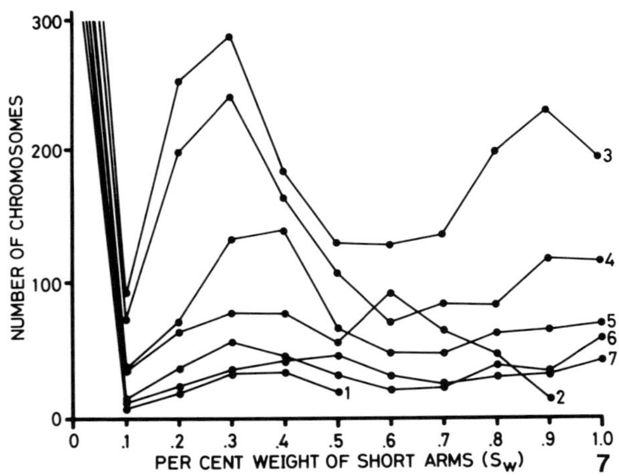

Fig. 6. Frequency distribution of the size of short arms with the antimodes at $S_w = 0.1$ and 0.6 (from Imai [60], with permission of the publisher.)

Fig. 7. Uneven frequency distributions of S_w in various sizes of chromosomes (C_w). The numerical orders show the distribution patterns of S_w when **(l)** $C_w < 1.2$, **(2)** $1.2 \leqslant C_w < 1.9$, **(3)** $1.9 \leqslant C_w < 3.0$, **(4)** $3.0 \leqslant C_w < 4.0$, **(5)** $4.0 \leqslant C_w < 5.0$, **(6)** $5.0 \leqslant C_w < 6.0$, and **(7)** $C_w \geqslant 6.0$. Note that two antimodes at $S_w = 0.1$ and 0.6 are found in all sized chromosomes having $C_w \geqslant 1.2$. (From Imai [60], with permission of the publisher.)

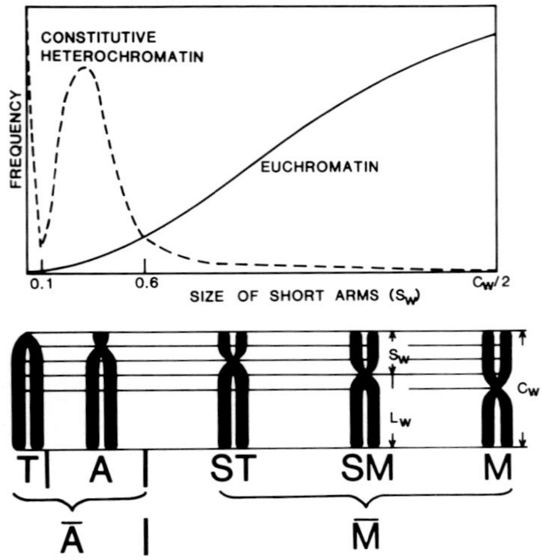

Fig. 8. Longitudinal differentiation of chromosome arms and classification of mammalian chromosome morphologies. The W-shaped distribution patterns of S_w (Fig. 6) are subdivided into three substantially different fractions (1) centromere with compound structure, (2) constitutive heterochromatin, and (3) euchromatin, which is defined roughly by S_w values as $S_w < 0.1$, $0.1 \leqslant S_w \leqslant 0.6$, and $S_w > 0.6$, respectively. With these definitions of short arms, the morphologies of mammalian chromosomes are classified into three biologically significant types (T, A, and $\overline{M}$) or simply two groups ($\overline{A}$ and $\overline{M}$). S_w, L_w, and C_w size of short and long arms and each chromosome is represented by percentage against the X-containing haploid set by the weight method combined with the photo cut-out technique. T (telocentric), A (acrocentric), and $\overline{M}$ (metacentric in a broad sense) are defined as chromosomes with $S_w < 0.1$, $0.1 \leqslant S_w \leqslant 0.6$, and $S_w > 0.6$. $\overline{M}$ includes the conventional M, SM, or ST.

THE FUSION AND FISSION CYCLES: CYCLICAL ALTERATION OF MAMMALIAN CHROMOSOME MORPHOLOGIES

Using the classification system (T, A, and $\overline{M}$) mentioned above, we can consider theoretically two alternative cycles of change in chromosome morphologies, termed here the fusion cycle and the fission cycle (Fig. 9). Chromosome morphologies change in the fusion cycle from A to $\overline{M}$ by centric fusion and the reverse by $\overline{M}$A inversion [p.i. ($\overline{M}$A)]. Alternatively morphological alterations proceed in the fission cycle as T→A→$\overline{M}$→T by tandem growth of constitutive heterochromatin (t.g.c.h.), A$\overline{M}$ inversion [p.i.(A$\overline{M}$)], and centric fission. The former just corresponds with the principles of the fusion hypothesis (see Fig. 2), and the latter is expected in the fission hypothesis (Fig. 3). Note that p.i.($\overline{M}$A) predominate p.i.(A$\overline{M}$) in the fusion hypothesis, whereas the reverse is true in the fission hypothesis. This means in other words that the directionality of pericentric inversion *in a*

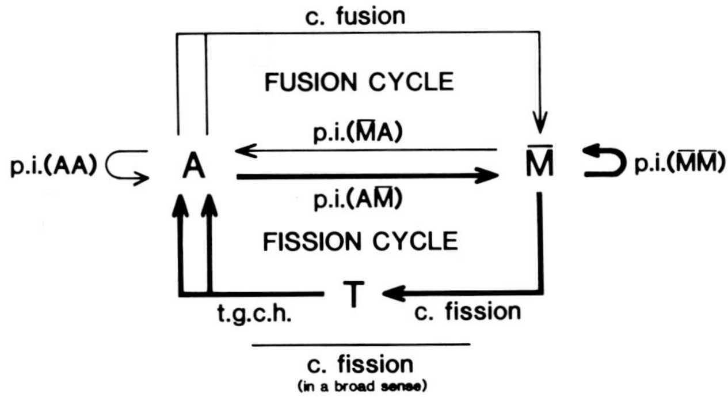

Fig. 9. Morphological alterations of mammalian chromosomes. When chromosomes are classified into T, A and $\overline{M}$ (for definitions see Fig. 8), two major cycles of chromosome morphology alteration (the fusion cycle and the fission cycle) are theoretically possible. In the fusion cycle, chromosomes change A —(fusion)→ $\overline{M}$ —(p.i.($\overline{M}$A))→ A, whereas A —(p.i.(A$\overline{M}$))→ $\overline{M}$—(fission)→ T —(t.g.c.h.)→ A occurs in the fission cycle. The fusion-fission controversy equates in the present context to the problem of which cycle (the fission or fusion) is the more dominant. Because of the strong unidirectionality of pericentric inversions (Fig. 10) and Robertsonian rearrangments (Fig. 14), the fusion cycle cannot dominate. In contrast, there is no such theoretical problem for the fission cycle, which can be reversed easily. The third cycle, i.e., fusion-fission cycle, is theoretically possible if the so-called centric fusion is dicentric fusion (Fig. 31). The fusion-fission cycle, more precisely fusion-fission fluctuation, may be important as a transitional phase or step of karyotype evolution.

quantitative sense may be essentially important for settling the fusion-fission controversy. The third possibility, i.e., fusion-fission cycle changing A→$\overline{M}$→T→A is discussed in a later section under the concept of fusion-fission fluctuation.

DIRECTIONALITY OF PERICENTRIC INVERSION

Pericentric inversions involve four different types that are named here $\overline{AA}$, $\overline{AM}$, $\overline{MM}$, $\overline{MA}$ inversions, and are symbolized as p.i.($\overline{AA}$), p.i.($\overline{AM}$), p.i.($\overline{MM}$), and p.i. ($\overline{MA}$) (Fig. 10). Pericentric inversion has in a quantative sense a strong directionality, i.e., p.i.($\overline{AM}$) and p.i.($\overline{MM}$) occur more frequently than p.i.($\overline{AA}$) or p.i.($\overline{MA}$). This directionality was suggested based on two different sources of experimental evidence: 1) Stochastic analysis of the occurrence probability of pericentric inversions using Monte Carlo simulation [64], and 2) well-known cytological evidence that most of all acrocentrics have totally heterochromatic short arms.

A rough outline of the simulation is as follows: A certain number of chromosomes with equal length and the centromeres at a fixed location are

10

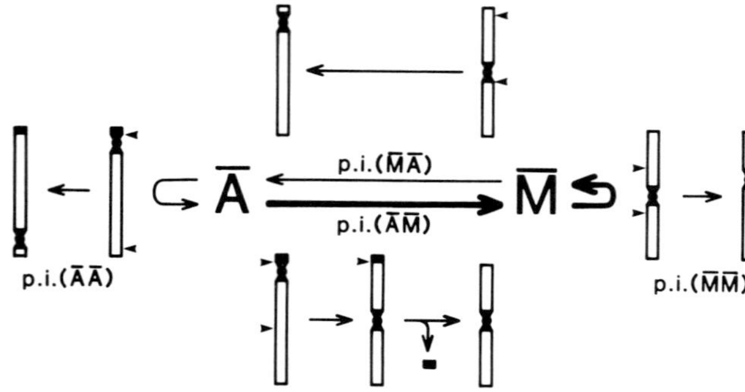

11

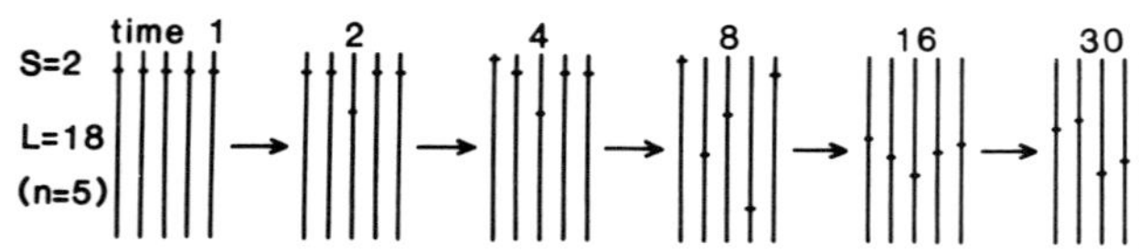

Fig. 10. Directionality of pericentric inversion. When chromosome morphologies are classified into $\overline{A}$ and $\overline{M}$ (for definitions see Fig. 8), pericentric inversions are subdivided into four types, p.i.($\overline{AA}$), p.i.($\overline{AM}$), p.i.($\overline{MM}$), and p.i.($\overline{MA}$), which changes $\overline{A}$ to $\overline{A}$, $\overline{A}$ to $\overline{M}$, $\overline{M}$ to $\overline{M}$, or $\overline{M}$ to $\overline{A}$, respectively. The Monte Carlo simulation experiments by Imai and Maruyama [64] revealed that p.i.($\overline{AM}$) and p.i.($\overline{MM}$) occur more frequently than p.i.($\overline{AA}$) or p.i. ($\overline{MA}$) (see Figs. 10 and 11). This directionality was supported by cytological observations that most of all $\overline{A}$s have totally heterochromatic short arms, but rarely euchromatic short arms, which are expected by p.i.($\overline{MA}$) or p.i.($\overline{AA}$).

Fig. 11. Karyotype alteration by pericentric inversion simulated by the Monte Carlo method. Small crossed lines are the centromeric positions on chromosomes. Once an acrocentric chromosome changed into a metacentric, the metacentric induced change again into another metacentric. By such stochastic characteristics, only $\overline{M}$s appear in $t > 16$ (see Fig. 12).

set in a computer (Fig. 11). Then each chromosome undergoes inversion by a chain of simulation processes, namely, two breaks occur randomly on a chromosome and the naked ends reunite inversely. In the exchange theory, inversions occur by randomly forming a loop and then exchanging chromatids at the intersecting point. The mechanisms for inducing inversions are different between the breakage and reunion theory and the exchange theory, but they are stochastically equivalent (for details see Imai [65].) After a certain estimation time (t), the locations of centromeres are estimated by means of S values, shown by a chain in Figure 11, in which the starting karyotype is K = 5A and karyotypes at t = 1, 2, 4, 8, and 30 are demonstrated. Note that all chromosomes at t = 16 and 30 are $\overline{M}$s ($K_{t=16} = K_{t=30} = 5\overline{M}$). The same phenomenon is found also at t = 200

(Fig. 13). In the figure, the frequency distribution of S_w that was obtained from 500 repeat simulation experiments indicates a clinal distribution pattern increasing from T to M (i.e., T < A < ST < SM < M). This clinal pattern resulted from a stochastic characteristic that one $\overline{A}$s change into $\overline{M}$s by p.i.($\overline{AM}$), then the derived $\overline{M}$s alter again into $\overline{M}$s by p.i.($\overline{MM}$), and it is very rare for p.i.($\overline{AA}$) or p.i.($\overline{MA}$) to occur. The clinical distribution pattern of S_w in Figure 8 is coincident with that predicted as a result of the simulation (Fig. 12), which was proposed as supporting evidence for the unidirectionality of pericentric inversion in mammals. Another line of evidence was found in accumulating C-banding data. If p.i.($\overline{MA}$) occurs frequently, the tip of the short arm of acrocentrics should be euchromatin (Figs. 3 and 10), whereas most observed acrocentrics have totally heterochromatic short arms. On the contrary, if p.i.($\overline{AM}$) occurs, we can expect a chromosome having a heterochromatic cap (Figs. 3 and 10); these kinds of chromosomes have actually been observed in ants [66] and a Japanese lung fluke [67]. C-bands at the telomeres of *Ammospheromophilus* may be such traces in mammals.

DIRECTIONALITY OF ROBERTSONIAN REARRANGEMENT

There are some pieces of indirect evidence suggesting unidirectionality of Robertsonian rearrangement as well as pericentric inversion. One piece of evidence is that the relative frequency of T chromosomes is about 30% in mammalian karyotypes, whereas there is no effective mechanism in the fusion hypothesis for inducing T chromosomes with $S_w < 0.1$[5]. The other pieces of evidence is that the size distribution of $\overline{M}$ chromosomes shows a unimodal pattern ranging between $C_w = 2$ and 18 with the modal point at $C_w = 5$, whereas $\overline{M}$ chromosomes resulting from Robertsonian polymorphisms ($\overline{M}_R$) are found only in large sized $\overline{M}$s with $C_w \geqslant 4$ (Fig. 13). This second piece of evidence may be puzzling for the fusion hypothesis particularly, because centric fusion is a reciprocal translocation occurring in pericentromeric heterochromatin of non-homologous acrocentrics, and thus there is no reason for fusion to occur only in large sized $\overline{M}$ chromosomes. The same argument is true if real fusions are dicentric translocations accompanying centromeric suppression (i.e., dicentric fusions; Fig. 3).

The fission hypothesis can interpret this evidence without any logical inconsistency. The high incidence of T chromosomes may be best interpreted if they are supplied unceasingly from $\overline{M}$ chromosomes by centric fission (Fig. 9). Note that karyotype evolution proceeds in the fission hypothesis towards increasing chromosome number and thus decreasing chromosome size. Therefore, the large sized chromosomes would be roughly older than

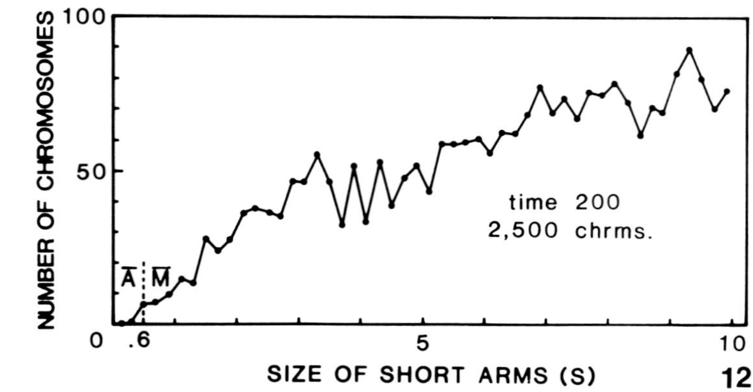

12

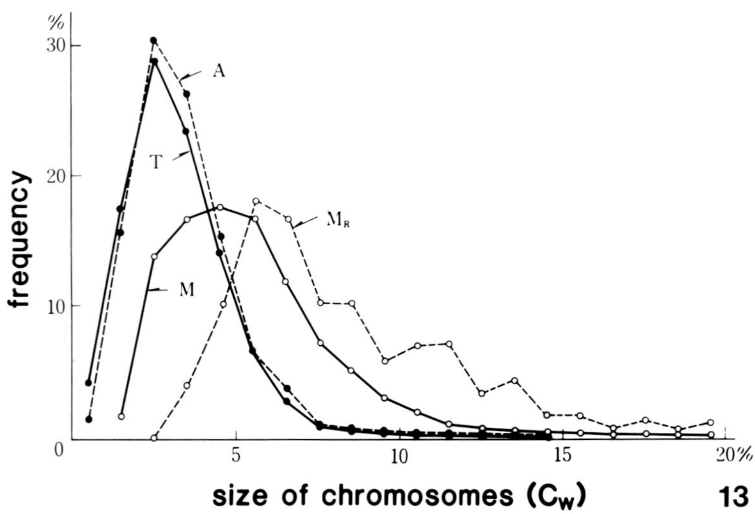

13

Fig. 12. Frequency distribution of S_w at t = 200. Starting from the K = $5\overline{A}$ shown in Figure 11, karyotype evolution by pericentric inversion is simulated up to t = 200. The clinal distribution pattern of S_w obtained from 500 repeated simulations is similar to that of euchromatin indicated in Figure 8, which was proposed as supporting evidence for the strong directionality of pericentric inversion in mammalian karyotype evolution (i.e., p.i.($\overline{AM}$) > p.i.($\overline{MA}$)) (see Fig. 11).

Fig. 13. Size distribution of T, A, $\overline{M}$, and $\overline{M}_R$ chromosomes observed in mammalian karyotypes. Each distribution totals 100%. $\overline{M}_R$ represents metacentrics involved in Robertsonian polymorphisms. It is noteworthy that Robertsonian rearrangements occur mostly in the large sized metacentrics with $C_w \geqslant 4$. This unexpected distribution pattern is difficult to interpret by the fusion hypothesis, but easy in the fission hypothesis.

the shorter ones; in other words, the latter would be induced recently by p.i.($\overline{AM}$). If so, it may be a reasonable assumption that the large sized $\overline{M}$s with $C_w > 4$ would have enough time to duplicate their centromeres but not yet in the small sized $\overline{M}$s $C_w < 4$. To confirm these assumptions, I artificially cut the $\overline{M}$s with $C_w \geqslant 4$ at the centromere, and examined the size distribution of T chromosomes resulting from such artificial fission (termed here as T′). If these series of assumptions are correct, the size distribution of T′ chromosomes should coincide with that of Ts actually observed in mammals. As a whole, the result shown in Figure 14 seems to support such a possibility.

Quantitative Analysis of Directionality in Mammalian Karyotype Evolution by the Karyograph Method

Let 2n and 2AN be the diploid chromosome number and the diploid arm number. The arm number is defined by counting one for each $\overline{A}$ and two for each $\overline{M}$ chromosomes. When a diploid karyotype (2K) having $2p\overline{A}$ and $2q\overline{M}$ is formulated as $2K = 2p\overline{A} + 2q\overline{M}$, 2n and 2AN of the karyotype is $2n = 2(p + q)$ and $2AN = 2(p + 2q)$. By these definitions, the karyotype can be represented by the point (2AN, 2n) on a two-dimensional graph of 2n against 2AN, which is named the karyograph [41]. The frequency distribution

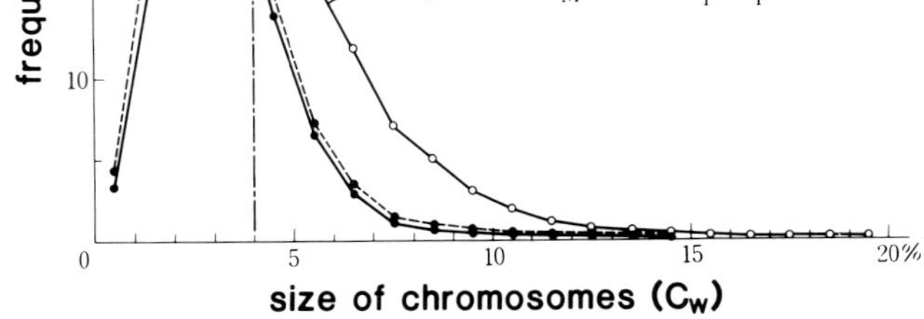

Fig. 14. Size distribution of T, T′, and $\overline{M}$ chromosomes with $C_w \geqslant 4$. T′ means hypothetical telocentrics induced artificially by cutting metacentrics with $C_w \geqslant 4$ at their centromeres (i.e., fission). $\overline{M}$s with $C_w \geqslant 4$ are assumed that they are induced recently by p.i.($\overline{AM}$) and thus they have not been exposed to centric fission. The frequency distribution of T′ coincides with that of T; this was proposed as evidence for the directionality of Robertsonian rearrangement (i.e., c. fission > c. fusion).

of 1,072 mammalian karyotypes and the distribution pattern of each order is represented in Figures 15 and 16, respectively.

The karyotypes can move on the karyograph vertically by centric fusion towards decreasing 2n, whereas the reverse occurs by centric fission. In these cases, 2AN values are conserved. On the other hand, the karyotypes can shift horizontally by p.i.($\overline{AM}$) or p.i. ($\overline{MA}$), i.e., 2AN increases by the former but decreases by the latter. Therefore, the density distribution of karyotypes on the karyograph has information about chromosomal mutations that have accumulated in each karyotype in terms of arm number or chromosome number alterations during the course of evoluton. As already discussed in the previous section, both pericentric inversion and Robertsonian rearrangement are in a *quantitative sense* unidirectional, i.e., p.i.($\overline{AM}$) > p.i.($\overline{MA}$) and centric fission > centric fusion. If so, each karyotype tends to move more easily towards increasing 2n and/or 2AN than the reverse (Fig. 17). Therefore, if we adopt these theoretical analyses (though the methods are not familiar to a majority of cytologists), the karyotype distributions shown in Figures 16 and 17 may be best interpreted by assuming the overall directionality of karyotype evolution as proceeding toward increasing 2n and/or 2AN mainly by centric fission and/or p.i.($\overline{AM}$).

If karyotype evolution started from 2K = 80–96$\bar{A}$ (or K = 40–48$\bar{A}$), as suggested by Wurster and Benirschke or Ohno (Fig. 4), the sequence of

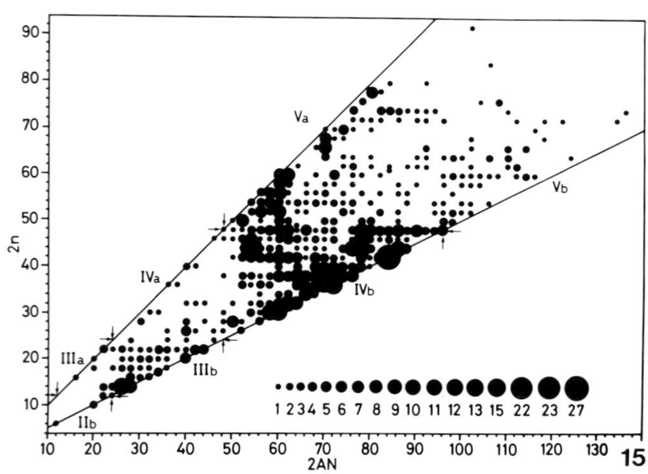

Fig. 15. Distribution of mammalian karyotypes on the karyograph. 2n: Diploid chromosome number. 2AN: Diploid arm number. Arrows and Roman numerals indicate distribution areas of mammalian karyotypes (for details see Fig. 16; from Imai and Crozier [41], with permission of the publisher).

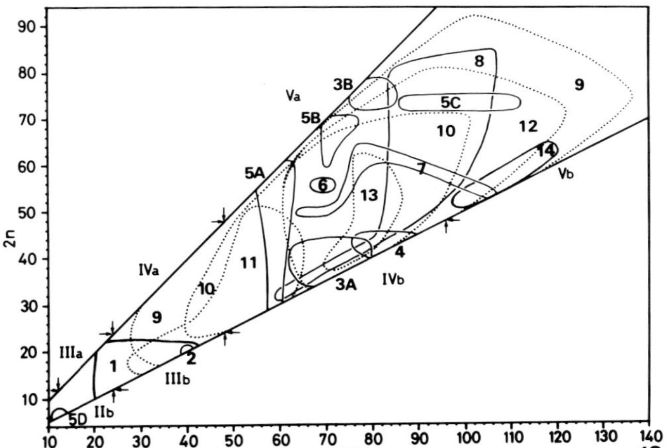

16

Fig. 16. Distribution patterns of mammalian karyotypes at order level. 1. Marsupialia. 2. Tubulidentata. 3A and 3B. Carnivora including Pinnipedia. 4. Cetacea. 5A-5D. Artiodactyla. 6. Proboscidae. 7. Edentata. 8. Perissodactyla. 9. Rodentia. 10. Insectivora. 11. Chiroptera. 12. Primates. 13. Lagomorpha. 14. Monotremata. (From Imai and Crozier [41], with permission of the publisher.)

karyotype evolution expected is 80–96$\overline{A}$—(fusion)→ 40–48$\overline{M}$ —(p.i.($\overline{MA}$))→ 40–48$\overline{A}$ —(fusion)→ 20–24$\overline{M}$—(p.i.($\overline{MA}$))→ 20–24$\overline{A}$ —(fusion)→ 10–12$\overline{M}$ —(p.i.($\overline{MA}$))→ 10–12$\overline{A}$ —(fusion)→ 5–6$\overline{M}$, where "p.i.($\overline{MA}$)" plays a significant role. The same problem also seems to negate the modal hypothesis of Matthey or Ohno (Fig. 4). Of course, I wish to emphasize that centric fusion still has a role as occasional "back eddies" in the main stream of centric fission. However, if this conclusion is correct, a simple question remains, that is, why do the chromosome numbers tend to increase in karyotype evolution? A possible answer for this question is discussed below.

A POSSIBLE BIOLOGICAL SIGNIFICANCE OF CENTRIC FISSION: THE MINIMUM INTERACTION HYPOTHESIS

The occurrence probability and the fixation probability are the two major guiding factors for evaluating chromosomal mutations that have contributed to karyotype evolution. The latter has been well documented in classic cytology by means of natural selection [68,69], but the former has rarely been investigated except the radiation-induced chromosomal mutations with somatic cells [2,34]. I wish to discuss here chromosomal configurations of interphase nuclei affecting the occurrence probability of spontaneous chromosomal mutations.

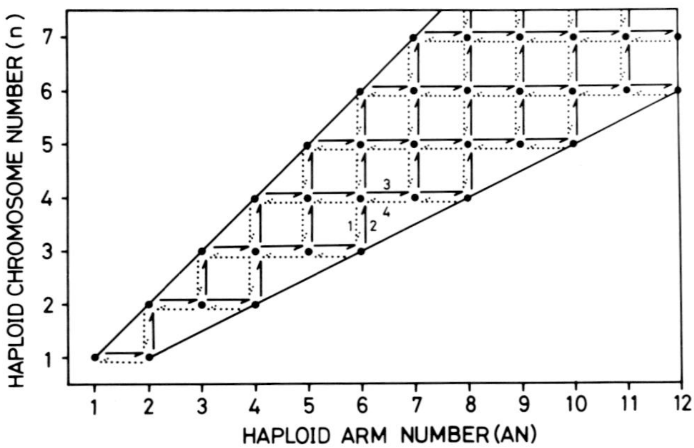

Fig. 17. Overall directionality of karyotype evolution on the karyograph. Mammalian karyotypes tend to evolve, in a statistical sense, towards increasing chromosome number and arm number mainly by the fission cycle shown in Figure 9. 1. Centric fusion. 2. Centric fission. 3. p.i. ($\overline{AM}$). 4. p.i. ($\overline{MA}$). Solid circles are haploid karyotypes plotted on the karyograph. I used 2n and 2AN in Figure 15 to make it possible to represent polymorphic karyotypes having odd 2n or 2AN numbers, but the haploid karyotype is convenient for theoretical analysis of karyotype evolution. The straight line n = AN and n = AN/2 is the borderline of the karyotype distribution due to the acrocentric karyotypes and the metacentric ones.

Theoretically speaking, chromosomal mutations that have contributed to karyotype evolution have to occur in either the germ cells or in early embryonic cells. The mis-resolution of interlocking and crossing-over may be the most plausible mechanism leading spontaneously to DNA strand breaks or exchanges in germ cells or early embryonic cells, because they are universally observed in eukaryotes and occur at much higher frequencies than others (e.g., spontaneous SCEs, transposons, or background radiation). Recent advances in electron microscopy have revealed that chromosomes are associated with their homologs laterally by the synaptonemal complex during zygotene and pachytene [70,71], and that the resulting bivalent is associated with the nuclear membrane at both termini by being fixed to it by the basal knob [19], or by the terminal attachment plaque [20] or telomere. Because of its shape, I term the resulting chromosome configuration at pachytene the "suspension arch structure" (Fig. 18). I examined the relative occurrence rates of reciprocal translocations under the suspension arch structure.

Let the nucleus of primary spermatocytes or primary oocytes be a sphere of radius r (Fig. 19). Let i and j be two randomly chosen chromosomes from a given karyotype, K, with haploid number n, and let C_i and C_j be their respective sizes in terms of a percentage of the combined length of all the

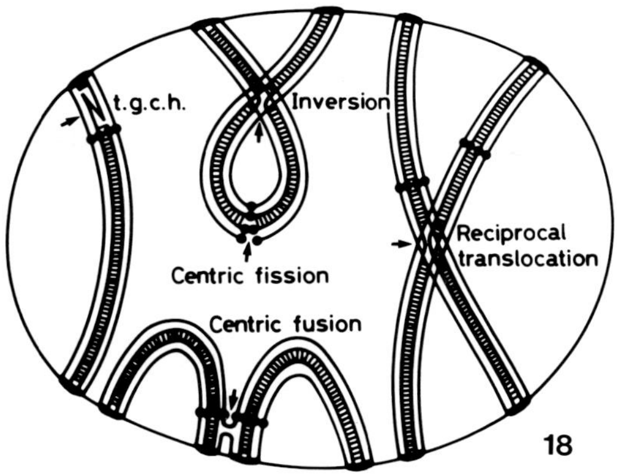

18

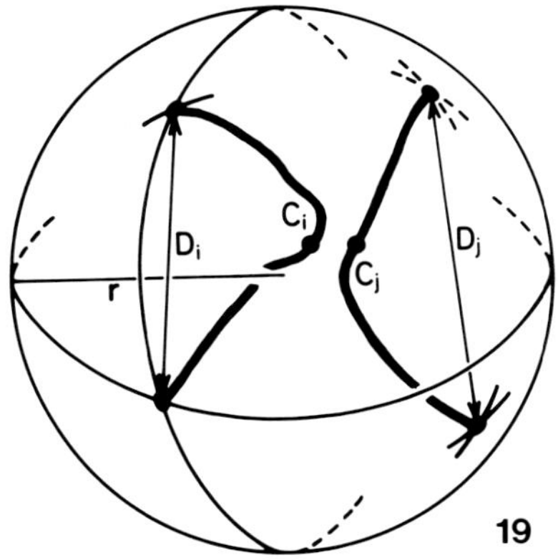

19

Fig. 18. Schematic representation of the suspension arch structure of pachytene chromosomes and chromosomal mutations that are theoretically expected by using the crossing-over mechanism named here "hetero-site crossing-over." Note that terminals of each bivalent are fixed on the nuclear membrane by the so-called telomere, which results in a configuration denoted here as the suspension arch structure, universal in eukaryotes. Short bars denote the synaptonemal complex (SC). The unit chromosome (sister chromatid or monad) comprising each tetrad or bivalent is represented by a solid line. t.g.c.h.: Tandem growth of constitutive heterochromatin in short arms. Arrows indicate the sites of chromosomal mutations. For reciprocal translocations, the mis-resolution of interlocking may be another important mechanism.

Fig. 19. Scheme for estimating the effect of the suspension arch structure affecting the occurrence probability of reciprocal translocation. r, Radius; C_i and C_j, chromosome i and j; D_i and D_j, distance between two terminals in chromosome i and j. The nucleus is schematically represented by a sphere with radius r containing two chromosomes i and j which were selected arbitrarily from a given haploid karyotype and which have their terminals fixed randomly onto the nuclear membrane. Chromosome i and j can move freely, except for their terminals. When we assume that if they contact (interact), they exchange their arms at those points (i.e., reciprocal translocation); the total frequency of interaction of chromosomes i and j estimated with all the combinations of chromosome numbers ($C_i = 1 \ldots n$, $C_j = 1 \ldots n$, $i \neq j$) corresponds to the occurrence probability of reciprocal translocation per cell.

chromosomes in a haploid set. Let the two ends of each chromosome be placed randomly on the inside surface of the sphere which is the interphase nucleus. Now, I ask whether the two chromosomes are close enough to come into physical contact. If they are in contact, a chromosomal exchange leading to a translocation can occur, but an exchange is impossible if they do not touch. This type of approach was already proposed by Revell [33,34] and is known as the exchange theory. This principle for simulation can be extended to all n chromosomes of the haploid set, enabling us to estimate the relative abundance of the number of chromosome pairs in close proximity in a nucleus. This abundance ratio is naturally a function of the nuclear radius, r, and the haploid chromosome number, n. Using five different species having n = 5, 11, 19, 20, and 41, I estimated the relative abundance for various r values (= 3, 5, 10, and 15) (Fig. 20). Inspection of the simulation results reveals two features: 1) When nuclear size (r) is fixed, the frequency of reciprocal translocation decreases as the chromosome number increases, and

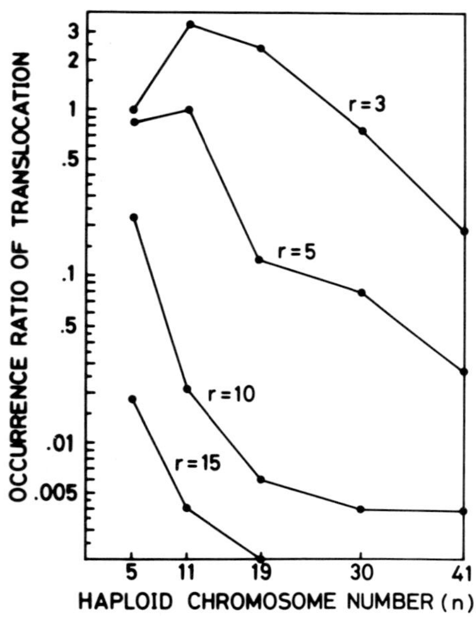

Fig. 20. Effects of nuclear volume and chromosome numbers affecting the occurrence probability of reciprocal translocation. r, Radius of nucleus (arbitrary value; see Fig. 19). The estimates were made using the karyotypes of Australian bulldog ants (*Myrmecia*) having n = 5, 11, 19, 30, and 41 (for details see Imai et al. [66]). The frequency of reciprocal translocation is represented as a proportion of the value for the case r = 3 and n = 5. Reciprocal translocations decrease with increasing nuclear volume or chromosome numbers.

2) when the chromosome number is fixed, the frequency decreases as nuclear size increases.

Based on these theoretical analyses, I propose that a major factor directing karyotype evolution may have been selection for reduced opportunity for spontaneous chromosomal mutations, especially reciprocal translocations. There has been a general tendency in eukaryotes for an increase of nuclear DNA (i.e., genome size) [72]. The increase of genome size has probably been instrumental in allowing various groups the opportunity for evolutionary advance by gene duplication and new gene creation [16], but it also increases the rate of the production of deleterious spontaneous chromosomal mutations such as reciprocal translocations. The risk-avoidance selection seems to have acted through at least five avenues: 1) improved DNA repair, 2) suspension arch structure, 3) increased nuclear volume in oocytes and spermatocytes, 4) increased chromosome contraction by higher-order helix formation of the constituent chromatin, and 5) asynchrony of the DNA replication cycle and chromatin contraction phases of sex chromosomes and autosomes. Chromosome number, if it links up with the suspension arch structure, also affects these probabilities as the sixth factor. The reason is as follows: the mean chromosome size will be larger in low-numbered karyotypes than in higher-numbered ones if the genome size is the same. Therefore, given the suspension arch structure, interaction between non-homologous chromosomes will be fewer in higher- than in lower-numbered karyothes (Fig. 21). This means in other words that high-numbered species are evolutionarily adapted for minimizing the genetic risks resulting from reciprocal translocations. Of course, the only available mechanism for increasing chromosome number is "centric fission." For these reasons, I propose this hypothesis as "the minimum interaction hypothesis" (for details see Imai [65]). The most important supporting evidence for this hypothesis was found in ant chromosomes [66,73]. As is shown in Figure 22 reciprocal translocation polymorphisms occur only in species with low chromosome numbers ($n \leqslant 12$), whereas Robertsonian polymorphisms occur exclusively in higher-numbered species ($n > 12$). This non-random distribution pattern of chromosomal polymorphisms may be best interpreted only under the fission hypothesis and the minimum interaction hypothesis.

MECHANISMS FOR THE HIGH INCIDENCE OF CENTRIC FUSION IN MAN AND OTHER MAMMALS

Centric fusion is an overwhelmingly abundant aberration in man [74], whereas centric fission is negligibly rare except the case by Sinha et al. [48],

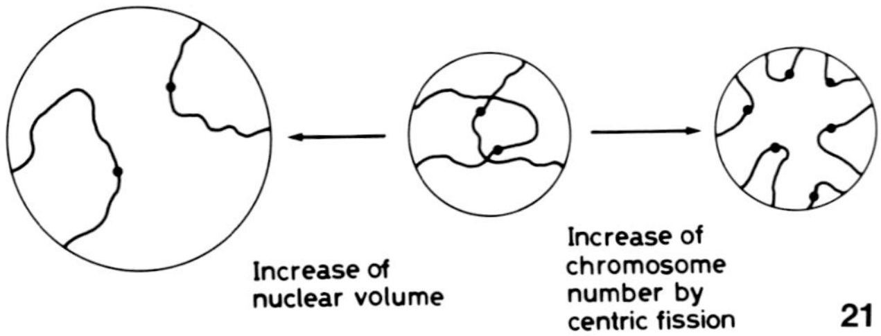

21

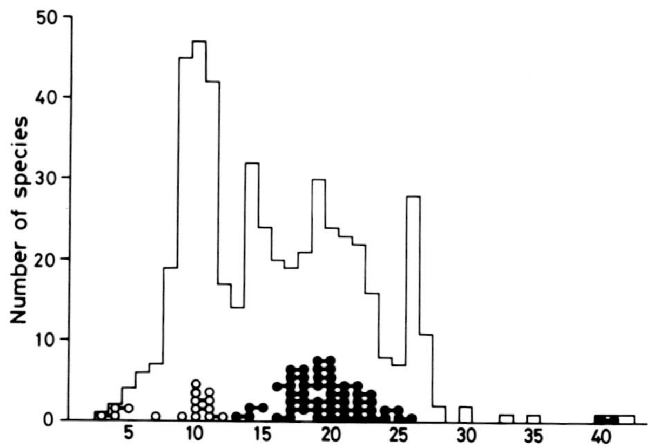

22

Fig. 21. Two possible solutions for reducing the occurrence probability of reciprocal translocation. If the genome size to nuclear volume ratio is high (center), increase of nuclear volume (left) or increase of chromosome number by centric fission (right) is adaptive in reducing the genetic risk due to reciprocal translocations.

Fig. 22. Frequency distribution of haploid chromosome numbers (n) in ants, reciprocal translocation polymorphisms (○), and Robertsonian polymorphisms (●). Chromosome numbers involved in the same polymorphic species are joined by short bars. The fraction with $n \leqslant 12$ is found commonly in hymenopterans other than ants (e.g., sawflies), whereas that with $n > 12$ is characteristic of ants. If reciprocal translocations occur by random breakage and reunion or random contact and exchange, the frequency increases as chromosome numbers increase. On the other hand, the occurrence probability of translocation increases greatly in low-numbered karyotypes under the suspension arch structure (Fig. 20). The non-random distribution of reciprocal translocation polymorphisms and Robertsonian polymorphisms in ants is consistently interpreted only assuming the minimum interaction hypothesis.

Hansen [49], Dallapiccola et al. [50], and Fryns et al. [51]. This indisputable evidence seems at a glance to be highly inconsistent with the fission hypothesis mentioned above. However, it may be too hasty to deny the fission hypothesis itself by such reasoning. The fission hypothesis as mentioned herein is more profound and flexible than the classic fusion hypothesis, because the fission hypothesis provides room for centric fusion as local eddies or a temporal transitional step of karyotype evolution.

According to the minimum interaction hypothesis, the physical proximity of chromosomes is an important factor for inducing chromosomal mutations (Fig. 18). Therefore, if there is a specific chromosome configuration, e.g., a tendency for the short arms of acrocentrics to associate, the occurrence probability of "dicentric fusion" will increase greatly (Fig. 3; also see Daniel and Lam-Po-Tang [22], Daniel [25]). Such a specific configuration is actually found in humans, which is the satellite association of D and G groups [75–77], though these observations were made with metaphase chromosomes of somatic cells. It is more essential for the present discussion that the short arms of D and G groups are associated at pachytene. This peculiar chromosome configuration was observed by three-dimensional reconstructions of meiotic chromosomes [78].

The case of mice may provide a more general model for the high incidence of centric fusion. Mice are characterized by an acrocentric karyotype ($2K = 40\overline{A}$). Their heterochromatic short arm ends and their euchromatic long arm ends are attached to the nuclear membrane at pachytene [19], i.e., forming the suspension arch structure. There is noteworthy evidence that heterochromatic short arms tend to associate with each other at pachytene (Fig. 23). If crossing-over occurs between short arms of non-homologous bivalents, we can expect "dicentric fusion" (Fig. 18). As such crossing-over is unusual in the classic context, I will temporarily refer to it as "hetero-site crossing-over" (for details see Imai et al. [65]).

In acrocentric karyotypes, since fission cannot occur, a high incidence of either centric fusion or pericentric inversion is to be expected.

Once acrocentric karyotypes appear by successive centric fissions, the minimum interaction hypothesis predicts that some acrocentrics will again become metacentrics by dicentric fusions. The occurrence probability of dicentric fusion will increase with increasing size of heterochromatic short arms, though not all acrocentrics will change into metacentrics because of the genetic risk due to such translocations. Indeed, heterozygotes for centric fusions are not always completely fertile, especially if more than two fusions are involved [79]. Some of dicentric $\overline{M}$s will change again into T by centric fission. By these means, "dicentric fusion" and centric fission will occur

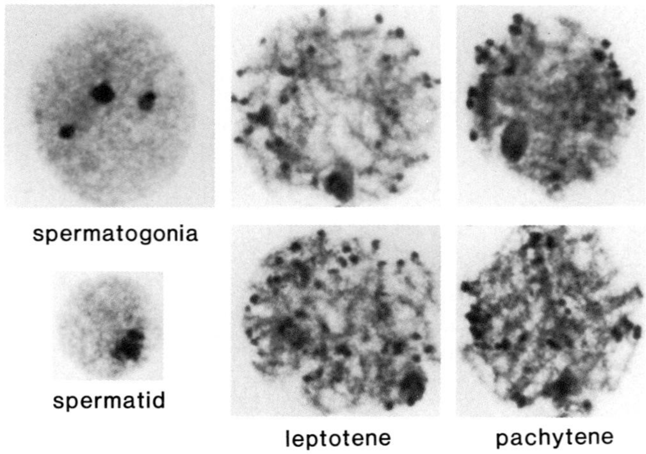

Fig. 23. Association of constitutive heterochromatin in mice spermatogonia, spermatocytes at leptotene or pachytene, and spermatids. Such association will accelerate the occurrence probability of centric fusion or tandem fusion if there is an effective mechanism leading to DNA strand breakages or exchanges.

repeatedly under some conditions, e.g., until acrocentrics change into metacentrics by p.i. ($\overline{AM}$). This phenomenon may be termed "fusion-fission fluctuation" and may be important as a transitional step of karyotype evolution.

Physical proximity of chromosomes is probably a necessary factor for inducing chromosome mutations but not a sufficient one. If we argue that physical proximity alone is important; then spermatogonia cells or spermatids may be the more likely area for centric fusion than pachytene cells, because of tight associations of constitutive heterochromatin (Fig. 23). However, such tight associations will not always induce centric fusions without some effective mechanisms leading to DNA strand breakages or exchanges. In this connection, hetero-site crossing-over and the "mis-resolution of interlocking" (discussed above) may be helpful mechanisms. On the other hand, radiation or chemical mutagens will induce centric fusions in spermatogonia or spermatids.

FINAL COMMENT ON THE FUSION-FISSION CONTROVERSY

In the minimum interaction hypothesis, the modes of karyotype evolution are largely determined by the ratio of genome size to nuclear volume (Fig.

24). If this ratio is low, reciprocal translocations occur rarely, and thus the increase of chromosome number will not always be advantageous for risk-avoidance selection. In such instances, low-numbered karyotypes may be conserved for a long evolutionary time. *Drosophila melanogaster* may be a case in point. There is a general tendency that nuclear volume increases with increasing genome size in eukaryotes [72], which may be important for minimizing the genetic risks of reciprocal translocation. The evidence that amphibians have rather low chromosome numbers in spite of their high cellular DNA content may also be interpreted in light of the large size of their oocyte nuclei.

In contrast, if the ratio is high, increase in chromosome number accompanied with the suspension arch structure may be favored for the risk-avoidance mechanism. The non-random distribution of translocation polymorphisms and Robertsonian polymorphisms found in ants (Fig. 22) may be an example of this phenomenon. The modal chromosome number in mammals (2n = 40–48; Fig. 4) may be best interpreted as a front of waves for increasing chromosome numbers by centric fission. I have concluded that centric fission on the whole predominates over centric fusion in mammals and ants, but there is still room for centric fusion. Consider the case of r = 10 in Figure 20, where the frequency of reciprocal translocations is reduced only very slightly (0.006→0.004) by a great increase of chromosome number (n = 19→ 41). In such a case, increase of chromosome number is not favored strongly by selection for risk-avoidance, and centric fusions may be fixed by chance. As mentioned above, the frequency of centric fusion will

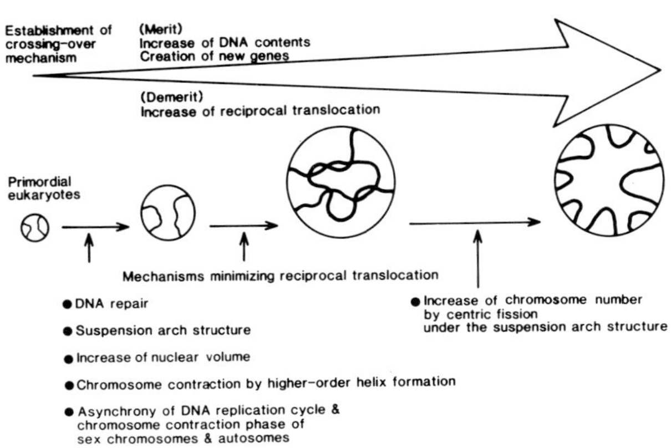

Fig. 24. Schematic representation of the minimum interaction hypothesis.

increase as increasing the size of heterochromatic short arms under the suspension arch structure model (Fig. 18). The clinal chromosome distribution laying the modal point near the highest number found for example, in fish, birds, butterflies, or grasshoppers may be interpreted consistently by the minimum interaction hypothesis as follows: their acrocentric modal karyotypes have evolved ancestrally by a series of centric fissions, by which they could succeed in minimizing reciprocal translocation, and then secondarily some acrocentrics changed into metacentrics by centric fusions. I suggested in the previous section that fission-fusion fluctuation is theoretically possible as a transitional phase or step of karyotype evolution. Dicentric translocation (Fig. 3j) as well as dicentric tandem fusion (Fig. 3k) can avoid the risk of chromosome bridge formation by invoking centromere suppression. However, these mechanisms would contribute little to karyotype evolution, because the minimum interaction hypothesis predicts that karyotype evolution proceeds as a whole toward reducing the occurrence probability of translocations (Fig. 24).

Many modern fusionists seem to accept centric fission itself, though they evaluate it as a minor component. The reason may be that they begin to consider karyotype evolution from acrocentric karyotypes. On the other hand, the fissionists question the origin of the acrocentric karyotypes. Pure experimentalists will say pessimistically that there is no experimental evidence to discuss the origin of acrocentric karyotypes. I wish to answer this criticism with an image of a firework exploding in the night sky. It is easy to trace back to the point of an explosion, but may be difficult to know its trajectory before the explosion. Is this yet reason enough to deny the assumption that the firework was shot from the ground before the explosion in the air?

ACKNOWLEDGMENTS

The Monte Carlo simulations for estimating the occurrence probability of reciprocal translocation were made by Dr. T. Maruyama (National Institute of Genetics, Mishima). I wish to thank to Dr. A. Daniel (Shodair Children's Hospital, USA) for his information on centromeric suppression. I also thank M.W.J. Crosland (The University of New South Wales, Australia) for kindly reading the manuscript. Contribution no. 1696 from the National Institute of Genetics, Mishima, Japan.

REFERENCES

1. Muller HJ: The remaking of chromosomes. Collect Net (Woods Hole) 8:182, 1938.

2. Muller HJ: Bearings of the Drosophila work on systematics. In Huxley J (ed): "The New Systematics." Oxford: Clarendon Press, 1940, p 185.
3. White MJD: "Animal Cytology and Evolution." 2nd ed. London: Cambridge University Press, 1954.
4. White MJD: "Animal Cytology and Evolution." 3rd ed. London: Cambridge University Press, 1973.
5. Imai HT: On the origin of telocentric chromosomes in mammals. J Theor Biol 61:195, 1978.
6. Lima-de-Faria A: The role of the kinetochore in chromosome organization. Hereditas 42:85, 1956.
7. Lima-de-Faria A: "Molecular Evolution and Organization of the Chromosome." Amsterdam: Elsevier, 1983, p 653.
8. John B, Hewitt GM: Karyotype stability and DNA variability in the Acrididae. Chromosoma (Berl) 20:155, 1966.
9. John B, Hewitt GM: Patterns and pathway of chromosome evolution within the Orthoptera. Chromosoma (Berl) 25:40, 1968.
10. Brinkley BR, Valdivia MM, Tousson A, Brenner SL: Compound kinetochores of the Indian muntjac. Evolution by linear fusion of unit kinetochores. Chromosome (Berl) 91:1, 1984.
11. Cavalier-Smith T: Palindromic base sequence and replication of eukaryote chromosome ends. Nature 250:467, 1974.
12. Bateman AJ: Simplification of palindromic telomere theory. Nature 253:379, 1975.
13. Cavalier-Smith T: Cloning chromosome ends. Nature 301:112, 1983.
14. Thomas CA, Pyeritz RE, Wilson DA, Dancis BM, Lee CS, Bick MD, Huang HL, Zimm BH: Cyclodromes and palindromes in chromosomes. Cold Spring Harber Symp Quant Biol 38:353, 1973.
15. Wilson DA, Thomas CA Jr: Palindromes in chromosomes. J Mol Biol 84:115, 1974.
16. Ohno S: "Evolution by Gene Duplication." New York: Springer-Verlag, 1970.
17. Britten RJ, Kohne DE: Implication of repeated nucleotide sequences. In Lima-de-Faria A (ed): "Handbook of Molecular Cytology." Amsterdam: North-Holland, 1969, p 37.
18. Bostock C: Repetitious DNA. In Prescott DM, Goldstein L, McConkey E (eds): "Advances in Cell Biology, Vol. 2." New York: Appleton, 1971, p 153.
19. Solari AJ: The spatial relationship of the X and Y chromosomes during meiotic prophase in mouse spermatocytes. Chromosoma (Berl) 29:217, 1970.
20. Moses MJ: Synaptonemal complex karyotyping in spermatocytes of the Chinese hamster (Cricetulus griseus). Chromosoma (Berl) 60:99, 1977.
21. Niebuhr E, Skovby F: Cytogenetic studies in seven individual with an i(Xq) karyotype. Hereditas 86:121, 1977.
22. Daniel A, Lam-Po-Tang PRL: Structure and inheritance of some heterozygous Robertsonian translocations in man. J Med Genet 13:381, 1976.
23. Pallister PD, Patau K, Inhorn SL, Optiz JM: A woman with congenital abnormalities, mental retardation and mosaicism for an unusual translocation t(6;19). Clin Genet 5:188, 1974.
24. Nakagome Y, Teramura F, Kataoka K, Hosono F: Mental retardation, malformation and partial 7p monosomy 45, XX, t dic(7;15)(p21;p11). Clin Genet 9:621, 1976.
25. Daniel A: Single Cd band in dicentric translocations with one suppressed centromere. Hum Genet 48:85, 1979.
26. Daniel A: Y isochromosomes and rings. In Sandberg AA (ed): "The Y Chromosome, Part B, Clinical Aspects of Y Chromosome Abnormalities." Alan R. Liss, 1985, p 105.

27. Wurster DH, Benirschke K: Indian muntjac, Muntiacus muntjak: A deer with a low diploid chromosome number. Science 168:1364, 1970.
28. Matthey R: The chromosome formulae of eutherian mammals. In Chiarelli AB, Capanna E (eds): "Cytotaxonomy and Vertebrate Evolution." London: Academic Press, 1973, p 531.
29. Shi L, Duax Ye Y: Comparative cytogenetic studies on the red muntjac, Chinese muntjac and their F_1 hybrids. Cytogenet Cell Genet 26:22, 1980.
30. Brinkley BR, Valdivia MM, Tousson A, Brenner SL: Compound kinetochores of the Indian muntjac. Evolution by linear fusion of unit kinetochores. Chromosoma (Berl) 91:1, 1984.
31. Elder FFB: Tandem fusion, centric fusion, and chromosomal evolution in the cotton rats, genus Sigmodon. Cytogenet Cell Genet 26:199, 1980.
32. Yosida TH: "Cytogenetics of the Black Rat. Karyotype Evolution and Species Differentiation." Tokyo: University of Tokyo Press, 1980.
33. Revell SH: The accurate estimation of chromatid breakage, and its relevance to a new interpretation of chromatid aberrations induced by ionizing radiations. Proc Roy Soc London Ser B 150:563, 1959.
34. Revell SH: The breakage-and-reunion theory and the exchange theory for chromosomal aberrations induced by ionizing radiations: A short history. Adv Rad Biol 4:367, 1974.
35. Hsu TC, Benirschke K: "An Atlas of Mammalian Chromosomes. Vol. 1–10". New York: Springer-Verlag, 1967–1977.
36. Ohno S: The mammalian genome in evolution and conservation of the original X-linkage group. In Benirschke K (ed): "Comparative Mammalian Cytogenetics." New York: Springer-Verlag, 1969.
37. Wurster DH, Benirschke K: Comparative cytogenetic studies in the order Carnivora. Chromosoma (Berl) 24:334, 1968.
38. Wurster DH, Benirschke K: Chromosome studies in the subfamily Bovoidea. Chromosoma (Berl) 25:151, 1968.
39. Todd NB: Karyotypic fissioning and canid phylogeny. J Theor Biol 26:445, 1970.
40. Todd NB: Chromosomal mechanisms in the evolution of artiodactyls. Paleobiology 1:175, 1975.
41. Imai HT, Crozier RH: Quantitative analysis of directionality in mammalian karyotype evolution. Am Nat 116:537, 1980.
42. Ohno S: Linkage group conservation and the notion of 24 primordial vertebrate linkage groups. Chromosomes Today 8:268, 1984.
43. Benirschke K: "Comparative Mammalian Cytogenetics." New York: Springer-Verlag, 1969.
44. Chiarelli AB, Capanna E: "Cytotaxonomy and Vertebrate Evolution." London: Academic Press, 1973.
45. Jacobs PA, Melville M, Ratcliffe S: A cytogenetic survey of 11,680 newborn infants. Am Hum Genet 37:359, 1974.
46. Gropp A, Winking H, Zech L, Muller H: Robertsonian chromosomal variation and identification of metacentric chromosomes in feral mice. Chromosoma (Berl) 39:265, 1972.
47. Kato H, Sagai T, Yosida TH: Stable telocentric chromosomes produced by centric fission in Chinese hamster cells in vitro. Chromosoma (Berl) 40:183, 1973.
48. Sinha AK, Pathak S, Nora JJ: A human family suggesting evidence for centric fission and stability of a telocentric chromosome. Hum Hered 22:423, 1972.

49. Hansen S: A case of centric fission in man. Humangenetik 26:257, 1975.
50. Dallapiccola B, Mastroiacovo P, Gandini E: Centric fission of chromosome No. 4 in the mother of two patients with trisomy 4p. Hum Genet 31:121, 1976.
51. Fryns JP, Bulcke J, Hens L, Van den Berghe H: Balanced transmission of centromeric fission products in man. Hum Genet 54:127, 1980
52. Yosida TH, Wada MY: Cytogenetical studies on the Japanese raccoon dog. VI. Distribution of B-chromosomes in 1,372 cells from 13 specimens, with special note on the frequency of the Robertsonian fission. Proc Jpn Acad B 60:301, 1984.
53. Whitehouse DB, Evans EP, Putt W, George AM: Karyotypes of the East African common zebra, Equus burchelli: Centric fission in a pedigree. Cytogenet Cell Genet 38:171, 1984.
54. Hsu TC: "Human and Mammalian Cytogenetics. An historical perspective." New York: Springer-Verlag, 1979.
55. Hayman DL, Martin PG: "Mammalia I: Monotremata and Marsupialia. Animal Cytogenetics 4 Chordata 4." Berlin: Gebruder Borntraeger, 1974.
56. Wurster DH, Snapper JR, Benirschke K: Unusually large sex chromosomes: New methods of measuring and description of karyotypes of six rodents (Myomorpha and Hystricomorpha) and one lagomorph (Ochotonidae). Cytogenetics 10:153, 1971.
57. Bachmann K: Genome size in mammals. Chromosoma (Berl) 37:85, 1972.
58. Manfredi Romanini MG: The DNA nuclear content and the evolution of vertebrates. In Chiarelli AB, Capanna E (eds): "Cytotaxonomy and Vertebrate Evolution." London: Academic Press, 1973.
59. Imai HT: Evidence for non-random localization of the centromere on mammalian chromosomes. J Theor Biol 49:111, 1975.
60. Imai HT: Further evidence and biological significance for non-random localization of the centromere on mammalian chromosomes. J Theor Biol 61:195, 1976.
61. Levan A, Fredga K, Sandverg AA: Nomenclature for centromeric position on chromosomes. Hereditas (Lund) 52:201, 1964.
62. Imai HT: Comparison between the length method and the weight method for formulating the centromeric position of mammalian chromosomes. Ann Rep Natl Inst Genet (Japan) 23:53, 1973.
63. Imai HT: Proposal of a new criterion for the classification of mammalian chromosomes. Ann Rep Natl Inst Genet (Japan) 23:50, 1973.
64. Imai HT, Maruyama T: Karyotype evolution by pericentric inversion as a stochastic process. J Theor Biol 70:253, 1978.
65. Imai HT: Theoretical bases for karyotype evolution. I. The minimum interaction hypothesis. Am Nat 128:900, 1986.
66. Imai HT, Crozier RH, Taylor RW: Karyotype evolution in Australian ants. Chromosoma (Berl) 59:341, 1977.
67. Hirai HY, Sakaguchi Y, Imai HT: C-band polymorphism in a Japanese lung fluke Paragonimus ohirai (Trematoda; Platyhelminthes). Heredity 47:249, 1981.
68. John B, Lewis KR: "The Meiotic System: Protoplasmatologia. Band VI F1." Wien: Springer-Verlag, 1965.
69. Sybenga J: "Meiotic Configuration." Berlin: Springer-Verlag, 1975.
70. Moses MJ: Synaptonemal complex. Ann Rev Genet 2:363, 1968.
71. Wettstein D von, Rasmussen SW, Holm PB: The synaptonemal complex in genetic segregation. Ann Rev Genet 18:331, 1984.
72. Cavalier-Smith T: Cell volume and the evolution of eukaryotic genome size. In Cavalier-Smith T (ed): "The Evolution of Genome Size." Chichester: John Wiley & Sons, 1985.

73. Imai HT, Baroni-Urbani C, Kubota M, Sharma GP, Narashimhanna MN, Das BC, Sharma AK, Sharma A, Deodikar GB, Vaidya VG, Rajasekarasetty MR: Karyological survey of Indian ants. Jpn J Genet 59:1, 1984.
74. Borgaonkar DS, Shaffer R, Reed WC, Jackson LG, Casey T, Levitsky K, Gonzalez I: Repository of chromosomal variants and anomalies in man. 8th Listing. Wilmington, Delaware: Wilmington Medical Center, 1981.
75. Nakagome Y: DNA replication studies of human D-group chromosomes in satellite association. Cytogenetics 8:296, 1969.
76. Nakagome Y: G-group chromosomes in satellite associations. Cytogenet Cell Genet 12:336, 1973.
77. Hemel JO van: ''Studies on Human Chromosome Satellite Association.'' Rotterdam: Bronder-offset N V, 1971.
78. Holm PB, Rasmussen SW: Three-dimensional reconstructions of meiotic chromosomes in human spermatocytes. Chromosomes Today 6:83, 1977.
79. Gropp A, Winking H, Redi C: Consequences of Robertsonian heterozygosity: Segregational impairment of fertility versus male-limited sterility. In Crosignani PG, Rubin BL, Fraccaro M (eds): "Genetic Control of Gamete Production and Function." London: Academic Press, 1982, p 115.
80. Rofe R, Hayman D: G-banding evidence for a conserved complement in the Marsupialia. Cytogenet Cell Genet 39:40, 1985.
81. Baker RJ, Qumsiyeh MB, Hood CS: Role of chromosomal banding patterns in understanding mammalian evolution. In Genoways HH (ed): "Current Mammalogy" Vol. 1. New York: Plenum Press, 1987, p 67.

ADDENDUM

To update the references, I have added two papers by Rofe and Hayman [80] and Baker et al. [81]. The former revealed by G–banding analyses that centric fission has played a dominant role in marsupials. In the latter, the preferential appearance of $\overline{AM}$ inversions (p.i.($A\overline{M}$)) in *Peromyscus* is highly consistent to the fission cycle in the present text (Fig. 9).

The Cytogenetics of Mammalian Autosomal Rearrangements, pages 583–599

Chapter 18
Preferential Paternal Origin of De Novo Structural Chromosome Rearrangements

Susan B. Olson and R. Ellen Magenis

Ninety-six patients were identified who had unique de novo structural chromosome rearrangements. Of these 96, 54 involved chromosomes with heterochromatic regions useful for parental origin studies. Using multiple staining techniques, establishment of origin was attempted in 41 of the 54 cases, and was successful in 32. Twenty-seven of the 32 cases (84.4%) were paternal in origin with a mean paternal age of 27.0 yr. Five cases were maternal in origin with a mean maternal age of 27.4 yr. Reports from the literature support this preferential paternal origin.

There was no evidence for an age-related effect associated with paternal errors. This is in contrast to the findings in cases of de novo trisomies and structural rearrangements involved in non-disjunction or segregation errors where there is a maternal age effect.

Exposure to radiation, drugs, or chemicals could not be demonstrated in this group of patients, although environmental insult to a vulnerably located testis undergoing continuous DNA replication and sperm production is an attractive explanation for the excess of paternally derived de novo rearrangements.

INTRODUCTION

Utilization of chromosome heteromorphisms observed by various staining techniques has allowed determination of the parental origin of different types of chromosome abnormalities. Approximately 80% of trisomy 21 (non-translocated) cases have been shown to be attributable to non-disjunction during maternal meiosis I [1–5]. This is consistent with the demonstrated maternal age effect [6,7]. Kajii and Niikawa [8] and Jacobs and Mayer [9] found that greater than two-thirds of human triploids examined resulted from

Department of Medical Genetics, Crippled Children's Division, and Clinical Cytogenetics Laboratory of University Hospitals, Oregon Health Sciences University, Portland, Oregon 97201 dispermy. Linder et al. [10] demonstrated the parthenogenic origin of an ovarian teratoma using similar marker studies.

Chamberlin and Magenis [11] studied 19 patients with de novo chromosome rearrangements and summarized data from 23 other cases in the literature. Five cases of a supernumerary bisatellited chromosome 15 were shown to be maternal in origin and associated with advanced maternal age. A maternal age effect has also been noted in cases of the cat eye syndrome [12], in which the origin of the extra bisatellited chromosome 22, present in the affected individuals, appears to be similar in mechanism to the bisatellited chromosome 15 [Magenis et al., unpublished data].

Of those de novo rearrangements that were neither Robertsonian translocations or supernumerary chromosomes, Chamberlin and Magenis [11] showed the majority to be paternal in origin, with no obvious age effect. Butler and Palmer [13] and Butler et al. [14] have shown paternal origin of de novo rearrangements in Prader-Willi patients. Using sequential Giemsa/quinacrine banding, they found that in the karyotypes of 13 of 13 patients, the de novo interstitial deletion in the proximal chromosome 15 long arm was paternal in origin. Further evidence for preferential paternal origin is shown in a compilation of data from the literature [15–35] (Table I).

Properties of Heterochromatin

The value of chromosome heteromorphisms in establishing the origin of chromosomes is dependent on their ability to satisfy the criteria for an informative genetic marker system. Through extensive population, family, and twin studies, chromosome heteromorphisms have been shown to be consistent from cell to cell in an individual, and inherited from one generation to the next in a Mendelian fashion [5,36–42]. Apparent inconsistencies in the inheritance pattern have been reported [5,43–45]. However, studies indicating a lack of chiasmata in centromeric regions [46,47] and absence of residual satellite DNA synthesis during pachytene thought to be necessary for recombination [48] suggest that crossing-over and subsequent recombination in heterochromatin is a rare event.

Chromosome Heteromorphism Analysis

Parental origin studies have utilized many staining protocols [49–52]. Table II illustrates five of the more useful staining techniques for demonstrating variant regions of eleven different chromosomes. For each region, there is a continuum of variation. C-banding [51] stains most heterochromatin heavily, the majority of which is seen in the juxtacentromeric regions. Particularly variable are the lengths of the secondary constriction regions of chromosomes 1, 9, 16, and the distal Y long arm.

TABLE I. Cases from the literature of parental origin of de novo chromosome rearrangements

Reference	Chromosome	Maternal age (yr)	Paternal age (yr)	Origin
Johnson et al. [15]	del(1q)	29	30	M
Pai et al. [16]	t(3;7;21;X)	24	30	P
Rosenfeld et al. [17]	dup(3)(q21→26)	42	42	P
Friedrich et al. [18]	del(7)(pter→q32:)	26	26	P
Deroover et al. [19]	del(9p)	21	21	P
Hoo et al. [20]	del(9p)	30	38	P
Ying et al. [21]	del(9)(q22q32)	21	30	M
Ikeuchi et al. [22]	t(13q21q)	23	26	P
Moedjano et al. [23]	t(13q18q)	22	24	M
Jacobs (personal comm.)	13p+	22	27	M
Imaizumi et al. [24]	13p+	33	38	P
Steinbach et al. [25]	r(13)	33	30	M
Steinbach et al. [25]	r(13)	20	23	P
Schwartz et al. [26]	psu dic(13)(q34q34)/ psu dic (13)(q22q11)	28	28	P
Kondo et al. [27]	del(13)(q14.1)	34	35	P
De los Cobos et al. [28]	t(13q15q)	22	—	M
Jacobs (personal comm.)	del(14q)	24	24	P
Hreidarsson et al. [29]	del(14)(q32.3)	35	35	M
Butler and Palmer [13]	del(15q)	27	27	P
Butler and Palmer [13]	del(15q)	34	36	P
Butler and Palmer [13]	del(15q)	23	22	P
Butler and Palmer [13]	del(15q)	26	26	P
Butler and Palmer [13]	del(15q)	30	39	P
Butler and Palmer [13]	del(15q)	21	23	P
Butler and Palmer [13]	del(15q)	40	40	P
Butler and Palmer [13]	del(15q)	21	22	P
Butler and Palmer [13]	del(15q)	21	36	P
Butler and Palmer [13]	del(15q)	31	32	P
Butler and Palmer [13]	del(15q)	21	24	P
Butler et al. [14]	del(15q)	—	—	P
Butler et al. [14]	del(15q)	—	—	P
Hoo et al. [30]	del(16q)	22	25	M
Hoo et al. [30]	del(16q)	27	30	M
Fryns et al. [31]	del(16)(q21)	17	17	P
Docherty et al. [32]	dir dup(17)(p12.00→q11.1)	—	—	P
Hecht (personal comm.)	t(18;21)	37	42	P
Reynolds et al. [33]	del(21)(q11.2q22.1)	39	44	P
Fowler et al. [34]	r(22)	27	30	M
Hamers et al. [35]	r(22)	—	—	P
Total				11M:28P

P, paternal; M, maternal.

TABLE II. Chromosome heteromorphisms observed using five banding techniques

	Banding																								
Chromosome	Q (49)											R [50]					C [51] G-11 [52]				DAPI [50]				
region — Chromosome	1	3	4	9	13	14	15	16	21	22	Y	13	14	15	21	22	1	9	16	Y	1	9	15	16	Y
Satellite																									
Size					+	+	+		+	+															
Shape					+	+	+		+	+															
Intensity					+	+	+		+	+															
Stalk																									
Size					o	o	o		o	o		+	+	+	+	+									
Intensity												+	+	+	+	+									
Short arm																									
Size					+	+	+		+	+													+		
Intensity					+	+	+		+	+													+		
Juxtacentromeric region																									
Size	o	+	+	o	+	+	+	o	+	+							+	+	+	+	+	+		+	+
Intensity		+	+		+	+	+		+	+															
Long arm																									
Size											+														

+, Stain reveals variability in this region; o, variants demarcated by absence of stain.

The Giemsa 11 (G-11) technique [52] stains a subset of the C-band region on chromosomes 1, 7, 9, 10, 13, 14, 15, 17, 20, 21, 22, Y, and sometimes 3, 4, and 5. There is not a correlation between size or presence of a G-11 band and the size of the C-band [53].

A similar pattern to that of C-banding can be achieved by first staining with the non-fluorescent antibiotic distamycin (DA) followed by the fluorochrome 4′-6-diamidino-2-phenylindole (DAPI) [54]. In addition, this staining procedure is specific among the acrocentrics for the chromosome 15 short arm. By staining the chromosomes with chromomycin A3 prior to the DA/DAPI treatment and excitation at a different wavelength (440–480 nm) than that used for viewing the DA/DAPI bands (360–400 nm), R-banding can be viewed on the same metaphases. This facilitates identification of the DA/DAPI-banded chromosomes, and provides information on length of stalk regions [50].

Quinacrine (Q-) banding [49] is the most useful single stain for visualizing human chromosome variation. Regions which are delineated by Q-banding are the short arm and satellites of the acrocentric chromosomes, the juxtacentromeric regions of chromosomes 3 and 4, and the distal Y long arm. Length of secondary constrictions in chromosomes 1, 9, and 16 and stalks on acrocentric chromosomes are demarcated by the lack of fluorescence. Visualization of certain variants is facilitated by a darkroom technique called serial printing, which is simply a range of printing exposures from very short (producing bright chromosomes in which the banding is obliterated, but pale variants are enhanced) to very long (producing pale chromosomes against which varying intensities of bright variants may be evaluated) [55]. Serial printing allows detection of variants not visible at routine exposures. In a study on amount of variation observed in a population of 39 unrelated people, serial printing revealed satellites on 6% of the acrocentric homologs that would otherwise have gone undetected [45].

MATERIALS AND METHODS

Chromosome studies were performed on 96 patients with unique de novo chromosome rearrangements and on their chromosomally normal parents. Fifty-four of these 96 involved chromosomes with potentially informative regions of heterochromatin. Parental origin studies were attempted in 41 of the 54 cases. Informative variant regions represented in the 32 cases in which determination of origin was possible were found in chromosomes 1 (two cases), 3 (three cases), 4 (seven cases), 9 (one case), 13 (seven cases), 15 (six cases), 16 (three cases), 21 (four cases), and 22 (four cases) (Table III).

Chromosomes were prepared from peripheral blood lymphocytes [56] or amniotic fluid cells [57]. Preparations were stained according to a protocol appropriate for the particular chromosomes involved in the abnormality. In some cases more than one chromosome with an informative variant region was involved in the rearrangement. Fifteen to twenty metaphases from each

TABLE III. Our cases of parental origin of de novo chromosome rearrangements

Case	Chromosome	Maternal age (yr)	Paternal age (yr)	Origin
1	1q+	24	24	P
2	t(1;21)(p31;q22)	40	43	P
3	t(2;4)(q31;q25)	27	29	P
4[a]	3p−	23	28	M
5	t(3;13)(q25.1;p13)	23	25	M
6	t(3;4;16)(p24.2;p15.2;q22)	22	24	P
7	4p+	31	32	P
8	del(4)(p16.2)	22	26	P
9	del(4)(q31.3)	26	39	P
10	del(4)(q12q21)	26	31	M
11	del(4q)	35	39	P
12	t(6;22)	—	—	P
13	t(9;12)(p24q11)	25	28	P
14[a]	t(10;13)	19	26	P
15	t(11;22)(q22.1;p13)	32	37	M
16[a]	t(12;15;21)	24	24	P
17	del(13)(q32)	28	29	P
18	del(13q)	19	19	P
19	t(13;15)q22;q25.2)	27	28	P
20[a]	t(13;18)/t(13;13)	17	17	P
21[a]	r(13)	25	23	P
22	del(15)(q12q14)	29	28	P
23	del(15q)	24	27	P
24[a]	15q+	21	23	P
25[a]	t(15;17)	27	28	P
26	del(16q)	33	34	M
27	del(16q)	21	21	P
28[a]	del(21q)	26	29	P
29[a]	21q+	—	—	P
30	dir dup(22)(q11.1→q11.2)	22	26	P
31	r(22)(p11;q13.3)	24	23	P
32[a]	t(X;21)	17	21	P
Total				5M:27P

P, paternal; M, maternal.
[a]Previously published, Chamberlin and Magenis [11].

person were photographed on a Zeiss photoscope III. Composites were constructed of the chromosomes involved in the abnormality from the proband and both parents. For those cases involving acrocentric chromosomes, serial printing was done [55].

Parental origin of the abnormal chromosome was established in one of the following ways: 1) The variant region of the abnormal chromosome was compared to the corresponding region of the normal parental chromosomes. 2) The variant region of the normal homolog of the aberrant chromosome in the proband was compared to the corresponding region of the normal parental chromosomes in cases where the heterochromatin was either deleted or altered due to the chromosome abnormality. By determining the origin of the normal homolog, parental origin of the de novo rearrangment was then established by exclusion.

RESULTS

Establishment of parental origin was successful for 32 of the 41 cases of de novo rearrangments. The results are summarized in Table III. Twenty-seven of the 32 (84.4%) were paternal in origin. The mean paternal age for these cases was 27.0 yr. For the five cases of maternal origin, the mean age was 27.4 yr. Three examples are shown of the clarity of interpretation possible in these cases (Figs. 1–3). Figure 1 illustrates the use of C-banding in demonstrating the paternal origin of the derivative 9 in a de novo t(9;12) (9qter→9p24::12q11→12qter;12pter →12q11::9p24→9pter) from a man whose wife has had multiple miscarriages. Figure 2 shows inheritance of a der(13), t(3;13) (3pter→3q25.1::13p13→13pter; 13qter→13p13::3q25.1→3qter). Figure 3 shows a chromosome 4 with extra material of unknown origin on the short arm. These Q-banded preparations (Figs. 2,3) are from two unrelated children with multiple dysmorphic features and their normal parents.

DISCUSSION

Our data on unselected de novo cases indicate a strong bias toward paternal origin. Table I lists results of published investigations of parental origin of de novo rearrangements. All data on origin and age of parent contributing the abnormal chromosome are summarized in Table IV.

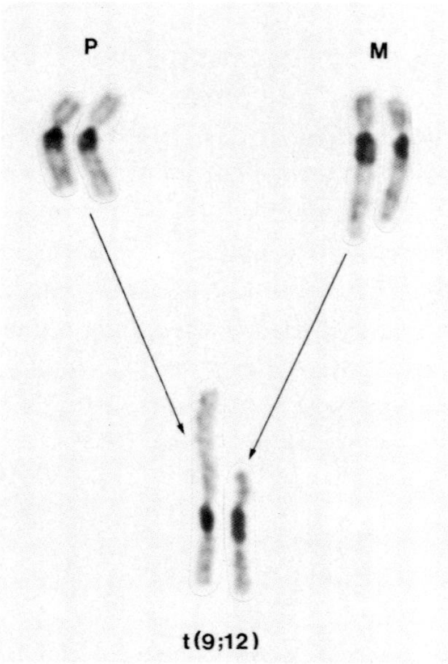

Fig. 1. C-banded paternal (P) and maternal (M) chromosomes 9, and the der(9) (left) and normal 9 (right) from balanced carrier of a de novo t(9p;12q). The normal 9 with the large block of heterochromatin is maternal in origin. Therefore, the abnormal 9 was the result of a paternal error.

In addition to cytological staining methods, parental origin has been possible for some de novo deletions by demonstration of the loss of alleles known to be in the deleted region. Using two-dimensional gel electrophoresis, we were able to show a missing paternal ESD gene to support the fluorescent variant results indicating the paternal origin of a 13q deletion (see case #18, Table III). Kondo et al. [27] reported similar deletion of the paternal ESD gene in a retinoblastoma patient with a de novo deletion of 13q.

Pai et al. [16] determined the origin of part of a complex translocation involving chromosomes 3, 7, 21, and X. Variant regions of the 3 and 21 were not informative with G-banding. They were able to show that the X chromosome involved in the translocation was always early replicating in lymphocyte cultures and, thus, active. Examination of G-6-PD phenotypes showed the mother to have the B variant and the father the A variant. The proband expressed only the A allele in red and white blood cells and fibroblasts rather than the predicted A/B. This lack of maternal G-6-PD expression combined with the X-inactivation results indicated that at least the

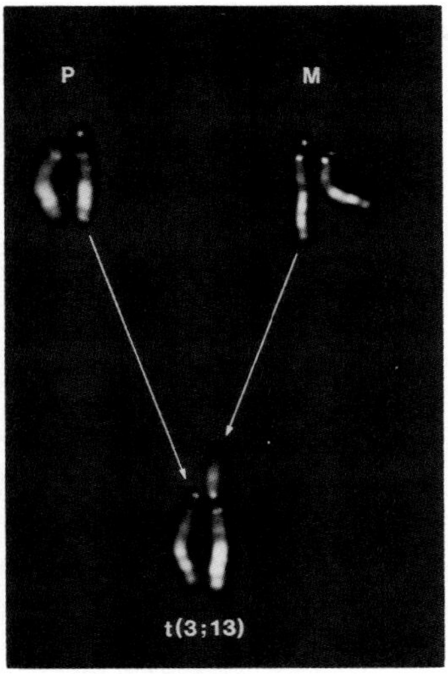

Fig. 2. Q-banded paternal (P) and maternal (M) chromosomes 13, and the normal 13 (left) and der(13) (right) from unbalanced offspring. The abnormal 13 with the brilliant short arm is maternal in origin.

portion of the rearrangement involving the X chromosome was paternal in origin.

Molecular analysis using multiple chromosome-specific probes is proving to be a valuable approach to the establishment of parental origin. This will potentially allow the origin of any chromosome to be determined even if it does not possess a cytologically observed variable region of heterochromatin. Stewart et al. [58] and Hassold [59] have demonstrated the usefulness of restriction fragment length polymorphisms (RFLPs) in determining the origin of the extra chromosome in Down syndrome. Hassold et al. [60] have used similar methods to confirm the previous conclusions from Xg blood group studies that the 45,X conceptus is most often the result of a missing paternal sex chromosome.

Using several X-linked RFLPs, Page et al. [61] examined the origin of the X chromosomes in XX males. Their data in most cases, indicating that one of the X chromosomes was paternal in origin, supported the X-Y interchange hypothesis of Ferguson-Smith [62] and the cytological evidence for presence of Y material on the X short arm presented by Magenis et al. [63]. Presence of Y-specific sequences in XX males has been further confirmed using DNA probes [64–66].

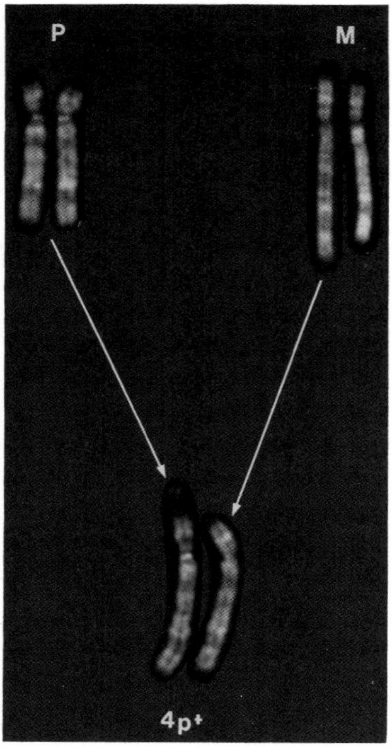

Fig. 3. Q-banded paternal (P) and maternal (M) chromosomes 4, and the der(4) (left) and normal 4 (right) from unbalanced offspring. The abnormal 4 with the bright juxtacentromeric region is paternal in origin.

TABLE IV. Summary of data

Total cases	Average maternal age (yr)	Average paternal age (yr)	Origin
Our cases			
32	27.4	27.0	5M:27P
Literature			
39	26.0	30.4	11M:28P
Combined			
71	26.7	28.7	16M:55P

M, maternal; P, paternal.

Studies were conducted on a patient with tyrosinemia, multiple dysmorphic features, and a partial deletion of 16q (see Table III, case #27). Paternal origin of the deletion was established from the loss of the paternal TAT and alpha haptoglobin genes shown by molecular hybridization techniques [67, 68].

The association between advanced maternal age and Down syndrome has been well documented long before the chromosomal defect was identified [6, 69–71]. A maternal age effect has also been implicated in other trisomies [72–75] and cases involving rearrangements associated with non-disjunction or segregation errors [11,76]. Possible relationships between advanced paternal age and de novo chromosome abnormalities have been investigated. Although 20–25 % of Trisomy 21 cases are due to a paternal error, there has been no evidence for a paternal age effect [6,77,78]. Neither has there been demonstrated a paternal age effect in the obligatory paternal error leading to the 47,XYY individual [73].

Martin and Rademaker [79] have recently reported a lack of correlation between paternal age and aneuploidy in the sperm of a group of 30 normal males studied using the human/hamster in vitro fertilization system for chromosome analysis. They did, however, demonstrate an apparent paternal age effect related to structural chromosome abnormalities. Our present data, as well as our previously published cases of obligate paternal de novo Y-chromosome rearrangements [64], have not supported a paternal age effect.

Warburton [80] reports the frequencies of de novo rearrangements in a newborn population of 59,452 to be 0.037 and 0.017 for balanced and unbalanced, respectively. Although she did not separate those involving autosomes from the sex chromosome abnormalities, in our experience, approximately half of the de novo rearrangements involve the Y-chromosome, an obligate paternal error. Different mechanisms are required to explain the formation of the dicentric Y, non-fluorescent Y, Y/autosome translocation, deleted Y, and the X-Y interchange involved in the origin of XX males. Postulated mechanisms have been reviewed in detail by Magenis et al. [64].

An attractive explanation for the preferential involvement of the paternal homolog in de novo rearrangements relates to differences between male and female gametogenesis. Females are born with all oocytes present in the ovary and arrested in the dictyotene stage of meiosis I. Males, on the other hand, begin sperm production at puberty, undergoing continued proliferation throughout adult life. The continual DNA replication and unprotected location of the testes may lend a vulnerability to environmental insult, with increased chromosome breakage and rearrangement.

Hook et al. [81] summarized data on structural chromosome rearrangements from 27,225 amniotic fluid studies reported in the New York State

Chromosome Registry and the United States Interregional Chromosome Register System. In comparing those studied due to mutagen exposure with those undergoing amniocentesis for some other reason, there was a significantly higher rate of non-familial structural rearrangements in the fetuses of the former. One case involved an inversion in the fetus of a woman whose husband had taken multiple licit and illicit drugs. The father of another fetus had been irradiated in the course of treatment for Hodgkin disease. Origin studies, however, were not reported for these cases. Caution must be exercised in drawing conclusions without such data. The father in our case #10 (Table III) had been treated with Dilantin, but the deleted chromosome was maternal in origin.

Martin et al. [82] studied 13 cancer patients before, and at regular intervals following, testicular radiation therapy. They examined chromosome preparations from the sperm of these men after penetration of zona-free hamster ova. There was a significantly higher incidence of both numerical and structural chromosome aberrations in the cancer group (20% at 36 months post-therapy) than in the non-irradiated controls (8.5%).

Information on suspected exposure to drugs, chemicals, and radiation is often difficult to retrieve from patients. Although we have no knowledge of paternal exposure in our group of 32 de novo rearrangements, completeness of family history is less than satisfactory for drawing conclusions. In general, better documentation of pre-pregnancy events is needed for both parents.

Our data and that of others support the preferential paternal origin of de novo rearrangements. A systematic investigation is now needed to provide a better understanding of the effect of environmental agents on the incidence of mutation and chromosome abnormalities, as well as to lend insight into mechanisms involved in breakage, reunion, and stability in rearrangements and deletions.

ACKNOWLEDGMENTS

We thank Douglas Hepburn, Catherine Olson, Marvin Yoshitomi, Michael Brown, Leslye Smith, and Elizabeth Clark from the Clinical Cytogenetics Laboratory who did some of the initial studies on the probands. We are also grateful to Jan Chamberlin, Lee Allen, Timothy Donlon, Diane Tomar, Sande Barton, Diane Waggoner, SuEllen Toth-Fejel, and Robert Sheehy for contributing to the origin data.

This work was supported by grants from NICHD (HD–07997) and Maternal and Child Health Services 920.

REFERENCES

1. Robinson JA: Origin of extra chromosome in trisomy 21. Lancet 1:131–133, 1973.
2. Schmidt R, Dar H, Nitowsky HM: Origin of extra 21 chromosome in patients with Down syndrome. Pediatr Res 9:318, 1975.
3. Magenis RE, Overton KM, Chamberlin J, Brady T, Lovrien E: Parental origin of the extra chromosome in Down syndrome. Hum Genet 37:7–16, 1977.
4. Mikkelsen M, Poulsen H, Grinsted J, Lange A: Non-disjunction in trisomy 21: Study of chromosomal heteromorphisms in 110 families. Ann Hum Genet 44:17–28, 1980.
5. Magenis RE, Chamberlin J: Parental origin of nondisjunction. In de la Cruz FF, Gerald PS (eds): "Trisomy 21 (Down Syndrome): Research Perspectives." Baltimore: University Park Press, 1981, pp 77–93.
6. Penrose LS: The relative effects of paternal and maternal age in mongolism. J Genet 27:219–224, 1933.
7. Hook EB: Down syndrome: Frequency in human populations and factors pertinent to variation in rates. In de la Cruz FF, Gerald PS (eds): "Trisomy 21 (Down Syndrome): Research Perspectives." Baltimore: University Park Press, 1981, pp 3–67.
8. Kajii T, Niikawa N: Origin of triploidy and tetraploidy in man: 11 cases with chromosome markers. Cytogenet Cell Genet 18:109–125, 1977.
9. Jacobs PA, Mayer M: The origin of human trisomy: A study of hetermorphisms and satellite associations. Ann Hum Genet 45:357–365, 1981.
10. Linder D, Kaiser McCaw B, Hecht F: Parthenogenic origin of benign ovarian teratomas. N Engl J Med 292:63–66, 1975.
11. Chamberlin J, Magenis RE: Parental origin of de novo chromosome rearrangements. Hum Genet 53:343–347, 1980.
12. Schinzel A, Schmit W, Fraccaro M, Tiepolo L, Zuffardi O, Opitz JM, Lindsten J, Zetterquist P, Enell H, Baccichetti C, Tenconi R, Pagon R: The "cat-eye-syndrome": Dicentric small marker chromosome probably derived from a no. 22 (tetrasomy 22 pter→q11) associated with a characteristic phenotype. Hum Genet 57:148–158, 1981.
13. Butler MG, Palmer CG: Parental origin of chromosome 15 deletion in Prader-Willi syndrome. Lancet 1:1285–1286, 1983.
14. Butler MG, Meany FJ, Palmer CG: Clinical and cytogenetic survey of 39 individuals with Prader-Labhart-Willi syndrome. Am J Med Genet 23:793–809, 1986.
15. Johnson VP, Heck LJ, Carter GA, Flom JO: Deletion of the distal long arm of chromosome 1: A definable syndrome. Am J Med Genet 22:685–694, 1985.
16. Pai GS, Thomas GH, Mahoney W, Migeon B: Complex chromosome rearrangements. Clin Genet 18:436–444, 1980.
17. Rosenfeld W, Verma RS, Jhaveri RC, Estrada R, Evans H, Dosik H: Duplication 3q: Severe manifestations in an infant with duplication of a short segment of 3q. Am J Med Genet 10:187–192, 1981.
18. Friedrich U, Osterballe O, Stengjerg S, Jorgensen J: A girl with karyotype 46,XX,del(7)(pter→q32:). Hum Genet 51:231–235, 1979.
19. Deroover J, Fryns JP, Parloir C, Haegeman J, van den Berghe H: Partial monosomy of the short arm of chromosome 9. Hum Genet 44:197–200, 1978.
20. Hoo JJ, Parslow MI, Shaw RL, Veale AMO: Complex de novo rearrangement of chromosome 9 with clinical features of monosomy 9p syndrome. Clin Genet 16:151–155, 1979.

21. Ying KL, Curry CJR, Rajani KB, Kassel SH, Sparkes RS: De novo interstitial deletion in the long arm of chromosome 9: A new chromosome syndrome. J Med Genet 19:68–76, 1982.
22. Ikeuchi T, Kondo I, Sasaki M, Kaneko Y, Kodama S, Hattori T: Unbalanced 13q/21q translocation: A revised study of the case previously reported as 21-monosomy. Hum Genet 33:327–330, 1976.
23. Moedjono SJ, Funderburk SJ, Sparkes RS: 18p− syndrome resulting from translocation (13q;18q) in a mildly affected adult male. J Med Genet 16:399–402, 1979.
24. Imaizumi K, Kajii T, Niikawa N: 13s+. Giant satellites or de novo rearrangement? Hum Genet 59:266–268, 1981.
25. Steinbach P, Drews K, Horstmann W, Barbi G, Scholz W: Two cases of ring chromosome 13. Ann Genet 24:152–157, 1981.
26. Schwartz S, Palmer CG, Weaver DD, Priest J: Dicentric chromosome 13 and centromere inactivation. Hum Genet 63:332–337, 1983.
27. Kondo I, Shin K, Honmura S, Nakajima H, Yamamura E, Satoh H, Terauchi M, Usuki Y, Takita H, Hamaguchi H: A case report of a patient with retinoblastoma and chromosome 13q deletion: Assignment of a new gene (gene for LCP1) on human chromosome 13. Hum Genet 71:263–266, 1985.
28. De Los Cobos LF: Origin of a paternal (13q:15q) translocation leading to dup(13q) in two half sibs. Am J Med Genet 14:617–623, 1983.
29. Hreidarsson SJ, Stamberg J: Distal monosomy 14 not associated with ring formation. J Med Genet 19:147–149, 1983.
30. Hoo JJ, Lowry RB, Lin CC, Haslam RHA: Recurrent de novo interstitial deletion of 16q in two mentally retarded sisters. Clin Genet 27:420–425, 1985.
31. Fryns JP, Bande-Knops J, Van den Berghe H: Partial monosomy of the long arm of chromosome 16: A distinct clinical entity? Hum Genet 46:115–120, 1979.
32. Docherty Z, Hultén MA, Honeyman MM: De ovo tandem duplication 17p11→cen. J Med Genet 19:138–142, 1983.
33. Reynolds JF, Wyandt HE, Kelly TE: De novo 21q interstitial deletion in a retarded boy with ulno-fibular dysostosis. Am J Med Genet 20:173–180, 1985.
34. Fowler G, Kaiser-McCaw B, Hecht T: The use of sequential silver and quinacrine staining to determine the parental origin and breakpoints of a ring-22 human chromosome. Clin Genet 18:274–279, 1980.
35. Hamers AJH, Klep-de-Pater JM, de France HF: Partial duplication in a ring chromosome 22. Ann Genet 26:34–37, 1983.
36. Madan K, Bobrow M: Structural variation in chromosome no. 9. Ann Genet 17:81–86, 1974.
37. Magenis RE, Chamberlin J, Overton K: Sequential Q- and C-band variants: Inheritance in four generations of a family. 13th Annual Somatic Cell Genetics Conference. US Virgin Islands, 1974.
38. McKenzie WH, Lubs HA: Human Q and C chromosomal variations: Distribution and incidence. Cytogenet Cell Genet 14:97–115, 1975.
39. Müller HJ, Klinger HP, Glasser M: Chromosome polymorphism in a human newborn population. II. Potentials of polymorphic chromosome variants for characterizing the ideogram of an individual. Cytogenet Cell Genet 15:239–255, 1975.
40. Magenis E, Palmer CG, Wang L, Brown M, Chamberlin J, Parks M. Merritt AD, Rivas ML, Yu PL: Heritability of chromosome banding variants. In Hook EB, Porter IH (eds): “Population Cytogenetics.” New York: Academic Press, 1977, pp 179–188.

41. Van Dyke DL, Palmer CG, Nance WE, Yu PL: Chromosome polymorphism and twin zygosity. Am J Hum Genet 29:431–447, 1977.
42. McCracken AA, Daly PA, Zolnick MR, Clark AM: Twins and Q-banded chromosome polymorphisms. Hum Genet 45:253–258, 1978.
43. Fitzgerald PH: The nature and inheritance of an elongated secondary constriction on chromosome 9 of man. Cytogenet Cell Genet 12:404–413, 1973.
44. Robinson JA, Buckton KE, Spowart G, Newton M, Jacobs P, Evans HJ, Hill R: The segregation of human chromosome polymorphisms. Ann Hum Genet 40:113–121, 1976.
45. Olson SB, Magenis RE, Lovrien EW: Human chromosome variation: The discriminatory power of Q band heteromorphism (variant) analysis in distinguishing between individuals with specific application to cases of questionable paternity. Am J Hum Genet 38:235–252, 1986.
46. Hultén M: Chiasma distribution at diakinesis in the normal human male. Hereditas 76:55–78, 1974.
47. Chandley AC, Seuanez H, Fletcher JM: Meiotic behavior of five human reciprocal translocations. Cytogenet Cell Genet 17:98–111, 1976.
48. Hotta Y, Stern H: Absence of satellite DNA synthesis during meiotic prophase in mouse and human spermatocytes. Chromosoma 69:323, 1978.
49. Caspersson T, Lomakka G, Zech L: The 24 fluorescence patterns of human metaphase chromosomes—distinguishing characters and variability. Hereditas 67:89–102, 1971.
50. Schweizer D: Simultaneous fluorescent staining of R bands and specific heterochromatic regions (DA-DAPI bands) in human chromosomes. Cytogenet Cell Genet 27:190–193, 1980.
51. Arrighi FE, Hsu TC: Localization of heterochromatin in human chromosomes. Cytogenetics 10:81–86, 1971.
52. Bobrow M, Madan K, Pearson PL: Staining of some specific regions of human chromosomes, particularly the secondary constriction of no. 9. Nature New Biol 238:122–124, 1972.
53. Magenis RE, Donlon TA, Wyandt HE: Giemsa-11 staining of chromosome 1: A newly described heteromorphism. Science 202:64–65, 1978.
54. Schweizer D, Ambros P, Andrle M: Modification of DAPI banding on human chromosomes by prestaining with a DNA binding oligopeptide antibiotic, distamycin A. Exp Cell Res 111:327–332, 1978.
55. Overton KM, Magenis RE, Brady T, Chamberlin J, Parks M: Cytogenetic darkroom magic: Now you see them, now you don't. Am J Hum Genet 28:417–419, 1976.
56. Moorhead PS, Nowell PC, Mellman WJ, Battips DM, Hungerford DA: Chromosome preparations of leukocytes cultured from human peripheral blood. Exp Cell Res 20:613, 1960.
57. Hoehn H, Bryant EM, Karp LE, Martin GM: Cultivated cells from diagnostic amniocentesis in second trimester pregnancies. I. Clonal morphology and growth potential. Pediatr Res 8:746–754.
58. Stewart GD, Harris P, Galt J, Ferguson-Smith MA: Cloned DNA probes regionally mapped to human chromosome 21 and their use in determining the origin of nondisjunction. Nucl Acids Res 13:4125–4132, 1982.
59. Hassold TJ: The origin of aneuploidy in humans. In Dellarco VL, Voytek PE, Hollaender H (eds): "Aneuploidy: Etiology and Mechanisms." New York: Plenum Press, 1985, pp 103–115.

60. Hassold T, Kumlin E, Takaesu N, Leppert M: Determination of the parental origin of sex-chromosome monosomy using restriction fragment length polymorphisms. Am J Hum Genet 37:965–972, 1985.
61. Page DC, de la Chapelle A, Weissenbach J: Chromosome Y-specific DNA in related human XX males. Nature 315:224–226, 1985.
62. Ferguson-Smith MA: X-Y chromosomal interchange in the aetiology of true hermaphroditism and of XX Klinefelter's syndrome. Lancet ii 475–476, 1966.
63. Magenis RE, Webb MJ, McKean RS, Tomar D, Allen LJ, Kammer H, Van Dyke DL, Lovrien E: Translocation (X;Y) (p22.33;p11.2) in XX males: Etiology of male phenotype. Hum Genet 62:271, 1982.
64. Magenis RE, Brown MG, Donlon T, Olson SB, Sheehy R, Tomar D: Structural aberrations of the Y chromosome, including the nonfluorescent Y: Cytologic origin and consequences. In Sandberg AA (ed): "The Y Chromosome, Part A: Basic Characteristics of the Y Chromosome." New York: Alan R. Liss, 1985, pp 537–574.
65. Guellaen G, Casanova M, Bishop C, Geldworth D, Andre G, Fellous M, Weissenbach J: Human XX males with Y single copy DNA fragments. Nature 307:172–173, 1984.
66. Page DC, de la Chapelle A, Weissenbach J: Chromosome Y-specific DNA in related human XX males. Nature 315:224–226, 1985.
67. Natt E, Koo F-T, Rettenmeier R, Scherer G: Assignment of the human tyrosine aminotransferase gene to chromosome 16. Hum Genet 72:225–228, 1986.
68. Toth-Fejel S, Magenis E, Scherer G, Olson S: Abnormal alpha haptoglobin findings in two patients with overlapping interstitial deletions of 16q one of whom also has tyrosinemia type II. Am J Hum Genet 39:A135, 1986.
69. Fraser J, Mitchell A: Kalmuc idiocy: Report of a case with autopsy with notes on sixty-two cases. J Ment Sci 22:168–179, 1876.
70. Penrose LS: Observations on the aetiology of mongolism. Lancet 2:505, 1954.
71. Penrose LS, Smith GF: "Down's Anomaly." Boston: Little, Brown and Co. 1966, pp 167–171.
72. Magenis RE, Hecht F, Milham S Jr: Trisomy 13 (D1) syndrome: Studies on parental age, sex ratio and survival. J Pediatr 73:222–228, 1968.
73. Carothers AD, Collyer S, de Mey R, Frackiewicz A: Parental age and birth order in the aetiology of some sex chromosome aneuploidies. Ann Hum Genet 41:277–287, 1978.
74. Hook EB: Rates of chromosome abnormalities at different maternal ages. Obstet Gynecol 58:282–285, 1981.
75. Hassold T, Chiu D: Maternal age-specific rates of numerical chromosome abnormalities with special reference to trisomy. Hum Genet 70:11–17, 1985.
76. Maraschio P, Zuffardi O, Bernardi F, Bozzola M, De Paoli C, Fonatsch C, Flatz SD, Ghersini L, Gimelli G, Loi M, Lorini R, Peretti D, Poloni L, Tonetti D, Vanni R, Zamboni G: Preferential maternal derivation in inv dup(15). Hum Genet 57:345–350, 1981.
77. Hook EB: Maternal age, paternal age, and human chromosome abnormality: Nature, magnitude, etiology, and mechanisms of effects. In Dellarco VL, Voytek PE, Hollaender A (eds): "Aneuploidy: Etiology and Mechanisms." New York: Plenum Press, 1985, pp 117–132.
78. Hook EB, Regal RR: A search for a paternal-age effect upon cases of 47, +21 in which the extra chromosome is of paternal origin. Am J Hum Genet 36:413–421, 1984.
79. Martin RH, Rademaker A: The effect of age on the frequency of sperm chromosome abnormalities in normal men. Am J Hum Genet 39:A123, 1986.

80. Warburton D: Outcome of cases of de novo structural rearrangements diagnosed at amniocentesis. Prenatal Diagnosis 4:69–70, 1984.
81. Hook EB, Schreinemachers DM, Willey AM, Cross PK: Rates of mutant structural chromosome rearrangements in human fetuses: Data from prenatal cytogenetic studies and associations with maternal age and parental mutagen exposure. Am J Hum Genet 35:96–109, 1983.
82. Martin RH, Burns K, Barnes M, Rademaker A, Douglas G, Arthur K, Ringrose T, Brown IS: An increased frequency of human sperm chromosomal abnormalities after radiotherapy. Am J Hum Genet 36:104S, 1984.

The Cytogenetics of Mammalian Autosomal
Rearrangements, pages 601–613

Chapter 19
Use of Recombinant DNA Probes in Determining the Origin of Chromosome Rearrangements

M.G. Mattei and J.F. Mattei

INTRODUCTION

Human cytogenetics is now 30 years old and in this short time has seen decisive technical progress occur which has each time brought about a considerable improvement in chromosome analysis.

The 1960s were the years of the standard technique of Giemsa chromosome staining which enabled the general morphology of the human karyotype to be described. Chromosomes appeared as small homogeneously stained structures which could only be differentiated by their relative size and centromeric index. The identification of certain chromosome pairs thus was uncertain, and numerous structural rearrangements remained undetected.

In the 1970s the chromosome banding techniques were developed, and these revealed a sequence of dark and light bands on the different chromosomes. Each pair could be identified by a characteristic pattern, and structural abnormalities could be detected by changes in the order or size of the bands. On average, 350 bands could be recognized on the haploid karyotype using these techniques.

In 1976, a new step forward in the precision of chromosome analysis was taken with the description of high resolution banding techniques, which have developed rapidly since [1]. These techniques were based on maximal despiralization of the chromosomes and revealed about a thousand bands on the human haploid karyotype. Microcytogenetics thus led to the precise identification of minor chromosome rearrangements corresponding to single subbands.

Finally, in 1981, using technical progress achieved in the development of in situ hybridization (ISH), molecular cytogenetics enabled a still finer degree

Centre de Génétique Médicale and INSERM U .242, Hôpital d'Enfants de la Timone, 13385
Marseille Cedex 5, France

of resolution. ISH permits any single copy DNA sequence of a few hundred base pairs to be detected and localized on metaphase chromosomes. This technique is now essential for cytogenetic diagnosis and detection of infra-microscopic chromosomal abnormalities.

IN SITU HYBRIDIZATION TECHNIQUE

Probe Labeling

The purified probe (generally a DNA fragment in a plasmid) is labeled by nick-translation with tritiated nucleotides (Fig. 1). The radioactive DNA is then dissolved in a solution containing different factors aimed at enhancing hybridization (formamide, dextran sulphate, etc.). The probe is denatured at 70°C before the hybridization reaction [2].

Chromosome Preparations

Chromosome preparations are carried out in accordance with normal procedures. The incorporation of 5-bromodeoxyuridine (5-BrdU) during the last 7 hr of culture enables good quality chromosome banding to be obtained after hybridization. The slides with the fixed metaphases are treated with ribonuclease A to avoid any non-specific hybridization with endogenous RNA. They are then immersed in formamide at an adequate temperature for the chromosomal DNA to be denatured and to make it accessible to the probe.

Hybridization and Autoradiography

The radioactive probe and chromosome preparations thus treated are brought into contact and incubated at 42°C overnight (Fig. 2). The hybridized slides are then carefully rinsed, taking into account the pH, temperature, and ionic concentration of the solutions so as to protect the chromosomal structure.

The chromosome spreads are coated with a photographic emulsion made up of a gelatin containing silver bromide crystals, reducible to silver metal by β radiation from the tritium. The slides are then exposed at 4°C, for 8–21 days in the dark, then developed and fixed.

Interpretation of Results

The slides simply Giemsa-stained are observed in a microscope and only the metaphases presenting a good chromosome dispersion and an average of 1 to 5 silver grains are photographed. The slides are then treated so as to reveal bands on the chromosomes, allowing them to be identified without

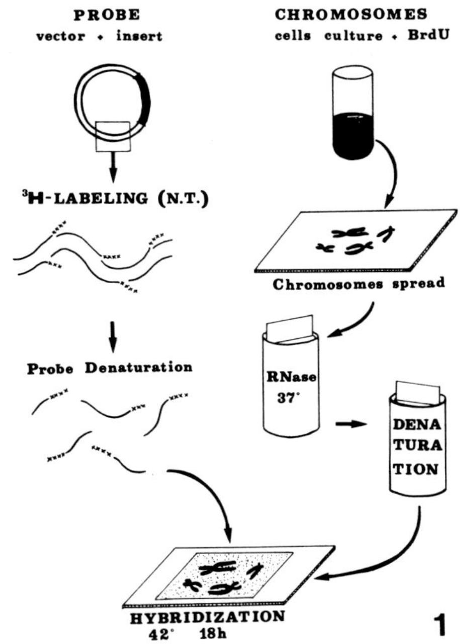

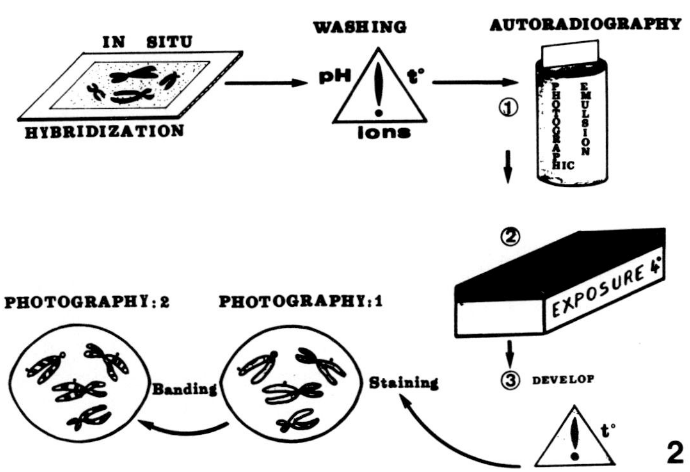

Fig. 1. In situ hybridization (ISH) technique: probe labeling and chromosome preparations.

Fig. 2. ISH technique: hybridization, autoradiography, and chromosome banding.

ambiguity. The preceding metaphases are again photographed and the photographs compared, enabling each silver grain to be attributed to a precise chromosome band. Grain accumulation on a given chromosome band reflects strong hybridization and enables the complementary chromosomal sequence to the probe to be identified.

ISH thus appears as a molecular biology tool perfectly adapted to the mapping of genes. Conversely, if the gene localization is known, ISH of given probes can be used to identify or to detect chromosome structural modifications whose reduced size does not enable them to be recognized with certainty, even with high resolution banding techniques. This is the case, for example, for the identification of small supernumerary chromosomes, or unbalanced rearrangements or for the precise localization of breakpoints in translocations. ISH can, last of all, enable inframicroscopic chromosome abnormalities to be detected, that is, involving chromosome segments of less than 3,000 kilobases (kb) (approximate size of a superband obtained after chromosomal despiralization).

IDENTIFICATION OF SMALL SUPERNUMERARY CHROMOSOMES

The presence of a small supernumerary chromosome in the human genome, not corresponding to any normal chromosome, poses a real problem for genetic counseling. Whether they are acrocentric or metacentric, satellited or non-satellited, these supernumerary chromosomes are generally too small to be precisely identified, even using chromosome despiralization techniques. It therefore proves absolutely necessary to be able to detect on these chromosomes the presence of a euchromatic region responsible for a pathologic phenotype. The in-depth clinical examination of the patient and the cytogenetic analysis first of all orientate the diagnosis towards a trisomy or a tetrasomy for a given chromosomal region. Then in situ hybridization with characteristic probes for this chromosome segment will confirm or invalidate the suggested hypothesis.

We have had the opportunity to study five subjects with encephalopathy, carriers of a small non-satellited supernumerary metacentric chromosome. In the five cases, cytogenetic techniques suggested a possible abnormality involving the chromosome 18 short arm. We had at our disposal, furthermore, the D18S3 probe (clone B74, gift from J.L. Mandel) previously localized at the distal extremity of the chromosome 18 short arm [2]. In situ hybridization of D18S3 with metaphase chromosomes of the five patients was performed. In each case, this probe hybridized quite specifically not only with the short arms of the two normal chromosomes 18, but also with the small supernumerary chromosome, proving that the latter is indeed derived from the short arm of chromosome 18. Moreover, in four of the five patients, we could observe bipolar hybridization pictures, that is, hybridization with the two extremities of the small supernumerary chromosome (Fig. 3). In these four subjects, the number of silver grains located on the small supernumerary chromosome was nearly equal to the number located on the two chromosomes 18 (Fig. 4a,b). We thus concluded, in these four cases, that the supernumerary chromosome was made up of two chromosome 18 short arms. It corresponded to an isochromosome 18p, and the subjects studied were tetrasomic for the chromosome 18 short arm. For the fifth subject, the D18S3 probe hybridized with a single extremity of the small supernumerary chromosome (Fig. 4c), and the grain number on this chromosome was equal to half those localized on the two normal chromosomes 18. This supernumerary chromosome thus included only one copy of the short arm of chromosome 18 and could be interpreted as a metacentric 18q−.

Patients with Cat eye syndrome generally have an extra bisatellited chromosome thought to be derived from chromosome 22. The most valuable evidence for its origin comes from cases with partial trisomy due to familial translocation. In situ hybridization with various probes derived from the 22q11 region of chromosome 22 confirmed this hypothesis [3–5].

DIAGNOSIS OF MINUTE CHROMOSOME REARRANGEMENTS

Small chromosome rearrangements, when they are detected, often remain difficult to identify despite the use of appropriate cytogenetic techniques. ISH using carefully chosen probes thus becomes an indispensable tool for precise cytogenetic diagnosis and, thus, for suitable genetic counseling. We have had the opportunity to apply this diagnostic procedure to the study of three significant observations.

An encephalopathic and dysmorphic child presented, in his karyotype, a chromosome 18 with a long arm including excess material and atypical banding. Since the C-banding technique revealed two centromeres, one in the normal position, the other at the distal long arm extremity, a partial duplication of the chromosome 18 itself, involving the short arm and the centromere, was suggested. ISH of the D18S3 probe with the metaphases of the patient enabled two hybridization peaks (Fig. 5) to be seen on the rearranged chromosome: one at the short arm extremity, the other at the long arm extremity, each having an intensity comparable with that of the hybridization peak observed on the normal chromosome 18 short arm. Partial duplication of chromosome 18 was thus confirmed.

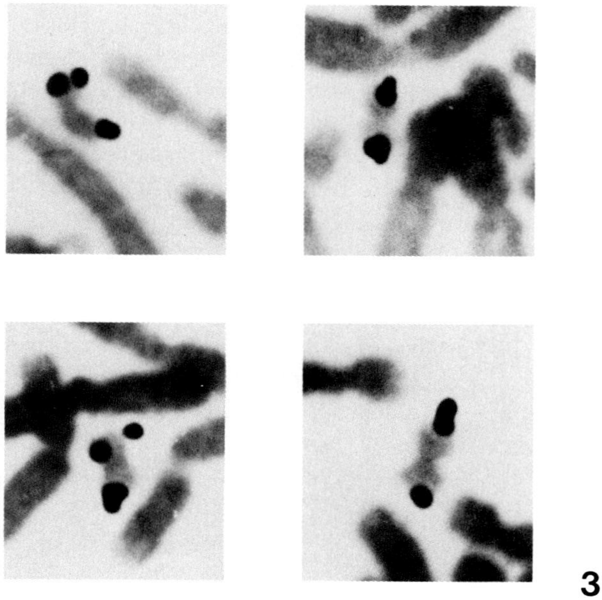

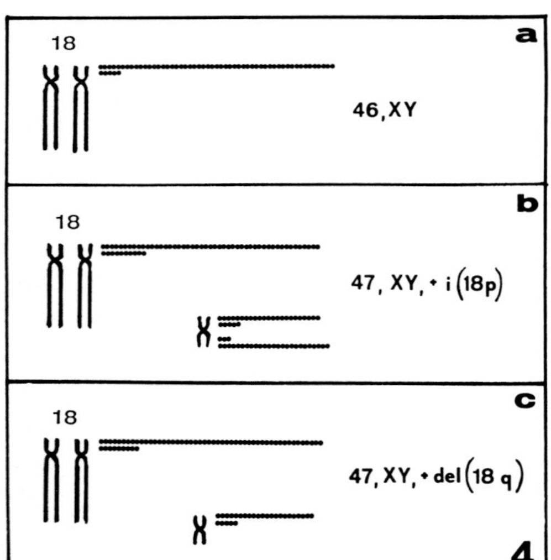

Fig. 3. ISH of the D18S3 probe with the two extremities of a small supernumerary chromosome (isochromosome 18p).

Fig. 4. Distribution of silver grains after ISH of the D18S3 probe with **a**: the metaphases of a normal subject; **b**: the metaphases of a patient with a supernumerary isochromosome (18p); **c**: the metaphases of a patient with a supernumerary deleted chromosome (18q−).

In the second observation, an abnormal karyotype was detected in a dysmorphic and mentally retarded subject, in whom one chromosome 18 had a short arm longer than normal. There again, the excess chromosomal material could not be precisely identified by the most accurate cytogenetic techniques. Two hypotheses could be formulated to explain this chromosome 18p+: either it resulted from a duplication of the chromosome 18 short arm, or it resulted from a translocation from any other chromosome fragment. Hybridization of the D18S3 probe with the metaphases of this subject enabled us to test the validity of the first hypothesis. The number of silver grains localized on the chromosome 18p+ was about twice as high as that localized on the normal chromosome 18 (Fig. 6), thus proving duplication of the chromosome 18 short arm. Furthermore, the presence of two hybridization peaks of comparable intensity on the chromosome 18p+ short arm, in the context of the very distal localization of probe D18S3 on the normal chromosome 18 short arm, suggests that a direct duplication is more likely than an inverted duplication (Fig. 7). In fact, in the latter case, a single hybridization peak of double intensity would have been obtained and this would have been situated in the middle of the chromosome 18p+ short arm.

A final example can be given, epitomizing the degree of analytical accuracy that can be obtained with chromosomal ISH. In this case, a young patient presented with slight mental retardation; the cytogenetic analysis revealed a partial deletion of one chromosome 18 short arm. The same deletion was observed in his phenotypically normal mother. Assuming the latter was a carrier of a balanced translocation, the problem was therefore to try to detect the distal region of the chromosome 18 short arm in the maternal metaphases. ISH of the D18S3 probe, with metaphases of the mother, revealed two equal intensity peak of hybridization: one on the normal chromosome 18 short arm, the other on the short arm of one chromosome 21 (Figs. 8,9). The mother of the proband was thus a carrier of a balanced translocation between the chromosome 18 and chromosome 21 short arms. Only ISH enabled a precise diagnosis in this case, since the slight lengthening of the chromosome 21 short arm could have been interpreted as a normal variant. Acrocentric chromosomes in fact have very polymorphic short arms. ISH of D18S3 with metaphases of the patient showed a single hybridization peak, situated on the normal chromosome 18, and it was concluded that an unbalanced rearrangement (i.e., an 18p partial deletion) was present.

TRANSLOCATIONS

Translocation Breakpoint Mapping

In certain cases, where a translocation has been provisionally identified by cytogenetic analysis, a determination of exactly where the breakpoint has

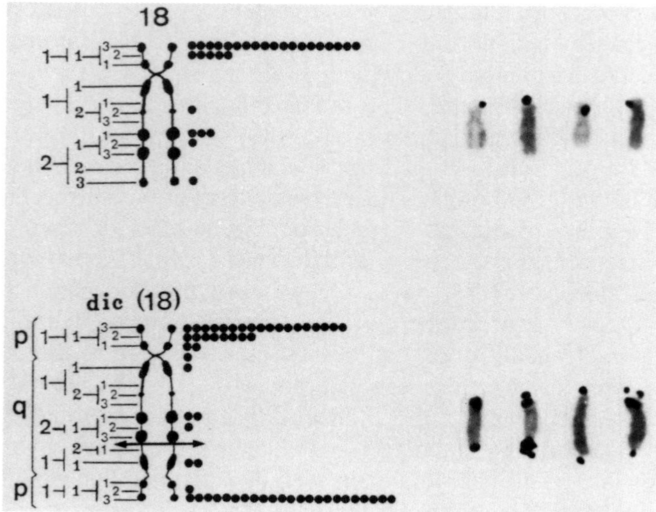

5

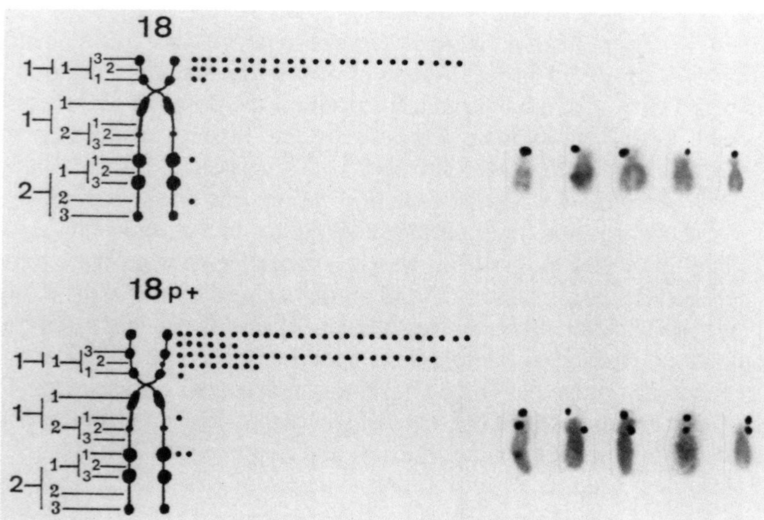

6

Fig. 5. Localization of silver grains after ISH of D18S3 probe with metaphases of a patient presenting a rearranged chromosome 18q+. **Top**: normal chromosome 18; **bottom**: rearranged chromosome 18—probable idic(18).

Fig. 6. ISH of the D18S3 probe with metaphases of a patient presenting an abnormal chromosome 18p+. **Top**: normal chromosome 18; **bottom**: rearranged chromosome 18—probable tandem duplication of 18p.

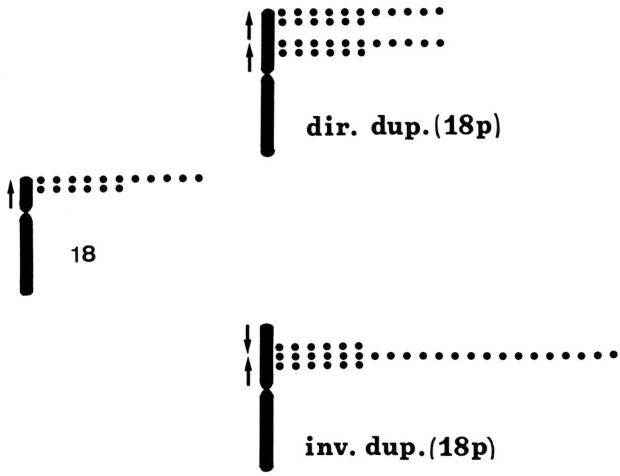

Fig. 7. Theoretical distribution of silver grains after ISH of D18S3 with direct duplication or inverted duplication of chromosome 18 short arm.

occurred can often be made by ISH. Let us consider the case of a subject who is a carrier of a balanced translocation between the X chromosome and chromosome 21, where the breakpoints are situated in the q27 band of the X chromosome long arm and in the q11 band of the chromosome 21 long arm. We have at our disposal two molecular probes which are very close to each other [6] in the X-chromosome region involved in the translocation: the coagulation Factor IX (F IX) gene and the St14 polymorphic probe. ISH of these two probes with the metaphases of the translocation (X;21) carrier gave the following results (Fig. 10): probe F IX hybridized with the translocated X-chromosome just above the breakpoint; probe St14 hybridized with derivative 21, i.e., with the X-chromosome fragment translocated onto chromosome 21. The translocation breakpoint could thus be very precisely localized on the X-chromosome between the loci of the F IX and St14 probes.

In the same way, Emanuel et al. [7] have demonstrated, using ISH of a probe for part of the lambda light chain constant region (Cλ), that the 22q11 breakpoints of chronic myelogenous leukemia t(9;22) and Burkitt lymphoma t(8;22) were not identical. For chronic myelogenous leukemia, the breakpoint was distal to the lambda light chain constant region, whereas for Burkitt lymphoma it was proximal.

Gene Mapping Using Translocations

By the physical separation of regions complementary to closely linked probes, chromosome translocations can increase ISH resolution and thus

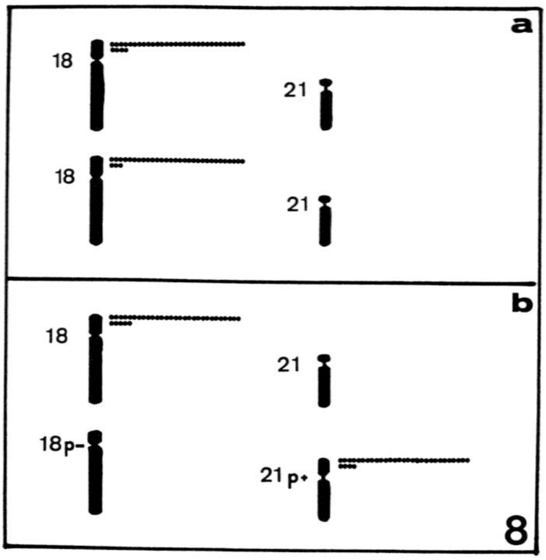

8

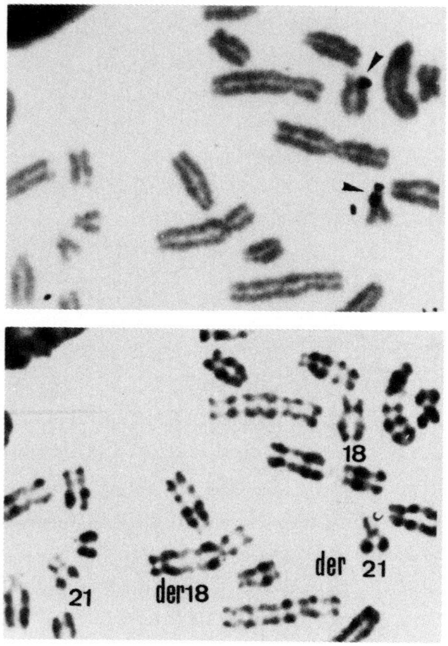

9

Fig. 8. Distribution of silver grains after ISH of D18S3 with metaphases of **a**: a normal subject; **b**: the patient with balanced translocation between chromosome 18 and chromosome 21 short arms.

Fig. 9. ISH of the D18S3 probe with patient affected by a balanced translocation (18p;21p). **Top**: partial metaphase with silver grains; **bottom**: the same metaphase after chromosomal banding enabling silver grains to be attributed to the normal chromosome 18 short arm and the short arm of a chromosome 21.

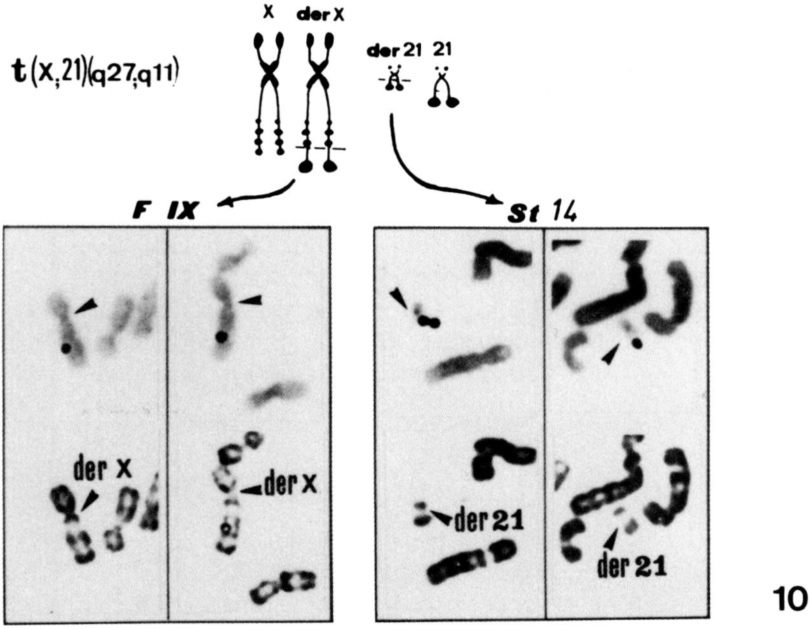

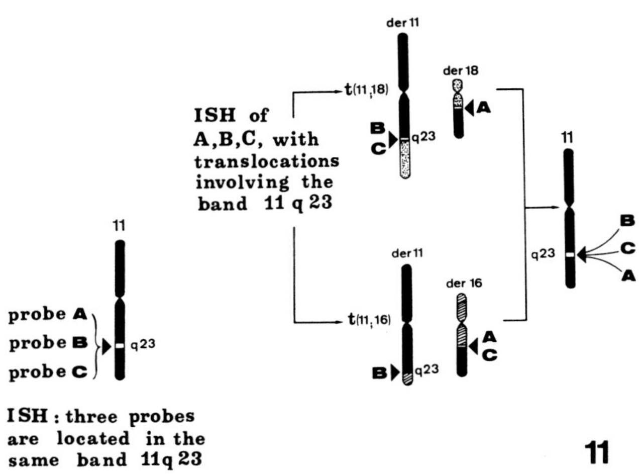

Fig. 10. ISH of the two probes F IX and St14 with partial metaphases of a subject with balanced translocation (X;21)(q27;q11): F IX hybridizes with the der(X) chromosome, and St14 with the der(21) chromosome.

Fig. 11. Example of the use of balanced translocations in precise mapping of molecular probes by ISH. The three probes A, B, C are mapped in proximal to distal order as B→C→A within band 11q23.

enable very precise gene maps to be established. For example (Fig. 11), let us consider three probes A, B, and C being localized in the same chromosome band 11q23; ISH does not enable these three probes to be ordered in the 11q23 band. We can nevertheless use chromosomal translocations having a breakpoint in the 11q23 band, such as the two translocations t(11;18) and t(11;16). The three probes, A, B, C, are first hybridized with the translocation (11;18) showing that A is translocated onto the derivative 18, whereas B and C remain on the derivative 11. The same three probes are then hybridized with the translocation (11;16) having a breakpoint at 11q23, and give slightly different results: probe B remains on the derivative 11, whereas probes C and A hybridize onto the derivative 16. We can thus conclude that B is the most proximal of the three probes and finally establish the following mapping on chromosome 11: cen11→B→C→A→tel11.

INFRAMICROSCOPIC CHROMOSOME ABNORMALITIES

ISH resolution being of the order of a few hundred base pairs can detect inframicroscopic chromosome abnormalities, i.e., less than 3,000 kb (the approximate size of a superband obtained after chromosome despiralization). The choice of probes to be hybridized is assisted by a very accurate clinical analysis of the subjects studied.

Studies on human sex chromosome aberrations reveal that male differentiation occurs only in the presence of a Y-chromosome, or part of the short arm of the Y-chromosome. The only exceptions to this rule are XX males (i.e., phenotypically male subjects having an apparent female karyotype 46,XX) and the rare cases of XX true hermaphroditism. In some of these cases, ISH has shown the presence of Y-specific sequences at the extremity of one of the X-chromosomes at the meiotic pairing region [8,9]. The X-chromosome including the Y sequences is of paternal origin and probably results from X-Y interchange mediated by accidental crossing-over during paternal meiosis.

CONCLUSIONS

The study of human chromosomes has benefitted in the last few years from progress realized by chromosome despiralization techniques enabling abnormalities of about 3,000 kb to be detected. Recent use of ISH of molecular probes whose maximum definition is 0.5 kb now constitutes a convergence between cytogenetics and molecular biology, i.e., between chromosome and gene.

ACKNOWLEDGMENTS

We thank J.L. Mandel (Strasbourg) for the generous gift of probes D18S3, F IX, and St14, and J.C. Kaplan (Paris), for the gift of 22q11 probes. Cell lines were kindly provided by F.D. Bricarelli (Genes), J. Couturier (Paris), M. Crawfurd (London), P. Jalbert (Grenoble), and M. Rolland (Toulouse). We also thank E. Passage and J. Arfi for their technical assistance.

REFERENCES

1. Yunis JJ: High resolution of human chromosomes. Science 191:1268, 1976.
2. Mattei MG, Philip N, Passage E, Moisan JP, Mandel JL, Mattei JF: DNA probe localization and identification of a small supernumerary chromosome. Hum Genet 69:268–271, 1985.
3. Magenis RE, McDermid H, White BN, Sheehy R: The extra chromosome in cat-eye syndrome (CES) is derived from chromosome 22; evidence from in situ hybridization of a chromosome 22 specific DNA probe. Cytogenet Cell Genet 40:685, 1985.
4. Mattei MG, Passage E, Julier C, Szajnert MF, Philip N, Jeanpierre M, Kaplan JC, Mattei JF: Chromosome 22 is involved in cat-eye syndrome as demonstrated by in situ hybridization. Cytogenet Cell Genet 40:693, 1985.
5. McDermid HE, Duncan AMV, Brasch KR, Holden JJA, Magenis E, Sheehy R, Burn J, Kardon N, Noel B, Schinzel A, Teshima I, White BN: Characterization of the supernumerary chromosome in cat-eye syndrome. Science 232:646–648, 1986.
6. Mattei MG, Baeteman-Voelckel MA, Heilig R, Oberlé I, Davies KE, Mandel JL, Mattei JF: Localization by in situ hybridization of the coagulation factor IX gene and of two polymorphic DNA probes with respect to the fragile X site. Hum Genet 69:327–331, 1985.
7. Emanuel BS, Selden JR, Wang E, Nowell PG, Croce CM: In situ hybridization and translocation breakpoint mapping. Cytogenet Cell Genet 38:127–131, 1984.
8. Buckle V, Boyd Y, Craig IW, Fraser N, Goodfellow PN, Wolfe J: Localization of Y chromosomal sequences in normal and "XX" males. Cytogenet Cell Genet 40:593, 1985.
9. Magenis RE, Sheehy R, Olsen S, Brown MG, Casanova M, Fellous M: Genes for maleness localized to distal one-half of Y chromosome short arm: Evidence from in situ hybridization of a Y-derived single copy DNA probe. Human Gene Mapping 8. Cytogenet Cell Genet 40:686, 1985.

The Cytogenetics of Mammalian Autosomal
Rearrangements, pages 615–634

Chapter 20
Origin and Clinical Significance of inv dup(15)

Paola Maraschio, Cristina Cuoco, Giorgio Gimelli, Orsetta Zuffardi, and Luciano Tiepolo

Six new cases of inv dup(15) are described, and all cases reported in the literature with this type of small extra chromosome are reviewed. Inv dup(15) appear to be morphologically heterogeneous and are found in subjects with different phenotypes. Our purpose was to find out whether inv dup(15) with different morphology correspond to specific phenotypes. Acrocentric inv dup(15) are either iso dic with presumably a breakpoint at band q12, or are derived from two homologous chromosomes; in the latter case the two breaks could be at the same site (e.g., in q12) or at different sites, i.e., in q11 and q13. These extra chromosomes are generally found in patients with mild to severe mental retardation. The metacentric inv dup(15) are smaller, almost entirely heterochromatic, and originate from breaks at q11, either in a single or in two homologous 15 chromosomes. They are associated either with normal phenotypes or with Prader-Willi syndrome (PWS). The inv dup(15) are usually de novo in subjects with mental retardation and PWS, and familial in normal individuals. Among the informative cases, a preferential maternal derivation is observed together with a clear maternal age effect.

INTRODUCTION

Supernumerary bisatellited chromosomes have been frequently reported in the literature both in normal and in phenotypically abnormal individuals: In a survey of 5,049 newborns Friedrich and Nielsen [1] found a frequency of 0.06% of these marker chromosomes. The introduction of techniques such as binding of anti-5-methylcytosine [2] and DAPI plus distamycin A staining [3], identified a large proportion of these markers as derived from chromosome 15. Among 64 small bisatellited additional chromosomes reported in the literature and studied with appropriate techniques, 46 were composed of chromosome 15 material and of these, 33 were interpreted as inverted duplications of chromosome 15 [4].

Biologia Generale e Genetica Medica, University of Pavia (P.M., L.T.), Instituto Gaslini, Genova (C.C., G.G.), and Biologia Generale, University of Florence (O.Z.), Italy

Buckton et al. [5] found one inv dup(15) among 3,993 liveborns, corresponding to a frequency of 0.025%. Inv dup(15) are extra, small chromosomes formed by an inverted duplication of satellites, short arm, and a variable amount of proximal long arm of chromosome 15. Because of inactivation of one centromere, they show two C positive regions but only one primary constriction.

Inv dup(15) chromosomes can be either smaller, or larger than G chromosomes, or equal to them, depending on the localization of the breakpoint and on the size of the satellites. When the breakpoint is close to the centromere their appearance is that of an almost entirely heterochromatic metacentric chromosome, but when the break point is more distal, the appearance is that of an acrocentric or sub-metacentric chromosome (Fig. 1).

Analysis of 40 inv dup(15) described in the literature and six unpublished personal cases shows that these extra chromosomes can be associated either with a normal phenotype, with more or less severe mental retardation, or with Prader-Willi syndrome.

In this review we take into account only bisatellited chromosomes showing at both extremities a banding morphology specific for chromosome 15, excluding therefore all the cases described as del(15q), some of which could indeed be inv dup(15) [6–9]. Our aim is to reveal a possible correlation between the various phenotypes observed and the inv dup(15) of different

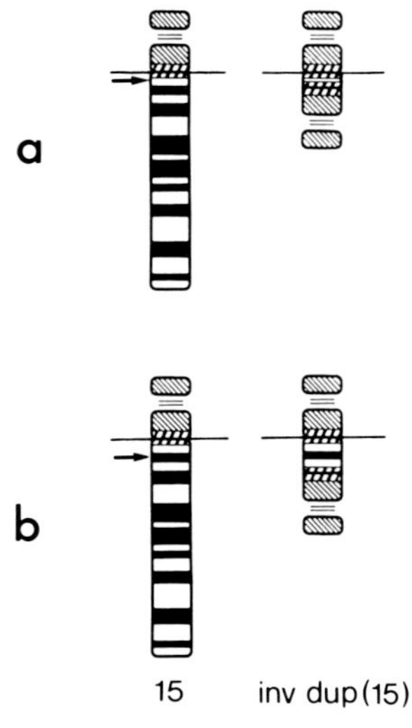

Fig. 1. Formation of a metacentric (**a**) and of an acrocentric or sub-metacentric (**b**) inv dup(15). The breakpoints are indicated by arrows. The horizontal line indicates the centromere.

appearance. We consider also the mechanisms of formation of these abnormal chromosomes, their parental derivation, and a possible correlation with parental ages.

Case Histories

Case 1

A male (M.M.) born at term on 30.8.76. The mother was 36 and the father 41 years old. He has one normal brother and normal sister. Examination at 5 years of age for psychomotor retardation revealed bilateral epicanthus and antimongoloid slant, low-set large ears, high forehead, macrocephaly, convergent strabismus, cafè-au-lait spots on the abdomen, kyphosis, and scoliosis (Fig. 2). I.Q. was 79.

Karyotype on cultured lymphocytes (Q-, C-, DA-DAPI, and NOR bands): 46,XY/47,XY,+inv dup(15)(pter→q12::q12→pter). The two cell lines were found in 8 and 92 cells, respectively. Both parents have normal chromosomes. The inv dup(15) is acrocentric, slightly larger than a G (Fig. 3A). Satellites at the two extremities are different, and the proximal ones (the nearest to the active centromere) are identical to those of chromosome 15 of maternal derivation. Familial analysis of the short arm polymorphisms of chromosome 15 after Q- and NOR-banding suggested a maternal derivation of the extra chromosome.

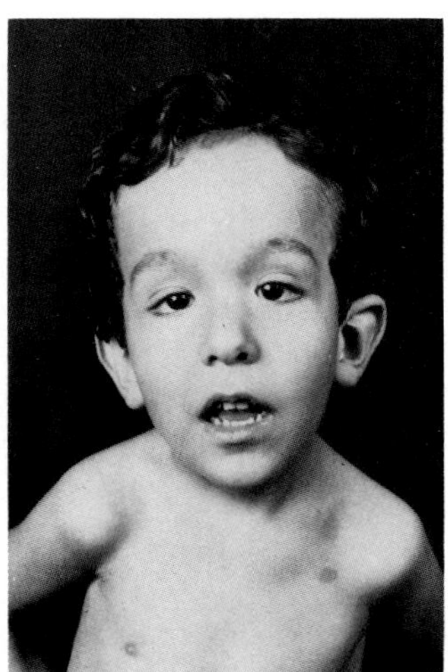

Fig. 2. Case 1 (M.M.) at 5 years.

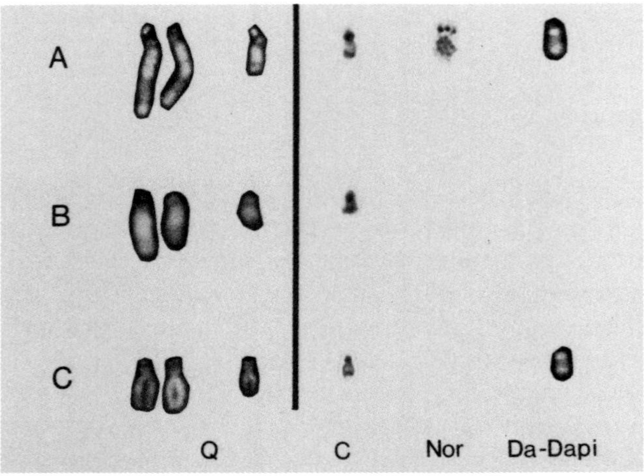

Fig. 3. **Left:** chromosomes 15 and inv dup(15) from the same cell after Q-banding; **Right:** inv dup(15) after C-, NOR-, and DA-DAPI banding. **A:** Case 1; **B:** case 2; **C:** case 3. Maternally derived normal 15 chromosome on left in case 1 and on right in case 2. In infants 1 and 2 the active centromere of the inv dup(15) is identical to that in the maternally derived normal 15. In case 3 the mother also had an inv dup(15).

Case 2

A male (V.F.) born at term on 20.4.78 when the mother was 40 and the father 44 years old. The mother had six pregnancies with two normal females, one normal male, two spontaneous abortions, and the propositus. Examination at 3 years and 7 months of age for severe psychomotor retardation revealed high forehead, flat occiput, low-seat ears, and kyphosis.

Karyotype on lymphocytes and fibroblasts (Q-, C-, DA-DAPI bands): 47,XY,inv dup(15)(pter→q12::q12→pter). Parents have normal chromosomes. The inv dup(15) is acrocentric, slightly larger than a G (Fig. 3B). It is of maternal derivation and has different satellites at the ends, with the proximal ones identical to those of the maternal 15.

Case 3

C.C. was born at term on 26.10.79 when the parents were both 37 years old; he was fourth-born with two normal brothers and one normal sister. Examination at 4 years and 10 months of age for psychomotor retardation revealed syndactyly of the second and third toes, hypotonia, and developmental retardation (Fig. 4). Karyotype on lymphocytes (G-, Q-, C-, DA-DAPI bands): 47,XY,+inv dup(15)(pter→q12::q12→pter).

The father has normal chromosomes, while the mother has in all cells the same inv dup(15). The mother is reported to be slightly mentally handicapped and "behaviourally disturbed." Because of poor banding results, we were unable to decide whether the extra acrocentric chromosome, which was larger than a G group chromosome, had at its ends similar or different satellites (Fig. 3C).

Case 4

A male (M.C.) born at term on 1.5.46 when the mother was 37 and the father 38 years old. He is fourth-born. He was referred for chromosome analysis when 37 years old because of olygozoospermia, but he was otherwise normal. Karyotype on lymphocytes (HRB, G-, Q-, C-, NOR-, DA-DAPI bands): 47,XY,+inv dup(15)(pter→q11.2::q11.2→pter). The parents and two brothers are dead. The third brother, unmarried, and with a normal phenotype, at karyotype analysis on lymphocytes revealed the same extra chromosome. This is an almost entirely heterochromatic metacentric, smaller than a G and with different satellites at the two ends (Fig. 5A).

Case 5

A female (M.A.) born at term on 10.12.59 when the mother was 24 and the father 29 years old. She is the second of a sibship of three. She was

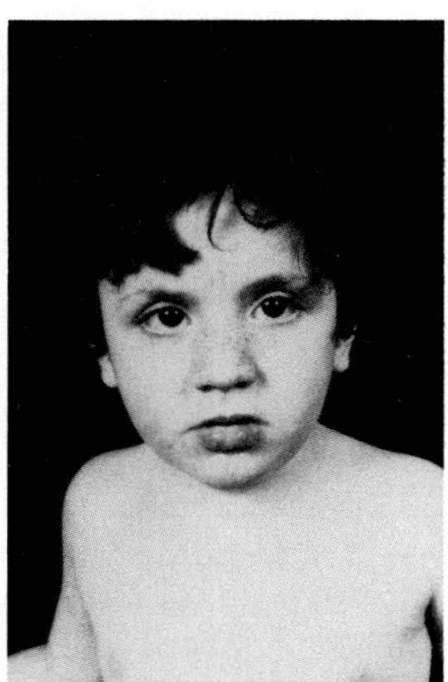

Fig. 4. Case 3 (C.C.) at 4 years and 10 months.

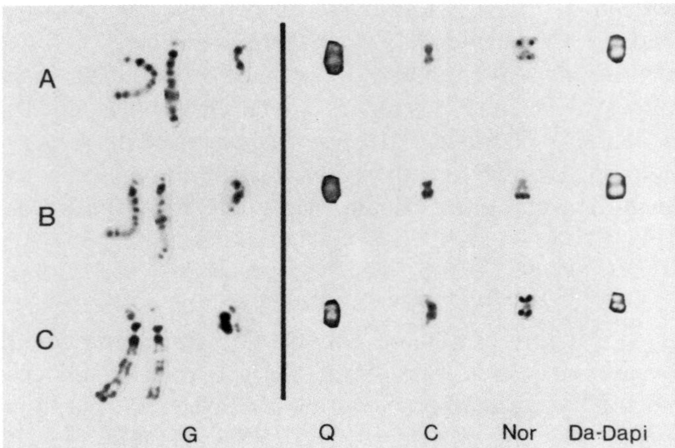

Fig. 5. **Left:** Chromosomes 15 and inv dup(15) from the same cell after HRB-G: **Right:** inv dup(15) after Q-, C-, NOR-, and DA-DAPI banding. **A:** Case 4; **B:** case 5; **C:** case 6.

referred for chromosome analysis after two spontaneous abortions; she has a normal phenotype. Karyotype on lymphocytes and fibroblasts (HRB, G-, Q-, C-, NOR-, DA-DAPI bands): 47,XX,+inv dup(15)(pter→q11.2::q11.2→pter). The same inv dup(15) was found in the mother, in the sister, and in the maternal grandfather, all phenotypically normal. No further information on chromosomal constitution of the family is available. The inv dup(15) chromosome is similar in size to a G and almost entirely heterochromatic. It appears as a sub-metacentric, deriving from two chromosomes 15 with short arms very different in size (Fig. 5B).

Case 6

A 41-year-old woman was referred to prenatal diagnosis for advanced maternal age. She had five previous pregnancies, four at term with normal children, and the fifth ended in a first trimester spontaneous abortion. Cultured amniocytes revealed a karyotype (HRB, G-, Q-, C-, NOR-, DA-DAPI bands): 47,XY,+inv dup(15)(pter)→q11::q11→pter). The pregnancy was interrupted, but the fetus could not be examined. The mother's karyotype was normal, while the father showed the same inv dup(15). No further information is available on the chromosomal constitution of the family. The inv dup(15) is an almost entirely heterochromatic metacentric, smaller than a G and with similar ends (Fig. 5C).

inv dup(15) IN MENTALLY RETARDED SUBJECTS

Table I shows the published cases of inv dup(15) in subjects with mental retardation, from slight to severe, but with no other major physical abnormalities [10–19]. We found a total of 33 such cases, 29 in the literature, and 4 in our sample. These represent the great majority of all cases of inv dup(15) described. Sex ratio in these cases is 20 males to 10 females ($\chi^2 = 3.33$ for 1 degree of freedom). Among minor phenotypic anomalies, behavioural disturbance and seizures are common to most patients. Low posterior hairline, epicanthus, low-set ears, and strabismus are all variably present, but these signs are common to several other chromosomal abnormalities and thus are of poor diagnostic value.

Inv dup(15) in all these patients are acrocentrics, similar in size or longer than the G group chromosomes. Only in case 2 of Yip et al. [16] is the extra chromosome metacentric and almost entirely heterochromatic. In 22 of the cases in Table I a different morphology for the two extremities of the inv dup(15) is reported and so the abnormal chromosome must be derived by a non-sister-chromatid exchange. In another six cases the chromosome ends are of similar morphology, and thus a sister chromatid fusion can be postulated.

A precise breakpoint is indicated only for some of the inv dup(15) in Table I, and in these cases, bands q12 and q13 are more frequently involved. In the case of Voss et al. [17] who was analyzed with high resolution banding, the breakpoint is indicated at q15, and this is one of the more distal.

We had the opportunity to re-examine with high resolution banding cases 6 and 7 previously studied by Maraschio et al. [15]. These inv dup(15) showing between the two centromeres a single G-positive band are probably 15pter→15q12::15q12→15pter. Only two cases with mosaicism were found: by Van Dyke et al. [11] in a patient with severe mental retardation and our case 1, moderately retarded. In the first, three cell lines are present: one with normal karyotype, one with an extra inv dup(15) of size similar to an E group chromosome (indicated as M1), and a third with 48 chromosomes and two inv dup(15). Of these one is identical to M1 and the other (M2) is smaller in size. Our case 1 presents two cell lines, one normal and one with an extra inv dup(15).

Twenty-four of the 25 cases in whom the parents were studied, are de novo. The only familiar case is our case 3, the same chromosome of the propositus being present in all the cells of the slightly mentally retarded mother.

In the case studied by Voss et al. [17], the propositus had two inv dup(15) with different morphology. The first, a sub-metacentric, is de novo and

TABLE I. inv dup(15) in mentally retarded subjects

Reference	Case identification	Sex	Breakpoints	Size[a]	Morphology[a]	Ends[a]	Derivation	Familiarity
Schreck et al. [10]	JF	M	q1::p1 or q1	G	Acrocentric	D	?	De novo
	CS	M	q1::p1 or q1	G	Acrocentric	D	?	De novo
	GK	M	q1::p1 or q1	G	Acrocentric	D	?	?
	JT	F	q1::p1 or q1	G	Acrocentric	E	?	?
	LV	F	q1::p1 or q1	G	Acrocentric	?	?	?
	RD	M	q1::p1 or q1	G	Acrocentric	E	?	De novo
Van Dyke et al. [11]		M	M_1q15	E	Acrocentric	D	?	De novo
			M_2q12	<G	Metacentric	E		
De Falco and Willey [12]	1	?	?	G	Acrocentric	?	?	De novo
	2	?	?	G	Acrocentric	?	?	De novo
	3	?	?	G	Acrocentric	?	?	?
Wisniewski et al. [13]	1	F	q1::p1 or q1	>G	Acrocentric	D	?	De novo
	2	M	q1::p1 or q1	>G	Acrocentric	D	Maternal	De novo
	3	M	q1::p1 or q1	>G	Acrocentric	D	Maternal	De novo
	4	F	q1::p1 or q1	>G	Acrocentric	D	?	?
	5	F	q1::p1 or q1	>G	Acrocentric	D	?	?
Chamberlin and Magenis [14]	17	F	?	>G G	?	D	Maternal	De novo
Maraschio et al. [15]	1	M	q12::q12 or q11::q13	>G	Acrocentric	D	Maternal	De novo
	2	F	q12::q12 or q11::q13	>G	Acrocentric	?	?	De novo

	3	M	q12::q12 or q11::q13	>G	Acrocentric	D	Maternal	De novo
	4	M	q12::q12 or q11::q13	>G	Acrocentric	D	Maternal	De novo
	5	M	q12::q12 or q11::q13	>G	Acrocentric	D	Maternal	De novo
	6	M	q12::q12 or q11::q13	>G	Acrocentric	E	Paternal	De novo
	7	F	q12::q12 or q11::q13	>G	Acrocentric	D	Maternal	De novo
	8	F	q12::q12 or q11::q13	>G	Acrocentric	D	Maternal	De novo
Yip et al. [16]	1	M	?	F	Sub-metacentric	D	?	De novo
	2	M	?	G	Metacentric	E	?	?
Voss et al. [17]	Propositus	M	I q15	G	?	D	Paternal	De novo
			II ?	?	Metacentric	E		Familial
Gilmore et al. [18]		M	q13::q13	G	?	D	?	De novo
Wisniewski and Doherty [19]	3	M	q1::q1	?	Sub-metacentric	D	?	De novo
Present study	1, Propositus	M	q12::q12	>G	Acrocentric	D	Maternal	De novo
	2, Propositus	M	q12::q12	>G	Acrocentric	D	Maternal	De novo
	3, Propositus	M	q12:q12	>G	Acrocentric	?	Maternal	Familial

[a]As defined by the authors. G, E, F refer to the size of the chromosomes. Chromosome ends are defined as equal (E) or different (D).

probably originated in the parental germinal line, while the second, a metacentric, was found also in the phenotypically normal father.

inv dup(15) AND PRADER-WILLI SYNDROME

Five cases of inv dup(15) found in patients with Prader-Willi syndrome are reported in the literature [20–24] (Table II). Other subjects with this syndrome have been described in whom a small extra chromosome is present but not identified [25,26]; in a further case the extra chromosome, showing a single C-positive region, was interpreted as 15q− [27].

All the inv dup(15) in Table II are de novo and smaller in size than those found in the patients with mental retardation (Table I). All are metacentric, similar in size to or smaller than an F or G group chromosome. In all these cases the authors indicate band q11 as involved in the breakpoint, identified in three cases [21,23,24] with high resolution banding. Four of these five inv dup(15) have identical ends and are iso-dicentrics [20,22–24], while in the case of Wisniewski et al. [21], the inv dup(15) has different satellites and thus seems to originate from two homologous chromosomes 15.

In the case of Wulsberg et al. [23], one of the two with mosaicism, the extra chromosome is present in 40% of lymphocytes but not in the fibroblasts. In all cells of this case there is a translocation between the distal portion of 1p and band q11 of a chromosome 15. The authors postulate that the iso-dic(15) originated after sister chromatid fusion in the small chromosome deriving from the translocation.

While PWS has been found frequently in patients with monosomy or transposition of region 15q11-13, its association with inv dup(15) is not easy to interpret. Fraccaro et al. [28], suggest the possibility of a gene disruption associated with a balanced translocation involving this region of a chromosome 15.

In patients with an extra inv dup(15), however, the possible effect of a gene disruption should be prevented by the presence of two normal chromosomes 15. Only in the patient described by Wulsberg et al. [23] in whom a translocation 1/15 is present together with an iso dic(15), there is only one normal chromosome 15 and so a gene disruption could explain the PWS.

Mattei et al. [29], in a patient with PWS and a 15/22 balanced translocation, found a small additional interstitial deletion of bands 15q11→q12. However, in two cases of Table II [22,24], high resolution techniques failed to show any concurrent deletion in the normal chromosomes 15.

De France et al. [30] describe a boy with PWS who had a 15q12–13→q15 duplication, and postulate a mutant gene suppressed by two normal chromo-

TABLE II. inv dup(15) in subjects with Prader-Willi syndrome

Reference	Case identification	Sex	Breakpoints	Size[a]	Morphology[a]	Ends[a]	Derivation	Familiarity
Fujita et al. [20]	—	F	q11::q11	< F	Metacentric	E	?	De novo
Wisniewski et al. [21]	—	M	q11 or q12::p11 or q11 or q12	< G	Metacentric	D	Paternal	De novo
Ledbetter et al. [22]	—	F	q11::q11	?	Metacentric	E	?	De novo
Wulfsberg et al. [23]	—	F	q11::q11	?	Metacentric	E	?	De novo
Mattei et al. [24]	9	M	q11::q11	G	Metacentric	E	?	De novo

[a]As defined by the authors. G, E, F refer to the size of the chromosomes. Chromosome ends are defined as equal (E) or different (D).

somes 15. This gene could be expressed both in the case of deficiency or duplication of the proximal 15q region. This interpretation does not fit with the inv dup(15) of size larger than a G of Table I because in these cases a tetrasomy of proximal 15q is present but is not expressed as PWS.

inv dup(15) IN NORMAL INDIVIDUALS

A third group of inv dup(15) is represented by those found in normal individuals [17,19,31–33] (Table III). The extra chromosomes are either metacentric smaller than a G or sub-metacentric similar in size to a G. Breakpoints, when indicated, are at 15q11 in all cases except one [33].

Two cases are de novo [19,32], while in all the others the inv dup(15) is found in two or more generations, for a total of 17 individuals. Three cases are mosaics. In the case of Voss et al. [17] the extra chromosome is present in 70% of lymphocytes while the same chromosome, together with a second larger inv dup(15), is found in 100% of the cells of his abnormal son.

In the case of Knight et al. [33], a mosaic inv dup(15) is present in the father and in the proposita, who in turn has a third cell line with two identical extra chromosomes. Two cell lines, a normal one and a 48,XY,+inv dup(15),+inv dup(15), are found in case 2 of Wisniewski and Doherty [19].

Inv dup(15) in three cases [17,19, our case 6] have similar ends and are probably iso-dicentrics, while in all the other cases the extra chromosome originates from two homologous chromosomes 15.

Indications for chromosomal analysis were oligospermia (our case 4), abortion (our case 5), and a child with inv dup(15) [17,33]. In five cases [31,32,19, and our case 6] the inv dup(15) was found in amniotic fluid cells, after an indication of advanced maternal age in our case, in those by Pecile et al. [32], and Wisniewski and Doherty [19]; indication for amniocentesis is not reported in two cases [31,19].

At autopsy, no abnormalities have been found in the fetuses in three cases [31,32,19]; in two (our cases 6 and 19), the autopsy results are not available.

POSSIBLE MECHANISMS OF inv dup(15) FORMATION

Schreck et al. [10] postulate five possible mechanisms for the origin of inv dup(15):

1) Pre-meiotic breakage in a single chromosome 15, followed by duplication and fusion of the two sister chromatids. The result is a dicentric inv dup(15) with identical ends, i.e., an iso dic(15).

2) Translocation between the proximal long arm of a chromosome 15 and the short arm of its homologue. The resulting inv dup(15) has a single centromere and different extremities.

TABLE III. inv. dup(15) in normal individuals

Reference	Case identification	Sex	Breakpoints	Size[a]	Morphology[a]	Ends[a]	Derivation	Familiarity
Stetten et al. [31]	Propositus	M	q1::q1	<G	?	D	Maternal	Familial
	Mother	F	q1::q1	<G	?	D	Maternal	Familial
Voss et al. [17]	Father	M	?	?	Metacentric	E	Maternal	Familial
Pecile et al. [32]	Fetus	M	?	G	?	D	?	De novo
Knight et al. [33]	Proposita	F	q12::q12	G	?	D	Paternal	Familial
	Father	M	q12::q12	G	?	D	Maternal	Familial
Wisniewski and Doherty [19]	1, Fetus	M	q1::q1	?	Sub-metacentric	D	Paternal	Familial
	1, Father	M	q1::q1	?	Sub-metacentric	D	?	Familial
	2, Fetus	M	q1::q1	?	Sub-Metacentric	E	?	De novo
Present case	4, Propositus	M	q11.2::q11.2	<G	Metacentric	D	?	Familial
	4, Brother	M	q11.1::q11.2	<G	Metacentric	D	?	Familial
Present case	5, Proposita	F	q11.2::q11.2	G	Sub-metacentric	D	Maternal	Familial
	5, Mother	F	q11.2::q11.2	G	Sub-metacentric	D	Paternal	Familial
	5, Sister	F	q11.2::q11.2	G	Sub-metacentric	D	Maternal	Familial
	5, Grandfather	M	q11.2::q11.2	G	Sub-Metacentric	D	Maternal	Familial
Present case	6, Fetus	M	q11::q11	<G	Metacentric	E	Paternal	Familial
	6, Father	M	q11::q11	<G	Metaentric	E	Paternal	Familial

[a]As defined by the authors. G, E, F refer to the size of the chromosomes. Chromosome ends are defined as equal (E) or different (D).

3) U-type exchange between two non-sister chromatids; a dicentric inv dup(15) with different ends will be formed.

4) A crossing-over within the inversion loop in a heterozygote for a paracentric inversion will result in an inv dup(15) similar to that formed with mechanism 3.

5) A crossing-over within the inversion loop in individuals heterozygous for a pericentric inversion; the inv dup(15) has single centromere and different extremities, as in mechanism 2.

None of the mechanisms take into account the possibility of a crossing-over in 15p, which is considered a highly improbable event [34]. Possibilities 2 and 5 are not likely to be involved in inv dup(15) formation because in fact they are expected to produce a monocentric abnormal chromosome, while all the inv dup(15) reported are dicentric.

Mechanism 1 is that solely consistent with iso dic(15) formation, while mechanisms 3 and 4 can both result in an inv dup(15) with different extremities. As regards mechanism 4, only De Falco and Willey [12] claim a probable inversion in one parental chromosome 15 in two of their three cases.

It should be noted that the U-type exchange postulated for mechanism 3 can take place both at the same locus in the two non-sister chromatids, but after breakage at different loci. In the first situation a tetrasomy for the short and for the proximal long arms would result. An asymmetrical exchange, on the other hand, will produce an inv dup(15) of similar size but with tetrasomy for the short arm and the proximal portion of the long arm, while for the more distal portion of the long arm there will only be a trisomy (Fig. 6). Of the 46 inv dup(15) considered, 41 have informative polymorphisms. Twenty-nine, having different extremities, originate from two homologous chromosomes 15 through mechanisms 3 or 4. Twelve inv dup(15) have similar ends and seven of these, obviously derived from one single chromosome 15, must be formed through mechanism 1.

DERIVATION OF inv dup(15)

Sixteen cases of de novo inv dup(15) were informative for maternal or paternal derivation: 13 originated in the maternal germinal line (12 from two homologues and one from a rearrangement of a single chromosome) and three in paternal gametogenesis (one from two and two from a single chromosome 15). The presence of an extra inv dup(15), independently from the various mechanisms of formation already discussed, requires in addition one non-disjunction and centromere inactivation. The timing of these events can be determined by polymorphism analysis of the inv dup(15) and the normal

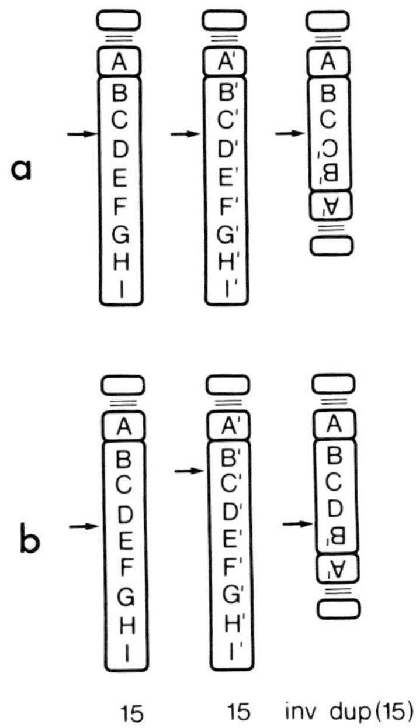

Fig. 6. U-type exchanges between non-sister chromatids of two chromosomes 15 result in inv dup(15) morphologically identical, but with different genetic content if the breakpoints are at the same locus (**a**) or at different loci (**b**).

chromosome 15. A first meiotic non-disjunction followed by a centromeric inactivation can be assumed whenever satellites of the short arm of the derivative chromosome and those of the normal chromosome 15 inherited from the parent in whom the anomaly took place are different (Fig. 7) Conversely, proximal satellites of the extra chromosome identical to those of the normal chromosome 15 indicate that the inactivation of one centromere at first meiotic division prevents a bridge formation between the two homologues and, after a second division non-disjunction, a gamete with inv dup(15) and a normal chromosome 15 is produced [13] (Fig. 8).

In five of the 14 informative cases, centromere inactivation followed first meiotic non-disjunction, while in nine the centromere inactivation was the primary event, followed by second meiotic non-disjunction.

PARENTAL AGE

Increased maternal age in inv dup(15) has been suggested by Wisniewski et al. [13] and by Maraschio et al. [15] who found a mean maternal age of

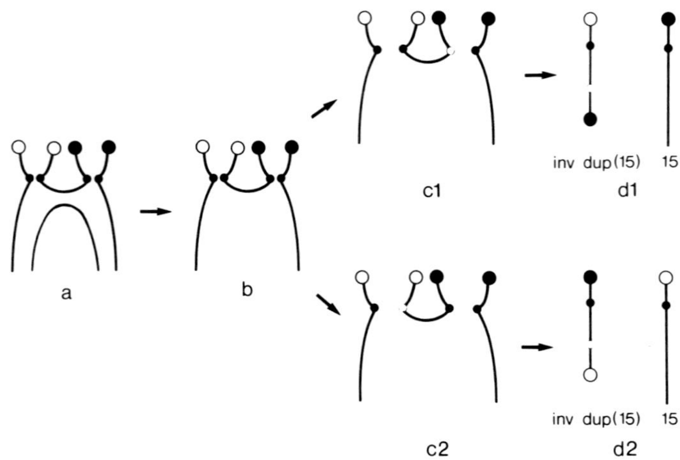

Fig. 7. U-type exchange (a), followed by a first meiotic non-disjunction (b) and inactivation (○) of one of the two centromeres in the derivative dicentric (c1 and c2), produces gametes with an extra inv dup(15) having proximal satellites different from those of the normal chromosome 15 (d1 and d2).

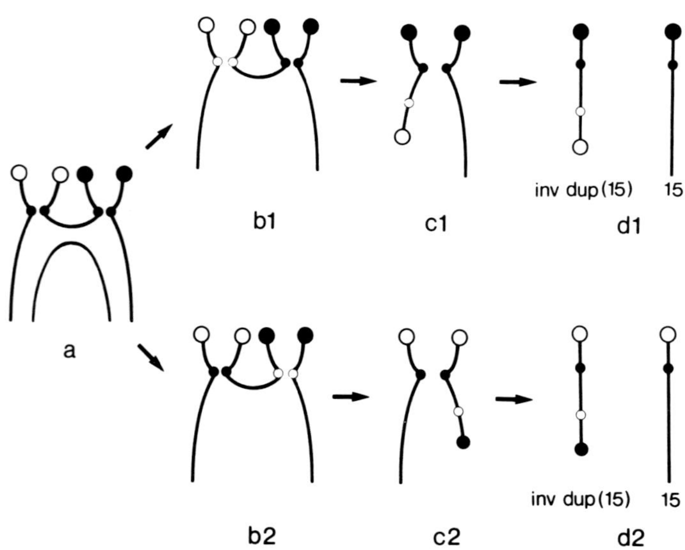

Fig. 8. U-type exchange (a) and inactivation (○) of one of the two centromeres (b1 and b2) can be followed by first meiotic disjunction (c1 and c2); a second meiotic non-disjunction will produce gametes with an extra inv dup(15) having proximal satellites identical to those of the normal chromosome 15 (d1 and d2).

35.7 and of 35.4 years, respectively. For all the 27 de novo cases considered in this review, the mean maternal age is 35.1 years and the mean paternal age 36.1 years. Connor and Gilmore [35] analyzed parental ages in their group of 16 sporadic inv dup(15) with a mean maternal and paternal age of 32.7 and 36.3 years, respectively, and found a strong direct maternal age effect, whereas the paternal age effect was not statistically significant.

If we consider separately the de novo inv dup(15) found in patients with 1) mental retardation only and 2) those with PWS, the two groups show a mean maternal age of 35.8 and 32.0 years, respectively. The two groups are, however, too different in size (24 cases with known maternal age in the first group and only 4 in the second) for an informative comparison.

CONCLUSIONS

All inv dup(15) reported in the literature from a morphologic point of view can be classified in two main groups, being reported either as (i) acrocentrics (or sub-metacentrics) of size similar or longer than a G group chromosome or as (ii) metacentrics, almost entirely heterochromatic and usually smaller than those of the previous group.

Inv dup(15) of the first type can be found after breaks at the same level or at different levels and can be therefore described as 15pter→15q12::15q12→15pter and as 15pter→15q11::15q13→15pter, respectively. The inv dup(15) of the second type are all a consequence of breaks at q11 (15pter→15q11::15q11→15pter).

The extra chromosomes of the first group are associated with a quite homogeneous phenotype, characterized by psychomotor retardation and no major physical abnormalities. In these cases mental retardation varies from mild to severe, and this variation could in turn be dependent on the different amounts of proximal 15q euchromatin.

The second type of inv dup(15), the small metacentrics, are instead found in two different groups of subjects, normal individuals and patients with PWS. In the PWS group, however, the inv dup(15) seem to originate mainly from one single chromosome 15, as indicated by the identical polymorphisms on their extremities. On the contrary most of the small metacentric inv dup(15) found in normal subjects, having different ends, derive from two homologous chromosomes 15. The same origin from a non-sister-chromatid exchange (i.e., from two homologues) appears to be predominant also in the acrocentric group of inv dup(15) found in patients with mental retardation.

Absence of phenotypic alterations in subjects with a small extra metacentric inv dup(15) can be due to the almost entire heterochromatic nature of these derivative chromosomes. However, it is not clear why PWS should also be associated with metacentric inv dup(15). The only obvious difference between the inv dup(15) found in the normal and PWS subjects is the mechanism of formation. Therefore, tentatively PWS could be related to a hypothetical regulatory system located on the proximal 15q portion, altered by the formation of an iso dic(15).

Preferential maternal derivation of inv dup(15) is evident, at least in the patients with mental retardation, the most informative group from this point of view. In the same group, there is also a maternal age effect.

ACKNOWLEDGMENTS

We are grateful to Professor M. Fraccaro for his help in the preparation of the paper.

REFERENCES

1. Friedrich U, Nielsen J: Bisatellited extra small metacentric chromosome in newborns. Clin Genet 6:23, 1974.
2. Miller OJ, Schnedl W, Allen J, Erlanger BF: 5-Methylcytosine localized in mammalian constitutive heterochromatin. Nature 251:636, 1974.
3. Schweizer D, Ambros P, Andrle M: Modification of DAPI banding on human chromosomes by prestaining with DNA-binding oligopeptide antibiotic, distamycin A. Exp Cell Res 111:327, 1978.
4. Mattei MG, Soviah N, Mattei JF: Chromosome 15 anomalies and the Prader-Willi syndrome: Cytogenetic analysis. Hum Genet 66:313, 1984.
5. Buckton KE, O'Riordan ML, Ratcliffe S, Slight J, Mitchell M, McBeath S, Keay AJ, Barr D, Short M: A G-band study of chromosomes in liveborn infants. Ann Hum Genet 43:227, 1980.
6. Watson EJ, Gordon RR: A case of partial trisomy 15. J Med Genet 11:400, 1974.
7. Power MM, Barry RG, Cannon DE, Masterson JG: Familial partial trisomy 15. Ann Genet 20:159, 1977.
8. Hongell K, Livanainen M: Partial trisomy 15 and temporal lobe syndrome in a retarded girl without gross malformation. Clin Genet 14:229, 1978.
9. Zannotti M, Preto A, Rossi Giovanardi P, Dallapiccola B: Extra dicentric 15pter→q21/22 chromosomes in five unrelated patients with a distinct syndrome of progressive psychomotor retardation, seizures, hyperreactivity and dermatoglyphic abnormalities. J Ment Defic Res 24:235, 1980.
10. Schreck RR, Breg WR, Erlanger BF, Miller OJ: Preferential derivation of abnormal human G-group-like chromosomes from chromosome 15. Hum Genet 36:1, 1977.
11. Van Dyke DL, Weiss L, Logan M, Pai GS: The origin and behavior of two isodicentric bisatellited chromosomes. Am J Hum Genet 29:294, 1977.
12. De Falco JE, Willey AM: Three cases of partial tetrasomy 15. Am J Hum Genet 30:77A, 1978.

13. Wisniewski L, Hassold T, Heffelfinger J, Higgins JV: Cytogenetic and clinical studies in five cases of inv dup(15). Hum Genet 50:259, 1979.
14. Chamberlin J, Magenis RE: Parental origin of de novo chromosome rearrangements. Hum Genet 53:343, 1980.
15. Maraschio P, Zuffardi O, Bernardi F, Bozzola M, De Paoli C, Fonatsch C, Flatz SD, Ghersini L, Gimelli G, Loi M, Lorini R, Peretti D, Poloni L, Tonetti D, Vanni R, Zamboni G: Preferential maternal derivation in inv dup(15). Analysis of eight new cases. Hum Genet 57:345, 1981.
16. Yip MY, Mark J, Hulten M: Supernumerary chromosomes in six patients. Clin Genet 21:397, 1982.
17. Voss R, Lerer I, Maftzir G, Sheinis M, Cohen MM: Partial trisomy 15 in a male with severe psychomotor retardation (48,XY,+15q−,+mar(15)). Am J Med Genet 12:131, 1982.
18. Gilmore DH, Boyd E, McClure JP, Batstone P, Connor JM: Inv dup(15) with mental retardation but few dysmorphic features. J Med Genet 21:221, 1984.
19. Wisniewski LP, Doherty RA: Supernumerary microchromosomes identified as inverted duplications of chromosome 15: A report of three cases. Hum Genet 69:161, 1985.
20. Fujita H, Sakamoto Y, Hamamoto Y: An extra idic (15p)(q11) chromosome in Prader-Willi syndrome. Hum Genet 55:409, 1980.
21. Wisniewski LP, Witt ME, Ginsberg-Fellner F, Wilner J, Desnick RJ: Prader-Willi syndrome and a bisatellited derivative of the chromosome 15. Clin Genet 18:42, 1980.
22. Ledbetter DH, Mascarello JT, Riccardi VM, Harper VD, Airhart SD, Strobel RJ: Chromosome 15 abnormalities and the Prader-Willi syndrome: A follow up report of 40 cases. Am J Hum Genet 34:278, 1982.
23. Wulfsberg EA, Sparkes RS, Klisak IJ, Gurfield WB: A (15→1) translocation in a patient mosaic for presence or absence of an isodici(15p)(q11). Am J Med Genet 13:417, 1982.
24. Mattei JF, Mattei MG, Giraud F, Prader-Willi syndrome and chromosome 15. A clinical discussion of 20 cases. Hum Genet 64:356, 1983.
25. Ridler MAC, Garrod O, Berg JM: A case of Prader-Willi syndrome in a girl with a small extra chromosome. Acta Paediatr Scand 60:222, 1971.
26. Michaelsen KF, Lundsteen C, Hansen FJ: Prader-Willi syndrome and chromosomal mosaicism 46,XY/47,XY,+mar in two cases. Clin Genet 16:147, 1979.
27. Kousseff BG: The cytogenetic controversy in the Prader-Labhort-Willi syndrome. Am J Med Genet 13:431, 1982.
28. Fraccaro M, Zuffardi O, Buhler E, Schinzel A, Simoni G, Witkowski R, Bonifaci E, Caufin D, Cignacco G, Delendi N, Gargantini L, Losanowa T, Marca L, Ullrich E, Vigi V: Deficiency, transposition, and duplication of one 15q region may be alternatively associated with Prader-Willi (or a similar) syndrome. Analysis of seven cases after varying ascertainment. Hum Genet 64:388, 1983.
29. Mattei MG, Mattei JF, Vidal I, Giraud G: Chromosome 15 abnormalities and the Prader-Willi syndrome. In: "Identification of Genetic Syndromes in Pediatrics. International Postgraduate Course." Firenze: October 14–17, 1982.
30. DeFrance HF, Beemer FA, Ippel PF: Duplication in chromosome 15q in a boy with the Prader-Willi syndrome; further cytogenetic confusion. Clin Genet 26:379, 1984.
31. Stetten G, Sroka-Zaczek B, Corson VL: Prenatal detection of an accessory chromosome identified as an inversion duplication (15). Hum Genet 57:357, 1981.
32. Pecile V, Archidiacono N, Rustico MA, Fiscella C, Rocchi M: A prenatally detected inv dup(15). Am J Med Genet 18:751, 1984.

33. Knight LA, Lipson M, Mann J, Bachman R: Mosaic inversion duplication of chromosome 15 without phenotypic effect: Occurrence in a father and daughter. Am J Med Genet 17:649, 1984.
34. Laurie DA, Hulten MA: Further studies on bivalent chiasma frequency in human males with normal karyotypes. Ann Hum Genet 49:189, 1985.
35. Connor JM, Gilmore DH: An analysis of the parental age effect of inv dup(15). J Med Genet 21:213, 1984.

The Cytogenetics of Mammalian Autosomal Rearrangements, pages 635–665

Chapter 21
Isochromosomes and Interstitial Tandem Direct and Inverted Duplications

Daniel L. Van Dyke

Tandem duplications of chromosome material can arise as direct or inverted intrachromosomal duplications and as inverted interchromosomal duplications (isochromosomes and isodicentrics). This chapter brings together the tandem duplications reported since chromosome banding techniques became available.

Direct intrachromosomal duplications appear to be more frequent than inverted, but both forms are quite uncommon and nearly every case has been unique in its breakpoints. Isochromosomes and isodicentrics are also quite rare, and for these there is non-random representation of only a few chromosome arms: 9p, 12p, 15q, 18p, 18q, 21q, 22q, Xq, Yp, and Yq. There is no parental age effect for direct duplications, but there is an advanced mean parental age for inverted duplications, isochromosomes, and isodicentrics.

A prezygotic origin for most tandem duplications is suggested by an absence of normal cell line mosaicism in most cases (except for secondary trisomy) and the advanced mean parental age. Direct duplications may arise by unequal crossing-over or unequal sister chromatid exchange. Inverted duplications may arise from a three-break rearrangement that includes a U-type exchange. Isodicentrics may arise from a two-break rearrangement with a U-type reunion. The origin of monocentric isochromosomes may be explained by centromere misdivision. The parental origin (maternal vs. paternal) is known in too few cases to have a meaningful interpretation.

INTRODUCTION

The purpose of this chapter is to discuss the types and origins of autosomal tandem duplications and isochromosomes. Most of the cases reported since the advent of chromosome banding techniques are cited. Inverted duplications involving chromosome 15 are fully discussed in Chapter 20 of this volume.

Medical Genetics and Birth Defects Center, Henry Ford Hospital, Detroit, Michigan 48202

A structural change in the karyotype that results in the doubling of a segment of chromosomal material is a duplication. If the duplicated segment is present within the same chromosome it is an intrachromosomal duplication. As it can be difficult to distinguish between a contiguous (tandem) and a non-contiguous duplication, common practice has assumed a tandem duplication whenever the duplicated segments are not separated by a cytogenetically visible non-duplicated segment [1–4]. A tandem duplication can be direct or inverted. There are two possible kinds of inverted tandem duplication within the arm, which can be represented diagrammatically as: centromere–ABCDE::EDC::F and centromereAB::EDC::CDEF. In humans, all of the cytogenetically visible inverted duplications appear to be of the first kind.

In an isochromosome, the arms are mirror images of one another: ABCDcenDCBA. In an isodicentric, the material between the centromeres is an extension of the mirror image from the arms: ABCDcenEF::FEcenDCBA. A stable isodicentric is usually pseudodicentric (there is only one primary constriction at metaphase); the term isopseudodicentric has been proposed [5]. In the short form of ISCN nomenciature [6], a monocentric isochromosome of 18p would be written i(18p), and an isodicentric 18p would be written dic(18) (q11). The classic definitions of isochromosome and isodicentric seem to require that the mirror images be genetically identical, originating from sister chromatids [7–9], but in general usage the terms have acquired broader definitions by not addressing their origins as identical (from sister chromatids) versus as merely homologous (from prezygotic or postzygotic exchange between homologs). Many mirror duplication chromosomes are called isochromosomes or isodicentrics but are not, sensu stricto, isochromosomes because they do not have genetically identical arms [10–14]. The terms isochromosome and isodicentric will be used here in the more general sense, not implying that the arms are genetically identical.

An isochromosome can be present as an extra element within an otherwise normal karyotype (secondary trisomy or so–called "partial tetrasomy"), and if so, one chromosomal segment is represented in four copies. Alternatively, a normal chromosome can be replaced by an isochromosome for one of its arms. This was termed "monoisodisomy" by Kimber and Sears [15–16], but I have not seen this term used in human cytogenetics as yet. Isochromosome X associated with gonadal dysgenesis and isochromosome 17q associated with leukemia are good examples of monoisodisomy. Rarely, a normal chromosome can be replaced by an isochromosome for each arm of the missing chromosome. This has been called "double monoisotrisomy" [15] or more simply "double isochromosome" [17].

LINKAGE AND PHENOTYPE-KARYOTYPE CORRELATION

It has been written that subjects with a tandem duplication are useful for study of the phenotype associated with specific chromosomal imbalances because there is no concurrent deletion of another segment, as is common in duplications derived from inversions and translocations. Hence these have been called "pure" duplications. The duplicated segment, especially when small, can be difficult to identify with confidence, even after multiple staining methods are employed. An ambiguous banding pattern and the absence of a parental chromosome rearrangement necessitates employment of other means to confirm the suspected identity of a small duplicated segment. It may be helpful to compare the phenotype of the subject with the phenotype of others who have a duplication of the same segment as that in the subject, such as in Down syndrome associated with dic(21) (q22) or iris colobomata with dup(22) (q11). Dosage studies of enzymes, other gene products, and DNA segments have been useful in some instances and need to be considered in others [18–31] (Table I). Conversely, confirmed tandem duplications can be useful in gene mapping. In one instance the concept worked both ways: a 10p duplication was confirmed by demonstrating increased phosphofructokinase activity (PFK was previously assigned to 10p), and the same duplication was employed to localize hexokinase I to the same segment [26].

Clinicians and geneticists are encouraged to bank cell lines in one of the mutant cell repositories since with few exceptions each tandem duplication has had unique chromosomal breakpoints. For the same reason, mutant cell repositories are encouraged to save these cell lines, even when there is uncertainty as to the exact segment that is duplicated.

TABLE I. Confirmatory genetic studies in tandem duplications

Duplicated segment	Assay	Reference
1q42→q43	5S rRNA	18
i(6p) vs i(17q)	Selection in cell hybrids	19
7q11→q22 ?	B-glucuronidase	20
8p12→p23	Glutathione reductase	21, 22
9p13→p24	Galactose-1-P uridyl transferase	23
+i(9p)	Galactose-1-P uridyl transferase	24, 25
10p11.1→p15.3	Phosphofructokinase and hexokinase I	26
+i(12p) vs +i(21q)	Lactate dehydrogenase B	27
+i(18p)	In situ hybridization of a probe	28
21q21→q22.2	Superoxide dismutase	29
dic(21)(q22)	Superoxide dismutase	30
22q11	In situ hybridization of a probe	31

A clinical reason for pursuing questions of apparent tandem duplications is that when the parents' chromosomes are normal, the empiric and the theoretical risk of recurrence in subsequent pregnancies appears to be very low. But in some cases, one parent has had a chromosome rearrangement. Rarely, a tandem duplication has been caused by an unusual recombination event within the meiotic inversion loop in a paracentric inversion heterozygote: e.g., dup 3q26.3 from an inv(3) (q12q29)pat [32], and dup 7q11.22→7q11.23 from an inv(7) (q11q22)mat [33]. An *apparent* inverted tandem duplication can arise as a stable dicentric duplication-deficiency recombinant from a paracentric inversion. Mules and Stamberg [34] reported a child with an "inverted tandem duplication" of chromosome 14, derived as a recombinant from a parental paracentric inversion inv(14) (q24.2q32.3). We have studied a family in which the malformed proband had what appeared to be a large dic(9) (q33) replacing a normal 9. Her mother and several relatives had a paracentric inv(9) (q22q34). Thus, the rec(14) and the rec(9) were inverted non-contiguous duplications, not inverted tandem duplications, since the segment within the parental inversion loop intervened and was not duplicated. Paracentric inversions in particular, and balanced rearrangements in general, are not always detected even with banding [35]. A small balanced paracentric inversion might not be detected in the parent of a patient with a dic(8)(p21), dic(13)(q3), dic(18)(p11), dic(21)(q22), or a small tandem duplication.

Among tandem duplications (Table II; refs. 36–97), there were three families with autosomal duplications which did not limit fertility but which did cause mild to moderate mental deficiency and mild dysmorphic features (Fig. 1, Table II). There are two published families with a direct duplication of an X-chromosome segment [36,37] and probably two familial isodicentric X-chromosomes [98,99]. Taylor et al. [100] described a child with secondary trisomy 18p whose phenotypically normal mother had a karyotype of 47,XX,18p−,+i(18p). Howell et al. [101] described a dicentric direct duplication, dup(21) (p13→q22), derived from a cross-over event in her ring 21 carrier mother.

DIRECT DUPLICATIONS

About 35 subjects have been reported with an autosomal direct duplication (Table II). Many more cases exist but have never been published, partly because of uncertainty regarding exactly what is duplicated. Direct duplications range in size from the two nucleotide duplication of Hemoglobin Cranston and Hemoglobin Tak, to Hemoglobin anti-Lepore mutants (the

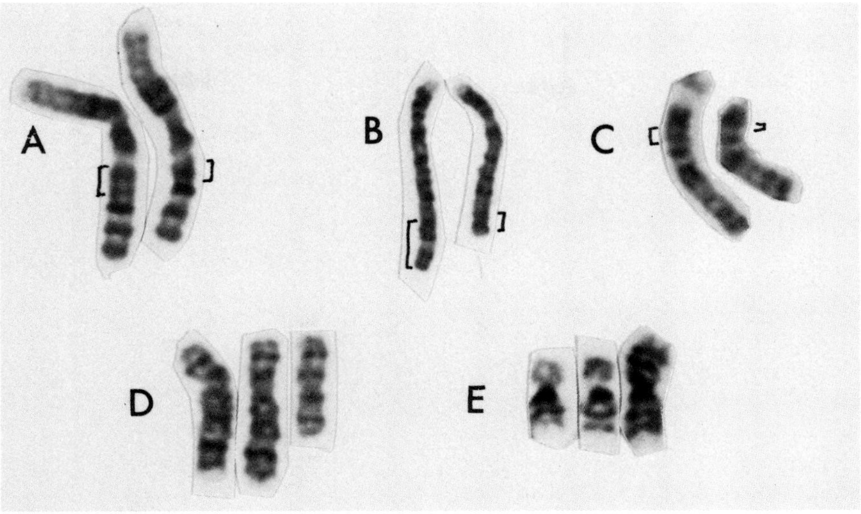

Fig. 1. Partial karyotypes of three familial duplications and two new secondary trisomies. **A:** 46,XX,dir dup(1)(q23q25). **B:** 46,XX,dir dup(4)(q31.22q33). **C:** 46,XX,dir dup(15)(q11q13). **D:** 46,XY/47,XY,+i(8p). **E:** 47,XY,+i(16q).

duplication counterparts of Hemoglobin Lepore deletions), to cytogenetically visible duplications.

A few patients with tandem duplications have been mosaic with a normal cell line: a dup(15) was interpreted as direct, but the direction of a dup(7) and two dup(12)s was uncertain [65, 84, 85, 88]. Postzygotic origin of the duplication is likely in these subjects. In the remainder, prezygotic origin is a good possibility because no normal cell line has been observed. Familial direct duplications have been observed and were mentioned above (Fig. 1, Table II). A direct duplication could originate in a meiotic cell or during G2 of mitosis in a somatic cell, by unequal crossing-over, a two-break reciprocal translocation, or a three-break insertion between sister or non-sister chromatids [49,58,89,102]. Theoretically, it could also arise during mitosis by unequal sister chromatid exchange. Direct duplication and triplication are rather common chromosome mutations in cancer and leukemia cells [38,102,103]. Unequal crossing-over in a female *Drosophila* homozygous for the tandem duplication "Bar" can create the tandem triplication "Ultra-bar" [104,105]. Multiple tandem repeats have been observed at the Bar locus, including a tandem quintuplication, Star [105], as well as sextuple-Bar and probable octuple-Bar [106].

TABLE II. Tandem duplications of euchromatic regions identified by banding techniques

Region	Maternal age	Paternal age	Reference
Direct duplications			
Familial cases			
Xp21→pter mat	23	25	36
Xq13→)q22 mat	23	40	37
1q23→)q25	22	23	Our case B86-85
1q23→)q25 mat	17	—	Our case B86-243
4q31.22→)q33	28	42	Our case R83-73
4q31.22→)q33 mat	30	34	—[a]
15q11→)q13	—	—	Our case B86-33
15q11→)q13 mat	20	—	Our case B85-671
Non-familial cases			
Xq13.3→)q24	—	—	38
Xq21→)q27	18	30	39
1p31→)p33	21	—	40, case 2276
1p31→)p35	26	—	41
1q32→)q42 or q43	20	25	42, 43
1q32→)q44	21	21	44
2q21→)p25	19	23	45
2p14→)p23	25	30	46
2q11.2→)q14.2	21	22	47
3q21→)q27	35	38	48
3q23→)q27	21	26	49
3q25→)q29	32	—	50, case 2
4q25→)q31	21	22	3
5q13→)q22	33	33	4
6q13→)q15	19	26	—[b]
6q21.1→)q24.2	28	33	Our case B80-437
8p21→)p23	27	25	51, case 1
8p11→)p22	28	38	52, case 7
9p13→)p22	30	29	1

Region	Maternal age	Paternal age	Reference
Inverted duplications			
Xq13→)q27.1	40	43	68
Yq	35	38	69
2q34→)q37	27	29	70
3q25→)q29	26	—	50, case 1
4p12→)p16	30	33	71, case 3
4q22→)q34	31	32	72, case 3
4q25→)q34	31	32	73, case 3
4q26→)q35	18	21	22, case 3
6p21→)p25	22	25	74
7q11→)q22?	31	43	20
8p21→)p23	27	28	21, 22, case 2
8p21→)p23	28	28	Our case B85-44
8p12→)p23	32	40	21, 22, case 1
8p11→)p22	37	40	52, case 8
9p13→)p24	36	39	23, case 1
9p21→)p24	29	33	75
9p21→)p24	23	23	76, case 1
10p11.1→)p15.3	30	34	26
11p15	24	28	77
12q24.2→)qter	27	27	78
14q130→)q243	24	26	79
19p13.2→)p13.3	23	23	Our case B85-829
Tandem duplications, direction is uncertain			
Xq13→)q24 ?	22	—	80
1q25→)q32	20	27	81
1q31→)q43 or 44	24	—	18
3q25→)q29	34	35	50, 82
3q26.3	26	31	32, case 2

9p12→)p24	26	35	53, case 2	4p14→)p16	27	32	83
9p12→)p24	31	35	54	7p1→)p2/normal	26	29	84, 85
9q11→)q33	28	29	55, case 1	7q11→)q22	29	29	86
9q12→)q32	27	29	56	7q11.22→)q11.23	22	26	33
9q12→)q32	33	38	57	10p12	34	29	2
10q11→)q22	27	28	58				
10q21→)q22	23	26	59	11p15	27	29	87
10q24.3→)q26.3	27	29	Our case R82-16	12q24.1→)q24.5/normal	28	—	88, case 1
12q14.3→)q21.1	—	—	60	12q24.1→)q24.5/normal	22	—	88, case 2
13q14→)qter	28	29	61	13q distal third	23	29	89
13q21→)qter	18	23	62	13q21→)q333	—	—	90
13q22→)q34	—	—	63	14q24→)q32	17	26	91
14q11.2→)q13	34	42	Our case B84-49	14q24→)q32	30	30	92
15q14→)q21	34	36	64	15q12→)q15 ?	35	36	93
15q21→)qter/ normal	18	29	65	15q14→)q15	—	—	94
				17q25	39	43	95
16p13.3→)p11.2	33	43	—[b]	18q12	18	21	96
dic 16p11→)q22 ?	—	—	14	21q223	29	34	97
16q11→)q13	—	—	66	22q11→)q13 ?	28	31	22, case 4
dic 17p12→)q11.1	—	—	67				
21q21→)q22.2	32	51	29				

In most instances the published interpretation was accepted. Except for the direct duplications inherited intact from one parent, and two inversion recombinants discussed in the text, all were de novo. All of the duplications cited were identified by banding techniques and except for the dic dup(16) and dic dup(17), all were monocentric. Mosaicism is indicated with a slash.

[a]S.A. Powell, personal communication, 1982.

[b]R. Stallard, personal communication, 1986.

INVERTED TANDEM DUPLICATIONS

Only about 20 autosomal inverted tandem duplications have been reported (Table II). Several mostly smaller duplications of uncertain origin have also been reported (Table II). A few chromosomal segments may be non-randomly represented among inverted duplications since half of the cases involved duplication of 4q, 8p, or 9p. A prezygotic origin in most cases is favored because the average parental age is advanced (Table III) and because none were determined to be mosaic with a normal cell line. Chiyo et al. [75] suggested that an inverted duplication can arise if a chromosome segment is transferred inverted to a homolog in gametogenesis of either parent. Taylor et al. [22] proposed that an inverted tandem duplication can arise by a three-break unequal exchange between homologous chromosomes or homologous chromatids. They also postulated unequal exchange between strands of a single DNA double helix, which would be far less likely because this would reverse the polarity of the DNA strands. A three-break exchange seems to be the minimal event, whether during meiosis or in a mitotic cell (Fig. 2) [68]. If one parent is a paracentric inversion heterozygote, a single U-type exchange during meiosis, within the inversion loop, can also create an

TABLE III. Mean parental ages and 95% confidence intervals for several classes of tandem duplications and isochromosomes

	Maternal age	Paternal age	No. cases	Reference
1970 Michigan	24.5	27.6	170,000	—[a]
1980 Michigan	25.0	28.1	145,000	—[a]
dir dup	25.8 (24.0 - 27.6)	30.6 (28.0 - 33.2)	33	
Monoisodisomy	26.2 (24.2 - 28.2)	28.2 (26.2 - 30.2)	19	
"t(X;X)"	27.4 (23.7 - 31.1)	30.2 (26.2 - 34.2)	15	—[b]
dic (X)(p1)	28.4 (25.7 - 31.1)	29.1 (26.5 - 31.7)	25	—[b]
inv dup	28.7 (26.3 - 31.1)	31.2 (28.1 - 34.3)	22	
Down syndrome 1970 - 1975	29.3	31.9	380	—[a]
Isochromosome	29.5 (27.7 - 31.3)	32.9 (30.8 - 35.0)	40	
Isodicentric	29.9 (28.0 - 31.8)	32.8 (30.4 - 35.2)	28	
dic(Y)(p or q)	—	32.9 (29.5 - 36.3)	14	—[b]
Secondary trisomy	31.0 (29.5 - 32.5)	34.6 (32.9 - 36.4)	49	
+dic(15)(q1)	34.5	36.4	35	—[b]

Inherited cases and those ascertained in a prenatal genetic study are excluded. For dic(X), t(X;X), and dic(Y) the ages of the parents were usually not published, so a reporting bias of the older parents is possible. The "secondary trisomy" classification does not include cases of +dic(15)(ql). The "isochromosome" classification includes the monocentrics and those not C-banded but apparently monocentric.

[a]Personal communication, Information Request Unit, Office of Vital Statistics, Michigan Department of Public Health.

[b]Van Dyke, unpublished.

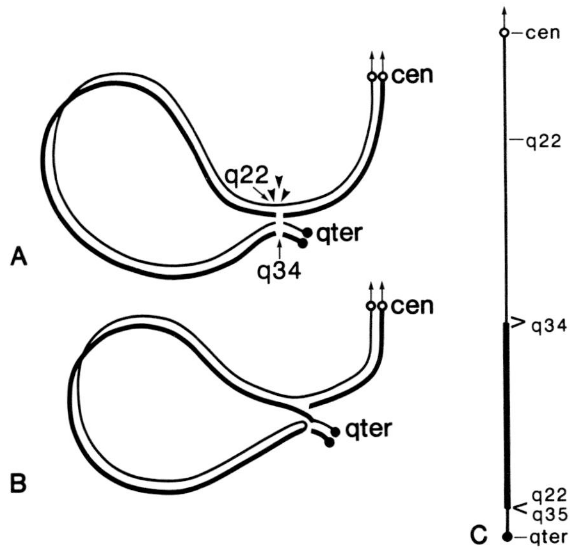

Fig. 2. Formation of an inverted duplication. Case 3 of Vogel et al. [72] is used as a typical example, with an inverted duplication of 4q22→4q34. Only two strands are necessary to effect the duplication, although two more strands would be in close promixity if the rearrangement were to occur during meiotic prophase I. The two strands can be sister chromatids or non-sister homologous chromatids or chromosomes. **A:** One strand is broken at band 4q22 and both strands are broken at the more distal band, 4q34. **B:** Normal repair processes create a U-type exchange at band 4q34 and ligation of the distal part of the 4q22 breakpoint with one of the two acentric telomeric fragments. **C:** Unfolding shows an inverted duplication. Inspection of the figure clarifies why inverted duplications tend to take the form: centromere-ABCDE::EDC::F rather than centromereAB:EDC::CDEF.

inverted tandem duplication (but may at the same time create a deletion distal to the duplication). Two possible examples were cited earlier [32,33]. It is plausible that among the other tandem duplications reported as de novo (Table II), a few of the parents carried an unrecognized paracentric inversion.

AUTOSOMAL ISOCHROMOSOMES

Like intrachromosomal duplications, isochromosomes are uncommon constitutional chromosome abnormalities [Table IV; refs. 107–176]. The nonrandom appearance of only a few of the many possible secondary trisomies and monoisodisomies among dysmorphic individuals (9p, 12p, 15q, 18p, 18q, 21q, and 22q) may be more the result of differential viability than of skewed origins, but could be a combination of the two. Probably the most common isochromosome is i(Xq), whereas i(Xp) is probably nonexistent [177,178]. The most common autosomal isochromosome, by a slim margin, appears to be i(18p). Metacentrics derived from acrocentric short arms (bisatellited markers) or homologous long arms (e.g., 13q;13q fusion) are usually classified as Robertsonian translocations, but the alternative origin as isochromosomes usually cannot be excluded.

Eleven patients with isochromosome 18q are described in the banding era literature. Monoisodisomy 18q has been divided into two clinical entities. If the isochromosome is monocentric, the patient is likely to have features of both trisomy 18 and 18p deletion, but the patient is likely to express only trisomy 18 features if the isochromosome is dicentric or if the patient has both i(18p) and i(18q) [113].

There have been several Down syndrome patients with a duplication of 21p13→q22 due to an isodicentric replacing one normal 21. That some subjects with this duplication have mild forms of Down syndrome underscores the importance of distal 21q in the pathogenesis of Down syndrome [125,126].

Secondary trisomy 9p has been reported in 13 patients, and we have observed it once in a prenatal genetic study (Table II). The karyotype can be confirmed by GALT assay [24,25,139]. In about half of the cases the extra chromosome is an isodicentric, and at least a few appear to be monocentric. The pattern of mosaicism is remarkable. All or most of the lymphocyte metaphase cells contain the isochromosome, whereas fibroblasts invariably have a karyotypically normal line ranging from 28 to 100% of metaphase cells. Remnant activity of the second centromere may partly account for this mosaicism; in one case, 4% of the isodicentrics had two primary constrictions, but there was no evidence of anaphase bridges or fragments [137]. The phenotype is quite variable [138].

Secondary trisomy 12p has been reported in about 12 cases (Table IV). It is also known as "Killian," 'Teschler-Nicola/Killian," "Pallister mosaic," and "mosaic tetrasomy 12p" syndrome. This diagnosis has doubtless been missed in some patients because all have been mosaic with the abnormal karyotype present in fewer than 1% of lymphocytes and more than 95% of fibroblasts [145]. Thus the pattern of mosaicism is quite the reverse of that in secondary trisomy 9p, most likely reflecting tissue-specific in vivo selection processes. The isochromosome 12p has also been mistakenly reported as isochromosome 21q more than once [143]. Biochemical studies have been helpful because secondary trisomy 12p has elevated LDH-B and normal SOD-1 activities (Table I) [27]. The i(12p) was apparently monocentric when C-banding was done. The syndrome seems characteristic [27,144]. Since isochromosome 12p is frequent in testicular tumors [179,180], it would be of interest to follow secondary trisomy 12p subjects for malignancies and to obtain testicular karyotypes when possible, and to obtain skin karyotypes on some patients with testicular tumors whose tumor cells have an isochromosome 12p.

Secondary trisomy 18p syndrome was described before banding [181], and about 20 cases have been reported with banding. This isochromosome has been almost always described as monocentric, but in some cases this is arguable because the C-band positive segment appears larger than than on the normal 18s in some published photographs, and on higher resolution may resolve into two C-bands. Peripheral blood karyotype mosaicism was observed in one case, and there was tissue-to-tissue variation in a case identified prenatally [166,168]. The other cases, including several with blood and fibroblast karyotype studies, had no evidence of mosaicism with a normal cell line. The syndrome seems characteristic [164,165]. All cases have been de novo, except one in which the mother had an 18p− and an i(18p) [100]. In situ hybridization of an 18p11-specific DNA probe confirmed the identity of the i(18p) in one case [28].

The cat eye syndrome has been caused by secondary trisomy 22pter→22q11 in more than 11 cases. In situ hybridization of a 22q11 probe in several cases has resolved the question of the origin of the extra chromosome [31]. Fibroblast karyotypes need to be studied on more of these patients to determine whether there is a tissue-dependent pattern of mosaicism similar to that observed in secondary trisomies 9p and 12p. One patient had normal cells in about 3% of lymphocytes, and 70% of fibroblasts [169].

I am aware of one case each of secondary trisomy 8p and secondary trisomy 16q (Fig. 1). Secondary trisomy 16q was observed in all of 23 metaphase cells from fibroblast cultures established from chorionic villi after a 10-week empty sac miscarriage. The parental karyotypes were normal. Secondary trisomy 8p was found in an infant with multiple malformations including patient ductus arteriosis, ventricular septal defect, abnormal thoracic vertebrae, flailed chest secondary to severely abnormal ribs, nevus flammeus on the forehead, prominent frontal bossing, large simple ears, anteverted nares, absent corpus callosum (by CAT scan), and grand mal seizures. At age 7 mo he had been on a ventilator since birth, had a head circumference of 16 in, weight 4,625 g, and developmental level of 6–8 wk. He died at age 8 mo. The peripheral blood karyotypes showed 27 of 89 cells 47,XY,+i(8p) and 62 cells (70%) normal. A skin fibroblast karyotype analysis showed 8 of 50 cells with the secondary trisomy, and 42 (84%) normal cells. A prenatal genetic study had shown 50 karyotypically normal amnio-

TABLE IV. Isochromosomes of euchromatic regions identified by banding techniques

Region	Maternal age	Paternal age	Reference
Monoisodisomy and double isochromosomes			
dic(8)(p21)/normal	32	32	107
i(9q)	35	34	10
dic(13)(q33)/r(13)/der dic	20	—	108
dic(13)(q34)/der dic	28	28	109
i(18q)	28	—	110
i(18q)	20	20	111
i(18q)	37 pgs	38	112
i(18q)	22	26	113
i(18q)/18p−	— pgs	—	114
dic(18)(p11)/r(18)/-18	25	31	5
dic(18)(p11)	—	—	115, case 2
dic(18)(p11)?	—	—	14
dic(18)(p11)	30	33	116
dic(18)(p11)	20	21	117
i(18p) and i(18q)	25	31	118
i(18p) and i(18q)	23	28	17
dic(21)(q2)	29	28	119, 120
dic(21)(q2)	—	—	121
dic(21)(q2)	25	26	122
dic(21)(q2)	—	—	123
dic(21)(q2)?	—	—	124
dic(21)(q22)	—	—	125
dic(21)(q22)	28	28	126
dic(21)(q223)	27	28	127
dic(21)(q22.3)	30	29	128
dic(21)(q22)/normal	26	25	129

Region	Maternal age	Paternal age	Reference
Secondary trisomy (continued)			
+i(12p)/normal	32	44	141
+i(12p)/normal	—	—	141, two cases
+i(12p)/normal	39	43	142, 143
+i(12p)/normal	32	32	143, 144, 145, case 2
+i(12p)/normal	25	—	146
+i(12p)/normal	—	—	147
+i(12p)/normal	30	32	148, 149
+i(12p)/normal	—	—	150, two cases
+i(12p)/normal	31	32	143, 151
+i(12p)/normal	— pgs	—	27
+i(12p)/normal	30	—	145, case 1
+i(12p)/normal	35 pgs	34	152
+i(12p)/normal	38	45	S.M. Gollin and F. Char, personal communication, 1986
+dic(15)(q15)/der dic	38	—	153
+dic(15)(q15)	33	33	Our case B79-303
+i(16q)	30	31	Our case T85-132
+(17p)?	26	32	154
+i(17q)?	29	30	155
+i(18p)	34	32	156
+i(18p)	31	43	157, case 1
+i(18p)	22	30	157, case 2
+i(18p)	32	41	157, case 3
+i(18p)	39	39	157, case 4
+i(18p)	31	36	158
+i(18p)	19	27	100
+i(18p)	27	25	159
+i(18p)	26	30	160

dic(21)(q22.3)	24	31	30
dic(21)(q223)	35 pgs	48	130
Secondary trisomy			
+i(8p)/normal	35	35	Our case T85-154, B85-460, S.A. Powell, personal communication
+i(9p)/normal	31	32	131
+i(9p)/normal	25	25	132
+i(9p)	19	24	24
+i(9p)	30	31	25
+i(9p)/normal	32	34	53, case 1
+i(9p)	27	31	133
+i(9p)	29 pgs	32	Our case A84-796
+dic(9)(q1)?/normal	—	—	134
+dic(9)(q21)	35	38	135
+dic(9)(q)/normal	—	—	136
+dic(9)(q22)	33	35	137
+dic(9)(q13)/normal	34	31	138
+dic(9)(q12)/normal	38	36	139
+dic(9)(q22.1)?	35	46	140

+i(18p)?	—	—	161, four cases
+i(18p)?	37	41	162
+i(18p)	35	35	163
+i(18p)	35	37	Our case B80-358
+i(18p)	20	22	164
+i(18p)	38	40	165
+i(18p)	36 pgs	34	Our case A84-356
+i(18p)/normal	30	36	166
+i(18p)	—	—	28
+i(18p)	28	—	167
+i(18p)/normal	36 pgs	—	168
+dic(22)(q1)/normal	38	46	169
+dic(22)(q11)?	32	36	170
+dic(22)(q11)	27	40	171, case 1
+dic(22)(q11)	36	39	171, case 5
+dic(22)(q11)	29	30	171, case 8
+dic(22)(q11)	35	38	171, twins, cases 9, 10
+dic(22)(q11)	28	31	172
+dic(22)(q11)	32	30	173
+dic(22)(q11)	27	30	174
+dic(22)(q11)	35	39	175
+dic(22)(q11)	—	—	176

In most instances the published interpretation was accepted. The parental origin of the abnormal chromosome, when known, is noted by "pat" or "mat." Dicentric cases are identified by "dic." A "pgs" after maternal age indicates ascertainment in a prenatal genetic study. Mosaicism is indicated with a slash. Most cases of inv dup(15) are not included.

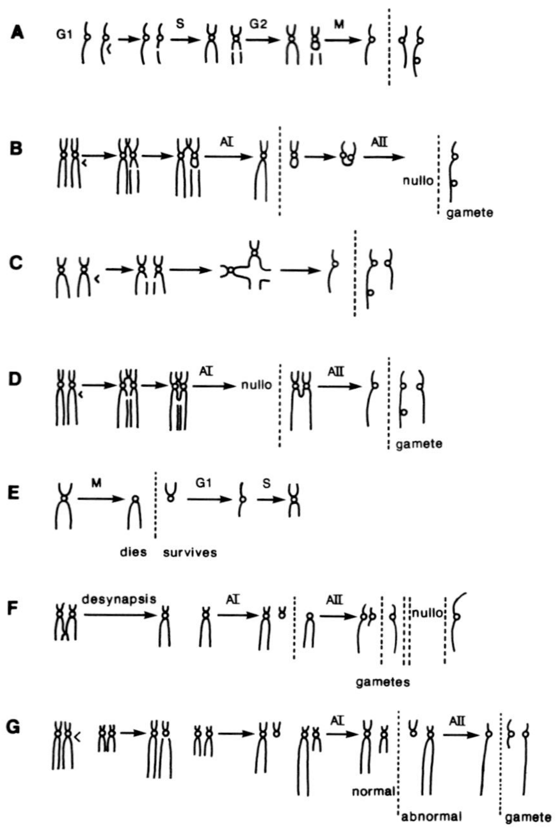

Fig. 3. Formation of isochromosomes. Open circle indicates a centromere; vertical dotted line indicates sister cells or gametes; G1, S, G2, M are mitotic stages; AI and AII are meiotic anaphase I and II. Acentric fragments are assumed to be lost. Sister chromatids can break and rejoin in mitosis (**A**) or meiosis (**B**), homologous non-sister chromatids can break and rejoin in mitosis (**C**) or meiosis (**D**), and centromere misdivision can occur in mitosis (**E**) or meiosis (**F**). In (**A**), a G1 break and G2 U-type reunion between sister chromatids in a mitotic cell can create monoisodisomy with an isodicentric. Under this model, the mirror images of the isodicentric would be genetically identical. In (**B**), a U-type exchange between sister chromatids during meiotic prophase I can create monoisodisomy with an isodicentric. There may be "centromere cooperation" at anaphase II. The arms are frequently non-identical due to genetic recombination. Fertilization results in monoisodisomy. In (**C**), a U-type exchange between homologs and adjacent segregation of the quadrivalent arms in mitotic anaphase results in monoisodisomy with an isodicentric. In (**D**), a U-type exchange between homologous non-sister chromatids during meiotic prophase I and "centromere cooperation" at anaphase I can create a gamete with one normal copy of the chromosome plus an isodicentric. Fertilization results in secondary trisomy (partial tetrasomy). In (**E**), centromere misdivision in a mitotic cell results in monocentric monoisodisomy. Under this model, the isochromosome would have

cyte metaphase cells (100% of cells normal). The parents had normal peripheral blood karyotypes (S. Powell, personal communication, 1986).

ISOCHROMOSOMES ARISING IN SOMATIC CELLS

Isochromosomes that replace normal chromosomes do arise in mitotic cells. Stallard et al. [182] found two cells with i(11q) among 14,710 leukocyte metaphases, and de la Chapelle [183] described an i(17p) and an i(17q) in adjacent leukocyte metaphase cells from a patient with an otherwise normal karyotype. Isodicentrics arise in Bloom syndrome lymphocytes as a result of exchange between homologues, or in endoreduplicated cells as a result of exchange between sister chromosomes [11,184]. Isochromosome formation appears to be a common means of karyotype evolution in neoplasia. Dicentric isochromosomes of the derived 17 have been observed in acute promyelocytic leukemia with t(15;17) [185–187], as have dicentric and monocentric i(17q) and iso-Philadelphia chromosomes in chronic myeloid leukemia [188–190]. Isochromosomes for 1p, 1q, 2p, 5q, 6p, 7q, 8p, 8q, 11q, 12p, 16q, 17q, 21q, and others have been reported in cancer [189,191–193]. We have observed isochromosomes 1q, 5p, 8q, 11q, 15q, and idic(3)(p25) in bladder tumor preparations [194], and isochromosome 5p may be the primary chromosome change in some bladder tumors [195]. Isochromosome 12p may be an important marker for malignant testicular tumors [179,180], as may i(6p) in retinoblastoma [19] and malignant melanoma [196].

ORIGIN OF DICENTRIC ISOCHROMOSOMES

An isochromosome's arms are homologous but may not be genetically identical. Both arms are probably identical if an isochromosome arises from centromere misdivision in a mitotic cell, from a mitotic exchange between sister chromatids, or from an asynaptic univalent during meiosis. However, meiotic recombination can result in all four chromatids being genetically distinct so that if an isochromosome arises during meiosis, the arms have a mirror image banding pattern but are not genetically identical. Likewise, an

---

genetically identical arms. In **(F)**, there is centromere misdivision involving a meiosis I univalent. The gamete on the left, if fertilized, results in secondary trisomy ("partial tetrasomy") for the short arm of the chromosome. The gamete on the right, if fertilized, results in monoisodisomy for the long arm of the chromosome. In **(G)**), there is a break at the centromere or proximal long arm, translocation of the long arm (whole arm translocation) to the distal long arm of a non-homolog, and subsequent formation of a short arm isochromosome.

isodicentric's centromeres and arms may be sisters (genetically identical) or homologous (not necessarily genetically identical).

Postzygotically, a U-type exchange [197] (or asymmetric reciprocal translocation [16]) could occur between sister chromatids or homologous chromosomes [11,12,198], and the resulting isochromosome would replace the normal homolog, causing monoisodisomy (Fig. 3A,C). Mosaicism with a normal cell line would be possible. Prezygotically, an exchange between sister chromatids in meiosis I or II could have the same outcome except that there would be no mosaicism with a normal cell line (Fig. 3B) [197]. Among the monoisodisomies, chromosomally normal cells have only been observed in one case [129], which suggests that in most cases the origin is prezygotic. This could be considered circumstantial evidence favoring prezygotic origin in most secondary trisomies ("partial tetrasomies") as well.

If the exchange occurs close to the centromere, the isodicentric can look similar to a monocentric isochromosome; i(18p) and dic(18) (q11) are best distinguished by C-banding. But if the exchange is distal to the centromere, the isodicentric can look quite unlike a classic isochromosome (e.g., dic(13)(q34), dic(15)(q1), and dic(21)(q22), Table IV).

For the origin of a secondary trisomy isodicentric, one of the simplest mechanisms is a meiosis I two-break U-type exchange between homologous non-sister chromatids (Fig. 3D). Extra isochromosomes or isodicentrics seem more likely to arise in meiosis I since a meiosis II or mitotic cell origin would require an independent non-disjunctional event in order to replace the homolog lost in formation of the isochromosome.

In the isodicentric, only one centromere forms a primary constriction at metaphase, and only that centromere is C-dot positive. However, the occasional rearrangement or loss of the isodicentric, plus the presence of anaphase pseudobridges and lagging if not a true bridge-break-reunion-break cycle, suggests residual spindle fiber function of the "latent," "inactivated," or "suppressed" centromere [109]. Vig [199] observed two primary constrictions at prometaphase and one at midmetaphase in a dicentric translocation; he suggested that out-of-phase separation of centromeres served to inhibit spindle microtubule attachment to the early-separating centromere.

Even when one centromere is not suppressed, there is evidence that centromeres in dicentrics "interact," or "cooperate," to increase the likelihood of parallel rather than criss-cross or interlocking orientation at anaphase [200–205]. The "cooperation" may be stronger if the centromeres are closer. In isodicentric X, the proportion of 45,X cells tends to be greater when the distance between the centromeres is greater [206]. The centromeres may have different relative strengths in moving toward the poles during anaphase.

Equal activity or strength of the centromeres in a dicentric would result in less stability. In human acrocentric/nonacrocentric translocation dicentrics, the centromere of the non-acrocentric usually predominates [207–210]. The interaction between the centromeres and the relative strengths of the centromeres may be influenced by heterochromatin or other chromosome material in the arms adjacent to the centromeres [211].

Mosaicism is commonly associated with isodicentrics, with some cells having new rearrangements derived from the original isodicentrics. The new rearrangements usually take the form of smaller isodicentrics. A normal karyotype is a common mosaicism in secondary trisomies; the assumption is that the isochromosome has been lost, rather than that this is evidence for postzygotic origin of the isochromosome.

A ring/isodicentric mosaicism was found in two cases involving chromosomes 13 and 18 [5,108], and in three other cases of ring/?isochromosome 14 and 21, C-banding was not performed [212,213]. It was postulated that the ring and isochromosome lines arose in an early mitotic cell, after a chromatid break in one arm and a chromosome break in the other arm. The strand with two breaks formed the ring and the strand with the one break formed the isodicentric after cell division, replication, and U-type fusion. In the chromosome 14 cases, the presence of normal cells confirms a postzygotic origin, but in the others, prezygotic origin is plausible for the isodicentric, with postzygotic origin of ring, perhaps derived from the isodicentric after fusion of one break between the centromeres and a second break in a free arm.

ORIGIN OF MONOCENTRIC ISOCHROMOSOMES

Darlington [8] proposed that monocentric isochromosomes in plants arise by centromere misdivision, thus forming a telocentric, and fusion of sister chromatids subsequent to the next replication. Many human isochromosomes appear to be monocentric, and their origin appears consistent with Darlington's centromere misdivision model. Monocentric monoisodisomy could arise by centromere misdivison in a meiotic or mitotic cell (Fig. 3E,F). Subsequent to the first mitotic division, centromere misdivision could result in mosaicism with a normal cell line. A normal line was not found in any of the patients with monocentric monoisodisomy (Table IV) and is very uncommon with isochromosome X [214,215], but is of course obligatory with isochromosome 17 in leukemia. Monocentric secondary trisomy may arise after a meiosis I centromere misdivision involving a univalent, followed by a U-type exchange (Fig. 3F) rather than by postzygotic centromere misdivision and subsequent non-disjunction of the normal homolog or by partial centromeres misdivision and non-disjunction involving the sister chromosomes. A monocentric "isochromosome" can also arise from a whole arm translocation or, for acrocentrics, from a Robertsonian translocation.

Since the DNA strand continues through the centromere, there is no reason why an intra-centromeric U-type exchange could not give rise to "centromere misdivision" and creation of a monocentric isochromosome. This would unify the mechanisms for the origins of mono- and dicentric isochromosomes. Lima-de-Faria [216,217] regards the centromere itself as "essentially a tandem reverse repeat" in several plant species. Comings and Okada [218] have suggested that the ultrastructure of quadripartite centromeres of metacentric chromosomes, including those of humans, is consistent with an origin by fusion of two telocentrics with bipartite centromeres. If the centromere is essentially a tandem reverse repeat, it could be easily reconstituted in its entirety by misdivision and U-type reunion.

Two cases have been described with both arms of chromosome 18 present as isochromosomes (Table IV) [17,118]. These "double isochromosomes" may have arisen by whole arm translocation or centromere misdivision and non-disjunction of the two isochromosomes in meiosis or in an early mitotic cell. Normal cells were not observed in either case.

There are seven patients who have 46 chromosomes, with the short arm of one chromosome present as an isochromosome, and the long arm translocated to the end of another chromosome (Table V) [219–226]. Three of these had i(9p) and two had i(5p). Apparently, all have been de novo. In at least one case the isochromosome was dicentric. Prezygotic origin seems likely since mosaicism with a normal cell line was not observed in any of the seven cases. The proposal that the end of the recipient chromosome is duplicated [220] needs to be confirmed. The origin appears compatible with centromeric misdivision or a break in the proximal long arm of one chromosome with

TABLE V. Isochromosomes associated with translocation of the other arm to another chromosome

Region	Maternal age (yr)	Paternal age (yr)	Reference
i(4p),t(1;4)(1qter→1p36::4cen→4qter)	26	27	219, 220
i(5p),t(2;5)(2pter→2q36::5q11→qter)	31	30	221
i(5p),t(5;22)(5qter→5q11::22p11→22qter)	28	26	222
i(9p),t(7;9)(7pter→7q36::9q11→9qter)	22	24	223
i(9p),t(9;18)(9qter→q12::18p11→18qter)	18	22	224
i(9p),t(9;15)(9qter→9q11::15p12→15qter)	22	28	225
dic(17)(q21),der(10)t(10;17)(p15;q21)	—	—	226

subsequent formation of an isochromosome of the short arm, and translocation of the telocentric or acentric long arm fragment to the end of some other chromosome (Fig. 3G).

NON-RANDOMNESS, PARENTAL AGE, AND PARENTAL ORIGINS

There is clear non-randomness in the frequency of isochromosomes of the autosomes in newborns and in cancers. Much of this non-randomness may be due to differences in viability of concepti (the constitutional isochromosomes involve small arms or small segments of chromosomal material) and individual cells (there is clear tissue-to-tissue variation in secondary trisomy mosaicism) with isochromosomes, but it is uncertain whether differential viability is the only reason for the non-randomness [177, 183]. For instance, chromosome 4 centromeric fission can result in two stable telocentric chromosomes (see Chapter 17, this volume), whereas telocentrics have not been reported for chromosome 5 as yet; perhaps centric fission of chromosome 5 preferentially results in formation of i(5p) and translocation of 5q to another chromosome (Table V).

Aside from the direction of the duplication relative to the centromere, there are two curious differences between direct and inverted tandem duplications. First, one breakpoint is terminal (or nearly so) in almost all inverted duplications, but is terminal in fewer than half of the direct duplications. Second, parental age is advanced for inverted duplications, but is apparently normal for direct duplications (Table III).

When the cases from Table IV are separated by number of centromeres, there is an apparent parental age effect for both monocentric isochromosomes and isodicentrics. However, whereas the parents of patients with secondary trisomy have advanced age (Table III), the parents of patients with monoisodisomy do not. This suggests that, although both secondary trisomy and monoisodisomy may arise by similar prezygotic mechanisms, the timing of their origins may differ. Thus, secondary trisomy may preferentially arise during meiosis I (Fig 3D,F) and monoisodisomy during meiosis II or early embryogenesis (Fig. 3A,C,E). A meiosis I univalent may be more likely to undergo centromere misdivision than a bivalent [7], and this may partly explain the advanced parental age, since the frequency of univalents increases with maternal age [227]. The absence of a parental age effect in monoisodisomy may relate to three models for its origin being postzygotic (Fig. 3A,C,E) and only two prezygotic (Fig. 3B,F), whereas a prezygotic origin of secondary trisomy seems nearly obligatory (Fig. 3D,F). Isodicentric X and Y chromosomes and large "dicentric" t(X;X) chromosomes may have origins similar to those suggested for autosomal isodicentric monoisodisomy (Fig. 3A,B,C), and a prezygotic origin is likely for at least some cases, since the parental age is slightly advanced and mosaicism with a normal cell line is very uncommon (Fig. 3B) [cf. 11,12,153,183,198]. Carothers et al. [214] have suggested a paternal age effect in i(X).

The advanced parental age associated with inverted duplication and secondary trisomy makes it relatively more likely that these will be found during prenatal genetic studies for advanced maternal age. Indeed several such cases have been observed (Table IV). The tissue-to-tissue variation in mosaicism for secondary trisomies will be problematic.

The parental origin of the abnormal chromosome is unknown for any de novo tandem duplication, except for a few isochromosomes [109, 137, 140, 175, 176] and an apparently direct dicentric duplication involving region 17q11.1→17p12, which appeared to be of maternal origin [67]. Investigation of parental origins would be useful to determine whether most arise as maternal or paternal errors and to compare the origins of monocentric duplications and isochromosomes. Such studies should be possible for many of the cases already published.

ACKNOWLEDGMENTS

I wish to thank Lester Weiss, Ramesh Babu, and Maria J. Worsham for helpful comments on the manuscript, and Susann M. Gollin, Sandra A. Powell, and Richard Stallard for sharing clinical data as personal communications.

REFERENCES

1. Fryns JP, Casaer P, Van den Berghe H: Partial duplication of the short arm of chromosome 9 (p13→p22) in a child with typical 9p trisomy phenotype. Hum Genet 46:231–235, 1979.
2. Fryns JP, Deroover J, Haegeman J, Van den Berghe H: Partial duplication of the short arm of chromosome 10. Karyotype: 46,XX,dup(10p) (pter→p12::p12::p12→qter). Hum Genet 47:217–220, 1979.
3. Fryns JP, Van den Berghe H: Partial duplication of the long arm of chromosome 4. Ann Genet 23:52–53, 1980.
4. Kessel E, Pfeiffer RA: Tandem duplication (5q13→22) in a mentally deficient girl. Hum Genet 52:217–220, 1979.
5. Madan K, Vlasveld L, Barth PG: Ring-18 and isopseudodicentric-18 in the same child: A hypothesis to account for common origin. Ann Genet 24:12–16, 1981.
6. Harnden DG, Klinger HP (eds): (1985): "An International System for Human Cytogenetic nomenclature." Birth Defects Original Article Series Vol. 21 No. 1. New York: March of Dimes Birth Defects Foundation.

7. Darlington CD: Misdivision and the genetics of the centromere. J Genet 37:341–364, 1939.
8. Darlington CD: The origin of isochromosomes. J Genet 39:351–361, 1940.
9. Darlington CD, Wylie AP: A dicentric cycle in Narcissus. Heredity (suppl) 6:197, 1953.
10. Gardner LI, Mitter N, Coplan J, Kalinowski P, Sanders KJ: Isochromosome 9q in an infant exposed to ethanol prenatally. N Engl J Med 312:1521, 1985.
11. Kuhn EM, Therman E: Cytogenetics of Bloom's syndrome. Cancer Genet Cytogenet 22:1–18, 1986.
12. Therman E, Trunca C, Kuhn EM, Sarto GE: Dicentric chromosomes and the inactivation of the centromere. Hum Genet 72:191–195, 1986.
13. Schmutz SM, Pinno E: Morphology alone does not make an isochromosome. Hum Genet 72:253–255, 1986.
14. Ward BE, Bradley CM, Cooper JB, Robinson A: Homodicentric chromosomes: A distinctive type of dicentric chromosome. J Med Genet 18:54–58, 1981.
15. Kimber G, Sears ER: Nomenclature for the description of aneuploids in the Triticinae. Proc 3rd Internat Wheat Genet Symp Canberra, The Griffin Press, 1968, pp 468–473.
16. Rieger R, Michaelis A, Green MM (1976): "Glossary of Genetics and Cytogenetics." 4th Ed. New York: Springer-Verlag.
17. Larson LM, Wasdahl WA, Saumur JH, Coleman ML, Jalal SM: Trisomy 18 syndrome with an unusual karyotype: Possible double isochromosome. J Med Genet 15:73–76, 1978.
18. Steffensen DM, Chu EHY, Speert DP, Wall PM, Meilinger K, Kelch RP: Partial trisomy of the long arm of human chromosome 1 as demonstrated by in situ hybridization with 5S ribosomal RNA. Hum Genet 36:25–33, 1977.
19. Squire J, Phillips RA, Boyce S, Godbout R, Rogers B, Gallie BL: Isochromosome 6p, a unique chromosomal abnormality in retinoblastoma: Verification by standard staining techniques, new densitometric methods, and somatic cell hybridization. Hum Genet 66:46–53, 1984.
20. Danesino C, Gimelli G, Cuoco C, Ciccione MO: Triplex gene dosage effect for B-glucuronidase and possible assignment to band q22 in a partial duplication 7q. Hum Genet 56:371–373, 1981.
21. George DL, Francke U: Gene dose effect: Regional mapping of human glutathione reductase on chromosome 8. Cytogenet Cell Genet 17:282–286, 1976.
22. Taylor KM, Francke U, Brown MG, George DL, Kaufhold M: Inverted tandem ("mirror") duplications in chromosomes: inv dup 8p, 4q, 22q. Am J Med Genet 1:3–19, 1977.
23. Zadeh TM, Funderburk SJ, Carrel R, Dumars KW: 9p duplication confirmed by gene dosage effect. Report of two patients. Ann Genet 24:242–244, 1981.
24. Moedjono SJ, Crandall BF, Sparkes RS: Tetrasomy 9p: Confirmation by enzyme analysis. J Med Genet 17:227–230, 1980.
25. Eydoux P, Junien C, Despoisse S, Chassevent J, Bibring C, Gregori C: Gene dosage effect for GALT in 9p trisomy and in 9p tetrasomy with an improved technique for GALT determination. Hum Genet 57:142–144, 1981.
26. Schwartz S, Cohen MM, Panny SR,Beisel JH, Vora S: Duplication of chromosome 10p: Confirmaton of regional assignments of platelet-type phosphofructokinase. Am J Med Genet 36:750–759, 1984.
27. Gilgenkrantz S, Droulle P, Schweitzer M, Foliguet N, Chadefaux B, Lombard M, Chery M, Prieur M: Mosaic tetrisomy 12p. Clin Genet 28:495–502,1985.

28. Mattei MG, Philip N. Passarge E, Moisan JP, Mandel JL, Mattei JF: DNA probe localization at 18p113 band by in situ hybridization and identification of a small supernumerary chromosome. Hum Genet 69:268–271, 1985.
29. Poissonnier M, Saint-Paul B, Dutrillaux B, Chassaigne M, Gruyer P, de Blignieres-Strouk G: Trisomie 21 partielle 21q21→21q22.2. Ann Genet 19:69–73, 1976.
30. Turleau C, Roubin M, Chavin-Colin F, de Grouchy J: Trisomie 21 par duplication en miroir 46,XX,psu dic(21)ter rea(21q21q). Ann Genet 23:187–189, 1980.
31. McDermid HE, Duncan AMV, Brasch KR, Holden JJA, Magenis E, Sheehy R, Burn J, Kardon N, Noel B, Schinzel A, Teshima I, White BN: Characterization of the supernumerary chromosome in cat eye syndrome. Science 232:646–648, 1986.
32. Kasai R, Narahara K, Kikkawa K, Takahashi Y, Wakita Y, Kimura S, Kataoka N, Kimoto H: Reproductive risk of paracentric inversion carriers : Report of two unrelated cases with paracentric inversion of the long arm of chromosome 3. Jpn J Hum Genet 30:57–67, 1985.
33. Hoo JJ, Lorenz R, Fischer A, Fuhrmann W: Tiny interstitial duplication of proximal 7q in association with a maternal paracentric inversion. Hum Genet 62:113–116, 1982.
34. Mules EH, Stamberg J: Reproductive outcomes of paracentric inversion carriers: Report of a liveborn dicentric recombinant and literature review. Hum Genet 67:126–131, 1984.
35. Van Dyke DL, Weiss L, Roberson JR, Babu VR: The frequency and mutation rate of balanced autosomal rearrangements in man estimated from prenatal genetic studies for advance maternal age. Am J Hum Genet 35:301–308, 1983.
36. Bernstein R, Jenkins T, Dawson B, Wagner J, Dewald G, Koo GC, Wachtel SS: Female phenotype and multiple abnormalities in sibs with a Y chromosome and partial X chromosome duplication: H-Y antigen and Xg blood group findings. J Med Genet 17:291–300, 1980.
37. Steinbach P, Horstmann W, Scholz W: Tandem duplication dup(X) (q13q22) in a male proband inherited from the mother showing mosaicism of X-inactivation. Hum Genet 54:309–313, 1980.
38. Knuutila S, Ruutu T, Partanen S, Vuopio P: Chromosome 1q+ in erythroid and granulocyte-monocyte precursors in a patient with essential thrombocythemia. Cancer Genet Cytogenet 9:245–249, 1983.
39. Varella-Garcia M, Tajara EH, Gagliardi ART: Structural aberration of the X chromosome in a patient with gonadal dysgenesis: An approach to karyotype-phenotype correlation. J Med Genet 18:228–231, 1981.
40. Verma RS, Peakman DC, Robinson A, Lubs HA: Comparison of G-, Q-, and R-banding in 28 cases of chromosomal abnormalities. Cytogenet Cell Genet 16:479–486, 1976.
41. Elejalde BR, Opitz JM, Elejalde MM, Gelbert EF, Abellera M, Meisner L, Lebel RR, Hartigan JM: Tandem dup (1p) within the short arm of chromosome 1 in a child with ambiguous genitalia and multiple congenital anomalies. Am J Med Genet 17:723–730, 1984.
42. Dallapiccola B, Lungarotti MS, Falorne A, Magnani M, Dacha M: Evidence for the assignment of GUK 1 gene locus to 1q32→q43 segment from gene dosage effect. Ann Genet 23:83–85, 1980.
43. Lungarotti MS, Falorini A, Calabro A, Passalacqua F, Dallapiccola B: De novo duplicaiton 1q31-q42: Variability of phenotypic features in partial 1q trisomics. J Med Genet 17:398–401, 1980.
44. Flatz S, Fonatsch C: Partial trisomy 1q due to tandem duplication. Clin Genet 15:541–542, 1979.

45. Say B, Carpenter NJ, Giacoia G, Jegathesan S: Agenesis of the lung associated with a chromosome abnormality (46,XX,2p+). J Med Genet 17:477–478, 1980.
46. Yunis E, Gonzalez J, Zuniga R, Torres de Caballero M, Mondragon A: Direct duplication 2p14→2p23. Hum Genet 48:241–244, 1979.
47. Mu Y, Van Dyke DL, Weiss L, Olgac S: De novo direct tandem duplication of the proximal long arm of chromosome 2: 46,XX,dir dup(2)(q11.2q14.2). J Med Genet 21:57–58, 1984.
48. Stengel-Rutkowski S, Murken JD, Pilar V, Dutrillaux B, Rodewald A, Goebel R, Bassermann R: New chromosomal dysmorphic syndromes. 3. Partial trisomy 3q. Eur J Pediatr 130:111–125, 1979.
49. Sciorra LJ, Bahng K, Lee M: Trisomy in the distal end of the long arm of chromosome 3. A condition clinically similar to the Cornelia de Lange syndrome. Am J Dis Child 133:727–730, 1979.
50. Wilson GN, Hieber VC, Schmickel RD: The association of chromosome 3 duplication and the Cornelia de Lange syndrome. J Pediatr 93:783–788, 1978.
51. Hongell K, Knuutila S, Westermarck T: Two cases of an abnormal short arm of chromosome 8 (8p+) associated with mental retardation. Clin Genet 13:237–240, 1978.
52. Rethore MO, Aurias A, Couturier J, Dutrillaux B, Prieur M, Lejeune J: Chromosome 8: Trisomie complete et trisomies segmentaires. Ann Genet 20:5–11, 1977.
53. Cuoco C, Gimelli G, Pasquali F, Poloni L, Zuffardi O, Alicata P, Battaglino G, Bernardi F, Cerone R, Cotellessa M, Ghidoni A, Motta S: Duplication of the short arm of chromosome 9. Analysis of five cases. Hum Genet 61:3–7, 1982.
54. Motegi T, Watanabe K, Nakamura N, Hasegawa T, Yanagawa Y: De novo tandem duplication 9p (p12→p24) with normal GALT activity in red cells. J Med Genet 22:64–80, 1985.
55. Turleau C, de Grouchy J, Chevin-Colin F, Roubin M, Brissaud PE, Repesse G, Safar A, Borniche P: Partial trisomy 9q: A new syndrome. Humangenetik 29:233–241, 1975.
56. Faed M, Robertson J, Brown S, Smail PJ, Muckhart RD: Pure partial trisomy for long arm of chromosome 9. J Med Genet 13:239–242, 1976.
57. Nakahori Y, Nakagome Y: A malformed girl with duplication of chromosome 9q. J Med Genet 21:387–388, 1984.
58. Vogel W, Back E, Imm W: Serial duplication of 10 (q11→q22) in a patient with minor congenital malformations. Clin Genet 13:159–163, 1978.
59. Koivisto M, Herva R, Linna SL: Serial duplication of 10 (q21→q22) in a mentally retarded boy with congenital malformations. Hum Genet 57:224–225, 1981.
60. Neville BGR, Berry AC, Stoddart Y: A case of malignant spinal cord ependymona in association with a duplication of part of the long arm of chromosome 12. J Med Genet 22:154, 1985.
61. Koske-Westphal T, Pruszak-Seel RE, Niss R, Passarge E: Partial trisomy 13 presumably due to recombination in an inversion heterozygote and by unequal crossing-over. Ann Hum Genet 41:315–322, 1978.
62. de Grouchy J, Turleau C, Danis F, Kohout G, Briard ML: Trisomie 13qter par duplication en tandem 46,XX,dir dup 13(q21→qter),9qh+. Ann Genet 21:247–251, 1978.
63. Kessel E, Pfeiffer RA, Baisch C: Postaxiale Polydaktylie als Kenzeichen einer partiellen Trisomie des langen Armes eines Chromosoms Nr. 13 (zwei falle mit 46,XX,t(22;13)(q13;q31) und 46,XY,dup(13)(pter-q34::q22-ater)). Klin Padiatr 192:85–90, 1980.
64. Herr HM, Scott CI, Horton SJ: De novo interstitial direct duplication of 15q: 46,XY,dir dup(15)(pter→q24::q14→q21.1::qter). J Med Genet 20:473–475, 1983.

65. Yip MY, Parsons A, Hulten M: A de novo tandem duplication 15(q21→qter) mosaic. Clin Genet 22:1–6, 1982.
66. Romain DR, Frazer AG, Columbano-Green LM, Parfitt RG, Smythe RH, Chapman CJ: Direct intrachromosomal duplication of 16q and heritable fragile site fra(10)(q25) in the same patient. Am J Med Genet 19:507–513, 1984.
67. Docherty Z, Hulten MA, Honeymann MM: De novo tandem duplication 17p11→cen. J Med Genet 20:138–142, 1983.
68. Van Dyke DL, Miller MJ, Weiss L: The origin of inverted tandem duplications, and phenotypic effects of tandem duplication of the X chromosome long arm. Am J Med Genet 15:441–450, 1983.
69. Schmid M, Gall H, Schempp W, Weber L, Schmidtke J: Characterization of a new aberration of the human Y chromosome by banding methods and DNA restriction endonuclease analysis. Hum Genet 59:26–35, 1981.
70. Yu CW, Chen H: De novo inverted tandem duplication of the long arm of chromosome 2(q34→q37). Birth defects 18(3B):311–320, 1982.
71. Giraud F, Mattei JF, Mattei MG, Ayme S, Bernard R: La trisomie 4p. A propos de 3 observations. Humangenetik 30:99–108, 1975.
72. Vogel W, Siebers JW, Gunkel J, Haas B, Knorr-Gartner H, Niethammer DG, Noel B: Uneinheitlicher Phanotyp bei partialtrisomie 4q. Humangenetik 28:103–112, 1975.
73. Dutrillaux B, Laurent C, Forabosco A, Noel B, Suerinc E, Biemont MC, Cotton JB: La trisomie 44q partielle. A propos de trois observations. Ann Genet 18:21–27, 1975.
74. Phelan MC, Albiez K, Stevenson RE: Trisomy 6p due to a tandem duplication. Proc Greenwood Genet Center 5:39–43, 1986.
75. Chiyo H, Furuyama J, Suehara N, Obashi Y, Kikkawa H, Ikoma F: Possible intrachromosomal duplication in a case of trisomy 9p. Hum Genet 34:317–221, 1976.
76. Baccichetti C, Lenzini E, Temperani P, Pallotta R, Giorgi PL, Tarantino E, Mengarda G, Dordi B: Partial trisomy 9: Clinical and cytogenetic correlations. Ann Genet 22:199–204, 1979.
77. Turleau C, de Grouchy J, Chavin-Colin F, Martelli H, Voyer M, Carlas R: Trisomy 11p15 and Beckwith-Wiedeman syndrome. A report of two cases. Hum Genet 67:219–221, 9184.
78. Ieshima A, Yorita T, Ohta S, Kuroki Y: A female infant with pure duplication 12q24.2→qter. Jpn J Hum Genet 29:391–397, 1984.
79. Strain JE, Smith ACM, Ward BE, Robinson A: Inverted tandem duplication of the middle segment of the long arm of chromosome 14. Pediatrics 67:273–275, 1981.
80. Sparkes RS, Salter WJ, Blaker RG, Muller HM: Insertional translocaiton into the X chromosome of a 46,XY male. Clin Genet 12:114–118, 1977.
81. Schinzel A: Possible trisomy 1q25 1q32 in a malformed girl with a de novo insertion in 1q. Hum Genet 49:167–173, 1979.
82. Falek A, Schmidt R, Jervis GA: Familial De Lange syndrome with chromosome abnormalities. Pediatrics 37:92–101, 1966.
83. Crisalli M, Monteverde R, Bonioli E, Vianello MG: Un nouveau cas de trisomie 4p partielle. J Genet Hum 23;187–196, 1975.
84. Eriksson B, Fraccaro M, Hulten M, Lindsten J, Tiepolo L: Unusual chromosomal mosaic (46,XX/46,XX,Cp+) in a girl with multiple malformations. Ann Genet 11:6–10, 1968.
85. Caspersson T, Hulten M, Lindsten J, Zech L: Chromatid interchange resulting in duplication of the short arm of chromosome number 7 in man. Ann Genet 14:143–144, 1971.

86. Kardon NB, Pollack L, Davis J, Broekman A, Krauss M: De novo duplication of the 7q11→q22 region. J Med Genet 20:471–473, 1983.
87. Waziri M, Patil SR, Hanson JW, Bartley JA: Abnormality of chromosome 11 in patients with features of Beckwith-Widemann syndrome. J Pediatr 102:873–8976, 1983.
88. Harrod MJE, Pyrne JB, Dev VG, Francke U: Duplication 12q mosaicism in two unrelated patients with a similar syndrome. Am J Med Genet 7:123–129, 1980.
89. Hoehn H, Wolf U, Schumacher H, Wehinger H: A chromosome 13q+ in a patient with characteristics of trisomy 13 syndrome. Humangenetik 13:34–42, 1971.
90. Fraccaro M: De novo duplication 13q (46,XX,dup(13)(q21→q333)). Hum Genet 58:444, 1981.
91. Nikolis J, Ivanonic K, Kosanovic M: Tandem duplication of chromosome 14 (q24→q32) in a male newborn with congenital malformations. Clin Genet 23:321–324, 1983.
92. Orye E, Van Bever H, Desimpel H: Distal trisomy 14q due to tandem duplication (q24→q32). Ann Genet 26:238–239, 1983.
93. de France HF, Beemer FA, Ippel PF: Duplication in chromosome 15q in boy with the Prader-Willi syndrome; further cytogenetic confusion. Clin Genet 26:379–382, 1984.
94. Veenema H, Beverstock GC, Zvelebil-Tarasevitch N, Doorn JL, van Parys JAP, van de Kamp JJP: Duplication in the proximal portion of the long arm of chromosome 15, in a girl without phenotypic features of the Prader-Willi syndrome. Clin Genet 26:65–68, 1984.
95. Orye E, Van Bever H: De novo distal trisomy 17q. Ann Genet 28:61–62, 1985.
96. Fryns JP, Vinken L, Marien J, Van den Berghe H: Partial trisomy 18q12, due to intrachromosomal duplication, is not associated with typical 18 trisomy phenotype. Hum Genet 46:341–344, 1979.
97. Mattei JF, Mattei MG, Bateman MA, Giraud F: Trisomy 21 for the region 21q223: Identification by high-resolution R-banding banding patterns. Hum Genet 56:409–411, 1981.
98. Kim HJ, Hsu LYF, Hirschhorn K: Familial X/X translocation: t(X;X)(p22;q13). Cytogenet Cell Genet 13:454–464, 1974.
99. Laurent C, Biemont M-C, Dutrillaux B: Sur quatre nouveaux cas de translocation du chromosome X chez l'homme. Humangenetik 26:35–46, 1975.
100. Taylor KM, Wolfinger HL, Brown MG, Chadwick DL: Origin of a small metacentric chromosome: Familial and cytogenetic evidence. Clin Genet 8:364–369, 1975.
101. Howell RT, McDermott A, Gardner A, Dickenson V: Down's syndrome with a recombinant tandem duplication of chromosome 21 derived from a maternal ring. J Med Genet 21:310–314, 1984.
102. Spriggs AI, Holt JM, Bedford J: Duplication of part of the long arm of chromosome 1 in narrow cells of a treated case of myelomatosis. Blood 48:595–599, 1976.
103. Papenhausen PR, Wolkin-Friedman E, Pekzar-Wissner C: Novel tandem triplication of 1q in a patient with a myelodysplastic syndrome. Cancer Genet Cytogenet 12:145–150, 1984.
104. Peterson HM, Laughnan JR: Intrachromosomal exchange at the Bar locus in Drosophila. Proc Natl Acad Sci 50:126–133, 1963.
105. Lindsley DL, Grell EH: "Genetic Variations of Drosophila melanogaster." Publ. No. 627. Washington D.C.: Carnegie Institution of Washington, 1968, pp 287, 308–321.
106. Rapoport JA: Multiple linear repetitions of the Bar gene. Drosophila Inform Serv 15:36–37, 1941.
107. Ray M, Hunter AGW: Partial trisomy 8 mosaicism with 46,XX/46,XX,-8,+dic(8). Ann Genet 23:100–102, 1980.

108. Hutchinson-Cole H, Aleck KA, Davis JR, Veomett IC, Hauck L: Mosaicism for ring and isopseudodicentric chromosome 13. Clin genet 29:88–91, 1986.
109. Schwartz S, Palmer CG, Weaver DD, Priest J: Dicentric chromosome 13 and centromere inactivation. Hum Genet 63:332–337, 1983.
110. Rodiere M, Donadio D, Emberger JM, Astruc J, Brunel D: Isochromosome 18: 46,XX,i(18q). Ann Genet 24:611–616, 1977.
111. Bass HN, Sparkes RS, Miller AA: Features of trisomy 18 and 18p− syndromes in an infant with 46,XY,i(18q). Clin Genet 16:163–168, 1979.
112. Froster-Iskenius U, Coerdt W, Rehder W, Schwinger E: Isochromosome 18q with karyotype 46,XX,i(18q). Cytogenetics and pathology. Clin Genet 26:549–554, 1984.
113. Ieschima A, Takashima S, Takada K, Akamatsu Y, Ohtani K, Ando G: Clinicopathological study in a female infant with 46,XX,i(18q) showing mixed features of both trisomy 18 and monosomy 18p. Jpn J Hum Genet 30:219–226, 1985.
114. Speed RM: Prophase pairing in a mosaic 18p−; iso 18q human female foetus studied by surface spreading. Hum Genet 72:256–259, 1986.
115. Laurent C, Biemont MC, Philip T, Dutruge J, Bellon G, Bethenod M, Gilly R, Ladreyt JF: Translocations termino-terminales de novo entre deux chromosomes 18. A propos de deux observations 46,XY,ter rea(18;18)(pter;pter). Ann Genet 21:78–82, 1978.
116. Fioretti G, Stabile M, Pagano L, Rinaldi A, Rolando D, Trapassi C, de Tollis G, Ventruto V: A case of Edward's syndrome with pseudodicentric isochromosome 18: 46,XY,i dic(18)(p11::p11). Ann Genet 25:116–118, 1982.
117. Wulfsberg EA, Sparkes RS, Klisak IJ, Teng A: Trisomy 18 phenotype in a patient with an isopseudodicentric 18 chromosome. J Med Genet 21:151–153, 1984.
118. Muller HJ, Buhler EM, Signer E, Egli F, Stalder GR: Trisomy-18 syndrome caused by translocation or isochromosome formation. J Med Genet 9:462–467, 1972.
119. Vogel W, Reinwein H, Engel E: Tandemchromosom (G/G) mit Satelliten am kurzen und langen Arm bei einem Patienten mit Translokationstriosomie G1. Humangenetik 9:361–371, 1970.
120. Vogel W: Identification of G-group chromosomes involved in a G/G tandem-translocation by the Giemsa-band technique. Humangenetik 14:255–256, 1972.
121. Sachdeva S, Wodnicki J, Smith GF: Fluorescent chromosomes of a tandem translocation in a mongol patient. J Ment Defic Res 15:181–184, 1971.
122. Bartsch-Sandhoff M, Schade H: Zwei subterminale Heterochromatinregionen bei einer seltenen Form einer 21/21-Translokation. Humangenetik 18:329–336, 1973.
123. Jernigan D, Curl N, Keeler C: Milledgeville mongoloid: A rare karyotype of Down's syndrome. J Hered 65:254–257, 1974.
124. Schuh BE, Korf BR, Salwen MJ: A 21/21 tandem translocation with satellites on both long and short arms. J Med Genet 11:297–299, 1974.
125. Niebuhr E: Down's syndrome. The possibility of a pathogenetic segment on chromosome no. 21. Humangenetik 21:99–104, 1974.
126. Hagemijer A, Smit EME: Partial trisomy 21. Further evidence that trisomy of band 21q22 is essential for Down's phenotype. Hum Genet 38:15–23, 1977.
127. Berg JM, Gardner HA, Gardner RJM, Goh EG, Markovic VD, Simpson NE, Worton RG: Dic(21;21) in a Down's syndrome child with an unusual chromosome 9 variant in the mother. J Med Genet 17:144–148, 1980.
128. Cantu JM, Hernandez A, Pascencia L, Vaca G, Moller M, Rivera H: Partial trisomy and monosomy 21 in an infant with an unusual de novo 21/21 translocation. Ann Genet 23:183–186, 1980.

129. Karukaya S, Yokoyama H: Tandem translocation of no. 21 chromosome in a patient with a typical Down syndrome. Proc Jpn Acad 56(B):514–518, 1980.
130. Cohen G, Manuel A, Cohen M, Endrey-Walder S, McBride W: Terminal rearrangement of chromosomes 21 detected in amniotic fluid, resulting in a trisomy 21. Hum Genet 64:203, 1983.
131. Rutten FJ, Scheres JMJC, Hustinx TWJ, ter Haar BGA: A presumptive tetrasomy for the short arm of chromosome 9. Humangenetik 25:163–170, 1974.
132. Orye E, Verhaaren H, Van Engmond H, Devloo-Blancquaert A: A new case of the trisomy 9p syndrome. Report of a patient with unusual chromosome findings (46,XX/47,XX, +i(9p)) and a peculiar congenital heart defect. Clin Genet 7:134–143, 1975.
133. Garcia-Cruz D, Vaca G, Ibarra B, Sanchez-Corona J, Ocampo-Campos R, Peregrina S, Moller M, Rivera H, Rivas F, Gonzalez-Angulo A, Cantu JM: Tetrasomy 9p: Clinical aspects and enzymatic gene dosage expression. Ann Genet 25:237–242, 1982.
134. Ghymers D, Hermann B, Disteche C, Frederic J: Tetrasomie partielle du chromosome 9, a l'etat de mosaique, chez un enfant porteur de malformations multiples. Humangenetik 20:273–282, 1973.
135. Abe T, Morita M, Kawai K, Misawa S, Takino T, Hashimoto H, Nakagome Y: Partial tetrasomy 9(9pter→9q2101) due to an extra iso-dicentric chromosome. Ann Genet 20:111–114, 1977.
136. Ferguson-Smith MA: Human Cytogenetics. In Littlefield JW, de Grouchy j, Ebling FJG (eds): "Birth Defects." Amsterdam: Excerpta Medica, 1978, pp. 349–363.
137. Wisniewski L, Politis GD, Higgins JV: Partial tetrasomy 9 in a liveborn infant. Clin Genet 14:147–153, 1978.
138. Peters J, Pehl C, Miller K, Sandlin CJ: Case report of mosaic partial tetrasomy 9 mimicking Klinefelter syndrome. Birth Defects 18(3B):287–293, 1982.
139. Balestrazzi P, Croci G, Frassi C, Franchi F, Giovannelli G: Tetrasomy 9p confirmed by GALT. J Med Genet 20:396–399, 1983.
140. Shapiro SD, Hansen KL, Littlefield CA: Non-mosaic partial tetrasomy and partial trisomy 9. Am J Med Genet 20:271–276, 1985.
141. Pallister PD, Meisner LF, Elejalde BR, Francke U, Herrmann J, Spranger J, Tiddy W, Inhorn SL, Opitz JM: The Pallister mosaic syndrome. Birth defects 8(3B):103–110, 1977.
142. Fryns JP, Petit P, Vinken J, Geutjens J, Marien J, van den Berghe H: Mosaic tetrasomy 21 in a severe mental handicap. Eur J Pediatr 139:87–89, 1982.
143. Hall BD: Mosaic tetrasomy 21 is mosaic tetrasomy 12p some of the time. Clin Genet 27:284–286, 1985.
144. Hunter AGW, Clifford B, Speevak M, MacMurray SB: Mosaic tetrasomy 21 in a liveborn male infant. Clin Genet 21:228–232, 1982.
145. Hunter AGW, Clifford B, Cox DM: The characteristic physiognomy and tissue specific karyotype distribution in the Pallister-Killian syndrome. Clin Genet 28:47–53, 1985.
146. Hersh JH, Graham JM Destrempes MM, Greenstein M: Teschler-Nicola/Killian syndrome: A case report. J Clin Dysmorphol 1:3:20–24, 1984.
147. Rimoin DL, Raffel L, Mohandas T: Clinical features of mosaic tetrasomy 12p. J Clin Dysmorphol 1:31, 1983.
148. Lubinsky M: A case report of Teschler-Nicola/Killian sydnrome. J Clin Dysmorphol 1:3:25–27, 1983.
149. Lubinsky M, Sanger W: "Killian syndrome" and mosaic tetrasomy 12p. J Clin Dysmorphol 2:1:5, 1984.

150. Vine DT, Kaffe S, Arrendondo-Vega F, Desnick RJ, Smith M: Teschler-Nicola/Killian syndrome and mosaic tetrasomy 12p. J Clin Dysmorphol 2:1:7, 1984.,
151. Kwee ML, Barth PG, Arwert F, Madan K: Mosaic tetrasomy 21 in a male child. Clin Genet 26:150–155, 1984.
152. Lopes V, Mak E, Wyatt PR: Prenattal diagnosis of tetrasomy 21. Prenat Diagn 5:233–235, 1985.
153. Van Dyke DL, Weiss L, Logan M, Pai GS: The origin and behavior of two isodicentric bisatellited chromosomes. Am J Hum Genet 29:294–300, 977.
154. Nielsen J, Homma A, Holm V: Inversion 19 and isochromosome short arm 17 or 18. Hum Genet 37:347–350, 1977.
155. Salamanca-Gomez F, Armendares S: Identification of isochromosome 17 in a girl with mental retardation and congenital malformations. Ann Genet 18:235–238, 1975.
156. Tangheroni W, Cao A, Furbetta M: Multiple anomalies associated with an extra small metacentric chromosome: Modified Giemsa stain results. Humangenetik 18:291–295, 1973.
157. Condron CJ, Cantwell RJ, Kaufman RL, Brown SB, Warren RJ: The supernumerary isochromosome 18 syndrome (+18pi). Birth Defects 10(10):36–42, 1974.
158. Nielsen J, Hreidarsson AB, Berggren S, Ried E, Tsuboi T, Saldana-Garcia P: A mentally retarded male with karyotype 47,XY,+mar=?i(18p). Ann Genet 17:129–133, 1974.
159. Balicek P, Zizka J, Lichy J: An isochromosome of the short arms of the no. 18 chromosome in a mentally retarded girl. Clin Genet 9:192–196, 1976.
160. Ogata K, Iinuma K, Kamimura K, Morinaga R, Kato J: A case report of a presumptive +1(18p) associated with serum IgA deficiency. Clin Genet 11:184–188,1977.
161. Nielsen KB, Dyggve H, Friedrich U, Hobolt N, Lyngbye T, Mikkelsen M: Small metacentric nonsatellited extra chromosome. Report of five mentally retarded invididuals and review of literature. Contribution to further delineation of a new syndrome. Hum Genet 44:59–69, 1978.
162. Cote GB, Petmezaki S, Bastaskis N: A gene for hypospadias in a child with presumed tetrasomy 18p. Am J Med Genet 4:141–146, 1979.
163. Rocchi M, Stormi M, Archidiacono N, Filippi G: Extra small metacentric chromosome identified as i(18p). J Med Genet 16:69–73, 1979.
164. Batista DAS, Vianna-Morgante AM, Richieri-Costa A: Tetrasomy 18p: Tentative delineation of a syndrome. J Med Genet 20:144–147, 1983.
165. Rivera H, Moller M, Hernandez A, Enriquez-Guerra MA, Arreola R, Cantu JM: Tetrasomy 18p: A distinctive syndrome. Ann Genet 27:187–189, 1984.
166. Fryns JP, Kleczkowska A, Marien P, van den berghe H: 18p tetrasomy. Further evidence for a distinctive clinical syndrome. Ann Genet 28:111–112, 1985.
167. Stamberg J, Thomas GH: Unusual supernumerary chromosomes: Types encountered in a referred population, and high incidence of associated maternal chromosome abnormalities. Hum Genet 72;140–144, 1986.
168. Gocke H, Muradow I, Zerres K, Hansmann M: Mosaicism of isochromosome 18p. Cytogenetic and morphological findings in a male fetus at 21 weeks. Prenat Diagn 6:151–157, 986.
169. Webb GC, Garson OM, Robson MK, Pitt DB: A partial D-trisomy/normal mosaic female. J Med Genet 8:522–527, 1971.
170. Petit P, Godart S, Fryns JP: Silver staining of the supernumerary chromosome in the cat-eye syndrome. Ann Genet 23:114–116, 1980.
171. Schinzel A, Schmid W, Fraccaro M, Tiepolo L, Zuffardi O, Opitz JM, Lindsten J, Zetterqvist P, Enell H, Baccichetti C, Tenconi R, Pagon RA: The "Cat Eye Syndrome":

Dicentric small marker chromosome probably derived from a no. 22 (tetrasomy 22pter→q11) associated with a characteristic phenotype. Hum Gen 57:148–158, 1981.
172. Smith A, Fraser IS, Shearman RP: A 13-year-old girl with karyotype 47,XX,i(22)(q11). J Med Genet 18:61–64, 1981.
173. Weilong G, Aide Y, Hongbao F, Shumou L, Ju Y: Bisatellited dicentric chromosome: A report on a case with karyotype 47,XY,+psu dic(22)(t(22;22)(22pter→cen→22q11::22q11→22pter). Hum Genet 61:325–328, 1982.
174. Wilson GN, Baker DL, Schau J, Parker J: Cat eye syndrome owing to tetrasomy 22pter→q11. J Med Genet 21:60–62, 1984.
175. Gabarron J, Glover G, Jimenez A, Lamata E: Pseudoisodicentric bisatellited extra marker chromosome (tetrasomy 22pter→q11, trisomy Yqh), derived from a maternal Y/22 translocation. Association between this tetrasomy and "cat eye" phenotypical features. Clin Genet 28:509–515, 1985.
176. Magenis RE, Zonana J, McDermid H, Weliky K, Sheehy RR: Origin of the extra chromosome in cat eye syndrome: Evidence from heteromorphisms and in situ hybridization analysis. An J Hum Genet 37:A105, 1985.
177. Therman E, Sarto GE, Patau K: Center for Barr body condensation on the proximal part of the human Xq: A hypothesis. Chromosoma 44:361–366, 1974.
178. Therman E, Sarto GE: Inactivation center on the human X chromosome. In Sandberg AA (ed): "Cytogenetics of the Mammalian X Chromosome, Part A." New York: Alan R. Liss, 1983, p 315–325.
179. Atkin NB, Baker MC: Chromosome analysis of three seminomas. Cancer Genet Cytogenet 17:315–323, 1985.
180. Delozier-Blanchet CD, Engel E, Walt H: Isochromosome 12p in malignant testicular tumors. Cancer Genet Cytogenet 15:375–376, 1985.
181. Abbo G, Zellweger H: The syndrome of the metacentric microchromosome. Helv Paediatr Acta 25:83–94, 1970.
182. Stallard R, Haney NR, Frank PA, Styron P, Juberg RC: Leukocyte chromosomes from parents of cytogenetically abnormal offspring: Preliminary observations. Cytogenet Cell Genet 30:50–53, 1981.
183. de la Chapelle A: How do human isochromosomes arise? Cancer Genet Cytogenet 5:173–179, 1982.
184. Therman E, Kuhn EM: Incidence and origin of symmetric and assymetric dicentrics in Bloom's syndrome. Cancer Genet Cytogenet 15:293–301, 1985.
185. Kondo K, Sasaki M: Cytogenetic studies in four cases of acute promyelocytic leukemia (APL). Cancer Genet Cytogenet 1:131–138, 1979.
186. de la Chapelle A: Chromosomal abnormalities in acute promyelocytic leukemia. Scand J Hematol 26:57–60, 1981.
187. Hagemeijer A, Lowenberg B, Abels J: Analysis of the breakpoints in translocation (15;17) observed in 4 patients with acute promyelocytic leukemia. Hum Genet 61:223–227, 1982.
188. McDermott A, Romain D, Fraser ID, Scott GL: Isochromosome 17q in two cases of acute blast transformation in myeloproliferative disorders. Hum Genet 45:215–218, 1978.
189. Mitelman F: "Catalog of Chromosome Aberrations in Cancer." New York: Alan R. Liss, 1985.
190. Kovacs G, Georgii A, Mainzer K: Three isodicentric Philadelphia chromosomes in acute phases of chronic myeloid leukemia: A case report. Cancer Genet Cytogenet 20:29–33, 1986.

191. Kakati S, Barcos M, Sandberg AA: Cytogenetic experience with non-Hodgkin, non-Burkitt lymphomas. Cancer Genet Cytogenet 2:199–220, 1980.
192. Oshimura M, Kakati S, Sandberg AA: Possible mechanisms for the genesis of common chromosome abnormalities, including isochromosomes and the 204 Philadelphia chromosome. Cancer Res 37:3501–3507, 1977.
193. Werner-Favre C, Cabrol C, Beris P, Engel E: Isochromosome 21 and other chromosomal abnormalities in a patient with erythroleukemia. Ann Genet 26:240–242, 1983.
194. Babu VR, Van Dyke DL, Lutz M, Miles B, Weiss L: Nonrandom chromosome changes in transitional cell carcinoma of the bladder. Clin Res 34:74A, 1986.
195. Gibas Z, Prout GR, Pontes JE, Connolly JG, Sandberg AA: A possible specific chromosome change in transitional cell carcinoma of the bladder. Cancer Genet Cytogenet 19:229–238, 1986.
196. Rey JA, Bello MJ, de Campos JM, Ramos MC, Benetiz J: Cytogenetic findings in a human malignant melanoma metastatic to the brain. Cancer Genet Cytogenet 16:179–183, 1985.
197. Jones GH: Further correlation between chiasmata and U-type exchanges in rye meiosis. Chromosoma 26;105–118, 1969.
198. Therman E, Denniston C, Sarto GE, Ulbert M: X chromosome constitution and the human female phenotype. Hum Genet 54:133–143, 1980.
199. Vig BK: Sequence of centromere separation: A possible mechanism for equational division of dicentrics in man. Mammalian Chromosome Newslett 5:59–65, 1984.
200. Sears ER, Camara A: Un chromosoma dicentrico transmisible en trigo. Genet Iberica 2:239–256, 1950.
201. Sears ER, Camara A: A transmissible dicentric chromosome. Genetics 37:125–135, 1952.
202. Steinitz-Sears LM, Sears ER: Chiasmata and crossing-over in a dicentric chromosome in wheat. Genetics 38:244–250, 1953.
203. Hair JB: The origin of new chromosomes in Agropyron. Heredity 6:215–233, 1953.
204. Niebuhr E: Dicentric and monocentric Robertsonian translocations in man. Humangenetik 16:217–226, 1972.
205. Niebuhr E: Unusual findings by fluorescence microscopy of a t(13q14q). Humangenetik 15:96–98, 1972.
206. Camargo M, Cervenka J: DNA replication and inactivation patterns in structural abnormality of sex chromosomes. I. X-A trnaslocations, rings, fragments, isochromosomes, and pseudo-isodicentrics. Hum Genet 67:37–47, 1984.
207. Roberts SH, Howell RT, Laurence KM, Heathcote ME: Stable dicentric autosome, tdic (8;22)(p23;p23), in a mentally retarded girl. J Med Genet 14:66–68, 1978.
208. Daniel A: Single Cd band in dicentric translocations with one suppressed centromere. Hum Genet 48:85–92, 1979.
209. Dewald GW, Boros SJ, Conroy MM, Dahl RJ, Spurbeck JL, Vitek HA: A tdic(5;15(p13;p11) chromosome showing variation for constriction in the centromeric regions in a patient with the cri du chat syndrome. Cytogenet Cell Genet 24:15–26, 1979.
210. Ing PS, Smith SD: Cytogenetic studies of a patient with mosaicism of isochromosome 13q and a dicentric (Y;13) translocation showing differentiation centrometric activity. Clin Genet 24:194–199, 1983.
211. Lindsley DL, Novitski E: Localization of the genetic factors responsible for the kinetic activity of X chromosomes of Drosophila melanogaster. Genetics 43:790–798, 1958.

212. Dallapiccola B, Bianco I. Brinchi V, Santulli B, Scarano G, Sicolo A, Stabile M, Ventruto V: t(21q21q)/r[t(21q21q)] mosaic in two unrelated patients with mild stigmata of Down's syndrome. Ann Genet 25:56–58, 1982.
213. Pangalos C, Velissariou V, Ghica M, Liacacos D: Ring-14 and trisomy 14q in the same child. Ann Genet 27:38–40, 1984.
214. Carothers AD, Frackiewicz A, de Mey R, Collyer S, Polani PE, Osztovics M, Horvath K, Papp Z, May HM, Ferguson-Smith MA: A collaborative study of the aetiology of Turner syndrome. Ann Hum Genet 43:355–368, 1980.
215. Palmer CG, Reichman A: Chromosomal and clinical findings in 110 females with Turner syndrome. Hum Genet 35:35–49, 1976.
216. Lima-de-Faria A: The role of the kinetochore in chromosome organization. Hereditas 42:85–160, 1956.
217. Lima-de-Faria A: "Molecular Evolution and Organization of the Chromosome." Amsterdam: Elsevier, 1983, pp 653–685.
218. Comings DE, Okada TA: Whole-mount electron microscopy of the centromere region of metacentric and telocentric mammalian chromosomes. Cytogenetics 9:436–449, 1970.
219. Andre MJ, Aurias A, de Berranger P, Gillot F, Lefranc G, Lejeune J: Trisomie 4p de novo par isochromosome 4p. Ann Genet 19:127–131, 1976.
220. Aurias A, Dutrillaux B: A possible new type of chromosome rearrangement: Telomere-centromere translocation (tct) followed by double duplication. Hum Genet 72:25–26, 1986.
221. Leschot NJ, Lim KS: "Complete" trisomy 5p: De novo translocation t(2;5)(q36;p11) [sic] with isochromosome 5p. Hum Genet 271–278, 1979.
222. Orye E, Benoit Y, Van Mele B: Complete trisomy 5p owing to de novo translocation t(5;22)(q11;11) with isochromosome 5p associated with a familial pericentric inversion of chromosome 2, inv 2(p21q11). J Med Genet 20:394–396, 1983.
223. Smith G, McCaa A, Kelly TE: Trisomy 9p with an isochromosome of 9p. Hum Genet 42:93–97, 1978.
224. Herva R, Koivisto M: Trisomy 9p with i(9p) and t(9q18p). Hum Gent 50:237–240, 1979.
225. Sandig KR, Mucke J, Veit H: Trisomy 9p resulting from de novo 9/15 translocation and a 9p isochromosome. Hum Genet 52:175–178, 1979.
226. Mascarello JT, Jones MC, Hoyme HE, Freebury MM: Duplication (17p) in a child with an isodicentric (17p) chromosome. Am J Med Genet 14:67–72, 1983.
227. Luthardt FW, Palmer CG, Yu PL: Chiasma and univalent frequencies in aging female mice. Cytogenet Cell Genet 12:68–79, 1973.

The Cytogenetics of Mammalian Autosomal
Rearrangements, pages 667–695

Chapter 22
Ring Autosomes: Identification, Familial Transmission, Causes of Phenotypic Effects and In Vitro Mosaicism

Herman E. Wyandt

INTRODUCTION

A ring chromosome was apparently first described in the plant *Crepis tectorum* by Nawashin [1]. Observations of ring chromosomes in a number of plants and animals subsequently followed. The most extensive studies of ring chromosome behavior were made by Barbara McClintock [2,3] in maize. Detailed genetic analysis of meiotic recombination in a ring X in *Drosophila*, which revealed its intrachromosomal origin, was made by L.V. Morgan [4].

The first human ring chromosomes were reported by Levan [5] in two human tumors. Human rings were also found, in vitro, in X-irradiated cultures by Puck [6], and the first constitutional rings in two patients with congenital anomalies were reported by Wang et al. [7]; both were thought to involve chromosome 18.

Ring chromosomes as a class of chromosome abnormality are relatively rare. Although a handful of cases of familial rings are now known, the majority are de novo rearrangements with no recurrence risk. Despite their relative rarity and lesser significance as a class of heritable chromosome abnormality, rings, nevertheless, provoke considerable curiosity and concern when they occur. Because of their instability, some authors have emphasized unpredictable phenotypic consequences of rings, particularly as they relate to phenotype–genotype correlation [8,9]. An additional intriguing aspect of ring chromosomes continues to be the implications of certain ring chromosome behavior for basic chromosome structure.

Cytogenetics Laboratory, Department of Laboratory Medicine, Danbury Hospital, Danbury, Connecticut 06810

FREQUENCY OF RING CHROMOSOMES

From pooled data of consecutive studies of 50,000 newborns from five countries, only one supernumerary ring chromosome was found [10]. In a study of structural rearrangements in 27,225 prenatal diagnoses from amniotic fluids in the New York Registry, one ring chromosome 13 was found [11], and in more recent European and United States collaborative studies of amniotic fluids, cases of ring 13 were found with frequencies of 1/52,965 and 1/62,279, respectively [12,13]. Despite these low frequencies, rings have been reported involving every human chromosome. Table I shows frequencies compiled from Borgaonkar's Catalog of Chromosomal Variation in Man [14], Repository [15] the Interregional Cytogenetics Registry (ICRS), and the more recent Registry of Cytogenetic Abnormalities and Phenylketonuria (RECAP). Notably frequent are rings involving chromosomes 13 and 18. Ring chromosome 13 is the most frequent, and comprises 17–20% of autosomal rings that have been recorded. Although a number of biases are inherent in each of the sources in Table I, the relative frequencies show a similar order.

Of interest with regard to the frequency of rings is the issue of supernumerary rings. For large chromosomes, identification is usually straightforward. For very small extra chromosomal fragments, however, identification may not be as simple. Not only may the identity of the chromosome be in question, but confirmation of a ring structure may be uncertain. One of the clues that a ring chromosome might be involved is mosaicism for cells that are missing the chromosome in question, have double-sized fragments, or have multiple copies. If chromosomes are examined in very elongated stages of the cell cycle, the ring structure may become evident [16]. In the case of very small supernumerary fragments that are not present in every cell, detection of the ring may be missed both cytologically and clinically.

RING CHROMOSOME BEHAVIOR

A durable model for ring chromosome behavior was provided by Barbara McClintock [3]. In very elegant studies of ring chromosomes in maize, she provided both cytological and genetic evidence that ring chromosomes were unstable. She further proposed that such instability was due to formation of a double-sized dicentric ring chromosome from a monocentric one by the mechanism of sister chromatid exchange, followed by a "bridge-breakage-fusion" cycle in subsequent cell divisions. She observed dicentric rings in anaphase lag as well as cells with variably sized, interlocking, and missing rings. Furthermore, by producing plants that were homozygous or heterozy-

TABLE I. Recorded cases of autosomal rings

Chr.	Chromosomal variants and anomalies in man			Chromosomal registries		
No.	CAT[a]	REP[b]	Total	ICRS[c]	RECAP[d]	Total
1	5	2	7	2		2
2	6		6			
3	4	1	5			
4	12	6	18	2	1	3
5	4	6	10	1		1
6	13	4	17	3		3
7	3	1	4	2		2
8	2	1	3			
9	13	7	20	1	1	2
10	12	1	13	1		1
11	7		7			
12	3	1	4			
13	33	34	67	9	2	11
14	16	11	27			
15	20	10	30	2	1	3
16	2	1	3			
17	10	2	12			
18	19	53	72	5	3	8
19	2	1	3			
20	6	2	8	1		1
21	21	14	35	6	2	8
22	21	15	36	5		5
Total	234	137	407	47	9	56

[a]Reference 14; on-line data provided 12/15/86 by Digamber S. Borgaonkar.
[b]Reference 15; on-line data provided 12/15/86 by Digamber S. Borgaonkar.
[c]The Interregional Cytogenetics Register System (ICRS) was initiated in 1973 and became operational in 1976. Project was funded by NIH contract NOI-HD-32713 (data courtesy of Dr. Linda Pasztor).
[d]Registry of Cytogenetic Abnormalities and Phenylketonuria (RECAP), the successor to ICRS was supported by NICHD contract NOI-HD-32819 (data courtesy of Dr. Jan Friedman).

gous for ring chromosomes carrying a gene for brows pigmentation formed during lignification, she was able to correlate variegation (i.e., uneven distribution of pigment) with alteration or loss of one or both rings. In plants with different size rings, the pattern of variegation changed, showing a correlation between ring size and stability.

In contrast, a ring X in *Drosophila* was regarded as highly stable. In the first ring X studied by L.V. Morgan [4], there was no evidence of variably sized rings such as seen in maize. Others have subsequently reported evidence of mitotic instability of rings in *Drosophila* [17–19]. The question of whether the mechanism of instability is the same as for maize, however, remains controversial [20].

The instability of rings in humans has been reported in very many cases. In contrast to rings in maize that continually generate new rings of varying size and genetic content, human rings tend to be less variable in size. Instability is typically reflected in the form of missing and double-sized dicentric rings and in changing ratios of centric, dicentric, and acentric fragments and pulverized rings.

It was early noted that some human rings were more unstable than others. Lejeune [21,22] suggested the instability of human rings, as in maize, was related to size. Kistenmacher and Punnet [23] prior to chromosome banding techniques, however, reviewed 56 cases of human rings and concluded that instability was unrelated to size, but was largely determined by structure and genetic content of the ring itself.

BANDING STUDIES OF RINGS AND RING PRODUCTS

As for chromosome abnormalities generally, the potential for making genotype-phenotype correlations and for understanding ring chromosome behavior more fully was greatly advanced with the advent of chromosome banding techniques. It not only became possible to identify the particular chromosome involved, but the genetic content of individual rings and ring products became more amenable to study. The following is a discussion of some of the most common forms of ring morphology and ring products in humans that have been observed by chromosome banding techniques. These structures are related by the mechanisms discussed and can be seen in the same individual.

Monocentric Rings

The percentage of cells with a monocentric ring (excluding cases where there is mosaicism for a normal cell line or some other stable structural rearrangement) ranges from 50–100%. In the majority of cases, the monocentric ring shows little or no variation in banding pattern. Such rings are presumably representative of their original form and typically would be expected to have arisen by terminal deletions in both chromosome arms. In many cases there appears to be little or no evidence of loss [24,25], or loss is apparent in only one arm. Such observations appear to be in conflict with the "two break" theory of ring formation [3] and the telomere hypothesis [26]. Just how minute loss can be is evident from rings involving the acrocentric chromosomes. In banding studies it has been evident that the nucleolar organizing region (Fig. 1) is often retained in the ring [27,28] and occasionally even the satellites are retained [29]. Studies by myself and colleagues (Rick Kasprzak and Ann Lamb, unpublished observations) of X-irradiated Indian muntjac cells in the first cell division after irradiation produced rings and compensating fragments that were clearly composed of the terminal ends of both chromosome arms (Figs. 2 and 3). In some instance these fragments were so small as to be composed of only the very tips of the chromosome arms.

It is conceivable that ring structures might be formed by other mechanisms such as a defect in normal telomere structure [30,31]. More likely, however, the deletions involved in most cases are simply below the resolution of the banding techniques, and significant or complex losses (especially involving both arms) are too deleterious to be viable [9].

Dicentric Rings

Dicentric rings typically comprise about 2–4% of cells, although as with other ring forms, the frequency may vary greatly from one case to the next. In many instances, cells with two dicentrics are present, and occasionally cells are even seen with three or four copies. The majority of these are of the type expected if a sister chromatid exchange occurred. That is, the centromeres are equidistant, and the chromosome is exactly duplicated in tandem with chromosome bands arranged in a trans-symmetrical pattern (Figs. 4 and 5). This pattern is the same irrespective of the location of the exchange. Theoretically any odd number of sister chromatid exchanges will result in the same pattern. An even number of exchanges will result in monocentric

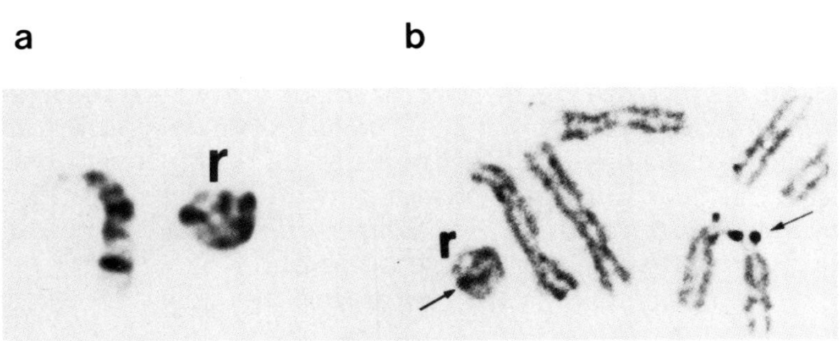

Fig. 1. a: Human ring 14 (r) and normal homolog by G-banding. **b:** Partial metaphase spread from the same patient, stained by silver staining technique of Denton and Howell [127]. Nucleolar organizing regions in the ring 14 (r) and in D group chromosome are indicated by the arrows (preparation by Rick Kasprzak).

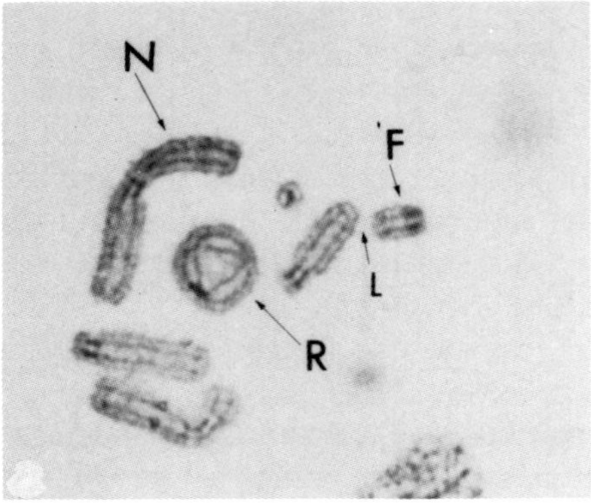

Fig. 2. Ring (R) and compensating fragment (F) induced in male Indian muntjac cell line by X-irradiation. Cells in monolayer were irradiated with 300 rads from a 250 kv X-ray machine filtered to an HVL of 1.3 mm at a dose rate of 103 rads per min. Cells were harvested 18 hr later so that all products of irradiation were present in the mitotic cells that were analyzed. Preparation was stained by the silver staining technqiue of Denton and Howell [127]. Axes of the chromatids are particularly accentuated as lines running the length of each chromatid. Particularly interesting is the size of the fragment and the lines within the fragment, both of which suggest that the fragment represents the acentric reconstituted ends of both arms of the chromosome from which the ring originated. As a comparison, the typical behavior of isochromatid breaks, which were not so reconstituted, was to form U-shaped unions (L), also easily visualized by this technique. The normal homolog to the ring is indicated (N) (preparation by Rick Kasprzak).

rings that are either independent or interlocking. Tetracentric and even octacentric rings are not infrequently seen in human ring cell lines, indicating that the cycle of ring doubling in size via sister chromatic exchange may be repeated in more than one cell division [32–35]. At least one case of a ring 12 has reported tricentrics and tetracentrics with latent centromeres [35].

Not all dicentric rings appear to arise in the same way. Some dicentric rings have a pattern not of the type described above, but rather a cis-symmetrical banding pattern with the centromeres non-equidistant apart. In the latter, the banding pattern of the two halves of the ring chromosome are identical (Fig. 5) as if made in the manner of an ink blot. Although rare, such dicentric rings appear to have arisen by the mechanism of isolocus joining between chromatids in a centric fragment [36]. Evidence for the formation of dicentric rings by this mechanism comes from sister chromatid exchange studies as well [34].

Rings of Various Size and Genetic Constitution

Rings of various size and genetic constitution are expected products of dicentric rings. Such rings presumably arise when a dicentric ring breaks at

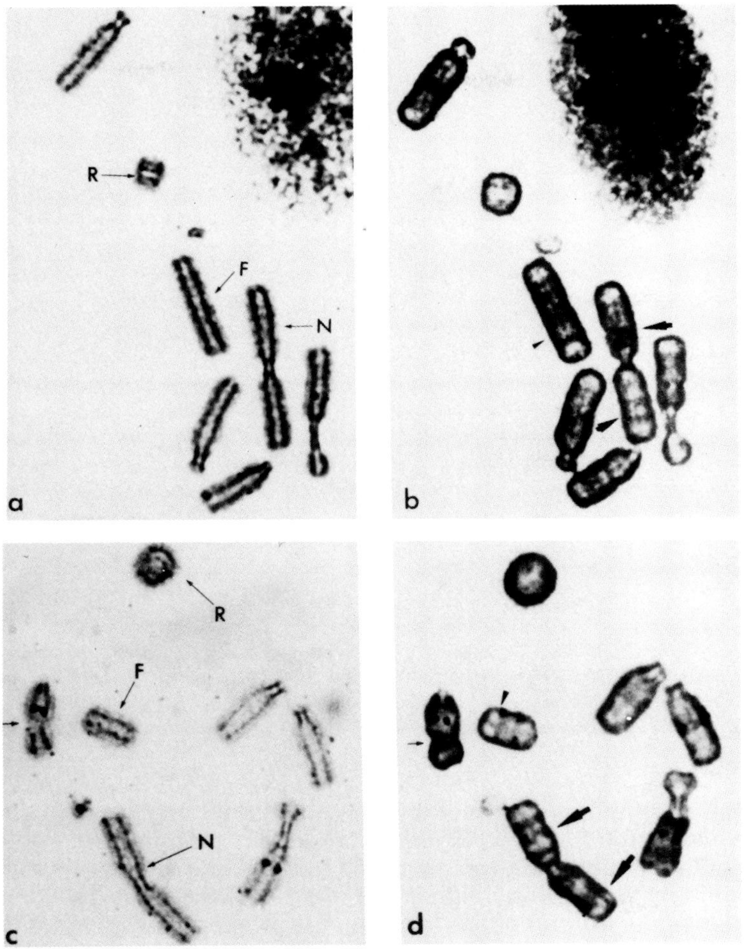

Fig. 3. Rings (R) and compensating fragments (F) observed by silver staining (**a, c**) and by G-banding (**b**, **d**) of the same metaphases. Rings were induced in an Indian muntjac cell line by X-irradiation (methods described in Fig. 2) and are observed in the first cell division following irradiation. Some 41 rings and fragments were found and studied by silver staining and/or G-banding. By silver staining, fragments typically had a parallel chromatid configuration (see Fig. 2), confirmed by G-banding, in many instances, to be composed of the ends of both chromosome arms. Large arrows above (b and d) indicate breakpoints in normal homolog; small arrow points to region of restitution of the two ends to form the fragment. Such observations are suggestive of a "one hit, two break mechanism" for ring formation (Wyandt, Kasprzak, and Lamb, unpublished observations).

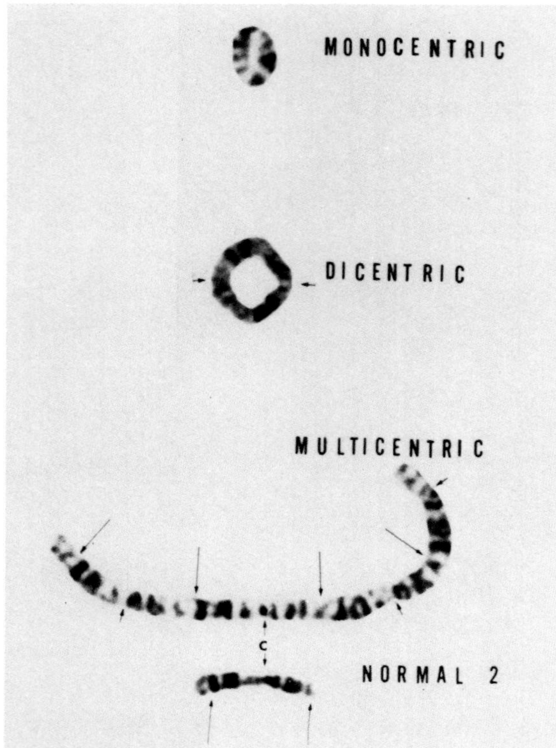

Fig. 4. Comparison of the banding patterns of monocentric and dicentric rings and a polycentric rod (open ring) with a normal chromosome 2 from a patient with a ring chromosome 2. Note the trans-symmetrical banding pattern of the dicentric ring. Short arrows indicate locations of centromeres. Long arrows indicate breakpoints or portion of the chromosome corresponding to that present in the monocentric ring. The structure of the multicentric is a (broken) tandem arrangement of four chromosome 2s. Reprinted from Wyandt et al. [33], with permission of S. Karger AG, Basel.

anaphase and the broken ends rejoin to form a new ring. A mathematical equation was formulated by Lejeune [21,22] to predict the probability that a particular gene could be deleted or duplicated, during this process, based on its distance from the centromere. In studies of human rings such variability is less evident in stable rings [37], but has been reported in a number of cases of unstable rings [29,32,36]. In a few reported cases, duplication and deletion within a ring that had to have arisen as mentioned above appears to be very stable. This was particularly evident in a case of ring 8 reported by Hamers and van Kempen [25], and in a ring 13 studied by myself and colleagues

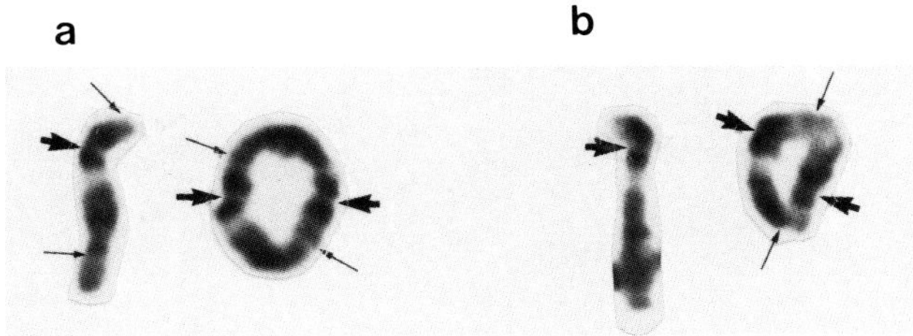

Fig. 5. Two cells from a patient with a ring 12 showing (**a**) a dicentric ring with a trans-symmetrical banding pattern, and (**b**) a dicentric ring with a cis-symmetrical banding pattern. Normal homolog from each cell is shown for comparison. Large arrows indicate the position of the centromeres; small arrows indicate approximate breakpoints. In the case of (b) an interstitial portion of 12 long arm appears to have been deleted, whereas the more distal segment of the long arm appears to have been duplicated. Such duplication and deletion (b) could occur if a dicentric ring (a) broke and the broken ends rejoined between sister chromatids (isolocus rejoining). Rings of type (b) seem to be extremely rare, compared to type (a). Presumably, a type (b) ring would have to survive at least two cell divisions as a dicentric chromosome. That and the fact that such a ring must originate from a dicentric ring of type (a) in the first place would seem to account for its relative rarity.

(Wyandt, Lamb, and Kasprzak, unpublished) that showed apparent duplication of material in a stable ring 13 (Fig. 6).

Other Ring Products

Detailed banding studies of more unusual products have been made in only a handful of cases of human rings. In many cases, even in the post-banding era, banding patterns of more unusual products have only been superficially studied. In the instances where such studies have been done, however, the variety of products is often complex and includes polycentric and interlocking rings, centric and polycentric rods, acentric fragments, incomplete rings, and chromosome pulverization [32,37].

Sister Chromatid Exchanges in Rings

Only a few studies have attempted to delineate sister chromatid exchange directly in a human ring chromosome [34,38]. Theoretically, monocentric rings that have undergone two rounds of DNA replication (Fig. 7) will form a dicentric ring if exchange has occurred during the first round, or form a pretzel configuration if exchange has occurred during the second round. The frequency and products of exchanges in a human ring 4 were studied by Bartram [34] after BrdU incorporation [39]. He found no difference in the rate of exchange in the ring compared to other chromosomes. Of 342 rings studied, most were monocentric. Of the monocentric rings, 18 had one exchange and were in the configuration of a pretzel, which at the next cell

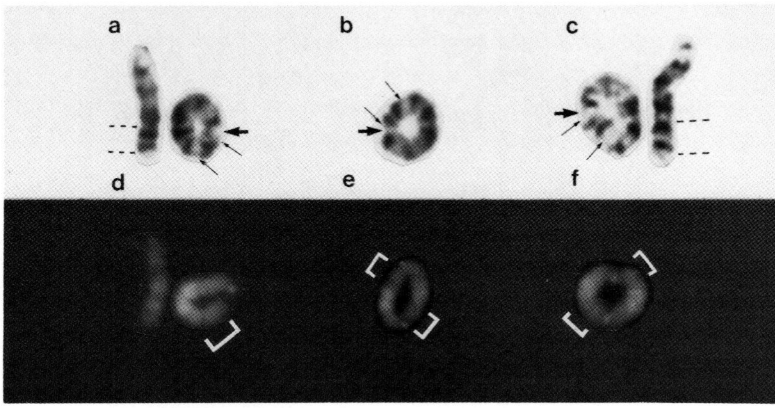

Fig. 6. Several monocentric (**a–d**) and dicentric (**e, f**) rings from a patient with ring chromosome 13. Of particular interest in this case was the apparent duplication of the most distal bands of 13q (between the broken lines, a and c) in both the monocentric and dicentric rings. Duplicated region in ring is indicated by small arrows (a–c) or brackets (d–f). Large arrows (a–c) represent position of centromeres in G-banded cells. In Q-banded preparations (d–f) centromeres are brightly fluorescent spots. Duplication in a monocentric ring presumably arises by breakage in a dicentric ring and rejoining of broken ends in a centric fragment to form a new ring. In this case the new ring must have been formed very early in development since all cells observed had the duplicated segment.

division would give rise to a dicentric ring with a trans-symmetrical pattern. Nine percent of rings were di- or polycentric. Almost half of these showed no exchanges, indicating that they must have originated prior to the addition of BrdU, at least three cell cycles earlier.

The data of Bartram are consistent with the interpretation that the majority of di- and polycentric rings are formed by the mechanism of sister chromatid exchange. Among the dicentric rings, there were also apparently examples of isolocus rejoinings.

RING CHROMOSOME MOSAICISM

In the previous sections, emphasis has been placed on chromosomal products that may arise as a result of ring instability. Of major clinical concern, however, is to what extent such instability contributes to chromosomal mosaicism. It is obvious that some products have a transient nature; others are more likely to become stable cell lines.

Ring Loss

One of the frequent products of dicentric bridge formation or interlocking rings is ring loss. Such loss can be a significant form of mosaicism depending

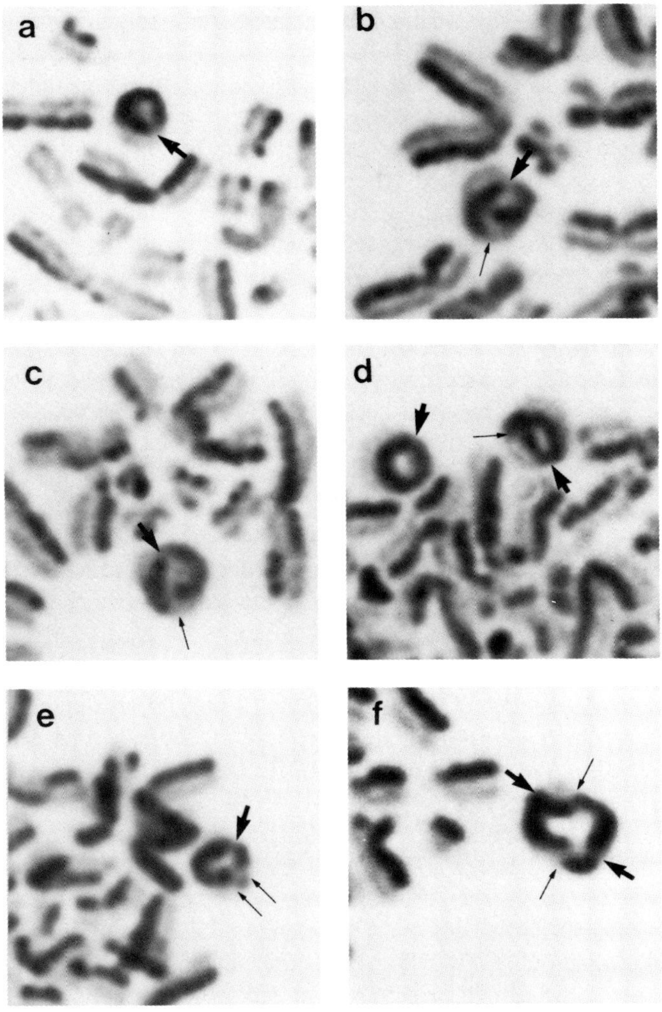

Fig. 7. Ring chromosomes that have incorporated 5′-bromo-deoxyuridine (BrdU) through two cycles of DNA replication. In the first cycle, BrdU is monofiliarly substituted; in the second cycle one chromatid (light staining) is bifiliarly substituted. **a:** Monocentric ring in which no sister chromatid exchange (SCE) has occurred. **b, c:** Monocentric rings in pretzel configurations with only one SCE present. **d:** Cell with two monocentric rings, one with a single exchange in a pretzel configuration and one without an exchange. **e:** Ring with two exchanges. **f:** Dicentric ring with a pair of symmetrical exchange points evident. Large arrows indicate position of centromere; small arrows indicate breakpoints.

on when in development and the extent to which ring loss occurs. The majority of cases with a ring chromosome show some degree of ring loss, ranging from an occasional cell up to 50% [40]. The extent to which ring loss represents in vivo mosaicism, however, needs to be studied more fully. In many cases, ring loss appears to be higher in fibroblast cultures than in lymphocyte cultures from the same patient [41,42]. However, numerous instances can also be cited where this is not the case [43–45]. Fibroblast cultures that have undergone numerous passages appear to gradually lose their ring chromosomes (Wyandt, unpublished results).

Ring Duplications

As with ring loss, the frequency of cells with double-sized (dicentric) rings or multiple copies of rings (both monocentric and dicentric) varies greatly from case to case. In most cases 5–10% of such cells are present. It is difficult to conceive of the dicentric rings in these cases becoming stable cell lines. In some cases, however, the frequency of dicentric rings may be higher [29]. In such cases there also seems to be a high frequency of cells with two or more monocentric rings. Arguments that duplications and loss of rings have a phenotypic affect seem particularly valid in cases where there is no apparent deletion of material from the monocentric ring [46,47].

Genetically Altered Rings

Some rings obviously must have arisen from the breakage of a prior ring and yet are stable. Very unstable rings may in fact have a tendency to evolve by selection to a more stable ring form. In these cases it is conceivable that a transient stage of mosaicism exists where multiple ring forms are present or that different proportions of various forms exist in different tissues or stages of development. The few longitudinal studies of ring chromosomes that have been done are of interest. In some cases where such studies have been done [33,48] there is surprisingly little evidence of significant changes in proportions of ring products over time. In other studies, significant changes have been observed [8,49,50]. Hecht [8] studied a patient with a ring 4 and found an increasing proportion of cells with a ring as the patient grew older. Studies were carried out before the advent of banding so that it is not known whether the ring itself may have been subtly altered. Sutherland and Carter [49] found a similar increase with age in the proportion of cells with a ring 2, and Parker et al [50] found an increase over time in tetraploid cells containing a ring in a patient with ring 4. In a more recent study of a ring 21 detected prenatally [29], only 25% of cells in amniotic fluid had a monocentric ring; 71% of cells had an asymmetrically dicentric ring; 2% had the usual symmetrical dicentric rings. Cultures of placenta, foreskin, and lymphocytes were obtained at birth, and at 14 months, a second lymphocyte culture was obtained. Of interest is that not only was the proportion of various ring forms different in the different tissues, but *two* monocentric ring populations were evident in the blood: one population was a ring that was smaller than a normal 21 and similar to the ring seen in the amniotic fluid; the second population had a larger ring with apparent duplication/deletion of 21. At 14 months the frequency of cells with the larger ring increased from 35% of cells to 73%; the frequency of the smaller ring had decreased from 40 to 14% of cells.

Stabilized Rod Configurations Arising from Rings

While the formation of stable rod configurations from rings is theoretically possible [3], in fact such occurrences seem not to be recorded for human rings. A number of authors report rod configurations that appear to be an open form of the ring without apparent loss of material [28,36,37]. It is not likely, however, that such configurations become stabilized with any significant frequency.

Complex Forms of Rod/Ring Mosaicism

A number of cases of reported rings have been associated with some form of complex rearrangement. In some instances these have been translocations with another chromosome that obviously preceded formation of the ring chromosome. A number of cases, however, have involved duplications and deletion of material in the rod and a corresponding deletion of material in a ring configuration. These configurations are such that they must have occurred simultaneously. At least four cases are now known that seem to have involved a similar mechanism [33,51,52]. Namely, three breaks involving both chromatids occurred in late S or G2 of the cell cycle such that one chromatid reconstituted to form a ring, while the second chromatid reconstituted in such a way that duplication of part of the chromosome occurred (Figs. 8 and 9).

PHENOTYPIC EFFECTS

The phenotypic effects of ring chromosomes have been much debated. For some chromosomes only a handful of cases have been reported. For others, numerous cases are known so that a distinctive phenotpye or spectrum of phenotypes has been elucidated. In the recent edition of Borgaonkar's *Chromosomal Variation in Man*, Appendix II [130], no less than 13 ring chromosomal syndromes are described.

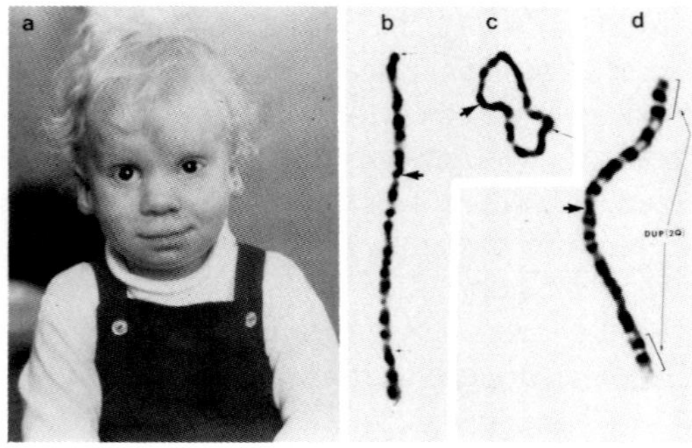

Fig. 8. **a:** Patient with two cell lines: 46,XX,r(2)(p25;q34) and 46,XX, dup(2)(qter→p25::q34→qter). The ring (**c**) was present in 30° of peripheral blood lymphocytes; the rod configuration (**d**) was present in the remaining 70° of lymphocytes and in all cells analyzed from skin fibroblasts. Normal homolog (**b**) is shown for comparison. Chromosomes are in prometaphase stage of elongation. Modified from Wyandt et al. [33]. Reprinted with permission from S. Karger AG, Basel.

The following is not intended to be an exhaustive discussion of the phenotypic features of known cases of rings, but rather is intended to touch on some of the most pertinent findings as they relate to phenotype-genotype correlation.

Clinical Findings in Reported Cases

Ring 1. Four cases of ring 1 are reported. All four cases seemed to show similar but non-specific clinical features: reduced birthweight, dwarfism, and mental retardation. Two were studied by banding [53,54] and appeared to be simple rings with little loss or no detectable loss of chromosomal material as could be determined by chromosome banding studies. A third case studied prior to banding by Bobrow et al. [55] is of interest in that a female, in addition to having the clinical features above, had acute myeloid leukemia and died at age nine.

Ring 2. Six cases of ring 2 have been reported [33,49,56,57]. Four cases involved simple rings with little or no evidence of genetic loss [49,57]. One case [57], involving a translocation between a no. 2 and a no. 6, showed deletion of the terminal bands of 2q and of 6q, but not of 2p. The clinical features in common for five cases included low birthweight, slow growth, microcephaly, epicanthus, flat nose, low-set ears, short neck, wide-spaced nipples, and psychomotor retardation. A non-specific finding of micrognathia was present in three cases. Maraschio et al. [57] point out that this is also a finding in cases with partial deletion of chromosome 6.

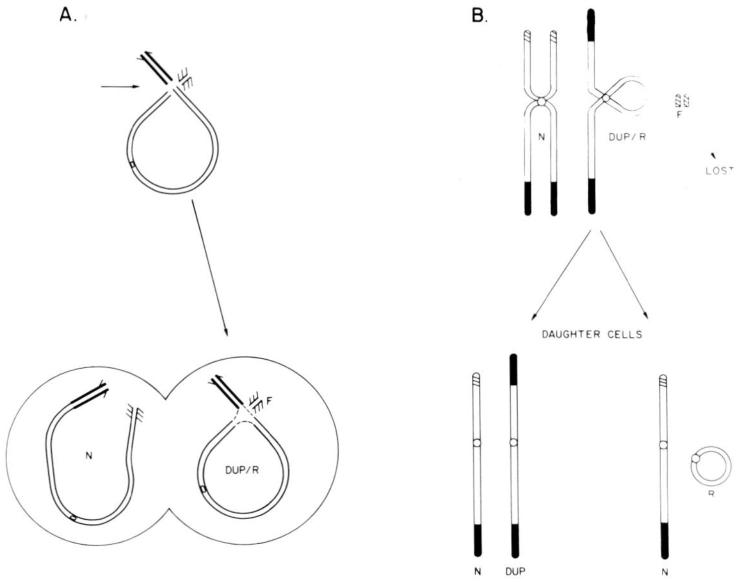

Fig. 9. Proposed mechanism for the origin of the two cell lines present in patient with rod/ring mosaicism (Fig. 8). Intrachromosomal exchange occurred either between complementary DNA strands (**A**) or between sister chromatids. Such exchange must have occurred just prior to or during the first zygotic cell division to yield cell lines depicted in daughter cells (**B**). N, normal homolog; DUP, duplication; R, ring; F, fragment. Solid areas represent segment of chromosome that was duplicated in the rod configuration; cross-hatched areas represent portion of the chromosome that deleted.

A sixth case of ring 2 was reported by myself and colleagues [33] involving complex mosaicism (Figs. 8 and 9). The primary cell line had an abnormal rod configuration showing duplication of a significant portion of the distal one-third of 2q (q33.2 to qter). The ring cell line constituting one-third of cells from peripheral blood showed a ring 2 missing precisely that segment that was duplicated in the rod. Clinical features of the child at 5 years of age were motor and mental retardation, growth retardation, malar hypoplasia, ptosis, epicanthal folds, and a narrow palate. He also had small hands with fifth finger clinodactyly and single palmar creases.

Ring 3. Two reported cases of ring 3 [58,59] were compared [59]. Details of specific breakpoints in the two cases are lacking, but according to Witowski et al. [59] there was little phenotypic correlation between the two cases.

Ring 4. Approximately 16 cases of ring 4 have been reported [47]. In a number of cases the ring 4 has been highly unstable [8,32,47]. The rings generally have been variable with respect to breakpoints and in the relative proportions of cells with various ring products, leading some authors to conclude that clinical features are due to ring-generated mosaicism rather than to deletion from the ring itself [46,47]. Despite cytogenetic variability, there has been considerable similarity in phenotypic findings, many of which overlap with those of 4p− (Wolf-Hirschorn) syndrome [60–62].

Ring 5. Despite the fact that there are well-established syndromes for deletions of both the long and short arms of chromosome 5, apparently only four cases of ring 5 have been reported [14,63,64]. Of these, all four involve deletion of bands p15 to pter. The clinical features are those associated with the well-established *cri-du-chat* syndrome [65].

Ring 6. For nine reported patients with ring 6 [42,66–70], breakpoints as well as phenotype were variable. Carnevale et al. [69] in a review of six cases suggested that phenotypic variability was related to the amount of material missing from the ring in the different patients. They also emphasized the apparent stability of the ring in different patients. Nishi et al. [70] who reviewed three additional cases also indicated the variability in phenotype and breakpoints in the different cases, but pointed out that two cases with similar beakpoints had quite different phenotypic manifestations, whereas two other cases with dissimilar breakpoints had similar phenotypes. The frequency of cells missing a ring also varied in these additional cases.

Ring 7. Of four reported cases of ring 7 [48,71,72], all appeared to show little if any loss of chromosomal material. Two cases described by Zachai and Breg [48] had very dissimilar phenotypes. In one case there was slow development and multiple congenital anomalies including proptosis, ptosis, microcornea, antimongoloid slant, small phallus, hypospadias, undescended testes, and a left simian crease. The second case had no abnormalities at birth, but later showed failure to thrive. Mental and motor development were normal. Phenotypic manifestations in two additional cases [71,72] were also apparently different. However features in common include short stature, skin pigmentation including nevus flammes, hemangioma, café-au-lait spots, and pigmented nevi. The case reported by DeLozier-Blanchet and Guenin [72] also had normal intelligence.

Ring 8. Two cases with ring 8 [25,73] had a normal cell line in addition to the ring cell line, i.e., 46,XY/46,XY,r(8). Both cases had low birthweight, short stature, dolichocephaly, high-arched palate, abnormal dentition, micrognathia, and mental retardation. One case reported by Hamers and van Kempen [25] is of interest in that ring size was constant, but the most distal bands of both arms appeared to be duplicated suggesting that the ring is a product of bridge-breakage-fusion in a dicentric ring. The most terminal band of the long arm, as revealed by R-banding, was retained.

Ring 9. The phenotypic features in ring 9 cases have been particularly variable causing several authors to reject the idea of an identifiable ring 9 syndrome. Of 14 cases of ring 9 [43,74–80] reported, one case was an inherited ring 9 present in a mother and her daughter [75]. The latter is of interest in that although the ring appeared identical in the mother and daughter, the mother had few if any phenotypic abnormalities, whereas the daughter had short stature, microcephaly, and mental retardation. Breakpoints for the remainder of cases are either non-specific or vary in location in the different cases. Recent reports of cases showing deletion of 9p, however, have apparently shown clinical features in common with cases of simple 9p deletion [130].

Ring 10. Nine cases of ring 10 have been reported [40,44,81–84]. One case [84] involved a complex translocation with chromosome 19 so that there was deletion of the short arm of no. 19 as well as of no. 10. A case reported by Lansky et al. [40] was missing more material than most of the other cases but had milder phenotypic features and reportedly fewer cells with an intact ring. Despite these complications, the following clinical features do appear in common: severe growth retardation, mild mental retardation, mild microcephaly, hypertelorism, ophthalmologic abnormalities, prominent nasal bridge, stubby nose, pointed chin, and undescended testes. Urinary tract anomalies also appear to be present in all cases studied.

Ring 11. Of approximately six cases of ring 11 [45,85–87], all have shown mental retardation, café-au-lait spots, short stature and microcephaly. Although little material is reportedly missing in any of the cases, two cases [85,86] show considerable overlap with cases of 11q− syndrome. One case is of interest in that there was development of Wilms tumor [87]. No history of aniridia was present, and bands p13 to p14 appeared to be intact in the ring.

Ring 12. Only four cases of ring 12 are known [14,35]. In one reported case [35], there was no obvious loss of chromatin material from the ring, and the child had rather mild phenotypic anomalies including growth retardation, mild motor retardation, peculiar facies, low-set ears, and prominent nose.

Ring 13. Among autosomal rings, ring 13 has been the most frequently reported. Some 75 cases are known. Breakpoints range from q21 to q34. Three distinctive syndromes are now recognized depending on the amount of material deleted from chromosome 13 [88–92]. Group I (loss of bands 13q33 and 13q34) is associated with mental retardation, microcephaly and craniofacial dysmorphism. Group II (loss of bands 13q34 through 13q32 and part of 13q31) have the above features with hypoplastic or absent thumbs. Group

III (interstitial deletion of band 13q14) has been associated with retinoblastoma. The phenotypic features associated with ring 13 and 13 deletion have been discussed in several reviews [14,88,89]. For the most part, the features seen in cases of ring chromosome 13 are similar to those seen in simple deletions of chromosome 13 [88].

Ring 14. As with ring 13, rings involving chromosomes 14 are emerging as a rather frequent anomaly [93,94,131]. According to Fryns et al. [93] who reviewed some 14 cases, a number of malformations and stigmata are seen in common and include moderate to severe mental retardation, microcephaly associated with central nervous system anomalies such as brain atrophy and porencephaly, failure to thrive, growth retardation, and severe generalized muscle hypotonia. Craniofacial dysmorphology is also typical and includes dolichocephaly, high forehead, epicanthal folds, hypertelorism, small palpebral fissures, broad nasal bridge, large dysplastic and low-set ears, macrostomia with thin lips, high-arched palate, micrognathia, and short neck. It is also of interest that at least two inherited ring 14s have been reported [95,122].

Ring 15. Approximately 23 cases of ring 15 have been reported [96–101]. As with rings of chromosomes 13 and 14, there appears to a definitive syndrome [100]. Low birthweight and growth retardation were present in all cases. Mild to moderate mental retardation is typically present [101]. Microcephaly is usually present and micrognathia is frequently present. There appears to be no constant facial dysmorphism, and two cases had a "bird headed profile" [97,101]. In most cases of ring 15 little or no detectable amount of material is missing from the long arm. However, Kosztolanyi and Pap [96] found considerable evidence of cell death in the form of micronuclei and nuclear projections in lymphocyte cultures and reduced viability of cells, by vital staining, in fibroblast cultures. Since their patient also had severe growth failure and atrophic intestinal mucosa, they speculate that these findings may be a manifestation of increased cell mortality rather than genetic loss per se.

Ring 16. Two cases are known; see Borgaonkar et al. [130].

Ring 17. Some nine cases of ring 17 have been reported [102,103]. Chromosomal breakpoints and phenotypic features have apparently been variable [130].

Ring 18. Ring 18 is of interest because not only is it one of the more frequent rings to be reported, but because of known cases of deletions involving both the short arm and the long arm of chromosome 18, ring 18 was one of the first rings for which any significant comparison of phenotype could be made [104,105]. The clinical features of ring 18 were shown by

Kunze et al. [104] to overlap with the clinical features of 18p− and 18q− syndromes. Two cases of inherited ring 18 are known [106,107].

Ring 19. Two cases are known [14].

Ring 20. Although based on a small number of cases, a definitive ring 20 syndrome was first recognized by Borgaonkar et al. [130,132]. Five cases compared by Jalpert et al. [108] revealed a ring 20 syndrome characterized by facial dwarfism, the absence of severe anomalies, and the late onset of encephalopathy and seizures.

Ring 21. Some 20 cases of ring 21 are known [109–115]. Phenotypic features associated with ring 21 appear to be similar to those associated with deletions of chromosome 21. These features have been described by some authors as the countertype of Down syndrome and include antimongoloid slant of eyes, prominent occiput, broad base of nose, microretrognathia, finger grasping with adducted thumbs, hypertonia, psychomotor retardation, and seizures [110]. Several cases of ring 21 are reported involving phenotypically normal individuals: one in a phenotypically normal but infertile man [109] and two in cases of normal adult females, one of whom was detected after a daughter was born with mosaicism for a ring 21. A tandemly duplicated 21 associated with Down syndrome has also been recently reported as being derived from a maternal carrier of a ring 21 [114]. Also, a 21q21q ring chromosome formed subsequent to a centric fusion translocation and associated with Down syndrome was reported by Orye and Craen [113]. A more recent case of a ring with a partially duplicated 21 in the majority of cells is of interest in that no stigmata of Down syndrome were present at birth. The ring in an unduplicated, monocentric form was initially detected prenatally [29].

Ring 22. Ring 22 also has been frequently described and is associated with a well-established syndrome. Distinction between ring 21 and ring 22 syndromes was first pointed out by Magenis et al. [116]. Since then some 23 patients with ring 22 have been reported [130]. Distinctive features have included mental retardation, verbal delay, microcephaly and growth failure, epicanthal folds, full eyebrows, dental malocclusions, thick and often fissured lips, large abnormally formed ears, second and third toe syndactyly, spine curvature anomalies, ataxic gait, hypotonia, and seizures [117]. Some variability in phenotype has been noted by Funderburk et al. [118], who reported two patients with similar rings but different phenotypes. They attributed the difference in phenotype as being due to the loss of the ring in a higher proportion of cells in one patient than in the other. Fowler et al. [27] reported a patient with a ring 22 which, on the other hand, was clinically almost identical to the first patient they had reported [16] and later identified as being ring 22 by banding [116]. There appeared to be little evidence of ring instability that had been observed in the first patient.

HERITABLE RINGS

Heritable rings have now been reported in at least eight different families [75,95,106,107,119–122]. Chromosomes involved have been 9, 14, 18, and 21, or G.

Ring 9

One case [75] of an inherited ring 9 has been reported. A female, delivered by Caesarian section, weighed 1,360 g. She sat at 8 mo, walked at 18 mo, and had a vocabulary of 3 to 4 words. She exhibited a failure to thrive. Her length, weight, and head circumference were all below the third percentile, and she had delayed bone age development. At 22 mo, further evaluation indicated she was alert and active but was not toilet-trained and still had a limited vocabulary. She had microcephaly, large ears, pointed teeth, prominent nose, wide down-turned mouth, and slight epicanthal folds. Skull bones were fused. The patient's mother was phenotypically normal with no remarkable medical history. Examination of chromosomes from both mother and child, however, revealed identical karyotypes, 46,XX, r(9)(p24q34).

Ring 14

Two cases of inherited ring 14 have been reported. One family [95] is described in which the mother, her two live offspring, and a therapeutically aborted fetus each had a ring 14 chromosome. The mother had epileptic fits from shortly after birth until she was 2.5 years old. Her intelligence was at the lower end of the normal range. From age 21 to 27 she had five pregnancies. The two liveborn children were mentally retarded. In addition, the mother had two spontaneous abortions, one of which was shown to be chromosomally normal. One liveborn child was apparently normal at birth. At age 6 months she had convulsions, followed by myoclonic convulsions at 16 mo. By age 2 yr there was regression in her developmental progress, with gradual loss of speech and ability to feed herself and of locomotion. I.Q. at 2.5 yr was below 50. She was microcephalic and had an abnormal electroencephalogram that was interpreted as generalized cerebral abnormality without striking epileptic features. Most cells from blood and skin showed a 46, XX, r(14)(p13q32). The ring was lost from 13% of peripheral blood cells.

The second child was delivered by breech. She failed to breathe spontaneously, muscle tone was poor, and she fed with difficulty. At 8 mo she was admitted to the hospital with grand mal convulsions. Prolonged convulsions were difficult to control medically. Electroencephalogram showed loss of activity in the right frontal area; tomography revealed cortical atrophy in that area. At 3 yr and 3 mo, she was an obese ataxic child without any congenital malformations. Her IQ was approximately 50.

The fifth pregnancy was therapeutically aborted because a prenatal chromosome analysis revealed the presence of the ring chromosome 14.

In a second family with ring 14 [122], a ring chromosome 14 was present in two brothers. The mother is mosaic for a ring 14 and a translocation t(14q;21q).

Ring 18

Two cases of inherited ring 18 are reported. In one case, a [107] 12-year-old girl and a 48-year-old mother were both institutionalized for severe mental retardation. Numerous congenital anomalies were present in both including antisocial and psychotic behavior. Of 52 metaphases analyzed from the mother, 50 had a ring; one cell was missing a ring and one cell had two rings. The ring was identified by autoradiography. Sixty-one cells were analyzed from the daughter: all cells had a ring chromosome replacing an E group; one cell with a 47 count had a small supernumerary acrocentric.

In the second case, a mother and her son are reported with a ring chromosome 18 [106]. The son had microcephaly, ptosis, short stature, and mental retardation. The mother was mentally retarded and had a similar facial appearance. Loss of genetic material appeared undetectable. Reportedly no mosaicism was present in either the mother or the child. Also there was no variability in the size and shape of the ring.

Ring 21

In one case [120], a family with G ring mosaicism in two generations is presented. The index patient had the chromosome constitution G ring/G monosomy. His mother, brother, and maternal uncle were mosaics with three cell lines: normal (predominantly)/G ring (few)/G monosomy (few). The index patient had small stature (below third percentile) and mental retardation. His mother, brothers and uncle were clinically normal except for some minor degenerative signs.

A second family of ring 21 was reported by Palmer et al. [119]. The mother, a 38-year-old white female, had three children; only the last was determined to be abnormal. Both the mother and child are reported with various clinical anomalies, including mild mental retardation, numerous skeletal and facial anomalies and prominent abdomen. Chromosome studies of the child and her mother revealed both to have a ring 21. In addition, the girl had an extra X-chromosome.

RINGS IN CANCER

A number of cases described above have also involved cancer or leukemia. In fact, the first rings in two tumor cell lines of lung and stomach [5]. Ring configurations have also apparently been seen in cases of erythroleukemia and acute myelogenous leukemia [123], in meningioma [124,128], in carcinoma of the cervix [125,128], and in bladder cancer [126]. As previously mentioned, a case of ring 1, in particular, was associated with acute myelogenous leukemia in conjunction with phenotypic anomalies [55]; and a case with ring chromosome 11 developed Wilms tumor [87]. The latter is of particular interest because of recent suggestions that loss of the genes on one chromosome may allow the mutant genes associated with conditions such as Wilms tumor and retinoblastoma on the remaining chromosome to be expressed.

CONCLUSIONS

Much emphasis has been placed on ring instability. This instability is more apparent in some cases than others. However, the problems of assessing such instability cannot be overemphasized. In many cases the details of ring products are only superficially described or even assumed. In only a few cases has more than one tissue been examined. Although a few studies have been attempted, the extent of in vitro vs. in vivo selection is, for the most part, unknown. The extent to which mosaicism seen in vitro in fact represents what is seen even in the particular tissue being sampled can be extensively debated. Hecht and Vlietinck [9] noted the following as complicating factors in attempting genotype-phenotype correlation in cases of ring chromosomes.

1) *Location of deletions:* Deletions involving metacentric vs. acrocentric chromosomes would provide twice as much variation since acrocentric involve loss of significant material from only one arm. 2) *Length of deletion:* Obviously the amount of material deleted will vary in different cases as for other chromosome rearrangement. 3) *Number of cell lines (mosaicism):* The tendency of rings to be lost or duplicated can lead to mosaicism. 4) *Similar looking but different rings.* 5) *Polycentric rings.*

In fact, the above list is incomplete, and the following factors can be added.

6) *Dynamic mosaicism:* The extent that deletion-duplication in a ring may be specific to certain tissues or stages in development and thus no longer

detectable in its original form. 7) *Genetic variability:* The expression of recessive alleles on normal homologs due to ring loss or deletion within the ring itself, i.e., unmasking of recessive genes.

Despite these complications, it is apparent that the phenotypic variability due to ring instability has in some cases been over-emphasized, or at the least, over-simplified. From the previous discussion it is evident that ring chromosome syndromes are gradually emerging that are at least as distinctive as for duplications and deletions arising from more conventional types of chromosome rearrangements. Some cases, on the other hand, show phenotypic variations that are not necessarily consistent with the apparent chromosomal imbalance. Cases of inherited rings have also demonstrated considerable phenotypic variability between parents and children indicating that one or more of the above factors must be operating.

Each case of ring chromosome should be treated as unique and informative only when studied as thoroughly as possible. Such thoroughness should include: examination of at least 100 cells, from more than one tissue, and with as little time in culture as possible; detailed banding studies of atypical ring products as well as an accurate assessment of mosaicism; a thorough clinical workup and detailed family history whenever possible.

ACKNOWLEDGMENTS

Certain previously unpublished data by Wyandt and colleagues was supported in part by NIH grant GM 25244 and by a Biomedical Research subgrant from the University of Virginia, BRS 5-23134. The author is indebted to David McGinnis for help in reviewing the literature, and to Linda Wyandt and Shivanana Patil for critically reading the manuscript.

REFERENCES

1. Nawashin M: Unbalanced somatic chromosomal variation in Crepis. Univ Calif Publ Agric Sci vi 3:95–106, 1930.
2. McClintock BA: Correlation of ring-shaped chromosomes with variegation in Zea mays. Proc Natl Acad Sci USA 18:677–681, 1932.
3. McClintock B: The production of homozygous deficient tissues with mutant characteristics by means of the aberrant mitotic behavior of ring-shaped chromosomes. Genetics 23:315–376, 1938.
4. Morgan LV: A closed X chromosome Drosophila melanogaster. Genetics 18:250–283, 1933.
5. Levan A: Self-perpetuating ring chromosomes in two human tumors. Hereditas 42:366–372, 1956.

6. Puck TT: The action of radiation on mammalian cells III: Relationship between reproductive death and induction of chromosome anomalies by X-irradiation of euploid cells in vitro. Proc Natl Acad Sci USA 44:772, 1958.
7. Wang HC, Melnyk J, McDonald LT, Uchida IA: Ring chromosomes in human beings. Nature 195:733–734, 1962.
8. Hecht F: Ring 4 chromosome: Ring autosomes, Lorelei of clinical-karyotype correlation and deletion mapping. Birth Defects 5(5):106–113, 1969.
9. Hecht F, Vlietinck RF: Autosomal rings and variable phenotype. Humangenetik 18:99–100, 1973.
10. Hook EB, Hamerton J: The frequency of chromosomal abnormalities detected in consecutive newborn studies—differences between studies—results by sex and by severity of phenotypic involvements. In Hook EB Porter IH (eds): "Population Cytogenetics, Studies in Humans." New York: Academic Press, 1977, pp 63–79.
11. Hook EB, Schreinemacher DM, Willey AM, Cross PK: Rates of mutant chromosome rearrangements in human fetuses: Data from prenatal cytogenetic studies and association with maternal age and parental mutagen exposure. Am J Hum Genet 35:96–109, 1983.
12. Ferguson-Smith MA, Yates JRW: Maternal age specific rate for chromosome aberrations and factors influencing them: Report of a collaborative European study on 52,965 amniocenteses. Prenat Diag 4:5–44, 1984.
13. Hsu LYF, Perlis TE: United States survey on chromosomal mosaicism and pseudomosaicism in prenatal diagnosis. Prenat Diag 4:97–120, 1984.
14. Borgaonkar DS: "Chromosomal Variation in Man." 4th Ed. New York: Alan R. Liss, 1984.
15. Borgaonkar DS, Shaffer R, Albright R, Reed WC, Jackson L: Repository of Chromosomal Variants and Anomalies In Man. 10th Listing. Wilmington Medical Center (Wilmington), University of Delaware (Newark), and Thomas Jefferson Univ. (Philadelphia), 1983.
16. Weleber RG, Hecht F, Giblett E: Ring G chromosome, a new G-deletion syndrome? Am Dis Child 115:489–493, 1968.
17. Griffin AB, Lindsley DL: The products of gynandromorphs through the use of unstable ring chromosomes in Drosophila melanogaster. Anat Rec 96:555–556, 1946.
18. Hinton CW: The analysis of rod derivatives of an unstable ring chromosome ring chromosome of Drosophila melanogaster. Genetics 42:55–65, 1957.
19. Pasztor LM: Unstable ring-X chromosome derived from a tandem metacentric compound in Drosophila melanogaster. Genetics 68:245–258, 1971.
20. Zalokar M, Erk I, Santamaria P: Distribution of ring X chromosomes in the blastoderm of gynandromorphic D. melanogaster. Cell 19:133–141, 1980.
21. Lejeune J: Modele theorique de la repartition de duplications et des deficiences dans le chromosomes en anneau. CR Acad Sci (Paris) 264:2586–2590, 1967.
22. Lejeune J: De la duplication de structures ciculaires. Ann Genet 11:71–77, 1968.
23. Kistenmacher ML, Punnett HH: Comparative behavior of ring chromosomes. Am J Hum Genet 22:304–318, 1970.
24. Wyandt HE, Vlietinck RF, Magenis RE, Hecht F: Colored reverse banding of human chromosomes with acridine orange following alkaline/formalin treatment: Densitometric validation and applications. Humangenetic 23:119–130, 1974.
25. Hamers AJH, van Kempen C: Casse report: Ring chromosome 8 in a boy with multiple congenital abnormalities and mental retardation. J Med Genet 14:451–455, 1977.
26. Muller HJ: An analysis of the process of structural changes in chromosomes in Drosophila. J Genet 40:1–66, 1940.

27. Fowler G, Kaiser-McGaw B, Hecht F: The use of sequential silver and quinacrine staining to determine the parental origin and breakpoints of a ring 22 human chromosome. Clin Genet 18:274–279, 1980.
28. Hernandez A, Garcia-Cruz D, Plascencia L, Nazara Z, Rivera H, Sanchez-Corona J, Cantu JM: Some clinical and cytogenetic observations on a ring chromosome 13 (p11q34). Ann Genet 22:221–224, 1979.
29. Stetten G, Sroka B, Corson VL, Boehm CD: Prenatal detection of an unstable ring 21. Hum Genet 68:310–313, 1984.
30. Dutrillaux B, Croquette MF, Viegas-Pequignot E, Aurias A, Cojet J, Couturier J, Lejeune J: Human somatic chains and rings. A preliminary note on end to end fusions. Cytogenet Cell Genet 20:70–77, 1978.
31. Morgan R, Jarzabek V, Jaffe JP, Hecht BK, Hecht F, Sandberg AA: Telomeric fusion in pre-T-cell acute lymphoblastic leukemia. Hum Genet 73:260–263, 1986.
32. Niss R, Passarge E: Derivative chromosomal structures from a ring 4. Humangenetic 28:9–23, 1975.
33. Wyandt HE, Kasprzak R, Lamb A, Willson K, Wilson WG, Kelly TE: Human chromosome 2 rod/ring mosaicism: Probable origin by prezygotic breakage and intrachromosomal exchange. Cytogenet Cell Genet 33:222–231, 1982.
34. Bartram CR: Sister chromatid exchange in a ring chromosome 4. Cytogenet Cell Genet 18:238–241, 1977.
35. Zuffardi O, Danesino C, Poloni L, Pavesi F, Bianchi C, Gargantini L: Ring chromosome 12 and latent centromeres. Cytogenet Cell Genet 28:151–157, 1980.
36. Pathak S, Sinha AK: An alternative model for ring chromosome perpetuation in a human subject. Ann Hum Genet (Lond) 35:471–476, 1972.
37. Hoo JJ, Obermann U, Cramer H: The behavior of a ring chromosome 13. Humangenetik 24:161–171, 1974.
38. Brewen JG, Peacock WJ: The effect of tritiated thymidine on sister-chromatid exchange in a ring chromosome. Mutat Res 7:433–440, 1969.
39. Latt SA: Microfluorometric detection of desoxyribonucleic replication in human metaphase chromosomes. Proc Natl Acad Sci USA 12:3395–3399, 1973.
40. Lansky S, Daniel W, Fleiszer K: Physical retardation associated with ring chromosome mosaicism: 46,XX,r(10)/45,XX,10−. J Med Genet 14:61–80, 1977.
41. Moore CM, Heller RH, Thomas GH: Developmental anomalies associated with a ring chromosome G. J Med Genet 10:299–303, 1973.
42. Van den Berghe H, Fryns J-P, Cassiman J-J, David G: Chromosome 6 en anneau caryotype 46,XY,r(6)/45,XY,-6. Ann Genet 17:29–35, 1974.
43. Jacobsen P, Mikkelsen M, Rosleff F: A ring chromosome, diagnosed by quinacrine fluorescence as no. 9, in a mentally retarded girl. Clin Genet 4:434–441, 1973.
44. Turleau C, Rethore M-O, Junien C, Lejeune J, de Grouchy J: Mosaique 46,XX/46,XX,10p−/47,XX,+r/47,XX,10p−, +r et phenotype de la trisomie partielle 10p. Ann Genet 22:178–181, 1979.
45. Velente M, Muller H, Sparkes R: Ring 11 chromosome (46,XX,r11(p15q25)). Hum Genet 36:345–350, 1977.
46. McDermott A, Voyce MA, Romain D: Ring chromosome 4. J Med Genet 14:228–232, 1977.
47. Kosztolanyi G: Ring chromosome 4: Wolf syndrome and unspecific developmental anomalies. Acta Pediatr Hung 26:157–165, 1985.
48. Zachai EH, Breg WR: Ring chromosome 7 with variable phenotypic expression. Cytogenet Cell Genet 12:40–48, 1973.

49. Sutherland GR, Carter RF: 46,XX/46,XX,r(2)(p25q37) mosaicism: Clinical and cytogenetic studies. Ann Genet 21:164–167, 1978.
50. Parker CE, Alfi OS, Derencsenyi A, Mavalwala J, Donnell G: A child with ring 4 chromosome (46,XX-46,XX,r4). Am J Dis Child 128:371–374, 1974.
51. Benn PA, Warburton D, Byrne JM, Rudelli R, Shonhaut A, Yeboa K, Mootabar H, Hsu LYF: A fetus with a chromosome 13 ring and placenta with chromosome rod/ring mosaicism. Prenat Diagn 3:297–302, 1983.
52. Hutchinson-Cole H, Aleck KA, Davis JR, Veomett IC, Hauck L: Mosaicism for a ring and isopseudodicentric chromosome 13. Clin Genet 29:88–91, 1986.
53. Kjessler B, Gustavson K-H, Wigertz A: Apparently non-deleted ring-1 chromosome and extreme growth failure in a mentally retarded girl. Clin Genet 14:8–15, 1978.
54. Wolf CB, Peterson JA, LoGrippo GA, Weiss L: Ring chromosome 1 and dwarfism—a possible syndrome. Pediatr 71:719–722, 1967.
55. Bobrow M, Emerson PM, Spriggs AI, Ellis HI: Ring-1 chromosome, microcephalic dwarfism, and acute myeloid leukemia. Am J Dis Child 126:257–260, 1973.
56. Vigfusson NV, Kapstafer KJ, Lloyd MA: Ring chromosome 2 in a child with growth failure and few congenital abnormalities. Am J Med Genet 7:383–389, 1980.
57. Maraschio P, Danesino C, Garau A, Saputo V, Vigi V, Volpatra S: Three cases of ring chromosome 2, one derived from a paternal 2/6 translocation. Hum Genet 48:157–167, 1979.
58. Picciano DJ, Berlin CM, Davenport SLH, Jacobsen CB: Human ring chromosomes: A report of five cases. Ann Genet 15:241–247, 1972.
59. Witowski R, Ullrich E, Piede U: Ring chromosome 3 in a retarded boy. Hum Genet 42:345–348, 1978.
60. Perez-Castillo A, Abrisqueta JA: Ring chromosome 4 and Wolf syndrome. Hum Genet 37:87–91, 1977.
61. Johnson VP, Mulder RD, Hosen R: The Wolf-Hirschorn (4p−) syndrome. Clin Genet 10:104–112, 1976.
62. Preus M, Ayme S, Kaplan P, Vekemans M: A taxonomic approach to the del (4) phenotype. Am J Med Genet 21:337–345, 1985.
63. Nakagome Y, Iinuma K, Taniguchi K: Points of exchange in a human no. 5 ring chromosome. Cytogenet Cell Genet 12:35–39, 1973.
64. Suerink E, Noel B, Rethore MO: Ring chromosome 5 in two malformed boys with cri du chat syndrome. Clin Genet 14:125–129, 1978.
65. Niebuhr E: The cri du chat syndrome. Hum Genet 44:227–275, 1978.
66. Wurster-Hill DH, Hoefnagel D: Banding identification of chromosomal abnormalities in four patients: Ring(6), translocation (2q−;15q+), translocation (21q;21q) and deletion (22q−). J Ment Defic Res 19:145–150, 1974.
67. Salmanca-Gonez F, Nava S, Armendares S: Ring chromosme 6 in a malformed boy. Clin Genet 8:370–375, 1975.
68. Sele B, Joannard A, Jalbert P, Boucharlat J: Chromosme 6 en anneau: Un tableau clinique peu specifique. Ann Genet 20:232–236, 1977.
69. Carnevale A, Blanco B, Catillo J, Castillo V, Dominguez D: Ring chromosome 6 in a child with minimal abnormalities. Am J Med Genet 4:271–277, 1979.
70. Nishi Y, Yoshimura O, Ohama K, Usui T: Ring chromosome 6: Case report and review. Am J Med Genet 12:109–114, 1982.
71. Nakano S, Miyamoto N: A ring C7 chromosome in a mentally and physically retarded male with various somatic abnormalities. J Hum Genet 22:90–98, 1977.

72. DeLozier-Blanchet CD, Guenin R: Cytogenetics of ring chromosome 7. Clin Genet 25:84–87, 1984.
73. Pfeiffer RA, Lenard HG: Ring chromosome 8 (46,XY,8r) bei einem Jungen. Klin Paediatr 185:187–191, 1973.
74. Fraisse J, Lauras B, Ooghe M-J, Freycon F, Rethore M-O: Apropos d'un cas de chromosome 9 en anneau. Ann Genet 17:175–180, 1974.
75. Caldwell R, Brown JA, Young RB: A report of a familial ring (9) chromosome. MCV Quarterly 13:179–182, 1977.
76. Kistenmacher ML, Punnett, HH: Comparative behavior of ring chromosomes. Am J Hum Genet 22:304–318, 1970.
77. Zdansky R, Andrle M, Buhler E, Tsuchimoto T, Mayr WR, Rett A: Irregular phenotypic expression of ring chromosomes. Humangenetik 26:193–198, 1975.
78. Metaxotou C, Kalpini-Mavrou A: Ring chromosome 9. 46,XY,r(9) in a male with ambiguous external genitalia. Hum Genet 37:351–354, 1977.
79. Nakajima S, Yanagisawa M, Kamoshita S, Nakagome Y: Mental retardation and congenital malformations associated with a ring chromosome 9. Hum Genet 32:289–293, 1976.
80. Inouye T, Matsuda H, Shimuira K, Hamazaki M, Kikuta I, Iinuma K, Nakagome Y: Ring chromosome 9 in an infant with malformations. Hum Genet 50:231–235, 1976.
81. Simoni G, Rossella F, Dalpra L, Visconti G, Piria-Schwarz C: Ring chromosome 10 associated with multiple congenital malformations. Hum Genet 51:117–121, 1979.
82. Michels V, Driscoll DJ, Ledbetter DH, Riccardi VM: Phenotype associated with ring 10 chromosome: Report of patient and review of literature. Am J Med Genet 9: 231–237, 1981.
83. Kondo I, Shimakura Y, Hirando T, Kaneko M, Yabuta, K: Ring chromosome 10 syndrome: Case report and possibility of clinical diagnosis. Clin Genet 25:196–200, 1984.
84. Fryns P, De Boeck K, Jaeken J, van den Berghe H: Malformation syndrome associated with a ring 10 chromosome and a translocated 10q/19 chromosome. Hum Genet 43:239–244, 1978.
85. Niikawa N, Jinno Y, Tomiyasu T, Fukushima Y, Kudo K: Ring chromosome 11 (46,XX,r(11)(p15q25)) associated with clinical features of the 11q− syndrome. Ann Genet 24:172–175, 1981.
86. Retbi JM, Raoul O, Allemon MC, Dayras JC: Chromosome 11 en anneau (46,XX,r11). Une nouvelle observation. La Nouvelle Press Medicale 10(3):181, 1981.
87. Romain DR, Gebbie OB, Parfitt RG, Columbano-Green LM, Smythe RH, Chapman CJ, Kerr A: Two cases of ring chromosome 11. J Med Genet 20:380–382, 1983.
88. Niebuhr E, Ottosen J: Ring chromosome D(13) associated with multiple congenital malformations. Ann Genet 16:199–202, 1973.
89. Parcheta B, Wisniewski L, Piontek E, Szymanska J, Skawinski W, Wermenski K: Clinical features in a case with ring chromosome 13. Eur J Pediatr 144:409–412, 1985.
90. Schmid W, Muhlethaler JP, Briner J, Knechtli H: Ring chromosome 13 in a polymalformed anencephalic. Humangenetik 27:63–66, 1975.
91. Magenis RE, Wyandt HE, Overton KM, Macfarlane J: Parental origin of a ring 13 chromosome in a female with multiple anomalies. Hum genet 33:181–186, 1976.
92. Couterier J, Morichon-Delvallez N, Dutrillaux B: Deletion of band 13q21 is compatible with normal phenotype. Hum Genet 70:87–91, 1985.
93. Fryns JP, Kubien E, Kleczkowska A, Nawrocka-Kanska B, Van den Berghe H: A ring chromosome 14. A distinct clinical entity. J Genet Hum 31:367–375, 1983.

94. Krawczun M, Melink G, Cervenka J: Ring chromosome 14 and immunoglobulin locus. Am J Med Genet 17:465–469, 1984.
95. Bowser Riley SB, Buckton KE, Ratcliffe SG, Syme J: Inheritance of a ring 14 chromosome. J Med Genet 18:209–213, 1981.
96. Kosztolanyi G, Pap M: Severe growth failure associated with atrophic intestinal mucosa and ring chromosome 15. Acta Paediatr Scand 75:326–331, 1986.
97. Wilson GN, Sauder SE, Bush M, Beitins IZ: Phenotypic delineation of ring chromosome 15 and Russell-Silver syndromes. J Med Genet 22(3):233–236, 1985.
98. Yunis E, Leibovici M, Quintero L: Ring(15) chromosome. Hum Genet 57:207–209, 1981.
99. Fryns JP, Timmermans J, Hondt FD, Francois B, Emmery L, van den Berghe H: Ring chromosome 15 syndrome. Hum Genet 51:43–48, 1979.
100. Schmid M, Henrichs I, Nestler H, Knorr-Gartner H, Teller WM, Krone W: Analysis of the banding pattern and mosaic configuration in a case of ring chromosome 15. Hum Genet 41:289–299, 1978.
101. Pfieferr RA, Dhadial R, Lenz W: 46,XX/46,XX,r(15) mosaicism. Report of a case. J Med Genet 14:63–65, 1977.
102. Carpenter NJ, Leichtman LG, Stamper S, Say B: An infant with ring 17 chromosome and unusual dermatoglyphics a new syndrome. J Med Genet 18:234–236, 1981.
103. Chudley AE, Pabello PD, McAlpine PJ, Nicol BE, Shokier HMK: Ring chromosome 17 in a mentally retarded young man—clinical, cytogenetic and biochemical investigations. Am J Hum Genet 12:219–225, 1982.
104. Kunze J, Stephan E, Tolksdorf M: Ring-chromosome 18. Ein 18p−/18q− Deletionssyndom. Humangenetik 15:289–318, 1972.
105. Hecht F, Vlietinck R: Autosomal rings and variable phenotypes. Humangenetik 18:99–100, 1973.
106. Donlan MA, Dolan CR: Ring chromosome 18 in a mother and son. Am J Med Genet 24:171–174, 1986.
107. Christensen KR, Friedrich U, Jacobsen P, Jensen K, Nielsen J, Tsuboi T: Ring chromosome 18 in mother and daughter J Ment Defic Res 14:49–67, 1970.
108. Jalpert P, Jalpert H, Sele B, Tachker D, Bost M, Meynard R, Pison H: Chromosme 20 en anneau: Un nouveau syndrome. Ann Genet 20:258–262, 1977.
109. Huret JL, Leonard C, Kanoui V: Ring chromosome 21 in a phenotypically normal but infertile man. Clin Genet 28:541–545, 1985.
110. Phillip N, Baeteman MA, Mattei MG, Mattei JF: Three new cases of partial monosomy 21 resulting from one ring 21 chromosome and two unbalanced reciprocal translocations. Eur J Pediatr 142:61–64, 1984.
111. Kazazian HH, Antonarakis SE, Wong C, Trusko S, Stetten G, Oliver M, Potter M, Gusella JF, Watkins PC: Ring chromosome 21: Characterization of DNA sequences at sites of breakage and reunion. Ann NY Acad Sci 450:33–42, 1985.
112. Ieshima A, Ogasawara N, Yamamoto Y, Kuroki Y: A case of r(21) with stigmata of atypical Down syndrome. Hum Genet 55:65–69, 1980.
113. Orye E, Craen M: A t(21q21q) ring chromosome. Hum Hered 24:253–258, 1974.
114. Howell RT, McDermott A, Gardner A, Dickinson V: Down's syndrome with recombinant tandem duplication of chromosome 21 dereived from a maternal ring. J Med Genet 21(4):310–314, 1984.
115. Cibis GW, Waeltermann J, Harris DJ: Peter's anomaly in association with ring 21 chromosome abnormality. Am J Ophthalmol 100:733–734, 1985.

116. Magenis R, Armendares S, Hecht F, Weleber RG, Overton K: Identification by fluorescence of two G rings: (46,XY,21r) G deletion syndrome I and (46,XX,r22) G deletion syndrome II. Ann Genet 15:265–266, 1972.
117. Aller V, Abrisqueta JA, de Torres ML, Martin-Lucas MA, Perez-Castillo A, Del Mazo J: An r(22)(p11 q13) in a moderately mentally retarded girl. Hum Genet 51:157–162, 1979.
118. Funderburk SJ, Sparkes RS, Klisak I: Phenotypic variation in two patients with a ring chromosome 22. Clin Genet 16:305–310, 1979.
119. Palmer CG, Hodes ME, Reed T, Kojetin J: Four new cases of ring 21 and 22 including familial transmission of ring 21. J Med Genet 14:54–60, 1977.
120. Zdansky R, Buhler EM, Vest M, Buhler UK, Stalder G: Familiares Mosaik mit G-Ring. Humangenetik 7:275–2867, 1969.
121. Hutchinson-Cole H, Aleck KA, Davis JR, Veomett IC, Hauck L: Mosaicism for ring and isopseudodicentric chromosome 13. Clin Genet 29:88–91, 1986.
122. Matalon R, Michals K, Thompson F, Rosenthal IM: Ring chromosome 14 in two brothers with mosaicism for ring 14 and translocation t(14q;21q) in the mother. (In Press).
123. Alimena G, Annino L, Dallapiccola B, Ricci N: Ring chromosomes and leukaemia. Experientia 15:1359–1360, 1975.
124. Mark J: Origin of the ring chromosome in a human recurrent meningioma studied with G-band technique. Acta Pathol Microbiol Scand 81:588–590, 1973.
125. Spriggs AI, Cowdell lRH: Ring chromosomes in carcinoma in situ of the cervix as evidence of clonal growth. J Obstet Gynaecol Br Commonw 79:833–840, 1972.
126. Falor WH, Ward RM: Fifty-three month persistence of ring chromosome in noninvasive bladder carcinoma. Acta Cytol 20:272–275, 1976.
127. Denton TE, Howell WM: The use of AgNO3 and ammoniacal silver to study NOR and satellite regions of chromosomes. Mammalian Chromosome Newsletter 16(4): 178–179, 1975.
128. Werner-Favre C, Beris P, Piguet D, Engel E: Ring chromosomes and hematologic disorders. Cancer Genet Cytogenet 23:265–267, 1986.
129. Hoo JJ, Forster C, Kindermann I, Zabel B, Hansen S: Supernumary small ring chromosome. Humangenetik 25:17–28, 1974.
130. Borgaonkar DS, Friedman J, Lewandowski RC, Eleuterio M: Chromosomal syndromes. In Borgaonkar, DS (ed.): "Chromosomal Variations in Man." 4th Ed. New York: Alan R. Liss, 1984, Appendix II.
131. Abe T, Misawa S, Nishioka K, Okuno T, Nakagome Y: Formation of a ring chromosome 14 subsequent to the de novo 13/14 reciprocal translocation. A new cytogenetic evidence obtained by the nucleolus-organizer staining. Ann Genet 21:109–112, 1978.
132. Borgaonkar DS, Lacassie YE, Stoll C: Usefulness of chromosome catalog in delineating new syndromes. Birth Defects: Original Article Series (The National Foundation), Volume XII(5):87–95, 1976.

The Cytogenetics of Mammalian Autosomal Rearrangements, pages 697–721

Chapter 23
Role of the Double Nucleolar Organizer Region (dNOR) Variant in the Etiology of Trisomy 21

C.K. Jackson-Cook, W.E. Nance, and J.A. Brown

For the vast majority of patients with trisomy 21 the cause(s) of meiotic non-disjunction have been unknown. One powerful approach for investigating the causes of non-disjunction is to identify and study individuals in whom such errors have in fact occurred. We have recently observed an apparent duplication of the nucleolar organizer region (a dNOR variant) on one of the acrocentric chromosomes in 13/41 Down syndrome parent couples. In every case in which it was possible to determine the parental origin of the extra chromosome, it came from the parent who carried the dNOR. In addition, we have found that at least one parent carried a dNOR variant in 9/10 families in whom two or more non-translocation aneuploid offspring were conceived. As in the previous cases, the parent carrying the dNOR variant was also the source of the non-disjunctional error in these families. In contrast, dNOR variants were noted in only 1/41 parents in whom non-disjunction did not occur and in 0/50 normal control individuals. Based on our studies, and on the work of other investigators, it seems possible that chromosome variants involving extra copies of the NOR may indeed be involved in the meiotic non-disjunction of chromosome 21. If the role of the dNOR can be confirmed, the use of this marker, along with advanced maternal age and low maternal serum alpha-fetoprotein levels, could ultimately permit the prior identification of most cases of trisomy 21.

INTRODUCTION

Down syndrome is the most common form of mental retardation in man. Its incidence has been estimated to range from 1 to 1.86/1,000 liveborn infants [1,2]. Cytogenetic studies of populations of affected individuals have revealed two main groups of chromosomally abnormal patients with Down syndrome [3]. The first subgroup, patients with primary trisomy 21, includes

Departments of Human Genetics (C.K. J.-C., W.E.N., J.A.B.), Pediatrics (W.E.N.), and Obstetrics and Gynecology (J.A.B.), Medical College of Virginia, Richmond, VA 23298 cases with an additional chromosome 21 that arises as the result of either a meiotic (47, XX or XY, +21) or mitotic (46/47, XX or XY, +21) non-disjunctional error. In the second group, which includes patients with secondary or tertiary trisomy 21, a trisomic dose of the genes present on the long arm of chromosome 21 is present as a consequence of the malsegregation of a translocation (most frequently Robertsonian) involving chromosome 21. Since many translocations involving chromosome 21 have been noted to be inherited, balanced translocation carriers represent one readily identifiable group of individuals who are at an increased risk of conceiving children with Down syndrome. However, collectively, translocations account for only 4–5% of all cases of Down syndrome [4]. Another high-risk subpopulation includes parents with cryptic mosaicism for a trisomy 21 cell line. However, parental mosaicism has been confirmed as a probable cause for offspring with Down syndrome in only a very few cases, and has been estimated to occur in less than 2% [5] of all parents of children with trisomy 21. Therefore, the vast majority of all cases of Down syndrome are attributable to the presence of an extra chromosome arising from a meiotic non-disjunctional error. However, the cause(s) of non-disjunction is still poorly understood.

Previously recognized subpopulations of chromosomally normal parents that are known to be at an increased risk for having a child with trisomy 21 include older mothers and couples with a previously affected child. More recently we have found that males or females who carry an unusual chromosomal variant of the nucleolar organizer region (NOR), a double NOR (dNOR), appear to represent an additional subgroup in the normal population who have an increased risk for having chromosome non-disjunctional errors [6]. The primary aim of this chapter is to review our data and previously reported information regarding the role of the dNOR in meiotic chromosome non-disjunction.

PREVIOUS PARENTAL ORIGIN STUDIES OF TRISOMY 21

One very powerful approach to assessing the significance of the putative cause(s) of non-disjunction is to identify and study individuals in whom non-disjunctional errors have in fact occurred. Through the utilization of variant or marker chromosomes (primarily quinacrine (QFQ) heteromorphisms), the parental and meiotic divisional origin of the extra chromosome in children with trisomy 21 has been determined in approximately one half (55%) of the families studied (Table I) prior to the present approach. Although non-disjunctional errors occur most frequently in mothers of liveborn Down syndrome children (79.7%), the errors have also been documented to arise in fathers (20.4%), thus suggesting that factors other than maternal age must be involved in the etiology of non-disjunction (Table I). A similar distribution of parental and meiotic errors has also been observed among aborted fetuses with trisomy 21 [50]. Therefore, no apparent association exists between the parental or meiotic origin of non-disjunction and the resulting viability of the trisomic fetuses.

Etiological Factors Examined Previously in Parental Origin Studies

When maternal age is examined with respect to the parental and meiotic origin of the extra chromosome, no significant differences are observed. Furthermore, in all groups, the mean maternal age is elevated compared to that of the general population [51]. These data provide valuable information about the role of advanced maternal age in the etiology of non-disjunction. It has long been conjectured that the increased risk among older mothers for Down syndrome was primarily attributable to meiosis I errors that occurred as a result of an aging of the ovum during the dictyotene stage of female gametogenesis. Additional hypotheses that have been suggested in an attempt to explain the advanced-maternal-age association include delayed fertilization [52,53] and/or ovulation [54]. However, none of these hypotheses are fully supported by the parental origin data, in which no significant difference was noted in the age of the mothers in whom non-disjunction occurred (regardless of a meiosis I vs. meiosis II error) compared to mothers in whom non-disjunction did not arise. Following the observation of an advanced age among all mothers of children with Down syndrome, including those cases in which a paternal origin of non-disjunction was determined, Mattei et al. [55] suggested that the advanced maternal age factor could act after fertilization, rather than during meiosis, by preferentially allowing the development of trisomic embryos. If this hypothesis were valid, one would expect 1) the mean age of mothers spontaneously aborting trisomy 21 conceptuses to be younger than that of mothers having liveborn children with Down syndrome; and 2) the mean age of mothers of liveborn children with secondary and tertiary trisomy to be comparable to that of mothers having liveborn children with primary trisomy. However, both younger and older age distributions have been reported among mothers spontaneously aborting conceptuses with Down syndrome [56,57]. In addition, no evidence of an advanced age distribution has been noted among mothers of liveborn children with Down syndrome resulting from either inherited or de novo translocations, regardless of the parental origin of the translocations [57,58]. Therefore, the hypothesis of a preferential survival of fetuses with Down syndrome conceived by older mothers remains controversial.

TABLE I. Summary of 42 investigations of the parental origin of the extra chromosome among liveborn patients with trisomy 21

Reference (year)	No. of families studied	Meiotic errors							Variants studied
		Maternal			Paternal				
		MI	MII	?	MI	MII	?	Unknown	
de Grouchy (1970) [7]	?	—	1	—	—	—	—	—	S
Juberg and Jones (1970) [8]	?	—	1	—	—	—	—	—	S
Licznerski and Linstein (1971) [9]	6	1	—	—	—	—	—	5	QFQC
Mutton (1973) [10]	?	—	1	—	—	—	—	—	S
Punnett and Kistenmacher (1973) [11]	10	—	1	—	—	1	—	8	S
Robinson (1973) [12]	12	4	—	—	—	—	—	8	QFQ
Sasaki and Hara (1973) [13]	?	—	—	—	—	1	—	—	QFQ
Smith and Sachdeva (1973) [14]	20	—	—	—	—	—	—	20	QFQ
Uchida (1973) [15] and Uchida (1977) [16]	?	3	2	3	—	2	—	—	QFQ
Bott et al. (1975) [17]	59	—	3	—	—	4	—	52	QFQ, C
Dutrillaux (1975) [18]	?	1	1	—	—	—	—	—	QFQ
Emberger and Taib (1975) [19]	?	—	1	—	—	—	—	30	QFQ
Giraud et al. (1975) [20]	32	1	1	—	—	—	—	30	QFQ
Hara and Sasaki (1975) [21]	33	1	2	—	—	1	—	29	QFQ
Schmidt et al. (1975) [22] and Schmidt et al. (1976) [23] and Schmidt et al. (1981) [24]	40	22	—	—	1	—	—	17	QFQ
Moore et al. (1976) [25]	1	1	—	—	—	—	—	—	QFQ
Wagenbichler (1976) [26] and Wagenbichler et al. (1976) [27]	70	12	6	4	6	3	3	36	QFQ

Mikkelsen et al. (1976) [28] and Hansson and Mikkelsen (1978) [29]	72	9	6	4	3	4	—	46	QFQ
Grass and Moore (1978) [30]	1	—	—	1	—	—	—	—	QFQ, NOR
Magenis et al. (1977) [31] and Magenis and Chamberlain (1981) [32]	54	31	2	—	6	3	—	12	QFQ
Verma and Dosik (1978) [33]	?	—	—	—	1	—	—	—	QFQ, RFA
Mattei et al. (1979) [34] and Mattei et al. (1980) [35]	70	33	8	—	7	3	—	19	QFQ, RHG
Calabro et al. (1980) [36]	1	—	1	—	—	—	—	—	QFQ
Mikkelsen et al. (1980) [37]	105	49	12	5	10	2	1	25	QFQ, NOR
Roberts and Callow (1980) [38]	20	4	1	3	1	—	—	11	QFQ
Hamers et al. (1981) [39] and Jongloet et al. (1981) [40] and Jongloet and Hamers (1982) [41]	115	40	16	10	7	8	2	32	QFQ
Jacobs and Mayer (1981) [42]	45	11	3	3	—	2	—	26	QFQ
Manning and Goodman (1981) [43]	15	8	—	—	2	2	—	3	QFQ
Moric-Petrovic and Laca (1981) [44]	?	2	—	—	1	—	—	?	QFQ
del Mazo et al. (1982) [45]	48	17	3	2	6	1	—	15	QFQ, C
Pilgard and Mikkelsen (1982) [46]	36	12	2	2	5	1	—	14	QFQ
Hatcher et al. (1983) [47]	?	24	—	—	3	—	—	—	QFQ
Pérez-Castillo et al. (1983) [48]	3	3	—	—	—	—	—	—	QFQ
Davies et al. (1984) [49]	6	—	—	1	—	—	—	5	RFLP
Total	917	289	74	38	59	38	6	413	

? No. of families studied was not reported; S, structural; QFQ, Q-band by fluorescence with quinacrine; C, C-bands; NOR, silver staining; RFA, R-bands by fluorescence with acridine orange; RHG, R-bands by heat denaturation with Giemsa; RFLP, restriction fragment length polymorphisms.

Other potential etiological agents that have been evaluated in conjunction with the knowledge of the parental origin of non-disjunction include 1) alpha-1-antitrypsin, 2) thyroid autoantibodies, 3) chromosome variants, and 4) satellite associations.

No direct correlation has been noted between the occurrence of non-disjunction and any specific alpha-1-antitrypsin variant phenotypes. Although Jongbloet et al. [40] reported a trend toward an increased frequency of Pi variants among mothers in whom non-disjunction occurred, Bufton et al. [59] reported concordance for the variant phenotype and the occurrence of non-disjunction in only four of seven informative families. In contrast, a correlation between the presence of thyroid autoantibodies and the occurrence of non-disjunction has been reported by Magenis and Chamberlin [32] and Flannery et al. [60]. However, since the number of informative cases in both studies (five and seven, respectively) is very small, more extensive evaluation will be required to confirm the existing evidence for an association between the presence of thyroid autoantibodies and non-disjunction.

Unfortunately, a paucity of data exists regarding the correlation between the presence of normal chromosome heteromorphisms and the parental origin of non-disjunction. In a case report by Calabro et al. [36], a bisatellited marker chromosome was observed in a proband with trisomy 21. Analyses of the parents' chromosomes revealed that the marker chromosome had been inherited from the father, while the non-disjunctional error occurred during the second meiotic division of the mother. In a second case report, the extra chromosome in a child with trisomy 21 was determined to originate as the result of a meiosis I error in a mother with a balanced Robertsonian translocation (t13ql4q) [25]. In contrast, no concordance regarding the source of the parental origin of non-disjunction and the presence of a balanced chromosomal rearrangement was noted in three Down syndrome families studied by Uchida and Freeman [61].

Two investigations on the interchromosomal effects of acrocentric associations have been conducted in reference to the parental origin of non-disjunction. In the first of these studies [29], a highly significant increase in the satellite association index for chromosomes 14 and 21 was noted among the parents in whom non-disjunction had occurred. However, although not as large, a significantly higher satellite association index was also found for chromosomes 14, 15, and 21 among the parents in whom non-disjunction did not occur when compared to the control group.

In the second study [42] a marginally significant increase in the overall frequency of satellite associations was noted in the parents in whom non-disjunction occurred. However, no significant differences in the association index for chromosome 21 (or any chromosome) were observed in these parents as compared to the parents in whom non-disjunction did not occur and controls. Jacobs and Mayer also observed that the frequency of satellite associations in uninformative parents was less than that of the parents producing normal gametes. Since the noninformative group is equally composed of individuals in whom non-disjunctional errors did and did not occur, one might have expected the association index of this cohort to be midway between those of the non-disjunction and normal parental subgroups had non-disjunction and satellite association been causally related.

Our Parental Origin Study Results

Using QFQ, NOR, and structural heteromorphisms, we were able to determine the parental origin of the extra chromosome 21 in approximately 80% of the families examined. When studying the NOR chromosome heteromorphisms we noted an increased incidence of what we termed "double nucleolar organizer region" (dNOR) variants among the parents in whom non-disjunction of chromosome 21 had occurred [6]. In 15 of the 50 families having children with trisomy 21 that were studied, a dNOR variant was observed on one of any of the ten acrocentric chromosomes in at least one of the parents (Fig. 1). The parental origin of non-disjunction could be determined in 13/15 dNOR(+) families; and, as shown in Table II, in each of the informative families, the parent carrying the dNOR was also determined to be the source of the non-disjunctional error (χ^2 = 10.08, 1 df, $P < 0.001$). Family 8 was uninformative with respect to this association, since both parents had a dNOR variant, even though the extra chromosome 21 in the child was shown to have been contributed by the mother in this family. Double NOR variants were observed in 13/41 (32%) of the parents in whom non-disjunction was known to have occurred and in 2/18 (11.1%) of the indeterminate parents (among one-half of whom non-disjunction had occurred). In contrast, only one of the 41 (2.4%) normal spouses and none of the 50 (0%) control subjects were noted to have a dNOR variant. The departure of the frequencies of dNORs from a random distribution among these two groups was highly significant (χ^2 = 27.95, $P < 0.005$).

Based on our observed frequencies of dNOR(+) variants in our experimental and control groups, we have derived estimates of the absolute and relative frequencies of dNOR mating types among families having two offspring with primary trisomy 21 (Table III). Based on these theoretical estimates, we would predict that families having at least one dNOR(+) parent would account for approximately 95% of all non-translocation/cryptic trisomy 21 mosaic multiplex Down syndrome families. Therefore, we initiated

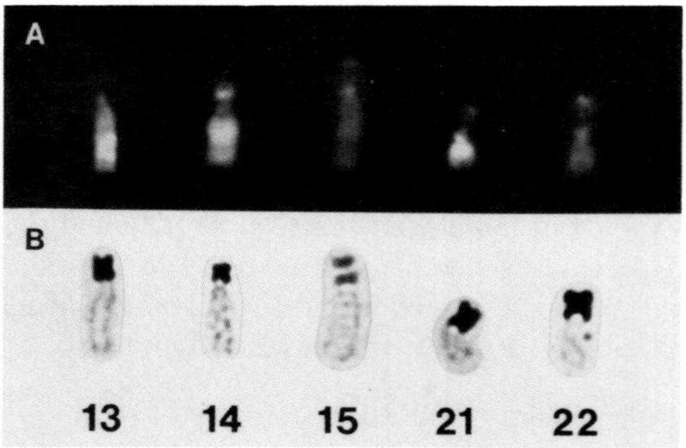

Fig. 1. Appearance of the double nucleolar organizer region (dNOR) variant chromosomes. The dNORS, which can be seen on any of the five acrocentric chromosomes, as they appear with quinacrine (QFQ staining) **(A)** and silver, or NOR staining (**B**). The dNORs may appear as two completely separated regions (as in the chromosome 15) or as two somewhat confluent regions (as seen in the other chromosomes) depending on chromosome morphology and the nature of the staining reaction. The dNOR variants are also characterized by elongated stalks as seen in the QFQ staining of these chromosomes.

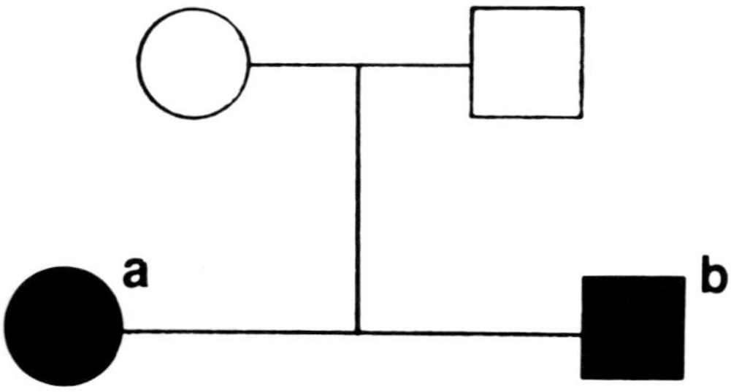

Fig. 2. Generalized pedigree illustrating the type of reproductive history noted in five couples who each conceived two offspring with trisomy 21.

a study to test the hypothesis that the vast majority of all families in which recurrent non-disjunctional events have been documented are dNOR(+). To date we have had an opportunity to study ten such families [62]. Three of these families were ascertained from our cytogenetic service records, and seven were ascertained through referrals from other genetic centers (see Acknowledgments). When possible, mid-metaphase chromosomes from the multiplex family members were stained and slides evaluated according to the same protocols followed in our initial study [6]. Five of the ten families were

TABLE II. Distribution of dNOR(+) phenotype and parental origin of non-disjunction

Family No.	dNOR(+) phenotype Chromosome	Frequency	dNOR(+) parent	Origin of non-disjunction	Inheritance of dNOR(+) by proband
1	13	10/10[a]	M	M	Yes
2	13	7/10	M	M	Yes
3	13	6/10	F	F	No
4	14	7/10	F	F	No
5	14	7/10	F	F	No
6	14	6/10	M	M	No[b]
7	15	6/10	M	M	No
8a	15	8/10	M	M	No
8b	21	5/10	F	M	No
9	21	5/10	M	M	Yes
10	21	5/10	M	M	Yes
11	21	6/10	M	?	Yes
12	22	8/10	M	?	No
13	22	6/10	F	F	No[b]
14	22	6/10	M	M	Yes
15	22	6/10	M	M	No

M, mother; F, father.
[a]No. of cells showing two silver-positive areas/total no. of cells examined.
[b]The proband did *not* inherit the dNOR. However, a second child with Down syndrome *did* inherit the dNOR.
Reproduced from Jackson-Cook et al. [6], with permission of the publisher.

TABLE III. Predicted distribution of dNOR parental mating types in families conceiving two offspring with trisomy 21

Parental mating type	Mating type frequency[a]	Absolute risk[b]	Frequency of families with two trisomy 21 offspring: Absolute	Relative (%)
N × N	$(0.98)^2$	$(0.0007)^2$	0.05×10^{-5}	4.6
N × dN	2(0.98) (0.02)	$(0.016)^2$	1.00×10^{-5}	91.7
dN × dN	$(0.02)^2$	$(0.031)^2$	0.04×10^{-5}	3.7
Total			1.09×10^{-5}	100.0

N, dNOR(−); dN, dNOR(+).
[a]Estimated frequencies in general population are based on initial study in which 1/91 dNOR(+) and 90/91 dNOR(−) controls were observed.
[b]Absolute risk estimated as described in Jackson-Cook, et al. [6].

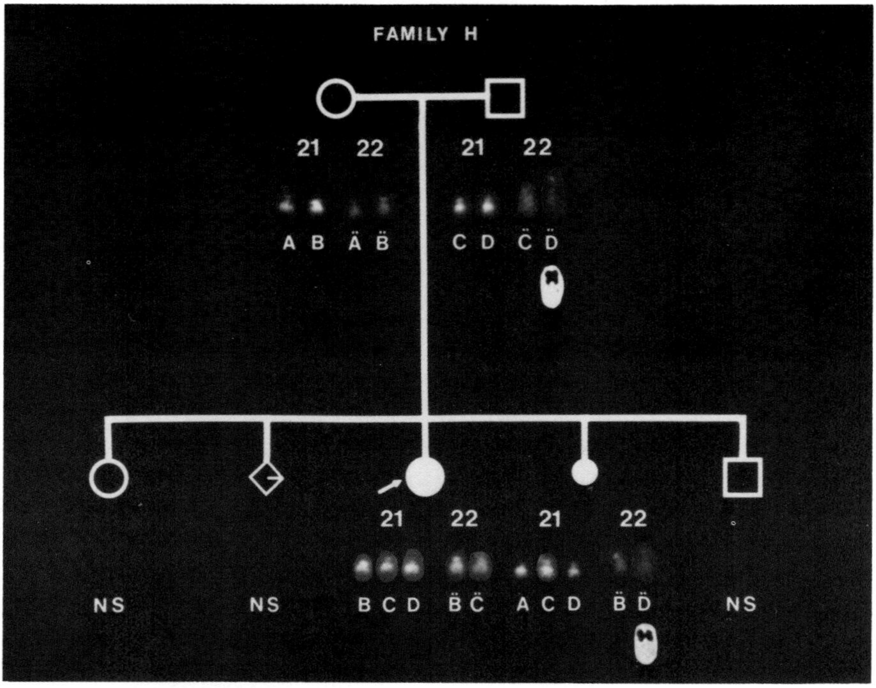

Fig. 3. Pedigree of family 1. The chromosomes 21 and 22 as they appear with QFQ staining are shown in this pedigree. The letters below the chromosomes refer to the heteromorphism assignments (based on QFQ and morphological variants) of the distinguishable chromosomes 21 (without dots) and 22 (with dots). The extra chromosome in both the proband (arrow) and the second affected offspring (the latter of whom was detected during antenatal screening) was contributed by the father, who has a dNOR on one of his chromosomes 22 (D) (see NOR-stained insert directly below chromosome). The dNOR was not inherited by the proband, but was inherited by the affected fetus (see insert below chromosome D). In addition to the two trisomic offspring, the couple also reported having a first-trimester abortion of unknown chromosomal constitution (NS, not studied cytogenetically) and two phenotypically normal children.

couples who had each conceived two offspring with trisomy 21 (Fig. 2–4). In addition to the ten probands, the reported reproductive histories included three spontaneous abortions and seven normal children (Table IV). No evidence of parental mosaicism for trisomy 21 was noted in the lymphocytes of any of these parents. A dNOR was seen in one parent in four of these five cases (Table V). In Family 5, access to slides stained only with ammoniacal

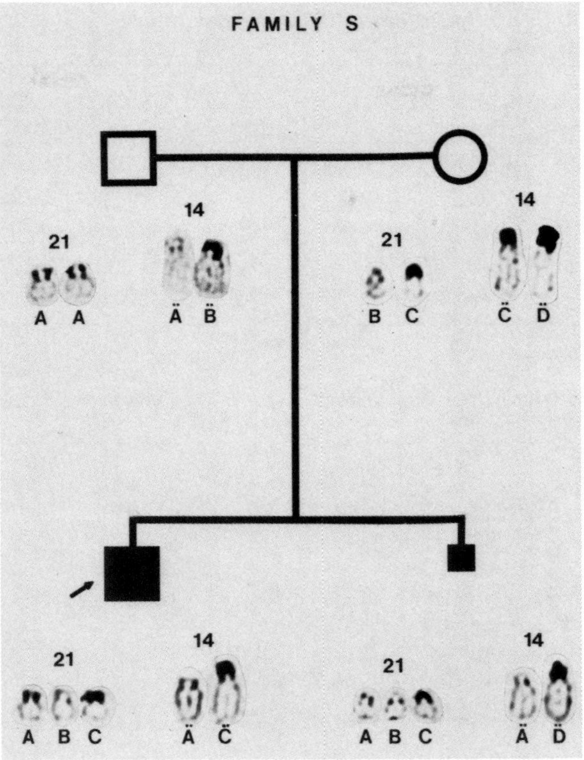

Fig. 4. Pedigree of family 2. The chromosomes 21 and 14 are shown in this pedigree as they appear with NOR or silver staining. The letters below the chromosomes refer to the heteromorphism assignments-based on NOR (shown) and QFQ (not shown) variants-of the distinguishable chromosomes 21 (without dots) and 14 (with dots). The mother, who has a dNOR variant (chromosome D), was determined to have been the source of the extra chromosome (meiosis I errors) in both the proband (arrow) and the second affected offspring (the latter of which was detected during antenatal screening subsequent to the dNOR finding in the mother). The dNOR was not inherited by the proband, but was inherited by the fetus (chromosome D).

silver and Giemsa were available for analysis. As a result, the definitive identification of the dNOR chromosome and the parental origin of non-disjunction could not be determined for this family. Of the remaining families, the parental origin could be determined for six of the eight probands, and in each case, the non-disjunctional error recurred in the parent carrying the dNOR. Also, in each case the non-disjunctional error recurred in the same parent during meiosis I.

TABLE IV. Reproductive histories of the five couples studied who each conceived two offspring with trisomy 21

Family No.	No. of offspring Trisomy 21	Abortus[a]	Normal	Total
1	2	1	2	5
2	2	0	0	2
3	2	1	2	5
4	2	0	1	3
5	2	1	2	5
Total	10	3	7	20

[a]Karyotype unknown.

TABLE V. dNOR and non-disjunctional status of the five couples conceiving two offspring with trisomy 21

Family	dNOR phenotype		Carrier	Origin
No.	dNOR(+)	Chromosome	parent	of non-disjunction
1	Yes	22	Father	a. Father b. Father
2	Yes	14	Mother	a. Mother b. Mother
3	Yes	21	Father	a. Father b. Father
4	No	—	—	a. Indeterminate b. Indeterminate
5*	Yes	D-group	Mother	a. Unknown b. Unknown

*Only slides stained with ammoniacal silver were available for analysis. Therefore, the specific chromosomal identification of the dNOR(+) variant and the parental origin of non-disjunction were indeterminate for this family.

In two of the ten multiplex families first-degree relatives each had a child with trisomy 21. In family 6 (Fig. 5), a brother and sister each had a child with Down syndrome. The same chromosome 21 dNOR was observed in the brother, sister, and their children. Although the parental origin of the extra chromosome 21 could not be determined for one of the affected offspring, the extra chromosome in the other affected child was noted to have been contributed during meiosis I by the father, who carried the familial dNOR variant.

In our seventh family a mother and her daughter each had a child with trisomy 21 (Fig. 6). In this case, a familial dNOR was observed on a chromosome 14 in a mother and her daughter. A dNOR variant was also observed in the daughter's spouse and in her stepfather. The parental origin

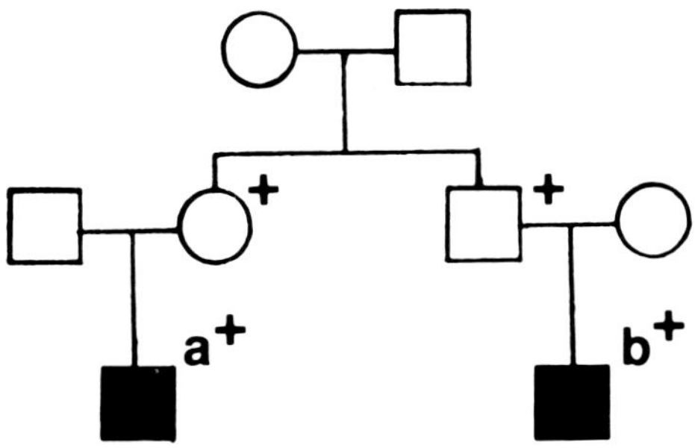

Fig. 5. Pedigree of family 6. In this family, siblings each had a child with trisomy 21. A dNOR (+) variant was noted on a chromosome 21 in the two siblings and both of the probands (a and b) with trisomy 21. The parental origin of the extra chromosome 21 could not be determined for proband (a) because the dNOR(−) chromosome 21 in the proband's mother was not distinguishable from a chromosome 21 in his father. The parental origin of the extra chromosome in proband (b) was contributed by the dNOR(+) father as a result of a meiosis I error.

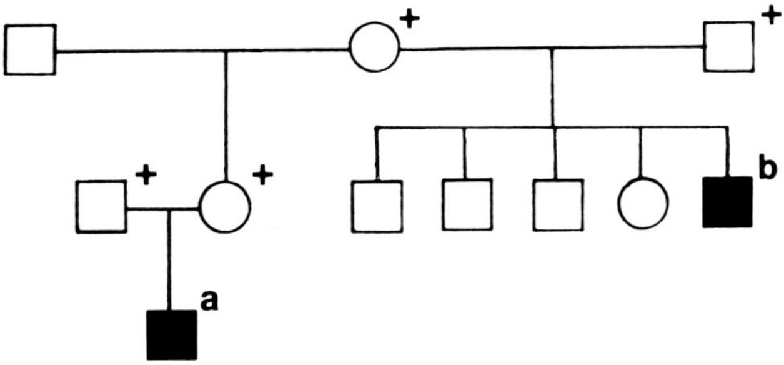

Fig 6. Pedgree of family 7. In this family a mother and her daugher each had a child with trisomy 21. The same dNOR(+) chromosome 14 was seen in the mother and her daughter. Interestingly, a dNOR(+) variant was also seen on a chromosome 13 in the father of proband (a). Similarly, a dNOR(+) variant was seen on a chromosome 22 in the father of proband (b) In both probands, the source of non-disjunction was determined to be attributable to a maternal meiosis I error.

of non-disjunction could be determined for both trisomic patients, and in each of these cases the extra chromosome was contributed by the mothers as the result of meiosis I errors.

Since the non-disjunctional events in our first-degree relative families are associated with a single familial dNOR variant chromosome, these families were pooled with our other five multiplex families to test our hypothesis of the distribution of NOR mating types in multiplex Down syndrome families (Table VI). Of the seven families studied, one involved a dNOR (−) couple,

7

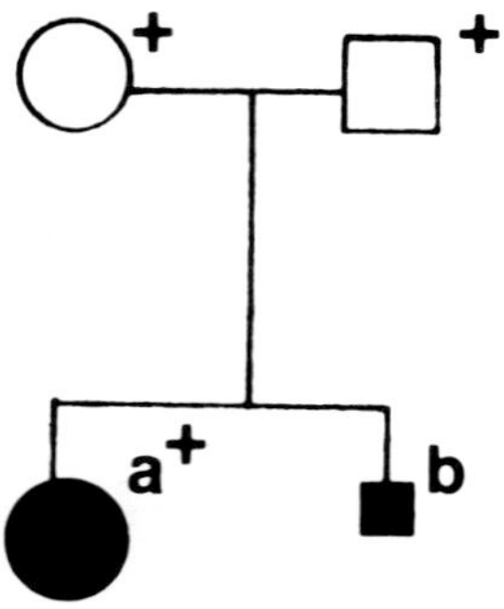

8

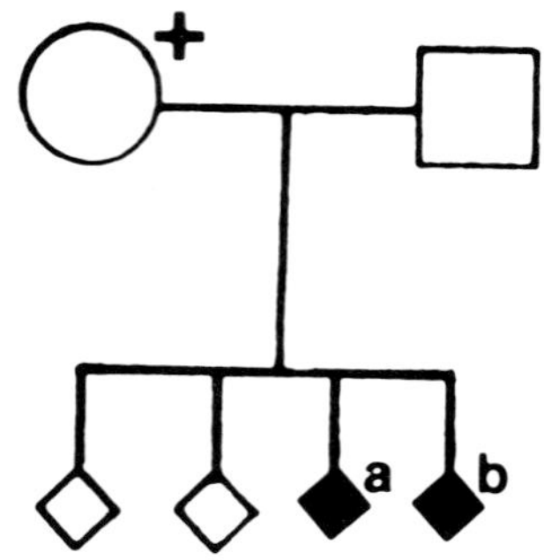

Fig. 7. Pedigree of family 8. In this family a child with trisomy 21 was born. In addition, during a second pregnancy, an abortus was determined to have trisomy 22. Analyses of the parents' chromosomes revealed a dNOR(+) variant in both the mother and father on a chromosome 14. The origin of the extra chromosome 21 in proband (a) was determined to be paternal, resulting from a meiosis I error. Chromosomes from proband (b) were not available for a parental-origin analysis.

Fig. 8. Pedigree from family 9. This family reported a reproductive history of four first-trimester spontaneous abortions. The chromosomal analyses, which were performed for only two of these abortuses (probands a and b), revealed that one abortus had trisomy 14 and one had trisomy 17. Since only the mother's chromosomes were available to us for study, no parental origin studies could be completed. However, a dNOR(+) variant was noted in the mother on a chromosome 15.

five were families in which one parent carried a dNOR, and one was a family in which both parents were dNOR (+). Because of the relatively low frequency of the dNOR variant in the general population we were at first surprised to observe the dNOR (+) by dNOR(+) mating within our sample. However, based on our predicted estimate (Table III), one would actually expect to find approximately as many dNOR(+) by dNOR(+) couples as dNOR(−) by dNOR(−) families.

Overall, our preliminary results are in good agreement with our predicted expectations and support our hypothesis that the majority of all non-translocation multiplex families have at least one dNOR(+) parent.

TABLE VI. Comparison of our observed vs. expected distributions of parental mating types among trisomy 21 multiplex families

	Parental mating types		
	N × N(%)	N × dN(%)	dN × dN(%)
Observed	1(14.3)	5(71.4)	1(14.3)
Expected	0.32(4.6)	6.42(91.7)	0.26(3.7)

N, = dNOR (−); dN, = dNOR(+).

Three additional families having a reproductive history of two aneuploid offspring were available for our analysis. In the first of these families a couple conceived a child with trisomy 21 and a first-trimester abortus with trisomy 22 (Fig. 7). The mother, father, and Down syndrome child all carried dNORs on a chromosome 14. The child inherited her dNOR variant from her father, who also contributed the extra chromosome 21 in this case.

Another couple ascertained in this study reported a reproductive history of four first-trimester spontaneous abortions (Fig. 8). Chromosomal analyses were obtained for two of these cases and revealed one abortus with trisomy 14 and one abortus with trisomy 17. Unfortunately, only the mother was available for study. However, an analysis of her chromosomes showed that she did indeed carry a dNOR variant on a chromosome 15.

Our tenth multiplex family studied had an interesting reproductive history of a chromosomally normal woman having two de novo translocation Down syndrome children [46,−21,+t (21;21)] who were fathered by different males. In this family, the mother, who showed no sign of mosaicism for a t(21;21) cell line in either lymphocytes or fibroblasts, was noted to carry two dNORs, one on a chromosome 14 and one on a chromosome 22 (Fig. 9). It is possible that this woman has a low level of gametic mosaicism for a t(21;21) cell line. However, one could also speculate that the dNOR variants were a factor in the formation of the apparently de novo Robertsonian translocations.

To summarize, in our preliminary study of multiplex kindreds, dNORs were observed in at least one parent from nine of the ten multiplex families we have studied.

DESCRIPTION OF THE dNOR VARIANT

The dNOR variant appears morphologically to consist of a linear doubling or duplication of the stalk or NOR area, which can be definitively seen following staining with ammoniacal silver (Fig. 1). However, the presence of this variant is often suggested by the characteristic elongated stalk region, as seen with quinacrine (QFQ) or Giemsa (GTG) banding, that is almost invariably associated with the variant and rarely seen with single NOR chromosomes. The dNOR can be distinguished from an intensely staining single NOR by the presence of two areas of silver staining (dots) stacked vertically, one on top of the other, rather than horizontally or side-by-side, as seen in some normal single variants. In our population, the dNORs varied in appearance from two completely separated areas of silver-grain deposit, as seen on the chromosome 15 in Figure 1, to two more-or-less confluent

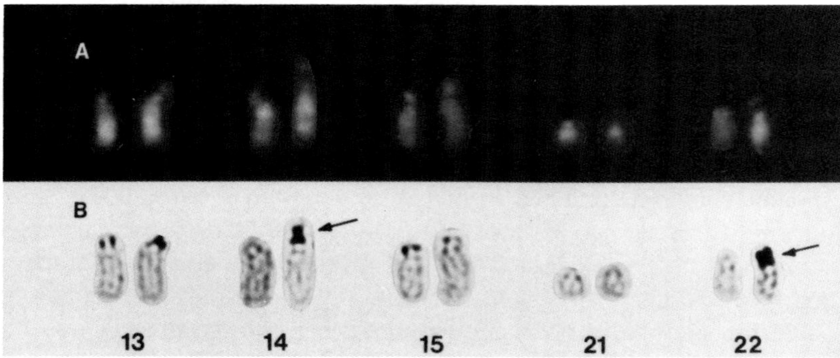

Fig. 9. QFQ and NOR heteromorphisms of the acrocentric chromosomes of a mother of two children with de novo trisomy translocation (21q21q). The acrocentric chromosomes are shown as they appear with QFQ **(A)** and NOR or silver **(B)** staining. Two dNORs (B), one on a chromosome 14 (arrow) and one on a chromosome 22 (arrow), were observed in this woman. One can note the presence of elongated stalks on these variant chromosomes as they appear with QFQ staining (A).

areas of silver staining (dumbbell shape) as seen in the other chromosomes. Both regions of silver staining on a dNOR variant were not seen in every cell. The observed frequency of cells in which double-stained regions were present ranged from 50% to 100% (Table II). The elongation of the stalks appeared to be a more consistent, if less obvious, feature of the dNOR variant. Finally, the specific morphology of the dNOR variant (two separate vs. two confluent regions) did appear to be consistent within an individual from cell to cell.

Only a small number of people known to have a dNOR variant chromosome have previously been reported (Table VII). Most of these individuals were described in case reports arising from diagnostic service laboratory surveys. The cytological appearance of the dNOR variants observed in our study are entirely consistent with descriptions in previously reported cases. As in our study, other investigators have described the dNOR as having two positively stained regions in only a portion of the metaphase spreads noted. However, to our knowledge only two groups of investigators have reported the distribution of silver-stained dNOR regions in individuals who carry the variant. Lau et al. [66] reported that at least two active areas on their multiple NOR variant could be seen in at least 57.8% of the cells, a proportion that is consistent with our observations in 15 subjects with the variant. In contrast, Perez-Castillo et al. [72] reported that the frequency of cells exhibiting at

TABLE VII. Previously reported dNOR(+) individuals

Reference	dNOR chromosome(s)	Reason for referral
Henderson and Atwood [63]	15	Unknown
Archidiacono et al. [64]	15	Unknown
	21	Unknown
Miller et al. [65]	14	Normal
Lau et al. [66]	14	Normal
Martin et al. [67]	21	Normal and 47, XXY
Balicek and Zizka [68]	22	Mental retardation (MR)
	21	47, +21
	21	Multiple aborter
	22	Anomalies
	14	Mother, 47, +21[a]
	22	Mental retardation with anomalies
	14	Father, FA[b]
	14	Mental retardation with anomalies
	13	Mental retardation with anomalies
	22	Father, NTD[c]
	13	Mother, 47, +21
	14	Mother, MR
	13	Mother, anomalies
	15	Mother, MR
Sofuni et al. [69]	14	Normal
	22	Children of
	13	parents exposed
	22	to A-bomb
Bernstein et al. [70]	22	Ambiguous genitalia
Jotterand-Bellamo and	13	Unknown
Van Melle [71]	15	Unknown
Pérez-Castillo, et al. [72]	15	Multiple aborter

[a]Mother of child with 47, +21.
[b]Father of child with Fanconi's anemia.
[c]Father of child with neural tube defect.
Reproduced from Jackson-Cook et al. [6], with permission of the publisher.

least two active areas in an inherited 15p+ variant chromosome that was observed in four siblings ranged from approximately 15% to 85%.

In situ hybridization studies have shown that rRNA-DNA hybridization occurs throughout the entire elongated stalk region of dNOR variants [63,65,70]. Based on the ratio of the number of grain counts in dNOR and normal NOR stalk areas, estimates of the multiplicity (or amplification) of the rDNA genes in these variants have ranged from a five-to sixfold increase [63,65]. Among the dNOR variants in which the two silver positive areas were completely separated by a silver negative area the rRNA-DNA in this negatively staining area was shown, by specific antibody binding, to be highly enriched in 5-methylcytosine [73]. These findings suggested that the entire elongated stalk was composed of rDNA genes, but that a portion of the extra genes may be inactivated, thus producing the intermittent silver-positive and -negative appearance. As with single NOR heteromorphisms, Miller noted the dNOR variant phenotype to be stably inherited between generations [65]. However, Pérez-Castillo et al. [72] reported considerable variability in the cytological appearance of an inherited 15p+ variant both within the cells of a single individual and between the cells of siblings carrying the same variant. These authors suggested that either compensatory interactions or threshold effect may influence the expression of NOR activity within human cells.

PREVIOUS REPORTS OF dNOR VARIANTS

It may not be fortuitous that four of the 28 previously reported individuals with dNORs were either aneuploid themselves, the parents of children with chromosomal aneuploidy, or individuals with a history of spontaneous multiple abortion (Table VII). For example, to illustrate the value of NOR heteromorphisms for determining the parental origin of non-disjunction, Mikkelsen published a photograph (her Fig. 1) of what appears to be a typical dNOR variant on a chromosome 21 in both a child with trisomy 21 and its mother [74]. As in our families, the non-disjunction in this case was noted to have occurred during meiosis I in the parent who carried the dNOR variant. Additional support for our findings is provided by many studies of Down syndrome families in whom acrocentric variants with elongated stalks have been noted [75–80]. Although silver staining was not performed in these studies, the morphology of these variants, as seen with QFQ and Giemsa banding, is quite consistent with that of our dNOR chromosomes. Of special interest is the report by Crandall and Ebbin [81] of a couple who had two spontaneous abortions, one offspring with trisomy 18 and one child with trisomy 21. Although, the father was not available for analysis, cytogenetic studies of the mother showed 46 chromosomes, with a variant chromosome 21 showing elongated stalks that closely resembled the morphology of the dNOR variants in our series. Giemsa stained or GTG-banded chromosomes that appear to have a morphology consistent with dNOR variants have also been observed in several families in whom a recurrence of Down syndrome was reported [82–84]. Reevaluation of the chromosomes of these family members with silver and quinacrine staining may allow for the assignment of the parental origin of the recurrent non-disjunctional events and would also provide additional information regarding the role of the dNOR in the non-disjunction of chromosome 21.

Another type of duplication of the NOR can be visualized in bisatellited marker chromosomes, a portion of which contain two silver-positive regions. Of the 70 separate cases of bisatellited marker chromosome reported, 11 were detected by chromosome analyses of probands with Down syndrome [85,86]. Of the 177 offspring of parents carrying this marker, 11 had trisomy 21, 9 of whom also carried the marker chromosome (48,XX or XY, +21, +mar) [84,86]. In addition, a child with secondary trisomy 21 [46,XX,t(21q21q)] was also noted among the offspring [85]. Because of the coincident occurrence of the bisatellited marker with trisomic individuals, it has been postulated that the marker chromosome may predispose carriers to meiotic non-disjunction [88–90]. However, Steinbach and Djalali [91] have suggested that the relatively high number of reported cases of accessory bisatellited marker chromosomes associated with Down syndrome is attributable to sample (and possibly publication) bias. Martin et al. [92] studied the rate of sperm chromosome aneuploidy in two men who each carried a small bisatellited marker chromosome. In one man the rate of aneuploidy was elevated, while in the other male no significant increase was observed. Clearly, more information is needed regarding the incidence of non-disjunction among carriers of bisatellited marker chromosomes.

A potential confounding factor to be considered in the apparent increased risk for having trisomic offspring for carriers of this bisatellited marker chromosome is the presence of other chromosome variants that may act interact to modify the risk of non-disjunction. For example, in one family studied by Leddet-Chevallier and Sparkes [87], the carrier mother (47,XX,+mar) also had a G-group chromosome with elongated short arms.

A RATIONALE FOR dNOR INVOLVEMENT IN MEIOTIC NON-DISJUNCTION

Miller [93] suggested that if the NOR of acrocentric chromosomes is involved in the occurrence of non-disjunction of chromosome 21, which now seems quite possible, then, based on our knowledge about NOR activity in meiotic and mitotic cells, one would expect 1) an excess of meiosis I compared to meiosis II errors (because of the lack of NOR activity in meiosis II) [94–97]; 2) an excess of maternal compared to paternal errors because of the rDNA amplification during oogenesis [98,99]; 3) no large maternal age effect (because of the somewhat questionable presence of NOR associations as ova enter dictyotene); 4) an increased risk in proportion to the silver deposit or rDNA cluster size; and 5) an excess of non-disjunctional errors among humans as compared to mice (because of the absence of association between ribosomal genes in the meiotic chromosomes of the mouse [94,100]).

In our study, we noted 86.8% meiosis I parental errors and 69.3% maternal non-disjunctional errors [6]. Furthermore, among our dNOR(+) individuals only meiosis I errors were observed. All of these results are in keeping with Miller's first two predictions regarding the expected excess of meiosis I and maternal errors. We found no difference in the mean age at the time of conception between mothers in whom non-disjunction did and did not occur, regardless of the dNOR status of the mother. This observation is consistent with his third prediction of a lack of a strong maternal-age effect in NOR associated non-disjunction.

Clearly, his fourth prediction regarding an increased risk to individuals with large rDNA clusters or silver-stained NORs is met by our observation of a significantly greater frequency of dNOR variants among parents in whom non-disjunction occurred in comparison with their normal spouses, and control individuals. Finally, in agreement with his fifth prediction, the incidence of chromosomal non-disjunction among mouse oocytes has indeed been shown to be much lower in mouse (<1%) than human oocytes [100].

In conclusion, on the basis of our study and Miller's predictions, it seems quite possible that the NOR, specifically the dNOR variant, may be involved in the etiology of the non-disjunction of chromosome 21. Clearly, if this causal role can be firmly established, testing for the dNOR variant would be of great value in the counseling and management not only of couples who have children with Down syndrome, but their extended families as well. The use of this marker along with advanced maternal age and possibly low maternal serum alpha-fetoprotein levels, would then permit the prenatal recognition of most cases of trisomy 21.

ACKNOWLEDGMENTS

We thank Carol Anderson, Ted Brown, Edward Jenkins, Daniel Van Dyke, Janice Edwards, Robert Young, David Flannery, Katy Phelan, Mark Steele, and Sharon Wenger for their assistance in ascertaining and collecting appropriate samples from the families used in this ongoing investigation.

REFERENCES

1. Hook E: Down Syndrome: Frequency in human populations and factors pertinent to variation in rates. In de la Cruz F, Gerald P (eds): "Trisomy 21 (Down Syndrome): Research Perspectives." Baltimore: University Park Press, 1981, pp 3–67.

2. McDonald A: Yearly and seasonal incidence of mongolism in Quebec. Teratology 61:1–4, 1972.
3. Hamerton J: Frequency of mosaicism, translocation, and other variants of trisomy 21. In de la Cruz F, Gerald P (eds): "Trisomy 21 (Down Syndrome): Research Perspectives." Baltimore: University Park Press, 1981, pp 99–107.
4. de Grouchy J, Turleau C: Trisomy 21. In "Clinical Atlas of Human Chromosomes." New York: John Wiley and Sons, 1984, pp 338-349.
5. Harris D, Begleiter M, Chamberlin J, Hankins L, Magenis R: Parental trisomy 21 mosaicism. Am J Hum Genet 34:125–133, 1982.
6. Jackson-Cook C, Flannery D, Corey L, Nance W, Brown J: Nucleolar organizer region variants as a risk factor for Down syndrome. Am J Hum Genet 37:1049–1061, 1985.
7. deGrouchy J: 21p− maternal en double exemplaire chez un trisomique 21. Ann Genet (Paris) 13:52-55, 1970.
8. Juberg RC, Jones B: The Christchurch chromosome (Gp-) mongolism, erythroleukemia, and an inherited Gp- chromosome (Christchurch). N Engl J Med 282:292-297, 1970.
9. Licznerski, G, Lindsten J: Trisomy 21 in man due to maternal non-disjunction during the first meiotic division. Hereditas 70:153–154, 1971.
10. Mutton DE: Origin of the trisomic 21 chromosome. Lancet 1:375, 1973.
11. Punnett HH, Kistenmacher ML: The origin of the extra chromosome in trisomy 21. Genetics 74 [Suppl]:5222, 1973.
12. Robinson JA: Origin of extra chromosome in trisomy 21. Lancet I:131–133, 1973.
13. Sasaki M, Hara Y: Paternal origin of the extra chromosome in Down's syndrome. Lancet 2:1257–1258, 1973.
14. Smith GF, Sachdeva S: Origin of extra chromosome in trisomy 21. Lancet I:487, 1973.
15. Uchida I: Paternal origin of the extra chromosome in Down's syndrome. Lancet 2:1258, 1973.
16. Uchida IA: Maternal radiation and trisomy 21. In Hook EB, Porter IH (eds): "Studies in Humans: Population Cytogenetics." New York: Academic Press, 1977, pp 285-299.
17. Bott CE, Sekhon GS, Lubs HA: Unexpected high frequency of paternal origin of trisomy 21. Am J Hum Genet 27:20A, 1975.
18. Dutrillaux B: Le caryotype humain normal et ses variations physiologiques. In "Sur la Nature et l'origine des Chromosomes Humains: l'Expansion Scientifique." Paris, 1975, pp 37-40.
19. Emberger JM, Taib J: 21p+ maternel en double exemplaire chez un trisomique 21. J Genet Hum [Suppl] 23:98–101, 1975.
20. Giraud F, Mattei JF, Mattei MG: Etude chromosomique chez les parents d'enfants trisomiques 21. Lyon Med 233:241-251, 1975.
21. Hara Y, Sasaki M: A note on the origin of extra chromosomes in trisomies 13 and 21. Proc Jpn Acad 51:295-299, 1975.
22. Schmidt R, Dar H, Nitowsky H: Origin of extra 21 chromosome in patients with Down syndrome. Pediatr Res 9:318, 1975.
23. Schmidt R, Dar H, Nitowsky H: Apparent parental mosaicism for 21 trisomy as a predisposing factor for children with Down's syndrome. In "International Congress of Human Genetics: Excerpta Medica," Amsterdam, 1976, Vol 5, p 151.
24. Schmidt R, Dar H, Nitowsky HM: Dermatoglyphic and cytogenetic studies in parents of children with trisomy 21. Clin Genet 20:203-210, 1981.
25. Moore CM, Arbisser AI, Simon FA, Morriss PH: Maternal first meiotic non-disjunction in a case of monozygous twins with Down's syndrome. In "International Congress Series Number 397: Excerpta Medica." Amsterdam, 1976, p 142.

26. Wagenbichler VP: Zur ätiologie meiotischer störungen unter besonderer berücksichtigung der trisomie 21. Wien Klin Wochenschr 88 (Suppl 63):3–23, 1976.
27. Wagenbichler P, Killian W, Rett A, Schnedl W: Origin of the extra chromosome no. 21 in Down's syndrome. Hum Genet 32:13–16, 1976.
28. Mikkelsen, Margareta, Hallberg, Anni, Poulsen H: Maternal and paternal origin of extra chromosome in trisomy 21. Hum Genet 32:17–21, 1976.
29. Hansson A, Mikkelsen M: The origin of the extra chromosome 21 in Down syndrome. Studies of fluorescent variants and satellite association in 26 informative families. Cytogenet Cell Genet 20:194–203, 1978.
30. Grass FS, Moore CM: Inheritance of chromosomes bearing active NOR's in a family with trisomy 21. Am J Hum Genet 29:50A, 1977.
31. Magenis RE, Overton KM, Chamberlin J, Brady T, Lovrien E: Parental origin of the extra chromosome in Down's syndrome. Hum Genet 37:7–16, 1977.
32. Magenis R, Chamberlin J: Parental origin of nondisjunction. In de la Cruz FF, Gerald PS (eds): "Trisomy 21 (Down Syndrome): Research Perspectives." Baltimore: University Park Press, 1981, pp 77–93.
33. Verma RS, Dosik H: Trisomy 21 in a child due to paternal nondisjunction as determined by RFA technique. Jpn J Hum Genet 23:17–21, 1978.
34. Mattei JF, Mattei MG, Ayme S, Giraud F: Origin of the extra chromosome in trisomy 21. Hum Genet 46:107–110, 1979.
35. Mattei JF, Ayme S, Mattei MG, Giraud F: Maternal age and origin of non-disjunction in trisomy 21. J Med Genet 17:368–372, 1980.
36. Calabro A, Lugarotti MS, Dallapiccola B: A comment on the paper: Recurrence of Down syndrome associated with microchromosome by C. Ramos, L. Rivera, J. Benitez, E. Tejedor, and A. Sanchez-Cascos. Hum Genet 53:287–288, 1980.
37. Mikkelsen, Margareta, Poulsen H, Grinsted J, Lange A: Non-disjunction in trisomy 21: Study of chromosomal heteromorphisms in 110 families. Ann Hum Genet 44:17–28, 1980.
38. Roberts DF, Callow MH: Origin of the additional chromosome in Down's syndrome: A study of 20 families. J Med Genet 17:363–367, 1980.
39. Hamers AJ, Heijnen GC, Jongbloet PH: Non-disjunction in de geslachtscellen bij ouders van patienten met net syndroom van Down. Ned T Geneesk, 1981.
40. Jongbloet PH, Frants RR, Hamers AJ: Parental a-1-antitrypsin (Pi) types and meiotic non-disjunction in the aetiology of Down Syndrome. Clin Genet 20:304–309, 1981.
41. Jongbloet PH, Hamers AJ: Personal communication to Mikkelsen. J Ment Defic Res 26:143–151, 1982.
42. Jacobs PA, Mayer M: The origin of human trisomy: a study of heteromorphisms and satellite associations. Ann Hum Genet 45:357–365, 1981.
43. Manning CH, Goodman HO: Parental origin of chromosomes in Down's syndrome. Hum Genet 59:101–103, 1981.
44. Morić-Petrović S, Laća Z: The origin of the extra chromosome 21 in Down's syndrome. Clin Genet 19:525–526, 1981.
45. Del Mazo J, Pérez-Castillo A, Abrisqueta JA: Trisomy 21: Origin of non-disjunction. Hum Genet 62:316–320, 1982.
46. Pilgaard, Mikkelsen M: Personal communication to Mikkelsen. J Ment Defic Res 26:143–151, 1982.
47. Hatcher NH: Unpublished observations cited in Hook E.B. (1983) Down syndrome rates and relaxed selection at older maternal ages. Am J Hum Genet 35:1307–1313, 1983.

48. Pérez-Castillo A, Del Mazo J, Abrisqueta JA: Three interesting cases of Down syndrome. Ann Gent (Paris) 26:123–128, 1983.
49. Davies KE, Harper K, Bonthron D, Krumlauf R, Polkey A, Pembrey ME, Williamson R: Use of a chromosome 21 cloned DNA probe for the analysis of non-disjunction in Down syndrome. Hum Genet 66:54–56, 1984.
50. Hassold T, Chiu D, Yamane J: Parental origin of autosomal trisomies. Ann Hum Genet 48:129–144, 1984.
51. Juberg R, Mowrey P: Origin of non-disjunction in trisomy 21 syndrome: All studies compiled, patental age analysis, and international comparisons. Am J Med Genet 16:111–116, 1983.
52. German J: Mongolism, delayed fertilization and human sexual behaviour. Nature 217:516–518, 1968.
53. Juberg R: Origin of chromosomal abnormalities: Evidence for delayed fertilization in meiotic non-disjunction. Hum Genet 64:122–127, 1983.
54. Jongbloet P, Vrieze O: Down syndrome: Increased frequency of maternal meiosis I non-disjunction during the transitional stages of the ovulatory seasons. Hum Genet 71:241–248, 1985.
55. Mattei J, Ayme S, Mattei M, Giraud F: Maternal age and origin of non-disjunction in trisomy 21. J Med Genet 17:368–372, 1980.
56. Ayme S, Lippman-Hand A: Maternal age effect in aneuploidy: Does altered embryonic selection play a role? Am J Hum Genet 34:558–565, 1982.
57. Hook E: Down syndrome rates and relaxed selection at older maternal ages. Am J Hum Genet 35:1307–1313, 1983.
58. Chamberlin J, Magenis R: Parental origin of de novo chromosome rearrangements. Hum Genet 53:343–347, 1980.
59. Bufton L, Magenis R, Lovrien E: Alpha-1-antitrypsin protease inhibitor (Pi) phenotypes in Down's syndrome patients and their parents. Clin Genet 21:14–18, 1982.
60. Flannery D, Jackson-Cook C: Thyroid antibodies are an independent risk factor for non-disjunctions of chromosome 21. Am J Hum Genet 39(3):A113, 1986.
61. Uchida IA, Freeman VCP: Trisomy 21 Down syndrome II. structural chromosome rearrangements in the parents. Hum Genet 72:118, 1986.
62. Jackson-Cook C, Brown J: Further evidence that the dNOR variant is a risk factor for trisomy 21. Am J Hum Genet 37:A99, 1985.
63. Henderson A, Atwood K: Satellite-association frequency and rDNA content of a double-satellited chromosome. Hum Genet 31:113–115, 1976.
64. Archidiacona N, deCapoa A, Ferraro M, Pelliccia F, Rocchis A, Rocchi M: Nucleolus organizer and N-band distribution in morphologic and fluorescence variants of human chromosomes. Hum Genet 37:285–289, 1977.
65. Miller A, Breg W, Warburton D, Dev G, Miller J: Regulation of rRNA gene expression in a human familial 14p+ marker chromosome. Hum Genet 43:289–297, 1978.
66. Lau Y, Wertelecki W, Pfeiffer R, Arrighi F: Cytological analyses of a 14p+ variant by means of N-banding and combinations of silver staining and chromosome bandings. Hum Genet 46:75–82, 1979.
67. Martin A, Miller L, Simpson J, Thomas C, Rzeszotarsk M, Elias S, Sarto G, Patel V: Localization of the nucleolar organizer by computer-aided analysis of a variant No. 21 in a human isolate. Hum Genet 48:211–219, 1979.
68. Balicek P, Zizka J: Intercalar satellites of human acrocentric chromosomes as a cytological manifestation of polymorphism in GC-rich material. Hum Genet 54:343–347, 1980.

69. Sofuni T, Tanabe K, Awa A: Chromosome heteromorphisms in the Japanese. II. Nucleolus organizer regions of variant chromosomes in D and G groups. Hum Genet 55:265–270, 1980.
70. Bernstein R, Dawson B, Griffiths JJ.: Human inherited marker chromosome 22 short-arm enlargement: Investigation of rDNA gene multiplicity, Ag-band size, and acrocentric association. Hum Genet 58:135–139, 1981.
71. Jotterand-Bellomo, Meele G: Is it always the same NOR that is more active in a pair of acrocentrics with distinct AG-staining? Hum Genet 59:185, 1981.
72. Pérez-Castillo A, Martin-Lucas MA, Abrisqueta JA: New insights into the effects of extra nucleolus organizer regions. Hum Genet 72:80–82, 1986.
73. Tantravahi U, Breg R, Wertelecki V, Erlanger BD, Miller Orlando J: Evidence for methylation of inaction human rRNA genes in amplified regions. Hum Genet 56:315–320, 1981.
74. Mikkelsen M: Parental origin of the extra chromosome in Down's syndrome. J Ment Defic Res 26:143–151, 1982.
75. Dekaban A, Bender M, Economos G: Chromosome studies in mongoloids and their families. Cytogenetics 2:61–75, 1963.
76. Therkelson A: Enlarged short arm of a small acrocentric chromosome in grandfather, mother and child, the latter with Down's syndrome. Cytogenet Cell Genet 3:441–451, 1964.
77. Sands V: Short arm enlargement in acrocentric chromosomes. Am J Hum Genet 21:293–304, 1969.
78. Gigliani F, Capoa deA, Rocchi, A: A marker chromosome number 14 with double satellite observed in two generations: An unbalanced chromosome constitution associated with normal phenotype. Humangenetik 15:191–195, 1972.
79. Hahnemann N, Eiberg H: Antenatal genetic diagnosis in a kindred with a 15p+ chromosome. Clin Genet 4:464–473, 1973.
80. Emberger J, Taib J: 21p+ Maternal en double exemplaire chez un trisomique 21. J Gen Hum [Suppl] 23:98–101, 1975.
81. Crandall B, Ebbin A: Trisomy 18 and 21 in two siblings. Clin Genet 4:517–519, 1973.
82. Shaw M: Familial mongolism. Cytogenetics 1:141–179, 1962.
83. Dhadial R, Pfeiffer R: Cytogenetic studies in families with two 47, +21-siblings. J Genet Hum 20:297–322, 1972.
84. Werner W, Gerrmann F, John B: Cytogenetic studies of a family with trisomy 21 mosaicism in two successive generations as the cause of Down's syndrome. Hum Genet 60:202–204, 1982.
85. Steinbach P, Djalali M, Hansmann I, Kattner E, Meisel-Stosiek M, Probeck HD, Schmidt A, Wolf M: The genetic significance of accessory bisatellited marker chromosomes. Hum Genet 65:155–164, 1983.
86. Pérez-Castillo A, Del Mazo J, Abrisqueta JA: Three interesting cases of Down syndrome. Ann Genet (Paris) 26:123–128, 1983.
87. Leddet-Chevallier I, Sparkes RS: Inherited 13/14 translocation and metacentric microchromosome associated with trisomy 21: Report of 2 cases. Clin Genet 23:386–390, 1983.
88. Friedrich U, Nielsen J: Bisatellited extra small metacentric chromosome in newborns. Clin Genet 6:23–31, 1974.
89. Ramos C, Rivera L, Benitez J, Tejedor E, Sanchez-Cascos A: Recurrence of Down syndrome associated with microchromosome. Hum Genet 49:7–10, 1979.

90. Bartsch HD, Azevedo-Moreira LM, de Rohrborn G, Claussen U, Gebauer HJ: Possible origin of small bisatellited additional chromosome. Hum Genet 54:319–322, 1980.
91. Steinbach P, Djalali M: Ineffectivity of accessory bisatellited marker chromosomes in inducing meiotic non-disjunction. Hum Genet 64:402–403, 1983.
92. Martin RH, Hildebrand KA, Yamamoto J, Peterson D, Rademaker AW, Taylor P, and Lin CC: The meiotic segregation of human sperm chromosomes in two men with accessory marker chromosomes. Am J Med Genet 25:381–388, 1986.
93. Miller OJ: Role of the nucleolus organizer in the etiology of Down syndrome. In Cruz FF, Gerald PS (eds):"Trisomy 21 (Down Syndrome): Research Perspectives." Baltimore: University Park Press, 1981, pp 163–176.
94. Mirre C, Hartung M, Stahl A: Association of ribsosomal genes in the fibrillar center of the nucleolus: A factor influencing translocation and non-disjunction in the human meiotic oocyte. Proc Natl Acad Sci USA 77:6017–6021, 1980.
95. Hartung M, Mirre C, Stahl A: Nucleolar organizers in human oocytes at meiotic prophase I, studied by the silver NOR method and electron microscopy. Hum Genet 52:295–308, 1979.
96. Schmid M, Miller H, Stasch S, Engel W: Silver staining of nucleolus organizer regions during human spermatogenesis. Hum Genet 64:363–370, 1983.
97. Schwarzacher H, Mikelsarr A, Schnedl W: The nature of the AG-staining of nucleolus organizer regions. Electron and light microscopic studies on human cells in interphase, mitosis, and meiosis. Cytogenet Cell Genet 20:24–39, 1978.
98. Hartung M, Keeling J, Patel C, Bobrow M, Stahl A: Nucleoli, micronucleoli, and nucleolus-like structures in human oocytes at meiotic prophase I studied by the silver-NOR technique. Cytogenet Cell Genet 35:2–8, 1983.
99. Wolgemuth D, Jagiello G, Henderson A: Quantification of ribosomal RNA genes in fetal human oocyte nuclei using rRNA:DNA hybridization in situ. Exp Cell Res 118:181–190, 1979.
100. Hansmann I, El-Nahass E: Incidence of non-disjunction in mouse oocytes. Cytogenet Cell Genet 24:115–121, 1979.

EDITOR'S NOTE

More recent evidence (e.g. Hassold T, Pettay D: Am J Hum Genet 39, A116, 1986) suggests that the parental dNOR may not be involved in the etiology of trisomy 21. Nevertheless early evidence suggested it was and Dr. Jackson-Cook and her colleagues deserve support for pursuing this opportunity to provide a counseling alternative to potential parents of trisomy 21 offspring.

SOME IMPLICATIONS OF CHROMOSOME BREAKPOINTS

The Cytogenetics of Mammalian Autosomal
Rearrangements, pages 725–738

Chapter 24
Phenotype in Autosomal Chromosome Aberrations: Distinctiveness, Variability, and Karyotype Correlations

Albert Schinzel

The phenotype of most individuals with autosomal chromosome aberrations is characterized by four major findings: mental deficiency, a specific pattern of minor dysmorphic features, major malformations, and (pre- and postnatal) growth retardation. Most consistent of all is mental deficiency. Its degree is often quite variable; however, within certain limits it is still a characteristic feature of any aberration. Growth in autosomal aberrations is often deficient and more often retarded, with at least some catching-up in late adolescence. A pattern of dysmorphic features is the most consistent finding and thus the one of most diagnostic importance in many autosomal aberrations. Major malformations are mostly facultatively associated with autosomal aberrations and are far less specific for a particular aberration. More specific than a single malformation is a particular pattern of malformations. There is virtually no malformation that is specifically associated with one chromosome aberration, and there is also no malformation that is found in one specific aberration only. If deletions and duplications of segments of varying length within the same chromosome region are considered, growth, mental deficiency, and the incidence and severity of major malformations correlate directly with the size of the (deleted or duplicated) segment. However, the pattern of dysmorphic features shows little correlation with the size of the imbalanced segment; it tends to be most distinctive in small duplications or deletions. The variability of clinical features in different patients with the same autosomal chromosome aberration, and the nonspecificity of malformations in them make it impossible to map congenital malformations to certain chromosome regions.

INTRODUCTION

Patients with autosomal chromosome aberrations are characterized by four major features (in order of their clinical importance): motor and mental

Institute of Medical Genetics, University of Zürich, CH-8001 Zürich, Switzerland

Abbreviations used: del, deletion, partial monosomy; dup, duplication, partial trisomy; trip, triplication, partial tetrasomy; r, ring chromosome; inv dup, inversion duplication (leading to partial tetrasomy of that segment).

deficit; a pattern of dysmorphic signs, especially of facies, distal limbs, and genitalia; congenital malformations, often multiple; and intrauterine and postnatal growth retardation. There is a great variability in clinical findings in most aberrations, and none of these four features can be considered obligatory for any aberration. Mental impairment is undoubtedly the feature with most impact on the individual and the society, and the pattern of dysmorphic signs is the by far most helpful feature for clinical diagnosis in order to prompt a chromosome examination [1].

THE PEDIGREE OF PATIENTS WITH AUTOSOMAL CHROMOSOME ABERRATIONS

In *de novo aberrations*, i.e., all triploidies, virtually all trisomies and the majority of structural aberrations, the pedigree tends to be unremarkable except for an increased mean parental (especially maternal) age for trisomies and, to a lesser extent, for a number of de novo deletions and other structural aberrations. By contrast, in many pedigrees ascertained by *unbalanced familial structural aberrations,* the following pedigree observations can be made: increase of fetal wastage, i.e., spontaneous abortions, stillborns, and premature or full-term newborns who die shortly after birth with multiple malformations and a specific phenotype. Often, larger pedigrees are seen with multiple balanced carriers and affected unbalanced offspring in several generations who do not fit into any of the monogenic modes of inheritance; in addition there is frequent reduced fertility, especially in male carriers. It is not unusual to find such (balanced or unbalanced) aberrations at prenatal diagnosis performed for increased maternal age because it took the mother a decade or more to conceive.

THE PREGNANCY WITH FETUSES WITH AUTOSOMAL CHROMOSOME ABERRATIONS: ULTRASONOGRAPHIC FINDINGS

As compared to most pregnancies, those with chromosomally unbalanced fetuses are more often complicated by bleeding episodes, threatened abortion, poly- or oligohydramnios, placental insufficiency, fetal growth retardation, abnormally high or low blood and amniotic fluid alpha-fetoprotein levels, and various abnormal ultrasonographic findings. With careful ultrasonographic examination during the early second trimester of any pregnancy (i.e., not only determination of the biparietal and amniotic sac diameters, localization of the placenta, and exclusion of a twin pregnancy), with the aid of certain abnormal parameters to prompt a fetal karyotope study, at least one-half of autosomal trisomies and one-third of structural chromosome aberrations could be detected at that stage. The finding of most prospective importance is fetal nuchal hygroma in autosomal trisomies and other aberrations [2]. Table I presents abnormal ultrasound findings in which fetal chromosome studies are indicated. Some major malformations amenable to ultrasound are, however, very rarely associated with autosomal chromosome aberrations, e.g., sirenomelia, iniencephalus, otocephalus, conjoint twinning, sacrococcygeal teratoma, and lobster-claw malformation (see Tables II, III). For later genetic counseling, it is important to perform chromosome studies in fetuses affected with one or a combination of the abnormalities listed in Table I even if the pregnancy will be terminated anyway. While omphalocele is a frequent

TABLE I. Abnormal fetal ultrasonographic findings which should prompt a prenatal chromosome examination

Hydrops; nuchal hygroma or swelling
Abnormal skull configuration
Abnormal facial morphology, e.g., hypoplastic nasal skeleton
Cleft lip and palate
Esophageal atresia, duodenal atresia
Gastroschisis, omphalocele
Many cardiovascular defects
Caudal regression, sacral agenesis
Intersexual external genitalia
Renal agenesis or malformations, hydronephrosis, bladder obstruction
Neural tube defects
Holoprosencephaly, midbrain defects
Microcephaly
Arthrogryposis, clubfeet
Limb defects, hexadactyly
Abnormal fetal movement patterns

TABLE II. Malformations common in autosomal chromosome aberrations

Microphthalmia, ocular coloboma
Cleft palate, cleft lip, or both
Brain malformations including particularly holoprosencephaly, agenesis of the corpus callosum and the Arnold Chiari malformation
Spina bifida, occipital or lumbar
Malformations of the heart and great vessels
Esophageal atresia, TE fistula, anal atresia with fistula
Malrotation of intestine, common mesentery; omphalocele
Malformation of the kidneys and urinary tract
Absence or hypoplasia of radius and thumb
Postaxial hexadactyly

TABLE III. Malformations uncommon in autosomal chromosome aberrations

Iniencephaly, otocephaly
Gastroschisis, body stalk defect
Atresia of the jejunum/ileum
Situs inversus totalis
Exstrophy of bladder or cloaca
Sirenomelia
Sacrococcygeal (and other) teratoma
Peromelia, amelia, phocomelia, ectrodactyly, split hand/foot
Ulnar or fibular ray defects
Congenital arthrogryposis
Acardiac and other severly disorganized twins, conjoined twins

finding in patients with autosomal chromosome aberrations, this is not the case for a gastrochisis; the reason that the latter malformation is listed in Table I is that a ruptured omphalocele cannot be distinguished at ultrasonic examination from a gastroschisis. In addition, as mentioned above, chronic placental insufficiency and intrauterine growth failure are also findings in which a prenatal chromosome study should be considered. The author's personal experience includes instances of fetuses with trisomy 18 with arthrogryposis, radial aplasia, esophageal atresia, omphalocele, severe intrauterine growth deficit, hydrops, and anencephalus; of trisomy 21 with nuchal swellings and with nonimmunological hydrops; of dup(1)(q32→qter) and del(7)q34→qter) with elevated serum alpha-fetoprotein, sacral agenesis, renal anomalies, and holoprosencephaly; of trisomy 13 with bilateral cleft lip and palate and bladder obstruction; of del(7)q32→qter) with intersexual genitalia; and of triploidy with severe growth deficit, large placenta with molar degeneration, and polyhydramnios.

GROWTH IN PATIENTS WITH AUTOSOMAL CHROMOSOME ABERRATIONS

Growth tends to be reduced in almost all autosomal aberrations. Although there is a natural variability influenced by genetic (parental) and other factors, each chromosome aberration has its own growth dynamics, and growth percentiles could be established for each of them. Extreme intrauterine growth retardation is found in some lethal chromosome aberrations, e.g., triploidy and nonmosaic trisomy 22, but it is quite unusual for other more viable aberrations. Extreme short stature, to the extent found in various skeletal dysplasias and in patient with growth hormone deficiency, is not at all common in autosomal aberrations. Growth hormone deficiency has been found as a rare, variable feature in several autosomal imbalances, and thus extreme dwarfism in a chromosome aberration for which it is not common should prompt a growth hormone study.

Patterns of growth in at least some autosomal aberrations deviate from the normal patterns. In general, many more patients are short during the period of growth than are adults with the same aberrations. This is in accordance with the often delayed bone maturation and delayed and diminished puberty in patients with many autosomal chromosome aberrations. There are limited data on this point, however duplication of 9p seems to be a typical example: during the period of growth, virtually all patients are very short, and bone age is almost always retarded, even up to 6 y or more; the patients continue to grow even up to the age of 22–25 y and a (yet unknown) proportion of adults are of normal height.

In a minority of autosomal chromosome aberrations growth tends to be normal, at least in the majority of the patients. These include mosaic trisomy 8, the cat-eye syndrome (tetrasomy 22 (pter→q11), additional isochromosome 18p (tetrasomy 18p), additional inv dup(15)(pter→q12), deletion of (9)(pter→p21), and duplication of (13)(q21→qter). In mosaic trisomy 8, patients with advanced growth and bone maturation have been repeatedly described; however, as adults these patients are in the upper normal range for parental and sibling heights. On the other extreme, there are some aberrations in which normal growth is never or virtually never observed, including trisomy 18 or 4p distal deletion, del (4)(pter→p16). Adult height of the latter patients are poorly known since survival to adulthood is exceptional. In summary, for autosomal chromosome aberrations in general there is a strong tendency towards growth retardation, but a marked individual variability as to the extent of this retardation.

PUBERTY AND FERTILITY

Most patients with autosomal chromosome aberrations do not undergo a normal puberty, but rather, experience a delayed and diminished puberty. This is more pronounced in males than in females. Fertility in *males* with autosomal aberrations is quite exceptional; it has so far only been recorded in mosaic trisomy 8, some autosomal rings, direct duplication of the terminal segment of 8p, and in the cat-eye syndrome with tetrasomy 22(pter→q11). This reduced male versus female fertility in autosomal aberrations is probably due to a higher sensitivity of male versus female meiosis to pairing anomalies and possibly also due to mental deficiency and lack of interest in sexuality. No proven case of fatherhood of a male with Down syndrome has so far been reported, although the results of meiotic studies, testicular histology, and spermiograms would not totally exclude such males from having, albeit reduced, fertility. In the future, DNA fingerprinting would be an excellent and easy method to confirm or exclude suspected paternity of males with trisomy 21.

Puberty in females with autosomal aberrations often starts later than in their sisters and is diminished, and menstrual cycles tend to be more irregular and to cease earlier. Gonadal dysgenesis is a facultative feature of a number of aberrations, including trisomies 13 and 18, and ovarian hypoplasia or agenesis is occasionally found. However, there are a number of aberrations for which fertility, be it normal or reduced, is known from reports of direct transmission of the aberration. Examples include cases with ring chromosome 18, del(18)(q21→qter), del(22)(pter→q11.2), mosaic trisomy 8, trisomy 21, the cat-eye syndrome, and duplications of various smaller autosomal segments [1]. Obviously, fertility of females with chromosome aberrations is easier to demonstrate than fertility of males.

As with virtually all features of autosomal chromosome aberrations, there is also variable manifestation of puberty and fertility: one patient may be fully fertile while for reasons completely unknown, another with the same aberration, even within the same family, may not be fertile at all. Data concerning puberty, fertility, and histologic findings in gonads of patients with autosomal aberrations are still scarce because most patients are diagnosed during childhood and generally are not further followed up.

PATTERNS OF DYSMORPHIC FEATURES

For many autosomal chromosome aberrations, a specific pattern of dysmorphic features is the most consistent finding and the one that is the most indicative for a specific aberration. This, however, does not hold true for the single dysmorphic feature: most of them are quite nonspecific, and it is only their *pattern* that makes many chromosome aberrations so phenotypically unique. Conversely, the pattern of dysmorphic features varies in specificity for known autosomal imbalances. For example, the pattern of trisomy 18, of deletion of (4)(pter→p16) and of duplication of 9p is very specific, while that of tetrasomy 15(pter→q12), additional isochromosome 18p, and of the cat-eye syndrome is less specific; but that of many autosomal rings is not at all specific. The absence of one or several elements of a specific pattern is usually not a major diagnostic problem given that there is no aberration in which one single dysmorphic sign would be a conditio sine qua non, and the corollary that there is also no aberration with specific pattern for which some or even exceptional cases would lack that specific pattern. The only exception to this rule would be the facultative presence of major malformations, which can overshadow the normal pattern in the particular aberration (see below).

A minor irritation to workers in this area is the fact that there is no commonly accepted delineation of dysmorphisms from malformations. For example, it is evident that downslanting palpebral fissures or low-set/protruding ears are dysmorphisms and not (even minor) malformations, but the definition of such features as preauricular pit or cleft uvula is more problematic. If a dysmorphic feature is defined as deviation from the usual structure with no clear-cut borderline from the normal, the latter two clearly would range among the malformations. If it would be that a malformation is a structure that is normally not present or the lack of a normal structure, than both would indeed be malformations. Another quality defining malformation is whether the function of an organ is adversely affected, in which case all of the above-mentioned would be dysmorphic features, but a pit could become a malformation in the case of repeated infections. In my opinion, all of these are examples of dysmorphisms, while I would insist on considering a very small atrial septal defect a malformation whether or not it has functional implications for the patient.

Dysmorphic features can be found anywhere over the body, but preferred locations are face, distal limbs and genitalia. Perhaps most striking are the facial dysmorphisms, which include abnormal size, shape, and position of the ears; aberrant palpebral fissure length and intraocular distance; nonhorizontal eye axes; unusual size and form of the nose and mandible; unusual length and shape of the upper and lower lip; deviant palatal configuration; and many others. Dysmorphic features often reflect a disproportionate growth of facial structures during fetogenesis, and in these instances there is often (at least a partial) catch-up growth during the late infancy or adolescence. Therefore, the pattern of characteristic facial dysmorphism is highly age dependent, and the age of a patient has to be considered if incidences of dysmorphic features are given for a special autosomal aberration. While, for example, most newborn and young children with the *cri-du-chat* syndrome show a round, "moon-shaped" face with full cheeks, adolescents and adults with this syndrome never show this finding; instead they mostly have a narrow, long, and asymmetric face. In patients with Down syndrome, upslanting palpebral fissure, epicanthic folds, and a depressed nasal bridge tend to become less striking or even to disappear with age, while the plicated tongue and radial lip furrows tend to show up in infancy and to become more obvious with advancing age. In most chromosome aberrations with a specific pattern of dysmorphisms, the best age for a clinical evaluation of a patient is between about 4 wk and 2 yr of age: deformations owing to intrauterine constraint have already resolved by that age, and the recognized dysmorphic patterns should still be present.

Dysmorphic patterns often reflect a disproportionate growth of skeletal and connective tissue structures. For example, undergrowth of the skeletal nasal base may lead to midface hypoplasia with depressed nasal bridge, flat nose, inner epicanthic folds, and the (not necessarily correct) impression of an increased intraocular distance. Dysmorphic patterns may also reflect malformations of underlying structures or organs or an abnormal fetal development not recognizable at clinical inspection. For example, the facial characteristics of holoprosencephaly reflect the associated brain malformation; similarly hypotelorism, a deficit of midbrain structures; mongoloid position of the eyes, a hypoplasia of the frontal cerebral lobes, and premature craniosynostosis with a prominent metopic suture, a lack of growth pressure from the forebrain. Other dysmorphisms may simply reflect a neurologic deficit during fetogenesis, e.g., protruding ears for lack of innervation of the posterior ear muscle, prominent lateral palatine ridges for lack of tongue thrust, ptosis and strabismus for lack of innervation of ocular muscles.

To become familiar with the clinical diagnosis of autosomal chromosome aberrations, it is necessary to know the characteristic clinical pattern from a number of uncomplicated and typical cases at different ages. The first reported cases are often not the most typical ones, but those which were recognized because they displayed particularly striking abnormal findings.

CONGENITAL MALFORMATIONS

In patients with chromosome aberrations that are characterized by a distinct pattern of dysmorphic features, major malformations are usually of minor diagnostic importance. This is not the case in patients with aberrations that do not commonly show a distinct dysmorphic pattern. The main reason is that the congenital malformations in individual aberrations are far less specific than dysmorphic *patterns*. There is virtually no malformation that is confined to one single aberration, but the incidence with which a specific malformation is present in an individual aberration is a feature of diagnostic value. Obviously, the latter is known only in aberrations of which a sufficient number of patients have been reported. Mean incidences of individual malformations in particular aberrations are usually in the range 5–10%. However, aberrations exist that show significantly higher incidences of certain malformations. For example, cardiovascular defects, if taken together, are found in over 90% of patients with trisomy 18; most patients with interstitial deletion of at least a part of 13q14 will develop retinoblastoma; and aniridia has so far been found in virtually all patients with interstitial deletion of 11p13. In the latter case, the association has probably resulted in a pronounced ascertainment bias since the deletion is far from obvious. A patient without aniridia may not prompt the high-resolution examination of chromosome 11p. On the other end of the spectrum, one finds isolated associations of virtually any malformation with a particular aberration, and causal relationships can only be inferred for very few of them.

As with dysmorphic features, a specific pattern of malformations is more characteristic than a single aberration. A not uncommon situation is that a patient may lack a component malformation and exhibit a further malformation that is less frequently seen. Of little diagnostic significance are combinations or rather nonspecific malformations, e.g., of heart defects with renal malformations or with facial clefts. Relatively specific combinations include postaxial hexadactyly, holoprosencephaly, microphthalmia, and facial clefts in trisomy 13; tracheo-esophageal fistula, radial aplasia, malrotation of the gut, cardiac, and renal malformations in trisomy 18; preauricular malformations (pits and/or tags), ocular coloboma, anal atresia, renal and cardiac malformations in the cat-eye syndrome with tetrasomy 22 (pter→q11); aniridia, ambiguous male genitalia with hypospadias, and hypernephroma in del(11)(p14→p12); anal atresia, ocular coloboma, absent thumbs, scrotum bifidum with penile hypoplasia and hypospadias, and holoprosencephaly in del(13)(q14→qter). In addition, these syndromes have particular associated dysmorphic patterns, and diagnostic skills should be focused more on the latter than on the malformations. The author remembers one of his first experiences in clinical cytogenetics: a blood sample was referred from a peripheral hospital from a newborn with "coloboma of the irises and anal atresia," and we expected to find the cat-eye chromosome; however, the karyotype revealed a r(18) chromosome. At a later clinical examination, we found the characteristic phenotype of del(18)(q21→qter) (the 18q− syndrome) with midface hypoplasia and narrow auditory canals, and if we had examined the patient's clinical features prior to his karyotype, we would most probably have considered an 18q deletion with additional malformations (anal atresia, and sacral hypoplasia later demonstrated from radiographs)and have certainly ruled out the cat eye syndrome.

In particular syndromes with a nondistinct pattern of dysmorphic features, there are a few malformations of high diagnostic value. In patients with the DiGeorge sequence, especially those with additional, nonrelated malformations, a deletion of 22q11 should be excluded no matter whether or not there is a distinct dysmorphic pattern; retinoblastoma is the most consistent feature of the interstitial deletions including a segment of 13q14; and aniridia is probably more diagnostic of the del(11p13) than any dysmorphic pattern. It is not by chance that these three are all examples of deletions and not of duplications: the deletions in the different patients are probably of quite different extent and overlap only for the segment "pathognomonic" of the malformation mentioned.

CAUSES FOR THE VARIABILITY IN OCCURRENCE OF CONGENITAL MALFORMATIONS IN MOST AUTOSOMAL CHROMOSOME ABERRATIONS: IMPLICATIONS FROM FINDINGS IN SIBLINGS AND TWINS

The reasons for the varying incidence of congenital malformations in autosomal chromosome aberrations are far from being understood. In siblings affected with the same inherited unbalanced rearrangement, there is a general tendency towards a higher concordance rate for congenital malformations than among all patients with the same aberration, and this might imply underlying genetic factors. However, such simple explanations have to be regarded with some reservation. Firstly, there is still considerable discordance between siblings in virtually all parameters of autosomal aberrations: most pronounced in malformations, less so in growth and psychomotor development, and least in dysmorphic patterns. Secondly, breakpoints of rearrangements are identical in siblings, but might vary considerably even on a submicroscopic level in other cases.

An impressive example for concordance of heart malformation in Down siblings was reported in a family from Italy [3]: In this case a 21/21 translocation heterozygote mother had eight unbalanced children with Down syndrome, all of whom had significant cardiovascular defects causing their demise, while less than one-third of all Down syndrome patients have life-threatening heart defects. A second case exhibiting this principle concerned two siblings with inherited duplication of 4p who were concordant for paraesophageal hernia, a finding otherwise quite rare in that aberration [4].

Monozygotic twins offer a chance to study the influence of intrauterine environmental factors on the occurrence of dysmorphic features, malformations, and growth retardation in autosomal aberrations [1]. It was found that monozygotic twins concordant for an autosomal aberration are highly concordant for dysmorphic features (not only for the pattern, but even for the individual features of which the pattern is composed). However, they tend to be almost as discordant for congenital malformations as nonrelatives. Concordance or discordance for growth and psychomotor development is influenced by their intrauterine growth and by the occurrence of malformations (especially of the brain)) in one or both of them. As with singletons, one can observe that the smaller twin will usually have more severe congenital malformations and a poorer outcome with respect to survival and mental development. These observations imply that the occurrence of a particular malformation to which patients with a specific aberration are more prone is, among other factors, determined by the placental blood supply during early embryogenesis. The same is probably true for growth and mental development, which, however, could be negatively influenced even long after termination of organogenesis. In other words, in monozygotic twins with autosomal aberrations who are genetically identical (except for cytoplasmic inheritance), are, because of their chromosome imbalance, prone to growth retardation and malformations. Yet they may have quite a different placental blood supply, or subtle differences in the blood supply and possibly in other intrauterine factors that might determine whether or not they will show the malformations and growth retardation to which they are prone.

KARYOTYPE—PHENOTYPE CORRELATIONS IN AUTOSOMAL CHROMOSOME ABERRATIONS

Patients with duplications of different lengths of the same chromosome arm (mostly translocated segments varying in length by more proximal breakpoints) are generally quite concordant for their dysmorphic patterns. The latter are already distinct in duplications of relatively small terminal segments and do not increase very much in richness and distinctiveness with increasing size of the segment. The situation is quite different for incidence and severity of congenital malformations and for growth and psychomotor retardation: the larger the duplicated segment, the more severe are these features. With increasing size of the segments of larger autosomal arms, the point is finally reached from where further increases produce no viable unbalanced offspring, but spontaneous abortion is still possible. Examples of this rule can be given for almost every autosome. Duplications of distal 7q and 13q are typical cases. All eight reported patients with familial duplication of (7)(q32→qter) survived for the period of observation, between 1 and 23 years; one had a heart malformation. By contrast, four out of seven reported cases with familial duplication of the larger segment (7)(q22→qter) died in early infancy, and in all families there was a high incidence of perinatal and early postnatal losses. There was a high incidence of various major malformations in the latter group, including heart defects, cleft palate, and renal and ocular anomalies. However, the patterns of dysmorphisms were not significantly different between these two groups [1]. In the 13q example, patients with full trisomy 13 show a distinct pattern of dysmorphisms and rather specific congenital malformations; their mean survival is short, and growth and particularly mental development are profoundly impaired. Patients with duplication of the distal half of 13q, dup(13)(q22→ qter), survive significantly longer, their growth is often normal, and mental impairment is severe, but certainly less than in the full trisomy. Dysmorphisms are quite similar apart from those caused by underlying holoprosencephaly, which are rare in the duplication and frequent in the full trisomy. If the duplicated segment comprises about two thirds of 13q, dup(13)(q13→qter), incidences of congenital malformations almost equal those in full trisomy, and survival becomes very limited, while in duplication of only 1/4 of distal 13q, dup(13)(q31→qter), mental retardation is only mild to moderate, malformations are rare except for hexadactyly, and survival is almost unrestricted. However, even patients with the latter, small duplication show a pattern of dysmorphisms quite similar to that of patients with larger duplications or even full trisomy [1]. Thus, for prognosis with respect to survival and mental development, it is important to know not only the origin, but also the exact size (breakpoints) of the duplicated segment.

MOTOR AND MENTAL DEVELOPMENT, PERSONALITY

The extent of mental retardation is a more-or-less characteristic feature of each autosomal chromosome aberration. The limits within which mental development may vary in a particular aberration are determined by the exact size and location of the aneuploid segment, additional genetic (familial) factors, environmental factors (stimulation or neglect), additional malformations, or handicaps acquired postnatally (e.g., congenital or acquired deafness, blindness, cerebral palsy, additional brain malformations), and others. Mosaicism (for a deletion, duplication, or trisomy) might mitigate the clinical picture as might attachment of the duplicated segment to an X chromosome with subsequent partial inactivation. As to the upper limit of possible achievements in a particular aberration, there are sharp borderlines, while the lower limits are quite variable. For example, there are a few aberrations that are compatible with normal or borderline normal mental development. These include mosaic trisomy 8, the cat-eye syndrome—tetrasomy 22(pter→q11) and deletion of (18)(pter→p11). The majority of patients with these three aberrations are moderately mentally retarded; however, there exist cases with severe or even profound mental deficiency. Those with 18p− will mostly have a form of holoprosencephaly that is facultatively associated with this aberration, and those with mosaic trisomy 8, agenesis of the corpus callosum with hydrocephalus. On the other hand, there is no patient with full trisomy 21 who is not developmentally delayed, and there is, so far, no case of deletion of the entire band 4p16 who is not at least severely mentally retarded.

Data concerning specific personality characteristics and specific disproportionate developmental delay in patients with autosomal aberrations are still very limited. However, many individual observations suggest that developmental delay is uneven in many aberrations. Language is disproportionally defective in various aberrations [5,6], as are higher intellectual functions required for combination puzzles and mathematics. Motor skills and social adaptation are often relatively unaffected, not only in patients with Down syndrome, but also in many other autosomal aberrations with which moderate mental deficiency is associated. In addition, a tendency towards specific behavioral difficulties is known from repeated observations in several chromosome aberrations, particularly in the tetrasomy 15(pter→q12) and in patients with ring chromosomes 18 with more terminal breakpoints, i.e., those who do not show the phenotype of either 18p− or 18q− syndrome. Most patients with the additional inv dup(15) chromosome are described as anxious, hyperactive, auto-aggressive, with a low frustration tolerance and autistic features. They are prone to psychosis or psychotic behavior and may have seizure disorders. Mental retardation is mostly of moderate degree, but the behavioral difficulties make learning very difficult for them. Aggressiveness, behavioral disturbance, and schizophrenic features are repeated findings in patients with ring 18 chromosomes [7], and schizophrenia has also been recorded in a mosaic trisomy 8 girl of normal intelligence [8].

MAPPING OF GENES RESPONSIBLE FOR SPECIFIC MALFORMATIONS THROUGH OBSERVATIONS OF UNBALANCED CHROMOSOME ABERRATIONS?

The nonspecificity of most malformations in autosomal chromosome aberrations excludes the possibility that one gene, in the hemizygous or duplicated (partially trisomic) state, could be responsible for them. One exception seems to be aniridia, which has been found in almost all reported cases of deletion of segment 11p13 [1]. In addition, the observation of a family with aniridia segregating as a dominant trait in carriers of a translocation involving a break point in 11p13 makes it likely that one gene for this malformation is localized at that band [9]. However, even aniridia is not entirely confined to 11p aberrations, it can be found as well in trisomy 13 and rarely in other autosomal aberrations. In addition, this association of deletion and malformation has been influenced by ascertainment bias (see above). The indication for such an examination is mostly aniridia in combination with Wilms tumor, ambiguous male external genitalia, and other abnormal findings. Thus, cases reported with this malformation and an 11p deletion are strongly biased towards the occurrence of aniridia. The situation is different in retinoblastoma and deletion of 13q14. A recessive retinoblastoma locus has been mapped to this chromosome segment, and thus a deletion of the latter would make an individual very prone to this tumor: its occurrence does not require several hypothetical steps of tumorigenesis, but only a second event, e.g., aneuploidy for chromosome 13, in a retina cell during early postnatal development.

A further connection concerns the mapping of specific malformations that frequently occur in a trisomy of one chromosome, to a segment of that chromosome. An example is postaxial hexadactyly in trisomy 13. At examination of patients with duplications of different regions of chromosome 13, it seems that hexadactyly is equally frequent (as in full trisomy) in duplications including segment (13)(q31→qter) but is not found in patients with duplications not comprising that segment. Thus, duplication of the segment (13)(q31→qter) is responsible for hexadactyly in trisomy 13. A second example is of agenesis of the corpus callosum in trisomy 8 and duplication of distal 8p[1].

REFERENCES

1. Schinzel A: "Catalogue of Unbalanced Chromosome Aberrations in Man." Berlin, New York: De Gruyter & Co., 1984.
2. Benacerraf BR, Frigoletto FD, Laboda LA: Sonographic diagnosis of Down syndrome in the second trimester. Am J Obstet Gynecol 153:45, 1985.
3. Furbetta M, Falorni A, Antignani P, Cao A: Sibship (21q21q) translocation Down's syndrome with maternal translocation. J Med Genet 10:371, 1973.
4. Darmady JM, Seabright M: A family with an inherited translocation involving the No.4/No.21 chromosomes. J Med Genet 12:408, 1975.
5. Debray-Ritzen P, Bursztejn C, Vivier R, Rethore MO, Prieur M, Lejeune J: Troubles graves d'integration du langue de l'enfant et aberrations chromosomiques à propos de trois cas. Rev Neurol Paris 130:357, 1974.
6. Mutton DE, Lea J: Chromosome studies of children with specific speech and language delay. Dev Med Child Neurol 22:588, 1980.
7. Schinzel A: Particular behavioral symptomatology in patients with rarer autosomal chromosome aberrations. In Schmid W, Niesen J (eds) : "Human Behavior and Genetics." Amsterdam: Elsevier/North Holland, 1981, pp 125–210.
8. Sperber MA: Schizophrenia and organic brain syndrome with trisomy 8(group C-trisomy 8 47,XX,+8). Biol Psychiatry 10:27, 1975.
9. Simola KOJ, Knuuttila S, Kaitila I, Pirkola A, Pohja P: Familial aniridia and translocation t(4;11)(q22;p13) without Wilms' tumor. Hum Genet 63:158, 1983.

The Cytogenetics of Mammalian Autosomal Rearrangements, pages 739–768

Chapter 25
Phenotypic Consequences of Chromosome Imbalance in the Mouse

K.M. Kirk and A.G. Searle

INTRODUCTION

The laboratory mouse usually has a haploid set of 20 acrocentric chromosomes, although many stocks with one or more metacentrics are now available. Over the last 5 years it has become apparent that mouse and man share a large number of homologous chromosome regions [1], some of which are known already to be quite extensive. Since human and mouse lineages must have separated very early in mammalian evolution, it can be surmised that their shared genetic regions contain the essential features of the mammalian genome. Thus, the study of what goes wrong in mouse development when all or part of a given chromosome is gained or lost may be helpful in unravelling the consequences of similar events in human chromosomes, as well as being of general interest for the elucidation of developmental processes in mammals and the organization of the mammalian genome.

Although Snell and colleagues carried out pioneer work over 50 yr ago on the developmental effects of chromosome imbalance generated by translocations in the mouse [2,3], it is only recently that the necessary knowledge and material for a comprehensive study has become available. Three major trends have been responsible for the changed situation:

i. The post war concern about genetic risks from radiation led to the induction and study of many translocations in the mouse.

ii. The development of banding methods for mammalian chromosomes allowed mouse chromosomes to be separately identified, followed by the successful assignment of linkage groups to these chromosomes.

iii. The discovery by Gropp and collaborators of an extraordinary diversity of Robertsonian (whole-arm) translocations in wild mice, with others arising

Medical Research Council Radiobiology Unit, Chilton, Didcot, Oxon OX11 ORD, England; present address (K.M.K.): Pitton Mill Cottage, Pitton Green, Rhosili, Gower, Wales spontaneously in laboratory stocks [4], led to improved methods for obtaining embryos with specific primary monosomies and trisomies.

Several Robertsonian translocations are now known for every mouse autosome (though not for the sex chromosome), so all autosomes have become available for study. The situation is more complex if we wish to study the effects of the gain or loss of a particular chromosome region.

MATERIALS AND METHODS

Robertsonian Translocations

The original method for obtaining primary monosomies and trisomies was to make use of the fact that heterozygotes for wild-derived Robertsonian translocations showed high frequencies of non-disjunction, especially after crossing to laboratory mice. Thus, reasonably high frequencies of zygotes monosomic and trisomic for the two chromosomes concerned would be expected on outcrossing such a heterozygote to wild type. A more efficient system [5] uses a combination of two Robertsonians with a common arm (monobrachial homology), since this leads to very high frequencies of non-disjunction for the arm concerned. Thus one specific type of monosomic and of the corresponding trisomic zygote (instead of two types when single Robertsonians are used) are produced at equally high frequencies. The only disadvantage of this method is that some male compounds with monobrachial homology are sterile. Where males are fertile, however, so that intercrosses are possible, these lead not only to monosomic and trisomic zygotes but to nullisomic and tetrasomic ones as well, which can throw further light on the consequences of chromosome imbalance [6,7].

Reciprocal Translocations

Normal meiotic segregation in heterozygotes for reciprocal translocations leads to the production of about 50% gametes with duplications and deficiencies for the parts of the chromosomes concerned distal to the translocation breakpoint. In nearly all mouse translocations, the resultant unbalanced zygotes die, usually around the time of implantation, which leads to the phenomenon known as "semisterility" of the heterozygous male or female parent. Sometimes, however, embryos with unbalanced karyotypes survive until later in gestation and may be malformed.

In order to study the phenotypic effects of this type of chromosomal imbalance male carriers for the translocation concerned are mated to wild-type females. Uterine contents are examined in middle to late pregnancy, i.e., 12–18 days after detection of a vaginal plug. Deciduomata (small moles) show where an embryo has died at or soon after implantation, while later deaths are shown by the presence of placentae alone (large moles) or dead embryos, and live embryos with malformations may also be found. The last three categories are rarely found in matings of normal untreated mice. Translocation heterozygosity can be diagnosed and translocations maintained by various genetic methods [8].

One interesting class of reciprocal translocation is that with one breakpoint near the centromere and the other distal on the chromosome [9]. This often leads to the generation of long and short marker chromosomes and is particularly associated with sterility of the male carrier [10]. Non-disjunction in the short marker arm, which is frequently achiasmate, leads to formation of tertiary monosomic or trisomic zygotes, in which effects of regional gains or losses of chromosomal material can be studied in isolation, without the complication of a simultaneous complementary loss or gain. Of course, these male-sterile translocations can only be maintained by breeding from female heterozygotes, partly diagnosed by small litter size.

Insertions

Only two insertions are known in the mouse, both of which involve chromosome 7 [4]. Like the tertiary aneusomies, they permit the effects of gains or losses of particular chromosome regions to be studied, namely those involved in the insertion.

Inversions

Paracentric inversions generate germ cells with dicentric and acentric chromosomes that normally do not survive to enter the zygote, although female carriers may show extra intrauterine mortality. Pericentric inversions, however, give gametes with duplications and deficiencies, like reciprocal translocations. In the normal mouse with 20 pairs of acrocentrics they can only arise if one breakpoint occurs beyond the centromere, which is most unusual. Very few have been reported yet in metacentric mouse stocks, but it is known that they can lead to congenital malformations in man [4,8].

EFFECTS OF PRIMARY MONOSOMY AND TRISOMY

It seems to be a general rule that loss of an autosome has much more severe zygotic consequences than the gain of one [12]. As already mentioned, the double Robertsonian method of generating aneuploids leads to equal frequencies of monosomic and trisomic zygotes, from which the monosomic component disappears earlier in development than the trisomic one.

Autosomal Monosomy

Baranov and Dyban [13–15] have reported on the effects of monosomy for chromosomes 1, 2, 3, 5, 6, 9, 15, 16, 17, and 19, while Magnuson et al. [16] have dealt with chromosomes 1, 4, 10, 11, 12, 14, 15, 16, 17, 18, and 19 monosomics. Thus it appears that only monosomics for chromosomes 7, 8, and 13 remain to be investigated. Monosomy 7 has been illustrated [13] as surviving only to the morula stage, but this is an inference from the very early lethality of embryos deficient for the distal end of chromosome 7 [17].

In some monosomics embryonic lethality is spread over a few days, but one group can be distinguished in which death is particularly early, i.e., almost entirely at the morula or early blastocyst stage (day 3 of gestation). This comprises chromosomes 2, 3, 5, 6, 11, and 15, and probably chromosome 7. Monosomics for chromosomes 1, 10, 14, and 16 may survive to be implanted, but very few live beyond the neurula stage; the other monosomics studied (chromosomes 4, 9, 12, 17–19) seem to have intermediate effects, with survival usually to the late blastocyst stage without implantation. Chromosome 1 monosomies seem to have an extended period of mortality, since Magnuson et al. [16] found a significant decline in numbers already on day 3 with no evidence for implantation, but Baranov [13] reported that some survived to the 8th and 9th day of gestation. Conversely, Baranov [13] found that all chromosome 17 monosomics were lethal at the morula stage, while Magnuson et al. [16] reported normal survival to the late blastocyst stage. These apparent discrepancies are probably the result of differences in genetic background, which seem important in determining the length of survival [16].

Studies on monosomy 19 have shown that its effects are not due to the expression of a recessive lethal gene [18] and that it is not a cell lethal, as judged by its survival in cells of 9-day chimaeric embryos [19], but clearly these cells are at a proliferative disadvantage, as in other monosomics [16]. Epstein and co-workers believe that the universal early lethality of the autosomal monosomies is the result of i) the decrease to half in the synthesis of specific gene products and ii) the existence of a large number of loci, scattered throughout the autosomes, that are vitally important in development and are involved in processes that are highly concentration-dependent. Baranov [13] has pointed out that haploid parthenogenotes survive about as well as embryos monosomic for single chromosomes despite having all chromosomes monosomic. Also, the earliest known lethals like c^{25H}, which are nullisomic for small stretches of chromosome, kill at about the same time as monosomics. He has suggested, therefore, that autosomal monosomy may lead to functional nullisomy, with a complete or partial inactivation of loci on the remaining chromosome. However, recent evidence on mouse chromosome 15 suggests that nullisomic embryos die earlier than monosomics; similarly, embryos tetrasomic for this chromosome die earlier than trisomics [7], the same holding for chromosome 16 [20].

Autosomal Trisomy

All autosomal trisomies have now been studied, some in considerable detail [11,21]. Although all survive longer than any of the monosomies, they resemble these in being roughly composed of a group which die early and another which survive until later. All, except possible trisomy 12, tend to be more or less retarded as compared with normal littermates. The early postimplantation lethal group comprises trisomies for chromosomes 2–5, 7–9, 11, 15, and 17 and is thus similar to the early-lethal monosomic group. All in the late-lethal group (except trisomy 6) tend to show specific malformations (Table I) that may affect all or nearly all members (e.g., in the exencephaly of trisomy 12 and the cardiovascular anomalies of trisomy 16) but are sometimes found in only a few. Sometimes, as with cleft palate in trisomy 19 [27], their presence or absence depends on the genetic background. As with monosomies, the length of survival of particular trisomics seems to depend on genetic background. It is now realised, as Table I shows, that at least four of the trisomies are compatible with survival to term, while trisomies for chromosome 19 may survive well into postnatal life. Thus the viability picture is not much different from that for human trisomies, in which trisomics for chromosome 21 frequently survive birth, while those for chromosomes 13 and 18 occasionally do so [29].

The discovery of a homologous region between human chromosome 21 and mouse chromosome 16 has stimulated much research on trisomy 16, which has been shown closely to resemble human trisomy 21 in its effects [11] and can be regarded, therefore, as a mouse model for Down syndrome. Other trisomies (e.g., that for chromosome 12) are being studied from the same point of view. Other lines of research on them include i) attempts to "rescue" particular trisomic cellular systems, either by the formation of tetraparental chimaeras or of radiation chimaeras (in the case of haemopoietic stem cells) in order to study their normality or otherwise and thus find out more about the involvement of particular chromosomes in developmental processes [21] and ii) the quantitative study of enzyme patterns in particular trisomics, so as to detect abnormal gene dosage effects that will help to reveal the chromosomal location of the gene concerned [30].

It should be pointed out that not all the phenotypic effects of monosomy and trisomy need necessarily be the direct result of the loss or gain of a

TABLE I. Congenital anomalies associated with specific autosomal trisomies in the mouse

Chromosome No.	Nature of malformation	Length of survival (days)	Reference
1	Holoprosencephaly with craniofacial hypoplasia, cyclopia, aprosopia, placental hypoplasia	14–15	[21]
10	Ventricular septal defect	12–15	[21]
12	Exencephaly (100%), anophthalmia, ventricular septal defect, placental hypoplasia	15–17	[21,22]
13	Pulmonary stenosis, double outlet right ventricle, cleft palate	16–term	[23]
14	Exencephaly (50%), ventricular septal defect, double outlet right ventricle, transposition of great arteries	16–term	[21]
16	Cardiovascular anomalies (96%) especially aortic arch anomalies, ventricular septal and endocardial cushion defects, severe thymic hypoplasia, hydronephrosis, etc.	14–term	[11,24]
17	Caudal necrosis and kinking	10–13	[25]
18	Cleft palate	To term[a]	[21,26]
19	Ventricular septal defect, double outlet right ventricle, cleft palate (in some stocks), placental hypoplasia	To term and beyond[b]	[21,27]

[a]Not found among liveborn.
[b]Usually lethal in utero or soon after birth, but one survived for 142 days [28] and produced one young.

chromosome. Some may be connected with disturbances in maternal imprinting. It has now been shown that there are regions on mouse chromosomes 2, 6–8, 11, and 17 that fail to show normal complementation when two doses of these regions come in from one parent and none from the other [31]. Usually, the defective complementation shows up when there are two doses of a chromosome region from the mother and none from the father, but two doses from the mother and one from the father might also cause trouble with imprinting, as well as one dose from the father and none at all from the mother.

EFFECTS OF RECIPROCAL TRANSLOCATIONS AND INSERTIONS

The usual method of generating autosomal trisomies, by the use of Robertsonian compounds with monobrachial homology, generates one easily detectable unbalanced product (the trisomy) at much greater frequencies than any other. On the other hand, the usual method of generating the unbalanced products or translocations—namely, the outcrossing of translocation heterozygotes, leads to two major unbalanced products at the same zygotic frequency and a number of other possible unbalanced products at a lower frequency. The two major products, and usually most of the others, will not be associated with any change in chromosome number, and their nature may be much more difficult to detect than a simple trisomy. Because the meiotic products of adjacent-1 and alternate disjunction are effectively equivalent in the mouse a translocation AB.CD/AD.CB (Fig. 1) will produce equal numbers of the following gametes: AB.CD, AD.CB, AB.CB, and AD.CD [32]. The last two are unbalanced, with duplication of one distal arm and deficiency of the other. The rarer adjacent-2 disjunction (non-disjunction of homologous centromeres) leads to AB.AB, AB.AD, AD.AD, CB.CB, CB.CD, and CD.CD gametes. These all have more extensive deficiencies than the products of normal segregation; so they are very unlikely to give rise to embryos that survive long enough to show specific malformations, in view of what has been discovered about the early embryonic lethality of all primary monosomies described so far in the mouse.

The main unbalanced gametic products of insertion heterozygotes will be tertiary trisomies and tertiary monosomies for the inserted segment, so

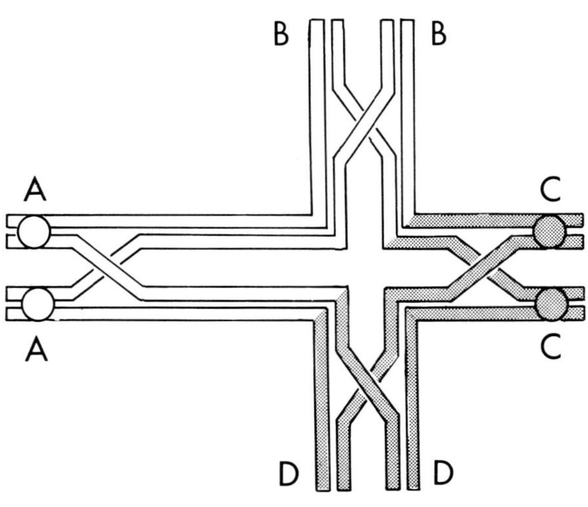

Fig. 1. Standard cross-configuration found at meiosis in a translocation heterozygote which involves normal chromosomes AB and CD (the latter stippled) with translocated chromosomes AD and CB. The centromeres are shown at A and C and there are single chiasmata in all four arms. (Reproduced from Searle et al. [32], with permission of Cambridge University Press.)

analysis is easier than where both a duplication and a deficiency are present. As already mentioned, the male-sterile group of translocations, with breaks near the ends of chromosomes, also give rise to tertiary monosomies and trisomies which may survive into postnatal life. Breakpoints of all the translocations considered in the following pages are given in Figure 2.

Early Embryonic Lethality

Many reciprocal translocations are known in which the amount of extra embryonic lethality beyond the implantation (small mole) stage is not increased above normal (Table II). As will be discussed later, they can provide valuable evidence on the relationship between size of deficiency and severity of lethality. The only detailed study on early embryonic lethality in this class of translocation is that on T(2;8)26H. A systematic study was made [33] of embryonic lethality on days 6 and 8 of pregnancy in outcrosses of T26H/+ males; no significant embryonic loss occurred after day 8. The expected proportions of embryos with different-sized deficiencies of chromosomes 2 and 8 were calculated on the basis of previous information on meiotic segregation. It was found that these proportions, grouped into three main classes, agreed well with the frequencies of different types of histological abnormality found in the implantation sites and with the supposition that embryos with a larger deficiency die first. Work on unbalanced embryos from outcrosses of T6Ca heterozygotes [34] supports the idea that large deficiencies die sooner. The problem that remains is the size of deficiency that is compatible with survival beyond the implantation stage.

Information on deficiency sizes associated with normal meiotic segregation in each translocation heterozygote is included in Table II. It should be remembered that in an unbalanced gamete each of these deficiencies will be associated with a duplication of the other translocated segment. The figures given have been calculated from known breakpoint positions [35] on the G-banded map of Nesbitt and Francke [36] by i) assuming that the breakpoint is in the middle of the band, ii) measuring what proportion of the chromosome concerned is included in the deficiency, and iii) calculating what proportion this is of the total haploid genome by using the figures given by the Nomenclature Committee [37] on chromosome lengths as a proportion of the haploid complement. The rarer adjacent-2 segregation will generate larger deficiencies than those from normal meiotic segregation. It can be seen from Table II that in all these translocations associated solely with extra lethality at or before embryonic implantation (except T145H) the zygotic imbalance would have included a deficiency of at least 1% of the haploid complement.

Later Lethality and Postnatal Viability

A number of translocations are now known in which one or more of the unbalanced products survive beyond the implantation period to later in pregnancy or possibly even into postnatal life. Table III lists those translocations in which there is later intrauterine lethality, again with estimates of the lengths of associated deficiencies. It will be seen that, with a few exceptions, the shorter of the two possible deficiencies involved in the imbalance contains less than 1% of the total haploid genome, the average being 0.8%. This contrasts with a larger average size of 1.8% for the smallest deficiency in translocations giving only early embryonic lethality.

The types of anomaly associated with tertiary monosomic or trisomic offspring, and with similar products of insertions, are summarized in Table IV. The tertiary trisomies, as well as the duplicated insertions, carry extra chromosomal material only. This is associated with a marked increase in viability as compared with the usual translocation products carrying both a duplication and deficiency. Thus, gains of up to 2% of the haploid genome may be compatible with postnatal survival to maturity, though duplications + deficiencies of this magnitude usually lead to embryonic death around the time of implantation (Table II). It is surprising that, in T31H, some tertiary monosomies manage to survive to maturity, despite a 2% loss of chromosomal material. The male infertility associated with a number of tertiary trisomies and monosomies [10,46] seems to be connected more with failures of pairing during meiosis than with the actual chromosome regions involved.

Specific Types of Developmental Anomaly

As Table III shows, many translocations have now been studied in which there is extra postimplantation lethality in mid- to late gestation. Although the causes are usually unknown, a number of specific malformations have been identified in particular translocation stocks, especially where uterine contents have been examined somewhat earlier than the usual 14½–17½ days post coitus (where day ½ p.c. is that on which the vaginal plug is found) so that the developmental defect can be picked out before death ensues.

Exencephaly and other CNS malformations. Snell and Picken [3] reported abnormal development leading to exencephaly in progeny of a translocation heterozygote (now extinct) and remarked that "at least eight out of ten translocations under investigation by Miss Bodemann have produced one or more embryos with similar brain deformities." More recent work has confirmed this finding, since exencephaly was observed in 12 out of 14 translocation stocks studied [38], although these were deliberately selected to exclude most of those with early-acting lethality only. We shall concentrate

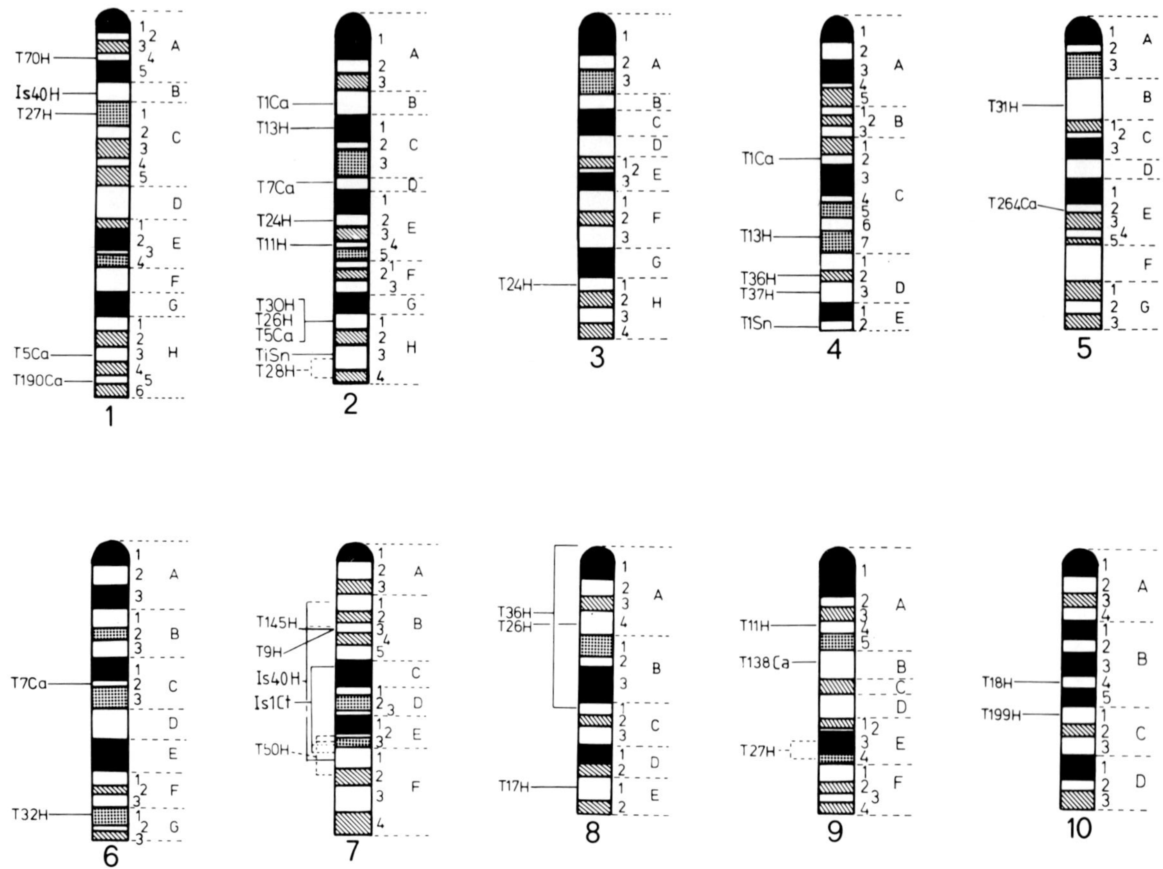

T70H
Is40H
T27H
T5Ca
T190Ca
1
T1Ca
T13H
T7Ca
T24H
T11H
T30H
T26H
T5Ca
TiSn
T28H
2
T24H
3
T1Ca
T13H
T36H
T37H
T1Sn
4
T31H
T264Ca
5
T7Ca
T32H
6
T145H
T9H
Is40H
Is1Ct
T50H
7
T36H
T26H
T17H
8
T11H
T138Ca
T27H
9
T18H
T199H
10

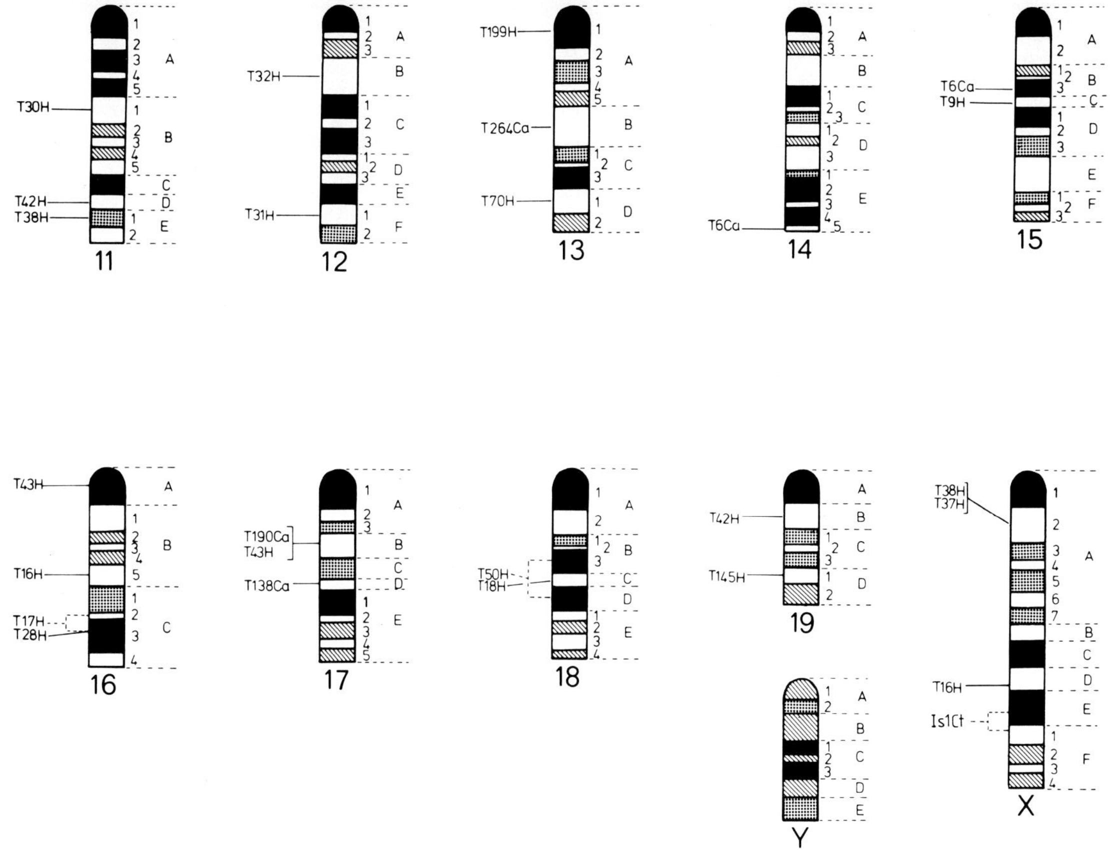

Fig. 2. Standard banded karyotype of the mouse [36] to show breakpoints of translocations under consideration.

TABLE II. Data on translocations giving early embryonic lethality of unbalanced products

Name and sex of carrier		No. females tested[a]	Extent of deficiency as % of haploid genome		Preimplantation loss (%)	% implants as	
						Deciduomata	Later deaths
Wild-type		35	—		8.0	3.2	2.6
T(1;9)27H	♂	41	5.3(1)	1.1(9)[b]	16.2	63.7	1.0
T(2;4)1Ca	♀	21	5.2(2)	3.2(4)	2.8	67.8	1.4
T(2;4)1Ca	♂	45			14.9	51.7	3.0
T(2;4)13H	♂	41	5.2(2)	1.7(4)	18.0	60.0	0.4
T(2;6)7Ca	♂	42	4.3(2)	3.0(6)	5.8	53.5	0.3
T(2;8)26H	♀	20	1.2(2)	3.6(8)	2.3	59.0	0.0
T(2;8)26H	♂	20			13.7	58.0	0.3
T(2;9)11H	♂	20	2.6(2)	3.5(9)	12.0	52.9	0.0
T(2;11)30H	♂	44	1.2(2)	2.6(11)	12.3	55.2	0.0
T(5;13)264Ca	♂	40	2.3(5)	2.1(13)	7.8	59.4	0.6
T(7;15)9H	♂	40	3.9(7)	2.4(15)	15.2	58.8	1.0
T(7;18)50H	♂	36	1.1(7)	1.2(18)[c]	12.9	61.6	0.4
T(7;19)145H	♀	41	3.9(7)	0.6(19)	20.7	54.4	1.6
T(9;17)138Ca	♂	39	2.8(9)	1.5(17)	—	59.1	0.6
T(10;18)18H	♂	30	2.3(10)	1.6(18)	11.2	58.0	1.5
T(X;16)16H	♀	41	1.9(X)	1.8(16)	15.0	66.4	2.8

Those for wild-type and T18H are from Kirk [38]; others are from previously unpublished analyses of data on Harwell translocations at M.R.C. Radiobiology Unit.

[a]Carrier females, or wild-type females mated to carrier males.

[b]Nos. in parenthesis are of the chromosomes carrying the two types of deficiency expected with normal disjunction. Each will be associated with a duplication of the size indicated in the other chromosome concerned.

[c]These are minimum sizes for the deficiencies, the range being 1.1–1.8% (chromosome 7) and 1.2–2.0% (chromosome 18).

TABLE III. Data on translocations giving later embryonic lethality of some unbalanced products

Name and sex of carrier		No. females tested	Extent of deficiency as % of haploid genome		Preimplantation loss (%)	% implants as Deciduomata	% implants as Later deaths[a]	Reference
T(1;2)5Ca	♂	42	1.4 (1)	1.2 (2)	—	44.1	4.8	[39]
T(1;13)70H	♂	15	6.4 (1)	0.6 (13)	18.1	52.5	12.7	[38]
	♀	24			13.6	39.3	24.8	
T(2;3)24H	♂	27	3.0 (2)	0.6 (3)	26.7	21.8	20.9	[38]
	♀	21			24.8	42.5	12.2	
T(2;4)1Sn	♂	42	0.6 (2)	0.1 (4)	12.4	25.7	17.5	[38]
	♀	14			38.1	33.9	17.4	
T(2;16)28H	♂	13	0.2 (2)	0.3 (16)	19.4	13.0	32.4	[38]
	♀	17			17.7	12.6	32.3	
T(5;12)31H	♂[b]	12	4.4 (5)	0.5 (12)	22.7	43.5	30.6	[38]
	♀	20			15.1	25.9	17.7	
T(8;16)17H	♂	29	0.5 (8)	1.0 (17)	26.3	42.1	16.2	[38]
	♀	23			15.6	17.6	17.0	
T(16;17)43H	♀[b]	14	3.5 (16)	2.3 (17)	11.1	44.5	13.3	[38]
T(11;19)42H	♀[b]	25	1.2 (11)	1.7 (19)	21.1	51.3	13.3	[38]
T(X;4)37H	♀[b]	34	5.1 (X)	1.0 (4)	26.6	36.7	19.0	[40]
T(X;11)38H	♀[b]	35	5.1 (X)	0.4 (11)	32.1	33.5	18.0	[40]
T(14;15)6Ca	♀	17	0.6 (14)	2.6 (15)	10.4	46.8	4.0	[41]
T(1;17)190Ca	♂	32	0.3 (1)	2.3 (17)	—	46.1	14.2	[39]
	♀	30			—	54.2	8.5	
T(10;13)199H	♀[b]	20	1.7 (10)	4.0 (13)	27.4	54.9	6.3	[39]
T(6;12)32H	♀[b]	21	0.5 (6)	3.3 (12)	—	56.7	9.9	[20]

See footnotes of Table II for further explanation.

[a]Including abnormal but still living embryos in the work of Kirk [38]. Some unbalanced products of the T1Sn heterozygote are known to survive birth [42].

[b]Translocation heterozygotes sterile or infertile in the male.

here on those translocations associated with quite high frequencies of exencephaly, anencephaly, and other failures of the neural tube to close, with omission of those in which only sporadic examples were observed.

Details of ten such translocations are summarized in Table V; in addition, the insertion Is(7;1)40H almost invariably led to exencephaly when the inserted segment of chromosome 7 was present in triplicate. This almost 100% penetrance of the condition is equivalent to an expectation of about 20% of implants being exencephalic in the translocations of Table V if the exencephaly was associated with one of the two unbalanced products of regular meiotic disjunction. If associated instead with products of nondisjunction (as in T42H and T70H), a much lower frequency would be expected. The 20% frequency expected with normal disjunction and full penetrance is not reached even in Snell's translocation; instead it appears that a particular chromosomal imbalance can lead to a range of different phenotypic effects with varying degrees of penetrance. The ancillary abnormalities shown in Table V are only those directly associated with the CNS defects in that they were found on occasion in the same animals; other abnormalities will be dealt with later.

The expression of exencephaly varied quite markedly between different translocations, as Figure 3 shows. In Is40H, for instance, the area of nonclosure always included the rhombencephalon (as in trisomy 12), with marked overgrowth of neural tissue (Figure 4), while in T1Sn its manifestation was very variable (sometimes with survival to term) and associated with a range of other anomalies (Table V). Exencephaly in T36H was severe and associated with cleft-face, while in T70H exencephaly was low-grade but associated with facial dysmorphy. Anencephaly was diagnosed occasionally in progeny from several other translocations but only in severely affected embryos. Exencephaly in rodents is generally considered analogous to human anencephaly.

Blood-stained amniotic fluid and occasional hydramnios were noted in some exencephalics, which suggested that there was a direct connection between the cerebrospinal and amniotic fluids. A detailed histological study on exencephalic T24H embryos [38] showed that the longitudinal axis was shortened and at an abnormal angle, reminiscent of that found in mouse embryos given excess vitamin A [22]. It is not surprising that exencephaly is often associated with eye abnormalities, but the reason for the connection with cardiac anomalies is not clear.

Cardiac anomalies. Five translocations (T1Sn, T17H, T28H, T36H, T38H) that had generated one or more embryos with severe generalized oedema (Fig. 5) were selected for a special study of associated cardiac

TABLE IV. Phenotypic effects of tertiary monosomy (Ms), trisomy (Ts), and insertional products

Translocation	Nature of product	Percent haploid genome gained (+) or lost (−)	Phenotype	Reference
T(1;7)190Ca	$Ts17^{1}$	+1.8	Retarded, postnatal lethals	[43]
T(1;13)70H	$Ts1^{13}$	+1.4	Males oligospermic, sometimes with dental and nasal anomalies	[44,45]
T(5;12)31H	$Ts5^{12}$	+2.1	Sterile in male, retarded	[46]
	$Ms5^{12}$	−2.1	Sterile in male, very retarded, sublethal	
T(6;12)32H	$Ts12^{6}$	+1.9	Sterile in male but female has bred	[20]
T(14;15)6Ca	$Ts15^{14}$	+2.1	Retarded, abnormal behaviour	[47]
	$Ms15^{14}$	−2.1	Early embryonic death	[34]
T(X;11)38H	TsX^{11}	+1.5	Some exencephalic, survivors retarded and male-sterile, but female has bred	[40]
T(7;19)145H	$Ts7^{19}$		Runted at birth	[48]
Is(7;1)40H	Dp(Is7)	+2.9	Exencephalic	[49]
Is(In7;X)1Ct	Dp(Is7)	+1.5	75% inviability	[50]
In(6.15LS)1Wa	Df[a]	−0.8	Viable and fertile	[51]

[a]Of 6Cen-6A2 and 15F2-15ter.

TABLE V. Translocations associated with high frequencies of exencephaly and related conditions in progeny of heterozygotes

Name	Nature of CNS abnormalities	% implants affected	Probable karyotype	Associated abnormalities	References
T(1;13)70H	Exencephaly (low-grade) ± facial dysmorphy	2	$1,13,13,13^{1}$?	—	[38]
T(2;4)1Sn (Snell's)	Exencephaly with cleft rhombencephalon or diencephalon; some with distended CNS (Snell)	12 (Snell) 9 (Kirk)	$2,2,4,4^{2}$ (Kirk)	Eye abnormalities, spine abnormalities, cleft face, some heart abnormalities (Kirk)	[2,38]
T(2;3)24H	Exencephaly, abnormal angle of skull base, coloboma	8	$2,2,3,3^{2}$	Some heart abnormalities	[38]
T(2;16)28H	Exencephaly (to variable extent) and anencephaly	4	$2,2^{16},16,16$	Retardation, open thorax, heart abnormalities, oedema	[38]
T(4;8)36H	Exencephaly ± cleft face	3	$4,4,8,8^{4}$?	Anophthalmia or microphthalmia, some heart anomalies	[38]
T(5;12)31H	Exencephaly, anencephaly	6	$5,5,12,12^{5}$ (mainly)	Severe retardation	[38]
T(8;16)17H	Exencephaly, occasional anencephaly, hydrocephaly	5	$8,8,16,16^{8}$	Agnathia (with anencephaly), spinal abnormalities	[38]
T(11;19)42H	Exencephaly, anencephaly, hydrocephaly	3	$11,19,19,19^{11}$ or Ts 19^{11}	Spinal abnormalities, severe retardation	[38]
T(X;4)37H	Exencephaly, hydrocephaly with spina bifida	3	Mainly X^{4}, 4^{X}	Eye abnormalities, hydramnios, oedema	[38,40]
T(X;11)38H	Exencephaly (low-grade), occasional hydrocephaly	8	XX^{11} or $N + X^{11}$	Eye abnormalities	[38,40]

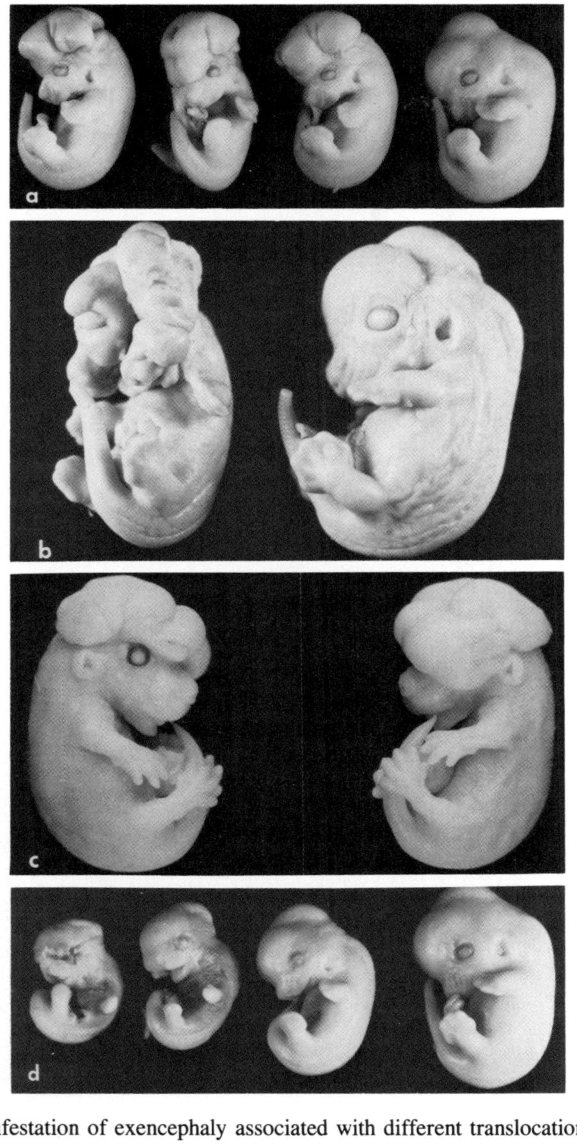

Fig. 3. Manifestation of exencephaly associated with different translocations. **a:** In Is40H, with inclusion of rhombencephalon (13½-day foetuses, normal on right). **b:** In T36H, with cleft face (13½-day foetuses, normal on right). **c:** In T1Sn, with asymmetric involvement of the telencephalon (14½-day foetus, right and left sides). **d:** In T70H with involvement of forebrain (extreme left), also facial dysmorphy (left two) and retardation compared with normal 13½-day littermate (right).

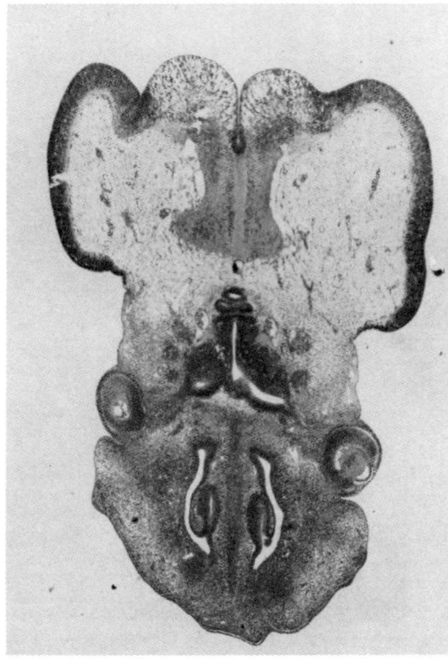

Fig. 4. Front section of head of 13½-day exencephalic foetus from Is40H outcross to show extent of overgrowth of neural tissue, with downward displacement of orbits.

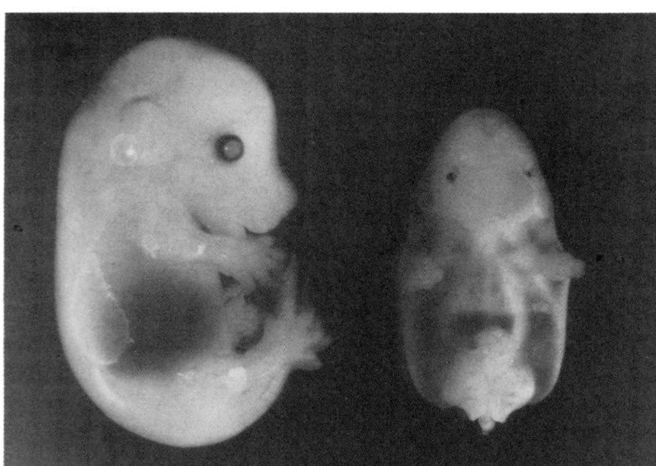

Fig. 5. 14½-day foetus from T36H outcross, to show severe oedema compared with normal littermate on right.

abnormalities in foetal progeny at 15½ days p.c. or later [38]. No cardiac anomalies were found in T(X;11)38H. Another of these translocations, T(8;16)17H, produced embryos with exencephaly and open thorax, also one very retarded embryo with a small right atrium and elongated aortic and pulmonary arches (Fig. 6). Table VI summarises the more frequent cardiac anomalies associated with the other three translocations. There was considerable variability between affected embryos in the manifestation of these anomalies, but ventricular hypertrophy was one of the commonest pathological conditions. Disproportionate reduction of the heart was noted in T28H and T36H progeny. Cardiac anomalies were also noted in 13% of the exencephalic embryos from T(2;3)24H but were not studied in detail.

Other anomalies. In the series of translocations examined by Kirk [38], retardation of embryos was a constant feature, although it was only slight in T(16;17)43H. Spine abnormalities, including spina bifida and kinked spine, were widespread, being found in T(8;16)17H (12% of abnormal embryos), T(2;3)24H (6%), T(X;11)38H (9%), T(11;19)42H (23%), T(1;13)70H (18%). In contrast, facial dysmorphy was found only in T70H (21% of abnormal embryos), where it was mainly associated with exencephaly (Fig. 3) and spinal abnormalities. Eye abnormalities (microphthalmia, anophthalmia, etc.) were found in some nonexencephalic embryos, particularly in T(X;4)37H (16% of abnormal embryos) and T(2;4)1Sn (4%). Limb abnormalities were remarkably rare, although adactyly (absence of digits) was noted in two out

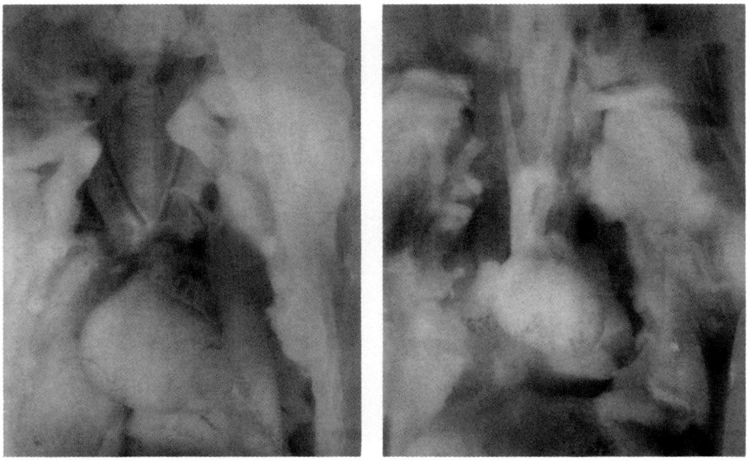

Fig. 6. Retarded heart with elongated aortic arch **(right)** compared with normal littermate **(left)** (17½ days) in T17H outcross.

TABLE VI. Translocations associated with cardiac anomalies

Translocation	Nature of anomaly	% implants affected	Probable karyotype	Associated abnormalities	References
T(2;4)1Sn	Oedema, hypertrophy of left ventricle	6	—	Exencephaly	[38]
T(2;16)28H	As above, also atresia of great vessels	12	$2,2^{16},16,16$	—	[38]
T(4;8)36H	Oedema, open thorax, persistent truncus arteriosus, tetralogy of Fallot etc.	8	$4,4,8,8^{4}$	—	[38]

of 47 abnormal embryos from outcrosses of T24H/+. Three other embryos from these crosses were noted as having a broad snout. Abnormalities of the skull and teeth are found in some tertiary trisomics from T(1;13)70H, with the extra small 1^{13} chromosome, while males are oligospermic. Male sterility is found in several other tertiary trisomics and in the tertiary monosomic for T(5;12)31H (see Table IV).

Apart from these tertiary aneusomics, unbalanced products of reciprocal translocations seem almost invariably to be lethal in the mouse. However, this may be because the usual "semisterility" test for translocations depends on the presence of about 50% intrauterine lethality in the progeny of the carrier male or female. The detection of viable imbalance is likely to be much more difficult. One probable example was reported [52] in the progeny of a male mouse treated with the chemical mutagen triethylenemelamine (TEM). An F_1 male, when tested for semisterility by litter size, produced normal-sized litters in which about 17% of offspring were retarded with domed heads and short faces. About 60% survived to weaning and developed circling and head-shaking behaviour, so the mutant phenotype was called "snaker." Breeding tests on these and on their normal sibs made it appear very probable that the original F_1 male was heterozygous for a balanced reciprocal translocation, with the snaker male one of its unbalanced products, while the other unbalanced product was apparently viable and normal. Other apparent dominant mutations may be of this type.

EVIDENCE FROM SIMPLE DELETIONS AND DUPLICATIONS

Deletions

Radiation mutagenesis has led to the recognition of a large number of deletions in the albino (*c*) region on mouse chromosome 7 and the dilute short ear (*d-se*) region on chromosome 9. These facilitate the study of the effects of homozygosity for partial chromosome loss as well as of heterozygosity.

At the albino locus, all the mutations that are lethal in the homozygote are deficiencies ranging in size from 2 centimorgans to 6–11 centimorgans [53]. Presumably, longer deficiencies have not been recovered because of deleterious effects in the heterozygote that prevent their transmission. Heterozygotes for the deficiencies studied seem to have normal viability, except for the two longest ones studied and a few others in which fewer heterozygotes than expected were produced [54]. Russell and colleagues have contrasted this situation with that in the *d-se* region, in which there was marked heterozygous viability depression in several shorter deficiencies, including one at the *se* locus which was less than 2 centimorgans long. They concluded that the length of deficiencies that can be tolerated without major adverse dominant effects is not constant throughout the mouse genome but depends on the content of specific chromosome regions. It is thought that the region immediately adjacent to the *c* locus contains regulatory genes concerned with liver function rather than structural genes [55]. In both the *c* and *d-se* regions complementation tests have revealed a large number of functional units associated with various phenotypic effects in the homozygote, especially neonatal and prenatal lethality, so that the larger the deficiency the more likely it is to lead to prenatal lethality in the homozygote.

More specific heterozygous effects of deletions are found in the t-complex, with effects on the tail in T^{hp} and T^{Or1}, for instance [56]. It seems probable that a number of other dominant mutations with homozygous lethality are really small deletions.

Duplication

A tandem duplication has been described on mouse chromosome 7 [57], that involves much the same region as the *c* locus deficiencies but also includes the *Hbb* locus. The length of chromosome 7 is increased by about 20%; thus the duplication is probably 15–20 centimorgans long. There are marked effects in the heterozygote, which weighs about 60% of normal at weaning, with survival to 10 days of age being only about 40% of normal. The duplication may also affect maternal functions. Homozygotes for it (i.e., tetrasomic for this region) die in prenatal life, probably around the time of implantation.

RELATIONSHIP BETWEEN PHENOTYPIC EFFECTS AND CHROMOSOME REGIONS

Research on phenotypic effects of primary trisomies in the mouse has revealed considerable variation between individual chromosomes in the degree of severity of the pathological effects and their actual nature, when these could be determined (Table I). The problem that now arises is whether, when all the data on effects of regional chromosome imbalance are analysed, it is possible to localize specific phenotypic effects to particular chromosome regions. The question of possible homology with human chromosome regions should also be taken into consideration.

Most of the work on phenotypic effects of chromosomal imbalance has been carried out on reciprocal translocations, which are associated with a combination of duplication and deficiency in their unbalanced gametes.

Therefore, it is often very difficult, if not impossible, to determine whether particular phenotypic anomalies are connected with the deficiency, the duplication or both, even in karyotyped foetuses. The location of one breakpoint very close to the end of the chromosome (so that duplication or deficiency of the terminal regional is likely to have little effect) or the comparison of results from different translocatins that have breakpoints on the same chromosome may help in the interpretation, as will studies on nonreciprocal insertions in which there is either a duplication or a deficiency of one particular chromosome region. Unfortunately, only two of these are known.

In Table VII, an attempt is made to pinpoint the chromosomal regions likely to be involved in the production of particular congenital anomalies. It should be remembered, however, that the frequencies of these anomalies are usually less than the expected frequency of zygotic imbalance of the relevant type, and that a range of different anomalies may result from one particular duplication-deficiency combination.

Trisomy for chromosome 1 leads to holoprosencephaly with cyclopia and facial dysmorphy. It is interesting to note that a number of neonates with similar abnormalities (Fig. 7) have been recovered from matings of homozygotes for the T(1;13)70H translocation, with a proximal breakpoint on chromosome 1. Facial dysmorphy with other cranial defects is also found in unbalanced products of this translocation which carry the long 13^{1} marker without the short 1^{13} one and seem to be deficient from the proximal part of chromosome 1. Duplication of this part of chromosome 1 and of the distal end of chromosome 13, in Ts(1^{13})70H, sometimes leads to cranial anomalies of a less severe kind (Table IV).

Duplication of a number of different chromosomal regions is associated with exencephaly to a variable extent. All of the chromosomes concerned belong to the early lethal group with respect to trisomic effects. In T31H exencephaly is also associated with deficiency of the distal end of chromosome 12, trisomy for which leads to 100% exencephaly. It is not clear, however, whether this deficiency or the associated duplication of most of chromosome 5 are chiefly responsible for maldevelopment of the neural tube in this instance. The exencephaly associated with Is40H is more clear-cut, since this insertion generates either duplications or deficiencies of a large part of chromosome 7 and "all embryos with duplications of the inserted segment were exencephalic" [49]. It also seems clear that the various anomalies associated with T36H imbalance (exencephaly with cleft face, eye abnormalities, oedema possibly resulting from cardiac anomalies) are the result of duplication for the distal end of chromosome 4, since the associated deficiency in the centromeric region of chromosome 8 must be very small

TABLE VII. Phenotypic effects of chromosome imbalance: Attempted localization of anomalies to specific chromosome regions

Chromosome	Translocation(s)	Nature of malformation(s)	Probable region involved
1	T(1;13)70H	Facial dysmorphy and other cranial defects	Cen-A4 (deficiency)
2	T(2;4)1Sn T(2;3)24H T(2;16)28H	Exencephaly, sometimes cardiac anomalies	H3-ter (duplication)
4	T(4;8S)In(8LS)36H	Cardiac anomalies, cranial defects, including exencephaly	D1-ter (duplication)
5	T(5;12)31H	Exencephaly	B-ter (duplication)
7	Is(7;1)40H	Exencephaly	B1-F1 (duplication)
8	T(8;16)17H	Hydrocephaly	E1-ter (deficiency)
11	T(X;11)38H T(11;19)42H	Exencephaly, some hydrocephaly	E1-ter (duplication) Cen-D (deficiency)
12	T(5;12)31H	Exencephaly	F1-ter (deficiency)
16	T(2;16)28H	Severe oedema, some exencephaly	C3-ter (duplication)

indeed. There is an extensive region of homology between this part of chromosome 4 and human chromosome 1q.

The anomalies associated with imbalance for T(2:16) 28H are of particular interest because of the homologies which exist between loci on mouse chromosome 16 and human chromosome 21 [11,58,59]. Present evidence suggests that the regions of man-mouse homology are in the distal parts of human chromosome 21q and mouse 16. Moreover, the loci of particular importance in the production of Down syndrome also seem to be located distally on human chromosome 21, at or beyond band 21q22 [60]. Thus, the fact that some unbalanced products of T28H, with duplication of the distal end of chromosome 16 and deficiency of a small distal part of chromosome 2 had severe oedema, perhaps the result of cardiac malformations, suggests that this translocation may prove useful in the search for a good mouse model for the trisomy 21 syndrome.

CONCLUSIONS AND SUMMARY

In the studies discussed in this paper, we can discern four categories of phenotypic effect that may arise as the result of chromosomal imbalance in the mouse. The most severe is very early embryonic death, even before implantation. This is associated particularly with autosomal monosomy but is also found to a lesser extent in reciprocal translocations which generate large deficiencies in particular chromosomes. The virtual absence of monosomic embryos from human abortions suggests that in human beings also these monosomies die at such an early stage as to prevent recognition.

The second category of effect is embryonic death at or soon after implantation. Dyban and Baranov [17,61] have dealt in detail with the few studies carried out on this and the previous category, which indicate that the distal segment of chromosome 7 and the central segment of chromosome 17 are very important for the control of early development. This category is associated with the early lethal group of primary trisomies and with most reciprocal translocations (Tables II, III), especially those with fairly central breakpoints so that both the main unbalanced products carry substantial chromosome deficiencies of at least 1% of the haploid component. If one of the breakpoints is more peripheral, then numbers of later deaths are also increased. This leads us into the third category, of deaths in the second half of gestation, or stillbirths, which are usually associated with visible malformations and which are therefore the main object of study. Nine primary trisomies are associated with this category—namely, eight of those shown in Table I and trisomy-6, which leads to foetal death without externally visible effects. As Tables IV and V show, there are also a number of translocations (including an insertion) that generate the type of chromosome imbalance which leads to this category of lethality.

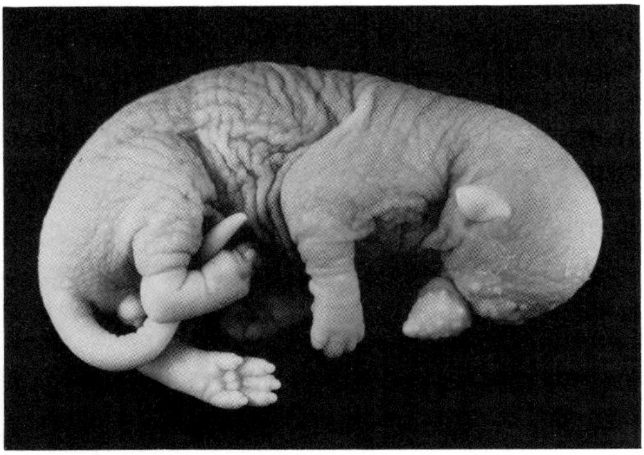

Fig. 7. Newborn mouse with holoprosencephaly from stock homozygous for T(1;13)70H.

The fourth category of effect consists of congenital and postnatal anomalies in survivors with chromosomal imbalance. These include mice with trisomy 19, which suffer from heart defects, etc., but may survive well beyond birth; several tertiary trisomies; and a tertiary monosomy. The commonest defect in these tertiary aneuploids is spermatogenic impairment, but this is also found in balanced heterozygotes for the translocations concerned and seems to be connected with meiotic pairing problems. However, surviving tertiary trisomics and monosomics also tend to be retarded and may have behaviour or skeletal defects.

Tertiary monosomies can also be classified as double deficiencies, and it seems likely that a number of small deficiencies in heterozygous condition may have specific postnatal phenotypic effects. These may be clearly connected with particular loci covered by the deficiency, such as with the hairpin tail (T^{hp}) deficiency in the t-complex [62] or the W^{19H} presumed deletion, with leads to low-grade depigmentation like *W*/+ but covers both the *W* and adjoining *Ph* loci and causes preimplantation lethality in the homozygote [63]. Other dominants with early embryonic lethal effects in the homozygote may well be small deficiencies, but these may be impossible to detect cytologically.

In mice, few of the unbalanced products of normal disjunction in recognised translocation heterozygotes seem to survive birth, one exception being some of the 4, 4^2 products of Snell's translocation, T(2;4)1Sn, sometimes lethal exencephalic foetuses but which may be runted liveborn without malformations [42].

The most studied category is that of late foetal death. The commonest malformation found was exencephaly, associated with trisomies 12 (100% incidence) and 14 (50% incidence), as well as at least ten reciprocal translocations and an insertion. With some chromosome anomalies the expression of exencephaly is very constant, e.g., in Ts 12 and in the duplicated form of Is40H it always included the rhombencephalon but in Ts14 segment is always closed. In others it is much more varied and may be associated with a range of facial, ocular, cardiac, and other defects. Thus there is a general nonspecificity of causation but evidence that the "window of opportunity" to close the neural folds in the cranial region may be missed in a number of different ways. These may depend partly on the nature of the chromosome imbalance and may have somewhat different consequences. In addition, there must be developmental thresholds, surmounted by some embryos of a particular unbalanced karyotype but not by others. Human anencephaly, usually regarded as analogous to mouse exencephaly, is associated with trisomy 18 and 21, but no homologies between these chromosomes and mouse chromosomes 12 and 14 have been reported.

Cardiac anomalies are probably just as common a category of defect but have been less studied because not so easily seen, although oedema may be an indication. They are associated with trisomies 10, 12–14, 16, and 19 and with several translocations in the mouse and also with a large number of human aberrations, again with considerable variability in expression [64]. Perhaps the most interesting is the endocardial cushion defect, which is frequently found with human trisomy 21 and with mouse trisomy 16, since these two chromosomes are known to share homologies in distal regions where the loci mainly responsible for Down syndrome are thought to be located.

In contrast to the high frequency of foetal malformations affecting the CNS and/or heart as the result of chromosome imbalance is the almost complete absence of detectable limb anomalies. Although a large number of mutants are known in the mouse that affect the digits and cause polydactyly, oligodactyly, syndactyly, etc., sometimes associated with hemimelia of the tibia or other long bones [56], rather few have the sorts of drastic effects on the limb that could be detected easily in utero. Perhaps the outgrowth of limb buds is less easily disturbed in development than closure of the neural folds or formation of the heart.

No doubt a more detailed examination of foetal progeny from the translocation and other stocks that generate chromosomal imbalance would reveal a wide range of other anomalies as well as those described. This is indeed a promising field for further study, the usefulness of which will continue to increase as further regions of man-mouse homology are recognised.

ACKNOWLEDGMENTS

We are grateful to Dr. Mary F. Lyon for her critical comments and for kindly allowing us to analyse data from a number of her translocation stocks. We are also indebted to Mr. Colin Beechey for his help with the illustrations, with data analysis, and in various other ways.

REFERENCES

1. Buckle VJ, Edwards JH, Evans EP, Jonasson JA, Lyon MF, Peters J, Searle AG, Wedd NS: Chromosome maps of man and mouse II. Clin Genet 26:1, 1984.
2. Snell GD, Bodemann E, Hollander W: A translocation in the house mouse and its effect on development. J Exp Zool 67:93, 1934.
3. Snell GD, Picken DI: Abnormal development in the mouse caused by chromosome unbalance. J Genet 31:213, 1935.
4. Searle AG: Numerical variants and structural rearrangements. In Green MC (ed): "Genetic Variants and Strains of the Laboratory Mouse." Stuttgart: Fischer, 1981, p 324.

5. Gropp A, Kolbus U, Giers D: Systematic approach to the study of trisomy in the mouse. II. Cytogenet Cell Genet 14:42, 1975.
6. Searle AG, Beechey CV: The use of Robertsonian translocations in the mouse for studies on non-disjunction. Cytogenet Cell Genet 33:81, 1982.
7. Beechey CV, Searle AG: Effects of 0–4 copies of chromosome 15 on mouse embryonic development. Cytogenet Cell Genet, 1987 (in press).
8. Searle AG: Chromosomal mutations: the genetic approach. In de Serres FJ (ed): "Chemical Mutagens," vol. 10. New York: Plenum, 1986, p 129.
9. Searle AG: Nature and consequences of induced chromosomal damage in mammals. Genetics 74:173, 1974.
10. Lyon MF, Meredith R: Autosomal translocation causing male sterility and viable aneuploidy in the mouse. Cytogenet 5:335, 1966.
11. Epstein CJ: Mouse monosomies and trisomies as experimental systems for studying mammalian aneuploidy. Trends Genet 1:129, 1985.
12. Epstein, CJ: "The Consequences of Chromosome Imbalance." Cambridge: Cambridge University Press, 1986.
13. Baranov VS: Chromosomal control of early embryonic development in mice. I. Experiments on embryos with autosomal monosomy. Genetika 61:165, 1983.
14. Baranov VS: Chromosomal control of the early development of mammals. Ontogenez 14:573, 1983.
15. Dyban AP, Baranov VS: "Cytogenetics of Mammalian Development." Moscow: Nauka, 1978 (in Russian).
16. Magnuson T, Debrot S, Dimpfl J, Zweig A, Zamora T, Epstein CJ: The early lethality of autosomal monosomy in the mouse. J Exp Zool 236:353, 1985.
17. Baranov VS: Chromosome control of early embryonic development in mice. II. Experiments on embryos with structural aberrations of autosomes 7, 9, 14 and 17. Genet Res, Camb 41:227, 1983.
18. Epstein CJ, Travis B: Preimplantation lethality of monosomy for mouse chromosome 19. Nature 280:144, 1979.
19. Magnuson T, Smith S, Epstein CJ: The development of monosomy 19 mouse embryo. J Embryol Exp Morphol 69:223, 1982.
20. Debrot S, Epstein CJ: Tetrasomy 16 in the mouse: a more severe condition than the corresponding trisomy. J Embryol Exp Morphol 91:169, 1986.
21. Gropp A: Value of an animal model for trisomy. Virchows Arch [A] 395:117, 1982.
22. Putz B, Krause G, Garde T, Gropp A: A comparison between trisomy 12 and vitamin A induced exencephaly and associated malformations in the mouse embryo. Virchows Arch [A] 368:65, 1980.
23. Hongell K, Gropp A: Trisomy 13 in the mouse. Teratology 26:95, 1982.
24. Miyabara S, Gropp A, Winking H: Trisomy 16 in the mouse fetus associated with generalized edema and cardiovascular and urinary tract anomalies. Teratology 25:369, 1982.
25. Gropp A, Winking H, Herbst EW, Claussen CP: Murine trisomy: developmental profiles of the embryos, and isolation of trisomic cellular systems. J Exp Zool 228:253, 1983.
26. Gropp D, Gropp A, Winking H: Trisomy 18 of the mouse embryo. Mouse News Lett 64:70, 1981.
27. White BJ, Tjio J-H, Van der Water LC, Crandall C: Trisomy 19 in the laboratory mouse. I. Frequency in different crosses at specific developmental stages and relationship of trisomy to cleft palate. Cytogenet Cell Genet 13:217, 1974.

28. Tease C, Fisher G: A long-lived trisomy 19 mouse. Mouse News Lett 72:105.
29. Sankaranaryanan K: "Genetic Effects of Ionizing Radiation in Multicellular Eukaryotes and the Assessment of Genetic Radiation Hazards in Man." Amsterdam: Elsevier, 1982, p 163.
30. Fundele R, Bucher T, Gropp A, Winking H: Enzyme patterns in trisomy 19 of the mouse. Dev Genet 2:291.
31. Searle AG, Beechey CV: Non-complementation phenomena and their bearing on non-disjunctional effects. In Dellarco VL, Voytek PE, Hollaender A (eds): "Aneuploidy." New York:Plenum, 1985, p 363.
32. Searle AG, Ford CE, Beechey CV: Meiotic disjunction in mouse translocations and the determination of centromere position. Genet Res, Camb 18:215, 1971.
33. Boer P de, Maar PHMD de: A histological study of embryonic death caused by heterozygosity for the T26H reciprocal mouse translocation. J Embryol Exp Morphol 35:595, 1976.
34. Oshimura M, Takagi N: Meiotic disjunction in T(14;15)6Ca heterozygotes and fate of chromosomally unbalanced gametes in embryonic development. Cytogenet Cell Genet 15:1, 1975.
35. Searle AG, Beechey CV: Map of reciprocal translocations and insertions. In Green MC (ed): "Genetic Variants and Strains of the Laboratory Mouse." Stuttgart: Fischer, 1981, p 360.
36. Nesbitt MN, Francke U: A system of nomenclature for band patterns of mouse chromosomes. Chromosoma 41:145, 1973.
37. Committee on Standardized Nomenclature for Mice: Standard karyotype of the mouse, *Mus musculus*. J Hered 63:69, 1972.
38. Kirk KM: "The Genesis and Effects of Unbalanced Genomes Resulting From Reciprocal Translocations in the Mouse." PhD Thesis, Reading University, Reading, 1984.
39. Lyon MF: unpublished data.
40. Searle AG, Beechey CV, Evans EP, Kirk M: Two new X-autosome translocations in the mouse. Cytogenet Cell Genet 35:279, 1983.
41. Kaufman MH: The incidence of chromosomally unbalanced gametes in T(14;15)6Ca heterozygote mice. Genet Res, Camb 27:77, 1976.
42. Eicher EM, Washburn LL: Personal communication. Mouse News Lett 56:43, 1977.
43. Lyon MF, Sayers I: Tertiary trisomics in T(1;17)190H. Mouse News Lett 57:17, 1977.
44. Boer P de: Fertile tertiary trisomy in the mouse. Cytogenet Cell Genet 12:435, 1973.
45. Boer P de, van der Hoeven FA, Chardon JAP: The production, morphology, karyotypes and transport of spermatozoa from tertiary trisomic mice and the consequences for egg fertilization. J Reprod Fertil 48:249, 1976.
46. Beechey CV, Kirk M, Searle AG: A reciprocal translocation induced in an oocyte and affecting fertility in male mice. Cytogenet Cell Genet 27:429, 1980.
47. Cattanach BM: A test of distributive pairing between two specific non-homologous chromosomes of the mouse. Cytogenet Cell Genet 6:67, 1967.
48. Eicher EM, Washburn LL: Assignment of genes to regions of mouse chromosomes. Proc Natl Acad Sci USA 75:946, 1978.
49. Searle AG, Beechey CV, Boer P de, Rooij DG de, Evans EP, Kirk M: A male-sterile insertion in the mouse. Cytogenet Cell Genet 36:617, 1983.
50. Cattanach BM: Position effect variegation in the mouse. Genet Res, Camb 23:291, 1974.
51. Beechey CV, Boer P de: Research news. Mouse News Lett 69:55, 1983.
52. Cattanach BM: Snaker: A dominant abnormality caused by chromosomal imbalance. Z Vererbungsl 96:275, 1965.

53. Russell LB, Montgomery CS, Raymer GD: Analysis of the albino-locus region of the mouse. IV. Characterisation of 34 deficiencies. Genetics 100:427, 1982.
54. Russell LB, Russell WL, Kelly EM: Analysis of the albino-locus region of the mouse. I. Origin and viability. Genetics 91:127, 1979.
55. Schmid W, Müller G, Schutz G, Gluecksohn-Waelsch S: Deletions near the albino locus on chromosome 7 of the mouse affect the level of tyrosine aminotransferase mRNA. Proc Natl Acad Sci USA 82:2866, 1985.
56. Green MC: Catalog of mutant genes and polymorphic loci. In Green MC (ed) "Genetic Variants and Strains of the Laboratory Mouse." Stuttgart: Fischer, 1981, p 8.
57. Russell LB, Russell WL, Cacheiro NLA, Vaughan CM, Popp RA, Jacobson KB: A tandem duplication in the mouse. Genetics 80:571, 1975.
58. Cox DR, Epstein LB, Epstein CJ: Genes coding for sensitivity to interferon (IfRec) and soluble superoxide dismutase (SOD-1) are linked in mouse and man and map to mouse chromosome 16. Proc Natl Acad Sci USA 77:2168, 1980.
59. Cox DR, Goldblatt D, Epstein CJ: Chromosomal assignment of mouse PRGS: further evidence for homology between mouse chromosome 16 and human chromosome 21. Am J Hum Genet 33:145A, 1981.
60. Hagermeijer A, Smit EME: Partial trisomy 21: Further evidence that trisomy of band 21q22 is essential for Down's phenotype. Hum Genet 38:15, 1977.
61. Dyban AP, Baranov VS: "Cytogenetics of Mammalian Embryonic Development." Oxford: Clarendon Press, 1987.
62. Bennett D: The T-locus of the mouse. Cell 6:441, 1975.
63. Lyon MF, Glenister PH, Loutit JF, Evans EP, Peters J: A presumed deletion covering the *W* and *Ph* loci of the mouse. Genet Res 44:161, 1984.
64. Schinzel A: "Catalogue of Unbalanced Chromosome Aberrations in Man." Berlin: de Gruyter, 1984.

The Cytogenetics of Mammalian Autosomal
Rearrangements, pages 769–801

Chapter 26
Mendelian Disorders and Autosomal Structural Rearrangements

Jean de Grouchy and Catherine Turleau

Autosomal disorders mapped through visible structural chromosomal rearrangements are reviewed. The term *Mendelian* refers to their indexing in McKusick's Catalogue. Their monogenic etiology is, however, not proven in most instances, and their mode of inheritance (if any) not well defined. They include retinoblastoma and deletion of band 13q14, nephroblastoma/aniridia and deletion 11p13, Prader-Willi syndrome and deletion 15q11, Langer-Giedion syndrome and deletion 8q24.1, DiGeorge syndrome and deletion 22q11, the cat-eye syndrome and proximal trisomy 22q, Miller-Dieker lissencephaly and deletion 17p13, and Beckwith-Wiedemann syndrome and trisomy 11p15. Lastly, we shall consider tentative but not confirmed assignments, tentative assignments secondarily disproven, and associations between a precise clinical feature and aneusomy for one discrete band.

INTRODUCTION

"The Morbid Anatomy of the Human Genome" by Victor McKusick [1] is the first attempt to prepare a map of genetic diseases in man. It is based essentially on mapping of the primary gene product, the deficiency of which is responsible for the disease. These primary products are most often enzymes, and less frequently structural proteins or other types of proteins that are neither enzymatic nor structural, sensu stricto. The corresponding diseases are enzymopathies, osteochondrodysplasias, and such disorders as the hemoglobinopathies, and immunological and endocrinological disorders. All of these are classical Mendelian traits transmitted in a recessive fashion for most enzymopathies, a dominant fashion for many osteoarticular disorders, and an intermediary fashion for, say, the hemoglobinopathies.

Laboratoire de Cytogénétique Humaine et Comparée, Hôpital Necker-Enfants-Malades, 75743 Paris Cedex 15, France

The aim of this chapter is to review a particular group of autosomal disorders mapped through visible structural chromosomal rearrangements. This group of diseases is characterized by several features. The primary gene product is not known. The monogenic etiology of the disorder is not proven in most cases, except for retinoblastoma. By way of consequence, the mode of inheritance (if any) is not well defined (except again for retinoblastoma). The term *Mendelian* used in the title refers solely to the indexing of the disorders under consideration in McKusick's Catalogue [2]. X-linked diseases are not considered in this volume devoted to autosomal rearrangements.

Precise mapping of these morbid entities has been possible owing to the refinement of cytogenetic techniques that are the basis of the new "microcytogenetics" [3]. The types of chromosomal rearrangements thus identified are diverse. The most frequent are interstitial deletions, which can be very subtle, at the limit of optic microscope definition. In fact, there is a great variability in the size of these deletions, and in some instances they had been observed even before the banding techniques era. A remarkable example is retinoblastoma, for which an association with a D group chromosome deletion was suggested as early as 1963 by Lele and co-workers [4]. It was only some 15 years later that the precise regional assignment was identified as 13q14 [5].

Other types of rearrangements are translocations resulting in a structural or functional lesion of a gene or group of genes. Ring chromosomes and small bisatellited supernumerary chromosomes and partial trisomies have also been observed.

In this review we shall first consider the most striking instances of specific rearrangements in Mendelian disorders, i.e., retinoblastoma and the deletion of band 13q14, nephroblastoma/aniridia and deletion 11p13, Prader-Willi syndrome and deletion 15q11, Langer-Giedion syndrome and deletion 8q24.1, DiGeorge syndrome and deletion 22q11, the cat-eye syndrome and proximal trisomy 22q, Miller-Dieker lissencephaly and deletion 17p13, and Beckwith-Wiedemann syndrome and trisomy 11p15. Lastly, we shall consider tentative but not confirmed assignments, tentative assignments secondarily disproven, and associations between a precise clinical feature and aneusomy for one discrete band.

TECHNICAL REMARKS

High-resolution banding is obtained by studying mitoses at prometaphase. This is done by synchronizing cell divisions, using an inhibitor such as thymidine or methotrexate, or bromodeoxyuridine (BrdU) [6,7]. The block is then released and the cells are harvested usually after 5–7 h to obtain a maximum number of mitoses at the proper stage. Banding is then obtained by classical methods or by BrdU incorporation followed by acridine-orange or fluorescence-plus-Giemsa-staining. These various procedures allow the identification of 850 or more discrete bands per haploid complement [8].

RETINOBLASTOMA AND DELETION 13q

Retinoblastoma (McKusick No. 18020) is an embryonic retinal tumor that occurs in infancy or early childhood. Its incidence is around 1 per 20,000 births. The majority of cases are phenocopies owing to nontransmissible somatic mutations. They are always unilateral and unifocal. Transmissible forms owing to germinal mutations are inherited in a dominant fashion or result from de novo mutations [9]. Lastly, a small number of cases are due to a chromosome rearrangement. In a number of these, retinoblastoma is associated with multiple congenital malformations and mental retardation, as was the case in the first reported large deletions [10]. But this is not always so, since, as we shall see, in very minute deletions retinoblastoma can be the only clinical feature. Conversely, mental retardation and malformations are not the rule in noncytogenetic forms of retinoblastoma.

The age of onset in cytogenetic forms is comparable to that observed in the other transmissible forms of the tumor [11]. One intriguing difference could be the lower incidence of bilateral tumors in cytogenetic forms as compared with genic forms [12]. A possible explanation will be discussed further on.

The frequency of detected cytogenetic forms of retinoblastoma has shown a threefold increase since the high-resolution banding techniques came into use. Their observed frequency may be as high as on the order of 10% of all retinoblastomas [11].

Different types of rearrangements can result in retinoblastoma through involvement of 13q14 (Table I) [13–31]. The most common are de novo interstitial deletions (Figs. 1,2). They are of variable length but always include 13q14.2 and the adjacent part of 13q14.1, which constitute the SRO (smallest region of overlapping) [5,27]. Their frequency is on the order of 80–90% of all cytogenetic forms.

De novo apparently balanced translocations with a breakpoint in or near 13q14 have also been observed (Fig. 2). They can involve another autosome or an X. Retinoblastoma may then result from a submicroscopic deletion or gene disruption. Another explanation for t(X;13) is position effect with inactivation spreading into 13q14 when the breakpoint is located in the vicinity of this band.

TABLE I. Cytogenetic findings in retinoblastoma patients

	Reference
De novo rearrangements	
Interstitial deletions	
De novo, apparently balanced translocations	
t(1;13) (p22q12)	Davison et al. (1979) [13]
t(13;18) (q14.1;q12.2)	Motegi et al. (1982) [14]
t(2;13) (p24.3;q14.2)	Turleau et al. (1985) [11]
t(X;13) (p22;q12)	Cross et al. (1977) [15]
	Nichols et al. (1980) [16]
	Mohandas et al. (1982) [17]
t(X;13) (p22;q12)	Hida et al. (1980) [18]
	Ejima et al. (1982) [19]
t(X;13) (p11.21;q12.3)	Kajii et al. (1985) [20]
Parental rearrangements	
Reciprocal translocation	
t(5;13) (p13;q14.3)	Noël et al. (1976) [21]
Maternal mosaicism	Michalova et al. (1982) [22]
Insertions	
inv ins(13) (q22q12q14)	Sparkes et al. (1979) [23]
	Hoegerman (1979) [24]
ins(12;13) (p11.21;q12.3q22.1)	Riccardi et al. (1979) [25]
ins(3;13) (p12;q13.1q14.3)	Strong et al. (1981) [26]
ins(20;13) (p12;q1307q14.3)	Rivera et al. (1981) [27]
	Turleau et al. (1985) [11]
ins(16;13) (q12.2;q13q21.1)	Turleau et al. (1983 [28], 1985 [11])
ins(5;13) (p11;q14.1q14.3)	Turleau et al. (1985) [11]
"Unrelated" translocations	
del(13) (q13q21.3),t(2;10)(q21;q24)	Grouchy et al. (1980) [29]
Without del(13)	
t(3;12) (q29;q15)pat	Bec et al. (1977) [30]
t(2;9) (q11;p11)de novo	Balestrazzi et al. (1984) [31]
t(8;15) (q24.2;q21.1)pat	Personal observation (unpublished)

Mosaicism has been reported in several instances.

A balanced parental rearrangement may lead to monosomy 13q14 either by malsegregation or by structural recombination (*aneusomie de recombinaison*). A reciprocal translocation and a maternal mosaicism have been observed, but the most frequent mechanism is insertion. Insertion may involve another chromosome or the same chromosome 13. As in the case of de novo deletions, breakpoints on 13q are variable.

Such insertions allow us new insight into the understanding of the familial occurrence of retinoblastoma. Indeed, pedigrees have been known with affected individuals related through unaffected parents. Until now these

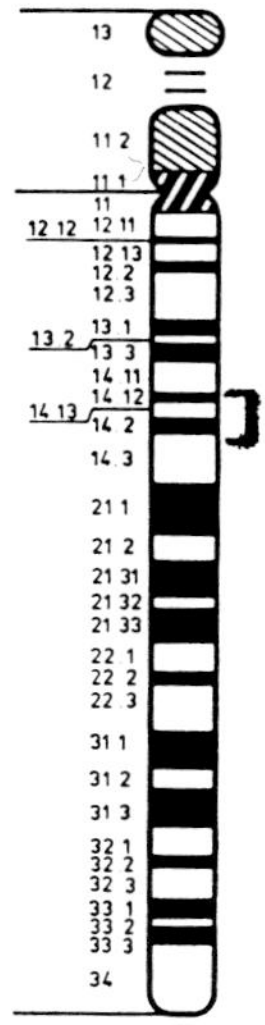

13

Fig. 1. Smallest region of overlapping (SRO) of deletions in retinoblastoma patients.

pedigrees had no convincing explanations, except for the notion of "delayed mutation." The demonstration of visible chromosomal insertions accounts for this mode of inheritance and it is probable that identical but submicroscopical insertions may also exist. It is foreseeable that in the near feature molecular tools will confirm this hypothesis.

It is noteworthy that in the reported families, individuals occur that are trisomic for the inserted segment and that their phenotype is most often normal or near normal [25–28].

Also of interest is an apparent increase of independent chromosomal imbalance in retinoblastoma patients: trisomy 21, balanced reciprocal translocations with and without deletion 13.

The identification of small deletions involving 13q14 allowed the localization of the gene responsible for an enzyme, esterase D, within this same band [32]. This was possible through gene-dosage effects in patients with retinoblastoma and short deletions of 13q14. Further studies confirmed the very strong linkage of both genes. A few exceptions, however, are known [33], which demonstrated that two different genes are indeed involved. They also showed that the gene for esterase D is proximal with regard to the retinoblastoma gene. In one family prenatal diagnosis was based on the gene-dosage

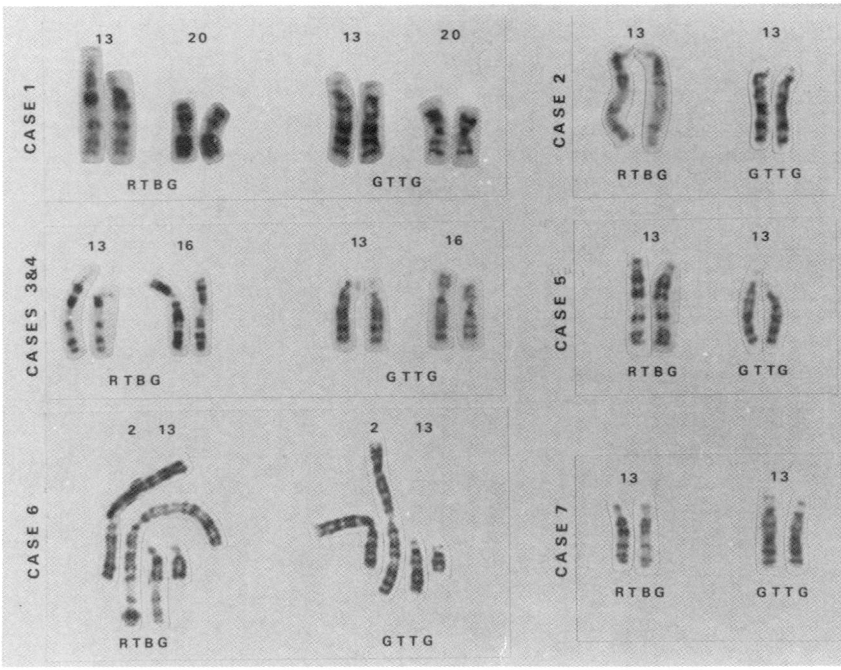

Fig. 2. Chromosomal rearrangements observed in retinoblastoma patients. Cases 2, 5, and 7, deletions; cases 1, 3, and 4, parental insertion; case 6, de novo translocation. (Reproduced from Turleau et al. [47], with permission of the publisher.)

effect of esterase D [34]. Low esterase D levels can lead to the detection of small deletion carriers [35].

Esterase D also shows an electrophoretic polymorphism. This allowed the demonstration of a very close linkage between the esterase D gene and the gene responsible for the dominant form of retinoblastoma, and thus localization of the latter within 13q14 [36]. It is currently conceived that this gene is unique and is involved in both the cytogenetic and "Mendelian" forms of the tumor.

The dosage of esterase D activity in the tumors first suggested that the deletion becomes homozygous in the tumor cell precursors [37]. This was further confirmed by using DNA probes, which showed that homozygosity is not restricted to the gene but also involves the entire chromosome 13 or large segments of it. This can arise by duplication of the deleted or mutated retinoblastoma gene bearing chromosome and loss of the normal homologue or by other mechanisms [38]. The deficit in bilateral forms can thus be explained by expression of recessive lethal genes, which limits the number of potential cancerous cells.

ANIRIDIA/WILMS TUMOR AND DELETION 11p3

Aniridia (McKusick No. 10620) is a rare congenital malformation that occurs most often as a sporadic condition, or as a dominant trait. Aniridia is frequently associated with Wilms tumor, which may itself be occasionally transmitted as a dominant trait [39]. Brusa and Torricelli [40] were the first to report an association between aniridia and nephroblastoma or Wilms tumor. This was confirmed by Miller et al. [41], who observed six cases of aniridia in a series of 440 patients with Wilms tumor, an incidence of 1/73. François et al. [42] reviewed some 50 cases of aniridia/Wilms tumor association (McKusick No. 19407). Francke et al. [43] and Riccardi et al. [44] demonstrated a short interstitial deletion of 11p in such patients. This chromosomal etiology was then confirmed by several reports (for review, see [45–47]) and the SRO of the deletion was pinpointed to band 11p13 (Figs. 3,4). The main features of the syndrome caused by deletion 11p13 are aniridia, a proneness to Wilms tumor, genital anomalies, and mental retardation. This complex syndrome has been variously defined as WAGR, AGR, or AWTA complex according to different authors.

As in the case of retinoblastoma the gene coding for an enzyme was shown to be located in the same band. Wieacker et al. [48] first mapped the catalase (CAT) gene to 11p by interspecific cellular hybridization, and Junien et al. [49] regionally assigned the gene to 11p13 by gene-dosage analysis in patients with imbalances for this region. This was confirmed by studies of series of patients [50,51].

In the case of retinoblastoma, however, the presence of the esterase D locus ultimately allowed the demonstration that the gene responsible for the dominant form of the tumor is the same as that in retinoblastoma owing to deletion 13q14. In the case of the WAGR complex it has not yet been possible to know with certainty whether the aniridia gene is the same as that responsible for the dominantly inherited form of aniridia, nor whether the nephroblastoma "locus" is the same as that responsible for the sporadic or dominant forms of the tumor.

Clinical Aspects [47]

Aniridia is present in all cases except one [52]. In practically all cases it is associated with nystagmus, cataracts, or glaucoma. As mentioned above, the question of two loci responsible for aniridia remains open: there is one

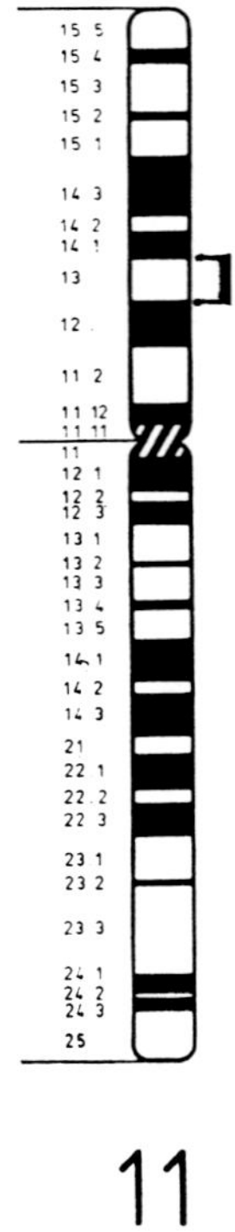

Fig. 3. SRO of deletions in WAGR patients.

suggestion of a gene assignment to 2p [53], but this has never been confirmed. The association of aniridia and deletion 11p13 demonstrates the existence of a locus in this band. That this locus is specific is shown by the observation of families with dominant transmission of isolated aniridia linked with translocations with a breakpoint in 11p13 [54,55].

In contrast with the retinoblastoma/deletion 13q14 situation, where the tumor is constant, in the aniridia/deletion 11p13 complex, nephroblastoma occurs only in approximately one-third of the patients. This may be related to incomplete penetrance of Wilms tumor, as shown by discordance in a pair of monozygotic twins [56]. Gonadoblastoma is occasionally observed but may be related to genital anomalies [57,45].

The sex-ratio is much in favor of males (2M/1F). Genital anomalies are practically constant in XY patients and vary in severity from cryptorchidism with or without hypospadias to male pseudohermaphroditism in approximately one-fifth of the cases. In girls, streak gonads were reported in only

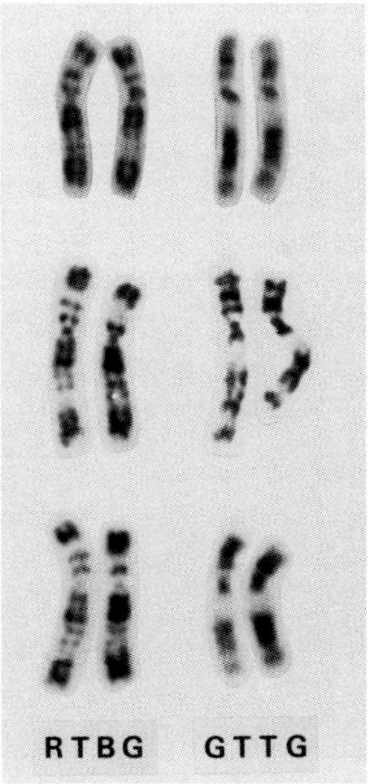

Fig. 4. Three examples of deletion 11p13 in WAGR complex. Deleted chromosomes on the left.

one case, but they are not systematically searched for. Two adolescent girls are known to have scanty, irregular menses [47,58].

Mental retardation is variable. Additional congenital malformations are occasionally present.

Cytogenetics (Table II) [59–70]

The most frequent rearrangement is, as in the case of retinoblastoma, a de novo deletion variable in size with an SRO in the midportion of 11p13. Complex de novo rearrangements can also be observed, with or without visible deletion. Parental rearrangements are possible, mostly insertions, the recipient chromosome being another autosome or the same chromosome 11. These parental rearrangements are responsible for familial cases of WAGR. Trisomy for the inserted segment is theoretically possible but does not seem to have been reported as yet. Unrelated de novo rearrangements are known to occur in addition to the 11p13 deletion, but there is no obvious explanation for their high incidence. An excess of various translocations not involving 11p13 has also been suggested in cases of nephroblastoma without deletion 11p13.

TABLE II. Cytogenetic findings in (W)AGR patients

	Reference
De novo rearrangements	
Interstitial deletions	See for review: Turleau et al. (1984) [47] Nakagome et al. (1984) [59]
Complex rearrangements	
t(8;11),del(11) (p12p14)	Gödde-Salz and Behnke (1981) [60]
dir ins(11;2) (p13;q13q23)	Niikawa et al. (1982) [61]
Associated rearrangements	
t(8;11),del(11) (p13p14)	Francke et al. (1979) [43]
t(4;7;15),del(11) (p13p14)	Turleau et al. (1981) [45]
inv(7)del(11)(p13p14)	Riccardi et al. (1980) [62]
t(1;2),del11p13	Waziri et al. (1981) [63]
t(2;17),del(11) (p11.2p13)	Narahara et al. (1984) [64]
Mosaicism	Nakagome et al. (1984) [59]
Translocations without del11p13	
t(8;5)	Giangiacomo et al. (1975) [65]
t(7;13)	Bernard et al. (1984) [66]
Parental rearrangements	
ins(3;11) (q22;p11.3p14.1)mat	Hittner et al. (1979) [67]
ins(2;11) (q32;p13p14.1)mat	Yunis and Ramsay (1980) [68]
ins(16;11) (p13;p11.3p14)mat	Kousseff and Agatucci (1981) [69]
ins(12;11) (p11.2;p12p14.2)mat	Nakagome and Nagahara (1985) [70]

Some rare observations are particularly noteworthy and allow a better understanding of the inner organization of the complex. One is the abovementioned observation of apparently balanced familial translocations with one breakpoint in 11p13 and associated with isolated aniridia [54,55]. Another observation is that of a male patient with deletion 11p13, nephroblastoma, and ambiguous genitalia but without aniridia [52]. Furthermore, associations of aniridia or iris dysplasia and Wilms tumor are known without visible deletion and normal catalase level [71]. These observations strongly suggest 1) the existence of several genes liable to be dissociated by chromosome breakage and 2) the existence of submicroscopic deletions.

The following order for these genes has been suggested: cen→ catalase→ Wilms tumor→ aniridia [47]. Another gene order largely based on lactate dehydrogenase (LDHA) regional assignment [64] can probably be discarded with the new, more distal assignment of this locus [72].

The locus responsible for genital ambiguity has not yet been identified. Yet, in one instance at least, deficiency of testosterone steroid receptors has been demonstrated [73]. Of possible significance is the observation that the gene coding for the subunit of follicle-stimulating hormone (FSH) is regionally assigned in close vicinity of the WAGR loci.

As for retinoblastoma and chromosome 13, homozygosity for chromosome 11 is observed in nephroblastoma tumor cells [74].

BECKWITH-WIEDEMANN SYNDROME

Beckwith-Wiedemann syndrome (BWS) (McKusick No. 13065) was described independently in 1963 by Beckwith [75] in three infants and in 1964 by Wiedemann [76] in three siblings. It has also been called the EMG syndrome because of its three major signs: exomphalos, macroglossia, and gigantism. Other signs are prematurity and hydramnios; neonatal hypoglycemia; microcephaly; facial naevus flammeus; fissures on the ear lobes; visceromegaly; gastrointestinal anomalies, mainly malrotation; posterior diaphragmatic eventration; cardiac anomalies; genitourinary anomalies; mental retardation; and increased frequency of neoplasms, notably nephroblastoma [77].

Trisomy 11p is a rare chromosomal disorder. Since the first report by Falk et al. [78], seven patients were described with this anomaly, however, there was no reference to BWS. Waziri et al. [79] were the first to relate specifically the clinical syndrome and trisomy for 11p15, by reporting two new patients with trisomy 11p15 and by emphasizing that these and the previously reported patients, indeed, had features of the syndrome. This was confirmed by Turleau et al. [80], who also reported on two patients with the syndrome (Fig. 5). One patient had a de novo unbalanced rearrangement, which later proved to be a trisomy 11p15 (Fig. 6). The other had trisomy 11p15 owing to a paternal translocation. A third report [81] concerned two sibs with partial trisomy 11p15 resulting from a paternal translocation.

Retrospectively, patients reported with trisomy 11p15, associated or not with other imbalance, exhibit signs of BWS (review in [47]). Conversely, cytogenetic surveys of BWS patients failed to show evidence of any chromosomal rearrangement, in particular of chromosome 11 [82,83]. A comparison of BWS patients with and without trisomy 11p15 shows that the main clinical differences are a higher incidence of cleft lip or palate, hypertelorism, congenital heart disease, and mainly mental retardation in chromosomal forms. Excessive growth advance is observed more frequently in BWS with normal karyotypes. It is noteworthy that six BWS patients with nephroblastoma had normal karyotypes [59,82] (personal unpublished observations).

In summary, BWS appears to be heterogeneous from a cytogenetic standpoint: most patients fail to show any chromosome imbalance; a minority are

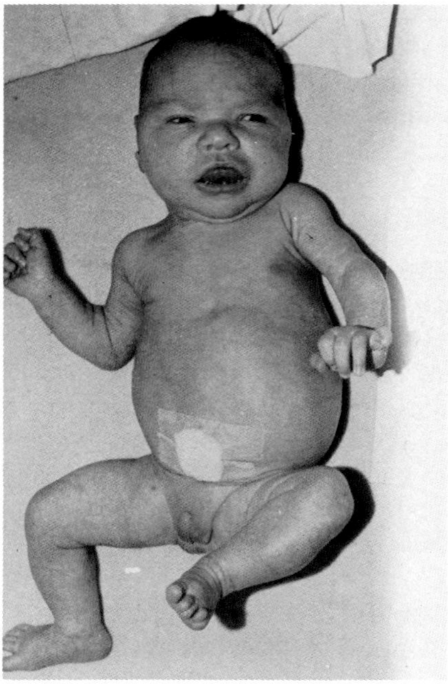

Fig. 5. Beckwith-Wiedemann patient (patient 1 in Turleau et al. [80]).

trisomic for at least part of 11p15, as a result of a parental pericentric inversion or translocation, or by a de novo rearrangement.

This situation remains most intriguing when considering the genic content of 11p, in particular the genes coding for insulin, insulin-like growth factor 2, and the HRAS1 oncogene, which all could be responsible for specific features of the syndrome. They were, however, present in two copies in patients with no visible chromosome imbalance [84]. There is no immediate explanation for this heterogeneity of BWS.

Proneness to malignancies is a common feature of BWS and deletion 11p13/aniridia. In this respect it must be emphasized that homozygosity for chromosome 11 is observed not only in nephroblastoma but also in other tumors that are liable to occur in BWS, namely, rhabdomyosarcoma and hepatoblastoma [85].

The genetic phenomena associated with the short arm of chromosome 11 are very provocative. The presence of the HRAS1 oncogene and the occurrence of tumors with duplication of 11p15 as well as with monosomy for 11p13 point to the possibility of a regulatory mechanism between an oncogene

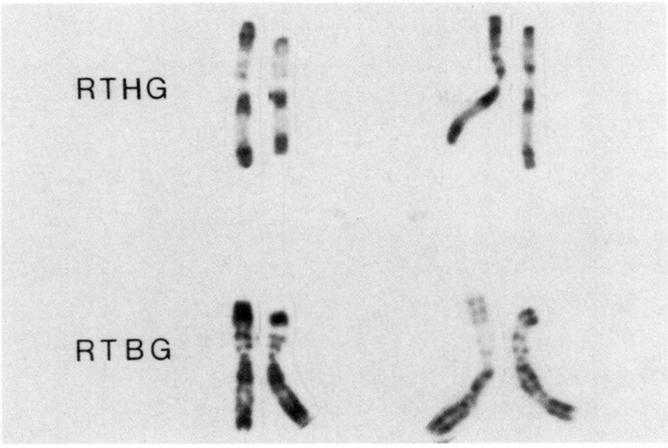

Fig. 6. Trisomy of 11p15 (see text) in a patient with Beckwith-Wiedemann syndrome. Abnormal chromosome on the left (patient 1 in Turleau et al. [80]).

in 11p15 and a supposed anti-oncogene in 11p13. The demonstration that the number of copies of human chromosome 11 controls tumorigenicity in somatic cell hybrids is of interest in this particular context [86].

PRADER-WILLI SYNDROME AND DELETION 15q11

The syndrome originally described in 1956 by Prader, Labhart, and Willi [87] in nine Swiss children is not a rare condition (McKusick No. 17627). The cardinal features are neonatal hypotonia with feeding difficulties, respiratory distress, diminished spontaneous movements, and weak or absent cry; hypogonadism and cryptorchidism; mental retardation; obesity with hyperphagia appearing around 2 yr of age; and short stature. Gestation is characterized by diminished intrauterine movements and frequent breech presentation. The facies is distinctive with a narrow bifrontal diameter, almond-shaped eyes with strabismus, and a triangular mouth. The hands and feet are small (acromicria) (Fig. 8).

In 1976, Hawkey and Smithies [88] reported the first t(15q15q) in a patient with Prader-Willi syndrome (PWS). A variety of structural rearrangements involving chromosome 15 have since been reported. They are present in around 60% of PWS patients. As shown in Table III these rearrangements include interstitial deletions, reciprocal and Robertsonian translocations, small additional bisatellited chromosomes, and pericentric inversions. They are extensively reviewed by Mattei et al. [89].

TABLE III. Cytogenetic findings in 94 patients with Prader-Willi syndrome reported in the literature

Interstitial deletions	52
Reciprocal translocations	
Unbalanced	17
Balanced	2
Robertsonian translocations	12
Small bisatellited additional chromosomes	9
Pericentric inversions	2
Total	94

After Mattei et al. (1984) [89].

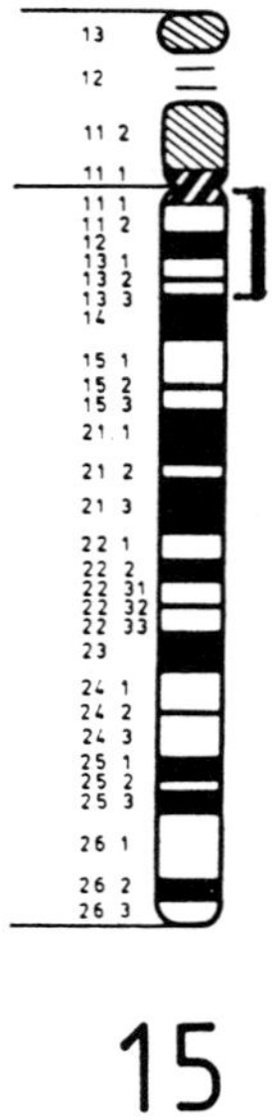

Fig. 7. SRO of deletions in Prader-Willi syndrome patients.

Interstitial deletions, first described by Ledbetter et al. [90] are the most frequent rearrangements. They involve 15q11 to 15q13 (Fig. 7,9). All have occurred de novo, except for one that was due to a t(14;15) [91] responsible for familial PWS.

Reciprocal translocations are usually unbalanced. They result from breakpoints, proximal in 15q and telomeric in the other chromosome. A 3:1 disjunction leads to a 45-chromosome karyotype and proximal monosomy 15q.

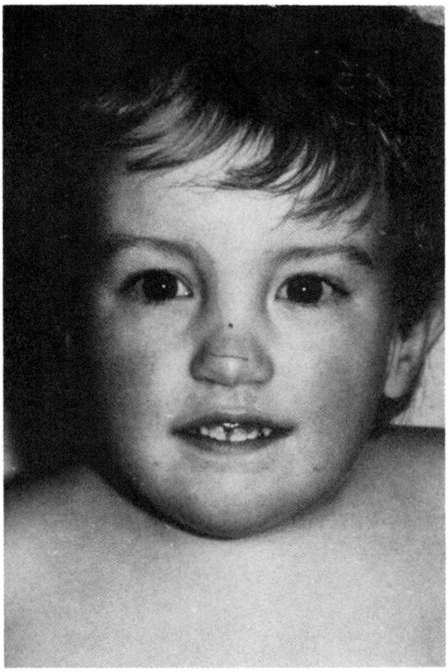

Fig. 8. Patient with deletion 15q11 and Prader-Willi syndrome.

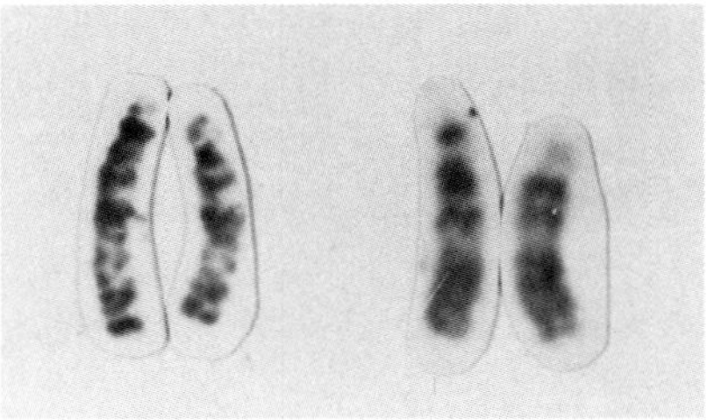

Fig. 9. Chromosomes 15 of patient from Figure 8. Deleted chromosome on the right.

Apparently balanced Robertsonian translocations are another frequent rearrangement in PWS. In fact, prometaphase analysis reveals associated tiny proximal deletions responsible for the clinical features. It is noteworthy that chromosome 15 is rarely involved in true, balanced Robertsonian translocations.

Small bisatellited additional chromosomes (SBACs) are intriguing, since they should apparently result in tetrasomy rather than monosomy. That they derive from chromosome 15 is proven by specific staining. Their size is variable and they may well result from complex rearrangements eventually

TABLE IV. Cytogenetic findings in Langer-Giedion syndrome

De novo rearrangements	Reference
Interstitial deletions	See for review: Bühler and Malik (1984) [102]
Complex rearrangement	
t(1;8;14) (p13.1;p23.3;q24.13;q12)	Schwartz et al. (1985) [105]
Associated rearrangements	
del(8) (q24.11q24.22),ins(8;1) (q22.1;p34.1p32.1)	Zaletajev and Marincheva (1983) [104]
del(8) (q24.11q24.12),t(2;9) (q21;p13)	Bowen et al. (1985) [103]

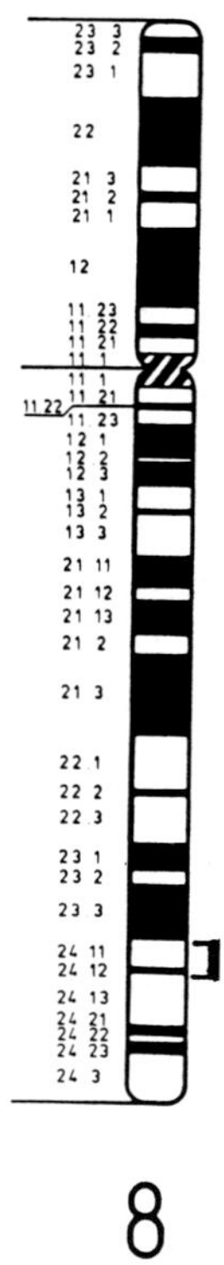

Fig. 10. SRO of deletions in Langer-Giedion syndrome patients.

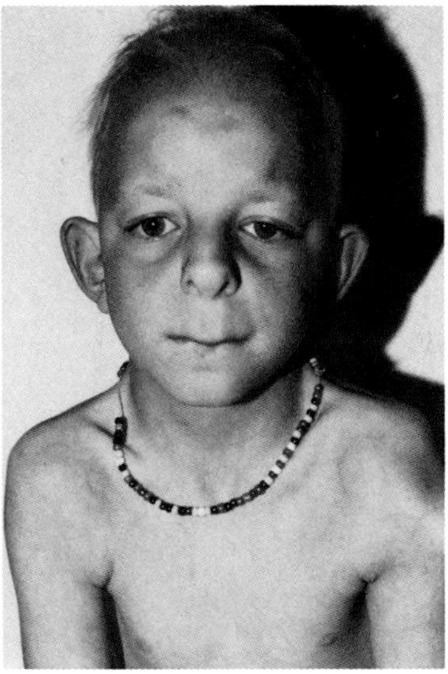

Fig. 11. Patient with deletion 8q and Langer-Giedion syndrome. (Reproduced from Turleau et al. [101], with permission of the publisher.)

concealing a functional monosomy. It must be emphasized, however, that if SBACs are observed in PWS, the majority of them are not observed in this condition but in others with normal or abnormal phenotypes.

The diversity and originality of chromosome rearrangements associated with PWS have led Mattei et al. [89] to suggest that the great lability of the proximal 15q region could be due to the existence of one or several palindromic sequencies.

Here again, as in other microcytogenetic conditions, the question of PWS heterogeneity is raised. In a recent review, Butler et al. [92] contend that patients with chromosomal rearrangements constitute a more homogeneous group than patients with normal karyotypes. They conform better to the classical syndrome described above. These authors further insist on the more frequent light hair and eye color and fair complexion in this group.

Some of the patients with normal karyotypes could correspond either to an erroneous clinical diagnosis or to a misinterpreted karyotype [93]. The difficulty encountered in analyzing the proximal 15q region owing to polymorphism of the short arm and variable degree of contraction of 15q1 must be underlined in this respect. Nonetheless, genuine cases of PWS are ob-

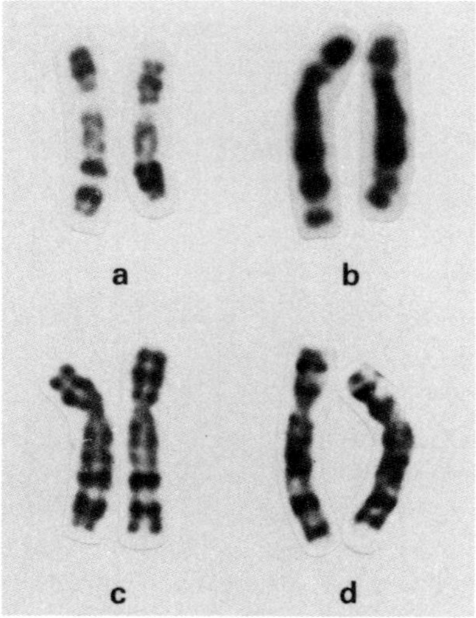

Fig. 12. Chromosomes 8 from two patients with Langer-Giedion syndrome. The largest deletion (top row) corresponds to patient in Figure 11. Patient 1: **(a)** RTBG; **(b)** GBTG.Patient 2: **(c)** RTBG, **(d)** GTTG.

TABLE V. Cytogenetic findings in Miller-Dieker lissencephaly

	Reference
De novo rearrangements	
Rings	
r(17) (p13q25)	Carpenter et al. (1981) [114]
r(17) (p13.3q25.2)	Dobyns et al. (1983) [110]
Deletions	
del(17) (p13.1)	Stratton et al. (1984) [111]
del(17) (p13)	Ketterer et al. (1985) [112]
Translocation	
17p+ (unbanded)	Jellinger and Rett (1976) [115]
Parental rearrangements	
t(7;17) (22.3;p13.2)pat	Dobyns et al. (1983) [110]
	Stratton et al. (1984) [111]
t(15;17) (q26.1;p13.1)pat	Stratton et al. (1984) [111]
t(12;17) (q24.31;p13.3)pat	Stratton et al. (1984) [111]
t(8;17) (q23.1;p13.1)mat	Stratton et al. (1984) [111]
inv(17) (p13.3;q25.1)mat	Greenberg et al. (1986) [113]

served with normal karyotypes, and, to date, it is not justified to rest the diagnosis on the existence of 15q1 monosomy or rearrangement.

LANGER-GIEDION SYNDROME AND DELETION 8q24.1

Giedion [94] reported in 1966 on the association of an abnormal facies, poor scalp hair growth, and specific radiological features. He named the syndrome "tricho-rhino-phalangeal." In 1969, he described a patient with this syndrome associated with multiple exostoses [95]. Independently, Langer [96] found the same association in another patient. The name *Langer-Giedion* or *tricho-rhino-phalangeal type II (TRP II)* was coined by Hall et al. [97], as opposed to type I, essentially without the multiple exostoses, as first delineated by Giedion [94].

Langer-Giedion syndrome (LGS) (McKusick No. 15023) is a very rare condition. The main clinical features are a bulbous, pear-shaped nose with tented alae, a prominent and long philtrum, a thin upper lip, micrognathia, and large protruding ears; sparse scalp hair; multiple exostoses; short stature; clinobrachydactyly; and cone-shaped epiphyses. There is mild to moderate mental deficiency with or without microcephaly (Fig. 11).

All cases are sporadic, except for one pair of monozygous (MZ) twins and an affected father and his daughter [98].

Since the first observation by Bühler et al. [99] of a rearrangement of chromosome 8, some 16 reports confirmed the existence of a deletion (Table IV). The breakpoints varied greatly in the first reports and were defined only when prometaphase analyses were done [100,101]. An agreement was finally reached to assign the critical segment to sub-band 8q24.1 [102,103] (Fig. 10). A visible deletion is observed in the majority of cases (Fig. 12). Only rare patients studied after applying high-resolution banding techniques were reported to have normal karyotypes [98]. The deletion can, however, be very minute at the limit of detection as in one of our two cases [101].

The size of the deletion varies greatly from patient to patient, from q22.3 to q24.2. The TRPII syndrome being considered constant, the larger deletions do not appreciably modify the phenotype, except for mental retardation, which may be absent in the tiny deletions.

As for other microdeletions, an abnormal frequency of complex rerrangements suggests a meiotic chromosome instability in the patients and/or their parents. They include one translocation [103], one insertion (into another site of the deleted 8) [104], and one circular three-chromosome translocation with breakpoints in 8q24.11 and 8q24.13 [105]. All rearrangements were de novo.

Exostoses are not malignancies; nonetheless, they represent abnormal proliferations. Their association with a deletion remains intriguing.

The question has been raised of a common etiology for TRPI and II [102]. There are two suggestions of a deletion of 8q23 or proximal 8q24 in TRPI [102,106], but in the majority of cases the karyotype is normal in this condition (unpublished personal observations).

MILLER-DIEKER SYNDROME AND DELETION 17p13

Lissencephaly means "smooth brain," without convolutions or gyri. In 1963, Miller [107] described this disorder in two sibs. Dieker et al. [108] reported in 1969 a family with three affected members and delineated a syndrome associating lissencephaly and malformations of the heart, kidneys, and other organs, as well as polydactyly and an unusual craniofacial appearance. Microcephaly is constant. The forehead is high and may be narrow in temporal areas. The remarkable ability to fold the forehead skin into many wrinkles is a striking feature of the syndrome. The occiput is prominent. There is hypertelorism. The palpebral fissures slant slightly downward. Corneal clouding can be observed. the nostrils are anteverted. The lips may have a puckered appearance. The ears are slightly rotated posteriorly with anomalies of differentiation. Hirsutism may be extensive. The neurological picture is severe with initial hypotonia, seizures, and progressive decerebration. Death invariably occurs in infancy or childhood (Fig. 14).

In 1980, Jones et al. [109] suggested the name "Miller-Dieker syndrome" (McKusick No. 24720) to distinguish this entity from other types of lissencephaly, of which at least five are known. It is a very rare disorder, considered to be an autosomal recessive trait. Dobyns et al. [110] were the first to report an association between the syndrome and 17p in two of three patients. One had a ring 17 and the other an unbalanced t(7p;17p) (Table V).

The same group further confirmed that Miller-Dieker lissencephaly is due to monosomy 17p13.3 (Figs. 13,15) by reporting one new patient with de novo 17p deletion and by reinvestigating four families already reported in the literature. In three of these they showed familial translocations involving 17p and a different autosome. In the fourth family the clinical diagnosis was later refuted [111]. A further case was briefly reported [112]. This case was associated with a deletion, but the parents' karyotypes were not mentioned. Lastly, the association of an inv[17] and Miller-Dieker syndrome was reported [113].

The chromosomal origin of Miller-Dieker syndrome appears well documented. It is, however, only one form of lissencephaly. This must be considered, in particular when giving genetic counseling. The majority of lissencephalies are recessive autosomal conditions not associated with chromosomal rearrangements permitting prenatal diagnosis.

Fig. 13. SRO of Miller-Dieker syndrome patients.

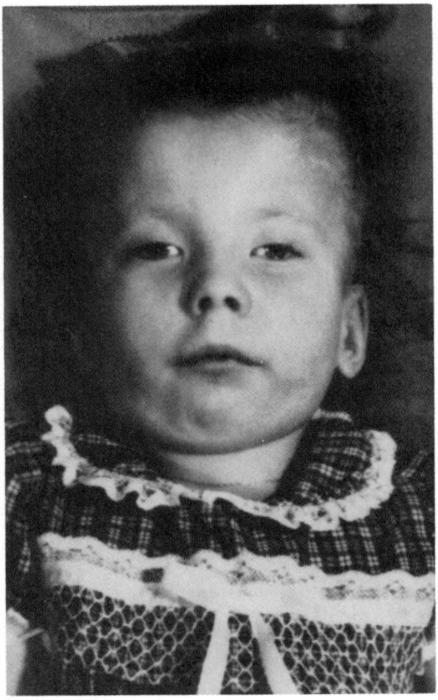

Fig. 14. Patient with deletion 17p13 and Miller-Dieker syndrome. (Reproduced from Stratton et al. [111], with permission of the publishers.)

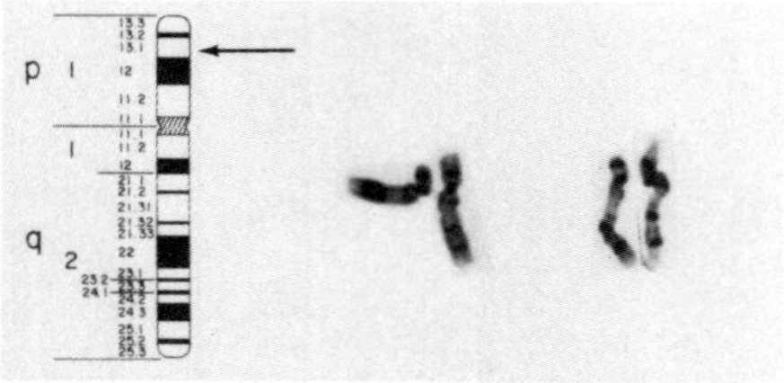

Fig. 15. Chromosomes of patient from Figure 14. Deleted chromosome on the left.

TABLE VI. Cytogenetic findings in DiGeorge syndrome

	Reference
De novo rearrangement	
Translocation	
45, XY,—22,t(3;22) (q29;q11)	Kelley et al. (1982) [118]
Parental rearrangements	
t(20;22) (q11;q11)	La Chapelle et al. (1981) [117]
t(10;22) (q25;q11)pat	Kelley et al. (1982) [118]
t(20;22) (q11;q11)mat	Kelley et al. (1982) [118]
t(18;22) (q12.2;p11.2)pat	Bowen et al. (1986) [120]
Unbalanced familial rearrangement	
45,XX,−22,t(4;22) (q35.2;q11.2)	Greenberg et al. (1984) [119]

DiGEORGE SYNDROME AND DELETION 22q11

The syndrome (McKusick No. 18840) described in 1968 by DiGeorge [116] consists of neonatal hypocalcemic tetany owing to aplasia or hypoplasia of the parathyroids; increased susceptibility to infection owing to aplasia or hypoplasia of the thymus; a characteristic facial dysmorphism with hypertelorism, downward-slanted palpebral fissures, short philtrum, low-set ears, and micrognathia; and cardiovascular malformations, mainly conotruncal and aortic arch anomalies. These malformations are considered to be due to a failure of embryonic differentiation of the third and fourth pharyngeal pouches.

La Chapelle et al. [117] were the first to report an association of partial monosomy 22 and DiGeorge syndrome (DGS). They studied a large family, including four affected children in different sibships, all of whom were carriers of an unbalanced t(20;22) (q11;q11). None of the six unaffected children had this translocation. Since none of the features of DGS is part of the trisomy 20p phenotype, the disease was attributed to monosomy 22q11. This hypothesis was supported by a previously reported child with complete monosomy 22 and several cardinal features of DGS (Table VI).

Kelley et al. [118] confirmed this pioneer observation by publishing the observations of three unrelated patients with DGS and carriers of a deletion of 22q11 owing to balanced parental or unbalanced de novo translocations (Fig. 16).

Greenberg et al. [119] described another family (briefly referred to by Kelley et al.) in which two of three sibs and their mother had complete or partial DGS associated with partial monosomy 22q owing to an unbalanced t(4;22) (q35.2;q11.2).

Bowen et al. [120] recently reported a patient with monosomy 22 resulting from a t(18;22) (q12.2;p11.2)pat.

Thus a deletion of 22q11 is now well documented in DGS. Molecular studies using a probe for the lambda immunoglobulin light chain constant region in two of the above patients suggest that the breakpoint of each rearrangement is proximal to the C lambda locus in 22q11 [121].

The deletion is, however, observed in a small number of cases. Greenberg and Ledbetter [122] consider that the incidence of a chromosomal rearrangement in DiGeorge syndrome is around 10%. In a series of 23 patients these authors observed one rearrangement of chromosome 22 as well as one

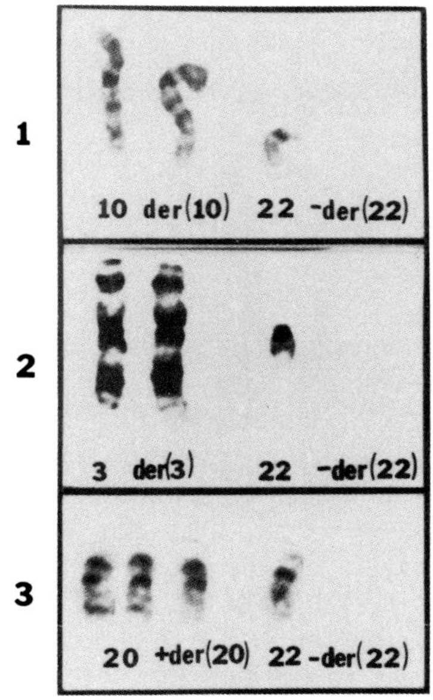

Fig. 16. Examples of translocations in DiGeorge syndrome. (Reproduced from Kelley et al. [118], with permission of the publisher.)

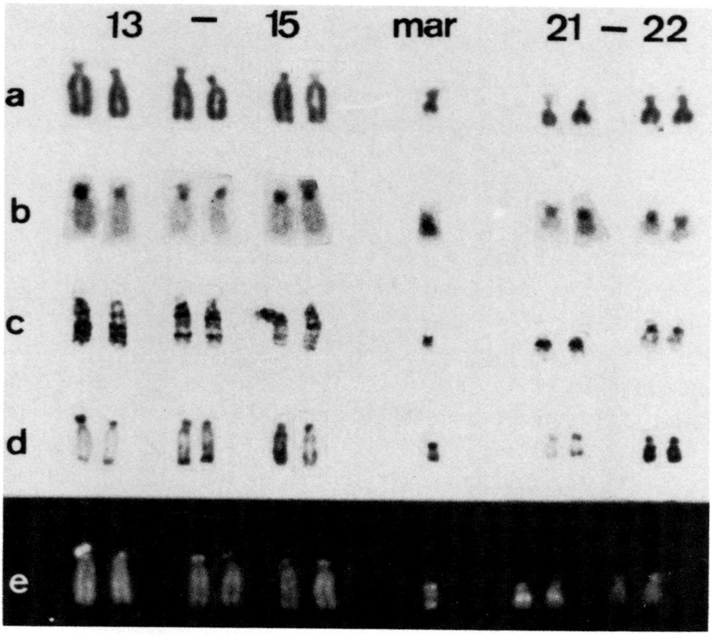

Fig. 17. Chromosomes from a patient with cat-eye syndrome. (Reproduced from Schinzel et al. [125], with permission of the publisher.)

monosomy 18q and one monosomy 10p. Another observation of monosomy 10p and partial DiGeorge sequence is known [123].

To return to chromosome 22, the only rearrangements to be observed are reciprocal translocations producing proximal monosomy 22q by either adjacent 2 or 3:1 type segregation (with a 45-chromosome karyotype as in PWS). De novo interstitial deletions appear to be extremely rare, if they exist. Failure to detect them may reflect the proximity of the centromere. The observation by Bowen et al. [120] is of interest in this respect, since these authors consider that the monosomy is in 22p, rather than 22q. Deletions may be overlooked because of the fact that the banding pattern of this particular region of 22q is relatively featureless. In other words, we might be in a situation similar to that of PWS for which true deletions were observed only some time after translocations.

THE CAT-EYE SYNDROME AND TETRASOMY 22pterq11

In 1965, Schachenmann et al. [124] reported four patients with a tiny metacentric supernumerary chromosome, and a similar phenotype that is now known as the cat-eye syndrome (McKusick No. 11547). The complete syndrome associates moderate mental retardation and congenital malformations including anal atresia, ocular coloboma, palpebral fissures slanting downward and outward, microphthalmia, pretragian tags and/or pits, CHD, anomalies of the urinary tract, and skeletal anomalies. In fact, the complete syndrome is observed in little more than a third of the cases. At least one major feature—coloboma or atresia of the anus—must be present in order to accept the diagnosis.

A chromosomal anomaly is observed in about five out of six cases. Schinzel et al. [125] suggested that the diagnosis of cat-eye syndrome should be restricted to cases with a chromosomal anomaly.

The characteristic chromosomal anomaly consists of a small supernumerary metacentric chromosome, whose size and morphology is variable (Fig. 17). It is generally shorter than, or the same size as, a chromosome 22, and it is usually present in mosaicism with normal cells. Several hypotheses have been advanced to explain its origin. The generally accepted view is that of a complex rearrangement of chromosome 22. Recently, cytogenetic observations [126,127] and molecular hybridization [72] have confirmed that tetrasomy of 22pter→q11 is implicated in cat-eye syndrome as proposed by Schinzel et al. [125].

MULTIPLE ENDOCRINE NEOPLASIA AND DELETION 20p12.2

Multiple endocrine neoplasia type 2A and 2B (MEN-2A and MEN-2B) are autosomal dominant conditions (McKusick Nos. 17140, 16230) in which medullary thyroid cancers are associated with adrenal pheochromocytomas. They differ by the type of other associated tumors (mucosal neuroma syndrome in type 2B). MEN-2A is also called Sipple syndrome.

An interstitial deletion of 20p12.2 was first suggested by Van Dyke et al. [128] and confirmed in further publications by the same group [129,130]. In all, these authors claimed to have found the deletion in 5 affected members of 3 of 5 MEN-2B families and 20 affected members of 10 of 11 MEN-2A families (60% of the MEN-2B families and 91% of the MEN-2A families).

In fact, this deletion has not been confirmed by other laboratories, and several groups failed to distinguish any deletion in MEN patients [131,132]. Molecular hybridization using various anonymous clones of the critical region did not show linkage with MEN-2A (HGM 8). The association of MEN and 20p12.2 deletion thus remains a matter of controversy, and, furthermore, deletions of 20p without MEN are known [133].

MISCELLANEOUS

An example of suggested assignments not confirmed is that of Cornelia de Lange syndrome (McKusick No. 12247). Falek et al. [134] reported a chromosomal rearrangement associated with a familial syndrome resembling Cornelia de Lange syndrome. The existence of common features in this syndrome and in trisomy 3q21→qter is now a classical notion. Both conditions can, however, be distinguished on clinical grounds. In many instances this distinction can be difficult and karyotyping is still indicated, particularly in cases of familial recurrence [135,136].

A number of microdeletions or duplications have been observed that offer the opportunity to map phenotypic features. And, indeed, proven but not exclusive associations are now known between microcephaly and deletion 1qter [136] and craniosynostosis and deletion 7p21. In the latter, the critical segment is of interest, since trisomy 7p21 is associated with delayed closure of the fontanels and cranial sutures, thus showing a type and countertype effect [136,137].

There are also associations between trigonocephaly and deletion 9p as well as deletion 11qter [136,137]. Finally, a variety of isolated reports concern the association of a microrearrangement in a chromosome arm and a Mendelian condition. For example, Crouzon syndrome and 2p [138], Franceschetti syndrome and 5q11 [139], and Holt-Oram syndrome and deletion 14q [140]. These reports remain isolated but are of great interest and must be pursued. It should be kept in mind that the Prader-Willi story began with isolated observations of apparently balanced Robertsonian translocations that could well have been discarded on the grounds that they were coincidental.

REFERENCES

1. McKusick VA: The human gene map, December 1984. Clin Genet 27:207, 1985.
2. McKusick VA: "Mendelian Inheritance in Man. Catalogs of Autosomal Dominant, Autosomal Recessive, and X-Linked Phenotypes." Baltimore and London: Johns Hopkins University Press, 1983, 6th Edition, p 1378.
3. Grouchy J de: Toward clinical microcytogenetics: The aniridia and the retinoblastoma stories. In Bonné-Tamir B (ed): "Human Genetics Part B: Medical Aspects." Proceedings of the Six International Congress of Human Genetics (Jerusalem, Sept 1981). New York: Alan R. Liss, Inc., 1982, p 359.

4. Lele KP, Penrose LS, Stallard HB: Chromosome deletion in a case of retinoblastoma. Am J Hum Genet 27:171, 1963.
5. Yunis JJ, Ramsay N: Retinoblastoma and subband deletion of chromosome 13. Am J Dis Child 132:161, 1978.
6. Yunis JJ: High resolution of human chromosomes. Science 191:1268, 1976.
7. Viegas-Péquignot E, Dutrillaux B: Une méthode simple pour obtenir des prophases et des prométaphases. Ann Génét (Paris) 21:122, 1978.
8. ISCN: Harnden DG, Klinger HP (eds): "An International System for Human Cytogenetic Nomenclature." Basel: Karger, 1985, also in Birth Defects: Original Article Series Vol 21 n°1, March of Dimes Birth Defects Foundation, New York, 1985.
9. Vogel F: Genetics of retinoblastoma. Hum Genet 52:1, 1979.
10. Howard RO: Chromosome errors in retinoblastoma. Birth Defects XVIII(6):703, 1982.
11. Turleau C, Grouchy J de, Chavin-Colin F, Junien C, Séger J, Schlienger P, Leblanc A, Haye C: Cytogenetic forms of retinoblastoma. Their incidence in a survey of 66 patients. Cancer Genet Cytogenet 16:321, 1985.
12. Matsunaga E: Retinoblastoma: Host resistance and 13q− chromosomal deletion. Hum Genet 56:53, 1980.
13. Davison EV, Gibbons B, Aherne GES, Roberts DF: Chromosomes in retinoblastoma patients. Clin Genet 15:505, 1979.
14. Motegi T, Komatsu M, Nakazato Y, Ohuchi M, Minoda K: Retinoblastoma in a boy with a de novo mutation of a 13/18 translocation: The assumption that the retinoblastoma locus is at 13q14.1, particularly at the distal portion of it. Hum Genet 60:193, 1982.
15. Cross HE, Hansen RC, Morrow G III, Davis JR: Retinoblastoma in a patient with a 13qXp translocation. Am J Ophthalmol 84:548, 1977.
16. Nichols WW, Miller RC, Sobel M, Hoffman RS, Sparkes RS, Mohandas T, Veomett I, Davis JR: Further observation on a 13qXp translocation associated with retinoblastoma. Am J Ophthalmol 89:621, 1980.
17. Mohandas T, Sparkes RS, Shapiro LJ: Genetic evidence for the inactivation of a human autosomal locus attached to an inactive X chromosome. Am J Hum Genet 34:811, 1982.
18. Hida TY, Kinoshita R, Matsumoto N, Susuki N, Tanaka H: Bilateral retinoblastoma with a 13qXp translocation. J Pediatr Ophthal Strabismus 17:144, 1980.
19. Ejima Y, Sasaki MS, Kaneko A, Tanooka H, Hara Y, Hida T, Kinoshita Y: Possible inactivation of part of chromosome 13 due to 13qXp translocation associated with retinoblastoma. Clin Genet 21:357, 1982.
20. Kajii T, Tsukahara M, Fukushima Y, Hata A, Matsuo K, Kuroki Y: Translocation (X;13) (p11.21;q12.3) in a girl with incontinentia pigmenti and bilateral retinoblastoma. Ann Génét (Paris) 28:219, 1985.
21. Noël B, Quack B, Rethoré MO: Partial deletions and trisomies of chromosome 13; mapping of bands associated with particular malformations. Clin Genet 9:593, 1976.
22. Michalova K, Kloucek F, Musilova J: Deletion of 13q in two patients with retinoblastoma, one probably due to 13q− mosaicism in the mother. Hum Genet 61:264, 1982.
23. Sparkes RS, Muller H, Klisak I, Abram JA: Retinoblastoma with 13q− chromosomal deletion associated with maternal paracentric inversion of 13q. Science 203:1027, 1979.
24. Hoegerman SF: Chromosome 13 long arm interstitial deletion may result from maternal inverted insertion. Science 205:1035, 1979.
25. Riccardi VM, Mintz-Hittner H, Francke U, Pippin S, Holmquist GP, Kretzer FL, Ferrell R: Partial triplication and deletion 13: Study of a family presenting with bilateral retinoblastomas. Clin Genet 15:332, 1979.

26. Strong LC, Riccardi VM, Ferrell RE, Sparkes RS: Familial retinoblastoma and chromosome 13 deletion transmitted via an insertional translocation. Science 213:1501, 1981.
27. Rivera H, Turleau C, Grouchy J de, Junien C, Despoisse S, Zucker JM: Retinoblastoma-del(13q14): Report of two patients, one with a trisomic sib due to maternal insertion. Gene-dosage effect for esterase D. Hum Genet 59:211, 1981.
28. Turleau C, Grouchy J de, Chavin-Colin F, Despoisse S, Leblanc A: Two cases of del(13q)-retinoblastoma and two cases of partial trisomy due to a familial insertion. Ann Génét (Paris) 26:158, 1983.
29. Grouchy J de, Turleau C, Cabanis MO, Richardet JM: Rétinoblastome et délétion intercalaire du chromosome 13. Arch Fr Pédiatr 37:531, 1980.
30. Bec P, Arne JL, Bourrouillou G, Colombiès P: Rétinoblastome et transmission de remaniements chromosomiques balancés. Mod Probl Ophthalmol 18:160, 1977.
31. Balestrazzi P, Mattei MG, Baeteman MA, Mattei JF, Giraud F: Bilateral retinoblastoma with de novo constitutional balanced translocation t(2;9)(q11;p11). Eur J Pediatr 141:250. 1984.
32. Sparkes RS, Sparkes MC, Wilson MG, Towner JW, Benedict W, Murphree AL, Yunis JJ: Regional assignment of genes for human esterase D and retinoblastoma to chromosome band 13q14. Science 208:1042, 1980.
33. Sparkes RS, Sparkes MC, Kalina RE, Pagon RA, Salk DJ, Disteche CM: Separation of retinoblastoma and esterase-D loci in a patient with sporadic retinoblastoma and del(15)(q14.1g22.3). Hum Genet 68:258, 1984.
34. Junien C, Despoisse S, Turleau C, Nicolas H, Picard F, Le Marec B, Kaplan JC, Grouchy J de: Retinoblastoma/del 13q14 and esterase D: application of gene dosage effect to prenatal diagnosis. Cancer Genet Cytogenet 6:281, 1982.
35. Cowell JK, Rutland P, Jay M, Hungerford J: Deletions of the esterase D locus from a survey of 200 retinoblastoma patients. Hum Genet 72:164, 1986.
36. Sparkes RS, Murphree AL, Lingua RW, Sparkes MC, Leigh Field L, Funderburk SJ, Benedict WF: Gene for hereditary retinoblastoma assigned to human chromosome 13 by linkage to esterase D. Science 219:971, 1983.
37. Benedict WF, Murphree AL, Banerjee A, Spiuna CA, Sparkes MC, Sparkes RS: Patient with 13 chromosome deletion: Evidence that the retinoblastoma gene is a recessive cancer gene. Science 219:973, 1983.
38. Cavenee WK, Dryja TP, Phillips RA, Benedict WF, Godbout R, Gallie BL, Murphree AL, Strong LC, White RL: Expression of recessive alleles by chromosomal mechanisms in retinoblastoma. Nature 305:779, (1983).
39. Matsunaga E: Genetics of Wilms' tumor. Hum Genet 57:231, 1981.
40. Brusa P, Torricelli C: Nefroblastoma di Wilms ed affezioni renali congenite nella casistica dell'IIPAI di Milano. Minerva Pediatr 5:457, 1953.
41. Miller RW, Fraumeni JF, Manning MD: Association of Wilms' tumor with aniridia, hemihypertrophy and other congenital malformations. N Engl J Med 270:922, 1964.
42. Francois J, Coucke D, Coppieters R: Aniridia-Wilms' tumour syndrome. Ophthalmologica 174:35, 1977.
43. Francke U, Holmes LB, Riccardi VM: Aniridia-Wilms' tumor association: Evidence for specific deletion of 11p13. Cytogenet Cell Genet 24:185, 1979.
44. Riccardi VM, Sujansky E, Smith AC, Francke U: Chromosomal imbalance in the aniridia-Wilms tumor association: 11p interstitial deletion. Pediatrics 61:604, 1978.
45. Turleau C, Grouchy J de, Dufier JL, Le Hong Phuc, Schmelck PH, Rappaport R, Nihoul-Fékété C, Diébold N: Aniridia-male pseudohermaphroditism-gonadoblastoma-mental retardation and del11p13. Hum Genet 57:300, 1981.

46. Gilgenkrantz S, Vigneron C, Gregoire MJ, Pernot C, Raspiller A: Association of del(11)(p15.1p12) aniridia, catalase deficiency and cardiomyopathy. Am J Med Genet 13:39, 1982.
47. Turleau C, Grouchy J de, Tournade MF, Gagnadoux MF, Junien C: Del 11p/aniridia complex. Report of three patient and review of 37 observations from the literature. Clink Genet 26:356, 1984.
48. Wieacker P, Mueller A, Mayerova KH, Grzeschik KH, Ropers HH: Assignment of the gene coding for human catalase to the short arm of chromosome 11. Ann Génét (Paris) 23:73, 1980.
49. Junien C, Turleau C, Grouchy J de, Said R, Rethoré MO, Tenconi R, Dufier JL: Regional assignment of catalase (CAT) gene to band 11p13. Association with the aniridia-Wilms' tumor gonadoblastoma (WAGR) complex. Ann Génét (Paris) 23:165, 1980.
50. Junien C, Turleau C, Lenoir GM, Philip T, Said R, Despoisse S, Laurent C, Rethoré MO, Kaplan JC, Grouchy J de: Catalase determination in various etiologic forms of Wilms' tumor and gonadoblastoma. Cancer Genet Cytogenet 7:51, 1983.
51. Ferrell RE, Riccardi VM: Catalase levels in patients with aniridia and/or Wilms' tumor: Utility and limitations. Cytogenet Cell Genet 31:120, 1981.
52. Turleau C, Grouchy J de, Nihoul-Fékété C, Dufier JL, Chavin-Colin F, Junien C: Del11p13/nephroblastoma without aniridia. Hum Genet 67:455, 1984.
53. Ferrel RE, Chakravarti A, Mintz Hittner N, Riccardi VM: Autosomal dominant aniridia: Probable linkage to acid phosphatase-1 locus on chromosome 2. Proc Natl Acad Sci USA 77:1580, 1980.
54. Simola KO, Knuutila S, Kaitila I, Pirkoia A, Pohja P: Familial aniridia and translocation t(4;11)(q22;p13) without Wilms' tumor. Hum Genet 63:158, 1983.
55. Moore JW, Hyman S, Antonavakis SE, Mules EH, Thomas GH: Familial isolated aniridia associated with a translocation involving chromosomes 11 and 22 t(11;22)(p13;q12.2). Hum Genet 72:297, 1986.
56. Maurer HS, Pendergrass TW, Borges W, Honig GR: The role of genetic factors in the etiology of Wilms' tumor. Two pairs of monozygous twins with congenital abnormalities (anirida; hemi-hypertrophy) and discordance for Wilms' tumor. Cancer 43:205, 1979.
57. Andersen SR, Geertinger P, Larsen HW, Mikkelsen M, Parving A, Vestermark S, Warburg M: Aniridia, cataract and gonadoblastoma in a mentally retarded girl with deletion of chromosome 11. Ophthalmologica 176:171, 1978.
58. Kaiser-Kupfer MI, White BJ, Papadopoulos N: Aniridia and mental retardation with deletion of the short arm of chromosome 11. Trans Am Ophthalmol Soc LXXIX:277, 1981.
59. Nakagome Y, Ise T, Sakurai M, Nakajo T, Okamoto E, Takano T, Nakahori Y, Tsuchida Y, Nagahara N, Takada Y, Ohsawa Y, Sawaguchi S, Toyosaka A, Kobayashi N, Matsunaga E, Saito S: High-resolution studies in patients with aniridia-Wilms' tumor association, Wilms' tumor or related congenital abnormalities. Hum Genet 67:245, 1984.
60. Gödde-Salz E, Behnke H: Aniridia, mental retardation and an unbalanced reciprocal translocation of chromosomes 8 and 11 with an interstitial deletion of 11p. Eur J Pediatr 136:93, 1981.
61. Niikawa N, Fukushima Y, Taniguchi N, Lizuka S, Kajii T: Chromosome abnormalities involving 11p13 and low erythrocyte catalase activity Hum Genet 60:373, 1982.

62. Riccardi VM, Hittner HM, Francke U, Yunis JJ, Ledbetter DB, Borges W: The aniridia-Wilms' tumor association: The critical role of chromosome band 11p13. Cancer Genet Cytogenet 2:131, 1980.
63. Waziri M, Patil SR, Hanson JW: Chromosomal and clinical findings in aniridia patients. Am J Hum Genet 33:125A, 1981.
64. Narahara K, Kikkawa K, Kimira S, Kimoto H, Ogata M, Kasai R, Hamawaki M, Matsuoka K: Regional mapping of catalase and Wilms tumor-aniridia, genitourinary abnormalities, and mental retardation triad loci to the chromosome segment 11p1305→p1306.
65. Giangiacomo J, Penchransky L, Monteleone PL, Thomson J: Bilateral-neonatal Wilms' tumor with BC chromosomal translocation. J. Pediatr 86:98, 1975.
66. Bernard JL, Baeteman MA, Mattei JF, Raybaud C, Giraud F: Wilms' tumor, malformative syndrome, mental retardation and de novo constitutional translocation, t(7;13)(q36;13). Eur J Pediatr 141:175, 1984.
67. Hittner HM, Riccardi VM, Francke U: Aniridia due to a heritable chromosome 11 deletion. Ophthalmology (Rochester) 86:1173, 1979.
68. Yunis JJ, Ramsay NKC: Familial occurrence of the aniridia-Wilms' tumor syndrome with deletion 11p13–14.1. J Pediatr 96:1027, 1980.
69. Koussef BG, Agatucci A: Aniridia-Wilms tumor association. J Pediatr 98:676, 1981.
70. Nakagome Y, Nagahara N: High-resolution studies in patients with aniridia-Wilms' tumor association. Hum Genet 70:289, 1985.
71. Riccardi VM, Hittner HM, Strong LC, Fernbach DJ, Lebo R, Ferrell RE: Wilms' tumor with aniridia/iris dysplasia and apparently normal chromosomes. J Pediatr 100:574, 1982.
72. HGM 8. Cytogenet Cell Genet 40, No. 1–4, 1985.
73. Malpuech G, Sultan C, Bertheas MF, Loire C, Renaud H, Francannet C, Vanlieferingen P: Male pseudohermaphroditism. Partial androgen receptors deficit. 11p13 deletion. Indication of gene localization. Am J Med Genet 24:679, 1986.
74. Koufos A, Hansen MF, Lampkin BC, Workiman ML, Lopeland NG, Jenkins NA, Cavenee WK: Loss of alleles at loci on human chromosome 11 during genesis of Wilms' tumour. Nature 309:170, 1984.
75. Beckwith JB: Extreme cytomegaly of the adrenal fetal cortex, omphalocele, hyerplasia of kidneys and pancreas, and Leydig-cell hyperplasia: Another syndrome? Western Society for Pediatric Research, Los Angeles, Nov. 11, 1963.
76. Wiedemann NR: Complexe malformatif familial avec hernie ombilicale et macroglossie: Un "syndrome nouveau"? J Génét Hum 13:223, 1964.
77. Sotelo-Avila C, Gonzalez-Crussi F, Fowler JW: Complete and incomplete forms of Beckwith-Wiedemann syndrome: Their oncogenic potential. J Pediatr 96:47, 1980.
78. Falk RE, Carrel RE, Valente M, Crandall BF, Sparkes RS: Partial trisomy of chromosome 11: A case report. Am J Ment Defic 77:383, 1973.
79. Waziri M, Patil SR, Hanson JW, Bartley IA: Abnormality of chromosome 11 in patients with features of Beckwith-Wiedemann syndrome. J Pediatr 102:873, 1983.
80. Turleau C, Grouchy J de, Chavin-Colin F, Martelli H, Voyer M, Charlas R: Trisomy 11p15 and Beckwith-Wiedemann syndrome. A report of two cases. Hum Genet 67:219, 1984.
81. Journel H, Lucas J, Allaire C, Le Mee F, Defawe G, Lecornu M, Jouan H, Roussey M, Le Marec B: Trisomy 11p15 and Beckwith-Wiedemann syndrome. Report of two new cases. Ann Génét (Paris) 28:97, 1985.

82. Turleau C, Grouchy J de: Beckwith-Wiedemann syndrome. Clinical comparison between patients with and without 11p15 trisomy. Ann Génét (Paris) 28:93, 1985.
83. Pettenati MJ, Higgins RR, Weaver DD: Cytogenetic investigation of Beckwith-Wiedemann syndrome (BWS) and its implications. Am J Hum Genet 37:A111, 1985.
84. Jeanpierre JM, Henry I, Turleau C, Ullrich A, Mallet J, Grouchy J de, Junien C: Beckwith-Wiedemann syndrome (BWS) and 11p15 markers. HGM 8, Cytogenet Cell Genet 40:661, 1985.
85. Koufos A, Hansen MF, Copeland NG, Jenkins NA, Lampkin BC, Cavenee WK: Loss of heterozygosity in three embryonal tumours suggests a common pathogenetic mechanism. Nature 316:330, 1985.
86. Kaelbling M, Klinger HP: Suppression of tumorigenicity in somatic cell hybrids. III. Cosegregation of human chromosome 11 of a normal cell and suppression of tumorigenicity in intraspecies hybrids of normal X malignant cells. Cytogenet Cell Genet 41:65, 1986.
87. Prader A, Labhart A, Willi H: Ein syndrome von Adipositas, Kleinwuchs, Kryptorchidismus und Oligophrenie nach Myatonie-artigen Zustand in Neugeborenenalter. Schweiz Med Wochenschr 86:1260, 1956.
88. Hawkey CJ, Smithies A: The Prader-Willi syndrome with a 15/15 translocation: Case report and review of the literature. J Med Genet 13:152, 1976.
89. Mattei MG, Souiah N, Mattei JF: Chromosome 15 anomalies and the Prader-Willi syndrome: Cytogenetic analysis. Hum Genet 66:313, 1984.
90. Ledbetter DH, Mascarello JT, Riccardi VM, Harper VD, Airhart SD, Strobel RJ: Chromosome 15 abnormalities and the Prader-Willi syndrome: A follow-up report of 40 cases. Am J Hum Genet 34:278, 1982.
91. Hasegawa T, Hara M, Ando M, Osawa M, Fukuyama Y, Takahashi M, Yamada K: Cytogenetic studies of familial Prader-Willi syndrome. Hum Genet 65:325, 1984.
92. Butler MG, Meaney FJ, Palmer CG: Clinical and cytogenetic survey of 3g individuals with Prader-Labhart-Willi syndrome. Am J Med Genet 23:793, 1986.
93. Niikawa N, Ishikiriyama S: Clinical and cytogenetic studies of the Prader-Willi syndrome: Evidence of phenotype-karyotype correlation. Hum Genet 69:22, 1985.
94. Giedion A: Das Tricho-Rhino-Phalangeale Syndrome. Helv Paediatr Acta 21:475, 1966.
95. Giedion A: Die Peripheren Dysotosen (PD)—ein Sammelbegriff. Fortschr Röntgenstr 110:507, 1969.
96. Langer LO: The thoraci-pelvi-phalangeal dystrophy. Clinical delineation of birth defects part IV, skeletal dysplasia, OAS V(4):55, 1969.
97. Hall BD, Langer LO, Giedion A, Smith DW, Cohen MM, Beals RK, Brandner M: Langer-Giedion syndrome. Birth Defects 10:147, 1974.
98. Langer LO, Krassikoff N, Laxova R, Sheerwilliams M, Lutter LD, Gorlin RJ, Jennings CG, Day DW: The tricho-rhino-phalangeal syndrome with exostoses (or Langer-Giedion syndrome): Four additional patients without mental retardation and review of the literature. Am J Med Genet 19:81, 1984.
99. Bühler EM, Bühler UK, Stalder GR, Jani L, Jurik LP: Chromosome deletion and multiple cartilagenous exostoses. Eur J Pediatr 133:163, 1980.
100. Zabel BU, Baumann WA: Langer-Giedion syndrome with interstitial 8q− deletion. Am J Med Genet 11:353, 1982.
101. Turleau C, Chavin-Colin F, Grouchy J de, Maroteaux P, Rivera H: Langer-Giedion syndrome with and without del 8q. Assignment of critical segment to 8q23. Hum Genet 62:183, 1982.

102. Bühler EM, Malik NJ: The tricho-rhino-phalangeal syndrome(s): chromosome 8 long arm deletion: Is there a shortest region of overlap between reported cases? TRP I and TRP II syndromes: Are they separate entities? Am J Med Genet 19:107, 1984.
103. Bowen P, Biederman B, Hoo JJ: The critical segment for the Langer-Giedion syndrome: 8q24. 11→q24.12. Ann Génét (Paris) 28:224, 1985.
104. Zaletajev DV, Marincheva GS: Langer-Giedion syndrome in a child with complex structural aberration of chromosome 8. Hum Genet 63:178, 1983.
105. Schwartz S, Beisel JH, Panny SR, Cohen MM: A complex rearrangement, including a deleted 8q, in a case of Langer-Giedion syndrome. Clin Genet 27:175, 1985.
106. Goldblatt J, Smart R: Evidence for homogeneity of the underlying defect in the tricho-rhino-phalangeal syndromes (TRPS). Am J Hum Genet 37:A94, 1985.
107. Miller JQ: Lissencephaly in two siblings. Neurology 13:841, 1963.
108. Dieker H, Edwards RH, ZuRhein G, Chou SM, Hartmasn HA, Optiz JM: The lissencephaly syndrome. Birth Defects 5:53, 1969.
109. Jones KL, Gilbert EF, Kaveggia EG, Opitz JM: The Miller-Dieker syndrome. Pediatrics 66:277, 1980.
110. Dobyns WB, Stratton RF, Parke JT, Greenberg F, Nussbaum RL, Ledbetter DH: The Miller-Dieker syndrome: lissencephaly and monosomy 17p. J Pediatr 102:552, 1983.
111. Stratton RF, Dobyns WB, Airhart SD, Ledbetter DH: New chromosomal syndrome: Miller-Dieker syndrome and monosomy 17p13. Hum Genet 67:193, 1984.
112. Ketterer DM, Katnik R, Phan D, Casey J, Giles HR: Monosomy 17p13 and Miller-Dieker (lissencephaly) syndrome: An underestimated entity? Am J Med Genet 37:A102, 1985.
113. Greenberg F, Stratton FR, Lockhart LH, Elder FFB, Dobyns WB, Ledbetter DH: Familial Miller-Dieker syndrome associated with pericentric inversion of chromosome 17. Am J Med Genet 23:853, 1986.
114. Carpenter NJ, Lechtman LG, Stamper S, Say B: An infant with ring 17 chromosome and unusual dermatoglyphics: A new syndrome? J Med Genet 18:234, 1981.
115. Jellinger K, Rett A: Agyria-Pachygyria (lissencephaly syndrome). Neuropediatrics 7:66, 1976.
116. DiGeorge AM: Congenital absence of the thymus and its immunologic consequences: concurrence with congenital hypoparathyroidism. In Good RA (ed): "Immunologic Deficiency Diseases." New York: National Foundation, 1968, p 116.
117. La Chapelle A de, Herva R, Koivisto M, Aula P: A deletion in chromosome 22 can cause DiGeorge syndrome. Hum Genet 57:253, 1981.
118. Kelley RI, Zakai EH, Emanuel BS, Kistenmarcher M, Greenberg F, Punnet N: The association of DiGeorge syndrome with partial monosomy of chromosome 22. J Pediatr 101:197, 1982.
119. Greenberg F, Crowder WE, Paschall V, Colon-Linares J, Lubianski B, Ledbetter DH: Familial DiGeorge syndrome and associated partial monosomy of chromosome 22. Hum Genet 65:317, 1984.
120. Bowen P, Pabst H, Berry D, Collins-Nakai R, Hoo JJ: Thymus deficiency in an infant with a chromosome t(18;22)(q12.2;p11.2) pat rearrangement. Clin Genet 29:174, 1986.
121. Cannizzaro LA, Emanuel BS: In situ hybridization and translocation breakpoint mapping. III. DiGeorge syndrome with partial monosomy of chromosome 22. Cytogenet Cell Genet 39:179, 1985.
122. Greenberg F, Ledbetter DH: Cytogenetic findings in a prospective series of patients with DiGeorge sequence. Am J Hum Genet 37:A95, 1985.

123. Hervé J, Warnek JF, Jeaneau-Bellego E, Portnoi MF, Taillemite JL, Hervé F: Monosomie partielle du bras court du chromosome 10 associée à un syndrome de Rieger et à un déficit partiel immunitaire du type DiGeorge. Ann Pédiatr (Paris) 31:77, 1984.
124. Schachenman G, Schmid W, Fraccaro M, Mannini A, Tiepolo L, Perona GP, Sartori E: Chromosomes in coloboma and anal atresia. Lancet 2:290, 1965.
125. Schinzel A, Schmid W, Fraccaro M, Tiepolo L, Zuffardi O, Opitz JM, Lindsten J, Zetterqvist P, Enell H, Bacchichetti C, Tenconi R, Pagon RA: The "cat eye syndrome." Dicentric small marker chromosome probably derived from a No. 22 (tetrasomy 22pter-q11) associated with a characteristic phenotype. Hum Genet 57:148,1981.
126. Wilson GN, Baker DL, Schau J, Parker J: Cat eye syndrome owing to tetrasomy 22pterq11. J Med Genet 21:60, 1984.
127. Gabarron J, Glover G, Jiminez A, Lamata E: Pseudoisodicentric bisatellited extra marker chromosome (tetrasomy 22pter→q11, trisomy Yqh), derived from a maternal Y/22 translocation. Association between this tetrasomy and "cat eye" phenotypical features. Clin Genet 28:509, 1985.
128. Van Dyke DL, Jackson CE, Babu VR: Localization of autosomal, dominant multiple endocrine neoplasia 2 syndrome (MEN1) to 20p12.2. HGM6, Cytogenet Cell Genet 32:324, 1982.
129. Babu VR, Van Dyke DL, Jackson CE: Chromsome 20p deletion in human multiple endocrine neoplasia types 2A and 2B: A double-blind study. Proc Natl Acad Sci USA 81:2525, 1984.
130. Jackson CE, Van Dyke DL, Babu VR: Chromosome 20 deletion-multiple endocrine neoplasia type 2 (MEN-2) up-date. Am J Hum Genet 37:A30, 1985.
131. Emmertsen K, Lamm LU, Rasmussen KZ Elbond O, Hansen HH, Henningsen K, Jorgensen J, Peterson GB: Linkage and chromosome study of multiple endocrine neoplasia IIA. Cancer Genet Cytogenet 9:251, 1983.
132. Hsu TC, Pathak S, Samaan N, Hickey RC: Chromosome instability in patients with medullary carcinoma of the thyroid. JAMA 246:2046, 1981.
133. Garcia-Cruz D, Rivera H, Barajas LO, Jimenez-Sainz M, et al.: Monosomy 20p due to a de novo del (20)(p12.2). Clinical and radiological delineation of the syndrome. Ann Génét (Paris) 28:231, 1985.
134. Falek A, Schmidt R, Jervis GA: Familial de Lange syndrome with chromosome abnormalities. Pediatrics 37:92, 1966.
135. Francke U: Clinical syndromes associated with partial duplications of chromosomes 2 and 3: dup(2p), dup(2q), dup(3p), dup(3q). Birth Defects XIV, No. 6C:191, 1978.
136. Grouchy J de, Turleau C: "Clinical Atlas of Human Chromosomes," 2nd Ed. New York: Wiley Medical, 1984, p 487.
137. Garcia-Esquivel L, Garcia-Cruz D, Rivera H, Plascencia ML, Cantu JM: De novo del (7)(pter→21.2::p15.2→qter) and craniosynostosis. Implications for critical segment in the 7p2 monosomy syndrome. Ann Génét (Paris) 29:36, 1986.
138. Aurias A, Prieur M, Dutrillaux B, Lejeune J: Systematic analysis of 95 reciprocal translocations of autosomes. Hum Genet 45:259, 1978.
139. Balestrazzi P, Baeteman MA, Mattei MG, Mattei JF: Franceschetti syndrome in a child with de novo balanced translocation (5;13) (q11;p11) and significant decrease of hexosaminidase B. Hum Genet 64:305, 1983.
140. Turleau C, Grouchy J de, Chavin-Colin F, Doré F, Séger J, Dautzenberg MD, Arthuis M, Jeanson C: Two patients with features of Holt-Oram syndrome. Exclusion mapping of PI (alpha-1-antitrypsin). Ann Génét (Paris) 27:237, 1984.

The Cytogenetics of Mammalian Autosomal Rearrangements, pages 803–833

Chapter 27
Mobile Genetic Elements in G-Band and R-Band DNA

Gerald P. Holmquist

INTRODUCTION

Mobile genetic elements represent chromosome rearrangements at the molecular level, and in many eukaryotes, movement is often associated with visible autosomal rearrangements. Almost all the interspersed repetitive sequences in the mammalian genome are in the class of mobile elements called retroposons, indicating they are transposed via reverse transcription of an RNA intermediate followed by insertion of the resulting DNA into a nicked site in the mammalian genome. Retroposons, such as Alu and L1, constitute about 20% of the human genome. Their movement appears both genetically and karyotypically innocuous. Retroviruses are a less common mobile element and exist in the genome as proviruses. They carry controlling sequences, and their mobility is often associated with mutation and oncogenesis. Classes of jumping genes with spectacular phenotypic effects, such as the Ty elements in yeast and controlling elements in corn, seem notably absent from mammals. Many families of retroposon sequences, and possibly retrovirus families, selectively reside in either G-band or R-band DNA. Where these genes jump seems directed by some property of the banded prepattern.

CLASSICAL STUDIES OF MOBILE ELEMENTS

Mobile or transposable genetic elements were first identified in *Zea maize* by McClintock (see review by Federoff [1]). They occurred as mobile "controlling elements," which, after inserting within or adjacent to a gene, inhibited the gene's expression. Excision of the element called "dissociation" occurred only when another element called "activator" was present in the

Department of Biology, Beckman Research Institute-City of Hope, Duarte, California 91010 genome. Excision of the dissociation element was detected as relief of the inhibition and occurred in both somatic and germline tissues. Restoration of one gene's activity was often concomitant with inhibition of a second gene's activity, as if the dissociation element had jumped from one gene to another. In the activator-dissociation system, the presence of the transposable dissociation element was first identified by its ability to act as a specific site of chromosome breakage [1]. The trans-acting product of the "activator" gene induced transposition of "dissociation" and chromatid breaks at the site harboring it. After replication, the broken chromosome ends, i.e., the sticky ends, often fused, producing dicentrics that subsequently underwent a breakage-fusion-bridge cycle. Thus, mobile elements can be responsible for cytologically visible chromosome rearrangements in addition to the submicroscopic rearrangements reflecting their mobility.

Ising and Ramel [2] identified transposing elements (TE) in *Drosophila.* TE1, for example, carried a copy of the *white* locus to other genomic loci. Some TE elements carried several hundred kilobases (kb) of *Drosophila* DNA and could be detected as visible insertions in the polytene chromosomes [3]. In rare strains wherein transposition is detected, transposition to a new location or loss of the TE element from the germline both occurred at frequencies up to 10^{-3}. Mobility of TE elements is attributed to small foldback elements that flank the transposable DNA [4]. Similar very high mobility systems, with the possible exception of retroviruses, have not been reported in mammals. However, molecular sleuthing reveals that most or possibly all of the interspersed repeat sequences in man are mobile elements. Only two classes of mobile elements, retroposons and retroviruses, have been documented in mammals. Readers should see [4,5] for in-depth reviews of mobile elements in eukaryotes or [6–12] for reviews of mobile elements in mammals.

INTERSPERSED REPEATS

A family of interspersed repetitive sequences consists of related but individually distinct members that either share sufficient homology to cross-hybridize or, when sequenced, appear similar but seldom identical. These repetitive sequences occur in two fairly distinct length classes—short (SINEs) and long (LINEs) interspersed repeated sequences; SINEs are less than 500 base pairs (bp), while LINEs are longer. Some families have many members. The human *Alu* sequence is a SINE 282 bp long, which probably originated by reverse transcription of 7 SL RNA (an RNA species involved in the transport of secreted proteins) [6,12]. The *Alu* family has about 500,000 members, which compose 5.5% of the human genome [10]. *Alu* seems well dispersed in that a human genomic library of 18 kb inserts in lambda DNA revealed over 90% of the clones hybridized to a cloned *Alu* sequence [14]. Mice have *Alu*-like familes called B1 and B2, which are 130 and 190 bp long, are present at 100,000 members/genome, and compose 1% of the genome each [15]. Both humans and mice have one major family of LINEs called L1 [6,8,11,16–18]. It was formerly called *KpnI* in primates, and various segments of it were called *Bam*HI, MIF-1, or R in rodents. The consensus sequence of L1 is about 7 kb long [17], but most of the copies in the genome are heterogeneously truncated from the 5′ end, so that the mouse genome has 85,000 members containing 3′ L1 sequences and only 10,000 members containing 5′ L1 sequences. Both humans [13] and mice [19] have 10% of their genomes (half of their interspersed repeat DNA) in small families of 1,000 members or less. These include processed pseudogenes complementary to processed mRNA transcripts, sequences complementary to a variety of t-RNA species or to small nuclear RNA species, and sequences complementary to retroviruses.

Retroposons

Alu, L1, processed pseudogenes, and most of the other families of interspersed repeat sequences excluding proviruses are now classified as retroposons [6]. They have arisen from reverse transcription of an RNA intermediate [6,7,8,12,20,21] by a common mechanism called retroposition (Fig. 1) as admirably reviewed by John H. Rogers [6]. This mechanism was inferred from the following common structure of a retroposon in genomic DNA.

genomic...... TR-RNA sequence-A_n-$(NA_x)_y$-(A)-rich sequence-
TR......genomic

Processed pseudogenes [7,20] were the first obvious candidates for originating from reverse transcription of an RNA intermediate. Many sequenced examples of processed pseudogenes are almost-perfect DNA copies of correctly processed (spliced, capped, and polyadenylated) mRNA. These were obviously not active genes because their cDNA sequences, when compared with that of the parental gene, contained frame-shift and stop-codon mutations that would obviate translation into an active gene product. The cDNA sequence (mRNA sequence-A_n) was usually imbedded in the common retroposon structure complete with terminal repeats. Subsequent sequencing studies showed that almost all interspersed repeats are of a similar general structure, implying they all arose from a common mechanism now called

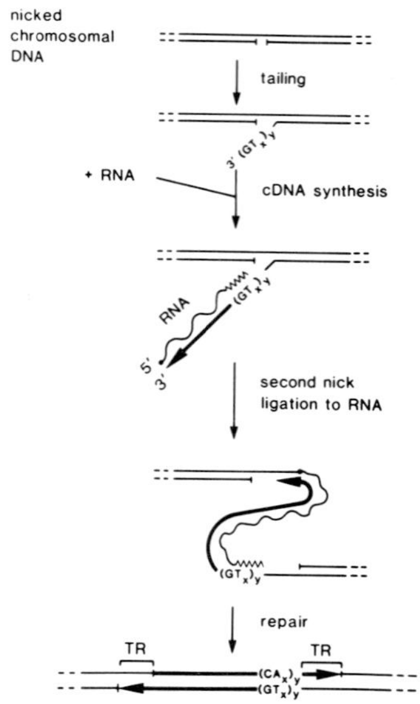

Fig. 1. One possible model of retroposition to explain the general structural features of most retroposons. The wavy line represents RNA; and the thick line, new DNA. Initial polymerization, 3′ tailing, of the nicked DNA by a process probably similar to the resolution of telomeres generates a $(GT_x)_y$ tail (its complement is the (A)-rich sequence) which both binds the poly(A) tail of mRNA and serves as a primer for reverse transcriptase. Ligation and gap filling generate the direct terminal repeats (TR) out of the insertion target site. (Reproduced from Rogers [6], with permission of the publisher.)

retroposition [6]. Processed pseudogenes, since they arose from the processed RNA polymerase II product mRNA, lack a transcription promotor. They are usually present as one to several copies/genome.

The more successful retroposons have family sizes of 1,000–500,000 and usually arose from a polymerase III transcript whose sequence includes an internal polymerase III promotor [6]. All the small nuclear RNA species U1–U6 are transcribed by RNA polymerase III, have 100–1,000 pseudogenes in the mammalian genome, and many of these copies are imbedded in the common retroposon structure [6]. The retroposons human *Alu,* murine B1, and murine B2 are the abundant SINE families in man and mouse. They share RNA polymerase III promotor sequences and many of the *Alu* and B2 sequences can function as polymerase III transcription units [6,22,23]. *Alu* seems distinctive not because of its sequence but because of its high copy number in primates. Various other short retroposon families and the homologous RNA species from which they supposedly originated were reviewed [6,12] as summarized (Table I).

The mammalian LINE family L1 has several unique features in addition to its consensus length of 7 kb and its retroposon structure. In primates, it has two overlapping open reading frames of 1,137 and 3,900 bp. In the sequenced family members studied, these show ratios (base substitutions which would alter an amino acid/silent substitutions) characteristic of DNA sequences that code for a functional protein [18]. The open reading frames are homologous to selected domains in several reverse transcriptases [9,17]. The success of L1 may be due to its ability to code for its own reverse transcriptase.

The genomic target site for retroposon insertion is duplicated during retroposition and appears as terminal repeat (TR) sequences flanking the retroposon (Fig. 1). This relation was deduced from sequenced genomic regions wherein the sequence of the target site before retroposition could be inferred. Such is the case for transposition into a repeat member of known consensus sequence [6] or a few known cases where a locus in the population is polymorphic for the presence of a retroposon [6,24–28]. The target site is usually a locally (A)-rich region and, in such cases, is often the (A)-rich sequence (Fig. 1) of another retroposon [6,25]. Despite much recent labor to find restriction length polymorphisms, especially in man, examples of retroposon insertion polymorphism in mammals are rare.

Retroviruses

Retroviruses, RNA tumor viruses (see general reviews [29,29a] and a review of the mouse mammary tumor virus [30]) are infectious, horizontally transmitted agents. Mouse mammary tumor virus (MMTV) is transmitted through mother's milk. Retroviral sequences are also integrated as proviruses into the genome wherein they are transmitted vertically in a Mendelian fashion and account for ca. 0.5% of the mouse genome [29a]. Retroviral RNA codes for viral-specific structural proteins and a reverse transcriptase for the synthesis of a covalently closed viral DNA replication intermediate. The DNA intermediate can become integrated into the mammalian genome by a process that differs from retroposition of an RNA intermediate. As in retroposition, however, a short region of the host target sequence is duplicated during integration and appears as direct repeats, each one forming an outer flanking region of the provirus. Proviruses almost always start with the sequence TG and end with CA [5]. The sequences of integration sites appear unrelated to each other [5], suggesting insertion along the host genome is random. Just inside the short flanking direct repeats are long (400–1,200 bp)

TABLE I. Properties and distributions of retroposons in mammals

Repetitive family	Base pairs	Species distribution	Parent gene	Divergence from parent (%)	Divergence from consensus (%)	Copy No.
Alu	282	All primates	7 SL RNA	18	14	3–5 × 10^5 (human)
Alu type IIa		Galago	Monomer + 7 SL RNA	27 + 24	6	
Alu type IIb		Galago	Monomer + 7 SL RNA	15 + 25	9	
Monomer		Galago	Met-tRNA	32	12	
B1	129	Rodents	7 SL RNA	25	8	1 × 10^5 (mouse)
B2a	180	Rodents	Ser-tRNA	34	10	8 × 10^4
B2b		Mouse	Ser-tRNA	39	4	
Identifier	87	Rat	Ala-tRNA	28	4	1 × 10^5 (rat)
C	300	Rabbits	Gly-tRNA			1.7 × 10^5 (rabbits)
Pol III SINE	162	Salmon	Lys-tRNA	16		
L1	7,000	Mammals	—			1 × 10^4 (mouse)

Data are from [12: Table I] and [6: Table V], excepting rabbit data [49], L1 data [17,18], and salmon data [23]. The exact species of parental t-RNA is sometimes in doubt (Sakamoto K, Okada N, J Mol Evol, in press).

terminal repeats (LTRs). LTRs provide gene expression functions such as promotion, initiation, and polyadenylation of transcripts. Unlike retroposons but similar to mobile "controlling elements" in *maize,* the provirus is capable of cis control of neighboring genes and c-oncogenes [5,9,29]. MMTV's LTR promoters are responsive to glucocorticoid stimulation, which determines in part the synthesis of infectious virions in the mammary epithelial cells during lactation. These LTRs can be used to regulate the activity of synthetic gene constructs [30].

Most of the ca. 20 copies of MMTV provirus [30] and 150 copies of VL30 provirus [31] are inactive and highly methylated. Some strains of mice have one or a few of the proviral loci in an activatable form. For example, only the Mtv-1 locus is expressed in C3Hf and DBAf strains, whereas both the Mtv-2 and Mtv-3 loci are active in GR mice [30]. Similarly, only the relatively hypomethylated subset of VL30 proviruses can be transmitted to rat cells with the aid of a leukemia helper virus [31]. Vertically transmissible viruses may arise from recombination between defective proviruses. Hybrid mice that inherit the relatively inefficient induction locus of strain C57BL/6 together with that of BALB/c can produce unusually high titers of virus, presumably owing to complementation and recombination between inactive (defective) and low-activity proviruses [29a]. Similarly, recombination during subsequent rounds of reverse transcription probably led to the formation of Kirsten and Harvey (v-Ras) sarcoma virus genomes from VL30 sequences and the c-Ras oncogene [31]. This summarizes a generally accepted relation: the set of cellular genes usually involved in the regulation of cell proliferation and called c-oncogenes are progenitors of retroviral transforming genes, v-oncogenes, which are mobile [29,29a,31].

High-ecotropic-virus inbred strains of mice such as AKR can have a new ecotropic (a retrovirus of species-specific host range) virus integrating into the genome at rates of once every ten generations [29a], similar to the transposition rate of TE in some inbred *Drosophila* strains [3]. However, genomes of feral mice harbor few ecotropic virus copies, and the two feral mouse populations known to harbor such ecotropic viruses have potent virus resistance genes that are absent from AKR; one such gene was derived from a retrovirus envelope glycoprotein gene [29a].

Intracisternal A particles (IAP) are noninfectious RNA-virus-like particles found in some types of mouse cells. These include oocytes, preimplantation embryos, tumors [5], and all mouse plasmacytomas [32]. Their DNA sequences are repeated about 1,000 times in the mouse genome. Like copia in *Drosophila,* and MMTV in mouse, IAP sequences have long terminal repeats characteristic of retroviruses. IAP insertion is associated with three mutations in mouse somatic cell lines. Two involve the insertional mutagenesis of kappa light chain genes in hybridomas [5,32] and a third involves activation of c-mos in a myeloma [5]. Mobility seems most evident in tissues wherein this retrovirus-like sequence is expressed.

In summary, feral mice have been depicted as retrovirus evolution factories containing thousands of proviruses and a multitude of mechanisms for the inactivation of same [29a]. Most provirus sequences are inactive because of methylation, repression by resistance genes, or defects such as deletions in the proviral gene itself. Activation of a provirus can produce neoplasia in the soma, insertional mutagenesis in the germline, recombination between two viral genomes, and recombination between a viral genome and a cellular gene. Recombination has lead to more virulent viruses of altered host range, viruses with v-oncogenes, and virus resistance genes. All these can integrate back into the mouse genome. Elucidating the extent of retroviral omnipresence in mice has, as in *Drosophila,* required many investigators to isolate many different retroviruses that usually could not be recovered and identified without the use of a variety of inbred strains that lack virus resistance factors. Similar knowledge of retroviruses in other mammalian species is less complete.

Other Mobility Mechanisms in Mammals

Up to half of the spontaneous mutations affecting phenotype in yeast [33] and *Drosophila* [5,34] are due to insertional mutagenesis or to gross rearrangements involving recombination between integrated mobile sequences located in different regions of the genome. In humans and mice, mutant alleles with repetitive elements inserted in or near the affected allele are rare. High mobility elements, such as the Ty elements in *Salmonella cerevisiae* [5,33], TE elements [4,5], foldback elements, and hybrid dysgenesis elements in *Drosophila melanogaster* [5], and controlling elements in *maize* [1,5], are different from retroposons and retroviruses in that the integrated sequences have terminal inverted repeats and move by a different set of mechanisms [5].

DNA-mediated transformation of mammalian cells involves the formation of covalently closed, circular DNA intermediates that stably integrate into the mammalian genome. Most of these insertions are due to nonhomologous recombination while under appropriate transformation conditions; up to 0.1% of these insertions can be due to homologous recombination [35]. Covalently closed circular DNA with sequences homologous to retroposons seem to be a common property of all cells [36]; this DNA is relatively abundant in thymocytes [37,38], presumably as a consequence of these cells' gene rearrangement mechanisms, but it also exists in fibroblasts [36] and embryonic cells [39]. The latter implies a possibility for vertical transmission of any DNA sequence by a mechanism unrelated to retroposition and involving circular DNA intermediates.

Gene amplification produces extrachromosomal DNA circles [40]. Double minute chromosomes seem to be very large examples of such circles [41]. In cultured somatic cells, amplification of the dihydrofolate reductase (DHFR) gene spontaneously occurs at the surprisingly high frequency of once every 1,000 divisions [42]. When this frequency is increased by the use of DNA synthesis inhibitors, one finds that increased amplification is associated with concomitant increases in both extrachromosomal DNA (increases of over a haploid amount were common), and chromosome rearrangements—i.e., sister chromatid exchanges (SCE), gaps, fragments, and breakage-bridge fusion chromosomes [40,43]. Quite convincingly, the chromosome rearrangements appear to be a consequence of recombination between extrachromosomal DNA and the remaining genome [40].

Homologous recombination between interspersed repeat sequences could result in a rearrangement. In yeast, Ty elements simply act as portable regions of homology, making it possible for the recombination system to promote inversions and translocations [33]. In mammals, unequal SCE between tandemly repeated sequences is a probable cause of satellite DNA amplification and C-band polymorphisms [44]. A deletion caused by recombination between two oppositely oriented *Alu* sequences was reported in the low-density-lipoprotein receptor gene from a patient with the internalization-defective form of familial hypercholesterolemia [45]. Such reports are extremely rare. The rarity may be due to the stringent homology requirements of homologous recombination [46,47]. In *Escherichia coli,* the frequency of homologous recombination depends on regions of *perfect* homology. Recombination requires at least 20 base pairs of perfect homology for a barely detectable frequency of recombination and a maximal recombination rate requires 74 base pairs of perfect homology [46]. Homologous recombination in mammals could possibly be even more demanding. *Alu* is present at $3\text{–}6 \times 10^5$ copies/genome, but the members of this family only share a 86% homology (Table I). This is sufficient for formation of intermolecular hybrids but probably not sufficient for homologous recombination. If 74 bp of perfect homology were required for homologous recombination between *Alu* sequences in mammals, then the average probability that one *Alu* sequence could recombine with another is $0.86^{74} = 10^{-5}$ times the recombination rate expected between identical *Alu* sequences. If gene conversion were very effective at homogenizing the sequences of repeat family members or if many *identical* members suddenly appeared in a genome, then a karyotypic catastrophe of interchromosomal recombination could possibly ensue but such catastrophes have not been reported.

Taxonomy of Retroposons

The recent evolution of mammalian SINE families from their respective RNA molecules (Table I) indicates that for many of these short retroposons, insertion of thousands of copies is occurring rather rapidly (see Table I and the review by Deininger and Daniels [12]). SINE families seem confined to the taxonomic order in which they originated. The *Alu* family is common to all primates and probably originated from 7 SL RNA 60 million years ago after primates diverged from rodents [12]. The "monomer" family of SINEs in the primitive primate *Galago* (African lemur or bush baby) originated from Met-tRNA and is not present in humans [12]. Similarly, the B2a family in rodents originated from ser-tRNA [6,12] and is absent from humans, while the rat identifier sequence, a repeat specifically transcribed in rat brain [48], originated from ala-tRNA and is found only in rats and not other rodents [12]. The major rabbit SINE seems specific to lagomorphs [49]. These rather recent additions to mammalian genomes, in contrast to processed pseudogenes, share an internal RNA polymerase III promoter and their polymerase III transcripts terminate in an oligo (U)-rich sequence that is capable of self-priming reverse transcription at an internal oligo (A) region [6,12]. Retroposon families are now being identified from the products of in vitro transcription of genomic DNA by RNA polymerase III [23]. Thus the most successful retroposons are ideally suited for self-propagation via an RNA intermediate (Fig. 1). In a variety of mouse cell types, about 20% of the transcripts complementary to the B2 family are discrete polymerase III products and not just polymerase II read-through products from adjacent structural genes [22]. Thus the mammalian genome is studded with hundreds of thousands of new additions that, like true selfish genes [50], have the ability to propagate themselves via the cell's molecular machinery. Taxonomic data indicates they have indeed propagated themselves.

While the mechanisms of gene mobility indicate that mobility could cause chromosome breaks, taxonomic data indicates that few breaks are ever fixed in the genome. *Alu* has presumably *fixed* an inserted copy of itself in the primate lineage leading to man at least 500,000 times in the last 65,000,000 years [12]. The actual insertion rate is probably much higher [51]. Yet only eight [52] (seventeen [56]) rearrangements, mostly pericentric inversions, have been fixed in the Hominidae (man and great apes). The G-banded patterns of several cat and human chromosomes are quite similar. This is apparent for about 20% of the genome when chromosome segments representing common syntenic linkage groups are compared [53]. During evolution, massive amounts of DNA have been inserted and deleted, yet few autosomal rearrangements have been fixed, and at least some banding patterns have remained remarkably unchanged. Gene mobility in mammals seems karyotypically innocuous.

Instances of karyotypic catastrophe, karyotypic megaevolution, have been documented in rats [54], muntjacs [55], and bats. The F_1 hybrid between an Indian (2n = 6) and a Chinese (2n = 46) muntjac typifies these catastrophes. The G-banded chromosome complements from the two almost phenotypically identical parental barking deer can be perfectly aligned. The Indian muntjac's chromosomes seem to have evolved by tandem fusions of the Chinese muntjac's chromosomes [55]. Some tandem and Robertsonian catastrophes (see Chapter 14 this volume), albeit visually spectacular, actually involve a limited number of strictly telic events and do not demonstrate the plethora of inversions and translocations one would expect from mobilized elements or recombination between homologous sequences interspersed throughout the genome. To this author's knowledge, no such instances have been reported. The retroposition mechanism of producing NA_4 tandem repeats (GTx)y in (Fig. 1) and the telomere resolution mechanism for generating similar tandemly repeated oligonucleotides show some similarities [6]; how this may relate to the propagation of interstitial telomeres or the silencing of centromeres and terminal telomeres during Robertsonian fission-fusion is quite speculative.

Each class of chromosome rearrangements is selectively found to be fixed only in certain groups of primates [56]. For example, out of 150 rearrangements distinguishing primates, Robertsonian translocations are preponderant (44 out of a total of 57 rearrangements) among the Lemuridae but are nonexistent among the Pongidae; fissions are very frequent (10/23) among the Cercopithecidae but not elsewhere; and pericentric inversions are especially preponderant (17/28) in Pongidae and man (Hominidae) [56]. Mobility of chromosomal elements may indeed provide part of the grist for chromosome rearrangements. However, the fraction of these rearrangements that is evident in individuals or becomes fixed in populations represents at best a very small and select subset of this grist.

The innocuous appearance of mobility could be misleading because the germline may *possibly* possess a mechanism to discard aberrant karyotypes. Schimke et al. [40] argue as follows: 1) Gene amplification via metabolic perturbation of established somatic cell lines generates extrachromosomal DNA, which induces chromosome breaks. 2) Gene amplification in the soma occurs in vivo in tumors. 3) Metabolic perturbation of proliferating primordial germ cells in female mice induces chromosome abnormalities. 4) In unperturbed established cell lines, spontaneous gene amplification rates are high. 5) Because of the amplification mechanism alone and its concomitant production of extrachromosomal DNA, the percentage of meiotic cells with chromosome breaks should be much higher than is detected in embryos. 6) Most, up to 98% but generally less, of the primordial germ cells in both the ovaries of chickens [57], rodents [58], and humans [59] and the testes of mice [58] degenerate before giving rise to viable meiotic products. While the possibility is unlikely, these cells could represent a selective trash bin for aberrant karyotypes. Data from neonates and spontaneous abortions on spontaneous chromosome rearrangements may not reflect the true germline rate of chromosome breakage.

MOBILE ELEMENTS AND THE BANDED PREPATTERN

Figure 2 exquisitely shows that the two major retroposon families in man, *Alu* and L1, are selectively restricted to (partitioned into) R- and G-bands, respectively, and are apparently absent from the C-bands [60]. This simple overview of retroposon distribution, readily absorbed by any cytogeneticist, has been ignored by reductionists in all the reviews and jumping gene evolution papers of which I am aware. The purpose of the remaining discussion is to present data from two methods for separating R-band DNA from G-band DNA. These data are concurrent in confirming the conclusions from Figure 2 and show that band partitioning is a common feature of retroposon and even retroviral families of sequences.

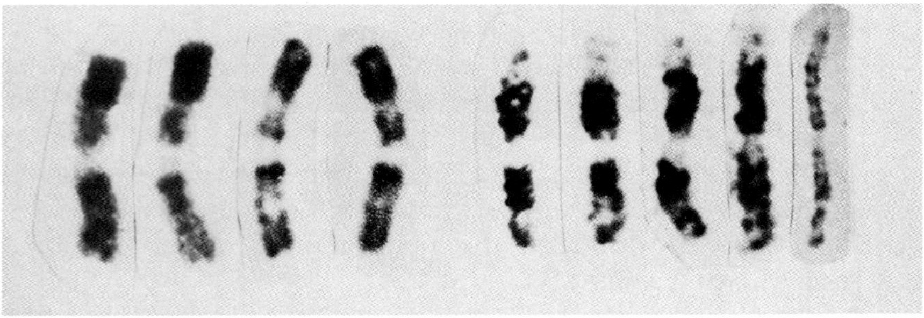

Fig. 2. In situ hybridization of biotinylated *Alu* (leftmost four chromosomes) or L1 interspersed repeat probes to human chromosome 1. The L1 probe was a cloned 1.9-kb Hind III fragment from human L1. Except for lack of signal in centric heterochromatin, *Alu* illuminates an R-banding pattern, while L1 illuminates a G-banding pattern. (Reproduced from Manuelidis and Ward [60], with permission of the publisher.)

Table II summarizes the differences between mammalian R-band and G-band DNA [61] as recently reviewed [62]. Separation of G-band DNA from R-band DNA utilizes two independent physical differences of DNA along the banded prepattern, base composition and replication time. Base composition differences are the basis of quinacrine banding [63–66]; replication time differences are the basis of replication banding [67–73]; and these procedures illuminate the same euchromatic prepattern of longitudinal bands as does trypsin G-banding.

Base Composition Fractionation

Euchromatic "mainband" DNA was fractionated according to base composition [74–82] into four major components (Fig. 3). When Southern transfers of these fractions were probed with *Alu* or L1, *Alu* selectively hybridized to the dG+dC-rich DNA fractions, while L1 selectively hybridized to the dA+dT-rich DNA fractions (Fig. 4). This is concordant with the in situ hybridization data of Figure 2.

Bernardi et al. [75,76,78,80] inferred that mammalian DNA consists of long base composition isochores. The existence of these long stretches of homogeneous-base-composition DNA was inferred from equilibrium density centrifugation data. Main band "euchromatic" DNA separated into discrete Gaussian components of various densities and average base compositions (Fig. 3). The densities and proportions of these components were independent of the initial size of the DNA molecules analyzed over the range of 5–200+ kb [76,78], 200–300 kb representing a practical upper limit to the size

TABLE II. Properties of R-band and G-band DNA

Housekeeping subgenome	Ontogenic subgenome
R-bands	G-bands
Early replicating	Late replicating
Constitutive genes	Tissue-specific genes
GGCGGG-rich promoters	CAAT, TATA box promoters
Quinacrine dull	Quinacrine bright
dG+dC-rich	dA+dT rich
Slow sequence divergence rate	Fast sequence divergence rate
	Concerted methylation
CCGG-rich	CCGG-poor
Shared SINEs/pseudogenes	
SINEs	LINEs
Alu, B1, B2	*Kpn*I, MIF-1
Bovine leukemia provirus	Mouse mammary tumor provirus
Hepatitis B provirus	

Reviewed in [62].

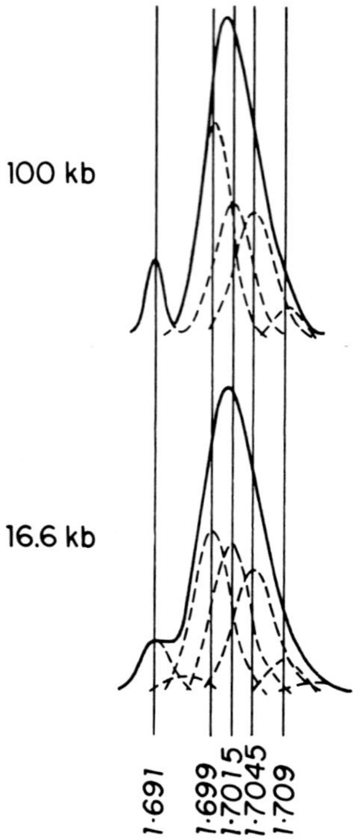

Fig. 3. Base composition fractionation of mouse DNA by analytical buoyant density centrifugation in CsCl gradients. Mouse satellite DNA bands as a light, very dA+dT-rich, satellite at a density of 1.691 g/cc. With the aid of a curve resolver, the main band was resolved as the sum of at least four Gaussian components of various densities and hence different base compositions. Note that the densities and proportions of the Gaussian components are similar for 100-kb DNA and for DNA sheared to 16.6 kb. Similar components from ligand-assisted preparative gradients were used in Figure 4. (Reproduced from Macaya et al. [78], with permission of the publisher.)

of DNA fragments they could analyze. They argued that the DNA in a chromosome band is a domain or isochore with a rather homogeneous base composition and with each isochore differing from the base composition of its neighboring isochores: they argued that they had fractionated DNA according to its G-band or R-band nature [75]. If there are at least 2,000 bands in the human genome [83], then a band represents 1,640 kb of DNA/haploid genome. The practical DNA size limit of 200–300 kb precluded a definite conclusion that DNA had actually been fractionated according to its banded nature [75].

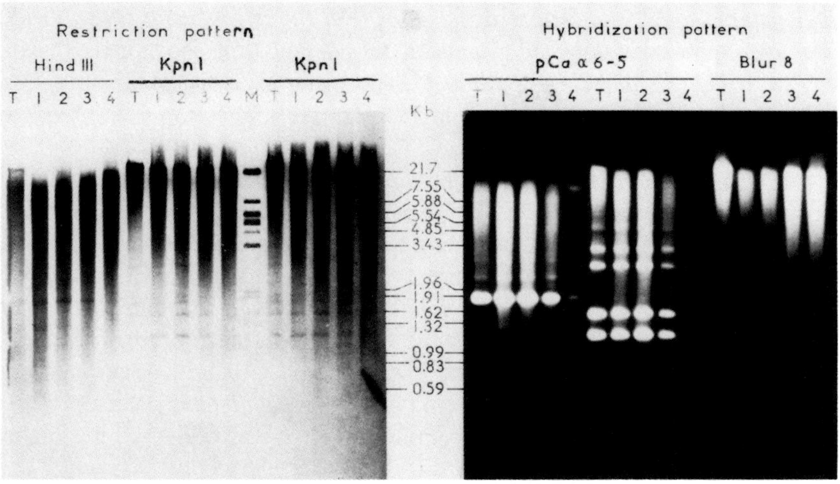

Fig. 4. Southern transfers of base composition fractionated DNA. Human DNA was fractionated by ligand-assisted equilibrium density centrifugation in Cs_2SO_4 into four fractions. T = total DNA; 1 = dA+dT-richest fraction...4 = dG+dC-richest fraction. Restriction endonuclease cleaved DNA was sized by gel electrophoresis (**left**) and transferred to nitrocellulose for hybridization and autoradiography (**right**). The probes used were pCaα 6-5 for the L1 family of interspersed repeats and Blur 8 for the *Alu* family. L1 hybridized preferentially to the dA+dT-rich fractions, while *Alu* hybridized preferentially to the dG+dC-rich fractions. The L1 family can be seen as a 1.9-kb fluorescent band of *Hind*III-digested DNA in the dA+dT-rich lanes of the ethidium-bromide-stained gel here and in the late-replicating DNA lanes in Figure 5. From this 1.9-kb band, the L1 probe used in Figure 2 was isolated (Reproduced from P. Soriano et al. [74], with permission of the Proceedings of the National Academy of Sciences, USA.)

Replication Time Fractionation

Bands are considered to be units of chromosome replication in addition to being units of chromosome structure [68,69,84–87]. When fiber autoradiography was first used to discover bidirectionally replicating replicons in Chinese hamster cells [88], the authors noted that replicons were temporally clustered; i.e., long stretches of DNA had multiple replication origins, all of which seemed to initiate DNA synthesis at the same time. At the termination of synthesis, the bidirectionally progressing replication forks finally came upon forks progressing in the opposite direction, and synthesis of the replicon cluster was complete. Replicons are heterogeneous in size, 50–330 kb [84], and a band is considered to represent a cluster of 10–25 replicons [69]. Since all [89] or almost all [68–70,73] autosomal R-bands replicate before G-bands and since R-bands probably finish synthesis before G-bands initiate synthesis

[89], so that their replication times never [89,90] or probably never [91,92] coincide, then bands should represent replication time isochores. R-bands should represent band size, 1,600 kb, domains of early replicating DNA whose replication times seldom coincide with the replication time of G-band DNA.

S-phase kinetics, replication banding studies, and PCC (premature chromosome condensation) studies are consistent with this isochore concept. S-phase kinetics, in at least several cell lines [61,69,90,93–100], are bimodal with R-band DNA replicating during the first, $S_{E,}$ mode, and G-band DNA replicating during the second, $S_{L,}$ mode (Fig. 6a) [69]. Replication banding [93,101] and S-phase PCC studies [86,87] indicate there is a short interval between the time R-bands complete synthesis and G-bands initiate synthesis during which little or no incorporation of DNA precursor is detected.

The minimum size of replication time isochores was measured in synchronized Chinese hamster cells by Kowalski and Cheevers [102]. The cells were pulsed with ^{14}C-thymidine in early S-phase (see Fig. 6a for temporal orientation): 2.5 h later, after the labeled replicons had finished replicating, bromodeoxyuridine (BrdU) was added for the duration of S-phase. DNA was isolated after the nascent strands had ligated to strands in neighboring contiguous replicons. Then very high molecular weight (515 kb) strands of single-stranded DNA, corresponding to about three or more contiguous replicons, were centrifuged to equilibrium in alkaline CsCl gradients to separate unsubstituted from BrdU-substituted DNA. The ^{3}H-radioactivity and optical density owing to BrdU-DNA were almost completely—95%—separated. At a resolution of almost 1,000 kb, DNA labeled during S_E is not covalently associated with DNA synthesized during S_L. These data strongly suggest that replication time isochores are the size of bands and definitely show, to a resolution of almost 1,000 kb, that early- and late-replicating DNA segments are not interspersed.

Figure 5 shows that a majority of *Alu* family members are in the early replicating DNA fraction and that majority of L1 family members are in the late replicating DNA fraction. This is consistent with Figures 2 and 4, showing *Alu* selectively partitioned into dG+dC-rich R-band DNA, and L1 selectively partitioned into dA+dT-rich G-band DNA. Table III shows that two-thirds of the interspersed repeat families in hamsters are selectively partitioned into early-replicating or into late-replicating DNA and that if partitioning is detected for a family of sequences, SINE families are selectively partitioned into the early-replicating (R-band) fraction, while LINE families are usually partitioned into the late-replicating (G-band) fraction. Figure 7 indicates that for one family of long interspersed repeats comple-

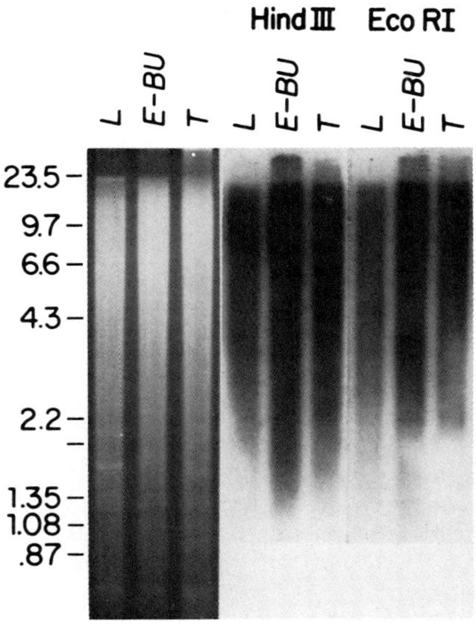

Fig. 5. Southern transfers of replication time fractionated DNA. Late-replicating (L), early-replicating (E-BU), and total (T) human HeLa cell DNA fractions were prepared much as described in Figure 6. DNA from an ethidium-stained gel (**left**) was transferred and probed with *Alu* (**right**), which selectively illuminated the early-replicating DNA fractions. We calculated that about 80% of the *Alu* copies are in the early-replicating 50% of the genome. Note that the 1.9-kb *Hind*III band representing the human L1 LINE family is almost absent from the early DNA fraction in the fluorogram **(left.)** Using our replication time fractionated hamster DNA, Vizard and Rosenberg reported similar results, mouse L1 probes selectively illuminated the late-replicating DNA lane [114].

mentary to the cloned sequence C1, partitioning is almost total in that all or almost all copies of this interspersed repeat family replicate in the last half of S-phase. In addition to the partitioning of retroposon families, proretrovirus families such as MMTV are also selectively partitioned into either the dA+dT-rich or the dG+dC-rich fractions [12,103] (Table II).

EVOLUTION AND CHROMOSOME REARRANGEMENTS

Individual bases show partitioning: R-bands are dC+dG-rich. CCGG sequences show partitioning: early-replicating DNA is terminally digested to a much lower molecular weight by *Msp*I than is late-replicating DNA (Table II) [61,62]. Partitioning of longer sequences such as mobile elements should therefore not be surprising. Selectively neutral mobile elements have been

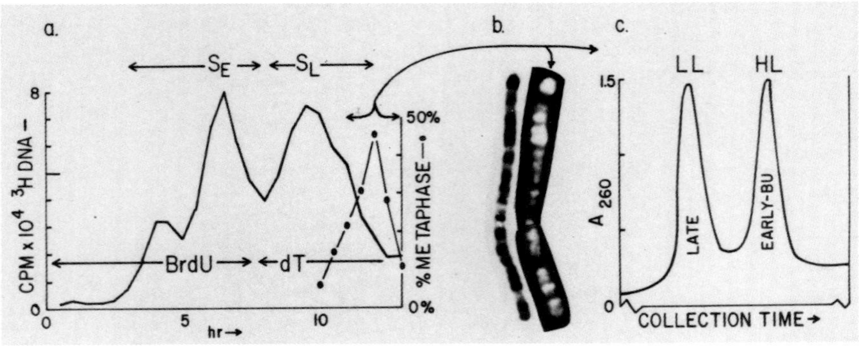

Fig. 6. Fractionation of early- and late-replicating V79 hamster cell DNA. **a:** Hamster V79 metaphase cells were transferred to 36°C at time zero and the amount of ^{3}H-labeled thymidine incorporated into DNA (solid line) was measured at intervals during S-phase. The two peaks of DNA synthesis correspond to the first (S_E) and second (S_L) halves of S-phase. The mitotic index is indicated by the broken line. (No subsequent bursts of mitoses were seen when observations ceased at 18 h. This discounts there being two populations of cells gated from G_1 at different times.) Another population of metaphase cells was grown in BrdU-substituted medium until the expected mid-S-phase and released into thymidine. At the time of the expected metaphase burst, chromosome spreads and metaphase DNA were prepared. **b:** A partial karyotype was prepared from a 33258 Hoechst-stained slide photographed in the fluorescence microsocope (rightmost chromosome). The pattern of fluorescence is equivalent to the trypsin Giemsa-banding pattern (leftmost). **c:** DNA was fractionated on a CsCl density gradient. The absorbancy at 260 nm of the fractions was monitored in a flow cell, and heavy (HL) and light (LL) peak fractions were collected and used in subsequent experiments. (Reproduced from Goldman et al. [61], with permission of the American Association of the Advancement of Science.)

subjected to the same diffusion equations as have neutral alleles. This initially prompted us to call R- and G-bands two separate subgenomes [61] because they seemed to represent two mutually isolated populations of sequences that were not at equilibrium for the diffusion of mobile elements.

Selection against organisms containing a mobile element of phenotypic consequence could act to partition mobile elements. Retro-proviruses are partitioned [103] (Table II). Partitioning means that a family of mobile elements has a majority of its members either near housekeeping genes or near tissue-specific genes. Since G-band DNA is transcriptionally repressed in most tissues, while most genes in R-band DNA are constitutively active in all tissues [61], partitioning of MMTV retrovirus family members into dA+dT-rich DNA fractions (Table II) [75] could conceivably reflect selection against organisms containing a provirus in a constitutively active domain. Selection against organisms could not act to partition retroposons such as *Alu*. *Alu* sequences have been fixed into the human lineage during primate evolution at the average rate of one copy every 130 yr (6.5×10^7 yr/5×10^5 copies [12]); selection against an almost equal frequency of individuals afflicted with *Alu* insertions into their G-bands would certainly represent an intolerably high genetic load.

Molecular drive mechanisms are more reasonable candidates for generating partitioning. Retroposons do show some target site preference [6], but this is in fact a dA+dT-rich target site preference for *Alu* and would not explain its partitioning into dG+dC-rich isochores unless they just happen to be rich in target site microenvironments. Partitioning would occur by a molecular drive mechanism if mobility was limited to a particular stage of S-phase, for example, by stage specific transcripts being retroposed only into nascent DNA. R-band DNA seems selectively mobilized by the amplification mechanism. With the possible exception of one positively C-banding homogenous staining region (HSR) [104], amplified regions such as HSRs or double minutes are both early-replicating and trypsin-Giemsa light regions, and amplified genes are house-keeping genes [104]. Genes involved in DNA synthesis, hypoxanthine phosphoribosyltransferase (HPRT), dihydrofolate reductase (DHFR), carbamyl phosphate synthetase, aspartate transcarbamylase, dihydroorotase (CAD), are particularly sensitive to amplification; and we noticed these genes are extremely early replicating [61]. The very early part of S-phase near the G1-S border is particularly sensitive to aphidicolin-induced DHFR amplification [40]. S-phase actually has three identifiable modes (Fig. 6a). These were independently identified with a series of DNA synthesis mutants and called S1, S2, and S3 [105]. In Figure 6a, S_E = S1+S2. In 1975, Klevecz et al. [98] called S1 catabolic DNA because much of the DNA synthesized during S1 was eventually degraded. Their data were immediately ignored but can now be more readily assimilated under the umbrella of amplification as reflecting the generation of short-lived extra-chromosomal DNA and as a mechanism that could result in the selective mobility of early replicating (R-band) DNA.

Partitioning of mobile elements could be a rather recently evolved addition to the banded prepattern. Frog and toad chromosomes show excellent replication bands [106] and G-bands [107,108], but they do not show banding with quinacrine or with mithramycin/chromomycin A_3 [106,109]. Consistent with the absence of cytologically detectable base composition differences between bands, *Xenopus* DNA does not show base composition isochores. Unlike mammalian DNA (Fig. 3), *Xenopus* DNA's density profiles are very dependent on the molecular weight of the starting material in the range 5-100 kb [76,80]. Thus, G-banding and replication banding represent primitive attributes of the banded prepattern, whereas base composition differences between bands reflect a more recent evolutionary event. When the remaining

TABLE III. Properties of randomly chosen repetitive DNA clones

Clone[a]	Insert size[c]	Copies early[b]	Copies late[b]	Class by spot[b]	Southern to V79[d]	Hybrid to Hela[e]
D3	4.5	120,000	18,000	E	E-SINE	+ E-SINE
A4	1.2	77,000	15,400	E	E-SINE	+ E-SINE
D4	2.1	55,000	4,970	E	E-SINE	?++
B4	1.0	3,360	252	E	vE-SINE[f]	+ E-SINE
B3	3.0	2,468	360	E	vE-SINE	++ E-SINE
C5	2.5	1,150	355	E	vE-SINE[f]	+E
A3		1,105	410	E		–
A7	3.1	995	220	E	vE-SINE	+
D10	3.9	985	80	E	vE-LINE	+ E(LINE?)
B6	1.1	660	160	E	E-SINE	–
B10		560	160	E		–
B9	1.0	1,820	3,460	S	E-SINE	–
C7	4.3	1,000	680	S	S–E SINE	++ E-SINE
A2	1.5	640	545	S		++
A9	.9	490	490	S	S-SINE[f]	–
D11		32	72	S		+
D6		44	70	S		–
A5		24	44	S	(SINE?)	–
C9	1.7	27	37	S	S–E	–
C2		32	32	S		–
C10		14	24	S		–
D7		14	16	S		–
C6		14	14	S		–

D2		17,000	44,000	L		–
C8	2.1	265	28,000	L	vL C-band	–
E2		560	4,000	L		–
B7	2.8	250	2,050	L	L-LINE[f]	+ L-(LINE?)
A1	8.2	605	1,820	L	L-LINE	–
C3	3.3	54	1,145	L	vL C-band[f]	–
C1	1.7	27	330	L	vL-LINE	–
D5		44	160	L		–
C8		14	60	L		–
D1		—	32	L		–
D12		—	14	L?		–
A36Fc						L-LINE
MIF-1					L-LINE[g]	
Alu						E-SINE[h]

E, early; S, shared; L, late; vE, very early; vL, very late.

[a]From a genomic bank of Chinese hamster *Pst*I-digested DNA in pBR322, clones containing hamster repeat DNA were screened by colony hybridization with total nick-translated hamster DNA.

[b]Copy No. in the two subgenomes was estimated by probing spot blots of serially diluted plasmid DNA with early- and late-replicating DNA from Chinese hamster V79 cells.

[c]Inserts from 19 different plasmids were sized.

[d]These inserts were used to probe Southern transfers of early- and late-replicating V79 DNA digested with *Eco*RI, *Msp*I, or *Hpa*II [67]. The replication time was qualitatively revaluated from the resulting autoradiograms. v, very; S-E shared to early. SINE or LINE nature was determined from the autoradiographic patterns and from *in situ* hybridization. Of the 19 inserts, only D3 and A4 cross-hybridized. D3, A4, and D4 probably represent the rodent *Alu* equivalent families B1 and B2.

[e]Human HeLa cell DNA cross-hybridized to some cloned plasmid DNAs under conditions that require 87% (++) or 77% (+) base-pairing homology. Inserts from these plasmids were used to probe Southern transfers of replication-time fractionated HeLa cell DNA.

[f][61].

[g][114].

[h]Figure 5.

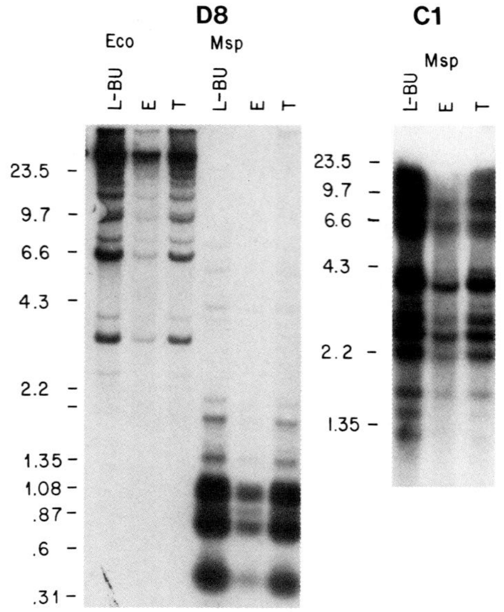

Fig. 7. Replication time-fractionated DNA. Autoradiograms of a Southern transfer that had been sequentially probed with several cloned hamster repeat sequences. Hamster V79 cell DNA was replication time fractionated as in Figure 6. Five micrograms of late-replicating-BrdU substituted (L-BU), early-replicating unsubstituted (E), and total (T) DNA were digested with *Eco*RI or *Msp*I, electrophoresed, Southern transferred, and probed with C5, an R-band partitioned repeat that selectively illuminated the E lanes (Fig. 5 from [61]). The transfer was reprobed with D8 and C1 (Table III), which selectively illuminate the L-BU lanes, as shown here. By ^{3}H-in situ hybridization, D8 hybridized to one C-band while C1 hybridized rather homogeneously to all chromosomes. D8 is a tandem repeat, illuminating an arithmetic *Eco*RI ladder of 3.3 kb, so we assume D8 represents late-replicating "satellite" DNA. The E/L-BU autoradiographic signal ratio for C1 is visually equal to that of D8, indicating that all or almost all C1 family members are partitioned into late-replicating DNA.

attributes (Table II), including mobile element partitioning, evolved to coincide with the banded prepattern is unknown.

Mobile Elements and the C Value Paradox

The C value (haploid DNA content) paradox [110] was deemed paradoxical because 1) eukaryotes have much more DNA than was thought necessary to act as genes and 2) many lower vertebrates have much more DNA than do higher vertebrates. While mammals have C values that within the narrow range of C = 2–3 pg (one picogram of DNA = 10^9 bp), amphibian C values vary from C = 1 pg to C = 100 pg [110]. Discoveries of satellite DNA and

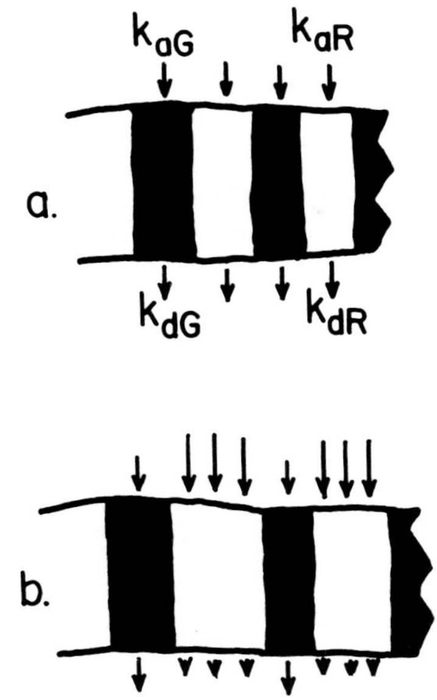

Fig. 8. A chromosome segment is depicted as a banded prepattern through which flows a karyotypically innocuous flux of mobile genetic elements. **a:** During evolutionary time, there is an influx of added elements with a rate constant k_a and an efflux of deleted elements with a rate constant k_d. Partitioning of a family of mobile elements means that the net flux, $k_a - k_d$, for that family during recent evolutionary time has been different for R-bands and G-bands. Therefore, the rate constants for G- vs. R-bands can be different, and an additional subscript has been used to indicate this. **b:** If, during recent evolutionary history, the net flux into R-bands had been high, the R-bands would have become enlarged.

introns only partly explain the paradox. Mizumo and Macgregor examined the intra and inter species reassociation kinetics of DNA within the salamander family Plethodontidae, wherein the C value varies from 18 to 69 pg and very little of the DNA is from C-banding heterochromatin [111]. The more rapidly reassociating (high copy number) fractions showed very low percentage homology to similar fractions from other plethodontid species while the slow-reassociating fractions were only slightly more taxonomically conservative. Only about 10% of the DNA seemed essential in that these sequences were common to all plethodontids. Since the unbanded karyotypes were, with the exception of chromosome length, almost identical for all species studied, they concluded that the presence of nonessential DNA, presumably DNA rather randomly inserted throughout the genome, was karyotypically innocuous. Since retroposons are found in fish [33], they are not restricted to mammals and may account for the nonessential DNA in salamanders. In conclusion, 90% of the plethodontid genome can be nonessential DNA, and this overwhelming mass of DNA has been mobilized in this salamander family while causing almost no chromosome rearrangements to become fixed.

The major primate and rodent short retroposon families are evolutionarily rather recent, having evolved from a parental RNA sequence 45–65 million years (my) ago [12] (Table II). Hundreds of thousands of copies of these sequences have been added to these genomes during this 45–65-my period. Since the C value of mammals is rather constant (C = 2–3 pg), the rate constant, k_a, for the addition of new elements must be almost identical with the rate constant, k_d, for the deletion of old elements (Fig. 8a). Chickens almost toally lack SINEs [112,113], and 87% of their DNA sequences are single-copy sequences [113]. Also, C values and their variance within Aves are low; C = 0.8–1.3 pg [110]. The data from the class Aves are consistent with $k_a < k_d$. With this condition, little extra DNA can accumulate. Conversely, C values and the variance of C values within the order Urodela, which includes salamanders, are consistent with $k_a >$ kd. With this condition, DNA simply accumulates with evolutionary time. Since the mechanism for deletion of retroposons most likely does not involve an RNA intermediate and therefore must be different from retroposition itself, one would expect k_a and k_d to usually be uncoupled and in groups where $k_a > k_d$, the variance of C values within the taxonomic group should be high. That k_a seems tightly coupled to and equal to k_d within the entire class Mammalia, is more unexpected than the initial paradox itself.

Partitioning of families of mobile elements into G- or R-bands can now be formally defined: for a partitioned family, $k_{aG} - k_{dG}$ and $k_{aR} - k_{dR}$ are unequal. This formality presents the banded karyotype as a pattern through which flows a (constant ?) flux of innocuous mobile elements. A banded karyotype displays a history of the net influx and efflux into and out of each band (Fig. 8). Twenty percent of the cat karyotype's being visually identical with that of the human [53] shows the net band per band flux can remain constant enough to allow visual band for band pairing of some homologous syntenic groups. After almost 100 my of evolution, syntenic groups that appeared visually different could have appeared so either because of rearrangements or because of differential ballooning (Fig. 8b) and contraction of the bands being visually compared.

CONCLUSIONS

The mechanisms by which mobile genetic elements move are now known or are at least partially known. Most of the interspersed repetitive DNA sequences in mammals are families of mobile elements that move by a common mechanism called retroposition. Here, reverse transcriptase acts upon an RNA intermediate, converts it into DNA, and inserts this sequence into the genome. Mobilization also seems to occur by a process involving gene amplification. Phenotypically dramatic classes of very high mobility elements, such as those described by McClintock in *maize,* have not been observed in mammals. While the mobility mechanisms provide ample indication that mobility and chromosome breaks could go hand-in-hand, mobility in mammals and urodeles has been karyotypically quite innocuous. While millions of repetitive sequences have been fixed during primate evolution and probably several orders of magnitude more of these events have occurred in primate germlines but failed to become fixed, only hundreds of autosomal rearrangement events have become fixed in primate genomes, and the basic banding pattern has remained visually intact. Speculation about how often a mobile event causes a chromosome rearrangement is limited by our present ignorance concerning possible selection against cells with chromosome rearrangements during germ cell atresia. While mechanisms for the addition of mobile elements to the genome are abundant, mechanisms for their elimination are still speculative. If the influx of mobile elements into the genome were greater than the efflux of mobile elements out, this would explain some aspects of the C value paradox. Without an efficient elimination mechanism, our genome would soon blow up like that of lungfish. Finally, most families of retroposons and retroviruses in mammals have become either partially or totally restricted to either G-bands or to R-bands. The net flux of a particular mobile family can somehow be directed by the chromosome's banded prepattern.

ACKNOWLEDGMENTS

We thank Laura Manuelidis for Figure 2 and Georgio Bernardi for Figures 3 and 4. Supported by NIH Public Health Service Grant GM 35645 and Career Development Award KO4-HD00323 to G.H.

REFERENCES

1. Federoff NV: Controlling elements in maize. In Shapiro JA (ed): "Mobile Genetic Elements." New York: Academic Press, 1983, p 1.
2. Ising G, Ramel C: The behaviour of a transposing element in Drosophila melanogaster. In Ashburner M, Novitski E (eds): "The Genetics and Biology of Drosophila." Vol 1b. New York and London: Academic Press, 1976, p 947.
3. Ising G, Block K: Derivation-dependent distribution of insertion sites for a Drosophila transposon. Cold Spring Harbor Symp Quant Biol 45:527, 1981.

4. Shapiro JA: "Mobile Genetic Elements." New York: Academic Press, 1983.
5. Finnegan DJ: Transposable elements in eukaryotes. Int Rev Cytol 93:281, 1985.
6. Rogers JH: The origin and evolution of retroposons. Int Rev Cytol 93:187, 1985.
7. Sharp PA: Conversion of RNA to DNA in mammals: Alu-like elements and pseudogenes. Nature 301:471, 1983.
8. Rogers J: The origin of retroposons. Nature 319:725, 1986.
9. Hardman N: Structure and function of repetitive DNA in eukaryotes. Biochem J 234:1, 1986.
10. Schmid CW, Jelinek WR: The Alu family of dispersed repetitive sequences. Science 216:1965, 1982.
11. Singer MF, Skowronski J: Making sense out of LINES: Long interspersed repeat sequences in mammalian genomes. Trends Biochem Sci 10:119, 1985.
12. Deininger PL, Daniels GR: The recent evolution of mammalian repetitive DNA elements. Trends Genet 2:76, 1986.
13. Sun L, Paulson KE, Schmid CW, Kadyk L, Leinwand L: Non-Alu family interspersed repeats in human DNA and their transcriptional activity. Nucleic Acids Res 12:2669, 1984.
14. Tashima M, Calabretta B, Torelli G, Scofield M, Maizel A: Presence of a highly repetitive and widely dispersed DNA sequence in the human genome. Proc Natl Acad Sci USA 78:1508, 1981.
15. Georgeiv GP, Kramerov DA, Ryskov AP, Skryabin KG, Lukanidin EM: Dispersed repetitive sequences in eukaryotic genomes and their possible biological significance. Cold Spring Harbor Symp Quant Biol 42:1109, 1982.
16. Voliva CF, Jahn CL, Comer MB, Hutchinson CA III, Edgell MH: The L1Md long interspersed repeat family in the mouse: Almost all examples are truncated at one end. Nucleic Acids Res 11:8847, 1983.
17. Loeb DD, Padgett RW, Hardies SC, Shehee WR, Comer MB, Edgell MH, Hutchison CA: The sequence of a large L1Md element reveals a tandemly repeated 5′ end and several features found in retrotransposons. Mol Cell Biol 6:168, 1986.
18. Burton FH, Loeb DD, Voliva CF, Martin SL, Edgell MH, Hutchison CA: Conservation throughout mammalia and extensive protein-encoding capacity of the highly repeated DNA long interspersed sequence one. J Mol Biol 187:291, 1986.
19. Bennett KL, Hill RE, Pietras DF, Woodworth-Gutai M, Kane-Hass C, Houston JM, Heath JK, Hastie ND: Most highly repeated dispersed DNA families in the mouse genome. Mol Cell Biol 4:1561, 1982.
20. Van Arsdell SW, Denison RA, Bernstein LB, Weiner AM: Direct repeats flank three small nuclear RNA pseudogenes in the human genome. Cell 26:11, 1981.
21. Jagadeeswaran P, Forget BG, Weissman SM: Short interspersed repetitive DNA elements in eucaryotes: Transposable DNA elements generated by reverse transcription of RNA Pol III transcripts. Cell 26:141, 1981.
22. Jackson M, Heller D, Leinwand L: Transcriptional measurements of mouse repeated DNA sequences. Nucleic Acids Res 13:3389, 1985.
23. Matsumoto KI, Murakami K, Okada N: Gene for lysine tRNA may be a progenitor of the highly repetitive and transcribable sequences present in the salmon genome. Proc Natl Acad Sci USA 83:3156, 1986.
24. Shyman S, Weaver S: Chromosomal rearrangements associated with LINE elements in the mouse genome. Nucleic Acids Res 13:5085, 1985.
25. Economou-Pachnis A, Tsichlis PN: Insertion of an Alu SINE in the human homologue of the M1v-2 locus. Nucleic Acids Res 13:8379, 1985.

26. Lakshmikumaran MS, D'Ambrosio ED, Laimins LA, Lin DT, Furano AV: Long interspersed repeated DNA (LINE) causes polymorphism at the rat insulin 1 locus. Mol Cell Biol 5:2197, 1985.
27. Kominami R, Muramatsu M, Moriwaki K: A mouse type 2 Alu sequence (M2) is mobile in the genome. Nature 301:87, 1983.
28. Heller D, Jackson M, Leinwand L: Organization and expression of non-Alu family interspersed repetitive DNA sequences in the mouse genome. J Mol Biol 173:419, 1982.
29. Varmus HE: Form and function of retroviral proviruses. Science 216:812, 1982.
29a. Kozak C, Silver J: The transmission and activation of endogenous mouse retroviral genomes. Trends Genet 1:331, 1985.
30. Ponta H, Gunzburg WH, Salmons B, Groner B, Herrlich P: Mouse mammary tumour virus: A proviral gene contributes to the understanding of eukaryotic gene expression and mammary tumorigenesis. J Gen Virol 66:931, 1985.
31. Carter AT, Norton JD, Gibson Y, Avery RJ: Expression and transmission of a rodent retrovirus-like (VL30) gene family. J Mol Biol 188:105, 1986.
32. Hawley RG, Shulman MJ, Murialdo H, Gibson DM, Hozumi N: Mutant immunoglobulin genes have repetitive DNA elements inserted into their intervening sequences. Proc Natl Acad Sci USA 79:7425, 1982.
33. Fink GR, Roeder GS: Transposable elements in yeast. In Shapiro JA (ed): "Mobile Genetic Elements." New York: Academic Press, 1983, p 300.
34. Cote B, Bender W, Curtis D, Chovnick A: Molecular mapping of the Rosy locus in Drosophila melanogaster. Genetics 112:769, 1986.
35. Smithies O, Gregg RG, Boggs SS, Koralewski MA, Kucherlapati RS: Insertion of DNA sequences into the human chromosomal beta-globin locus by homologous recombination. Nature 317:230, 1985.
36. Riabowol K, Schmookler-Reis RJ, Goldstein S: Interspersed repetitive and tandemly repetitive sequences are differentially represented in extrachromosomal covalently closed circular DNA of human diploid fibroblasts. Nucleic Acids Res 13:5563, 1985.
37. Fujimoto S, Tsuda T, Toda M, Yamagishi H: Transposon-like sequences in extrachromosomal circular DNA from mouse thymocytes. Proc Natl Acad Sci USA 82:2072, 1985.
38. Paulson KE, Deka N, Schmid CW, Misra R, Schindler CW, Rush MG, Kadyk L, Leinwand L: A transposon-like element in human DNA. Nature 316:359, 1985.
39. Alberts B, Bray D, Lewis J, Raff M, Roberts K, Watson JD: "Molecular Biology of the Cell." New York: Garland Publications, 1983.
40. Schimke RT, Sherwood SW, Hill AB, Johnston RN: Overreplication and recombination of DNA in higher eukaryotes: Potential consequences and biological implications. Proc Natl Acad Sci USA 83:2157, 1986.
41. Hamkalo BA, Farnham PJ, Johnson R, Schimke RT: Ultrastructural features of minute chromosomes in a methotrexate-resistant mouse 3T3 cell line. Proc Natl Acad Sci USA 82:1126, 1985.
42. Johnston RN, Beverley SM, Schimke RT: Rapid spontaneous dihydrofolate reductase gene amplification shown by fluorescence-activated cell sorting. Proc Natl Acad Sci USA 80:3711, 1983.
43. Sager R, Gadi IK, Stephens L, Grabowy CT: Gene amplification: An example of accelerated evolution in tumorigenic cells. Proc Natl Acad Sci USA 82:7015, 1985.
44. Smith GP: Evolution of repeated DNA sequences by unequal crossover. Science 191:528, 1976.

45. Lehrman MA, Schneider WJ, Sudhof TC, Brown MS, Goldstein JL, Russell DW: Mutation in LDL receptor: Alu-Alu recombination deletes exons encoding transmembrane and cytoplasmic domains. Science 227:140, 1985.
46. Watt VM, Ingles CJ, Urdea MS, Rutter WJ: Homology requirements for recombination in Escherichia coli. Proc Natl Acad Sci USA 82:4768, 1985.
47. Shen P, Huang HV: Homologous recombination in Escherichia coli: Dependence on substrate length and homology. Genetics 112:441, 1986.
48. Sutcliffe JG, Milner RJ, Bloom FE, Lerner RA: Common 82-nucleotide sequence unique to brain RNA. Proc Natl Acad Sci USA 79:4942, 1982.
49. Hardison RC, Printz R: Variability within the rabbit C repeats and sequences shared with other SINES. Nucleic Acids Res 13:1073, 1985.
50. Orgel LE, Crick FHC: Selfish DNA: The ultimate parasite. Nature 284:604, 1980.
51. Walsh JB: How many processed pseudogenes are accumulated in a gene family? Genetics 110:345, 1985.
52. Seuanez HN: "The Phylogeny of Human Chromosomes." Berlin, Heidelberg, New York: Springer-Verlag, 1979.
53. Nash WG, O'Brien SJ: Conserved regions of homologous G-banded chromosomes between orders in mammalian evolution: Carnivores and primates. Proc Natl Acad Sci USA 79:6631, 1982.
54. Elder FFB: Tandem fusion, centric fusion, and chromosomal evolution in the cotton rats, genus Sigmodon. Cytogenet Cell Genet 26:199, 1980.
55. Liming S, Yingying Y, Xingsheng D: Comparative cytogenetic studies on the red muntjac, Chinese muntjac, and their F1 hybrids. Cytogenet Cell Genet 26:22, 1980.
56. Dutrillaux B: Chromosomal evolution in primates: Tentative phylogeny from Microcebus murinus (Prosimian) to man. Hum Genet 48:251, 1979.
57. Hughes GC: The population of germ cells in the developing female chick. J Embryol Exp Morphol 11:513, 1963.
58. Whittingham DG, Wood MJ: Reproductive physiology. In Foster HL, Small JD, Fox JG (eds): "The Mouse in Biological Research, Vol. 3." New York: Academic Press, 1983, pp 138–165.
59. Baker TG: A quantitative and cytological study of germ cells in human ovaries. Proc R Soc Lond [Biol]158:417, 1963.
60. Manuelidis L, Ward DC: Chromosomal and nuclear distribution of the HindIII 1.9-kb human DNA repeat segment. Chromosoma 91:28, 1984.
61. Goldman MA, Holmquist GP, Gray MC, Caston LA, Nag A: Replication timing of mammalian genes and middle repetitive sequences. Science 224:686, 1984.
62. Holmquist G: DNA sequences in G-bands and R-bands. In Adolph KR (ed): "Chromosome and Chromatin Structure." Boca Raton, FL: CRC Press, In Press.
63. Comings DE, Kovacs BW, Avelino E, Harris DC: Mechanisms of chromosome banding. V. Quinacrine Banding. Chromosoma 50:111, 1975.
64. Latt SA, Brodie S, Munroe SH: Optical studies of complexes of quinacrine with DNA and chromatin: Implications for the fluorescence of cytological chromosome preparations. Chromosoma 49:17, 1974.
65. Weisblum B, De Haseth PL: Quinacrine, a chromosome stain specific for deoxyadenylate-deoxythymidylate-rich regions in DNA. Proc Natl Acad Sci USA 69:1972.
66. Korenberg JR, Engels WR: Base ratio, DNA content, and quinacrine-brightness of human chromosomes. Proc Natl Acad Sci USA 75:3382, 1978.
67. Latt SA: Microfluorometric detection of deoxyribonucleic acid replication in human metaphase chromosome. Proc Natl Acad Sci USA 70:3395, 1973.

68. Latt SA: Fluorescence analysis of late DNA replication in human metaphase chromosomes. Somatic Cell Mol Genet 1:293, 1975.
69. Holmquist G, Gray M, Porter T, Jordan J: Characterization of Giemsa dark- and light-band DNA. Cell 31:121, 1982.
70. Grzeschik K.-H, Kim MA, Johannsmann R: Late replicating bands of human chromosomes demonstrated by fluorochrome and Giemsa staining. Humangenetik 29:41, 1975.
71. Epplen JT, Siebers J.-W, Vogel W: DNA replication patterns of human chromosomes from fibroblasts and amniotic fluid cells revealed by a Giemsa staining technique. Cytogenet Cell Genet 15:177, 1975.
72. Epplen JT, Vogel W: DNA replication patterns of human C group chromosomes from fibroblasts and aminiotic fluid cells revealed by a Giemsa staining technique. Humangenetik 30:337, 1975.
73. Meer B, Hameister H, Cerrillo M: Early and late replication patterns of increased resolution in human lymphocyte chromosomes. Chromosoma 82:315, 1981.
74. Soriano P, Meunier-Rotival M, Bernardi G: The distribution of interspersed repeats is nonuniform and conserved in the mouse and human genomes. Proc Natl Acad Sci USA 80:1816, 1983.
75. Bernardi G, Olofsson B, Filipski J, Zerial M, Salinas J, Cuny G, Meunier-Rotival M, Rodier M: The mosaic genome of warm-blooded vertebrates. Science 228:953, 1985.
76. Thiery JP, Macaya G, Bernardi G: An analysis of eukaryotic genomes by density gradient centrifugation. J Mol Biol 108:219, 1976.
77. Cuny G, Soriano P, Macaya G, Bernardi G: The major components of the mouse and human genomes. 1. Preparation, basic properties and compositional heterogeneity. Eur J Biochem 115:227, 1981.
78. Macaya G, Thiery JP, Bernardi G: An approach to the organization of eukaryotic genomes at a macromolecular level. J Mol Biol 108:237, 1976.
79. Soriano P, Macaya G, Bernardi G: The major components of the mouse and human genomes. 2. Reassociation kinetics. Eur J Biochem 115:235, 1981.
80. Bernardi G, Olofsson B, Filipski J, Zerial M, Salinas J, Cuny G, Meunier-Rotival M, Rodier F: The mosaic genome of warm-blooded vertebrates. Science 228:953, 1985.
81. Olofsson B, Bernardi G: Organization of nucleotide sequences in the chicken genome. Eur J Biochem 130:241, 1983.
82. Meunier-Rotival M, Soriano P, Cuny G, Strauss F, Bernardi G: Sequence organization and genomic distribution of the major family of interspersed repeats of mouse DNA, Proc Natl Acad Sci USA 79:355, 1982.
83. Yunis JJ: Mid-prophase human chromosomes. The attainment of 2,000 bands. Hum Genet 56:293, 1981.
84. Hand R: Eucaryotic DNA: Organization of the genome for replication. Cell 15:317, 1978.
85. Stubblefield E, Analysis of the replication pattern of Chinese hamster chromosomes during 5-bromodeoxyuridine suppression of 33258 Hoechst fluorescence. Chromosoma 53:209, 1975.
86. Lau Y.-F, Arrighi FE: Studies of mammalian chromosome replication. II. Evidence for the existence of defined chromosome replicating units. Chromosoma 83:721, 1981.
87. Hameister H, Sperling K: Description of a chromosome replication unit in individual prematurely condensed human S-phase chromosomes. Chromosoma 90:389, 1984.
88. Huberman JA, Riggs AD: On the mechanism of DNA replication in mammalian chromosomes. J Mol Biol 32:327, 1968.

89. Camargo M, Cervenka J: Patterns of DNA replication of human chromosmes. II. Replication map and replication model. Am J Hum Genet 34:757, 1982.
90. Schmidt M: Two phases of DNA replication in human cells. Chromosoma 76:101, 1980.
91. Dutrillaux B, Couturier J, Richer CL, Viegas-Pequignot E: Sequence of DNA replication in 277 R- and Q-bands of human chromosomes using a BrdU treatment. Chromosoma 58:51, 1976.
92. Kondra RM, Ray M: Analysis of DNA replication patterns of human fibroblast chromosomes. The replication map. Hum Genet 43:139, 1978.
93. Schempp W, Vogel W: Difference between diploid and aneuploid Chinese hamster cells and replication at mid-S-phase. Chromosoma 73:109, 1979.
94. Klevecz RR, Kapp LN: Intermittent DNA synthesis and periodic expression of enzyme activity in the cell cycle of WI-38. J Cell Biol 58:564, 1973.
95. Kapp LN, Klevecz RR: The cell cycle of low passage and high passage human diploid fibroblasts. Exp Cell Res 101:154, 1976.
96. Klevecz RR: Temporal structure of S phase analyzed using an automated synchrony system. Biotechnology Bioengineering XVII:675, 1975.
97. Klevecz RR, Keniston BA, Deaven LL: The temporal structure of S phase. Cell 5:195, 1975.
98. Klevecz RR, Kapp LN, Remington JA: Intermittent amplification and catabolism of DNA and its correlation with gene expression. In "Control of Proliferation in Animal Cells." New York: Cold Spring Harbor Laboratories, 1974, p 817.
99. Remington JA, Klevecz RR: Families of replicating units in cultured hamster fibroblasts. Exp Cell Res 76:410, 1973.
100. Clausen OPF, Kirkhus B, Elgjo K, Pedersen S, Bolund L: DNA synthesis rate changes during the S phase in mouse epidermis. Cell Tissue Kinet 18:445, 1985.
101. Schempp W, Vogel W: Decrease of DNA synthesis in amniotic fluid cells during the middle part of S-phase revealed by differential chromosome staining after incorporation of BrdU. Chromosoma 67:193, 1978.
102. Kowalski J, Cheevers WP: Synthesis of high molecular weight DNA strands during S phase. J Mol Biol 104:603, 1976.
103. Kettman R, Meunier-Rotival M, Cortadas J, Cuny G, Ghysdael J, Mammerick M, Burny A, Bernardi G: Integration of bovine leukemia virus DNA in the bovine genome. Proc Natl Acad Sci USA 76:4822, 1979.
104. Schimke RT: "Gene Amplification." New York: Cold Spring Harbor Laboratories, 1982.
105. Marcus M, Fainsod A, Diamond G: The genetic analysis of cell-cycle mutants. Annu. Rev Genet 19:389, 1985.
106. Schempp W, Schmid M: Chromosome banding in amphibia. VI. BrdU-replication patterns in Anura and demonstration of XX/XY sex chromosomes in Rana esculenta. Chromosoma 83:697, 1981.
107. Stock AD, Mengden GA: Chromosome banding pattern conservatism in birds and nonhomology of chromosome banding patterns between birds, turtles, snakes and amphibians, Chromosoma 50:69, 1975.
108. Stock AD: The occurrence of G-bands in the mitotic chromosomes of the amphibian Xenopus muelleri, Genetika 64:225, 1984.
109. Schmid M: Chromosome banding in amphibia. IV. Differentiation of GC- and AT-rich chromosome regions in Anura. Chromosoma 77:83, 1980.
110. Lewin B: "Gene Expression 2, Eucaryotic Chromosomes." New York: John Wiley and Sons, 1980.

111. Mizuno S, Macgregor HC: Chromosomes, DNA sequences, and evolution in salamanders of the genus Plethodon. Chromosoma 48:239, 1974.
112. Epplen JT, Leipoldt M, Engel W, Schmidtke J: DNA sequence organisation in avian genomes. Chromosoma 69:307, 1978.
113. Eden FC, Hendrick JP: Unusual organization of DNA sequences in the chicken. Biochemistry 17:5835, 1978.
114. Vizard DL, Rosenberg NL: Temporal replication of an interspersed repeated sequence of mouse DNA. Biochim Biophys Acta 782:402, 1984.

The Cytogenetics of Mammalian Autosomal
Rearrangements, pages 835–854

Chapter 28
Chromosome Rearrangements, Protooncogenes, and Human Neoplasia

Avery A. Sandberg

The specific (primary) chromosome changes, particularly rearrangements (translocations, inversions), in human neoplasia are tabulated. The utilization of these specific events in the recognition and classification of subsets and subtypes within presently assumed entities is discussed. In lymphoid disorders the frequent juxtaposition of protooncogenes c-*myc* and *bcl*-1 and *bcl*-2 to proximity with Ig and TcR loci is summarized and correlated with commonly associated rearrangements. The known variability of breakpoints in CML on 22q11 in the *bcr* region is compared with breakpoint variability in other disorders. The involvement in acute leukemias of genes such as those for ferritin, metallothionein, interferon, and myeloperoxidase is investigated, and the possibility is considered that some oncogenes and other genes in proximity to neoplastic breakpoints are not actually involved in oncogenesis. The future application of molecular approaches in the diagnosis and classification of various neoplasia entities as a further extension of cytogenetics is presented.

INTRODUCTION

Cytogenetic (chromosomal, karyotypic) analyses, particularly those based on banding techniques, have revealed a large number of chromosome rearrangements characterizing various human neoplastic disease states [1–4]. These rearrangements have generally consisted of reciprocal and usually balanced exchanges of chromosomal material between two chromosomes. In Tables I–XI are listed those translocations or rearrangements as well as other morphologic chromosome changes thought to be specific (primary) in nature and to characterize each of the neoplastic diseases listed. In fact, the specificity of these translocations and structural changes has been utilized as a

Roswell Park Memorial Institute, Department of Genetics and Endocrinology, Buffalo, New York 14263 and The Genetics Center of Southwest Biomedical Research Institute, Scottsdale, Arizona 85251

TABLE I. Common chromosomal rearrangements in acute nonlymphocytic leukemia (ANLL)

t(1;16)(q21;q22) M7	t(8;21)(q22.1;q22.3) M2
inv(3)(q21q25–27) M1, M2	t(9;11)(p22;q23) M5a
t(3;9)(p14;q34) M7	t(9;22)(q34.1;q11.2) M1, M2
t(6;9)(p22.2;q34) M1, M2	t(15;17)(q22;q11.2) M3
t(8;16)(p11;p13) M5b	inv(16)(p13.1q22.1) M4

French-American-British (FAB) classification of ANLL: M1, acute myeloblastic leukemia; M2, acute myeloblastic leukemia (with differentiation); M3, acute promyelocytic leukemia; M4, acute myelomonocytic leukemia; M5, acute monoblastic leukemia; M6, acute erythroleukemia; M7, acute megakaryoblastic leukemia.

TABLE II. Common morphologic chromosome changes in acute nonlymphocytic leukemia (ANLL)

del(5)(q13q31)	del(16)(q23)
7q−	del(3p) or translocation with 3
Translocation or del (12)(p11–13)	i(17q)
Translocation or del (11)(q23)	

TABLE III. Common chromosomal rearrangements in acute lymphoblastic leukemia

t(1;19)(q21–q23; p13.3) L1	Pre-B cell
t(2;8)(p11–13;q24.1) L3	B-cell
t(4;11)(q21.1;q23) L2	Early pre-B precursor
t(8;14)(q24;q11) L1	T-cell
t(8;14)(q24.1;q32.3) L3	B-cell
t(8;22)(q24.1;q11) L3	B-cell
t(9;22)(q34.1;q11.2) L1	Early pre-B cell
t(11;14)(p13–14;q11.2–13) L2	T-cell
t(11;14)(q13;q32) L1	B-cell

TABLE IV. Common morphologic changes in acute lymphoblastic leukemia

del(6)(q21q25)	11q−
9p−	12p−
14q+(q32)	i(17q)
del(14)(q11)	

TABLE V. Chromosome rearrangements and other morphologic changes in chronic myelocytic leukemia

t(9;22)(q34;q11) chronic phase	t(3;21)(q26;q22) blastic phase[a]	i(17q)

[a]21q22 band may be involved in changes related to previous chemotherapy.

TABLE VI. Common chromosomal rearrangements in myelodysplastic disorders myeloproliferative disease, and preleukemia

t(1;3)(p36;q21)	t(2;15)(p13 or p21;q12 or q23)
t(1;7)(p11;p11)	t(3;17)(q26;q22)
t(2;11)(p11;q23)	t(11;21)(q22;q21)

TABLE VII. Morphologic chromosome changes (other than translocations) in myelodysplastic disorders, myeloproliferative diseases, and preleukemia

del(9)(q13q22)	del(5)(q31q35)
del(20)(q12q13)	7q−
del(21)(q22)	12p−

TABLE VIII. Common chromosome rearrangements in lymphoid disorders

Rearrangement	Disorder	Juxtaposed protooncogene	Other involved loci[a]
t(1;5)(p36;q13-14)	multiple myeloma	?	?
t(2;8)(p11-13;q24)	B-cell lymphoma and leukemia (Burkitt type)	c-*myc* at 8q24	Igk at 2p11-13
t(8;14)(q24;q11)	T-cell leukemia	c-*myc* at 8q24	αTcR at 14q11
t(8;14)(q24;q32)	B-cell lymphoma and leukemia (Burkitt type)	c-*myc* at 8q24	Igh at 14q32
t(8;22)(q24;q11)	B-cell lymphoma and leukemia (Burkitt type)	c-*myc* at 8q24	Igl at 22q11
t(11;14)(p13;q11)	T-cell leukemia	?	αTcR at 14q11
t(11;14)(q13;q32)	B-cell lymphoma and leukemia	*bcl*-1 at 11q13	Igh at 14q32
t(14;14)(q32;q11)	T-cell leukemia	—	Igh at 11q32 αTcR at 14q11
t(14;18)(q32;q21)	B-cell lymphoma	*bcl*-2 at 18q21	Igh at 14q32
inv(14)(q11q32)	T-cell leukemia, T-cell CLL	—	Igh at 14q32 αTcR at 14q11

[a]Ig, immunoglobulin loci; Igk, Igl, kappa and lamba light chains; Igh, heavy chains; TcR, T-cell receptor loci.

diagnostic criterion in some of these diseases, e.g., differentiation of myxoid liposarcoma from other myxoid tumors [5], as well as cytogenetically "dissecting" previously defined (cytologically, histologically, cytochemically, immunologically) entities into a number of subsets, e.g., acute myeloblastic leukemia (AML), in which several unique subsets are characterized by specific chromosome changes such as t(8;21), t(9;22), and t(6;9) [6]. Thus, in each of these states the rearrangement or other change, consisting primarily of a translocation, has been shown to be a sine-qua-non of the disease, the karyotypic change possibly being an essential part, if not directly neces-

TABLE IX. Specific (primary) chromosome rearrangements in tumors

Benign tumors	
t(3;8)(p21;q12)	Mixed tumors of salivary gland
t(3;12)(q28;q14)	Lipoma
t(9;12)(p13-22;q13-15)	Mixed tumors of salivary gland
Adenocarcinomas	
t(5;14)(q13;q22)	Kidney
t(6;14)(q21;q24)	Ovary
Sarcomas	
t(X;18)(p11.2;q11.2)	Synovial
t(2;13)(q37;q14)	Rhabdomyosarcoma (alveolar)
t(9;22)(q31;q12.2)	Chondrosarcoma (extraskeletal, myxoid)
t(12;16)(q13;p11)	Liposarcoma (myxoid)
Embryonal and other tumors	
t(1;19)(q12;q13)	Malignant melanoma
t(11;22)(q24;q12)	Ewing sarcoma and neuroectodermal tumors

TABLE X. Morphologic (non-translocational) chromosome changes in tumors

Benign tumors	
12q−	Colon polyps
22q−	Meningioma
Adenocarcinomas	
1q−	Uterus
del(3)(p14p23)	Lung
i(5p),9q−,11p−	Bladder
6q−	Ovary
del(10)(q24)	Prostate
12q−,17p−	Colon
Sarcomas	
1q−	Rhabdomyosarcoma (embryonal)
Embryonal and other tumors	
del(1)(p32p36)	Neuroblastoma
del(6)(q11q27),i(6p),del(1)(p11p22)	Malignant melanoma
del(11)(p13)[a]	Wilms tumor
i(12p)	Testes (germ cell tumors)
del(13)(q14)[a],i(6p)	Retinoblastoma

[a]May be associated with constitutional chromosome change.

sary, for the development of the malignant transformation [7–9]. Though the bulk of the states shown in Tables I–XI are characterized by translocations, other rearrangements such as inversions, e.g., of chromosomes 3 and 16, also appear to characterize a small number of disease states. In the case of translocations, material is exchanged between two or more chromosomes, whereas in inversions the rearrangement occurs within a single chromosome, thus changing the sequence of the DNA within this chromosome.

TABLE XI. Chromosome oncogene locations in the human genome

Chromosome	Location	Oncogene
1	1p32	Blym-1
1	1p32	L-myc
1	1p11-p13	N-ras
1	1p32-35	c-lck
1	1p34-p36	c-src-2 (pseudo?)
1	1p36.1-p36.2	c-fgr
1	1q22-q24	c-ski
2	2p11-2q14	c-rel
2	2p21-p23	N-myc
3	3p25	c-raf-1
4		c-raf-2 (pseudo)
4		c-kit
5	5q33-34	c-fms
6	6p21	c-pim-1
6	6p11-p12	c-K-ras-1 (pseudo)
6	6p23-q12	c-yes-2
6	6q16-q22	mcf-3 (c-ros)
6	6q21-q23	c-myb
7	7p11-p13	c-erb-B
7	7q21-q31	c-met
7		HA-raf-2
8	8q11	c-mos
8	8q24	c-myc
9	9q34.1	c-abl
11	11p14.1	c-H-ras-1
11	11q13	bc1-1
11	11q23-q24	c-ets-1
12	12p11.1-p12.1	c-K-ras-2
12	12q12-q13	c-int
12	12q13-q14.3	c-gli
14	14q21-q31	c-fos
14	14q32	H-akt-1
15	15q26.1	c-fes
17	17q11.2	c-erb-A1(neu)
17	17q21.3	c-erb-A2
18	18q21	bcl-2
18	18q21.3	c-yes-1
18		c-erv-1
19	19p13.2-q13.2	c-mel
20	20q12-q13	c-src-1
21	21q22.1-q22.3	c-ets-2
22	22q13.1	c-sis
X	Xpter-q28	c-H-ras-2 (pseudo)
X	Xp21-cen	HA-raf-1

The other morphologic changes shown in Tables II, IV, VII, and X consist primarily of partial deletions of various chromosomes, as well as a few isochromosomes, with each abnormal chromosome constituting a specific or nonrandom karyotypic change for the neoplastic conditions in which the cytogenetic change is observed. In all of these cases the missing chromosomal segment has not been found to be translocated to another chromosome; hence, it is assumed to have been lost from the genome of the affected cells.

Molecularly, loss of chromosomal material has been related to possible removal of an anti-oncogene from the genome with possible abnormal expression of a contralateral abnormal gene possibly being responsible for the malignant transformation [9,10]. Examples of such situations are the 11p− and 13q− observed in Wilms tumor and retinoblastoma, respectively [11,12].

The translocations, inversions, and deletions shown in Tables I–XI are thought to constitute the specific (primary) karyotypic event in each of these neoplastic states, an event that may affect protooncogenes (or other genes) thought to participate in the initiation and/or maintenance of malignant transformation in these diseases [7–9]. The mechanisms by which such protooncogenes are converted to oncogenes appear to vary from disease state to disease state.

REARRANGEMENTS INVOLVING PROTOONCOGENES AND LOCI Ig (IN B-CELL) AND TcR (IN T-CELL) LYMPHOMAS AND LEUKEMIAS

Mistakes by the enzyme system involved in the arrangement of variable, heavy chain, and joining (V-D-J) segments have been implicated in certain translocations in B-cell malignancies, e.g., joining of separate segments of DNA on different chromosomes rather than on the same chromosome [13,16]. Thus, most of the Burkitt lymphomas with an early-B-cell phenotype are the result of mistakes in V-D-J joining, where the more differentiated Burkitt lymphomas with rearrangements within a switch region may be the result of mistakes during heavy chain isotype formation. Thus, in the translocation t(8;14) (Fig. 1) the break on chromosome 8 at q24 is followed by translocation of exons of the oncogene *myc* to chromosome 14 where it is juxtaposed to genes associated with the production of heavy chain immunoglobulins [17]. In this case no abnormal gene products have been shown to exist, though in all probability quantitative differences from the normal in the production of mRNA and hence protein products apparently may be present. One could postulate that an imbalance between the products of the immuno-

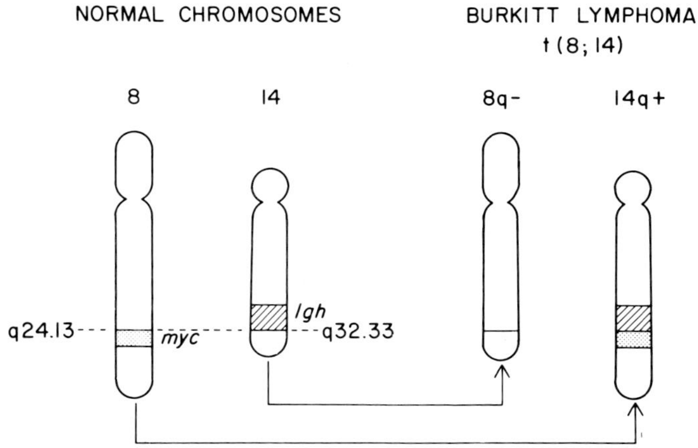

Fig. 1. Schematic presentation of the common translocation seen in Burkitt lymphoma, i.e., t(8;14)(q24;q32). The break on chromosome 8 generally affects the *myc* protooncogene, which is then translocated to chromosome 14 at q32. This juxtaposes the oncogene c-*myc* to the gene for heavy chain immunoglobulins (Igh), the latter apparently activating expression of the c-*myc* leading to the genesis of the lymphomas.

globulin genes and of the *myc* oncogene may possibly be responsible, at least in part, for the malignant transformation observed in Burkitt lymphoma. The translocation in Burkitt lymphoma or leukemia, though most often involving chromosomes 8 and 14, may also show variant forms, such as translocations between chromosome 8 with either chromosome 2 or 22 (Table VIII). In each of these variant translocations, there appears to be juxtaposition of the oncogene *myc* to genes responsible for the production of light chain immunoglobulins. A major difference between t(8;14) and t(8;22) as well as t(2;8) in Burkitt lymphoma is that in t(8;14) the oncogene *myc* is translocated to chromosome 14, whereas in t(8;22) and t(2;8) *myc* remains on chromosome 8 with genes from the loci for the light chain immunoglobulins on chromosomes 2 (kappa light chain) and 22 (lambda light chain) being translocated and juxtaposed to the *myc* gene on chromosome 8. Furthermore, the protooncogene *myc* is not involved by breakage in these variant translocations; i.e., *myc* remains intact, though it is activated by the events [13]. Since the disease associated with these variant translocations appears to be for all practical purposes clinically and histologically the same, it would appear that in the genesis of Burkitt lymphoma (BL) or leukemia the participation of the oncogene c-*myc* is a crucial event resulting from its juxtaposition to genes responsible for heavy or light chain immunoglobulin production. Breaks in 14q differ between endemic and sporadic BL, as well as the break in chromosome 8, which occurs 5′ to the oncogene c-*myc* (i.e., more proximal to the centromere) in endemic BL in contrast to the break involving c-*myc* directly in sporadic BL [13].

In follicular lymphoma with t(14;18)(q32;q21) an oncogene, *bcl*-2, in chromosome 18 is thought to be activated by its proximity to the translocated material of the heavy-chain locus from chromosome 14 to chromosome 18 [13]. In the t(14;18) of follicular lymphoma most breaks affecting the *bcl*-2 gene lead to rearrangement of this gene and others occurring 3′ or 5′ to the gene. The break on chromosome 14 is in the J-locus (5′ region) of the heavy chain immunoglobulins. In a manner similar to that in follicular lymphoma, the translocation t(11;14)(q13;q32) in lymphomas and leukemias of B-cell origin appears to involve the juxtaposition of heavy chain immunoglobulin regions to the gene *bcl*-1. This event either activates or deregulates *bcl*-1 with the possibility that this is directly involved in the genesis of the disease.

The chromosomal location of the alpha chain of the T-cell receptor (αTcR) has been mapped to band 14q11.2 [13]. This is a region that is frequently associated with translocations and inversions seen in T-cell neoplasia, and this has led to the speculation that the αTcR locus may function in T-cell translocations in a manner similar to that of the Igh genes B-cell translocations [18,19].

Transcriptional deregulation of c-*myc* by the translocation of the α-locus of the T-cell receptor has been postulated to occur in T-cell leukemias [13–15, 18, 19]. In one form, T-cell acute lymphoblastic leukemia (ALL) with t(8;14)(q24;q11), the α-chain of the T-cell receptor (located at 14q11.2) is involved by breakage as a result of the translocation, thus affecting the genes between the variable (V_α) and the constant (C_α) regions of the α-chain of the T-cell receptor, with the V_α genes being proximal and the C_α genes distal to the 14q11 breakpoint [20]. Translocation of this region of chromosome 14 to the long arm of 8 leads to deregulation of c-*myc* with this possibly playing a key role in the genesis of the T-cell malignancy. The involvement of c-*myc* in the translocation and the association of joining signals at the breakpoints provides a parallel to the situation observed in the translocation involving c-*myc* and the immunoglobulin loci in B-cell neoplasms and suggests their common mechanisms of translocation and oncogene deregulation seen in B-cell and T-cell malignancies [18].

In T-cell diseases with inversion of chromosome 14, inv(14)(q11.2q32), the constant region of the T-cell receptor (αTcR) is juxtaposed to the immunoglobulin region V_H and thus the change possibly affects a gene related to the genesis of the disease [14,15,]. Involvement of the β-chain of the T-cell receptor (βTcR) gene located at 7q34 [21,22] has been found in a number of translocations in T-cell disease, though the molecular events related to neoplasia have not been defined as much as those of αTcR.

THE PHILADELPHIA CHROMOSOME AND MOLECULAR EVENTS IN LEUKEMIA

In chronic myelocytic leukemia (CML) the translocation (Fig. 2) between chromosomes 9 and 22, t(9;22)(q34;q11), appears to affect the protooncogene *abl* on chromosome 9 and the gene *bcr* on chromosome 22, respectively [24]. As a result of the translocation the "breakpoint cluster region" (*bcr*) gene is split with the exact location of the break within this gene showing some variability though occurring within a region of DNA consisting of 5.8 kilobases (Fig. 3). The protooncogene *abl* is juxtaposed to the abbreviated *bcr* gene, creating a new gene, *bcr-c-abl*, of which the messenger RNA (mRNA) and the resulting protein product are of different size and nature as compared to the normal products of each one of these genes. Thus, the

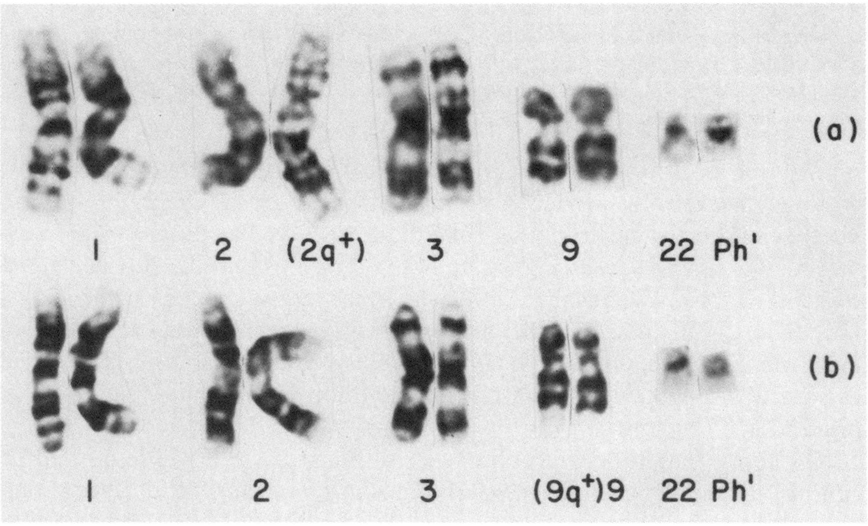

Fig. 2. Ph-translocation (**b**) of the standard type showing the abbreviated chromosome 22 (Ph) and extra material on one chromosome 9 (9q+) resulting from a reciprocal translocation t(9;22)(q34;q11). A variant (simple) Ph-translocation is shown in (**a**), with chromosome 2 (2q+) participating in an exchange with chromosome 22; chromosome 9 appears essentially normal and, hence, it was assumed that neither chromosome 9 participated in the Ph-translocation. However, molecular studies in similar cases have revealed a translocation of the oncogene c-*abl* to the Ph and, hence, chromosome 9 did participate in the genesis of the Ph, though cytologically it may be difficult to show the changes involving chromosome 9.

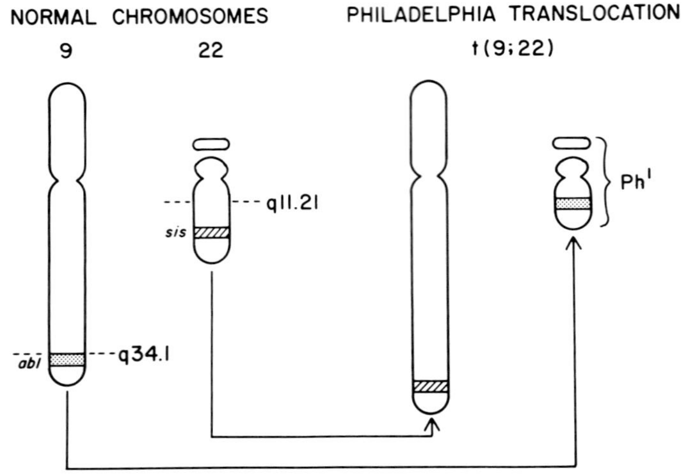

Fig. 3. Schematic presentation of the Ph-translocation and the main events associated with it. The break on chromosome 9 affects the protooncogene *abl*, which is translocated to the abbreviated chromosome 22 constituting the Ph chromosome. The break on chromosome 22 affects the *bcr* gene, part of which is translocated to chromosome 9. The oncogene c-*sis* is too distal to be affected by the events of the Ph chromosome, though it is translocated in toto to chromosome 9 as part of the reciprocal translocation t(9;22).

normal mRNA for c-*abl* is about 6–7 kilobases and that for *bcr* 4–5 kilobases, whereas the mRNA produced by the abnormal gene, *bcr-c-abl*, is 7–8 kilobases. The proteins resulting from these mRNAs also differ in size, with that for the normal *abl* being about 145,000 daltons and that for *bcr* about 190,000 daltons, whereas that for the abnormal gene is about 210,000 daltons [25]. Thus, in the case of CML the abnormal protein may play a key role in the genesis of the disease; in all probability attempts at neutralizing the activity of this protein will constitute a key effort in the control and possible prevention of the disease.

The translocation between chromosome 9 and 22 in CML results in the genesis of the Philadelphia (Ph) chromosome, a cytogenetic feature that characterizes this disease. In all probability diseases in which these karyotypic events do not occur may not constitute CML and may belong to other myeloproliferative or myelodysplastic disorders. In a rather significant number of cases of CML (ca. 5%) the translocation between chromosomes 9 and 22 leading to the genesis of the Ph involves a chromosome other than chromosome 9 (simple-variant Ph) (Figs. 1, 4) or the participation of chromosomes additional to 9 and 22 (complex-variant Ph). In almost all cases an abbreviated form of chromosome 22 constitutes the pH. One of the important

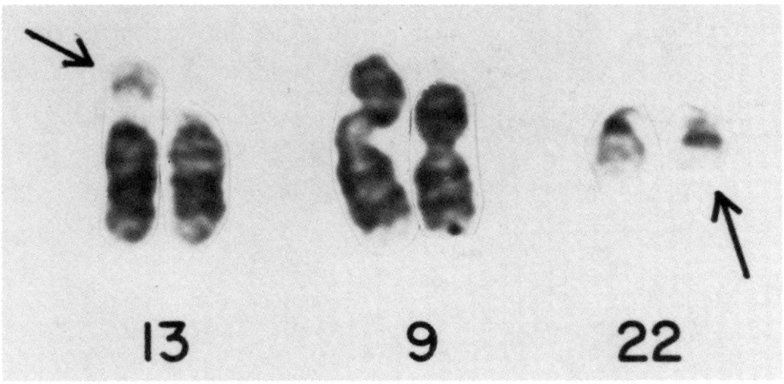

Fig. 4. Variant (simple) Ph-translocation in which the Ph appeared to be generated by an exchange between chromosome 13 and 22. As in the case shown in Figure 2, it was assumed on a cytologic basis that chromosome 9 was not involved; molecularly, such cases have been shown to involve chromosome 9 through translocation of the oncogene c-*abl* to the Ph, the morphology and size of the chromosome 9 being not sufficiently affected to be readily discernible in cytogenetic preparations. In this case, as in that of Figure 2, molecular studies have expanded and revealed the cytogenetic events associated with the genesis of the Ph which otherwise would not be ascertained.

contributions of molecular biology to our understading of CML and of the Ph in particular has been the demonstration that regardless of the cytologic evidence for a variant Ph, a translocation of c-*abl* from chromosome 9 to the *bcr* region of chromosome 22 occurs in each one of these cases [24]. In other words, whether the Ph chromosome is masked by the presence of a segment of another chromosome resulting in a marker chromosome larger than the normal chromosome 22 or whether a chromosome other than 9 is involved in the genesis of the Ph chromosome, the translocation of c-*abl* to *bcr* probably occurs in all cases. Hence, the basic molecular events associated with the Ph in CML appear to be the same though some variations may be present. On the other hand, in some cases of acute lymphoblastic leukemia (ALL), molecular studies have revealed that the Ph chromosome, however similar its morphology, may not be associated with a break in the *bcr* region though accompanied morphologically by an abbreviated chromosome with features that microscopically cannot be differentiated from those of the Ph in CML. Thus, in some cases of Ph-positive ALL the juxtaposition of the translocated c-*abl* occurs to a gene 3′ (proximal) to the *bcr* gene [26–29]. In other words, the *bcr* region may not be involved in the chromosome rearrangement of the Ph with the chromosome breakpoints being proximal (5′) to the *bcr* region at band 22q11. In such cases of Ph-positive ALL, c-*abl* is inappropriately expressed in ALL cells without *bcr* rearrangement, and the genetic mechanism of activation must be different from that reported for Ph-positive CML. Not all cases of ALL with a Ph show this deviation from events in CML and hence it appears that Ph-positive ALL may consist of at least two entities, one in which the molecular events related to the Ph are similar to those in CML and another entity in which the Ph chromosome is generated without a break in the *bcr* region but with the translocation of c-*abl* to chromosome 22.

HETEROGENEITY OF BREAKPOINTS WITH INVOLVED ONCOGENES

As in the case of the cytogenetic observations, the molecular studies in CML, ALL, Burkitt lymphoma, and other diseases are revealing a heterogeneity of the underlying molecular events characterizing these states. Thus, the breakpoint within *bcr* shows variability within the 5.8-kilobase region of this gene, the size of the *abl* oncogene material translocated to the Ph differs among cases, and the size or involvement of the *myc* oncogene involved in translocations of the Burkitt type varies not only with the type of translocation—e.g., t(8;14) *vs* t(8;22) or t(2;8)—but also within the same type of translocation. The reflections of these molecular nuances in the biology and clinical manifestations of these diseases are yet to be fully determined.

The diseases with chromosomal rearrangements discussed above—i.e., Ph-positive CML and ALL and Burkitt lymphoma—have been those in which protooncogenes have been shown to be involved and to be associated with molecular events that may play an important role in the malignant transformation. Though as shown in the Tables I–X other chromosomal rearrangements established as specific events in human neoplasia may affect protooncogenes, in most cases the demonstration of their involvement remains to be established. It must be kept in mind that a chromosomal band or even a sub-band contains a relatively large amount of DNA, an average band containing millions of base pairs of DNA, in which an oncogene (or any other gene) could be readily accommodated. Furthermore, the breaks described for chromosomal rearrangements in various neoplastic states may or may not involve an oncogene, since the breakpoints shown in Tables I–X, however accurate they are in cytogenetic terms, may not reflect events at the much higher level of (molecular) resolution possible within a band than is afforded by microscopic examination. In many instances in which specific rearrangements have been shown to exist, no protooncogenes have been described to be located within these chromosomal areas. This does not mean that key genes that could function as protooncogenes (oncogenes represent active forms of protooncogenes capable of inducing malignant transformation) are not located within such bands, but their description remains to be established and their characteristics to be demonstrated.

REARRANGEMENTS OF NONONCOGENE LOCI IN ACUTE LEUKEMIAS

Attempts at demonstrating the involvement of genes through chromosomal translocations and other rearrangements have led to the concept that a gene may be split by these cytogenetic events, and this derangement may possibly be responsible for, or related to, the development of the neoplasia. Thus, it has been shown that genes for ferritin and metallothionein may be affected in cytogenetic events occurring within chromosome 3 and 16 in two forms of acute nonlymphocytic leukemia (ANLL) [30,31]. In each case apparently the break leading to the chromosomal change occurs within these genes and one could postulate that the resulting abnormal situtation may result in abnormal products' leading to the neoplastic state or that a combination of the abbreviated genes with another gene or their parts may lead to events similar to those described above for CML or lymphoma.

It has been shown that in the translocation t(9;11)(p22;q23) seen in acute monoblastic leukemia the beta-1 interferon gene is split on chromosome 9 and that the oncogene *ets* is translocated from chromosome 11 to 9, thus juxtaposing the oncogene to the abbreviated interferon gene [30]. It has also been shown that in the inversion, inv(3)(q21q26), as well as in the translocation, t(3;3)(q21;q26), seen in some forms of ANLL, the chromosomal event affects the tranferrin gene located at 3q21 and/or the transferrin receptor located at 3q26 [32]. In a similar approach, the same authors indicated that the metallothionein gene is split by the rearrangement resulting from the inversion occurring within chromosome 16 in ANLL of the M4 type [31]. Thus, the molecular events shown to occur by these authors may point to the possibility that the changes in the genes described may serve the same function as those previously described for protooncogenes in CML and BL; i.e., the molecular events associated with the chromosomal changes may be sufficient to initiate malignant transformation in each case, resulting in a form of ANLL with characteristic clinical, cytologic, and other aspects. In other cases it has been shown that in translocations putatively involving protooncogenes, the oncogenes are conserved or are merely transferred from one chromosome to another whether they are within or outside the chromosomal breakage point (Fig. 6) [33–37]. Several potentially interesting but invalid associations come to mind.

The human *erb*-A gene is on the long arm of chromosome 17 but is proximal to the breakpoint of the characteristic t(15;17)(q22;q21) (Fig. 5)

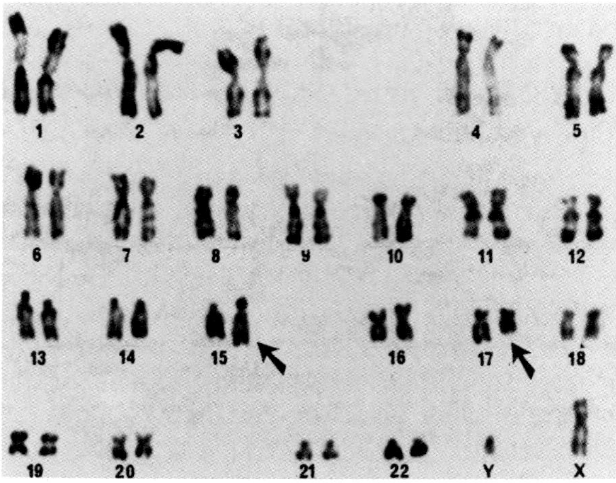

Fig. 5. Translocation, t(15;17), a characteristic cytogenetic event in acute promyelocytic leukemia (APL) encountered in most of the cases with this disease. It has not been established whether any known oncogenes or other genes are involved in or affected by this translocation.

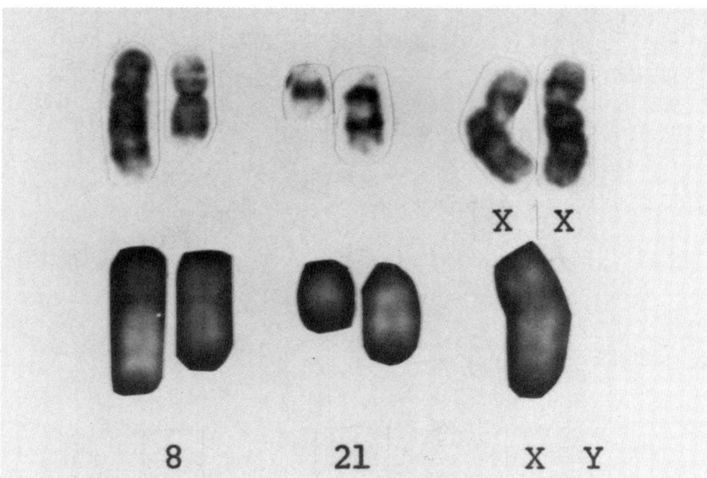

Fig. 6. Specific translocation t(8;21)(q22;q22) seen in M2 type of AML in which a reciprocal exchange occurs between chromosomes 8 and 21. To date, no oncogene has been shown to be involved by this translocation; the oncogenes *mos* and *myb* are located on chromosome 8, but they are not affected by the cytogenetic events, though *myb* is translocated to chromosome 21. Loss of a sex chromosome (Y in males, X in females) is a common additional event in these cases and is usually associated with progression of the disease, as is true of other additional karyotypic changes.

seen in acute promyelocytic leukemia (APL). The p53 gene that was originally reported to be closely distal to the same breakpoint on chromosome 17 [38] has been subsequently shown to actually be located on the short arm of this chromosome [39,40]. The location of the human gene *neu* (the homolog of the rat transforming gene *erb*-A), which most likely codes for a cell surface receptor [41], maps to the chromosome region 17q12→17q22. However, this gene is distal to the 17q11 breakpoint of APL and thus is translocated to chromosome 15 from the involved chromosome 17. The gene encoding the nerve growth factor (NGF) is at region 17q21 [42], which is also distal to the characteristic breakpoint in t(15;17). It remains to be determined whether regulatory sequences of the myeloperoxidase gene [42] are involved in the t(15;17) of APL. Thus, it is clear from the aforementioned that in APL with t(15;17) cogent molecular events associated with the translocation have not been described to date.

Since the human *pim* oncogene has been mapped to region 6p21 [43], which is near the breakpoint affected in the translocation t(6;9)(p21;q33) seen in some ANLL cases, it will be of interest to ascertain whether the *pim* locus is directly affected in this translocation. The oncogene *abl* is apparently not involved in this translocation [35].

A c-*raf*-related gene, A-*raf*-1 [44], and the gene for a tissue inhibitor of metalloproteins (TIMP) [45] have been mapped to the Xp11 region, a band involved in a specific translocation seen in synovial sarcoma, i.e., t(X;18)(p11.2;q11.2). It will be of interest to establish whether either or both of these genes are rearranged or otherwise affected.

In the case of the 5q- hematopoietic syndrome (refractory sideroblastic anemia) with an interstitial deletion of variable length of the long arm of chromosome 5, it has been shown that the oncogene *fms* may or may not be lost [46], but that the locus for the granulocyte/monocyte colony stimulating factor (GM-CSF) gene is invariably missing as a result of this deletion [46]. Whether this results in a situation where the loss of one allele, e.g., akin to 11p− in Wilms tumor and 13q− in retinoblastoma [10–12], leads to an expression of a mutated gene present on the intact chromosome which may possibly be responsible for the malignant transformation is yet to be established. This may apply to conditions associated with 11q− and affecting the locus for the oncogene *ets*-1 [47,48].

SPECIFIC CHROMOSOME CHANGES AND ONCOGENESIS—THE FUTURE

More and more human leukemias and cancers in which specific chromosomal events, particularly translocations, not only characterize disease entities and subsets but may also be responsible for their genesis are being described. These specific cytogenetic changes as summarized and appended (Tables I–X) have not only been used diagnostically and prognostically but also present key information for the molecular biologist in determining the nature of the genes involved and possible mechanisms underlying the genesis of neoplasia. Since in all probability a significant proportion of human neoplastic states will be shown to be accompanied by such rearrangements, especially translocations, though it must be stated here that some of these states are accompanied by either numerical or other morphologic changes not apparently associated with chromosomal rearrangements (deletions or insertions), the number of oncogenes (or other genes) responsible for human neoplasia will greatly exceed the three dozen or so whose chromosomal locations have been described to date (Table XI). In other words, each chromosomal rearrangement in human neoplasia will be shown to affect a gene (protooncogene?) with the possibility that the involvement of such a gene is responsible for the genesis or maintenance of the neoplasia. If that is so, the genes affected in human neoplasia will greatly exceed in number those that have been located on the various chromosomes to date and may, in fact, number in the hundreds (one for each specific neoplastic entity of the several hundred known).

Almost all cancers and the preponderant percentage of leukemias are associated with chromosome changes, which in most cases represent rearrangements discussed above. In all probability these chromosomal events lead to some sort of deregulation of protooncogenes and other genes whose role in the development of the malignant state is thought to be crucial if not paramount. However, abnormalities (particularly of expression) of other oncogenes and other genes are also present in these states even in the presence of the specific (primary) karyotypic change as the sole cytogenetic anomaly in the involved cells [49,50]. Apparently, the specific event and the subsequent events associated with it can trigger other gene activations and expressions. What compounds the situation further is the almost invariable presence of additional (secondary) chromosome changes in epithelial tumors (adenocarcinomas), often in sarcomas and in a high percentage of the leukemias [1]. Such additional chromosome changes can be very complex and may represent an array of rearrangements and numerical anomalies. There is little doubt that these additional cytogenetic events must be expressed molecularly, thus leading to an array of metabolic and physiologic abnormalities which affect (negatively usually) the biology of the neoplasia. The complexity of the additional cytogenetic and molecular events has defied meaningful analysis in most cases, though without an understanding of these events we will probably make little progress in our search for key clues in cancer biology and control.

One informative aspect emerging from molecular studies of chromosomal rearrangements is the finding that hidden within what appear to be homogeneous karyotypic events is a diversity of molecular events occurring within the region affected cytogenetically. Thus, molecular approaches, akin to the introduction of banding techniques, will undoubtedly expand the cytogenetic information and horizons and yield more details regarding chromosomal rearrangements in leukemia and cancer.

It should be pointed out that activation or expression of protooncogenes may occur in human leukemias and cancers in the absence of additional chromosomal rearrangements possibly owing to a triggering of a cascade of gene activations resulting from the primary karyotypic event and/or the additional cytogenetic changes not infrequently seen in leukemic and cancer cells.

It should be pointed out that mechanisms other than those already discussed that affect genes (including oncogenes) leading to neoplasia are possible, e.g., abnormal mRNA resulting from translocation of a total protooncogene to a new locus or the interaction of several normal or abnormal products of genes not necessarily modified by the cytogenetic events.

In all probability there exists a specificity not only for the type of cell affected by a chromosome (and consequent gene or oncogene) change and the ability of the cell to undergo malignant transformation, but also for the stage of the cell (particularly in hematopoietic disorders) in the differentiation pathway. In other words, a specific chromosomal rearrangement must affect a particular (and susceptible) cell type for the cell to become malignant, though such a change may have to occur at a propitious stage of the cell. Whether rearrangements affecting nonsusceptible cells cause the death of such cells or have no effect on the cells remains to be established.

Cancer cytogenetics, in concert with the general field of cytogenetics, has taken occasional giant steps if not leaps that have led to an explosion of findings related to human neoplasia. One has only to mention the effects of the introduction of banding techniques which made possible not only the identification of individual chromosomes, both normal and abnormal, but also the recognition of chromosomal bands and sub-bands affected by rearrangements in a large number of human cancers and leukemias. The application of molecular approaches to the cytogenetic findings in cancer has already led to the recognition of karyotypic changes at the submicroscopic level that could not be or are not likely to be visualized microscopically. Thus, the resolution of cytogenetic findings has been vastly expanded by these approaches and will undoubtedly lead to a new explosion of genetic information on chromosomal events underlying, if not responsible for, the genesis of many more neoplastic states than are presently recognized.

REFERENCES

1. Sandberg AA: "The chromosomes in Human Cancer and Leukemia." Elsevier New York: North-Holland, 1980.
2. Sandberg AA: Chromosomal changes in human cancers: Specificity and heterogeneity. In Owens AH, Jr, Coffey DS, Baylin SB (eds): "Tumor Cell Heterogeneity: Origins and Implications." New York: Academic Press, 1982, pp 367–397.
3. Sandberg AA: Cytogenetics of the leukemias and lymphomas. In Luderer AA, Weetall HH (eds): "The Human Oncogenic Viruses." Clifton, NJ: Humana Press, Inc., 1986, pp 1–41.
4. Sandberg AA: The chromosomes in human leukemia. Semin Hematol 23:201–217, 1986.
5. Sandberg AA, Turc-Carel C: The cytogenetics of solid tumors: Relation to diagnosis, classification and pathology. Cancer 59:387–395, 1987.
6. Second MIC Cooperative Study Group: Morphologic, immunologic and cytogenetic (MIC) working classification of the acute myeloid leukemias. Cancer Genet Cytogenet 30:301–315, 1988.
7. Yunis JJ: The chromosomal basis of human neoplasia. Science 221:227–236, 1983.
8. Rowley JD: Biological implications of consistent chromosome rearrangements in leukemia and lymphoma. Cancer Res 44:3159–3168, 1984.
9. Sandberg AA: A chromosomal hypothesis of oncogenesis. Cancer Genet Cytogenet 8:277–285, 1983.
10. Knudson, AG: Modal hereditary cancers of man. Prog Nucleic Acid Res Mol Biol 29:17–25, 1983.
11. Koufos A, Hansen MF, Lampkin BC, Workman ML, Copeland NG, Jenkins NA, Cavenee WK: Loss of alleles at loci on human chromosome 11 during genesis of Wilms' tumour. Nature 309:170–172, 1984.
12. Cavenee WK, Dryja TP, Phillips RA, Benedict WF, Godbout R, Gallie BL, Murphree AL, Strong LC, White RL: Expression of recessive alleles by chromosomal mechanisms in retinoblastoma. Nature 305:779–784, 1983.
13. Croce CM: Chromosome translocations and human cancer. Cancer Res 46:6019–6023, 1986.
14. Showe LC, Croce CM: Chromosome translocations in B- and T-cell neoplasias. Semin Hematol 23:237–243, 1986.
15. Showe LC, Croce CM: Translocation mechanisms in B- and T-cell leukemias. Cancer Treat Rev 2:18–33, 1986.
16. Leder P, Battey J, Lenoir G, Maulding G, Murphy W, Potter H, Stewart T, Taub R: Translocations among antibody genes in human cancer. Science 222:765–767, 1983.
17. Haluska FG, Sheldo F, Tsujimoto Y, Croce CM: The t(8;14) chromosomal translocation occurring in B-cell malignancies result from mistakes in V-D-J joining. Nature 324:158–161, 1986.
18. Finger LR, Harvey RC, Moore RCA, Showe LC, Croce CM: A common mechanism of chromosomal translocation in T- and B-cell neoplasia. Science 234:982–985, 1986.

19. Erikson J, Finger L, Sun L, ar-Rushdi A, Nishikura K, Minowada J, Finan J, Emanuel BS, Nowell PC, Croce CM: Deregulation of c-myc by translocation of the α-locus of the T-cell receptor in T-cell leukemias. Science 232:884–886, 1986.
20. McKeithan TW, Shima EA, Le Beau MM, Minowada J, Rowley JD, Diaz MO: Molecular cloning of the breakpoint junction of a human chromosomal 8;14 translocation involving the T-cell receptor α-chain gene and sequences of the 3′ side of MYC. Proc Natl Acad Sci USA 83:6636–6640.
21. Isobe M, Erickson J, Emanuel BS, Nowell PC, Croce CM: Location of gene for β subunit of human T-cell receptor at band 7q35, a region prone to rearrangements in T cells. Science 228:580–582, 1985.
22. Morton CC, Duby AD, Eddy RL, Shows TB, Seidman JG: Genes for β-chain of human T-cell antigen receptor map to regions of chromosomal rearrangements in T cells. Science 228:582–585, 1985.
23. Hecht F, Morgan R, Gemmill RM, Hecht BK: Translocations in T-cell leukemia and lymphoma. N Engl J Med 313:758, 1985.
24. Sandberg AA, Gemmill RM, Hecht BK, Hecht F: The Philadelphia chromosome: A model of cancer and molecular cytogenetics. Cancer Genet Cytogenet 21:129–146, 1986.
25. Ben-Neriah Y, Daley GQ, Mes-Masson AM, Witte ON, Baltimore D: The chronic myelogenous leukemia-specific P210 protein is the product of the bcr/abl hybrid gene. Science 233:212–214, 1985.
26. Cannizzaro LA, Nowell PC, Belasco JB, Croce CM, Emanuel BS: The breakpoint in 22q11 in a case of Ph′-positive acute lymphocytic leukemia interrupts the immunoglobulin light chain gene cluster. Cancer Genet Cytogenet 18:173, 1985.
27. Rodenhuis S, Smets LA, Slater RM, Behrendt H, Veerman A: Distinguishing the Philadelphia chromosome of acute lymphoblastic leukemia from its counterpart in chronic myelogenous leukemia. N Engl J Med 313:51, 1985.
28. Erikson J, Griffi CA, ar-Rushdi A, Valtieri M, Hoxie J, Finan J, Emanuel BS, Rovera G, Nowell RC, Croce CM: Heterogeneity of chromosome 22 breakpoint in Philadelphia-positive (Ph+) acute lymphocytic leukemia. Proc Natl Acad Sci USA 83:1807–1811, 1986.
29. de Klein A, Hagemeijer A, Bartram CR, Houwen R, Hoefsloot L, Carbonell F, Chan L, Barnett M, Greaves M, Kleihauer E, Heisterkamp N, Groffen J, Grosveld G: bcr Rearrangement and translocation of the c-abl oncogene in Philadelphia positive acute lymphoblastic leukemia. Blood 68:1369–1375.
30. Diaz MO, Le Beau MM, Pitha P, Rowley JD: Interferon and c-*ets*-1 genes in the translocation (9;11)(p22;q23) in human acute monocytic leukemia. Science 231:265–267, 1986.
31. Le Beau MM, Diaz MO, Karin M, Rowley JD: Metallothionein gene cluster is split by chromosome 16 rearrangements in myelomonocytic leukemia. Nature 313:709–711, 1985.
32. LeBeau MM, Espinoza R III, Yang F, Schneider C, Larson RA, Rowley JD, Diaz MO: Juxtaposition of transferrin and transferrin receptor genes in the inv(3) and t(3;3) in nonlymphocytic leukaemia, (in press).
33. Diaz MO, Le Beau MM, Rowley JD, Drabkin HA, Patterson D: The role of the c-mos gene in the 8;21 translocation in human acute myeloblastic leukemia. Science 229:767–769, 1985.
34. Sacchi N, Watson DK, Geurts van Kessel AHM, Hagemeijer A, Kersey J, Drabkin HD, Patterson D, Papas SP: Hu-ets-1 and Hu-ets-2 gene are transposed in leukemias with (4;11) and (8;21) translocations. Science 231:379–382, 1986.

35. Wesbrook BA, Le Beau MM, Diaz MO, Groffen J, Rowley JD: Chromosomal localization and characterization of c-abl in the t(6;9) of acute nonlymphocytic leukemia. Proc Natl Acad Sci USA 82:8742–8746, 1985.
36. Groffen J, Heisterkamp N, Stephenson JR, Geurts van Kesel A, de Klein A, Grosveld G, Bootsma D: C-sis is translocated from chromosome 22 to chromosome 9 in chronic myelocytic leukemia. J Exp Med 158:9–15, 1983.
37. Le Beau MM, Westbrook CA, Diaz MO, Rowley JD: C-*src* is consistently conserved in the chromosomal deletion (20q) observed in myeloid disorders. Proc Natl Acad Sci USA 82:6692–6696, 1985.
38. Le Beau MM, Westbrook CA, Diaz MO, Rowley JD: Translocation of the p53 gene in t(15;17) in acute promyelocytic leukemia. Nature 316:826–828,; 1985.
39. Benchimal S, Lamb P, Crawford LV, Sheer D, Shows TB, Bruns GA, Peacock J: Transformation associated p53 is encoded by a gene in human chromosome 17. Somatic Cell Mol Genet II:505–510, 1985.
40. Isobe M, Emanuel BS, Givol D, Owen M, Croce CM: The gene for the human p53 tumor antigen is located at band 17p13. Nature 320:84–85, 1986.
41. Schechter AL, Hung M-C, Vaidyanathan L, Weinberg RA, Yang-Feng TL, Francke U, Ullrich A, Coussens L: The neu gene: An erbB-homologous gene distinct from and unlinked to the gene encoding the EGF receptor. Science 229:976–978, 1985.
42. Huebner K, Isobe M, Chao M, Bothwell M, Ross AH, Finan J, Hoxie JA, Sehgal A, Buck CR, Lanahan A, Nowell PC, Koprowski H, Croce CM: The nerve growth factor receptor gene is at human chromosome region 17q12-17q22, distal to the chromosome 17 breakpoint in acute leukemias. Proc Natl Acad Sci USA 83:1403–1407, 1986.
43. Nagarajan L, Louie E, Tsujimoto Y, ar-Rushdi A, Huebner K, Croce CM: Localization of the human pim oncogene (PIM) to a region of chromosome 6 involved in translocations in acute leukemias. Proc Natl Acad Sci USA 83:2556–2560, 1986.
44. Huebner K, ar-Rushdi A, Griffin CA, Isobe M, Kozak LC, Emanuel BS, Nagarajan L, Cleveland JL, Bonner TI, Goldsborough MD, Croce CM, Rapp U: New active genes in the *raf* oncogene group located on the X chromosome in mouse and man. Proc Natl Acad Sci USA 83:3934–3938, 1986.
45. Huebner K, Isobe M, Gasson JC, Golde DW, Croce CM: Localization of the gene encoding human erythroid-potentiating activity to chromosome region Xp11.1→p11.4. Am J Hum Genet 38:819–826, 1986.
46. Le Beau MM, Westbrook CA, Diaz MO, Larson RA, Rowley JD, Gasson JC, Golde DW, Sherr CJ: Evidence for the deletion (5q) in myeloid disorders. Science 231:984–987, 1986.
47. Ohyashiki K, Ohyashiki J, Sandberg AA, Minowada J, Kinniburgh AJ: Loss of a Hu-*ets*-1 allele in human leukemia cell liver ML-1, -2, -3 is associated with a chromosome change at 11q24. Cancer Res (submitted).
48. Rovigatti U, Watson DK, Yunis JJ: Amplification and rearrangement of Hu-ets-1 in leukemia and lymphoma with involvement of 11q23. Science 232:398–400, 1986.
49. Slamon DJ, deKernion JB, Verma IM, Cline MJ: Expression of cellular oncogenes in human malignancies. Science 224:256–262, 1984.
50. Blick M, Westin E, Gutterman J, Wong-Staal F, Gallow R, McCredie K, Keating M, Murphy E: Oncogene expression in human leukemia. Blood 64:1234–1239, 1984.

The Cytogenetics of Mammalian Autosomal
Rearrangements, pages 855–893

Chapter 29
Application of Structural Rearrangements and DNA Probes to Gene Mapping

R. Ellen Magenis

INTRODUCTION

Of the estimated 10,000 to 50,000 genes in the human genome [1,2], about 374 have been firmly assigned to autosomes and 124 to the X chromosome [3]. In addition, there are 300 anonymous DNA segments localized to autosomes and 186 to the X chromosome, as well as antigens, protein spots, pseudogenes, and fragile sites for a total of approximately 1,650 assignments [3]. This leaves an overwhelming number still to be localized. Though the number of genes assigned is small, the rate of progress is accelerating, and already significant contributions to basic knowledge as well as clinical medicine have been made.

It has recently become possible to clone genes based on precise map position. Those cloned include the genes causing chronic granulomatous disease [4], Duchenne muscular dystrophy [5], and retinoblastoma [6]. This molecular cloning will allow identification of the abnormal protein product of these genes and greater understanding of the disease process and treatment potential. Further application of gene mapping studies has included the ability to make a diagnosis of disease otherwise not amenable to such diagnosis, by linkage of gene loci to marker loci. Examples include the prenatal diagnosis of cystic fibrosis [7–10] and the detection of Huntington disease before the manifestation of symptoms [11]. Mapping data has also allowed identification of marker chromosomes [12–14], of chromosome rearrangement breakpoints [15–19], and of small to cryptic deletions [20–22]. A detailed chromosome map will aid in the understanding of genetic regulatory mechanisms, of cellular differentiation, of neoplastic transformation, and of the pathogenesis

Departments of Medical Genetics, Pediatrics, and Crippled Children's Division, Oregon Health Sciences University, Portland, Oregon 97201 and embryogenesis of genetic defects and chromosomal malformation syndromes.

LINKAGE AND MAPPING PRIOR TO CHROMOSOME BANDING

A number of genetic traits were assigned to the X chromosome by the X-linked pattern of inheritance prior to specific knowledge of human cytogenetics, the first, color blindness, as early as 1911 [23]. The first human autosomal linkage group was defined by Mohr in 1954, between the Lutheran blood group and secretor [24]; pedigree analysis indicated nonindependent segregation of these two loci. The recombination fraction was .13 instead of .5, as would be expected for free assortment. Although assignment to the same human autosome was made, the particular chromosome could not be identified. Human autosome assignments awaited the cytogenetic developments of lymphocyte and fibroblast culture methods, and of harvest and staining techniques [25,26], so that numbers of cells from many individuals could be examined and chromosomes (though unbanded) identified.

In the same year, 1968, two genes were assigned to specific autosomes, the Duffy blood group to the centromeric region of chromosome 1 [27,28] and thymidine kinase to chromosome 17 [29,30]. Donahue et al. [27] and Ying and Yves [28] independently reported families with a greatly lengthened chromosome 1, described by Donahue et al. [27] as a region that extends "from the centromere along part of the long arm in which the chromatids may be thinner, with, occasionally, an alternating pattern of light and dark staining, suggestive of coils." The region was inherited in these families for several generations without phenotypic effect. A number of polymorphic marker loci were examined in these families; of these, only the Duffy blood group showed nonrandom segregation with the elongated chromosome, suggestive of linkage, and hence, assignment of the Duffy locus to chromosome 1. It was later recognized that the increased length of chromosome 1 was due to a large region of C-band-positive heterochromatin and that there was a range of heterochromatin size in the population (Fig. 1).

The assignment of the thymidine kinase gene was through the use of the then novel technique of somatic cell hybridization. During co-cultivation of two chromosomally different mouse cell lines, Barski et al. in 1960 [31] noted proliferating cells arising from fusion of the two lines, which was documented by chromosome analysis, the fused cells containing the chromosomes of both cell lines. These results were soon confirmed by others, and Littlefield [32] added a selective system, hypoxanthine-aminopterin-thymidine (HAT), in which neither parent line could grow, to isolate hybrids

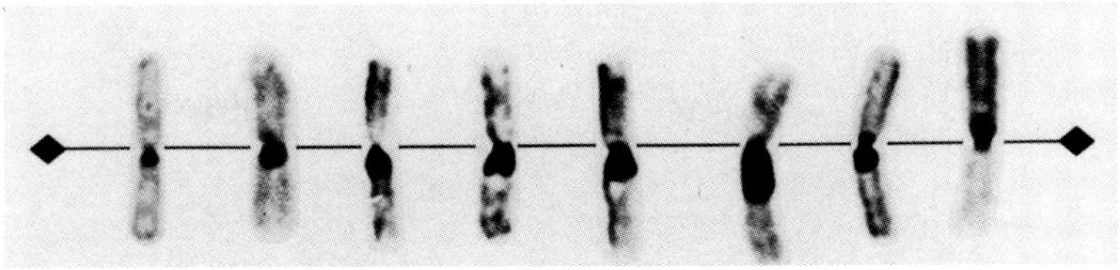

Fig. 1. C-banded chromosomes 1 from eight individuals showing varying amounts of genetically inactive centromeric heterochromatin. The last two chromosomes on the right illustrate a partial inversion and a complete inversion of the heterochromatin, respectively.

between two drug-resistant cell lines. Ephrussi and Weiss [33] were able to fuse rat and mouse cells and isolate hybrids, and Weiss and Green [29] showed that human-rodent hybrids could also be produced and isolated in a selective system. Weiss and Green [29] used a thymidine-kinase-deficient mouse cell line that could not grow in HAT medium and required the human thymidine kinase to proliferate. In these hybrids there was rapid and preferential loss of human chromosomes until after several generations in culture, only one human chromosome remained, this chromosome presumably containing the thymidine kinase gene. They misidentified the chromosome as a C-group chromosome, which was later shown to be an E-group [30] and still later identified as chromosome 17 [34].

Translocation chromosomes segregating in families were first utilized as markers for gene mapping studies in 1969. Robson et al. [35] studied two families with rearrangements involving chromosome 16 for linkage with a number of blood group markers. The translocation in one family was between chromosomes 2 and 16; since the chromosomes were unbanded, it was not certain which chromosome arms were involved. There were seven certain nonrecombinants and four probable nonrecombinants in eleven opportunities for recombination between α-haptogoblin (Hp) and the rearranged chromosomes, suggesting linkage. Study of the second family, which carried a 1/16 rearrangement, favored linkage to chromosome 16, the chromosome in common rather than chromosome 1 or 2; there were several probable nonrecombinants and no certain recombinants between the rearrangement and Hp. Combining this information with that from studies of segregation with chromosome 16 long arm marker heterocromatin and data on chromosome lengths gave a probability of 0.97 that Hp was somewhere on chromosome 16. This localization was confirmed by studies of a very large family in which a recurrent break of chromosome 16 (fragile site) was segregating [36]. The fragile site was located at about the junction of the middle and distal third of the long arm; later studies using banding techniques showed the break to be at the junction of bands 16q21 and 16q22.1 (Fig. 2) [48]. The fragile site was

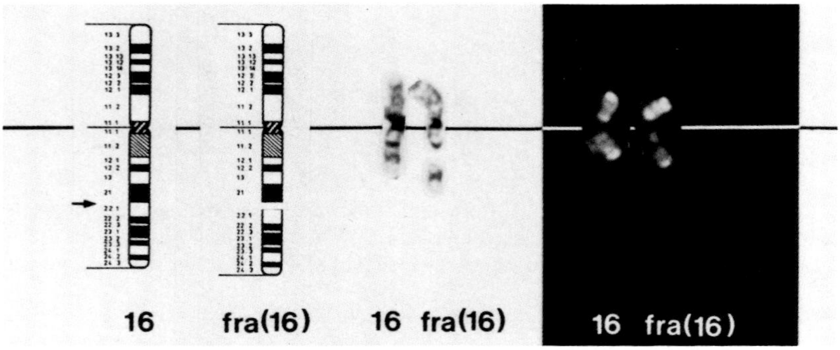

Fig. 2. G-banded (**center**) and R-banded (**right**) chromosomes 16; the chromosome 16 on the right in each panel shows a fragile site on the long arm. **Left:** The breakpoint (q22.1) is shown in the idiogram [48].

shown to segregate as an autosomal dominant trait, though its presence could not be detected in every cell. In 127 family members studied (Fig. 3), there were 33 opportunities for recombination between the fragile site and the α-haptoglobin locus; 30 were nonrecombinant, for a lod (logarithim of the odds) score of 5 at a recombination fraction of .1, indicating that the α-haptoglobin gene is located near the fragile site on chromosome 16. In addition to confirming the assignment of α-haptoglobin to chromosome 16, this work placed the gene at a point on the long arm two-thirds of the distance from the centromere to the end.

Gene dose effects were sought by a number of investigators soon after the discovery that Down syndrome was associated with an extra G-sized chromosome. Elevated levels of a number of red cell enzymes were reported [37–41] and the genes for these enzymes provisionally assigned to the extra chromosome. However, elevated levels of X-linked enzymes also were reported [42] and the gene dosage method of gene mapping fell into disrepute. It became apparent that the mean red cell age is reduced in Down syndrome and that there is a concomitant increase in red cell enzyme levels. One report, however, among those assigning red cell enzyme genes to the extra chromosome stood out as likely to be valid. This report by Baikie et al. in 1965 [39] described increased red cell activity of phosphofructokinase (PFK) in Down syndrome over that of other red cell enzymes. Increased activity of PFK was also found by Layzer and Epstein in 1972 [43]; and the assignment of PFK, liver type, to chromosome 21 was later confirmed by Vora and Francke in 1981 [44] using somatic cell hybrids.

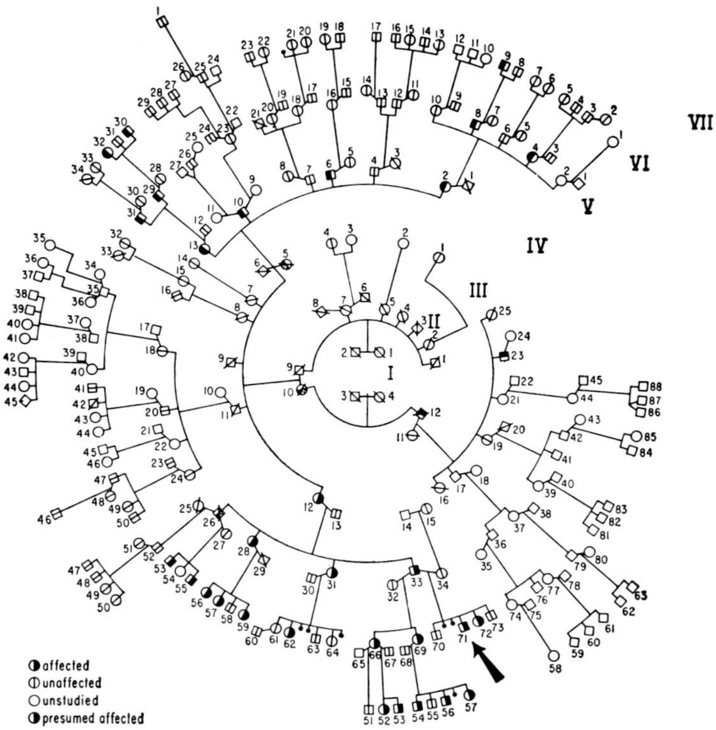

Fig. 3. Pedigree of family with an inherited fragile 16. Arrow points to the proband. Only a portion of cells from each expressing [the fra (16)] individual exhibited the fragile site. However, every person with the fragile site whose parents were alive was shown to have an expressing parent. Unexpressing persons had only unexpressing children. Thus, the fragile site appeared to be fully penetrant.

LINKAGE AND MAPPING FOLLOWING INTRODUCTION OF CHROMOSOME BANDING TECHNIQUES

With the advent of chromosome banding techniques in the early 1970s such that every human chromosome could be identified and aberrations of even small portions of chromosomes documented [45,46], there was a rapid increase in the numbers of genes mapped to chromosomes and in the precision of the techniques used. A continuing increase in precision of gene mapping has occurred, owing primarily to the use of balanced and unbalanced chromosome aberrations, with breakpoints defined by high-resolution techniques [47,48] together with a number of gene mapping approaches [49].

Despite the disfavor of gene dosage mapping as an appropriate technique, some researchers continued to investigate chromosomally unbalanced individuals for evidence of loss or gain of an allele and or quantitative changes in levels of gene products. Ferguson-Smith et al. in 1973 [50] first conclusively showed a concurrent loss of an allele, half-normal enzyme levels and a deletion of part of a chromosome. The patient studied had a deletion of the distal end of the short arm of chromosome 2 (2p23→pter). He was missing a red cell acid phosphatase allele (ACP1), which he should have inherited from his mother, and his activity level of ACP1 was 45% less than expected. This was consistent with hemizygosity at the ACP1 locus, and assignment of the gene to distal 2p. Further evidence for this assignment was provided by Magenis et al. [51] also using gene dosage. They presented an infant with an imbalanced product of a translocation between chromosomes 2 and 18 inherited from her father (Fig. 4). The child was ACP1 type BA, her mother type BA, and her father B. Activity levels of ACP1 have been shown to be type

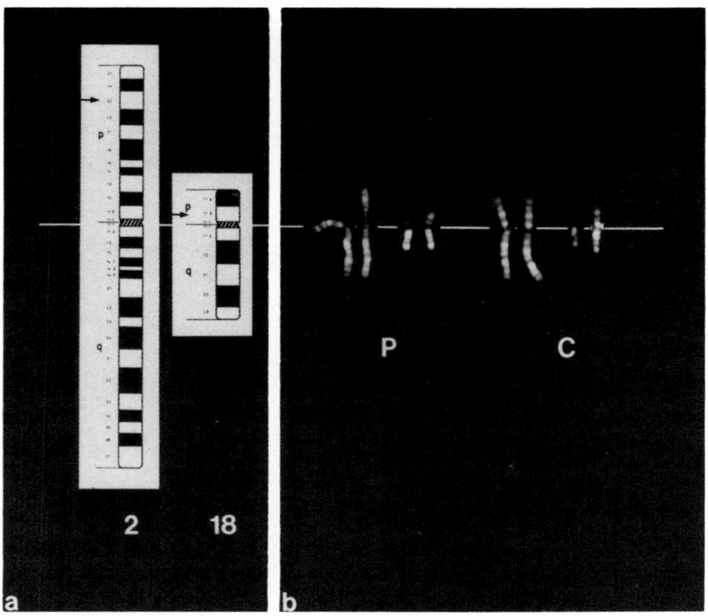

Fig. 4. Panel a: Idiogram [48] at the 400 band stage of chromosomes 2 and 18. The arrow indicates the breakpoint in chromosome 2. There does not appear to be reciprocal exchange. **Panel b:** Representative Q-banded chromosomes 2 and 18 from the father (P) carrying a balanced 2/18 translocation and from the child (C) who has the unbalanced product, a duplication of the distal three bands of the short arm of chromosome 2.

dependent, with a ratio of activity for the different alleles of 2:3:4 (pA:pB:pC) [52]. For example, age-matched controls in this study showed type AA activity levels to be 114.9 units/g of hemoglobin, type BB 160.9 ± 15.4. The heterozygote appears to have activity commensurate with adding the activity levels for the respective alleles (e.g., BA 138. ± 10.9). The level in the child, 201.8 ± 10.4, corresponded to two doses of type B and one of type A. This assigned ACP1 to bands p23→pter [48] of chromosome 2. Still further documentation of this gene assignment was provided by somatic cell hybrid studies [53] and by family studies of chromosomal rearrangements [54].

Gene dosage has also been unequivocally demonstrated by the presence of three different markers in patients with partial trisomy. Four patients with trisomy for 6p21→6pter who had three different major histocompatibility complex class I antigen (HLA) haplotypes each have been reported [55–57], and another patient with duplication of the region p31.2→22.1 of chromosome 1 had three different phosphoglucomutase 1 types [58]. These results gave credence to dosage evaluations of chromosome abnormalities that had been reported in malignancies. Marsh et al. [59] found a cell line in a patient with preleukemia that had a partial deletion of chromosome 1; they found both Rh-positive and Rh-negative red cells and concluded that the Rh locus is on the distal short arm of chromosome 1. Elevated glutathione reductase activity was found by de la Chapelle et al. [60] in five patients with trisomy 8 mosaicism in bone marrow; the extra chromosome was evident in over 95% of cells examined in the subjects, and the mean activity of glutathione reductase was approximately 1.5-fold. Even greater precision in dosage mapping was achieved by the use of multiple individuals with duplications or deletions of relevant regions of chromosomes to determine a smallest region of overlap (SRO). Ferguson-Smith et al. used a panel of chromosome 9 aberrations to regionally localize adenylate kinase 1 to 9q34 [61], and later galactase-1-phosphate uridyltransferase [62] to the proximal short arm of chromosome 9.

This same approach has been applied to somatic cell hybrid studies. The use of a human cell line carrying a translocation aids in regional localization, and the use of multiple overlapping rearrangements allows a smallest region of overlap to be determined, as well as gene order. For example, Carritt et al. [63] analyzed Chinese hamster-human lymphocyte hybrids containing rearrangements of chromosome 1 and were able to conclude that glucose dehydrogenase and enolase 1 are distal to phosphogluconate dehydrogenase but all three are located distal to 1p36.13. They further concluded that fucosidase 1 and adenylate kinase 2 are in band 1p34.

METHODS OF GENE MAPPING

At least 18 different methods have now been successfully utilized both independently and in different combinations to map genes on chromosomes [63], included those already discussed:

1 and 2. Two of the methods, classical family linkage study and somatic cell hybridization account for over 60% of the 1,096 loci mapped to autosomes, most of these the result of somatic cell hybrid study. During the years 1970–1980, following advances in techniques, there was an explosion of effort and results in somatic cell genetics. Although until recently only traits expressed in cultured cells could be utilized for gene localization, results were generally accurate and quickly obtained. The speed was increased further by the development of hybrid clone panels of known partial sets of human chromosomes for preliminary assignment [64]. In contrast, family studies have been time-consuming and slow. Family studies have also been essentially limited to segregation analysis of blood marker phenotype and rare disease trait, or of chromosome variant or rearrangement and rare disease. Most families have been small in size and have required sophisticated mathematical formulae and computer programs to determine the likelihood that traits under study are linked [65,66]. The study of families for linkage and mapping purposes has recently increased in importance with the advent of molecular techniques which detect DNA polymorphism. The use of bacterial restriction enzymes to cut human DNA at specific points [67] combined with the use of gel electrophoresis to detect size differences (restriction fragment length polymorphism) has allowed an estimate of genomic variation far greater than that inferred from protein marker data. Much of the variation, if not present in proteins, must be in noncoding sequences. Restriction enzyme digestion can result in length variation whether the DNA alteration is a single-base-pair change at a recognition site or sequence deletion, addition, or rearrangement; most appear to be due to a single-base-pair substitution. The number of such polymorphisms already detected is large; a summary list compiled in 1985 included 559 polymorphisms detected by arbitrary cloned segments [68]. It is anticipated that when the number is large enough and appropriately spaced throughout the genome that any disease gene can be mapped, even when the action of the gene is unknown [69] such as the mapping of the Huntington [11] and cystic fibrosis genes [7–10].

To map such genes onto specific chromosomal regions using restriction fragment length polymorphisms (RFLP) as markers requires that the location of the linked polymorphic segment be known. The same arbitrary DNA probes detecting RFLPs, as well as cloned known genes, can be utilized for mapping purposes with the DNA of a somatic cell hybrid clone panel on Southern blots or more precisely with in situ hybridization techniques, and even more precisely when in situ hybridization is accomplished using multiple overlapping chromosomal arrangements (See No. 3 below). A diagram of the process of construction of a cloned DNA probe and of Southern blot analysis is presented in Figure 5.

3. A third method, the in situ hybridization technique of gene localization, has contributed 159 of the 1,096 autosomal loci mapped (14.5%) [3,68], almost all of these since 1981. The use of this technique is rapidly increasing as it allows direct visualization of the location of a cloned probe on a chromosome.

A method for forming molecular hybrids between RNA in solution and the DNA of intact cytological preparations was first described by Gall and Pardue [70] and independently by John et al. [71] in 1969. Gall and Pardue [70] studied *Xenopus* cells using slide preparations treated with alkali to denature the DNA, incubated the slides in a solution of tritium-labeled *Xenopus* ribosomal RNA and detected the hybrids by autoradiography. In these preparations, silver grains were found primarily over the nucleolus. The in situ hybridization technique was utilized by a number of investigators to localize different molecular species of repetitive RNA or DNA to chromosomal regions, but until the work of Harper and Saunders in 1981 [72] there was limited success in assignment of single copy DNA or RNA segments to chromosomes. Their modifications, which allowed assignment of unique DNA sequences, were the addition of 10% dextran sulfate to the hybridization reaction and the use of cloned probes. The dextran sulfate was used to accelerate the hybridization rate of DNA; the use of a cloned probe allowed the attached vector molecules, which cannot hybridize to the DNA of the chromosome, to participate in network formation, increasing the radioactive deposit at the specific hybridized site. Notable chromosomal assignments achieved by this method include insulin [73], β-globin [74], α-globin [75], amylase, propiomelanocortin, and somatostatin genes [76] and the clotting factor IX gene [77, 78]. The in situ hybridization procedure is illustrated in diagram form in Figure 6. Detailed in situ hybridization methodology has been published by Donlon [79] and Buckle and Craig [80].

Several different approaches using cytogenetic aberrations for in situ hybridization procedures have been successful in attaining exact map positions. Such approaches are necessary for precise mapping because of inherent scattering or shift of the grains from the site of hybridization owing to path length of radioactive particles from the site and technical problems with emulsion [81]. One of the most definitive approaches is the use of a translo-

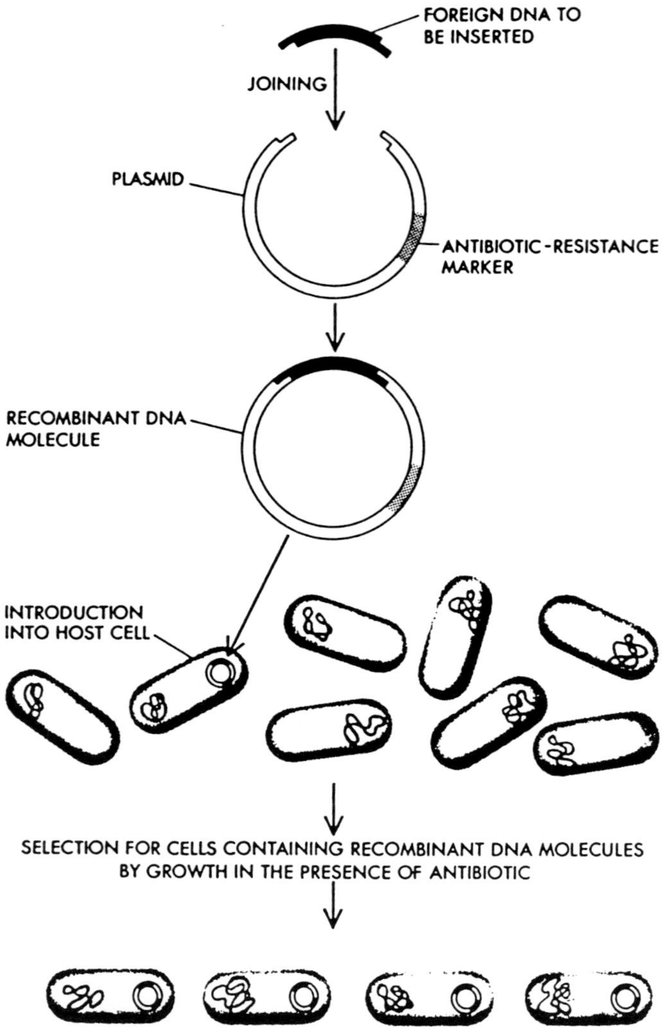

Fig. 5. **a.** Preparation of a cloned probe. A segment of desired double-stranded DNA cut with a restriction enzyme is spliced into a similarly cut plasmid that contains a gene coding for antibiotic resistance. The recombinant plasmid is introduced into a bacterium (usually *Escherichia coli*), which is then grown on appropriate media in the presence of an antibiotic. Only bacteria containing the plasmid confering resistance to the antibiotic will grow. Thus the plasmid-containing bacteria will be selected for; and large clones of such bacteria prepared. Plasmid may then be isolated by lysis of the bacteria followed by isopycnic centrifugation.

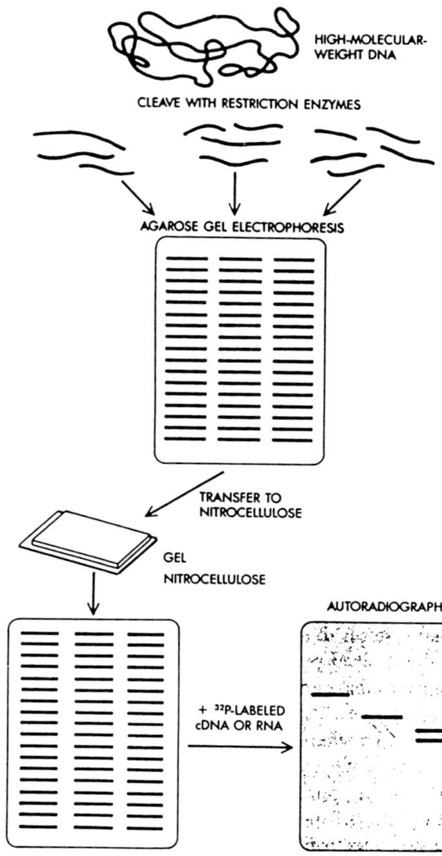

Fig. 5. b. Southern blot analysis. Double-stranded genomic DNA is cut with a restriction enzyme and applied to an agarose gel. Electrophoresis is performed to separate the cleaved segments of DNA by size. The DNA is transferred to a nitrocellulose filter by blotting. It is denatured and then hybridized to select denatured cloned probes which have been radiactively labeled. Autoradiography is done, and the sites of hybridization visualized on film.

cation in which only a very small chromosomal region is moved to another chromosome and the hybridization grains follow the changed position. The choice of translocation may be based on preliminary hybridization to normal chromosomes alone, or include use of larger rearrangements, as in the following example, modified from Magenis et al. [82].

Initial chromosome preparations were from four subjects, one normal 46,XY male and three patients with different rearrangements of chromo-

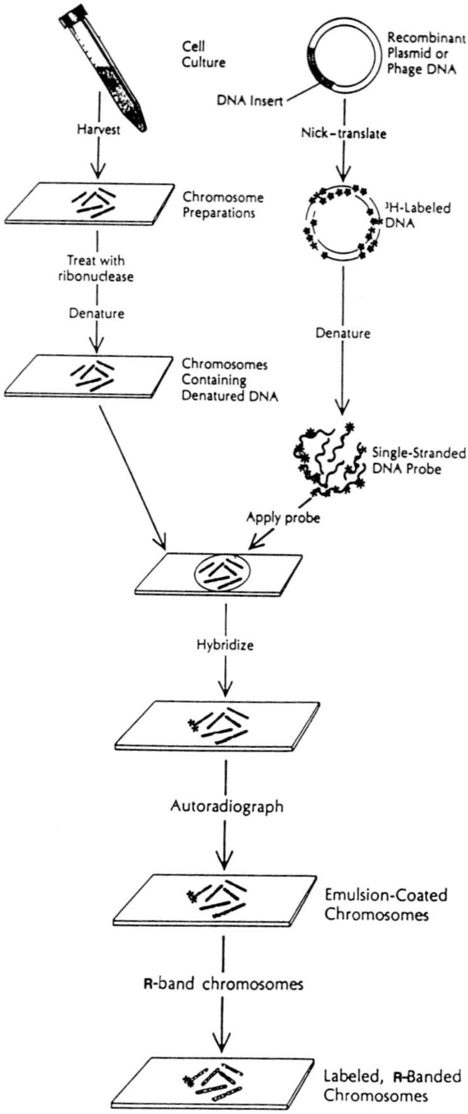

Fig. 6. The in situ hybridization procedure. White blood cells from selected individuals are grown in a routine manner for chromosome study. Cells are harvested and slides made. The chromosomes are treated for 1 h with RNAse at a concentration of 200 μg/ml in 2 × SSC at 37°C, denatured in 70% formamide (MC/B, Norwood, OH) 2 × SSC at 70°C for 2 min, then immersed in 70%, 80%, 95%, ethanol for 1 min each. The probe is prepared by nick-translation of the plasmid, which contains the insert using tritiated thymidine. It is then denaturated and diluted to a final concentration of 0.1 μ/ml in pH 7 hybridization buffer

(continued)

somes 11: 46,XX, del(11)(p11.2→p15.1); 46,XY, inv (11)(p13q24.2), (Fig. 7), and 46,XX,rec(11)inv(11)(p13q24.2)pat. These were utilized for in situ hybridization studies with a tritium-labeled cDNA probe containing a β-globin insert. With the hybridization technique described by Harper and Saunders [72], there were 1–2 grains over each labeled metaphase. Of 360 cells scored, 88 were labeled over chromosome 11, band p15 (24%). Approximately half of the chromosomes 11 labeled from the abnormal patients were the del(11) or inv(11). These results exclude the β-globin locus from 11p11→p14, since these bands were not present in the deleted 11, and assign it to 11p15. (Figs. 8,9).

With the β-globin gene assignment clearly made to 11p15, rearrangement within band p15 was sought; a chromosome rearrangement was identified in a phenotypically normal person which divides the distal short arm band p15 of chromosome 11. This rearrangement, t(7;11), was used for in situ hybridization with the β-globin probe as described above (Fig. 10). The breakpoint in chromosome 7 is in band 7p11.2, while the breakpoint in 11 is either distal 11p15.3 or p15.4 [48]. One hundred cells were scored for silver grains; seven of these cells had label over the distal short arm of the derivative 7 which includes the distal 11, bands p15.4→pter, while 9 cells were labeled on the distal short arm of the normal 11. No cells had label over the derivative 11 in the region of the breakpoint, but three cells had label over other parts of the chromosome (Figs. 11,12). These results place the β-globin gene on the distal tip of chromosome 11, bands p15.4→pter.

The chromosome abnormalities utilized were chosen as best suited to exclude early suggested assignments of the β-globin gene [74,83] as well as to localize the gene firmly. The chromosome abnormalities are constitutional with good correlations of phenotype abnormalities with the chromosome aberrations, which reduce the problems of karyotype interpretation. Also, lymphocyte cultures can be manipulated to produce large numbers of high-

---

Fig. 6 *(continued)*
containing 50% formamide/2 × SSCP, 10% dextran sulfate, and 0.1 mg/ml sonicated denatured salmon sperm DNA. Chromosomes are hybridized for 12 h at 37°C; excess probe is then removed with three changes of 50% formamide/2 × SSC for 3 min each, followed by five changes of 2 × SCC for 2 min each, all at 39°C. Chromosome preparations are again dehydrated with an alcohol series, and Kodak autoradiographic NTB-2 liquid emulsion applied to the slides for a 4–6-day exposure. After the slides are developed, they are R-banded, examined and photographed using a Zeiss photoscope III with a KP 490 excitor filter, and FT 510 dichroic mirror, and an LP 520 barrier filter. Silver grains are analyzed over fluorescent R-banded preparation by a double-illumination system. For illustrative purposes some of these preparations were destained in methanol and restained with Wrights's stain.

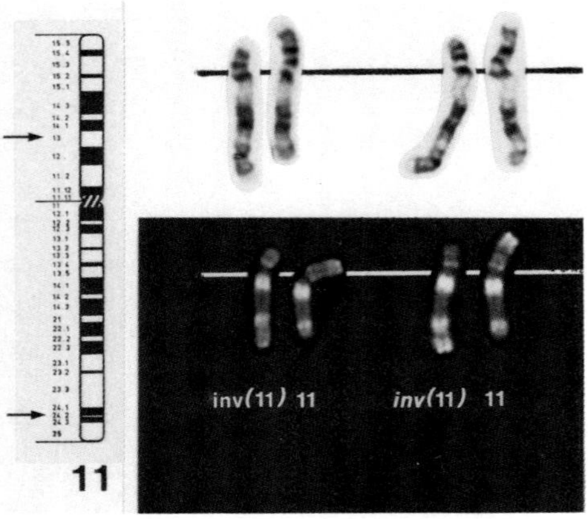

Fig. 7. G- (**top**) and R-banded (**bottom**) chromosomes from a carrier of a balanced pericentric inversion of chromosome 11 [46,XY,inv11(p13q24.2)]. Breakpoints are shown in the idiogram on the left [48].

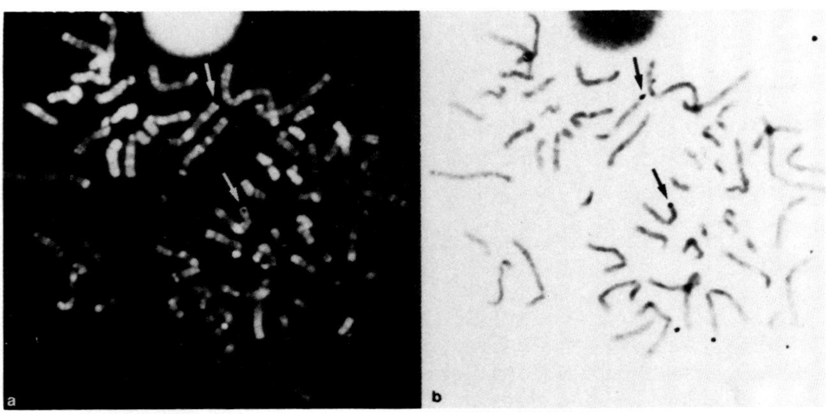

Fig. 8. Lymphocyte metaphase spread sequentially stained with **a.** an R-band technique and **b.** Wright's stain from a balanced carrier of an inversion 11 [46,XY,inv(11)p13q24.2)], hybridized with a 4.4-kb Pstl fragment containing the β-globin sequence. A grain is present on the distal short arm of the normal 11 and on the terminal long arm of the inverted chromosome 11 (arrows).

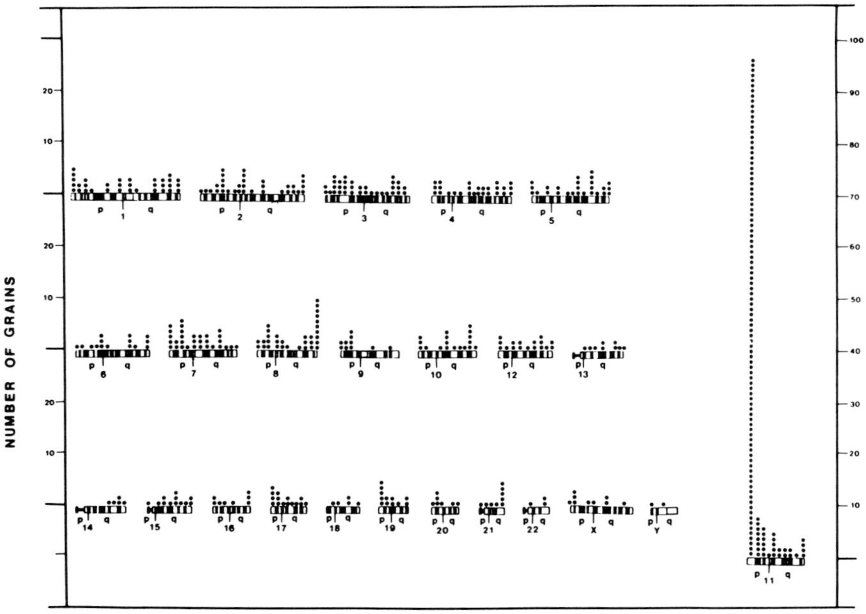

Fig. 9. In-situ hybridization of a 4.4-kb Pst1 fragment containing the β-globin sequence. Compilation of grains distributed over the chromosomes of the four individuals studied. Eighty-eight of the 360 cells scored were labeled on chromosome 11, band 15p, thus localising the β-globin gene.

resolution chromosomes for breakpoint analysis, accurate breakpoints being critical for precise mapping.

Since the insulin gene had also been assigned to chromosome 11p15 [73], the t(7;11) was utilized with an insulin probe, phins 321, provided by Bell for within-band placement and possible ordering with the β-globin gene. An excess of grains was found on the derivative 7, but also on the derivative 11. These findings are yet unexplained.

Another approach to increasing accuracy of mapping is the use of multiple overlapping deletions to determine the smallest region of overlap (SRO), as shown by the following study, modified from Magenis et al. [84]. A 5.5-kilobase (kb) single-sequence DNA fragment (G8) that reveals the Huntington disease-linked DNA polymorphic locus D4S10 on Southern blot analysis [85] was hybridized to the metaphase chromosomes of five patients. Two of them had different interstitial deletions of chromosome 4 short arm (T.A. and B.A.), two had different chromosome 4 short arm terminal deletions (R.G. and D.Q.) (breakpoints and overlapping segments are illustrated in

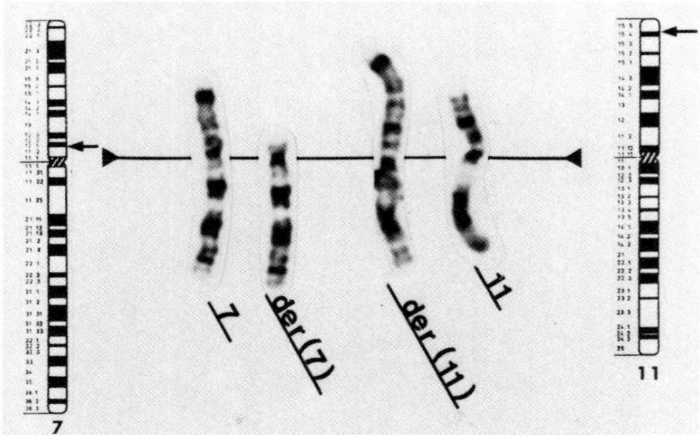

Fig. 10. G-banded chromosomes 7 and 11 from a phenotypically normal carrier of a balanced translocation between the short arms of chromosome 7 and 11 (46,XY, t(7;11)(p11.2;p 15.4) [48]. Arrows on idiograms point to breakpoints.

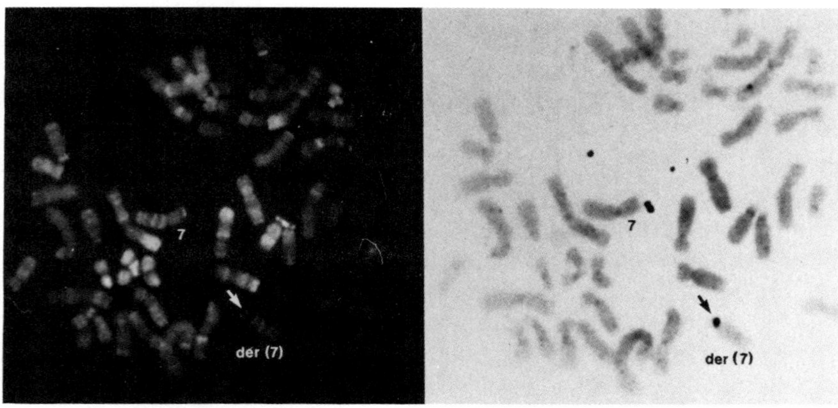

Fig. 11. Sequentially stained lymphocyte metaphase spread hybridized with a 4.4.-kb Pstl fragment containing the β-globin gene. The arrows point to the derivative 7. The other 7 is normal. On the left, the spread is R-banded to identify the chromosomes; on the right, the same spread has been destained and restained with Wright's stain to show grains. A grain is present on the short arm of the derivative 7, in the region of translocated chromosome 11 material.

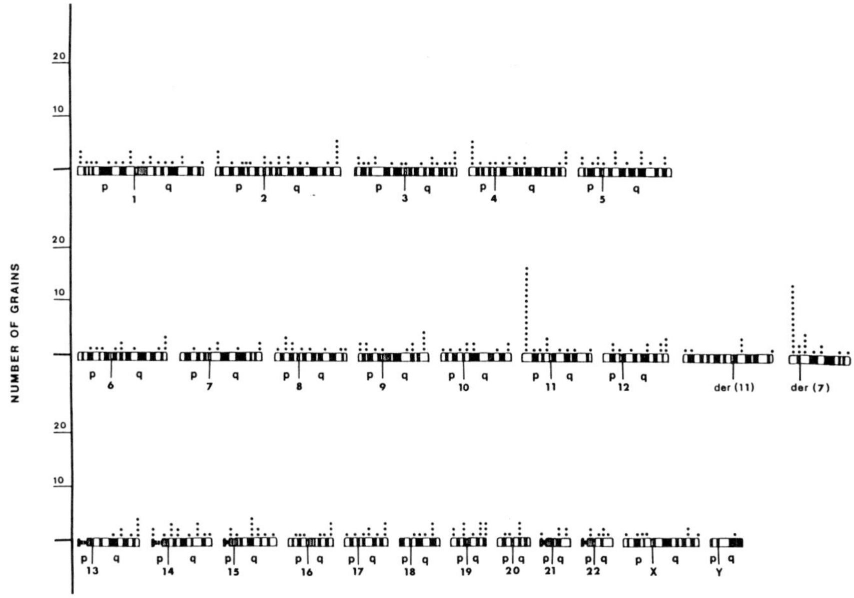

Fig. 12. Distribution of grains in 100 cells scored following in situ hybridization with a β-globin probe, from a patient with a balanced translocation between chromosomes 7 and 11. Note excess of grains on the distal short arm of the normal chromosome 11 and on the derivative 7 at the site of the translocated chromosome 11 material. Grains are not in excess on the short arm of the derivative chromosome 11. This indicates that the β-globin gene is located in the most distal pale band p15.5 of chromosome 11.

Fig. 13), and one had a normal male karyotype. Chromosomes of D.Q. and T.A. are showing in Figure 14. Results are summarized in Table I and in Figure 15. In all the deletions except B.A., the locus D4S10 homologous to probe G8 is missing. The smallest region of overlap is that between the deletions of T.A. and D.Q. In Figure 16, the deleted segments from T.A. and D.Q., both of which must contain D4S10, are differentially shaded so that the overlap region is clear. The combined data suggest that the D4S10 site must be located in the overlap region, within the distal half of band p16.1. The tightly linked Huntington disease locus is then also mapped either in or close to this segment.

Combinations of techniques may also provide a means to pinpoint a specific region and/or an SRO for fine gene localization, along with in situ hybridization procedures. For example, family studies of a fragile site on the long arm of chromosome 16 first placed the α-Hp gene on the long arm of 16 at a point two-thirds of the distance from the centromere to the end [36].

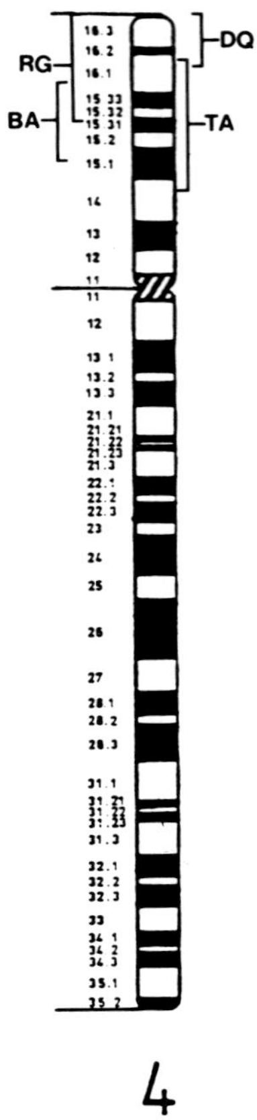

Fig. 13. Approach to fine mapping of Huntington disease by studying the smallest region of overlap in deletion 4p patients. Idiogram of chromosome 4 [48], illustrating by brackets the breakpoints and extent of the deletions present in patients RG, BA, DQ, and TA.

In situ hybridization studies of chromosomes bearing a fragile 16 site with a haptoglobin probe (pULB 1148 and pULB 1741 provided by Bollen), showed consistent labeling of the portion of 16 distal to the fragile site (Figs. 17, 18) in band q22 as has been reported by Simmers et al. [86]. In situ hybridization studies (Figs. 19, 20) of a mentally retarded male with multiple anomalies and tyrosinemia (owing to tyrosine aminotransferase deficiency) and a very small interstitial deletion of chromosome 16 long arm (q22.1→q.22.3) showed grain numbers approximately half those of controls on the long arm of 16 (Fig. 20). It was not possible to differentiate the normal and deleted chromosomes 16 owing to the quality of preparations achieved with the in situ

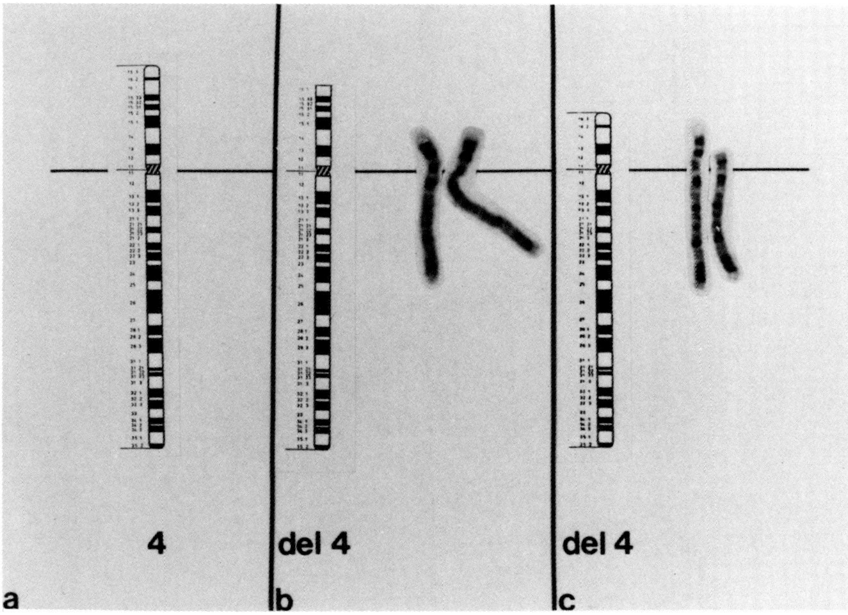

Fig. 14. The deletions in $4p^-$ patients for the fine mapping of Huntington disease. **a:** Idiogram of chromosome 4 at the 850 band stage [48]. **b:** G-banded chromosomes 4 from DQ. The chromosome 4 on the right has a small terminal deletion. This deletion is shown in the idiogram, breakpoint in band p16.1. **c:** G-banded chromosome 4 of TA; the right-hand chromosome 4 has a large deletion of the short arm, extending from p16.1→p14.

TABLE I. Results of hybridization of G8 to 4p15.3→4pter

		Percentage of cells with hybridization		
Patient	No. cells scored	Normal 4	Deleted 4	Total
R.G.	200	12	<1	12
D.Q.	200	12	<1	12
T.A.	200	11	0	11
B.A.	240	8	8	16
Normal male	200	22	—	22

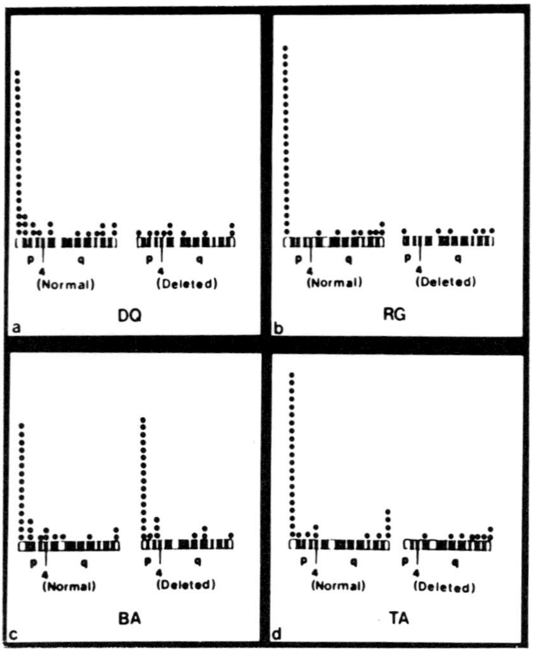

Fig. 15. Distribution of grains following hybridization with the Huntington-disease-linked probe G8 over chromosomes 4 from four individuals with different deletions of chromosome 4 short arm. At least 200 cells were scored from each individual. An excess of grains is present on the normal chromosome 4 distal short arm in each case. The deleted chromosome 4 in each case did not show excess grains, except for B.A. (**c**), indicating the probable absence of the G8 probe homologous site in D.O. (**a**), R.G. (**b**), and T.A. (**d**), and presence in BA.

hybridization process and the smallness of the deletion. Dosage studies of the α-Hp phenotype level showed a weak α-Hp type 2 compared with parents (Fig. 21), and Laurell electroimmunoassay of the α-Hp protein showed approximately half-normal levels (Fig. 22). All these results together support the assignment of α-Hp to a small, discrete region of chromosome 16q bands 22.1→22.3. In situ hybridization with a probe pchTAT2-16 (provided by Scherer) for the tyrosine aminotransferase (TAT) gene, which had been previously assigned to chromosome 16 [87], to chromosomes of the same subject did not show above-background grain numbers, whereas a control showed heavy labeling over 16q (Fig. 23). These results are consistent with loss of both TAT loci; this supports the work of Scherer et al., who determined on Southern blots that both TAT genes had been lost [88]. They also showed that one allele present in the patient's father was lost in the deletion

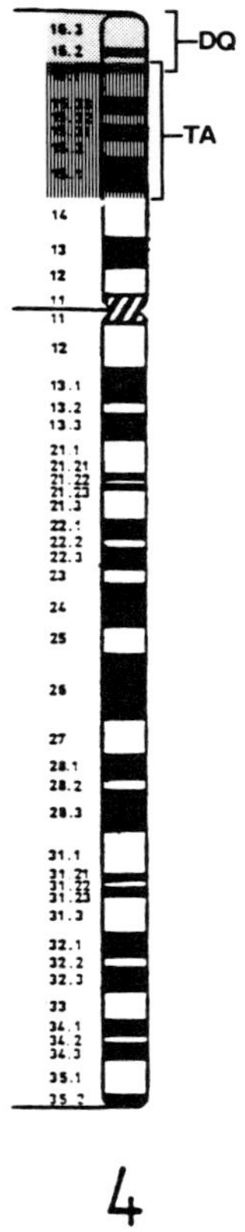

Fig. 16. Idiogram of chromosome 4 showing the deletions of D.O. and T.A. in brackets. The deleted regions are differentially shaded so that the small overlap region between the two deletions is clear. The G8 probe does not significantly hybridize with the chromosomes of either case. The homologous segment to G8 therefore must lie within the overlap region of 4p16.1.

visible at the cytologic level, and the other allele, deleted at the molecular level, was also deleted in his mother.

A still different approach to enhance precision of in situ hybridization results using rearrangements is the use of the technique of somatic cell hybrids containing human chromosome rearrangements. Utilizing such methods, Munke et al. [89] were able to determine precise positions of the protooncogene neuroblastoma RAS viral oncogene homolog (NRAS), the genes for the β-subunit of nerve growth factor (NGFB) and the gene for

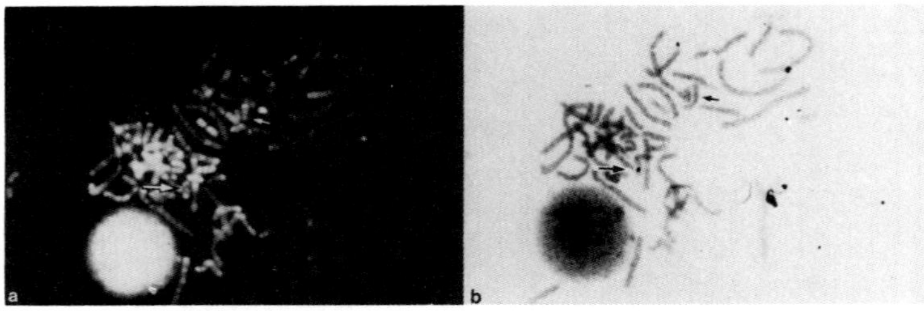

Fig. 17. **a:** R-banded metaphase spread from a patient with a fragile 16 that has been hybridized with a probe containing a haptoglobin sequence. The normal 16 is indicated by the small arrow; the large arrow points to the fragile site present in the other chromosome 16. **b:** The same metaphase in panel a. after destaining and restaining with Wright's stain so that the grains produced by the autoradiographic process can be seen. A grain is present on the long arm of the chromosome 16 with the fragile site; it is located distal to the break site.

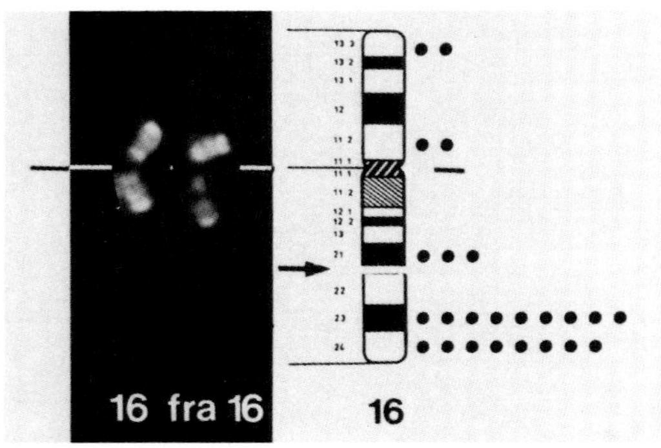

Fig. 18. **Left:** R-banded chromosomes 16 from a subject with a fragile 16. The right-hand 16 has a fragile site in proximal q22.1 **Right:** Distribution of grains on chromosome 16 following hybridization with the haptoglobin probe in 75 cells scored. Note excess of grains distal to the fragile site.

amylase (AMY). These genes had been previously assigned to chromosome 1 short arm, but precise positions had not been established. By in situ hybridization of cloned gene probes they were able to locate the complementary sequences in mouse-human somatic cell hybrids containing translocations of chromosome 1. The results agreed with the presence or absence of the human sequences on Southern blots made from hybrid cells, and, further, confirmed that the genes were present on the rearranged chromosome. They found that both NRAS and NGBF map to band 1p22, and AMY to band 1p21.

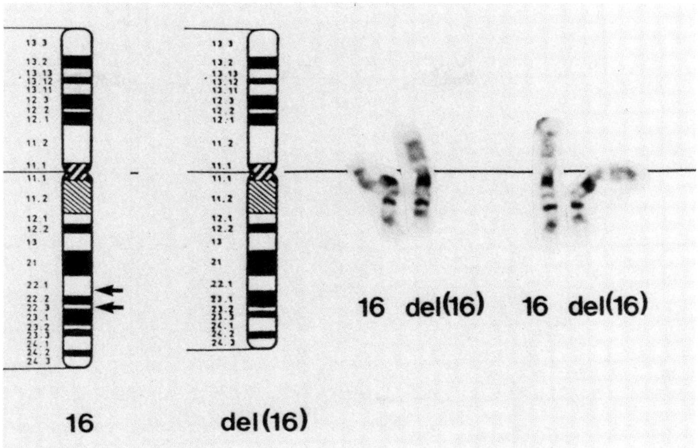

Fig. 19. Chromosomes 16 from a male (T.H.) with multiple anomalies, mental retardation, and tyrosinemia. The chromosomes are G-banded and show a small deletion of the long arm of the chromosome 16 on the right. The breakpoints are shown by arrows on the left hand idiogram and the chromosome 16 deletion is illustrated in the idiogram on the right.

4. Gene dose effects (deletion mapping, trisomy mapping) utilizing individuals with unbalanced chromosome complements have contributed approximately 7% of the total loci mapped. Methodology and limitations have already been discussed. Gene dosage mapping together with family linkage studies, somatic cell hybrid studies, and in situ hybridization procedures have been instrumental in over 80% of loci mapped to specific chromosomes. In each of these methods use of chromosome aberrations increases the precision of localization.

5. Chromosome changes associated with particular phenotypes have provided 42 of the 1,096 autosomal localizations [3]. These changes include the fragile Xq27 site and the associated fragile X syndrome [90], deletion of band q14 of chromosome 13 and retinoblastoma [91], deletion of band q11 of chromosome 15 and Prader-Willi syndrome [92], deletion of terminal 17p and Miller-Dieker lissencephaly syndrome [93], among others. Some of these deleted segments also include specific missing gene loci such as esterase D deletion accompanying 13q14 deletion and retinoblastoma [91], and overlap with gene dosage mapping.

Some changes may also segregate in families, such as the X-linked form of chondrodysplasia punctata (Conradi syndrome) [94] described below. The proband, an adolescent male, had short stature, very flat facies, ichthyosis, and mental retardation (Fig. 24). His young nephew through his maternal half-sister appeared very similar to the proband, having similar facies, ich-

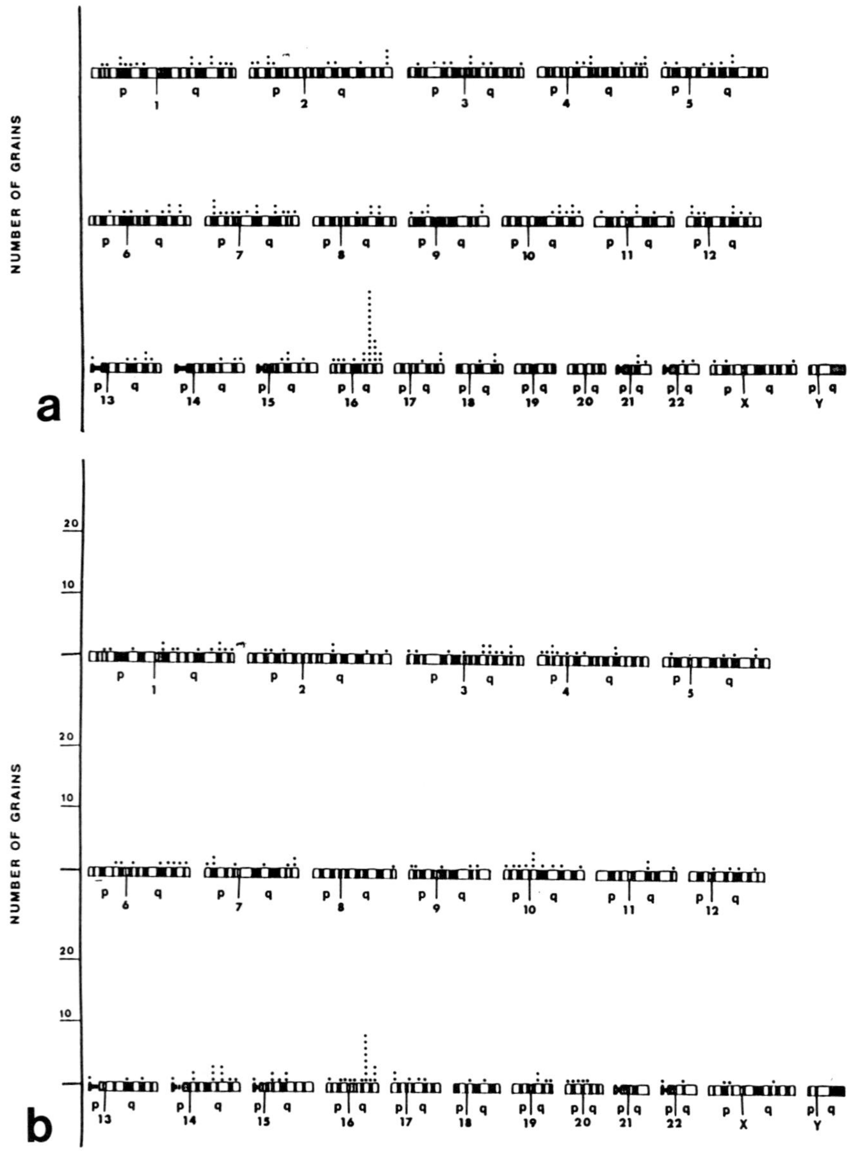

Fig. 20. **a:** Distribution of grains in 100 cells scored from a normal female following hybridization with a haptoglobin probe. There is a marked excess of grains over the long arm of 16. **b:** Distribution of grains in 100 cells scored from patient T.H., who has a small interstital deletion of 16 long arm. Grain numbers on the chromosome 16 long arm are approximately one-half of those of the control, an indication of loss of the gene due to the deletion.

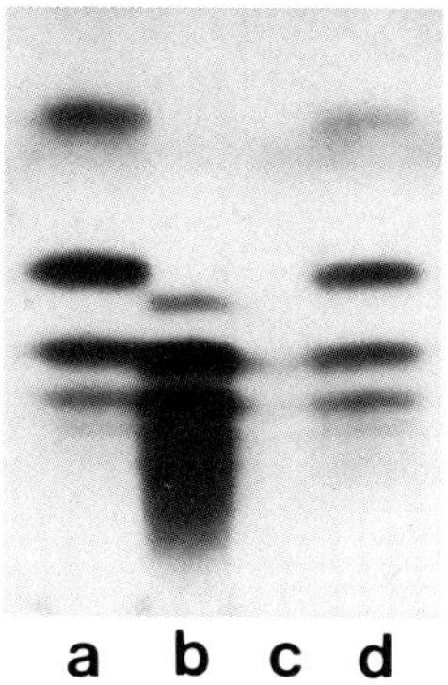

Fig. 21. Protein electrophoresis of serum stained for haptoglobin from father of T.H. (**a**), mother (**b**), patient T.H. (**c**), and brother (**d**). Though sample size was the same, the c lane (T.H.) is very weak.

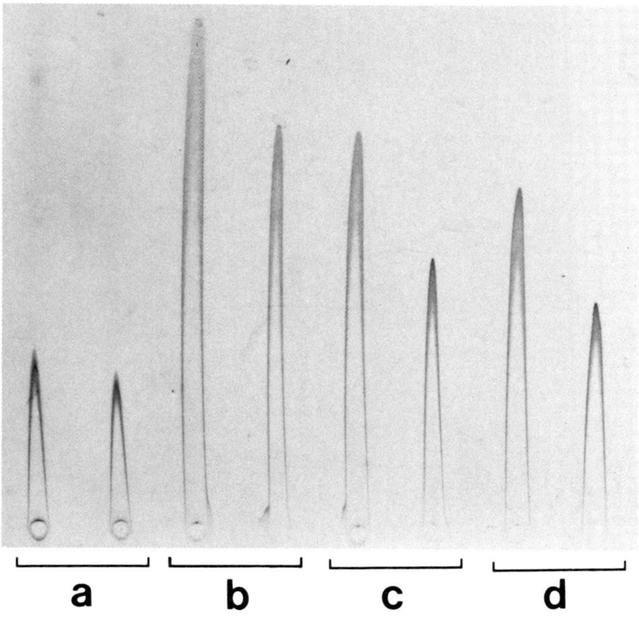

Fig. 22. Laurell electroimmunoassay of haptoglobin. **a:** Patient T.H. **b:** Mother of T.H. **c:** Father of T.H. **d:** brother of T.H. Each individual was run at two different dilutions. Levels in T.H. were less than half those of his brother and parents.

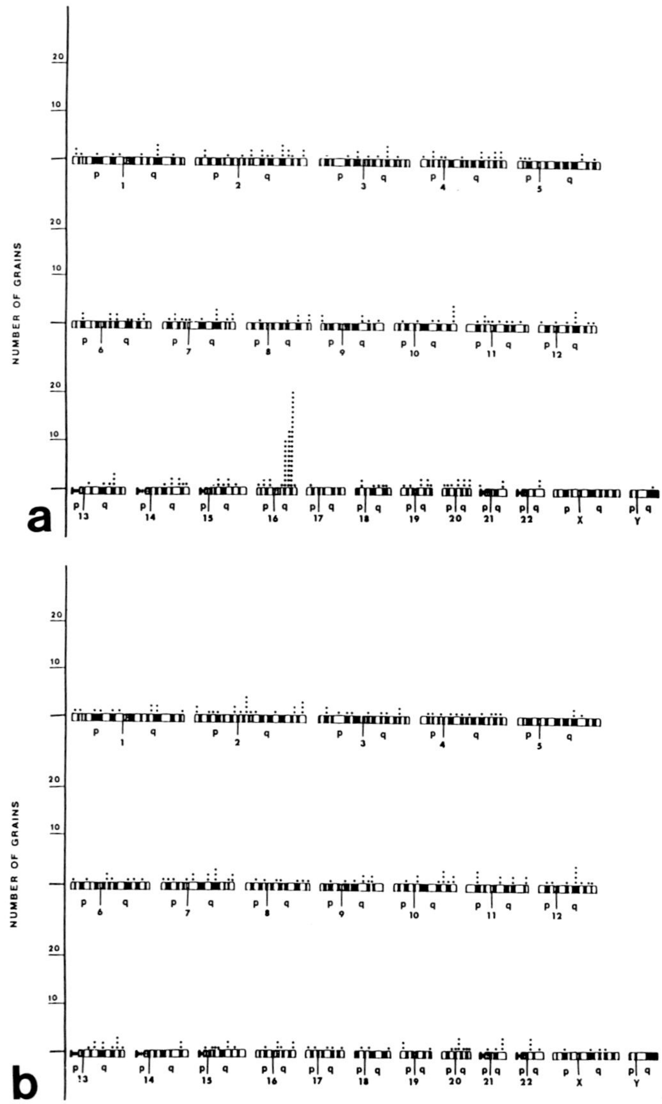

Fig. 23. **a:** Summarized grain counts in 100 cells from a normal control hybridized with a probe (pchTAT2-16) containing the tyrosine aminotransferase gene (TAT). There are excess grains on the long arm of 16 in the same region of hybridization as the haptoglobin probe. **b:** Distribution of grains in 100 cells from patient T.H. hybridized with the probe containing the TAT gene. No specific hybridization region is evident, indicating probable absence of the TAT gene in both chromosomes 16 in the patient.

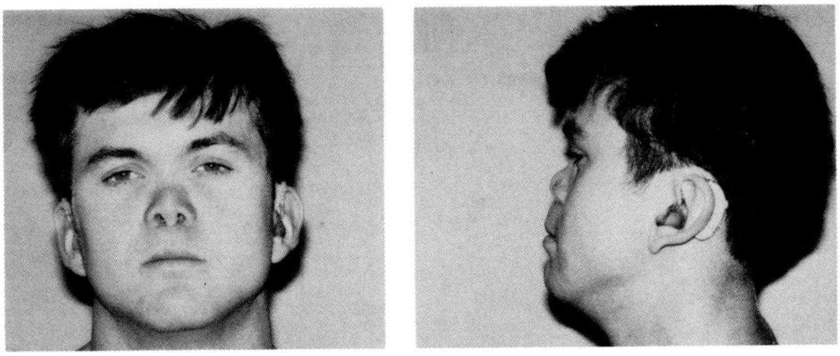

Fig. 24. Photograph of face of patient with X-linked chondrodysplasia punctata (Conradi syndrome).

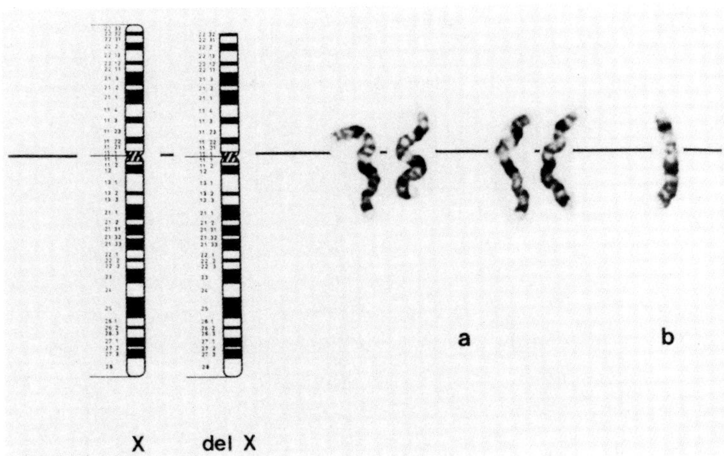

Fig. 25. **a:** G-banded X chromosomes at approximately the 750 band level from the mother of the proband with X-linked Conradi disease. The right-hand chromosome in each pair has a small terminal deletion. **b:** X chromosome from proband showing the deletion inherited from his mother. The breakpoint, p22.32, is shown in the idiogram.

thyosis and suspected developmental delay. In addition, the nephew on X-ray had punctate calcification in some joints. Chromosomal studies of the boys showed a small terminal deletion of the short arm of the X, and a similar deletion of the short arm of one X in their mothers (Fig. 25). Steroid sulfatase assay of the fibroblasts of the proband showed no activity, mapping the steroid sulfatase locus to the deleted region p.22.32→pter of the X chromosome as well as this X-linked form of chondrodysplasia punctata.

6. Exclusion mapping provides information on chromosomal regions in which a locus cannot reside. The information is usually ascertained by the study of chromosomally unbalanced patients who by heterozygosity at a locus or normal values on dosage analysis can be shown not to be unbalanced at the locus studied. Ferguson-Smith in 1975 demonstrated that useful mapping information can be obtained in this way [95] and later summarized the available exclusions [49]. As an example, by combining reports of exclusion of the MNS blood group (MNS) from various chromosomal segments [96], a total of 75% of the genome could be excluded as the site of the MNS gene. These excluded segments may provide information critical for determining the SRO for accurate mapping of a locus.

7. X-autosome translocations in a female affected with an X-linked disorder provide information on the regional localization of the disorder on the X. Most notable is the assignment of the Duchenne muscular distrophy locus to Xp21 by this method [97]. Localization of the gene for incontinentia pigmenti to the proximal short arm of the X is also suspected by this method [98].

8. A corollary to method 7 is the use of Lyonization of the X chromosome to assign a locus to the X. This method was utilized to assign the gene for the testicular feminization syndrome to the X [99].

9. Forty-two autosomal loci have thus far been finely mapped using restriction endonuclease techniques [3]. Fine mapping refers to determination at the molecular level of the transcriptional unit and usually requires that genomic and cDNA clones for the gene in question be available. A restriction map is constructed using a number of different restriction enzymes in conjuction with Southern blotting for both the cDNA and the genomic DNA clones. The results provide information for locations of introns and exons. This fine structure mapping further may provide information on adjacent loci such as in the β-globin cluster [100] and allow determination of order and orientation. The technique also allows precise breakpoints at the molecular level to be determined in the chromosome translocations associated with certain malignances. For example, Haluska et al. [101] cloned and sequenced the t(8;14) breakpoints of a Burkitt's lymphoma cell line and a pre-B-cell acute lymphoblastic leukemia (ALL) cell line. They found immunoglobulin heavy chain variable segment-diversity segment-joining segment-(V-D-J-) joining in each case, which has implications for causation of malignancy.

A new approach to molecular gene mapping that has been developed [102] is the result of the ability to resolve large DNA fragments on gel electrophoresis. Conventional gel electrophoresis permits analysis of small DNA molecules, such as cloned genes, but does not resolve DNA fragments larger than 20 kilobases. This is apparently because long molecules align themselves end on; thus mobility is independent of length. Schwartz and Cantor [103] and Carle and Olson [104] in 1984 described a technique called orthogonal field alternation gel electrophoresis (OFAGE or pulse field electrophoresis), in which the DNA molecules are subjected to two alternating fields at 90°–180° angle to each other, which force the molecules to turn each time the field is changed. The speed of turning is dependent on length, and on time between change of field direction. This resolution of large molecules should allow cloning and analysis of DNA from breakpoint regions of any chromosomal deletion and translocation since the breakpoint should be detected as a band of abnormal size and mobility on an OFAGE large molecule restriction map. Such a map can be constructed with the use of restriction enzymes such as notI, which cut genomic DNA at rare intervals and create large DNA fragments. Each large fragment should be identifiable by a cloned probe for any gene or small fragment along its length. This technique then should also facilitate chromosome "walking" and the identification of linked marker and disease genes.

The remaining nine techniques together have resulted in only a few more than 100 localizations. These methods, in brief, are:

10. Deductions from the amino acid sequence of proteins particularly those that are combined as Lepore hemoglobin which allowed deduction of close linkage of δ- and β- globins [105].

11. Chromosome-mediated gene transfer. McBride and Ozer in 1973 [106] first showed transfer of functional genes into cells by uptake of isolated metaphase chromosomes. Transfer of an intact human X chromosome was accomplished by this method [107], and also cotransfer of human syntenic genes thymidine kinase and galactokinase [108]. This method is limited by the requirement of appropriate selective systems, which are few in number.

12. Microcell-mediated gene transfer. Micronuclei, induced by treatment with colchicine, containing only a portion of cell DNA, can be isolated as microcells by centrifugation using cytochalasin B. The microcells then can be introduced into somatic cells by fusion techniques such that the somatic cells gain small but varying amounts of foreign genetic information.

Fournier and Ruddle in 1977 [109] demonstrated the presence of 1–6 mouse chromosomes in human epithelial carcinoma (HeLa) cells and Chinese hamster cells following microcell transfer. Mouse enzymes were also identified in some of the stable cell lines produced. Tourian et al. [110] introduced human chromosomes into hypoxanthine-guanine phosphoribosyl transferase-negative ($HGPRT^-$) mouse cells and showed the presence of HGPRT following transfer. The assignment of human type I procollagen to chromosome 17 was made in this way [111].

13. Homology of synteny. This method is the result of comparative chromosome mapping of genes in other organisms, particularly the great apes, whose chromosomes are very similar to those of humans. A number of human genes on several different chromosomes are known to be located on the morphologically similar chromosomes in the gorilla, orangutan, and chimpanzee. Thus, a gene mapped to the morphologically homologous chromosome in orangutan might also be mapped to that chromosome in the human. This is true for even more remote organisms as mouse and dog for genes on the X chromosome (e.g., hemophilia).

14. cDNA/DNA solution hybridization (Cot analysis). This is a method to detect the presence of specific DNA sequences, using rate of hybridization in solution [112]. It requires either an aneuploid complement of human chromosomes, deletions/duplication, or segregation of chromosomes, in somatic cell hybrids. For example, Deisseroth showed, using cDNA made from hemoglobin mRNA, that α- and β- hemoglobin sequences could be detected in hybrid cells [113]. He further showed by Cot analysis utilizing a somatic cell hybrid panel, that the β-globin gene segregated with chromosome 11, localizing the gene to that chromosome [114].

15. Linkage disequilibrium. This method is exemplified by the close linkage of major histocompatibility complex class I antigens A & B (HLA-A and -B) genes such that some specific haplotypes appear to be preserved in the population. This also has been shown between HLA and properdin factor B (BF), complement components 2 and 4 (C_2, C_4), and 21-hydroxylase deficiency [115].

16. Ovarian teratoma centromere mapping. Ovarian teratomas are generally the result of lack of division at second meiosis (or failure of second polar body formation) of germ cells. Thus the centromeric region is essentially identical for the homologous chromosomes, while crossing-over has taken

TABLE II. Publication of proceedings of human gene mapping (HGM) workshops

	Birth Defects: Original Article Series	Cytogenetics and Cell Genetics (Basel)
HGM1 New Haven (1973)	X(3):1–216, 1974	13:1–216, 1974
HGM2 Rotterdam (1974)	XI(3):1–310, 1975	14:162–480, 1975
HGM3 Baltimore (1975)	XII(7):1–452, 1976	16:1–452, 1976
HGM4 Winnipeg (1977)	XIV(4):1–730, 1978	22:1–730, 1978
HGM5 Edinburgh (1979)	XV(11):1–236, 1979	25:2–236, 1979
HGM6 Oslo (1981)	XVIII(2):1–343, 1982	32:1–343, 1982
HGM7 Los Angeles (1983)	XX(2):1–666, 1984	37:1–666, 1984
HGM8 Helsinki (1985)	XXI(4):1–832, 1985	40:1–832, 1985

TABLE III. Number of loci assigned—update of April 15, 1986 [3]

Gene mapping workshop	No. of autosomal assignments[a] Confirmed	Provisional	Tentative ("in limbo"; inconsistent)	Total	X-chromosome assignments[b] Confirmed	In limbo
New Haven (1973)	31	28	5	64	88	67
Rotterdam (1974)	48	32	6	86	91	70
Baltimore (1975)	72	46	7	125	95	80
Winnipeg (1977)	83	82	11	176	102	96
Edinburgh (1979)	123	87	20	230	112	101
Oslo (1981)	180	120	45	345	116	118
Los Angeles (1983)	247	161	80	488	118	136
Helsinki (1985)	351	216	66	633	123	157
April 15, 1986	374	232	37	641	124	157 [3]

[a]Plus 300 anonymous DNA segments; 82 antigens; 19 O'Farrell protein spots, all of unknown function; 19 pseudogenes; 18 "like" genes; and 63 fragile sites, for a grand total, for autosomes, of 1,134.
[b]Plus 186 DNA segments, 4 protein spots, 4 antigens, 3 fragile sites, 25 pseudogenes and 20 "like" genes, for a grand total, for the X, of 522.

place at further distances. The amount of crossing-over reflects distance from the centromere [116], at least for short distances.

17. Radiation-induced gene segregation [117] uses quantitated radiation of normal human cells to break chromosomes and then hybridization with nonirradiated, nonhuman cells. This may result in segments of human chromosomes attaching to the nonhuman host chromosomes and segregating with them as well as human chromosome deletions and rearrangements. Loci consistently segregating together, then, are likely to be syntenic. If the rodent cells are deficient in a specific enzyme, a selective medium may be used to retain the normal human counterpart. Other linked genes would be retained according to distance from the selected gene and radiation dose.

18. Adenovirus-induced chromosome change. For example, adenovirus 12 produces a specific break or gap region in the long arm of chromosome 17; it also induces thymidine kinase (TK) to be produced, suggesting that the gap may be an uncondensed functioning segment, containing the TK gene. McDougall also showed that galactokinase was induced in the same way. [118]

GENE MAPPING WORKSHOPS

The progress of chromosome mapping has been documented in the chronicles of Gene Mapping Workshops. The first was organized by F.H. Ruddle and V. McKusick and held in New Haven, Connecticut in 1973, five years after the first genes were mapped to autosomes. By the end of the workshop, 31 genes were confirmed on autosomes, 16 of these on chromosome 1, and 88 on the X. No genes had been assigned to chromosomes 3, 8, 9, and the Y; and only ribosomal genes assigned to chromosomes 13, 15, and 22. Thymidine kinase was still the only gene assigned to chromosome 17. Most of the assignments were made using somatic cell hybrids.

By the Los Angeles meeting, Gene Mapping Workshop 7, 1983, the character of the workshops had changed. They were larger, and molecular geneticists were strongly represented. At least half of the assignments used recombinant DNA techniques. This was also the case at the Helsinki meeting, Human Gene Workshop 8, 1985, but there was in addition a shift back to the use of family linkage studies in conjunction with molecular techniques, the linkage markers being restriction fragment linked polymorphisms.

The Human Gene Workshops are referenced in Table II and numbers of loci assigned are summarized in Table III [3]. Throughout these workshops it is evident that advances in cytogenetic technology, including high-resolution techniques, and use of well-delineated chromosome aberrations have been the basis for increased precision of gene localization.

ACKNOWLEDGMENTS

I thank SuEllen Toth-Fejel, Susan Olson, and Bob Sheehy for their contributions to the studies described in this chapter. I am also grateful to Cathi Schmidt for typing the manuscript.

This work was supported by grants from NICHD (HD-07997) and Maternal and Child Health Services 920.

REFERENCES

1. Strickberger M: Gene mutation. In "Genetics." New York: The MacMillan Company, pp 515–535, 1968.
2. Laird CD: Chromatid structure: Relationship between DNA content and nucleotide sequence diversity. Chromosoma 32:378–406, 1971.
3. McKusick VA: The human gene map 15 April 1986. Clin Genet 29:545–588, 1986.
4. Royer-Pokora B, Kunkel LM, Monaco AP, Goff SC, Newburger PE, Baehner RL, Cole FS, Curnutte JT, Orkin SH: Cloning the gene for an inherited human disorder—chronic granulomatous disease—on the basis of its chromosomal location. Nature 322:32–38, 1986.
5. Monaco AP, Neve RL, Colletti-Feener C, Bertelson CJ, Kurnit DM, Kunkel LM: Isolation of candidate cDNAs for portions of the Duchenne muscular dystrophy gene. Nature 323:646–650, 1986.

6. Friend SH, Bernards R, Rogelj S, Weinberg RA, Rapaport JM, Albert DM, Dryja TP: A human DNA segment with properties of the gene that predisposes to retinoblastoma and osteosarcoma. Nature 323:643–646, 1986.
7. Scrambler PJ, Wainwright BJ, Farrell M, Bell J, Stanier P, Lench NJ, Bell G, Kruyer H, Ramirez F, Williamson R: Linkage of COL1A2 collagen gene to cystic fibrosis and its clinical implications. Lancet ii:1241–1242, 1985.
8. Tsui L-C, Buchwald M, Barker D, Braman JC, Knowlton R, Schumm JW, Eiberg H, Mohr J, Kennedy D, Plasvic N, Zsiga M, Markiewicz D, Akots G, Brown V, Helms C, Gravius T, Parker C, Rediker K, Donis-Keller H: Cystic fibrosis locus defined by a genetically linked polymorphic marker. Science 230:1054–1057, 1985.
9. Wainwright BJ, Scrambler PJ, Schmidtke J, Watson EA, Law H-Y, Farrell M, Cooke HJ, Eiberg H, Williamson R: Localization of cystic fibrosis locus to human chromosome 7 cen→q22. Nature 318:384–385, 1985.
10. White R, Woodward S, Leppert M, O'Connell P, Hoff M, Herbst J, Lalouel J-M, Dean M, Vande Woude G: A closely linked genetic marker for cystic fibrosis. Nature 318:382–384, 1985.
11. Gusella JF, Tanzi RE, Anderson MA, Hobbs W, Gibbons K, Raschtchian R, Gilliam TC, Wallace MR, Wexler NS, Conneally PM: DNA markers for nervous system diseases. Science 225:1320–1326, 1984.
12. McDermid HE, Duncan A.M.V., Brasch KR, Holden JAJ, Magenis RE, Sheehy R, Burn J, Kardon N, Noel B, Schinzel A, Teshima I, White BN: Characterization of the supernumerary chromosome in cat eye syndrome. Science 232:646–648, 1986.
13. Magenis RE, Sheehy RR, Brown MG, McDermid HE, White BN, Zonana J, Weleber R: Parental origin of the extra chromosome in cat eye syndrome: Evidence from heteromorphism and in situ hybridization analysis. Am J Med Genet (in press).
14. Gilgenkrantz S, Droulle P, Seboun E, Casanova M, Schweitzer M, Goosens M, Fellous M: Prenatal identification of a Y-chromosome deletion by Y-specific single copy DNA probes. Prenat Diagn 6:307–311, 1986.
15. Emanuel BS, Seldon JR, Wang E, Nowell PC, Croce CM: In situ hybridization and translocation breakpoint mapping I. Nonidentical 22q11 breakpoints for the t(9;22) of CML and the t(8;22) of Burkitt lymphoma. Cytogenet Cell Genet 38:127–131, 1984.
16. Emanuel BS, Nowell PC, McKeon C, Croce CM, Israel MA: Translocation breakpoint mapping: Molecular and cytogenic studies of chromosome 22. Cancer Genet Cytogenet 19:81–92, 1986.
17. LeBeau MM, Diaz MO, Karin M, Rowley JD: Metallothionein gene cluster is split by chromosome 16 rearrangements in myelomonocytic leukemia. Nature 313:709–711, 1985.
18. Boyd Y, Buckle V, Holt S, Munro E, Hunter D, Craig I: Muscular dystrophy in girls with X; autosome translocations. J Med Genet 23:484–490, 1986.
19. Magenis RE, Brown MG, Donlon T, Olson SB, Sheehy R, Tomar D: Structural aberrations of the Y chromosome, including the non-fluorescent Y: Cytological origin and consequences. In Sandberg A (ed): "The Y Chromosome, Part A: Basic Characteristics of the Y Chromosome." Progress and Topics in Cytogenetics, Volume 6A. New York: Alan R, Liss, Inc., 1985, pp 537–574.
20. Magenis RE, Brown M, Chamberlin J, Donlon T, Hepburn D, Lamvik N, Yoshitomi M: Resolution of breakpoints in a complex de novo rearrangement by use of multiple staining techniques: 45,XY,t(12;15;21)(12qter→12p12.3::21q11.1→21qter;15qter→15p11::12p12.3→12pter): Confirmation of suspected 12p12.3 intraband deletion by dosage effect of LDH-B. Amer J Med Genet 9:95–109, 1981.

21. Francke U, Ochs HD, de Martinville B, Giacalone J, Lindgren V, Disteche C, Pagon RA, Hofker MH, van Omenn G-J B, Pearson PL, Wedgwood RJ: Minor Xp21 chromosome deletion in a male associated with expression of Duchenne muscular dystrophy, chronic granulomatous disease, retinitis pigmentosa and MacLeod syndrome. Am J Hum Genet 37:250–267, 1985.
22. Benedict WF, Murphree AL, Banerjee A, Spina CA, Sparkes MC, Sparkes RS: Patient with 13 chromosome deletion: Evidence that the retinoblastoma gene is a recessive cancer gene. Science 219:973–975, 1983.
23. Wilson EB: The sex chromosomes. Arch Mikrosk Anat 77:249–271, 1911.
24. Mohr J: "A study of Linkage in Man." Copenhagen: Munksgaard, 1954.
25. Tjio JH, Levan A: The chromosome number of man. Hereditas 42:1–6, 1956.
26. Moorhead PS, Nowell PC, Mellman WJ, Battips DM, Hungerford DA: Chromosome preparations of leukocytes cultured from peripheral blood. Exp Cell Res 20:613–616, 1960.
27. Donahue RP, Bias WB, Renwick JH, McKusick VA: Probable assignment of the Duffy blood group locus to chromosome 1 in man. Proc Natl Acad Sci USA 61: 949–955, 1968.
28. Ying K-L, Ives EJ: Asymmetry of chromosome number 1 pair in three generations of a phenotypically normal family. Can J Genet Cytol 10:575–589, 1968.
29. Weiss MC, Green H: Human-mouse hybrid cell lines containing partial complements of human chromosomes and functioning human genes. Proc Natl Acad Sci USA 58:1104–1111, 1967.
30. Migeon BR, Miller CS: Human-mouse somatic cell hybrids with single human chromosome (Group E): Link with thymidine kinase activity. Science 162:1005–1006, 1968.
31. Barski G, Sorieul S, Cornefert F: "Hybrid" type cells in combined cultures of two different mammalian cell strains. J Natl Cancer Inst 26:1269–1291, 1961.
32. Littlefield, J: Selection of hybrids from mating of fibroblasts in vitro and their presumed recombinants. Science 145:709–710, 1964.
33. Ephrussi B, Weiss MC: Interspecific hybridization of somatic cells. Proc Natl Acad Sci USA 53:1040–1042, 1965.
34. Miller OJ, Allderdice PW, Miller DA, Breg WR, Migeon BR: Human thymidine gene locus: Assignment to chromosome 17 in a hybrid of man and mouse cells. Science 173:244–245, 1971.
35. Robson EB, Polani PE, Dart SJ, Jacobs PA, Renwick JH: Probable assignment of the alpha locus of haptoglobin to chromosome 16 in man. Nature 223:1163–1165, 1969.
36. Magenis RE, Hecht F, Lovrien EW: Heritable fragile site on chromosome 16: Probable localization of haptoglobin locus in man.Science 170:85–87, 1970.
37. Brandt NJ, Frøland A, Mikkleson M, Nielson A, Tolstrup N: Galactosemia locus and the Down's syndrome chromosome. Lancet 2:700–703, 1963.
38. Donnell GN, Ng WG, Bergren WR, Melnyk J, Koch R: Enhancement of erythrocyte-galactokinase activity in Langdon-Down trisomy. Lancet 1:533, 1965.
39. Baikie AG, Loder PB, de Gruchy GC, Pitt DB: Phosphohexokinase activity of erythrocytes in mongolism, another possible marker for chromosome 21. Lancet 1:412–414, 1965.
40. Bartels H, Kruse K: Enzymbestimmungen in Erythrocyten bei Kindern mit Down-Syndrome. Humangenetik 5:305–309, 1968.
41. Pantelakis SN, Karaklis AC, Alexiou D, Vardas E, Valaes T: Red cell enzyme in trisomy 21. Am J Hum Genet 22:184–193, 1970.

42. Mellman WJ, Oski FA, Tedesco TA, Maciera-Coelho A, Harris H: Leucocytes enzymes in Down's syndrome. Lancet 2:674–675, 1964.
43. Layzer RB, Epstein CJ: Phosphofructokinase and chromosome 21. Am J Hum Genet 24:533–543, 1972.
44. Vora S, Francke U: Assignment of the human gene for liver-type 6-phosphofructokinase isozyme (PFKL) to chromosome 21 using somatic cell hybrids and monoclonal anti-L antibody. Proc Natl Acad Sci USA 78:3738–3742, 1981.
45. Caspersson T, Zech L, Johansson C, Modest EJ: Identification of human chromosomes by DNA reacting fluorescing agents. Chromosoma 30:215–227, 1970.
46. Seabright M: A rapid banding technique for human chromosomes. Lancet 2:971–972, 1971.
47. Yunis JJ: High resolution of human chromosomes. Science 191:1268–1271, 1976.
48. ISCN: (1985): "An International System for Human Cytogenetic Nomenclature," Harnden DG, Klinger HP (eds). Published in collaboration with Cytogenet Cell Genet (Karger, Basel, 1985). Also in Birth Defects: Original Article Series, Vol. 21, No. 1 (March of Dimes Birth Defects Foundation, New York, 1985).
49. Ferguson-Smith MA, Aitken DA: The contribution of chromosome aberrations to the precision of human gene mapping. Cytogenet Cell Genet 32:24–42, 1982.
50. Ferguson-Smith MA, Newman BF, Ellis PM, Thompson DMG, Riley ID: Assignment by deletion of human red cell acid phosphatase gene locus to the short arm of chromosome 2. Nature new Biol 243:271–274, 1973.
51. Magenis RE, Koler RD, Lovrien EW, Bigley RW, DuVal M, Overton K: Gene dosage: Evidence for assignment of erythrocyte acid phosphatase locus to chromosome 2. Proc Natl Acad Sci USA 72:4526–4530, 1975.
52. Hopkinson DA, Spencer N, Harris H: Genetical studies on human red cell acid phosphatase. Am J Hum Genet 16:141–154, 1964.
53. Povey S, Swallow DM, Brobow M, Craig I, van Heymingen V: Probable assignment of locus determining human red-cell acid-phosphatase ACP1 to chromosome 2 using somatic cell hybrids. Ann Hum Genet 38:1–5, 1974.
54. Mace MA, Noades J, Robson EB, Hultén M, Lindsten J, Polani PE, Jacobs PA, Buckton E: Segregation of ACP_1 and MNSs in families with structural rearrangements involving chromosome 2. Ann Hum Genet 38:479–484, 1975.
55. McGillavray BC, Dill FJ, Lowry RB: HLA investigation and phenotype correlation of two siblings with trisomy 6p12pter. Am J Hum Genet 30:59A, 1978.
56. Pearson G, Mann JD, Bensen J, Bull RW: Inversion duplication of chromosome 6 with trisomic codominant expression of HLA antigens. Am J Hum Genet 31:29–34, 1979.
57. Berger R, Bernheim A, Sasportes M, Hauptmann G, Hors J, Legrand L, Fellous M: Regional mapping of the HLA on the short arm of chromosome 6. Clin Genet 15:245–251, 1979.
58. Cousineau AJ, Higgins JV, Hackel E, Waterman DF, Toriello H, Carlile PA, Cook PJL: Cytogenetic recognition of chromosomal duplication [dup(1)(p31.4→p22.1)] and the detection of three different alleles at the PGM_1 locus. Ann Hum Genet 45:337–340, 1981.
59. Marsh WL, Chaganti RSK, Gardner FH, Mayer K, Nowell PC, German J: Mapping human autosomes—Evidence supporting assignment of Rhesus to short arm of chromosome No. 1. Science 183(4/28):966–968, 1974.
60. de la Chapelle A, Vuopio P, Icen A: Red cells glutathione reductase activity in patients with bone-marrow trisomy 8. HGM3 Baltimore (1975). Birth Defects: XII (7):1–452, 1952; also in Cytogenet Cell Genet 16:277–279, 1976.

61. Ferguson-Smith MA, Aitken DA, Turleau C, Grouchy J de: Localization of the human ABO:Np-1:Ak-1 linkage group by regional assignment of Ak-1 to 9q34. Hum Genet 34:35–43, 1976.
62. Aitken DA, Ferguson-Smith MA: Intrachromosomal assignment of the structural gene for GALT to the short arm of chromosome 9 by gene dosage studies. Cytogenet Cell Genet 25:131, 1979.
63. Carritt B, King J, Welch, HM: Gene order and localization of enzyme loci on the short arm of chromosome 1. Ann Hum Genet 46:329–335, 1982.
64. Creagan RP, Ruddle FH: The clone panel: A systematic approach to gene mapping using interspecific somatic cell hybrids. Cytogenet Cell Genet 14:282–286, 1975.
65. Ott J: Estimation of the recombination fraction in human pedigrees: Efficient computation of the likelihood for human linkage studies. Am J Hum Genet 26:588–597, 1974.
66. Conneally PM, Rivas ML: Linkage analysis in man. Adv Hum Genet 10:209–266, 1980.
67. Smith HO: Nucleotide sequence specificity of restriction endonucleases. Science 205:455–462, 1979.
68. Human Gene Mapping 8 (1985). "Eighth International Workshop on Human Gene Mapping." Cytogenet Cell Genet 40(Nos. 1–4), 1985.
69. Botstein D, White RL, Skolnick M, Davis RW: Construction of a genetic linkage map in man using restriction fragment length polymorphisms. Am J Hum Genet 32:314–331, 1980.
70. Gall JG, Pardue ML: Formation and detection of RNA-DNA hybrid molecules in cytological preparations. Proc Natl Acad Sci USA 63:378–383, 1969.
71. John HA, Birnstiel ML, Jones KW: RNA-DNA hybrids at the cytological level. Nature 223:582–587, 1969.
72. Harper ME, Saunders GF: Localization of a single copy DNA sequence on G-banded human chromosomes by in situ hybridization. Chromosoma 83:431–439, 1981.
73. Harper ME, Ullrich A, Saunders GF: Localization of the human insulin gene to the distal end of the short arm chromosome 11. Proc Natl Acad Sci USA 78:4458–4460, 1981.
74. Malcom S, Barton P, Murphy C, Ferguson-Smith MA: Chromosomal localization of a single copy gene by in *situ* hybridization-human beta globin on the short arm of chromosome 11. Ann Hum Genet 45:135–141, 1981.
75. Gerhard DS, Kawasaki E, Bancroft FC, Szabo P: Localization of a unique gene by directed hybridization *in situ*. Proc Natl Acad Sci USA 78:3755–3579, 1981.
76. Zabel BU, Naylor SL, Sakaguchi AY, Bell GI, Shows TB: High resolution chromosomal localization of human genes for amylase, proopiomelanocortin, somatostatin, and a DNA fragment (D3S1) by in situ hybridization. Proc Natl Acad Sci USA 80:6932–6936, 1983.
77. Buckle VJ, Craig IW, Edwards JH: Localization of the factor IX structural gene by in situ hybridization. Cytogenet Cell Genet: 37:430, 1984.
78. Chance PF, Dyer KA, Kurachi K, Yoshitake S, Ropers HH, Wieacker P, Gartler SM: Regional assignment of the human factor IX gene by molecular hybridization. Cytogen Cell Genet 37:435, 1984.
79. Donlon T: Practical approach to in situ hybridization. Karyogram 12:3–10, 1986.
80. Buckle VJ, Craig IW: In situ hybridization. In Davis KE (ed): "Human Genetic Diseases, a Practical Approach." Washington, DC: Oxford (IRL Press), 1986, pp 85–100.
81. Rogers AW: The resolution of autoradiographs. In "Techniques of autoradiography." New York: Amsterdam: Elsevier, 1967, pp 49–67.

82. Magenis RE, Donlon TA, Tomar DR: Localization of the β-globin gene to 11p15 by in situ hybridization: Utilization of chromosome 11 rearrangements. Hum Genet 69:300–303, 1985.
83. Gusella J, Varsanyi-Breiner A, Kao F-T, Jones C, Puck TT, Keys C, Orkin S, Housman D: Precise localization of human gene complex on chromosome 11. Proc Natl Acad Sci USA 76:5239–5243, 1979.
84. Magenis RE, Gusella J, Weliky K, Olson S, Haight G, Toth-Fejel S, Sheehy R: Huntington disease-linked restriction fragment length polymorphism localized within band p16.1 of chromosome 4 by in situ hybridization. Am J Hum Genet 39:383–391, 1986.
85. Gusella JF, Wexler NS, Conneally PM, Naylor SL, Anderson MA, Tanzi RE, Watkins PC, Ottina K, Wallace MR, Sakaguchi AY, Young AB, Shouldson I, Bonilla E, Martin JB: A polymorphic DNA marker genetically linked to Huntington's disease. Nature 306:234–238, 1983.
86. Simmers RN, Stupans I, Sutherland GR: Localization of the human haptoglobin genes distal to the fragile site at 16q22 using in situ hybridization. Cytogenet Cell Genet 41:38–41, 1986.
87. Natt E, Kao F-T, Rettenmeier R, Scherer G: Assignment of the human tyrosine aminotransferase gene to chromosome 16. Hum Genet 72:225–228, 1986.
88. Natt E, Westphal E-M, Toth-Fejel S, Magenis RE, Buist NRM, Rettenmeier R, Scherer G: Inherited and *de novo* deletion of the tyrosine aminotransferase gene locus at 16q22.1→22.3 in a patient with tyrosinemia type II. Hum Genet (in press).
89. Münke M, Lindgren V, deMartinville B, Francke U: Comparative analysis of mouse-human hybrids with rearranged chromosomes 1 by in situ hybridization and Southern blotting: High resolution mapping of NRAS, NGFB, and Amy on human chromosome 1. Somatic Cell Mol Genet 10:589–599, 1984.
90. Lubs HA: A marker X chromosome. Am J Hum Genet 21:231–244, 1969.
91. Sparkes RS, Sparkes MC, Wilson MG, Towner JW, Benedict W, Murphree AL, Yunis JJ: Regional assignment of genes for human esterase D and retinoblastoma to chromosome band 13q14. Science 208:1042–1044, 1980.
92. Ledbetter DH, Riccardi VM, Airhart SD, Strobel RJ, Keenan BS, Crawford JD: Deletions of chromosome 15 as a cause of Prader-Willi syndrome. N Engl J Med 304:325–329, 1981.
93. Stratton RF, Dobyns WB, Airhart SD, Ledbetter DH: New chromosomal syndrome: Miller-Dieker syndrome and monosomy 17p13. Hum Genet 67:193–200, 1984.
94. Curry CJR, Magenis RE, Brown MG, Lanman JT, Tsai T, O'Lague P, Goodfellow P, Mohandas T, Bergner EA, Shapiro LJ: Inherited chondrodysplasia punctata due to a deletion of the terminal short arm of an X chromosome. N Engl J Med 311:1010–1015, 1984.
95. Ferguson-Smith MA: Gene mapping by exclusion. Cytogenet Cell Genet 14:296–299, 1975.
96. Cook PJL, Noades JE, Buckton KE, Lindenbaum RH: Exclusion mapping of MNS and GPT. Cytogenet Cell Genet 25:143, 1979.
97. Greenstein RM, Reardon MP, Chan TS: An X/autosome translocation in a girl with Duchenne Muscular Dystrophy (DMD): Evidence for DMD gene localization. Pediatr Res 11:457A, 1977.
98. Gilgenkrantz S, Tridon P, Pinel-Briquel N, Beurey J, Weber M: Translocation (X;9)(p11;q34) in a girl with incontinenti pigmenti (IP): Implications for the regional assignment of the IP locus to Xp11? Ann Genet 28:90–92, 1985.

99. Meyer WJ, III, Migeon BR, Migeon CJ: Locus on human X chromosome for dihydrotestosterone receptor and androgen insensitivity. Proc Natl Acad Sci 72:1469–1472, 1975.
100. Lawn RM, Fritsch EF, Parker RC, Blake G, Maniatis T: The isolation and characterization of linked δ- and β-globin genes from a cloned library of human DNA. Cell 15:1157–1174, 1978.
101. Haluska FG, Finver S, Tsujimoto Y, Croce CM: The t(8;14) chromosomal translocation occurring in B-cell malignancies results from mistakes in V-D-J joining. Nature 324:158–161, 1986.
102. Shaw DJ: A new strategy for mapping the human genome. J Med Genet 23:421–424, 1986.
103. Schwartz DC, Cantor CR: Separation of yeast chromosome-sized DNAs by pulsed field gradient gel electrophoresis. Cell 37:67–75, 1984.
104. Carle GF, Olson MV: Separation of chromosomal DNA molecules from yeast by orthogonal-field-alternation gel electrophoresis. Nucleic Acids Res 12:5647–5664, 1984.
105. Baglioni C: The fusion of two polypeptide chains in hemoglobin Lepore and its interpretation as a genetic deletion. Proc. Natl Acad Sci USA 48:1880–1885, 1962.
106. McBride OW, Ozer HL: Transfer of genetic information by purified metaphase chromosomes. Proc Natl Acad Sci USA 70:1258–1262, 1973.
107. Wullems GJ, Van der Horst J, Bootsma D: Transfer fo the human X chromosome to human Chinese hamster cell hybrids via isolated Hela metaphase chromosomes somatic. Cell Genet 2:359–371, 1976.
108. Willecke K, Lange R, Krüger A, Reber T: Cotransfer of two linked human genes into cultured mouse cells. Proc Natl Acad Sci 73:1274–1278, 1976.
109. Fournier REK, Ruddle F: Microcell mediated transfer of murine chromosomes into mouse, Chinese hamster, and human somatic cells. Proc Natl Acad Sci USA 74:319–323, 1977.
110. Tourian A, Johnson RT, Burg K, Nicolson SW, Sperling K: Transfer of human chromosomes via human minisegregant cells into mouse cells and the quantitation of the expression of HGPRT. J Cell Sci 30: 193–210, 1978.
111. Sundar Raj CV, Church RL, Klobutcher LA, Ruddle FH: Genetics of the connective tissue proteins: assignment of the gene for human type I procollagen to chromosome 17 by analysis of cell hybrids and microcell hybrids. Proc Natl Acad Sci USA 74:4444–4448, 1977.
112. Britten RJ, Kohne DE: Repeated sequences in DNA. Science 161:529–540, 1968.
113. Deisseroth A: Isolation of hybrid cells that exhibit markers of erythroid differentiation. N Engl J Med 148–152, 1976.
114. Deisseroth A, Nienhuis A, Lawrence J, Giles R, Turner P, Ruddle F: Chromosomal localization of human β-globin gene on human chromosome 11 in somatic cell hybrids. Proc Natl Acad Sci USA 75:1456–1460, 1978.
115. O'Neill GJ, Pollack MS, Yang SY, Levine LS, New MI, Dupont B: Gametic associations of the HLA linked genes: BF, C2, C4S, C4F, glyoxalase I and 21-hydroxylase deficiency. "Human Gene Mapping 5." Birth Defects XV(11): 1979; also in Cytogenet Cell Genet 25:195, 1979.
116. Linder D, McCaw BK, Hecht F: Parthenogenic origin of benign ovarian teratomas. N Engl J Med 292:63–66, 1975.

117. Goss SJ, Harris H: New method for mapping genes in human chromosomes. Nature 255:680–684, 1975.
118. McDougall JK, Elsevier SM, Kucherlapati RS, Ruddle FH: Regional localization of human genes in virus-induced uncoiler regions. "Rotterdam Conference (1974): Second International Workshop on Human Gene Mapping." Birth Defects XI (3):202–204, 1975; also in Cytogenet Cell Genet 14:372–374, 1975.

Index